EXTENDED SURFACE HEAT TRANSFER

EXTENDED SURFACE HEAT TRANSFER

ALLAN D. KRAUS
University of Akron

ABDUL AZIZ
Gonzaga University

JAMES WELTY
Oregon State University

WILEY

Extended Surface Heat Transfer

Authorized reprint by Wiley India Pvt. Ltd., 4435-36/7, Ansari Road, Daryaganj, New Delhi - 110002.

Reprint: 2013

Printed at: *Lalit Printer & Binder, Delhi*

ISBN: 978-81-265-4053-2

TO OUR WIVES

Ruth, Ayesha, and Sharon

TO OUR WIVES

Ruth, Ayesha, and Sharon

CONTENTS

PREFACE

Extended surfaces, in the forms of longitudinal or radial fins or spines are ubiquitous in applications where the need exists to enhance heat transfer between a surface and an adjacent fluid. Applications range from very large scale, as with tubes in heat exchangers, to the very small, as is the case for the temperature control of electronic components.

At the fundamental level, the analysis of heat transfer from finned surfaces involves solving second-order differential equations in a variety of coordinate systems. The subject of extended surface heat transfer is one where analytical methods have been very successful in providing design information for a variety of geometries, some of which are very complex. As both primary and extended surfaces involve convective exchange as a boundary condition, the convective heat transfer coefficient h, which appears as a parameter in the solution, must be evaluated using standard analysis or empirical correlations.

This coefficient can be modeled, most simply, as a constant, in which the governing second-order differential equation is linear. When the more realistic definition of h as a function of temperature is employed, the problem becomes nonlinear and is considerably more difficult to solve. This nonlinearty is exaggerated when the solid–fluid interface encounters a phase change in the form of evaporation or condensation. The subject of convective heat transfer is included as a separate chapter in this work, and this chapter provides some direction in evaluating the coefficient h. Two later chapters are devoted entirely to the areas of boiling and condensation.

A rather extensive treatise on the subject of extended surface heat transfer is a book by Donald Q. Kern and Allan D. Kraus published in 1972. One might have thought this subject area sufficiently mature that only modest developments, probably of an incremental nature, would occur in the ensuing 28 years following its publication. The reality is that considerable additional work has been done and that the field is much in need of a contemporary treatment which includes both the fundamental material addressed earlier and the substantial additional contributions that have been made during the interim. Indicative of the advances made is the fact that the 1972 work of Kern and Kraus contained about 150 references to the technical literature, most applicable to the design of heat exchangers. The present work represents about a tenfold increase over the number of fundamental citations listed in 1972.

When examining the table of contents, the reader will observe that several additional subjects are included in the present work. The first eight chapters treat the fundamentals of extended surface analysis, including the implications of the Murray–Gardner assumptions, the analysis of single fins, and analysis and optimization of finned arrays. Subsequently, application areas considered include finned passages, heat exchangers of various configurations, radiation effects, optimum designs, heat transfer with change of phase, and augmentation. Three of the later chapters treat two-dimensional effects, time-dependent and periodic heat transfer. A common feature of this book is the extensive employment of example problems to illustrate the use of expressions derived and/or discussed in the text.

During the course of the preparation of this work, many obligations have been incurred that must be acknowledged. First and foremost are two rather basic acknowledgments to two people who did not partake in the preparation of the manuscript. The first is to Donald Kern, who died in 1971 during the preparation of the first *Extended Surface Heat Transfer*. Dr. Kern's well-received *Process Heat Transfer* was published in 1949 and remained in print for almost 50 years. Kern, known as a master of material presentation, delineated a procedure for establishing the impact of heat exchanger fouling on fin efficiency. It was at his suggestion that the first edition of *Extended Surface Heat Transfer* was written, and his wife, Natalie, has remained one of the dearest friends of the first author. The second basic obligation, which is common to all three authors, is to our first editor, Frank Cerra. While with John Wiley & Sons, he was able to provide considerable guidance on publishing economics and production.

We are indebted to Detlev Kroger at the University of Stellenbosch in South Africa, who provided a catalog of heat transfer correlations for airflow on the outside of finned tubes, and to Peter Heggs of the University of Manchester in the United Kingdom, for his review of the chapters on heat exchangers. We also acknowledge, with thanks, the help and general guidance of Avram Bar-Cohen (University of Minnesota), Dave Snider (University of South Florida), Raj Manglik (University of Cincinnati), Virgil Lunardini (U.S. Army Corps of Engineers, Cold Regions Research and Engineering Laboratory), Paul Marto (U.S. Naval Postgraduate School), and Adrian Bejan (Duke University). Bob Argentieri and Bob Hilbert at Wiley handled all editorial and production details. Finally, the second author wishes to acknowledge the support of his children, Fahad, Sheza, and Kashif.

Allan D. Kraus
Abdul Aziz
James Welty

CHAPTER 1

CONVECTION WITH SIMPLIFIED CONSTRAINTS

1.1 INTRODUCTION

Three quarters of a century ago, a paper by Harper and Brown (1922) appeared as an NACA report. It was an elegant piece of work and appears to be the first really significant attempt to provide a mathematical analysis of the interesting interplay between convection and conduction in and upon a single extended surface. Harper and Brown called this a *cooling fin*, which later became known merely as a *fin*. It is most probable that Harper and Brown were the pioneers even though Jakob (1949) pointed out that published mathematical analyses of extended surfaces can be traced all the way back to 1789. At that time, Ingenhouss demonstrated the differences in thermal conductivity of several metals by fabricating rods, coating them with wax, and then observing the melting pattern when the bases of the rods were heated. Jakob also pointed out that Fourier (1822) and Despretz (1822, 1828a,b) published mathematical analyses of the temperature variation of the metal bars or rods.

Although these ancient endeavors may have been quite significant at the time they were written, it appears that the Harper–Brown work should be considered as the forerunner of what has become a burgeoning literature that pertains to a very significant subject area in the general field of heat transfer.

The NACA report of Harper and Brown was inspired by a request from the Engineering Division of the U.S. Army and the U.S. Bureau of Standards in connection with the heat-dissipating features of air-cooled aircraft engines. It is interesting to note that this request came less than halfway through the time period between the first flight of the Wright Brothers at Kitty Hawk and the actual establishment of the U.S. Air Force. The work considered longitudinal fins of rectangular profile and trapezoidal profile (which Harper and Brown called *wedge-shaped fins*) and radial fins of rectangular profile (which Harper and Brown called *circumferential fins*). It also introduced the concept of *fin efficiency*, although the expression employed by Harper and Brown was called the *fin effectiveness*. From this modest, yet masterful beginning, the analysis and evaluation of the performance of individual components of extended surface and arrays of extended surface where individual components are assembled into complicated configurations has become an art.

Harper and Brown (1922) provided thorough analytical solutions for the two-dimensional model for both rectangular and wedge-shaped longitudinal fins and the circumferential fin of uniform thickness. The solutions culminated in expressions for the fin efficiency (called the *effectiveness*) or in correction factors that adjusted the efficiency of the rectangular profile longitudinal fin. They concluded that the use of a one-dimensional model was sufficent and they proposed that the tip heat loss could be accounted for through the use of a corrected fin height, which increases the fin height by a value equal to half of the fin thickness. Lost in the shuffle, however, was the interesting observation that with dx as the differential element of fin height, the differential face surface area of the element is $dx/\cos\kappa$, where κ is the taper angle, which is 90° for rectangular profile straight and circumferential fins as well as for spines of constant cross section.

Schmidt (1926) covered the three profiles considered by Harper and Brown from the standpoint of material economy. He stated that the least material is required for given conditions if the fin temperature gradient (from base to tip) is linear, and he showed how the fin thickness of each type of fin must vary to produce this result. Finding, in general, that the calculated shapes were impractical to manufacture, he proceeded to show the optimum dimensions for longitudinal and radial fins of constant thickness (rectangular profile) and the longitudinal fin of trapezoidal profile. He also considered the longitudinal fin of triangular profile as the case of the longitudinal fin of trapezoidal profile with zero tip thickness.

The case of integral pin fins of different profiles was considered by Bueche and Schau (1936). They determined for conical fins that the heat dissipation was a function of the Biot modulus based on the base radius and aspect ratio of fin height to base radius. They also showed that a weight optimization could be effected.

Murray (1938) considered the problem of the radial fin of uniform thickness (the radial fin of rectangular profile) presenting equations for the temperature gradient and effectiveness under conditions of a symmetrical temperature distribution around the base of the fin. He also proposed that the analysis of extended surface should be based on a set of assumptions that have been known since 1945 as the Murray–Gardner assumptions. These assumptions are deemed to be of considerable importance because their elimination, either one at a time or in combination, provided a series of paths for subsequent investigators to follow.

A stepwise procedure for calculating the temperature gradient and efficiency of fins whose thickness varies in any manner was presented by Hausen (1940). The temperature gradient in conical and cylindrical spines was determined by Focke (1942) who, like Schmidt, showed how the spine thickness must vary to keep the material required to a minimum. He too, found the resulting shape impractical and went on to determine the optimum cylindrical and conical spine dimensions.

Avrami and Little (1942) derived equations for the temperature gradient in thick bar fins and showed under what conditions fins might act as insulators on the base or prime surface. Carrier and Anderson (1944) discussed straight fins of constant thickness, radial fins of constant thickness, and radial fins of constant cross-sectional area, presenting equations for the fin efficiency of each. However, in the latter two cases, the efficiencies are given in the form of an infinite series.

Gardner (1945), in a giant leap forward, derived general equations for the temperature excess profile and fin efficiency for any form of extended surface for which the Murray–Gardner assumptions are applicable and whose thickness varies as some power of the distance measured along an axis normal to the base or prime surface (the fin height). He proposed the profile function (his nomenclature)

$$y = y_b \left(\frac{x}{x_b} \right)^{(1-2n)/(1-n)}$$

for the straight or longitudinal fins,

$$y = y_b \left(\frac{x}{x_b} \right)^{(1-2n)/(2-n)}$$

for spines and

$$y = y_b \left(\frac{x}{x_b} \right)^{-2n/(1-n)}$$

for radial or circumferential (radial) fins.

These equations depend on the assignment of some number to n; for example, the straight fin of rectangular profiles results when $n = 0$ in the first of these. This also serves to show that in Gardner's profile functions, the positive sense of the height coordinate x is in a direction from fin tip to fin base.

With the foregoing profile functions in hand and working with a general differential equation that he derived, Gardner was able to provide solutions for the temperature excess profile in terms of modified Bessel functions. For n equal to zero or an integer,

$$\theta = \theta_b \left(\frac{u}{u_b} \right)^n \frac{I_n(u) + \beta K_n(u)}{I_n(u_b) + \beta K_n(u_b)}$$

and for n equal to a fraction,

$$\theta = \theta_b \left(\frac{u}{u_b} \right)^n \frac{I_n(u) + \beta I_{-n}(u)}{I_n(u_b) + \beta I_{-n}(u_b)}$$

where u depends on the type of fin or spine and where

$$\beta = \frac{I_{n-1}(u_a)}{K_{n-1}(u_a)}$$

if n is equal to zero or an integer and

$$\beta = -\frac{I_{n-1}(u_a)}{I_{1-n}(u_a)}$$

if n is equal to a fraction.

For straight (longitudinal) fins

$$u = 2(1-n)\left(\frac{x}{x_b}\right)^{1/2(1-n)}\sqrt{\frac{h}{ky_b}}x_b$$

for spines

$$u = \frac{2\sqrt{2}\,(2-n)}{3}\left(\frac{x}{x_b}\right)^{3/2(2-n)}\sqrt{\frac{h}{ky_b}}x_b$$

and for radial fins

$$u = (1-n)\left(\frac{x}{x_b}\right)^{1/(1-n)}\sqrt{\frac{h}{ky_b}}x_b$$

The fin efficiency, defined as the ratio of the heat transferred from the fin to the heat that would be transferred by the fin if its thermal conductivity were infinite (if the entire fin were to operate at the base temperature excess), was provided by Gardner for all the fins that he considered. Gardner designated the fin efficiency η, and for n equal to zero or an integer,

$$\eta = \frac{2(1-n)}{u_b[1-(u_a/u_b)^{2(1-n)}]}\,\frac{I_{n-1}(u_b)-\beta K_{n-1}(u_b)}{I_n(u_b)-\beta K_n(u_b)}$$

and for n equal to a fraction,

$$\eta = \frac{2(1-n)}{u_b[1-(u_a/u_b)^{2(1-n)}]}\,\frac{I_{n-1}(u_b)-\beta I_{1-n}(u_b)}{I_n(u_b)-\beta I_{-n}(u_b)}$$

Graphs were provided that plotted the efficiency as a function of a parameter that embraced the fin dimensions and thermal properties. Two of the graphs (for straight fins and spines) are reproduced here as Figs. 1.1 and 1.2.

Gardner also pointed out that the terms *fin efficiency* and *fin effectiveness* had not been used consistently in the English literature. He redefined the fin effectiveness as the ratio of the heat transferred through the base of the fin to the heat transferred through the same prime or base surface area if the fin were not present. He also provided a relationship to permit the conversion from fin efficiency to fin effectiveness.

It is felt that the Gardner paper is remarkable for several reasons. First and probably foremost is the fact that he reemphasized the concept of the fin efficiency, thereby creating an itch that literally thousands of equipment designers have been scratching ever since. Moreover, it appears that Gardner was one of the first to demonstrate the use of applied mathematics to yield concepts that engineers could use to build

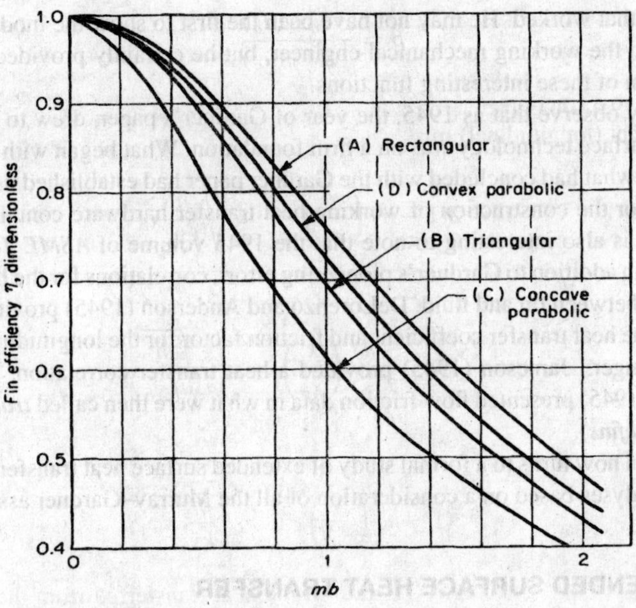

Figure 1.1 Gardner's graph for the efficiency of straight (longitudinal) fins. (Reproduced from *Trans. ASME*, 67, 1945.)

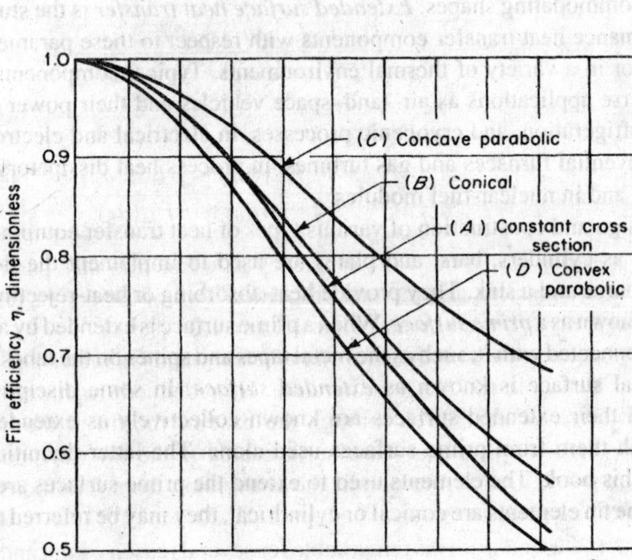

Figure 1.2 Gardner's graph for the efficiency of spines. (Reproduced from *Trans. ASME*, 67, 1945.)

equipment that worked. He may not have been the first to show the modified Bessel functions to the working mechanical engineer, but he certainly provided an intense reexposition of these interesting functions.

One may observe that as 1945, the year of Gardner's paper, drew to a close, the extended surface technology was on a firm foundation. What began with Harper and Brown and what had concluded with the Gardner paper had established useful design equations for the construction of working heat transfer hardware containing finned surfaces. It is also interesting to note that the 1945 volume of *ASME Transactions* contained, in addition to Gardner's pioneering effort, correlations for the heat transfer coefficient between fin and fluid. DeLorenzo and Anderson (1945) provided a correlation for the heat transfer coefficient and friction factor for the longitudinal fin–axial flow exchanger.[1] Jameson (1945) provided a heat transfer correlation, and Gunter and Shaw (1945) presented flow friction data in what were then called *transverse fins* (now *radial fins*).

Attention now turns to a formal study of extended surface heat transfer and begins with fin analyses based on a consideration of all the Murray–Gardner assumptions.

1.2 EXTENDED SURFACE HEAT TRANSFER

A growing number of engineering disciplines are concerned with energy transitions requiring the rapid movement of heat. They produce an expanding demand for high-performance heat transfer components with progressively smaller weights, volumes, costs, or accommodating shapes. *Extended surface heat transfer* is the study of these high-performance heat transfer components with respect to these parameters and of their behavior in a variety of thermal environments. Typical components are found in such diverse applications as air–land–space vehicles and their power sources, in chemical, refrigeration, and cryogenic processes, in electrical and electronic equipment, in convential furnaces and gas turbines, in process heat dissipators and waste heat boilers, and in nuclear-fuel modules.

In the design and construction of various types of heat transfer equipment, simple shapes such as cylinders, bars, and plates are used to implement the flow of heat between a source and a sink. They provide heat-absorbing or heat-rejecting surfaces, and each is known as a *prime surface*. When a prime surface is extended by appendages intimately connected with it, such as the metal tapes and spines on the tubes in Fig. 1.3, the additional surface is known as *extended surface*. In some disciplines, prime surfaces and their extended surfaces are known collectively as extended surfaces to distinguish them from prime surfaces used alone. The latter definition prevails throughout this book. The elements used to extend the prime surfaces are known as *fins*. When the fin elements are conical or cylindrical, they may be referred to as *spines* or *pegs*.

The demands for aircraft, aerospace, gas turbine, air conditioning, and cryogenic auxilliaries have placed particular emphasis on the compactness of the heat exchanger

[1]What Kern (1950) referred to as the *double pipe*.

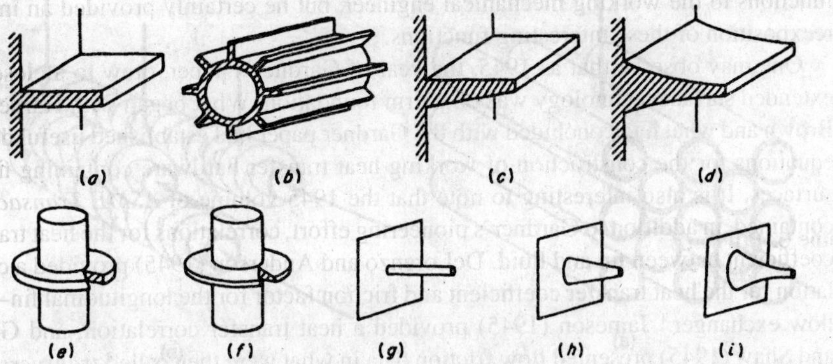

Figure 1.3 Some typical examples of extended surfaces: *(a)* longitudinal fin of rectangular profile; *(b)* cylindrical tube equipped with fins of rectangular profile; *(c)* longitudinal fin of trapezoidal profile; *(d)* longitudinal fin of parabolic profile; *(e)* cylindrical tube equipped with radial fin of rectangular profile; *(f)* cylindrical tube equipped with radial fin of trapezoidal profile; *(g)* cylindrical spine; *(h)* truncated conical spine; *(i)* truncated parabolic spine.

surface, particularly on those surfaces that induce small pressure gradients in the fluids circulated through them. Several are shown in Fig. 1.4. *Compactness* refers to the ratio of heat transfer surface per unit of exchanger volume.

An early definition by Kays and London (1950) established a compact exchanger element as one containing in excess of 245 m^2 per cubic meter of exchanger. Compact exchanger elements have been available with over 4100 m^2 per cubic meter compared with 65 to 130 m^2 per cubic meter for conventional heat exhangers with $\frac{5}{8}$- to 1-in. tubes. Many compact heat exchanger elements consist of prime surface plates or tubes separated by plates, bars, or spines, which also act as fins. As shown in Fig. 1.4*d*, each of the fins may be treated as a single fin with fin height equal to half of the separation plate spacing and with the separation plate acting as the prime surface. Thus, the compact heat exhanger is considered as another form of extended surface.

1.2.1 Fin Efficiency

It can be shown quite readily that when a fin and its prime surface are exposed to a uniform thermal environment, a unit of fin surface will be less effective than a unit of prime surface. Consider the plate with a longitudinal fin of rectangular cross section shown in Fig. 1.5. Let the inner plate surface remove heat from a source with a uniform heat transfer coefficient and temperature T_1, and let the outer plate and fin surfaces reject it to colder surroundings with a uniform heat transfer coefficient and temperature T_s. The colder surface of the plate is at some intermediate temperature T_p, and the heat from the source leaves the plate because of the temperature potential, $T_p - T_s$. Similarly, the fin surface is at some temperature, T, and the heat leaves the fin because of the temperature potential, $T - T_s$. The heat enters the fin through its base, where it joins the plate and moves continuously through it by conduction. In

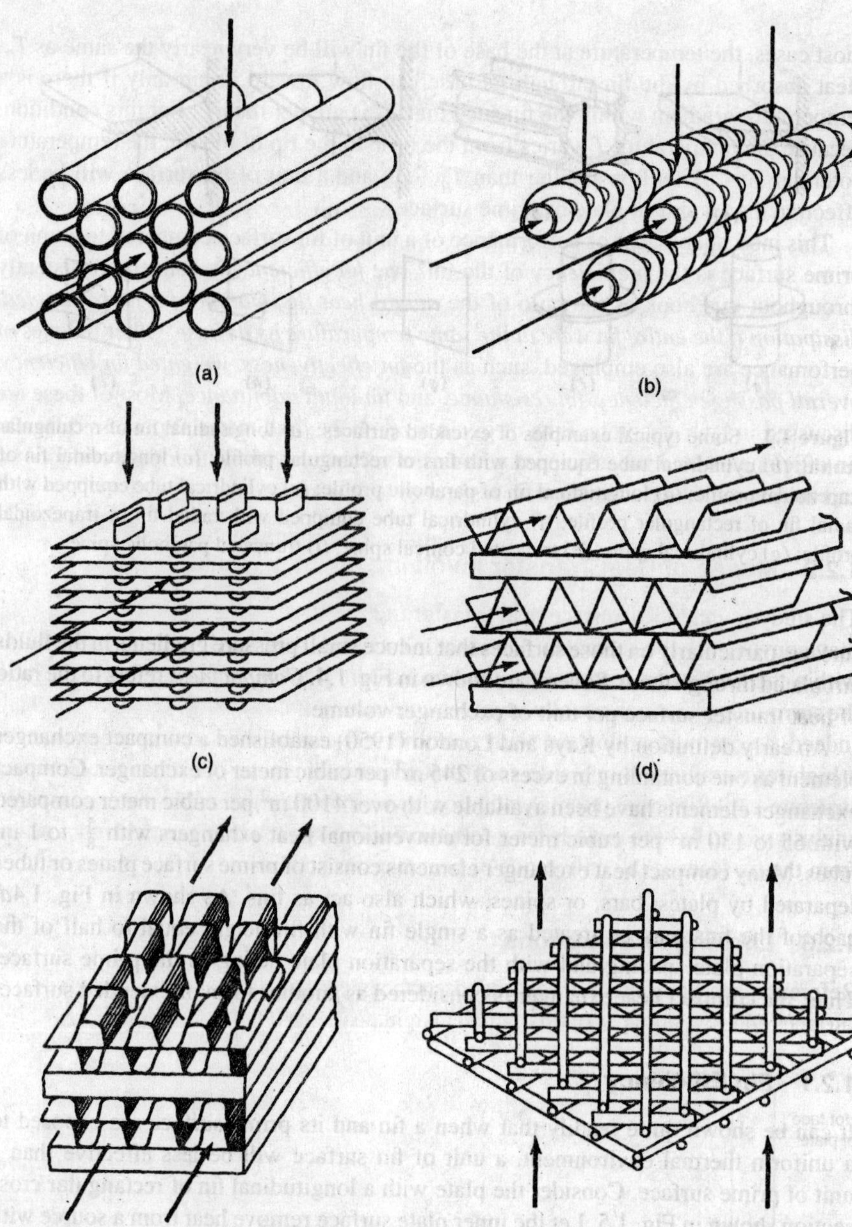

Figure 1.4 Some typical examples of compact heat exchanger surfaces: *(a)* cylindrical tube; *(b)* cylindrical tube with radial fins; *(c)* flat tube with continuous fins; *(d)* plate fin; *(e)* offset plate fin; *(f)* crossed rod matrix. (From W. M. Kays and A. L. London, *Compact Heat Exchangers*, 3rd ed., McGraw-Hill, New York, 1984, by permission.)

most cases, the temperature at the base of the fin will be very nearly the same as T_p. Heat absorbed by the fin through its base can flow toward its tip only if there is a temperature gradient within the fin such that T_p is greater than T. For this condition, because the temperature T varies from the base to the tip of the fin, the temperature potential $T - T_s$ will be smaller than $T_p - T_s$ and a unit of fin surface will be less effective than a unit of plate or prime surface.

This inescapable loss of performance of a unit of fin surface compared to a unit of prime surface is the *inefficiency* of the fin. The *fin efficiency* is defined consistently throughout this book as the ratio of the *actual heat dissipation of a fin to its ideal dissipation if the entire fin were at the same temperature as its base*. Other indexes of perfomance are also employed, such as the *fin effectiveness, weighted fin efficiency, overall passage efficiency, fin resistance*, and *fin input admittance*. Most of these are discussed in later chapters. Fins of given size, shape, and material possess different fin efficiencies, and the efficiency of any fin will vary with its thermal conductivity and the mode of heat transfer with respect to its environment.

1.2.2 Modes of Heat Transfer Involving Fins and Surroundings

The study of extended surface heat transfer in most cases comprises two factors that may conveniently be separated. One factor considers only the movement of the heat within the fin by conduction. The other considers how the fin exchanges heat with the surroundings, which usually involves convection and radiation singly or together. Indeed, a contrived hollow fin with poor thermal conductivity could be exposed to a high temperature on one side such that the movement of the heat within the fin structure could involve internal radiation as well as conduction. In such a case, the analysis would have to include the internal radiation as well. In this chapter convection from the fin faces is the only mode of heat transfer considered.

1.2.3 Limiting Assumptions

Reference has been made to extended surfaces comprising several types of prime surfaces and several types of fins. Much insight on the heat flows, temperature profiles,

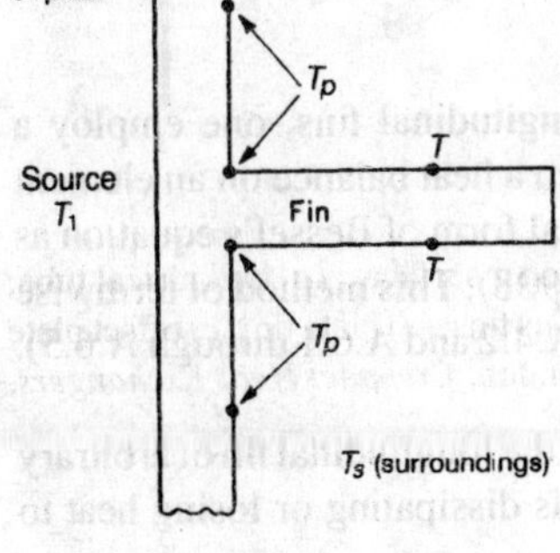

Figure 1.5 Temperature–potential differences in fins.

efficiencies, and optimization of fin parameters can be obtained from analysis of the three fundamental fin geometries shown in Fig. 1.3: longitudinal fins, radial fins, and spines.

Fins of various geometries and thermal conductivities respond differently to identical and uniform heat sources and sinks. Similarly, there are numerous ways in which the temperatures and heat transfer coefficients of sources and sinks may vary. Important to the analysis of fin geometries are the constraints or assumptions that are employed to define and limit the problem and often to simplify its solution. The analysis of the three fundamental fin geometries provided in this chapter employ the assumptions proposed by Murray (1938) and Gardner (1945). These limiting assumptions, which are almost always referred to as the *Murray–Gardner assumptions*, are:

1. The heat flow in the fin and its temperatures remain constant with time.
2. The fin material is homogeneous, its thermal conductivity is the same in all directions, and it remains constant.
3. The convective heat transfer coefficient on the faces of the fin is constant and uniform over the entire surface of the fin.
4. The temperature of the medium surrounding the fin is uniform.
5. The fin thickness is small, compared with its height and length, so that temperature gradients *across* the fin thickness and heat transfer from the edges of the fin may be neglected.
6. The temperature at the base of the fin is uniform.
7. There is no contact resistance where the base of the fin joins the prime surface.
8. There are no heat sources within the fin itself.
9. The heat transferred through the tip of the fin is negligible compared with the heat leaving its lateral surface.
10. Heat transfer to or from the fin is proportional to the temperature excess between the fin and the surrounding medium.

1.3 LONGITUDINAL FINS

1.3.1 Generalized Differential Equation

Gardner (1945) proposed that for the analysis of longitudinal fins, one employ a generalized fin. The differential equation resulting from a heat balance on an element of fin height can be compared termwise with the general form of Bessel's equation as given originally by Douglass in Sherwood and Reed (1938). This method of termwise comparison is demonstrated in Appendix A (Sections A.4.2 and A.6.1 through A.6.5).

Terminology and Coordinate System. Consider the longitudinal fin of arbitrary profile displayed in Fig. 1.6*a* and assume that the fin is dissipating or losing heat to its surroundings. Note that the dimension x pertains to the *height coordinate* which has its origin at the *fin tip* and has a positive orientation from fin tip to *fin base*. The fin profile shown in Fig. 1.6*b* is confined by two curves which are almost always symmetrical, $y = f_2(x)$ and $y = -f_2(x)$, so that the *fin thickness* is $\delta(x) = 2f_2(x)$. The *fin cross section* shown in Fig. 1.6*c* is $A(x) = f_1(x) = 2Lf_2(x)$, where L is the *fin length*, which is directed into the plane of Fig. 1.6. The *edges* of the fin are bounded by the fin profile curves, $\pm f_2(x)$, and the *fin faces* are the *lateral surfaces* of the fin bounded by one of the fin profile curves and the fin length. The fin base is shown by the crosshatched area in Fig. 1.6*c*.

Properties at the fin base, located at $x = b$, are designated by a subscript b. For example, θ_b, q_b, and T_b represent the temperature excess, heat flow, and temperature at the fin base, respectively. Properties at the fin tip are designated by a subscript a (θ_a, q_a, and T_a). In general, the tip is located at $x = a$, but for the longitudinal fins and spines discussed in this chapter, the tip and the origin of the height coordinate is specifically located at $x = a = 0$.

It is customary in extended surface analysis to deal with a *temperature excess*. Let $\theta(x)$, a function of the height coordinate x, be the temperature difference or excess between a point on the fin surface and the surroundings. Thus

$$\theta(x) = T(x) - T_s$$

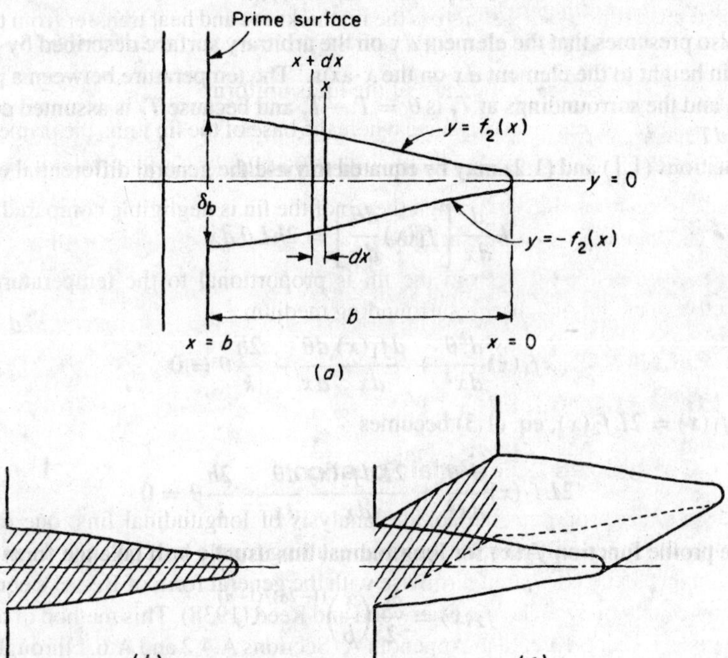

Figure 1.6 Longitudinal fin with arbitrary profile: *(a)* coordinate system; *(b)* fin profile area; *(c)* fin cross-sectional area.

Generalized Differential Equation. The differential equation for the fin temperature profile is formulated from a consideration of the steady-state heat balance over the differential element of height dx. This differential element is bounded by planes parallel to the fin base at x and $x+dx$ and by the confining profile curves, $y = \pm f_2(x)$.

If the fin surface temperature is $T(x)$, so that at dx the temperature is T and k is the fin thermal conductivity, the difference between the heat entering the element by conduction at $x + dx$ and the heat leaving the element by conduction at x is

$$dq = k\frac{d}{dx}\left[f_1(x)\frac{dT}{dx}\right] \tag{1.1}$$

To comply with the assumption of a steady state, the difference in heat conduction into and out of the element dx, as described by eq. (1.1), must be offset by some mode of heat dissipation from the exposed lateral surface of the fin. If the heat is dissipated by convection to the surrounding medium, P is the fin perimeter and h is the convective coefficient,

$$dq = hP(T - T_s)\,dx = 2h[L + f_2(x)](T - T_s)$$

However, by Murray–Gardner assumption 5, which states that the fin thickness must be small in comparison to its height and length, $L >> 2f_2(x)$. Hence

$$dq = 2hL(T - T_s)\,dx \tag{1.2}$$

This also presumes that the element dx on the arbitrary surface described by $f_2(x)$ is equal in height to the element dx on the x-axis.[2] The temperature between a point on the fin and the surroundings at T_s is $\theta = T - T_s$ and because T_s is assumed constant, $d\theta = dT$.

Equations (1.1) and (1.2) may be equated to yield the general differential equation

$$k\frac{d}{dx}\left[f_1(x)\frac{d\theta}{dx}\right] = 2hL\theta\,dx$$

or

$$f_1(x)\frac{d^2\theta}{dx^2} + \frac{df_1(x)}{dx}\frac{d\theta}{dx} - \frac{2h}{k}\theta = 0 \tag{1.3}$$

With $f_1(x) = 2Lf_2(x)$, eq. (1.3) becomes

$$2Lf_2(x)\frac{d^2\theta}{dx^2} + \frac{2Ldf_2(x)}{dx}\frac{d\theta}{dx} - \frac{2h}{k}\theta = 0 \tag{1.4}$$

The profile function $f_2(x)$ for longitudinal fins usually will take the form

$$f_2(x) = \frac{\delta_b}{2}\left(\frac{x}{b}\right)^{(1-2n)/(1-n)} \tag{1.5}$$

[2] Gardner (1945) pointed out that this is generally valid for thin fins and spines because the square of the slope of the fin sides is negligible compared with unity. This is what has come to be called the *length of arc assumption.*

where δ_b is the fin thickness at its base. The particular solution may be obtained by substituting the *boundary conditions* into the general solution to eliminate the arbitrary constants

$$\theta(x = b) = \theta_b \tag{1.6a}$$

and

$$\left.\frac{d\theta}{dx}\right|_{x=0} = 0 \tag{1.6b}$$

The particular solution may also be obtained by substituting the *initial conditions*

$$\theta(x = b) = \theta_b \tag{1.6c}$$

and

$$q(x = b) = q_b = -kA\left.\frac{d\theta}{dx}\right|_{x=b} \tag{1.6d}$$

into the general solution to eliminate the arbitrary constants.

1.3.2 Longitudinal Fin of Rectangular Profile

For the longitudinal fin of rectangular profile displayed with its terminology and coordinate system in Fig. 1.7, the exponent on the general fin profile of eq. (1.5) satisfies the geometry when $n = \frac{1}{2}$. The profile function for this fin then becomes

$$f_2(x) = \frac{\delta_b}{2} = \frac{\delta}{2}$$

because $\delta_b = \delta$ and

$$\frac{df_2(x)}{dx} = 0$$

When these are substituted into eq. (1.4), the governing differential equation becomes

$$\frac{d^2\theta}{dx^2} - \frac{2h}{k\delta}\theta = 0 \tag{1.7}$$

which is an ordinary second-order differential equation with constant coefficients. The general solution is

$$\theta = C_1 e^{mx} + C_2 e^{-mx} \tag{1.8}$$

where m is referred to as the *fin performance factor*,

$$m = \left(\frac{2h}{k\delta}\right)^{1/2}$$

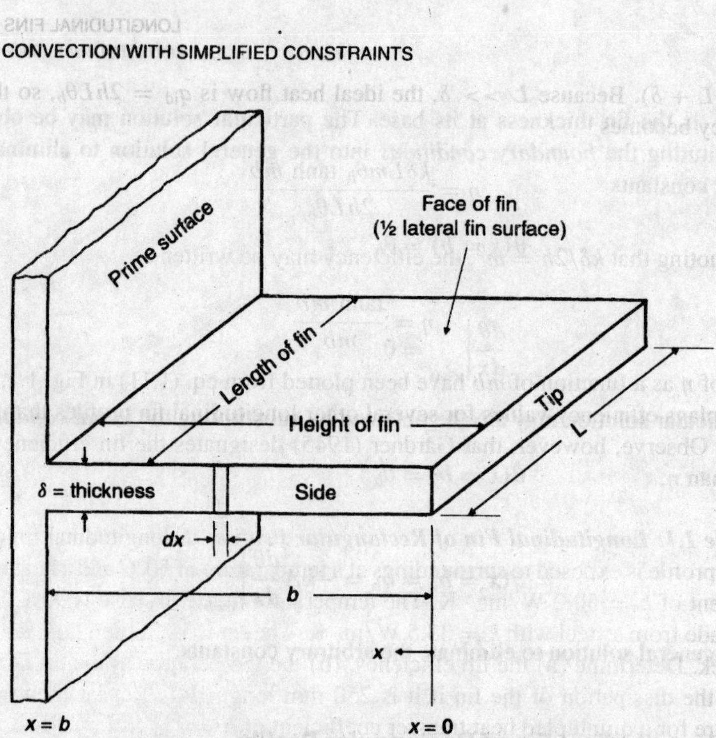

Figure 1.7 Terminology and coordinate system for the longitudinal fin of rectangular profile.

Application of the boundary conditions of eqs. (1.6*a*) and (1.6*b*) permits evaluation of the arbitrary constants C_1 and C_2. This evaluation yields the point-to-point temperature excess, $\theta(x) = T(x) - T_s$:

$$\theta(x) = \frac{\theta_b \cosh mx}{\cosh mb} \tag{1.9}$$

Because the orientation of the height coordinate x is opposed to the direction of the heat flow in the fin, the heat flow through the base of the fin q_b is obtained from

$$q_b = kA\frac{d\theta}{dx}\bigg|_{x=b}$$

The fin cross-sectional area is equal to $A = \delta L$, so that in using the derivative of eq. (1.9) evaluated at $x = b$,

$$q_b = \frac{k\delta L m\theta_b \sinh mb}{\cosh mb}$$

or

$$q_b = k\delta L m\theta_b \tanh mb \tag{1.10}$$

For the longitudinal fin of rectangular profile, the actual heat flow is given by eq. (1.10). The ideal heat flow is $q_{id} = hP\theta_b$, where P is the perimeter of the fin,

$P = 2(L + \delta)$. Because $L >> \delta$, the ideal heat flow is $q_{id} = 2hL\theta_b$, so that the efficiency becomes

$$\eta = \frac{k\delta L m\theta_b \tanh mb}{2hL\theta_b}$$

and by noting that $k\delta/2h = m^2$, the efficiency may be written as

$$\eta = \frac{\tanh mb}{mb} \tag{1.11}$$

Values of η as a function of mb have been plotted from eq. (1.11) in Fig. 1.1, which also displays efficiency values for several other longitudinal fin profiles that will be studied. Observe, however, that Gardner (1945) designates the fin efficiency by ϕ rather than η.

Example 1.1: Longitudinal Fin of Rectangular Profile. A longitudinal fin of rectangular profile is exposed to surroundings at a temperature of 50°C and a heat transfer coefficient of $h = 50.2\ \mathrm{W/m^2 \cdot K}$. The temperature at the fin base is 90°C and the fin is made from a steel with $k = 33.5\ \mathrm{W/m \cdot K}$. The fin is 101.6 mm high and 9.525 mm thick. Determine (**a**) the fin efficiency, (**b**) the temperature at the tip of the fin, and (**c**) the dissipation of the fin if it is 250 mm long. (**d–f**) Repeat the foregoing procedure for a quintupled heat transfer coefficient of $h = 251\ \mathrm{W/m^2 \cdot K}$.

SOLUTION. For the surroundings with $h = 50.2\ \mathrm{W/m^2 \cdot K}$,

$$\theta_b = 90 - 50 = 40°\mathrm{C} \qquad \delta = 9.525/1000 = 9.525 \times 10^{-3}\ \mathrm{m}$$

$$m = \left(\frac{2h}{k\delta}\right)^{1/2} = \left[\frac{(2)(50.2)}{(33.5)(9.525 \times 10^{-3})}\right]^{1/2} = 17.738\ \mathrm{m^{-1}}$$

$$b = 101.6/1000 = 0.1016\ \mathrm{m} \qquad mb = (17.738)(0.1016) = 1.802$$

(**a**) From eq. (e.11),

$$\eta = \frac{\tanh mb}{mb} = \frac{\tanh 1.802}{1.802} = \frac{0.947}{1.802} = 0.526$$

(**b**) The tip temperature of the fin is determined from eq. (1.9) at $x = 0$, where $\theta(x = 0) = \theta_a$:

$$\theta_a = \frac{\theta_b \cosh mx}{\cosh mb} = \frac{40(\cosh 0)}{\cosh 1.802} = \frac{(40)(1.00)}{3.114} = 12.8°\mathrm{C}$$

so that at $x = 0$, where $T(x = 0) = T_a$,

$$T_a = \theta_a + T_s = 12.8 + 50 = 62.8°\mathrm{C}$$

(**c**) The heat dissipated by the fin is calculated from eq. (1.10):

$$q_b = k\delta L m\theta_b \tanh mb$$

$$= (33.5)(9.525 \times 10^{-3})(0.25)(17.738)(40)(0.947)$$

$$= 53.6 \text{ W}$$

The heat transferred can also be obtained from the ideal heat dissipation and the efficiency determined in part (a). With the surface area of the fin, $S = 2bL$,

$$q_{\text{id}} = 2bLh\theta_b = (2)(0.1016)(0.25)(50.2)(40) = 102.0 \text{ W}$$

the heat dissipated is

$$q_b = \eta q_{\text{id}} = (0.526)(102.0) = 53.6 \text{ W}$$

(d) With $h = 251 \text{ W/m}^2 \cdot \text{K}$

$$m = \left(\frac{2h}{k\delta}\right)^{1/2} = \left[\frac{(2)(251)}{(33.5)(9.525 \times 10^{-3})}\right]^{1/2} = 39.664 \text{ m}^{-1}$$

$$mb = (39.664)(0.1016) = 4.030$$

and from eq. (1.11),

$$\eta = \frac{\tanh mb}{mb} = \frac{\tanh 4.030}{4.030} = \frac{0.999}{4.030} = 0.248$$

(e) The tip temperature is determined from eq. (1.9):

$$\theta_a = \frac{\theta_b \cosh mx}{\cosh mb} = \frac{40(\cosh 0)}{\cosh 4.030} = \frac{40}{28.139} = 1.4°\text{C}$$

so that at $x = 0$, where $T(x = 0) = T_a$,

$$T_a = \theta_a + T_s = 1.4 + 50 = 51.4°\text{C}$$

(f) The heat dissipated by the fin is calculated from eq. (1.10) from parts (a), (c), and (d):

$$q_b = k\delta Lm\theta_b \tanh mb$$

$$= (33.5)(9.525 \times 10^{-3})(0.25)(39.664)(40)(0.999)$$

$$= 126.5 \text{ W}$$

The heat transferred can also be obtained from the ideal heat dissipation and t[illegible] efficiency determined in part (d). With

$$q_{\text{id}} = 2bLh\theta_b = (2)(0 \text{[illegible]} 25)(251)(40) = 510.0 \text{ W}$$

the heat dissipated is

$$q_b = \eta q_{\text{id}} = (0.248)\text{[illegible]} 6.5 \text{ W}$$

1.3.3 Longitudinal Fin of Triangular Profile

For the longitudinal fin of triangular profile shown in Fig. 1.8, it is noted that the exponent on the general fin profile of eq. (1.5) satisfies the geometry when $n = 0$. The profile function for this fin then becomes

$$f_2(x) = \frac{\delta_b}{2}\frac{x}{b}$$

and

$$\frac{df_2(x)}{dx} = \frac{\delta_b}{2b}$$

When these are substituted into eq. (1.4), the governing differential equation for the temperature excess, $\theta(x) = T(x) - T_s$, becomes

$$x\frac{d^2\theta}{dx^2} + \frac{d\theta}{dx} - m^2 b\theta = 0 \tag{1.12}$$

where again, $m = (2h/k\delta)^{1/2}$.

Equation (1.12) is an ordinary second-order differential equation with variable coefficients. As shown in Section A.6.1, its general solution is

$$\theta(x) = C_1 I_0(2m\sqrt{bx}) + C_2 K_0(2m\sqrt{bx}) \tag{1.13}$$

and it can be observed that to have a finite temperature excess at the fin tip where $x = 0$, the arbitrary constant C_2 must equal zero because $K_0(0)$ is unbounded. This leaves

$$\theta(x) = C_1 I_0(2m\sqrt{bx})$$

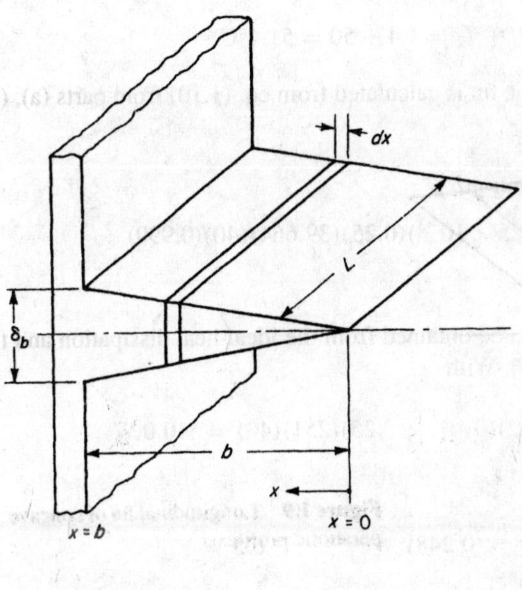

Figure 1.8 Longitudinal fin of triangular profile.

C_1 is evaluated using the boundary condition of eq (1.6a). The particular solution is

$$\theta(x) = \frac{\theta_b I_0(2m\sqrt{bx})}{I_0(2mb)} \tag{1.14}$$

The heat disspated by the fin must equal the heat flow through the base of the fin and is obtained using eq. (1.14), noting that $A = \delta_b L$, writing the Bessel function series expansion for $I_0(2m\sqrt{bx})$, differentiating term by term, and evaluating the derivative at $x = b$. The result is

$$q_b = kA\frac{dT}{dx}\bigg|_{x=b} = \frac{2hL\theta_b I_1(2mb)}{mI_0(2mb)} \tag{1.15}$$

The fin efficiency is the ratio of the actual heat flow given by eq. (1.15) to the ideal heat flow, $q_{id} = 2hbL\theta_b$:

$$\eta = \frac{2hL\theta_b[I_1(2mb)/mI_0(2mb)]}{2hbL\theta_b} = \frac{I_1(2mb)}{(mb)I_0(2mb)} \tag{1.16}$$

Values of η as a function of mb have been plotted from eq. (1.16) in Fig. 1.1.

1.3.4 Longitudinal Fin of Concave Parabolic Profile

For the longitudinal fin of concave parabolic profile shown in Fig. 1.9, it is noted that the exponent on the general fin profile of eq. (1.5) satisfies the geometry when $n = \infty$. The profile function for this fin then becomes

$$f_2(x) = \frac{\delta_b}{2}\left(\frac{x}{b}\right)^2$$

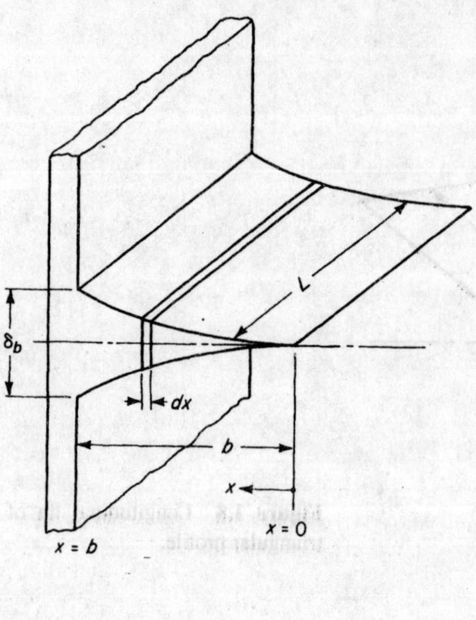

Figure 1.9 Longitudinal fin of concave parabolic profile.

and

$$\frac{df_2(x)}{dx} = \frac{\delta_b}{b}\frac{x}{b}$$

When these are substituted into eq. (1.4), the governing differential equation for the temperature excess, $\theta(x) = T(x) - T_s$, becomes

$$x^2\frac{d^2\theta}{dx^2} + 2x\frac{d\theta}{dx} - m^2b^2\theta = 0 \tag{1.17}$$

where again, $m = (2h/k\delta)^{1/2}$.

Equation (1.17) is an ordinary second-order differential equation with variable coefficients. It is known as an *Euler equation* and its general solution is obtained by making the transformation $x = e^v$ or $v = \ln x$. Then

$$\frac{d\theta}{dx} = \frac{d\theta}{dv}\frac{dv}{dx} = \frac{1}{x}\frac{d\theta}{dv}$$

and

$$\frac{d^2\theta}{dx^2} = \frac{d\,[(1/x)(d\theta/dv)]}{dx} = -\frac{1}{x^2}\frac{d\theta}{dv} + \frac{1}{x}\frac{d(d\theta/dv)}{dx}$$

and after simplification,

$$\frac{d^2\theta}{dx^2} = -\frac{1}{x^2}\frac{d\theta}{dv} + \frac{1}{x^2}\frac{d^2\theta}{dv^2}$$

With these transformations in hand, eq. (1.17) becomes

$$x^2\left(\frac{1}{x^2}\frac{d^2\theta}{dv^2} - \frac{1}{x^2}\frac{d\theta}{dv}\right) + 2x\left(\frac{1}{x}\frac{d\theta}{dv}\right) - m^2b^2\theta = 0$$

Canceling common terms gives an ordinary differential equation with constant coefficients:

$$\frac{d^2\theta}{dv^2} + \frac{d\theta}{dv} - m^2b^2\theta = 0$$

which has as its solution

$$\theta = C_1e^{\alpha v} + C_2e^{\beta v}$$

or in terms of the independent variable x,

$$\theta(x) = C_1x^\alpha + C_2x^\beta \tag{1.18}$$

where

$$\alpha, \beta = -\tfrac{1}{2} \pm \tfrac{1}{2}(1 + 4m^2b^2)^{1/2}$$

The general solution may be written

$$\theta(x) = C_1 x^{\alpha} + \frac{C_2}{x^{1/\beta}}$$

and it can be observed that at $x = 0$, the temperature excess, $T - T_s$, will be unbounded unless $C_2 = 0$. Therefore,

$$\theta(x) = C_1 x^{\alpha}$$

and from a consideration of the temperature excess at the fin base where $x = b$, the particular solution is obtained as

$$\theta(x) = \theta_b \left(\frac{x}{b}\right)^{\alpha} \tag{1.19}$$

Heat flow through the base of the fin is obtained by differentiating eq. (1.19) and evaluating the derivative at $x = b$. Noting that $A = \delta_b L$, the result is

$$q_b = kA \left.\frac{d\theta}{dx}\right|_{x=b} = \frac{k\delta_b L\theta_b \alpha}{b}$$

or

$$q_b = \frac{k\delta_b L\theta_b}{2b}\left[-1 + \sqrt{1 + (2mb)^2}\right] \tag{1.20}$$

The expression for the fin efficiency results when eq. (1.20) is divided by the ideal heat flow, $q_{\text{id}} = 2hbL\theta_b$:

$$\eta = \frac{k\delta_b L\theta_b\left[-1 + \sqrt{1 + (2mb)^2}\right]}{(2b)(2hbL\theta_b)}$$

This may be simplified by multiplying the numerator and denominator by $-1 - \sqrt{1 + (2mb)^2}$ and noting that as $m^2 = 2h/k\delta_b$,

$$\eta = \frac{-1 + \sqrt{1 + (2mb)^2}}{2(mb)^2} \frac{-1 - \sqrt{1 + (2mb)^2}}{-1 - \sqrt{1 + (2mb)^2}}$$

so that

$$\eta = \frac{2}{1 + \sqrt{1 + (2mb)^2}} \tag{1.21}$$

Values of η as a function of mb have been plotted from eq. (1.21) in Fig. 1.1.

1.3.5 Longitudinal Fin of Convex Parabolic Profile

For the longitudinal fin of convex parabolic profile shown in Fig. 1.10, it is noted that the exponent on the general fin profile of eq. (1.5) satisfies the geometry when $n = \frac{1}{3}$. The profile function for this fin then becomes

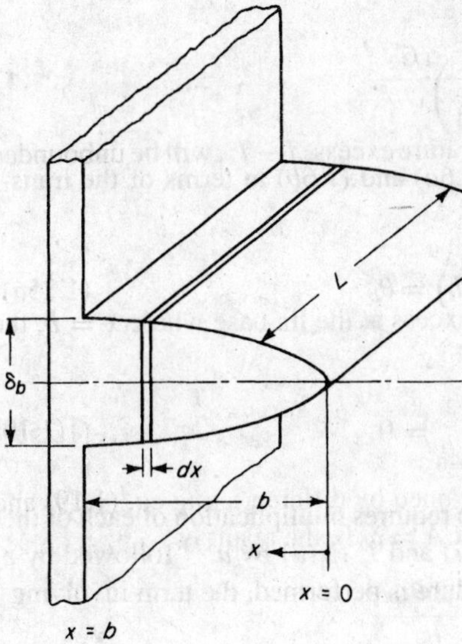

Figure 1.10 Longitudinal fin of convex parabolic profile.

$$f_2(x) = \frac{\delta_b}{2}\left(\frac{x}{b}\right)^{1/2}$$

and

$$\frac{df_2(x)}{dx} = \frac{\delta_b}{4\sqrt{bx}}$$

When these are substituted into eq. (1.4), the governing differential equation for the temperature excess, $\theta(x) = T(x) - T_s$, becomes

$$\sqrt{x}\frac{d^2\theta}{dx^2} + \frac{1}{2\sqrt{x}}\frac{d\theta}{dx} - m^2\sqrt{b}\,\theta = 0 \qquad (1.22)$$

where again, $m = (2h/k\delta)^{1/2}$.

A termwise comparison (see Section A.6.2) with the general Bessel equation leads to the general solution for the temperature excess, $\theta(x) = T(x) - T_s$:

$$\theta(x) = x^{1/4}\left[C_1 I_{1/3}\left(\tfrac{4}{3}mb^{1/4}x^{3/4}\right) + C_2 I_{-1/3}\left(\tfrac{4}{3}mb^{1/4}x^{3/4}\right)\right] \qquad (1.23)$$

Evaluation of the arbitrary constants in eq. (1.23) requires careful consideration of the infinite series expansions of the two Bessel functions.

Define a transformed variable u:

$$u \equiv \tfrac{4}{3}mb^{1/4}x^{3/4}$$

so that eq. (1.23) may be rewritten as

$$\theta(u) = \Omega u^{1/3}[C_1 I_{1/3}(u) + C_2 I_{-1/3}(u)] \qquad (1.24)$$

where

$$\Omega \equiv \left(\frac{3}{4mb^{1/4}}\right)^{1/3}$$

and where the boundary conditions of eqs. (1.6a) and (1.6b) in terms of the transformed variable u are

$$\theta\left(u = u_b = \tfrac{4}{3}mb\right) = \theta_b \tag{1.25a}$$

and

$$\left.\frac{d\theta}{du}\right|_{u=0} = 0 \tag{1.25b}$$

Use of the boundary condition of eq. (1.25b) requires multiplication of each of the terms of the infinite series expansion for $I_{1/3}(u)$ and $I_{-1/3}(u)$ by $u^{1/3}$ followed by a term-by-term differentiation. When this procedure is performed, the term involving

$$\frac{d}{du}\left[u^{1/3}I_{1/3}(u)\right]$$

becomes unbounded at $x = 0$. This requires that $C_1 = 0$. Then application of the boundary condition of eq. (1.25a) yields a value for C_2 such that the particular solution of eq. (1.24) in terms of u becomes

$$\theta(u) = \frac{\Omega u^{1/3}\theta_b I_{-1/3}(u)}{\Omega u_b^{1/3} I_{-1/3}(u_b)} = \theta_b\left(\frac{u}{u_b}\right)^{1/3}\frac{I_{-1/3}(u)}{I_{-1/3}(u_b)}$$

and in terms of x,

$$\theta(x) = \theta_b\left(\frac{x}{b}\right)^{1/4}\frac{I_{-1/3}\left(\frac{4}{3}mb^{1/4}x^{3/4}\right)}{I_{-1/3}\left(\frac{4}{3}mb\right)} \tag{1.26}$$

The heat flow through the base of the fin is obtained by differentiating eq. (1.26) term by term and evaluating the derivative at $x = b$. Again noting that $A = \delta bL$,

$$q_b = k\delta_b Lm\theta_b\frac{I_{2/3}\left(\frac{4}{3}mb\right)}{I_{-1/3}\left(\frac{4}{3}mb\right)} \tag{1.27}$$

The fin efficiency can be obtained by taking the ratio of eq. (1.27) to the ideal heat flow, $q_{\text{id}} = 2hbL\theta_b$:

$$\eta = \frac{q_b}{2hbL\theta_b} = \frac{mk\delta_b\theta_b I_{2/3}\left(\frac{4}{3}mb\right)}{2hb\theta_b I_{-1/3}\left(\frac{4}{3}mb\right)}$$

or

$$\eta = \frac{1}{mb}\frac{I_{2/3}\left(\frac{4}{3}mb\right)}{I_{-1/3}\left(\frac{4}{3}mb\right)} \tag{1.28}$$

Values of η as a function of mb have been plotted from eq. (1.28) in Fig. 1.1.

Example 1.2: Comparison of Longitudinal Fins of Different Profiles. Longitudinal fins of different profiles are exposed to surroundings at a temperature of 20°C and a heat transfer coefficient of $h = 40\ \text{W/m}^2 \cdot \text{K}$. In all cases, the temperature at the fin base is 90°C and the fins are made from a steel with $k = 30\ \text{W/m} \cdot \text{K}$. All fins are 10 cm high with bases 0.80 cm thick. Compare the fin efficiencies, the dissipation of the fin per unit length, and the tip temperatures if the profiles are (**a**) rectangular, (**b**) triangular, (**c**) concave parabolic, and (**d**) convex parabolic.

SOLUTION. For all fins

$$\theta_b = 90 - 20 = 70°\text{C} \qquad \delta = 0.80/1000 = 0.008\ \text{m}$$

$$m = \left(\frac{2h}{k\delta}\right)^{1/2} = \left[\frac{(2)(40)}{(30)(0.008)}\right]^{1/2} = 18.257\ \text{m}^{-1}$$

$$b = 10/100 = 0.100\ \text{m} \qquad mb = (18.257)(0.100) = 1.8257$$

(**a**) For the rectangular profile, by eq. (1.11),

$$\eta = \frac{\tanh mb}{mb} = \frac{\tanh 1.8257}{1.8257} = \frac{0.949}{1.8257} = 0.520$$

by eq. (1.10),

$$\begin{aligned} q_b &= k\delta_b L m\theta_b \tanh mb \\ &= (30)(0.008)(1.00)(18.257)(70)(0.949) \\ &= 291.1\ \text{W} \end{aligned}$$

and by eq. (1.9),

$$\theta_a = \frac{\theta_b \cosh mx}{\cosh mb} = \frac{70 \cosh 0}{\cosh 1.8257} = \frac{70}{3.184} = 22.0°\text{C}$$

so that

$$T_a = \theta_a + 20 = 22 + 20 = 44°\text{C}$$

(**b**) For the triangular profile,[3]

$$2mb = (2)(1.8257) = 3.6514$$

[3]Numerical values for all of the modified Bessel functions in this example can be obtained from tables or computer codes.

By eq. (1.16),

$$\eta = \frac{I_1(2mb)}{mbI_0(2mb)} = \frac{I_1(3.6514)}{1.8257I_0(3.6514)} = \frac{7.1133}{(1.8257)(8.3327)} = 0.468$$

by eq. (1.15),

$$\begin{aligned} q_b &= \frac{2hL\theta_b I_1(2mb)}{mI_0(2mb)} \\ &= \frac{(2)(40)(1.00)(70)(7.1133)}{(18.257)(8.3327)} \\ &= 261.8 \text{ W} \end{aligned}$$

and by eq. (1.14),

$$\theta_a = \frac{\theta_b I_0(2m\sqrt{bx})}{I_0(2mb)} = \frac{70I_0(0)}{8.3327} = \frac{(70)(1.00)}{8.3327} = 8.4°\text{C}$$

so that

$$T_a = \theta_a + 20 = 8.4 + 20 = 28.4°\text{C}$$

(c) For the concave parabolic profile, by eq. (1.21),

$$\eta = \frac{2}{1 + \sqrt{1 + (2mb)^2}} = \frac{2}{1 + \sqrt{1 + (3.6514)^2}} = 0.418$$

and by eq. (1.20),

$$\begin{aligned} q_b &= \frac{k\delta_b L\theta_b}{2b}\left[-1 + \sqrt{1 + (2mb)^2}\right] \\ &= \frac{(30)((0.008)(1.00)(70)}{(2)(0.100)}\left[-1 + \sqrt{1 + (3.6514)^2}\right] \\ &= 234.0 \text{ W} \end{aligned}$$

Equation (1.19) shows that at $x = 0$, $\theta(x = 0) = 0$ and the tip temperature will approximate the temperature of the surroundings. Thus,

$$T_a \approx 20°\text{C}$$

(d) For the convex parabolic profile, by eq. (1.28),

$$\eta = \frac{1}{mb}\frac{I_{2/3}\left(\frac{4}{3}mb\right)}{I_{-1/3}\left(\frac{4}{3}mb\right)}$$

Here

$$\tfrac{4}{3}mb = 2.4343$$

and

$$\eta = \frac{1}{1.826} \frac{I_{2/3}(2.434)}{I_{-1/3}(2.434)} = \frac{2.7419}{(1.826)(3.0512)} = 0.492$$

and by eq. (1.27),

$$\begin{aligned} q_b &= k\delta_b L m \theta_b \frac{I_{2/3}\left(\frac{4}{3}mb\right)}{I_{-1/3}\left(\frac{4}{3}mb\right)} \\ &= (30)(0.008)(1.00)(70)(18.257)\left(\frac{2.7419}{3.0512}\right) \\ &= 275.6 \text{ W} \end{aligned}$$

Here, too, the temperature at the tip of the fin will approximate the temperature of the surroundings, This can be deduced from eq. (1.26), where the presence of the $(x/b)^{1/4}$ term causes θ_b to reduce to zero at $x = 0$. Thus $T_a \approx 20°C$.

Following is a summary of the performance of the four fins for the conditions imposed:

Fin Profile	η	q_b (W)	T_a (°C)
Rectangular	0.520	291.1	44.0
Triangular	0.468	261.8	28.4
Concave parabolic	0.418	234.0	≈ 20
Convex parabolic	0.492	275.6	≈ 20

1.3.6 Longitudinal Fin of Least Material

A discussion of the longitudinal fin that yields minimum weight is presented in Chapter 3.

1.4 RADIAL FINS

1.4.1 Generalized Differential Equation

Consider the radial fin of arbitrary profile shown in Fig. 1.11. A generalized differential equation can be developed for any radial fin of arbitrary profile function by a procedure similar to that used for the longitudinal fin whose profile is confined by two symmetrical curves that are functions of *the radial coordinate* r, $y = f_2(r)$ and $y = -f_2(r)$. Observe that the fin height is in the direction of the radial coordinate and that the origin of this coordinate is taken at the point where $r = 0$.

The difference in heat conducted into the differential element at r and that leaving the element a $r + dr$ in terms of the temperature excess $\theta = T - T_s$ is

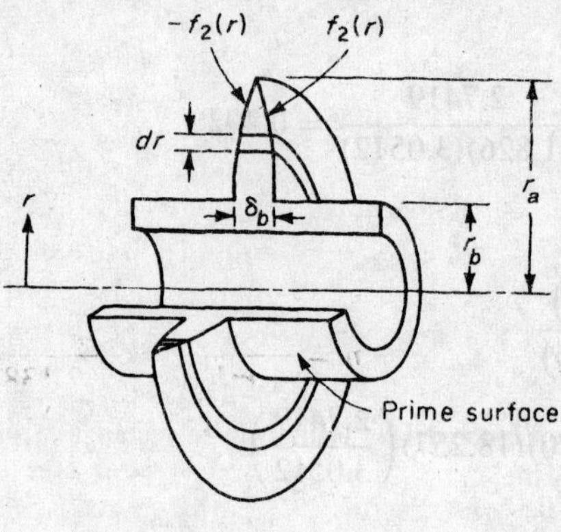

Figure 1.11 Radial fin of arbitrary profile.

$$dq = k\frac{d}{dr}\left[(2\pi r)2f_2(r)\frac{d\theta}{dr}\right]dr$$

This is the equation for a time-invariant steady-state system, and in accordance with an energy balance, it can be equated to the heat leaving the element dr by convection:

$$dq = 2h(2\pi r\,dr)\theta$$

The heat balance is

$$4\pi k\frac{d}{dr}\left[rf_2(r)\frac{d\theta}{dr}\right]dr = 4\pi h\theta r\,dr$$

or

$$k\left[f_2(r)\frac{d^2\theta}{dr^2} + f_2(r)\frac{d\theta}{dr} + r\frac{df_2(r)}{dr}\frac{d\theta}{dr}\right] = h\theta r$$

which, upon rearrangement, yields the generalized differential equation

$$f_2(r)\frac{d^2\theta}{dr^2} + \frac{f_2(r)}{r}\frac{d\theta}{dr} + \frac{df_2(r)}{dr}\frac{d\theta}{dr} - \frac{h}{k}\theta = 0 \tag{1.29}$$

1.4.2 Radial Fin of Rectangular Profile

For the radial fin of rectangular profile shown in Fig. 1.12, the profile function is

$$f_2(r) = \frac{\delta}{2}$$

and its derivative

$$\frac{df_2(r)}{dr} = 0$$

With these substituted, eq. (1.29) becomes

$$r^2\frac{d^2\theta}{dr^2} + r\frac{d\theta}{dr} - m^2r^2\theta \tag{1.30}$$

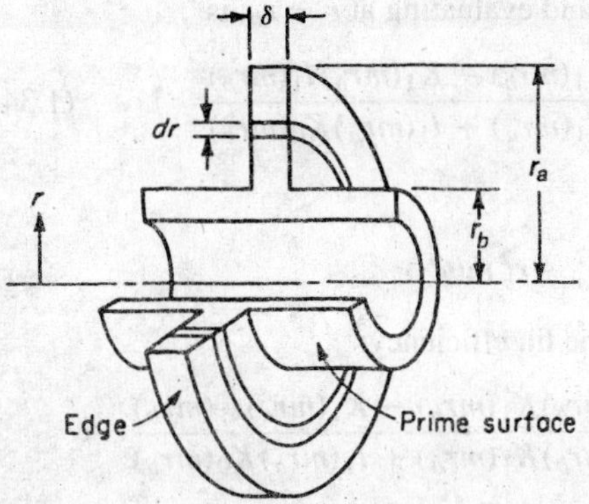

Figure 1.12 Radial fin of rectangular profile.

where $m = (2h/k\delta)^{1/2}$. Equation (1.30) is Bessel's modified equation, which has a solution in terms of the modified Bessel functions:

$$\theta(r) = C_1 I_0(mr) + C_2 K_0(mr) \tag{1.31}$$

The arbitrary constants are evaluated using the boundary conditions

$$\theta(r = r_b) = \theta_b \tag{1.32a}$$

and

$$\left.\frac{d\theta}{dr}\right|_{r=r_a} = 0 \tag{1.32b}$$

When these boundary conditions are used with eq. (1.31), two simultaneous equations in C_1 and C_2 result

$$\begin{aligned} \theta_b &= C_1 I_0(mr_b) + C_2 K_0(mr_b) \\ 0 &= C_1 I_1(mr_a) - C_2 K_1(mr_a) \end{aligned}$$

When C_1 and C_2 are evaluated and inserted into eq. (1.31), the equation for the temperature excess becomes

$$\theta(r) = \frac{\theta_b[K_1(mr_a)I_0(mr) + I_1(mr_a)K_0(mr)]}{I_0(mr_b)K_1(mr_a) + I_1(mr_a)K_0(mr_b)} \tag{1.33}$$

Note that when $r = r_b$, eq. (1.33) reduces to $\theta = \theta_b$, as it should.

The heat flow through the base is determined[4] from the general relationship

$$q_b = -2\pi r_b k \delta \left.\frac{d\theta}{dr}\right|_{r=r_b}$$

[4]Here the minus sign must be used because the temperature gradient decreases with increasing height coordinate r.

The result, after differentiating eq. (1.33) and evaluating at $r = r_b$, is

$$q_b = 2\pi r_b \delta k m \theta_b \frac{I_1(mr_a)K_1(mr_b) - K_1(mr_a)I_1(mr_b)}{I_0(mr_b)K_1(mr_a) + I_1(mr_a)K_0(mr_b)} \tag{1.34}$$

The ideal heat flow

$$q_{\text{id}} = 2\pi(r_a^2 - r_b^2)h\theta_b$$

can be used with eq. (1.34) to determine the fin efficiency:

$$\eta = \frac{q_b}{q_{\text{id}}} = \frac{2\pi r_b \delta k m \theta_b}{2\pi(r_a^2 - r_b^2)h\theta_b} \frac{I_1(mr_a)K_1(mr_b) - K_1(mr_a)I_1(mr_b)}{I_0(mr_b)K_1(mr_a) + I_1(mr_a)K_0(mr_b)}$$

and by noting that $m^2 = 2h/k\delta$, an alternative form is obtained:

$$\eta = \frac{2r_b}{m(r_a^2 - r_b^2)} \frac{I_1(mr_a)K_1(mr_b) - K_1(mr_a)I_1(mr_b)}{I_0(mr_b)K_1(mr_a) + I_1(mr_a)K_0(mr_b)} \tag{1.35}$$

The fin efficiency expressed by eq. (1.35) does not lend itself to comparison with the efficiencies of fins of other radial profiles. However, eq. (1.35) can be adjusted by expressing the efficiency in terms of the radius ratio,

$$\rho \equiv \frac{r_b}{r_a} \tag{1.36}$$

and a parameter ϕ defined by

$$\phi \equiv (r_a - r_b)^{3/2} \left(\frac{2h}{kA_p} \right)^{1/2} \tag{1.37}$$

where A_p is the profile area of the fin:

$$A_p = \delta(r_a - r_b)$$

The arguments of the Bessel functions in eq. (1.35) can be expressed in terms of the profile area:

$$mr_a = \left(\frac{2h}{k\delta} \right)^{1/2} r_a = r_a \left[\frac{2h(r_a - r_b)}{kA_p} \right]^{1/2}$$

or

$$mr_a = r_a(r_a - r_b)^{1/2} \left(\frac{2h}{kA_p} \right)^{1/2} \tag{1.38}$$

In a similar fashion,

$$mr_b = r_b(r_a - r_b)^{1/2} \left(\frac{2h}{kA_p} \right)^{1/2} \tag{1.39}$$

The fin height is the difference between the fin radii, $b = r_a - r_b$. If both numerators and denominators of eqs. (1.38) and (1.39) are multiplied by $b = r_a - r_b$, the result is

$$mr_a = \frac{r_a(r_a - r_b)^{3/2}(2h/kA_p)^{1/2}}{r_a - r_b} = \frac{r_a\phi}{r_a - r_b} \tag{1.40}$$

and an identical procedure gives

$$mr_b = \frac{r_b\phi}{r_a - r_b} \tag{1.41}$$

Now define two additional radius functions,

$$R_a \equiv \frac{1}{1 - r_b/r_a} = \frac{1}{1-\rho}$$

and

$$R_b \equiv \rho R_a = \frac{\rho}{1-\rho}$$

Substituting these into eqs. (1.40) and (1.41) yields

$$mr_a = \frac{\phi}{1-\rho} = \phi R_a$$

and

$$mr_b = \frac{\rho\phi}{1-\rho} = \phi R_b$$

Finally, the portion of eq. (1.35) preceding the final term may be expressed in terms of ϕ and ρ so that the efficiency of the radial fin of rectangular profile becomes

$$\eta = \frac{2\rho}{\phi(1+\rho)} \frac{I_1(\phi R_a)K_1(\phi R_b) - I_1(\phi R_b)K_1(\phi R_a)}{I_0(\phi R_b)K_1(\phi R_a) + I_1(\phi R_a)K_0(\phi R_b)} \tag{1.42}$$

Gardner's (1945) plot of the efficiencies of radial fins of rectangular profile is provided in Fig. 1.13. The curves are for values of x_e/x_b (Gardner's nomenclature), which is the reciprocal of ρ.

Example 1.3: Radial Fin of Rectangular Profile. A radial fin of rectangular profile is exposed to surroundings at a temperature of 35°C and a heat transfer coefficient of $h = 40\ \text{W/m}^2 \cdot \text{K}$. The temperature at the fin base is 110°C and the fin is made from a steel with $k = 40\ \text{W/m} \cdot \text{K}$ with outer and inner diameters of 25 and 10 cm. The fin thickness is 0.25 cm. Determine (**a**) the fin efficiency, (**b**) the temperature at the tip of the fin, and (**c**) the dissipation of the fin.

SOLUTION. For this particular radial fin,

$$\theta_b = 110 - 35 = 75°\text{C} \qquad \delta = 0.0025\ \text{m}$$

$$r_a = 0.125\ \text{m} \qquad r_b = 0.050\ \text{m}$$

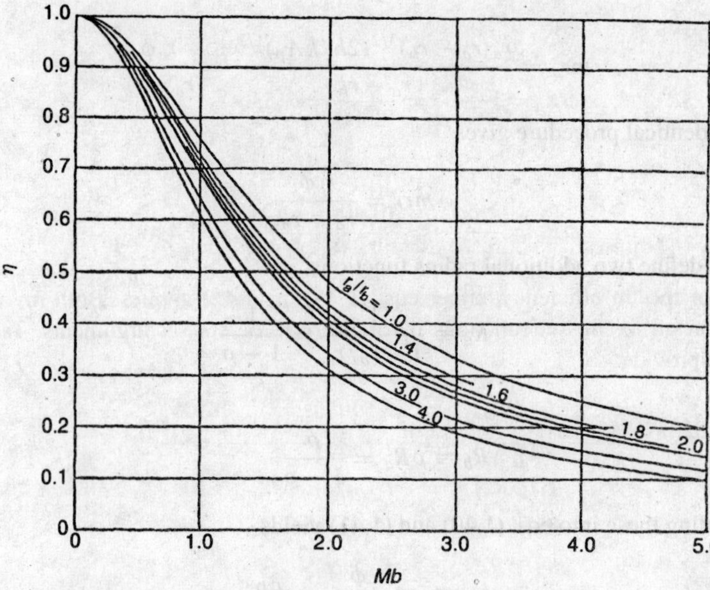

Figure 1.13 Gardner's graph for the efficiency of radial fins. (Reproduced from *Trans. ASME*, 67, 1945.)

$$\rho = \frac{r_a}{r_b} = \frac{0.050}{0.125} = 0.400$$

$$m = \left(\frac{2h}{k\delta}\right)^{1/2}$$

$$= \left[\frac{(2)(40)}{(40)(0.0025)}\right]^{1/2}$$

$$= 28.284 \text{ m}^{-1}$$

$$A_p = \delta(r_a - r_b)$$

$$= (0.0025)(0.125 - 0.050)$$

$$= 1.875 \times 10^{-4} \text{ m}^2$$

and

$$\phi = (r_a - r_b)^{3/2}\left(\frac{2h}{kA_p}\right)^{1/2}$$

$$= (0.125 - 0.050)^{3/2}\left[\frac{(2)(40)}{(40)(1.875 \times 10^{-4})}\right]^{1/2}$$

$$= (0.0205)(103.28)$$

$$= 2.121$$

Then

$$\phi R_a = \frac{\phi}{1-\rho} = \frac{2.121}{0.600} = 3.536$$

and

$$\phi R_b = \frac{\rho\phi}{1-\rho} = \frac{(0.400)(2.121)}{0.600} = 1.414$$

(**a**) For the fin efficiency, either eqs. (1.35) or (1.42) applies. Both involve the evaluation of six Bessel functions involving mr_a or mr_b as arguments. Tables of software provide

$$\begin{aligned}
I_0(\phi R_b) &= I_0(mr_b) &= 1.5661 \\
I_1(\phi R_b) &= I_1(mr_b) &= 0.8992 \\
I_1(\phi R_a) &= I_1(mr_a) &= 6.4081 \\
K_0(\phi R_b) &= K_0(mr_b) &= 0.2387 \\
K_1(\phi R_b) &= K_1(mr_b) &= 0.3136 \\
K_1(\phi R_a) &= K_1(mr_a) &= 0.0213
\end{aligned}$$

The final term in eq. (1.35) or (1.42) is evaluated as

$$\frac{(6.4081)(0.3136) - (0.8992)(0.0213)}{(1.5661)(0.0213) + (6.4081)(0.2387)} = 1.2733$$

so that by eq. (1.42),

$$\eta = \frac{2\rho}{\phi(1+\rho)}(1.2733) = \frac{(2)(0.400)}{(2.121)(1.400)}(1.2733) = 0.343$$

(**b**) For the tip temperature, use eq. (1.33) with $r = r_a = 0.125$ m. This requires the evaluation of two more modified Bessel functions. Tables or software provide

$$\begin{aligned}
I_0(\phi R_a) &= I_0(mr_a) &= 7.5897 \\
K_0(\phi R_a) &= K_0(mr_a) &= 0.0189
\end{aligned}$$

and the combination of modified Bessel functions in the bracketed term of eq. (1.33) becomes 0.1804.

$$\theta(r = r_a) = \theta_a = 0.1804\theta_b = 0.1804(75) = 13.5°\text{C}$$

and

$$T_a = \theta_a + T_s = 13.5 + 35 = 48.5°\text{C}$$

(**c**) The heat dissipation can be obtained from eq. (1.34) or the efficiency. With the fin surface

$$S = 2\pi(r_a^2 - r_b^2) = 2\pi(0.0131) = 8.247 \times 10^{-2} \text{ m}^2$$

the heat dissipation is

$$q_b = hS\eta\theta_b = 40(8.247 \times 10^{-2})(0.343)(75) = 84.9 \text{ W}$$

1.4.3 Radial Fin of Hyperbolic Profile

For the radial fin of hyperbolic profile shown in Fig. 1.14, the profile function is

$$f_2(r) = \frac{C}{r}$$

where C is a constant. The derivative of the profile function is

$$\frac{df_2(r)}{dr} = -\frac{C}{r^2}$$

and the cross section will be

$$f_1(r) = (2\pi r)2f_2(r) = 4\pi r\frac{C}{r} = 4\pi C$$

which shows that the cross section of this fin is a constant. Moreover, $f_2(r_b) = \delta_b/2$ and the constant becomes $C = \delta_b r_b/2$. Substitution of the profile function into eq. (1.29) gives, upon rearrangement, the governing differential equation for the temperature excess

$$\frac{d^2\theta}{dr^2} - \frac{m^2}{r_b}r\theta = 0 \tag{1.43}$$

where again $m^2 = 2h/k\delta_b$.

With

$$M^2 \equiv \frac{m^2}{r_b}$$

the general solution can be obtained as

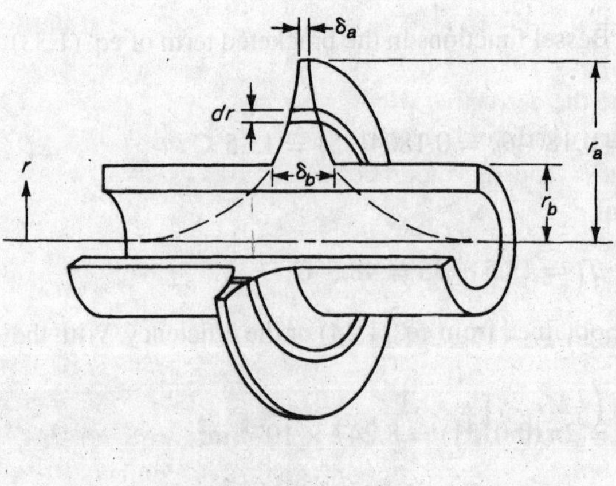

Figure 1.14 Radial fin of hyperbolic profile.

$$\theta(r) = r^{1/2}\left[C_1 I_{1/3}\left(\tfrac{2}{3}Mr^{3/2}\right) + C_2 I_{-1/3}\left(\tfrac{2}{3}Mr^{3/2}\right)\right] \tag{1.44}$$

where the arbitrary constants C_1 and C_2 are evaluated from the boundary conditions defined by eqs. (1.32) and where the details for obtaining this general solution may be found in Section A.6.3.

As in the case of the longitudinal fin of convex parabolic profile, it is helpful to make a transformation of variables. Let

$$u \equiv \tfrac{2}{3}Mr^{3/2}$$

so that

$$\frac{du}{dr} = Mr^{1/2}$$

and

$$r^{1/2} = \left(\frac{3u}{2M}\right)^{1/3} = \Omega u^{1/3}$$

where

$$\Omega = \left(\frac{3}{2M}\right)^{1/3}$$

The use of the foregoing permits the representation

$$\theta(r) = \Omega u^{1/3}[C_1 I_{1/3}(u) + C_2 I_{-1/3}(u)] \tag{1.45}$$

Equation (1.45) is identical in form with the general solution for the longitudinal fin of convex parabolic profile. However, the transformed boundary conditions in this case become

$$\theta\left(u = u_b = \tfrac{2}{3}Mr_b^{3/2}\right) = \theta_b \tag{1.46a}$$

and

$$\theta'\left(u = u_a = \frac{2}{3}Mr_a^{3/2}\right) = 0 \tag{1.46b}$$

where the prime is used to indicate the derivative with respect to u.

The use of the second boundary condition requires term-by-term differentiation of the product terms involving $u^{1/3}$ and the modified Bessel functions. In using eq. (1.46*b*), one obtains in terms of r:

$$0 = r_a^{1/2}\left[C_1 I_{-2/3}\left(\tfrac{2}{3}Mr_a^{3/2}\right) + C_2 I_{2/3}\left(\tfrac{2}{3}Mr_a^{3/2}\right)\right]$$

When this is coupled with eq. (1.32a), eq. (1.44) gives

$$\theta_b = r_b^{1/2}\left[C_1 I_{1/3}\left(\tfrac{2}{3}Mr_b^{3/2}\right) + C_2 I_{-1/3}\left(\tfrac{2}{3}Mr_b^{3/2}\right)\right]$$

These equations permit evaluation of the arbitrary constants C_1 and C_2. The particular solution for the temperature excess in the radial fin of hyperbolic profile, incorporating the solutions for C_1 and C_2, is

$$\theta(r) = \theta_b \Lambda \left(\frac{r}{r_b} \right)^{1/2} \tag{1.47}$$

where

$$\Lambda = \frac{I_{2/3}\left(\frac{2}{3}Mr_a^{3/2}\right) I_{1/3}\left(\frac{2}{3}Mr^{3/2}\right) - I_{-2/3}\left(\frac{2}{3}Mr_a^{3/2}\right) I_{-1/3}\left(\frac{2}{3}Mr^{3/2}\right)}{I_{2/3}\left(\frac{2}{3}Mr_a^{3/2}\right) I_{1/3}\left(\frac{2}{3}Mr_b^{3/2}\right) - I_{-2/3}\left(\frac{2}{3}Mr_a^{3/2}\right) I_{-1/3}\left(\frac{2}{3}Mr_b^{3/2}\right)}$$

Equation (1.47) is seen to reduce to θ_b when $r = r_b$, as it should.

The heat flow through the fin base is determined from

$$q_b = -2\pi r_b \delta_b k \frac{d\theta}{dr}$$

using the derivative of eq. (1.47) evaluated at $r = r_b$. The differentiation involves two product terms:

$$I_{1/3}\left(\tfrac{2}{3}Mr^{3/2}\right)$$

and

$$I_{-1/3}\left(\tfrac{2}{3}Mr^{3/2}\right)$$

and is most readily accomplished using the transformed variable u and the condition at the base given by eq. (1.46*a*). Termwise differentiation and subsequent use of eq. (1.46*a*) lead to the expression for the fin heat dissipation, which is the heat flow through the base:

$$q_b = 2\pi k \delta_b \theta_b M \psi r_b^{3/2} \tag{1.48}$$

where

$$\psi = \frac{I_{2/3}(u_a) I_{-2/3}(u_b) - I_{-2/3}(u_a) I_{2/3}(u_b)}{I_{-2/3}(u_a) I_{-1/3}(u_b) - I_{2/3}(u_a) I_{1/3}(u_b)}$$

The ideal heat flow is

$$q_{\text{id}} = 2\pi (r_a^2 - r_b^2) h \theta_b$$

Hence, the efficiency will be given by the ratio of eq. (1.48) to this ideal heat flow:

$$\eta = \frac{2\pi k \delta_b \theta_b M r_b^{3/2} \psi}{2\pi (r_a^2 - r_b^2) h \theta_b} = \frac{2 r_b \psi}{m (r_a^2 - r_b^2)} \tag{1.49}$$

The fin efficiency given by eq. (1.49) does not lend itself to comparison with the efficiencies of other radial profiles but can be adjusted by expressing the efficiency in terms of the radius ratio ρ. To do this, it is necessary to evaluate the fin profile area,

$$A_p = \int_{r_b}^{r_a} 2 f_2(r)\, dr = \int_{r_b}^{r_a} 2\left(\frac{\delta_b r_b}{2r}\right) dr = \delta_b r_b \ln \frac{r_a}{r_b} = \delta_b r_b \ln \frac{1}{\rho}$$

This profile area may be used in the efficiency relationship expressed by eq. (1.49). The base thickness δ_b is evaluated in terms of the profile area:

$$\delta_b = \frac{A_p}{r_b \ln(1/\rho)}$$

so that

$$m = \left(\frac{2h}{k\delta_b}\right)^{1/2} = \left[\frac{2hr_b \ln(1/\rho)}{kA_p}\right]^{1/2}$$

and

$$M = \frac{m}{r_b^{1/2}} = \left[\frac{2h \ln(1/\rho)}{kA_p}\right]^{1/2}$$

The terms exclusive of ψ in eq. (1.49) can be represented by

$$\frac{2r_b}{m(r_a^2 - r_b^2)} = \frac{2r_b^{1/2}}{(2h/kA_p)^{1/2}[\ln(1/\rho)]^{1/2}(r_a - r_b)(r_a + r_b)}$$

$$= \frac{2r_b^{1/2}(r_a - r_b)^{1/2}}{\phi[\ln(1/\rho)]^{1/2}(r_a + r_b)}$$

As in the case of the radial fin of rectangular profile,

$$\phi = (r_a - r_b)^{3/2}\left(\frac{2h}{kA_p}\right)^{1/2}$$

and a further simplification through the use of the radius ratio ρ provides

$$\frac{2r_b^{1/2}(r_a - r_b)^{1/2}}{\phi[\ln(1/\rho)]^{1/2} r_a(1 + r_b/r_a)} = \frac{2r_b^{1/2} r_a^{1/2}(1 - r_b/r_a)^{1/2}}{\phi[\ln(1/\rho)]^{1/2} r_a(1 - r_b/r_a)}$$

$$= \frac{1}{\phi}\left[\frac{4\rho(1-\rho)}{(1+\rho)^2 \ln(1/\rho)}\right]^{1/2}$$

Now define

$$R_b \equiv \tfrac{2}{3} M r_b^{3/2}$$

and

$$R_a \equiv \tfrac{2}{3} M r_a^{3/2}$$

These may be expressed in terms of ϕ and ρ:

$$R_b = \frac{2}{3} M r_b^{3/2}$$

$$= \frac{2}{3}\left(\frac{2h}{kA_p}\right)^{1/2}\left(\ln\frac{1}{\rho}\right)^{1/2} r_b^{3/2}$$

$$= \frac{2}{3}\phi\left(\ln\frac{1}{\rho}\right)^{1/2}\left(\frac{\rho}{1-\rho}\right)^{3/2}$$

and

$$R_a = \frac{2}{3} M r_a^{3/2}$$

$$= \frac{2}{3}\left(\frac{2h}{kA_p}\right)^{1/2}\left(\ln\frac{1}{\rho}\right)^{1/2} r_a^{3/2}$$

$$= \frac{2}{3}\phi\left(\ln\frac{1}{\rho}\right)^{1/2}\left(\frac{1}{1-\rho}\right)^{3/2}$$

The fin efficiency of the radial fin of hyperbolic profile with all the foregoing terms substituted into eq. (1.49) becomes

$$\eta = \frac{1}{\phi}\zeta\left[\frac{4\rho(1-\rho)}{(1+\rho)^2(\ln 1/\rho)}\right]^{1/2} \tag{1.50}$$

where

$$\zeta = \frac{I_{2/3}(R_a)I_{-2/3}(R_b) - I_{-2/3}(R_a)I_{2/3}(R_b)}{I_{-2/3}(R_a)I_{-1/3}(R_b) - I_{2/3}(R_a)I_{1/3}(R_b)}$$

an expression that is a function only of ϕ and ρ.

The efficiencies of radial fins with hyperbolic profile are plotted in Fig. 1.15 for values of the radius ratio, $\rho = 0.8$ and 0.4. These efficiencies may be compared with the radial fins of rectangular profile, which are also displayed. By inspection, the hyperbolic profile fins yield a higher efficiency because equal profile areas and fin heights provide a greater proportion of both profile and cross-sectional areas near the fin base.

1.4.4 Radial Fin of Triangular Profile

An exact solution for the temperature excess of the radial fin of triangular profile was provided by Smith and Sucec (1969), who followed the lead provided by Bert (1963). The configuration is displayed in Fig. 1.16, where it can be observed that the profile function and its derivative are

$$f_2(r) = \frac{\delta_b}{2b}(r_a - r)$$

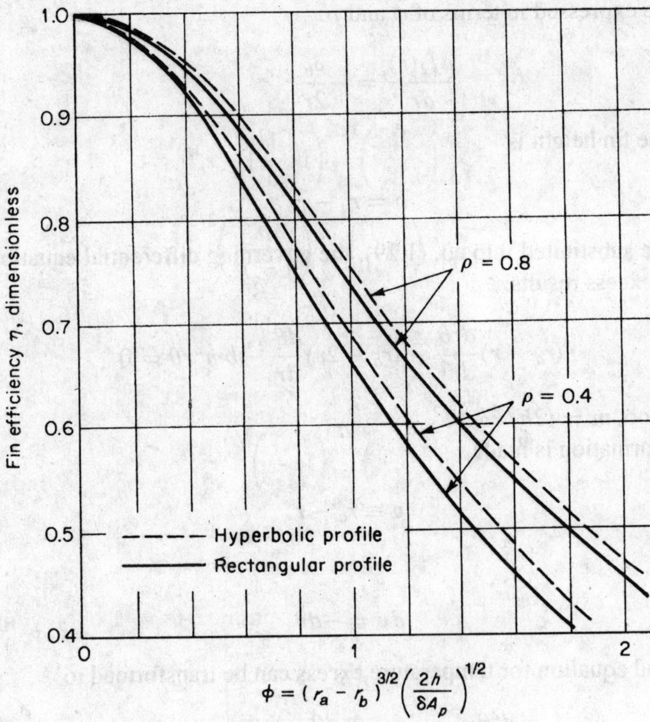

Figure 1.15 Comparison of fin efficiences: radial fins of rectangular and hyperbolic profile.

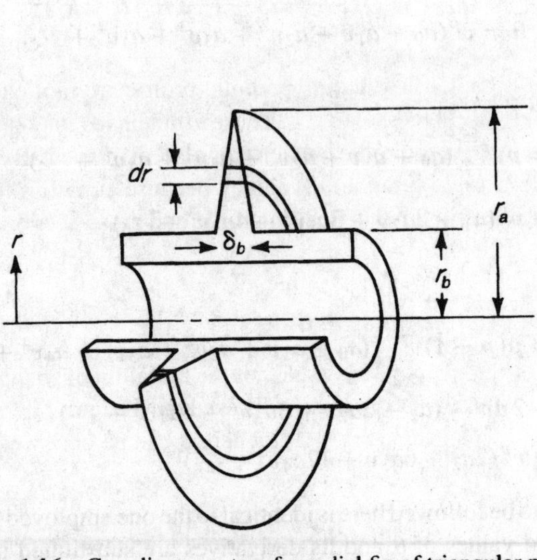

Figure 1.16 Coordinate system for radial fin of triangular profile.

and

$$\frac{df_2(r)}{dr} = -\frac{\delta_b}{2r}$$

and where the fin height is

$$b = r_a - r_b$$

With these substituted into eq. (1.29), the governing differential equation for the temperature excess results:

$$r(r_a - r)\frac{d^2\theta}{dr^2} - (r_a - 2r)\frac{d\theta}{dr} - bm^2 r\theta = 0$$

where, here too, $m = (2h/k\delta_b)^{1/2}$.

If a transformation is made,

$$v \equiv r_a - r$$

so that

$$dv = -dr$$

the differential equation for temperature excess can be transformed to

$$v(r_a - v)\frac{d^2\theta}{dv^2} - (r_a - 2v)\frac{d\theta}{dv} - bm^2(r_a - v)\theta = 0 \tag{1.51}$$

and this equation can be solved by the method of Frobenius.

Assume that

$$\theta = v^p(a_0 + a_1 v + a_2 v^2 + a_3 v^3 + a_4 v^4 + \cdots)$$

so that

$$\frac{d\theta}{dv} = pv^{p-1}(a_0 + a_1 v + a_2 v^2 + a_3 v^3 + a_4 v^4 + \cdots)$$
$$+ v^p(a_1 + 2a_2 v + 3a_3 v^2 + 4a_4 v^3 + \cdots)$$

and

$$\frac{d^2\theta}{dv^2} = p(p-1)v^{p-2}(a_0 + a_1 v + a_2 v^2 + a_3 v^3 + a_4 v^4 + \cdots)$$
$$+ 2pv^{p-1}(a_1 + 2a_2 v + 3a_3 v^2 + 4a_4 v^3 + \cdots)$$
$$+ v^p(2a_2 + 6a_3 v + 12a_4 v^2 + \cdots)$$

The procedure to be followed here is identical to the one employed in Section A.4.3. When the assumed values of θ and its derivatives are substituted into eq. (1.51), a series involving v to various powers of p results:

$$A_1 v^{p-1} + A_2 v^p + A_3 v^{p+1} + A_4 v^{p+2} + \cdots = 0$$

where A_1, A_2, and A_3 are

$$A_1 = p^2 r_a a_0$$

$$A_2 = [(p^2 + 2p + 1)r_a]a_1 - (p^2 + p + bm^2 r_a)a_0$$

$$A_3 = [(p^2 + 4p + 4)r_a]a_2 - (p^2 + 3p + 2 + bm^2 r_a)a_1 + bm_2 a_0$$

The indicial equation derives from the sum of the entries in the column headed by the lowest power of v. Because eq. (1.51) demands that the coefficients A_k be identically equal to zero, it is observed from A_1 that

$$p^2 r_a a_0 = 0$$

and because r_a is a physical dimension that cannot equal zero, a trivial solution would result if $a_0 = 0$. Thus the only alternative is that $p = 0$. The theory than says that the solution for θ will be

$$\theta = C_1 \sum_{k=0}^{\infty} a_k v^k + C_2 \ln v \left(\sum_{k=0}^{\infty} a_k v^k + \sum_{k=0}^{\infty} b_k v^k \right)$$

But to keep θ finite at $r = r_a$, where $v = r_a - r_a = 0$, C_2 must be zero, and after application of the boundary condition of eq. (1.32*a*), the particular solution will be

$$\theta = C_1 \sum_{k=0}^{\infty} a_k v^k \tag{1.52}$$

The coefficients, $a_0, a_1, a_2, a_3, \ldots$ are all related by a recurrence relationship, and it is determined by looking at the coefficients, $A_1, A_2, A_3, \ldots$ with $p = 0$. For A_1 with $p = 0$,

$$r_a a_1 = bm^2 r_a a_0$$

or

$$a_1 = bm^2 a_o \tag{1.53}$$

It can be shown after a rather laborious procedure that for $k \geq 2$,

$$a_k = \frac{[k(k-1) + br_a m^2]a_{k-1} - bm^2 a_{k-2}}{k^2 r_a} \tag{1.54}$$

with

$$a_o = \frac{\theta_b}{1 + (mb)^2 + \sum_{k=2}^{\infty} (a_k/a_0) b^k} \tag{1.55}$$

Use of eqs. (1.54) and (1.55) in eq. (1.52) allows a computation of the heat flow and the fin efficiency. A plot of the efficiency is shown in Fig. 1.17.

1.4.5 Radial Fin of Least Material

A discussion of the longitudinal fin that yields minimum weight is presented in Chapter 3.

1.5 SPINES

1.5.1 Generalized Differential Equation

Gardner (1945) also proposed a profile function for spines:

$$f_2(x) = \frac{\delta_b}{2}\left(\frac{x}{b}\right)^{(1-2n)/(2-n)} \tag{1.56}$$

With the appropriate value of n, eq. (1.56) may be used for the development of a generalized differential equation for spines.

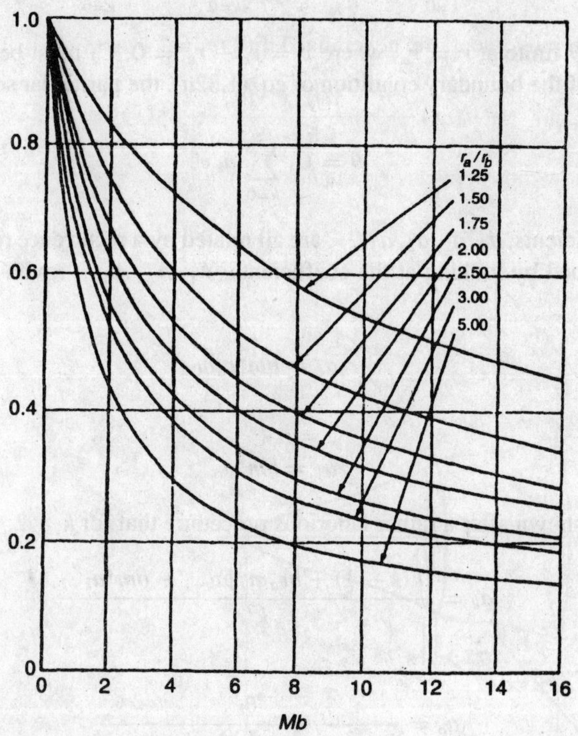

Figure 1.17 Efficiency of radial fin of triangular profile. (Reproduced from *Trans. ASME*, 91, 1969.)

Figure 1.18 shows a spine of arbitrary profile. It can be seen that the spine cross-sectional normal to the flow of heat, the confining profile, and the perimeter of the spine are all perfectly arbitrary functions of the distance x from the tip of the spine. A differential equation for the temperature excess, $\theta(x) = T(x) - T_s$, may be written in a manner similar to that for the longitudinal and radial fins by considering the heat conduction into and out of the element dx through the cross section $f_1(x)$:

$$dq = k\frac{d}{dx}\left[f_1(x)\frac{d\theta}{dx}\right]dx$$

This difference in heat flow into and out of the element must equal that dissipated by the surface of the spine. If dissipation occurs solely by convection and h is the convection coefficient,

$$dq = hf_3(x)\theta\, dx$$

where $f_3(x)$ defines a perimeter function $P(x)$ which depends on the distance x from the origin of the coordinate system. The energy balance involving conduction and convection with regard to the element dx yields

$$k\frac{d}{dx}\left[f_1(x)\frac{d\theta}{dx}\right] = hf_3(x)\theta$$

and upon rearrangement, the generalized differential equation becomes

$$f_1(x)\frac{d^2\theta}{dx^2} + \frac{df_1(x)}{dx}\frac{d\theta}{dx} - \frac{k}{k}f_3(x)\theta = 0 \tag{1.57}$$

The relationships between $f_1(x)$ and $f_2(x)$ and $f_3(x)$ and $f_2(x)$ are

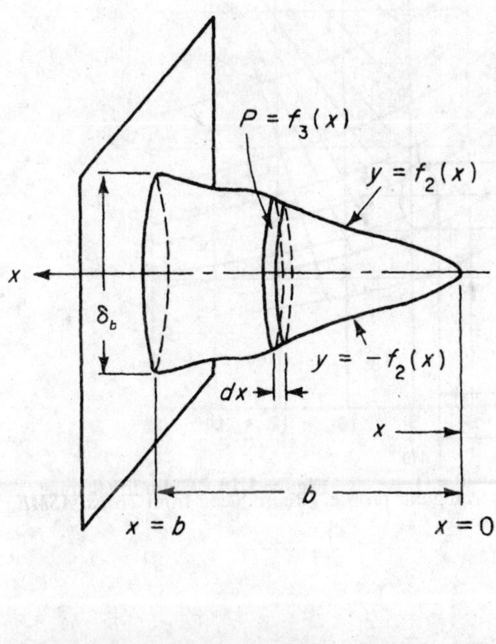

Figure 1.18 Spine of arbitrary profile.

$$f_1(x) = \pi[f_2(x)]^2 \quad \text{and} \quad f_3(x) = 2\pi f_2(x)$$

Thus eq. (1.57) may also be written as

$$[f_2(x)]^2 \frac{d^2\theta}{dx^2} + \frac{d}{dx}[f_2(x)]^2 \frac{d\theta}{dx} - \frac{2k}{k} f_2(x)\theta = 0 \tag{1.58}$$

Equation (1.58) is a second-order differential equation with variable coefficients except where the spine cross section normal to the heat flow is constant. It may be solved via termwise comparison with the general Bessel equation and is identical to the procedure used for the generalized longitudinal fin. Moreover, a comparison of Figs 1.6 and 1.18 indicates that the boundary conditions are the same for both cases and the general solution of eq. (1.58) will have two arbitrary constants evaluated using the boundary conditions of eqs. (1.6):

$$\theta(x = b) = \theta_b \tag{1.6a}$$

and

$$\left.\frac{d\theta}{dx}\right|_{x=0} = 0 \tag{1.6b}$$

1.5.2 Spines of Constant Cross Section

Cylindrical Spine. For the cylindrical spine shown in Fig. 1.19, the profile function of eq. (1.56) matches the configuration when $n = \frac{1}{2}$. Actually, δ_b in eq. (1.56) may be replaced by the spine diameter d. Thus

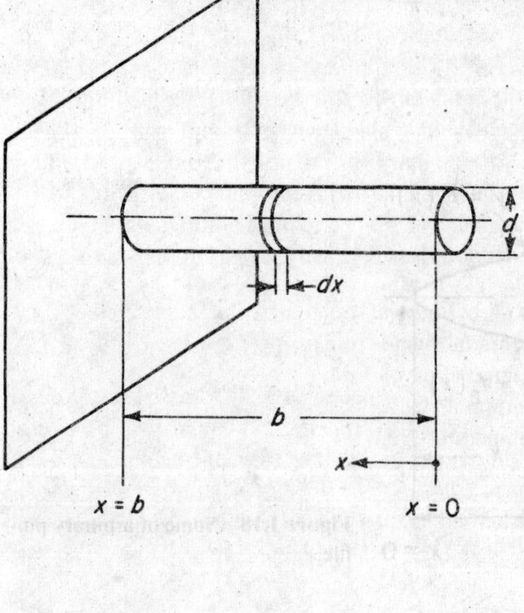

Figure 1.19 Cylindrical spine.

$$f_2(x) = \frac{d}{2}$$

$$[f_2(x)]^2 = \frac{d^2}{4}$$

and

$$\frac{d}{dx}[f_2(x)]^2 = 0$$

These values may be substituted into eq. (1.58) to obtain

$$\frac{d^2\theta}{dx^2} - m^2\theta = 0 \tag{1.59}$$

where

$$m = \left(\frac{4h}{kd}\right)^{1/2}$$

Equation (1.59) is identical in form to eq. (1.7). The general solution, the boundary conditions, the particular solution, the heat flow through the fin base, and the fin efficiency all have the same form as those for the longitudinal fin of rectangular profile. The exceptions involve the use of the spine diameter instead of the fin thickness and the performance factor $m = (4h/kd)^{1/2}$ instead of $m = (2h/k\delta_b)^{1/2}$. These comparisons yield for the temperature excess,

$$\theta(x) = \theta_b \frac{\cosh mx}{\cosh mb} \tag{1.60}$$

for the heat flow through the spine base and the heat dissipation,

$$q_b = \frac{\pi}{4} kd^2 m\theta_b \tanh mb \tag{1.61}$$

and for the fin efficiency,

$$\eta = \frac{\tanh mb}{mb} \tag{1.62}$$

Values of η using eq. (1.62) have been plotted in Fig. 1.2.

Example 1.4: Cylindrical Spine. A cylindrical rod is used as a spine. Its diameter is 0.875 cm and its height is 8 cm. It is fabricated of a steel with $k = 32$ W/m · K and is exposed to surroundings at a temperature of 30°C via a heat transfer coefficient of $h = 50$ W/m^2 · K. The temperature at the fin base of the spine is 85°. Determine (**a**) the fin efficiency, (**b**) the tip temperature, and (**c**) the heat dissipation. (**d**–**f**) Repeat the foregoing procedure for a thermal conductivity of $k = 200$ W/m · K.

SOLUTION. For the surroundings, with $h = 50$ W/m^2 · K,

$$\theta_b = 85 - 30 = 55°\text{C} \qquad d = 8.75/1000 = 8.75 \times 10^{-3}\ \text{m}$$

$$m = \left(\frac{4h}{kd}\right)^{1/2} = \left[\frac{(4)(50)}{(32)(8.75 \times 10^{-3})}\right]^{1/2} = 26.726\ \text{m}^{-1}$$

$$b = 8/100 = 0.08\ \text{m} \qquad mb = (26.726)(0.08) = 2.138$$

(a) From eq. (1.62),

$$\eta = \frac{\tanh mb}{mb} = \frac{\tanh 2.138}{2.138} = \frac{0.973}{2.138} = 0.455$$

(b) The tip temperature of the fin is determined from eq. (1.60) at $x = 0$, where $\theta(x = 0) = \theta_a$:

$$\theta_a = \frac{\theta_b \cosh mx}{\cosh mb} = \frac{55(\cosh 0)}{\cosh 2.138} = \frac{(55)(1.00)}{4.301} = 12.8°\text{C}$$

so that at $x = 0$, where $T(x = 0) = T_a$,

$$T_a = \theta_a + T_s = 12.8 + 30 = 42.8°\text{C}$$

(c) The heat dissipated by the fin is calculated from eq. (1.61):

$$\begin{aligned} q_b &= \frac{\pi}{4} k d^2 m \theta_b \tanh mb \\ &= \frac{\pi}{4}(32)(8.75 \times 10^{-3})^2(26.726)(55)(0.973) \\ &= 2.75\ \text{W} \end{aligned}$$

The heat transferred can also be obtained from the ideal heat dissipation and the efficiency determined in part (a). With the surface area of the fin,

$$S = \pi d b = \pi(8.75 \times 10^{-3})(0.08) = 2.199 \times 10^{-3}\ \text{m}^2$$

$$q_{\text{id}} = hS\theta_b = 50(2.199 \times 10^{-3})(55) = 6.045\ \text{W}$$

the heat dissipated is

$$q_b = \eta q_{\text{id}} = (0.455)(6.045) = 2.75\ \text{W}$$

(d) With $k = 200\ \text{W/m} \cdot \text{K}$,

$$m = \left(\frac{4h}{kd}\right)^{1/2} = \left[\frac{(4)(50)}{(200)(8.75 \times 10^{-3})}\right]^{1/2} = 10.690\ \text{m}^{-1}$$

$$mb = (10.690)(0.08) = 0.855$$

and from eq. (1.11),

$$\eta = \frac{\tanh mb}{mb} = \frac{\tanh 0.855}{0.855} = \frac{0.694}{0.855} = 0.811$$

(e) The tip temperature is determined from eq. (1.9):

$$\theta_a = \frac{\theta_b \cosh mx}{\cosh mb} = \frac{55(\cosh 0)}{\cosh 0.855} = \frac{55}{1.389} = 39.6°\text{C}$$

so that at $x = 0$, where $T(x = 0) = T_a$,

$$T_a = \theta_a + T_s = 39.6 + 30 = 69.6°\text{C}$$

(f) The heat transferred can also be obtained from the ideal heat dissipation and the efficiency. With

$$S = \pi db = \pi(8.75 \times 10^{-3})(0.008) = 2.199 \times 10^{-3}\ \text{m}^2$$

and

$$q_{\text{id}} = hS\theta_b = (50)(2.199 \times 10^{-3})(55) = 6.045\ \text{W}$$

the heat dissipated is

$$q_b = \eta q_{\text{id}} = (0.811)(6.045) = 4.91\ \text{W}$$

Rectangular Spine. For the rectangular spine shown in Fig. 1.20, the profile function of eq. (1.56) also matches the configuration when $n = \frac{1}{2}$. Here, however, the sides are designated by δ_1 and δ_2. With

$$f_1(x) = \delta_1\delta_2$$

$$f_3(x) = 2(\delta_1 + \delta_2)$$

and

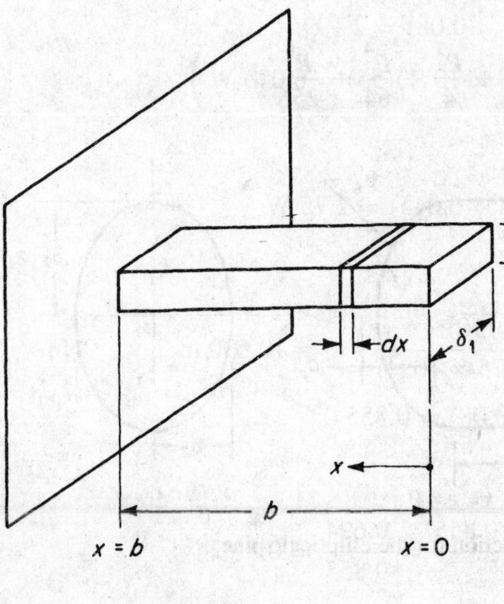

Figure 1.20 Rectangular spine.

$$\frac{df_1(x)}{dx} = 0$$

substituted into eq. (1.57), the differential equation for temperature excess is obtained:

$$\frac{d^2\theta}{dx^2} - m^2\theta = 0 \tag{1.63}$$

Here m has a special form involving the perimeter of the spine:

$$m = \left(\frac{hP}{kA}\right)^{1/2} = \left[\frac{2h(\delta_1 + \delta_2)}{k\delta_1\delta_2}\right]^{1/2}$$

which is a general expression for m for spines and all values given previously for spines of constant cross section are specific values of this expression. For the special case of the square cross section, $\delta = \delta_1 = \delta^2$, and eq. (1.63) applies with $m = (4h/k\delta)^{1/2}$.

Equation (1.63) is identical with eq. (1.59). Hence the temperature excess and the efficiency will be given by eqs. (1.60) and (1.62), respectively, as long as the proper value of m is employed. The heat flow through the base is

$$q_b = k\delta_1\delta_2 m\theta_b \tanh mb \tag{1.64}$$

Elliptical Spine. The cross section for the elliptical spine is shown in Fig. 1.21, where it may be noted that the semimajor and semiminor axes are designated, respectively, by δ_1 and δ_2. Here

$$f_1(x) = \pi\delta_1\delta_2$$

$$\frac{df_1(x)}{dx} = 0$$

and

$$f_3(x) = \pi(\delta_1 + \delta_2)\left(1 + \frac{p^2}{4} + \frac{p^4}{64} + \frac{p^6}{256} + \cdots\right)$$

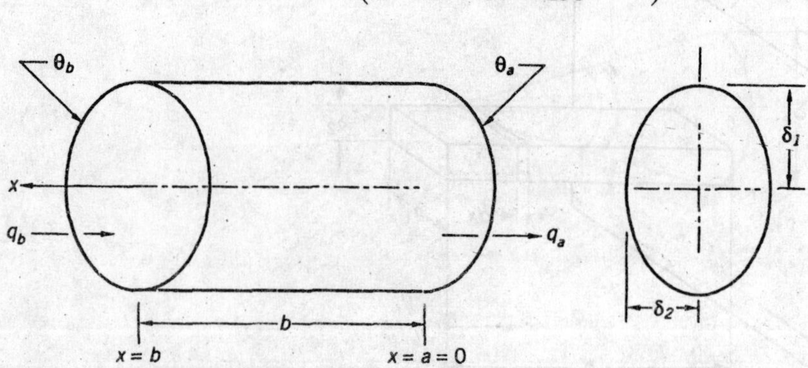

Figure 1.21 Cross section for the elliptical spine.

where

$$p = \frac{\delta_1 - \delta_2}{\delta_1 + \delta_2}$$

Here, too, the differential equation for the temperature excess is given by

$$\frac{d^2\theta}{dx^2} - m^2\theta = 0 \tag{1.59}$$

but with

$$m = \left(\frac{hP}{kA}\right)^{1/2} = \left[\frac{h(\delta_1 + \delta_2)}{k(\delta_1\delta_2)}\left(1 + \frac{p^2}{4} + \frac{p^4}{64} + \frac{p^6}{256} + \cdots\right)\right]^{1/2}$$

The temperature excess and the fin efficiency are given by eqs. (1.60) and (1.62), respectively, but the heat dissipated is given by

$$q_b = \left[hk\pi^2\delta_1\delta_2(\delta_1\delta_2)\left(1 + \frac{p^2}{4} + \frac{p^4}{64} + \frac{p^6}{256} + \cdots\right)^{1/2}\theta_b \tanh mb\right] \tag{1.65}$$

1.5.3 Conical Spine

For the conical spine shown in Fig. 1.22, the profile function is defined by eq. (1.56) with $n = -1$. Hence

$$f_2(x) = \frac{\delta_b}{2}\frac{x}{b}$$

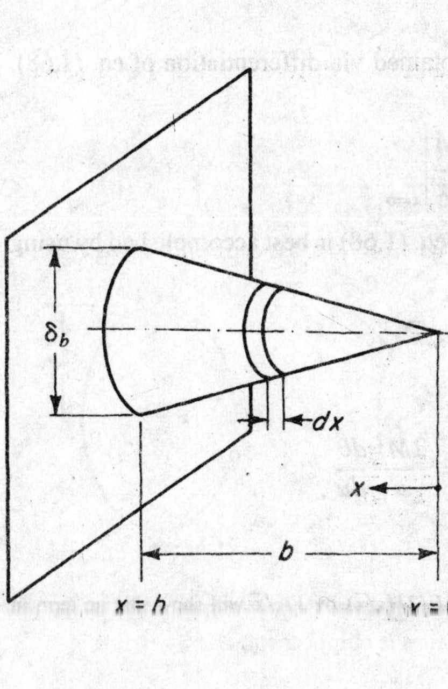

Figure 1.22 Conical spine.

and

$$\frac{df_2(x)}{dx} = \frac{\delta_b}{2b}$$

Use of these in eq. (1.58) gives the differential equation for the temperature excess, $\theta(x) = T(x) - T_s$:

$$x^2\frac{d^2\theta}{dx^2} + 2x\frac{d\theta}{dx} - M^2 x\theta = 0 \tag{1.66}$$

where

$$M = (2m^2 b)^{1/2}$$

and $m = (2h/k\delta_b)^{1/2}$.

The general solution of eq. (1.66) is shown in Section A.6.5 to be

$$\theta(x) = x^{-1/2}[C_1 I_1(2M\sqrt{x}) + C_2 K_1(2M\sqrt{x})] \tag{1.67}$$

where C_1 and C_2 are arbitrary constants to be evaluated from the boundary conditions of eqs. (1.6). However, it can be noted immediately that in order to have a finite temperature excess at $x = 0$, C_2 must equal zero because $K_1(2M\sqrt{x})/\sqrt{x}$ is unbounded[5] at $x = 0$. Hence, only C_1 need be evaluated. This evaluation is carried out at $x = b$, and when the result is substituted into eq. (1.67), the particular solution for the temperature excess is obtained:

$$\theta(x) = \theta_b\left(\frac{b}{x}\right)^{1/2}\frac{I_1(2M\sqrt{x})}{I_1(2M\sqrt{b})} \tag{1.68}$$

which reduces, as it should, to θ_b at $x = b$.

The heat flow through the base can be obtained via differentiation of eq. (1.68), evaluation at $x = b$, and substitution into

$$q_b = kA\left.\frac{d\theta}{dx}\right|_{x=b}$$

As in many previous cases, differentiation of eq. (1.68) is best accomplished by using a transformation of variable

$$u \equiv 2M\sqrt{x}$$

so that

$$\frac{d\theta}{dx} = \frac{du}{dx}\frac{d\theta}{du} = \frac{2M^2}{u}\frac{d\theta}{du}$$

and with $d\theta/du$ evaluated at

[5]Multiplication of each term of the infinite series for $I_1(2M\sqrt{x})$ by $1/\sqrt{x}$ will show that no term in $I_1(2M\sqrt{x})/\sqrt{x}$ is unbounded at $x = 0$.

$$u_b = 2M\sqrt{b}$$

the heat flow through the base in terms of u is

$$q_b = \frac{\pi k \delta_b^2 \theta_b M^3 \sqrt{b}}{u_b^2} \frac{I_2(u_b)}{I_1(u_b)}$$

In terms of x it is

$$q_b = \frac{\pi k \delta_b^2 \theta_b M}{4\sqrt{b}} \frac{I_2(2M\sqrt{b})}{I_1(2M\sqrt{b})} \tag{1.69}$$

The surface area of the conical spine is the integral of the perimeter function evaluated between the limits $x = 0$ and $x = b$:

$$S = \int_{x=0}^{x=b} f_3(x)\,dx = \int_{x=0}^{x=b} \pi \frac{\delta_b}{b} x\,dx = \frac{\pi}{2}\delta_b b$$

The ideal heat flow is obtained from this surface with operation at the base temperature excess:

$$q_{id} = h\left(\frac{\pi}{2}\delta_b b\right)\theta_b$$

Then the efficiency with $M = m\sqrt{2b}$ is

$$\eta = \frac{q_b}{q_{id}} = \frac{\pi k \delta_b \theta_b M I_2(2M\sqrt{b})}{(\pi/2)\, h \delta_b b \theta_b I_1(2M\sqrt{b})}$$

or

$$\eta = \frac{q_b}{q_{id}} = \frac{\sqrt{2} I_2(2\sqrt{2}\,mb)}{(mb) I_1(2\sqrt{2}\,mb)} \tag{1.70}$$

Values of η as a function of mb employing eq. (1.70) have been plotted in Fig. 1.2.[6]

1.5.4 Spine of Concave Parabolic Profile

The coordinate system for the spine of concave parabolic profile is shown in Fig. 1.23. It requires that the exponent on the profile function of eq. (1.56) be derived from $n = \infty$. The profile function for this spine then becomes

$$f_2(x) = \frac{\delta_b}{2}\left(\frac{x}{b}\right)^2$$

and

[6] A typographical error appears in the original Gardner (1945) paper. It shows a difference by the $\sqrt{2}$ in the modified Bessel function argument. Correspondence with Gardner confirmed eq. (1.70), and Gardner provided an errata in 1976.

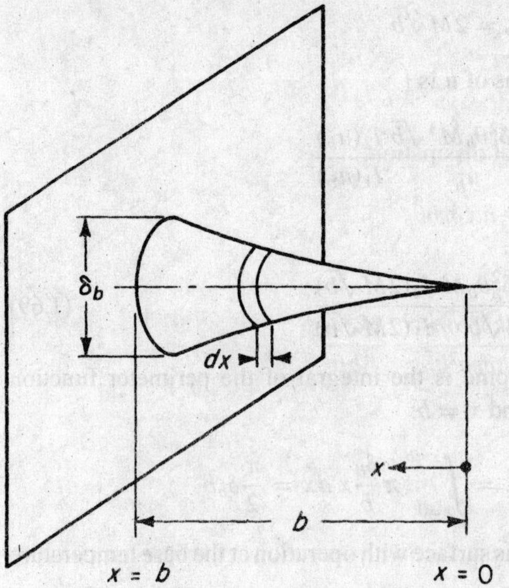

Figure 1.23 Spine of concave parabolic profile.

$$\frac{df_2(x)}{dx} = \frac{\delta_b x}{b^2}$$

Use of these in eq. (1.58) gives the governing differential equation for the temperature excess, $\theta(x) = T(x) - T_s$:

$$x^4 \frac{d^2\theta}{dx^2} + 4x^3 \frac{d\theta}{dx} - M^2 x^2 \theta = 0 \tag{1.71}$$

where $M = \sqrt{2}\, mb$ and $m = (2h/k\delta)^{1/2}$.

Equation (1.71) is recognized as an *Euler equation*, and by a procedure similar to that used for the longitudinal fin of concave parabolic profile, a particular solution is obtained:

$$\theta(x) = \theta_b \left(\frac{x}{b}\right)^{\alpha} \tag{1.72}$$

where

$$\alpha = -\tfrac{3}{2} + \tfrac{1}{2}\sqrt{(9 + 4M^2)^{1/2}}$$

The heat flow through the base of the fin is obtained by evaluating

$$q_b = k\left(\frac{\pi}{4}\delta_b^2\right) \frac{d\theta}{dx}\bigg|_{x=b}$$

and using eq. (1.72):

$$q_b = \frac{\pi k \delta_b^2 \theta_b[-3 + (9 + 4M^2)^{1/2}]}{8b} \tag{1.73}$$

The spine surface area is

$$S = \int_{x=0}^{x=b} f_3(x)\,dx = \int_{x=0}^{x=b} \pi\,\delta_b \left(\frac{x}{b}\right)^2 dx = \frac{1}{3}\pi\,\delta_b b$$

which can be used to obtain the ideal heat dissipation,

$$q_{\text{id}} = \tfrac{1}{3} h\pi\,\delta_b b\theta_b$$

The efficiency is

$$\eta = \frac{q_b}{q_{\text{id}}} = \frac{3\pi k\delta_b^2 \alpha\theta_b}{4\pi h\delta_b b^2\theta_b} = \frac{3k\delta_b\alpha}{4hb^2}$$

which may be adjusted to give

$$\eta = \frac{3}{2m^2b^2}\left[-\frac{3}{2} + \frac{1}{2}\sqrt{(9+4M^2)^{1/2}}\right]$$

or after suitable algebraic readjustment, the final and simplest form for the longitudinal fin of concave parabolic profile:

$$\eta = \frac{2}{1 + \left(1 + \frac{8}{9}m^2b^2\right)^{1/2}} \tag{1.74}$$

Values of η as a function of mb have been plotted from eq. (1.74) in Fig. 1.2.

1.5.5 Spine of Convex Parabolic Profile

The coordinate system for the spine of convex parabolic profile shown in Fig. 1.24. It requires that the exponent on the general fin profile of eq. (1.56) be obtained when $n = 0$. The profile function for this spine then becomes

$$f_2(x) = \frac{\delta_b}{2}\left(\frac{x}{b}\right)^{1/2}$$

and

$$\frac{df_2(x)}{dx} = \frac{\delta_b}{4}\left(\frac{1}{bx}\right)^{1/2}$$

When these are substituted into eq. (1.58), the governing differential equation for the temperature excess, $\theta(x) = T(x) - T_s$, becomes

$$x\frac{d^2\theta}{dx^2} + \frac{d\theta}{dx} - M^2\sqrt{x}\,\theta = 0 \tag{1.75}$$

where

$$M = (2m^2 b^{1/2})^{1/2}$$

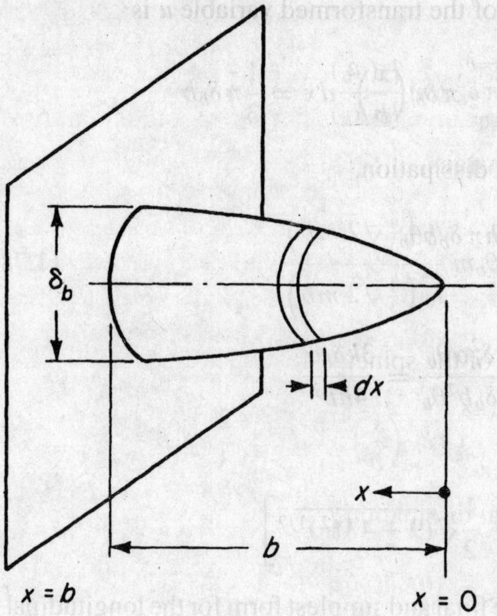

Figure 1.24 Spine of convex parabolic profile.

and $m = (2h/k\delta)^{1/2}$

As indicated in Section A.6.4, the general solution of eq. (1.75) is

$$\theta(x) = C_1 I_0\left(\tfrac{4}{3}Mx^{3/4}\right) + C_2 K_0\left(\tfrac{4}{3}Mx^{3/4}\right) \tag{1.76}$$

and it may be noted that because $K_0(0)$ is unbounded, in order to maintain a finite temperature excess at $x = 0$, C_2 must be zero. Thus eq. (1.76) reduces to

$$\theta(x) = C_1 I_0\left(\tfrac{4}{3}Mx^{3/4}\right)$$

Then C_1 may be evaluated by using the boundary condition of eq. (1.6*a*), and this leads to the particular solution

$$\theta(x) = \frac{\theta_b I_0\left(\frac{4}{3}\sqrt{2}\, mb^{1/4}x^{3/4}\right)}{I_0\left(\frac{4}{3}\sqrt{2}\, mb\right)} \tag{1.77}$$

The heat flow through the base of the fin is obtained by making the transformation

$$u \equiv \tfrac{4}{3}\sqrt{2}\, mb^{1/4}x^{3/4}$$

Then, because at $x = b$

$$u(x = b) = u_b = \tfrac{4}{3}\sqrt{2}\, mb$$

after transformation of eq. (1.76),

$$q_b = kA\left.\frac{d\theta}{du}\right|_{u=u_b} = \frac{\pi}{4}k\delta_b^2\theta_b\left(\frac{16m^4b}{3u}\right)^{1/3}\frac{d}{du}\left[\frac{I_0(u)}{I_0(u_b)}\right]_{u=u_b}$$

The heat flow through the base in terms of the transformed variable u is

$$q_b = \frac{\sqrt{2}}{4} k\pi \delta_b^2 \theta_b m \frac{I_1(u_b)}{I_0(u_b)}$$

and in terms of the specified spine dimensions,

$$q_b = \frac{\sqrt{2}}{4} k\pi \delta_b^2 \theta_b m \frac{I_1\left(\frac{4}{3}\sqrt{2}\, mb\right)}{I_0\left(\frac{4}{3}\sqrt{2}\, mb\right)} \tag{1.78}$$

Once more, consider the surface area of the spine:

$$S = \int_{x=0}^{x=b} f_3(x)\, dx = \int_{x=0}^{x=b} \pi \delta_b \left(\frac{x}{b}\right)^{1/2} dx = \frac{2}{3}\pi \delta_b b$$

which can be used to obtain the ideal heat dissipation,

$$q_{\text{id}} = hS\theta_b = \tfrac{2}{3} h\pi \delta_b b \theta_b$$

The spine efficiency then becomes

$$\eta = \frac{q_b}{q_{\text{id}}} = \frac{\left(\sqrt{2}/4\right) k\pi \delta_b^2 m \theta_b I_1\left(\frac{4}{3}\sqrt{2}\, mb\right)}{\left(\frac{2}{3}\right) \pi \delta_b h b \theta_b I_0\left(\frac{4}{3}\sqrt{2} mb\right)}$$

or

$$\eta = \frac{3\sqrt{2}}{4} \frac{I_1\left(\frac{4}{3}\sqrt{2} mb\right)}{(mb) I_0\left(\frac{4}{3}\sqrt{2} mb\right)} \tag{1.79}$$

Values of η as a function of mb have been plotted from eq. (1.79)[7] in Fig. 1.2.

Example 1.5: Spines of Different Profiles. **Spines of cylindrical, conical, concave parabolic and convex parabolic, profiles are exposed to surroundings at a temperature of 25°C via a heat transfer coefficient of $h = 40\ \text{W/m}^2\cdot\text{K}$. In all cases, the spine base temperature is 100°, the spine thermal conductivity is, $k = 100\ \text{W/m}\cdot\text{K}$, the spine base diameter is 0.92 cm, and the spine height is 10 cm. Compare the fin efficiencies and heat dissipations of the four spines.**

SOLUTION. For all the spines

$$\theta_b = 100 - 25 = 75°\text{C}$$

$$\delta_b = 0.0092\ \text{m} \quad \text{and} \quad b = 0.10\ \text{m}$$

[7] Equation (1.79) is at variance with the original Gardner (1945) reference. Correspondence with Gardner confirmed eq. (1.78) and Gardner published an errata in 1976.

(a) For the cylindrical spine with $\delta_b = d$,

$$m = \left(\frac{4h}{kd}\right)^{1/2} = \left[\frac{(4)(40)}{(100)(0.0092)}\right]^{1/2} = 13.188 \text{ m}^{-1}$$

and

$$mb = (13.188)(0.100) = 1.319$$

By eq. (1.62),

$$\eta = \frac{\tanh mb}{mb} = \frac{\tanh 1.319}{1.319} = \frac{0.866}{1.319} = 0.657$$

and by eq. (1.61),

$$\begin{aligned} q_b &= \frac{\pi}{4} k d^2 m \theta_b \tanh mb \\ &= \frac{\pi}{4}(100)(0.0092)^2(13.188)(75)(0.866) \\ &= 5.70 \text{ W} \end{aligned}$$

(b) For the conical spine with $\delta_b = 0.0092$ m,

$$m = \left(\frac{2h}{k\delta_b}\right)^{1/2} = \left[\frac{(2)(40)}{(100)(0.0092)}\right]^{1/2} = 9.325 \text{ m}^{-1}$$

and

$$mb = (9.325)(0.100) = 0.9325$$

By eq. (1.70).

$$\frac{\sqrt{2}\, I_2(2\sqrt{2}\, mb)}{(mb) I_1(2\sqrt{2}\, mb)}$$

It is interesting to note that some software and mathematical tables may not include values for $I_2(x)$. The user may have to resort to a computation using the infinite series or recognize that the recurrence relationships may help. Employment of entry 24 in Table A.1.1 gives

$$I_{n+1}(x) = I_{n-1} - \frac{2n}{x} I_n(x)$$

so that with $n = 1$ and

$$x = 2\sqrt{2}\, mb = 2\sqrt{2}\,(0.9325) = 2.638$$

one finds that

$$I_0(2.638) = 3.6585$$

$$I_1(2.638) = 2.8505$$

and then

$$I_2(2.638) = 1.4969$$

This makes the efficiency

$$\frac{\sqrt{2}\,I_2(2\sqrt{2}\,mb)}{(mb)I_1(2\sqrt{2}\,mb)} = \frac{\sqrt{2}\,(1.4969)}{(0.933)(2.8505)} = 0.796$$

The heat dissipation is calculated from

$$q_b = \frac{\pi k \delta_b^2 \theta_b M}{4\sqrt{b}} \frac{I_2(2M\sqrt{b})}{I_1(2M\sqrt{b})} \tag{1.69}$$

where

$$M = m\sqrt{2b} = 9.325\sqrt{(2)(0.100)} = 4.170$$

and where

$$2M\sqrt{b} = 2\sqrt{2}\,mb = 2\sqrt{2}\,(0.9325) = 2.638$$

Thus

$$\begin{aligned} q_b &= \frac{\pi k \delta_b^2 \theta_b M}{4\sqrt{b}} \frac{I_2(2M\sqrt{b})}{I_1(2M\sqrt{b})} \\ &= \frac{\pi(100)(0.0092)^2(75)(4.170)}{4\sqrt{0.100}} \left(\frac{1.4969}{2.8505}\right) \\ &= 3.45 \text{ W} \end{aligned}$$

Observe that the conical spine operates at a higher efficiency than the cylindrical spine ($0.796 > 0.657$), but it dissipates substantially less heat (3.45 W < 5.70 W).

(c) For the concave parabolic spine with $\delta_b = 0.0092$ m, $mb = 0.9325$, and by eq. (1.74),

$$\begin{aligned} \eta &= \frac{2}{1 + \left(1 + \frac{8}{9}m^2b^2\right)^{1/2}} \\ &= \frac{2}{1 + \left[1 + \frac{8}{9}(0.9325)^2\right]^{1/2}} \\ &= 0.858 \end{aligned}$$

By eq. (1.73),

$$q_b = \frac{\pi k \delta_b^2 \theta_b [-3 + (9 + 4M^2)^{1/2}]}{8b} \tag{1.73}$$

In this case,

$$M = \sqrt{2}\, mb = \sqrt{2}\,(0.9325) = 1.319$$

and

$$q_b = \frac{\pi(100)(0.0092)^2(75)\{-3 + [9 + 4(1.319)^2]^{1/2}\}}{(8)(0.10)}$$

$$= 2.48 \text{ W}$$

(d) For the convex parabolic spine with $\delta_b = 0.0092$ m, $mb = 0.9325$ and by eq. (1.79),

$$\eta = \frac{3\sqrt{2}}{4} \frac{I_1\left(\frac{4}{3}\sqrt{2}\, mb\right)}{(mb) I_0\left(\frac{4}{3}\sqrt{2}\, mb\right)}$$

With

$$\tfrac{4}{3}\sqrt{2}\, mb = \tfrac{4}{3}\sqrt{2}\,(0.9325) = 1.758$$

then

$$I_0(1.758) = 1.9354$$

$$I_1(1.758) = 1.2652$$

and the efficiency is computed as

$$\eta = \frac{3\sqrt{2}}{4} \frac{1.2652}{(0.9325)(1.9354)} = 0.744$$

By eq. (1.78),

$$q_b = \frac{\sqrt{2}}{4} k\pi \delta_b^2 \theta_b m \frac{I_1\left(\frac{4}{3}\sqrt{2}\, mb\right)}{I_0\left(\frac{4}{3}\sqrt{2}\, mb\right)}$$

$$= \frac{\sqrt{2}}{4}(100)\pi(0.0092)^2(75)(9.325)\left(\frac{1.2652}{1.9354}\right)$$

$$= 4.30 \text{ W}$$

The efficiencies and heat dissipations of all four spines in the surroundings imposed are summarized as follows:

Spine Profile	η	q_b (W)
Cylindrical	0.657	5.70
Conical	0.796	3.45
Concave parabolic	0.858	2.48
Convex parabolic	0.744	4.30

1.6 NOMENCLATURE

Roman Letter Symbols

- A cross-sectional or profile area, m^2; combination of coefficients in Frobenius analysis
- a coefficient in Frobenius analysis, dimensions vary
- b fin height, m
- C arbitrary constant, dimensionless
- d diameter, m; derivative or differential, dimensions vary
- f function, dimensions vary
- h heat transfer coefficient, $W/m^2 \cdot K$
- I modified Bessel function of the first kind
- K modified Bessel function of the second kind
- k thermal conductivity, $W/m \cdot K$
- L fin length, m
- M modified fin performance parameter, dimensions vary
- m fin performance parameter, m^{-1}
- P fin perimeter, m
- p ratio of elliptical spine diameters, dimensionless
- q heat flow, W
- R radius function, dimensionless
- r radial coordinate, m; radius, m
- S surface area, m^2
- T temperature, K
- u transformed variable, dimensionless
- v transformed variable, dimensionless; variable in Frobenius expansion, dimensionless
- x height coordinate, m
- y fin thickness function, m

Greek Letter Symbols

- β ratio of Bessel functions, dimensionless
- Δ change in, dimensionless
- δ fin thickness, m; side of rectangular spine, m; diameter of elliptical spine, m
- ζ combination of Bessel functions, dimensionless
- η fin efficiency, dimensionless

θ temperature excess, K
Λ combination of Bessel functions, dimensionless
ρ radius ratio, dimensionless
ϕ combination of terms, dimensionless
ψ combination of Bessel functions, dimensionless
Ω combination of terms, dimensionless

Roman Letter Subscripts

a tip of fin
b base of fin
id ideal
n order of Bessel function
p profile area
s surroundings

Roman Letter Superscripts

k exponent in Frobenius solution
n indicates type of profile
p exponent in Frobenius solution

Greek Letter Superscipts

α exponent in Euler equation general solution, dimensionless
β exponent in Euler equation general solution, dimensionless

CHAPTER 2

CONVECTION WITH REAL CONSTRAINTS

2.1 INTRODUCTION

Attention now turns to the further historical advances of the extended surface technology where the limitations of the Murray–Gardner assumptions are revealed. The elimination of these assumptions, either individually or in combination, has been the self-imposed task of many investigators. As these assumptions are removed, the mathematical models used for analysis come closer to the real-world situation.

The fourth Murray–Gardner assumption calls for a uniform heat transfer coefficient over the fin faces. Although this assumption makes for a simpler mathematical treatment of the fin conduction–convection interplay, it is certainly one of the assumptions that is most vulnerable. Most of the attention focused on elimination of the Murray–Gardner assumptions has been devoted to the proposition that the convection coefficient must be constant and uniform over the faces of the fin.

The assumption that the heat transfer between the fin and the surrounding fluid must be proportional to the temperature difference is the tenth Murray–Gardner assumption. The strict sense of this assumption precludes an analysis of fin heat dissipation by radiation. Clearly, this is not an acceptable restriction during the space age. The removal of some of the Murray–Gardner assumptions is considered in this chapter. It begins with the mathematical development for the temperature excess, heat flow, and efficiency for the longitudinal fin of rectangular and trapezoidal profile and the radial fin of rectangular profile. It continues with fin and spine analysis when the coefficient of heat transfer on the fin faces is variable. It then goes on to consider the performance of the longitudinal fin of rectangular profile with internal heat generation, and it concludes with some numerical studies that show the effect of removal of more of the other Murray–Gardner assumptions.

2.2 FINS WITH TIP HEAT LOSS

In removing the assumption that the tip of the fin is completely insulated, either of two methods of analysis is suggested.[1] In the first method, the analytical solution adjusts the tip boundary condition to include the heat flow to the surroundings, and a particular solution is developed for the adjusted boundary condition. The second method, proposed by Harper and Brown (1922) and referred to as the *Harper–Brown approximation*, fictitiously adjusts the fin height to accommodate the heat flow from the tip while considering the tip to be completely insulated.

2.2.1 Longitudinal Fin of Rectangular Profile

Analytical Solution. The terminology and coordinate system for the longitudinal fin of rectangular profile with tip heat loss is shown in Fig. 2.1. Observe that the positive sense of the height coordinate x is from fin base to fin tip and that the fin height is still designated by b. Let the fin transfer heat from the fin faces via a heat transfer coefficient h and from the fin tip via a heat transfer coefficient h_a. This permits consideration of different heat transfer coefficients at the faces and tip of the fin because they probably will not be equal. The heat flow from the tip will be

$$q_a = h_a \delta L \theta_a$$

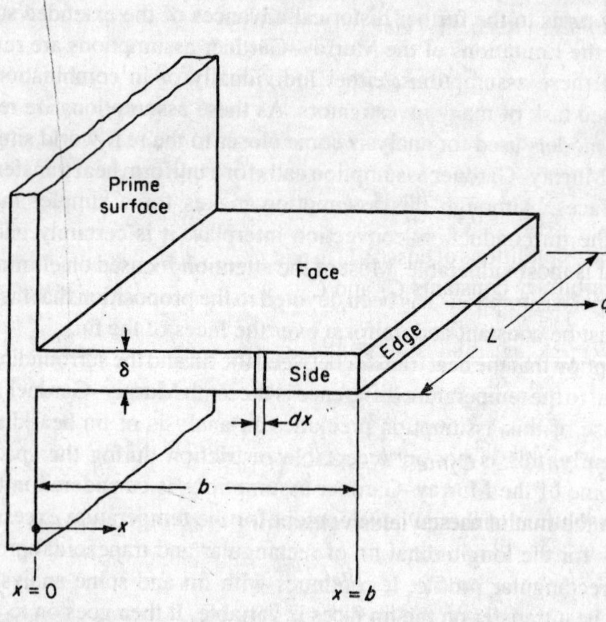

Figure 2.1 Terminology and coordinate system for the longitudinal fin of rectangular profile with tip heat loss.

[1]A third method that involves a linear transformation is decribed in Chapter 7.

The differential equation for temperature excess, eq. (1.7), also applies when the orientation of the height coordinate is reversed:

$$\frac{d^2\theta}{dx^2} - m^2\frac{d\theta}{dx} = 0 \tag{1.7}$$

where

$$m = \left(\frac{2h}{k\delta}\right)^{/12}$$

This differential equation is to be solved using one boundary condition at the fin base:

$$\theta(x = 0) = \theta_b \tag{2.1a}$$

At the fin tip, continuity demands that

$$q(x = b) = q_a = -k\delta L\frac{d\theta}{dx}\bigg|_{x=b} = h_a\delta L\theta_a$$

so that the second boundary condition becomes

$$\frac{d\theta}{dx}\bigg|_{x=b} = -\frac{h_a}{k}\theta_a \tag{2.1b}$$

The general solution to eq. (1.7) is

$$\theta = C_1e^{mx} + C_2e^{-mx} \tag{2.2a}$$

and its derivative is

$$\frac{d\theta}{dx} = m\left(C_1e^{mx} - C_2e^{-mx}\right) \tag{2.2b}$$

Use of the boundary conditions of eqs. (2.1) in eqs. (2.2) yields the pair of simultaneous equations in the arbitrary constants C_1 and C_2:

$$C_1 + C_2 = 0$$

and

$$C_1me^{mb} - C_2me^{-mb} = -\frac{h_a}{k}\left(C_1e^{mb} + C_2e^{-mb}\right)$$

Simultaneous solution of these yields values for the arbitrary constants:

$$C_1 = \frac{\theta_be^{-2mb}}{\alpha + e^{-2mb}}$$

and

$$C_2 = \frac{\alpha\theta_b}{\alpha + e^{2mb}}$$

where

$$\alpha = \frac{m + h_a/k}{m - h_a/k}$$

and the particular solution for the temperature excess may be written as

$$\theta(x) = \frac{\theta_b\{e^{mb[(x/b)-2]} + \alpha e^{-mb(x/b)}\}}{\alpha + e^{-2mb}} \tag{2.3}$$

The particular solution for the case of no heat flow from the tip for the configuration of Fig. 2.1 can be shown to be

$$\theta = \frac{\theta_b \cosh m(b-x)}{\cosh mb} \tag{2.4}$$

and if the tip heat transfer coefficient h_a is equal to zero, no heat will flow from the fin tip, $\alpha = 1$, and eq. (2.3) reduces to eq. (2.4)

Equation (2.3) represents the exact solution and eq. (2.4) the approximate solution when the heat flow from the fin tip is assumed negligible. The difference is shown in Fig. 2.2, where the ordinate represents the percent temperature excess over the base temperature excess as a function of the fraction x/b and α.

The heat flow through the base of the fin derives from

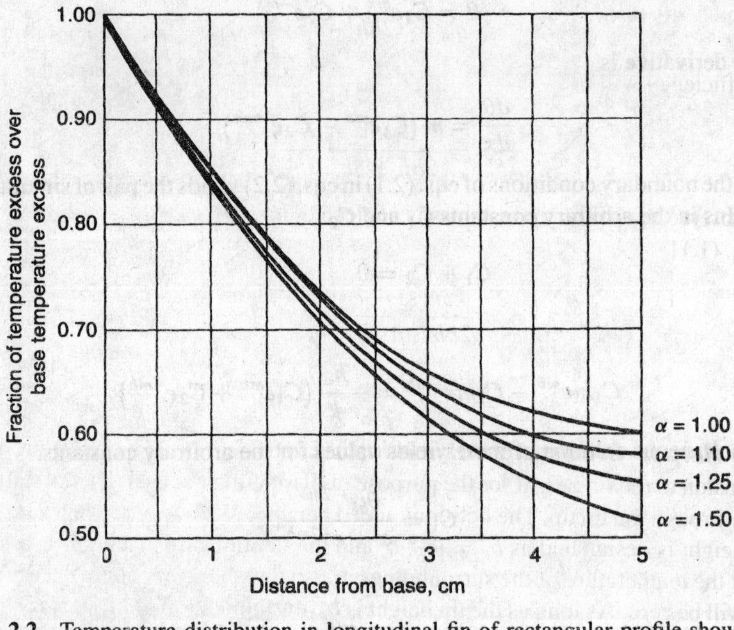

Figure 2.2 Temperature distribution in longitudinal fin of rectangular profile showing the effect of tip heat loss. The presentation is for a copper fin with $\delta = 0.1588$ cm, $b = 5.08$ cm. $k = 389.4$ W/m · K, and a heat transfer coefficient on the fin faces of 142 W/m^2 · K.

$$q = -k\delta L \left.\frac{d\theta}{dx}\right|_{x=0}$$

From eq. (2.3),

$$\frac{d\theta}{dx} = \frac{d}{dx} \frac{\theta_b\{e^{mb[(x/b)-2]} + \alpha e^{-mb(x/b)}\}}{\alpha + e^{-2mb}}$$

and at $x = 0$,

$$\left.\frac{d\theta}{dx}\right|_{x=0} = \frac{m\theta_b(e^{-2mb} - \alpha)}{\alpha + e^{-2mb}}$$

so that

$$q_b = -k\delta L m \left.\frac{d\theta}{dx}\right|_{x=0} = k\delta L m\theta_b \frac{\alpha - e^{-2mb}}{\alpha + e^{-2mb}} \tag{2.5}$$

or more conveniently,

$$q_b = \frac{k\delta L m\theta_b\,[(h_a/mk) + \tanh mb]}{1 + (h_a/mk)\tanh mb} \tag{2.6}$$

The fin efficiency is based on the actual heat flow given by eq. (2.5) and the ideal heat flow from the fin faces and the fin tip if all surfaces were to operate at the base temperature excess. Thus, with the ideal heat flow given by

$$q_{\text{id}} = (2bh + \delta h_a)L\theta_b$$

the efficiency will be

$$\eta = \frac{k\delta m}{2bh + \delta h_a} \frac{\alpha - e^{-2mb}}{\alpha + e^{-2mb}} \tag{2.7}$$

Observe that if the tip heat loss is neglected, $h_a = 0$, $\alpha = 1$ and eq. (2.7) reduces to eq. (1.11):

$$\eta = \frac{k\delta m}{2bhL} \frac{1 - e^{-2mb}}{1 + e^{2mb}} = \frac{\tanh mb}{mb} \tag{1.11}$$

The Harper–Brown Approximation. Harper and Brown (1922) proposed an extension to the fin height for the purpose of dissipating the heat that normally would pass through the fin tip. The fictitious added height b' is sketched in Fig. 2.3. The total fin height is designated as $b_c = b + b'$ and the fictitious tip at $x = b_c$ is assumed to be at the temperature of the surroundings, b_s, so that the temperature excess at the fin tip will be zero. As long as the fin height is b_c, an application of a boundary condition entailing no heat loss at the fin tip is permitted. Thus the differential equation for temperature excess, given by

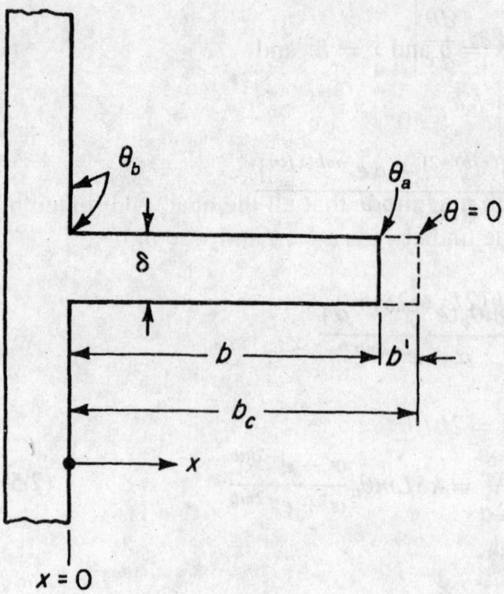

Figure 2.3 Coordinate system for the Harper–Brown approximation.

$$\frac{d^2\theta}{dx^2} - m^2\frac{d\theta}{dx} = 0 \tag{1.7}$$

and its general solution, given by

$$\theta = C_1 e^{mx} + C_2 e^{-mx} \tag{2.2a}$$

still apply. However, the boundary conditions are

$$\theta(x = 0) = \theta_b$$

and the heat flow at the adjusted fin tip

$$q(x = b_c) = 0$$

which requires that

$$\left.\frac{d\theta}{dx}\right|_{x=b_c} = 0 = mC_1 e^{mb_c} - mC_2 e^{-mb_c}$$

so that

$$C_1 e^{mb_c} = C_2 e^{-mb_c}$$

When the corrected fin height is not used,[2] the heat loss from the tip is

$$q = h\delta L\theta_a \tag{2.8}$$

[2]The Harper–Brown approximation is based on h, the heat transfer coefficient on the faces of the fin and not on any special value of h.

If this heat is to be dissipated between $x = b$ and $x = b_c$ and

$$\left.\frac{d\theta}{dx}\right|_{x=b_c} = 0$$

this requires that $b' = b_c - b$ be of such magnitude that all the heat represented by eq. (2.8) be dissipated by the faces of the fin between $x = b$ and $x = b_c$ or

$$h\delta L\theta_a = h(2L + 2\delta)b'\theta_a$$

If $L >> \delta$,

$$h\delta L\theta_a = 2hLb'\theta_a$$

The value of b' is now obtained as

$$b' = \frac{\delta}{2}$$

and the corrected fin height becomes

$$b_c = b + b' = b + \frac{b}{2} \tag{2.9}$$

Thus b_c is the sum of the original, uncorrected fin height and half of the fin thickness and it appears as though the tip of the fin were hinged and one-half of the fin thickness folded and added to the fin height.

The temperature excess is given by eq. (2.4) with a modification that employs b_c:

$$\theta = \frac{\theta_b \cosh m(b_c - x)}{\cosh mb_c} \tag{2.10}$$

The heat flow through the base of the fin can be obtained from eq. (1.10) with b_c substituted for b:

$$q_b = k\delta Lm\theta_b \tanh mb_c \tag{2.11}$$

However,

$$m = \left(\frac{2h}{k\delta}\right)^{1/2}$$

and

$$\delta = \frac{2h}{km^2}$$

The corrected fin height then becomes

$$b_c = b + \frac{\delta}{2} = b + \frac{h}{km^2}$$

so that an alternative expression to eq. (2.11) can be written as

$$q_b = k\delta L m\theta_b \tanh\left[mb + \left(\frac{h\delta}{2k}\right)^{1/2}\right] \tag{2.12}$$

because

$$\frac{h}{km} = \frac{h}{k(2h/k\delta)^{1/2}} = \left(\frac{h\delta}{2k}\right)^{1/2}$$

The fin efficiency is merely eq. (1.11) with b_c used instead of b:

$$\eta = \frac{\tanh mb_c}{mb_c} \tag{2.13}$$

Example 2.1: Longitudinal Fin of Rectangular Profile with Tip Heat Loss. A longitudinal fin of rectangular profile is exposed to surroundings at a temperature of 50°C and a heat transfer coefficient of $h = 50\ \text{W/m}^2 \cdot \text{K}$. The temperature at the fin base is 100°C and the fin is made from a steel with $k = 30\ \text{W/m} \cdot \text{K}$. The fin is 10 cm high and 0.80 mm thick and 1 m long. Using (1) the case of no heat flow from the tip, (2) the ideal case *with* heat flow from the tip, and (3) the Harper–Brown approximation, compare (**a**) the fin efficiency, (**b**) the tip temperature excess, and (**c**) the heat dissipated.

SOLUTION. For this fin

$$\theta_b = 100 - 50 = 50°\text{C}$$

$$m = \left(\frac{2h}{k\delta}\right)^{1/2} = \left[\frac{(2)(50)}{(30)(0.008)}\right]^{1/2} = 20.412\ \text{m}^{-1}$$

$$mb = (20.412)(0.10) = 2.041$$

and

$$\alpha = \frac{m + h_a/k}{m - h_a/k} = \frac{20.412 + 50/30}{20.412 - 50/30} = 1.178$$

(**a**) In the fin efficiency comparison, for the simplified case, use eq. (1.11):

$$\eta = \frac{\tanh mb}{mb} = \frac{\tanh 2.041}{2.041} = \frac{0.967}{2.041} = 0.474$$

For the ideal case where $h_a = 50\ \text{W/m}^2 \cdot \text{K}$, use eq. (2.7):

$$\eta = \frac{k\delta m}{2bh + \delta h_a} \frac{\alpha - e^{-2mb}}{\alpha + e^{-2mb}}$$

and with $e^{-2mb} = 0.0169$,

$$\eta = \frac{(30)(0.008)((20.412))}{(2)(0.10)(50) + (0.008)(50)} \left(\frac{1.178 - 0.0169}{1.178 + 0.0169} \right) = 0.458$$

For the Harper–Brown approximation, obtain b_c from eq. (2.9):

$$b_c = b + \frac{\delta}{2} = 0.100 + \frac{0.008}{2} = 0.104 \text{ m}$$

Then

$$mb_c = (20.412)(0.104) = 2.123$$

and by eq. (2.13),

$$\eta = \frac{\tanh mb_c}{mb_c} = \frac{\tanh 2.123}{2.123} = \frac{0.972}{2.123} = 0.458$$

(**b**) In the tip temperature excess comparison, for the simplified case, use eq. (2.4):

$$\theta(x) = \frac{\theta_b \cosh m(b - x)}{\cosh mb}$$

At $x = b$,

$$\theta(x = b) = \theta_a = \frac{50 \cosh 0}{\cosh 2.041} = \frac{(50)(1.00)}{3.915} = 12.8°\text{C}$$

For the ideal case at $x = b$ where $h_a = 50 \text{ W/m}^2 \cdot \text{K}$, use eq. (2.3) with $\alpha = 1.178$ and $e^{-2mb} = 0.0169$:

$$\begin{aligned}
\theta_a &= \left. \frac{\theta_b\{e^{mb[(x/b)-2]} + \alpha e^{-mb(x/b)}\}}{\alpha + e^{-2mb}} \right|_{x=b} \\
&= \frac{50[e^{2.041(1-2)} + 1.178e^{-2.041}]}{1.178 + 0.0169} \\
&= \frac{(50)(0.2828)}{1.1949} \\
&= 11.8°\text{C}
\end{aligned}$$

For the Harper–Brown approximation, use eq. (2.10) with $b_c = 0.104$ m

$$\theta_a = \frac{\theta_b \cosh m(b_c - b_c)}{\cosh mb_c} = \frac{50 \cosh 0}{\cosh 2.123} = \frac{50(1.0)}{4.238} = 11.8°\text{C}$$

(**c**) In the heat dissipation comparison, for the simplified case, use eq. (1.10):

$$\begin{aligned}
q_b &= k\delta Lm\theta_b \tanh mb \\
&= (30)(0.008)(1.000)(20.412)(50) \tanh 2.041 \\
&= (244.45)(0.967) \\
&= 236.8 \text{ W}
\end{aligned}$$

For the ideal case, use eq. (2.6) with

$$\frac{h_a}{mk} = \frac{50}{20.412(30)} = 0.0816$$

so that

$$\begin{aligned} q_b &= \frac{k\delta L m\theta_b\,[(h_a/mk) + \tanh mb]}{1 + (h_a/mk)\tanh mb} \\ &= \frac{(30)(0.008)(1.00)(20.412)(50)(0.0816 + \tanh 2.041)}{1.000 + 0.0816 \tanh 2.041} \\ &= \frac{(244.95)(1.049)}{1.079} \\ &= 238.0 \text{ W} \end{aligned}$$

For the Harper–Brown approximation, use eq. (2.11) with $b_c = 0.104$ m:

$$\begin{aligned} q_b &= k\delta L m\theta_b \tanh mb_c \\ &= (30)(0.008)(1.000)(20.412)(50) \tanh 2.123 \\ &= (244.45)(0.972) \\ &= 238.0 \text{ W} \end{aligned}$$

The results may be summarized as follows:

	η	θ_a (°C)	q_b W
Simplified case	0.474	12.8	236.8
Ideal case	0.458	11.8	238.0
Harper–Brown	0.458	11.8	238.0

Usefulness Criteria for Longitudinal Fins of Rectangular Profile. Once again, consider eq. (2.6), which gives the actual heat flow through the fin base when the tip heat loss is considered:

$$q_b = \frac{k\delta L m\theta_b\,[(h_a/mk) + \tanh mb]}{1 + (h_a/mk)\tanh mb} \tag{2.6}$$

The form of this equation indicates that when all variables but the fin height b are held constant, a single value of the fin height corresponds to a maximum heat flow through the fin base. This maximum or optimum may be obtained in the customary manner by finding where $\partial q_b/\partial b$ vanishes. Recalling that

$$\frac{d}{dx}\tanh ax = a\,\text{sech}^2 ax$$

the derivative is

$$\frac{dq_b}{dx} = \frac{k\delta L m\theta_b\,[(h_a/mk)\tanh mb + 1]\,m\,\text{sech}^2 mb}{[1 + (h_a/mk)\tanh mb]^2}$$

$$-\frac{k\delta L m\theta_b\,[(h_a/mk) + \tanh mb]\,(h_a/mk)\,\text{sech}^2 mb}{[1 + (h_a/mk)\tanh mb]^2}$$

When this is set equal to zero, it is observed that

$$m\left(\frac{h_a}{mk}\tanh mb + 1\right) = \frac{h_a}{k}\left(\frac{h_a}{mk} + \tanh mb\right)$$

$$\frac{h_a}{k}\tanh mb + m = \frac{h_a^2}{mk^2} + \frac{h_a}{k}\tanh mb$$

or

$$\frac{h_a^2}{mk^2} = \frac{\delta h_a^2}{2hk} = 1$$

The foregoing indicates that the fin heat dissipation will be at a maximum when the dimensionless parameter $\delta h_a^2/2hk$ equals unity. For values of this parameter less than unity, the fin will have a favorable effect and will assist the flow of heat from the prime or base surface to the surrounding environment. For values greater than unity, the fin will have a retarding effect on the heat flow. When the tip coefficient h_a and the face coefficient are equal,

$$\frac{h_a^2}{m^2k^2} = \frac{\delta h}{2k} = 1$$

or

$$h = \frac{2k}{\delta} \tag{2.14}$$

and it is observed that the maximum occurs when the surface coefficient just equals twice the internal conductance of the fin.

Example 2.2: Usefulness of a Longitudinal Fin of Rectangular Profile. Determine whether a longitudinal fin of rectangular profile 1 m in length with a height of 5.08 cm and a thickness of 0.3175 cm can be employed advantageously in surroundings in which water boils with a heat transfer coefficient of $h = 14{,}250\ \text{W/m}^2\cdot\text{K}$. The fin is made of a steel with $k = 33.5\ \text{W/m}\cdot\text{K}$. The base temperature is 105°C and water boils at 100°C.

SOLUTION. Neither of the temperatures at the fin base or environment or the fin length have anything to do with whether or not the fin should be used. The criterion is given by eq. (2.14):

$$h = \frac{2k}{\delta} \tag{2.14}$$

Using the given values of k and δ yields

$$h = \frac{(2)(33.5)}{3.175 \times 10^{-3}} = 21{,}102\ \mathrm{W/m^2 \cdot K}$$

The value of h for this application was stated as $h = 14,250\ \mathrm{W/m^2 \cdot K}$. Hence the fin could be used advantageously. Alternatively,

$$\frac{\delta h}{2k} = \frac{3.175 \times 10^{-3}(14{,}250)}{(2)(33.5)} = 0.675 < 1$$

The fin efficiency is obtained from eq. (1.11):

$$m = \left(\frac{2h}{k\delta}\right)^{1/2} = \left[\frac{(2)(14{,}250)}{(33.5)(3.175 \times 10^{-3})}\right]^{1/2} = 517.64\ \mathrm{m^{-1}}$$

$$mb = (517.64)(0.0508) = 26.296$$

and

$$\eta = \frac{\tanh mb}{mb} = \frac{\tanh 26.296}{26.296} = \frac{1.000}{26.296} = 0.038$$

The boiling coefficient, $h = 14,250\ \mathrm{W/m^2 \cdot K}$, is quite high but is typical of boiling applications. This value leads to the conclusion that even at a very low value of fin effiicency ($\eta = 0.038$), the fin will dissipate some heat. With $\theta_b = 105 - 100 = 5°\mathrm{C}$,

$$q_b = 2hLb\eta\theta_b = (2)(14{,}250)(1.00)(0.0508)(0.038)(5) = 275.3\ \mathrm{W}$$

This is more heat than that which would be dissipated through the footprint of the fin on the base or prime surface:

$$q = hL\delta\theta_b = (14{,}250)(1.00)(3.175 \times 10^{-3})(5) = 226.3\ \mathrm{W}$$

2.2.2 Longitudinal Fin of Trapezoidal Profile

Temperature Excess, Heat Flow, and Efficiency. Harper and Brown (1922) suggested that the longitudinal fin of trapezoidal profile could be treated using the configuration of Fig. 2.4, where it may be observed that the tip heat loss is accounted for by extending the fin height. As in the case of the longitudinal fin of rectangular profile, the fin is extended $\delta_a/2$ (half the fin thickness), so that the corrected fin height is $b_c = b + \delta_a/2$. Notice that the height coordinate has its origin at the fin tip and its positive sense is in a direction from fin tip to fin base.

The generalized differential equation of eq. (1.3),

$$f_1(x)\frac{d^2\theta}{dx^2} + \frac{df_1(x)}{dx}\frac{d\theta}{dx} - \frac{2h}{k}\theta = 0 \qquad (1.3)$$

may be used in this case. The cross-sectional area function for unit length and its derivative in terms of the fin taper angle κ are

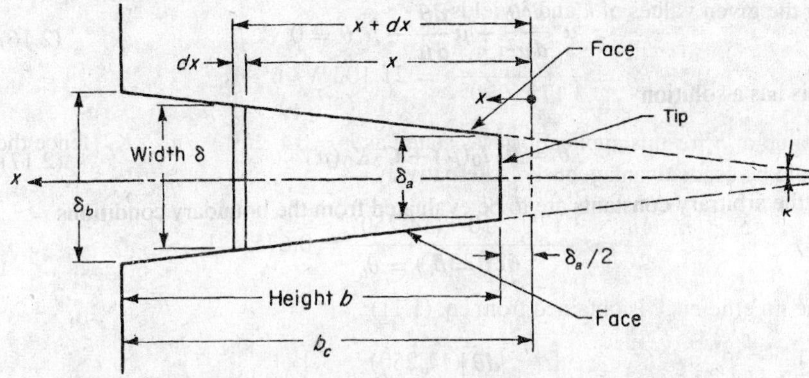

Figure 2.4 Terminology and coordinate system for the longitudinal fin of trapezoidal profile with tip heat loss.

$$f_1(x) = \delta_a + 2\left(x - \frac{\delta_a}{2}\right) \tan \kappa$$

and

$$\frac{df_1(x)}{dx} = 2 \tan \kappa$$

When these are substituted into eq. (1.3), the result is

$$\left[\delta_a + 2\left(x - \frac{\delta_a}{2}\right) \tan \kappa\right] \frac{d^2\theta}{dx^2} + 2 \tan \kappa \frac{d\theta}{dx} - \frac{2h}{k \cos \kappa} \theta = 0$$

and rearrangement yields[3]

$$\frac{d^2\theta}{dx^2} + \frac{2 \tan \kappa}{\delta_a(1 - \tan \kappa) + 2x \tan \kappa} \frac{d\theta}{dx} - \frac{2h}{k \cos \kappa [\delta_a(1 - \tan \kappa) + 2x \tan \kappa]} \theta = 0$$

Harper and Brown (1922) suggested a rather involved tranformation of variable from x to u:

$$\mu^2 \equiv 4K^2 \left[x + \frac{\delta_a(1 - \tan \kappa)}{2 \tan \kappa}\right] \tag{2.15}$$

where

$$K^2 = \frac{h}{k \sin \kappa}$$

With this transformation and appropriate substitution, Bessel's modified equation results:

[3]Harper and Brown acknowledged that the element of face surface area is not dx but $dx/\cos\kappa$. This accounts for the appearance of the $\cos\kappa$ in the denominator of the last term.

$$\mu^2 \frac{d^2\theta}{d\mu^2} + \mu \frac{d\theta}{d\mu} - \mu^2\theta = 0 \tag{2.16}$$

and this has a solution

$$\theta = C_1 I_0(\mu) + C_2 K_0(\mu) \tag{2.17}$$

where the arbitrary constants are to be evaluated from the boundary conditions

$$\theta(x = b_c) = \theta_b$$

and

$$\left.\frac{d\theta}{dx}\right|_{x=0} = 0$$

However, these boundary conditions must be transformed to the system where μ is the independent variable. When eq. (2.15) is differentiated,

$$2\mu\, d\mu = 4K^2\, dx$$

so that

$$\frac{d\mu}{dx} = \frac{2K^2}{\mu}$$

and because

$$\frac{d\theta}{d\mu} = \frac{d\theta}{dx}\frac{dx}{d\mu}$$

then

$$\frac{d\theta}{d\mu} = \frac{\mu}{2K^2}\frac{d\theta}{dx}$$

The latter shows that for all finite values of μ, the derivative $d\theta/d\mu$ vanishes when $d\theta/dx$ vanishes. This leads to

$$\left.\frac{d\theta}{d\mu}\right|_{x=0} = 0 \tag{2.18a}$$

$$\mu(x = 0) = \mu_a = 2K\left[\frac{\delta_a(1 - \tan\kappa)}{2\tan\kappa}\right]^{1/2} \tag{2.18b}$$

and

$$\mu(x = b_c) = \mu_b = 2K\left[b_c + \frac{\delta_a(1 - \tan\kappa)}{2\tan\kappa}\right]^{1/2} \tag{2.18c}$$

Working now with with the general solution of eq. (2.17) and the transformed boundary conditons of eqs. (2.18),

$$\left.\frac{d\theta}{d\mu}\right|_{x=0} = [C_1 I_1(\mu) - C_2 K_1(\mu)]_{\mu=\mu_a} = 0$$

is obtained. Hence

$$C_1 I_1(\mu_a) = C_2 K_1(\mu_a)$$

and in addition,

$$\theta(\mu = \mu_b) = C_1 I_0(\mu_b) + C_2 K_0(\mu_b) = \theta_b$$

The arbitrary constants are evaluated by simultaneous solution of the foregoing two equations:

$$C_1 = \frac{\theta_b K_1(\mu_a)}{I_0(\mu_b)K_1(\mu_a) + I_1(\mu_a)K_0(\mu_b)}$$

and

$$C_2 = \frac{\theta_b I_1(\mu_a)}{I_0(\mu_b)K_1(\mu_a) + I_1(\mu_a)K_0(\mu_b)}$$

The particular solution for the temperature excess is

$$\theta(\mu) = \theta_b \frac{K_1(\mu_a)I_0(\mu) + I_1(\mu_a)K_0(\mu)}{I_0(\mu_b)K_1(\mu_a) + I_1(\mu_a)K_0(\mu_b)} \tag{2.19}$$

In the steady state, the heat flow through the fin base must equal the heat dissipated by both faces of the fin. For unit length,

$$q_b = 2h \int_0^{b_c} \theta \frac{dx}{\cos \kappa}$$

But dx in terms of $d\mu$ is known:

$$dx = \frac{\mu}{2K^2} d\mu$$

so that

$$q_b = \frac{h}{K^2 \cos \kappa} \int_{\mu_a}^{\mu_b} \theta \mu \, d\mu$$

The governing differential equation of eq. (2.16) may be modified to the form

$$\mu \frac{d^2\theta}{d\mu^2} + \frac{d\theta}{d\mu} - \mu\theta = 0$$

from which it is observed that

$$\mu\theta = \mu\frac{d^2\theta}{d\mu^2} + \frac{d\theta}{d\mu} = \frac{d}{d\mu}\left(\mu\frac{d\theta}{d\mu}\right)$$

With this relationship in hand, the integration can be performed:

$$\int_{\mu_a}^{\mu_b} \theta\mu\, d\mu = \int_{\mu_a}^{\mu_b} \frac{d}{d\mu}\left(\mu\frac{d\theta}{d\mu}\right) d\mu = \mu\frac{d\theta}{d\mu}\bigg|_{u_a}^{\mu_b}$$

Noting that $d\theta/d\mu$ vanishes for $\mu = \mu_a$,

$$q_b = \frac{h\mu_b\theta_b}{K^2\cos\kappa}\frac{K_1(\mu_a)I_1(\mu_b) - I_1(\mu_a)K_1(\mu_b)}{I_0(\mu_b)K_1(\mu_a) + I_1(\mu_a)K_0(\mu_b)} \tag{2.20}$$

or with $K^2 = h/k \sin\kappa$, as defined previously, the heat flow through the fin base may also be expressed as

$$q_b = k\theta_b\mu_b \tan\kappa \frac{K_1(\mu_a)I_1(\mu_b) - I_1(\mu_a)K_1(\mu_b)}{I_0(\mu_b)K_1(\mu_a) + I_1(\mu_a)K_0(\mu_b)} \tag{2.21}$$

The fin efficiency is the actual heat flow of eq. (2.20) divided by the ideal heat flow that would occur if the entire fin operated at the base temperature excess

$$q_{\text{id}} = \frac{2hb_c\theta_b}{\cos\kappa}$$

Thus

$$\eta = \frac{q_b}{q_{\text{id}}} = \frac{q_b}{2hb_c\theta_b/\cos\kappa}$$

or

$$\eta = \frac{\mu_b}{2K^2b_c}\frac{K_1(\mu_a)I_1(\mu_b) - I_1(\mu_a)K_1(\mu_b)}{I_0(\mu_b)K_1(\mu_a) + I_1(\mu_a)K_0(\mu_b)} \tag{2.22}$$

2.2.3 Radial Fin of Rectangular Profile with Tip Heat Loss

Ideal Case. The radial fin of rectangular profile with tip heat loss is displayed in Fig. 2.5. When no heat flows through the fin tip, the differential equation for the temperature excess and its general solution are as developed in Chapter 1:

$$r^2\frac{d^2\theta}{dr^2} + r\frac{d\theta}{dr} - m^2r^2\theta = 0 \tag{1.30}$$

where $m^2 = 2h/k\delta$ and

$$\theta(r) = C_1 I_0(mr) + C_2 K_0(mr) \tag{1.31}$$

For the ideal case that considers heat flow through the fin tip, the two boundary conditions are

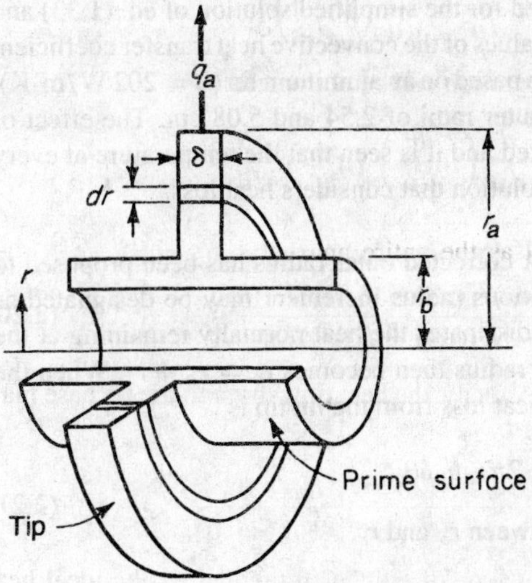

Figure 2.5 Radial fin of rectangular profile with tip heat loss.

$$\theta(r = r_b) = \theta_b \tag{2.23a}$$

and

$$\left.\frac{d\theta}{dr}\right|_{r=r_a} = -\frac{h_a}{k}\theta_a = -\frac{h_a}{k}[C_1 I_0(mr_a) + C_2 K_0(mr_a)] \tag{2.23b}$$

The arbitrary constants are obtained from simultaneous solution of eqs. (2.23), and this solution leads to the particular solution for the temperature excess:

$$\theta(r) = \theta_b \frac{I_0(mr) + \gamma K_0(mr)}{I_0(mr_b) + \gamma K_0(mr_b)} \tag{2.24}$$

where

$$\gamma = \frac{(h_a/mk)I_0(mr_a) + I_1(mr_a)}{K_1(mr_a) - (h_a/mk)K_0(mr_a)}$$

It is apparent that if the heat transfer coefficient at the tip of the fin is zero, the simplified solution results. For this condition

$$\gamma = \frac{I_1(mr_a)}{K_1(mr_a)}$$

and eq. (2.24) reduces, as it should, to eq. (1.33)[4]:

$$\theta(r) = \theta_b \left[\frac{K_1(mr_a)I_0(mr) + I_1(mr_a)K_0(mr)}{I_0(mr_b)K_1(mr_a) + K_0(mr_b)I_1(mr_a)}\right]$$

[4]Keller and Somers (1959) have shown the effects of heat flow through the tips of relatively thick fins. However, their approach focused on the effects of fin thickness rather than the effects of tip heat loss.

Temperature excess profiles obtained for the simplified solution of eq. (1.33) and the ideal solution of eq. (2.24) for two values of the convective heat transfer coefficient are compared in Fig. 2.6. The curves are based on an aluminum fin ($k = 202$ W/m·K), 0.1588 cm thick and with inner and outer radii of 2.54 and 5.08 cm. The effect of tip heat transfer coefficient may be noted and it is seen that the temperature at every point on the fin is lower for the ideal solution that considers heat loss.

Harper–Brown Approximation. A corrected outer radius has been proposed to account for the tip heat loss. The fictitious radius increment may be designated as and it is this radius increment that dissipates the heat normally remaining at the tip, where $r = r_a$. The fictitious outer radius then becomes $r_c = r_a + r'$. When the corrected radius is not employed, the heat loss from the fin tip is

$$q_a = 2\pi r_a h_a \delta \theta_a$$

If all of this heat is to be dissipated between r_a and r_c,

$$q_a = 2\pi (r_c^2 - r_a^2) h_a \theta_a$$

These dissipations may then be equated:

$$2\pi r_a h_a \delta \theta_a = 2\pi (r_c^2 - r_a^2) h_a \theta_a$$

or

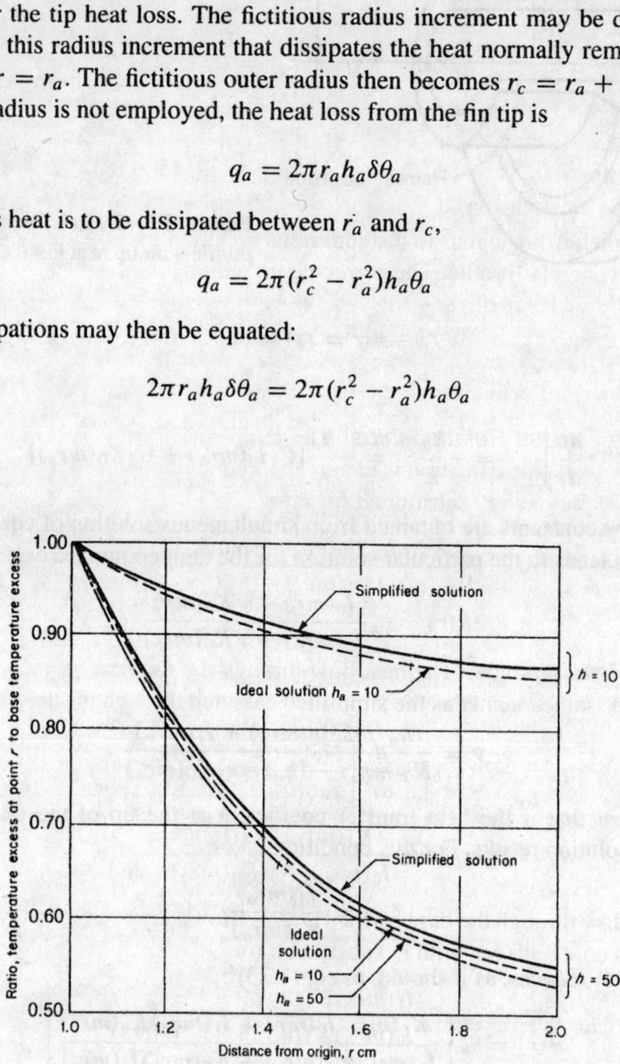

Figure 2.6 Temperature distribution for a radial fin of rectangular profile showing the effect of tip heat loss. The fin is aluminum 0.1588 cm thick with inner and outer radii of 2.54 and 5.08 cm.

$$r_a\delta = r_c^2 - r_a^2$$
$$= (r_c - r_a)(r_c + r_a)$$
$$= r'(r_c + r_a)$$

It is assumed that r' is small and that r_a and r_c are approximately equal. Should this not be the case, the heat loss from the fin tip will be significant and the approximation, as well as the entire analysis, will not be valid. With this in mind, observe that $r_c + r_a = 2r_a$ and that

$$r_a\delta = r'(r_c + r_a) = 2r_a$$

so that[5]

$$r' = \frac{\delta}{2}$$

which is completely analogous to the additional height in the analysis of the longitudinal fin of rectangular profile. The corrected tip radius becomes

$$r_c = r_c + r' = r_a + \frac{\delta}{2}$$

Temperature Excess Relationships. The temperature excess relationship using the Harper–Brown approximation is identical with the relationship for the simplified case, eq. (1.33), but with r_c substituted for r_a:

$$\theta(r) = \theta_b \frac{K_1(mr_c)I_0(mr) + I_1(mr_c)K_0(mr)}{I_0(mr_b)K_1(mr_c) + K_0(mr_b)I_1(mr_c)} \tag{2.25}$$

Heat Flow Relationships. The heat flow through the fin base for the ideal case is obtained in the same manner as the simplified case and through the use of eq. (2.24):

$$q_b = -2\pi k r_b \delta\theta_b \frac{d}{dr}\left[\frac{I_0(mr) + \gamma K_0(mr)}{I_0(mr_b) + \gamma K_0(mr_b)}\right]_{r=r_b}$$

$$q_b = 2\pi k r_b \delta m\theta_b \frac{\gamma K_1(mr_b) - I_1(mr_b)}{I_0(mr_b) + \gamma K_0(mr_b)} \tag{2.26}$$

The heat flow through the base by the Harper–Brown approximation will be in the same form as eq. (1.34) but with r_c substituted for r_a:

$$q_b = 2\pi k r_b \delta m\theta_b \frac{I_1(mr_c)K_1(mr_b) - K_1(mr_c)I_1(mr_b)}{I_0(mr_b)K_1(mr_c) + K_0(mr_b)I_1(mr_c)} \tag{2.27}$$

[5]The exact solution would be $r' = (\delta/2)[1 - 4(\delta/4r_a) + \cdots]$. However, the sum of the terms following the 1 will amount to about 1%, and omission of these terms causes no appreciable error, particularly because r' is itself a very small correction.

with

$$r_c = r_a + \frac{\delta}{2} = r_a + \frac{h}{km^2}$$

Efficiency Relationships. For the Harper–Brown approximation, a development similar to that given in Chapter 1 leads to the fin efficiency in terms of the corrected radius r_c:

$$\eta = \frac{2\rho}{\phi(1+\rho)} \frac{I_1(\phi R_c)K_1(\phi R_b) - K_1(\phi R_c)I_1(\phi R_b)}{I_0(\phi R_b)K_1(\phi R_c) + K_0(\phi R_b)I_1(\phi R_c)} \tag{2.28}$$

where here

$$\rho = \frac{r_b}{r_c}$$

$$\phi = (r_c - r_b)^{1/2}\left(\frac{2h}{k\delta}\right)^{1/2}$$

$$\phi R_c = \frac{\phi}{1-\rho}$$

and

$$\phi R_b = \frac{\rho\phi}{1-\rho}$$

Example 2.3: Radial Fin of Rectangular Profile with Tip Heat Loss. A longitudinal fin of radial profile is exposed to surroundings at 35°C via a heat transfer coefficient of $h = 40\ \text{W/m}^2 \cdot \text{K}$. The temperature at the fin base is 110°C and the fin is made from a steel with $k = 40\ \text{W/m} \cdot \text{K}$. The inner and outer diameters are 10 and 25 cm, respectively, and the fin thickness is 0.25 cm. Using (1) the case of no heat flow from the tip, (2) the ideal case *with* heat flow from the tip, and (3) the Harper–Brown approximation, compare (**a**) the fin tip temperature excess and (**b**) the heat dissipation.

SOLUTION. For this particular radial fin

$$\theta_b = 110 - 35 = 75°\text{C}$$

$$r_a = 0.125\ \text{m} \qquad r_b = 0.050\ \text{m}$$

$$\rho = \frac{r_a}{r_b} = \frac{0.050}{0.125} = 0.400 \qquad \delta = 0.0025\ \text{m}$$

$$m = \left(\frac{2h}{k\delta}\right)^{1/2}\left[\frac{(2)(40)}{(40)(0.0025)}\right]^{1/2} = 28.2843\ \text{m}^{-1}$$

$$mr_a = (28.2843)(0.125) = 3.5355$$

$$mr_b = (28.2843)(0.050) = 1.4142$$

and

$$\frac{h_a}{mk} = \frac{40}{(28.284)(40)} = 0.0354$$

Tables or software provide

$$\begin{aligned}
I_0(mr_b) &= I_0(1.4142) &= 1.5661\\
I_0(mr_a) &= I_0(3.5355) &= 7.5897\\
I_1(mr_b) &= I_1(1.4142) &= 0.8992\\
I_1(mr_a) &= I_1(3.5355) &= 6.4081\\
K_0(mr_b) &= K_0(1.4142) &= 0.2387\\
K_0(mr_a) &= K_0(3.5355) &= 0.0189\\
K_1(mr_b) &= K_1(1.4142) &= 0.3136\\
K_1(mr_a) &= K_1(3.5355) &= 0.0213
\end{aligned}$$

(a) In the tip temperature excess comparison, for the simplified case, the tip temperature excess derives from eq. (1.33) and was found in Example 1.3 to be

$$\theta_b = 13.5°\text{C}$$

For the ideal case eq. (2.24) is used with $h_a = 40\ \text{W/m}^2 \cdot \text{K}$. The parameter γ is evaluated first:

$$\begin{aligned}
\gamma &= \frac{(h_a/mk)I_0(mr_a) + I_1(mr_a)}{K_1(mr_a) - (h_a/mk)K_0(mr_a)}\\
&= \frac{(0.0354)(7.5897) + 6.4081}{0.0213 - (0.0354)(0.0189)}\\
&= \frac{6.6765}{0.0207}\\
&= 322.96
\end{aligned}$$

Then

$$\theta_a = \theta_b \frac{I_0(mr_a) + \gamma K_0(mr_a)}{I_0(mr_b) + \gamma K_0(mr_b)}$$

$$= 75\left[\frac{7.5897 + (322.96)(0.0189)}{1.5661 + (322.96)(0.2387)}\right] = 75\left(\frac{13.6936}{78.6692}\right) = 13.0°\text{C}$$

For the Harper–Brown approximation, use eq. (2.25) with

$$r_c = r_a + \frac{\delta}{2} = 0.1250 + 0.00125 = 0.1263$$

This makes

$$mr_c = (28.2843)(0.1263) = 3.5709$$

and tables or software provide

$$\begin{aligned} I_0(mr_c) &= I_0(3.5709) = 7.8324 \\ I_1(mr_c) &= I_1(3.5709) = 6.6164 \\ K_0(mr_c) &= K_0(3.5709) = 0.0181 \\ K_1(mr_c) &= K_1(3.5709) = 0.0205 \end{aligned}$$

Then

$$\begin{aligned} \theta_a &= \theta_b \frac{K_1(mr_c)I_0(mr_c) + I_1(mr_c)K_0(mr_c)}{I_0(mr_b)K_1(mr_c) + K_0(mr_b)I_1(mr_c)} \\ &= 75\left[\frac{(0.0205)(7.8324) + (6.6164)(0.0181)}{(1.5661)(0.0205) + (0.2387)(6.6164)}\right] \\ &= 75\left(\frac{0.2800}{1.6117}\right) \\ &= 13.0°C \end{aligned}$$

(b) In the heat dissipation comparison, for the simplified case with no tip heat loss, the heat dissipation was obtained in Example 1.3 as

$$q_b = 84.9 \text{ W}$$

For the ideal case, eq. (2.26) is used with $\gamma = 322.96$:

$$\begin{aligned} q_b &= 2\pi k r_b \delta m \theta_b \frac{\gamma K_1(mr_b) - I_1(mr_b)}{I_0(mr_b) + \gamma K_0(mr_b)} \\ &= 2\pi(40)(0.05)(0.0025)(28.243)(75)\left[\frac{(322.96)(0.3136) - 0.8992}{1.5661 + (322.96)(0.2387)}\right] \\ &= 66.643\left(\frac{100.3838}{78.6692}\right) \\ &= 85.0 \text{ W} \end{aligned}$$

For the Harper–Brown approximation, use eq. (2.27):

$$q_b = 2\pi k r_b \delta m \theta_b \frac{I_1(mr_c)K_1(mr_b) - K_1(mr_c)I_1(mr_b)}{I_0(mr_b)K_1(mr_c) + K_0(mr_b)I_1(mr_c)}$$

The term outside the brackets is the same as in eq. (2.26):

$$2\pi k r_b \delta m \theta_b = 2\pi(40)(0.05)(0.0025)(28.243)(75) = 66.643$$

so that

$$q_b = 66.643\left[\frac{(6.6164)(0.3136) - (0.0205)(0.8992)}{(1.5661)(0.0205) + (0.2387)(6.6164)}\right]$$

$$= 66.643\left(\frac{2.0565}{1.6114}\right)$$

$$= 85.0 \text{ W}$$

2.3 NONUNIFORM HEAT TRANSFER COEFFICIENTS

Gardner's 1945 paper brought forth a great deal of printed discussion which tended to trigger the investigation of the validity of assuming a constant value for the heat transfer coefficient. Apparently, the first item of interest here was a paper by Ghai (1951), who presented an experimental investigation of straight rectangular fins. This investigation revealed large differences in the heat transfer coefficient both from fin base to fin tip and along the fin in the direction of the airflow. The evaluation of the variation of the point-to-point heat transfer coefficients was made possible by the construction of special hollow fins of stainless steel with as many as 50 thermocouples embedded in an area of 2 by 4 in. without disturbing the airflow. Ghai's results showed that the heat transfer coefficient was minimal at the point where the fin was attached to the prime surface and that the coefiicient increased as the fin tip was approached.

In a discussion of Ghai's paper, Gardner (1951), showed that Ghai's data could be represented by an equation for the heat transfer coefficient of the form

$$h = h_o \frac{(1+a)[(x/b) - c]^a}{(1+c)^{a+1} - c^{a+1}}$$

where h_o is the mean value of h, a and c are constants, b the fin height, and x/b the fractional distance from the fin base. Gardner then worked from his generalized differential equation using the foregoing value of h. He was able to adjust his results for the fin efficiency contained in his 1945 paper to accommodate the fin efficiency based on this heat transfer coefficient variation. Here, too, the fin efficiency expressions were in terms of the modified Bessel functions.

Melese (1957) examined the problem of the longitudinal fins of rectangular, trapezoidal, and triangular profiles with a heat transfer coefficient that varied linearly over the fin height. In general, the fin efficiencies that he obtained were lower for the linear variation than for a constant value equal to the mean of the linear variation. In this, and many subsequent works, little or no reasoning was provided to justify why the coefficient should vary in the manner assumed. In the Melese (1957) study, the thermal conductivity of the fin and the environmental temperature were assumed constant and internal heat generation in the fin was not included.

It was Fortescue (1957) who suggested that the heat transfer correlations should relate the *Stanton number* to the *Reynolds number* and a thermal conductance ratio to account for the variation in fin efficiency between the variable h and the constant h cases. This method was later adopted by Hughes and Slack (1958) in an empirical investigation involving spiral fuel elements.

Han and Lefkowitz (1960) assumed a power law variation of h over the height of the fin:

$$h(x) = (\gamma + 1)h_o \left(\frac{x}{b}\right)^{\gamma}$$

where γ is a number, h_o the mean value of h, and b the height of the fin. It may be noted that $h = 0$ at the base of the fin, h is contant when $\gamma = 0$, and linear and parabolic variations occur when γ is set equal to 1 and 2, respectively. Here, too, no attempt was made to relate this type of variation to the real world. Chen and Zyskowski (1963) considered an exponential variation for the heat transfer coefficient:

$$h(x) = h_o \frac{1 - ae^{-c(x/b)}}{1 - (c/a)(1 - e^{-c})}$$

where again, h_o is the mean value of the heat transfer coefficient and a, b, and c are constants. The works of Han and Lefkowitz (1960) and Chen and Zyskowski (1963) is discussed in subsequent sections.

Migai (1963) applied linear and hyperbolic variations to h and produced curves that were based on the maximum value of h at both the fin base and the fin tip, giving the error in the fin efficiency resulting from the use of a constant value of h.

Stynes and Myers (1964) took advantage of the similarity between the transfer of heat and the transfer of mass. The quantity of benzoic acid to dissolve locally in water in cross flow was investigated for both upstream- and downstream-facing sides of the fins. Although no mathematical correlation was attempted, plots of what is known as the *j-factor* were produced for varying fin spacings, fin heights, channel widths, and Reynolds numbers. Cumo et al. (1964) gave a rather brief theoretical analysis of the variation of the convective coefficient on the walls of channels in parallel flow.

Ueda and Harada (1964) performed an extensive numerical study of the heat transfer coefficient and the friction factor in banks of straight rectangular profile fins in cross flow. They also considered the effects of fin height and spacing, channel height, and the thermal properties of the fins themselves. One of the graphs produced showed a variation of the fluid temperature in the gaps between the fins and in the direction of the fin height. A large variation in the surrounding fluid temperature was demonstrated. This appears to be the first published case in which a constant surrounding or environmental temperature was put into question. It pointed out that a systematic study of the variation of the environmental temperature was needed.

It appears that the first two-dimensional study of a straight fin of general profile to incorporate a variable heat transfer coefficient on the fin faces and to allow for internal heat generation within the fin was conducted by Cumo et al. (1965a). This treatment employed a numerical solution using the Gauss–Seidel iteration method which is computationally efficient when the coefficient matrix obtained by taking first-law energy balances on a large number of presumed isothermal subvolumes[6] has a dominant principal diagonal. This work, in Italian, showed that the one-dimensional formulation of the conduction within the fin could be misleading under certain circumstances. However, the guidance provided with regard to the variable heat transfer coefficient is significant and a two-dimensional variation of the temperature at the fin base was also included. The Fortran listing of the computer code used is provided in the work of Cumo et al. (1965b).

Roizen (1966) published additional developments with regard to the variation of the heat transfer coefficient on the faces of a longitudinal fin of rectangular profile when the fin was in cross flow to the fluid stream. Graphs were presented that showed the variation of h over the upstream and downstream faces of the fin. Correction curves were presented to enable the true one-dimensional fin efficiency to be calculated by the designer when the fin efficiency based on a constant h had been determined. The error in assuming a constant h was significant, but no mathematical correlation between the heat transfer coefficient and the flow geometry was attempted.

Melese and Wilkins (1966) treated longitudinal fins of arbitrary profile with variable heat transfer coefficients on the fin faces, internal heat generation, and variable thermal conductivity. The shapes considered were rectangular, trapezoidal, and triangular profiles. They also provided an optimized profile based on all three of the foregoing constraints.

Joyce (1967) gave a procedure for optimization of heat exchangers with finned tubes containing longitudinal fins of rectangular profile. Here, with a variation of h on the fin faces, two-dimensional temperature variations were taken into account, and Evreinova (1967) conducted an empirical study of heat transfer and friction of longitudinal finned surfaces.

Sturman (1973) confirmed for a rather limited number of cases that very large variations of the temperature within the gaps between fins in cross flow do exist. He also showed that the constant-h and constant-environmental-temperature models are in rather considerable disagreement with experimental results.

An investigation of the variation of h over the surface of straight fins of rectangular profile and with parallel flow was conducted by Stachiewicz (1969). The results showed a general increase in h from fin base to fin tip but with a marked dip at about 75% of the fin height. It was observed that the same general shape of the h curve seemed to apply over a wide range of Reynolds numbers.

Attention will now be focused on the Han and Leflowitz (1960) and Chen and Zyskowski (1963) studies. The treatments contained in these studies are interesting because they indicate that mathematical analyses of this rather complicated problem are possible.

[6] These subvolumes are called *nodes* or *cells*.

2.3.1 Han and Lefkowitz (1960)

Han and Lefkowitz (1960) treated the fin shown in Fig. 2.7. The general differential equation for the temperature excess, $\theta(x) = T(x) - T_s$, with any heat transfer function taken as a function of the distance from the fin base is

$$\frac{d^2\theta}{dx^2} - \frac{2h(x)}{k\delta}\theta = 0 \tag{2.29}$$

Han and Lefkowitz assumed the power series form for the heat transfer coefficient

$$h(x) = (\gamma + 1)h_o\left(\frac{x}{b}\right)^\gamma \tag{2.30}$$

where h_o is the average coefficient. Substitution of eq. (2.30) into eq. (2.29) yields

$$\frac{d^2\theta}{dx^2} - \frac{2(\gamma + 1)h_o}{k\delta}\left(\frac{x}{b}\right)^\gamma \theta = 0 \tag{2.31}$$

When $\gamma = 0$, the coefficient $h(x)$ is constant over the faces of the fin. When $\gamma = 1$, the coefficient increases linearly from $x = 0$ to $x = b$. Values of $\gamma \geq 2$ produce parabolic distributions. In all cases of $\gamma \geq 1$, the magnitude of the heat transfer coefficient is zero at the fin base.

The method of Section A.4 may be applied to eq. (2.31) because it is in the form

$$\frac{d}{dx}\left(x^p \frac{d\theta}{dx}\right) - ax^j = 0$$

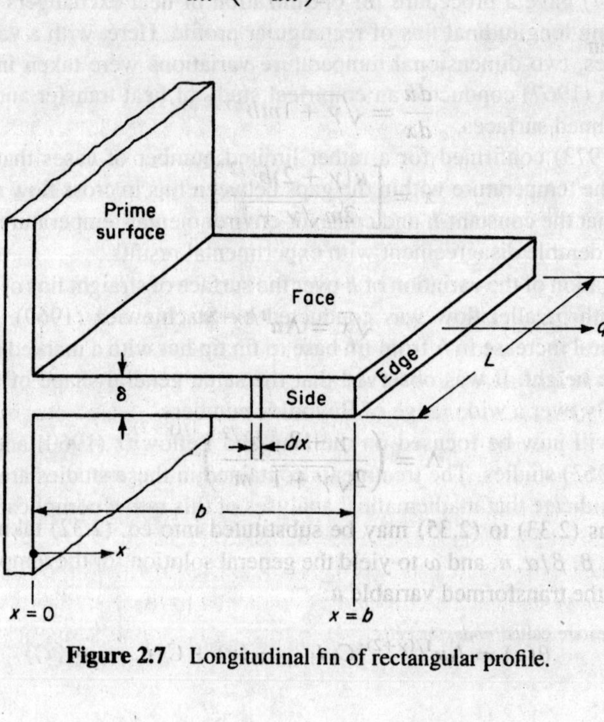

Figure 2.7 Longitudinal fin of rectangular profile.

where $p = 0,\ j = \gamma$ and

$$a = (\gamma + 1)b^{-\gamma}\left(\frac{2h_o}{k\delta}\right) = (\gamma + 1)b^{-\gamma}m^2$$

The general solution takes the form

$$\theta(x) = x^{\beta/\alpha}[C_1 I_n(\omega x^{1/n}) + C_2 I_{-n}(\omega x^{1/n})] \tag{2.32}$$

where

$$\alpha = \frac{2}{\gamma + 2} \qquad \beta = \frac{1}{\gamma + 2}$$

$$\frac{\beta}{\alpha} = \frac{1}{2} \qquad n = \frac{1}{\gamma + 2}$$

and

$$\omega = \alpha\sqrt{a} = \frac{2\sqrt{\gamma + 1}}{\gamma + 2}mb^{-\gamma/2}$$

If the transformation

$$u \equiv \frac{2\sqrt{\gamma + 1}}{\gamma + 2}mb^{-\gamma/2}x^{(\gamma+2)/2} \tag{2.33}$$

is made, then

$$\frac{du}{dx} = \sqrt{\gamma + 1}mb^{-\gamma/2}x^{\gamma/2}$$

$$x = \left[\frac{u(\gamma + 2)b^{\gamma/2}}{2m\sqrt{\gamma + 1}}\right]^{2/(\gamma+2)}$$

and

$$\sqrt{x} = \Lambda u^{1/(\gamma+2)} \tag{2.34}$$

where

$$\Lambda = \left(\frac{\gamma + 2}{2\sqrt{\gamma + 1}}\frac{b^{\gamma/2}}{m}\right)^{1/(\gamma+2)} \tag{2.35}$$

Equations (2.33) to (2.35) may be substituted into eq. (2.32) taking note of the values of α, β, β/α, n, and ω to yield the general solution for the temperature excess in terms of the transformed variable u:

$$\theta(u) = \Lambda u^{1/(\gamma+2)}[C_1 I_{1/(\gamma+2)}(u) + C_2 I_{-1/(\gamma+2)}(u)] \tag{2.36}$$

The arbitrary constants C_1 and C_1 are evaluated by means of the boundary conditions. At $u = 0$,

$$\theta(u = 0) = \theta_b$$

and at $u = u_b = \left[2\sqrt{\gamma + 1}/(\gamma + 1)\right] mb$

$$\left.\frac{d\theta}{du}\right|_{u=u_b} = 0$$

Use of these boundary conditions in eq. (2.36) permits the arbitrary constants to be evaluated and leads to the particular solution for the temperature excess:

$$\theta(u) = \frac{\theta_b \left\{\Gamma_{[(\gamma+1)/(\gamma+2)]}\right\}}{2^{1/(\gamma+2)}} u^{1/(\gamma+2)} \Upsilon \tag{2.37}$$

where Γ represents the *gamma function* and where

$$\Upsilon = I_{-1/(\gamma+2)}(u) + \frac{I_{(\gamma+1)/(\gamma+2)}(u_b)}{I_{-(\gamma+1)/(\gamma+2)}(u_b)} I_{1/(\gamma+2)}(u)$$

The fin efficiency is defined by

$$\eta = \frac{q_b}{q_{\text{id}}} = \frac{-k\delta L \left.\frac{d\theta}{du}\right|_{x=0}}{2h_o L b \theta_b}$$

and after termwise differentiation and evaluation at $u = 0$

$$\eta = \left[\frac{(\gamma + 2)^{\gamma}(\gamma + 1)}{(mb)^{2(\gamma+1)}}\right]^{1/(\gamma+2)} \frac{I_{(\gamma+1)/(\gamma+2)}(u_b)}{I_{-(\gamma+1)/(\gamma+2)}(u_b)} \frac{\Gamma(\gamma + 1)/(\gamma + 2)}{\Gamma(1/(\gamma + 2))} \tag{2.38}$$

where, as already noted,

$$u_b = \frac{2\sqrt{\gamma + 1}}{\gamma + 2} mb$$

Example 2.4: Longitudinal Fin of Rectangular Profile in Surroundings with Nonuniform Heat Transfer Coefficient. A longitudinal fin of rectangular profile is 7.62 cm high and 0.3226 cm thick. The fin is made from a steel with $k = 30\ \text{W/m} \cdot \text{K}$. Compare the fin efficiencies for (**a**) a constant heat transfer coefficient of $h = 15\ \text{W/m}^2 \cdot \text{K}$ and (**b**) a heat transfer coefficient that varies as the fourth power of the distance from the base, $h = 75(x/b)^4\ \text{W/m}^2 \cdot \text{K}$.

SOLUTION. The efficiency of the longitudinal fin of rectangular profile with a uniform heat transfer coefficient is given by eq. (1.11). With $h = 15\ \text{W/m}^2 \cdot \text{K}$,

$$m = \left(\frac{2h}{k\delta}\right)^{1/2} = \left[\frac{(2)(15)}{(30)(3.226 \times 10^{-3})}\right]^{1/2} = 17.607 \text{ m}^{-1}$$

and

$$mb = (17.607)(0.0762) = 1.342$$

Then

$$\eta = \frac{\tanh mb}{mb} = \frac{\tanh 1.342}{1.342} = \frac{0.871}{1.342} = 0.650$$

(**b**) For the fin with

$$h = 75\left(\frac{x}{b}\right)^4$$

eq. (2.30) indicates that $\gamma = 4$ and $h_o = 15 \text{ W/m}^2 \cdot \text{K}$. Thus $mb = 1.342$ and

$$u_b = \frac{2\sqrt{\gamma+1}}{\gamma+2} mb = \frac{2\sqrt{4+1}}{4+2}(1.342) = 1.000$$

In the fin efficiency given by eq. (2.38),

$$\eta = \left[\frac{(\gamma+2)^{\gamma}(\gamma+1)}{(mb)^{2(\gamma+1)}}\right]^{1/(\gamma+2)} \times \frac{I_{(\gamma+1)/(\gamma+2)}(u_b)}{I_{-(\gamma+1)/(\gamma+2)}(u_b)} \frac{\Gamma(\gamma+1)/(\gamma+2)}{\Gamma(1/(\gamma+2))} \tag{2.38}$$

the term in the first line can be evaluated as

$$\left[\frac{(6)^4(5)}{(1.342)^{(2)(5)}}\right]^{1/6} = 2.646$$

so that

$$\eta = (2.646)\frac{I_{5/6}(1.000)}{I_{-5/6}(1.000)} \frac{\Gamma(5/6)}{\Gamma(1/6)}$$

For the gamma functions, working from tables with the fundamental relationship

$$\Gamma(n+1) = n\Gamma(n)$$

or from software,

$$\Gamma(5/6) = 1.1282$$

and

$$\Gamma(1/6) = 5.5662$$

The modified Bessel functions can be evaluated from software, the polynomial approximations of Abramowitz and Stegun (1964), or the defining equations [eqs. (A.87)],

$$I_{5/6}(1.000) = 0.6811$$

and

$$I_{-5/6}(1.000) = 0.8541$$

Thus the fin efficiency is

$$\eta = 2.646 \left(\frac{1.1282}{5.5662} \right) \left(\frac{0.6811}{0.8541} \right) = 0.429$$

The marked decrease in fin efficiency over the case where h is uniform is due to the fact that the fourth power coefficient of 4 leads to values of the heat transfer coefficient of zero at the fin base and close to zero at points on the fin where the temperature excess, θ, has its highest values.

2.3.2 Chen and Zyskowski (1963)

Chen and Zyskowski (1963) took eq. (2.29) and applied an exponential variation of h:

$$h(x) = h_o \frac{1 - ae^{-c(x/b)}}{1 - (a/c)(1 - e^{-c})} \tag{2.39}$$

where a, b, and c are constants. They also employed a coefficient, h_∞, to define the asymptotic value of the heat transfer coefficient as the fin becomes infinitely high. The particular solution for the temperature excess with the assumed variation of the heat transfer coefficient where

$$n = \frac{2mb}{c}$$

is not an integer is

$$\theta(u) = \theta_b[C_1 J_n(u) + C_2 J_{-n}(u)] \tag{2.40}$$

where

$$u = n(ae^{-cx/b})^{1/2}$$

In eq. (2.40), the arbitrary constants in terms of

$$\psi = n \left(\frac{a}{e^c} \right)^{1/2} \tag{2.41a}$$

and

$$\phi = n\sqrt{a} \tag{2.41b}$$

are given by

$$C_1 = -\frac{nJ_{-n}(\psi) + \psi J_{1-n}(\psi)}{J_n(\phi)[-nJ_n(\psi) - \psi J_{1-n}(\psi)] - J_{-n}(\phi)[-nJ_n(\psi) + \psi J_{n-1}(\phi)]} \tag{2.42a}$$

and

$$C_2 = \frac{nJ_n(\psi) - \psi J_{n-1}(\psi)}{J_n(\phi)[-nJ_{-n}(\psi) - \psi J_{1-n}(\psi)] - J_{-n}(\phi)[-nJ_n(\psi) + \psi J_{n-1}(\psi)]} \tag{2.42b}$$

If n is an integer, the particular solution is

$$\theta(u) = \theta_b[C_3 J_n(u) + C_4 Y_n(u)] \tag{2.43}$$

and the arbitrary constants are given by

$$C_3 = \frac{Y_{1-n}(\psi) - Y_{n+1}(\psi)}{J_n(\phi)[Y_{n-1}(\psi) - Y_{n+1}(\psi)] - Y_n(\phi)[J_{n-1}(\psi) - J_{n+1}(\psi)]} \tag{2.44a}$$

and

$$C_4 = \frac{-J_{n-1}(\psi) + J_{n+1}(\psi)}{J_n(\phi)[Y_{n-1}(\psi) - Y_{n+1}(\psi)] - Y_n(\phi)[J_{n-1}(\psi) - J_{n+1}(\psi)]} \tag{2.44b}$$

Two fin efficiency relationships evolve from the definition

$$\eta = \frac{q_b}{q_{\mathrm{id}}} = \frac{-k\delta L \left.\dfrac{d\theta}{dx}\right|_{x=0}}{2\theta_b \displaystyle\int_0^b h(x)\,dx}$$

If n is not an integer,

$$\eta = \frac{c(A_1 - B_1)}{2(mb)^2\,[1 - (a/c)(1 - e^{-c})]} \tag{2.45}$$

where

$$A_1 = \frac{[-nJ_{-n}(\psi) - \psi J_{1-n}(\psi)][-nJ_n(\phi) + \phi J_{n-1}(\phi)]}{J_n(\phi)[-nJ_{-n}(\psi) - \psi J_{1-n}(\psi)] - J_{-n}(\phi)[-nJ_n(\psi) + \psi J_{n-1}(\psi)]}$$

and

$$B_1 = \frac{[-J_n(\psi) + \psi J_{n-1}(\psi)][-nJ_{-n}(\phi) - \phi J_{1-n}(\phi)]}{J_n(\phi)[-nJ_{-n}(\psi) - \psi J_{1-n}(\psi)] - J_{-n}(\phi)[-nJ_n(\psi) + \psi J_{n-1}(\psi)]}$$

If n is an integer,

$$\eta = \frac{\sqrt{a}\,(A_2 - B_2)}{2mb\,[1 - (a/c)(1 - e^{-c})]} \tag{2.46}$$

where

$$A_2 = \frac{[Y_{n-1}(\psi) - Y_{n+1}(\psi)][J_{n-1}(\phi) - J_{n+1}(\phi)]}{J_n(\phi)[Y_{n-1}(\psi) - Y_{n+1}(\psi)] - Y_n(\phi)[J_{n-1}(\psi) - J_{n+1}(\psi)]}$$

and

$$B_2 = \frac{[J_{n-1}(\psi) - J_{n+1}(\psi)][Y_{n-1}(\phi) - Y_{n+1}(\phi)]}{J_n(\phi)[Y_{n-1}(\psi) - Y_{n+1}(\psi)] - Y_n(\phi)[J_{n-1}(\psi) - J_{n+1}(\psi)]}$$

2.3.3 Gardner (1951)

Gardner (1951) pointed out that for the longitudinal fin of rectangular profile with origin of the height coordinate at the fin base, a heat transfer coefficient variation of the form

$$h(x) = h_o \frac{(1+a)[(x/b)+c]^a}{(1+c)^{a+1} - c^{a+1}} \tag{2.47}$$

yields a form of the Bessel equation

$$\frac{d^2\theta}{d[(x/b)+a]^2} - \left[(mb)^2 \frac{1+a}{(1+c)^{a+1} - c^{a+1}}\right]\left(\frac{x}{b}+a\right)^a \theta = 0 \tag{2.48}$$

where

$$m = \left(\frac{2ha}{k\delta}\right)^{1/2}$$

Define

$$u \equiv 2n\left(\frac{1-n}{n}\right)^{1/2} \frac{[(x/b)+c]^{1/n}}{(1+c)^{(1-n)/n} - c^{(1-n)/n}} mb$$

and

$$n = \frac{1}{2+a}$$

Then at $x = 0$, where $u = u_b$,

$$u_b = 2n\left(\frac{1-n}{n}\right)^{1/2} \frac{c^{1/n}}{(1+c)^{(1-n)/n} - c^{(1-n)/n}} mb$$

and at $x = b$, where $u = u_a$,

$$u_a = 2n\left(\frac{1-n}{n}\right)^{1/2} \frac{1+c^{1/n}}{(1+c)^{(1-n)/n} - c^{(1-n)/n}} mb$$

the solution of eq. (2.48) becomes

$$\theta(x) = \theta_b \left(\frac{u}{u_b}\right)^n \frac{I_{1-n}(u_a)I_n(u) - I_{n-1}(u_a)I_{-n}(u)}{I_n(u_b)I_{1-n}(u_a) - I_{-n}(u_b)I_{n-1}(u_a)} \tag{2.49}$$

and the fin efficiency is

$$\eta = \frac{2(1-n)}{u_b\left[1-(u_a/u_b)^{2(1-n)}\right]} \frac{I_{n-1}(u_b)I_{1-n}(u_a) - I_{1-n}(u_b)I_{n-1}(u_a)}{I_n(u_b)I_{1-n}(u_a) - I_{-n}(u_b)I_{n-1}(u_a)} \tag{2.50}$$

2.4 FINS WITH INTERNAL HEAT GENERATION

Minkler and Rouleau (1960) considered the effect of heat sources in longitudinal fins of rectangular and triangular profiles with a constant heat transfer coefficient on the fin faces and with no tip heat loss. Consider a longitudinal fin of arbitrary profile and let the origin of the height coordinate be at the fin base and let b be the height of the fin. The fin profile is confined by two symmetrical curves, $y = \pm f_2(x)$. The fin cross-sectional area for unit length ($L = 1$ m) is $A = f_1(x) = 2f_2(x)$ and the temperature excess at any point on the fin is $\theta = T - T_s$, where T is the temperature on the fin and T_s is the temperature of the surroundings. The differential equation for the fin temperature profile is formulated from a steady-state energy balance over a differential element of fin height, dx.

The difference between the heat entering the element by conduction at x and that leaving the element by conduction at $x + dx$ is

$$2\frac{d}{dx}\left[kf_2(x)\frac{dT}{dx}\right]dx$$

minus the heat losses through the faces of the element,

$$2h(T - T_s)\,dx$$

plus the heat generated within the element,

$$2q_i f_2(x)$$

provide the basis for the energy balance over dx, which leads to the general differential equation

$$2\frac{d}{dx}\left[kf_2(x)\frac{d\theta}{dx}\right] - 2h\theta - 2q_i f_2(x) = 0 \tag{2.51}$$

where $d\theta = dT$ and q_i is the uniform internal heat generation in W/m^3.

For the longitudinal fin of rectangular profile, $f_2(x) = \delta/2$, and with a constant thermal conductivity, k, eq. (2.51) becomes

$$\frac{d^2\theta}{dx^2} - m^2\theta = \frac{q_i}{k} \tag{2.52}$$

where $m^2 = 2h/k\delta$. This second-order *nonhomogeneous* equation possesses a complementary function

$$\theta_c = C_1 e^{mx} + C_2 e^{-mx}$$

and a particular integral

$$\theta_{\text{pi}} = \frac{q_i \delta}{2h}$$

The general solution is

$$\theta = \theta_c + \theta_{\text{pi}} = C_1 e^{mx} + C_2 e^{-mx} + \frac{q_i \delta}{2h} \tag{2.53}$$

where the arbitrary constants C_1 and C_2 are evaluated from the boundary conditions

$$-k\delta \left.\frac{d\theta}{dx}\right|_{x=0} = q_b \tag{2.54a}$$

and

$$\left.\frac{d\theta}{dx}\right|_{x=b} = 0 \tag{2.54b}$$

Use of these boundary conditions in eq. (2.53) yields the particular solution of eq. (2.52):

$$\theta(x) = \frac{q_b}{(2hk\delta)^{1/2}} \left(\frac{\cosh mx}{\tanh mb} - \sinh mx \right) + \frac{q_i \delta}{2h} \tag{2.55}$$

It is convenient to express eq. (2.55) in terms of three dimensionless parameters which pertain to the longitudinal fin of rectangular profile.

The *removal number* or *fin effectiveness* is the Gardner (1945) definition. It is the ratio of the fin dissipation (equal, in the steady state, to the heat passing through the base of the fin by conduction) to the heat passing through the fin footprint of the base or prime surface if the fin were not present. Thus, for unit fin length,

$$N_R = \frac{q_b}{h\delta\theta_b} \tag{2.56}$$

The value of N_R must exceed unity for the employment of the fin to be worthwhile. In fact, economic considerations may dictate that N_R must be considerably greater than unity. If $N_R \le 1$, the fin acts as an insulator.

The *generation number* is the ratio of the heat generated to the heat dissipated by the fin if the entire fin were to operate at the base temperature excess. For two fin faces with unit length,

$$N_G = \frac{q_i \delta b}{2h\theta_b b} = \frac{q_i \delta}{2h\theta_b} \tag{2.57}$$

The generation number is related to the fin efficiency. The numerator and denominator of eq. (2.57) can be adjusted:

$$N_G = \frac{q_i \delta b}{2h \int_0^b \theta\, dx} \frac{2h \int_0^b \theta\, dx}{2h\theta_b b} \tag{2.58}$$

The first term of eq. (2.58) is the ratio of the heat generated to the heat actually dissipated by the fin. The second term is the fin efficiency when there is no heat generation. When eq. (2.58) is written as

$$N_G = \eta \frac{q_i b \delta}{2h \int_0^b \theta\, dx} \tag{2.59}$$

it is observed that the generation number is a measure of the *fin inefficiency* due to the heat generation. When $N_G = 0$ there is no heat generation and no fin inefficiency. When $N_G = 1$ the fin is generating heat precisely as fast as it can be removed by convection. When $N_G > 1$ the heat flows into the fin base and, as given by eq. (2.57), N_R is negative.

The *Biot number*, Bi, measures the ratio of the surface conductance to the internal conductance of the fin:

$$\text{Bi} = \frac{h\delta}{2k} \tag{2.60}$$

and as indicated in Section 2.2, it is often used as a usefulness criterion for the longitudinal fin of rectangular profile.

Returning now to eq. (2.55), it is seen that the first term on the right may be adjusted to

$$\frac{q_b}{(2hk\delta)^{1/2}} = \frac{q_b \theta_b (h\delta_b)^{1/2}}{h\delta\theta_b (2k)^{1/2}} = \theta_b N_R (\text{Bi})^{1/2}$$

and the last term becomes $\theta_b N_G$. Thus, eq. (2.55) may be written as

$$\theta(x) = \theta_b N_R (\text{Bi})^{1/2} \left(\frac{\cosh mx}{\tanh mb} - \sinh mx \right) + \theta_b N_G \tag{2.61}$$

and at $x = 0$ where $\theta = \theta_b$, eq. (2.61) reduces to

$$N_R (\text{Bi})^{1/2} = (1 - N_G) \tanh mb \tag{2.62}$$

Figure 2.8 is a plot of eq. (2.62). Here it is seen that as mb increases, the value of $N_R(\text{Bi})^{1/2}$ becomes asymptotic. At $N_G = 0$, the value of $N_R(\text{Bi})^{1/2}$ approaches unity, and for $N_G = 1$, the value of $N_R(\text{Bi})^{1/2}$ is always zero. The latter can be deduced from eq. (2.62) and represents the case where the heat generated by the fin just equals the heat dissipated by it. Figure 2.9 is a plot of the temperature excess ratio, θ/ϑ_b, for a copper fin. The copper fin is 3.08 cm high and 0.1588 cm wide, with $h = 285\ \mathrm{W/m^2 \cdot K}$. This figure is based on eq. (2.62) and shows that the tip temperature of the fin increases significantly as the tip temperature increases.

2.5 POLYGONAL FINS

A recurring arrangement of extended surface is that of a single sheet of metal pierced by round tubes in either a square or an equilateral triangular arrangement. Zabronsky (1955) obtained the efficiency of the square fin shown with its coordinate system in Fig. 2.10. Observe that the tube has a radius r_b and that the tubes are on pitch $2p$.

The difference between the heat flow entering and leaving the element of fin surface in the radial direction is

$$dq = k\delta \int \nabla\theta\, ds \tag{2.63}$$

where $\theta = T - T_s$ is the temperature excess over the surroundings at T_s, ds is the arc length at the differential element, and ∇T is the temperature gradient. The heat dissipated by the two fin faces is

$$dq = 2h\theta \iint dx\, dy \tag{2.64}$$

An application of the divergence theorem allows eq. (2.63) to be written as

$$dq = k\delta \iint \nabla^2\theta\, dx\, dy \tag{2.65}$$

where ∇^2 is the Laplacian operator. Equations (2.63) and (2.65) may be equated to produce

$$\nabla^2\theta = m^2\theta \tag{2.66}$$

where $m^2 = 2h/k\delta$. Equation (2.66) applies at any point on the fin surface.

If it is assumed that the temperature excess is constant at the tube surface,

$$\theta = \theta_b \tag{2.66a}$$

four additional boundary conditions can be obtained

$$\frac{\partial\theta}{\partial x} = \begin{cases} 0 & \text{at } x = 0 \text{ and } r \le y \le p \quad (2.66b) \\ 0 & \text{at } x = p \text{ and } 0 \le y \le p \quad (2.66c) \end{cases}$$

$$\frac{\partial\theta}{\partial y} = \begin{cases} 0 & \text{at } y = 0 \text{ and } r \le x \le p \quad (2.66d) \\ 0 & \text{at } y = 0 \text{ and } 0 \le x \le p \quad (2.66e) \end{cases}$$

Because the same conditions hold for any of the squares, the solution to eq. (2.66) must be an *even doubly periodic function* which is constant on the circles

$$(x - 2jp)^2 + (y - 2kp)^2 = r^2$$

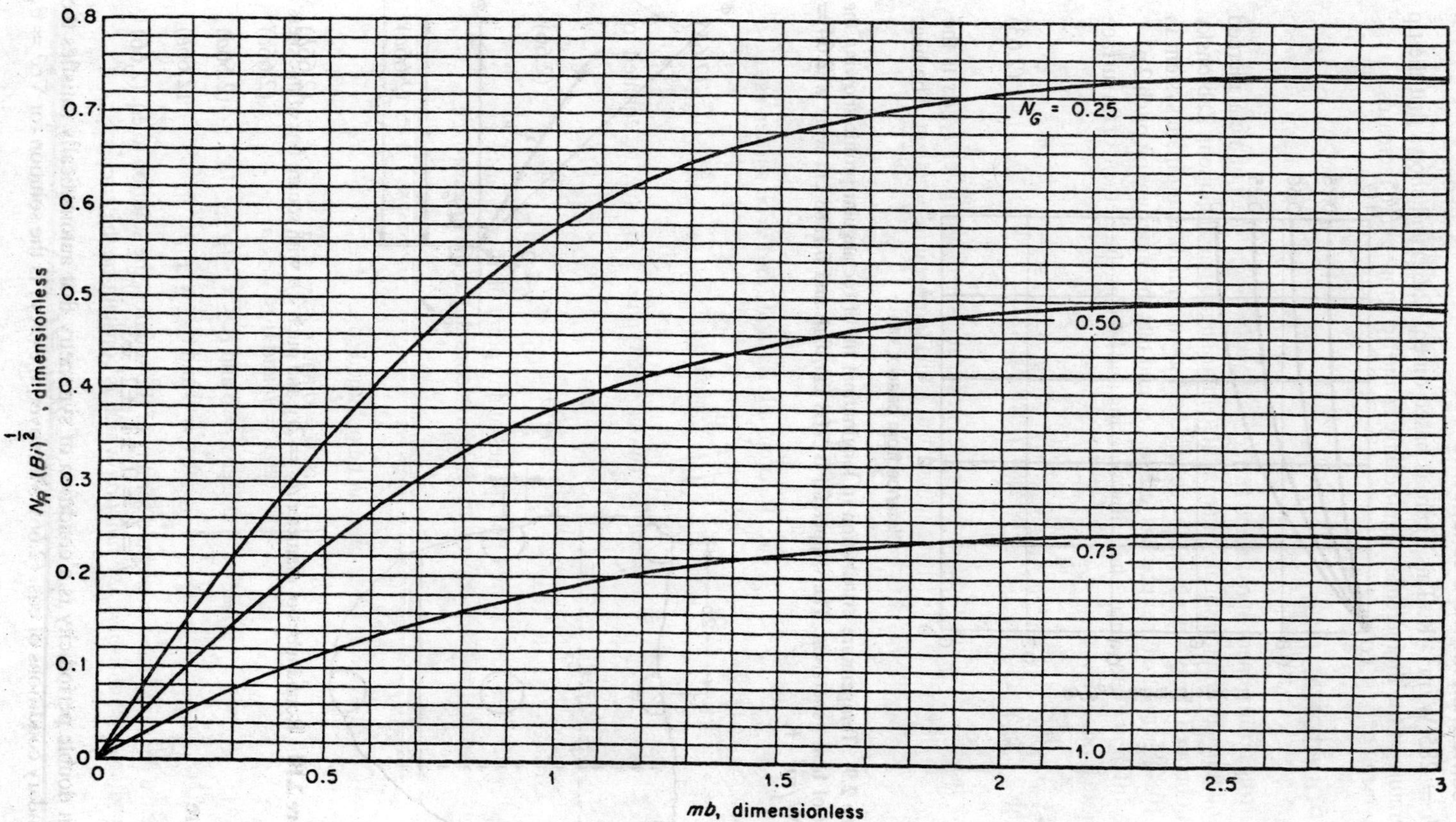

Figure 2.8 Plot of $N_R(\mathrm{Bi})^{1/2}$ for various values of the generation number N_G.

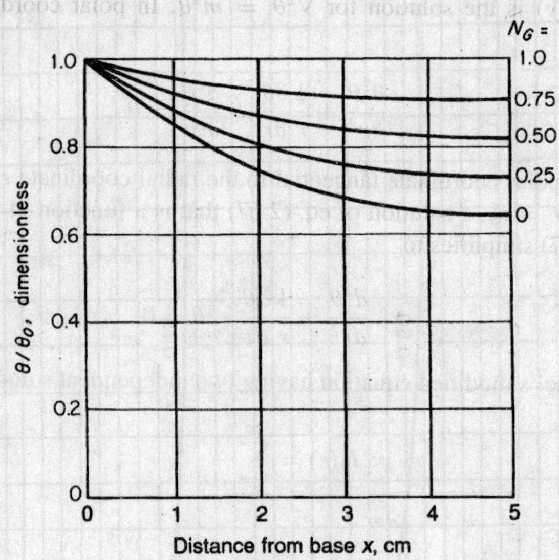

Figure 2.9 Temperature distribution in longitudinal fin of rectangular profile showing the effect of heat generation. The copper fin is 5.08 cm high and 0.1588 cm wide with $h = 285\ \text{W/m}^2 \cdot \text{K}$.

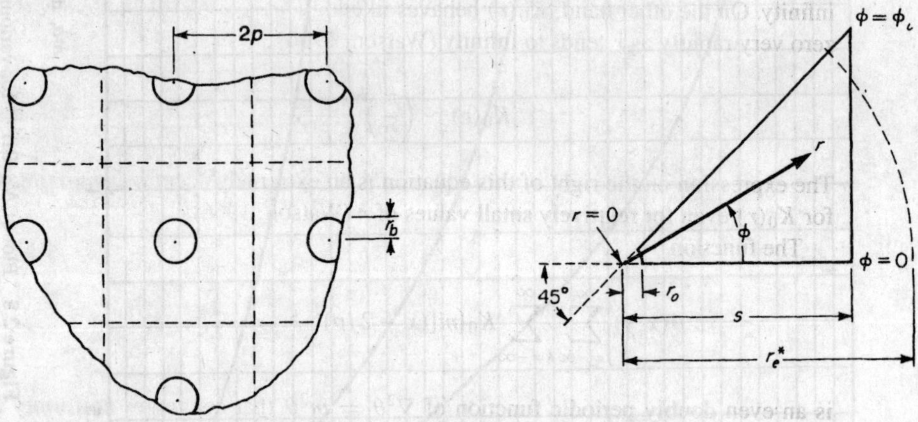

Figure 2.10 Round tubes of diameter $d_b = 2r_b$ and pitch $2p$ with square fins of thickness δ.

where

$$j = k = 0, \pm 1, \pm 2, \pm 3, \ldots$$

Even double periodicity is a condition of symmetry that automatically satisfies the boundary conditions of eqs. (2.66). Moreover, if $\theta(x, y)$ is the solution for $\nabla^2\theta = \theta$, then $\theta(mx,my)$ is the solution for $\nabla^2\theta = m^2\theta$. In polar coordinates, $\nabla^2\theta = \theta$ becomes

$$\frac{\partial^2\theta}{\partial r^2} + \frac{1}{r}\frac{\partial\theta}{\partial r} + \frac{\partial^2\theta}{\partial\phi^2} = \theta \tag{2.67}$$

where ϕ is a polar coordinate tangential to the radial coordinate r. Because $\theta = \theta_b$ on the circle $r = r_b$, a solution of eq. (2.67) that is a function of r alone is sought. Thus eq. (2.67) simplifies to

$$\frac{d^2\theta}{dr^2} + \frac{1}{r}\frac{d\theta}{dr} - \theta = 0 \tag{2.68}$$

which is Bessel's modified equation having two independent solutions

$$I_0(r) = \sum_{n=0}^{\infty}\frac{(r/2)^{2n}}{(n!)^2} \tag{2.69}$$

and

$$K_o(r) = \int_1^{\infty} e^{-rz}(z^2-1)^{-1/2}\,dz \qquad (r > 0) \tag{2.70}$$

Here, $I_0(r)$ is an entire function that behaves somewhat like e^r as r tends toward infinity. On the other hand, $K_0(r)$ behaves like $\log 1/r$ near $r = 0$ and tends toward zero very rapidly as r tends to infinity (Watson, 1966). In fact,

$$K_0(r) \sim \left(\frac{\pi}{2}\right)^{1/2}\frac{e^{-r}}{r^{1/2}} \tag{2.71}$$

The expression on the right of this equation is an extremely accurate approximation for $K_0(r)$ even for relatively small values of r (Watson, 1966).

The function

$$\theta(x,y)\sum_{j=-\infty}^{\infty}\sum_{k=-\infty}^{\infty} K_0\{m[(x-2jp)^2+(y-2pk)^2]^{1/2}\} \tag{2.72}$$

is an even doubly periodic function of $\nabla^2\theta = m^2\theta$ that converges uniformly and absolutely in any closed region, excluding the points $(2jp,2kp)$, and behaves like

$$-\ln\{m[(x-2jp)^2+(y-2pk)^2]^{1/2}\}$$

in the neighborhood of point $(2jp,2kp)$. Because of the extremely rapid convergence of $K_0(r)$ to the limit zero as r tends to infinity, $\theta(x,y)$ is nearly constant on the circles

$$(x-2jp)^2+(y-2pk)^2 = r_b^2$$

Therefore, an accurate solution is given by the function

$$\frac{\theta(x,y)}{M}$$

where M is the mean value of $\theta(x,y)$ on the circle

$$x^2 + y^2 = r_b^2$$

The addition theorem for the Bessel functions that is employed to find the mean value of $\theta(x,y)$ is

$$K_0(w) = K_0(Z)I_0(z) + 2\sum_{j=1}^{\infty} K_m(Z)I_m(z) \cos m\varphi \tag{2.73}$$

where

$$w = (Z^2 + z^2 - 2Zz \cos u)^{1/2}$$

$$Z = 2pm(j^2 + k^2)^{1/2}$$

$$z = mr$$

$$u = \phi - \psi$$

and

$$\psi = \arctan \frac{k}{j}$$

Now

$$\theta(x,y) = \sum_{j=-\infty}^{\infty} \sum_{k=-\infty}^{\infty} K_0\{m[(x - 2jp)^2 + (y - 2pk)^2]^{1/2}\}$$

or

$$\theta(x,y) = \sum_{j=-\infty}^{\infty} \sum_{k=-\infty}^{\infty} K_0\{m[r^2 - 4rp(j^2 + k^2)^{1/2} \cos(\phi - \psi)4p^2(j^2 + k^2)]^{1/2}\} \tag{2.74}$$

Therefore, the mean value of $\theta(x,y)$ on the circle $x^2 + y^2 = r_b^2$ is

$$K_0(2mr_b) + I_0(2mr_b)\theta(x,y) = {\sum_{j=-\infty}^{\infty} \sum_{k=-\infty}^{\infty}}' K_0[2mp(j^2 + k^2)^{1/2}] \tag{2.75}$$

where the prime in the double summation indicates that the term for $j = k = 0$ is omitted. Therefore, the relationship for $\theta(x,y)$ at any point on the fin is given by

$$\theta(x, y) = \theta_b \frac{\Sigma_{j=-\infty}^{\infty} \Sigma_{k=-\infty}^{\infty} K_0\{m[(x - 2jp)^2 + (y - 2pk)^2]^{1/2}\}}{I_0(mr_b)\Sigma_{j=-\infty}^{\infty} \Sigma_{k=-\infty}^{\infty}{}' K_0[2pm(j^2 + k^2)^{1/2}] + K_0(mr_b)} \tag{2.76}$$

It must be emphasized that the transition from eq. (2.67) to eq. (2.68) is a simplification that results in a very satisfactory approximation. Equation (2.72) does not represent the boundary condition, $\theta = \theta_b$ on $r = r_b$ exactly. Equation (2.76) is a smoothing of eq. (2.72) that fulfills the foregoing boundary condition even more exactly. Thus Zabronsky's solution satisfies the adiabatic condition at the fin edge exactly but satisfies the condition at the fin base only approximately. Sparrow and Lin (1964) employed an entirely different method of analysis that satisfies the isothermal boundary condition at the fin base exactly and fulfills the boundary condition at the fin edge approximately to any desired accuracy.

The fin efficiency is obtained directly from eq. (2.76) using Green's theorem and is given by

$$\eta = \frac{2\pi r_b}{m[1 - \pi(r_b)^2]} \frac{K_1(mr_b) - \Sigma_{j=-\infty}^{\infty}\Sigma_{k=-\infty}^{\infty} I_1(mr_b)K_0[2pm[((j^2 + k^2)^{1/2}]}{K_0(mr_b) - \Sigma_{j=-\infty}^{\infty}\Sigma_{k=-\infty}^{\infty} I_0(mr_b)K_0[2pm[((j^2 + k^2)^{1/2}]} \tag{2.77}$$

2.6 CLOSURE

The preliminary study of extended surface that has encompassed Chapters 1 and 2 has been concluded. The study proceeds to Chapter 3 which considers optimization of individual fins.

2.7 NOMENCLATURE

Roman Letter Symbols

- A combination of Bessel functions, dimensionless
- a coefficient in variable heat transfer coefficient analysis, dimensions vary; coefficient in general Bessel equation, dimensions vary; parameter, dimensionless
- B combination of Bessel functions, dimensionless
- Bi Biot modulus, dimensionless
- b fin height, m; coefficient in variable heat transfer coefficient analysis, m
- C arbitrary constant, dimensionless
- c coefficient in variable heat transfer coefficient analysis, dimensionless
- d derivative or differential, dimensions vary
- f function, dimensions vary
- h heat transfer coefficient, $W/m^2 \cdot K$
- I modified Bessel function of the first kind
- J Bessel function of the first kind
- j counter, dimensionless

K modified Bessel function of the second kind
k thermal conductivity, W/m · K; counter, dimensionless
L fin length, m
m fin performance parameter, m^{-1}
N number, dimensionless
n combination of terms, dimensionless
p fin pitch, dimensionless
q heat flow, W
R radius function, dimensionless
r radial coordinate, m; radius, m
S surface area, m^2
s element of arc length, m
T temperature, K
u transformed variable, dimensionless
w transformed variable, dimensionless
x height coordinate, m
Y Bessel function of the second kind
y length coordinate, m
Z transformed variable, dimensionless
z transformed variable, dimensionless

Greek Letter Symbols

α combination of terms, dimensionless
Γ Gamma function, dimensionless
γ combination of terms, dimensionless; coefficient in variable heat transfer coefficient, dimensionless relationship, dimensionless
δ fin thickness, m
ζ combination of terms, dimensionless
η fin efficiency, dimensionless
θ temperature excess, K
κ taper angle, rad
Λ combination of terms, dimensionless
μ transformed variable, dimensionless
ρ radius ratio, dimensionless
υ combination of Bessel functions, dimensionless
ϕ combination of terms, dimensionless; circumferential polar coordinate, rad
ψ angle, rad
ω coefficient in reduced Bessel equation, dimensionless
∇ differential operator, dimensions vary

Roman Letter Subscripts

a tip of fin
b base of fin; base (unit fin length)
c corrected fin height; complementary function

G generation
i internal generation (unit fin length)
id ideal
n order of Bessel function
o nominal or average value
p profile
pi particular integral
R removal
s surroundings

Roman Letter Superscripts

j exponent in general Bessel equation
k exponent in general Bessel equation
p exponent in general Bessel equation

Greek Letter Superscripts

α exponent in reduced Bessel equation
β exponent in reduced Bessel equation

Symbolic Superscript

$'$ added fin height

CHAPTER 3

CONVECTIVE OPTIMIZATIONS

3.1 INTRODUCTION

There are three types of optimizations that pertain to extended surface design and analysis. All have received considerable attention in the literature and are treated here. The first type of optimization involves for longitudinal fins, radial fins, and spines the profile that yields minimum weight for a specified heat flow or weight allocation.

The second type of optimization is intimately linked to the first and concerns the *synthesis* of a *particular* fin or spine profile based on a set of specified conditions. The conditions may cover a wide range and may include the mode of dissipation from the fin faces and whether or not internal heat dissipation is present. Included in this type of optimization is the interesting problem of the fin dimensions for a prescribed fin mass. Indeed, an optimization for minimum mass can be obtained from an optimization for minimum volume as well as minimum profile area. This was demonstrated by Sonn and Bar-Cohen (1981) in an optimumization for the minimum mass of the cylindrical spine.

The third type of optimization is in the placement of individual fins to form channels. This leads to the treatment of an *array of fins*[1] in which each fin is operating in an optimum manner. All of these optimization types are treated in this chapter to a fair amount of detail.

3.1.1 Longitudinal Fin Optimizations

Snider and Kraus (1986) traced the history of the quest for the optimum longitudinal fin profile for a fin with uniform and constant heat transfer coefficient on its faces,

[1]Not a *finned array*, which is treated in Chapter 7.

constant thermal conductivity, and heat dissipation only by convection. They noted in this interesting historical survey that a fascinating case of the intermarriage of heuristics, numerics, and rigor in the formulation of the mathematical model could be obtained.

Schmidt (1926) gave a heuristic solution which acknowledged that the true differential equation governing the temperaure at all points on the fin was two-dimensional. He then proposed a distribution of isolated *heat flux lines* within the fin which would form *flux tubes*. Then he reasoned that if a flux tube or flux pipe were replaced by one of uniform cross section but of the same length and volume, its thermal resistance would be lowered. For an optimal fin containing these flux pipes, all of the pipes would be uniform and parallel and, as a result, the temperature gradient would be horizontal everywhere and the fin temperature would be a function of only the fin height coordinate x. This indicated that the temperature was, in fact, a linear function of x and that the heat flux was uniform over the fin. Moreover, this implied that the heat flow in the fin was one-dimensional. The substitution of the linear temperature distribution into the governing differential equation led to the fact that the rate of change of the fin profile function is also linear and that the optimum longitudinal profile is concave parabolic.

Duffin (1959) filled in the logical gaps of the Schmidt (1926) endeavor by publishing a rigorous proof of the optimality of the parabolic profile fin. He employed some powerful formulations of variational techniques that he had developed earlier.

Both Schmidt (1926) and Duffin (1959) made the customary assumption that the differential surface element per unit fin length was equal to twice[2] the differential element of fin height. Of course, the only longitudinal profile that possesses this property is the rectangular profile, which is by no means the optimum profile. In fact, Gardner (1945) observed that the *area-cosine principle* demanded that the surface element for the rectangular profile fin should be $dS = 2dx/\cos\kappa$, where κ is the taper angle. More recently it has been pointed out, notably by Maday (1974), that the universal relationship for unit fin length is

$$dS = (dx^2 + dy^2)^{1/2}$$

where $y = y(x)$ is the fin profile function.

Using $dS = 2dx$ has been known as making the *length of arc idealization*, and with this idealization, Schmidt's fin is indeed the optimum profile. But the universal relationship for dS reopened the question as to whether the Schmidt fin is really the optimal fin.

It is not clear when the elimination of the length of arc idealization was given serious consideration, but the paper by Maday (1974) certainly is a landmark. In it, Maday carefully stated the correct formulation without the idealization, and he derived the Euler–Lagrange conditions parameterized in a form suitable for numerical integration. He also pointed out that the Schmidt (1926) parabolic fin does not satisfy these conditions, and even more significantly, he reported simulations

[2]Twice because of the two fin faces.

demonstrating that the performance of the parabolic fin could be surpassed by fins having a wavy profile.

Snider and Kraus (1986) proposed a ripple surface that had the aim of placing lots of exposed surface near the fin base. But as more and more ripples were introduced, it was observed that the coolant would experience greater difficulty in penetrating the surfaces between the ripples. The effect was observed to be extremely difficult to model, and experimentation was suggested as being necessary to determine how far the benefits of the ripple concept could go.

3.1.2 Optimum Dimensions of Individual Fins

Brown (1965) treated the radial fin of rectangular profile and provided a graphical presentation for the optimum dimensions in a practical working range, and Cobble (1971) optimized the total volume of the longitudinal fin with respect to a temperature distribution parameter. His solution yielded the Schmidt parabolic profile for the simple convective case with constant heat transfer coefficient on the fin faces. His solution also yielded the result obtained by Wilkins (1961) for the simple radiating fin. Cobble also provided the profile for a fin that exchanges heat with the surroundings by convection and radiation.

Eder (1973) provided an analytic treatment of rectangular and triangular fins and gave closed-form solutions for the fin temperature, dissipated heat flux, and optimized dimensions with respect to the fin mass. Bhargava and Duffin (1973) investigated dual extrumum principles relating to cooling fins, and Ahmadi and Razani (1973) gave an approximate solution for the optimum dimensions of a convecting–radiating fin. Guceri and Maday (1975) followed the Maday (1974) effort and found the dimensions for the least-weight circular (radial) fin through an application of the minimum principle. The fin temperature, thickness, and heat flux were considered to be functions only of the radius, and solutions were obtained for the exact one-dimensional representation. This was extended to the approximate case where the curvature is neglected in the convection calculations. It is to be noted that this work essentially looked at the problem with and without the length of arc idealization.

Mikk (1980) looked at, as he put it, "the problem of reducing to a minimum the mass of a convective annular/circular fin" with a prescribed heat transfer requirement. He showed that the limiting case of the minimization was obtained when the temperature of the tip achieved the temperature of the environment. He also showed that the mass of the "ideal" fin differed little from the fin of triangular profile.

Sonn and Bar-Cohen found the optimum dimensions of a pin fin (cylindrical spine) by optimizing on the basis of spine volume rather than profile area. Poulikakos and Bejan (1982) established a theoretical framework for the minimization of entropy generation in extended surfaces. They first derived an entropy generation rate relationship and then, based on this general result, developed analytical methods and gave graphical presentations for the selection of optimum dimensions for pin fins, rectangular plate fins, plate fins with a trapezoidal cross section, and triangular plate fins with a rectangular cross section.

Razelos (1983) looked at spines with circular cross sections that vary along their height in a specified manner. He found the optimum dimensions of base radius and spine height that minimized the volume for a given dissipation. His results were expresssed in a dimensionless form such that certain properties of the optimized pin fin could be derived immediately and, in addition, a meaningful comparison with other optimum configurations could be made.

Razelos and Imre (1980) looked at the optimization requirements for circular (radial) fins with variable thermal parameters, and Razelos (1986) provided optimum dimensions of convective pin fins (cylindrical spines) with internal heat generation. Hati and Rao (1983) studied the optimum dimensions of cooling fins disspating by convection and radiation, and Hrymak et al. (1985) presented an efficient numerical method to discover the optimum shape for a fin subject to both convective and radiative heat loss.

3.2 QUEST FOR THE OPTIMUM PROFILE

As indicated in the foregoing remarks, Schmidt (1926), proposed that the radial fin of least material should also possess a linear temperature gradient, and he derived an equation that would yield the profile function to accomplish this. However, this profile is of little interest because of presumed manufacturing difficulties. But Schmidt's analyses of both the longitudinal and radial fins of rectangular profile are of historical interest and details of these analyses are provided in the subsections that follow.

3.2.1 Longitudinal Fin of Least Material

All of the relationships for temperature excess, heat dissipation, and efficiency derived in Chapter 1 were based on the general differential equation

$$2Lf_2(x)\frac{d^2\theta}{dx^2} + \frac{2Ldf_2(x)}{dx}\frac{d\theta}{dx} - \frac{2h}{k}\theta = 0 \qquad (1.4)$$

where $f_2(x)$ is the profile function. A fin is frequently made of highly conducting or corrosion-resistant metal, which may be expensive. In other instances, the value of a fin must be related to its weight as payload. For either reason, it is desirable to learn the proportions of various fins that can perform specific services with the least weight of metal. This reduces to determining the profile function yielding the minimum profile area A_p, where

$$A_p = 2\int_0^b f_2(x)\,dx$$

Consider a longitudinal fin with its faces perfectly insulated so that it can conduct heat only *along* its longitudinal axis. If the fin has a constant cross section from $x = 0$ to $x = b$, each unit of fin cross section is equally effective in conducting heat because the heat flux q/A must be constant by definition. Then

$$\frac{q}{A_1} = \frac{q}{A_2} = \frac{q}{A_3} = \cdots = \frac{q}{A}$$

However, if the insulation is removed from the fin faces, heat will flow from its lateral surfaces to the colder surroundings. The constant fin cross section will no longer possess constant heat flux at every point in the fin. Indeed, for constant heat flux and equal effectiveness in conducting heat, the cross section of the fin must decrease as some function of its distance from the fin base but in such a manner that q/A allows for heat flow from the lateral surfaces.

Next, consider the heat flow at some point in a fin of any profile

$$q = -kA\frac{d\theta}{dx}$$

where A is a function of x. Then

$$\frac{d\theta}{dx} = -\frac{q}{kA}$$

For a constant heat flux

$$\frac{d\theta}{dx} = C_1 \qquad \text{(a constant)}$$

and hence $\theta = C_1 x + C_2$, which is a linear relationship for the temperature excess of the lateral surface and the temperature distribution in the fin. The only longitudinal fin possessing a linear temperature distribution is the fin of concave parabolic profile, where the temperature excess is given by

$$\theta = \theta_b \left(\frac{x}{b}\right)^{\alpha} \qquad \text{for } \alpha = 1 \tag{3.1}$$

The concave parabolic profile fin is therefore a fin of least material only under the condition that $\alpha = 1$. Note also that for this condition,

$$\alpha = -\frac{1}{2} + \frac{1}{2}\sqrt{1 + (2mb)^2} = 1$$

$$1 + (2mb)^2 = 9$$

$$4(mb)^2 = 8$$

and

$$mb = \sqrt{2}$$

3.2.2 Radial Fin of Least Material

Schmidt (1926) also proposed that the radial fin of least material is the one whose temperature excess is linear. For the fin whose radial coordinate increases in the outward direction, this linear temperature excess relationship can be written as

$$\theta = \theta_b \left(1 - \frac{r - r_b}{r_a - r_b}\right) \tag{3.2}$$

Two differentiations yield

$$\frac{d\theta}{dr} = \frac{-\theta_b}{r_a - r_b}$$

and

$$\frac{d^2\theta}{dr^2} = 0$$

When eq. (3.2) and its derivatives are substituted into the general differential equation for all radial fins given by

$$f_2(r)\frac{d^2\theta}{dr^2} + \frac{f_2(r)}{r}\frac{d\theta}{dr} + \frac{df_2(r)}{dr}\frac{d\theta}{dr} - \frac{h}{k}\theta = 0 \tag{1.29}$$

one obtains

$$-\frac{f_2(r)}{r}\frac{\theta_b}{r_a - r_b} - \frac{df_2(r)}{dr}\frac{\theta_b}{r_a - r_b} - \frac{h}{k}\theta_b\left(1 - \frac{r - r_b}{r_a - r_b}\right) = 0$$

or

$$\frac{\theta_b}{r_a - r_b}\left[f_2(r) + r\frac{df_2(r)}{dr}\right] + \frac{h}{k}\theta_b\left(1 - \frac{r - r_b}{r_a - r_b}\right) = 0 \tag{3.3}$$

In the first term of eq. (3.3), it can be seen that

$$f_2(r) + r\frac{df_2(r)}{dr} = \frac{d}{dr}[rf_2(r)]$$

Integration of eq. (3.3) gives

$$\frac{\theta_b}{r_a - r_b}rf_2(r) + \frac{h}{2k}\theta_b r^2 - \frac{h}{3k}\theta_b\frac{r^3}{r_a - r_b} + \frac{h}{2k}\theta_b\frac{r_b r^2}{r_a - r_b} = C \tag{3.4}$$

where C is an arbitrary constant of integration determined at $r = r_a$ where $f_2(r_a) = 0$. Hence

$$C = \frac{h}{k}\theta_b\left[\frac{r_a^2}{2} - \frac{r_a^3}{3(r_a - r_b)} + \frac{r_b r_a^2}{2(r_a - r_b)}\right]$$

The profile equation is obtained when this value of C is put into eq. (3.4):

$$\frac{rf_2(r)}{r_a - r_b} = \frac{h}{k}\left[\frac{r^3 - r_a^3}{3(r_a - r_b)} - \frac{r^2 - r_a^2}{2} - \frac{r_b(r^2 - r_a^2)}{2(r_a - r_b)}\right]$$

or, after adjustment,

$$\frac{kf_2(r)}{hr_a^2} = \frac{1}{3}\left(\frac{r}{r_a}\right)^2 - \frac{1}{2}\left(\frac{r}{r_a}\right) + \frac{1}{6}\left(\frac{r_a}{r}\right) \tag{3.5}$$

A plot of eq. (3.5) appears in Fig. 3.1.

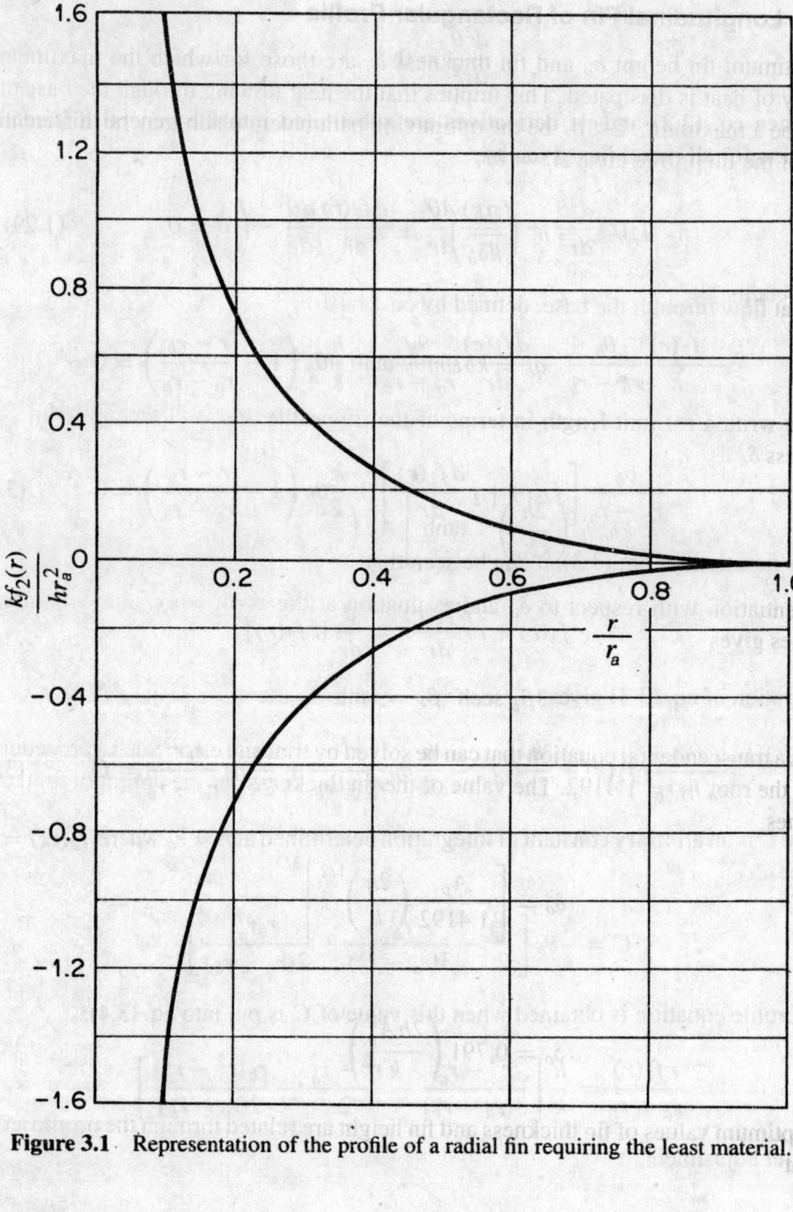

Figure 3.1 Representation of the profile of a radial fin requiring the least material.

3.3 OPTIMUM SHAPES OF LONGITUDINAL FINS

Attention now turns to the optimum dimensions for the longitudinal fins of rectangular, triangular, and concave parabolic profiles when either a weight specification or a heat dissipation is provided. These are but two of the more commmon optimizing parameters.

3.3.1 Longitudinal Fin of Rectangular Profile

The optimum fin height b_o and fin thickness δ_o are those for which the maximum quantity of heat is dissipated. This implies that the heat flowing through the base of the fin be a maximum. Let β_L be the optimizing parameter, which may be defined in terms of the fin profile area, $A_p = b\delta_o$:

$$\beta_L = mb_o = b_o \left(\frac{2h}{k\delta_o}\right)^{1/2} = A_p \left(\frac{2h}{k}\right)^{1/2} \left(\frac{1}{\delta_o}\right)^{3/2}$$

The heat flow through the base, defined by eq. (1.10),

$$q_b = k\delta L m\theta_b \tanh mb \tag{1.10}$$

may be written for unit length in terms of the fin profile area and the optimum fin thickness δ_o as

$$q_b = k\delta_o\theta_b \left(\frac{2h}{k\delta_o}\right)^{1/2} \tanh \left[A_p \left(\frac{2h}{k}\right)^{1/2} \left(\frac{1}{\delta_o}\right)^{3/2}\right]$$

Differentiation with respect to δ_o and evaluation at the point where the derivative vanishes gives

$$3\beta_L \operatorname{sech}^2 \beta_L = \tanh \beta_L$$

This is a transcendental equation that can be solved by trial and error. Such a procedure yields the root $\beta_L = 1.4192$. The value of the fin thickness for the optimum fin then becomes

$$\delta_o = \left[\frac{A_p}{1.4192}\left(\frac{2h}{k}\right)^{1/2}\right]^{2/3}$$

or

$$\delta_o = 0.791 \left(\frac{2hA_p^2}{k}\right)^{1/3} \tag{3.6}$$

The optimum values of fin thickness and fin height are related through the profile area so that

$$b_o = \frac{A_p}{\delta_o} = 1.262\left(\frac{kA_p}{2h}\right)^{1/3} \tag{3.7}$$

Aziz (1992) has provided expressions for the optimum dimensions in terms of a specified heat-dissipating requirement q_b:

$$\delta_o = \frac{0.632}{hk}\left(\frac{q_b}{\theta_b}\right)^2 \tag{3.8}$$

and

$$b_o = 0.798\left(\frac{q_b}{h\theta_b}\right) \tag{3.9}$$

3.3.2 Longitudinal Fin of Triangular Profile

The heat flow through the base of the longitudinal fin of triangular profile is given by eq. (1.15):

$$q_b = \frac{2hL\theta_b I_1(2mb)}{mI_0(2mb)} \tag{1.15}$$

and for unit length ($L = 1$ m),

$$q_b = \frac{2h\theta_b I_1(2mb)}{mI_0(2mb)}$$

With the optimum height b_o and the optimum fin base thickness δ_o related by the profile area, $b_o = 2A_p/\delta_o$, define β_T as

$$\beta_T = 2mb_o = 2\left(\frac{2h}{k\delta_o}\right)^{1/2}\frac{2A_p}{\delta_o} = 4A_p\left(\frac{2h}{k}\right)^{1/2}\left(\frac{1}{\delta_o}\right)^{3/2}$$

The fin thickness is

$$\delta_o = \left[\frac{4A_p(2h/k)^{1/2}}{\beta_T}\right]^{2/3}$$

The heat flow per unit length with these values of β_T and δ_o is

$$q_b = \left[4A_p(2h)^2k\right]^{1/3}\theta_b\beta_T^{-1/3}\frac{I_1(\beta_T)}{I_0(\beta_T)}$$

which may be differentiated with respect to β_T and the result equated to zero. Then

$$I_0(\beta_T)I_2(\beta_T) + \frac{2}{3}\frac{I_0(\beta_T)I_1(\beta_T)}{\beta_T} = I_1^2(\beta_T)$$

A trial-and-error solution, apparently first performed by Schmidt (1926), yields a root, $\beta_T = 2.6188$, which fixes the base thickness for the optimum fin as

$$\delta_o = \left[\frac{4A_p(2h/k)^{1/2}}{2.6188}\right]^{2/3}$$

or

$$\delta_o = 1.328\left(\frac{2hA_p^2}{k}\right)^{1/3} \tag{3.10}$$

and the optimum fin height as

$$b_o = \frac{2A_p}{\delta_o} = 1.506\left(\frac{kA_p}{2h}\right)^{1/3} \tag{3.11}$$

Aziz (1992) has provided expressions for the optimum dimensions in terms of a specified heat-dissipating requirement q_b:

$$\delta_o = \frac{0.827}{hk}\left(\frac{q_b}{\theta_b}\right)^2 \tag{3.12}$$

and

$$b_o = 0.842\left(\frac{q_b}{h\theta_b}\right) \tag{3.13}$$

3.3.3 Longitudinal Fin of Concave Parabolic Profile

The search for the optimum dimensions for the longitudinal fin of concave parabolic profile begins with the equation for the heat flow through the base given by

$$q_b = \frac{k\delta_b L\theta_b}{2b}\left[-1 + \sqrt{1 + (2mb)^2}\right] \tag{1.20}$$

which for unit length ($L = 1$ m) is

$$q_b = \frac{k\delta_b\theta_b}{2b}\left[-1 + \sqrt{1 + (2mb)^2}\right]$$

This time, define the parameter β_P as

$$\beta_P = mb_o = \left(\frac{2h}{k\delta_o}\right)^{1/2} b_o$$

where b_o is the optimum fin height. With the optimum fin base thickness δ_o the fin profile area is

$$A_p = \frac{\delta_o b_o}{3}$$

In terms of the profile area and the base thickness,

$$\beta_P = \left(\frac{2h}{k\delta_o}\right)^{1/2} \frac{3A_p}{\delta_o} = 3A_p \left(\frac{2h}{k}\right)^{1/2} \left(\frac{1}{\delta_o}\right)^{3/2}$$

The fin thickness is

$$\delta_o = \left[\frac{3A_p(2h/k)^{1/2}}{\beta_P}\right]^{2/3}$$

Substitution of this fin thickness into the heat flow relationship gives

$$q_b = \frac{k\theta_b}{6A_p}\left[3A_p\left(\frac{2h}{k}\right)^{1/2}\right]^{4/3} \beta_P^{-4/3}\left[-1 + (1 + 4\beta_P^2)^{1/2}\right]$$

Differentiating with respect to β_P and setting the result equal to zero provides

$$-\frac{4}{3}\beta_P^{-7/3}\left[-1 + (1 + 4\beta_P^2)^{1/2}\right] + \frac{1}{2}\beta_P^{-4/3}\frac{8\beta_P}{(1 + 4\beta_P^2)^{1/2}} = 0$$

After simplification this yields $\beta_P^2 = 2$, which has a positive root at $\beta_P = \sqrt{2}$. The optimum base thickness is therefore

$$\delta_o = \left[\frac{3A_p(2h/k)^{1/2}}{\sqrt{2}}\right]^{2/3}$$

or

$$\delta_o = 1.651\left(\frac{2hA_p^2}{k}\right)^{1/3} \tag{3.14}$$

and the optimum fin height is

$$b_o = \frac{3A_p}{\delta_o} = 1.817\left(\frac{kA_p}{2h}\right)^{1/3} \tag{3.15}$$

Aziz (1992) has provided expressions for the optimum dimensions in terms of a specified heat-dissipating requirement q_b:

$$\delta_o = \frac{1}{hk}\left(\frac{q_b}{\theta_b}\right)^2 \tag{3.16}$$

and

$$b_o = \frac{q_b}{h\theta_b} \tag{3.17}$$

3.3.4 Comparison of Fins

Longitudinal fins of rectangular, triangular, and concave parabolic profile may be compared to determine which profile requires the least profile area for dissipation of a given amount of heat. The comparison is accomplished by substitution of the optimum fin thickness into the expression for the heat flow through the base of each fin.

Rectangular Fin. The heat flow through the base (for unit length) may be written

$$q_b = k\delta_b m\theta_b \tanh mb = (2hk)^{1/2}\delta_b^{1/2}\theta_b \tanh mb$$

and by noting that $mb = 1.4192$ for the optimum

$$q_b = (2hk)^{1/2}\left[0.791\left(\frac{2hA_p^2}{k}\right)^{1/3}\right]^{1/2}\theta_b \tanh 1.4192$$

or

$$q_b = 1.258(h^2A_pk)^{1/3}\theta_b$$

from which

$$A_p = \frac{1}{h^2k}\left(\frac{1}{1.258}\frac{q_b}{\theta_b}\right)^3$$

or

$$A_p = \frac{0.500}{h^2k}\left(\frac{q_b}{\theta_b}\right)^3 \tag{3.18}$$

Triangular Fin. The same procedure may be followed. The heat flow through the base (for unit length) is

$$q_b = \frac{2h\theta_b I_1(2mb)}{mI_0(2mb)} = (2hk)^{1/2}\delta_o^{1/2}\theta_b\frac{I_1(2mb)}{I_0(2mb)}$$

For the optimum, $2mb = 2.6188$ and

$$q_b = (2hk)^{1/2}\left[1.328\left(\frac{2hA_p^2}{k}\right)^{1/3}\right]^{1/2}\theta_b\frac{I_1(2.6188)}{I_0(2.6188)}$$

or

$$q_b = 1.422(h^2A_pk)^{1/3}\theta_b$$

from which

$$A_p = \frac{1}{h^2k}\left(\frac{1}{1.422}\frac{q_b}{\theta_b}\right)^3$$

or

$$A_p = \frac{0.347}{h^2 k}\left(\frac{q_b}{\theta_b}\right)^3 \tag{3.19}$$

Concave Parabolic Fin. The heat flow through the base for unit length is

$$q_b = \frac{k\delta_o^2\theta_b}{6A_p}\left[-1+\sqrt{1+(2mb)^2}\right]$$

With $mb = \sqrt{2}$ and using eq. (3.14), this becomes

$$q_b = \frac{k\theta_b}{6A_p}\left[1.651\left(\frac{2hA_p^2}{k}\right)^{1/3}\right]^2\left\{-1+\left[1+\left(2\sqrt{2}\right)^2\right]^{1/2}\right\}$$

or

$$q_b = 1.450(h^2 A_p k)^{1/3}\theta_b$$

from which

$$A_p = \frac{1}{h^2 k}\left(\frac{1}{1.450}\frac{q_b}{\theta_b}\right)^3$$

or

$$A_p = \frac{0.327}{h^2 k}\left(\frac{q_b}{\theta_b}\right)^3 \tag{3.20}$$

Discussion. Equations (3.18) through (3.20) give the fin profile area required for rectangular, triangular, and concave parabolic fins, respectively. The profile area in each case is seen to be a function of the base heat flow to temperature excess ratio q_b/θ_b, These profile areas are also seen to be inversely proportional to the thermal conductivity of the fin and to the square of the heat transfer coefficient between the fin and its surroundings.

Three significant conclusions can be drawn from these equations. The first conclusion is that for the same material, surrounding conditions, and the ratio of base heat flow to base temperature excess, the optimum concave parabolic fin requires only about 65% as much material as the optimum rectangular fin. The triangular profile requires about 69% as much material as the rectangular profile and about 6% more than the comparable concave parabolic fin.

The second conclusion concerns the choice of material for any particular profile. The equations show that the profile area is inversely proportional to the thermal conductivity of the fin material. The total weight of the fin is proportional to the specific weight of the material used. Hence, the fin weight is directly proportional to the specific weight γ and inversely proportional to the thermal conductivity k.

Consider, for example, three materials:

Copper:	$\gamma = 8937$ N/m^3	$k = 401$ W/m · K
Aluminum:	$\gamma = 2702$ N/m^3	$k = 235$ W/m · K
Steel:	$\gamma = 7817$ N/m^3	$k = 41$ W/m · K

For a given heat flow, temperature excess and heat transfer coefficient, the aluminum fin requires only (2702)(41)/(7817)(235) = 0.0603 or 6% as much weight as the steel fin. From the same point of view, the copper fin requires (8937)(41)/(7817)(401) = 0.1169 or 11.7% as the steel fin, but requires only (8937)(237)/(2702)(401) = 1.955 or 95.5% more weight than that of the comparable aluminum fin.

Finally, it is seen that the profile area and fin volume increase as the cube of the base heat flow to base temperature exess ratio. If it is desired that the heat flow be doubled,[3] there is a choice of using two identical fins or making one fin eight times as large. It is obvious from a design standpoint that larger numbers of short stubby fins are more desirable than fewer higher ones.

Example 3.1: Optimum Longitudinal Fins. A prime surface is to be fitted with longitudinal fins of a material with $k = 100$ W/m · K and a specific weight of $\gamma = 2700$ N/m^3. The fins are to dissipate heat to the surroundings with a heat transfer coefficient of $h = 400$ W/m^2 · K. If the maximum tolerable weight per linear meter is $w = 0.0372$ N/m, determine the optimum dimensions for longitudinal fins of (**a**) rectangular, (**b**) triangular and (**c**) concave parabolic profiles.

SOLUTION. The maximum possible profile area of any fin is

$$A_p = \frac{w}{L\gamma} = \frac{0.0372}{(1)(2700)} = 1.378 \times 10^{-5} \text{ m}^2$$

(**a**) For the rectangular fin, use eqs. (3.6) and (3.7):

$$\delta_o = 0.791\left(\frac{2hA_p^2}{k}\right)^{1/3}$$

$$= 0.791\left[\frac{(2)(400)(1.378 \times 10^{-5})^2}{100}\right]^{1/3}$$

$$= 9.09 \times 10^{-4} \text{ m}$$

or

$$\delta_o = 0.0909 \text{ cm}$$

[3]For the same base temperature excess.

and

$$b_o = \frac{A_p}{\delta_o}$$

$$= \frac{1.378 \times 10^{-5}}{9.09 \times 10^{-4}}$$

$$b_o = 0.01515 \text{ m}$$

or

$$b_o = 1.515 \text{ cm}$$

(b) For the triangular fin, use eqs. (3.10) and (3.11):

$$\delta_o = 1.328 \left(\frac{2hA_p^2}{k} \right)^{1/3}$$

$$= 1.328 \left[\frac{(2)(400)(1.378 \times 10^{-5})^2}{100} \right]^{1/3}$$

$$= 1.527 \times 10^{-3} \text{ m}$$

or

$$\delta_o = 0.1527 \text{ cm}$$

and

$$b_o = \frac{2A_p}{\delta_o}$$

$$= \frac{(2)(1.378 \times 10^{-5})}{1.527 \times 10^{-3}}$$

$$b_o = 0.01805 \text{ m}$$

or

$$b_o = 1.805 \text{ cm}$$

(c) For the concave parabolic fin, use eqs. (3.14) and (3.15)

$$\delta_o = 1.651 \left(\frac{2hA_p^2}{k} \right)^{1/3}$$

$$= 1.651 \left[\frac{(2)(400)(1.378 \times 10^{-5})^2}{100} \right]^{1/3}$$

$$= 1.898 \times 10^{-3} \text{ m}$$

or

$$\delta_o = 0.1898 \text{ cm}$$

and

$$b_o = \frac{3A_p}{\delta_o}$$

$$= \frac{(3)(1.378 \times 10^{-5})}{1.898 \times 10^{-3}}$$

$$b_o = 0.0218 \text{ m}$$

or

$$b_o = 2.18 \text{ cm}$$

These dimensions may be summarized

Profile	Base Width (cm)	Height (cm)
Rectangular	0.0909	1.515
Triangular	0.1527	1.805
Concave parabolic	0.1898	2.180

3.4 OPTIMUM SHAPES OF RADIAL FINS

3.4.1 Radial Fin of Rectangular Profile

Equation (3.5) gives the profile function for the radial fin of least material. This fin is of interest, but because of manufacturing difficulties, it has had little, if any, practical application. Brown (1965) has treated the radial fin of rectangular profile shown in Fig. 3.2*a* and has developed a graphical presentation of the optimum dimensions for a practical working range.

The heat dissipated by the radial fin of rectangular profile is given by

$$q_b = 2\pi r_b \delta k m \theta_b \frac{I_1(mr_a)K_1(mr_b) - K_1(mr_a)I_1(mr_b)}{I_0(mr_b)K_1(mr_a) + I_1(mr_a)K_o(mr_b)} \tag{1.34}$$

Using $m^2 = 2h/k\delta$ and the fin volume,

$$V = \pi\delta(r_a^2 - r_b^2) \tag{3.21}$$

eq. (1.34) may be written as

$$q_b = 2\pi k r_b \theta_b \left(\frac{2h\delta}{k}\right)^{1/2} \frac{I_1(Z_a)K_1(Z_b) - K_1(Z_a)I_1(Z_b)}{I_0(Z_b)K_1(Z_a) + I_1(Z_a)K_0(Z_b)} \tag{3.22}$$

where

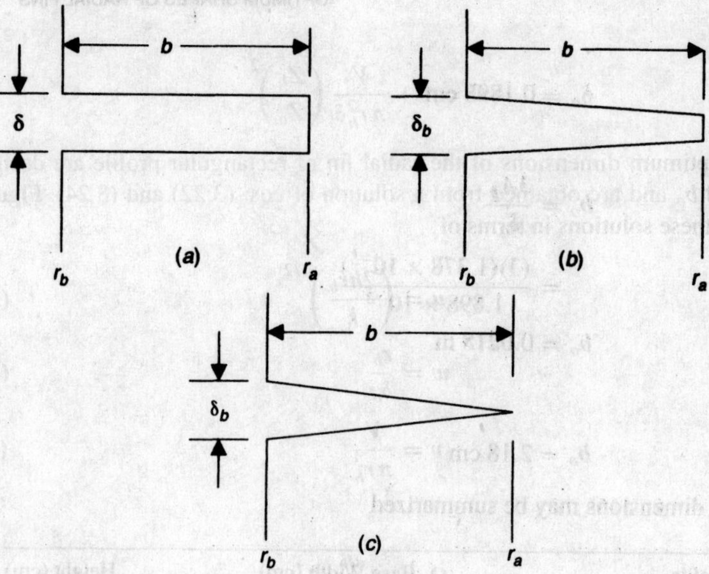

Figure 3.2 Radial fins of (*a*) rectangular, (*b*) trapezoidal and (*c*) triangular profile showing optimum dimensions. [Adapted from Brown (1965).]

$$Z_b = \left(\frac{2hr_b^2}{k\delta} \right)^{1/2} \tag{3.23a}$$

and

$$Z_a = \left[\frac{2hr_b^2 \left(1 + V/\pi \delta r_b^2\right)}{k\delta} \right]^{1/2} \tag{3.23b}$$

Equation (3.22) represents the heat flow as a function of fin thickness, δ. Keeping h, k, θ_b, r_b, and V constant, the heat dissipated by the fin will be a maximum when $dq_b/d\delta$ vanishes. Differentiating eq. (3.22) with respect to δ, equating to zero, and rearranging gives

$$2 + Z_b \Lambda_1 + Z_a C \Lambda_2 \Lambda_3 = 0 \tag{3.24}$$

where, with $b = r_a - r_b$,

$$\Lambda_1 = \frac{I_1(Z_b)K_1(Z_a) - K_1(Z_b)I_1(Z_a)}{I_0(Z_b)K_1(Z_a) + K_0(Z_b)I_1(Z_a)} - \frac{I_0(Z_b)K_1(Z_a) + K_0(Z_b)I_1(Z_a)}{I_1(Z_b)K_1(Z_a) - K_1(Z_b)I_1(Z_a)}$$

$$\Lambda_2 = \frac{I_0(Z_b)K_1(Z_b) + I_1(Z_b)K_0(Z_b)}{I_0(Z_b)K_1(Z_a) + I_1(Z_a)K_0(Z_b)}$$

$$\Lambda_3 = \frac{I_0(Z_a)K_1(Z_a) + I_1(Z_a)K_0(Z_a)}{I_1(Z_b)K_1(Z_a) - K_1(Z_b)I_1(Z_a)}$$

and

$$C = 1 + \frac{V}{\pi r_b^2 \delta} \left(\frac{Z_b}{Z_a} \right)^2$$

The optimum dimensions of the radial fin of rectangular profile are designated as δ_o and b_o and are obtained from a solution of eqs. (3.22) and (3.24). Figure 3.3 presents these solutions in terms of

$$u = \left(\frac{2hr_b}{k} \right)^{1/2} \tag{3.25a}$$

$$w = \frac{r_b}{\delta_o} \tag{3.25b}$$

$$v = \frac{V}{\pi r_b^3} \tag{3.25c}$$

and

$$Q = \frac{q_b}{2\pi k r_b \theta_b} \tag{3.25d}$$

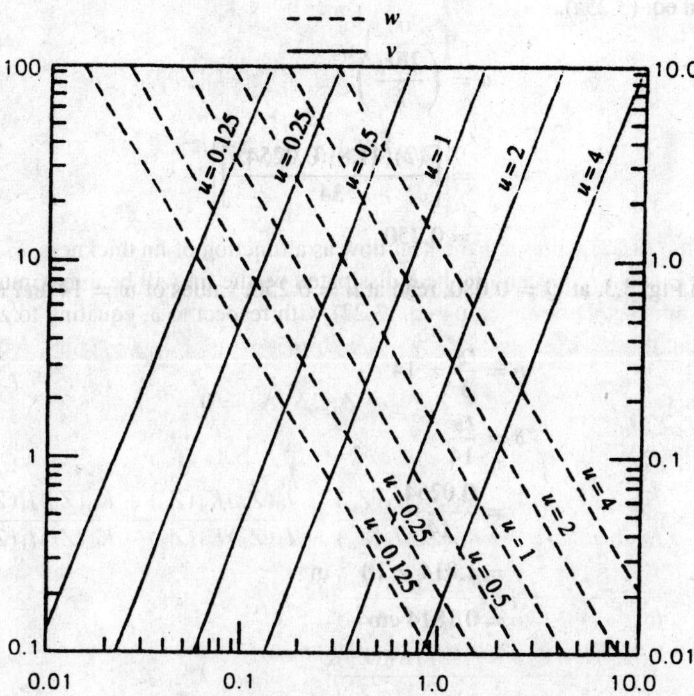

Figure 3.3 Chart for the determination of the optimum dimensions of the radial of rectangular profile.

In a sense, Fig. 3.3 is a design chart. If q_b and r_b are specified, the optimum thickness δ_o can be found via w, and the tip radius r_a can be found via v because

$$V = \pi\delta(r_a^2 - r_b^2)$$

Example 3.2: Optimum Radial Fin of Rectangular Profile. Determine the thickness of a radial fin of rectangular profile which is required to dissipate 15.625 W in an application where the base temperature excess is 36°C with a heat transfer coefficient of $h = 41.8\ \text{W/m}^2 \cdot \text{K}$. The fin is fabricated of a steel with $k = 34\ \text{W/m} \cdot \text{K}$ and the fin has a base diameter of 5.08 cm.

SOLUTION. From eq. (3.25*d*), with $r_b = 5.08/2 = 2.54$ cm,

$$Q = \frac{q_b}{2\pi k r_b \theta_b}$$

$$= \frac{15.625}{2\pi(34)(0.0254)(36)}$$

$$= 0.080$$

and from eq. (3.25a),

$$u = \left(\frac{2hr_b}{k}\right)^{1/2}$$

$$= \left[\frac{(2)(41.8)(0.0254)}{34}\right]^{1/2}$$

$$= 0.250$$

From Fig. 3.3, at $Q = 0.080$, read at $u = 0.250$, values of $w = 14$ and $v = 0.31$. Then

$$w = \frac{r_b}{\delta} = 14$$

$$\delta = \frac{r_b}{14}$$

$$= \frac{0.0254}{14}$$

$$= 1.814 \times 10^{-3}\ \text{m}$$

$$= 0.1814\ \text{cm}$$

and

$$v = \frac{V}{\pi r_b^3} = 0.31$$

$$
\begin{aligned}
V &= 0.31\pi r_b^3 \\
&= 0.31\pi(0.0254)^3 \\
&= 1.6387 \times 10^{-5}\ \text{m}^3 \\
&= \pi\delta(r_a^2 - r_b^2) \\
r_a^2 &= \frac{V}{\pi\delta} + r_b^2 \\
&= \frac{1.6387 \times 10^{-5}\ \text{m}^3}{\pi(1.814 \times 10^{-3})} + (0.0254)^2 \\
&= 3.5202 \times 10^{-3} \\
r_a &= 5.933 \times 10^{-2}\ \text{m} \\
&= 5.933\ \text{cm}
\end{aligned}
$$

Mikk (1980), Razelos and Imre (1980), and Ullman and Kalman (1989) have all considered the dimensions of optimal radial fins. Mikk's work concerns rectangular and triangular profiles, and the Razelos and Imre work pertains to rectangular, triangular, and trapezoidal profiles. The most recent work, that of Ullman and Kalman, includes four profiles: rectangular, triangular, hyperbolic, and parabolic.

Mikk (1980) presents his results for the fin of rectangular profile in the form of three dimensionless parameters:

$$\phi = \frac{r_a}{r_b} \tag{3.26a}$$

$$\sigma = \frac{h(r_a - r_b)^2}{k\delta_o} \tag{3.26b}$$

and

$$\nu = 8\pi^3 r_b^2 h^2 k\delta_o(r_a^2 - r_b^2)\left(\frac{\theta_b}{q_b}\right)^3 \tag{3.26c}$$

where δ_o is the optimum fin thickness. The optimum values of these parameters are shown in Table 3.1, which also gives the efficiency η of the fin, defined as

$$\eta = \frac{q_b}{2\pi(r_a^2 - r_b^2)h\theta_b} \tag{3.26d}$$

For a given q_b and r_b, Mikk's results require trial and error to determine r_a and δ_o after an assumption of r_a, which fixes the value of ϕ. Table 3.1 then gives σ and ν. The former gives δ_o and the latter q_b. If the q_b calculated does not match the q_b given, the calculation must be repeated with a new value of r_a.

Another study of the optimization of the radial fin of rectangular profile is that of Razelos and Imre (1980), who employed a Runge–Kutta scheme in conjunction with a quasi-Newton algorithm to locate the optimum. Their results are plotted in Fig. 3.4.

TABLE 3.1 Optimum Values of ϕ, σ, and ν for a Radial Fin of Rectangular Profile

ϕ	σ	ν	η
1.0	1.0071	1.0085	0.627
1.5	0.7657	0.8115	0.636
2.0	0.6328	0.6618	0.643
2.5	0.5507	0.5493	0.646
3.0	0.4944	0.4628	0.649
4.0	0.4208	0.3426	0.652
5.0	0.3741	0.2646	0.655
6.0	0.3410	0.2116	0.656
∞	0.0000	0.0000	0.667

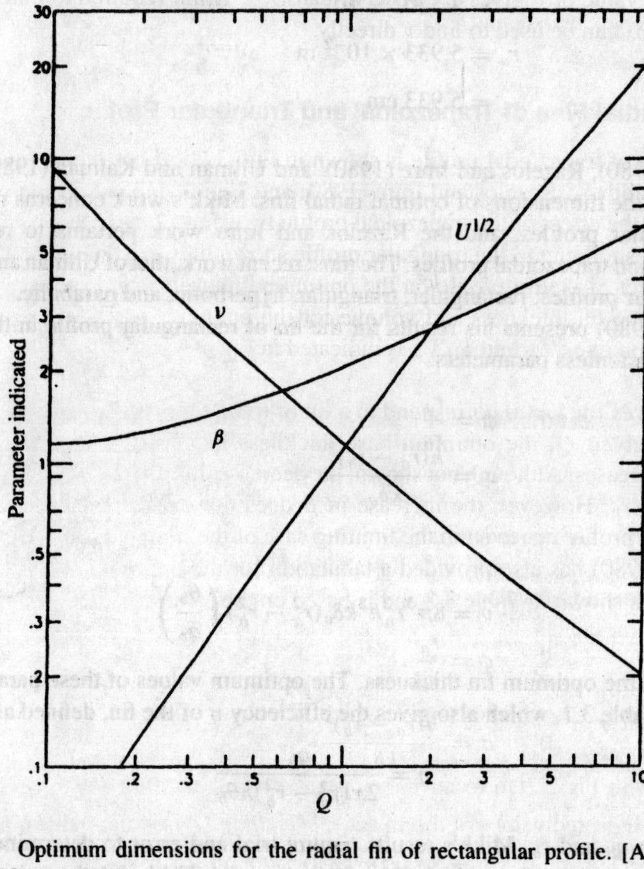

Figure 3.4 Optimum dimensions for the radial fin of rectangular profile. [Adapted from Razelos and Imre (1980)].

The parameters ν, β, and $U^{1/2}$ are plotted against q, where the optimum fin thickness is δ_o and

$$\nu = r_b \left(\frac{2h}{k\delta_o} \right)^{1/2} \tag{3.27a}$$

$$U^{1/2} = \left(\frac{kV}{4\pi r_b^4 h}\right)^{1/2} \tag{3.27b}$$

$$\beta = \frac{r_a}{r_b} \tag{3.27c}$$

and

$$Q = \frac{q_b}{4\pi h r_b^2 \theta_b} \tag{3.27d}$$

The calculation procedure in this case begins with the determination of Q from given values of q_b and r_b. Figure 3.4 is then employed to find β and ν. The value of β gives r_a, and the value of ν gives δ_o. Observe that $U^{1/2}$ is not required to find V because r_a, r_b, and δ_o can be used to find ν directly.

3.4.2 Radial Fins of Trapezoidal and Triangular Profile

A radial fin of trapezoidal profile is shown in Fig. 3.2b. It has a base width $\delta_b = \delta_o$ and a tip width δ_a. Razelos and Imre (1980) and Netrakanti and Huang (1985) have considered the effect of the trapezoidal profile by using a taper ratio $\lambda = \delta_a/\delta_b$, with $\lambda = 0$ corresponding to the triangular profile shown in Fig. 3.2c.

The effect of the fin profile on the optimum dimensions is based on δ_R and V_R, which are the fin thickness and volume for the optimum fin of rectangular profile. The variation of δ_b/δ_R and V/V_R is indicated in Fig. 3.5 for three values of the taper ratio λ.

The curves for $\lambda = 0$ correspond to a fin of triangular profile, and Fig. 3.5 shows that for a given Q, the optimum base thickness increases as λ increases but the volume decreases. Although not shown, the decrease in λ also causes a slight increase in $\beta = r_a/r_b$. However, the increase in β does not exceed 3% (compared to the rectangular profile fin) even in the limiting case of the triangular fin ($\lambda = 0$).

Mikk (1980) has also provided a tabulation for the optimum triangular fin. This tabulation is shown in Table 3.2 and is based on the same parameters as Table 3.1:

$$\phi = \frac{r_a}{r_b} \tag{3.26a}$$

$$\sigma = \frac{h(r_a - r_b)^2}{k\delta_o} \tag{3.26b}$$

and

$$\nu = 8\pi^3 r_b^2 h^2 k \delta_o (r_a^2 - r_b^2) \left(\frac{\theta_b}{q_b}\right)^3 \tag{3.26c}$$

and

$$\eta = \frac{q_b}{2\pi (r_a^2 - r_b^2) h \theta_b} \tag{3.26d}$$

Mikk's results require an iterative approach if q_b and r_b are specified and r_a and $\delta_o = \delta_b$ are sought.

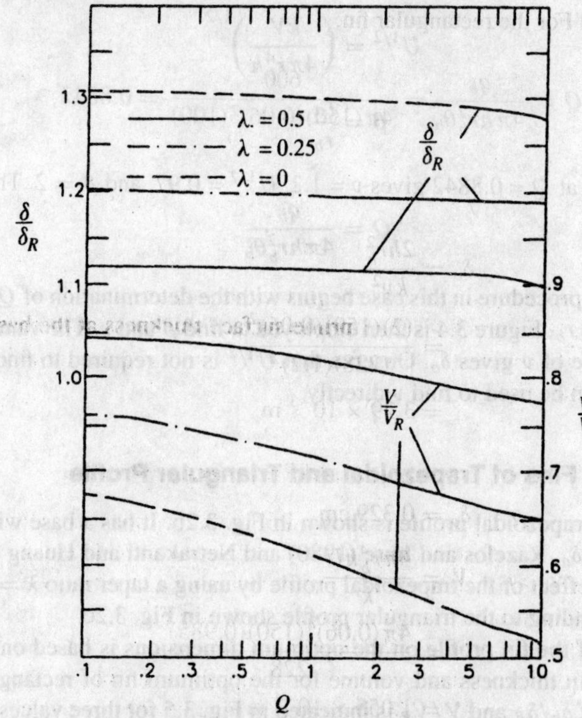

Figure 3.5 Optimum dimensions of radial fins of trapezoidal and triangular profiles. [Adapted from Razelos and Imre (1980) and Netrakanti and Huang (1985).]

TABLE 3.2 Optimum Values of ϕ, σ, and ν for a Radial Fin of Triangular Profile

ϕ	σ	ν	η
1.0	0.8574	0.7037	0.592
1.5	0.6143	0.5312	0.612
2.0	0.4941	0.4143	0.623
2.5	0.4232	0.3312	0.630
3.0	0.3758	0.2708	0.635
4.0	0.3169	0.1911	0.642
5.0	0.2814	0.1426	0.646
6.0	0.2571	0.1109	0.648
∞	0.0000	0.0000	0.667

Example 3.3: Optimum Radial Fins. Design radial fins of optimum dimensions for (**a**) rectangular, (**b**) triangular, and (**c**) trapezoidal ($\lambda = 0.50$) profiles to dissipate 600 W. The base of the fin has a radius of 6 cm and the base temperature excess is 100°C. The average convective heat transfer coefficient is $h = 150\ \text{W/m}^2 \cdot \text{K}$ and the thermal conductivity of the aluminum fin is 228 W/m · K.

SOLUTION. (a) For the rectangular fin,

$$Q = \frac{q_b}{4\pi h r_b^2 \theta_b} = \frac{600}{4\pi(150)(0.06)^2(100)} = 0.8842$$

Reading Fig. 3.4 at $Q = 0.8842$ gives $\nu = 1.2$, $U^{1/2} = 0.97$, and $\beta = 2$. Thus

$$\delta_o = \frac{2hr_b^2}{k\nu^2}$$

$$= \frac{(2)(150)(0.06)^2}{(228)(1.2)^2}$$

$$= 3.29 \times 10^{-3} \text{ m}$$

or

$$\delta_o = 0.329 \text{ cm}$$

$$V = \frac{4\pi r_b^4 hU}{k}$$

$$= \frac{4\pi(0.06)^4(150)(0.985)}{228}$$

$$= 1.055 \times 10^{-4} \text{ m}^3$$

and

$$r_a = \beta r_b$$

$$= (2)(0.06) = 0.12 \text{ m}$$

or

$$r_a = 12 \text{ cm}$$

(b) For the triangular fin, for $Q = 0.8842$, Fig. 3.5 gives $\delta_o/\delta_R = 1.3$, $V/V_R = 0.61$ and $r_a/r_b = 2$. Thus

$$\delta_o = 1.3\delta_R$$

$$= (1.3)(3.29 \times 10^{-3})$$

$$= 4.28 \times 10^{-3} \text{ m}$$

or

$$\delta_o = 0.428 \text{ cm}$$

and

$$V = 0.61 V_R$$

$$= (0.61)(1.055 \times 10^{-4})$$

$$= 6.44 \times 10^{-5}\ \text{m}^3$$

and

$$r_a = 2r_b$$

$$= (2)(0.06) = 0.12\ \text{m}$$

or

$$r_a = 12\ \text{cm}$$

(c) For the trapezoidal fin with $\lambda = 0.50$, for $Q = 0.8842$, Fig. 3.5 gives $\delta_o/\delta_R = 1.12$, $V/V_R = 0.83$, and $r_a/r_b = 2$. Thus

$$\delta_o = 1.12\delta_R$$

$$= (1.12)(3.29 \times 10^{-3})$$

$$= 3.68 \times 10^{-3}\ \text{m}$$

or

$$\delta_o = 0.368\ \text{cm}$$

and

$$V = 0.83V_R$$

$$= (0.83)(1.055 \times 10^{-4})$$

$$= 8.76 \times 10^{-5}\ \text{m}^3$$

and

$$r_a = 2r_b$$

$$= (2)(0.06) = 0.12\ \text{m}$$

or

$$r_a = 12\ \text{cm}$$

Additional computations may be made for copper ($k = 382$ W/m · K) and steel ($k = 47$ W/m · K) fins. For the same conditions of heat loading, Q will remain at a value of $Q = 0.8842$ because Q is not a function of the material thermal conductivity. Hence ν, $U^{1/2}$, β, δ_o/δ_R, and V/V_R read from Figs. 3.4 and 3.5 will remain at the values used in Example 3.3. The reader may verify the results shown in the following tabulations:

For aluminum:

Profile	δ_o (cm)	V (m^3)	r_a (cm)
Rectangular	0.329	1.01×10^{-4}	12
Triangular	0.428	6.15×10^{-5}	12
Trapezoidal	0.368	8.37×10^{-5}	12

For copper:

Profile	δ_o (cm)	V (m^3)	r_a (cm)
Rectangular	0.196	6.00×10^{-5}	12
Triangular	0.255	3.67×10^{-5}	12
Trapezoidal	0.220	4.99×10^{-5}	12

For carbon steel:

Profile	δ_o (cm)	V (m^3)	r_a (cm)
Rectangular	1.596	4.90×10^{-4}	12
Triangular	2.075	2.98×10^{-4}	12
Trapezoidal	1.788	4.06×10^{-4}	12

Discussion. Five significant observations can be made:

1. For any given profile, the volume for a specified q_b is inversely proportional to the thermal conductivity of the material. For example, the rectangular carbon steel fin requires about eight times the volume required for the corresponding copper fin and about five times that for the aluminum fin.

2. As the taper ratio λ decreases, the optimum base thickness δ_o increases. Comparing the three aluminum fins, the base thicknesses are 0.329, 0.368, and 0.428 cm for the rectangular ($\lambda = 1$), the trapezoidal ($\lambda = 0.5$), and the triangular ($\lambda = 0$) profiles, respectively.

3. The volume decreases as λ decreases. Comparison of the tabulated values shows that for the aluminum fins, $V = 1.01 \times 10^{-4}$ m^3 for $\lambda = 1$ (rectangular profile), $V = 8.37 \times 10^{-5}$ m^3 for $\lambda = 0.5$ (trapezoidal profile), and $V = 6.15 \times 10^{-5}$ m^3 for $\lambda = 0$ (triangular profile). Thus for the same q, the triangular profile takes about 38% less material than the corresponding rectangular profile.

4. The height of the fin, $r_a - r_b$, is virtually independent of the fin shape. More accurate calculations by Razelos and Imre (1980) and Netrakanti and Huang (1985) reveal an increase in length of about 3% from the rectangular to the triangular shape.

5. For the rectangular geometry, the Brown (1965) solution (Fig. 3.3) or the Mikk (1980) solution (Table 3.1) could have been used. Moreover, for the triangular geometry, the alternative is to use Table 3.2. For the trapezoidal geometry, the results of Netrakanti and Huang (1985) provide an alternative to Fig. 3.5. Such calculations have revealed that the final results are essentially the same.

3.4.3 Radial Fins of Parabolic and Hyperbolic Profile

Ullman and Kalman (1989) have presented solutions for the optimum dimensions of the radial fins of rectangular, triangular, parabolic (Fig. 3.6a) and hyperbolic (Fig. 3.6b) profile. Their results are shown in Fig. 3.7, where the parameters are

$$Q = \frac{q_b(n+1)(n+2)}{2\pi^2 k\delta_o\theta_b\,[(r_a/r_b)-1]\,[(r_a/r_b)+n+1]} \qquad (n = 0, 1, 2) \qquad (3.28a)$$

and

$$Q = \frac{q_b}{2\pi^2 k\delta_o\theta_b\,[(r_a/r_b)-1]} \qquad (n = -1) \qquad (3.28b)$$

and $k\delta_o/2hr_b^2$ are plotted against r_a/r_b for

$$n = \begin{cases} 2 & \text{(parabolic)} \\ 1 & \text{(triangular)} \\ 0 & \text{(rectangular)} \\ -1 & \text{(hyperbolic)} \end{cases}$$

If the fin is to be designed for a given q_b, the procedure is to select a trial value of the optimum fin base thickness δ_o which fixes the parameter $k\delta_o/2hr_b^2$. Figure 3.7 is then read to obtain r_a/r_b and Q. In the event that r_a is specified, Q can be used to establish q_b, which could then be compared to the q_b specified. This procedure could then be repeated until a match is obtained.

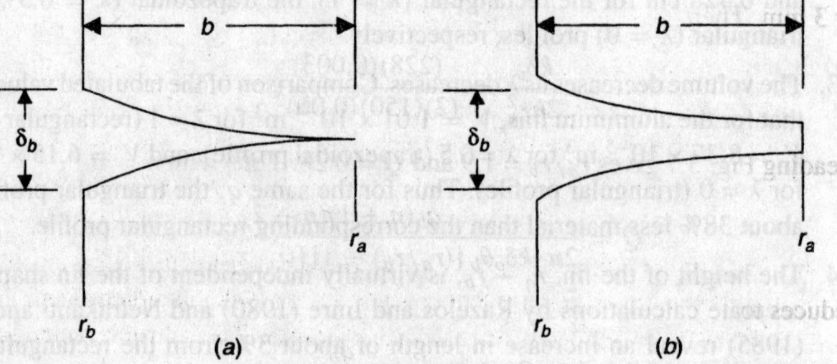

Figure 3.6 Optimum radial fins of (*a*) parabolic and (*b*) hyperbolic profile.

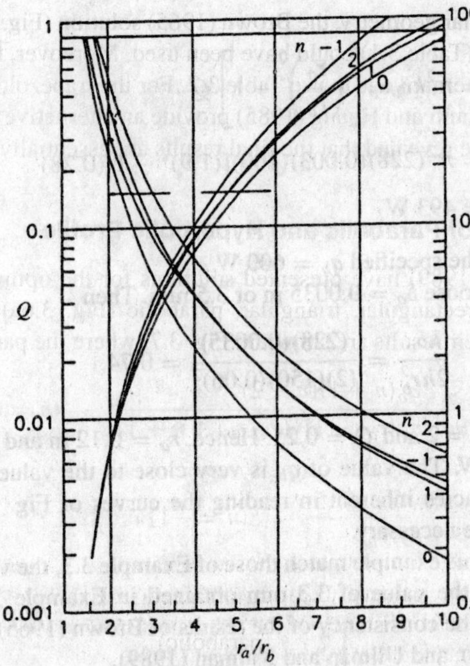

Figure 3.7 Optimum dimensions of radial fins. [Adapted from Ullman and Kalman (1989)].

Example 3.4: Optimum Radial Fins. Design radial fins of optimum dimensions for **(a)** rectangular, **(b)** triangular,**(c)** hyperbolic, and **(d)** parabolic shapes to dissipate 600 W. The base of the fin has a radius of 6 cm and the base temperature excess is 100°C. The average convective heat transfer coefficient is $h = 150$ W/m$^2 \cdot$ K and the thermal conductivity of the aluminum fin is 228 W/m · K. The design is to be based on Fig. 3.7.

SOLUTION. **(a)** For the rectangular fin ($n = 0$), as a first trial, select $\delta_o = 0.003$ m or 3 mm. Then

$$\frac{k\delta_o}{2hr_b^2} = \frac{(228)(0.003)}{(2)(150)(0.06)^2} = 0.63$$

Reading Fig. 3.7 gives $r_a/r_b = 1.9$ and $Q = 0.28$. Thus with $n = 0$,

$$Q = \frac{q_b(n+1)(n+2)}{2\pi^2 k\delta_o\theta_b\,[(r_a/r_b) - 1]\,[(r_a/r_b) + n + 1]} \tag{3.28a}$$

reduces to

$$Q = \frac{q_b}{\pi^2 k\delta_o\theta_b\left[(r_a/r_b)^2 - 1\right]}$$

so that

$$q_b = \pi^2 k\delta_o\theta_b \left[\left(\frac{r_a}{r_b}\right)^2 - 1\right] Q$$

$$= \pi^2(228)(0.003)(100)[(1.9)^2 - 1](0.28)$$

$$= 493 \text{ W}$$

which does not match the specified $q_b = 600$ W.

As a second trial, choose $\delta_o = 0.0035$ m or 3.5 mm. Then

$$\frac{k\delta_o}{2hr_b^2} = \frac{(228)(0.0035)}{(2)(150)(0.06)^2} = 0.74$$

and Fig. 3.7 gives $r_a/r_b = 2$ and $Q = 0.25$. Hence, $r_a = 0.12$ m and q_b is evaluated from Q as $q_b = 591$ W. The value of q_b is very close to the value specified, and because of the inaccuracies inherent in reading the curves of Fig. 3.7, no further refinement appears to be necessary.

Because the data in this example match those of Example 3.3, the value of 3.5 mm can be compared with the value of 3.3 mm obtained in Example 3.3. This close agreement vouches for the consistency of the results of Brown (1965), Mikk (1980), Razelos and Imre (1980), and Ullman and Kalman (1989).

(**b**) For the triangular fin ($n = 1$), as a first trial select $\delta_o = 0.003$ m or 3 mm. Then

$$\frac{k\delta_o}{2hr_b^2} = \frac{(228)(0.003)}{(2)(150)(0.06)^2} = 0.63$$

Reading Fig. 3.7 gives $r_a/r_b = 1.9$ and $Q = 0.39$. Thus with $n = 1$,

$$r_a = 1.9r_a = (1.9)(0.06) = 0.114 \text{ m} \quad \text{or} \quad 11.4 \text{ cm}$$

and

$$q_b = \frac{1}{3}\pi^2 k\delta_o\theta_b \left(\frac{r_a}{r_b} - 1\right)\left(\frac{r_a}{r_b} + 2\right) Q$$

$$= \frac{1}{3}\pi^2(228)(0.003)(100)(1.90 - 1)(1.90 + 2)(0.54)$$

$$= 308 \text{ W}$$

which is less than the specified 600 W.

As a second trial, choose $\delta_o = 4$ mm. Then

$$\frac{k\delta_o}{2hr_b^2} = \frac{(228)(0.004)}{(2)(150)(0.06)^2} = 0.84$$

and Fig. 3.7 gives $r_a/r_b = 2.1$ and $Q = 0.4$. Hence $r_a = 0.126$ m or 12.6 cm and q_b is evaluated from Q as $q_b = 541$ W.

Finally, try $\delta_o = 4.3$ mm. Then

$$\frac{k\delta_o}{2hr_b^2} = \frac{(228)(0.0043)}{(2)(150)(0.06)^2} = 0.86$$

and Fig. 3.7 gives $r_a/r_b = 2.1$ and $Q = 0.41$. Hence $r_a = 0.126$ m or 12.6 cm and q_b is evaluated from Q as $q_b = 601$ W, which is very close.

The value of $\delta_o = 4.3$ mm is very close to the value of $\delta_o = 4.28$ mm obtained in Example 3.3.

(**c**) For the parabolic fin ($n = 2$) as a first trial, select $\delta_o = 0.004$ m (4 mm). Then

$$\frac{k\delta_o}{2hr_b^2} = \frac{(228)(0.004)}{(2)(150)(0.06)^2} = 0.84$$

and Fig. 3.7 gives $r_a/r_b = 2.1$ and $Q = 0.54$. Thus with $n = 2$,

$$r_a = 2.1r_a = (2.1)(0.06) = 0.126 \text{ m} \quad \text{or} \quad 12.6 \text{ cm}$$

and

$$q_b = \frac{1}{6}\pi^2 k\delta_o\theta_b\left(\frac{r_a}{r_b} - 1\right)\left(\frac{r_a}{r_b} + 3\right)Q$$

$$= \tfrac{1}{6}\pi^2(228)(0.004)(100)(2.10 - 1)(2.10 + 3)(0.54)$$

$$= 454 \text{ W}$$

which is less than the specified 600 W.

As a second trial, choose $\delta_o = 4.3$ mm. Then

$$\frac{k\delta_o}{2hr_b^2} = \frac{(228)(0.0043)}{(2)(150)(0.06)^2} = 0.90$$

and Fig. 3.7 gives $r_a/r_b = 2.2$ and $Q = 0.5$. Hence $r_a = 0.132$ m or 13.2 cm and q_b is evaluated from Q as $q_b = 503$ W. The final value is $\delta_o = 5.8$ mm. Then

$$\frac{k\delta_o}{2hr_b^2} = \frac{(228)(0.0058)}{(2)(150)(0.06)^2} = 1.22$$

and Fig. 3.7 gives $r_a/r_b = 2.3$ and $Q = 0.40$. Hence $r_a = 0.138$ m or 13.8 cm and q_b is evaluated from Q as $q_b = 599$ W, which is very close.

These values cannot be checked against other optimization data because no such other data appear to exist for the radial fin of parabolic profile.

(**d**) For the hyperbolic fin ($n = -1$), as a first trial select $\delta_o = 0.005$ m (5 mm). Then

$$\frac{k\delta_o}{2hr_b^2} = \frac{(228)(0.005)}{(2)(150)(0.06)^2} = 1.06$$

and Fig. 3.7 gives $r_a/r_b = 2.2$ and $Q = 0.27$. Thus with $n = -1$,

$$r_a = 2.2r_a = (2.2)(0.06) = 0.132 \text{ m} \quad \text{or} \quad 13.2 \text{ cm}$$

and

$$\begin{aligned} q_b &= 2\pi^2 k\delta_o\theta_b\left(\frac{r_a}{r_b} - 1\right)Q \\ &= 2\pi^2(228)(0.005)(100)(2.20 - 1)(0.27) \\ &= 729 \text{ W} \end{aligned}$$

which is more than the specified 600 W. The final trial value is $\delta_o = 4$ mm. Then

$$\frac{k\delta_o}{2hr_b^2} = \frac{(228)(0.004)}{(2)(150)(0.06)^2} = 0.84$$

and Fig. 3.7 gives $r_a/r_b = 2.1$ and $Q = 0.3$. Hence $r_a = 0.126$ m or 12.6 cm and q_b is evaluated from Q as $q_b = 594$ W, which is close enough.

Here, too, these values cannot be checked against other optimization data because no such other data appear to exist for the radial fin of hyperbolic profile.

3.5 OPTIMUM SPINE DIMENSIONS

3.5.1 Cylindrical Spine

The Sonn and Bar-Cohen (1981) optimization for the cylindrical spine shown in Fig. 3.8*a* is based on the minimization of the spine volume. Working with the heat flow through the base of the spine given by

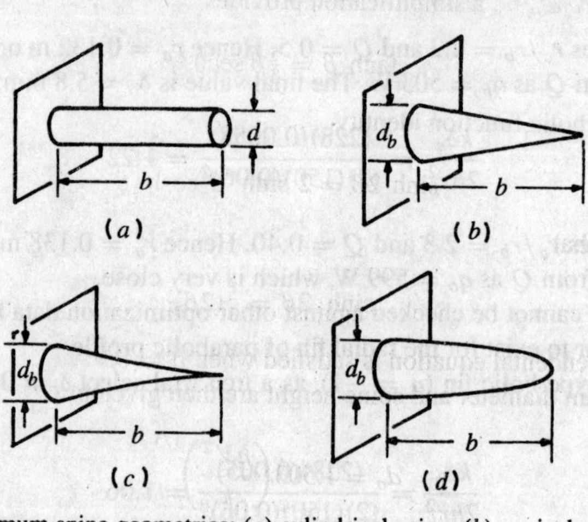

Figure 3.8 Optimum spine geometries: (*a*) cylindrical spine; (*b*) conical spine; (*c*) concave parabolic spine; (*d*) convex parabolic spine.

$$\dot{q}_b = \frac{\pi}{4} k d^2 m \theta_b \tanh mb \tag{1.61}$$

they defined a volume parameter

$$V = d_o^2 b_o$$

where d_o and b_o are the optimum spine dimensions and where $m^2 = 4h/kd_o$. With $b_o = V/d_o^2$, they expressed eq. (1.61) in terms of d_o and V:

$$q_b = \frac{\pi}{2}\theta_b (hk)^{1/2} d_o^{3/2} \tanh\left[2V\left(\frac{h}{k}\right)^{1/2}\left(\frac{1}{d_o}\right)^{5/2}\right]$$

If

$$\Lambda_1 = \frac{\pi}{2}\theta_b (hk)^{1/2} \quad \text{and} \quad \Lambda_2 = 2V\left(\frac{h}{k}\right)^{1/2}$$

then

$$\beta = mb_o = \left(\frac{4h}{kd_o}\right)^{1/2} b_o = 2\left(\frac{h}{kd_o}\right)^{1/2}\frac{V}{d_o^2} = 2V\left(\frac{h}{k}\right)^{1/2} d_o^{-5/2} = \Lambda_2 d_o^{-5/2}$$

and

$$q_b = \Lambda_1 d_o^{3/2} \tanh \Lambda_2 d_o^{-5/2}$$

The point at which $\partial q_b/\partial d_o$ vanishes yields the desired maximum

$$\frac{\partial q_b}{\partial d_o} = \frac{3}{2}\Lambda_1 d_o^{1/2} \tanh \Lambda_2 d_o^{-5/2} - \left(\Lambda_1 d_o^{3/2}\right)\left(\frac{5}{2}\Lambda_2 d_o^{-7/2}\right) \operatorname{sech}^2 \Lambda_2 d_o^{-5/2} = 0$$

and with $\beta = \Lambda_2 d_o^{-5/2}$, a simplification provides

$$\tanh \beta = \tfrac{5}{3}\beta \operatorname{sech}^2\beta \tag{3.29}$$

With the hyperbolic function identity

$$\sinh 2\beta = 2 \sinh \beta \cosh \beta$$

it is observed that

$$\sinh 2\beta = \tfrac{5}{3}(2\beta)$$

and this transcendental equation is satisfied when $\beta = 0.9193$.

The optimum diameter and spine height are then given by

$$d_o = 1.503\left(\frac{hV^2}{k}\right)^{1/5} \tag{3.30}$$

and

and

$$b_o = 0.564\left(\frac{Vk^2}{h^2}\right)^{1/5} \tag{3.31}$$

Aziz (1992) has provided expressions for the optimum diameter and fin height in terms of a specified heat-dissipating requirement.

$$d_o = 0.917\left(\frac{q_b^2}{hk\theta_b^2}\right)^{1/3} \tag{3.32}$$

and

$$b_o = 0.440\left(\frac{q_b k}{h^2\theta_b}\right)^{1/3} \tag{3.33}$$

3.5.2 Other Spines

Aziz (1992) has provided expressions for the optimum dimensions δ_o and d_o of the conical, concave parabolic, and convex parabolic spines in terms of a specified volume V or a specified heat-dissipating requirement q_b.

Conical Spine. For the conical spine (Fig. 3.8*b*)

$$d_o = 1.954\left(\frac{hV^2}{k}\right)^{1/5} \tag{3.34}$$

$$b_o = 1.001\left(\frac{k^2V}{h^2}\right)^{1/5} \tag{3.35}$$

$$d_o = 1.099\left(\frac{q_b^2}{hk\theta_b^2}\right)^{1/3} \tag{3.36}$$

and

$$b_o = 0.751\left(\frac{q_b k}{h^2\theta_b}\right)^{1/3} \tag{3.37}$$

Concave Parabolic Spine. For the concave parabolic spine (Fig. 3.8*c*)

$$d_o = 2.097\left(\frac{hV^2}{k}\right)^{1/5} \tag{3.38}$$

$$b_o = 1.448\left(\frac{k^2V}{h^2}\right)^{1/5} \tag{3.39}$$

$$d_o = 1.175\left(\frac{q_b^2}{hk\theta_b^2}\right)^{1/3} \tag{3.40}$$

and

$$b_o = 1.084\left(\frac{q_b k}{h^2 \theta_b}\right)^{1/3} \tag{3.41}$$

Convex Parabolic Spine. For the convex parabolic spine (Fig. 3.8*d*)

$$d_o = 1.798\left(\frac{hV^2}{k}\right)^{1/5} \tag{3.42}$$

$$b_o = 0.788\left(\frac{k^2 V}{h^2}\right)^{1/5} \tag{3.43}$$

$$d_o = 1.026\left(\frac{q_b^2}{hk\theta_b^2}\right)^{1/3} \tag{3.44}$$

and

$$b_o = 0.595\left(\frac{q_b k}{h^2 \theta_b}\right)^{1/3} \tag{3.45}$$

Example 3.5: Optimum Spines. A spine is to be employed to dissipate 10 W from a primary surface at 90°C by convection. Determine the optimum dimensions d_o and b_o for (**a**) cylindrical, (**b**) conical, (**c**) concave parabolic, and (**d**) convex parabolic spines which are fabricated from a material having a thermal conductivity $k = 65$ W/m · K. The environmental temperature is 30°C and the heat transfer coefficient is $h = 80$ W/m^2 · K.

SOLUTION. For all spines

$$\theta_b = 90 - 30 = 60°\text{C}$$

(**a**) For the cylindrical spine, use eqs. (3.32) and (3.33):

$$d_o = 0.917\left(\frac{q_b^2}{hk\theta_b^2}\right)^{1/3}$$

$$= 0.917\left[\frac{(10)^2}{(80)(65)(60)^2}\right]^{1/3}$$

$$= 0.0160 \text{ m}$$

or

$$= 1.60 \text{ cm}$$

$$b_o = 0.440\left(\frac{q_b k}{h^2\theta_b}\right)^{1/3}$$

$$= 0.440\left[\frac{(10)(65)}{(80)^2(60)}\right]^{1/3}$$

$$= 0.0524 \text{ m}$$

or

$$= 5.24 \text{ cm}$$

(b) For the conical spine, use eqs. (3.36) and (3.37):

$$d_o = 1.099\left(\frac{q_b^2}{hk\theta_b^2}\right)^{1/3}$$

$$= 1.099\left[\frac{(10)^2}{(80)(65)(60)^2}\right]^{1/3}$$

$$= 0.0192 \text{ m}$$

or

$$= 1.92 \text{ cm}$$

and

$$b_o = 0.751\left(\frac{q_b k}{h^2\theta_b}\right)^{1/3}$$

$$= 0.751\left[\frac{(10)(65)}{(80)^2(60)}\right]^{1/3}$$

$$= 0.0895 \text{ m}$$

or

$$= 8.95 \text{ cm}$$

(c) For the concave parabolic spine, use eqs. (3.40) and (3.41):

$$d_o = 1.175\left(\frac{q_b^2}{hk\theta_b^2}\right)^{1/3}$$

$$= 1.175\left[\frac{(10)^2}{(80)(65)(60)^2}\right]^{1/3}$$

$$= 0.0205 \text{ m}$$

or

$$= 2.05 \text{ cm}$$

and

$$b_o = 1.084\left(\frac{q_b k}{h^2\theta_b}\right)^{1/3}$$

$$= 1.084\left[\frac{(10)(65)}{(80)^2(60)}\right]^{1/3}$$

$$= 0.1292 \text{ m}$$

or

$$= 12.92 \text{ cm}$$

(d) For the convex parabolic spine, use eqs. (3.44) and (3.45):

$$d_o = 1.026\left(\frac{q_b^2}{hk\theta_b^2}\right)^{1/3}$$

$$= 1.026\left[\frac{(10)^2}{(80)(65)(60)^2}\right]^{1/3}$$

$$= 0.0179 \text{ m}$$

or

$$= 1.79 \text{ cm}$$

and

$$b_o = 0.595\left(\frac{q_b k}{h^2\theta_b}\right)^{1/3}$$

$$= 0.595\left[\frac{(10)(65)}{(80)^2(60)}\right]^{1/3}$$

$$= 0.0709 \text{ m}$$

or

$$= 7.09 \text{ cm}$$

Discussion. If a single spine is employed, the design results in "chubby" spines. However, if two spines are used, each dissipating 5 W, the dimensions for the cylindrical spine become $d_o = 1.01$ cm and $b_o = 4.16$ cm (for 10 W, $d_o = 1.60$ cm and $b_o = 5.24$ cm). The spine is shorter and slimmer. Moreover, the volume used for a single spine is 1.054×10^{-5} m^3, but with two spines, only two-thirds of the volume is needed, 3.333×10^{-6} m^3. If four spines are used, the dimensions become $d_o = 0.64$ cm and $b_o = 3.30$ cm and the volume needed for the four spines is 5.407×10^{-6} m^3, which is about one-half of the single spine design.

If further reduction in volume is necessary, a conical geometry can be considered. For four conical spines, the optimum dimensions of each would be $d_o = 0.76$ cm and $b_o = 5.63$ cm, but the total volume of the four would be 1.084×10^{-6} m^3.

3.6 IMPROVED ANALYSES FOR CONVECTING FINS

The optimization analyses of previous sections have been improved to include the effects of tip heat loss, internal heat generation, variable heat transfer coefficient, temperature-dependent thermal properties, and a convective boundary condition and prime surface thickness at the base of the fin. These analyses are considered in the following subsections.

3.6.1 Effect of Tip Heat Loss

The heat flow through the base of a longitudinal fin of rectangular profile with tip heat loss is given by

$$q_b = \frac{k\delta Lm\theta_b\,[(h_a/mk) + \tanh mb]}{1 + (h_a/mk)\tanh mb} \tag{2.6}$$

where $m^2 = 2h/k\delta$ and h_a is the heat transfer coefficient at the fin tip. This may be rewritten for unit fin length ($L = 1$) and by noting that $\tanh mb = \sinh mb/\cosh mb$:

$$q_b = (2hk\delta)^{1/2}\theta_b \frac{\sinh mb + \Upsilon \cosh mb}{\cosh mb + \Upsilon \sinh mb} \tag{3.46}$$

where

$$\Upsilon = h_a\left(\frac{\delta}{2hk}\right)^{1/2}$$

Aziz (1978) has shown that for a fixed profile area with optimum dimensions, $A_p = \delta_o b_o$, b_o can be replaced by $b_o = A_p/\delta_o$ and eq. (3.46) can be used to set $\partial q_b/\partial\delta_o$ equal to zero. This procedure leads to an optimizing relationship between mb and Υ for the optimum conditions

$$(1 + \Upsilon^2)\sinh 2mb + 2\Upsilon(1 + \cosh 2mb) = 6mb(1 - \Upsilon^2) \tag{3.47}$$

The solution of eq. (3.47) is shown in Fig. 3.9, which provides a starting point for the trial-and-error procedure necessary to fix the optimum fin dimensions.

Example 3.6: Optimum Fin with Tip Heat Loss. Design an optimum longitudinal fin of rectangular profile with the following specifications:

$$q_b = 170 \text{ W/m} \qquad \theta_b = 50°\text{C}$$

$$k = 31 \text{ W/m}\cdot\text{K} \qquad h = 34\ W/m^2\cdot K$$

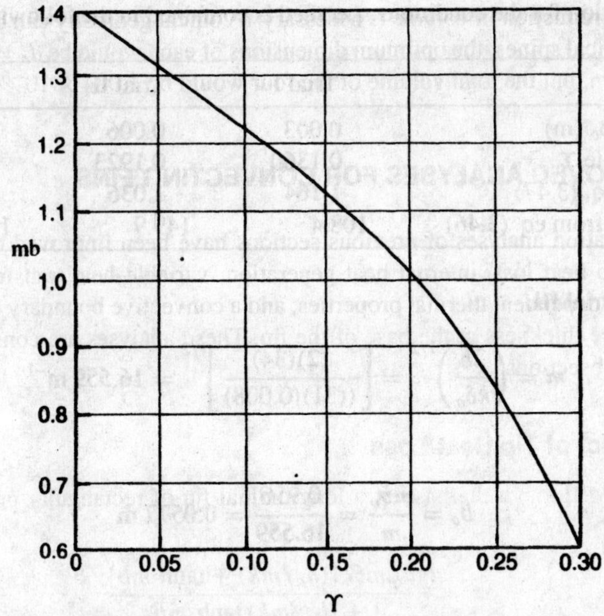

Figure 3.9 Effect of tip heat loss on the optimum dimensions of a longitudinal fin of rectangular profile. [Adapted from Aziz (1978).]

and with a tip heat transfer coefficient of

$$h_a = 114 \text{ W/m}^2 \cdot \text{K}$$

SOLUTION. The procedure is:

1. Assume a trial value of δ_o.
2. Compute Υ:

$$\Upsilon = h_a \left(\frac{\delta}{2hk} \right)^{1/2}$$

3. Consult Fig. 3.9 to locate an approximate value of mb.
4. Use repeated values of mb in eq. (3.47) until a particular value of mb satisfies eq. (3.47).
5. Use eq. (3.46):

$$q_b = (2hk\delta)^{1/2} \theta_b \frac{\sinh \, mb + \Upsilon \cosh \, mb}{\cosh \, mb + \Upsilon \, \sinh \, mb}$$

to determine q_b. If this value does not equal the specified value of q_b, the procedure must begin again with a different value of δ_o.

The solution for the conditions specified is contained in the following tabulation:

	Trial I	Trial II	Trial III
Select δ_o (m)	0.003	0.006	0.008
Compute Υ	0.1360	0.1923	0.2220
mb in eq. (3.47)	1.164	1.036	0.956
q_b (W) from eq. (3.46)	108.4	149.9	170.2

Then, from trial III,

$$m = \left(\frac{2h}{k\delta_o}\right)^{1/2} = \left[\frac{(2)(34)}{((31)(0.008)}\right]^{1/2} = 16.559 \text{ m}^{-1}$$

and

$$b_o = \frac{mb_o}{m} = \frac{0.956}{16.559} = 0.0577 \text{ m}$$

In summary,

$$\delta_o = 0.800 \text{ cm} \quad \text{and} \quad b_o = 5.77 \text{ cm}$$

3.6.2 Effect of Variable Heat Transfer Coefficient

The optimization analysis for a single cylindrical spine presented in Section 3.5.1 was based on the assumption that the heat transfer coefficient h was constant. Li (1983) modified this analysis by arguing that h was actually inversely proportional to the diameter; that is, $h = CD^{-n}$, where n and C are constants. The value of n, however, depends on the cross-flow Reynolds number Re_D and on whether the spine is single or part of an array. Table 3.3 gives the value of n for different configurations and flow conditions.

TABLE 3.3 Values of n in $h = CD^{-n}$

Configuration	Re_D	n
Single	10^2–10^3	0.50
	10^3–2×10^5	0.40
	2×10^5–10^6	0.30
Aligned array	10^2–10^3	0.50
	10^3–2×10^5	0.37
	2×10^5–10^6	0.20
Staggered array	10^2–10^3	0.50
	10^3–2×10^5	0.40
	2×10^5–10^6	0.20

The optimization analysis, which allows for a variable h of the form $h = CD^{-n}$, leads to a relation between mb_o and n:

$$\sinh\ 2mb_o = \frac{5+n}{3-n} mb_o \tag{3.48}$$

where b_o is the optimum fin height. Observe that the constant C is of no consequence with regard to the optimum dimensions. Equation (3.48) reduces to eq. (3.29) when $n = 0$ (constant h). The solutions of eq. (3.48) for various values of n are displayed in Table 3.4. Once the value of n has been obtained, the solutions for optimum fin height b_o optimum diameter d_o, heat dissipated q_b, the fin efficiency η can be derived. Table 3.4 also furnishes this information for various values of n, but here it should be noted that because h is a function of d, a trial-and-error solution is necessary to fix the dimensions of the optimum spine.

3.6.3 Effect of Temperature Dependent Thermal Conductivity

The effect of temperature-dependent thermal conductivity on the optimum dimensions of a longitudinal fin of rectangular profile has been studied by Aziz and Enamul-Huq (1975) and Krane (1976). If the temperature dependence of the thermal conductivity is of the form

$$k = k_s[1 + \beta(T - T_s)]$$

where T_s designates the temperature of the surroundings, then for the longitudinal fin of rectangular profile, a regular perturbation analysis gives the heat dissipated as

$$q_b = \frac{k_s \delta_o \theta_b}{b_o}(mb_o \tanh mb_o)\left[1 + \left(\frac{1}{3}\right)\epsilon \tanh^2 mb_o\right] \tag{3.49}$$

where

$$\epsilon = \frac{k_b}{k_s} - 1$$

The transcendental equation defining the optimum point is

TABLE 3.4 Optimum Parameters for a Cylindrical Spine with Variable Heat Transfer Coefficient

n	mb_o	$d_o\left(\frac{k}{hV^2}\right)^{0.2}$	$d_o\frac{k}{hb_o^2}$	$\frac{q_b k}{\theta_b h^2 b_o^3}$	η
0.50	1.1709	1.2388	2.9175	6.4545	0.704
0.40	1.1217	1.2603	3.1794	7.1965	0.721
0.37	1.1068	1.2670	3.2651	7.4414	0.725
0.30	1.0721	1.2833	3.4801	8.0589	0.737
0.20	1.0220	1.3081	3.8296	9.0724	0.754
0.00	0.9193	1.3660	4.7300	11.736	0.789

$$6mb_o(1 + \epsilon \tanh^2 mb_o) = \sinh 2mb_o + \tfrac{2}{3}\epsilon \tanh mb_o \sinh^2 mb_o \tag{3.50}$$

The solution of eq. (3.50) is shown in Fig. 3.10. Once the thermal conductivity variation is known, ϵ can be found. The use of Fig. 3.10 then gives mb_o. If the profile area, A_p, is specified, the optimum dimensions are given by

$$\delta_o = \left[\frac{2hA_p^2}{k_s(mb_o)^2}\right]^{1/3} \tag{3.51}$$

and

$$b_o = \frac{A_p}{\delta_o} \tag{3.52}$$

The corresponding expressions in terms of q_b are

$$\delta_o = \frac{(q_b/\theta_b)^2}{hk_s\left\{2\tanh^2 mb_o\left[1 + \left(\frac{1}{3}\right)\epsilon \tanh^2 mb_o\right]\right\}} \tag{3.53}$$

and

$$b_o = \frac{mb_o(q_b/\theta_b)}{2h\left[1 + (\epsilon/3)\tanh^2 mb_o\right]\tanh mb_o} \tag{3.54}$$

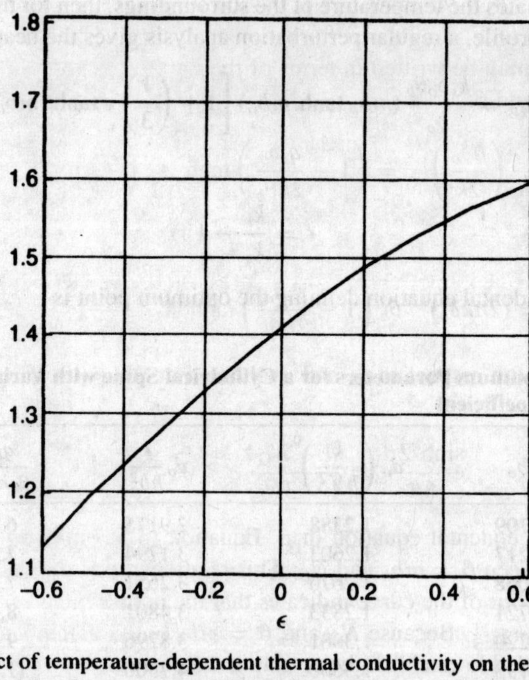

Figure 3.10 Effect of temperature-dependent thermal conductivity on the optimum dimensions of a longitudinal fin of rectangular profile. [Adapted from Aziz and Enamul-Huq (1975).]

3.6.4 Effect of Internal Heat Generation

The effect of internal heat generation on the optimum dimensions of a longitudinal fin of rectangular profile was first studied by Minkler and Rouleau (1960) and the relationship between q_i, q_b, θ_b and the dimensions of the fin is given by

$$N_R(\text{Bi})^{1/2} = (1 - N_G) \tanh mb \tag{2.62}$$

where N_R is the removal number or fin effectiveness,

$$N_R = \frac{q_b}{h\delta\theta_b}$$

N_G is the generation number,

$$N_G = \frac{q_i \delta b}{2h\theta_b b} = \frac{q_i \delta}{2h\theta_b}$$

and Bi is the Biot number,

$$\text{Bi} = \frac{h\delta}{2k}$$

For a given profile area, a relationship between the optimum dimensions b_o and δ_o can be determined. Let β be the optimizing parameter which may be defined in terms of the profile area, $A_p = b_o \delta_o$:

$$\beta = mb_o = \left(\frac{2h}{k\delta_o}\right)^{1/2} = A_p \left(\frac{2h}{k}\right)^{1/2} \left(\frac{1}{\delta_o}\right)^{3/2}$$

Equation (2.62) may be written in terms of the fin profile area using the definitions for N_R, N_G, Bi, and β:

$$\frac{q_b}{h\delta_o\theta_b}\left(\frac{h\delta_o}{2k}\right)^{1/2} = \left(1 - \frac{q_i\delta_o}{2h\theta_b}\right) \tanh A_p \left(\frac{2h}{k}\right)^{1/2} \left(\frac{1}{\delta_o}\right)^{3/2}$$

or with $\beta = mb_o$

$$q_b = (2hk\delta_o)^{1/2}\theta_b \left(1 - \frac{q_i\delta_o}{2h\theta_b}\right) \tanh A_p \left(\frac{2h}{k}\right)^{1/2} \left(\frac{1}{\delta_o}\right)^{3/2} \tag{3.55}$$

Differentiation with respect to δ_b and evaluation at the point where the derivative vanishes yields

$$\frac{\sinh 2\beta}{6\beta} = \frac{1 - N_G}{1 - 3N_G} \qquad 0 \le N_G \le \frac{1}{3} \tag{3.56}$$

which is a transcendental equation in β. Equation (3.56) may be used to obtain a relationship between $\beta = mb_o$ and N_G. This relationship is shown in Fig. 3.11. The asymptotic behavior of the curve indicates that the optimum value, $\beta = mb_o$, exists only for $0 \le N_G \le \frac{1}{3}$. Because N_G and $\beta = mb_o$ both contain δ_o, a trial-and-error procedure is needed to establish the optimum dimensions.

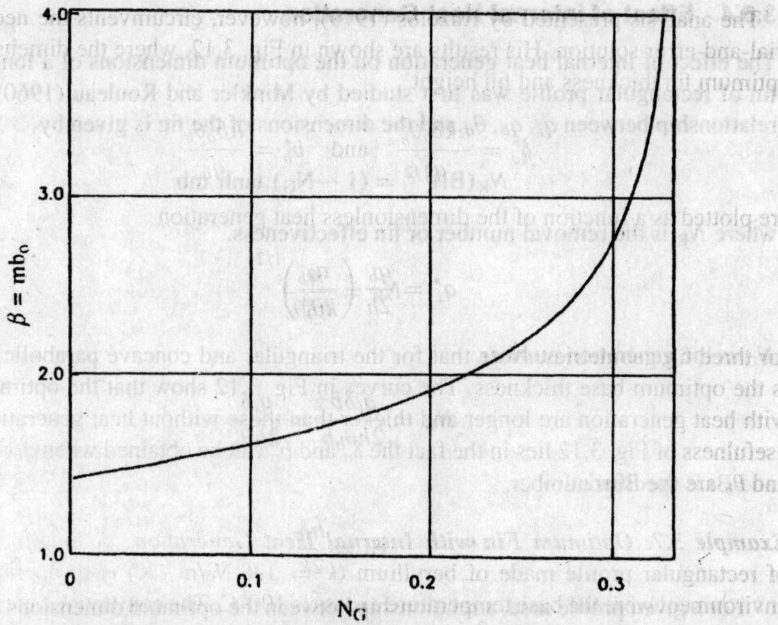

Figure 3.11 Relationship between the generation number N_G and $\beta = mb_o$ for the optimum longitudinal fin of rectangular profile with heat generation. [Adapted from Minkler and Rouleau (1960).]

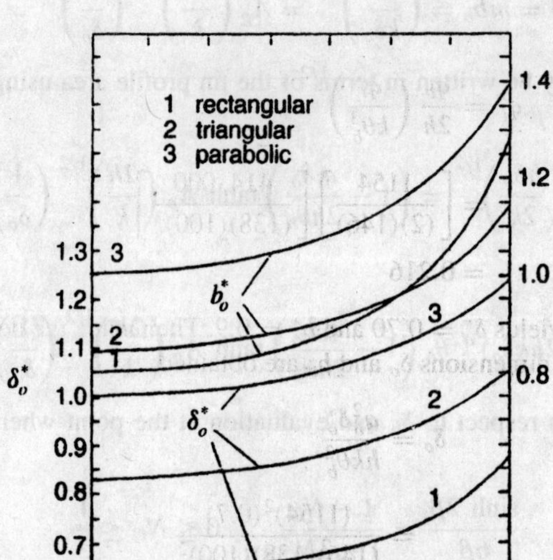

Figure 3.12 Optimum base thicknesses and heights for three longitudinal fins with heat generation. [Adapted from Razelos (1979).]

The analysis presented by Razelos (1979), however, circumvents the need for a trial-and-error solution. His results are shown in Fig. 3.12, where the dimensionless optimum fin thickness and fin height

$$\delta_o^* = \frac{hk\delta_o\theta_b^2}{q_b^2} \quad \text{and} \quad b_o^* = \frac{hb_o\theta_b}{q_b}$$

are plotted as a function of the dimensionless heat generation

$$q_i^* = \frac{q_b}{2h}\left(\frac{q_i}{k\theta_b^3}\right)^{1/2}$$

for three fin geometries. Note that for the triangular and concave parabolic fins, δ_o is the optimum base thickness. The curves in Fig. 3.12 show that the optimum fins with heat generation are longer and thicker than those without heat generation. The usefulness of Fig. 3.12 lies in the fact the δ_o and b_o can be obtained when q_b, q_i, h, k, and θ_b are specified.

Example 3.7: Optimum Fin with Internal Heat Generation. A longitudinal fin of rectangular profile made of beryllium ($k = 138$ W/m · K) is to operate in an environment were the base temperature excess is 100°C. The convective heat transfer coefficient is $h = 146$ W/m^2 · K and because of gamma-ray absorption, the heat generation in the fin is 414 kW/m^3. In addition to the internal heat generation, the fin must dissipate 1154 W/m. Design a fin with optimum dimensions.

SOLUTION. With

$$\begin{aligned} q_i^* &= \frac{q_b}{2h}\left(\frac{q_i}{k\theta_b^3}\right)^{1/2} \\ &= \left[\frac{1154}{(2)(146)}\right]\left[\frac{414,000}{(138)(100)^3}\right]^{1/2} \\ &= 0.216 \end{aligned}$$

curve 1 in Fig. 3.12 yields $\delta_o^* = 0.70$ and $b_o^* = 0.9$. Then using the definitions for δ_o^* and b_o^*, the optimum dimensions δ_o and b_o are obtained:

$$\begin{aligned} \delta_o &= \frac{q_b^2\delta_o^*}{hk\theta_b^2} \\ &= \frac{(1154)^2(0.7)}{(146)(138)(100)^2} \\ &= 4.63 \times 10^{-3}\ \text{m} \end{aligned}$$

or

$$\delta_o = 0.463\ \text{cm}$$

and

$$b_o = \frac{q_b b_o^*}{h\theta_b}$$

$$= \frac{(1154)(0.9)}{(146)(100)}$$

$$= 0.071 \text{ m}$$

or

$$= 7.1 \text{ cm}$$

In certain situations, the maximum permissible fin height may be specified rather than the heat dissipation. Although Fig. 3.12 can still be used, the process becomes one of trial and error. To avoid this, Razelos (1979) has provided an alternative shown here as Fig. 3.13. Here the dimensionless thickness and fin efficiency

$$\delta_o^* = \frac{k\delta_o}{hb_o^2} \quad \text{and} \quad \eta = \frac{q_b}{2hb_o\theta_b}$$

are plotted against the parameter

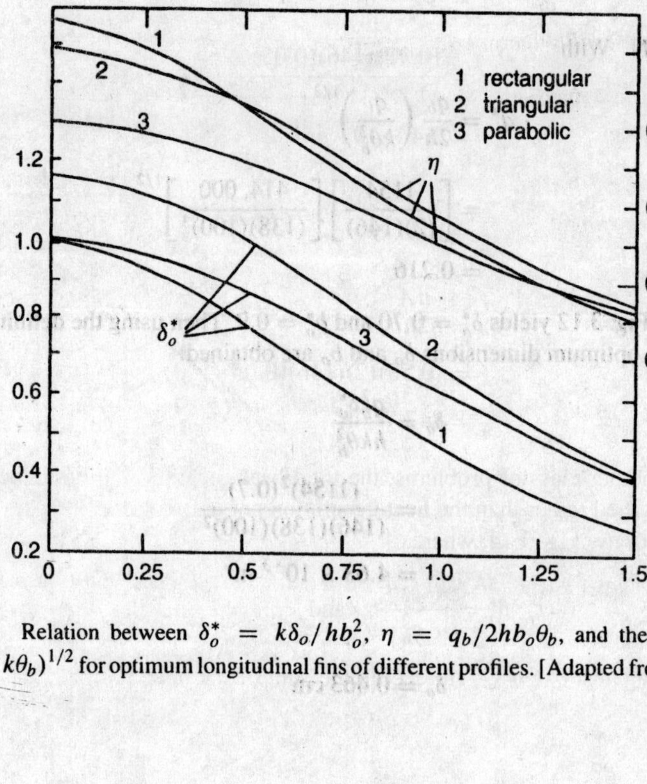

Figure 3.13 Relation between $\delta_o^* = k\delta_o/hb_o^2$, $\eta = q_b/2hb_o\theta_b$, and the parameter $\zeta = b_o\,(q_i/k\theta_b)^{1/2}$ for optimum longitudinal fins of different profiles. [Adapted from Razelos (1979).]

$$\zeta = b_o \left(\frac{q_i}{k\theta_b} \right)^{1/2}$$

It is interesting to note once more that the efficiency of the optimum fin with heat generation is lower than that for a fin without heat generation.

Example 3.8: Optimum Fin with Internal Heat Generation. Consider the data of Example 3.7 except that q_b is not specified. Instead, the maximum permissible fin height is specified as 5 cm. Determine the optimum fin thickness and the heat dissipated.

SOLUTION. Here

$$\zeta = b_o \left(\frac{q_i}{k\theta_b} \right)^{1/2}$$

$$= 0.05 \left[\frac{441,000}{(138)(100)} \right]^{1/2}$$

$$= 0.28$$

Figure 3.13 gives $\delta_o^* = 0.92$ and $\eta = 0.58$, so that

$$\delta_o = \frac{\delta_o^* h b_o^2}{k}$$

$$= \frac{(0.92)(146)(0.05)^2}{138}$$

$$= 2.4 \times 10^{-3} \text{ m}$$

or

$$= 0.24 \text{ cm}$$

and

$$q_b = \eta 2 h b_o \theta_b$$

$$= (0.58)(2)(146)(0.05)(100)$$

$$= 847 \text{ W/m}$$

In yet another class of problems, the total heat generation to be accommodated may be specified rather than the heat generation per unit volume. If q_b is given, it is convenient to use Fig. 3.14, where

$$\delta_o^* = \frac{hk\delta_o\theta_b^2}{q_b^2} \quad \text{and} \quad b_o^* = \frac{hb_o\theta_b}{q_b}$$

are plotted against $q_b C_1/Q$. Here Q is the sum of the heat dissipated and the heat generated:

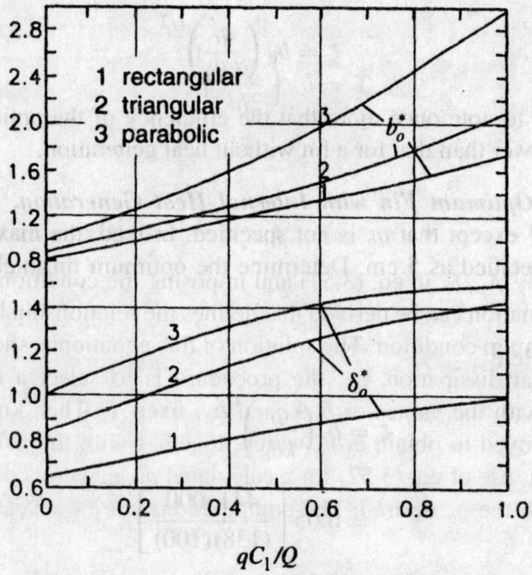

Figure 3.14 Optimum base thickness $hk\delta_o\theta_b^2/q_b^2$ and the optimum height $hb_o\theta_b/q_b$ plotted as a function of the parameter q_bC_1/Q. [Adapted from Razelos (1979).]

$$Q = q_b + q_i$$

and $C_1 = 1$ for the rectangular profile, $C_1 = \frac{1}{2}$ for the triangular profile and $C_1 = \frac{1}{3}$ for the concave parabolic profile.

3.6.5 Effect of Tip Heat Loss and Temperature-Dependent Thermal Conductivity

The combined effect of tip heat loss and temperature-dependent thermal conductivity on the optimum dimensions of a longitudinal fin of rectangular profile has been investigated by Aziz (1978). The linear thermal conductivity–temperature relationship of the form given in Section 3.6.3 was assumed and a two-term regular perturbation solution for the heat dissipation rate was derived. It is of the form

$$\frac{q_b b_o}{k_s \delta_o \theta_b} = (1+\epsilon)mb_o \frac{\sinh mb_o + \Upsilon \cosh mb_o}{\cosh mb_o + \Upsilon \sinh mb_o} + \frac{\epsilon mb_o}{3(\cosh mb_o + \Upsilon \sinh mb_o)^3}\psi \tag{3.57}$$

where

$$\psi = [(1-3\Upsilon^2)\sinh mb_o + \Upsilon(\Upsilon^2 - 3)\cosh mb_o]\cosh 2mb_o - 2[\cosh mb_o + \Upsilon^3 \sinh mb_o]\sinh 2mb_o - \Upsilon^3$$

where

$$\Upsilon = h_a \left(\frac{\delta}{2hk} \right)^{1/2}$$

and

$$\epsilon = \frac{k_b}{k_s} - 1$$

Replacing b_o by A_p/δ_o in eq. (3.57) and imposing the condition $\partial q_b/\partial \delta_o = 0$, a transcendental equation can be derived that defines the relationship between mb_o, Υ, and ϵ for the optimum condition. The solution of this equation is shown in Fig. 3.15. For a desired heat dissipation, q_b, the procedure is to select a trial value of δ_o which, together with the values of h, k, and h_a, fixes Υ. Then knowing Υ and ϵ, Fig. 3.16 is employed to obtain mb_o, which, together with the values of h, k, and δ_o, gives b_o. Next, use of eq. (3.57) for a calculated q_b leads to a comparison of the calculated q_b with the q_b desired. The computations can be repeated until the two values match.

3.6.6 Effect of Internal Heat Generation and Temperature-Dependent Thermal Conductivity

Aziz (1977) used a perturbation approach to establish the optimum dimensions δ_o and

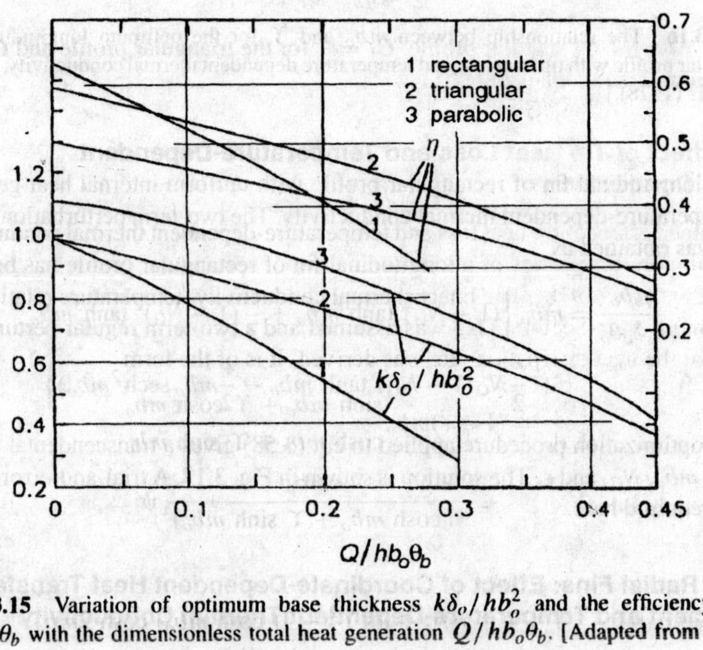

Figure 3.15 Variation of optimum base thickness $k\delta_o/hb_o^2$ and the efficiency $\eta = q_b/2hb_o\theta_b$ with the dimensionless total heat generation $Q/hb_o\theta_b$. [Adapted from Razelos (1979).]

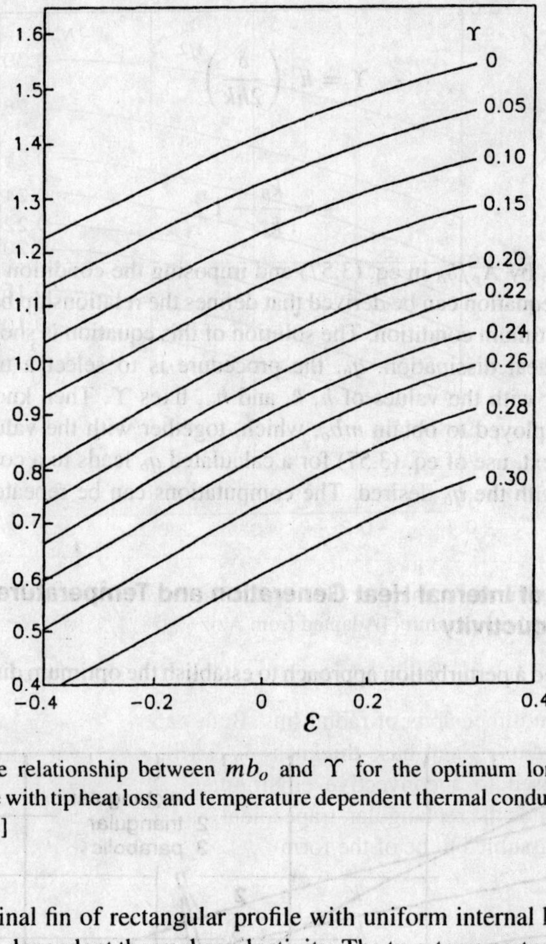

Figure 3.16 The relationship between mb_o and Υ for the optimum longitudinal fin of rectangular profile with tip heat loss and temperature dependent thermal conductivity. [Adapted from Aziz (1978).]

b_o of a longitudinal fin of rectangular profile with uniform internal heat generation and temperature-dependent thermal conductivity. The two-term perturbation solution for q_b was obtained as

$$\frac{q_b b_o}{k_s \delta_o \theta_b} = mb_o \left\{ (1 - N_G) \tanh mb_o + \frac{\epsilon}{3}(1 - N_G)^2 \tanh^3 mb_o + \frac{\epsilon}{2} N_G (1 - N_G) \tanh mb_o - \frac{\epsilon}{2} mb_o \operatorname{sech}^2 mb_o) \right\} \quad (3.58)$$

The optimization procedure applied to eq. (3.58) gives a transcendental equation relating mb_o, N_G, and ϵ. The solution is shown in Fig. 3.17. A trial-and-error solution is also required here.

3.6.7 Radial Fins: Effect of Coordinate-Dependent Heat Transfer Coefficient and Temperature-Dependent Thermal Conductivity

Razelos and Imre (1980) and Netrakanti and Huang (1985) have considered the combined effect of variable heat transfer coefficient and variable thermal conductivity

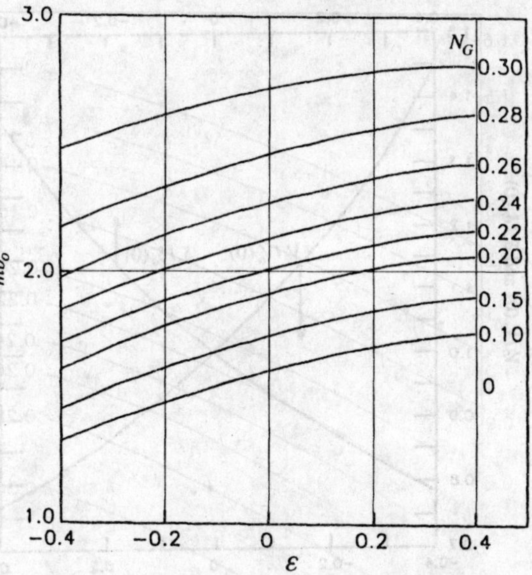

Figure 3.17 Optimization chart for the longitudinal fin of rectangular profile with uniform heat generation and temperature. [Adapted from Aziz (1978).]

on the optimum dimensions of radial fins. Both papers consider radial fins of base radius r_b tip radius r_a and base thickness δ_b (in the case of the rectangular profile $\delta_b = \delta$) exposed to a convective environment at temperature T_s. The profiles considered include the rectangular, trapezoidal, and triangular shapes. The variations of k and h are assumed to be of the form

$$\frac{k}{k_s} = 1 + \epsilon \frac{T - T_s}{T_b - T_s} \tag{3.59}$$

and

$$\frac{h}{h_s} = \frac{[1 + (r_b/r_a)](m+1)(m+2)}{2\,[1 + (m+1)(r_b/r_a)]} \left(\frac{r - r_b}{r_a - r_b} \right)^m \tag{3.60}$$

where ϵ and m are constants to be specified. The analysis assumes a constant base temperature T_b and no heat loss from the fin tip.

The optimization problem is posed as follows: Given the fin volume and the base radius, find the base thickness and the tip radius so that the heat flow through the base is maximized. Razelos and Imre (1980) solved the problem using a quasi-Newton technique to seek the optimum. Netrakanti and Huang (1985), on the other hand, employed a method of invariant embedding to achieve the solution.

The effect of variable thermal conductivity on the optimum volume V_o and optimum base thickness δ_o is shown in Fig. 3.18 for $m = 0$ (constant heat transfer coefficient). The values $V_o(0)$ and $\delta_o(0)$ denote the values corresponding to $\epsilon = 0$ (constant thermal conductivity). Because the corresponding results for the trapezoidal

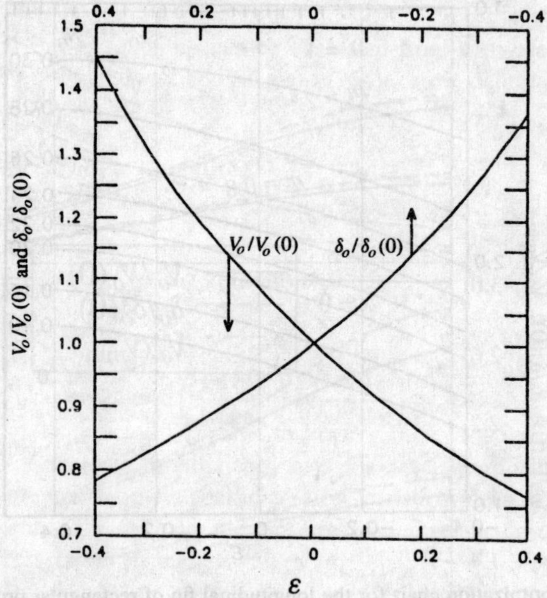

Figure 3.18 Effect of variable thermal conductivity on the optimum thickness and optimum volume of a radial fin of rectangular profile. [Adapted from Razelos and Imre (1980).]

and triangular profiles are virtually identical to those for the rectangular profile, Fig. 3.18 is applicable to all three profiles. Figure 3.18 shows that if the thermal conductivity increases with temperature ($\epsilon > 0$), both the volume and base thickness decrease compared to their values for $\epsilon = 0$. However, if the thermal conductivity decreases with temperature ($\epsilon < 0$), both the volume and the base thickness increase. Unlike V_o and δ_o, the fin height $r_a - r_b$, was found to be hardly affected by the variation of the thermal conductivity.

Figure 3.19 shows how the optimum volume and optimum base thickness are affected by variation of the heat transfer coefficient. The lower portion of the figure is for the triangular fin ($\lambda = 0$) and the upper portion is for the rectangular fin ($\lambda = 1$). For each geometry, curves are shown for $m = 0$ (constant heat transfer coefficient), $m = 0.8$, and $m = 2$. Here $V_o(0)$ and $\delta_o(0)$ denote the optimum volume and optimum base thickness for $m = 0$. The quantity $Q = q_b/4\pi h_s r_b^2 \theta_b$ is the dimensionless dissipation rate. The results show that both V_o and δ_o increase significantly. For example, at $Q = 0.10$, as m changes from 0 to 2, the base thickness of a rectangular fin is almost doubled. For the same change in m, the base thickness of a triangular fin is almost tripled. The ratio of optimum volume V_o to the least material fin volume $V_{\min}$, $V_o/V_{\min}$, is also shown in Fig. 3.19. This ratio increases with Q for both rectangular and triangular fins. As m increases, the ratio decreases quite significantly. Razelos and Imre (1980) have concluded that the effect of m on the fin height, $r_a - r_b$, is quite small. The following example

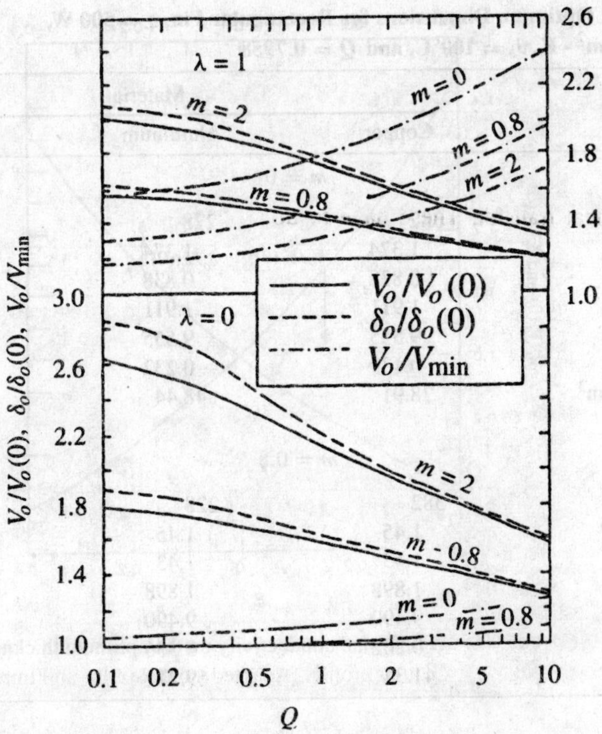

Figure 3.19 Effect of variable heat transfer coefficient on optimum base thickness and optimum volume of radial fins of rectangular ($\lambda = 1$) and triangular ($\lambda = 0$) profiles.

taken from Razelos and Imre (1980) illustrates the effect of variable heat transfer coefficient.

Example 3.9: Optimum Fin Radial Fins with Variable Heat Transfer Coefficient. Determine the optimum dimensions of a radial fin with base radius 5 cm which will dissipate 500 W when the average heat transfer coefficient is 200 W/m^2 · K and the temperature difference between the base and the environment is 100°C. The fin materials to be considered are copper, aluminum, and carbon steel. The profile shape is either rectangular or triangular. Calculations are to be performed for $m = 0$ and $m = 0.8$. Discuss the effect of the variable heat transfer coefficient, $m \neq 0$, on the optimum dimensions. Assume no heat loss from the tip.

SOLUTION. The results obtained using Figs. 3.4, 3.5, and 3.19 are shown in Tables 3.5 and 3.6. Table 3.5 shows that rectangular copper fins with variable heat transfer coefficient ($m = 0.8$) require 43% more volume and are thicker by 45% compared to those with constant heat transfer coefficient ($m = 0$). The corresponsing figures for triangular fins (Table 3.6) are 60 and 65%. However, for both rectangular and triangular fins, the tip radius r_a is affected only slightly by the change in m.

TABLE 3.5 Optimum Dimensions for Rectangular Fin, $q = 500$ W, $h_s = 200$ W/m$^2 \cdot$ K, $\theta_b = 100$°C, and $Q = 0.7958$

Property	Copper	Aluminum	Carbon Steel
	Material		
	$m = 0$		
k	382	228	47
V	1.374	1.374	1.374
$U^{1/2}$	0.838	0.838	0.939
β	1.911	1.911	1.911
r_a cm	9.555	9.555	9.555
δ_R cm	0.139	0.232	1.127
$V_R = V(0)$ cm^3	28.91	48.44	234.79
	$m = 0.8$		
k	382	228	47
$\delta_o/\delta_o(0)$	1.45	1.45	1.45
$V_o/V(0)$	1.43	1.43	1.43
β	1.898	1.898	1.898
r_a cm	9.490	9.490	9.490
δ_o cm	0.202	0.336	1.634
V_o cm^3	41.34	69.27	335.75

3.6.8 Effect of Base Convection and Primary Surface Thickness

Aziz (1985a,b) has considered the optimization of longitudinal fins in which the base is heated by convection though the primary surface wall. The fluid at the inside of the primary surface is at temperature T_f, and the heat transfer coefficient between the interior fluid and the primary surface is h_f. The thickness of the primary surface wall is δ_p and the wall has a thermal conductivity, k_p.

Aziz found that the optimum mb_o is a function of the parameter

$$\mathrm{Bi}^{-1} + \frac{k_f \delta_p}{k_p b_o}$$

where Bi is the Biot modulus, $\mathrm{Bi} = h_f b_o / k_p$, and k_f is the thermal conductivity of the fluid. The functional relationship is shown in Fig. 3.20 for three profile shapes: the longitudinal fins of rectangular, triangular, and concave parabolic profiles. For a desired heat dissipation q_b, the design procedure is to select a trial value of b_o, which together with the other data, fixes the abscissa value in Fig. 3.20 as well as the value of mb_o. The definition of mb_o can then be used to find δ_o, and then q_b can be computed as

$$q_b = \frac{T_f - T_s}{R_T} \tag{3.61}$$

where R_T is the thermal resistance and where

TABLE 3.6 Optimum Dimensions for Triangular Fin, $q = 500$ W, $h_s = 200$ W/m^2 · K, $\theta_b = 100$°C, and $Q = 0.7958$

	Material		
Property	Copper	Aluminum	Carbon Steel
		$m = 0$	
k	382	228	47
V	1.202	1.202	1.202
$U^{1/2}$	0.661	0.661	0.661
β	1.958	1.958	1.958
r_a cm	9.790	9.790	9.790
δ_R cm	0.181	0.304	1.474
$V_R = V(0)$ cm^3	17.98	30.13	146.15
		$m = 0.8$	
k	382	228	47
$\delta_o/\delta_o(0)$	1.65	1.65	1.65
$V_o/V(0)$	1.60	1.60	1.60
β	1.935	1.935	1.935
r_a cm	9.680	9.680	9.680
δ_o cm	0.299	0.502	2.432
V_o cm^3	41.34	69.27	233.84

$$R_T = R_f + \frac{1}{h_f \delta_o} + \frac{\delta_p}{k_p \delta_o} \tag{3.62}$$

and for the rectangular profile fin

$$R_f = \frac{b_o}{kmb\delta_o \tanh mb_o} \tag{3.63}$$

for the triangular fin

$$R_f = \frac{b_o I_0(2mb_o)}{kmb_o \delta_o I_1(2mb_o)} \tag{3.64}$$

and for the concave parabolic fin

$$R_f = \frac{2b_o}{k\delta_o[-1 + [1 + 4(mb_o)^2]^{1/2}} \tag{3.65}$$

3.6.9 Effect of Temperature-Dependent Heat Transfer Coefficient and Thermal Conductivity

When fins are cooled by natural convection, the heat transfer coefficient is a nonlinear function of temperature difference. Work by Hagen (1988) considers the optimization

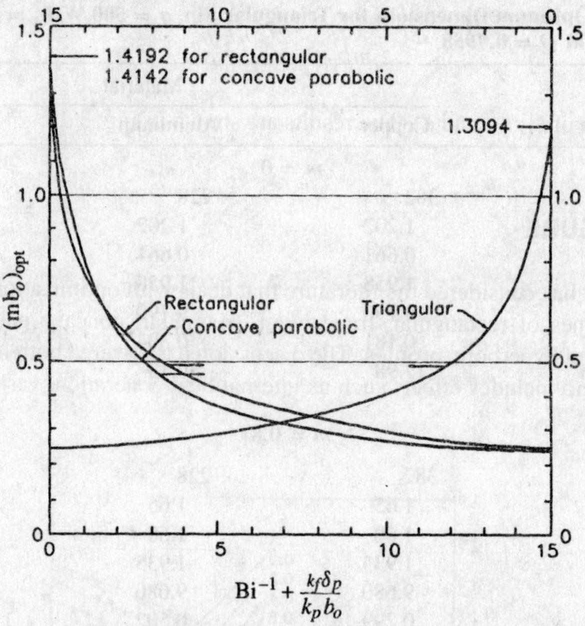

Figure 3.20 Relationship between mb_o and $\mathrm{Bi}^{-1} + k_f \delta_p / k_p b_o$ for the optimum dimensions of longitudinal fins with base convection and finite wall resistance.

of a tapered fin when both the heat transfer coefficient and the thermal conductivity are temperature dependent. The specific forms of the variations considered here are

$$k = k_s[1 + \beta(T - T_s)] \tag{3.66}$$

and

$$h = C(T - T_s)^n \tag{3.67}$$

where C is a constant and $n = 0.25$ for laminar free convection and $n = 0.33$ for turbulent free convection, respectively. For the tapered fin of optimum height b_o, optimum base thickness $\delta_b = \delta_o$, and tip thickness δ_a, the formulation leads to the identification of three parameters,

$$\xi_1 = \beta\theta_b \tag{3.68a}$$

$$\xi_2 = \frac{2\delta_p}{\delta_o} \tag{3.68b}$$

and

$$\xi_3 = n \tag{3.68c}$$

Using a perturbation solution, Hagen (1988) found the optimum mb_o,

$$mb_o = \frac{2C\theta_b \xi_3 b_o^2}{k_s \delta_o} \tag{3.69}$$

as a function of ξ_1, ξ_2, and ξ_3. The results are shown in Fig. 3.21.

3.7 CLOSURE

This chapter has considered the literature that deals with optimization of convecting fins and spines of rectangular, trapezoidal, triangular, concave parabolic, convex parabolic, and hyperbolic profiles. The discussion has covered both longitudinal and radial fins and includes effects such as internal heat generation, variable heat trans-

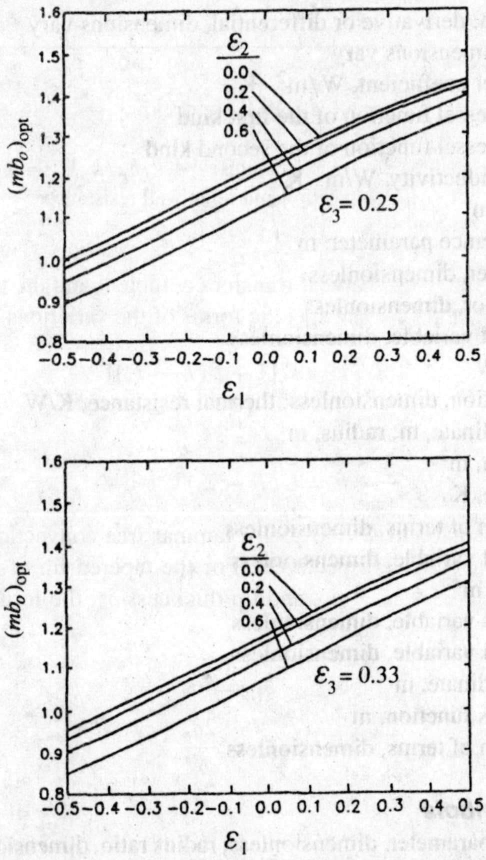

Figure 3.21 Optimum mb_o as a function of ξ_1 and ξ_2 for laminar and turbulent natural convection.

fer coefficient (both coordinate and temperature dependent), temperature-dependent thermal conductivity, base convection, and wall resistance.

A treatment of studies relating to arrays of convecting fins, fins in boiling and condensing applications, and fins that are designed to minimize entropy generation rather than meet the criterion of maximum heat transfer for a given amount of material is left to subsequent chapters. Representative examples of such studies are the works of Bar-Cohen and Jelinek (1985), Unal (1986), and Poulikakos and Bejan (1982).

3.8 NOMENCLATURE

Roman Letter Symbols

- A cross-sectional or profile area, m^2
- Bi Biot number, dimensionless
- b fin height, m
- C arbitrary constant, dimensionless
- d diameter, m; derivative or differential, dimensions vary
- f function, dimensions vary
- h heat transfer coefficient, W/m^2 · K
- I modified Bessel function of the first kind
- K modified Bessel function of the second kind
- k thermal conductivity, W/m · K
- L fin length, m
- m fin performance parameter, m^{-1}
- N fin parameter, dimensionless
- n fin designator, dimensionless
- Q transformed variable, dimensionless
- q heat flow, W
- R radius function, dimensionless; thermal resistance, K/W
- r radial coordinate, m; radius, m
- S surface area, m^2
- T temperature, K
- U combination of terms, dimensionless
- u transformed variable, dimensionless
- V fin volume, m^3
- v transformed variable, dimensionless
- w transformed variable, dimensionless
- x height coordinate, m
- y fin thickness function, m
- Z combination of terms, dimensionless

Greek Letter Symbols

- β optimizing parameter, dimensionless; radius ratio, dimensionless; coefficient of thermal conductivity, K^{-1}
- Δ change in, dimensionless

- δ fin thickness, m
- ϵ thermal conductivity ratio, dimensionless; perturbation parameter, dimensionless; parameter, dimensionless
- ζ combination of terms, dimensionless
- η fin efficiency, dimensionless
- θ temperature excess, K
- κ taper angle, rad
- Λ combination of terms, dimensions vary
- λ profile indicator, dimensionless
- ν combination of terms, dimensionless
- ξ parameter, dimensions vary
- σ combination of terms, dimensionless
- Υ combination of terms, dimensionless
- ϕ radius ratio, dimensionless
- ∂ partial derivative, dimensionless

Roman Letter Subscripts

- a tip of fin
- b base of fin
- f fin; fluid
- G generation number
- i internal
- L longitudinal fin
- m parameter in variable heat transfer coefficient analysis
- o optimum value
- P concave parabolic fin
- p profile area
- R removal number; rectangular profile
- s surroundings
- T triangular fin; thermal

Roman Letter Superscripts

- m parameter in variable heat transfer coefficient analysis
- n natural convection exponent

Greek Letter Superscript

- α exponent in Euler equation solution, dimensionless

Symbolic Superscript

- $*$ dimensionless variable

CHAPTER 4

CONVECTION COEFFICIENTS

4.1 INTRODUCTION

In evaluating the heat transfer performance of surfaces enhanced by the presence of fins or fin arrays, the convective heat transfer coefficient h is clearly an important quantity. It has already been demonstrated that fins are particularly effective under conditions where the controlling thermal resistance is due to convection (where the coefficient h is relatively small).

This chapter is devoted to discussing the phenomenon of convective heat transfer and to cataloging the relationships that are available for evaluating the convective heat transfer coefficient h for use in the analysis of extended surfaces.

There are three principal approaches for determining the convective heat transfer coefficient. These are:

1. Analysis of the thermal and hydrodynamic boundary layers by analytical or numerical means
2. Analogies between fluid flow, heat transfer, and mass transfer
3. Experimentation

With the first of these approaches, the analytical complexity of boundary layer analysis is generally formidable. There are a few classic cases that have been solved, and these are reported in virtually every handbook and textbook on this subject.

The second approach, utilizing analogies among fluid flow (often referred to as *momentum transfer*), heat transfer, and mass transfer, has proven very useful. There are numerous situations where quantitative knowledge of one transfer mode is useful in providing information on another.

In the case of experimental studies that yield empirical relationships for determining the heat transfer coefficient h, the technical literature includes large numbers of publications dealing with convective heat transport. The combination of surface geometry, fluid flow type, fluid composition, and thermal boundary conditions has led to hundreds of publications, each dealing with a specific case.

Convective heat transfer equations, achieved by whatever means, generally relate important variables in terms of dimensionless parameters. These can be generated, formally, using well-accepted dimensional analysis techniques, which are described in all textbooks on this subject [see, e.g., Welty et al. (1986), Incropera and DeWitt (1996), and Holman (1990)].

The dimensionless parameters used for correlating convective heat transfer are:

Nusselt number, Nu	$\frac{hL}{k}, \quad \frac{hd}{k}$
Stanton number, St	$\frac{h}{\rho c_p v}$
Reynolds number, Re	$\frac{Lv\rho}{\mu}, \quad \frac{dv\rho}{\mu}$
Prandtl number, Pr	$\frac{\mu c_p}{k}, \quad \frac{\nu}{\alpha}$
Grashof number, Gr	$\frac{g\beta L^3\theta}{\nu^2}, \quad \frac{g\beta d^3\theta}{\nu^2}$
Rayleigh number, Ra	$\frac{g\beta L^3\theta}{\nu\alpha}, \quad \frac{\rho g d^3\theta}{\nu\alpha}$

The first two parameters, Nu and St, contain the desired quantity, h, and they are always the *dependent* parameter. St is used only in the context of forced convection. The Reynolds number plays an important role in forced convection, while the Rayleigh number[1] is important in natural convection. The Prandtl number is observed to be a ratio of fluid properties and thus may be considered a fluid property in its own right.

Some fundamental concepts are introduced in the following section to assist in an understanding of the physical processes that occur when heat is transferred by convection. Because the convection phenomenon is intimately involved with fluid motion, flow considerations are treated first.

4.2 FLUID FLOW CONSIDERATIONS

The flow of a fluid past a solid surface gives rise to certain phenomena. If the solid body represents an obstruction, as with flow normal to a pin fin, the fluid flow path is altered to flow around the solid. If flow is parallel to a plane solid surface, the presence of the solid–fluid interface will be transmitted some distance into the fluid.

A fundamental distinction to be made regarding surface–fluid interaction is whether the flow is internal or external. For *internal flow* the fluid is confined to

[1]The Rayleigh number is the product of the Grashof and Prandtl numbers.

flow in a passage such as a circular tube or a rectangular channel. The walls of the flow passage comprise the surface of interest in this case. With *external flow*, the fluid is not confined and flows over the exterior surface of the solid body. Flow may be parallel to the walls of a channel, normal to a cylinder, or with any orientation relative to a solid of any geometric configuration.

4.2.1 Boundary Layer Flow Past a Plane Wall

The situation shown in Fig. 4.1 is one of steady (time-independent) uniform flow parallel to a plane wall. Velocity profiles in the fluid are illustrated at various locations along the surface. In Fig. 4.1 the x-direction is measured along the surface measured from the point of initial contact, *the leading edge*, and the y-direction is normal to the surface. The *free-stream velocity*, designated by v_∞, remains unchanged for large values of y. For relatively small values of y, however, the velocity is seen to vary from a value of zero at the surface, $y = 0$, to the free-stream velocity, v_∞, at a value of y that increases with the distance x.

The region close to the surface, where the velocity gradient dv/dy is nonzero, is the *boundary layer*. The interaction between the flowing fluid and the boundary is a function of the velocity gradient at the surface

$$\left.\frac{dv}{dy}\right|_{y=0}$$

The relationship between the force per unit area of contact (the shear stress) and the velocity gradient at the wall is expressed as

$$\tau_o = \mu \left.\frac{dv}{dy}\right|_{y=0} \tag{4.1}$$

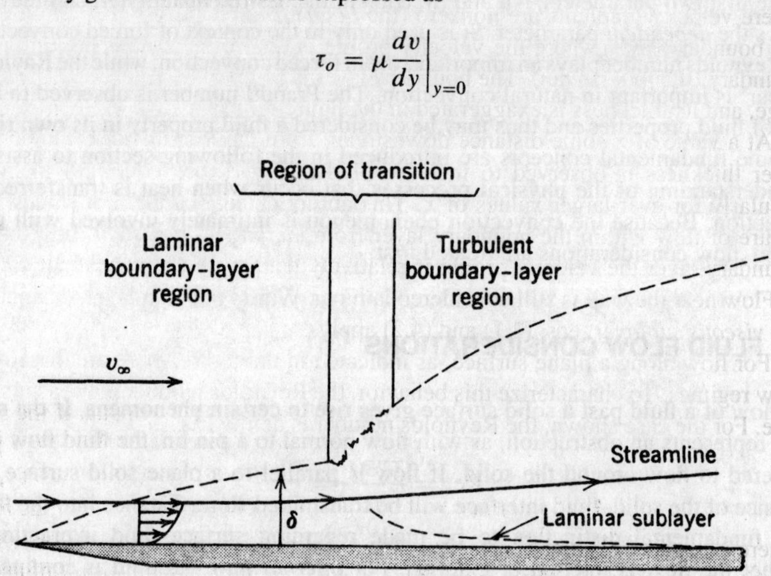

Figure 4.1 Flow parallel to a plane wall.

where the coefficient μ is the viscosity of the fluid. This equation, originally proposed by Sir Isaac Newton, is the defining relationship for the viscosity, which is an important fluid property.

An obvious similarity exists between eq. (4.1) and the Fourier relation for one-dimensional conduction:

$$\frac{q}{A} = -k\frac{dT}{dy} \tag{4.2}$$

Analogous quantities in these two relationships are shear stress and heat flux, velocity gradient and temperature gradient, and the transport properties of viscosity and thermal conductivity.

In relatively low velocity flows there is a distinct layerlike effect where layers (*lamina*) of fluid particles move past one another with exchange between adjacent layers at a molecular level only. This very orderly, predictable flow is designated *laminar* and is described by eq. (4.1). Because there is no bulk motion of particles between fluid layers, the transfer of heat is also by molecular means. Thus, in laminar flow, heat transfer normal to the flow is by conduction and eq. (4.2) applies.

With an increase in flow velocity there is an increasing tendency for bulk transfer of fluid particles normal to the flow direction. Above some threshold velocity this bulk transport becomes pronounced and the laminar flow structure disappears. The resulting flow is characterized by irregular, chaotic behavior which is termed *turbulent*. For turbulent flows the dynamic nature of fluid motion enhances the rate of heat transfer when a temperature difference exists between a surface and adjacent fluid.

Referring again to Fig. 4.1, the dashed line divides the region close to the surface where velocity gradients are nonzero (the *boundary layer*) from the region outside the boundary layer where the velocity profile is unaffected by the presence of the boundary (the *free stream*). The boundary layer is actually very thin near the leading edge, and its thickness is exaggerated in the figure for clarity.

At a value of x some distance downstream from the leading edge, the boundary layer thickness is observed to increase markedly and then continues to increase regularly for ever-larger values of x. This abrupt change is due to a change in the nature of flow within the boundary layer from *laminar* to *turbulent*. In a turbulent boundary layer the velocity profile is relatively flat for all but very small values of y. Flow near the wall is still considered laminar. Within this thin layer, designated as the *viscous sublayer*, eqs. (4.1) and (4.2) apply.

For flow along a plane surface, as indicated in the figure, there are three distinct flow regimes. To characterize this behavior, the Reynolds number plays a significant role. For the case shown, the Reynolds number is defined as

$$\text{Re}_x = \frac{x v_\infty \rho}{\mu} = \frac{x v_\infty}{\nu} \tag{4.3}$$

where v_∞ is the free-stream velocity, x the distance from the leading edge, ρ the fluid density, μ the fluid viscosity, and $\nu = \mu/\rho$ the fluid kinematic viscosity.

The Reynolds number is dimensionless and is generally regarded as the most important parameter in fluid mechanics. For a more complete discussion of the

Reynolds number, the reader may consult any of a large number of fluid mechanics textbooks and treatises [see White (1994)].

For the case of flow parallel to a plane wall, the regimes are characterized as follows:

(a) $0 < \mathrm{Re}_x < 2 \times 10^5$ The boundary layer flow is laminar
(b) $2 \times 10^5 < \mathrm{Re}_x < 3 \times 10^6$ The boundary layer flow may be laminar or turbulent
(c) $3 \times 10^6 < \mathrm{Re}_x$ The boundary layer flow is turbulent

Case (b) is often referred to as the *transition regime.*

Example 4.1. Air at atmospheric pressure and a temperature of 300 K flows with a velocity of 30 m/s along a plane surface. Determine the distance from the leading edge to where (a) the transition begins to occur and (b) the boundary layer becomes fully turbulent.

SOLUTION. The key parameter involved in this solution is Reynolds number defined by eq. (4.3):

$$\mathrm{Re}_x = \frac{x v_\infty}{\nu} \tag{4.3}$$

For air at 300 K and atmospheric pressure, $\nu = 1.570 \times 10^{-5}\ \mathrm{m^2/s}$.

(a) At transition, $\mathrm{Re}_x = 2 \times 10^5$. Thus, the distance from the leading edge is determined as

$$x = \frac{\mathrm{Re}_x \nu}{v_\infty} = \frac{(2 \times 10^5)(1.570 \times 10^{-5})}{30} = 0.105\ \mathrm{m}$$

(b) The calculation is similar to the one in part (a). Here $\mathrm{Re}_x = 3 \times 10^6$. The distance from the leading edge is given by

$$x = \frac{\mathrm{Re}_x \nu}{v_\infty} = \frac{(3 \times 10^6)(1.570 \times 10^{-5})}{30} = 1.57\ \mathrm{m}$$

It is clear that for air at this velocity, the laminar boundary layer occupies a small portion of the flow surface.

Example 4.2. Evaluate the same quantities as in Example 4.1 for the case of water at 300 K flowing at 30 m/s along the plate.

SOLUTION. For water at 300 K, $\nu = 0.879 \times 10^{-6}\ \text{m}^2/\text{s}$. The corresponding lengths at transition become

$$x = \frac{(2 \times 10^5)(0.879 \times 10^{-6})}{30} = 0.0059 \text{ m}$$

and for turbulence

$$x = \frac{(3 \times 10^6)(0.879 \times 10^{-5})}{30} = 0.0879 \text{ m}$$

Observe that the boundary layer enters the transition regime after less than 1 cm of contact with the surface.

4.2.2 External Flow Past Bluff Bodies

In Section 4.2.1 the free-stream velocity was parallel to the plane surface. Flow conditions are much different when the free stream is directed normal to, or at an oblique angle to, the surface. Consideration will now be given to those cases where the body over which flow occurs is not streamlined, designated *bluff*, and the associated flow is irregular.

In Fig. 4.2 the case is examined where external flow is normal to a cylinder, as with the case of a pin fin. Six conditions are illustrated for various values of Reynolds number, defined for this situation as

$$\text{Re}_d = \frac{d v_\infty \rho}{\mu} = \frac{d v_\infty}{\nu} \tag{4.4}$$

This is a similar form to the Reynolds number for flat-late flow except that the length scale in this case is the cylinder diameter.

For streamlines that conform to the shape of the cylinder, the pressure and free-stream velocity are smoothly varying functions of the angle θ. For $0 < \theta < 90$ deg, the pressure decreases and v_∞ increases to minimum and maximum values, respectively, at 90 deg. For 90 deg $< \theta <$ 180 deg, the pressure increases and the velocity decreases until conditions at $\theta = 180$ deg are equal to those at $\theta = 0$ deg. This case, indicated in Fig. 4.2*a*, is associated with fully attached streamlines. This will occur only at very small values of v_∞ and is aptly referred to as *creeping flow*.

The other cases illustrated in Fig. 4.2 are for increasing values of Re_d, and in every instance, there is some degree of boundary layer separation. The occurrence of separation has direct implications on both the force exerted on the cylinder by the fluid and on heat transfer between the solid surface and the fluid.

4.2.3 Internal Flow

In Fig. 4.3, the velocity profiles and boundary layer thickness are shown for an internal flow situation: namely, flow in a circular conduit such as a pipe or tube. In both parts of Fig. 4.3, conditions near the entrance of the pipe are those of laminar boundary layer flow and regular boundary layer growth. As the flow proceeds through the pipe the

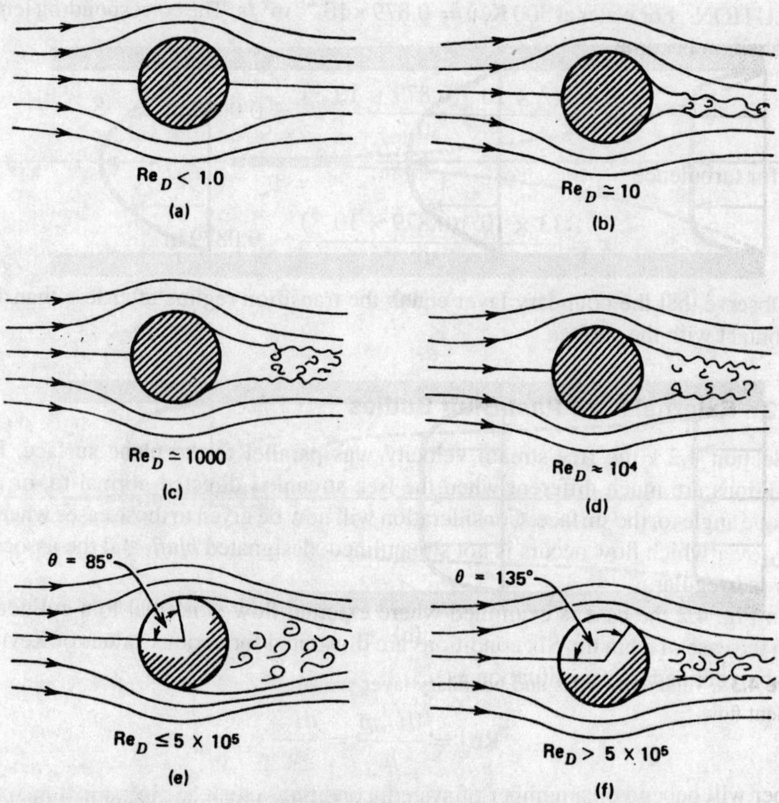

Figure 4.2 Streamline patterns for a cylinder in cross flow.

boundary layer grows to occupy an ever-increasing portion of the flow cross section. Eventually, the condition is reached where the boundary layer fills the total flow area. Physically, this means that the presence of the pipe wall is felt throughout the cross section.

For distances along the flow beyond where the boundary layers meet, the velocity profiles no longer change along the flow path. This is the region of *fully developed* flow. The region nearer the entrance where velocity profiles are changing is designated the entrance region, and the distance x from the entrance to the point when fully developed flow is reached is designated the *entrance length.*

The indicator of flow type for pipe flow is the Reynolds number, defined as

$$\mathrm{Re}_d = \frac{dv\rho}{\mu} = \frac{dv}{\nu} \tag{4.5}$$

where the length scale d is the pipe diameter and where v is the average velocity in the tube or pipe. Flow in a pipe is laminar for $\mathrm{Re}_d \leq 2300$. Transition to turbulent flow will occur at values of Re_d greater than this value and the specific transition Reynolds

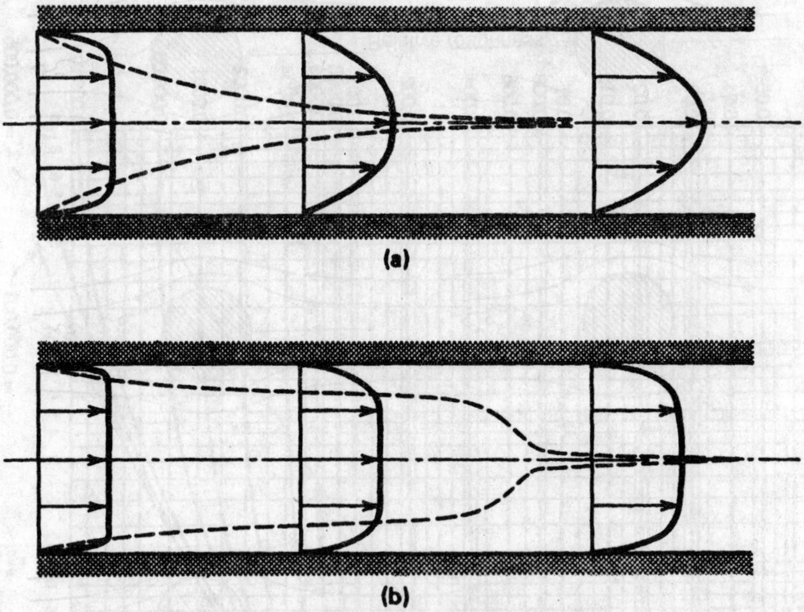

Figure 4.3 Velocity profiles and boundary layer growth in pipe flow: (*a*) laminar flow; (*b*) turbulent flow.

number will depend on a number of system conditions, such as pipe smoothness (or roughness), system vibration, flow fluctuations, and direction changes. Most sources suggest that flow may be safely considered turbulent for values of Re_d greater than 10^4.

The frictional pressure drop for flow in a pipe is expressed as

$$\frac{\Delta p}{\rho} = f \frac{L}{d} \frac{v^2}{2} \tag{4.6}$$

where ρ is the fluid density, L and d the length and diameter of the pipe, v the average flow velocity, and f the friction factor. Equation (4.6) is, in reality, a definition of the friction factor, f.

The friction factor is shown as a function of Reynolds number in Fig. 4.4. There are three distinct regions in this plot. In the laminar region, f varies with Re_d in a regular manner according to

$$f = \frac{64}{\mathrm{Re}_d} \tag{4.7}$$

The derivation of this relationship, from first principles, is presented in most textbooks on fluid mechanics [e.g., Welty et al. (1986) and White (1994)]. Recall that laminar flow will exist for $\mathrm{Re}_d < 2300$.

With large values of Re_d, the flow is termed *fully turbulent* and the values for f in Fig. 4.4 do not vary with Re_d. Thus f is a function only of the relative roughness

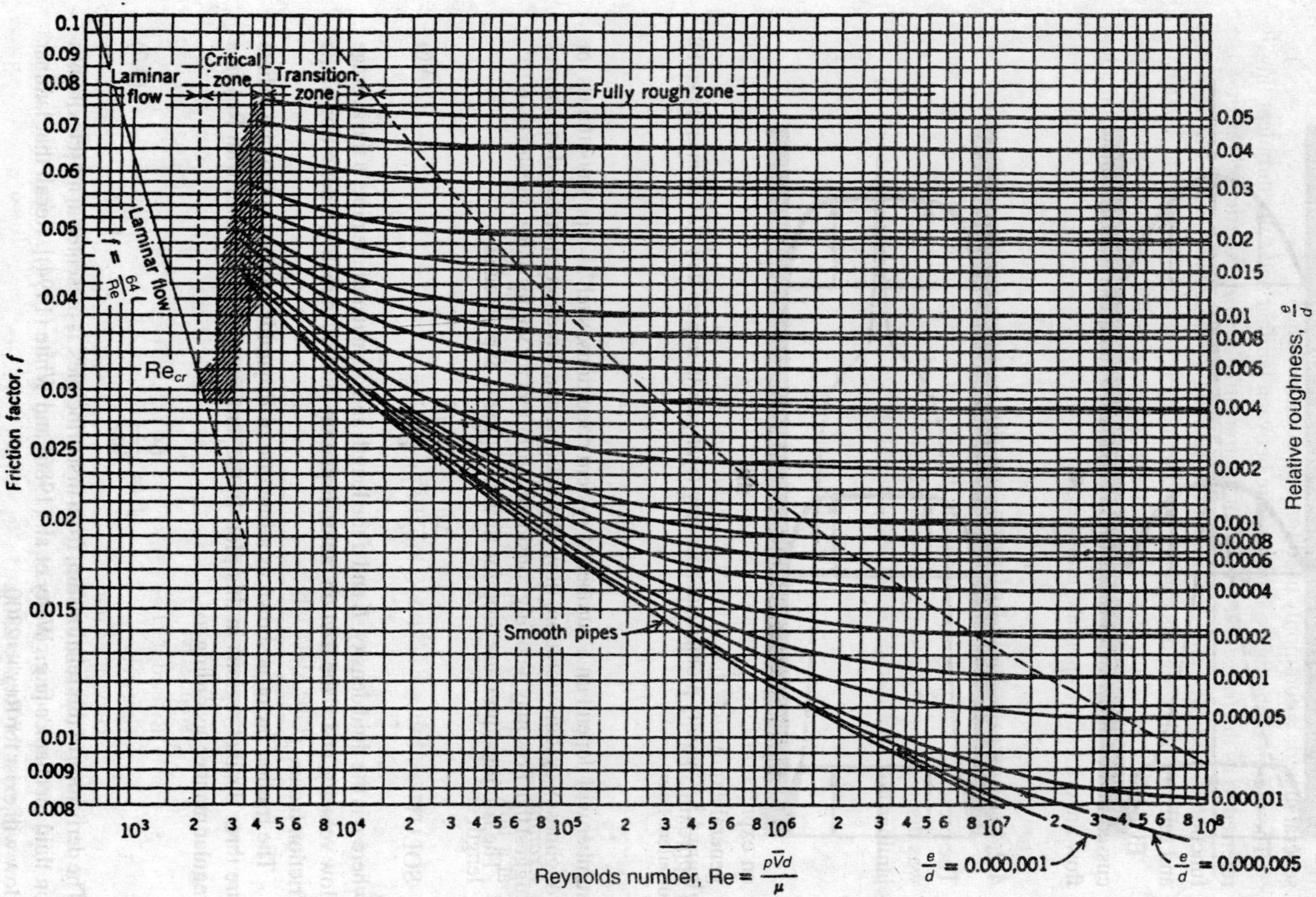

Figure 4.4 Friction factor as a function of Reynolds number for pipe flow.

ϵ/d of the pipe. Average values for the pipe roughness ϵ are listed in Table 4.1 for several materials commonly used for pipe construction.

The region in the figure intermediate between laminar and fully turbulent flow is referred to as the *transition region*. In the transition region, the friction factor f is a function of both Re_d and ϵ/d. Equations relating f, Re_d, and ϵ/d for the transition and fully turbulent regions can be found in many references (Welty et al. 1986).

Evaluation of heat transfer coefficients associated with extended surfaces is discussed in the following portions of this chapter. Fundamental ideas concerning fluid flow are critically important in this regard.

4.2.4 Hydrodynamic Entrance Length

The presence of an entrance length that precedes the region of fully developed flow was discussed in the preceding section. This concept is also shown in Fig. 4.3. For laminar flow, with $\text{Re}_d < 2300$, the entrance length is given by

$$\left.\frac{x_e}{d}\right|_{\text{lam}} = 0.05\text{Re}_d \tag{4.8}$$

an expression first suggested by Langhaar (1942). With turbulent flow, there is no general expression that is totally satisfactory. Kays and Crawford (1993) suggest an approximation that is independent of Reynolds number,

$$10 \le \left.\frac{x_e}{d}\right|_{\text{turb}} \le 60 \tag{4.9}$$

Example 4.3. Water, at 300 K, flows through a 0.15-m-inside diameter cast iron pipe at a flow rate of 3.52 kg/s. For a pipe that is 600 m long, evaluate (**a**) the entrance length and (**b**) the pressure drop.

SOLUTION. Water, at 300 K, has the properties:

TABLE 4.1 Average Roughness Values for Commercial Pipe Materials

Material	ϵ (mm)
Concrete	0.30–3.00
Wood stave	0.18–0.90
Riveted steel	0.90–9.00
Cast iron	0.26
Galvanized iron	0.15
Asphalted cast iron	0.12
Commercial steel	0.046
Wrought iron	0.046
Drawn tubiing	0.0015
Glass	Smooth

$$\rho = 997 \text{ kg/m}^3 \qquad \mu = 8.55 \times 10^{-4} \text{ Pa} \cdot \text{s} \qquad \nu = 8.58 \times 10^{-7} \text{ m}^2/\text{s}$$

The Reynolds number for this flow is determined as

$$\text{Re}_d = \frac{dv}{\nu} = \frac{d\dot{m}}{\mu A} = \frac{4\dot{m}}{(\pi d)\mu}$$

$$= \frac{(4)(3.52)}{\pi(0.15)(8.55 \times 10^{-4})}$$

$$= 34{,}900 \qquad \text{(the flow is turbulent)}$$

(**a**) According to eq. (4.9), the entrance length lies between the limits

$$10 \le \left.\frac{x_e}{d}\right|_{\text{turb}} \le 60$$

or

$$1.5 \text{ m} \le x_e \le 9 \text{ m}$$

(**b**) The pressure drop, as given by eq. (4.6), requires that the friction factor f be determined. The friction factor is plotted in Fig. 4.4 as a function of ϵ/d and Re_d.

For cast iron, from Table 4,

$$\epsilon = 0.26 \text{ mm} \quad \text{or} \quad 0.26 \times 10^{-3} \text{ m}$$

giving

$$\frac{\epsilon}{d} = \frac{0.26 \times 10^{-3}}{0.15} = 0.00173$$

and the friction factor for $\text{Re}_d = 34{,}900$ and $\epsilon/d = 0.00173$ is

$$f = 0.0028$$

The pressure drop is now evaluated as

$$\frac{\Delta p}{\rho} = f\frac{L}{d}\frac{v^2}{2}$$

and in terms of the mass flow

$$\Delta p = f\frac{8fL\dot{m}^2}{\rho\pi^2 d^5}$$

or

$$= \frac{(8)(0.0028)(600)(3.52)^2}{997\pi^2(0.15)^5}$$

$$= 223 \text{ Pa}$$

It is worth noting that even at the largest value of the entrance length, the fully developed region occupies 591 m or 98.5% of the total pipe length.

4.3 HEAT TRANSFER CONSIDERATIONS

There are two main classifications of convective heat transfer. These have to do with the driving force causing the fluid to flow. *Natural* or *free convection* occurs when fluid motion is a result of the heat exchange. When a fluid is heated or cooled, the associated density change and buoyant effect produce a natural circulation in which the affected fluid moves of its own accord past the solid surface. The fluid that replaces it is similarly affected by the energy transfer and the process continues. *Forced convection* is the term used to describe those convective situations in which fluid circulation is produced by external means, such as a fan, pump, or blower.

4.3.1 Forced Convection

The hydrodynamic boundary layer examined earlier in this chapter plays a major role in this process. There is, in addition, a *thermal boundary layer*, which is of primary importance in heat transfer. The thermal boundary layer, for flow along a flat plate, is illustrated in Fig. 4.5. In this case there is flow along the plate, producing boundary layer development as discussed earlier and illustrated in Fig. 4.1. For the situation shown in Fig. 4.5, there is, in addition, a temperature difference between the surface, at T_s, and the free stream, at T_∞.

For other cases of external flow the thermal boundary layer is not as regular or as well defined. Flow around bluff bodies involves boundary layer separation, as shown in Fig. 4.2, which affects heat transfer quite significantly.

4.3.2 Natural Convection

The mechanism of natural convection involves the movement of a fluid past a solid boundary due to the density effects caused by the energy exchange. In this book, those

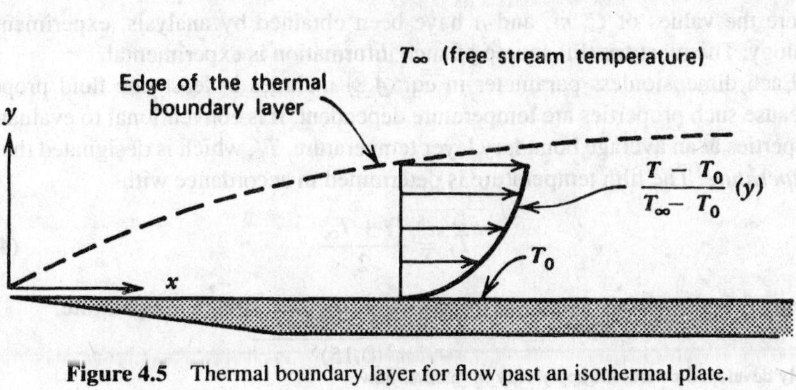

Figure 4.5 Thermal boundary layer for flow past an isothermal plate.

natural-convection flows classified as free boundary flows, which involve plumes or buoyant jets, are not discussed. Heat transfer involving extended surfaces, by definition, involves a bounding surface.

Natural convection is treated extensively by a number of authors [see Gebhart et al. (1988) and Kakac et al. (1985)]. The present discussion is restricted to presenting relationships for the coefficient h likely to be encountered in heat exchange processes involving extended surface.

4.4 CORRELATIONS FOR FORCED CONVECTION HEAT TRANSFER

In this section the relevant correlations for forced convection that apply to extended surface analysis are cataloged. The correlating relationships involve either the Nusselt number or the Stanton number, parameters that include the heat transfer coefficient h as a function of other dimensionless parameters, most often the Reynolds and Prandtl numbers.

The principal distinction to be made is whether flow is internal (confined) or external (unconfined) relative to the heat transfer surface. Other considerations concern whether flow is laminar or turbulent, whether fully developed conditions have been achieved,[2] the magnitude of the Prandtl number, and the temperature used in evaluating fluid properties.

4.4.1 External Flow

Earlier in this chapter some fundamental ideas regarding external flow were presented. An important concern is whether the boundary layer remains in contact with the surface, as with flow along a plane wall, or if the boundary layer separates from the surface, as for flow normal to a cylinder.

In all external forced flows, correlating expressions are presented in the general form

$$\mathrm{Nu} = C\mathrm{Re}^m \mathrm{Pr}^n \tag{4.10}$$

where the values of C, m, and n have been obtained by analysis, experiment, or analogy. The most fruitful source of such information is experimental.

Each dimensionless parameter in eq. (4.8) includes at least one fluid property. Because such properties are temperature dependent, it is conventional to evaluate all properties at an average boundary layer temperature, T_f, which is designated the *film temperature*. The film temperature is determined in accordance with

$$T_f = \frac{T_s + T_\infty}{2} \tag{4.11}$$

where T_s is the surface temperature and T_∞ is the free-stream temperature.

[2]Fully developed conditions apply only to internal flow.

An alternative to the use of T_f is to evaluate all properties at the free-stream temperature and to multiply the right-hand side of eq. (4.8) by a factor that accounts for property variation. These factors are commonly of the form

$$\left(\frac{\mu_\infty}{\mu_s}\right)^r \quad \text{and} \quad \left(\frac{\text{Pr}_\infty}{\text{Pr}_s}\right)^r$$

where the subscripts ∞ and s refer to free-stream and surface conditions, respectively.

Flow Along a Plane Wall. This is the case depicted in Fig. 4.5. The region close to the solid surface where the temperature varies between T_s and T is the *thermal boundary layer*. This region is directly analogous to the hydrodynamic boundary layer, discussed in Section 4.2.1.

The thickness of the hydrodynamic and thermal boundary layers for laminar flow is

$$\frac{\delta}{\delta_t} \approx \text{Pr}^{1/3} \tag{4.12}$$

It is evident that for cases with a Prandtl number close to unity, the two boundary layers are nearly equal in thickness. This condition is approached for gases that have $\text{Pr} \sim 0.7$ and $\text{Pr}^{1/3} \sim 0.89$.

For *laminar flow parallel to a flat plate*, the mean Nusselt number is given by[3]

$$\text{Nu}_L = 0.664\text{Re}_L\text{Pr}^{1/3} \qquad (\text{Pr} > 0.6) \tag{4.13}$$

In the case of fluids with $\text{Pr} < 0.6$, which is characteristic of liquid metals, the Nusselt number expression is

$$\text{Nu}_L = 1.13\text{Re}_L^{1/2}\ \text{Pr}^{1/2} \qquad (\text{Pr} < 0.05) \tag{4.14}$$

Churchill and Ozoe (1972) have presented a single correlation, applicable for all Prandtl numbers for laminar flat-plate flow. Their expression is

$$\text{Nu}_L = \frac{0.6774\,\text{Re}_L^{1/2}\ \text{Pr}^{1/3}}{\left[1 + (0.0468/\text{Pr})^{2/3}\right]^{1/4}} \tag{4.15}$$

Experiments with turbulent boundary layer flow parallel to a flat plate have yielded the relationship

$$\text{Nu}_L = \frac{hL}{k} = 0.037\text{Re}_L^{4/5}\text{Pr}^{1/3} \tag{4.16}$$

where

$$0.60 < \text{Pr} < 60 \quad \text{and} \quad 10^6 < \text{Re}_L < 10^8$$

[3]Unless otherwise indicated, all Nusselt numbers are to be considered as mean Nusselt numbers yielding an average heat transfer coefficient.

It is clear that for some region near the leading edge, laminary boundary layer flow will always exist. If the Reynolds number, based on total plate length, is very large, the contribution of the laminar region to the total heat transfer may be negligible. Conversely, if the laminar flow region occupies a substantial portion of the surface, the contributions of both the laminar and turbulent flow regions must be taken into account. A discussion of these mixed boundary layer conditions is given by Incropera and DeWitt (1996).

A useful expression that accounts for these mixed conditions is given, assuming that transition occurs abruptly at $\text{Re}_x = 5 \times 10^5$, by

$$\text{Nu}_L = (0.037\text{Re}_L^{4/5} - 871)\text{Pr}^{1/3} \tag{4.17}$$

where

$$0.60 < \text{Pr} < 60 \quad \text{and} \quad 5 \times 10^5 < \text{Re}_L < 10^8$$

Observe that with $\text{Re}_L >> \text{Re}_c$, a reasonable approximation for plane wall heat transfer is

$$\text{Nu}_L = 0.037\text{Re}_L^{4/5}\,\text{Pr}^{1/3} \tag{4.18}$$

Cylinders in Cross Flow. Flow normal to circular cylinder exhibits a number of complex features that were discussed at some length in Section 4.2.2. Of interest in this section is the case where, for such a situation, a temperature difference exists between the surface and the flowing fluid.

Figures 4.6 and 4.7 show how the local Nusselt number varies from point to point around the surface of a cylinder with fluid flowing normal to its axis. In Fig. 4.6 the variation of Nu_d is observed to be relatively smooth for Reynolds numbers up to 600. Figure 4.7, with Re_θ varying from 70,000 up to 220,000, shows Nu_d to vary in a much more complex fashion. The effects of boundary layer separation are evident. These representations of Nu_θ show clearly that local heat transfer rates vary quite irregularly. From the standpoint of engineering calculations the overall average value of is of greater interest.

McAdams (1954) has plotted the experimental results of 13 separate investigators for the flow of air normal to single cylinders and found good agreement when plotted as Nu_d versus Re_d. His plot is reproduced as Fig. 4.8. A correlation of the form

$$\text{Nu}_d = B\text{Re}^n\text{Pr}^{1/3} \tag{4.19}$$

has been used to fit McAdams' plot piecewise over the entire Reynolds number range. The constants B and n in eq. (4.17) vary with Reynolds number. Values of these constants are given in Table 4.2. Fluid properties should be evaluated at the film temperature.

Churchill and Bernstein (1977) recommended a single correlating equation covering the entire range of Re. This correlation is expressed as

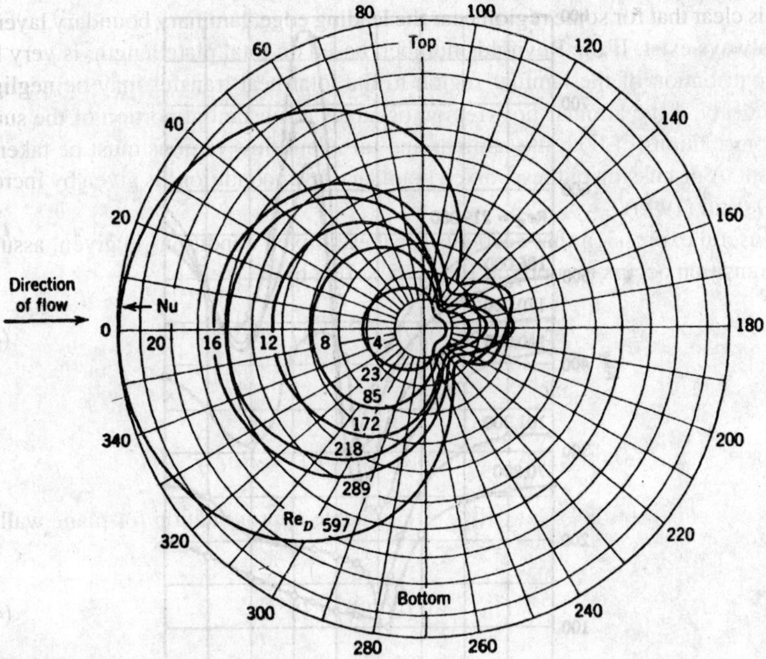

Figure 4.6 Local Nusselt numbers for cross flow about a circular cylinder at low Reynolds numbers. [Adapted from Eckert and Soehngen (1952).]

$$Nu_d = 0.30 + \frac{0.62Re_d^{1/2}\,Pr^{1/3}}{[1 + (0.40/Pr)^{2/3}]^{1/4}}\left[1 + \left(\frac{Re_d}{282,000}\right)^{5/8}\right]^{4/5} \qquad (4.20)$$

and is recommended as long as $Re_d Pr > 0.20$.

Example 4.4. A stainless steel pin fin, cylindrical in shape, is attached to a plane wall to increase the heat transfer to air at 300 K. The plane wall (which is the fin base) is maintained at 400 K. Air flows normal to the 5-mm-diameter pin with a velocity of 5 m/s. For this case the maximum possible heat removal from the fin is to be determined.

SOLUTION. It will be assumed that the convective coefficient h is uniform over the length of the fin. The maximum heat transfer rate will occur for a pin of infinite length, the value being

$$q = (hPkA)^{1/2}(T_s - T_\infty)$$

The value of h can be determined using either eqs. (4.19) or (4.20). For air at

$$T_f = \frac{T_s + T_\infty}{2} = \frac{400 + 300}{2} = 350\text{ K}$$

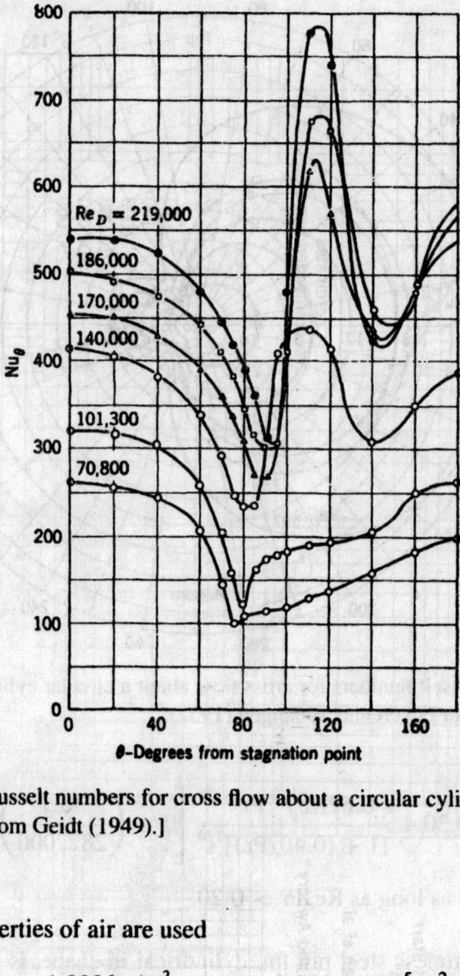

Figure 4.7 Local Nusselt numbers for cross flow about a circular cylinder at high Reynolds numbers. [Adapted from Geidt (1949).]

The following properties of air are used

$$\rho = 1.000\,\text{kg/m}^3 \qquad \nu = 2.057 \times 10^{-5}\,\text{m}^2/\text{s}$$

$$k = 3.003 \times 10^{-2}\,\text{W/m}\cdot\text{K} \qquad \text{Pr} = 0.697$$

and for stainless steel

$$k_{\text{st}} = 16\,\text{W/m}\cdot\text{K}$$

The value of Re_d is calculated as

$$\begin{aligned}\text{Re}_d &= \frac{dv}{\nu}\\ &= \frac{(0.005)(5)}{2.057 \times 10^{-5}}\\ &= 1215\end{aligned}$$

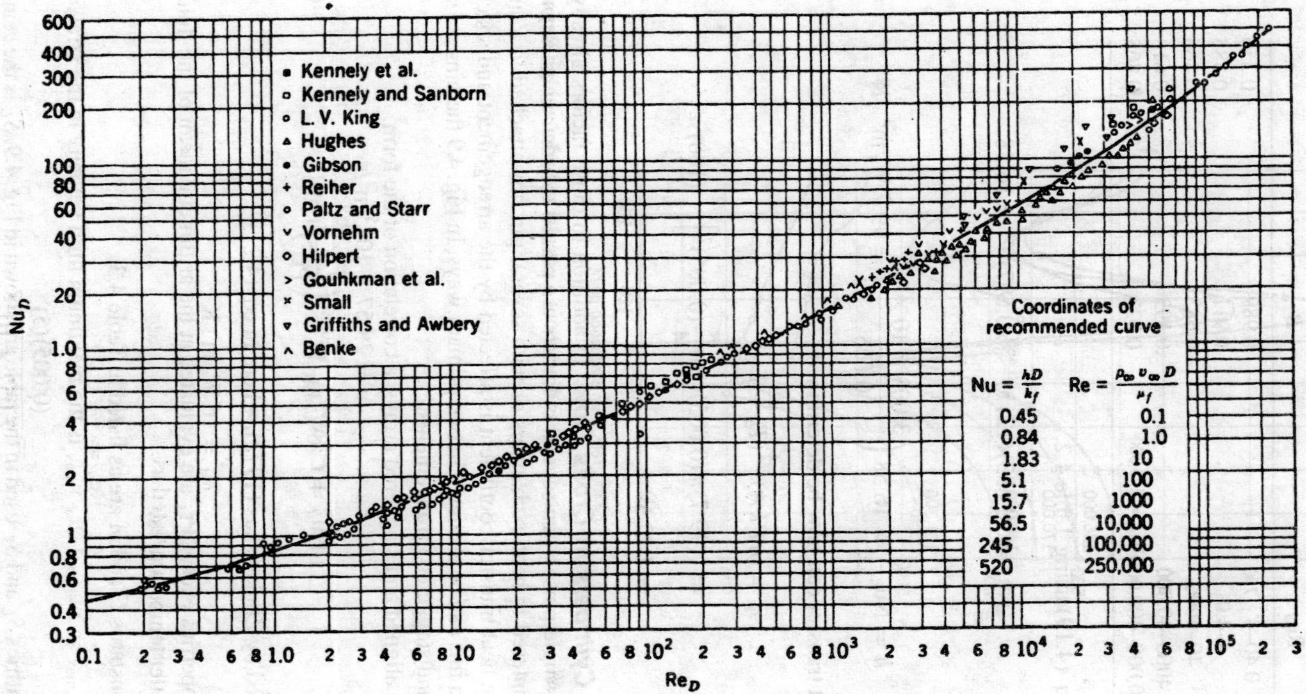

Figure 4.8 Nu_d as a function of Re_d for the flow of air normal to single cylinders. [Adapted from McAdams (1954).]

TABLE 4.2 Values of *B* and *n* for Use in Eq. (4.17)

Re_d	B	n
0.40–4	0.989	0.330
4–4.0	0.911	0.385
40–4000	0.683	0.466
4000–40,000	0.193	0.618
40,000–400,000	0.027	0.805

and from eq. (4.19) using Table 4.2,

$$\text{Nu}_d = (0.683)(1215)^{0.466}(0.697)^{1/3} = 16.58$$

giving

$$h = \text{Nu}_d \frac{k}{d} = 16.58 \left(\frac{3.003 \times 10^{-2}}{0.005} \right) = 99.7\ \text{W/m}^2 \cdot \text{K}$$

The heat transfer can now be determined

$$\begin{aligned} q &= (hPkA)^{1/2}(T_b - T_\infty) \\ &= \left[(99.7)\pi(0.005)(16)\frac{\pi}{4}(0.005)^2 \right]^{1/2} (100) \\ &= 2.22\ \text{W} \end{aligned}$$

Banks of Cylinders in Cross Flow. In addition to those factors already considered for single pins, when several cylinders are placed together in an array, as is often encountered in heat exchangers or with several pin fins attached to a surface, the effective heat transfer coefficient is affected by the arrangement and spacing of the pins. Pin fins can be arranged in two primary ways. In Fig. 4.9 the pins are shown in aligned and staggered arrangements.

For both aligned and staggered arrays, a correlation of the form

$$\text{Nu}_d = C\,\text{Re}_{d,\max}^m\ \text{Pr}^{0.36} \left(\frac{\text{Pr}}{\text{Pr}_s} \right)^{1/4} \tag{4.21}$$

was proposed by Zhukauskas (1972), where in eq. (4.21):

- All properties except Pr_s are evaluated at the arithmetic mean of the fluid inlet and outlet temperatures.
- The constants C and m are as listed in Table 4.3.
- In $\text{Re}_{d,\max} = \rho v_{\max} d/\mu$, $v_{\max}$ is the maximum fluid velocity in the array.

The quantities S_L and S_T used in the table are shown in Fig. 4.9. S_L is the center-to-center distance between the fins *along* the flow direction. S_T is this distance measured normal to the flow direction.

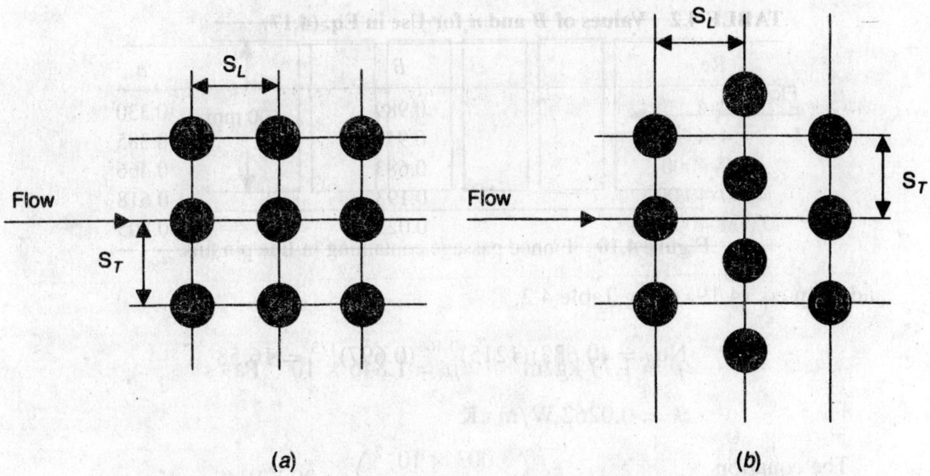

Figure 4.9 Pin arrangements in (*a*) aligned and (*b*) staggered arrangements.

TABLE 4.3 Values of Constants *C* and *m* for Eq. (4.21)

Configuration	$Re_{d,max}$	C	m
Aligned	$10–10^2$	0.80	0.40
Staggered	$10–10^2$	0.90	0.40
Aligned	$10^2–10^3$	Use single	Use single
Staggered	$10^2–10^3$	cylinder	cylinder
Aligned $\left(\frac{S_T}{S_L} > 0.7\right)$	$10^3–2 \times 10^5$	0.27	0.63
Staggered $\left(\frac{S_T}{S_L} < 2\right)$	$10^3–2 \times 10^5$	$0.35\left(\frac{S_T}{S_L}\right)^{1/5}$	0.60
Staggered $\left(\frac{S_T}{S_L} > 2\right)$	$10^3–2 \times 10^5$	0.40	0.60
Aligned	$2 \times 10^5–2 \times 10^6$	0.021	0.84
Staggered	$2 \times 10^5–2 \times 10^6$	0.022	0.84

Example 4.5. Atmospheric air is used to cool electronic components that are mounted on two plates (considered isothermal) which are attached to an in-line array of aluminum pin fins arranged as shown in Fig. 4.10. The pins are arranged in a square pattern such that $S_L = S_T = 4$ mm. Each pin is 2 mm in diameter and 100 mm high. There are a total of 625 pins. Air enters the array at 300 K with a velocity of 10 m/s. The thermal conductivity for aluminum is 240 W/m · K. Using air properties at 300 K, estimate the value of h for this array.

SOLUTION. For air at 300 K and 1 atmosphere,

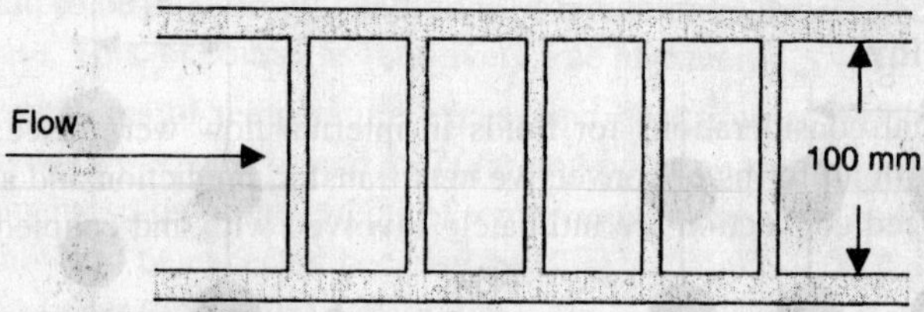

Figure 4.10 Finned passage containing in-line pin fins.

$$\rho = 1.77 \text{ kg/m}^3 \qquad \mu = 1.846 \times 10^{-5} \text{ Pa} \cdot \text{s}$$

$$k = 0.0262 \text{ W/m} \cdot \text{K}$$

The equation

$$\text{Nu}_d = C\,\text{Re}_{d,\max}^m\,\text{Pr}^{0.36}\left(\frac{\text{Pr}}{\text{Pr}_s}\right)^{1/4} \tag{4.21}$$

requires the use of the maximum velocity, which has a value

$$v_{\max} = \frac{S_T}{S_T - d}v = \frac{4}{4-2}(10) = 20 \text{ m/s}$$

giving

$$\begin{aligned}\text{Re}_{d,\max} &= \frac{\rho v_{\max} d}{\mu} \\ &= \frac{(1.77)(20)(0.002)}{1.846 \times 10^{-5}} \\ &= 3835\end{aligned}$$

The values of C and m can be read from Table 4.3 as

$$C = 0.27 \quad \text{and} \quad m = 0.63$$

The solution can now be completed. Nu_d is evaluated as

$$\begin{aligned}\text{Nu}_d &= C\,\text{Re}_{d,\max}^m\,\text{Pr}^{0.36}\left(\frac{\text{Pr}}{\text{Pr}_s}\right)^{1/4} \\ &= (0.27)(3835)^{0.63}(0.707)^{0.36}(1)^{1/4} \\ &= 43.14\end{aligned}$$

and h has the value

$$h = \text{Nu}_d\frac{k}{d} = 43.14\left(\frac{0.0262}{0.002}\right) = 565 \text{ W/m}^2 \cdot \text{K}$$

4.4.2 Internal Flow

The fluid mechanical considerations for fluids in internal flow were discussed in Section 4.2.2. As with all forms of convective heat transfer, prediction and analysis of internal flow forced convection are intimately involved with and coupled to the fluid flow.

A first consideration to be made in examining forced convection inside passages is the type of flow, that is, whether it is laminar or turbulent. Recall that the criterion is the magnitude of Re_d based on a form of eq. (4.5):

$$\mathrm{Re}_d = \frac{\rho v d}{\mu} = \frac{v d}{\nu}$$

where v is the mean velocity of flow and d is the diameter of a circular flow passage or the equivalent diameter of a noncircular passage. As discussed earlier, for $\mathrm{Re}_d < 2300$ the flow is laminar, for $\mathrm{Re}_d > 10^4$ the flow is turbulent, and between these values there is a transition regime where the flow tends to fluctuate between the two.

Entrance effects are important considerations for heat transfer, just as was the case for fluid flow. Criteria for the extent of the entrance region are different for laminar and turbulent flows. These criteria are discussed next in conjunction with each flow type.

Laminar Flow. A classic analytical solution exists for the case of steady laminar incompressible flow in a circular tube. The Graetz solution (Graetz, 1885) established development of the velocity field for this problem as well as the criteria for fully developed flow. This work presumed that the fluid and pipe walls were at the same temperature.

The analysis of heat transfer in laminar internal flow involves an extension to the Graetz solution to include boundaries that are at a different temperature than the fluid. Clearly, there are an infinite number of ways in which wall heating or cooling can occur. The two primary boundary conditions are those of *constant wall temperature* and *constant wall heat flux*. Any more complicated boundary conditions may be treated mathematically as combinations and variations of these basic forms. Methods for doing this are discussed by Kays and Crawford (1993).

Figure 4.11 shows the variation in Re_d with a nondimensional axial location for these two types of wall conditions. It is interesting to note that asymptotic solutions exist when both the hydrodynamic and thermal entrance lengths have been exceeded. These fully developed results are

$$\mathrm{Re}_d = 3.66 \qquad (T_s = \text{constant}) \tag{4.22}$$

and

$$\mathrm{Re}_d = 4.36 \qquad (q'' = \text{constant}) \tag{4.23}$$

In Fig. 4.11 it is also evident that there is a criterion for fully developed conditions to be reached (or the extent of the thermal entry length). This criterion is for a fully developed temperature profile,

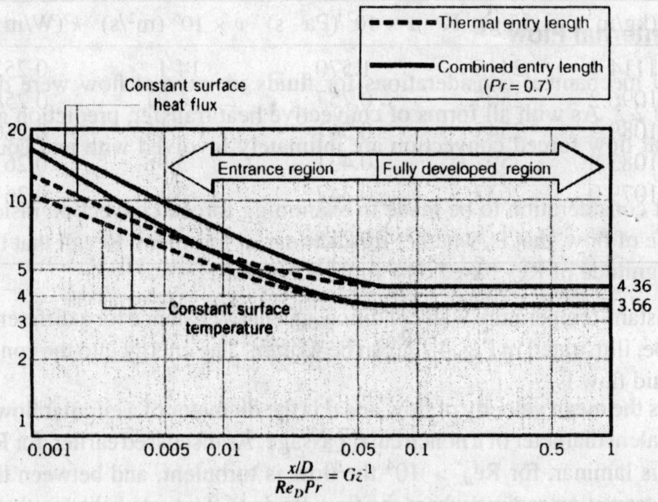

Figure 4.11 Local Nusselt number variation for laminar flow in a tube.

$$\frac{x}{d} = 0.05\,\text{Re}_d\,\text{Pr} \tag{4.24}$$

For values of x/d less than the fully developed condition, Nu_d will be greater than the values shown in eqs. (4.22) and (4.23).

A popular empirical expression for laminar flow in a circular tube has been developed by Sieder and Tate (1936),

$$\text{Nu}_d = 1.86\left(\text{Re}_d\text{Pr}\frac{d}{L}\right)^{1/3}\left(\frac{\mu}{\mu_s}\right)^{0.14} \tag{4.25}$$

where T_s is constant. All fluid properties, except μ_s, are evaluated at the bulk mean temperature.

The qualifications for using eq. (4.25) indicate that fluid properties, with the exception of μ_s, are to be evaluated at the mean bulk temperature of the fluid, which is the average between entering and exiting fluid temperatures. Normally, the entering temperature is known but the temperature of the fluid leaving the tube is not; thus the bulk temperature T_b is unknown. This leads to a trial-and-error problem, illustrated in the following example.

Example 4.6. Ethylene glycol, with a fully developed velocity profile, flows through a 0.75-m-long tube with an inside diameter of 2.5 cm, and an average velocity of 0.05 m/s. The liquid enters at 21°C. The tube surface is maintained constant at 99°C by condensing steam on its outside surface. Find the rate of heat transfer to the glycol.

SOLUTION. Properties of ethylene glycol are temperature dependent:

T (K)	ρ(kg/m^3)	c (J/kg) · K	$\mu \times 10^2$ (Pa · s)	$\nu \times 10^6$ (m^2/s)	k (W/m · K)	Pr
300	1114.4	2415	1.570	14.1	0.252	151
320	1096.2	2505	0.757	6.91	0.258	73.5
330	1089.5	2549	0.561	5.15	0.260	55.0
340	1083.8	2592	0.431	3.98	0.261	42.8
350	1079.0	2637	0.342	3.17	0.261	34.6
373			0.215			

For a constant-temperature wall condition, an energy balance for a differential portion of the tube, illustrated in Fig. 4.12, can be written. The energy into the control volume due to fluid flow is

$$\frac{\rho\pi d^2}{4} cvT$$

where c is the specific heat of the ethylene glycol at the bulk temperature. The energy into the control volume due to convection from the wall is

$$\pi dh\,dx(T_s - T)$$

These are balanced by the energy leaving the control volume due to flow:

$$\frac{\rho\pi d^2}{4} cv(T + dT)$$

and the overall energy balance may be written as

$$\frac{\rho\pi d^2}{4} cvT + hd\pi\,dx(T_s - T) - \frac{\rho\pi d^2}{4} pv(T + dT) = 0$$

This expression can be simplified to yield

$$\frac{dT}{dx} + \frac{4}{d}\frac{h}{\rho cv}(T_s - T) = 0$$

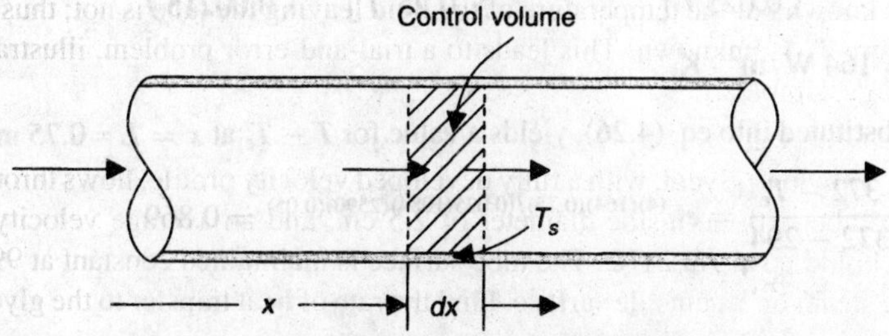

Figure 4.12 Control volume for analyis of flow for Example 4.6.

or, letting $\theta = T_s - T$,

$$\frac{d\theta}{dx} = -\frac{4}{d}\frac{h}{\rho c v}\theta$$

The foregoing equation can be solved to yield the fluid temperature as a function of x by separation of the variables and integration:

$$\int_{\theta_i}^{\theta} \frac{d\theta}{\theta} = -\frac{4}{d}\frac{h}{\rho c v}\int_0^x dx$$

and

$$\ln\frac{\theta}{\theta_i} = -\frac{4x}{d}\frac{h}{\rho c v}$$

and with T_i taken as the fluid temperature at the entrance or inlet of the tube,

$$\frac{\theta}{\theta_i} = \frac{T_s - T}{T_s - T_i} = e^{-(4hx/d\rho c v)} \tag{4.26}$$

The total heat transfer to the liquid is given by an energy balance

$$q = \frac{\rho \pi d^2}{4} c v (T_e - T_i) \tag{4.27}$$

where T_e is the fluid temperature at the exit or outlet of the tube. Equations (4.26) and (4.27) include three unknown quantities: q, T_e, and h.

The initial step is to determine the type of flow. Assuming the bulk temperature $T_b = 340$ K, and obtaining the value of ν from the table of properties,

$$\text{Re}_d = \frac{dv}{\nu} = \frac{(0.025)(0.05)}{3.98 \times 10^{-6}} = 314 << 2300$$

The flow is clearly laminar.

The Sieder–Tate equation is used to evaluate h:

$$h = \text{Nu}_d \frac{k}{d} = 1.86\frac{k}{d}\left(\text{Re}_d \, \text{Pr}\, \frac{d}{L}\right)^{1/3}\left(\frac{\mu}{\mu_s}\right)^{0.14}$$

$$= 1.86\left(\frac{0.261}{0.025}\right)\left[(314)(42.8)\left(\frac{0.025}{0.75}\right)\right]^{1/3}\left(\frac{0.00431}{0.00215}\right)^{0.14}$$

$$= 164 \text{ W/m}^2 \cdot \text{K}$$

This result, substituted into eq. (4.26), yields a value for $T - T_e$ at $x = L = 0.75$ m of

$$\frac{372 - T_e}{372 - 294} = e^{-(4)(164)(0.75)/(0.025(1084)(2590)(0.05))} = 0.869$$

This gives

$$T_e = T_s - 0.869(T_s - T_i)$$

$$= 372 - (0.869)(78)$$
$$= 304.2 \text{ K}$$

The mean bulk temperature with this value of T_e is

$$T_b = \frac{294 + 304.2}{2} = 299 \text{ K}$$

This result differs by 41° from the original assumption. The new T_b value must now be used to evaluate fluid properties, and the solution must be completed at least one more time. With the revised fluid properties used, the resulting values are

$$\text{Re}_d = 88.6$$
$$h = 189.2 \text{ W/m}^2 \cdot \text{K}$$
$$T_e = 306.1 \text{ K}$$

This agreement is very good and the value for heat flow rate can now be determined using eq. (4.27):

$$q = \frac{\rho \pi d^2}{4} c v (T_e - T_i)$$
$$= \frac{(1114.4)\pi(0.025)^2}{4}(2.415)(0.05)(306.1 - 294)$$
$$= 0.799 \text{ kW}$$

For those cases involving laminar flow in noncircular passages such as ducts and annuli, heat transfer calculations can be made with reasonable accuracy by using an equivalent diameter defined as

$$d_e = 4\frac{A_F}{P_W} \tag{4.28}$$

where A_F is the flow area (or cross-sectional area) and P_W is the wetted perimeter. The replacement of d by d_e in correlations for both flow and heat transfer is common practice and generally yields acceptable results, particularly in turbulent flow. Better accuracy can be achieved in laminar flow by using the analytical results as reported by Kays and Crawford (1993) for noncircular flow passages. Table 4.4 gives the asymptotic values for Nu_d for several configurations.

Transition Flow. For the transition region ($2300 \le \text{Re}_d \le 10,000$), the Nusselt number, based on the work of Hausen (1943), is given by

$$\text{Nu}_d = \frac{hd}{k} = 0.116\left[\left(\frac{\rho v d}{\mu}\right)^{2/3} - 125\right]\text{Pr}^{1/3}\left(\frac{\mu}{\mu_s}\right)^{0.14}\left[1 + \left(\frac{d}{L}\right)^{2/3}\right] \tag{4.29}$$

TABLE 4.4 Nusselt Numbers for Fully Developed Conditions with Laminar Flow in Noncircular Passages ($Nu_d = hd_e/k$)

Cross Section	$\frac{b}{a}$	Uniform q_s''	Uniform T_s
	—	4.36	3.66
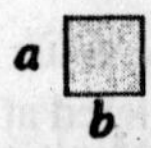	1.00	3.61	2.98
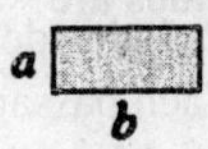	1.43	3.73	3.08
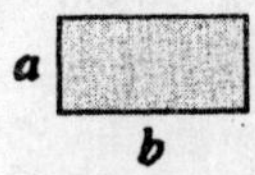	2.00	4.12	3.39
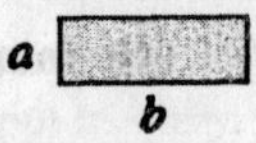	3.00	4.79	3.96
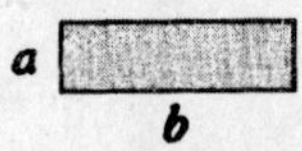	4.00	5.33	4.44
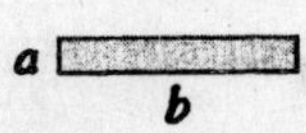	8.00	6.49	5.60
	∞	8.23	7.54
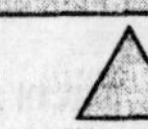	–	3.11	2.47

Turbulent Flows. Forced convection heat transfer with turbulent flows is universally evaluated using experimentally determined correlations. Probably the most used expression for internal, turbulent flow heat transfer is the Dittus–Boelter equation (Dittus and Boelter, 1930):

$$Nu_d = 0.023\, Re_d^{0.80}\, Pr^n \tag{4.30}$$

where $n = 0.3$ for $T_s < T_m$ and 0.4 for $T_s > T_m$ and is valid when $0.7 < Pr < 160$, $Re_d > 10,000$, and $L/d > 10$. Fluid properties are evaluated at the average bulk fluid temperature.

A modification of eq. (4.30) is recommended by Sieder and Tate (1936):

$$Nu_d = 0.027\, Re_d^{0.80}\, Pr^{1/3} \left(\frac{\mu}{\mu_s}\right)^{0.14} \tag{4.31}$$

where all fluid properties except μ_s are evaluated at T_m. This expression is valid when $0.7 < Pr < 16,700$, $Re_d > 10,000$, and $L/d > 10$.

A reportedly more accurate and clearly, more complex relationship is attributed to Petukhov (1970):

$$\mathrm{Nu}_d = \frac{(f/8)\mathrm{Re}_d \mathrm{Pr}}{1.07 + 12.7(f/8)^{1/2}(\mathrm{Pr}^{2/3} - 1)} \tag{4.32}$$

which is valid for $0.5 < \mathrm{Pr}_d < 2000$ and $10^4 < \mathrm{Re}_d < 5 \times 10^6$. In this expression the quantity f is the friction factor, evaluated using Fig. 4.4. The use of a turbulent flow correlation is illustrated in the following example.

Example 4.7. An electronic circuit board is cooled by contact with a heat sink, which consists of a large number of flow passages. Air at 300 K is drawn through this device at a volumetric flow rate of 0.003 m^3/s per each rectangular passage. Each passage is 5 mm wide with a height of 25 mm and dissipates 2 W. The heat sink is 15 cm long. Estimate (**a**) the operating temperature of this unit and (**b**) the pressure drop.

SOLUTION. (**a**) An energy balance for a portion of the rectangular passage will follow a procedure similar to that of Example 4.6. For the differential control volume shown in Fig. 4.13, the energy balance consists of three terms: the energy in to the control volume by fluid flow is

$$\rho A v c_p T$$

where c_p is the specific heat of the air at constant pressure, the energy into the control volume via convection from the wall is

$$hP(T_s - T)\,dx$$

and the energy leaving the control volume due to fluid flow is

$$\rho A v c_p (T + dT)$$

Thus the energy balance is

$$\rho A v c_p T + hP(T_s - T)\,dx - \rho A v c_p (T + dT) = 0$$

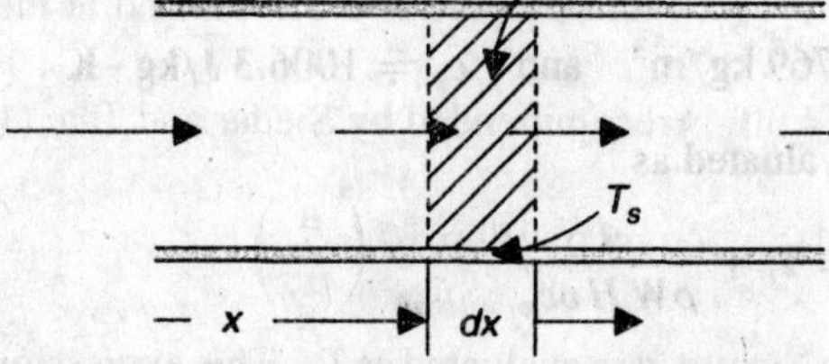

Figure 4.13 Control volume for analyis of flow for Example 4.7.

which can be simplified to yield

$$\frac{dT}{dx} = \frac{hP}{\rho A v c_p}(T_s - T)$$

Letting $\theta = T_s - T$, this expression becomes

$$\frac{d\theta}{dx} = -\frac{hP}{\rho A v c_p}\theta \qquad (4.33)$$

In this example the flow area A and perimeter P can be expressed in terms of the channel dimensions W and H as

$$A = WH \quad \text{and} \quad P = 2(W + H)$$

and eq. (4.33) takes the form

$$\frac{d\theta}{dx} = -\frac{2(W + H)h}{\rho v W H c_p}\theta$$

which can be solved for the desired temperature via separation of the variables and integration if the value for h is known:

$$\int_{\theta_i}^{\theta_e} \frac{d\theta}{\theta} = -\frac{2(W + H)}{\rho v W H c_p} h \int_0^L dx$$

and

$$\ln \frac{\theta_e}{\theta_i} = -\frac{2(W + H)hL}{\rho \dot{V} c_p}$$

where $\dot{V}$ is the volumetric flow rate (m^3/s). Then

$$\frac{\theta_e}{\theta_i} = e^{-2(W+H)hL/\rho \dot{V} c_p} \qquad (4.34)$$

An energy balance for the entire flow passage is written as

$$q = \rho A v c_p (T_e - T_i)$$

which can be solved for the unknown exit temperature of the air, T_e.

Choosing a value of 300 K as an initial guess for the mean air temperature, the values of ρ and c_p are

$$\rho = 1.1769 \text{ kg/m}^3 \quad \text{and} \quad c_p = 1006.3 \text{ J/kg} \cdot \text{K}$$

The value for T_e is thus evaluated as

$$T_e = T_i + \frac{q}{\rho W H v c_p}$$

$$= T_i + \frac{q}{\rho \dot{V} c_p}$$

$$= 300\text{ K} + \frac{2}{(1.1769)(0.003)(1006.3)}$$

$$= 300.6\text{ K}$$

This result suggests that air properties can be evaluated at a mean temperature, between entrance and exit, of 300 K. The properties of interest are

$$\rho = 1.1769\text{ kg/m}^3 \qquad \nu = 1.569 \times 10^{-5}\text{ m}^2/\text{s}$$

$$k = 0.0262\text{ W/m}\cdot\text{K} \qquad \text{Pr} = 0.708$$

The equivalent diameter for this passage is

$$d_e = 4\frac{WH}{2(W+H)}$$

$$= \frac{2(0.005)(0.025)}{0.005 + 0.025}$$

$$= 0.00833\text{ m}$$

and the Reynolds number is evaluated as

$$\text{Re}_d = \frac{d_e v}{\nu} = \frac{d_e}{\nu}\frac{\dot{V}}{WH}$$

$$= \frac{0.00833}{1.569 \times 10^{-5}}\frac{0.003}{(0.005)(0.025)}$$

$$= 12{,}747 \qquad \text{(the flow is turbulent)}$$

Equation (4.30) can now be used to evaluate h:

$$h = \text{Nu}_d\frac{k}{d_e} = 0.023\text{Re}_d^{0.80}\text{Pr}^{0.40}\frac{k}{d_e}$$

$$= \frac{(0.023)(12,747)^{4/5}(0.708)^{0.40}(0.0262)}{0.00833}$$

$$= 121.3\text{ W/m}^2\cdot\text{K}$$

The value of the surface temperature T_s can now be obtained from eq. (4.34):

$$\frac{\theta_e}{\theta_i} = e^{-(2)(0.005+0.025)(121.3)(0.15)/(1.1769)(0.003)(1006.3)}$$

$$\frac{T_s - 300.6}{T_s - 300} = 0.735$$

and this yields

$$T_s = 302\text{ K}$$

(b) The pressure drop through the channel is evaluated by applying eq. (4.6) using the equivalent diameter d_e:

$$\frac{\Delta p}{\rho} = f \frac{L}{d_e} \frac{v^2}{2}$$

and because Re_d has already been determined, the value of f can be read immediately from Fig. 4.4. Assuming smooth channel walls, the value is

$$f = 0.028$$

Thus the pressure drop becomes

$$\frac{\Delta p}{\rho} = f \frac{L}{d_e} \frac{v^2}{2}$$

$$\Delta p = \frac{1}{2} \frac{(1.1769)(0.028)(0.15)}{0.00833} \left[\frac{0.003}{(0.005)(0.025)} \right]^2$$

$$= 171 \text{ Pa}$$

4.5 NATURAL CONVECTION

The mechanism of natural convection involves the movement of a fluid past a solid boundary due to the density effects caused by the energy exchange. In this book, those natural convection flows classified as *free boundary flows*, which involve plumes or buoyant jets, are not discussed. Heat transfer involving extended surfaces, by definition, involves a bounding surface.

Natural convection is treated extensively by a number of authors [e.g., Kakac et al. (1985) and Gebhart et al. (1988)]. The present discussion is restricted to presenting relationships for the heat transfer coefficient h likely to be encountered in heat exchange processes involving extended surface.

Correlations for natural convection normally involve three parameters: the Nusselt number Nu, the Rayleigh number Ra, and the Prandtl number Pr:

$$\text{Nu} = f(\text{Ra}, \text{Pr}) \tag{4.35}$$

The Nusselt and Prandtl numbers were discussed at the beginning of this chapter. They are defined as

$$\text{Nu}_L = \frac{hL}{k} \tag{4.36}$$

and

$$\text{Pr} = \frac{\mu c_p}{k} = \frac{\nu}{\alpha} \tag{4.37}$$

The new parameter, Ra, which arises in natural convection correlations is the product of the Grashof and Prandtl numbers:

$$\mathrm{Ra}_L = \frac{\beta g L^3 \,\Delta T}{\nu \alpha} \tag{4.38}$$

where β is the coefficient of thermal expansion for the fluid,[4] g the acceleration of gravity, L the significant length of the system geometry involved, ΔT the temperature difference between the surface and the ambient fluid, and ν and α the kinematic viscosity and thermal diffusivity of the fluid.

Equation (4.38) defines the Rayleigh number in terms of ΔT, which implies that the surface temperature has a single value. If the boundary condition is one of known heat flux, a modified Rayleigh number is commonly used in correlations such as eq. (4.35). This modified form is defined as

$$\mathrm{Ra}_L^* = \frac{\beta g L^4 q''}{k \nu \alpha} \tag{4.39}$$

The Rayleigh number plays a role in natural convection that is similar to that of the Reynolds number in forced convection. Among its uses is the indication of whether a flow is laminar or turbulent. For instance, with natural convection adjacent to a vertical plane wall, the critical value for the Rayleigh number, $\mathrm{Ra}_{L,\mathrm{c}}$, is

$$\mathrm{Ra}_{L,\mathrm{c}} \simeq 10^9 \tag{4.40}$$

When $\mathrm{Ra}_{L,\mathrm{c}} < 10^9$, such flows are treated as laminar.

Empirical correlations in this section apply to simple geometries in both external and internal flow. There are an enormous number of geometric arrangements that involve extended surfaces, particularly those associated with cooling electronic equipment. For an extended discussion of this subject the interested reader may examine the review article by Incropera (1988).

4.5.1 Vertical Plane Wall

The classic case of natural convection heat transfer is that of the heated (or cooled) vertical plate adjacent to a stagnant ambient fluid. Buoyancy forces resulting from density changes in the fluid give rise to natural-convective flow: upward for a heated surface and downward if the surface is cooled.

A single correlation, applicable to both laminar and turbulent flows (i.e., all values of Ra), has been proposed by Churchill and Chu (1975b). This expression is

$$\mathrm{Nu}_L = \frac{hL}{k} = \left\{ 0.825 + \frac{0.387\mathrm{Ra}_L^{1/6}}{\left[1 + (0.492/\mathrm{Pr})^{9/16}\right]^{4/9}} \right\}^2 \tag{4.41}$$

Equation (4.41) is strictly applicable for an isothermal plate. If the wall heating (or cooling) condition is one of constant heat flux, this same equation can be used with

[4] For ideal gases, β is the reciprocal of the absolute temperature in Kelvin, $1/T$.

the plate temperature taken as that at its midpoint. Churchill (1983) has provided a more detailed discussion of this case.

Equation (4.41) can also be used for vertical cylinders provided that they satisfy the condition given by Sparrow and Gregg (1956),

$$\frac{d}{L} > \frac{35}{(\mathrm{Ra}/\mathrm{Pr})^{1/4}} \tag{4.42}$$

This condition establishes that the boundary layer thickness will be substantially less than the cylinder diameter. When this condition is not met, the heat transfer rate is enhanced. Results for small-diameter cylinders have been reported by Minkowycz and Sparrow (1974).

4.5.2 Horizontal Plates and Cylinders

The following correlations apply to *horizontal plates* for the specific cases specified. For a heated plate facing up or a cooled plate facing down,

$$\mathrm{Nu}_L = 0.54\,\mathrm{Ra}_L^{1/4} \qquad (10^4 < \mathrm{Ra}_L < 10^7) \tag{4.43}$$

and

$$\mathrm{Nu}_L = 0.15\,\mathrm{Ra}_L^{1/3} \qquad (10^7 < \mathrm{Ra}_L < 10^{11}) \tag{4.44}$$

For a heated plate facing down or a cooled plate facing up,

$$\mathrm{Nu}_L = 0.27\,\mathrm{Ra}_L^{1/4} \qquad (10^5 < \mathrm{Ra}_L < 10^{10}) \tag{4.45}$$

Another geometry of interest is that of a very long horizontal cylinder. This case has been described by the correlation of Churchill and Chu (1975a):

$$\mathrm{Nu}_d = \left\{0.60 + \frac{0.387\mathrm{Ra}_d^{1/6}}{\left[1 + (0.559/\mathrm{Pr})^{9/16}\right]^{8/27}}\right\}^2 \tag{4.46}$$

which applies for values of $\mathrm{Ra}_d < 10^{12}$

4.5.3 Horizontal Pin-Fin Arrays

Sparrow and Vemuri (1986) and Liebermann and Gebhart (1969) provide insight regarding the dependence of heat transfer coefficients in pin-fin arrays and along horizontal cylinders on array configuration and geometric parameters. The optimal spacing between horizontal isothermal cylinders configured in an equilateral triangular arrangement was found by Bejan et al. (1995) and Aihara et al. (1990) provided a correlation for the average heat transfer coefficient along the surface of staggered cylindrical pin fins with a vertical base plate.

The Aihara et al. (1990) correlation gives the Nusselt number, $\mathrm{Nu} = hS_h/k$:

$$\mathrm{Nu} = \frac{\pi d}{2S_v}\left[\frac{1}{20}\eta \mathrm{Ra}_h^{3/4}\left(1 - e^{-120/\eta \mathrm{Ra}_h}\right)^{1/2} + \frac{1}{200}\eta \mathrm{Ra}_h^{1/4}\right] \tag{4.47}$$

where S_h is the center-to-center pin-fin spacing in the horizontal and vertical directions (Fig. 4.14), η the pin-fin efficiency, and Ra_h the modified Rayleigh number based on S_h, respectively.

4.5.4 Parallel-Plate Channels

A case of practical interest is that of natural convection in the channel formed by two parallel-plane walls. The vertical channel is considered first.

The benchmark paper involving natural convection in a vertical channel was published by Elenbaas (1942). For symmetrically heated vertical plates Elenbaas obtained the expression

$$\mathrm{Nu}_z = \frac{1}{24}\mathrm{Ra}_z \frac{z}{L}\left[1 - e^{-35/\mathrm{Ra}_z(z/L)}\right]^{3/4} \tag{4.48}$$

where $\mathrm{Nu}_z = hz/k$, z, and L represent the channel width and height, respectively, and $\mathrm{Ra}_z = g\beta z^3\,\Delta T/\nu\alpha$ is the Rayleigh number based on channel width.

It is convenient to write the foregoing in a slightly more compact form by defining the *Elenbaas number* as

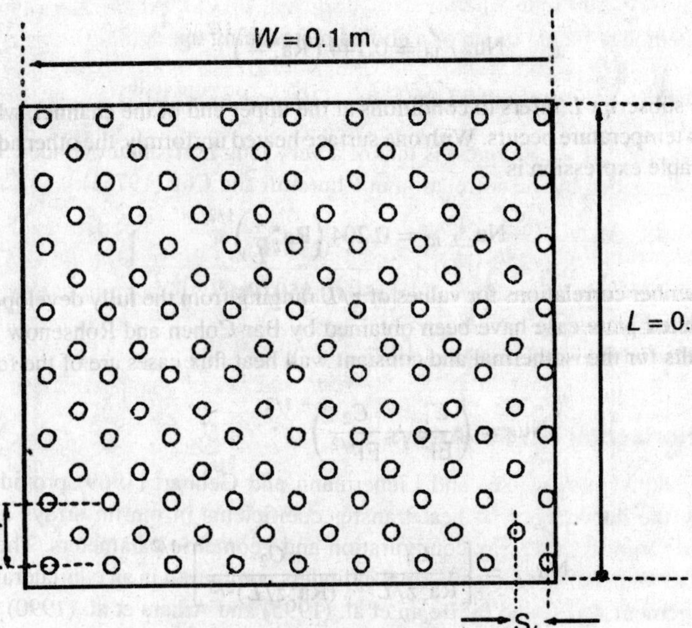

Figure 4.14 Pin-fin array configuration.

$$\text{El} \equiv \text{Ra}_z \frac{z}{L} = \frac{g\beta\alpha z^4 \, \Delta T}{\nu L}$$

so that an alternative expression for Nu_z can be written as

$$\text{Nu}_z = \frac{\text{El}}{24}\left(1 - e^{-35/\text{El}}\right)^{3/4} \tag{4.49}$$

In the (fully developed) limit as $z/L \to 0$, eq. (4.49) reduces to

$$\text{Nu}_{z,\text{fd}} = \frac{\text{El}}{24} \tag{4.50}$$

A modification of this result applies for the case of only one isothermal plate and the other plate adiabatic. For this case

$$\text{Nu}_{z,\text{fd}} = \frac{\text{El}}{12} \tag{4.51}$$

If the wall heating condition is one of *uniform heat flux* rather than constant temperature, the Rayleigh number is modified to the form

$$\text{Ra}_z^* = \frac{g\beta}{\nu\alpha}\frac{z^4}{k}q'' \tag{4.52}$$

For both surfaces heated uniformly, the fully developed limit is given by Bar-Cohen and Rohsenow (1984):

$$\text{Nu}_{z,L,\text{fd}} = 0.144\left(\text{Ra}_z^* \frac{z}{L}\right)^{1/2} \tag{4.53}$$

where the subscript L refers to conditions at the upper end of the channel, where the maximum temperature occurs. With one surface heated uniformly, the other adiabatic, the applicable expression is

$$\text{Nu}_{z,L,\text{fd}} = 0.204\left(\text{Ra}_z^* \frac{z}{L}\right)^{1/2} \tag{4.54}$$

Nusselt number correlations for values of z/L ranging from the fully developed limit to the isolated plate case have been obtained by Bar-Cohen and Rohsenow (1984). Their results for the isothermal and constant wall heat flux cases are of the form

$$\text{Nu}_z = \left(\frac{C_1}{\text{El}^2} + \frac{C_2}{\text{El}^{1/2}}\right)^{-1/2} \tag{4.55}$$

and

$$\text{Nu}_{z,L} = \left[\frac{C_1}{\text{Ra}_s^* z/L} + \frac{C_2}{(\text{Ra}_z^* z/L)^{2/5}}\right]^{-1/2} \tag{4.56}$$

with constants C_1 and C_2 given in Table 4.5. In the cases considered, the fully developed limit and isolated plate limit correspond to $\text{El} \le 10$ or $\text{Ra}_z^* z/L \le 10$ and $\text{El} \ge 100$ or $\text{Ra}_z^* z/L \ge 100$. When employing the relationships in Table 4.5, the

TABLE 4.5 Heat Transfer Parameters for Natural Convection between Parallel Plates

Surface Condition	C_1	C_2	z_{opt}	z_{max}/z_{opt}
Symmetric isothermal plates, $T_{s,1} = T_{s,2}$	576	2.87	$2.71(\mathrm{Ra}_z/z^3L)^{-1/4}$	1.71
Symmetric isoflux plates, $q''_{s,1} = q''_{s,2}$	48	2.51	$2.12(\mathrm{Ra}^*_z/z^4L)^{-1/5}$	4.77
Isothermal/adiabatic plates, $T_{s,1} = q''_{s,2}$	144	2.87	$2.15(\mathrm{Ra}_z/z^3L)^{-1/4}$	1.71
Isoflux/adiabatic plates, $q''_{s,1} = q''_{s,2} = 0$	24	2.52	$2.69(\mathrm{Ra}^*_z/z^4L)^{-1/5}$	4.77

fluid properties should be evaluated at $T = (T_s + T_\infty)/2$ for isothermal surfaces and $T = (T_{s,L} + T_\infty)/2$ for constant-flux surfaces.

Example 4.8. An electronic circuit board is to be cooled using a heat sink arrangement as shown in Fig. 4.15. The fin surfaces are considered isothermal at the temperature of the circuit board, and the ambient air is at 300 K. The assembly has a total length L of 30 cm and consists of 20 fin pairs. (**a**) If the circuit board temperature is not to exceed 325 K, determine the maximum power that can be dissipated through the heat sink, and (**b**) determine the optimum spacing of these fins for maximum power dissipation and maximum heat removal rate.

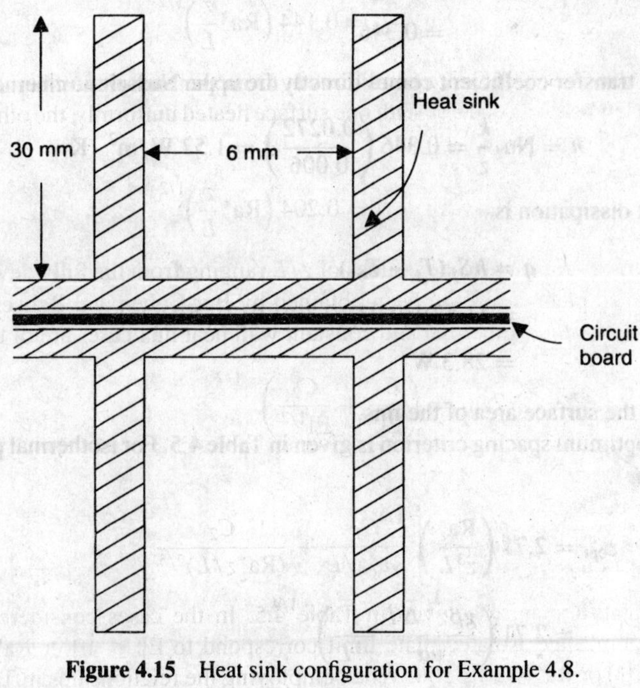

Figure 4.15 Heat sink configuration for Example 4.8.

SOLUTION. The properties of air evaluated at

$$T_f = \frac{T_s + T_\infty}{2} = \frac{325 + 300}{2} = 312.5\ K$$

are

$$k = 27.2 \times 10^{-3}\ \text{W/m}\cdot\text{K} \qquad \nu = 1.682 \times 10^{-5}\ \text{m}^2/\text{s}$$

$$\alpha = 2.386 \times 10^{-5}\ \text{m}^2/\text{s} \qquad \beta = \frac{1}{312.5} = 3.200 \times 10^{-3}\ \text{K}^{-1}$$

(a) The rate of power dissipation can be evaluated using eq. (4.49) to determine the heat transfer coefficient. The Elenbaas number is evaluated first :

$$\begin{aligned}
\text{El} &= \frac{g\beta\,\Delta T z^4}{\nu\alpha L} \\
&= \frac{(9.81)(0.0032)(25)(0.006)^4}{(1.682 \times 10^{-5})(2.386 \times 10^{-5})(0.30)} \\
&= 8.448
\end{aligned}$$

Equation (4.48) can now be solved:

$$\begin{aligned}
\text{Nu}_z &= \frac{\text{El}}{24}\left(1 - e^{-35/\text{El}}\right)^{3/4} \\
&= \frac{8.448}{24}\left(1 - e^{-35/8.448}\right)^{3/4} \\
&= 0.346
\end{aligned}$$

The heat transfer coefficient comes directly from the Nusselt number:

$$h = \text{Nu}_z \frac{k}{z} = 0.346\left(\frac{0.0272}{0.006}\right) = 1.57\ \text{W/m}^2\cdot\text{K}$$

and the heat dissipation is

$$\begin{aligned}
q &= hS_f(T_s - T_\infty) \\
&= (1.57)[(0.03)(0.30)(2)(2)(20)](25) \\
&= 28.3\ \text{W}
\end{aligned}$$

where S_f is the surface area of the fins.

(b) The optimum spacing criterion is given in Table 4.5. For isothermal plates with $L = 0.30$ m,

$$\begin{aligned}
z_{opt} &= 2.71\left(\frac{\text{Ra}_z}{z^3 L}\right)^{-1/4} \\
&= 2.71\left(\frac{g\beta z^3\,\Delta T}{\nu\alpha}\frac{1}{z^3 L}\right)^{-1/4}
\end{aligned}$$

$$= 2.71\left(\frac{g\beta\,\Delta T}{\nu\alpha L}\right)^{-1/4}$$

$$= 2.71\left[\frac{(9.81)(0.0032)(25)}{(1.682\times10^{-5})(2.386\times10^{-5})(0.30)}\right]^{-1/4}$$

$$= 0.00953 \text{ m} \quad \text{or} \quad 9.53 \text{ mm}$$

The value of the Elenbaas number is

$$\text{El} = \frac{g\beta z_{\text{opt}}^4\,\Delta T}{\nu\alpha L}$$

$$= \frac{(9.81)(0.0032)(0.00953)^4(25)}{(1.682\times10^{-3})(2.386\times10^{-3})(0.30)}$$

$$= 53.76$$

Equation (4.55) can now be used to evaluate Nu_z for this condition :

$$\text{Nu}_z = \left[\frac{C_1}{(\text{El})^2} + \frac{C_2}{(\text{El})^{1/2}}\right]^{-1/2}$$

with values of C_1 and C_2 taken from Table 4.5. The value of Nu_z is determined as

$$\text{Nu}_z = \left[\frac{576}{(53.76)^2} + \frac{2.87}{(53.76)^{1/2}}\right]^{-1/2}$$

$$= 1.301$$

The new value for the heat dissipation is

$$q = hS_{f,n}(T_s - T_\infty)$$

where $S_{f,n}$ is the surface area corresponding to the new fin arrangement at the optimum spacing. Because the new spacing is larger than the original configuration, there will be fewer fins in the optimum arrangement. For the same-size circuit board the number of fins spaced the distance z_{opt} apart,

$$n_{\text{opt}} = 20\left(\frac{0.006}{0.00953}\right) \simeq 12$$

and for 12 fin pairs, optimally spaced, the heat dissipation rate will be

$$q = \text{Nu}_z\frac{k}{z}S_{f,n}\,\Delta T$$

$$= 1.301\left(\frac{0.0272}{0.00953}\right)(0.03)(0.30)(2)(2)(12)(25)$$

$$= 40.1 \text{ W}$$

This is an increase of 42% relative to the initial design in part (a) and is accomplished with eight fewer fin pairs.

4.5.5 Parallel-Plate Arrays

The correlation of natural convection heat transfer from arbitrarily shaped bodies has been the subject of considerable research. Of particular importance are the studies by Bilitsky (1986), Culham et al. (1994), Karagiozis (1991), Lee et al. (1991), Starner and McManus (1963), Van de Pol and Tierney (1973), Welling and Wooldridge (1965), and Yovanovich (1987a,b).

Work of Yovanovich and Co-workers. The efforts of Yovanovich (1987a,b) and Lee et al. (1991) have revealed that in the domain of developing flow (i.e., for relatively large values of the Rayleigh number), the Nusselt number for natural convection cooling of isothermal bodies can be obtained by superposition of the diffusive and convective limits. When the characteristic length is taken as the square root of the wetted area, the Nusselt and Rayleigh numbers are defined by

$$\mathrm{Nu}_{\sqrt{S}} \equiv \frac{h\sqrt{S}}{k}$$

and

$$\mathrm{Ra}_{\sqrt{S}} \equiv \frac{g\beta \mathrm{Pr}\, \Delta T(\sqrt{S})^3}{\nu^2}$$

so that

$$\mathrm{Nu}_{\sqrt{S}} = \mathrm{Nu}^{\infty}_{\sqrt{S}} + f(\mathrm{Pr})G_{\sqrt{S}}\mathrm{Ra}^{0.25}_{\sqrt{S}} \tag{4.57}$$

In this equation, the "universal" Prandtl number function takes the form

$$f(\mathrm{Pr}) = \frac{0.67}{[1 + (0.5/\mathrm{Pr})^{9/16}]^{4/9}} \tag{4.58}$$

where $\mathrm{Nu}^{\infty}_{\sqrt{S}}$ is the diffusive limit, $\mathrm{Ra}_{\sqrt{S}}$ the Rayleigh number, and G a body gravity function that is based on the square root of the wetted surface area, S. Following Culham et al. (1994), eq. (4.57) can be applied to an array of widely spaced fins by combining the diffusive limit $\mathrm{Nu}^{\infty}_{\sqrt{S}}$ and the body gravity function $G_{\sqrt{S}}$ for the wetted area. Thus, with this nomenclature and viewing just a vertical base plate, the diffusive limit for a cuboid, with dimensions $L \times W \times H$, takes the form

$$\mathrm{Nu}^{\infty}_{\sqrt{S}} = \frac{3.192 + 1.868(H/L)^{0.76}}{\sqrt{1 + 1.189(H/L)}} \tag{4.59}$$

Similarly, for an arbitrary array of longitudinal fins of rectangular profile with base plate dimensions $L \times W \times \delta_p$ and n fins, each with dimensions $b \times \delta \times L$, the body function $G_{\sqrt{S}}$ can be written as

$$G_{\sqrt{S}} = 1.0904 \left[\frac{L\Lambda^2}{(n\delta_b b + \delta_b W + L\Lambda)^{3/2}} \right]^{1/4} \tag{4.60}$$

where

$$\Lambda = nb + \delta_p + W$$

In this equation, the base plate and fin height L are parallel to the gravity vector. The thickness of the base plate and the fin are δ_p and δ_b, respectively, and it has been implicitly assumed that the wetted area of the heat sink surface is isothermal.

The results of applying this modeling approach to convective finned arrays with widely spaced fins can be seen in Figs. 4.16 and 4.17, which show the Nusselt number variation with Rayleigh number for four arrays of fins (Karagiozis, 1991). All four arrays have an identical base plate 22 cm wide, 15 cm high, and 0.95 mm thick, but differ in the fin geometry. In the first and second arrays, fins 5 cm high and 0.95 cm thick were spaced 3 and 3.5 cm apart (Fig. 4.16), while in the third and fourth arrays, fins 1.5 cm high and 0.15 cm thick were spaced 1.5 and 3.0 cm apart (Fig. 4.17). Examination of these figures reveals that for relatively large Rayleigh numbers where developing flow prevails, excellent agreement is found between the correlation and the data. However, as the Rayleigh number decreases, signifying an approach to fully developed flow associated with smaller interfin spaces and lower base heat fluxes, the measured values fall below the correlation. It may also be noted that in keeping with the fundamental isothermal fin assumptions used in developing this correlation, all four arrays examined contain fins that are relatively thick and widely spaced.

Example 4.9. Determine the effective natural convection heat transfer coefficient and the total heat transfer rate from a vertically oriented sand-cast aluminum array of fins on a base 10 mm thick, 225 mm wide, and 200 mm high. There are five fins 35 mm high, 10 mm thick, and spaced 35 mm apart, and the array can be assumed isothermal at a temperature of 350 K, with the ambient air at 300 K.

SOLUTION. This description suggests that the superposition relation of eq. (4.57) can be used to determine the thermal performance of this isothermal array of fins with thick, widely spaced fins. Using eq. (4.59), the diffusive limit Nusselt number can be calculated. First determine the dimensions of the "cuboid." The base area, $L \times W$, is 200 mm × 225 mm, and the extension H is the base thickness plus the fin height, or 10 mm + 35 mm = 45 mm. Using these values in eq. (4.59) yields

$$\begin{aligned} \text{Nu}^{\infty}_{\sqrt{S}} &= \frac{3.192 + 1.868(H/L)^{0.76}}{\sqrt{1 + 1.189(H/L)}} \\ &= \frac{3.192 + (1.868)(0.045/0.2)^{0.76}}{[1 + (1.189)(0.045/0.2)]^{1/2}} \\ &= \frac{3.793}{1.126} = 3.369 \end{aligned}$$

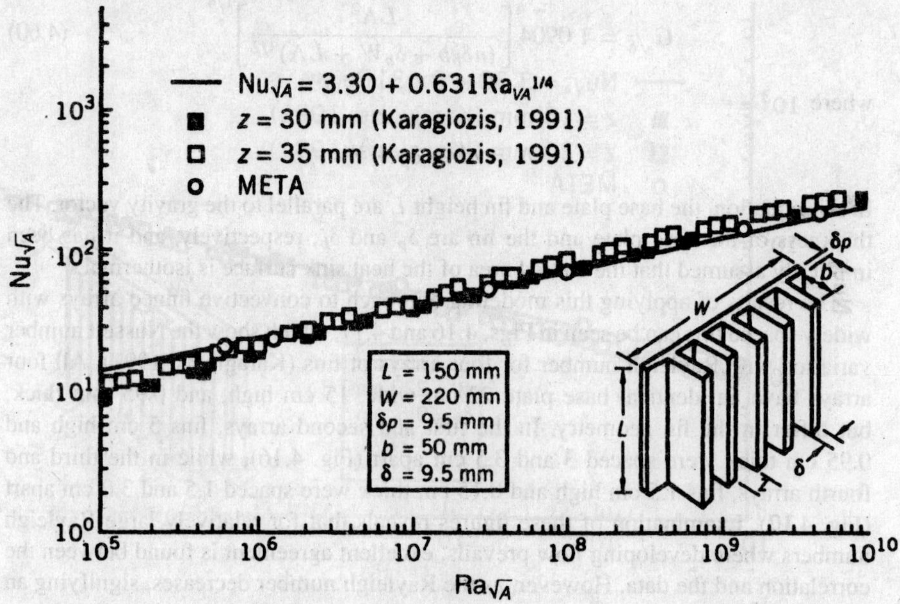

Figure 4.16 Rectangular fin assembly heat transfer for widely spaced high fins. [Adapted from Culham et al. (1994).] *META* is a computer code used at the University of Waterloo.

Next, find the body gravity function $G_{\sqrt{S}}$ using eq. (4.60) with the specified dimensions:

$$G_{\sqrt{S}} = 1.0904\left[\frac{L\Lambda^2}{(n\delta_b b + \delta_b W + L\Lambda)^{3/2}}\right]^{1/4}$$

Here

$$\begin{aligned}\Lambda &= nb + \delta_b + W\\ &= (5)(0.035) + 0.010 + 0.225\\ &= 0.410 \text{ mm}\end{aligned}$$

so that

$$\begin{aligned}G_{\sqrt{S}} &= 1.0904\left\{\frac{(0.20)(0.410)^2}{[(5)(0.01)(0.035) + (0.01)(0.225) + (0.20)(0.410)]^{3/2}}\right\}^{1/4}\\ &= 1.0904\left(\frac{0.03362}{0.02522}\right)^{1/4}\\ &= 1.173\end{aligned}$$

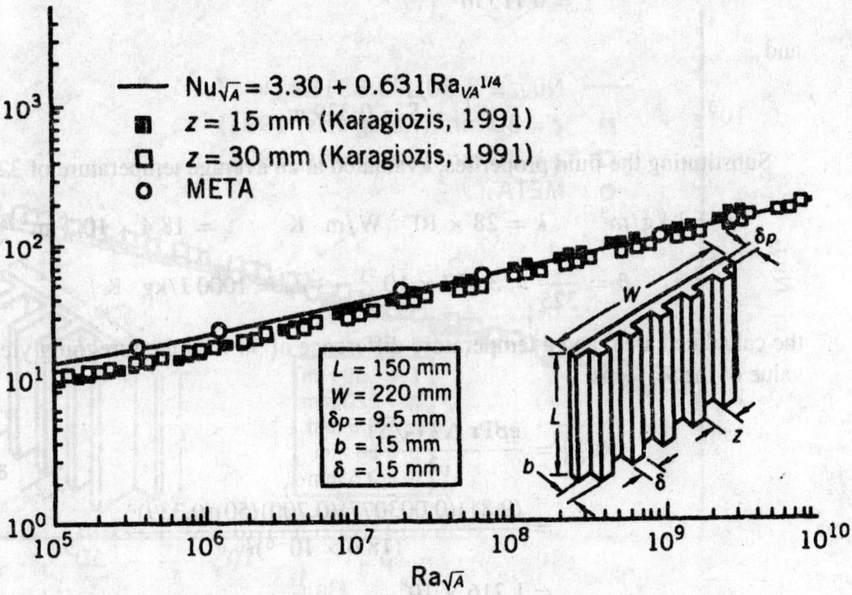

Figure 4.17 Rectangular fin assembly heat transfer for widely spaced low fins/ [Adapted from Culham et al. (1994).] *META* is a computer code used at the University of Waterloo.

For an average fluid temperature of 325 K, the Prandtl number is 0.70 and the "universal" Prandtl number function $f(\text{Pr})$ using eq. (4.58) is found to be 0.513. Consequently, the superposition relation of eq. (4.57) takes the form

$$\begin{aligned}\text{Nu}_{\sqrt{S}} &= \text{Nu}^{\infty}_{\sqrt{S}} + f(\text{Pr})G_{\sqrt{S}}\text{Ra}^{0.25}_{\sqrt{S}} \\ &= 3.369 + (0.513)(1.173)\text{Ra}^{0.25}_{\sqrt{S}} \\ &= 3.369 + 0.601\text{Ra}^{0.25}_{\sqrt{S}}\end{aligned}$$

The Rayleigh number $\text{Ra}_{\sqrt{S}}$, based on the wetted surface area S, is expressed as

$$\text{Ra}_{\sqrt{S}} = \frac{g\beta\,\Delta T\,\text{Pr}(\sqrt{S})^3}{\nu^2}$$

and the total surface area of the heat sink is equal to the base surface area plus the product of the number of fins and the surface area of each fin:

$$\begin{aligned}S &= LW + n(2bL) \\ &= (0.200)(0.225) + (5)(2)(0.035)(0.200) \\ &= 0.045 + 0.070\end{aligned}$$

$$= 0.115 \text{ m}^2$$

and

$$\sqrt{S} = 0.339 \text{ m}$$

Substituting the fluid properties, evaluated at an average temperature of 325 K,

$$\rho = 1 \text{ kg/m}^3 \qquad k = 28 \times 10^{-3} \text{ W/m}\cdot\text{K} \qquad \nu = 18.4 \times 10^{-6} \text{ m}^2/\text{s}$$

$$\beta = \frac{1}{325} = 3.077 \times 10^{-3} \qquad c_p = 1000 \text{ J/kg}\cdot\text{K}$$

the calculated area, and a temperature difference of 50 K in the foregoing yields the value of the $\text{Ra}_{\sqrt{S}}$ as

$$\text{Ra}_{\sqrt{S}} = \frac{g\beta \text{Pr}\, \Delta T (\sqrt{S})^3}{\nu^2}$$

$$= \frac{(9.81)(0.003077)(0.700)(50)(0.339)^3}{(18.4 \times 10^{-6})^2}$$

$$= 1.216 \times 10^8$$

Returning to the basic relation, the Nusselt number can now be found to equal

$$\text{Nu} = 3.369 + 0.602\text{Ra}^{1/4}$$

$$= 3.369 + 0.602(1.216 \times 10^8)^{1/4}$$

$$= 3.369 + (0.602)(105.0)$$

$$= 66.37$$

so that

$$h = \text{Nu}\frac{k}{\sqrt{S}}$$

$$= \frac{(66.37)(0.028)}{0.339}$$

$$= 5.48 \text{ W/m}^2\cdot\text{K}$$

Finally, using this heat transfer coefficient, it is possible to determine the heat dissipation of the array of fins, which is assumed to be isothermal:

$$q = hS(T_b - T_{\text{air}}) = (5.48)(0.115)(50) = 31.5 \text{ W}$$

As a first-order correction for a nonisothermal heat sink, this value could be multiplied by an overall surface efficiency.

Observe that even this relatively "coarse" array with thick, widely spaced fins offers approximately 2.5 times the heat transfer capability of the unfinned base area.

Work of Van de Pol and Tierney. A careful review of the Starner and McManus (1963) and the Welling and Wooldridge (1965) studies, as well as other data available in the literature at the time, led Van de Pol and Tierney (1973) to suggest that successful application of the Elenbaas correlation to arrays of fins would require channel aspect ratio corrections in eq. (4.49):

$$\text{Nu}_z = \frac{\text{El}}{24}\left(1 - e^{-35/\text{El}}\right)^{3/4} \tag{4.49}$$

They proposed that the differences observed, between actual finned arrays and the relatively wide fins studied by Elenbaas be addressed by the introduction of a geometric parameter, ψ, into the Elenbaas equation, in the manner suggested by

$$\text{Nu}_r = \frac{\text{El}_r}{\psi}\left[1 - e^{-\psi(0.50/\text{El}_r)^{3/4}}\right] \tag{4.61}$$

Here the Nusselt and channel Rayleigh numbers are based on a hydraulic radius, $r = 2A/P$, which is *twice* the cross-sectional area divided by the wetted perimeter. The geometric parameter ψ takes the form

$$\psi = \frac{(24)(1 - 0.483e^{-0.17/a})}{\{(1 + a/2)[1 + (1 - e^{-0.83a})(9.14a^{1/2}e^{-0.4646S} - 0.61)]\}^3} \tag{4.62}$$

where a is the channel aspect ratio z/b.

The Van de Pol and Tierney (1973) relationship was shown to fit the data of Welling and Wooldridge with a standard error of just 0.039 in the Nusselt number (Fig. 4.18). Moreover, the form of this correlation assures that at very small aspect ratios, eq. (4.62) approaches the original Elenbaas correlation, and that at very large values it merges smoothly into the isolated plate limit.

It should be noted that in keeping with the data of Van de Pol and Tierney (1973), all fluid properties in eq. (4.61) are to be evaluated at the wall temperature, with the exception of the volumetric expansion coefficient β, which is taken at the fluid temperature.

Work of Bilitzky. Analysis of data for thermal transport from his vertically oriented arrays of fins led Bilitzky (1986) to recognize that closely spaced narrow fins display substantially higher heat transfer coefficients than predicted by the Elenbaas correlation. Echoing earlier results by Sparrow and Bahrami (1980), the empirical data for the Nusselt number do not appear to approach the anticipated fully developed limit at small-channel Rayleigh numbers; appearing instead to approach an asymptote nearly parallel to, although lower than, the isolated plate limit. Flow visualization studies of these arrays revealed the presence of a significant inflow from the open front edge of the interfin channels, which increased in importance as the fin width and fin spacing decreased.

Based on these observations, Bilitzky modified the Van de Pol and Tierney (1973) geometric factor ψ in eq. (4.62) by replacing the product $0.4646S$ by a purely geometric factor B, reflecting the contribution of fin-tip inflow to the overall channel

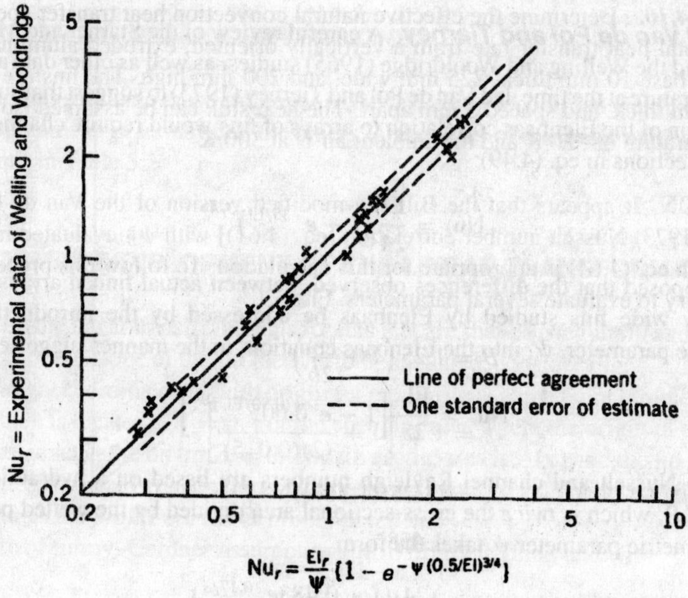

Figure 4.18 Comparison of predicted array Nusselt numbers. [Adapted from Van De Pol and Tierney (1973).]

flow. This parameter is equivalent to 1.25 times the ratio of the hydraulic radius associated with the fin-tip opening to that for the vertical channel (with attention paid to the actual wetted perimeter)

$$B = 1.25\left(1 + \frac{z}{2b}\right) \tag{4.63}$$

Insertion of this additional term into the Van de Pol and Tierney (1973) geometric parameter of eq. (4.62) yields

$$\psi_B = \frac{24\Lambda_1}{[(1 + a/2)(1 + \Lambda_2\Lambda_3)]^3} \tag{4.64}$$

where, with $a = z/b$,

$$\Lambda_1 = 1 - 0.483e^{-0.17/a}$$

$$\Lambda_2 = 1 - e^{-0.83a}$$

and

$$\Lambda_3 = 9.14a^{1/2}e^{-B} - 0.61$$

Use of this correction was reported by Bilitsky (1986) to yield agreement within 5% for his data, approximately 10% for the Welling and Wooldridge (1965) data, and similar agreement with the results of Starner and McManus (1963).

Example 4.10. **Determine the effective natural convection heat transfer coefficient and the total heat transfer rate from a vertically oriented, extruded aluminum heat sink on a base 10 mm thick, 225 mm wide, and 200 mm high. The fins are 15 mm high, 2 mm thick, and spaced 5 mm apart. The heat sink can be assumed isothermal at a temperature of 350 K and the ambient air is at 300 K.**

SOLUTION. It appears that the Bilitzky-modified version of the Van de Pol and Tierney (1973) Nusselt number correlation [eq. (4.61)] with ψ_B evaluated in accordance with eq. (4.64) is appropriate for this calculation. To follow this procedure, it is necessary to evaluate several parameters. First

$$B = 1.25\left(1 + \frac{z}{2b}\right)$$

$$= 1.25\left[1 + \frac{0.005}{(2)(0.015)}\right]$$

$$= (1.25)(1.167)$$

$$= 1.458$$

Next,

$$a = \frac{z}{b}$$

$$= \frac{0.005}{0.015}$$

$$= 0.333$$

Then, noting that the wetted perimeter of the channel is $2b + z$,

$$r = \frac{2bz}{2b + z}$$

$$= \frac{(2)(0.015)(0.005)}{(2)(0.015) + 0.005}$$

$$= 0.00429 \text{ m}$$

and finally,

$$\text{El}_r = \text{Ra}_r \frac{r}{L} = \frac{\rho^2 g \beta c_p \Delta T r^4}{\mu k L}$$

At the average temperature of 325 K, the fluid properties are

$$\rho = 1 \text{ kg/m}^3 \qquad k = 28 \times 10^{-3} \text{ W/m} \cdot \text{K} \qquad \mu = 2 \times 10^{-5} \text{ N} \cdot \text{s/m}^2$$

$$\beta = \frac{1}{325} = 3.077 \times 10^{-3} \text{ K}^{-1} \qquad c_p = 1000 \text{ J/kg} \cdot \text{K}$$

Here Ra_z is evaluated at 325 K, where the temperature difference is 50 K, the hydraulic

radius is 4.29 mm, and the channel length is $L = 0.2$ m. A partial evaluation using the listed properties gives

$$\mathrm{El}_r = 5.385 \times 10^7 \frac{\Delta T\, r^4}{L}$$

and then

$$\mathrm{El}_r = (5.385 \times 10^7)\frac{(50)(0.00429)^4}{0.20}$$
$$= 4.55$$

Insertion of these values into eq. (4.64) will give the modified ψ parameter, ψ_B. However, the Λ values are needed:

$$\Lambda_1 = 1 - 0.483e^{-0.17/a}$$
$$= 1 - 0.483e^{-0.17/0.333}$$
$$= 1 - (0.483)(0.600)$$
$$= 0.710$$
$$\Lambda_2 = 1 - e^{-0.83a}$$
$$= 1 - e^{-(0.83)(0.333)}$$
$$= 1 - 0.758$$
$$= 0.242$$

and

$$\Lambda_3 = 9.14a^{1/2}e^{-B} - 0.61$$
$$= (9.14)(0.333)^{1/2}e^{-1.458} - 0.61$$
$$= 1.227 - 0.61$$
$$= 0.617$$

Then, by eq. (4.64),

$$\psi_B = \frac{24\Lambda_1}{[(1 + a/2)(1 + \Lambda_2\Lambda_3)]^3}$$
$$= \frac{(24)(0.710)}{\{(1 + 0.333/2)\,[1 + (0.242)(0.618)]\}^3}$$
$$= \frac{17.040}{[(1.167)(1.150)]^3}$$
$$= \frac{17.040}{2.411}$$
$$= 7.066$$

Returning now to eq. (4.61) with $\psi = \psi_B = 7.066$,

$$\begin{aligned}\text{Nu}_r &= \frac{\text{El}_r}{\psi}\left[1 - e^{-\psi(0.50/\text{El}_r)^{3/4}}\right]\\ &= \frac{4.55}{7.066}\left[1 - e^{-[(7.066)(0.50/4.55)^{3/4}]}\right]\\ &= 0.648\,(1 - e^{-1.348})\\ &= 0.479\end{aligned}$$

It is now possible to determine the heat transfer coefficient for the finned surface of the array from the Nusselt number as

$$h = \frac{k}{r}\text{Nu}_r = \left(\frac{0.028}{0.00429}\right)(0.479) = 3.11\ \text{W/m}^2\cdot\text{K}$$

The total heat dissipation from all the wetted surfaces in the array is based on the assumption that a single value of h can be employed. With $h = 3.13\ \text{W/m}^2\cdot\text{K}$ and assuming a fin efficiency near unity,

$$\begin{aligned}q_T &= hS(T_b - T_{\text{air}})\\ &= (3.11)[(0.225)(0.200) + (32)(2)(0.015)(0.200)](50)\\ &= (3.11)(0.045 + 0.192)(50)\\ &= 36.9\ \text{W}\end{aligned}$$

It is interesting to note that this heat sink with relatively thin, narrow, and closely spaced fins is thus capable of dissipating nearly as much heat as the heat sink with the relatively massive fins examined in Example 4.9.

4.6 CLOSURE

The information presented in this chapter was chosen to provide the reader with two kinds of information. Fundamental concepts associated with fluid flow and convective heat transfer were presented to provide a general understanding of the physical principles that are involved. Correlations were next presented for representative conditions involving extended surface heat transfer. Examples were included to illustrate the use of certain correlations to representative applications.

There is a substantial body of technical literature dealing with convective heat transfer, and considerable research is currently under way dealing, in particular, with applications to cooling electronic equipment. A brief summary of this literature follows.

The ability to determine the heat transfer coefficient in physical configurations that employ extended surfaces has come from a wealth of experimental data. Starner and McManus (1963) presented heat transfer coefficients for four arrays of fins of fins positioned with their bases vertical and inclined at an angle of 45° with the horizontal. They analyzed the fins as isothermal surfaces, claiming that the lowest fin efficiency encountered was over 98% and that this indicated that the temperature drop from base to tip was negligible. They claimed that the heat transfer coefficients for the vertical arrays fell 10 to 30% below those of similarly placed parallel plates and that the arrays inclined at an angle of 45° yielded results 5 to 20% below those of the vertical arrays. They looked at two flow patterns for the horizontal arrays, and they claimed that the heat transfer coefficients could be reduced sharply by preventing three-dimensional flow.

An extensive numerical study of the overall heat transfer coefficient and friction factors for banks of straight (longitudinal) profile fins in cross flow was undertaken by Ueda and Harada (1964). The effect of fin height and spacing, channel height, and fin material, all as a function of Reynolds number, were considered. One plot of the local value of the heat transfer coefficient averaged between the local values for the two fin sides was included. The Ueda–Harada (1964) study may well have been the first to show the large variation in the surrounding fluid temperature, which had, heretofore, been assumed constant. This was revealed by a plot which showed the variation of the fluid temperature in the gaps between the fins and in the direction of the fin height.

Hilding and Coogan (1964) conducted a study of 10 internal fin configurations using air. The air was heated inside the finned tubes, all of which had an inside diameter of 1.397 cm and a length of 45.72 cm. Their results have limited applicability because the measurements were taken with air as the only fluid in tubes with the same internal diameter. However, their conclusions are characteristic of nonviscous fluids and they arrived at several generalizations involving comparisons between finned and unfinned tubes.

Theoclitus and Eckrich (1964) described an experimental technique for the determination of the efffectiveness (efficiency) of extended surface. The technique depends on a correlation between steady-state and transient testing methods. Karasev (1967) investigated the effect of rectangular turbulizers on the fluid resistance, the structure of the flow, and the local heat transfer, and Zozulya and Korunzhii (1968) considered the effect of low-frequency high-amplitude oscillations on the rate of heat transfer from smooth and finned tubes to transformer oil. Ackerman and Brunsvold (1969) provided heat transfer and draft loss (pressure drop) for five staggered and one in-line tube bank arrangement. In this experimental study, 4.673-cm-outside diameter tubes with 2.54-cm-long elliptical studs were investigated. The tube banks, which were in a clean condition, were eight or 10 rows deep and possessed various transverse and longitudinal spacing-to-tube diameter ratios. The performance was described adequately by suitable arrangement factors and the relative performance of all arrangements was shown to depend on the basis of comparison.

Horn and Atherton (1969) pointed out that manufacturing differences in mass-produced fin elements forming heat transfer surfaces could affect the heat transfer and pressure drop performance. They took performance data on production finned tubes in axial flow and used the result to illustrate statistical techniques for determining whether observed performance differences are real when assessed against the magnitude of experimental error.

Shcherbakov and Semena (1969) reported on studies of tubes with longitudinal internal fins, and Ornatskii et al. (1970) showed that in internally finned tubes, the velocity at the boundary between the interfin space and the main flow depends on the hydrodynamic and geometrical characteristics of the main flow and the interfin space. Lipets and Zholudov (1970) provided a method for calculating the effectiveness of internally finned tubes and Jones (1970) studied the use of integral fins inside tubes to enhance boiling heat transfer.

Chaddock (1970) conducted an experimental study to effect optimum design procedures for finned heat transfer surfaces in the free convection mode. The study was prompted by the desire to apply finned surfaces to thermoelectric refrigerators, and a versatile heat exchanger that could be changed to any one of 12 rectangular fin arrays was constructed. An equation for the overall free convection heat transfer coefficient was developed and a numerical example showing its use for a vertical fin array was provided.

Thomas and Hayes (1970) provided a study that gave heat transfer coefficients on surfaces used in distillation plant heat exchangers, and Aihara (1970a,b), in a pair of papers, reported on experimental natural convection studies in a finned array. The first of these considered the effect of fin geometry and temperature, and an empirical relationship for the average Nusselt number in the vertical open channel between the fins was derived. The second paper was considered to be a sequel to the first, and in it, the author considered the fin edge as one of the component surfaces in the array. After having made the observation that surface roughening produces a considerable increase in heat transfer, Dalle Donne and Meerwald (1970), measured heat transfer and friction coefficients of rods with various thread-type ribs.

Zappa and Geiger (1971) conducted an experimental investigation on the effect of disturbance fins on heat transfer and pressure drop of transformer oil in laminar flow. The test section was a horizontal annulus with circumferential (radial) fins. The fin height was varied for each test, and the results indicated that the fins almost doubled the heat transfer performance. An optimum fin spacing-to-fin height ratio was observed and reported. Mantle et al. (1971) considered heat transfer from rib-roughened surfaces, particularly in gas-cooled reactor fuel elements and the electrically heated scaled-up dummy elements used in the laboratory, to establish their performance. They showed that important corrections were required based on the thermal conductivities of the test element and coolant. They also correlated computed temperature distributions within the ribbed surfaces.

Forslund (1972) reported on a successful technique for the measurement of thermal conductance in thin radiating fins. The work was inspired by the desire to obtain a lightweight radiator–condenser configuration for the mercury-Rankine cycle of a space power plant. Taranyan et al. (1972) showed the dependence of heat transfer on the relative depth of the interfin cavity in staggered tube bundles with staggered transverse (radial) fins.

Rich (1973) took air-side heat transfer and pressure drop measurements on nine multirow plate fin and tube arrangements. The typical dimensions of the test cores were identical, with the exception of the fin pitch. Sturman (1973) provided a brief description of methods for measuring the heat transfer coefficient on fin surfaces. The method, which depends on the measurement of the first and second derivatives of the temperature distribution within the fin, was examined in more detail and refined in its applications. A method was described that enabled errors to be estimated and confidence intervals to be obtained.

Zozulya et al. (1973) presented results of an experimental study of heat transfer from a bundle of staggered finned tubes when the incoming flow was turbulized by means of metal grids. Okamoto and Kameoka (1973) analyzed the temperature distribution in a plate fin normal to the direction of flow for various fin sizes, fluid velocities, and material thermal conductivities. The variation in local heat transfer coefficient along the fin height was measured using a local heater element. Normalized equations for the Nusselt number, fin effectiveness, and fin efficiency were provided and it was clearly shown that the local heat transfer coefficient was largely influenced by the vortex flow between the fin spaces.

Hu and Chang (1973) investigated analytically the heat transfer in fully developed flow in internally finned tubes. They observed that the highest value of the Nusselt number was obtained for a tube containing 22 fins, with the fin height at about 80% of the tube radius. The value of this Nusselt number was almost 20 times that of the finless tube under the same flow conditions.

Saboya and Sparrow (1974a,b) observed on the basis of the range of Reynolds numbers encountered in air-conditioning equipment of the 1970s, that the local heat transfer coefficients were very low on the portion of the fin that is washed by the wake downstream of the tube. Consequently, that portion contributes very little to the overall heat transfer capabilities of the exchanger. They considered candidate approaches for eliminating the deleterious effects of the wake. Among these was an investigation of tube relocation, and it was shown that the overall heat transfer capability of the system was increased moderately by rearward relocation of the tube.

Cowan et al. (1974) made aerodynamic and heat transfer performance measurements on an industrial air-cooled heat exchanger. Results were presented for the airflow and air temperature distribution above the finned tube bundle and for the static pressure distribution beneath the bundle. Michiyoshi et al. (1974) tested one plain rod and 28 transverse finned rods in an annulus. They were able to obtain correlations for both radiative and convective heat transfer as well as pressure drop. Their correlations included factors to account for fin height and fin spacing.

Stasiulevicius and Survila (1974) reported experimental results of heat transfer and what they referred to as *hydraulic drag* with regard to staggered banks of tubes with helical annular (radial) fins. Kuntysh et al. (1974) established a quantitative relationship between the rate of heat transfer from staggered tubes with circular (radial) fins in bundles as a function of interfin spacing. Legkii et al. (1974) used arrays of miniature heat flux sensors for measuring local heat transfer coefficients on the surface of a tube with constant-thickness annular (radial) fins of rectangular profile in cross flow. They showed that two radial heat transfer maxima occurred on the fin surfaces, one at the fin base and the other at the fin tip, and they claimed that the rate of heat transfer from the fin tip and the tube surface exceeded average heat transfer levels.

Zozulya and Khavin (1974) presented experimental data for heat transfer and friction loss in single and multirow bundles of large-diameter tubes with transverse (radial) welded fins. Senshu et al. (1979) presented a method for measuring the heat transfer coefficient with high accuracy for several types of fins. Their method involved unsteady heat flow in an air-conditioning apparatus between the airstream and cooler fins and tube.

Rabas and Eckels (1975) presented heat transfer and pressure drop data taken on seven segmented tube bundle arrangements. All tube bundles were three rows deep, and comparisons between pitch arrangements showed that the heat transfer and pressure drop performances were almost identical for staggered equilateral and diamond pitch. However, both the heat transfer and pressure drop decreased when changing from a staggered diamond to a square in-line pitch and from a square in-line pitch to a rectangular in-line pitch.

Watkinson et al. (1975) took heat transfer and pressure drop measurements on integral inner-fin tubes of several different designs using oil in laminar flow. They showed that at a Reynolds number of 500, heat transfer was enhanced over smooth tube values by 8 to 22%, depending on the tube geometry. At constant pumping power and the same Reynolds number, the increase in heat transfer ranged from 1 to 18%. Correlating equations for heat transfer and friction factor were presented for both straight and spiral arrangements.

Saboya and Sparrow (1976) performed experiments on a three-row finned tube exchanger, and Goldstein and Sparrow (1976) reported on experiments conducted on a corrugated fin-and-tube heat exchanger configuration. Legkii et al. (1976) looked at the distribution of the heat transfer coefficient on the lateral faces and ends of a circular (radial) fin using heat flux sensors.

Masliyah and Nandakumar (1976, 1977) also studied heat transfer in internally finned tubes. They provided numerical solutions of the energy equation and the equations of motion for fully develped Newtonian fluid flow in an internally finned helical coil for the case of uniform heat flux with peripherally uniform temperature for low Reynolds numbers. Plots of the axial velocity, secondary stream function, and fluid temperature were presented. It was shown that the Reynolds number–friction factor product was a function of the Dean number[5] and the radius of curvature, whereas the average Nusselt number is a function of the Dean number only. They also claimed that for a given fluid at a constant pressure drop, a finned tube gave lower values for the average Nusselt number than a finless tube with the same inside diameter and radius of curvature.

Agrawal and Bhavsar (1977) conducted an experimental investigation on the heat transfer characteristics for airflow across single finned cylinders. Their apparatus consisted of two physically identical finned cylinders of different materials: brass and aluminum. The fins were of the transverse rectangular spiral type and the results,

[5]The *Dean number* is equal to the Reynolds number multiplied by the square root of the ratio of the tube or pipe radius to its radius of curvature.

which were given in graphical form, indicated the influence of the airflow on the heat transfer rate.

Sparrow et al. (1978) studied the heat transfer in forced convection in a shrouded fin array and experimentally determined the effect of the shroud and shroud clearance. Gentry and Small (1978) obtained results for heat transfer and power expenditure (friction loss) for an agitated vessel equipped with bare tubes, low-fin tubes, and medium-fin tubes. Both heating and cooling data were obtained using water, ethylene glycol, and two petroleum oils as test fluids. The mixing Reynolds number range was from 55 to 200,000. Jones and Russell (1978) took measurements of the local fin temperature on an annular (radial) finned tube in cross flow. Various techniques were used and indicated large variations in the local fin temperature. Later, measurements of local heat transfer coefficients employed transient techniques. Patankar et al. (1979) conducted an experimental study pertaining to heat transfer in turbulent flow in internally finned tubes and channels.

Soliman (1979) gave an analytical solution for the temperature distribution and Nusselt number during fully developed laminar flow in internally finned tubes. It was obtained assuming axially uniform heat flux and circumferentially uniform temperature at the outer surface of the tube. Kubanek and Miletti (1979) took heat transfer and pressure drop measurments on three integral spiraled inner-fin tubes with two-phase flow of refrigerant R-22 under evaporating conditions.

A great deal of experimental work pertaining to heat transfer from and to extended surfaces has been conducted at the University of Minnesota by Sparrow and his colleagues (Patankar, Goldstein, and Bar-Cohen). Sparrow and Charmchi (1980) conducted experimental studies of heat transfer in laminar flow on externally finned tubes. This was followed by Sparrow et al. (1980), who used the napthalene sublimation technique to determine the row-by-row distribution of the heat transfer coefficient in experiments on in-line pin-fin arrays. It was shown that fully developed conditions prevailed for the fourth row and beyond, and in a comparison to staggered pin-fin arrays, it was shown that the heat transfer coefficients were, in general, lower for in-line array. However, the pressure drop for the in-line array was lower, as expected. The napthalene sublimation technique was also employed by Sparrow and Bahrami (1980) in an experimental study to determine the natural convection heat flow distribution on the faces of isothermal circumferential (radial) fins affixed to a horizontal tube. The mass transfer results so obtained were transformed to heat transfer results using the analogy between heat and mass transfer.

Sparrow and Semle (1981) measured heat transfer coefficients at and adjacent to the tip of a wall-attached cylinder in cross flow with application to fins and Webb (1981a,b) took heat transfer and flow friction measurements for finned tubes having plain fins. Sparrow and Becky (1981) pointed out that the design of energy-efficient heat exchange devices requires a knowledge of both pressure drop (and/or pumping power) and heat transfer characteristics. This inspired an experimental study on a configuration consisting of longitudinal fins attached to one of the proncipal walls of a flat duct.

Carnavos (1981) experimentally obtained the heat transfer performance in heating water using 11 different tubes having integral internal spiral and longitudinal fins.

Individual fin performance was presented and compared with a smooth tube datum at constant pumping power. Correlating equations were provided for heat transfer and friction data that were based on air, water, and ethylene glycol data. Webb and Scott (1980) performed a parametric analyis of the performance of internally finned tubes for heat exchanger applications, and Scott and Webb (1981) reported on the analytic prediction of the friction factor for internally finned channels. Their model assumed the validity of the "law of the wall" and applied a logarithmic velocity distribution to the interfin and core regions of the flow. Theoretically based friction factor equations were developed for internally finned circular tubes and rectangular channels. Their model predicts the Carnavos (1981) friction data for 21 internally finned tubes within a spread of 10%.

Van Fossen (1982) measured heat transfer coefficients on pin and end wall surfaces of several staggered arrays of short pin fins attached to the trailing edges of turbine blades. The study was conducted because short pin fins were often used to increase the heat transfer to the coolant, and due to the limitations of the casting technology, it was not possible expeditiously to manufacture fins of optimum height for fins in the trailing-edge region. In fact, in many cases, the fins were so short that the fins in this area suffered a decrease in heat transfer compared to a plane wall. When plotted against the Reynolds number, the measured Nusselt numbers fell on a single curve for all surfaces tested.

Sparrow and Larsen (1982) performed experiments to determine "per fin" heat transfer coefficients for pin-fin arrays exposed to a novel airflow arrangement. The flow entered the array longitudinally, but owing to the blocking action of the base surface, was forced to turn and exit from the configuration in a predominantly cross-flow configuration. Geometric parameters such as the fin height-to-diameter and the fin pitch-to-diameter ratios were varied and results were obtained for Reynolds numbers that ranged over an order of magnitude. Metzger and Haley (1982) also conducted and reported on experiments as well as flow visualization studies for arrays of short pin fins, and Patankar and Murthy (1982) reported on work pertaining to heat transfer from rotating cylinders equipped with circumferential (radial) fins.

Metzger et al. (1982a) examined and reported on an experimental study of the problem of developing heat transfer in rectangular ducts with staggered arrays of short pin fins. Metzger et al. (1982b) also reported on pressure loss and heat transfer through multiple rows of short pin fins.

Nakayama and Xu (1983) provided correlations for heat transfer and friction factors for what they claimed were enhanced fins for air coolers, and Said and Trupp (1983) published heat transfer data for turbulent flow in internally finned tubes. An analysis of laminar flow and heat transfer in tubes with internal circumferential (radial fins) was published by Rowley and Patankar (1983), and Ping (1983) provided additional insight into the heat transfer and pressure drop characteristics of a pin-fin cooling configuration. Arora and Messeh (1983) conducted heat transfer experiments in high-aspect-ratio rectangular channels with epoxied short pin fins, and Krause and Peters (1983) reported on a study of heat transfer from horizontally finned tubes in an air-fluidized bed of uniformly sized particles. A primary objective of this work was to gather heat transfer coefficient data for the finned case, which were then compared to data obtained from the bare horizontal tube in a similar flow environment.

Zhukauskas et al. (1983) reported on heat transfer and hydraulic resistance meaurements taken on spirally finned staggered (checkerboard) tube banks in transverse arrays. The empirical data obtained allowed for the influence of thermal properties of the fluid as a function of temperature and the geometry of the tube bank. Ning and Qilang (1983) conducted an experimental investigation on heat transfer in helically welded tube banks, and Metzger et al. (1983) reported on the heat transfer and flow friction characteristics of very rough transverse ribbed surfaces with and without pin fins.

Cooper and Buford (1984) made a comparison of 11 friction factor correlations on triangular pitch tube banks to determine the best method for calculating the pressure drop across the convection section of a fired heater. Sparrow and Kadle (1984) investigated the heat transfer of spadelike rods that extended outward from a rotating shaft. The experiment encompassed a number of geometrical parameters, including the length and diameter of the rods, the number of rods deployed around the circumference of the shaft, and the distance between the rods and the free end of the shaft. The rotational Reynolds nunber was also varied, and the Nusselt numbers were obtained on a rod-to-rod basis.

Bemrose and Bott (1984) deliberately fouled banks of spiral-wound fin tubes with calcium carbonate dust. They obtained friction factors as a function of time and gave curves of the friction factor to the unfouled friction factor ratio that were normalized on the basis of a constant Reynolds number. Brigham (1984) presented array-averaged heat transfer coefficients that were obtained for two configurations of short pin fins and for two flat-plate configurations in a converging channel. The purpose of the work was the determination of the effect of flow acceleration due to channel convergence and the effect of varying pin length on the heat transfer. Results were presented in the form of Nusselt number as a function of Reynolds number.

Brigham and Van Fossen (1984) measured array-averaged heat transfer coefficients on pin and end wall surfaces for two configurations of staggered pin fins. The work was found to be necessary because prior work concerning pin fins indicated that the amount of heat transfer from short pin fins was lower than from higher pin fins found in tube banks and other configurations. The study assessed the effect of the number of pin rows and the row geometry.

Pis'mennyi and Legkii (1984) provided results of experiments made for the purpose of determining heat transfer from staggered multirow bundles of tubes containing radial fins, and Metzger et al. (1984) experimentally studied two families of pin-fin array geometries that had potential for improving air foil internal cooling performance. One family utilized pins of cylindrical cross section (cylindrical spines) with various orientations of the array with respect to the mean flow direction. The second family was composed of fins with oblong cross section with various pin orientations with respect to the mean flow direction. Both heat transfer and pressure loss characteristics were presented, and it was observed that the use of circular pins with an array orientation somewhere between in-line and staggered can, in some cases, enhance the heat transfer while decreasing the pressure loss.

Yao (1984) also reported on the heat transfer and friction loss characteristics of pin-fin cooling configurations, and Hatada and Senshu (1984) performed an experimental study on the heat transfer characteritics of convex louvered fins for air-conditioning heat exchangers. The performance of one- and two-row plate fin heat exchangers was investigated by Rosman et al. (1984), and Baliga and Azrak (1984) reported on their studies of laminar fully developed flow and heat transfer in triangular plate fin ducts.

Azer and Sivakumar (1984) reported results of an experimental investigation in which heat transfer and pressure drop data were taken during saturated flow boiling inside four electrically heated tubes. One tube was smooth and the remaining three were internally finned with different diameters and fin geometries. The smooth tube was employed as the basis of comparison with the data for the finned tubes. It was found that over the mass flux range tested, an enhancement of the heat transfer coefficient as high as 146% over the smooth tube results on a nominal area basis was obtained. Correlation equations were proposed for predicting the overall heat transfer and pressure drop during saturated flow boiling inside internally finned tubes.

Prakash and Ronzoni (1985) numerically analyzed laminar fully developed flow in an internally finned vertical concentric circullar annular duct. The fins were radial and were affixed on the outside of the inner tube. The outer wall was insulated, while a uniform heat flux was provided on the inner wall. It was found that the resulting bouyancy increased both the heat transfer and friction. This effect, compared with respect to the no-buoyancy case, is stronger when the number of fins was small and when the fins were short.

Sparrow and Vemuri (1985) conducted experiments to determine the combined-mode natural convection/radiation heat transfer of highly populated arrays of rodlike cylindrical fins or spines. The fins were oriented with their axes horizontal and were attached to a heated base plate. The parameters investigated included the number of fins in the array, fin height and diameter, base plate-to-ambient temperature difference, and presence or absence of adcavent shrouded surfaces. Calculations showed that the contribution to the total heat transfer was substantial and was greatest for the more populous arrays, for higher fins, and at small temperature differences.

Eckels and Rabas (1985a,b) gave data on heat transfer and pressure drop in typical air-cooler finned tubes, and Leung et al. (1985a) reported on an experimental study that attempted to determine the optimum spacing of vertical rectangular fins in an array under natural convection conditions. This was followed by Leung et al. (1985b), who then examined the effect of protrusions on the thermal performance of vertically finned arrays.

Sparrow and Kang (1985) performed heat transfer and pressure drop experiments on cross-flow tube banks in which the individual tubes were equipped with longitudinal fins. The parameters investigated included placement of the fin at the front of the tube, at the rear, and at both the front and rear. Also included were the fin-tip shape, whether blunt or contoured, and the fin thickness. For each tube bank geometry, the Reynolds number was varied by nearly an order of magnitude. The results showed that a high degree of heat transfer enhancement can indeed be obtained with longitudinal fins in cross flow.

Sparrow and Myrum (1985) made the point that commercially available interrupted annular finned tubes possessed highly specific and often irregular geometries. They fabricated an annular (radially) finned tube of regular and unambiguous geometry and made several tests. The first used the finned tube without radial interruptions to determine a set of baseline heat transfer coefficients. Subsequently, radial interruptions were cut into the fins and the effect of interruptions on the heat transfer were quantified.

Kadle and Sparrow (1986) investigated heat transfer from an array of parallel longitudinal fins to a turbulent airstream passing through the interfin spaces analytically, numerically, and experimentally. The fins were integrally attached to a heated base plate, while the fin tips were shrouded to avoid leakage. In the analytical/numerical work, a conjugate problem was solved. The solution encompassed turbulent flow and heat transfer in the airstream and heat conduction in the fin and base plate. The turbulent model and the computational scheme wre verified by comparison with experimental data. It was found that the local heat transfer coefficient varied along the fin and along the surface of the base plate, with the lowest values in the corners formed by the fin–base plate and fin–shroud intersections.

Sparrow et al. (1986) performed experiments to determine the heat transfer characteristics of a horizontal finned tube situated in a vertical channel that was open to the ambient at the top and bottom. The heat transfer from the finned tube was by natural convection and radiation, and the performance of the configuration was determined as a function of three geometric parameters: the vertical position of the tube in the channel, the clearance between the fin tips and channel walls, and the height of the channel. Experiments were also carried out with the finned tube situated in free space, and it was found that the in-channel positioning of the tube gives substantially higher heat transfer rates.

A method for handling the variation of a heat transfer coefficient using the cascade algorithm of Kraus et al. (1978) is presented in Chapter 7. There it was confirmed that the cascade algorithm could be used to verify tha analytical method proposed by Han and Lefkowitz (1960). There are, of course, significantly more references to the case of the variable heat transfer coefficient in the literature.

Harris and Wilson (1961) considered the variation of the heat flux (from which the variation of h could easily be deduced) along the height of the longitudinal fin of rectangular profile with cross-flow forced convection. Although this study was for a diffferent situtation than the one described by Ghai (1951), it was shown that the assumption of a constant h value was substantially in error.

Lubny-Gerczyk and Bobkova (1963) looked at various geometries of fins and spines with varying distributions of h, and Advani and Sukhatme (1973) looked at the problem of a vertical fin losing heat by laminar free convection. In their study they employed a simultaneous solution of the flow and energy equations at the surface and the heat conduction equation inside the fin. This procedure eliminated the need for assumption of a constant heat transfer coefficient on the fin faces.

Garg and Velusamy (1986) described a method for simultaneous solution of the convective heat transfer problem to a flowing fluid and the conduction problem within a plate fin. Their motivation came from a desire to enhance an analysis carried out for a Prandtl number equal to 0.7.

The technical literature dealing, in particular, with electronic equipment cooling is growing rapidly at the present time. For users of this book with special needs that were not addressed here, the current technical literature must be examined. A good starting point is the review article by Incropera (1988), which includes a comprehensive review of the pertinent literature through the mid-1980s. Moreover, Chapters 11 and 12 in this book deal with finned heat exchangers, and additional correlations for heat transfer and pressure drop that are germaine to heat exchangers are considered.

In Chapter 10, which deals with compact heat exchangers, Kays and London (1984) are cited for providing heat transfer and friction data for 85 plate, louvered, strip, wavy fin, and perforated fin surfaces as well as circular tubes with circular and continuous fins and flat tubes with continuous fins. The heat transfer data are presented in terms of a j-factor that represents a Stanton number correlation, and the friction data provides a friction factor to be used in a pressure loss equation that includes entrance, exit, acceleration, and friction losses. Additional compact heat exhanger correlations are provided in Chapter 20.

Chapter 11 considers longitudinal fin double–pipe heat exchangers. Once again, the Sieder–Tate (1936), Hausen (1943) and Dittus–Boelter correlations are cited although the examples make use of a correlation due to Gnielinski (1976) which is employed for both tubes and the annular region for both heat transfer and friction loss.

Radial fin heat exchangers with transverse high fins are considered in Chapter 12. The correlations of Briggs and Young (1963), Vampola (1966), and Ganguli et al. (1985) are provided for flow over the outside of the high–fin tubes. In an example pertaining to the design of an air–fin cooler, the Briggs and Young (1963) and Gnielinski (1985) correlations are employed for the air-fin side and the tube side respectively and the Robinson and Briggs (1966) correlation is employed for the pressure loss on the air-fin side.

4.7 NOMENCLATURE

Roman Letter Symbols

A	cross-sectional area, m^2
a	channel aspect ratio, dimensionless
B	constant, dimensionless
b	fin height, m
C	constant, dimensionless
c	specific heat, J/kg · K
d	diameter, m; derivative or differential, dimensionless
El	Elenbaas number, dimensionless
f	function, dimensions vary
$f(\text{Pr})$	universal Prandtl number function, dimensionless
G	body gravity function, dimensionless
g	acceleration of gravity, dimensionless
H	channel or cuboid dimension, m
h	heat transfer coefficient, $W/m^2 \cdot K$

k	thermal conductivity, W/m · K
L	length, cuboid or channel height, m
Nu	Nusselt number, dimensionless
P	perimeter, m
Pr	Prandtl number, dimensionless
p	pressure, N/m^2 (Pa)
q	heat flow, W
q''	heat flux, W/m^2
R	radius function, dimensionless
Ra	Rayleigh number, dimensionless
Re	Reynolds number, dimensionless
r	radial coordinate, m; radius, m; hydraulic radius, m
S	surface area, m^2
$\sqrt{S}$	square root of surface area, m
St	Stanton number, dimensionless
T	temperature, K
$\dot{V}$	volumetric flow rate, m^3/s
v	velocity, m/s
W	channel or cuboid dimension, m
x	length dimension, m
y	length dimension, m
z	clear space between fins, m

Greek Letter Symbols

α	thermal diffusivity, m^2/s
β	coefficient of volumetric thermal expansion, K^{-1}
Δ	change in, dimensionless
θ	temperature excess, K
Λ	combination of terms, dimensions vary
λ	profile indicator, dimensionless
μ	dynamic viscosity, kg/m ·s
ν	kinematic viscosity, m^2/s
ρ	hspace.25in & density, kg/m^3
τ	shear stress, N/m^2
ψ	geometric parameter, dimensionless

Roman Letter Subscripts

d	diameter basis
e	entrance length; exit
fd	fully developed
i	inlet
L	based on length
lam	laminar
max	maximum
r	based on hydraulic radius

$\sqrt{S}$	square root of surface
s	surface
T	transverse
t	thermal
turb	turbulent
x	based on x
z	based on fin spacing

Symbolic Subscript

∞	free stream

Roman Letter Superscripts

m	exponent
n	exponent

Symbolic Superscripts

$*$	modified
$''$	heat flux
∞	diffusive limit

CHAPTER 5

LINEAR TRANSFORMATIONS

5.1 INTRODUCTION

Interest in the efficiency of finned passages (which, in reality, are finned arrays) for convection heat transfer was stimulated by the availability of the double-pipe heat exchanger, with its low cost per unit heat transfer surface. This exchanger consists of two concentric pipes with longitudinal fins of rectangular profile bonded to the outside of the inner pipe. This produces a finned annular passage with heat flow from the inner pipe and fins. It was apparent that if the fins were bonded to both the outside of the inner pipe and the inside of the outer pipe, the inside surface of the outer pipe could also contribute to the total heat transfer surface. The heat loss from the outer pipe to the surroundings could be made negligible by the application of insulation to the exterior of the outer pipe.

Kraus (1960) developed an expression for the efficiency of such a configuration, which was later adapted by Kraus (1961) for a forced-convection-cooled electronic chassis or cold plate containing longitudinal fins of rectangular profile. The cross sections of the fins in the array were optimized by Kraus (1962).

The growing availability of compact heat exchanger cores, as disclosed by Kays and London (1984, and in their earlier editions in 1955 and 1964), stimulated interest in the fin efficiency of stacked heat exchanger cores. Kays (1960) treated as stack components two- and three-layer cores, termed a *double* and *triple sandwich*, and developed the efficiency for equal input on both exterior plates. Kern and Kraus (1972) looked at single, double, and triple stacks[1] with heat input on one side and heat input on two sides.

[1] *Stacks*, rather than *sandwiches* was the nomenclature used.

Kraus et al. (1978) showed that for longitudinal fins of rectangular, trapezoidal, and triangular profiles, and in *finned arrays* (not arrays of fins) composed of these fins, conditions of heat flow and temperature excess (relative to the presumed constant and uniform temperature environment) at any point on a fin are induced by similar conditions at the fin base. In particular, it was shown that there was a *linear transformation* that mapped conditions from the fin base to conditions at the fin tip, and vice versa. They defined a **T**-matrix that mapped the heat flow and temperature excess at the fin tip to those conditions at the fin base and claimed that this was a new parameterization that would completely describe the performance of an individual fin. The conventional fin efficiency was abandoned and it was proposed that single fins could be characterized by a single, yet important parameter, the *thermal transmission ratio* or *input admittance*.

Development of the thermal transmission matrix **T** and thermal transmission ratio μ is the subject of this chapter. These two quantities are the parameters to be used in the analysis of individual fins and finned arrays. Algorithms for the assembly of individual fins into finned arrays are treated in Chapter 7. Here, study of linear transformations for individual fins begins with a consideration of the limitations of the fin efficiency.

5.2 LIMITATIONS OF FIN EFFICIENCY

There is a delicate interplay between conduction and convection in a single extended surface or fin and, as shown in Section 1.2.1, an inescapable loss of performance due to the negative temperature gradient that occurs within the fin. This loss of performance is handled analytically through use of a design parameter called the *fin efficiency*, defined as the ratio of actual heat dissipated by the fin to the ideal heat dissipated if the entire fin were to operate at the fin base temperature:

$$\eta = \frac{q}{q_{\text{id}}}$$

For example, for the longitudinal fin of rectangular profile shown in Fig. 5.1, the fin efficiency, as derived in Section 1.3.2, is given by eq. (1.11), repeated here:

$$\eta = \frac{\tanh mb}{mb} \tag{5.1}$$

The general idea of an efficiency as a performance parameter is sound; it is a dimensionless ratio comparing performance with a certain standard. However, the particular way that efficiency has been defined for fins compares every fin with a different standard; the actual fin performance as compared with what a geometrically identical fin could do if conditions were perfect. Two fins of different dimensions in the same environment may have the same efficiency but they may transmit different quantities of heat. A simple example, based on a real-world optimization, can show that one fin can transmit more heat than another under identical environmental

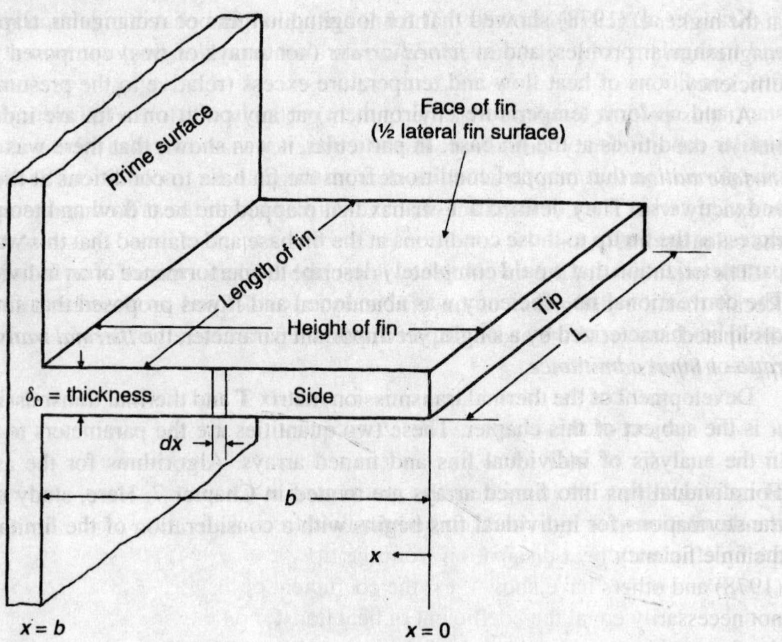

Figure 5.1 Terminology and coordinate system, longitudinal fin of rectangular profile.

conditions and operate at a lower value of fin efficiency. This is vividly demonstrated in the following paragraphs.

Using the terminology in Fig. 5.1, one may consider an aluminum longitudinal fin of rectangular profile ($k = 202$ W/m · K) with a base temperature of 200°C, dissipating to an environment at 100°C under natural convection conditions where h is taken as $h = 10$ W/m^2 · K. The fin dimensions are height, $b = 10$ cm; thickness, $\delta = 0.2286$ cm; and length $L = 25$ cm. Using eq. (1.10), repeated, yields

$$q_b = (2hk\delta)^{1/2} L\theta_b \tanh mb \tag{5.2}$$

with $\theta_b = 200 - 100 = 100$°C, the heat dissipation may be computed as 43.85 W. Then, using eq. (5.1), the efficiency may be computed as $\eta = 0.877$.

In a desire to save mass, it is proposed that a magnesium fin ($k = 148$ W/m·K) with the same length and thickness dimensions operating at otherwise identical thermal conditions be employed. Using a dissipation of 43.85 W, one may solve eq. (5.2) to obtain $b = 10.64$ cm, which shows that because of its poorer conductivity, the height of the magnesium fin must be increased in order to accommodate an identical dissipation requirement. For the magnesium fin, the efficiency, computed from eq. (5.1). is $\eta = 0.824$. Observe that the magnesium fin. which has a huge advantage in a weight optimization, dissipates the *same* quantity of heat but at a *lower* fin efficiency.

If the heat transfer coefficient is reduced to, say, $h = 8$ W/m^2 · K, eqs. (5.1) and (5.2) show that the magnesium fin (with the greater fin height) will dissipate 36.30 W at an efficiency of 0.853, whereas the aluminum fin will dissipate less heat, 35.94 W, at the higher efficiency of 0.899. The fins are no longer behaving identically, and the magnesium fin is clearly outperforming its aluminum *design-point twin* at a lower efficiency.

At this point it is noted that the fault is probably not in the efficiency concept but in the efficiency definition. If the efficiency had originally been called the *surface utilization factor* (which as demonstrated in Example 1.1 in Section 1.3.2 is exactly what it is), it might never have entered either the preceding discussion or the calculations.

The usefulness of the fin efficiency is debatable when a finned array is considered. Recall that the definition of fin efficiency, also defined as the ratio of the heat dissipated, say q_1, to the heat dissipated if the fin were entirely at the base temperature $q_{id} = hS\theta_b$, is

$$\eta = \frac{q_1}{q_{\text{id}}} \tag{5.3}$$

For a single fin without tip heat dissipation, $q_1 = q_b$, and eq. (5.3), consistent with the development that is based on no heat flow from the fin tip, results. However, for a single fin with heat dissipation from the tip, $q_a = h_a A_a$, where as Sparrow et al. (1978) and others have shown, h_a, the coefficient of heat transfer at the fin tip, does not necessarily equal the coefficient of heat transfer on the fin faces. In this case, A_a, the fin cross-sectional area at the fin tip, is equal to the fin surface area at the fin tip. But to be sure, the entire convective dissipation (dissipation from the fin faces and the fin tip) passes through the base of the fin. Thus, with $q_1 = q_b$, the efficiency of the fin is also given by eq. (5.3).

But in a finned array, heat loss from an individual fin includes both convection to the surrounding environment and conduction from its tip to the contiguous fin or a cluster of fins located downstream, q_a. Consequently, the convective heat dissipation is $q_1 = q_b - q_a$, and the fin efficiency becomes

$$\eta = \frac{q_1}{q_{\text{id}}} = \frac{q_b - q_a}{q_{\text{id}}}$$

It is observed that the efficiency of a fin in a finned array depends not only on the heat transfer condition h, the fin thermal conductivity k, and the fin dimensions L, b, and δ, but also on where the fin is mounted in the array. Thus, while the use of the fin efficiency in obtaining an overall passage efficiency is very useful in finned passage analysis and heat exchanger design, it is felt that this is a major inadequacy of the concept of fin efficiency when dealing with finned arrays and justifies development of an alternative approach.

5.3 LONGITUDINAL FIN OF RECTANGULAR PROFILE REVISITED

Once again, consider Fig. 5.1, which displays a longitudinal fin of rectangular profile. The governing differential equation for the point-to-point temperature excess along the fin given in Section 1.3.2 as eq. (1.7) is repeated here:

$$\frac{d^2\theta}{dx^2} - m^2\theta = 0 \tag{5.4}$$

This equation was solved using *boundary value data*, that is, the temperature excess at the fin base,

$$\theta(x = b) = \theta_b$$

and the heat flow at the fin tip,

$$q(x = 0) = 0$$

which led to

$$\left.\frac{d\theta}{dx}\right|_{x=0} = 0$$

The general differential equation may also be solved using *initial value data*, which involve conditions of temperature excess and heat flow at the same point on the fin. If this point is taken at the fin base, the initial conditions are

$$\theta(x = b) = \theta_b$$

and

$$q(x = b) = q_b$$

Use of these conditions with the general solution of eq. (5.4),

$$\theta = C_1 e^{mx} + C_2 e^{-mx}$$

and its derivative,

$$\frac{d\theta}{dx} = m\left(C_1 e^{mx} - C_2 e^{-mx}\right)$$

yields a pair of simultaneous linear algebraic equations in the arbitrary constants C_1 and C_2 which can be written in matrix form as

$$\begin{bmatrix} e^{mb} & e^{-mb} \\ kAme^{mb} & -kAme^{-mb} \end{bmatrix} \begin{bmatrix} C_1 \\ C_2 \end{bmatrix} = \begin{bmatrix} \theta_b \\ q_b \end{bmatrix}$$

The solution for C_1 and C_2 is obtained expeditiously using a matrix inversion.[2] With the fin cross-sectional area $A = \delta L$,

$$C_1 = \frac{k\delta L e^{-mb}\theta_b + e^{-mb} q_b}{2k\delta L m}$$

[2]Recall that the inverse of a 2 by 2 matrix is easily obtained by swapping the elements of the principal diagonal, changing the sign of the off-diagonal elements, and then dividing all elements by the determinant of the matrix. Appendix B is a brief introduction to matrices and some of the basic operations concerning matrices.

and

$$C_2 = \frac{k\delta L e^{mb}\theta_b - e^{mb} q_b}{2k\delta Lm}$$

Substitution of these values into the general solution gives

$$\theta(x) = \left(\frac{e^{-mb}}{2}\theta_b + \frac{e^{-mb}}{2k\delta Lm} q_b\right) e^{mx} + \left(\frac{e^{mb}}{2}\theta_b - \frac{e^{mb}}{2k\delta Lm} q_b\right) e^{-mx}$$

or upon rearrangement,

$$\theta(x) = \frac{e^{m(b-x)} + e^{-m(b-x)}}{2}\theta_b - \frac{1}{k\delta Lm}\frac{e^{m(b-x)} - e^{m(b-x)}}{2} q_b$$

Observe that it has been possible to separate the effects of base temperature excess and base heat flow on the fin temperature excess at any point x. Noting the importance of the fin parameter, $k\delta Lm = (2hk\delta)^{1/2}L$, it is convenient to define the *characteristic admittance of the fin* as

$$Y_o \equiv (2hk\delta)^{1/2} L \tag{5.5}$$

Thus the temperature excess at any point on the fin may be written in terms of the hyperbolic functions

$$\theta(x) = [\cosh m(b-x)]\theta_b - \frac{1}{Y_o}\sinh m(b-x) q_b \tag{5.6}$$

and a simple differentiation yields the heat flow,

$$q(x) = k\delta L \frac{d\theta(x)}{dx} = -[Y_o \sinh m(b-x)]\theta_b + [\cosh m(b-x)]q_b \tag{5.7}$$

Observe in the foregoing that at $x = b$, eq. (5.6) gives, as it should, $\theta(x = b) = \theta_b$, and eq. (5.7) reduces, as it should, to $q = q_b$.

The temperature excess and heat flow at the tip of the fin can be determined by merely letting $x = 0$ in eqs. (5.6) and (5.7):

$$\theta(x = 0) = \theta_a = (\cosh mb)\,\theta_b - \left(\frac{1}{Y_o}\sinh mb\right) q_b \tag{5.8}$$

and

$$q(x = 0) = q_a = -(Y_o \sinh mb)\,\theta_b + (\cosh mb)\, q_b \tag{5.9}$$

Equations (5.8) and (5.9) are very significant to an understanding of fin analysis because they lead to three important conclusions, one of which is intuitively physical and two of which are mathematical:

- Aside from the physical dimensions and the two important heat transfer properties h and k, conditions of temperature excess and heat flow at the fin tip can be said to be induced by conditions of temperature excess and heat flow at the fin base. This, of course, is intuitively true physically.

- Conditions of temperature excess and heat flow at the fin tip and indeed, as shown by eqs. (5.6) and (5.7), at any point on the fin can be determined as a superposition of the effects of temperature excess and heat flow at the fin base. This is to be expected because the differential equation for temperature excess is a linear differential equation.
- There is a linear transformation relating conditions of temperature excess and heat flow at the fin tip and indeed at any point on the fin with conditions of temperature excess and heat flow at the fin base.

5.4 LINEAR TRANSFORMATION

The linear transformation suggested by eqs. (5.8) and (5.9) may be written in matrix form as

$$\begin{bmatrix} \theta_a \\ q_a \end{bmatrix} = \Gamma \begin{bmatrix} \theta_b \\ q_b \end{bmatrix} = \begin{bmatrix} \cosh mb & -Z_o \sinh mb \\ -Y_o \sinh mb & \cosh mb \end{bmatrix} \begin{bmatrix} \theta_b \\ q_b \end{bmatrix} \tag{5.10}$$

where, as defined by eq. (5.5), $Y_o = (2hk\delta)^{1/2}L$. The *characteristic impedance* Z_o is the reciprocal of the characteristic admittance:

$$Z_o = \frac{1}{Y_o} = \frac{1}{(2hk\delta)^{1/2}L}$$

The Γ matrix in eq. (5.10) containing the hyperbolic functions is the linear transformation matrix that results from the solution of eq. (5.4) as an initial value problem. In eq. (5.10) the independent variables are θ_b and q_b, and the dependent variables are θ_a and q_a. Attention is often focused on a different set of dependent and independent variables. Thus, other linear transformations are useful and can readily be obtained.

5.5 OTHER LINEAR TRANSFORMATIONS

The general problem concerns the linear transformation from two variables to another two and it is seen that four variables taken two at a time yields six combinations. Thus, six linear transformations are required to characterize completely *all* of the combinations of temperature excesses and heat flows at both fin tip and fin base. The general formulation involves the matrix representation

$$\begin{bmatrix} \text{dep}_1 \\ \text{dep}_2 \end{bmatrix} = \begin{bmatrix} \alpha_{11} & \alpha_{12} \\ \alpha_{21} & \alpha_{22} \end{bmatrix} \begin{bmatrix} \text{indep}_1 \\ \text{indep}_2 \end{bmatrix}$$

where *dep* and *indep* stand for dependent and independent variables, respectively.

Here, the α-matrix represents a linear combination of variables, as a simple matrix multiplication clearly shows:

$$\text{dep}_1 = \alpha_{11}[\text{indep}_1] + \alpha_{12}[\text{indep}_2]$$

and

$$\text{dep}_2 = \alpha_{21}[\text{indep}_1] + \alpha_{22}[\text{indep}_2]$$

Here the α's may take different forms, depending on which two variables are selected as independent variables.

Table 5.1 presents a summary of the six possible combinations of dependent and independent variables. Observe that the choice of parameters shown in the table is clearly a function of the variables chosen as independent variables. This, in turn, may depend on which variables are easily measured or calculated and possibly on how individual fins may be connected with other fins to form a finned array.

5.5.1 Z and Y Matrices

The z-parameters are assembled into an impedance matrix $\mathbf{Z}$, and these z-parameters relate the dependent variables of temperature excess at fin base and fin tip, θ_b and θ_a, to the independent variables of base and tip heat flows. This is case 1 in Table 5.1 and the matrix representation is

$$\begin{bmatrix} \theta_b \\ \theta_a \end{bmatrix} = \mathbf{Z} \begin{bmatrix} q_b \\ q_a \end{bmatrix} = \begin{bmatrix} z_{11} & z_{12} \\ z_{21} & z_{22} \end{bmatrix} \begin{bmatrix} q_b \\ q_a \end{bmatrix} \tag{5.11}$$

The y-parameters are assembled into an admittance matrix $\mathbf{Y}$, and these y-parameters, also shown in Table 5.1 (case 2), relate the base and tip heat flows, q_b and q_a, as dependent variables to the base and tip temperature excesses, θ_b and θ_a, as independent variables. The matrix representation is

$$\begin{bmatrix} q_b \\ q_a \end{bmatrix} = \mathbf{Y} \begin{bmatrix} \theta_b \\ \theta_a \end{bmatrix} = \begin{bmatrix} y_{11} & y_{12} \\ y_{21} & y_{22} \end{bmatrix} \begin{bmatrix} \theta_b \\ \theta_a \end{bmatrix} \tag{5.12}$$

If eq. (5.12) is put into eq. (5.11), the result is

$$\begin{bmatrix} \theta_b \\ \theta_a \end{bmatrix} = \begin{bmatrix} z_{11} & z_{12} \\ z_{21} & z_{22} \end{bmatrix} \begin{bmatrix} y_{11} & y_{12} \\ y_{21} & y_{22} \end{bmatrix} \begin{bmatrix} \theta_b \\ \theta_a \end{bmatrix}$$

which shows that the z-parameter matrix is the inverse of the y-parameter matrix

TABLE 5.1 Summary of Linear Transformations

Case	dep_1	dep_2	indep_1	indep_2	α_{11}	α_{12}	α_{21}	α_{22}	Matrix
1	θ_b	θ_a	q_b	q_a	z_{11}	z_{12}	z_{21}	z_{22}	$\mathbf{Z}$
2	q_b	q_a	θ_b	θ_a	y_{11}	y_{12}	y_{21}	y_{22}	$\mathbf{Y}$
3	θ_b	q_b	θ_a	q_a	τ_{11}	τ_{12}	τ_{21}	τ_{22}	$\mathbf{T}$
4	θ_a	q_a	θ_b	q_b	γ_{11}	γ_{12}	γ_{21}	γ_{22}	$\mathbf{\Gamma}$
5	θ_b	q_a	q_b	θ_a	h_{11}	h_{12}	h_{21}	h_{22}	$\mathbf{H}$
6	q_b	θ_a	θ_b	q_a	g_{11}	g_{12}	g_{21}	g_{22}	$\mathbf{G}$

$$\mathbf{Z} = \mathbf{Y}^{-1} \tag{5.13a}$$

and

$$\mathbf{Y} = \mathbf{Z}^{-1} \tag{5.13b}$$

Equations (5.13) form the basis for conversion from the z-parameters to the y-parameters, and vice versa. With the determinants of $\mathbf{Z}$ and $\mathbf{Y}$ designated as Δ_Z and Δ_Y, respectively, this procedure yields

$$z_{11} = \frac{y_{22}}{\Delta_Y} \tag{5.14a}$$

$$z_{12} = -\frac{y_{12}}{\Delta_Y} \tag{5.14b}$$

$$z_{21} = -\frac{y_{21}}{\Delta_Y} \tag{5.14c}$$

and

$$z_{22} = \frac{y_{11}}{\Delta_Y} \tag{5.14d}$$

and a similar procedure gives the y's in terms of the z's:

$$y_{11} = \frac{z_{22}}{\Delta_Z} \tag{5.15a}$$

$$y_{12} = -\frac{z_{12}}{\Delta_Z} \tag{5.15b}$$

$$y_{21} = -\frac{z_{21}}{\Delta_Z} \tag{5.15c}$$

and

$$y_{22} = \frac{z_{11}}{\Delta_Z} \tag{5.15d}$$

5.5.2 T and Γ Matrices

The τ- and γ-parameters differ from the z- and y-parameters in that they relate the temperature excess and heat flows at either the base or tip to the temperature excesses and heat flow at the tip or base. These are represented as cases 3 and 4 in Table 5.1, and their matrix representations are

$$\begin{bmatrix} \theta_b \\ q_b \end{bmatrix} = \mathbf{T} \begin{bmatrix} \theta_a \\ q_a \end{bmatrix} = \begin{bmatrix} \tau_{11} & \tau_{12} \\ \tau_{21} & \tau_{22} \end{bmatrix} \begin{bmatrix} \theta_a \\ q_a \end{bmatrix} \tag{5.16}$$

and

$$\begin{bmatrix} \theta_a \\ q_a \end{bmatrix} = \mathbf{\Gamma} \begin{bmatrix} \theta_b \\ q_b \end{bmatrix} = \begin{bmatrix} \gamma_{11} & \gamma_{12} \\ \gamma_{21} & \gamma_{22} \end{bmatrix} \begin{bmatrix} \theta_b \\ q_b \end{bmatrix} \tag{5.17}$$

Equation (5.16) is very significant and it forms an important cornerstone for the combination of individual fins into finned arrays. It is the *matrix of the linear transformation* that maps fin-tip conditions into fin-base conditions.

If eq. (5.17) is put into eq. (5.16), the result is

$$\begin{bmatrix} \theta_b \\ q_b \end{bmatrix} = \begin{bmatrix} \tau_{11} & \tau_{12} \\ \tau_{21} & \tau_{22} \end{bmatrix} \begin{bmatrix} \gamma_{11} & \gamma_{12} \\ \gamma_{21} & \gamma_{22} \end{bmatrix} \begin{bmatrix} \theta_b \\ q_b \end{bmatrix}$$

which shows that the τ-parameter matrix is the inverse of the γ-parameter matrix

$$\mathbf{T} = \mathbf{\Gamma}^{-1} \tag{5.18a}$$

and

$$\mathbf{\Gamma} = \mathbf{T}^{-1} \tag{5.18b}$$

Equations (5.18) form the basis for the conversion from the τ-parameters to the γ-parameters, and vice versa. With the determinants of **T** and **Γ** designated as Δ_T and Δ_Γ, respectively,

$$\tau_{11} = \frac{\gamma_{22}}{\Delta_\Gamma} \tag{5.19a}$$

$$\tau_{12} = -\frac{\gamma_{12}}{\Delta_\Gamma} \tag{5.19b}$$

$$\tau_{21} = -\frac{\gamma_{21}}{\Delta_\Gamma} \tag{5.19c}$$

and

$$\tau_{22} = \frac{\gamma_{11}}{\Delta_\Gamma} \tag{5.19d}$$

one may work in the opposite direction to obtain

$$\gamma_{11} = \frac{\tau_{22}}{\Delta_T} \tag{5.20a}$$

$$\gamma_{12} = -\frac{\tau_{12}}{\Delta_T} \tag{5.20b}$$

$$\gamma_{21} = -\frac{\tau_{21}}{\Delta_T} \tag{5.20c}$$

and

$$\gamma_{22} = \frac{\tau_{11}}{\Delta_T} \tag{5.20d}$$

5.5.3 H and G Matrices

The *h*- and *g*-parameters shown in Table 5.1 as cases 5 and 6 bear no resemblance to cases 1 through 4. They are called *hybrids* because they freely mix both heat flows,

q_b and q_a, and both temperature excesses, θ_b and θ_a, without regard to consistency at either fin base or fin tip. Their matrix representations are

$$\begin{bmatrix} \theta_b \\ q_a \end{bmatrix} = \mathbf{H} \begin{bmatrix} q_b \\ \theta_a \end{bmatrix} = \begin{bmatrix} h_{11} & h_{12} \\ h_{21} & h_{22} \end{bmatrix} \begin{bmatrix} q_b \\ \theta_a \end{bmatrix} \tag{5.21}$$

and

$$\begin{bmatrix} q_b \\ \theta_a \end{bmatrix} = \mathbf{G} \begin{bmatrix} \theta_b \\ q_a \end{bmatrix} = \begin{bmatrix} g_{11} & g_{12} \\ g_{21} & g_{22} \end{bmatrix} \begin{bmatrix} \theta_b \\ q_a \end{bmatrix} \tag{5.22}$$

If eq. (5.22) is put into eq. (5.21), the result is

$$\begin{bmatrix} \theta_b \\ q_a \end{bmatrix} = \begin{bmatrix} h_{11} & h_{12} \\ h_{21} & h_{22} \end{bmatrix} \begin{bmatrix} g_{11} & g_{12} \\ g_{21} & g_{22} \end{bmatrix} \begin{bmatrix} \theta_b \\ q_a \end{bmatrix}$$

which shows that the h-parameter matrix is the inverse of the g-parameter matrix,

$$\mathbf{H} = \mathbf{G}^{-1} \tag{5.23a}$$

and

$$\mathbf{G} = \mathbf{H}^{-1} \tag{5.23b}$$

Equations (5.24) form the basis for the conversion from h-parameters to g-parameters, and vice versa. With the determinants of **Z** and **Y** designated as Δ_{H} and Δ_{G}, respectively,

$$h_{11} = \frac{g_{22}}{\Delta_{\mathrm{G}}} \tag{5.24a}$$

$$h_{12} = -\frac{g_{12}}{\Delta_{\mathrm{G}}} \tag{5.24b}$$

$$h_{21} = -\frac{g_{21}}{\Delta_{\mathrm{G}}} \tag{5.24c}$$

and

$$h_{22} = \frac{g_{11}}{\Delta_{\mathrm{G}}} \tag{5.24d}$$

and working in reverse,

$$g_{11} = \frac{h_{22}}{\Delta_{\mathrm{H}}} \tag{5.25a}$$

$$g_{12} = -\frac{h_{12}}{\Delta_{\mathrm{H}}} \tag{5.25b}$$

$$g_{21} = -\frac{h_{21}}{\Delta_{\mathrm{H}}} \tag{5.25c}$$

and

$$g_{22} = \frac{h_{11}}{\Delta_H} \tag{5.25d}$$

5.5.4 Relationships between Parameters

Each of the six sets of parameters can be represented in terms of the other five. This section will show how all parameters may be represented in terms of the z-parameters and then to provide a tabulation of all possible conversions.

Begin with the defining equations for the z, τ and h-parameters

$$\theta_b = z_{11} q_b + z_{12} q_a \tag{a}$$

$$\theta_a = z_{21} q_b + z_{22} q_a \tag{b}$$

$$\theta_b = \tau_{11} \theta_a + \tau_{12} q_a \tag{c}$$

$$q_b = \tau_{21} \theta_a + \tau_{22} q_a \tag{d}$$

$$\theta_b = h_{11} q_b + h_{12} \theta_a \tag{e}$$

$$q_a = h_{21} q_b + h_{22} \theta_a \tag{f}$$

The y-parameters have already been established in terms of the z-parameters by eqs. (2.13). These equations may be summarized by an equality of two matrices:

$$\mathbf{Y} = \begin{bmatrix} y_{11} & y_{12} \\ y_{21} & y_{22} \end{bmatrix} = \begin{bmatrix} z_{22}/\Delta_Z & -z_{12}/\Delta_Z \\ -z_{21}/\Delta_Z & z_{11}/\Delta_Z \end{bmatrix} \tag{5.26}$$

To find the τ-parameters in terms of the z-parameters, begin with eq. (b) and solve for q_b:

$$q_b = \frac{1}{z_{21}} \theta_a - \frac{z_{22}}{z_{21}} q_a \tag{g}$$

Then make a term-by-term comparison of eqs. (d) and (g) to obtain

$$\tau_{21} = \frac{1}{z_{21}}$$

and

$$\tau_{22} = -\frac{z_{22}}{z_{21}}$$

Next, put eq. (g) into eq. (a) to obtain

$$\theta_b = z_{11} \left(\frac{1}{z_{21}} \theta_a - \frac{z_{22}}{z_{21}} q_a \right) + z_{12} q_a$$

$$= \frac{z_{11}}{z_{21}} \theta_a + \left(z_{12} - \frac{z_{11} z_{22}}{z_{21}} \right) q_a$$

$$= \frac{z_{11}}{z_{21}}\theta_a - \frac{\Delta_Z}{z_{21}}q_a$$

where $\Delta_Z = z_{11}z_{22} - z_{12}z_{21}$. Then a term-by-term comparison with eq. (c) shows that

$$\tau_{11} = \frac{z_{11}}{z_{21}}$$

and

$$\tau_{12} = -\frac{\Delta_Z}{z_{21}}$$

These may be summarized by the matrix representation

$$\mathbf{T} = \begin{bmatrix} \tau_{11} & \tau_{12} \\ \tau_{21} & \tau_{22} \end{bmatrix} = \begin{bmatrix} z_{11}/z_{21} & -\Delta_Z/z_{21} \\ 1/z_{21} & -z_{22}/z_{11} \end{bmatrix} \tag{5.27}$$

The elements of $\mathbf{\Gamma}$ in terms of the z-parameters may be obtained by taking the inverse of eq. (5.27). First obtain det $\mathbf{T}$:

$$\det \mathbf{T} = \frac{1}{(z_{21})^2}(\Delta_Z - z_{11}z_{22})$$

$$= \frac{1}{(z_{21})^2}(z_{11}z_{22} - z_{12}z_{21} - z_{11}z_{22})$$

$$= -\frac{z_{12}z_{21}}{(z_{21})^2} = -\frac{z_{12}}{z_{21}}$$

Then

$$\mathbf{\Gamma} = -\frac{z_{21}}{z_{12}}\begin{bmatrix} -z_{22}/z_{21} & \Delta_Z/z_{21} \\ -1/z_{21} & z_{11}/z_{21} \end{bmatrix}$$

or

$$\mathbf{\Gamma} = \begin{bmatrix} \gamma_{11} & \gamma_{12} \\ \gamma_{21} & \gamma_{22} \end{bmatrix} = \begin{bmatrix} z_{22}/z_{12} & -\Delta_Z/z_{12} \\ 1/z_{12} & -z_{11}/z_{12} \end{bmatrix} \tag{5.28}$$

To find the h-parameters in terms of the z-parameters, begin with eq. (b) and solve for q_a:

$$q_a = -\frac{z_{21}}{z_{22}}q_b + \frac{1}{z_{22}}\theta_a \tag{h}$$

Then make a term-by-term comparison of eqs. (f) and (h) to obtain

$$h_{21} = -\frac{z_{21}}{z_{22}}$$

and

$$h_{22} = \frac{1}{z_{22}}$$

Next, put eq. (h) into eq. (a) to obtain

$$\theta_b = z_{11}q_b + z_{12}\left(-\frac{z_{21}}{z_{22}}q_b + \frac{1}{z_{22}}\theta_a\right)$$

$$= \left(z_{11} - \frac{z_{12}z_{21}}{z_{22}}\right)q_b + \frac{z_{12}}{z_{22}}\theta_a$$

$$= \frac{\Delta_Z}{z_{22}}q_b + \frac{z_{12}}{z_{22}}\theta_a$$

where $\Delta_Z = z_{11}z_{22} - z_{12}z_{21}$. Then a term-by-term comparison with eq. (e) shows that

$$h_{11} = \frac{\Delta_Z}{z_{22}}$$

and

$$h_{12} = \frac{z_{12}}{z_{22}}$$

In summary,

$$\mathbf{H} = \begin{bmatrix} h_{11} & h_{12} \\ h_{21} & h_{22} \end{bmatrix} = \begin{bmatrix} \Delta_Z/z_{22} & z_{12}/z_{22} \\ -z_{21}/z_{22} & 1/z_{22} \end{bmatrix} \tag{5.29}$$

The elements of $\mathbf{G}$ in terms of the z-parameters may be obtained by taking the inverse of eq. (5.29). First, obtain det $\mathbf{H}$:

$$\det \mathbf{H} = \frac{\Delta_Z + z_{12}z_{21}}{(z_{22})^2}$$

$$= \frac{z_{11}z_{22} - z_{12}z_{21} + z_{12}z_{21}}{(z_{22})^2}$$

$$= \frac{z_{11}}{z_{22}}$$

Then

$$\mathbf{G} = \frac{z_{22}}{z_{11}}\begin{bmatrix} 1/z_{22} & -z_{12}/z_{22} \\ z_{21}/z_{22} & \Delta_Z/z_{22} \end{bmatrix}$$

or

$$\mathbf{G} = \begin{bmatrix} g_{11} & g_{12} \\ g_{21} & g_{22} \end{bmatrix} = \begin{bmatrix} 1/z_{11} & -z_{12}/z_{11} \\ z_{21}/z_{11} & \Delta_Z/z_{11} \end{bmatrix} \tag{5.30}$$

5.6 SUMMARY OF ALL CONVERSIONS

Machinations similar to those employed in the preceding section can yield expressions for each parameter in terms of all the others. A summary of these is provided (without derivation) in Table 5.2.

5.7 FORMAL DEVELOPMENT OF THE LINEAR TRANSFORMATIONS

To develop the linear transformations for fins that are shaped irregularly and in which the effects of the slope of the fin faces is to be included, consider Fig. 5.2. Here, too, x is the height coordinate but with origin set arbitrarily at some point remote from the fin tip which is located at $x = a$. For the specific cases where the origin of the height coordinate is set at the fin tip, $a = 0$. The fin profile area is $A(x)$ and the fin thickness $\delta(x)$ is the difference between the two curves $y_1(x)$ and $y_2(x)$. These curves are often referred to as the *profile curves* or *profile functions*.

With the Murray–Gardner assumptions[3] still firmly in mind and with

TABLE 5.2 Summary of Parameter Conversions

z_{11}	=	y_{22}/Δ_Y	τ_{11}/τ_{21}	γ_{22}/γ_{21}	Δ_H/h_{22}	$1/g_{11}$
z_{12}	=	$-y_{12}/\Delta_Y$	$-\Delta_T/\tau_{21}$	$1/\gamma_{21}$	h_{12}/h_{22}	$-g_{12}/g_{11}$
z_{21}	=	$-y_{21}/\Delta_Y$	$1/\tau_{21}$	$-\Delta_\Gamma/\gamma_{21}$	$-h_{21}/h_{22}$	g_{21}/g_{11}
z_{22}	=	y_{11}/Δ_Y	$-\tau_{22}/\tau_{21}$	γ_{11}/γ_{21}	$1/h_{22}$	Δ_G/g_{11}
y_{11}	=	z_{22}/Δ_Z	τ_{22}/τ_{12}	$-\gamma_{11}/\gamma_{12}$	$1/h_{11}$	Δ_G/g_{22}
y_{12}	=	$-z_{12}/\Delta_Z$	$-\Delta_T/\tau_{12}$	$1/\gamma_{12}$	$-h_{12}/h_{11}$	g_{12}/g_{22}
y_{21}	=	$-z_{21}/\Delta_Z$	$1/\tau_{12}$	$-\Delta_\Gamma/\gamma_{12}$	h_{21}/h_{11}	$-g_{21}/g_{22}$
y_{22}	=	z_{11}/Δ_Z	$-\tau_{11}/\tau_{12}$	γ_{22}/γ_{12}	Δ_H/h_{11}	$1/g_{22}$
τ_{11}	=	z_{11}/z_{21}	$-y_{22}/y_{21}$	$\gamma_{22}/\Delta_\Gamma$	$-\Delta_H/h_{21}$	$1/g_{21}$
τ_{12}	=	$-\Delta_Z/z_{21}$	$1/y_{21}$	$-\gamma_{12}/\Delta_\Gamma$	h_{11}/h_{21}	$-g_{22}/g_{21}$
τ_{21}	=	$1/z_{21}$	$-\Delta_Y/y_{21}$	$-\gamma_{21}/\Delta_\Gamma$	$-h_{22}/h_{21}$	g_{11}/g_{21}
τ_{22}	=	$-z_{22}/z_{21}$	y_{11}/y_{21}	$\gamma_{11}/\Delta_\Gamma$	$1/h_{21}$	$-\Delta_G/g_{21}$
γ_{11}	=	z_{22}/z_{12}	$-y_{11}/y_{12}$	τ_{22}/Δ_T	$1/h_{12}$	$-\Delta_G/g_{12}$
γ_{12}	=	$-\Delta_Z/z_{12}$	$1/y_{12}$	$-\tau_{12}/\Delta_T$	$-h_{11}/h_{12}$	g_{22}/g_{12}
γ_{21}	=	$1/z_{12}$	$-\Delta_Y/y_{12}$	$-\tau_{21}/\Delta_T$	h_{22}/h_{12}	$-g_{11}/g_{12}$
γ_{22}	=	$-z_{11}/z_{12}$	y_{22}/y_{11}	τ_{12}/Δ_T	$-\Delta_H/h_{12}$	$1/g_{12}$
h_{11}	=	Δ_Z/z_{22}	$1/y_{11}$	τ_{12}/τ_{22}	$-\gamma_{12}/\gamma_{11}$	g_{22}/Δ_G
h_{12}	=	z_{12}/z_{22}	$-y_{12}/y_{11}$	Δ_T/τ_{22}	$1/\gamma_{11}$	$-g_{12}/\Delta_G$
h_{21}	=	$-z_{21}/z_{22}$	y_{21}/y_{11}	$1/\tau_{22}$	$\Delta_\Gamma/\gamma_{11}$	$-g_{21}/\Delta_G$
h_{22}	=	$1/z_{22}$	Δ_Y/y_{11}	$-\tau_{21}/\tau_{22}$	γ_{21}/γ_{11}	g_{11}/Δ_G
g_{11}	=	$1/z_{11}$	Δ_Y/y_{22}	τ_{21}/τ_{12}	$-\gamma_{21}/\gamma_{22}$	h_{22}/Δ_H
g_{12}	=	$-z_{12}/z_{11}$	y_{12}/y_{22}	Δ_T/τ_{11}	$1/\gamma_{22}$	$-h_{12}/\Delta_H$
g_{21}	=	z_{21}/z_{11}	$-y_{21}/y_{22}$	$1/\tau_{11}$	$\Delta_\Gamma/\gamma_{22}$	$-h_{21}/\Delta_H$
g_{22}	=	Δ_Z/z_{11}	$1/y_{22}$	$-\tau_{12}/\tau_{11}$	γ_{12}/γ_{22}	h_{11}/Δ_H

[3]Presented in Section 1.2.3.

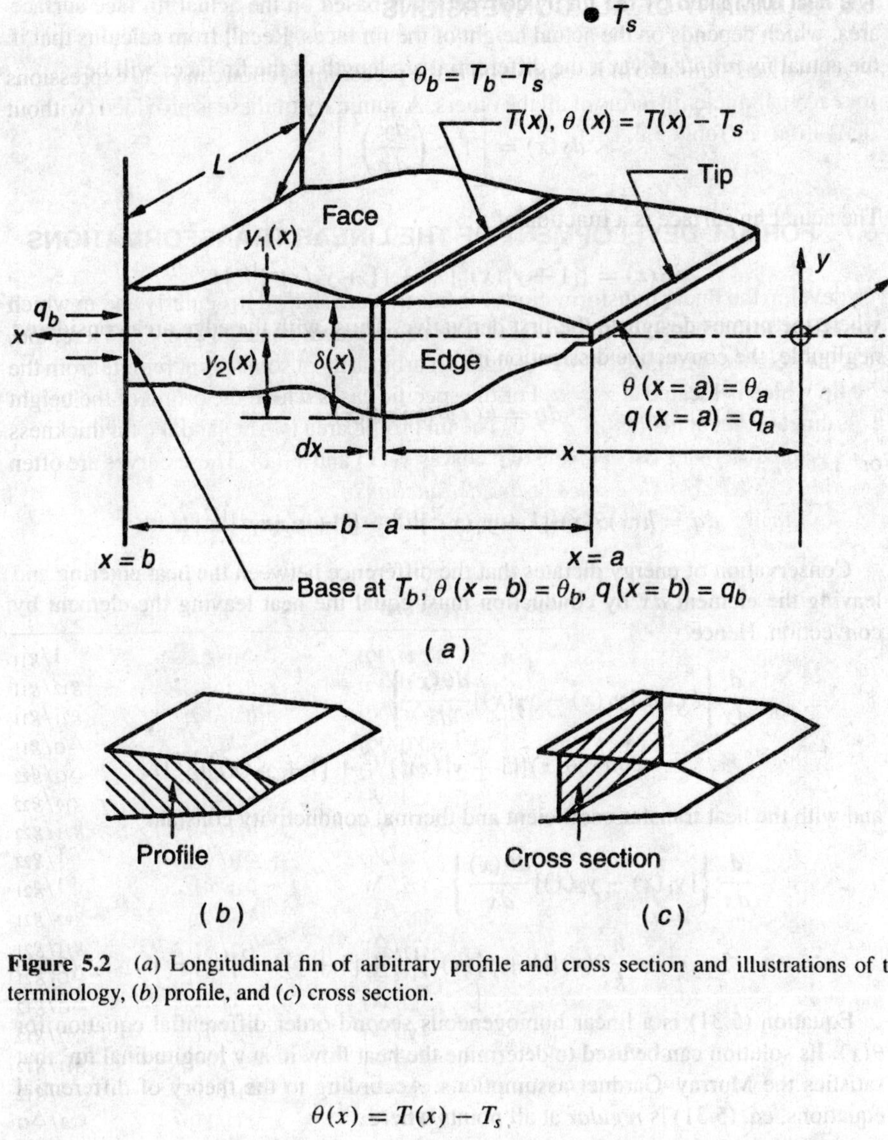

Figure 5.2 (*a*) Longitudinal fin of arbitrary profile and cross section and illustrations of the terminology, (*b*) profile, and (*c*) cross section.

$$\theta(x) = T(x) - T_s$$

and

$$A(x) = [y_1(x) - y_2(x)]L$$

the difference between the heat entering and leaving the element dx by conduction will be

$$dq = \frac{d}{dx}\left\{k(x)L[y_1(x) - y_2(x)]\frac{d\theta(x)}{dx}\right\}$$

The heat dissipated by the fin by convection is based on the actual fin face surface area, which depends on the actual height of the fin faces. Recall from calculus that if the actual *fin profile* is $y(x)$, the differential arc length of the fin faces will be

$$ds(x) = \left[1 + \left(\frac{dy}{dx}\right)^2\right]^{1/2}$$

The actual fin surface is a function of x:

$$dS(x) = \{[1 + y_1'(x)^2]^{1/2} + [1 + y_2'(x)^2]^{1/2}\}L$$

where the primes designate the first derivative. Thus, with the edge area considered negligible, the convective dissipation is

$$dq = h(x)\theta(x)\,dS(x)$$

or[4]

$$dq = h(x)\theta(x)\{[1 + y_1'(x)^2]^{1/2} + [1 + y_2'(x)^2]^{1/2}\}L$$

Conservation of energy dictates that the difference between the heat entering and leaving the element dx by conduction must equal the heat leaving the element by convection. Hence

$$\frac{d}{dx}\left\{k(x)L[y_1(x) - y_2(x)]\frac{d\theta(x)}{dx}\right\}$$
$$= h(x)\theta(x)\{[1 + y_1'(x)^2]^{1/2} + [1 + y_2'(x)^2]^{1/2}\}L$$

and with the heat transfer coefficient and thermal conductivity constant

$$\frac{d}{dx}\left\{[y_1(x) - y_2(x)]\frac{d\theta(x)}{dx}\right\}$$
$$-\frac{h}{k}\theta(x)\{[1 + y_1'(x)^2]^{1/2} + [1 + y_2'(x)^2]^{1/2}\} = 0 \qquad (5.31)$$

Equation (5.31) is a linear homogeneous second-order differential equation for $\theta(x)$. Its solution can be used to determine the heat flow in any longitudinal fin[5] that satisfies the Murray–Gardner assumptions. According to the theory of differential equations, eq. (5.31) is *regular* at all points where

$$k(x)A(x) = k(x)L[y_1(x) - y_2(x)] \neq 0 \qquad (5.32)$$

and *singular* when $k(x)A(x)$ vanishes. The latter situation can occur physically only when $y_1(x) = y_2(x)$, that is, when the fin has zero thickness. This happens at the

[4]If the height b rather than the actual arc length s is used to compute the surface area of the fin, the procedure is known as *making the length of arc assumption.*

[5]Or spine or, indeed, any radial fin as long as it is noted that the radial height coordinate has its positive sense from fin base to fin tip.

tips of certain fins or spines, such as the longitudinal fins of triangular and concave parabolic profiles and the conical and concave parabolic spines.

The theory also guarantees that eq. (5.32) possesses two solutions, $\lambda_1(x)$ and $\lambda_2(x)$, satisfying the *initial conditions* (the conditions at $x = b$)

$$\lambda_1(b) = 1; \qquad \frac{d\lambda_1(b)}{dx} = 0 \tag{5.33a}$$

and

$$\lambda_2(b) = 0; \qquad \frac{d\lambda_2(b)}{dx} = \frac{1}{k(b)A(b)} \tag{5.33b}$$

subject to certain continuity conditions that must be imposed on k and A. If k and A happen to have discontinuities, one merely breaks the fin into shorter segments. The temperature excess and heat flow are then given in terms of the values at the base of the fin,

$$\theta(x) = \theta_b \lambda_1(x) + q_b \lambda_2(x) \tag{5.34}$$

and

$$q(x) = k(x)A(x)[\theta_b \lambda_1'(x) + q_b \lambda_2'(x)] \tag{5.35}$$

where the primes denote derivatives with respect to x.

The linear transformation that maps the temperature excess and heat flow at any point on the fin in terms of conditions at the fin base is

$$\begin{bmatrix} \theta(x) \\ q(x) \end{bmatrix} = \begin{bmatrix} \lambda_1(x) & \lambda_2(x) \\ k(x)A(x)\lambda_1'(x) & k(x)A(x)\lambda_2'(x) \end{bmatrix} \begin{bmatrix} \theta_b \\ q_b \end{bmatrix} \tag{5.36}$$

which is used in

$$\begin{bmatrix} \theta(x) \\ q(x) \end{bmatrix} = \mathbf{\Lambda} \begin{bmatrix} \theta_b \\ q_b \end{bmatrix} \tag{5.37}$$

where Λ may be written as

$$\mathbf{\Lambda} = \begin{bmatrix} 1 & 0 \\ 0 & k(x)A(x) \end{bmatrix} \begin{bmatrix} \lambda_1(x) & \lambda_2(x) \\ \lambda_1'(x) & \lambda_2'(x) \end{bmatrix} \tag{5.38}$$

The determinant of the second matrix is the familiar *Wronskian*:

$$w\{\lambda_1, \lambda_2 : x\} = \det \begin{bmatrix} \lambda_1(x) & \lambda_2(x) \\ \lambda_1'(x) & \lambda_2'(x) \end{bmatrix}$$

The $\mathbf{\Gamma}$-matrix is generated when x is set equal to a (a may equal zero) in eq. (5.38):

$$\mathbf{\Gamma} = \begin{bmatrix} \gamma_{11} & \gamma_{12} \\ \gamma_{21} & \gamma_{22} \end{bmatrix}$$

so that the actual linear transformation from base conditions to tip conditions is

$$\begin{bmatrix} \theta_a \\ q_a \end{bmatrix} = \mathbf{\Gamma} \begin{bmatrix} \theta_b \\ q_b \end{bmatrix} = \begin{bmatrix} \gamma_{11} & \gamma_{12} \\ \gamma_{21} & \gamma_{22} \end{bmatrix} \begin{bmatrix} \theta_b \\ q_b \end{bmatrix} \tag{5.39}$$

Here

$$\gamma_{11} = \lambda_1(a) \tag{5.40a}$$

$$\gamma_{12} = \lambda_2(a) \tag{5.40b}$$

$$\gamma_{21} = k(a)A(a)\lambda_1'(a) \tag{5.40c}$$

and

$$\gamma_{22} = k(a)A(a)\lambda_2'(a) \tag{5.40d}$$

Observe that if $A(a) = 0$, $\gamma_{21} = \gamma_{22} = 0$ and

$$\mathbf{\Gamma} = \begin{bmatrix} \lambda_1(0) & \lambda_2(0) \\ 0 & 0 \end{bmatrix} = \begin{bmatrix} \gamma_{11} & \gamma_{12} \\ 0 & 0 \end{bmatrix}$$

The determinant of this matrix is equal to zero and the matrix is termed *singular* because it has no inverse. This is why fins and spines that taper to a zero cross section are called singular and why a different approach (considered in Chapter 6) is necessary for their parameterization.

An example of the determination of the parameters follows. It concerns a familiar configuration, the longitudinal fin of rectangular profile.

5.8 EXAMPLE OF FINDING THE PARAMETERS

For the longitudinal fin of rectangular profile shown in Fig. 5.3, the profile functions are $y_1(x) = \delta/2$ and $y_2(x) = -\delta/2$. With these in eq. (5.31), the differential equation for temperature excess of eq. (5.4),

$$\frac{d^2\theta}{dx^2} - m^2\theta = 0 \tag{5.4}$$

is obtained where

$$m = \left(\frac{2h}{k\delta}\right)^{1/2}$$

With the initial conditions given by

$$\theta(x = b) = \theta_b$$

and

$$q(x = b) = q_b$$

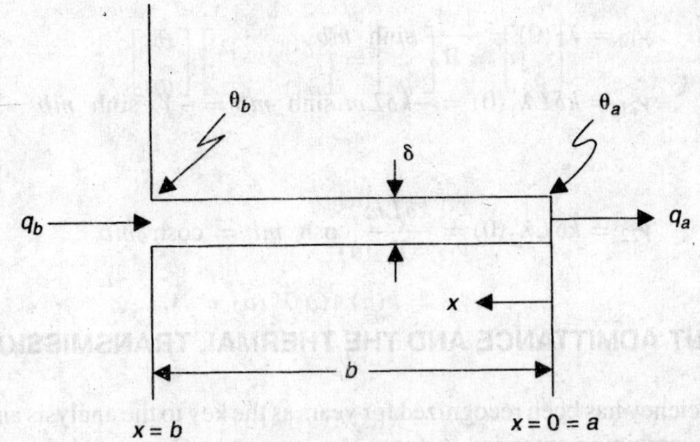

Figure 5.3 Longitudinal fin of rectangular profile.

the development follows that employed in Section 5.3 to arrive at the temperature excess at any point on the fin:

$$\theta(x) = [\cosh m(b-x)]\theta_b - \frac{1}{Y_o}[\sinh m(b-x)]q_b \tag{5.6}$$

and the heat flow at any point in the fin

$$\begin{aligned} q(x) &= k\delta L \frac{d\theta(x)}{dx} \\ &= -[Y_o \sinh m(b-x)]\theta_b + [\cosh m(b-x)]q_b \end{aligned} \tag{5.7}$$

where Y_o is the characteristic admittance of the fin defined by

$$Y_o = (2hk\delta)^{1/2} L \tag{5.5}$$

Reference to eqs. (5.33) shows that

$$\lambda_1(x) = \cosh\ m(b-x); \qquad \lambda_1(b) = 1 \tag{5.41a}$$

$$\lambda_2(x) = -\frac{1}{Y_o} \sinh\ m(b-x); \qquad \lambda_2(b) = 0 \tag{5.41b}$$

$$\lambda_1'(x) = -m \sinh\ m(b-x); \qquad \lambda_1'(b) = 0 \tag{5.41c}$$

and

$$\lambda_2'(x) = \frac{m}{Y_o} \cosh\ m(b-x); \qquad \lambda_2'(b) = \frac{m}{Y_o} = \frac{1}{k\delta L} \tag{5.41d}$$

and at $x = a = 0$, the γ-parameters are obtained from eqs. (5.40):

$$\gamma_{11} = \lambda_1(0) = \cosh\ mb \tag{5.42a}$$

$$\gamma_{12} = \lambda_2(0) = -\frac{1}{Y_o}\sinh\ mb \tag{5.42b}$$

$$\gamma_{21} = k\delta L\lambda_1'(0) = -k\delta Lm\sinh\ mb = -Y_o\sinh\ mb \tag{5.42c}$$

and

$$\gamma_{22} = k\delta L\lambda_2'(0) = \frac{k\delta Lm}{Y_o}\cosh\ mb = \cosh\ mb \tag{5.42d}$$

5.9 INPUT ADMITTANCE AND THE THERMAL TRANSMISSION RATIO

The fin efficiency has been recognized for years as the key to the analysis and design of equipment containing extended surfaces. An attempt has been made here (in Section 5.2) to show that the fin efficiency has its limitations. Indeed, it must be boldly emphasized that the fin efficiency is certainly not a design parameter that can be used in the design and analysis of the intricate arrays because, for complicated shapes, the fin efficiency of individual fins segments may be difficult to obtain.

Two completely new parameterizations were developed by Kraus et al. (1978). One of them that pertains to *regular fins* is the *input admittance*, and the second, which involves *singular fins*, is the *thermal transmission ratio*. Both parameters are based on the Murray–Gardner limiting assumptions.

5.9.1 Input Admittance

The input admittance applies to an individual fin or a finned array and is defined by

$$Y_{\text{in}} = \frac{q_b}{\theta_b} \tag{5.43}$$

It is the ratio of the base heat flow to the base temperature excess, and it is related to the fin efficiency η by

$$Y_{\text{in}} = \eta h S \tag{5.44}$$

where S is the total convecting surface in either the single fin under consideration or the finned array containing many fins. It should be emphasized that use of the input admittance in conjunction with the base or prime surface is additive; the total input admittance, $Y_{in,T}$, is the sum of the input admittance of the base or prime surface, $Y_{\text{in},b} = hS_b$ and the fin or array input admittance, Y_{in}, so that

$$q_T = Y_{\text{in},T}\theta_b = (Y_{\text{in},b} + Y_{\text{in}})\theta_b$$

Recall that the fin or array input admittance is easily formulated from the linear transformation involving $\mathbf{T}$:

$$\begin{bmatrix} \theta_b \\ q_b \end{bmatrix} = \mathbf{T}\begin{bmatrix} \theta_a \\ q_a \end{bmatrix} = \begin{bmatrix} \tau_{11} & \tau_{12} \\ \tau_{21} & \tau_{22} \end{bmatrix}\begin{bmatrix} \theta_a \\ q_a \end{bmatrix} \tag{5.45}$$

A matrix multiplication yields

$$\theta_b = \tau_{11}\theta_a + \tau_{12}q_a$$
$$q_b = \tau_{21}\theta_a + \tau_{22}q_a$$

and now the input admittance may be defined using its defininition [eq. (5.43)]:

$$Y_{\text{in}} = \frac{q_b}{\theta_b} = \frac{\tau_{22}\theta_a + \tau_{21}q_a}{\tau_{11}\theta_a + \tau_{12}q_a}$$

or

$$Y_{\text{in}} = \frac{q_b}{\theta_b} = \frac{\tau_{21} + \tau_{22}(q_a/\theta_a)}{\tau_{11} + \tau_{12}(q_a/\theta_a)} \tag{5.46}$$

Observe that the tip heat flow may be zero or negligible, in which case

$$Y_{\text{in}} = \frac{q_b}{\theta_b} = \frac{\tau_{22}}{\tau_{11}} \tag{5.47}$$

5.9.2 Thermal Transmission Ratio

As indicated by eq. (5.46), the input admittance is easily obtained from the elements of **T**. However, an attempt to follow the same procedure for the singular fin is foolhardy because, as observed at the end of Section 5.7, **Γ** for the singular fin,

$$\mathbf{\Gamma} = \begin{bmatrix} \lambda_1(0) & \lambda_2(0) \\ 0 & 0 \end{bmatrix} = \begin{bmatrix} \gamma_{11} & \gamma_{12} \\ 0 & 0 \end{bmatrix}$$

possesses a zero determinant. Thus **T**, which is the inverse of **Γ**, does not exist.

The use of the input admittance for a regular fin or finned array suggests that because no heat can ever leave the tip of a singular fin, it is sufficent to parameterize the singular fin with a single parameter,

$$\mu = \frac{q_b}{\theta_b} \tag{5.48}$$

called the *thermal transmission ratio*, which, like the input admittance, is seen to be the ratio of the base heat flow to the base temperature excess. While both are ratios, the input admittance is a ratio reserved for regular fins or the base of an array of fins, and the thermal transmission ratio is reserved for use with singular fins. This is but a subtle difference and use of the terminology will be quite apparent as the reader proceeds.

Attention now turns to development of the six linear transformations for several longitudinal fins and spines and the radial fin of rectangular profile. The thermal transmission ratios for four singular fins and spines are also developed, and this is the task for Chapter 6.

5.10 NOMENCLATURE

Roman Letter Symbols

A	cross-sectional or profile are, m^2
a	fin tip dimension, m
b	fin height, m
C	arbitrary constant, dimensionless
d	diameter, m; derivative or differential, dimensions vary
f	function, dimensions vary
G	thermal conductance, W/K
$\mathbf{G}$	inverse hybrid matrix, dimensions vary
g	elements of inverse hybrid matrix, dimensions vary
$\mathbf{H}$	hybrid matrix, dimensions vary
h	heat transfer coefficient, $W/m^2 \cdot K$, elements of hybrid matrix, dimensions vary
k	thermal conductivity, $W/m \cdot K$
L	fin length, m
m	fin performance parameter, m^{-1}
q	heat flow, W
S	surface area, m^2
T	temperature, K
w	Wronskian, m
x	height coordinate, m
Y	thermal admittance, W/K
$\mathbf{Y}$	thermal admittance matrix, W/K
y	elements of thermal admittance matrix, W/K profile, m; profile function, m
$\mathbf{Z}$	thermal impedance matrix, W/K
z	elements of thermal impedance matrix, W/K

Greek Letter Symbols

α	elements of a matrix, dimensions vary
$\mathbf{\Gamma}$	thermal transmission matrix, dimensions vary
γ	elements of thermal transmission matrix, dimensions vary
Δ	change in, dimensionless; determinant, dimensions vary
δ	fin thickness, m
η	fin efficiency, dimensionless
θ	temperature excess, K
κ	taper angle, dimensionless
$\mathbf{\Lambda}$	matrix, dimensions vary
λ	elements of a matrix, dimensions vary
μ	thermal transmission ratio, W/K
$\mathbf{T}$	linear transmission matrix, dimensions vary
τ	elements of linear transmission matrix, dimensions vary

Roman Letter Subscripts

a	tip of fin
b	base of fin
G	inverse hybrid matrix
H	hybrid matrix
id	ideal
in	input condition
o	characteristic
s	surroundings
Y	thermal admittance matrix
Z	thermal impedance matrix

Greek Letter Subscripts

Γ	matrix of a linear transformation
T	transmission matrix

Symbolic Superscript

′	first derivative

CHAPTER 6

ELEMENTS OF LINEAR TRANSFORMATIONS

6.1 INTRODUCTION

It has been noted that those fins that possess linear transformations relating two of the four physical quantities of base and tip heat flow and base and tip temperature excess to the other two are termed *regular fins*. Several of the individual useful shapes are regular, and the following types are treated in this chapter:

1. Longitudinal fin of rectangular profile
2. Longitudinal fin of rectangular profile with one face insulated
3. Longitudinal fin of trapezoidal profile
4. Longitudinal fin of trapezoidal profile with one face insulated
5. Longitudinal fin of truncated parabolic profile
6. Radial fin of rectangular profile
7. Spines of constant cross section
8. Truncated conical spine
9. Truncated concave parabolic spine

Moreover, in contrast to regular fins, there are several useful fins that cannot be characterized by any of the linear transformations. These are the *singular fins* and those considered are

1. Longitudinal fin of triangular profile
2. Longitudinal fin of concave parabolic profile
3. Conical spine

4. Concave parabolic spin

In addition, it is often useful to represent a bond resistance and a fin or spine tip heat dissipation as a linear transformation. The linear transformations in these two cases can be represented by a **T**-matrix, and these are also developed in this chapter.

6.2 REGULAR FINS AND SPINES

Because interest is most often focused on fin base conditions (temperature excess and heat flow), the differential equations for the individual fins and spines treated in this chapter are solved using initial value data. This guarantees that solutions of the equations are analytically well posed (as opposed to solutions obtained from boundary value data), easily computed numerically, and convenient to tabulate. The procedure here involves use of the generalized differential equation derived in Section 5.7 and development of the **Γ**-matrix, which maps base conditions of temperature excess and heat flow into similar conditions at the tip. Then the other five sets of parameters are developed through the use of Table 5.2.

It was shown in Section 5.7 that a generalized differential equation for the point-to-point temperature excess on a regular longitudinal fin could be developed. The result was eq. (5.31) which is repeated here:

$$\frac{d}{dx}\left\{[y_1(x) - y_2(x)]\frac{d\theta(x)}{dx}\right\} - \frac{h}{k}\theta(x)\{[1 + y_1'(x)^2]^{1/2} + [1 + y_2'(x)^2]^{1/2}\} = 0 \qquad (6.1)$$

where the primes represent derivatives.

Equation (6.1) is a linear homogeneous second-order differential equation for $\theta(x)$, and its solution can be used to determine the heat flow in any longitudinal fin that satisfies the Murray–Gardner assumptions. Moreover, it was shown that the theory guarantees two solutions to eq. (6.1), $\lambda_1(x)$ and $\lambda_2(x)$, that satisfy the *initial conditions*

$$\lambda_1(b) = 1; \qquad \frac{d\lambda_1(b)}{dx} = 0 \qquad (6.2a)$$

and

$$\lambda_2(b) = 0; \qquad \frac{d\lambda_2(b)}{dx} = \frac{1}{k(b)A(b)} \qquad (6.2b)$$

subject to certain continuity conditions that must be imposed on k and A.

The solutions are

$$\theta(x) = \theta_b\lambda_1(x) + q_b\lambda_2(x) \qquad (6.3)$$

and

$$q(x) = k(x)A(x)[\theta_b\lambda_1'(x) + q_b\lambda_2'(x)] \qquad (6.4)$$

Equations (6.3) and (6.4) produce a linear transformation that maps the temperature excess and heat flow at the fin tip from conditions at the fin base. This is frequently referred to as the $\mathbf{\Gamma}$-matrix, which is generated when x is set equal to the tip location, a (a may equal zero):

$$\mathbf{\Gamma} = \begin{bmatrix} \gamma_{11} & \gamma_{12} \\ \gamma_{21} & \gamma_{22} \end{bmatrix}$$

so that the actual linear transformation from base conditions to tip conditions is

$$\begin{bmatrix} \theta_a \\ q_a \end{bmatrix} = \mathbf{\Gamma} \begin{bmatrix} \theta_b \\ q_b \end{bmatrix} = \begin{bmatrix} \gamma_{11} & \gamma_{12} \\ \gamma_{21} & \gamma_{22} \end{bmatrix} \begin{bmatrix} \theta_b \\ q_b \end{bmatrix} \tag{6.5}$$

Here, because the fin tip is located at $x = a$,

$$\gamma_{11} = \lambda_1(a) \tag{6.6a}$$

$$\gamma_{12} = \lambda_2(a) \tag{6.6b}$$

$$\gamma_{21} = k(a)A(a)\lambda_1'(a) \tag{6.6c}$$

and

$$\gamma_{22} = k(a)A(a)\lambda_2'(a) \tag{6.6d}$$

and if $a = 0$, these become

$$\gamma_{11} = \lambda_1(0) \tag{6.7a}$$

$$\gamma_{12} = \lambda_2(0) \tag{6.7b}$$

$$\gamma_{21} = k(0)A(0)\lambda_1'(0) \tag{6.7c}$$

and

$$\gamma_{22} = k(0)A(0)\lambda_2'(0) \tag{6.7d}$$

This theory will now be employed to develop the six linear transformations for the regular fins of interest.

6.2.1 Longitudinal Fin of Rectangular Profile

The* $\mathbf{\Gamma}$*-Matrix. The $\mathbf{\Gamma}$-matrix for the longitudinal fin of rectangular profile, displayed in Fig. 6.1, was developed in Section 5.3 and summarized in Section 5.4. The matrix representation is

$$\begin{bmatrix} \theta_a \\ q_a \end{bmatrix} = \mathbf{\Gamma} \begin{bmatrix} \theta_b \\ q_b \end{bmatrix} = \begin{bmatrix} \cosh mb & -Z_o \sinh mb \\ -Y_o \sinh mb & \cosh mb \end{bmatrix} \begin{bmatrix} \theta_b \\ q_b \end{bmatrix} \tag{6.8}$$

where

$$Y_o = (2hk\delta)^{1/2} L$$

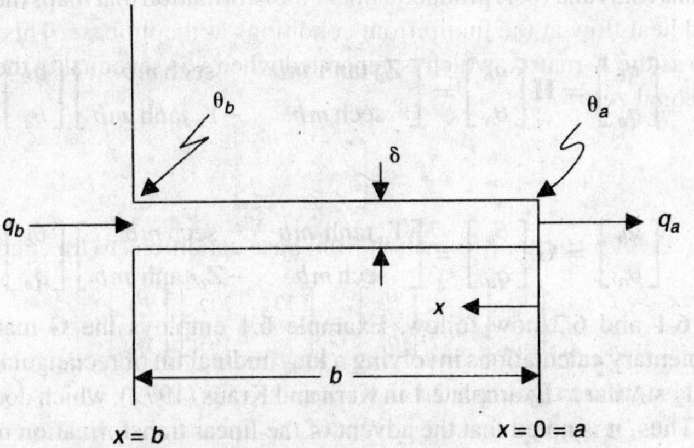

Figure 6.1 Longitudinal fin of rectangular profile.

is the *characteristic admittance*, its reciprocal,

$$Z_o = \frac{1}{Y_o} = \frac{1}{(2hk\delta)^{1/2} L}$$

is the *characteristic impedance*, and

$$m = \left(\frac{2h}{k\delta}\right)^{1/2}$$

is the fin performance parameter. The use of Table 5.2 provides the other five matrices.

T-*Matrix*

$$\begin{bmatrix} \theta_b \\ q_b \end{bmatrix} = \mathbf{T} \begin{bmatrix} \theta_a \\ q_a \end{bmatrix} = \begin{bmatrix} \cosh mb & Z_o \sinh mb \\ Y_o \sinh mb & \cosh mb \end{bmatrix} \begin{bmatrix} \theta_a \\ q_a \end{bmatrix} \tag{6.9}$$

Z-*Matrix*

$$\begin{bmatrix} \theta_b \\ \theta_a \end{bmatrix} = \mathbf{Z} \begin{bmatrix} q_b \\ q_a \end{bmatrix} = \begin{bmatrix} Z_o \coth mb & -Z_o \operatorname{csch} mb \\ Z_o \operatorname{csch} mb & -Z_o \coth mb \end{bmatrix} \begin{bmatrix} q_b \\ q_a \end{bmatrix} \tag{6.10}$$

Y-*Matrix*

$$\begin{bmatrix} q_b \\ q_a \end{bmatrix} = \mathbf{Y} \begin{bmatrix} \theta_b \\ \theta_a \end{bmatrix} = \begin{bmatrix} Y_o \coth mb & -Y_o \operatorname{csch} mb \\ Y_o \operatorname{csch} mb & -Y_o \coth mb \end{bmatrix} \begin{bmatrix} \theta_b \\ \theta_a \end{bmatrix} \tag{6.11}$$

H-*Matrix*

$$\begin{bmatrix} \theta_b \\ q_a \end{bmatrix} = \mathbf{H} \begin{bmatrix} q_b \\ \theta_a \end{bmatrix} = \begin{bmatrix} Z_o \tanh mb & \operatorname{sech} mb \\ \operatorname{sech} mb & -Y_o \tanh mb \end{bmatrix} \begin{bmatrix} q_b \\ \theta_a \end{bmatrix} \tag{6.12}$$

G-*Matrix*

$$\begin{bmatrix} q_b \\ \theta_a \end{bmatrix} = \mathbf{G} \begin{bmatrix} \theta_b \\ q_a \end{bmatrix} = \begin{bmatrix} Y_o \tanh mb & \operatorname{sech} mb \\ \operatorname{sech} mb & -Z_o \tanh mb \end{bmatrix} \begin{bmatrix} \theta_b \\ q_a \end{bmatrix} \tag{6.13}$$

Examples 6.1 and 6.2 now follow. Example 6.1 employs the **G**-matrix to make some rudimentary calculations involving a longitudinal fin of rectangular profile. It is substantially similar to Example 2.1 in Kern and Kraus (1972), which does not use the **G**-matrix. Thus, it is noted that the advent of the linear transformation matrices does not imply that the simple fin temperature excess, heat flow, and efficiency relationships cannot be used. It means only that there is now another alternative. Example 6.2 concerns the heat loss from the tip of a longitudinal fin of rectangular profile. It is similar to Example 6.1 in Kern and Kraus (1972) and involves the **T**-matrix. The solution contained here seems to be less involved that the solution in Kern and Kraus (1972), but nevertheless, the Kern and Kraus solution has validity.

The point to be made is that solution methods that do not involve the linear transformations developed here have validity, and the linear transformations do not replace methods that heretofore have been customary. But there will be cases where the linear transformations provide the only possible expeditious solutions, and these facts will be demonstrated as this study proceeds.

Example 6.1. A longitudinal fin of rectangular profile is exposed to surroundings with a temperature of 50°C and a heat transfer coefficient of 50 W/m^2 · K. The temperature at the fin base is 100°C and the fin is made of a material with thermal conductivity $k = 160$ W/m · K. The fin is 10 cm high, 40 cm long, and 0.9525 cm thick. Determine the tip temperature and the base heat flow, assuming that the fin tip is insulated.

SOLUTION. Here θ_b is a given and $q_a = 0$ because the tip of the fin is stated as being insulated. The values of θ_a and q_b are to be found. This suggests use of the **G**-matrix, which maps conditions of tip heat flow and base temperature excess to these sought-after conditions. The necessary computations now follow using four significant figures[1]:

$$m = \left(\frac{2h}{k\delta} \right)^{1/2}$$

$$= \left[\frac{(2)(50)}{(160)(0.009525)} \right]^{1/2}$$

[1] At least four decimal places are customary in computations involving matrices in order to fend off the phenomena known as *catastrophic cancellation* and *round-off error*.

$$= 8.1004 \text{ m}^{-1}$$

$$mb = (8.1004)(0.1000) = 0.8100$$

$$Y_o = (2hk\delta)^{1/2} L$$

$$= [(2)(50)(160)(0.009525)]^{1/2}(0.40)$$

$$= 4.9380 \text{ W/K}$$

$$Z_o = 1/Y_o = 0.2025 \text{ K/W}$$

and from eqs. (6.13),

$$g_{11} = Y_o \tanh mb$$

$$= 4.9380 \tanh 0.8100$$

$$= (4.9380)(0.6696) = 3.3066 \text{ K/W}$$

$$g_{12} = \operatorname{sech} mb$$

$$= \operatorname{sech} 0.8100$$

$$= 0.7427$$

$$g_{21} = \operatorname{sech} mb$$

$$= \operatorname{sech} 0.8100$$

$$= 0.7427$$

$$g_{22} = -Z_o \tanh mb$$

$$= -0.2025 \tanh 0.8100$$

$$= -(0.2025)(0.6696) = -0.1356 \text{ K/W}$$

The linear transformation may be written as

$$\begin{bmatrix} q_b \\ \theta_a \end{bmatrix} = \mathbf{G} \begin{bmatrix} \theta_b \\ q_a \end{bmatrix} = \begin{bmatrix} 3.3066 & 0.7427 \\ 0.7427 & -0.1356 \end{bmatrix} \begin{bmatrix} \theta_b \\ q_a \end{bmatrix}$$

and with

$$\theta_b = 100 - 50 = 50°\text{C}$$

and

$$q_a = 0 \text{ W}$$

then

$$\begin{bmatrix} q_b \\ \theta_a \end{bmatrix} = \mathbf{G} \begin{bmatrix} \theta_b \\ q_a \end{bmatrix} = \begin{bmatrix} 3.3066 & 0.7427 \\ 0.7427 & -0.1356 \end{bmatrix} \begin{bmatrix} 50 \\ q_a = 0 \end{bmatrix}$$

or

$$\begin{bmatrix} q_b \\ \theta_a \end{bmatrix} = \begin{bmatrix} 165.3 \text{ W} \\ 37.1°\text{C} \end{bmatrix}$$

Thus

$$\theta_b = 37.1 + 50 = 87.1°\text{C}$$

and

$$q_b = 165.3 \text{ W}$$

Example 6.2. Show that if the tip heat transfer coefficient is h_a, the heat flowing through the base of a longitudinal fin of rectangular profile is given by

$$q_b = \frac{k\delta L m \theta_b \left[(h_a/mk) + \tanh mb\right]}{1 + (h_a/mk) \tanh mb}$$

where all symbols are as defined previously.

SOLUTION. The given expression contains q_b and θ_b, and the strategy is to obtain these base conditions in terms of the tip conditions q_a and θ_a. This leads to use of the **T**-matrix with tip heat loss, $q_a = h_a \delta L \theta_a$:

$$\begin{bmatrix} \theta_b \\ q_b \end{bmatrix} = \mathbf{T} \begin{bmatrix} \theta_a \\ q_a \end{bmatrix} = \begin{bmatrix} \cosh mb & Z_o \sinh mb \\ Y_c \sinh mb & \cosh mb \end{bmatrix} \begin{bmatrix} \theta_a \\ h_a \delta L \theta_a \end{bmatrix}$$

After expansion,

$$\theta_b = \cosh mb\, \theta_a + (h_a \delta L) Z_o \sinh mb\, \theta_a$$

$$q_b = Y_o \sinh mb\, \theta_a + (h_a \delta L) \cosh mb\, \theta_a$$

Rearrangement provides

$$\theta_b = [(\cosh mb + h_a \delta L Z_o \sinh mb)]\theta_a$$

$$q_b = [(Y_o \sinh mb + h_a \delta L \cosh mb)]\theta_a$$

Division of the second by the first yields

$$\frac{q_b}{\theta_b} = \frac{Y_o \sinh mb + h_a \delta L \cosh mb}{\cosh mb + h_a \delta L Z_o \sinh mb}$$

The Y_o in the numerator and the $\cosh mb$ in both the numerator and denominator may be factored. With $Z_o = 1/Y_o$, this leads to

$$\frac{q_b}{\theta_b} = \frac{Y_o(\tanh mb + h_a \delta L Z_o)}{1 + h_a \delta L Z_o \tanh mb}$$

and when it is noted that

$$Y_o = (2hk\delta)^{1/2} L = k\delta L m$$

so that

$$h_a \delta L Z_o = \frac{h_a \delta L}{k \delta L m} = \frac{h_a}{mk}$$

the result is obtained:

$$q_b = \frac{k\delta L m \theta_b \left[(h_a/mk) + \tanh mb\right]}{1 + (h_a/mk)\tanh mb}$$

6.2.2 Longitudinal Fin of Rectangular Profile with One Face Insulated

In the event that one face of the longitudinal fin of rectangular profile (see Fig. 6.1) is insulated, all of eqs. (6.8) to (6.13) apply but with

$$Y_o = (hk\delta)^{1/2} L$$

$$Z_o = \frac{1}{Y_o} = \frac{1}{L}(hk\delta)^{-1/2}$$

and

$$m = \left(\frac{h}{k\delta}\right)^{1/2}$$

6.2.3 Longitudinal Fin of Trapezoidal Profile

For the longitudinal fin of trapezoidal profile shown in Fig. 6.2, the origin of the x-coordinate system is taken at the fictitious point of intersection of the two fin faces. This puts the tip of the fin at $x = a$. The fin profile is confined by the two smooth planes $y_1(x)$ and $-\ y_2(x)$. The fin taper angle is defined by

$$\kappa = \arctan \frac{\delta_b}{2b}$$

and a consideration of similar triangles provides a relationship between the base and tip coordinates and thicknesses:

$$\frac{\delta_a}{\delta_b} = \frac{a}{b}$$

The differential equation for the temperature excess can be obtained from eq. (6.1) as

$$2x \tan \kappa \frac{d^2\theta}{dx^2} + 2 \tan \kappa \frac{d\theta}{dx} - \frac{2h}{k \cos \kappa}\theta = 0 \tag{6.14}$$

The presence of the $\cos\kappa$ in the last term accounts for the fact that the fin face surface element is not dx but $dx/\cos\kappa$. This accounts for the taper at the fin faces, and a slight adjustment of eq. (6.14) then gives

$$x\frac{d^2\theta}{dx^2} + \frac{d\theta}{dx} - n^2\theta = 0 \tag{6.15}$$

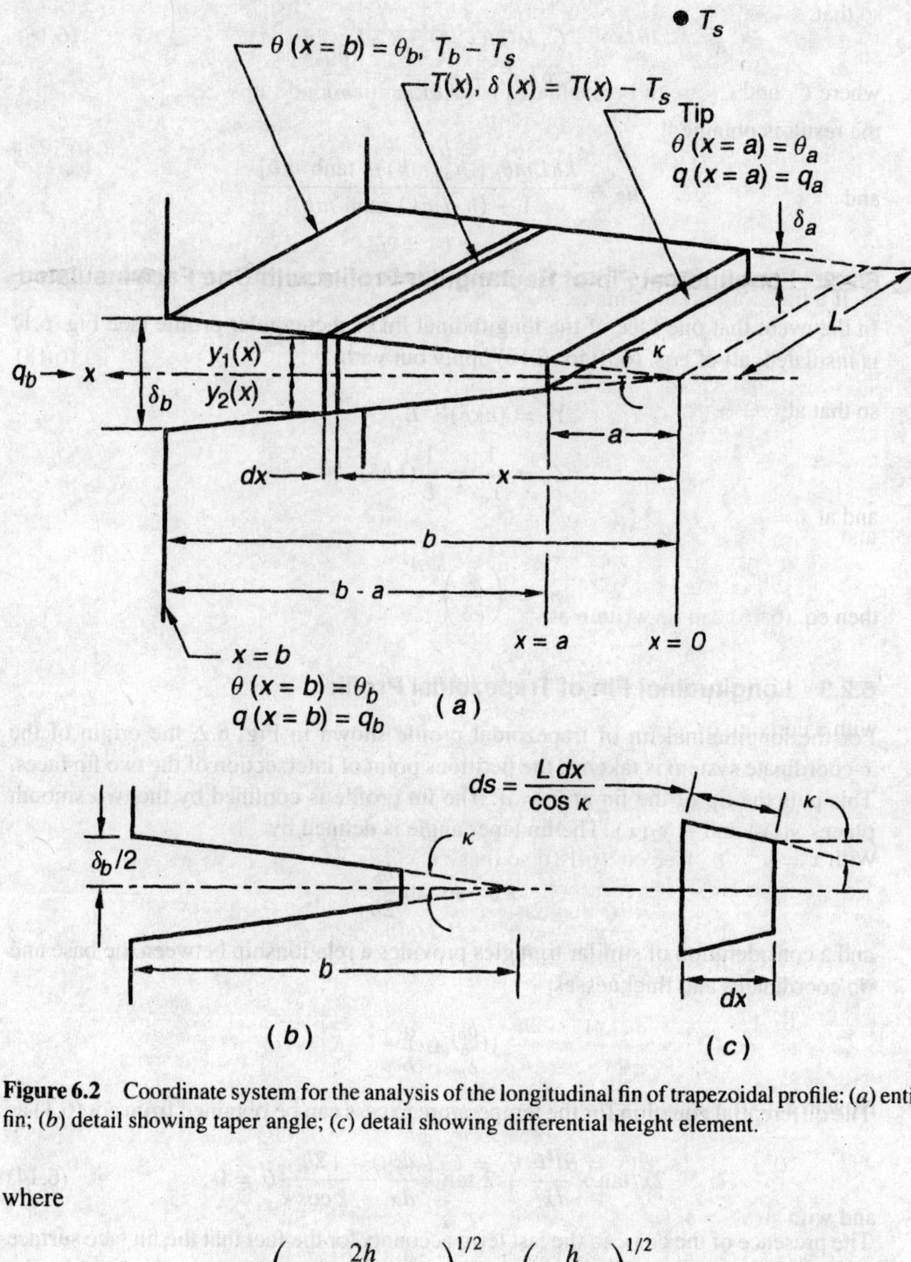

Figure 6.2 Coordinate system for the analysis of the longitudinal fin of trapezoidal profile: (*a*) entire fin; (*b*) detail showing taper angle; (*c*) detail showing differential height element.

where

$$n=\left(\frac{2h}{2k\cos\kappa\ \tan\kappa}\right)^{1/2}=\left(\frac{h}{k\sin\kappa}\right)^{1/2}$$

Equation (6.15) is a linear homogeneous second-order differential equation with variable coefficients that possesses a general solution in terms of the modified Bessel functions

$$\theta(x) = C_1 I_0(2nx^{1/2}) + C_2 K_0(2nx^{1/2}) \tag{6.16}$$

where C_1 and C_2 are to be evaluated from the initial conditions

$$\theta(x = b) = \theta_b \tag{6.17a}$$

and

$$q(x = b) = q_b \tag{6.17b}$$

If a transformation is made,

$$u = 2nx^{1/2} \longrightarrow x = u^2/4n^2 \tag{6.18}$$

so that at $x = b$,

$$u_b = 2nb^{1/2}$$

and at $x = a$

$$u_a = 2na^{1/2}$$

then eq. (6.16) can be written as

$$\theta(u) = C_1 I_0(u) + C_2 K_0(u) \tag{6.19}$$

with a derivative

$$\frac{d\theta(x)}{dx} = \frac{d}{dx}[C_1 I_0(u) + C_2 K_0(u)]$$

With $x = u^2/4n^2$ [see eq. (6.18)] so that $dx = (u/2n^2)\,du$, this becomes

$$\frac{d\theta(x)}{dx} = \frac{2n^2}{u}\frac{d}{du}[C_1 I_0(u) + C_2 K_0(u)]$$

or

$$\frac{d\theta(x)}{dx} = \frac{2n^2}{u}[C_1 I_1(u) - C_2 K_1(u)]$$

Now at $x = b$ where $u = u_b$,

$$\theta(x = b) = \theta_b = C_1 I_0(u_b) + C_2 K_0(u_b) \tag{6.20a}$$

and with $A(b) = \delta_b L$,

$$q(x = b) = q_b = kA(b)\left.\frac{d\theta(x)}{dx}\right|_{x=b}$$

$$= \frac{2k\delta_b L n^2}{u_b}[C_1 I_1(u_b) - C_2 K_1(u_b)] \tag{6.20b}$$

When solved simultaneously for C_1 and C_2, eqs. (6.20) provide

$$C_1 = \frac{K_1(u_b)\theta_b + (u_b/2k\delta_b Ln^2)K_0(u_b)q_b}{I_0(u_b)K_1(u_b) + I_1(u_b)K_0(u_b)}$$

and

$$C_2 = \frac{I_1(u_b)\theta_b - (u_b/2k\delta_b Ln^2)I_0(u_b)q_b}{I_0(u_b)K_1(u_b) + I_1(u_b)K_0(u_b)}$$

Use of the Bessel function identity

$$I_0(z)K_1(z) + I_1(z)K_0(z) = \frac{1}{z}$$

permits the arbitrary constants to be written as

$$C_1 = u_b\left[K_1(u_b)\theta_b + \frac{u_b}{2k\delta_b Ln^2}K_0(u_b)q_b\right]$$

and

$$C_2 = u_b\left[I_1(u_b)\theta_b - \frac{u_b}{2k\delta_b Ln^2}I_0(u_b)q_b\right]$$

Now, with C_1 and C_2 inserted into eq. (6.19),

$$\theta(u) = u_b[K_1(u_b)I_0(u) + I_1(u_b)K_0(u)]\theta_b + \frac{u_b^2}{2k\delta_b Ln^2}[K_0(u_b)I_0(u) - I_0(u_b)K_0(u)]q_b \tag{6.21}$$

which shows that

$$\lambda_1(u) = u_b[K_1(u_b)I_0(u) + I_1(u_b)K_0(u)] \tag{6.22a}$$

and

$$\lambda_2(u) = \frac{u_b^2}{2k\delta_b Ln^2}[K_0(u_b)I_0(u) - I_0(u_b)K_0(u)] \tag{6.22b}$$

Observe that $\lambda_1(u_b) = 1$ and $\lambda_2(u_b) = 0$, as required by eq. (6.2*a*).

To determine $q(x)$, it is necessary to obtain expressions for $\lambda_1'(u)$ and $\lambda_2'(u)$. First,

$$\lambda_1'(u) = \frac{d}{dx}\{u_b[K_1(u_b)I_0(u) + I_1(u_b)K_0(u)]\}$$

and with $dx = (u/2n^2)\,du$,

$$\lambda_1'(u) = 2n^2\frac{u_b}{u}[K_1(u_b)I_1(u) - I_1(u_b)K_1(u)] \tag{6.22c}$$

Then

$$\lambda_2'(u) = \frac{d}{dx}\left\{\frac{u_b^2}{2k\delta_b Ln^2}[K_0(u_b)I_0(u) - I_0(u_b)K_0(u)]\right\}$$

and again with $dx = (u/2n^2)\,du$,

$$\lambda_2'(u) = \frac{u_b}{k\delta_b L}\frac{u_b}{u}[K_0(u_b)I_1(u) + I_0(u_b)K_1(u)] \tag{6.22d}$$

Observe that at $x = b$ where $u = u_b$, $\lambda_1'(u_b) = 0$ and $\lambda_2'(u_b) = 1/k\delta_b L = 1/kA(b)$, both of which are required by eq. (6.2b).

Then with

$$q(x) = k(x)A(x)[\theta_b\lambda_1'(x) + q_b\lambda_2'(x)] \tag{6.4}$$

the $\mathbf{\Gamma}$ matrix can now be formulated. Equations (6.22) can be evaluated at $x = a$ to form

$$\begin{bmatrix}\theta_a\\ q_a\end{bmatrix} = \mathbf{\Gamma}\begin{bmatrix}\theta_b\\ q_b\end{bmatrix} = \begin{bmatrix}\gamma_{11} & \gamma_{12}\\ \gamma_{21} & \gamma_{22}\end{bmatrix}\begin{bmatrix}\theta_b\\ q_b\end{bmatrix}$$

and in accordance with eqs. (6.6),

$$\gamma_{11} = \lambda_1(u_a) = u_b[I_0(u_a)K_1(u_b) + I_1(u_b)K_0(u_a)] \tag{6.23a}$$

$$\gamma_{12} = \lambda_2(u_a) = \frac{u_b^2}{2k\delta_b Ln^2}[I_0(u_a)K_0(u_b) - I_0(u_b)K_0(u_a)] \tag{6.23b}$$

$$\gamma_{21} = kA(a)\lambda_1'(u_a)$$

or

$$\gamma_{21} = 2k\delta_a Ln^2\frac{u_b}{u_a}[I_1(u_a)K_1(u_b) - I_1(u_b)K_1(u_a)]\} \tag{6.23c}$$

and with $\delta_a/\delta_b = a/b$,

$$\frac{\delta_a u_b^2}{\delta_b u_a} = \frac{a(4n^2b)}{b(2na^{1/2})} = 2na^{1/2} = u_a$$

and

$$\gamma_{22} = kA(a)\lambda_2'(u_a) = u_a[I_1(u_a)K_0(u_b) + I_0(u_b)K_1(u_a)] \tag{6.23d}$$

The elements of the other linear tranformations can now be formed using Table 5.2. However, before employing this procedure, it is necessary to obtain the value of the determinant of $\mathbf{\Gamma}$. If the γ's are written as

$$\gamma_{11} = u_b\gamma_1$$

$$\gamma_{12} = \frac{u_b^2}{2k\delta_b Ln^2}\gamma_2$$

$$\gamma_{21} = 2k\delta_a L n^2 \frac{u_b}{u_a}\gamma_3$$

$$\gamma_{22} = u_a\gamma_4$$

where

$$\gamma_1 = I_0(u_a)K_1(u_b) + I_1(u_b)K_0(u_a)$$

$$\gamma_2 = I_0(u_a)K_0(u_b) - I_0(u_b)K_0(u_a)$$

$$\gamma_3 = I_1(u_a)K_1(u_b) - I_1(u_b)K_1(u_a)$$

and

$$\gamma_4 = [I_1(u_a)K_0(u_b) + I_0(u_b)K_1(u_a)$$

then

$$\begin{aligned}\det \mathbf{\Gamma} &= \det \begin{bmatrix} \gamma_{11} & \gamma_{12} \\ \gamma_{21} & \gamma_{22} \end{bmatrix} \\ &= \gamma_{11}\gamma_{22} - \gamma_{12}\gamma_{21} \\ &= u_a u_b \gamma_1\gamma_4 - \frac{\delta_a}{\delta_b}\frac{u_b^3}{u_a}\gamma_2\gamma_3 \\ &= u_a u_b[\gamma_1\gamma_4 - \gamma_2\gamma_3]\end{aligned}$$

with

$$\begin{aligned}\gamma_1\gamma_4 &= [I_0(u_a)K_1(u_b) + I_1(u_b)K_0(u_a)][I_1(u_a)K_0(u_b) + I_0(u_b)K_1(u_a)] \\ &= I_0(u_a)I_1(u_a)K_0(u_b)K_1(u_b) + I_0(u_a)I_0(u_b)K_1(u_a)K_1(u_b) \\ &\quad + I_1(u_a)I_1(u_b)K_0(u_a)K_0(u_b) + I_0(u_b)I_1(u_b)K_0(u_a)K_1(u_a)\end{aligned}$$

and

$$\begin{aligned}\gamma_2\gamma_3 &= [I_0(u_a)K_0(u_b) - I_0(u_b)K_0(u_a)][I_1(u_a)K_1(u_b) - I_1(u_b)K_1(u_a)] \\ &= I_0(u_a)I_1(u_a)K_0(u_b)K_1(u_b) - I_0(u_a)I_1(u_b)K_0(u_a)K_1(u_a) \\ &\quad - I_0(u_b)I_1(u_a)K_0(u_a)K_1(u_b) + I_0(u_b)I_1(u_b)K_0(u_a)K_1(u_a)\end{aligned}$$

Subtraction provides

$$\begin{aligned}\gamma_1\gamma_4 - \gamma_2\gamma_4 &= I_0(u_a)I_0(u_b)K_1(u_a)K_1(u_b) + I_0(u_a)I_1(u_b)K_0(u_b)K_1(u_a) \\ &\quad + I_1(u_a)I_1(u_b)K_0(u_a)K_0(u_b) + I_0(u_b)I_1(u_a)K_0(u_a)K_1(u_b) \\ &= I_0(u_a)K_1(u_a)[I_0(u_b)K_1(u_b) + I_1(u_b)K_0(u_b)] \\ &\quad + I_0(u_a)K_0(u_a)[I_0(u_b)K_1(u_b) + I_1(u_b)K_0(u_b)] \\ &= [I_0(u_a)K_1(u_a) + I_1(u_a)K_0(u_a)][I_0(u_b)K_1(u_b) + I_1(u_b)K_0(u_b)] \\ &= \frac{1}{u_a}\frac{1}{u_b}\end{aligned}$$

so that

$$\det \boldsymbol{\Gamma} = u_a u_b \frac{1}{u_a} \frac{1}{u_b} = 1 \tag{6.24}$$

T-*Matrix*

$$\tau_{11} = u_a[I_1(u_a)K_0(u_b) + I_0(u_b)K_1(u_a)] \tag{6.25a}$$

$$\tau_{12} = \frac{u_b^2}{2k\delta_b Ln^2}[I_0(u_b)K_0(u_a) - I_0(u_a)K_0(u_b)] \tag{6.25b}$$

$$\tau_{21} = 2k\delta_a Ln^2 \frac{u_b}{u_a}[I_1(u_b)K_1(u_a) - I_1(u_a)K_1(u_b)] \tag{6.25c}$$

and

$$\tau_{22} = u_b[I_0(u_a)K_1(u_b) + I_1(u_b)K_0(u_a)] \tag{6.25d}$$

Z-*Matrix*

$$z_{11} = \frac{u_b}{2k\delta_b Ln^2} \frac{I_1(u_a)K_0(u_b) + I_0(u_b)K_1(u_a)}{I_1(u_b)K_1(u_a) - I_1(u_a)K_1(u_b)} \tag{6.26a}$$

$$z_{12} = \frac{(1/2k\delta_a Ln^2)(u_a/u_b)}{I_1(u_a)K_1(u_b) - I_1(u_b)K_1(u_a)} \tag{6.26b}$$

$$z_{21} = \frac{(1/2k\delta_a Ln^2)(u_a/u_b)}{I_1(u_b)K_1(u_a) - I_1(u_a)K_1(u_b)} \tag{6.26c}$$

and

$$z_{22} = \frac{u_a}{2k\delta_a Ln^2} \frac{I_0(u_a)K_1(u_b) + I_1(u_b)K_0(u_a)}{I_1(u_a)K_1(u_b) - I_1(u_b)K_1(u_a)} \tag{6.26d}$$

Y-*Matrix*

$$y_{11} = \frac{2k\delta_b Ln^2}{u_b} \frac{I_0(u_a)K_1(u_b) + I_1(u_b)K_0(u_a)}{I_0(u_b)K_0(u_a) - I_0(u_a)K_0(u_b)} \tag{6.27a}$$

$$y_{12} = \frac{2k\delta_b Ln^2/u_b^2}{I_0(u_a)K_0(u_b) - I_0(u_b)K_0(u_a)} \tag{6.27b}$$

$$y_{21} = \frac{2k\delta_b Ln^2/u_b^2}{I_0(u_b)K_0(u_a) - I_0(u_a)K_0(u_b)} \tag{6.27c}$$

and

$$y_{22} = \frac{2k\delta_a Ln^2}{u_a} \frac{I_1(u_a)K_0(u_b) + I_0(u_b)K_1(u_a)}{I_0(u_a)K_0(u_b) - I_0(u_b)K_0(u_a)} \tag{6.27d}$$

H-*Matrix*

$$h_{11} = \frac{u_b}{2k\delta_b Ln^2} \frac{I_0(u_b)K_0(u_a) - I_0(u_a)K_0(u_b)}{I_0(u_a)K_1(u_b) + I_1(u_b)K_0(u_a)} \tag{6.28a}$$

$$h_{12} = \frac{1}{u_b[I_0(u_a)K_1(u_b) + I_1(u_b)K_0(u_a)]} \tag{6.28b}$$

$$h_{21} = \frac{1}{u_b[I_0(u_a)K_1(u_b) + I_1(u_b)K_0(u_a)]} \tag{6.28c}$$

and

$$h_{22} = \frac{2k\delta_a Ln^2}{u_a} \frac{I_1(u_a)K_1(u_b) - I_1(u_b)K_1(u_a)}{I_0(u_a)K_1(u_b) + I_1(u_b)K_0(u_a)} \tag{6.28d}$$

G-*Matrix*

$$g_{11} = \frac{2k\delta_b Ln^2}{u_b} \frac{I_1(u_b)K_1(u_a) - I_1(u_a)K_1(u_b)}{I_1(u_a)K_0(u_b) + I_0(u_b)K_1(u_a)} \tag{6.29a}$$

$$g_{12} = \frac{1}{u_a[I_1(u_a)K_0(u_b) + I_0(u_b)K_1(u_a)]} \tag{6.29b}$$

$$g_{21} = \frac{1}{u_a[I_1(u_a)K_0(u_b) + I_0(u_b)K_1(u_a)]} \tag{6.29c}$$

and

$$g_{22} = \frac{u_a}{2k\delta_a Ln^2} \frac{I_0(u_a)K_0(u_b) - I_0(u_b)K_0(u_a)}{I_1(u_a)K_0(u_b) + I_0(u_b)K_1(u_a)} \tag{6.29d}$$

Example 6.3. A longitudinal fin of trapezoidal profile with a thermal conductivity of 217 W/m · K is dissipating heat in a convective environment with a coefficient of heat transfer of 100 W/m^2 · K. The fin is 7.62 cm high, 25 cm long, and has base and tip thicknesses of 0.9525 cm (0.375 in.) and 0.2381 cm (0.0938 in.), respectively. Measurements indicate a tip temperature excess of $\theta_a = 27.6$°C and a base heat flow of $q_b = 138.8$ W. Evaluate the base temperature excess and the tip heat flow.

SOLUTION. Here θ_a and q_b are given and θ_b and q_a are to be found. This suggests use of the **H**-matrix. Observe that $b - a = 7.62$ cm and that by similar triangles, $a/b = \delta_a/\delta_b = 0.2381/0.9525 = \frac{1}{4}$. Use of these two relationships shows that

$$b = 10.16 \text{ cm} \quad \text{and} \quad a = 2.54 \text{ cm}$$

The taper angle is

$$\kappa = \arctan \frac{\delta_b}{2b} = \arctan \frac{0.009525}{(2)(0.1016)} = 2.68°$$

so that

$$\sin\kappa = \sin 2.68° = 0.0468$$

Then

$$n = \left(\frac{h}{k\sin\kappa}\right)^{1/2} = \left[\frac{100}{(217)(0.0468)}\right]^{1/2} = 3.1372\ \text{m}^{-1/2}$$

and

$$u_a = 2na^{1/2} = (2)(3.1372)(0.0254)^{1/2} = 1.0000$$

$$u_b = 2nb^{1/2} = (2)(3.1372)(0.1016)^{1/2} = 2.0000$$

The result will come from use of the **H**-matrix. The elements of this matrix are given in eqs. (6.28) and the values of eight modified Bessel functions will be needed. A table of Bessel functions or a software routine provides for $u_b = 2.0000$ and $u_a = 1.0000$:

$$\begin{aligned} I_0(u_b) &= 2.27959 & I_0(u_a) &= 1.26607\\ I_1(u_b) &= 1.59064 & I_1(u_a) &= 0.56516\\ K_0(u_b) &= 0.11389 & K_0(u_a) &= 0.42102\\ K_1(u_b) &= 0.13987 & K_1(u_a) &= 0.60191 \end{aligned}$$

From eqs. (6.28),

$$h_{11} = \frac{u_b}{2k\delta_b Ln^2}\,\frac{I_0(u_b)K_0(u_a) - I_0(u_a)K_0(u_b)}{I_0(u_a)K_1(u_b) + I_1(u_b)K_0(u_a)}$$

and with

$$\frac{u_b}{2k\delta_b Ln^2} = \frac{2.0000}{(2)(217)(.009525)(0.25)(3.1372)^2} = 0.1966$$

this becomes

$$h_{11} = 0.1966\left[\frac{(2.27959)(0.42102) - (1.26607)(0.11389)}{(1.26607)(0.13987) + (1.59064)(0.42102)}\right]$$

or

$$h_{11} = 0.1894\ \text{K/W}$$

$$h_{12} = h_{21} = \frac{1}{u_b[I_0(u_a)K_1(u_b) + I_1(u_b)K_0(u_a)]}$$

$$= \frac{1}{(2.0000)[1.26607(0.13987) + (1.59064)(0.42102)]} = 0.5905$$

and

$$h_{22} = \frac{2k\delta_a Ln^2}{u_a}\,\frac{I_1(u_a)K_1(u_b) - I_1(u_b)K_1(u_a)}{I_0(u_a)K_1(u_b) + I_1(u_b)K_0(u_a)}$$

With

$$\frac{2k\delta_a L n^2}{u_a} = \frac{(2)(217)(0.002381)(0.25)(3.1372)^2}{1.0000} = 2.5428$$

this results in

$$h_{22} = 2.5428\left[\frac{(0.56516)(0.13987) - (1.59064)(0.60191)}{(1.26607)(0.13987) + (1.59064)(0.42102)}\right]$$

or

$$h_{22} = -2.6377 \text{ W/K}$$

The conclusion of the example comes from a straightforward matrix multiplication:

$$\begin{bmatrix} \theta_a \\ q_b \end{bmatrix} = \mathbf{H}\begin{bmatrix} q_b \\ \theta_a \end{bmatrix}$$

$$= \begin{bmatrix} 0.1894 & 0.5905 \\ 0.5905 & -2.6377 \end{bmatrix}\begin{bmatrix} q_b = 138.8 \text{ W} \\ \theta_a = 27.6°\text{C} \end{bmatrix}$$

$$= \begin{bmatrix} 42.6°\text{C} \\ 9.1 \text{ W} \end{bmatrix}$$

Observe that a tip heat flow of 9.1 W when the base heat flow is 138.8 W is *not* negligible.

6.2.4 Longitudinal Fin of Half Trapezoidal Profile

The configuration for the longitudinal fin of half-trapezoidal profile is shown in Fig. 6.3. It may be observed that this fin is just half of the longitudinal fin of trapezoidal profile of Fig. 6.2 taken along the x-coordinate axis in Fig. 6.2. In both fins, the taper angle κ and the profile function $y_1(x)$ are given by identical expressions:

$$\kappa = \arctan\frac{\delta_b}{2b}$$

and

$$y_1(x) = x \tan\kappa$$

However, in Fig. 6.3, the profile function $y_2(x) = 0$, so that

$$y_1'(x) = \tan\kappa$$

and

$$y_2'(x) = 0$$

Insertion of these into eq. (6.1) with $h_1 = h_2 = h$ yields

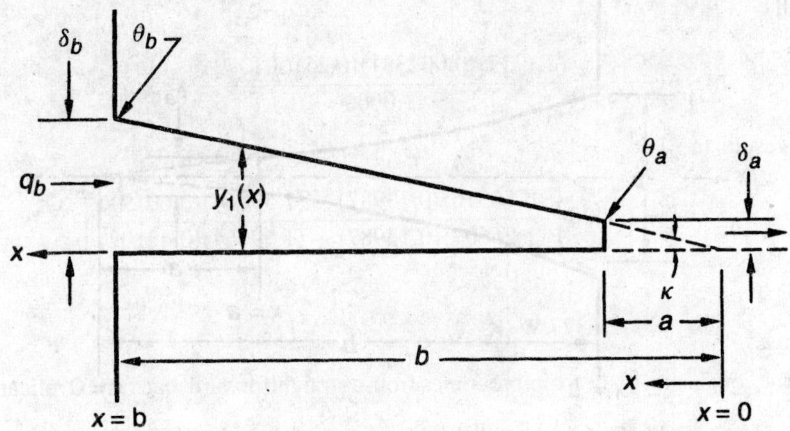

Figure 6.3 Longitudinal fin of half-trapezoidal profile.

$$x \tan \kappa \frac{d^2\theta}{dx^2} + \tan \kappa \frac{d\theta}{dx} - \frac{h}{k}(\sec \kappa + 1)\theta = 0$$

and further simplification gives

$$x \frac{d^2\theta}{dx^2} + \frac{d\theta}{dx} - n^2\theta = 0 \tag{6.30}$$

where n for this case is given by

$$n = \left[\frac{h(\sec \kappa + 1)}{k \tan \kappa} \right]^{1/2}$$

Equations (6.15) and (6.30) are identical in form and must therefore have identical solutions. As long as it is recognized that the value of n differs in the two cases, eqs. (6.23) and (6.25) through (6.29) may be used to evaluate the elements of all the linear transformations for the longitudinal fin of half trapezoidal profile.

6.2.5 Longitudinal Fin of Truncated Concave Parabolic Profile

The configuration and coordinate system for the longitudinal fin of truncated concave parabolic profile is displayed in Fig. 6.4. It may be noted that the origin of the x-coordinate axis is *not* at the fin tip and that the fin *does not* taper to zero thickness at the tip. The profile function for this fin is given by

$$y_1(x) = -y_2(x) = \frac{\delta_b}{2} \left(\frac{x}{b} \right)^2$$

and with this in eq. (6.1), with $h_1 = h_2 = h$ and with no provision made for the length of arc of the fin faces, the differential equation for the temperature excess becomes

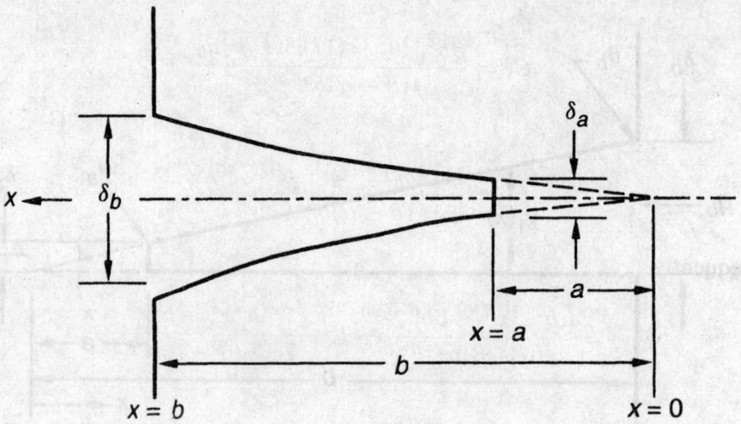

Figure 6.4 Longitudinal fin of truncated parabolic profile.

$$x^2\frac{d^2\theta}{dx^2} + 2x\frac{d\theta}{dx} - m^2b^2\theta = 0 \tag{6.31}$$

where again

$$m = \left(\frac{2h}{k\delta_b}\right)^{1/2}$$

Equation (6.31) is an Euler equation and by appropriate transformation, its general solution can be shown to be

$$\theta(x) = C_1x^\alpha + C_2x^\beta \tag{6.32}$$

where

$$\alpha, \beta = -\frac{1}{2} \pm \frac{(1+4m^2b^2)^{1/2}}{2}$$

Application of the initial conditions of eqs. (6.17) leads to the temperature excess and heat flow at the fin base in terms of the arbitrary constants

$$\theta(b) = \theta_b = C_1b^\alpha + C_2b^\beta$$

$$q(b) = q_b = k\delta_bL(\alpha C_1b^{\alpha-1} + \beta C_2b^{\beta-1})$$

because

$$A(b) = 2y_1(b)L = 2\frac{\delta_b}{2}\left(\frac{b}{b}\right)^2 L = \delta_bL$$

Simultaneous solution for the arbitrary constants C_1 and C_2 yields

$$C_1 = \frac{\beta b^{\beta-1}\theta_b - (1/k\delta_b L)b^{\beta} q_b}{(\beta - \alpha)b^{\alpha+\beta-1}}$$

and

$$C_2 = \frac{(1/k\delta_b L b^{\alpha} q_b - \alpha b^{\alpha-1}\theta_b}{(\beta - \alpha)b^{\alpha+\beta-1}}$$

Consequently,

$$\theta(x) = \frac{\alpha b^{\alpha-1}x^{\beta} - \beta b^{\beta-1}x^{\alpha}}{(\alpha - \beta)b^{\alpha+\beta-1}}\theta_b + \frac{b^{\beta}x^{\alpha} - b^{\alpha}x^{\beta}}{k\delta_b L(\alpha - \beta)b^{\alpha+\beta-1}}q_b \tag{6.33a}$$

and

$$q(x) = \frac{k\delta_b L\alpha\beta[b^{\alpha-1}x^{\beta-1} - b^{\beta-1}x^{\alpha-1}]}{(\alpha - \beta)b^{\alpha+\beta-1}}\theta_b$$

$$+ \frac{\alpha b^{\beta}x^{\alpha-1} - \beta b^{\alpha}x^{\beta-1}}{(\alpha - \beta)b^{\alpha+\beta-1}}q_b \tag{6.33b}$$

Then

$$\lambda_1(x) = \frac{\alpha b^{\alpha-1}x^{\beta} - \beta b^{\beta-1}x^{\alpha}}{(\alpha - \beta)b^{\alpha+\beta-1}}$$

$$= \frac{1}{\alpha - \beta}\left[\alpha\left(\frac{x}{b}\right)^{\beta} - \beta\left(\frac{x}{b}\right)^{\alpha}\right] \tag{6.34a}$$

and

$$\lambda_2(x) = \frac{b^{\beta}x^{\alpha} - b^{\alpha}x^{\beta}}{k\delta_b L(\alpha - \beta)b^{\alpha+\beta-1}}$$

$$= \frac{b}{k\delta_b L(\alpha - \beta)}\left[\left(\frac{x}{b}\right)^{\alpha} - \left(\frac{x}{b}\right)^{\beta}\right] \tag{6.34b}$$

and, as required by eq. (6.2*a*), $\lambda_1(b) = 1$ and $\lambda_2(b) = 0$. The derivatives of eqs. (6.34) are

$$\lambda_1'(x) = \frac{\alpha\beta(b^{\alpha-1}x^{\beta-1} - b^{\beta-1}x^{\alpha-1})}{(\alpha - \beta)b^{\alpha+\beta-1}} \tag{6.35a}$$

and

$$\lambda_2'(x) = \frac{\alpha b^{\beta}x^{\alpha-1} - \beta b^{\alpha}x^{\beta-1}}{k\delta_b L(\alpha - \beta)b^{\alpha+\beta-1}} \tag{6.35b}$$

and as required by eq. (6.2*b*), $\lambda_1'(b) = 0$ and $\lambda_2'(b) = 1/k\delta_b L$.

The elements of the **Γ**-matrix can now be set down from an inspection of eqs. (6.33) using $x = a$ at the tip of the fin:

$$\gamma_{11} = \lambda_1(a) = \frac{1}{\alpha - \beta}\left[\alpha\left(\frac{a}{b}\right)^{\beta} - \beta\left(\frac{a}{b}\right)^{\alpha}\right] \tag{6.36a}$$

$$\gamma_{12} = \lambda_2(a) = \frac{b}{k\delta_b L(\alpha - \beta)}\left[\left(\frac{a}{b}\right)^{\alpha} - \left(\frac{a}{b}\right)^{\beta}\right] \tag{6.36b}$$

and with $A(a) = a^2\delta_b L/b^2$,

$$\gamma_{21} = kA(a)\lambda_1'(a) = \frac{k\delta_b L\alpha\beta}{b(\alpha - \beta)}\left[\left(\frac{a}{b}\right)^{\beta+1} - \left(\frac{a}{b}\right)^{\alpha+1}\right] \tag{6.36c}$$

and

$$\gamma_{22} = kA(a)\lambda_2'(a) = \frac{1}{\alpha - \beta}\left[\alpha\left(\frac{a}{b}\right)^{\alpha+1} - \beta\left(\frac{a}{b}\right)^{\beta+1}\right] \tag{6.36d}$$

The determinant of $\mathbf{\Gamma}$ is

$$\begin{aligned}
\det\begin{bmatrix}\gamma_{11} & \gamma_{12}\\ \gamma_{21} & \gamma_{22}\end{bmatrix} &= \gamma_{11}\gamma_{22} - \gamma_{12}\gamma_{21}\\
&= \frac{1}{(\alpha-\beta)^2}\left[\alpha\left(\frac{a}{b}\right)^{\beta} - \beta\left(\frac{a}{b}\right)^{\alpha}\right]\left[\alpha\left(\frac{a}{b}\right)^{\alpha+1} - \beta\left(\frac{a}{b}\right)^{\beta+1}\right]\\
&\quad - \frac{\alpha\beta}{(\alpha-\beta)^2}\left[\left(\frac{a}{b}\right)^{\alpha} - \left(\frac{a}{b}\right)^{\beta}\right]\left[\left(\frac{a}{b}\right)^{\beta+1} - \left(\frac{a}{b}\right)^{\alpha+1}\right]
\end{aligned}$$

The algebra is really not that horrendous:

$$\begin{aligned}
\det\mathbf{\Gamma} &= \frac{1}{(\alpha-\beta)^2}\left[\alpha^2\left(\frac{a}{b}\right)^{\alpha+\beta+1} - \alpha\beta\left(\frac{a}{b}\right)^{2\alpha+1} - \alpha\beta\left(\frac{a}{b}\right)^{2\beta+1} + \beta^2\left(\frac{a}{b}\right)^{\alpha+\beta+1}\right]\\
&\quad - \frac{1}{(\alpha-\beta)^2}\left[-\alpha\beta\left(\frac{a}{b}\right)^{\alpha+\beta+1} + \alpha\beta\left(\frac{a}{b}\right)^{2\alpha+1}\right.\\
&\quad\left. + \alpha\beta\left(\frac{a}{b}\right)^{2\beta+1} - \alpha\beta\left(\frac{a}{b}\right)^{\alpha+\beta+1}\right]\\
&= \frac{1}{(\alpha-\beta)^2}(\alpha^2 - 2\alpha\beta + \beta^2)\left(\frac{a}{b}\right)^{\alpha+\beta+1}\\
&= \frac{(\alpha-\beta)^2}{(\alpha-\beta)^2}\left(\frac{a}{b}\right)^{\alpha+\beta+1}\\
&= \left(\frac{a}{b}\right)^{\alpha+\beta+1}
\end{aligned}$$

and because

$$\alpha + \beta + 1 = \left[-\frac{1}{2} + \frac{(1+4m^2b^2)^{1/2}}{2}\right] + \left[-\frac{1}{2} - \frac{(1+4m^2b^2)^{1/2}}{2}\right] + 1 = 0$$

the determinant reduces to unity:

$$\Delta_\Gamma = \left(\frac{a}{b}\right)^{\alpha+\beta+1} = 1$$

The other five linear transformations can be determined through the use of Table 5.2. The elements of these matrices follow.

T-*Matrix*

$$\tau_{11} = \frac{1}{\alpha - \beta}\left[\alpha\left(\frac{a}{b}\right)^{\alpha+1} - \beta\left(\frac{a}{b}\right)^{\beta+1}\right] \tag{6.37a}$$

$$\tau_{12} = \frac{b}{k\delta_b L(\alpha - \beta)}\left[\left(\frac{a}{b}\right)^{\beta} - \left(\frac{a}{b}\right)^{\alpha}\right] \tag{6.37b}$$

$$\tau_{21} = \frac{k\delta_b L\alpha\beta}{b(\alpha - \beta)}\left[\left(\frac{a}{b}\right)^{\alpha+1} - \left(\frac{a}{b}\right)^{\beta+1}\right] \tag{6.37c}$$

and

$$\tau_{22} = \frac{1}{\alpha - \beta}\left[\alpha\left(\frac{a}{b}\right)^{\beta} - \beta\left(\frac{a}{b}\right)^{\alpha}\right] \tag{6.37d}$$

Z-*Matrix*

$$z_{11} = \frac{b}{k\delta_b L\alpha\beta}\left[\frac{\beta(a/b)^{\beta+1} - \alpha(a/b)^{\alpha+1}}{(a/b)^{\beta+1} - (a/b)^{\alpha+1}}\right] \tag{6.38a}$$

$$z_{12} = \frac{b(\alpha - \beta)}{k\delta_b L\alpha\beta}\left[\left(\frac{a}{b}\right)^{\beta+1} - \left(\frac{a}{b}\right)^{\alpha+1}\right]^{-1} \tag{6.38b}$$

$$z_{21} = \frac{b(\beta - \alpha)}{k\delta_b L\alpha\beta}\left[\left(\frac{a}{b}\right)^{\beta+1} - \left(\frac{a}{b}\right)^{\alpha+1}\right]^{-1} \tag{6.38c}$$

and

$$z_{22} = \frac{b}{k\delta_b L\alpha\beta}\left[\frac{\alpha(a/b)^{\beta} - \beta(a/b)^{\alpha}}{(a/b)^{\beta+1} - (a/b)^{\alpha+1}}\right] \tag{6.38d}$$

Y-*Matrix*

$$y_{11} = \frac{k\delta_b L}{b}\left[\frac{\beta(a/b)^{\alpha} - \alpha(a/b)^{\beta}}{(a/b)^{\alpha} - (a/b)^{\beta}}\right] \tag{6.39a}$$

$$y_{12} = \frac{k\delta_b L(\alpha - \beta)}{b}\left[\left(\frac{a}{b}\right)^{\alpha} - \left(\frac{a}{b}\right)^{\beta}\right]^{-1} \tag{6.39b}$$

$$y_{21} = \frac{k\delta_b L(\beta - \alpha)}{b}\left[\left(\frac{a}{b}\right)^{\alpha} - \left(\frac{a}{b}\right)^{\beta}\right]^{-1} \tag{6.39c}$$

and

$$y_{22} = \frac{k\delta_b L}{b} \frac{\alpha(a/b)^{\alpha+1} - \beta(a/b)^{\beta+1}}{(a/b)^{\alpha} - (a/b)^{\beta}} \tag{6.39d}$$

H-*Matrix*

$$h_{11} = \frac{b}{k\delta_b L} \frac{(a/b)^{\beta} - (a/b)^{\alpha}}{\alpha(a/b)^{\beta} - \beta(a/b)^{\alpha}} \tag{6.40a}$$

$$h_{12} = (\alpha - \beta)\left[\alpha\left(\frac{a}{b}\right)^{\beta} - \beta\left(\frac{a}{b}\right)^{\alpha}\right]^{-1} \tag{6.40b}$$

$$h_{21} = (\alpha - \beta)\left[\alpha\left(\frac{a}{b}\right)^{\beta} - \beta\left(\frac{a}{b}\right)^{\alpha}\right]^{-1} \tag{6.40c}$$

and

$$h_{22} = \frac{k\delta_b L\alpha\beta}{b} \frac{(a/b)^{\beta+1} - (a/b)^{\alpha+1}}{\alpha(a/b)^{\beta} - \beta(a/b)^{\alpha}} \tag{6.40d}$$

G-*Matrix*

$$g_{11} = \frac{k\delta_b L\alpha\beta}{b} \frac{(a/b)^{\alpha+1} - (a/b)^{\beta+1}}{\alpha(a/b)^{\alpha+1} - \beta(a/b)^{\beta+1}} \tag{6.41a}$$

$$g_{12} = (\alpha - \beta)\left[\alpha\left(\frac{a}{b}\right)^{\alpha+1} - \beta\left(\frac{a}{b}\right)^{\beta+1}\right]^{-1} \tag{6.41b}$$

$$g_{21} = (\alpha - \beta)\left[\alpha\left(\frac{a}{b}\right)^{\alpha+1} - \beta\left(\frac{a}{b}\right)^{\beta+1}\right]^{-1} \tag{6.41c}$$

and

$$g_{22} = \frac{b}{k\delta_b L} \frac{(a/b)^{\alpha} - (a/b)^{\beta}}{\alpha(a/b)^{\alpha+1} - \beta(a/b)^{\beta+1}} \tag{6.41d}$$

6.2.6 Radial Fin of Rectangular Profile

The radial fin of rectangular profile is the only radial fin to be treated in this work. Thus there is no need to develop a general differential equation for radial fins.

For the radial fin of rectangular profile displayed in Fig. 6.5, observe that the origin of the height coordinate system is taken at the center of curvature of the fin base and fin tip. The height coordinate r has a positive orientation from fin base to fin tip, and it is to be noted that $r_a > r_b$.

Refer to Fig. 6.5 and consider the differential element dr. The difference in heat conducted into the differential element at dr and that leaving at $r + dr$ is

$$dq = k\frac{d}{dr}\left[(2\pi r)\delta\frac{d\theta}{dr}\right]dr$$

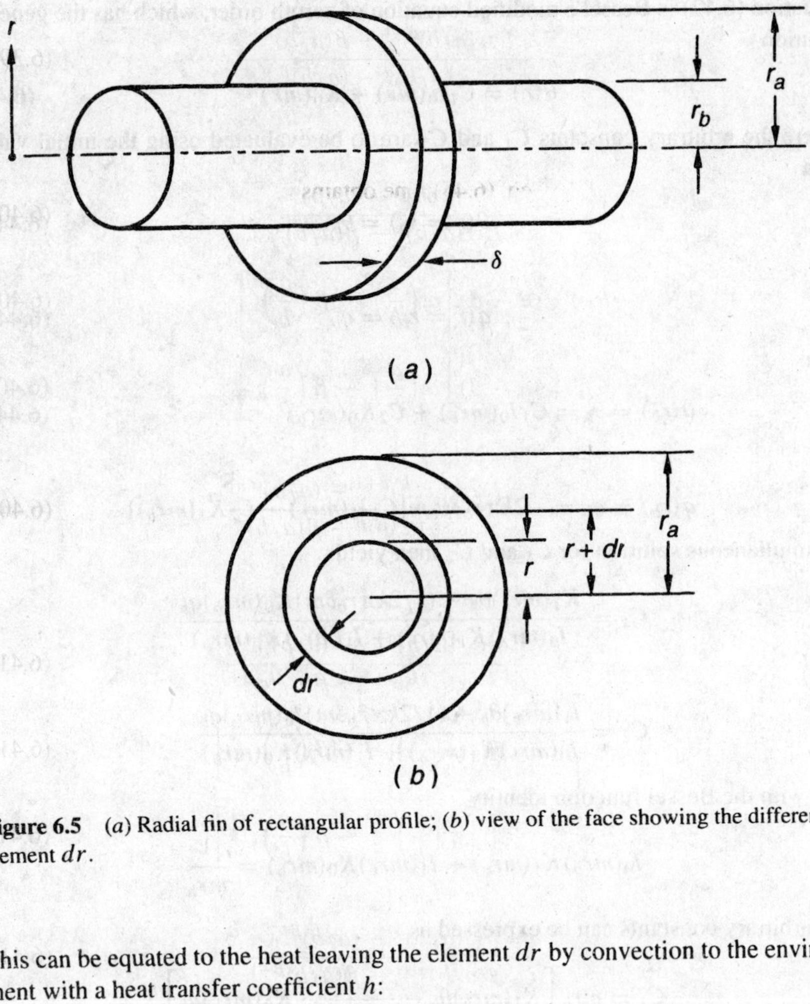

Figure 6.5 (*a*) Radial fin of rectangular profile; (*b*) view of the face showing the differential element dr.

This can be equated to the heat leaving the element dr by convection to the environment with a heat transfer coefficient h:

$$dq = 2h(2\pi r\, dr)\theta$$

so that the heat balance is

$$2k\pi\delta \frac{d}{dr}\left(r\frac{d\theta}{dr}\right) dr = 4h\pi\theta r\, dr$$

This leads to the differential equation for temperature excess:

$$r^2\frac{d^2\theta}{dr^2} + r\frac{d\theta}{dr} - m^2r^2\theta = 0 \tag{6.42}$$

where

$$m = \left(\frac{2h}{k\delta}\right)^{1/2}$$

Equation (6.42) is Bessel's modified equation of zeroth order, which has the general solution

$$\theta(r) = C_1 I_0(mr) + K_0(mr) \tag{6.43}$$

where the arbitrary constants C_1 and C_2 are to be evaluated using the initial value data

$$\theta(r = r_b) = \theta_b \tag{6.44a}$$

and

$$q(r = r_b) = q_b \tag{6.44b}$$

Thus

$$\theta(r_b) = \theta_b = C_1 I_0(mr_b) + C_2 K_0(mr_b) \tag{6.44c}$$

and

$$q(r_b) = q_b = -2k\pi r_b \delta \theta_b m[C_1 I_1(mr_b) - C_2 K_1(mr_b)] \tag{6.44d}$$

A simultaneous solution for C_1 and C_2 then yields

$$C_1 = \frac{K_1(mr_b)\theta_b - (1/2k\pi r_b \delta m)K_0(mr_b)q_b}{I_0(mr_b)K_1(mr_b) + I_1(mr_b)K_0(mr_b)}$$

and

$$C_2 = \frac{I_1(mr_b)\theta_b + (1/2k\pi r_b \delta m)I_0(mr_b)q_b}{I_0(mr_b)K_1(mr_b) + I_1(mr_b)K_0(mr_b)}$$

and with the Bessel function identity

$$I_0(mr_b)K_1(mr_b) + I_1(mr_b)K_0(mr_b) = \frac{1}{mr_b}$$

the arbitrary constants can be expressed as

$$C_1 = mr_b \left[K_1(mr_b)\theta_b - \frac{1}{2k\pi r_b \delta m} K_0(mr_b)q_b \right]$$

$$C_2 = mr_b \left[I_1(mr_b)\theta_b + \frac{1}{2k\pi r_b \delta m} I_0(mr_b)q_b \right]$$

These make

$$\theta(r) = mr_b[K_1(mr_b)I_0(mr) + I_1(mr_b)K_0(mr)]\theta_b + \frac{1}{2k\pi\delta}[I_0(mr_b)K_0(mr) - K_0(mr_b)I_0(mr)]\,q_b \tag{6.45}$$

which indicates that

$$\lambda_1(r) = mr_b[K_1(mr_b)I_0(mr) + I_1(mr_b)K_0(mr)] \tag{6.46a}$$

and

$$\lambda_2(r) = \frac{1}{2k\pi\delta}\left[I_0(mr_b)K_0(mr) - K_0(mr_b)I_0(mr)\right] \tag{6.46b}$$

and that $\lambda_1(r_b) = 1$ and $\lambda_2(r_b) = 0$, as required by eqs. (6.2*a*).

The heat flow at any point r is $q(r) = -2k\pi\delta r\theta'(r)$, and after a differentiation of eq. (6.45), one obtains

$$q(r) = 2\pi k\delta m^2 r_b r[I_1(mr_b)K_1(mr) - K_1(mr_b)I_1(mr)]\theta_b + mr[I_0(mr_b)K_1(mr) + K_0(mr_b)I_1(mr)]q_b$$

and this makes

$$\lambda_1'(r) = m^2 r_b[I_1(mr_b)K_1(mr) - K_1(mr_b)I_1(mr)] \tag{6.47a}$$

and

$$\lambda_2'(r) = \frac{m}{2k\pi\delta}\left[I_0(mr_b)K_1(mr) + K_0(mr_b)I_1(mr)\right] \tag{6.47b}$$

Here it is to be noted that $\lambda_1'(r_b) = 0$ and $\lambda_2'(r_b) = -m/2k\pi\delta r_b = -1/kA(r_b)$, as required by eqs. (6.2*b*), where the minus sign appears because the heat flow is in the direction of the positive sense of the coordinate system.

Let a radius ratio be proposed:

$$\rho \equiv \frac{r_b}{r_a} \qquad (\rho < 1) \tag{6.48}$$

Moreover, consider the profile area

$$A_p = \delta(r_a - r_b) \tag{6.49}$$

and note that

$$\delta = \frac{A_p}{r_a - r_b}$$

so that

$$m = \left(\frac{2h}{k\delta}\right)^{1/2} = (r_a - r_b)^{1/2}\left(\frac{2h}{kA_p}\right)^{1/2} \tag{6.50}$$

Then

$$mr_b = r_b(r_a - r_b)^{1/2}\left(\frac{2h}{kA_p}\right)^{1/2}$$

and if both numerator and denominator are multiplied by $(r_a - r_b)$,

$$mr_b = r_b\frac{(r_a - r_b)^{3/2}}{r_a - r_b}\left(\frac{2h}{kA_p}\right)^{1/2} \tag{6.51}$$

An identical procedure yields

$$mr_a = r_a \frac{(r_a - r_b)^{3/2}}{r_a - r_b} \left(\frac{2h}{kA_p} \right)^{1/2} \tag{6.52}$$

Now define

$$\phi \equiv \frac{(r_a - r_b)^{3/2}}{r_a - r_b} \left(\frac{2h}{kA_p} \right)^{1/2} \tag{6.53}$$

and write

$$\alpha = mr_b = \frac{r_b \phi}{r_a - r_b} = \frac{\rho\phi}{1 - \phi} \tag{6.54a}$$

and

$$\beta = mr_a = \frac{r_a \phi}{r_a - r_b} = \frac{\phi}{1 - \phi} \tag{6.54b}$$

The parameters α and β given by eqs. (6.54) allow the Bessel function arguments to be written in terms of the parameter ϕ given by eq. (6.53) and the radius ratio ρ.

With eqs. (6.46) and (6.47) in hand, the $\mathbf{\Gamma}$-matrix can be set up in accordance with eqs. (6.6) using α and β as defined by eqs. (6.54).

$$\gamma_{11} = \lambda_1(r_a) = \alpha[I_0(\beta)K_1(\alpha) + I_1(\alpha)K_0(\beta)] \tag{6.55a}$$

$$\gamma_{12} = \lambda_2(r_a) = \frac{1}{2k\pi\delta}[I_0(\alpha)K_0(\beta) - I_0(\beta)K_0(\alpha)] \tag{6.55b}$$

$$\gamma_{21} = kA(r_a)\lambda_1'(r_a) = 2k\pi\delta m r_a \alpha[I_1(\alpha)K_1(\beta) - I_1(\beta)K_1(\alpha)]$$

or with $mr_a = \beta$,

$$\gamma_{21} = 2k\pi\delta\alpha\beta[I_1(\alpha)K_1(\beta) - I_1(\beta)K_1(\alpha)] \tag{6.55c}$$

and

$$\gamma_{22} = kA(r_a)\lambda_2'(r_a) = 2k\pi\delta r_a \frac{m}{2k\pi\delta}[I_0(\alpha)K_1(\beta) + I_1(\beta)K_0(\alpha)]$$

and again, with $mr_a = \beta$,

$$\gamma_{22} = \beta[I_0(\alpha)K_1(\beta) + I_1(\beta)K_0(\alpha)] \tag{6.55d}$$

The determinant of the $\mathbf{\Gamma}$-matrix is

$$\det \mathbf{\Gamma} = \gamma_{11}\gamma_{22} - \gamma_{12}\gamma_{21}$$

and using eqs. (6.55) it can be shown that

$$\det \mathbf{\Gamma} = 1$$

so that the elements of the other linear transformations can be developed from Table 5.2 using the **Γ**-matrix.

T-*Matrix*

$$\tau_{11} = \beta[I_0(\alpha)K_1(\beta) + I_1(\beta)K_0(\alpha) \tag{6.56a}$$

$$\tau_{12} = \frac{1}{2k\pi\delta}[I_0(\beta)K_0(\alpha) - I_0(\alpha)K_0(\beta)] \tag{6.56b}$$

$$\tau_{21} = 2k\pi\delta\alpha\beta[I_1(\beta)K_1(\alpha) - I_1(\alpha)K_1(\beta)] \tag{6.56c}$$

and

$$\tau_{22} = \alpha[I_0(\beta)K_1(\alpha) + I_1(\alpha)K_0(\beta)] \tag{6.56d}$$

Z-*Matrix*

$$z_{11} = \frac{1}{2k\pi\delta\alpha}\frac{I_0(\alpha)K_1(\beta) + I_1(\beta)K_0(\alpha)}{I_1(\beta)K_1(\alpha) - I_1(\alpha)K_1(\beta)} \tag{6.57a}$$

$$z_{12} = \frac{1}{2k\pi\delta\alpha\beta}\frac{1}{I_1(\alpha)K_1(\beta) - I_1(\beta)K_1(\alpha)} \tag{6.57b}$$

$$z_{21} = \frac{1}{2k\pi\delta\alpha\beta}\frac{1}{I_1(\beta)K_1(\alpha) - I_1(\alpha)K_1(\beta)} \tag{6.57c}$$

and

$$z_{22} = \frac{1}{2k\pi\delta\beta}\frac{I_0(\beta)K_1(\alpha) + I_1(\alpha)K_0(\beta)}{I_1(\alpha)K_1(\beta) - I_1(\beta)K_1(\alpha)} \tag{6.57d}$$

Y-*Matrix*

$$y_{11} = 2k\pi\delta\alpha\frac{I_0(\beta)K_1(\alpha) + I_1(\alpha)K_0(\beta)}{I_0(\beta)K_0(\alpha) - I_0(\alpha)K_0(\beta)} \tag{6.58a}$$

$$y_{12} = \frac{2k\pi\delta}{I_0(\alpha)K_0(\beta) - I_0(\beta)K_0(\alpha)} \tag{6.58b}$$

$$y_{21} = \frac{2k\pi\delta}{I_0(\beta)K_0(\alpha) - I_0(\alpha)K_0(\beta)} \tag{6.58c}$$

and

$$y_{22} = 2k\pi\delta\beta\frac{I_0(\alpha)K_1(\beta) + I_1(\beta)K_0(\alpha)}{I_0(\alpha)K_0(\beta) - I_0(\beta)K_0(\alpha)} \tag{6.58d}$$

H-*Matrix*

$$h_{11} = \frac{1}{2k\pi\delta} \frac{I_0(\beta)K_0(\alpha) - I_0(\alpha)K_0(\beta)}{I_0(\beta)K_1(\alpha) + I_1(\alpha)K_0(\beta)} \tag{6.59a}$$

$$h_{12} = \frac{1}{\alpha[I_0(\beta)K_1(\alpha) + I_1(\alpha)K_0]} \tag{6.59b}$$

$$h_{21} = \frac{1}{\alpha[I_0(\beta)K_1(\alpha) + I_1(\alpha)K_0(\beta)]} \tag{6.59c}$$

and

$$h_{22} = 2k\pi\delta\beta \frac{I_1(\alpha)K_1(\beta) - I_1(\beta)K_1(\alpha)}{I_0(\beta)K_1(\alpha) + I_1(\alpha)K_0(\beta)} \tag{6.59d}$$

G-*Matrix*

$$g_{11} = 2k\pi\delta\beta \frac{I_1(\beta)K_1(\alpha) - I_1(\alpha)K_1(\beta)}{I_0(\alpha)K_1(\beta) + I_1(\beta)K_0(\alpha)} \tag{6.60a}$$

$$g_{12} = \frac{1}{\beta[I_0(\alpha)K_1(\beta) + I_1(\beta)K_0(\alpha)]} \tag{6.60b}$$

$$g_{21} = \frac{1}{\beta[I_0(\alpha)K_1(\beta) + I_1(\beta)K_0(\alpha)]} \tag{6.60c}$$

and

$$g_{22} = \frac{1}{2k\pi\delta\beta} \frac{I_0(\alpha)K_0(\beta) - I_0(\beta)K_0(\alpha)}{I_0(\alpha)K_1(\beta) + I_1(\beta)K_0(\alpha)} \tag{6.60d}$$

Example 6.4. A temperature measurement at the tip of a radial fin of rectangular profile indicates that the temperature excess is 18.4°C. The fin is fabricated from a material with a thermal conductivity $k = 328\ \mathrm{W/m \cdot K}$ and the inner and outer radii are 2 and 4 cm, respectively. The fin is 0.3175 cm (0.125 in.) thick and the heat transfer coefficient between the fin faces and the surrounding environment is known to be $h = 133.3\ \mathrm{W/m^2 \cdot K}$. What are θ_b and q_b if the tip heat transfer coefficient is $h_a = 52\ \mathrm{W/m^2 \cdot K}$?

SOLUTION. The **T**-matrix can be employed after the value of q_a is computed. Here

$$S_a = 2\pi r_a \delta = 2\pi(0.04)(0.003175) = 7.980 \times 10^{-4}\ \mathrm{m^2}$$

so that

$$q_a = h_a S_a \theta_a = (52)(7.980 \times 10^{-4})(18.4) = 0.763\ \mathrm{W}$$

Then

$$m = \left(\frac{2h}{k\delta}\right)^{1/2} = \left[\frac{(2)(133.3)}{(328)(0.003175)}\right]^{1/2} = 16.0000\ \mathrm{m^{-1}}$$

so that

$$\alpha = mr_b = (16)(0.02) = 0.320$$

and

$$\beta = mr_a = (16)(0.04) = 0.640$$

The values of eight modified Bessel functions will be needed:

$$\begin{aligned} \beta = mr_a &= 0.640 & \alpha = mr_b &= 0.320 \\ I_0(\beta) &= 1.10505 & I_0(\alpha) &= 1.02576 \\ I_1(\beta) &= 0.33667 & I_1(\alpha) &= 0.16206 \\ K_0(\beta) &= 0.72767 & K_0(\alpha) &= 1.31356 \\ K_1(\beta) &= 1.19227 & K_1(\alpha) &= 2.83899 \end{aligned}$$

By eqs. (6.56),

$$\begin{aligned} \tau_{11} &= \beta[I_0(\alpha)K_1(\beta) + I_1(\beta)K_0(\alpha)] \\ &= (0.64)[1.02576(1.19227) + (0.33667)(1.31356)] \\ &= (0.64)[1.2230 + 0.4422] = 1.0657 \\ \tau_{12} &= (1/2k\pi\delta)\,[I_0(\beta)K_0(\alpha) - I_0(\alpha)K_0(\beta)] \end{aligned}$$

With

$$\frac{1}{2k\pi\delta} = \frac{1}{(2)(328)\pi(0.003175)} = 0.1528$$

$$\begin{aligned} \tau_{12} &= (0.1528)[(1.10505)(1.31356) - (1.02576)(0.72767)] \\ &= (0.1528)[1.4515 - 0.7464] = 0.1078 \text{ K/W} \\ \tau_{21} &= 2k\pi\delta\alpha\beta[I_1(\beta)K_1(\alpha) - I_1(\alpha)K_1(\beta)] \end{aligned}$$

and with

$$2k\pi\delta\alpha\beta = (2)(328)\pi(0.003175)(0.32)(0.64) = 1.3401$$

$$\begin{aligned} \tau_{21} &= (1.3401)[(0.33667)(2.83899) - (0.16206)(1.19227)] \\ &= (1.3401)(0.9558 - 0.1932) = 1.0219 \text{ W/K} \end{aligned}$$

and

$$\begin{aligned} \tau_{22} &= \alpha[I_0(\beta)K_1(\alpha) + I_1(\alpha)K_0(\beta)] \\ &= (0.32)[(1.10505)(2.83899) - (0.16206)(0.72767)] \\ &= (0.32)(3.1372 - 0.1179) = 0.9662 \end{aligned}$$

Now with $\theta_b = 18.4$°C and $q_a = 0.763$ W,

$$\begin{bmatrix} \theta_b \\ q_b \end{bmatrix} = \mathbf{T}\begin{bmatrix} \theta_a \\ q_a \end{bmatrix} = \begin{bmatrix} 1.6652 & 0.1078 \\ 1.0219 & 0.9662 \end{bmatrix}\begin{bmatrix} 18.4 \\ 0.763 \end{bmatrix} = \begin{bmatrix} 30.7^\circ\text{C} \\ 19.54 \text{ W} \end{bmatrix}$$

6.2.7 Generalized Differential Equation for Spines

The first consideration concerns spines of circular cross section, and Fig. 6.6 presents the terminology and coordinate system. Observe that the x-axis is along the axis of symmetry and its positive sense is directed from the spine tip at $x = a$ to the spine base at $x = b$. The coordinate r denotes the radius of the circular cross section and is a function of x, but in accordance with the Murray–Gardner assumptions, the steady-state temperature excess, $\theta = T - T_s$, is a function only of x.

The heat flow q is taken positive from spine base to spine tip and the difference in heat conducted into the differential element at x and that leaving at $x + dx$ is

$$dq(x) = \frac{d}{dx}\left[k(x)(\pi r^2)\frac{d\theta(x)}{dx}\right] dx$$

or, for the case of constant thermal conductivity,

$$dq(x) = k\pi \frac{d}{dx}\left[r^2 \frac{d\theta(x)}{dx}\right] dx \tag{6.61a}$$

The heat dissipated through the differential element of spine surface is given by

$$dq(x) = h(x)\, dS(x)\theta(x)$$

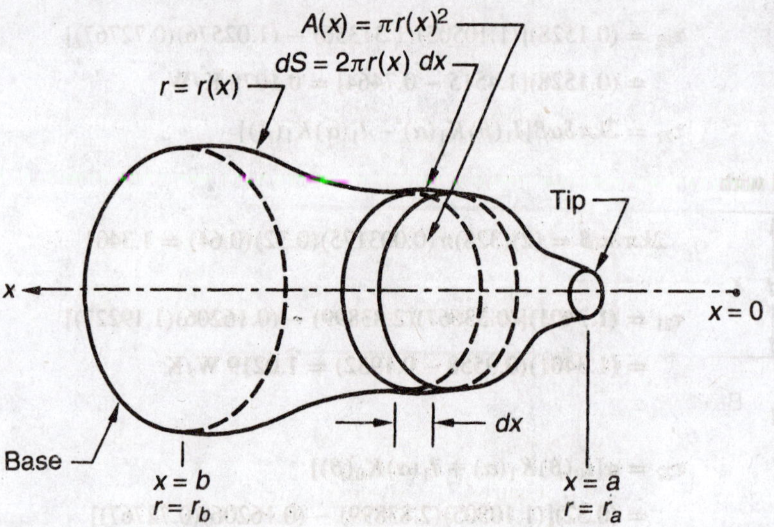

Figure 6.6 Coordinate system for the generalized spine of circular cross section.

and with the spine surface area given by

$$dS(x) = 2\pi r \left[1 + \left(\frac{dr}{dx}\right)^2\right]^{1/2} dx$$

and with $h(x) = h$ (a constant),[2] one obtains

$$dq(x) = 2h\pi r \left[1 + \left(\frac{dr}{dx}\right)^2\right]^{1/2} \theta(x)\, dx \tag{6.61b}$$

When eqs. (6.61) are equated, the result is

$$k\pi \frac{d}{dx}\left[r^2 \frac{d\theta(x)}{dx}\right] dx = 2h\pi r \left[1 + \left(\frac{dr}{dx}\right)^2\right]^{1/2}$$

and simple adjustment yields the general equation for the temperature excess of the spine of circular cross section:

$$r\frac{d^2\theta}{dx^2} + 2\frac{dr}{dx}\frac{d\theta}{dx} - \frac{2h}{k}\left[1 + \left(\frac{dr}{dx}\right)^2\right]^{1/2} \theta = 0 \tag{6.62}$$

6.2.8 Cylindrical Spine

In Fig. 6.7 the cylindrical spine of diameter d has its origin of the height coordinate x, taken at the fin tip ($x = a = 0$). Here $r(x) = d/2$, and $dr/dx = 0$ and eq. (6.62) reduces to

$$\frac{d^2\theta}{dx^2} - m^2\theta = 0 \tag{6.63}$$

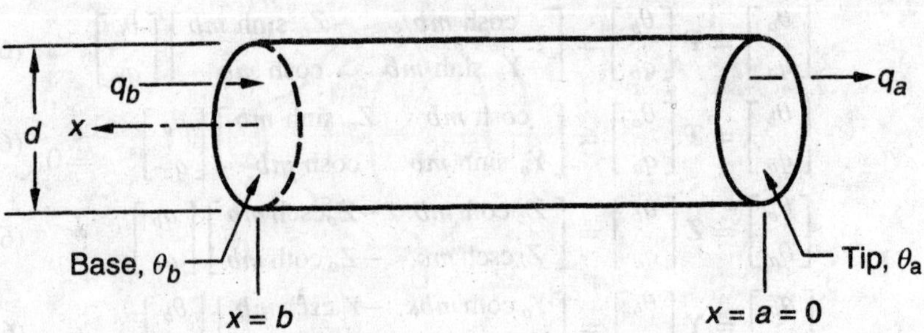

Figure 6.7 Cylindrical spine.

[2]A case where h is not a constant is treated in Chapter 7.

where

$$m = \left(\frac{4h}{kd}\right)^{1/2}$$

Equation (6.63) is identical in form to eq (1.59), with the solution for the temperature excess given by eq. (1.60):

$$\theta = \theta_b \frac{\cosh mx}{\cosh mb}$$

The heat flow through the base of the spine is given by

$$q_b = kA \left.\frac{d\theta}{dx}\right|_{x=b}$$

and with $A = \pi d^2/4$,

$$q_b = \frac{\pi}{4} kd^2 m \left.\frac{\sinh mx}{\cosh mb}\right|_{x=b}$$

or with $m = (4h/kd)^{1/2}$,

$$q_b = \frac{1}{2}(\pi^2 hkd^3)^{1/2} \tanh mb$$

which makes

$$Y_o = \frac{(\pi^2 hkd^3)^{1/2}}{2}$$

and

$$Z_o = \frac{2}{(\pi^2 hkd^3)^{1/2}}$$

The linear transformations of eqs. (6.8) through (6.13) apply as long as the proper values of m, Y_o, and Z_o are employed. These equations are repeated here:

$$\begin{bmatrix} \theta_a \\ q_a \end{bmatrix} = \mathbf{\Gamma} \begin{bmatrix} \theta_b \\ q_b \end{bmatrix} = \begin{bmatrix} \cosh mb & -Z_o \sinh mb \\ -Y_o \sinh mb & \cosh mb \end{bmatrix} \begin{bmatrix} \theta_b \\ q_b \end{bmatrix} \tag{6.64}$$

$$\begin{bmatrix} \theta_b \\ q_b \end{bmatrix} = \mathbf{T} \begin{bmatrix} \theta_a \\ q_a \end{bmatrix} = \begin{bmatrix} \cosh mb & Z_o \sinh mb \\ Y_o \sinh mb & \cosh mb \end{bmatrix} \begin{bmatrix} \theta_a \\ q_a \end{bmatrix} \tag{6.65}$$

$$\begin{bmatrix} \theta_b \\ \theta_a \end{bmatrix} = \mathbf{Z} \begin{bmatrix} q_b \\ q_a \end{bmatrix} = \begin{bmatrix} Z_o \coth mb & -Z_o \operatorname{csch} mb \\ Z_o \operatorname{csch} mb & -Z_o \coth mb \end{bmatrix} \begin{bmatrix} q_b \\ q_a \end{bmatrix} \tag{6.66}$$

$$\begin{bmatrix} q_b \\ q_a \end{bmatrix} = \mathbf{Y} \begin{bmatrix} \theta_b \\ \theta_a \end{bmatrix} = \begin{bmatrix} Y_o \coth mb & -Y_o \operatorname{csch} mb \\ Y_o \operatorname{csch} mb & -Y_o \coth mb \end{bmatrix} \begin{bmatrix} \theta_b \\ \theta_a \end{bmatrix} \tag{6.67}$$

$$\begin{bmatrix} \theta_b \\ q_a \end{bmatrix} = \mathbf{H} \begin{bmatrix} q_b \\ \theta_a \end{bmatrix} = \begin{bmatrix} Z_o \tanh mb & \operatorname{sech} mb \\ \operatorname{sech} mb & -Y_o \tanh mb \end{bmatrix} \begin{bmatrix} q_b \\ \theta_a \end{bmatrix} \tag{6.68}$$

$$\begin{bmatrix} q_b \\ \theta_a \end{bmatrix} = \mathbf{G}\begin{bmatrix} \theta_b \\ q_a \end{bmatrix} = \begin{bmatrix} Y_o \tanh mb & \operatorname{sech} mb \\ \operatorname{sech} mb & -Z_o \tanh mb \end{bmatrix}\begin{bmatrix} \theta_b \\ q_a \end{bmatrix} \tag{6.69}$$

6.2.9 Spine of Rectangular Cross Section

A spine of rectangular cross section is shown in Fig. 6.8,[3] where it may be noted that the sides are designated by δ_1 and δ_2 and where, again, the origin of the x-coordinate is taken at the spine tip where $x = a = 0$. Of course, there is no radius function because the spine is rectangular but the cross-sectional area is constant, $A = \delta_1\delta_2$, and there is no spine taper.

For an element of spine with surface area $dS = 2(\delta_1 + \delta_2)\,dx$, the theory leading to eq. (6.62) yields

$$\frac{d^2\theta}{dx^2} - m^2\theta = 0 \tag{6.63}$$

as the differential equation for temperature excess but with

$$m = \left[\frac{2h(\delta_1 + \delta_2)}{k\delta_1\delta_2}\right]^{1/2}$$

$$Y_o = [2hk\delta_1\delta_2(\delta_1 + \delta_2)]^{1/2}$$

and

$$Z_o = [2hk\delta_1\delta_2(\delta_1 + \delta_2)]^{-1/2}$$

As long as these values of m, Y_o, and Z_o are used, eqs. (6.64) through (6.69) provide the 24 elements of the linear transformations.

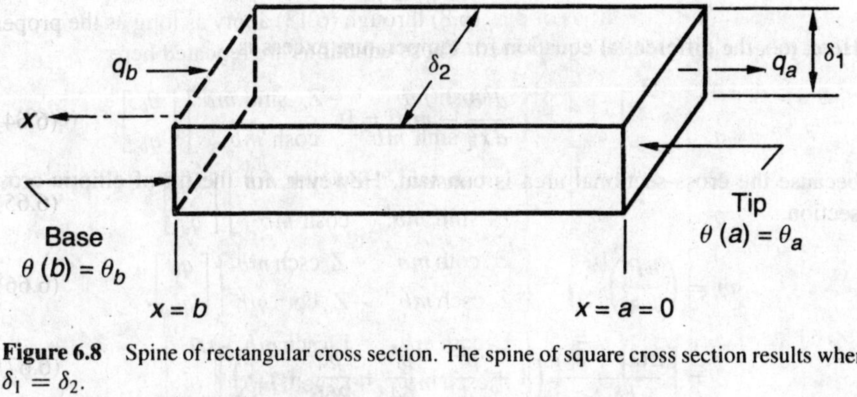

Figure 6.8 Spine of rectangular cross section. The spine of square cross section results when $\delta_1 = \delta_2$.

[3]Be aware that if one of the δ's is very much greater than the other, this spine of rectangular cross section becomes a longitudinal fin of rectangular profile.

In the event that $\delta_1 = \delta_2 = \delta$, the spine is square and eqs. (6.64) through (6.69) may be used but with

$$m = \left(\frac{4h}{k\delta}\right)^{1/2}$$

$$Y_o = (4hk\delta^3)^{1/2}$$

and

$$Z_o = (4hk\delta^3)^{-1/2}$$

6.2.10 Elliptical Spine

The spine of elliptical cross section is displayed in Fig. 6.9. The semimajor axis is designated by δ_1 and the semiminor axis is designated by δ_2. The origin of the height coordinate is taken at the tip of the spine, where $x = a = 0$, and the positive sense of the coordinate system is toward the spine base at $x = b$. The cross-sectional area is constant and is given by

$$A = \pi\delta_1\delta_2$$

The differential surface area is given by $dS = P\,dx$ with P to three-decimal-accuracy given by

$$P = \pi(\delta_1 - \delta_2)\left(1 + \frac{p^2}{4} + \frac{p^4}{64} + \frac{p^6}{256} + \cdots\right)$$

where

$$p = \frac{\delta_1 - \delta_2}{\delta_1 + \delta_2}$$

Here, too, the differential equation for temperature excess is

$$\frac{d^2\theta}{dx^2} - m^2\theta = 0 \tag{6.63}$$

because the cross-sectional area is constant. However, for the fin of elliptic cross section,

$$m = \left(\frac{hP}{kA}\right)^{1/2}$$

$$= \left[\frac{h(\delta_1 + \delta_2)}{k\delta_1\delta_2}\left(1 + \frac{p^2}{4} + \frac{p^4}{64} + \frac{p^6}{256} + \cdots\right)\right]^{1/2}$$

and

$$Y_o = kAm$$

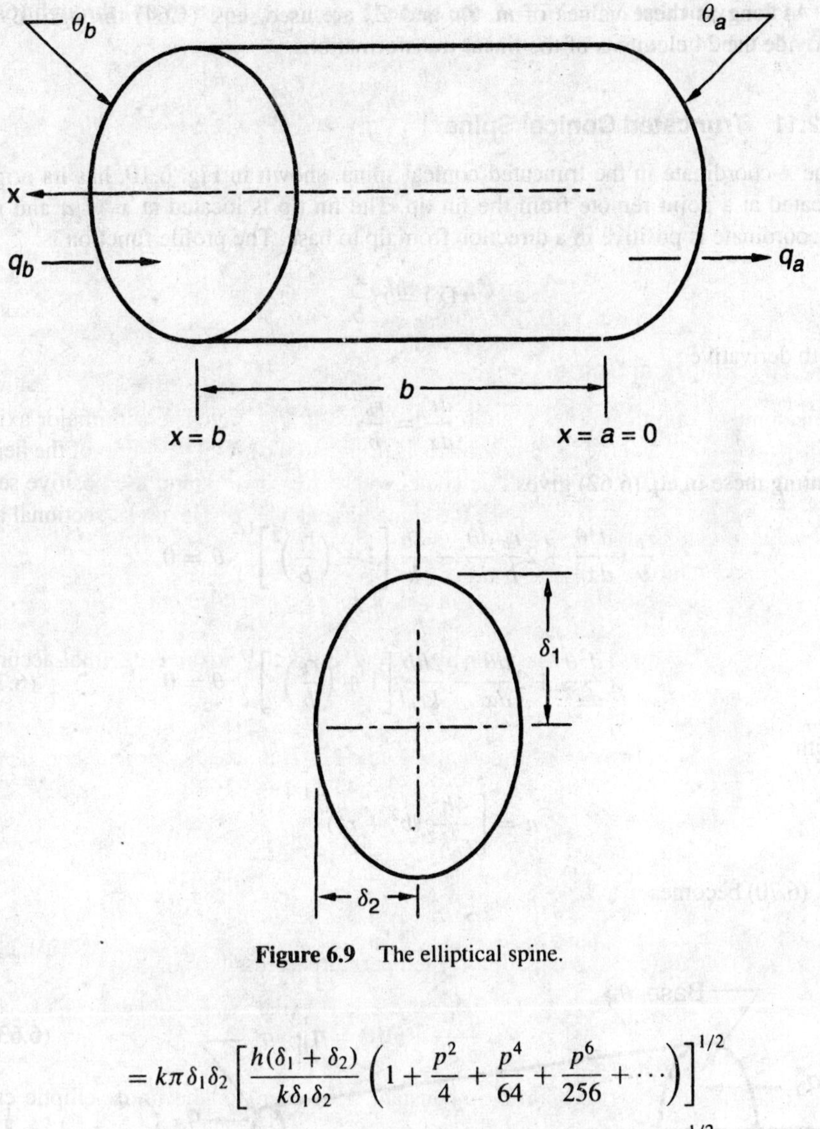

Figure 6.9 The elliptical spine.

$$= k\pi\delta_1\delta_2\left[\frac{h(\delta_1+\delta_2)}{k\delta_1\delta_2}\left(1+\frac{p^2}{4}+\frac{p^4}{64}+\frac{p^6}{256}+\cdots\right)\right]^{1/2}$$

$$= \left[hk\pi\delta_1\delta_2(\delta_1+\delta_2)\left(1+\frac{p^2}{4}+\frac{p^4}{64}+\frac{p^6}{256}+\cdots\right)\right]^{1/2}$$

so that

$$Z_o = \frac{1}{Y_o}$$

$$= \left[hk\pi\delta_1\delta_2(\delta_1+\delta_2)\left(1+\frac{p^2}{4}+\frac{p^4}{64}+\frac{p^6}{256}+\cdots\right)\right]^{-1/2}$$

As long as these values of m, Y_o, and Z_o are used, eqs. (6.64) through (6.69) provide the 24 elements of the linear transformations.

6.2.11 Truncated Conical Spine

The x-coordinate in the truncated conical spine. shown in Fig. 6.10, has its origin located at a point remote from the fin tip. The fin tip is located at $x = a$ and the x-coordinate is positive in a direction from tip to base. The profile function is

$$r(x) = r_b \frac{x}{b}$$

with derivative

$$\frac{dr}{dx} = \frac{r_b}{b}$$

Putting these in eq. (6.62) gives

$$\frac{r_b}{b} x \frac{d^2\theta}{dx^2} + 2\frac{r_b}{b}\frac{d\theta}{dx} - \frac{2h}{k}\left[1 + \left(\frac{r_b}{b}\right)^2\right]^{1/2} \theta = 0$$

or

$$x\frac{d^2\theta}{dx^2} + 2\frac{d\theta}{dx} - \frac{2hb}{kr_b}\left[1 + \left(\frac{r_b}{b}\right)^2\right]^{1/2} \theta = 0 \qquad (6.70)$$

With

$$n = \left[\frac{4h^2}{r_b^2 k^2}(b^2 + r_b^2)\right]^{1/4}$$

eq. (6.70) becomes

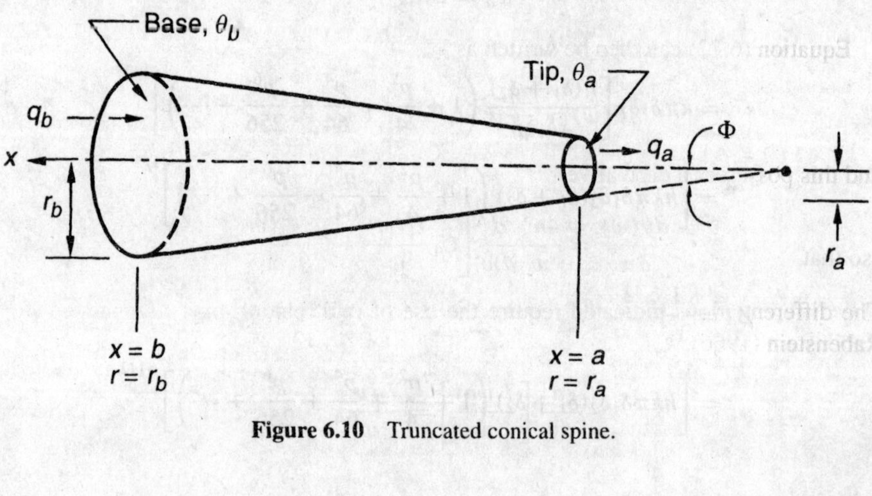

Figure 6.10 Truncated conical spine.

$$x^2 \frac{d^2\theta}{dx^2} + 2x \frac{d\theta}{dx} - n^2 x\theta = 0 \tag{6.71}$$

which has a solution

$$\theta(x) = x^{-1/2}[C_1 I_1(2nx^{1/2}) + C_2 K_1(2nx^{1/2})] \tag{6.72}$$

where C_1 and C_2 are to be evaluated from the initial conditions

$$\theta(x = b) = \theta_b$$

and

$$q(x = b) = q_b$$

If a transformation is made,

$$w = 2nx^{1/2}$$

so that

$$x = \frac{w^2}{4n^2}$$

and

$$dx = \frac{w}{2n^2} dw$$

then at $x = b$,

$$w_b = 2nb^{1/2}$$

and at $x = a$,

$$w_a = 2na^{1/2}$$

Equation (6.72) can then be written as

$$\theta(w) = \frac{2n}{w}[C_1 I_1(w) + C_2 K_1(w)] \tag{6.73}$$

and this possesses a derivative

$$\frac{d\theta(w)}{dw} = \frac{4n^3}{w} \frac{d}{dw}\left[C_1 \frac{I_1(w)}{w} + C_2 \frac{k_1(w)}{w}\right] \tag{6.74}$$

The differentiations indicated require the use of two relationships summarized by Rabenstein (1966):

$$\frac{d}{dw}\left[\frac{I_1(w)}{w}\right] = \frac{I_2(w)}{w}$$

and

$$\frac{d}{dw}\left[\frac{K_1(w)}{w}\right] = -\frac{K_2(w)}{w}$$

and with these in eq. (6.74) one obtains

$$\frac{d\theta(w)}{dw} = \frac{4n^3}{w}\left[C_1\frac{I_2(w)}{w} - C_2\frac{K_2(w)}{w}\right]$$

and with

$$A(x) = \pi r_b^2\left(\frac{x}{b}\right)^2 = \frac{\pi r_b^2 w^4}{16b^2n^4}$$

the heat flow at any point will be

$$q(w) = \frac{k\pi r_b^2 w^2}{4b^2 n}[C_1 I_2(w) - C_2 K_2(w)] \tag{6.75}$$

Now, at $x = b$ where $w = w_b$,

$$\theta_b = \frac{2n}{w_b}[C_1 I_1(w_b) + C_2 K_1(w_b)]$$

and

$$q_b = \frac{k\pi r_b^2 w_b^2}{4b^2 n}[C_1 I_2(w_b) - C_2 K_2(w_b)]$$

Solution of this set yields

$$C_1 = \frac{(w_b/2n)K_2(w_b)\theta_b + (4nb^2/k\pi r_b^2 w_b^2)K_1(w_b)q_b}{I_1(w_b)K_2(w_b) + I_2(w_b)K_1(w_b)}$$

and

$$C_2 = \frac{(w_b/2n)I_2(w_b)\theta_b - (4nb^2/k\pi r_b^2 w_b^2)I_1(w_b)q_b}{I_1(w_b)K_2(w_b) + I_2(w_b)K_1(w_b)}$$

With the Bessel function identity

$$I_1(w_b)K_2(w_b) + I_2(w_b)K_1(w_b) = \frac{1}{w_b}$$

these relationships for C_1 and C_2 may be inserted into eq. (6.75) to provide

$$\theta(x) = \frac{w_b^2}{w}[K_2(w_b)I_1(w) + I_2(w_b)K_1(w)]\theta_b$$
$$+ \frac{8n^2b^2}{k\pi r_b^2 w_b w}K_1(w_b)I_1(w) - I_1(w_b)K_1(w)]q_b \tag{6.76}$$

which shows that

$$\lambda_1(w) = \frac{w_b^2}{w}[K_2(w_b)I_1(w) + I_2(w_b)K_1(w)] \tag{6.77a}$$

and

$$\lambda_2(w) = \frac{8n^2b^2}{k\pi r_b^2 w_b w}[K_1(w_b)I_1(w) - I_1(w_b)K_1(w)] \tag{6.77b}$$

which at $w = w_b$ reduce to $\lambda_1(w_b) = 1$ and $\lambda_2(w_b) = 0$ as required by eqs. (6.2).

Expressions for $\lambda_1'(w)$ and $\lambda_2'(w)$ can be obtained by substituting the expressions for C_1 and C_2 into eq. (6.74) or by differentiating eqs. (6.77a) and (6.77b). Adopting the latter idea yields

$$\lambda_1'(w) = \frac{2n^2}{w}w_b^2[K_2(w_b)I_2(w) - I_2(w_b)K_2(w)] \tag{6.77c}$$

and

$$\lambda_2'(w) = \frac{16n^4b^2}{k\pi r_b^2 w_b w^2}[K_1(w_b)I_2(w) + I_1(w_b)K_2(w)] \tag{6.77d}$$

which at $w = w_b$ reduce to

$$\lambda_1'(w_b) = 0 \quad \text{and} \quad \lambda_2'(w_b) = 16n^4b^2/k\pi r_b^2 w_b^4$$

By noting that $16n^4b^2 = w_b^4$, it is seen that

$$\lambda_2'(w_b) = 1/k\pi r_b^2 = 1/kA(b)$$

Both of these results are required by eqs. (6.2*b*).

A listing of the **Γ**-elements now follows. Their formulation requires the use of eqs. (6.77) evaluated at $x = a$ where $w = w_a$ and the multiplication of $\lambda_1'(w)$ and $\lambda_2'(w)$ by $kA(a) = k\pi r_b^2(a/b)^2 = k\pi r_b^2(w_a/w_b)^4$:

$$\gamma_{11} = \lambda_1(w_a) = \frac{w_b^2}{w_a}[I_1(w_a)K_2(w_b) + I_2(w_b)K_1(w_a)] \tag{6.78a}$$

$$\gamma_{12} = \lambda_2(w_a) = \frac{8n^2b^2}{k\pi r_b^2 w_b w_a}[I_1(w_a)K_1(w_b) - I_1(w_b)K_1(w_a)] \tag{6.78b}$$

$$\begin{aligned}\gamma_{21} &= kA(a)\lambda_1'(w_a) \\ &= 2k\pi n^2 r_b^2 \left(\frac{w_a}{w_b}\right)^2 [I_2(w_a)K_2(w_b) - I_2(w_b)K_2(w_a)]\end{aligned} \tag{6.78c}$$

and

$$\gamma_{22} = kA(a)\lambda_2'(w_a) = \frac{16n^4b^2w_a^2}{w_b^5}[I_2(w_a)K_1(w_b) + I_1(w_b)K_2(w_a)]$$

With $16n^4b^2 = w_b^4$, this becomes

$$\gamma_{22} = \frac{w_a^2}{w_b}[I_2(w_a)K_1(w_b) + I_1(w_b)K_2(w_a)] \tag{6.78d}$$

It can be shown that[4]

$$\det \mathbf{\Gamma} = 1$$

and use of this fact and eqs. (6.78) in conjunction with Table 5.2 will provide the elements of all the other linear transformations.

T-*Matrix*

$$\tau_{11} = \frac{w_a^2}{w_b}[I_2(w_a)K_1(w_b) + I_1(w_b)K_2(w_a)] \tag{6.79a}$$

$$\tau_{12} = \frac{8n^2b^2}{k\pi r_b^2 w_b w_a}[I_1(w_b)K_1(w_a) - I_1(w_a)K_1(w_b)] \tag{6.79b}$$

$$\tau_{21} = 2k\pi n^2 r_b^2\left(\frac{w_a}{w_b}\right)^2[I_2(w_b)K_2(w_a) - I_2(w_a)K_2(w_b)] \tag{6.79c}$$

and

$$\tau_{22} = \frac{w_b^2}{w_a}[I_1(w_a)K_2(w_b) + I_2(w_b)K_1(w_a)] \tag{6.79d}$$

Z-*Matrix*

$$z_{11} = \frac{w_b}{2k\pi n^2 r_b^2}\frac{I_2(w_a)K_1(w_b) + I_1(w_b)K_2(w_a)}{I_2(w_b)K_2(w_a) - I_2(w_a)K_2(w_b)} \tag{6.80a}$$

$$z_{12} = \frac{(a/b)}{2k\pi n^2 r_b^2[I_2(w_a)K_2(w_b) - I_2(w_b)K_2(w_a)]} \tag{6.80b}$$

$$z_{21} = \frac{(a/b)}{2k\pi n^2 r_b^2[I_2(w_b)K_2(w_a) - I_2(w_a)K_2(w_b)]} \tag{6.80c}$$

and

$$z_{22} = \frac{w_a}{2k\pi n^2 r_a^2}\frac{I_1(w_a)K_2(w_b) + I_2(w_b)K_1(w_a)}{I_2(w_a)K_2(w_b) - I_2(w_b)K_2(w_a)} \tag{6.80d}$$

Y-*Matrix*

$$y_{11} = \frac{k\pi r_b^2 w_b}{2b}\frac{I_1(w_a)K_2(w_b) + I_2(w_b)K_1(w_a)}{I_1(w_b)K_1(w_a) - I_1(w_a)K_1(w_b)} \tag{6.81a}$$

[4]In a manner that is substantially similar to the method used to show the same result for the longitudinal fins of trapezoidal and truncated concave parabolic profile.

$$y_{12} = \frac{k\pi r_b^2 w_a w_b}{8n^2b^2[I_1(w_a)K_1(w_b) - I_1(w_b)K_1(w_a)} \tag{6.81b}$$

$$y_{21} = \frac{k\pi r_b^2 w_a w_b}{8n^2b^2[I_1(w_b)K_1(w_a) - I_1(w_a)K_1(w_b)]} \tag{6.81c}$$

and

$$y_{22} = \frac{k\pi r_a^2 w_a}{2a} \frac{I_2(w_a)K_1(w_b) + I_1(w_b)K_2(w_a)}{I_1(w_a)K_1(w_b) - I_1(w_b)K_1(w_a)} \tag{6.81d}$$

H-*Matrix*

$$h_{11} = \frac{2b}{k\pi r_b^2 w_b} \frac{I_1(w_b)K_1(w_a) - I_1(w_a)K_1(w_b)}{I_1(w_a)K_2(w_b) + I_2(w_b)K_1(w_a)} \tag{6.82a}$$

$$h_{12} = \frac{w_a/w_b^2}{I_1(w_a)K_2(w_b) + I_2(w_b)K_1(w_a)} \tag{6.82b}$$

$$h_{21} = \frac{w_a/w_b^2}{I_1(w_a)K_2(w_b) + I_2(w_b)K_1(w_a)} \tag{6.82c}$$

and

$$h_{22} = \frac{2k\pi n^2 r_a^2}{w_a} \frac{I_2(w_a)K_2(w_b) - I_2(w_b)K_2(w_a)}{I_1(w_a)K_2(w_b) + I_2(w_b)K_1(w_a)} \tag{6.82d}$$

G-*Matrix*

$$g_{11} = \frac{2k\pi n^2 r_b^2}{w_b} \frac{I_2(w_b)K_2(w_a) - I_2(w_a)K_2(w_b)}{I_2(w_a)K_1(w_b) + I_1(w_b)K_2(w_a)} \tag{6.83a}$$

$$g_{12} = \frac{w_b/w_a^2}{I_2(w_a)K_1(w_b) + I_1(w_b)K_2(w_a)} \tag{6.83b}$$

$$g_{21} = \frac{w_b/w_a^2}{I_2(w_a)K_1(w_b) + I_1(w_b)K_2(w_a)} \tag{6.83c}$$

and

$$g_{22} = \frac{2a}{k\pi r_a^2 w_a} \frac{I_1(w_a)K_1(w_b) - I_1(w_b)K_1(w_a)}{I_2(w_a)K_1(w_b) + I_1(w_b)K_2(w_a)} \tag{6.83d}$$

6.2.12 Truncated Concave Parabolic Spine

For the truncated concave parabolic spine displayed in Fig. 6.11, the profile function is

$$r(x) = r_b \left(\frac{x}{b}\right)^2$$

with derivative

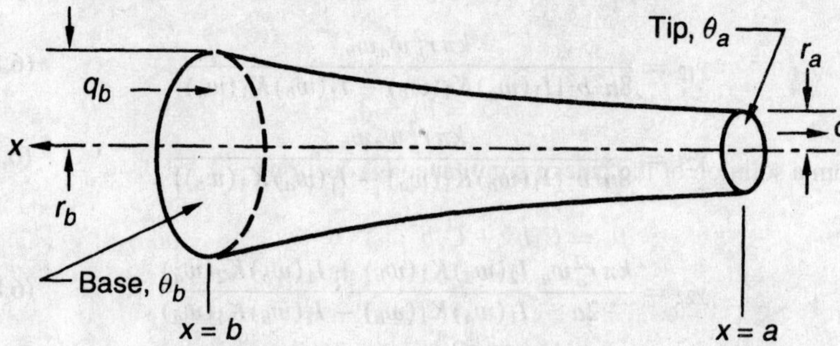

Figure 6.11 Truncated parabolic spine.

$$\frac{dr(x)}{dx} = \frac{2r_b}{b^2}x$$

Insertion of these into eq. (6.62) gives

$$r_b\left(\frac{x}{b}\right)^2\frac{d^2\theta}{dx^2} + \frac{4r_b}{b^2}x\frac{d\theta}{dx} - \frac{2h}{k}\left(1 + \frac{4r_b^2}{b^4}x^2\right)^{1/2}\theta = 0$$

This complicated differential equation can be simplified significantly if one recognizes that spines are usually long and thin so that the ratio $4r_b^2/b^2$ is small ($4r_b^2/b^2 << 1$). In this case it reduces to

$$r_b\left(\frac{x}{b}\right)^2\frac{d^2\theta}{dx^2} + \frac{4r_b}{b^2}x\frac{d\theta}{dx} - \frac{2h}{k}\theta = 0$$

or

$$x^2\frac{d^2\theta}{dx^2} + 4x\frac{d\theta}{dx} - m^2b^2\theta = 0 \tag{6.84}$$

where

$$m = \left(\frac{2h}{kr_b}\right)^{1/2}$$

Equation (6.84) is recognized as an Euler equation and its general solution is

$$\theta = C_1x^\alpha + C_2x^\beta \tag{6.85}$$

where

$$\alpha, \beta = -\frac{3}{2} \pm \frac{(9 + 4m^2b^2)^{1/2}}{2}$$

The arbitrary constants C_1 and C_2 must be evaluated from the initial conditions

$$\theta(x = b) = \theta_b$$

and

$$q(x = b) = q_b$$

and from a solution of the pair of simultaneous equations

$$\theta_b = C_1 b^{\alpha} + C_2 b^{\beta}$$

and

$$q_b = k\pi r_b^2(\alpha C_1 b^{\alpha-1} + \beta C_2 b^{\beta-1})$$

Let

$$\psi = (9 + 4m^2b^2)^{1/2}$$

so that the solutions for C_1 and C_2 can be written as

$$C_1 = \frac{(b^{\beta}/k\pi r_b^2)q_b - \beta b^{\beta-1}\theta_b}{\psi b^{\alpha+\beta-1}}$$

and

$$C_2 = \frac{\alpha b^{\alpha-1}\theta_b - (b^{\alpha}/k\pi r_b^2)q_b}{\psi b^{\alpha+\beta-1}}$$

With C_1 and C_2 inserted into eq. (6.85),

$$\theta(x) = \frac{1}{\psi}\left[\alpha\left(\frac{x}{b}\right)^{\beta} - \beta\left(\frac{x}{b}\right)^{\alpha}\right]\theta_b + \frac{b}{k\pi r_b^2\psi}\left[\left(\frac{x}{b}\right)^{\alpha} - \left(\frac{x}{b}\right)^{\beta}\right]q_b \tag{6.86}$$

and one can observe immediately that

$$\lambda_1(x) = \frac{1}{\psi}\left[\alpha\left(\frac{x}{b}\right)^{\beta} - \beta\left(\frac{x}{b}\right)^{\alpha}\right]$$

and

$$\lambda_2(x) = \frac{b}{k\pi r_b^2\psi}\left[\left(\frac{x}{b}\right)^{\alpha} - \left(\frac{x}{b}\right)^{\beta}\right]$$

When $x = b$ it is noted that $\lambda_1(b) = 1$ because $\alpha - \beta = \psi$ and that $\lambda_2(b) = 0$, both of which are required by eqs. (6.2a).

Further, with $\alpha\beta = -(mb)^2$,

$$\lambda_1'(x) = \frac{2hb}{kr_b\psi}\left[\left(\frac{x}{b}\right)^{\alpha-1} - \left(\frac{x}{b}\right)^{\beta-1}\right]$$

and

$$\lambda_2'(x) = \frac{1}{k\pi r_b^2 \psi}\left[\alpha\left(\frac{x}{b}\right)^{\alpha-1} - \beta\left(\frac{x}{b}\right)^{\beta-1}\right]$$

and one can observe that at $x = b$, $\lambda_1'(b) = 0$ and that

$$\lambda_2'(b) = (\alpha - \beta)/k\pi r_b^2 \psi = 1/k\pi r_b^2 = 1/kA(b)$$

which also satisfies the requirements of eqs. (6.2*b*).

The elements of the $\boldsymbol{\Gamma}$-matrix may be set down from the foregoing evaluated at $x = a$ and with appropriate multiplication of $\lambda_1'(a)$ and $\lambda_2'(a)$ by $kA(a) = k\pi r_a^2$.

$$\gamma_{11} = \lambda_1(a) = \frac{1}{\psi}\left[\alpha\left(\frac{a}{b}\right)^{\beta} - \beta\left(\frac{a}{b}\right)^{\alpha}\right] \tag{6.87a}$$

$$\gamma_{12} = \lambda_2(a) = \frac{b}{k\pi r_b^2 \psi}\left[\left(\frac{a}{b}\right)^{\alpha} - \left(\frac{a}{b}\right)^{\beta}\right] \tag{6.87b}$$

$$\gamma_{21} = kA(a)\lambda_1'(a) = \frac{2\pi r_b a h}{\psi}\left[\left(\frac{a}{b}\right)^{\alpha+2} - \left(\frac{a}{b}\right)^{\beta+2}\right] \tag{6.87c}$$

and

$$\gamma_{22} = kA(a)\lambda_2'(a) = \frac{1}{\psi}\left[\alpha\left(\frac{a}{b}\right)^{\alpha+3} - \beta\left(\frac{a}{b}\right)^{\beta+3}\right] \tag{6.87d}$$

It can be shown that

$$\det \boldsymbol{\Gamma} = 1$$

and use of this fact and eqs. (6.87) in conjunction with Table 5.2 will generate the elements of all the other linear transformations.

T-*Matrix*

$$\tau_{11} = \frac{1}{\psi}\left[\alpha\left(\frac{a}{b}\right)^{\alpha+3} - \beta\left(\frac{a}{b}\right)^{\beta+3}\right] \tag{6.88a}$$

$$\tau_{12} = \frac{b}{k\pi r_b^2 \psi}\left[\left(\frac{a}{b}\right)^{\beta} - \left(\frac{a}{b}\right)^{\alpha}\right] \tag{6.88b}$$

$$\tau_{21} = \frac{2\pi r_b a h}{\psi}\left[\left(\frac{a}{b}\right)^{\beta+2} - \left(\frac{a}{b}\right)^{\alpha+2}\right] \tag{6.88c}$$

and

$$\tau_{22} = \frac{1}{\psi}\left[\alpha\left(\frac{a}{b}\right)^{\beta} - \beta\left(\frac{a}{b}\right)^{\alpha}\right] \tag{6.88d}$$

Z-Matrix

$$z_{11} = \frac{1}{2\pi r_b a h}\left[\frac{\beta(a/b)^{\beta+3} - \alpha(a/b)^{\alpha+3}}{(a/b)^{\alpha+2} - (a/b)^{\beta+2}}\right] \tag{6.89a}$$

$$z_{12} = \frac{\psi}{2\pi r_b a h\left[(a/b)^{\alpha+2} - (a/b)^{\beta+2}\right]} \tag{6.89b}$$

$$z_{21} = \frac{\psi}{2\pi r_b a h\left[(a/b)^{\beta+2} - (a/b)^{\alpha+2}\right]} \tag{6.89c}$$

and

$$z_{22} = \frac{1}{2\pi r_b a h}\left[\frac{\alpha(a/b)^{\beta} - \beta(a/b)^{\alpha}}{(a/b)^{\alpha+2} - (a/b)^{\beta+2}}\right] \tag{6.89d}$$

Y-Matrix

$$y_{11} = \frac{k\pi r_b^2}{b}\left[\frac{\beta(a/b)^{\alpha} - \alpha(a/b)^{\beta}}{(a/b)^{\alpha} - (a/b)^{\beta}}\right] \tag{6.90a}$$

$$y_{12} = \frac{k\pi r_b^2 \psi}{b\left[(a/b)^{\alpha} - (a/b)^{\beta}\right]} \tag{6.90b}$$

$$y_{21} = \frac{k\pi r_b^2 \psi}{b\left[(a/b)^{\beta} - (a/b)^{\alpha}\right]} \tag{6.90c}$$

and

$$y_{22} = \frac{k\pi r_b^2}{b}\left[\frac{\alpha(a/b)^{\alpha+3} - \beta(a/b)^{\beta+3}}{(a/b)^{\alpha} - (a/b)^{\beta}}\right] \tag{6.90d}$$

H-Matrix

$$h_{11} = \frac{b}{k\pi r_b^2}\left[\frac{(a/b)^{\beta} - (a/b)^{\alpha}}{\alpha(a/b)^{\beta} - \beta(a/b)^{\alpha}}\right] \tag{6.91a}$$

$$h_{12} = \frac{\psi}{\left[\alpha(a/b)^{\beta} - \beta(a/b)^{\alpha}\right]} \tag{6.91b}$$

$$h_{21} = \frac{\psi}{\left[\alpha(a/b)^{\beta} - \beta(a/b)^{\alpha}\right]} \tag{6.91c}$$

and

$$h_{22} = 2\pi r_b a h\left[\frac{(a/b)^{\alpha+2} - (a/b)^{\beta+2}}{\alpha(a/b)^{\alpha} - \beta(a/b)^{\beta}}\right] \tag{6.91d}$$

G-*Matrix*

$$g_{11} = 2\pi r_b a h \left[\frac{(a/b)^{\beta+2} \ (a/b)^{\alpha+2}}{\alpha(a/b)^{\alpha+3} - \beta^{\beta+3}} \right] \tag{6.92a}$$

$$g_{12} = \frac{\psi}{\alpha(a/b)^{\alpha+3} - \beta(a/b)^{\beta+3}} \tag{6.92b}$$

$$g_{21} = \frac{\psi}{\alpha(a/b)^{\alpha+3} - \beta(a/b)^{\beta+3}} \tag{6.92c}$$

and

$$g_{22} = \frac{b}{k\pi r_b^2} \left[\frac{(a/b)^{\alpha} - (a/b)^{\beta}}{\alpha(a/b)^{\alpha+3} - \beta(a/b)^{\beta+3}} \right] \tag{6.92d}$$

6.3 SINGULAR FINS AND SPINES

As shown in Section 6.2, for any fin or spine, the temperature excess and heat flow at any point may be given as a superposition of base temperature excess and base heat flow

$$\theta(x) = \theta_b \lambda_1(x) + q_b \lambda_2(x) \tag{6.3}$$

and

$$q(x) = k(x)A(x)[\theta_b \lambda_1'(x) + q_b \lambda_2'(x)] \tag{6.4}$$

These equations produce a linear transformation that maps the temperature excess and heat flow at the fin tip where $x = a$ (a may equal zero) from similar conditions at the fin base where $x = b$. This linear transformation is referred to as the $\mathbf{\Gamma}$-matrix, so that the linear transformation is represented by

$$\begin{bmatrix} \theta_a \\ q_a \end{bmatrix} = \mathbf{\Gamma} \begin{bmatrix} \theta_b \\ q_b \end{bmatrix} = \begin{bmatrix} \gamma_{11} & \gamma_{12} \\ \gamma_{21} & \gamma_{22} \end{bmatrix} \begin{bmatrix} \theta_b \\ q_b \end{bmatrix} \tag{6.5}$$

where because the fin tip is located at $x = a$,

$$\gamma_{11} = \lambda_1(a) \tag{6.6a}$$

$$\gamma_{12} = \lambda_2(a) \tag{6.6b}$$

$$\gamma_{21} = k(a)A(a)\lambda_1'(a) \tag{6.6c}$$

and

$$\gamma_{22} = k(a)A(a)\lambda_2'(a) \tag{6.6d}$$

and if $a = 0$, these become

$$\gamma_{11} = \lambda_1(0) \tag{6.7a}$$

$$\gamma_{12} = \lambda_2(0) \tag{6.7b}$$

$$\gamma_{21} = k(0)A(0)\lambda_1'(0) \tag{6.7c}$$

and

$$\gamma_{22} = k(0)A(0)\lambda_2'(0) \tag{6.7d}$$

If the fin or spine tapers to a point at $x = 0$, where $A(0) = 0$, $\mathbf{\Gamma}$ becomes singular[5]:

$$\begin{bmatrix} \gamma_{11} & \gamma_{12} \\ \gamma_{21} & \gamma_{22} \end{bmatrix} = \begin{bmatrix} \lambda_1(0) & \lambda_2(0) \\ 0 & 0 \end{bmatrix} \tag{6.93}$$

It must be stressed that when regular as shown by eq. (6.5) or when singular as indicated by eq. (6.93), the $\mathbf{\Gamma}$-matrix depends only on the fin or spine geometry and the heat transfer parameters of h and k. Observe that $\mathbf{\Gamma}$ is singular when the product $k(a)A(a)$ or $k(0)A(0)$ vanishes and that this condition does indeed occur for the longitudinal fins of triangular and concave parabolic profile and for the conical and concave parabolic spines. These four shapes require special interpretation through the use of a *thermal transmission ratio* defined on the basis of the heat flow and temperature excess at the fin or spine base:

$$\mu \equiv \frac{q_b}{\theta_b} \tag{6.94}$$

which will permit inclusion of these fins and spines at remote locations in arrays of extended surface which are commonly referred to as *heat sinks*. Figure 6.12 shows a sketch of a finned array containing nine elements (fins). It may be observed that six of the remote fins are *singular* longitudinal fins of triangular profile. Moreover, two of the fins are *regular* longitudinal fins of rectangular profile, and the fin containing the base of the array is the *regular* longitudinal fin of trapezoidal profile.

6.3.1 Longitudinal Fin of Triangular Profile

The longitudinal fin of triangular profile shown, with its coordinate system in Fig. 6.13, is seen to be but a specific case of the longitudinal fin of trapezoidal profile (Fig. 6.2) with the origin located at $x = a = 0$ and with tip thickness $\delta_a = 0$. Observe in Fig. 6.2 that the origin of the height coordinate is at the tip of the fin with positive orientation toward the fin base and that fin height is b.

In this case, the differential equation for the temperature excess is identical to the differential equation for the temperature excess of the longitudinal fin of trapezoidal profile

$$x\frac{d^2\theta}{dx^2} + \frac{d\theta}{dx} - n^2\theta = 0 \tag{6.15}$$

with

[5]Recall that a singular matrix is a matrix whose determinant is equal to zero. This is the origin of the terminology *singular fin*.

Figure 6.12 Finned array containing nine fins.

$$n = \left(\frac{h}{k \sin \kappa}\right)^{1/2}$$

where κ is the taper angle

$$\kappa = \arctan \frac{\delta_b}{2b}$$

With the initial conditions given by

$$\theta(x = b) = \theta_b$$

and

$$q(x = b) = q_b$$

the particular solution for the temperature excess in terms of a transformed variable

$$u = 2nx^{1/2}$$

is

$$\theta(u) = u_b[K_1(u_b)I_0(u) + I_1(u_b)K_0(u)]\theta_b + \frac{u_b^2}{2k\delta_b L n^2}[K_0(u_b)I_0(u) - I_0(u_b)K_0(u)]\,q_b \qquad (6.21)$$

where

$$u_b = 2nb^{1/2}$$

However, as $x \to 0$, $u \to 0$ and $\theta(0) \to \theta_a$. This temperature difference must be finite. With $I_0(u \to 0) = 1$, eq. (6.21) becomes

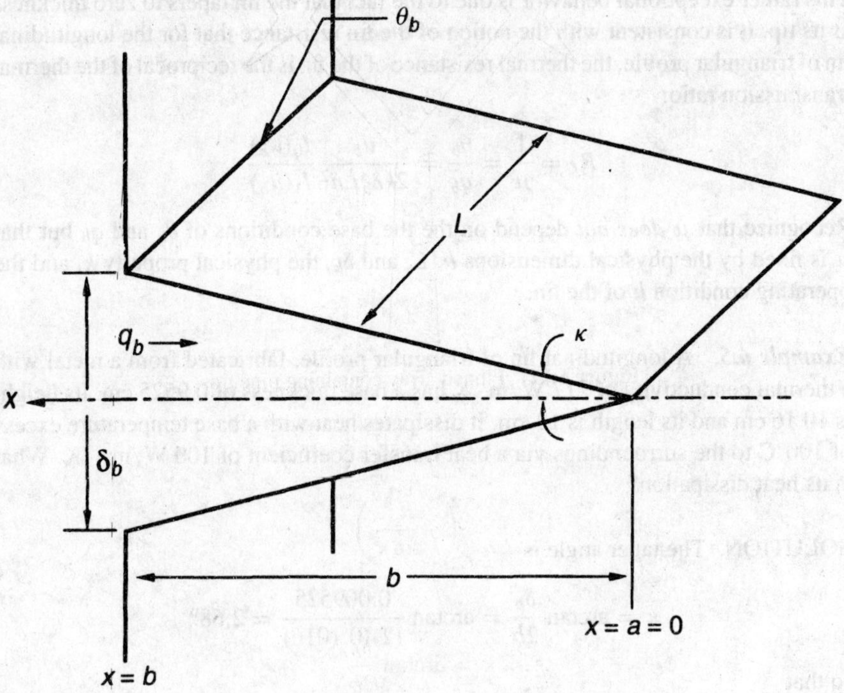

Figure 6.13 Longitudinal fin of triangular profile.

$$\theta(u \to 0) = \theta_a = u_b[K_1(u_b) + I_1(u_b)K_0(u \to 0)]\theta_b$$
$$+ \frac{u_b^2}{2k\delta_b L n^2}[K_0(u_b) - I_0(u_b)K_0(u \to 0)]\, q_b \qquad (6.95)$$

Because $K_0(u)$ becomes unbounded as $u \to 0$, the terms involving $K_0(u)$ in eq. (6.95) must cancel one another. This implies a relationship between θ_b and q_b:

$$u_b I_1(u_b)K_0(u \to 0)]\theta_b = \frac{u_b^2}{2k\delta_b L n^2} I_0(u_b)K_0(u \to 0)q_b$$

so that after cancellation of the common $K_0(u \to 0)$ terms, one can obtain

$$q_b = \frac{2k\delta_b L n^2}{u_b}\frac{I_1(u_b)}{I_0(u_b)}\theta_b$$

and this leads to the thermal transmission ratio μ:

$$\mu = \frac{q_b}{\theta_b} = \frac{2k\delta_b L n^2}{u_b}\frac{I_1(u_b)}{I_0(u_b)} \qquad (6.96)$$

which is all that is needed to characterize the longitudinal fin of triangular profile. This rather exceptional behavior is due to the fact that the fin tapers to zero thickness at its tip. It is consistent with the notion of the fin resistance that for the longitudinal fin of triangular profile, the thermal resistance of the fin is the reciprocal of the thermal transmission ratio:

$$R_f = \frac{1}{\mu} = \frac{\theta_b}{q_b} = \frac{u_b}{2k\delta_b L n^2} \frac{I_0(u_b)}{I_1(u_b)}$$

Recognize that μ *does not* depend on the the base conditions of θ_b and q_b but that μ is fixed by the physical dimensions b, L, and δ_b, the physical property k, and the operating condition h of the fin.

Example 6.5. A longitudinal fin of triangular profile, fabricated from a metal with a thermal conductivity of 217 W/m · K has a base thickness of 0.9525 cm. Its height is 10.16 cm and its length is 12 cm. It dissipates heat with a base temperature excess of 100°C to the surroundings via a heat transfer coefficient of 100 W/m² · K. What is its heat dissipation?

SOLUTION. The taper angle is

$$\kappa = \arctan \frac{\delta_b}{2b} = \arctan \frac{0.009525}{(2)(0.1016)} = 2.68^\circ$$

so that

$$\sin \kappa = 0.0468$$

Then

$$n = \left(\frac{h}{k \sin \kappa} \right)^{1/2} = \left[\frac{100}{(217)(0.0468)} \right]^{1/2} = 3.1372 \text{ m}^{-1}$$

and

$$u_b = 2nb^{1/2} = (2)(3.1372)(0.1016)^{1/2} = 2.0000$$

With

$$I_1(u_b) = I_1(2.0000) = 1.59064$$

and

$$I_2(u_b) = I_2(2.0000) = 2.27959$$

by eq. (6.96),

$$\mu = \frac{q_b}{\theta_b} = \frac{2k\delta_b L n^2}{u_b} \frac{I_1(u_b)}{I_2(u_b)}$$

$$= \frac{(2)(217)(0.009525)(0.12)(3.1372)^2}{2.0000}\left(\frac{1.59064}{2.27959}\right)$$

$$= (2.4411)(0.6978) = 1.7033 \text{ W/K}$$

and if $\theta_b = 100°\text{C}$,

$$q_b = (100)(1.7033) = 170.3 \text{ W/K}$$

6.3.2 Longitudinal Fin of Concave Parabolic Profile

The coordinate system for the longitudinal fin of concave parabolic profile is shown in Fig. 6.14. Notice that the height coordinate is taken with origin at the fin tip where $x = a = 0$ and $\delta_a = 0$ and that the fin height is b. The height coordinate has a positive sense from tip to base.

Here, too, the differential equation for temperature excess is identical to the differential equation for temperature excess when the fin is truncated and $\delta_a \neq 0$. This was shown in Section 6.2 to be

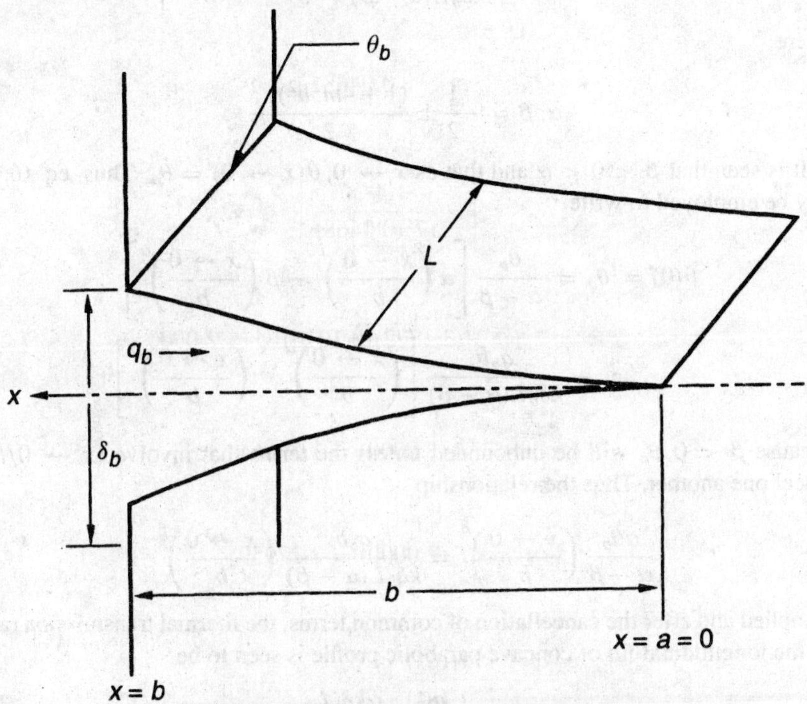

Figure 6.14 Longitudinal fin of concave parabolic profile.

$$x^2 \frac{d^2\theta}{dx^2} + 2x \frac{d\theta}{dx} - m^2 b^2 \theta = 0 \qquad (6.31)$$

where

$$m = \left(\frac{2h}{k\delta_b} \right)^{1/2}$$

After an appropriate transformation, the particular solution for $\theta(x)$ is obtained using the initial value data

$$\theta(x = b) = \theta_b$$

and

$$q(x = b) = q_b$$

and is

$$\theta(x) = \frac{\theta_b}{\alpha - \beta} \left[\alpha \left(\frac{x}{b} \right)^{\beta} - \beta \left(\frac{x}{b} \right)^{\alpha} \right] + \frac{q_b b}{k\delta_b L(\alpha - \beta)} \left[\left(\frac{x}{b} \right)^{\alpha} - \left(\frac{x}{b} \right)^{\beta} \right] \qquad (6.97)$$

where

$$\alpha, \beta = -\frac{1}{2} \pm \frac{(1 + 4m^2 b^2)^{1/2}}{2}$$

It is seen that $\beta < 0 < \alpha$ and that as $x \to 0, \theta(x \to 0) = \theta_a$. Thus, eq. (6.97) may be employed to write

$$\theta(0) = \theta_a = \frac{\theta_b}{\alpha - \beta} \left[\alpha \left(\frac{x \to 0}{b} \right)^{\beta} - \beta \left(\frac{x \to 0}{b} \right)^{\alpha} \right] + \frac{q_b b}{k\delta_b L(\alpha - \beta)} \left[\left(\frac{x \to 0}{b} \right)^{\alpha} - \left(\frac{x \to 0}{b} \right)^{\beta} \right]$$

Because $\beta < 0, \theta_a$ will be unbounded unless the terms that involve $(x \to 0/b)^{\beta}$ cancel one another. Thus the relationship

$$\frac{\alpha\theta_b}{\alpha - \beta} \left(\frac{x \to 0}{b} \right)^{\beta} = \frac{q_b b}{k\delta_b L(\alpha - \beta)} \left(\frac{x \to 0}{b} \right)^{\beta}$$

is implied and after the cancellation of common terms, the thermal transmission ratio for the longitudinal fin of concave parabolic profile is seen to be

$$\mu = \frac{q_b}{\theta_b} = \frac{\alpha k\delta_b L}{b} \qquad (6.98)$$

Example 6.6. A longitudinal fin of concave parabolic profile is to have the same dimensions as the longitudinal fin of triangular profile in Example 6.5. It is fabricated from the same metal ($k = 217$ W/m · K) and dissipates heat at the same base temperature excess ($\theta_b = 100$°C) to the surroundings with an identical heat transfer coefficient of $h = 100$ W/m^2 · K. Compare its heat dissipation to that of the triangular fin.

SOLUTION. With

$$m = \left(\frac{2h}{k\delta_b}\right)^{1/2} = \left[\frac{(2)(100)}{(217)(0.009525)}\right]^{1/2} = 9.8368 \text{ m}^{-1}$$

and $b = 0.1016$ m, the value of α becomes

$$\begin{aligned}\alpha &= -\frac{1}{2} + \frac{(1+4m^2b^2)^{1/2}}{2}\\ &= -\frac{1}{2} + \frac{[1+(4)(9.8368)^2(0.1016)^2]^{1/2}}{2}\\ &= -\frac{1}{2} + \frac{(1+3.9954)^{1/2}}{2}\\ &= 0.6175\end{aligned}$$

Then by eq. (6.98),

$$\begin{aligned}\mu &= \frac{q_b}{\theta_b} = \frac{\alpha k \delta_b L}{b}\\ &= \frac{(0.6175)(217)(0.009525)(0.12)}{0.1016}\\ &= 1.5074 \text{ W/K}\end{aligned}$$

and if $\theta_b = 100$°C,

$$q_b = (100)(1.5074) = 150.74 \text{ W}$$

Here the thermal resistance is

$$R_f = \frac{\theta_b}{q_b} = \frac{1}{\mu} = \frac{1}{1.5074} = 0.6634 \text{ K/W}$$

which shows that the longitudinal fin of concave parabolic profile has a higher thermal resistance than the longitudinal fin of triangular profile for the same height, length, and base thickness, thermal properties, and operating conditions. This means that the fin of concave parabolic profile is, basically, not as good a heat dissipator as the longitudinal fin of triangular profile. However, the parabolic profile fin possesses only two-thirds of the profile area as the triangular fin and hence only two-thirds of the mass. This suggests that *a fin of least material* is more than just an item of interest, and Chapter 3 in its entirety, was devoted to this very important consideration.

6.3.3 Conical Spine

The conical spine shown in Fig. 6.15 is a special case of the truncated conical spine where the origin of the height coordinate is located at $x = a = 0$. Thus, the origin is at the tip of the spine, which tapers to a point and the height coordinate is positive toward the fin base. Again, the height is designated as b.

The differential equation for temperature excess is the same as that for the truncated conical spine

$$x^2 \frac{d^2\theta}{dx^2} + 2x \frac{d\theta}{dx} - n^2 x\theta = 0 \tag{6.71}$$

where

$$n = \left[\frac{4h^2}{r_b^2 k^2} (b^2 + r_b^2) \right]^{1/4}$$

The particular solution in terms of the transformed variable,

$$w = 2nx^{1/2}$$

is obtained using the initial value data

$$\theta(x = b) = \theta_b$$

and

$$q(x = b) = q_b$$

and is

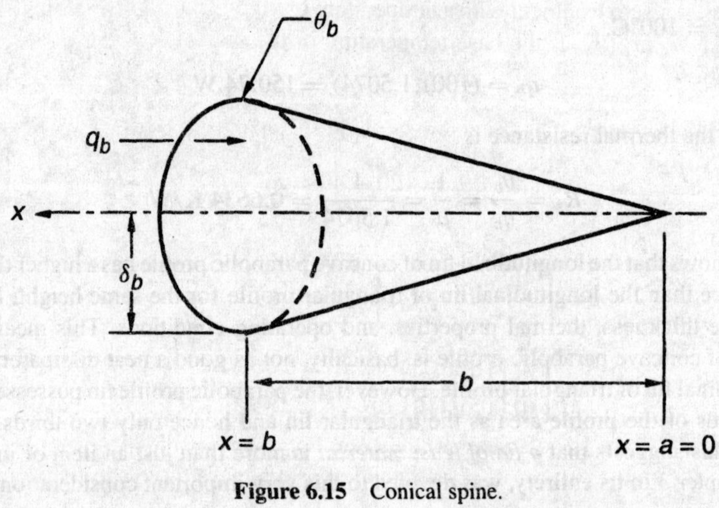

Figure 6.15 Conical spine.

$$\theta(w) = \frac{w_b^2}{w}[K_2(w_b)I_1(w) + I_2(w_b)K_1(w)]\theta_b + \frac{8n^2b^2}{k\pi r_b^2 w_b w}[K_1(w_b)I_1(w) + I_1(w_b)K_1(w)]q_b \tag{6.99}$$

where

$$w_b = 2nb^{1/2}$$

and

$$w_a = 2na^{1/2}$$

As $x \to 0$, $w \to 0$ and $w_a \to 0$. In eq. (6.99), $K_1(w)/w$ is indeterminant, but an application of the L'Hôpital rule shows that $K_1(w)/w$ is unbounded. For the tip temperature to remain finite, terms involving $K_1(w)/w$ in eq. (6.99) must cancel. Hence

$$\frac{w_b^2}{w}I_2(w_b)K_1(w)\theta_b = \frac{8n^2b^2}{k\pi r_b^2 w_b w}I_1(w_b)K_1(w)q_b$$

or

$$\frac{q_b}{\theta_b} = \frac{k\pi r_b^2 w_b^3}{8n^2b^2}\frac{I_2(w_b)}{I_1(w_b)}$$

But $w_b^3 = 8n^4b^{3/2}$, so that

$$\mu = \frac{q_b}{\theta_b} = \frac{k\pi n r_b^2}{b^{1/2}}\frac{I_2(w_b)}{I_1(w_b)} \tag{6.100}$$

Example 6.7. A conical spine possesses a base diameter of 1 cm and is 4 cm high. It is fabricated from a metal with a thermal conductivity of 101.6 W/m · K and it operates in an environment with a temperature of 40 °C and a heat transfer coefficient of 56.71 W/m² · K. If the base temperature of the spine is 120°C, what is its heat dissipation?

SOLUTION. Here

$$\theta_b = 120 - 40 = 80°\text{C}$$

and

$$n = \left[\frac{4h^2}{r_b^2k^2}(b^2 + r_b^2)\right]^{1/4}$$

$$= \left\{\frac{(4)(56.71)^2}{(0.005)^2(101.6)^2}[(0.04)^2 + (0.005)^2]\right\}^{1/4}$$

$$= (80.975)^{1/4} = 2.9998 \quad \text{or} \quad 3.0000$$

and

$$\begin{aligned} w_b &= 2nb^{1/2} \\ &= (2)(3.000)(0.04)^{1/2} \\ &= 1.2000 \end{aligned}$$

Then with

$$I_1(w_b) = 0.71648$$

and

$$I_2(w_b) = 0.20260$$

eq. (6.100) can be employed to give

$$\begin{aligned} \mu &= \frac{q_b}{\theta_b} \\ &= \frac{k\pi n r_b^2}{b^{1/2}} \frac{I_2(w_b)}{I_1(w_b)} \\ &= \frac{101.6\pi(3.0000)(0.005)^2}{(0.04)^{1/2}} \left(\frac{0.20260}{0.71648}\right) \\ &= 0.0339 \text{ W/K} \end{aligned}$$

and

$$q_b = \mu\theta_b = (0.0339)(80) = 2.71 \text{ W}$$

6.3.4 Concave Parabolic Spine

The concave parabolic spine shown in Fig. 6.16 is a special case of the truncated concave parabolic spine where the origin of the height coordinate is located at $x = a = 0$. Here, too, the origin is at the tip of the spine, which tapers to a point, and the height coordinate is positive toward the fin base located at $x = b$.

The differential equation for temperature excess is that for the truncated concave parabolic spine:

$$x^2 \frac{d^2\theta}{dx^2} + 4x \frac{d\theta}{dx} - m^2 b^2 \theta = 0 \qquad (6.84)$$

where

$$m = \left(\frac{2h}{kr_b}\right)^{1/2}$$

Here the particular solution for $\theta(x)$ is also obtained using the initial value data

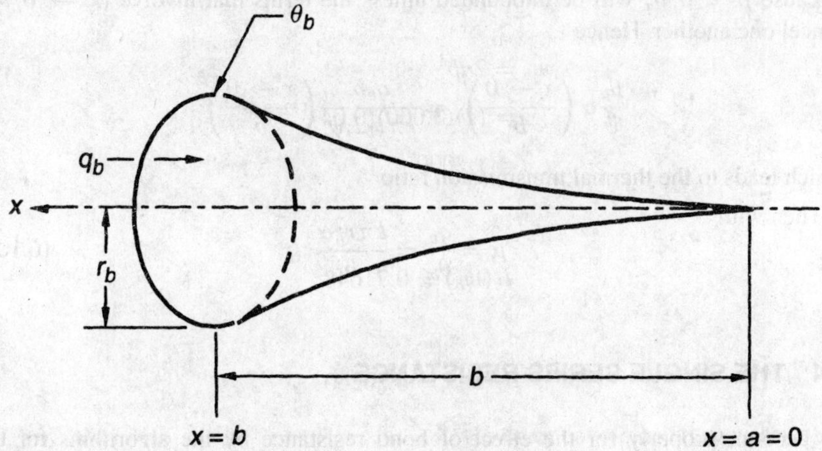

Figure 6.16 Concave parabolic spine.

$$\theta(x = b) = \theta_b$$

and

$$q(x = b) = q_b$$

and is

$$\theta(x) = \frac{1}{\psi}\left[\alpha\left(\frac{x}{b}\right)^{\beta} - \beta\left(\frac{x}{b}\right)^{\alpha}\right]\theta_b + \frac{b}{k\pi r_b L\psi}\left[\left(\frac{x}{b}\right)^{\alpha} - \left(\frac{x}{b}\right)^{\beta}\right]q_b \quad (6.101)$$

where

$$\alpha, \beta = -\frac{3}{2} \pm \frac{(1+4m^2b^2)^{1/2}}{2}$$

and

$$\psi = \alpha\beta = -(mb)^2 = -\frac{2hb^2}{kr_b}$$

It is seen that $\beta < 0 < \alpha$ and that as $x \to 0$, $\theta_a = \theta(0)$, and eq. (6.101) can be written in the form

$$\theta(0) = \theta_a = \frac{\theta_b}{\psi}\left[\alpha\left(\frac{x \to 0}{b}\right)^{\beta} - \beta\left(\frac{x \to 0}{b}\right)^{\alpha}\right]$$

$$+ \frac{q_b b}{k\pi r_b L\psi}\left[\left(\frac{x \to 0}{b}\right)^{\alpha} - \left(\frac{x \to 0}{b}\right)^{\beta}\right]$$

Because $\beta < 0$, θ_a will be unbounded unless the terms that involve $(x \rightarrow 0/b)^\beta$ cancel one another. Hence

$$\frac{\theta_b}{\psi}\alpha\left(\frac{x \rightarrow 0}{b}\right)^\beta = \frac{q_b b}{k\pi r_b L\psi}\left(\frac{x \rightarrow 0}{b}\right)^\beta$$

which leads to the thermal transmission ratio

$$\mu = \frac{q_b}{\theta_b} = \frac{k\pi r_b^2 \alpha}{b} \tag{6.102}$$

6.4 THE SINGLE SERIES RESISTANCE

To account properly for the effect of bond resistance by the algorithms for the combination of extended surface that will be developed in Chapter 7, a linear transformation for a single series thermal resistance must be formulated. Figure 6.17 shows such a representation and it is proposed that the temperature drop across the thermal resistance is equal to the difference between the temperatures at the two ends. Notice that T_b is taken at the end of the resistance where the heat flow enters and that this thermal resistance is in K/W.

Continuity demands that

$$q_b = q_a = q \tag{6.103a}$$

and the temperature drop will be equal to

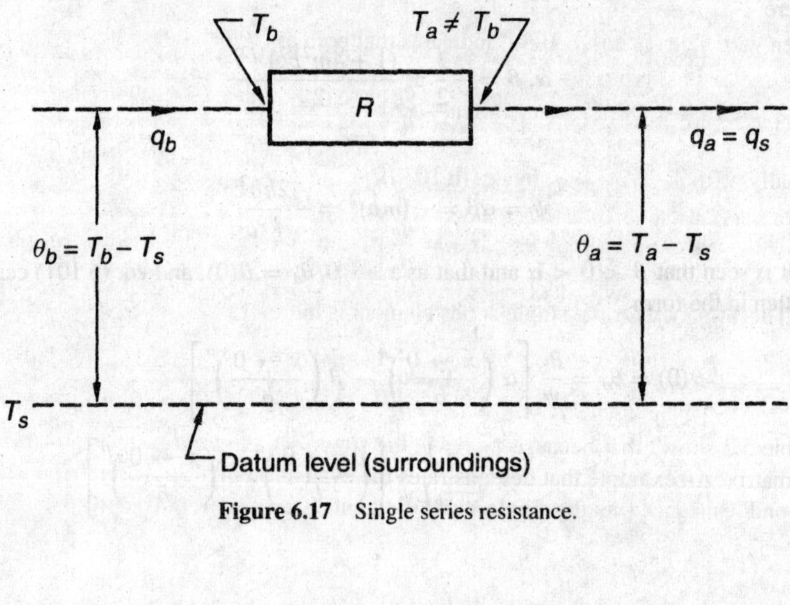

Figure 6.17 Single series resistance.

$$T_b - T_a = qR$$

If the surrounding temperature is both added and subtracted to the left side of the latter, the result is

$$T_b - T_s - (T_a - T_s) = qR$$

and this shows that

$$\theta_b - \theta_a = qR \tag{6.103b}$$

The **T**-matrix is defined by

$$\begin{bmatrix} \theta_b \\ q_b \end{bmatrix} = \mathbf{T} \begin{bmatrix} \theta_a \\ q_a \end{bmatrix} = \begin{bmatrix} \tau_{11} & \tau_{12} \\ \tau_{21} & \tau_{22} \end{bmatrix} \begin{bmatrix} \theta_a \\ q_a \end{bmatrix} \tag{5.16}$$

A matrix multiplication yields

$$\theta_b = \tau_{11}\theta_a + \tau_{12}q_a$$
$$q_b = \tau_{21}\theta_a + \tau_{22}q_a$$

and the τ's can be found by an application of superposition. If $q = q_a = 0$, eq. (6.103b) shows that $\theta_a = \theta_b$, so that

$$\tau_{11} = \left.\frac{\theta_b}{\theta_a}\right|_{q_a=0} = 1 \tag{6.104a}$$

Next, with $\theta_a = 0$, eq. (6.103b) shows that $\theta_b = Rq_b = Rq_a$. Hence

$$\tau_{12} = \left.\frac{\theta_b}{q_a}\right|_{\theta_a=0} = R \tag{6.104b}$$

Then with $q_a = 0$, eq. (6.103a) indicates that both q's are zero and

$$\tau_{21} = \left.\frac{q_b}{\theta_a}\right|_{q_a=0} = 0 \tag{6.104c}$$

Finally, with $q_b = q_a = q$, by eq. (6.103a),

$$\tau_{22} = \left.\frac{q_b}{q_a}\right|_{\theta_a=0} = 1 \tag{6.104d}$$

The **T**-matrix for the single series element is therefore

$$\mathbf{T} = \begin{bmatrix} 1 & R \\ 0 & 1 \end{bmatrix} \tag{6.105}$$

Table 5.2 shows that because $\tau_{21} = 0$, the single series element does not possess a **Z**-matrix. An example that demonstrates the evaluation of a single fin that possesses a bond resistance (via the single series element) is presented in Chapter 7.

6.5 THE SINGLE SHUNT CONDUCTANCE

Of equal importance is the single shunt conductance, which can be used to account for the fin tip heat loss. Because the single shunt conductance is somewhat more difficult to visualize, the details of the development of its **T**-matrix are somewhat more difficult to grasp. The single shunt element is the dual of the single series resistance and is displayed in Fig. 6.18. Notice that here the temperatures at the extremes are equal; that is,

$$T_b = T_a$$

and if T_s is subtracted from each side of the equation,

$$T_b - T_s = T_a - T_s$$

or

$$\theta_b = \theta_a \tag{6.106a}$$

Of course, the shunt element is merely an artifice to account for the tip heat loss. For a tip surface area $S(a)$, the shunt conductance will be $G = h_a S(a)$. Notice in Fig. 6.18 that this is *not* q_a.[6]

Continuity can be employed to show that with reference to Fig. 6.18,

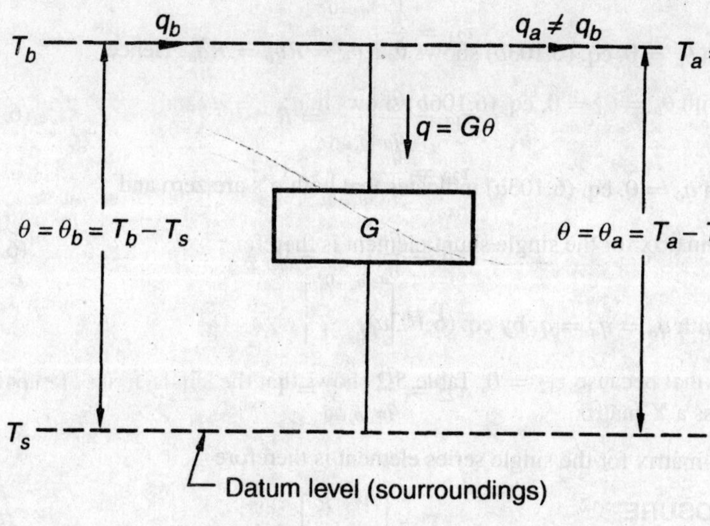

Figure 6.18 Single shunt conductance.

[6]The reason for the presence of the q_a will become more apparent in Chapter 7, where it is shown that use of the single shunt element permits the tip to be treated as adiabatic with $q_a = 0$.

$$q_b = G\theta_b + q_a = G\theta_a + q_a \tag{6.106b}$$

where G is the reciprocal of the thermal resistance

$$G = \frac{1}{R}$$

and is, therefore, a thermal conductance in W/K. The **T**-matrix, defined by eq. (5.16), can be used here as well. Again, the matrix multiplication yields

$$\theta_b = \tau_{11}\theta_a + \tau_{12}q_a$$

$$q_b = \tau_{21}\theta_a + \tau_{22}q_a$$

and the τ's can be found by another application of superposition. If $\theta_a = \theta_b = 0$ as required by eq. (6.106*a*), regardless of q_a,

$$\tau_{11} = \left.\frac{\theta_b}{\theta_a}\right|_{q_a=0} = 1 \tag{6.107a}$$

Next, if $\theta_a = 0$, eq. (6.106*a*) requires that $\theta_b = 0$. Hence

$$\tau_{12} = \left.\frac{\theta_b}{q_a}\right|_{\theta_a=0} = 0 \tag{6.107b}$$

Then with $q_a = 0$, eq. (6.106b) indicates that $q_b = G\theta_b$ and

$$\tau_{21} = \left.\frac{q_b}{\theta_a}\right|_{q_a=0} = G \tag{6.107c}$$

Finally, with $\theta_b = \theta_a = 0$, eq. (6.106*b*) shows that $q_a = q_b$ and

$$\tau_{22} = \left.\frac{q_b}{q_a}\right|_{\theta_a=0} = 1 \tag{6.107d}$$

The **T**-matrix for the single shunt element is therefore

$$\mathbf{T} = \begin{bmatrix} 1 & 0 \\ G & 1 \end{bmatrix} \tag{6.108}$$

Notice that because $\tau_{12} = 0$, Table 5.2 shows that the single series element does not possess a **Y**-matrix.

6.6 CLOSURE

With the elements of the linear transformations for regular fins and spines and the thermal transmission ratio for singular fins and spines in hand, attention can now turn to the algorithms for finned array assembly in Chapter 7.

6.7 NOMENCLATURE

Roman Letter Symbols

A cross-sectional or profile area, m^2
a fin-tip dimension, m
b fin height, m
C arbitrary constant, dimensionless
d diameter, m; derivative or differential, dimensions vary
G thermal conductance, W/K
$\mathbf{G}$ inverse hybrid matrix, dimensions vary
g elements of inverse hybrid matrix, dimensions vary
$\mathbf{H}$ hybrid matrix, dimensions vary
h heat transfer coefficient, W/m^2 · K; elements of hybrid matrix, dimensions vary
I modified Bessel function of the first kind
K modified Bessel function of the second kind
k thermal conductivity, W/m · K
L fin length, m
m fin performance parameter, m^{-1}
n combination of terms, m$^{-1/2}$
q heat flow, W
R thermal resistance, K/W
r radial coordinate, m; radius, m
S surface area, m^2
T temperature, K
u combination of terms, dimensionless
w combination of terms, dimensionless
x height coordinate, m
Y thermal admittance, W/K
$\mathbf{Y}$ thermal admittance matrix, W/K
y elements of thermal admittance matrix, W/K; profile, m; profile function, m
$\mathbf{Z}$ thermal impedance matrix, W/K
z elements of thermal impedance matrix, W/K; combination of Bessel functions

Greek Letter Symbols

α exponent in Euler equation general solution, dimensionless; combination of terms, dimensionless
β exponent in Euler equation general solution, dimensionless; combination of terms, dimensionless
$\mathbf{\Gamma}$ thermal transmission matrix, dimensions vary
γ elements of thermal transmission matrix, dimensions vary
Δ change in, dimensionless; determinant, dimensions vary
δ fin thickness, m; side of rectangular spine, m; diameter of elliptical spine, m
η fin efficiency, dimensionless
θ temperature excess, K

κ taper angle, rad
Λ matrix, dimensions vary
λ elements of a matrix, dimensions vary
μ thermal transmission ratio, W/K
ρ radius ratio, dimensionless
$\mathbf{T}$ linear transmission matrix, dimensions vary
τ elements of linear transmission matrix, dimensions vary
ϕ combination of terms, m^{-1}
ψ combination of terms, m^{-1}

Roman Letter Subscripts

a tip of fin
b base of fin
f fin
id ideal
in input
o characteristic value
p profile area
s surroundings

Symbolic Superscript

$'$ first derivative

CHAPTER 7

ALGORITHMS FOR FINNED ARRAY ASSEMBLY

7.1 INTRODUCTION

The thermal transmission ratio developed in the previous chapter is defined as the ratio of the heat entering the base of a single fin to the temperature excess at the base of the fin. This was extended by Kraus et al. (1978) to finned arrays in the form of the fin or finned array *input admittance*, and algorithms were provided for what were called the *cascade, cluster*, and *parallel connections*. Kraus et al. (1978) also showed that a modification of the procedure developed therein led to an exact solution for the double stack with unequal heat distribution on the two stack faces at opposite sides.

The significant result of Chapters 5 and 6 was the development of a new parameterization for heat transfer in indiviual fins or spines in the form of a **T**-matrix for fins that were regular, those fins and spines whose cross section is never zero. Those fins that did taper to zero cross section, termed *singular fins*, were shown to be characterized by the aforementioned thermal transmission ratio. Use of the **T**-matrix for individual regular fins and the thermal transmission ratio for singular fins can both be employed via the *cascade, cluster*, and *parallel algorithms* to yield an input admittance for an entire finned array.

Kraus and Snider (1980) elaborated on the status, validity, and virtue of the foregoing new parameters by discussing their apparent superiority over the classic notion of the fin efficiency with regard to adaptability and versatility. They also provided many of the parameters that were summarized in Chapter 6.

Attention now turns to a detailed development of the cascade, cluster, and parallel algorithms and the consideration of several examples that illustrate their use.

7.2 ALGORITHMS FOR THE COMBINATION OF FINS

Figure 7.1 displays two finned arrays. Each of them contains three fin profiles, and in each, the fin of triangular profile is the most remote fin in the array. Algorithms for the combination of fins that does not contain loops in the graph-theoretical sense (such as the array shown in Fig. 7.1*a*, which contains nine fins) are developed in this section.

7.2.1 Cascade Algorithm

Consider the simplest case of two fins in cascade shown in Fig. 7.2. Here, continuity dictates that the heat leaving fin 2 must enter fin 1 and that at the point of intersection, the temperature excesses must match. Thus, a matrix equality that embraces both

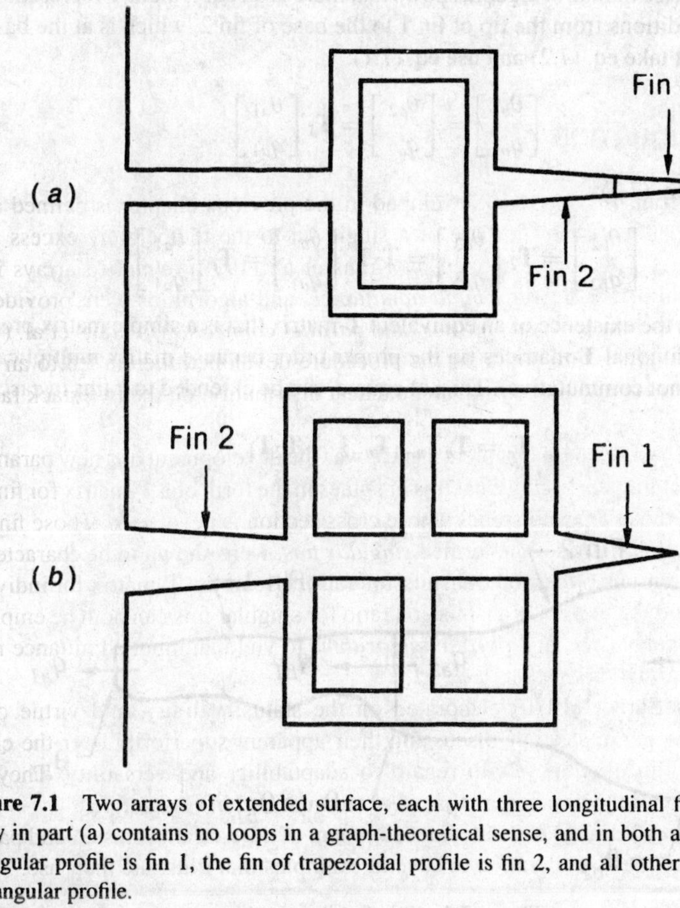

Figure 7.1 Two arrays of extended surface, each with three longitudinal fin profiles. The array in part (a) contains no loops in a graph-theoretical sense, and in both arrays, the fin of triangular profile is fin 1, the fin of trapezoidal profile is fin 2, and all other fins are fins of rectangular profile.

continuity and compatability at the intersection point between the fins may be set down as

$$\begin{bmatrix} \theta_{a2} \\ q_{a2} \end{bmatrix} = \begin{bmatrix} \theta_{b1} \\ q_{b1} \end{bmatrix} \tag{7.1}$$

In addition, each fin has a **T** parameterization, which is the linear transformation from tip conditions to base conditions. For fin 1,

$$\begin{bmatrix} \theta_{b1} \\ q_{b1} \end{bmatrix} = \mathbf{T}_1 \begin{bmatrix} \theta_{a1} \\ q_{a1} \end{bmatrix} \tag{7.2}$$

and for fin 2,

$$\begin{bmatrix} \theta_{b2} \\ q_{b2} \end{bmatrix} = \mathbf{T}_2 \begin{bmatrix} \theta_{a2} \\ q_{a2} \end{bmatrix} \tag{7.3}$$

A simple exercise in matrix algebra shows that there is an equivalent **T** representation that maps conditions from the tip of fin 1 to the base of fin 2, which is at the base of the array. First take eq. (7.2) and use eq. (7.1):

$$\begin{bmatrix} \theta_{b1} \\ q_{b1} \end{bmatrix} = \begin{bmatrix} \theta_{a2} \\ q_{a2} \end{bmatrix} = \mathbf{T}_1 \begin{bmatrix} \theta_{a1} \\ q_{a1} \end{bmatrix}$$

and then use eq. (7.3):

$$\begin{bmatrix} \theta_{b2} \\ q_{b2} \end{bmatrix} = \mathbf{T}_2 \begin{bmatrix} \theta_{a2} \\ q_{a2} \end{bmatrix} = \mathbf{T}_2\mathbf{T}_1 \begin{bmatrix} \theta_{a1} \\ q_{a1} \end{bmatrix} = \mathbf{T}_e \begin{bmatrix} \theta_{a1} \\ q_{a1} \end{bmatrix} \tag{7.4}$$

This shows the existence of an equivalent **T**-matrix that is a simple matrix product of the two individual **T**-matrices (in the proper order because matrix multiplication is, in general, not commutative). This, of course, can be extended to n fins in cascade:

$$\mathbf{T}_e = \mathbf{T}_n\mathbf{T}_{n-1}\mathbf{T}_{n-2}\cdots\mathbf{T}_2\mathbf{T}_1 \tag{7.5}$$

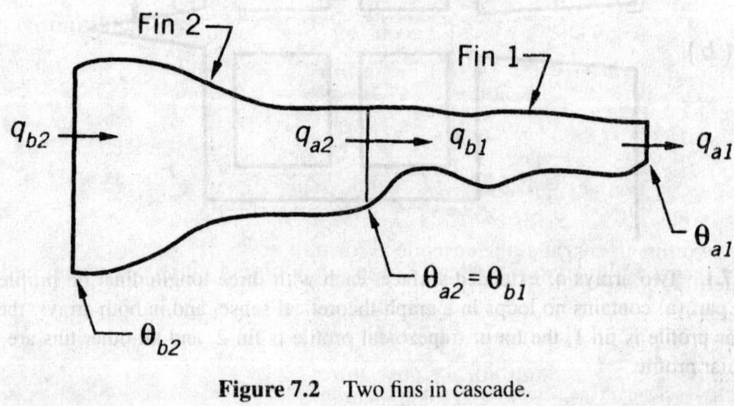

Figure 7.2 Two fins in cascade.

and eq. (7.5) is the *cascade algorithm.*

Observe that the foregoing development is based on the assignment of fin 1 as the most remote fin in the array. This applies to the development pertaining to just the two fins and the extension to the general case of n fins given by the cascade algorithm of eq. (7.5).

Array Input Admittance. The array input admittance can be obtained using eq. (5.46), called the *reflection relationship* and is repeated here:

$$Y_{\text{in}} = \frac{\tau_{21} + \tau_{22}(q_a/\theta_a)}{\tau_{11} + \tau_{12}(q_a/\theta_a)} \tag{7.6}$$

where the τ elements here can pertain either to an individual fin or an equivelent **T**-matrix. The ratio $q_a/\theta_a = 0$ for the case of a fin or finned array with an adiabatic tip or $q_a/\theta_a = h_{a1}A_{a1}$ for an array with remote fin 1 not adiabatic. Alternatively, the input admittance my be determined for n fins from repeated applications of eq. (7.6) by first using $q_a\theta_a = 0$ or $q_a/\theta_a = h_{a1}A_{a1}$ with the elements of **T** to find the input admittance at the base of fin 1. By the continuity and compatability conditions of eq. (7.1), this input admittance is the value of q_a/θ_a for fin 2, and another application of eq. (7.6), this time using the elements of $\mathbf{T_2}$, can find the value of the input admittance for fin 2. The procedure then repeats until the input admittance for fin n is obtained.

Example 7.1: The Cascade Algorithm and the Reflection Relationship for the Longitudinal Fin of Rectangular Profile. Consider the two longitudinal fins of rectangular profile placed end to end (or in cascade) as shown in Fig. 7.3. The two fins are identical with heights, $b/2$, and no heat flows from the tip of fin 1. Show that the two fins in cascade are formally equivalent to a single fin of height b by using (**a**) an equivalent **T**-matrix and (**b**) by determining the input admittance at fin 2 by the reflection relationship.

SOLUTION. (**a**) For an equivalent **T**-matrix, observe that this is the simplest possible cascade and two **T**-matrices will be required. Each of them have a **T**-matrix with elements given by eq. (6.9). With $u = b/2$,

$$\mathbf{T}_2 = \mathbf{T}_1 = \begin{bmatrix} \cosh mu & Z_o \sinh mu \\ Y_o \sinh mu & \cosh mu \end{bmatrix}$$

where

$$Y_o = (2hk\delta)^{1/2}L \quad \text{and} \quad Z_o = \frac{1}{(2hk\delta)^{1/2}}L$$

For the two fins in cascade, the cascade algorithm of eq. (7.5) is employed to give an equivalent $\mathbf{T}_e$:

$$\mathbf{T}_e = \begin{bmatrix} \cosh mu & Z_o \sinh mu \\ Y_o \sinh mu & \cosh mu \end{bmatrix}^2$$

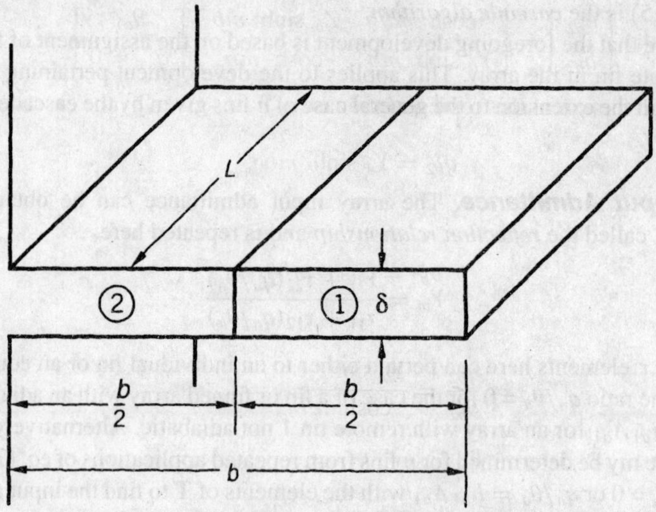

Figure 7.3 Two longitudinal fins of rectangular profile in cascade.

or

$$\mathbf{T}_e = \begin{bmatrix} \cosh^2 mu + \sinh^2 mu & 2Z_o \cosh mu \sinh mu \\ 2Y_o \cosh mu \sinh mu & \cosh^2 mu + \sinh^2 mu \end{bmatrix}$$

There are two hyperbolic function identities that can be employed here:

$$\cosh 2mu = \cosh^2 mu + \sinh^2 mu$$

and

$$\sinh 2mu = 2 \cosh mu \sinh mu$$

Thus

$$\mathbf{T}_e = \begin{bmatrix} \cosh 2mu & Z_o \sinh 2mu \\ Y_o \sinh 2mu & \cosh 2mu \end{bmatrix}$$

and because $mb = 2mu$,

$$\mathbf{T}_e = \begin{bmatrix} \cosh\ mb & Z_o \sinh\ mb \\ Y_o \sinh\ mb & \cosh\ mb \end{bmatrix}$$

Because

$$\begin{bmatrix} \theta_{b2} \\ q_{b2} \end{bmatrix} = \mathbf{T}_e \begin{bmatrix} \theta_{a1} \\ q_{a1} \end{bmatrix} = \begin{bmatrix} \cosh\ mb & Z_o \sinh\ mb \\ Y_o \sinh\ mb & \cosh\ mb \end{bmatrix} \begin{bmatrix} \theta_{a1} \\ q_{a1} \end{bmatrix}$$

and because $q_{a1} = 0$ (the tip of fin 1 is presumed to be adiabatic),

$$\begin{bmatrix} \theta_{b2} \\ q_{b2} \end{bmatrix} = \begin{bmatrix} \cosh\ mb & Z_o \sinh\ mb \\ Y_o \sinh\ mb & \cosh\ mb \end{bmatrix} \begin{bmatrix} \theta_{a1} \\ q_{a1} = 0 \end{bmatrix}$$

An expanded version of this representation is

$$q_{b2} = Y_o \sinh\ mb\theta_{a1}$$

and

$$\theta_{b2} = \cosh\ mb\theta_{a1}$$

so that

$$Y_{\text{in},2} = \frac{q_{b2}}{\theta_{b2}} = \frac{Y_o \sinh\ mb}{\cosh\ mb} = Y_o \tanh\ mb$$

and because $Y_o = (2hk\delta)^{1/2} L$,

$$Y_{\text{in},2} = \frac{q_{b2}}{\theta_{b2}} = Y_o \tanh\ mb = (2hk\delta)^{1/2} L \tanh\ mb$$

Notice that this is eq. (1.10):

$$q_b = (2hk\delta)^{1/2} L\theta_b \tanh\ mb \tag{1.10}$$

(**b**) Use of the reflection relationship requires two steps. The first step pertains to fin 1 with $u = b/2$:

$$\begin{bmatrix} \theta_{b1} \\ q_{b1} \end{bmatrix} = \mathbf{T}_1 \begin{bmatrix} \theta_{a1} \\ q_{a1} \end{bmatrix} = \begin{bmatrix} \cosh\ mu & Z_o \sinh\ mu \\ Y_o \sinh\ mu & \cosh\ mu \end{bmatrix} \begin{bmatrix} \theta_{a1} \\ q_{a1} \end{bmatrix}$$

Because the tip of fin 1 is adiabatic,

$$Y_{\text{in},1} = \frac{\tau_{21,1}}{\tau_{11,1}} = \frac{Y_o \sinh\ mu}{\cosh\ mu}$$

and the reflection relationship of eq. (7.6) gives with $q_{a1}/\theta_{a1} = Y_{\text{in},1}$:

$$\begin{aligned} Y_{\text{in},2} &= \frac{\tau_{21,2} + Y_{\text{in},1}\tau_{22,2}}{\tau_{11,2} + Y_{\text{in},1}\tau_{12,2}} \\ &= \frac{Y_o \sinh\ mu + (Y_o \sinh\ mu/\cosh\ mu)\cosh\ mu}{\cosh\ mu + (Y_o \sinh\ mu/\cosh\ mu) Z_o \sinh\ mu} \\ &= \frac{2Y_o \sinh\ mu \cosh\ mu}{\cosh^2 mu + \sinh^2 mu} \end{aligned}$$

and with the same two hyperbolic function identities,

$$Y_{\text{in},2} = \frac{Y_o \sinh\ 2mu}{\cosh\ 2mu}$$

With $u = b/2$,

$$Y_{\text{in},2} = \frac{Y_o \sinh\ mb}{\cosh\ mb} = Y_o \tanh mb$$

which is the same result as in part (a).

7.2.2 Cluster Algorithm

Figure 7.4 shows n fins that are appended to fin m. These fins are said to be in a *cluster arrangement* and the tip of fin m is called a *cluster point.* Here, continuity dictates that the heat that leaves fin m must divide among fins 1 through n.[1] Moreover, compatability dictates that the temperature excesses of all fins involved in the cluster must match. Thus, with the use of subscripts a and b which has become customary, it is observed that

$$q_{am} = q_{b1} + q_{b2} + q_{b3} + \cdots + q_{bn} \tag{7.7a}$$

and

$$\theta_{am} = \theta_{b1} = \theta_{b2} = \theta_{b3} = \cdots = \theta_{bn} \tag{7.7b}$$

Consider the admittance at the tip of fin m, which is termed the *cluster admittance* Y_C:

$$Y_{am} = Y_C = \frac{q_{am}}{\theta_{am}}$$

and employ eqs. (7.7):

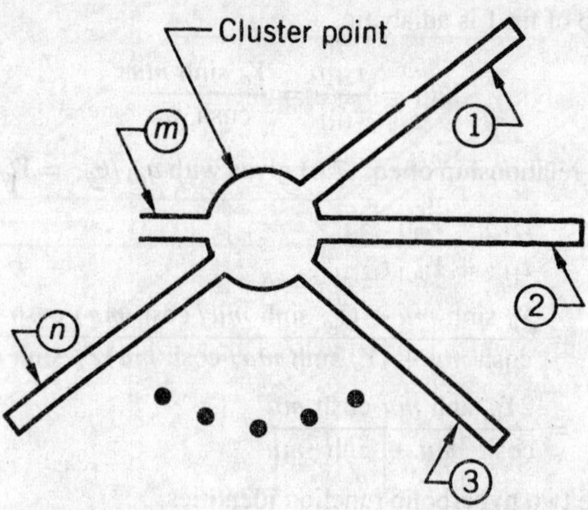

Figure 7.4 Fins in cluster. There are n fins attached to the tip of fin m and the tip of fin m is designated as the cluster point.

[1]But equally only if fins 1 though n all have exactly the same dimensions.

$$Y_{am} = Y_C = \frac{q_{am}}{\theta_{am}} = \frac{q_{b1} + q_{b2} + q_{b3} + \cdots + q_{bn}}{\theta_{am}}$$

indent or

$$Y_{am} = Y_C = \frac{q_{b1}}{\theta_{b1}} + \frac{q_{b2}}{\theta_{b2}} + \frac{q_{b3}}{\theta_{b3}} + \cdots + \frac{q_{bn}}{\theta_{bn}} \tag{7.8}$$

Each term on the right-hand side of eq. (7.8) represents an individual input admittance or thermal transmission ratio. Thus a cluster admittance may be defined for r-regular fins and s-singular fins as

$$Y_C = \sum_{i=1}^{r} Y_{\text{in},i} + \sum_{j=1}^{s} \mu_j \tag{7.9}$$

This addition of input admittances and thermal transmission ratios to form the cluster admittance is the *cluster algorithm*.

7.2.3 Parallel Algorithm

A parallel arrangement of n fins is shown in Fig. 7.5. Notice that a single fin, designated as fin m, is delivering heat through its tip to the parallel combination, and that a single fin, designated as fin p, is picking up, at its base, all the heat leaving the parallel combination. In this case, continuity demands that

$$q_{am} = q_{b1} + q_{b2} + q_{b3} + \cdots + q_{bn} \tag{7.10a}$$

and

$$q_{bp} = q_{a1} + q_{a2} + q_{a3} + \cdots + q_{an} \tag{7.10b}$$

In addition, compatability requires that at point m,

$$\theta_{am} = \theta_{b1} = \theta_{b2} = \theta_{b3} = \cdots = \theta_{bn} \tag{7.11a}$$

and at point p,

$$\theta_{bp} = \theta_{a1} = \theta_{a2} = \theta_{a3} = \cdots = \theta_{an} \tag{7.11b}$$

Use of the **Y**-matrix, which is referred to as the *thermal admittance matrix*, defined by eq. (5.12), and reproduced here:

$$\begin{bmatrix} q_b \\ q_a \end{bmatrix} = \mathbf{Y} \begin{bmatrix} \theta_b \\ \theta_a \end{bmatrix} = \begin{bmatrix} y_{11} & y_{12} \\ y_{21} & y_{22} \end{bmatrix} \begin{bmatrix} \theta_b \\ \theta_a \end{bmatrix} \tag{7.12}$$

permits a linear transformation from temperature excesses to heat flows for any individual fin. Thus, from eqs. (7.10) using fins 1, 2, ..., n,

$$\begin{bmatrix} q_{am} \\ q_{bp} \end{bmatrix} = \begin{bmatrix} q_{b1} \\ q_{a1} \end{bmatrix} + \begin{bmatrix} q_{b2} \\ q_{a2} \end{bmatrix} + \cdots + \begin{bmatrix} q_{bn} \\ q_{an} \end{bmatrix}$$

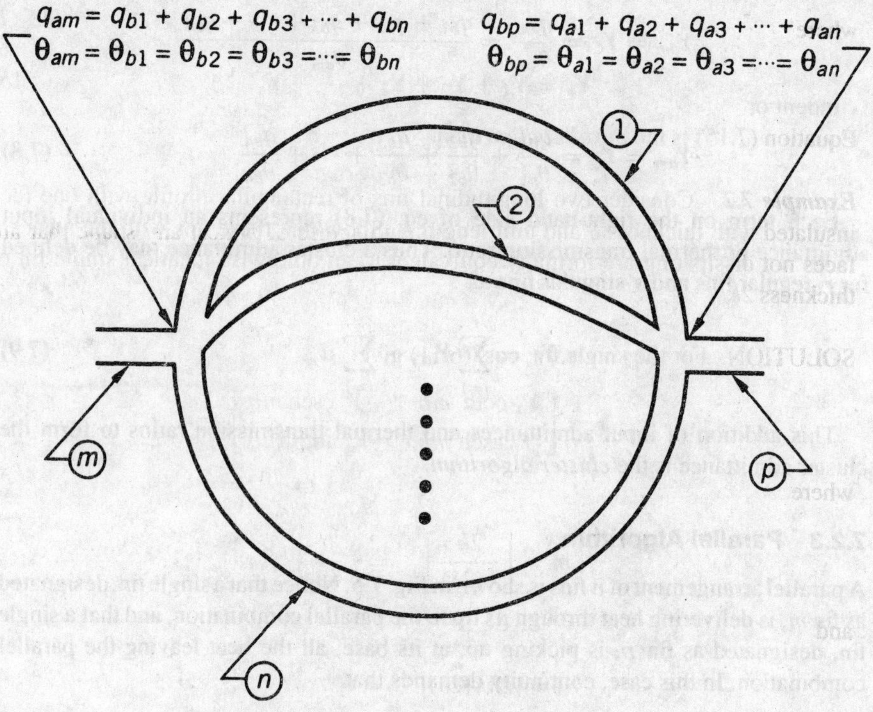

Figure 7.5 Parallel combination of n fins.

and the use of eq. (7.12) provides

$$\begin{bmatrix} q_{am} \\ q_{bp} \end{bmatrix} = \mathbf{Y}_1 \begin{bmatrix} \theta_{b1} \\ \theta_{a1} \end{bmatrix} + \mathbf{Y}_2 \begin{bmatrix} \theta_{b2} \\ \theta_{a2} \end{bmatrix} + \cdots + \mathbf{Y}_n \begin{bmatrix} \theta_{bn} \\ \theta_{an} \end{bmatrix} \tag{7.13}$$

But, the compatability conditions of eqs. (7.11),

$$\begin{bmatrix} \theta_{am} \\ \theta_{bp} \end{bmatrix} = \begin{bmatrix} \theta_{b1} \\ \theta_{a1} \end{bmatrix} = \begin{bmatrix} \theta_{b2} \\ \theta_{a2} \end{bmatrix} = \cdots = \begin{bmatrix} \theta_{bn} \\ \theta_{an} \end{bmatrix}$$

can be incorporated into eq. (7.13) to show that

$$\begin{bmatrix} q_{am} \\ q_{bp} \end{bmatrix} = [\mathbf{Y}_1 + \mathbf{Y}_2 + \cdots + \mathbf{Y}_n] \begin{bmatrix} \theta_{am} \\ \theta_{bp} \end{bmatrix} \tag{7.14}$$

This shows that there is an equivalent thermal admittance matrix $\mathbf{Y}_e$ that relates the total heat fows to the temperature excesses at the base and tip of a parallel combination of fins:

$$\begin{bmatrix} q_b \\ q_a \end{bmatrix} = \mathbf{Y}_e \begin{bmatrix} \theta_b \\ \theta_a \end{bmatrix}$$

where

$$\mathbf{Y}_e = \mathbf{Y}_1 + \mathbf{Y}_2 + \mathbf{Y}_3 + \cdots + \mathbf{Y}_n \tag{7.15}$$

Equation (7.15) is the *parallel algorithm.*

Example 7.2. Consider two longitudinal fins of rectangular profile with one face insulated with thickness δ and unit length L. Show that these fins with the adjacent faces not dissipating are formally equivalent to an otherwise identical single fin of thickness 2δ.

SOLUTION. For the single fin, eqs. (6.11) give

$$\mathbf{Y} = \begin{bmatrix} Y_o \coth mb & -Y_o \operatorname{csch} mb \\ Y_o \operatorname{csch} mb & -Y_o \coth mb \end{bmatrix}$$

where

$$m = \left[\frac{2h}{k(2\delta)}\right]^{1/2} = \left(\frac{h}{k\delta}\right)^{1/2}$$

and

$$Y_o = [2hk(2\delta)]^{1/2} = (4hk\delta)^{1/2}$$

For the two fins with one face insulated, eqs. (6.11) also give

$$\mathbf{Y} = \begin{bmatrix} Y_o \coth mb & -Y_o \operatorname{csch} mb \\ Y_o \operatorname{csch} mb & -Y_o \coth mb \end{bmatrix}$$

but as indicated in Section 6.2.2,

$$m = \left(\frac{h}{k\delta}\right)^{1/2}$$

which is the same as for the single fin and

$$Y_o = (hk\delta)^{1/2}$$

By the parallel algorithm, an equivalent can be formed for the two fins

$$\mathbf{Y}_e = 2\begin{bmatrix} Y_o \coth mb & -Y_o \operatorname{csch} mb \\ Y_o \operatorname{csch} mb & -Y_o \coth mb \end{bmatrix}$$

For equivalence, this $\mathbf{Y}_e$ for the two fins must equal $\mathbf{Y}$ for the single fin. Because m is the same for both, the equivalence is proven if $\mathbf{Y}_e = \mathbf{Y}$ and

$$\begin{bmatrix} Y_o \coth mb & -Y_o \operatorname{csch} mb \\ Y_o \operatorname{csch} mb & -Y_o \coth mb \end{bmatrix} = 2\begin{bmatrix} Y_o \coth mb & -Y_o \operatorname{csch} mb \\ Y_o \operatorname{csch} mb & -Y_o \coth mb \end{bmatrix}$$

or if

$$(4hk\delta)^{1/2}\begin{bmatrix}\coth mb & -\operatorname{csch} mb\\ \operatorname{csch} mb & -\coth mb\end{bmatrix} = 2(hk\delta)^{1/2}\begin{bmatrix}\coth mb & -\operatorname{csch} mb\\ \operatorname{csch} mb & -\coth mb\end{bmatrix}$$

Because

$$(4hk\delta)^{1/2} = 2(hk\delta)^{1/2}$$

the equivalence is proven.

Example 7.3: Analysis of a Finned Array. The array shown in Fig. 7.6 contains 12 fins. It is fabricated of aluminum ($k = 200$ W/m · K) and is used in an application where $h = 50$ W/m^2 · K. All fins are longitudinal fins of rectangular profile with a length (into the plane of the paper) of 8 cm. All other dimensions are shown in Table 7.1. The tips of fins 1, 4, 5, 7, and 9 are adiabatic, and a design decision as to whether this array can dissipate 70 W at a temperature excess between the base of fin 12 and the environment of 50°C.

SOLUTION. The first step is to determine the elements of the **T**-matrix for all the fins in the array. This can be done by using eqs. (6.9), where it is seen that

$$\tau_{11} = \cosh mb \tag{a}$$

$$\tau_{12} = Z_o \sinh mb \tag{b}$$

$$\tau_{21} = Y_o \sinh mb \tag{c}$$

$$\tau_{22} = \cosh mb \tag{d}$$

The parameters for all fins in Fig. 7.6 may be generated by hand calculation or via the use of commercially available software. These parameters, which require the use of the characteristic admittance

TABLE 7.1 Heights and Thicknesses of Fins in Fig. 7.6

Fin	b (cm)	δ (cm)
1	1.400	0.3048
2	2.500	0.3048
3	4.000	0.3048
4	1.400	0.3048
5	1.400	0.3048
6	4.828	0.3048
7	1.400	0.3048
8	1.200	0.3048
9	1.400	0.3048
10	1.200	0.3048
11	2.828	0.3048
12	4.000	0.4572

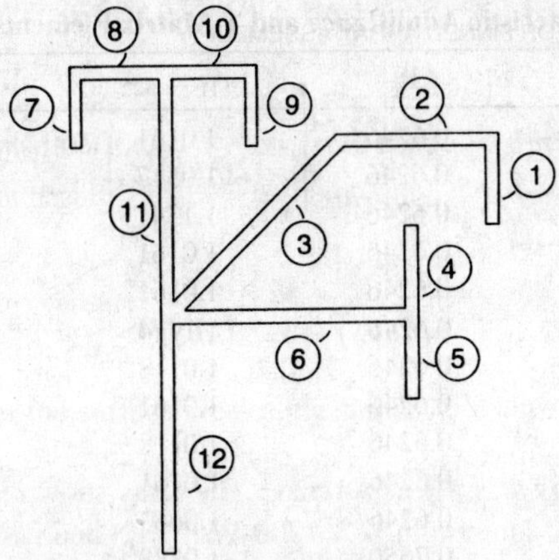

Figure 7.6 Finned array for Example 7.3.

$$Y_o = (2hk\delta)^{1/2} L$$

the characteristic impedance

$$Z_o = \frac{1}{(2hk\delta)^{1/2}} L$$

and

$$m = \left(\frac{2h}{k\delta}\right)^{1/2}$$

are shown in Table 7.2.

The detailed procedure involves repeated use of the cascade algorithm, eq. (7.5), the reflection relationship, eq. (7.6), and the cluster algorithm, eq. (7.9).

(1) Begin with fins 1, 2, and 3. Fin 1 is remote and has an adiabatic tip. The cascade algorithm gives an equivalent matrix, $\mathbf{T}_e$:

$$\begin{aligned}
\mathbf{T}_e &= \mathbf{T}_3\mathbf{T}_2\mathbf{T}_1 \\
&= \begin{bmatrix} 1.1341 & 0.8566 \\ 0.3342 & 1.1341 \end{bmatrix} \begin{bmatrix} 1.0517 & 0.5214 \\ 0.2034 & 1.0517 \end{bmatrix} \begin{bmatrix} 1.0161 & 0.2886 \\ 0.1126 & 1.0162 \end{bmatrix} \\
&= \begin{bmatrix} 1.1341 & 0.8566 \\ 0.3342 & 1.1341 \end{bmatrix} \begin{bmatrix} 1.1274 & 0.8334 \\ 0.3251 & 1.1274 \end{bmatrix} \\
&= \begin{bmatrix} 1.5571 & 1.9109 \\ 0.7455 & 1.5571 \end{bmatrix}
\end{aligned}$$

Then, by the reflection relationship

TABLE 7.2 Characteristic Admittance and T-Matrix Elements for All Fins in Fig. 7.6

Fin	mb	Y_o	$\tau_{11} = \tau_{22}$	τ_{12}	τ_{21}
1	0.1793	0.6246	1.0161	0.2886	0.1126
2	0.3202	0.6246	1.0517	0.5214	0.2034
3	0.5123	0.6246	1.1341	0.8566	0.3342
4	0.1793	0.6246	1.0161	0.2886	0.1126
5	0.1793	0.6246	1.0161	0.2886	0.1126
6	0.6184	0.6246	1.1974	1.0544	0.4113
7	0.1537	0.6246	1.0118	0.2470	0.0964
8	0.1793	0.6246	1.0161	0.2886	0.1126
9	0.1537	0.6246	1.0118	0.2470	0.0964
10	0.1793	0.6246	1.0161	0.2886	0.1126
11	0.3622	0.6246	1.0663	0.5926	0.2312
12	0.4183	0.7650	1.0888	0.5633	0.3292

$$Y_{\text{in},3} = \frac{\tau_{21,e}}{\tau_{11,e}} = \frac{0.7455}{1.5571} = 0.4788 \text{ W/K}$$

This may be confirmed by treating the entity of fins 1, 2, and 3 as a single fin of height

$$b = 1.40 + 2.50 + 4.00 = 7.90 \text{ cm}$$

In this case

$$m = \left(\frac{2h}{k\delta}\right)^{1/2} = \left[\frac{2(50)}{(200)(0.003048)}\right]^{1/2} = 12.8079$$

$$mb = (12.8079)(0.0790) = 1.0118$$

and

$$Y_o = (2hk\delta)^{1/2} L = [(2)(50)(200)(0.003048)]^{1/2}(0.080) = 0.6246$$

For the equivalent fin, 7.90 cm high,

$$\begin{aligned} \tau_{21} &= Y_o \sinh\ mb \\ &= 0.6246(\sinh 1.0118) \\ &= 0.6246(1.1935) \\ &= 0.7455 \text{ W/K} \end{aligned}$$

and

$$\begin{aligned} \tau_{11} &= \cosh\ mb \\ &= \cosh 1.0118 \end{aligned}$$

$$= 1.5571$$

This time the reflection relationship gives for this equivalent

$$Y_{\text{in},3} = \frac{\tau_{21,e}}{\tau_{11,e}} = \frac{0.7455}{1.5571} = 0.4788 \text{ W/K}$$

which checks.

(2) Fins 4 and 5 have indentical dimensions and have adiabatic tips. For both

$$Y_{\text{in},4} = Y_{\text{in},5} = \frac{\tau_{21,4}}{\tau_{11,4}} = \frac{0.1126}{1.0161} = 0.1108 \text{ W/K}$$

The cluster algorithm says that at the tip of fin 6, the cluster admittance is the sum of the input admittances of fins 4 and 5. Thus

$$Y_{a,6} = Y_{C,6} = \frac{q_{a,6}}{\theta_{a,6}} = Y_{\text{in},4} + Y_{\text{in},5} = (2)(0.1108) = 0.2216 \text{ W/K}$$

(3) The reflection relationship may now be used to find the input admittance at the base of fin 6. With $q_{a,6}/\theta_{a,6} = Y_{C,6}$,

$$\begin{aligned} Y_{\text{in},6} &= \frac{\tau_{21,6} + Y_{C,6}\tau_{22,6}}{\tau_{11,6} + Y_{C,6}\tau_{12,6}} \\ &= \frac{0.4113 + 0.2216(1.1974)}{1.1974 + 0.2216(1.0544)} \\ &= \frac{0.6767}{1.4311} \\ &= 0.4728 \text{ W/K} \end{aligned}$$

(4) Fins 7 and 8 and fins 9 and 10 are identical combinations and fins 7 and 9 have adiabatic tips. Hence

$$\begin{aligned} \mathbf{T}_e &= \mathbf{T}_8\mathbf{T}_7 = \mathbf{T}_{10}\mathbf{T}_9 \\ &= \begin{bmatrix} 1.0161 & 0.2886 \\ 0.1126 & 1.0161 \end{bmatrix} \begin{bmatrix} 1.0118 & 0.2470 \\ 0.0964 & 1.0118 \end{bmatrix} \\ &= \begin{bmatrix} 1.0560 & 0.5430 \\ 0.2119 & 1.0560 \end{bmatrix} \end{aligned}$$

The input admittance for these entities is obtained from the reflection relationship

$$Y_{\text{in},8} = Y_{\text{in},10} = \frac{\tau_{21,e}}{\tau_{11,e}} = \frac{0.2119}{1.0560} = 0.2006 \text{ W/K}$$

(5) The cluster admittance at the tip of fin 11 is found from the cluster algorithm

$$Y_{a,11} = Y_{C,11} = \frac{q_{a,11}}{\theta_{a,11}} = Y_{\text{in},8} + Y_{\text{in},10} = (2)(0.2006) = 0.4012 \text{ W/K}$$

(6) The input admittance at the base of fin 11 is found from the reflection relationship.

$$Y_{\text{in},11} = \frac{\tau_{21,11} + Y_{C,11}\tau_{22,11}}{\tau_{11,11} + Y_{C,11}\tau_{12,11}}$$

$$= \frac{0.2312 + (0.4012)(1.0663)}{1.0663 + (0.4012)(0.5926)}$$

$$= \frac{0.6591}{1.3061}$$

$$= 0.5046 \text{ W/K}$$

(7) At the tip of fin 12, by the cluster algorithm

$$Y_{a,12} = Y_{C,12} = \frac{q_{a,12}}{\theta_{a,12}}$$

$$= Y_{\text{in},3} + Y_{\text{in},6} + Y_{\text{in},11}$$

$$= 0.4788 + 0.4728 + 0.5046$$

$$= 1.4562 \text{ W/K}$$

(8) The input admittance at the base of the array which is the base of fin 12 is now found via the reflection relationship:

$$Y_{\text{in},12} = \frac{\tau_{21,12} + Y_{C,12}\tau_{22,12}}{\tau_{11,12} + Y_{C,12}\tau_{12,12}}$$

$$= \frac{0.3292 + (1.4562)(1.0888)}{1.0888 + (1.4562)(0.5633)}$$

$$= \frac{1.9147}{1.9091}$$

$$= 1.0030 \text{ W/K}$$

Observe that the performance of the array is better at the tip of fin 12 than at the base of fin 12. Indeed, the performance of the finned array at the tip of fin 12

$$q = Y_{\text{in}}\theta = (1.4562)(50) = 72.81 \text{ W} > 70.0 \text{ W}$$

meets the specification for the finned array but that at the base of fin 12,

$$q = Y_{\text{in}}\theta = (1.0030)(50) = 50.15 \text{ W} < 70.0 \text{ W}$$

which is a dismal failure.

Fin 12 in this array has caused a *heat flow choke* because its dimensions, placement, or arrangement does not permit the transfer of a sufficient amount of heat to its tip. This is a major limitation in the use of finned arrays where the input admittance of one of the fins is less than the cluster admittance at its tip. A discussion of the *choking phenomenon* now follows and a return to this example with suggested remedies for the elimination of the heat flow choke will take place after this discussion.

7.2.4 Choking Phenomenon

A fin or spine that accommodates a cluster at its tip may exhibit a heat flow choke because the fin or spine dimensions may not permit, in a cooling application, the transfer of a sufficient amount of heat to the cluster at its tip. This phenomenon is dramatically evident in Example 7.3.

In the case of a longitudinal fin of rectangular profile, eq. (7.6) can be written, with the tip admittance designated by Y_a, as

$$Y_{\text{in}} = \frac{Y_o \sinh mb + Y_a \cosh mb}{\cosh mb + (Y_a/Y_o) \sinh mb} \tag{7.16}$$

If numerator and denominator are divided by cosh mb, the result is

$$Y_{\text{in}} = \frac{Y_o \tanh mb + Y_a}{1 + (Y_a/Y_o) \tanh mb}$$

and it may be observed that

$$\lim_{b \to 0} Y_{\text{in}} = Y_a$$

and

$$\lim_{b \to \infty} Y_{\text{in}} = Y_o$$

These provide the extremes for the input admittance Y_{in} for the extreme values of the fin height b.

Kraus and Snider (1985) show that to find the fin height that makes Y_{in} a maximum, the customary method of finding where $\partial Y_{\text{in}}/\partial b$ vanishes is employed. In eq. (7.16), once again observe that neither Y_o nor Y_a is a function of b, and let the numerator and denominator be designated by u and v:

$$u = Y_o \sinh mb + Y_a \cosh mb$$

and

$$v = \cosh mb + Y \sinh mb$$

where

$$Y = \frac{Y_a}{Y_o}$$

Then

$$\frac{\partial Y_{\text{in}}}{\partial b} = \frac{vm(Y_o \cosh mb + Y_a \sinh mb) - um(\sinh mb + Y \cosh mb)}{v^2} = 0$$

The m's in the numerator and the v^2 in the denominator may be canceled and hence

$$v(Y_o \cosh\ mb + Y_a \sinh\ mb) = u(\sinh\ mb + Y \cosh\ mb)$$

and with u and v substituted, the result after considerable algebraic tedium is

$$Y_o = YY_a = \frac{Y_a^2}{Y_o}$$

or

$$Y_o = Y_a \tag{7.17}$$

Because eq. (7.17) does not involve b, it implies that the slope of $Y_{\text{in}} = f(b)$ is never zero unless $Y_o = Y_a$, in which case it is always zero. Thus, there is no question of finding the maximum Y_{in} with respect to fin height, and three cases are evident.

- *Case I.* If $Y_o < Y_a$, the characteristic admittance is less than the tip admittance, and the fin is choking the heat flow regardless of its height. This is the case for fin 12 in Example 7.3, and any fix cannot involve fin 12. However, in this case, b should be as low as possible ($b \to 0$).
- *Case II.* If $Y_o > Y_a$, no choking occurs and b can be made as high as possible because a benefit exists for high b ($b \to \infty$) consistent with mass or volume limitations.
- *Case III.* If $Y_o = Y_a$, the value of b has no effect on the heat transfer process.

The foregoing alternatives boil down to a consideration of case I. To remedy a choke, three things can be attempted

- Remove the offending fin by merely cutting it away. This may not be possible if the offending fin is used as a supporting structure.
- Because adjustment of the height is futile, attempt to adjust the characteristic admittance by making the fin thicker if a longitudinal fin or making the diameter larger if a spine. Because $Y_o = (2hk\delta)^{1/2}L$ for the longitudinal fin of rectangular profile, a break-even point exists when $Y_o = Y_a$:

$$\delta = \frac{Y_a^2}{2hkL^2} \tag{7.18}$$

and for the cylindrical spine where $Y_o = (\pi^2 hkd^3/4)^{1/2}$, the break-even point will occur when

$$d = \left(\frac{4Y_a^2}{\pi^2 hk}\right)^{1/3} \tag{7.19}$$

- Replace the single offending fin with a parallel combination of fins.

A return to Example 7.3 is beneficial so that these alternatives can be examined in detail.

Example 7.3 (Continued): Analysis of a Finned Array

SOLUTION. There are three choices for the elimination of the heat flow choke, but only the second and third need be investigated, as fin 12 is a suppporting structure and probably cannot be removed. A modification of the fin thickness via eq. (7.18) can be investigated with a value of $Y_a = 1.4562$ W/K, to yield

$$\delta = \frac{Y_a^2}{2hkL^2} = \frac{(1.4562)^2}{(2)(50)(200)(0.08)^2} = 0.01657 \text{ m} \quad \text{or} \quad 1.657 \text{ cm}$$

This gives an aspect ratio of

$$\text{AR} = \frac{4.000}{1.668} = 2.415$$

a value that leads to a violation of Murray–Gardner assumption 5, which states that the fin thickness must be small compared with its height and length, so that temperature gradients across the fin and heat transfer from the edges of the fin may be neglected.

Instead, use two fins in parallel to replace fin 12. These fins, designated as fins 12 and 13, are each 4.00 cm high with $mb = 0.4183$ and $Y_o = 0.7650$ W/K. These are the values used previously for the original fin 12, and by eq. (6.11), the **Y**-matrix will be

$$\mathbf{Y} = \begin{bmatrix} Y_o \coth mb & -Y_o \operatorname{csch} mb \\ Y_o \operatorname{csch} mb & -Y_o \coth mb \end{bmatrix} = \begin{bmatrix} 1.9343 & -1.7765 \\ 1.7765 & -1.9343 \end{bmatrix}$$

The parallel algorithm of eq. (7.15) says that there is an equivalent **Y**-matrix that is the sum of the two individual **Y**-matrices for the fins in parallel. Hence

$$\mathbf{Y}_e = 2\begin{bmatrix} 1.9343 & -1.7765 \\ 1.7765 & -1.9343 \end{bmatrix} = \begin{bmatrix} -3.8686 & -3.5530 \\ 3.5530 & -3.8686 \end{bmatrix}$$

This matrix has a determinant

$$\Delta_Y = -2.3423$$

A return to the **T**-representation can be accomplished by following Table 5.2.

$$\tau_{11} = -\frac{y_{22}}{y_{21}} = -\frac{3.8686}{3.5530} = 1.0888$$

$$\tau_{12} = -\frac{1}{y_{21}} = \frac{1}{3.5530} = 0.2815$$

$$\tau_{21} = -\frac{\Delta_Y}{y_{21}} = -\frac{-2.3423}{3.5530} = 0.6592$$

and

$$\tau_{22} = \frac{y_{11}}{y_{21}} = \frac{3.8686}{3.5530} = 1.0888$$

Thus

$$\mathbf{T}_e = \begin{bmatrix} 1.0888 & 0.2815 \\ 0.6592 & 1.0888 \end{bmatrix}$$

and with $Y_C = 1.4562$ W/K for this parallel entity, the reflection relationship of eq. (7.6) gives

$$\begin{aligned} Y_{\text{in}} &= \frac{\tau_{21} + Y_C \tau_{22}}{\tau_{11} + Y_C \tau_{12}} \\ &= \frac{0.6592 + (1.4562)(1.0888)}{1.0888 + (1.4562)(0.2815)} \\ &= \frac{2.2447}{1.4987} \\ &= 1.4978 \text{ W/K} \end{aligned}$$

Because $Y_{\text{in}} = 1.4987 > Y_{\text{in}} = 1.4611$, the heat flow choke has been eliminated and the finned array with

$$q_b = Y_{\text{in}} \theta_b = (1.4978)(50) = 74.89 \text{ W} > 70 \text{ W}$$

is now meeting the specification.

7.2.5 Single Elements

The effects of bond resistance and tip heat loss can be handled in the cascade procedure described by eq. (7.5) or the reflection relationship of eq. (7.6). For the case of the series resistance, the single series element is placed in a manner such that it premultiplies either the **T**-matrix for a single fin or the equivalent **T**-matrix for an array of fins. For the case of tip heat loss, the single shunt element is placed in a manner such that it postmultiplies either the **T**-matrix for a single fin or the **T**-matrix for an array of fins.

7.2.6 Single Shunt Element

The single shunt admittance matrix can be used effectively in a computationally efficient procedure to account for the fin-tip heat loss, as Example 7.4 will demonstrate.

Example 7.4: The Use of a Single Shunt Element to Account for Fin-Tip Heat Loss. Consider a single longitudinal fin of rectangular profile of height $b = 15$ cm, length $L = 32$ cm, and thickness $\delta = 0.2286$ cm. It is fabricated of a material with $k = 180$ W/m · K, and the coefficients of heat transfer on the fin faces and the fin tip are $h = 24$ W/m^2 · K and $h_a = 16$ W/m^2 · K, respectively. Determine the input admittance using (**a**) the reflection relationship of eq. (7.6) and (**b**) the single shunt admittance matrix of eq. (6.13).

SOLUTION. For this fin

$$m = \left(\frac{2h}{k\delta}\right)^{1/2} = \left[\frac{(2)(24)}{(180)(0.002286)}\right]^{1/2} = 10.8006 \text{ m}^{-1}$$

$$mb = (10.8006)(0.15) = 1.6201$$

and

$$Y_o = (2hk\delta)^{1/2}L = [(2)(24)(180)(0.002286)]^{1/2}(0.32) = 1.4221 \text{ W/K}$$

The elements of the **T**-matrix are obtained from eqs. (6.9), so that

$$\mathbf{T} = \begin{bmatrix} 2.6257 & 1.7071 \\ 3.4527 & 2.6257 \end{bmatrix}$$

(**a**) The tip heat loss is given by

$$q_a = h_a S\theta_a$$

and this can be converted to a tip admittance:

$$Y_a = h_a S = h_a \delta L = (16)(0.002286)(0.32) = 0.0117 \text{ W/K}$$

Then using eq. (7.6) with $Y_a = q_a/\theta_a$,

$$Y_{\text{in}} = \frac{\tau_{21} + \tau_{22}(q_a/\theta_a)}{\tau_{11} + \tau_{12}(q_a/\theta_a)}$$

$$= \frac{3.4527 + (2.6257)(0.0117)}{2.6257 + (1.7071)(0.1117)}$$

$$= 1.3167 \text{ W/K}$$

Notice that this is a slightly higher input admittance that would be obtained if the heat loss were neglected:

$$Y_{\text{in}} = \frac{\tau_{21}}{\tau_{11}} = \frac{3.4517}{2.6257} = 1.3146 \text{ W/K}$$

(**b**)

$$\mathbf{T}_{sh} = \begin{bmatrix} 1.0000 & 0 \\ 0.0117 & 1.0000 \end{bmatrix}$$

and the cascade algorithm of eq. (7.5) will provide $\mathbf{T}_e$:

$$\mathbf{T}_e = \begin{bmatrix} 2.6257 & 1.7071 \\ 3.4527 & 2.6257 \end{bmatrix} \begin{bmatrix} 1.0000 & 0 \\ 0.0117 & 1.0000 \end{bmatrix}$$

or

$$\mathbf{T}_e = \begin{bmatrix} 2.6457 & 1.7071 \\ 3.4834 & 2.6257 \end{bmatrix}$$

The input admittance in this case is $Y_{in} = \tau_{21}/\tau_{11}$ because use of the single shunt admittance matrix puts $Y_a = 0$ at the tip of the fin. Thus

$$Y_{in} = \frac{3.4834}{2.6457} = 1.3167 \text{ W/K}$$

This is the same result as in part (a).

7.2.7 Single Series Element

The bond resistance of several types of interference-fit high fin tubes[2] was studied by Gardner and Carnavos (1960). They considered an edge-wound ribbon fin to be a thin circular disk concentric with the tube outside diameter and determined whether or not a radial gap would exist between the fin and the tube. This determination was made on the basis of contact pressure, which will be developed by the difference in temperature between the fin and the tube. This is, in reality, a consideration of the thermal stress problem, and with the magnitude of the gap in hand, Gardner and Carnavos were able to evaluate the thermal resistance. This was followed by another study by Gardner (1964) in a symposium volume that covered air-cooled heat exchangers exclusively.

Kraus and Specht (1974) developed equations that permitted an adjustment of the fin efficiency when the thermal bond resistance between a longitudinal fin of rectangular profile and its base or prime surface was known. Manzoor et al. (1984) investigated the validity of the perfect contact assumption in the context of fin assembly heat transfer, and Serebryanyi (1985) gave recommendations on the evaluation of the contact resistance for a bundle of rectangular fins attached to a module containing dissipating semiconductors.

Sheffield et al. (1985) examined the surface characteristics of finned tube heat exchangers to help predict the contact geometry between the metal surfaces. Primary characteristics evaluated include surface roughness, surface waviness, and the material hardness. A scanning electron microscope was employed to observe the surface topography of the fins as well as the actual contact with the tube. The results were subsequently used to develop a mathematical model for prediction of the heat transfer between tube and fin.

If the bond resistance is known, use of the single series element matrix permits an efficient evaluation of its effect. This is demonstrated in Example 7.5

Example 7.5: Use of a Single Series Element to Account for Bond Resistance between Fin and Prime Surface. The fin in Example 7.4, operating with tip heat loss, has been bonded to the prime surface in a manner such that a 0.325-K/W bond resistance must be included. Evaluate the effect of the bond resistance.

[2]These tubes have radial fins of pronounced rectangular profile.

SOLUTION. First, recall that the input admittance for the fin with tip heat loss was found to be

$$Y_{\text{in}} = \frac{3.4834}{2.6457} = 1.3167 \text{ W/K}$$

This will be used as the reference for the determination of the effect of the bond resistance.

To determine the effect of the bond resistance, use the cascade algorithm of eq. (7.5) with the single series element **T**-matrix of eq. (6.105) acting as a premultiplier of the **T**-matrix, which is

$$\mathbf{T} = \begin{bmatrix} 2.6257 & 1.7071 \\ 3.4527 & 2.6257 \end{bmatrix}$$

With the bond resistance of 0.3250 K/W,

$$\mathbf{T}_{se} = \begin{bmatrix} 1.0000 & 0.3250 \\ 0.0000 & 1.0000 \end{bmatrix}$$

the cascade algorithm gives

$$\mathbf{T}_e = \begin{bmatrix} 1.0000 & 0.3250 \\ 0.0000 & 1.0000 \end{bmatrix} \begin{bmatrix} 2.6257 & 1.7071 \\ 3.4527 & 2.6257 \end{bmatrix}$$

or

$$\mathbf{T}_e = \begin{bmatrix} 3.7478 & 2.5605 \\ 3.4527 & 2.6257 \end{bmatrix}$$

The input resistance with the bond resistance is

$$Y_{\text{in}} = \frac{3.7478}{3.4527} = 1.0855 \text{ W/K}$$

Clearly, the bond resistance is destroying the performance of the fin.

7.3 CASCADE ALGORITHM AND THE VARIABLE HEAT TRANSFER COEFFICIENT

It has been noted in Chapter 2 that ever since publication of the limiting assumptions for ideal fin analysis by Murray (1938) and Gardner (1945), investigators have been considering the effect of relaxing the requirement that the heat transfer coefficient between the fin faces and the surrounding environment be uniform and constant. Ghai (1951) was apparently the first to take exception to this assumption, and this was followed in short order by Gardner (1951) in a discussion of Ghai's work, Melese (1957), Fortescue (1957), and Hughes and Slack (1958). Han and Lefkowitz (1960) assumed a power law variation of h over the height of the fin, Chen and Zyskowski

(1963) considered an exponential variation and Cumo et al. (1965a) published a two-dimensional study that allowed for variation of the heat transfer coefficient as well as internal heat generation. Snider and Kraus (1982) demonstrated how perturbation theory could be used to estimate the effect of a variation in the heat transfer coefficient.

Although Snider and Kraus (1982) provided some insight into a correction for variability of the heat transfer coefficient, the **T**-matrix and the cascade algorithm of eq. (7.5) can be used to evaluate the performance of a combination of individual fins and spines of almost any shape that are attached end to end to form an array. The ultimate goal is, of course, the computation of a numerical value for the input admittance, Y_{in}. A fin with a variation of the heat transfer coefficient on its faces may be decomposed into n subfins which are attached end to end. The **T**-matrix can be developed for each subfin, each with its own local heat transfer coefficient and a computationally efficient procedure involving the cascade algorithm can then be employed to determine the overall equivalent **T**-matrix for the single fin composed of the n subfins. Then, from this equivalent **T**-matrix, the input admittance or fin efficiency can be computed.

7.3.1 Han and Lefkowitz (1960) Solution

As indicated in Section 2.3, Han and Lefkowitz (1960) considered a heat transfer coefficient variation of the form given by eq. (2.30) and repeated here as eq. (7.20):

$$h(x) = h_o(\gamma + 1)\left(\frac{x}{b}\right)^{\gamma} \tag{7.20}$$

on the faces of the longitudinal fin of rectangular profile shown with its terminology and coordinate system in Fig. 7.7, where it should be noted that the height coordinate has its origin at the fin base with its positive orientation from base to tip and that h_o is the average heat transfer coefficient.

The differential equation for the temperature excess is given by

$$\frac{d^2\theta}{dx^2} - (\gamma + 1)m^2\left(\frac{x}{b}\right)^{\gamma}\theta = 0 \tag{7.21}$$

where

$$m = \left(\frac{2h_o}{k\delta}\right)^{1/2}$$

so that

$$\frac{d^2\theta}{dx^2} = \frac{2(\gamma + 1)h_o}{k\delta}\left(\frac{x}{b}\right)^{\gamma}\theta = 0$$

which is eq. (2.31).

After a rather involved transformation of variables (see Section 2.3.1), Han and Lefkowitz showed that the solution to the differential equation of eq. (7.21) with the boundary conditions of fin base temperature excess:

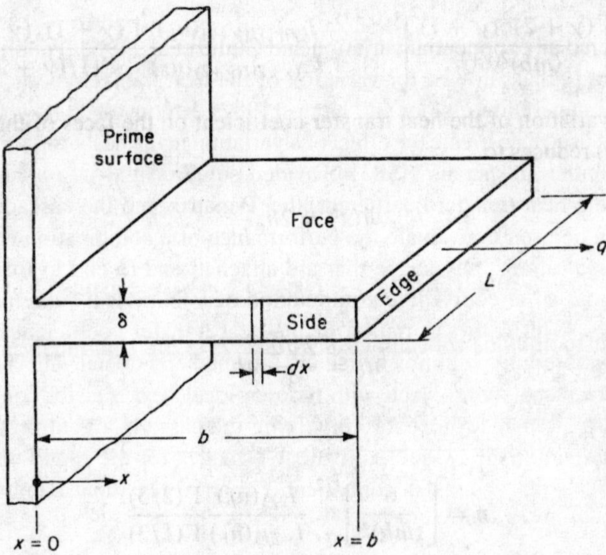

Figure 7.7 Terminology and coordinate system for the longitudinal fin of rectangular profile.

$$\theta(x = 0) = \theta_b$$

and the condition for the insulated tip is eq. (2.37), repeated here:

$$\theta(u) = \frac{\theta_b \left\{\Gamma_{[(\gamma+1)/(\gamma+2)]}\right\}}{2^{1/(\gamma+2)}} u^{1/(\gamma+2)} \Upsilon \tag{7.22}$$

where

$$u = \frac{2\sqrt{\gamma+1}}{\gamma+2} mb^{-\gamma/2} x^{(\gamma+2)/2} \tag{7.23}$$

$$u_b = \frac{2\sqrt{\gamma+1}}{\gamma+2} mb \tag{7.24}$$

and where

$$\Upsilon = \left[I_{-1/(\gamma+2)}(u) + \frac{I_{(\gamma+1)/(\gamma+2)}(u_b)}{I_{-(\gamma+1)/(\gamma+2)}(u_b)} I_{1/(\gamma+2)}(u) \right]$$

With the fin efficiency defined by

$$\eta = \frac{q_b}{q_{\text{id}}} = \frac{-k\delta L \left.\dfrac{d\theta}{dx}\right|_{x=0}}{2h_o Lb\theta_b}$$

a termwise differentiation of eq. (7.22) and subsequent evaluation at $x = 0$ gives

$$\eta = \left[\frac{(\gamma+2)^{\gamma}(\gamma+1)}{(mb)^{2(\gamma+1)}}\right]^{1/(\gamma+1)} \frac{I_{(\gamma+1)/(\gamma+2)}(u_b)}{I_{-(\gamma+1)/(\gamma+2)}(u_b)} \frac{\Gamma(\gamma+1)/(\gamma+2)}{\Gamma(1/(\gamma+2))} \tag{7.25}$$

For a linear variation of the heat transfer coefficient on the faces of the fin, $\gamma = 1$, and eq. (7.20) reduces to

$$h(x) = 2h_o \frac{x}{b} \tag{7.26}$$

With

$$u_b = \frac{2\sqrt{2}}{3} mb$$

the efficiency is

$$\eta = \left[\frac{6}{(mb)^4}\right]^{1/2} \frac{I_{2/3}(u_b)}{I_{-2/3}(u_b)} \frac{\Gamma(2/3)}{\Gamma(1/3)} \tag{7.27}$$

7.3.2 Use of the Cascade Algorithm

The connection of fins in cascade pertains to the configuration where at least two fins are connected tip to base. For n fins in cascade, the n fins may be assembled by the cascade algorithm:

$$\mathbf{T}_e = \mathbf{T}_n \mathbf{T}_{n-1} \mathbf{T}_{n-2} \cdots \mathbf{T}_2 \mathbf{T}_1 \tag{7.5}$$

Clearly, care must be exercised in selection of the fin numbering system and in the order of multiplication because matrix multiplication in general is not commutative. Every effort should be made to number the fins in sequence from the tip up through the fin base, where the base is defined as the point where the heat to be dissipated by the fin or the entity of fins is injected.

The thrust here is to consider that any single fin with any arbitrary piecewise continuous and single-valued heat transfer coefficient distribution on its faces can be modeled by n fins in cascade. Each of the n fins will possess its own heat transfer coefficient, which may be assumed constant and uniform over its faces. The cascade algorithm guarantees that the equivalent $\mathbf{T}$-matrix will represent the entire configuration. Only the number of subfins needed to guarantee the required accuracy is in question.

7.3.3 Number of Subfins Required

An example has been contrived to provide an idea as to the number of subfins required to match the classical solution of Han and Lefkowitz (1960). The example pertains to a longitudinal fin of rectangular profile operating with the thermal and physical data

$$b = 10 \text{ cm} \qquad L = 10 \text{ cm} \qquad \delta = 0.1975 \text{ cm}$$

$$h_o = 50.5 \text{ W/m}^2 \cdot \text{K} \qquad k = 202 \text{ W/m} \cdot \text{K} \qquad \gamma = 1$$

These numbers, which give an aspect ratio for the fin profile of approximately 50, are for a fin that is unusually high for its thickness and will yield a value for the fin efficiency which is quite low. However, the numbers may be used to provide a point of comparison between the classical and the cascade methods.

With the foregoing numbers, and with the linear variation of the heat transfer coefficient ($\gamma = 1$) over the fin faces,

$$h(x) = 2h_o \frac{x}{b} = (2)(50.5)\left(\frac{x}{0.100}\right) = 1010x \text{ W/m}^2 \cdot \text{K}$$

the fin efficiency given by eq. (7.27) is

$$\eta = 0.4474$$

and with the fin face area equal to

$$S = 2bL = (2)(0.100)(0.100) = 0.0200 \text{ m}^2$$

the input admittance is

$$Y_{\text{in}} = h_o S \eta = (50.5)(0.0200)(0.4474) = 0.4519 \text{ W/K}$$

Now, in what follows, let it be stressed that how a discretized analysis converges to a continuous analysis depends on the specific nature of the problem and not on the nature of the variation of one of the parameters. Thus, the following discussion represents an example and not a generalization.

The number of subfins required to yield an input admittance of $Y_{\text{in}} = 0.4519$ W/K has been determined in a computationally efficient procedure using a hand calculator with a huge memory. To be sure, use of the hand calculator is not relevant here. The point is that all numbers computed were stored in memory and four-figure accuracy was assured. One program took the thermal and physical data as input and, using eq. (7.6), quickly determined Y_{in} as a function of n, the total number of sub fins. The results are shown in Table 7.3, where it is seen that 50 subfins will provide the sought-after accuracy.

Table 7.3 shows that if three-decimal-place accuracy is all that is required because the value of $h(x)$ does not have nearly this accuracy, as few as 20 subfins may be employed. But if 50 subfins are to be used, Table 7.4 provides an insight as to just what is happening along the fin surface as the base of the fin is approached. While Table 7.4 shows data for every even-numbered subfin, data for subfin 31 are also included. It is at subfin 31 that the heat flow through the fin begins to choke, and this indicates that close to 40% of the fin is actually acting as a partial insulator.

TABLE 7.3 Number of Subfins Required to Yield $Y_{in} - 0.4519$ W/K When $b = L = 10$ cm, $\delta = 0.1975$ cm, $h_o = 50.5$ W/m^2 · K, and $k = 202$ W/m · K

Number of Subfins	y_{in} (W/K)
10	0.4534
20	0.4522
30	0.4520
40	0.4520
50	0.4519
75	0.4519

TABLE 7.4 Cascade Calculations for 50 Subfins[a]

i	x	h	Y_o	mb	$\tau_{11} = \tau_{22}$	τ_{12}	τ_{21}	Y_{in}
2	0.097	97.97	0.8842	0.0443	1.0010	0.0501	0.0392	0.0790
4	0.093	93.93	0.8658	0.0434	1.0009	0.0501	0.0376	0.1535
6	0.089	89.89	0.8470	0.0425	1.0009	0.0501	0.0360	0.2226
8	0.085	85.85	0.8277	0.0415	1.0009	0.0501	0.0343	0.2586
10	0.081	81.81	0.8080	0.0405	1.0008	0.0501	0.0327	0.3419
12	0.077	77.77	0.7877	0.0395	1.0008	0.0501	0.0311	0.3914
14	0.073	73.73	0.7670	0.0385	1.0007	0.0501	0.0295	0.4341
16	0.069	69.69	0.7457	0.0374	1.0007	0.0501	0.0279	0.4701
18	0.065	65.65	0.7238	0.0363	1.0007	0.0501	0.0263	0.4998
20	0.061	61.61	0.7011	0.0351	1.0006	0.0501	0.0246	0.5236
22	0.057	57.57	0.6778	0.0340	1.0006	0.0501	0.0230	0.5419
24	0.053	53.53	0.6535	0.0328	1.0005	0.0501	0.0214	0.5534
26	0.049	49.49	0.6284	0.0315	1.0005	0.0501	0.0198	0.5643
28	0.045	45.45	0.6022	0.0302	1.0005	0.0501	0.0182	0.5693
30	0.041	41.41	0.5748	0.0288	1.0004	0.0501	0.0166	0.5706
31	0.039	39.39	0.5606	0.0281	1.0004	0.0501	0.0158	0.5700
32	0.037	37.37	0.5461	0.0274	1.0004	0.0501	0.0149	0.5687
34	0.033	33.33	0.5157	0.0259	1.0003	0.0501	0.0133	0.5640
36	0.029	29.29	0.4834	0.0242	1.0003	0.0501	0.0117	0.5568
38	0.025	25.25	0.4489	0.0225	1.0003	0.0501	0.0101	0.5472
40	0.021	21.21	0.4114	0.0206	1.0002	0.0501	0.0085	0.5356
42	0.017	17.17	0.3701	0.0186	1.0002	0.0501	0.0069	0.5221
44	0.013	13.13	0.3237	0.0162	1.0001	0.0501	0.0053	0.5068
46	0.009	9.09	0.2693	0.0135	1.0001	0.0501	0.0036	0.4900
48	0.005	5.05	0.2007	0.0101	1.0001	0.0501	0.0020	0.4717
50	0.001	1.01	0.0898	0.0045	1.0000	0.0501	0.0004	0.4519

[a] Thermal and physical data are the same as for Table 7.3. Only data for even-numbered fins are presented.

7.4 NOMENCLATURE

Roman Letter Symbols

b	fin height, m
d	diameter, m; derivative or differential, dimensions vary
h	heat transfer coefficient, $W/m^2 \cdot K$
I	modified Bessel function of the first kind
k	thermal conductivity, $W/m \cdot K$
L	fin length, m
m	fin performance parameter, m^{-1}
n	number of fins, dimensionless
q	heat flow, W
S	surface area, m^2
T	temperature, K
u	combination of terms, dimensionless
v	combination of terms, dimensionless
x	height coordinate, m
Y	thermal admittance, W/K
$\mathbf{Y}$	thermal admittance matrix, W/K
y	elements of thermal admittance matrix, W/K
Z	thermal impedance, W/K

Greek Letter Symbols

Γ	gamma function, dimensionless
γ	paramater, dimensionless
δ	fin thickness, m
∂	partial derivative, dimensionless
η	fin efficiency, dimensionless
θ	temperature excess, K
μ	thermal transmission ratio, W/K
$\mathbf{T}$	linear transmission matrix, dimensions vary
τ	elements of linear transmission matrix, dimensions vary

Roman Letter Subscripts

a	tip of fin
b	base of fin
e	equivalent
i	counter
in	input condition
j	counter
m	fin designator
n	number of fins; fin designator
o	characteristic value; average or nominal value

p	fin designator
s	surroundings
sh	shunt

Greek Letter Superscript

γ	exponent

CHAPTER 8

ADVANCED ARRAY METHODS AND ARRAY OPTIMIZATION

8.1 INTRODUCTION

The algorithms provided by Kraus et al. (1978) were insufficient for the analysis of finned arrays that contained loops in the graph-theoretical sense. This limitation was removed by Snider and Kraus (1981), who identified the subclass of arrays that possessed these loops and then extended their prior technique by deriving general equations for the efficient analysis of a perfectly arbitrary configuration of fins.

Ingham et al. (1981) reported on a numerical two-dimensionsal analysis of fin assembly heat transfer. This was followed by Heggs et al. (1982), who provided a series truncation method for the analysis of finned assemblies. Kraus (1982) showed that the algorithms provided by Kraus et al. (1978) could be employed effectively in a manner that relied on the fact that each individual fin in an array was analogous to an electrical two-port network. This was an extension of the transmission line analogy proposed by Kern and Kraus (1972). Two-port matrices were also developed to accommodate a single series element to account for bond resistance and the single shunt element to be used for tip heat loss, and these have been developed in Chapter 6 with examples provided in Chapter 7.

Kraus et al. (1982), working with these two-ports, showed that all of the regular fins could be represented by a single π-network, and they showed that these π-networks could be assembled in a manner to permit a nodal analysis that would yield a temperature excess at all the fin intersection points.

The cascade algorithm described in Chapter 7 is applicable to arrays of extended surface with multiple heat inputs. In such cases, while the q's and θ's may be known, the input admittance at the tips of the fins where heat is injected may not be known a priori. Consider, for example, the repeating section of the unevenly loaded *double stack* shown in Fig. 8.1. Here, two base surfaces are dissipating heat to the coolant fluid. The quantities of interest are the left- and right-side base temperature excesses, θ_l and θ_r and the associated heat flows, q_l and q_r. Observe that there are *two* array bases, and if two unequal temperature excesses are prescribed, $\theta_l \neq \theta_r$, and this temperature asymmetry forbids an a priori assessment of the input admittances at the two array bases. Details on the analysis of such an array are provided in Section 8.2, and a more formal method is provided in Sections 8.3 and 8.4.

One could, in fact, follow the cascade algorithm of eqs. (7.5), the reflection relationship of eq. (7.6), and the cluster algorithm of eq. (7.13) by treating $Y_{\text{in},r} = q_{ar}/\theta_{ar}$ as an unknown and formally carry it through the cluster-cascade process from right to left. However, the "special" procedure described in Section 8.2 is somewhat more illuminating.

Finned arrays with multiple heat input as well as finned arrays containing loops in the graph-theoretical sense, such as the array shown in Fig. 8.2, may be analyzed by a method of node analysis proposed by Kraus et al. (1982) or by the general array algorithm proposed by Snider and Kraus (1981). These methods are described in detail in Sections 8.3 and 8.4. In Sections 8.5 and 8.6 optimization of arrays of fins and finned arrays are considered.

8.2 MULTIPLE HEAT INPUTS

Consider Fig. 8.1 and fin 4 which is at its extreme right where the heat q_r is injected at the *tip* of fin 4. Keeping in mind that a positive q indicates a heat flow from right to left, the $\mathbf{\Gamma}$-representations for fins 1 and 4 are

$$\begin{bmatrix} \theta_{a1} \\ q_{a1} \end{bmatrix} = \mathbf{\Gamma}_1 \begin{bmatrix} \theta_l \\ q_l \end{bmatrix} \tag{8.1a}$$

and

$$\begin{bmatrix} \theta_r \\ q_r \end{bmatrix} = \mathbf{\Gamma}_4 \begin{bmatrix} \theta_{b4} \\ q_{b4} \end{bmatrix} \tag{8.1b}$$

Observe that the bases of fins 2, 3 and 4 are in cluster at the tip of fin 1.

Compatability and continuity dictate that

$$\theta_{a1} = \theta_{b2} = \theta_{b3} = \theta_{b4} \tag{8.2a}$$

and if $\theta_{b1} > \theta_{a4}$

$$q_{b4} = q_{a1} - q_{b2} - q_{b3} \tag{8.2b}$$

Whether or not the tips of fins 2 and 3 dissipate, they will possess input admittances

$$Y_{\text{in},2} = \frac{q_{b2}}{\theta_{b2}} \tag{8.3a}$$

and

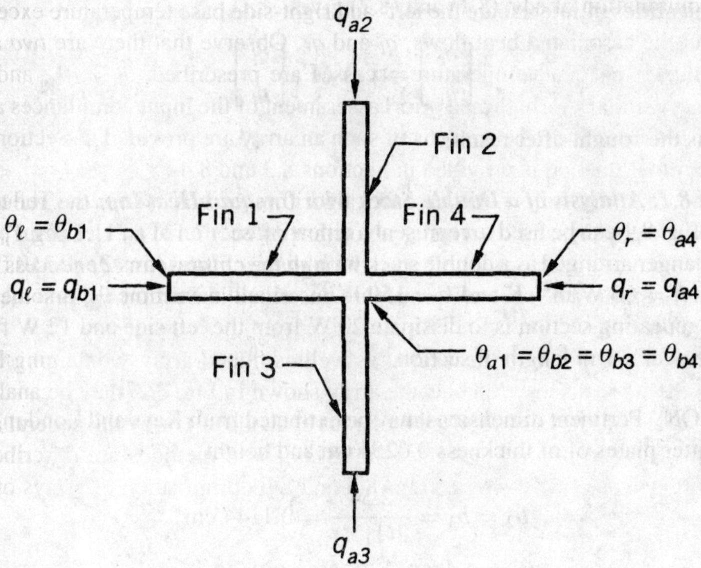

Figure 8.1 Repeating section of an unevenly loaded double stack. The subscripts l and r represent "left" and "right."

$$Y_{\text{in},3} = \frac{q_{b3}}{\theta_{b3}} \tag{8.3b}$$

and a relationship that links the base of fin 4 to the tip of fin 1 can now be established via the continuity and compatability conditions of eqs. (8.2) using a mathematical artifice:

$$\begin{bmatrix} \theta_{b4} \\ q_{b4} \end{bmatrix} = \begin{bmatrix} \theta_{a1} \\ q_{a1} \end{bmatrix} - \begin{bmatrix} 0 \\ q_{b2} \end{bmatrix} - \begin{bmatrix} 0 \\ q_{b3} \end{bmatrix}$$

or, using eqs. (8.3)

$$\begin{bmatrix} \theta_{b4} \\ q_{b4} \end{bmatrix} = \begin{bmatrix} \theta_{a1} \\ q_{a1} \end{bmatrix} - \begin{bmatrix} 0 \\ Y_{\text{in},2}\theta_{b2} \end{bmatrix} - \begin{bmatrix} 0 \\ Y_{\text{in},3}\theta_{b3} \end{bmatrix}$$

Again, invoking the continuity and compatability conditions of eqs. (8.2) yields

$$\begin{bmatrix} \theta_{b4} \\ q_{b4} \end{bmatrix} = \begin{bmatrix} \theta_{a1} \\ q_{a1} \end{bmatrix} - \begin{bmatrix} 0 \\ Y_{\text{in},2}\theta_{a1} \end{bmatrix} - \begin{bmatrix} 0 \\ Y_{\text{in},3}\theta_{a1} \end{bmatrix}$$

The relationship between the conditions at the base of fin 4 and the tip of fin 1 is in the form of an equivalent $\mathbf{Y}_{\text{in}}$-matrix:

$$\begin{bmatrix} \theta_{b4} \\ q_{b4} \end{bmatrix} = \mathbf{Y}_{\text{in}} \begin{bmatrix} \theta_{a1} \\ q_{a1} \end{bmatrix} = \begin{bmatrix} 1 & 0 \\ -(Y_{\text{in},2} + Y_{\text{in},3}) & 1 \end{bmatrix} \begin{bmatrix} \theta_{a1} \\ q_{a1} \end{bmatrix} \tag{8.4}$$

Now, a combination of eqs. (8.1) and (8.2) gives

$$\begin{bmatrix} \theta_r \\ q_r \end{bmatrix} = \mathbf{\Gamma}_4 \mathbf{Y}_{\text{in}} \mathbf{\Gamma}_1 \begin{bmatrix} \theta_l \\ q_l \end{bmatrix} \tag{8.5}$$

and this is the sought-after result.

Example 8.1: Analysis of a Double Stack with Unequal Heat Inputs. The configuration of Fig. 8.1 can be used to represent a repeating section of an 11.1 plain plate fin[1] heat exchanger arranged as a double stack with *splitter plates* (fins 2 and 3) 0.0254 cm thick. Let $h = 60\ \text{W/m}^2 \cdot \text{K}$ and $k = 150\ \text{W/m} \cdot \text{K}$ and determine θ_{b1} and θ_{a4} (θ_l and θ_r) if the repeating section is to dissipate 20 W from the left side and 12 W from the right side over a 1-m length of section.

SOLUTION. Pertinent dimensions may be extracted from Kays and London (1984). With splitter plates of of thickness 0.0254 cm and height

$$b_2 = b_3 = \frac{2.54}{2(11.1)} = 0.1144\ \text{cm}$$

$$b_1 = 0.6350\ \text{cm} \quad \delta_1 = 0.0152\ \text{cm}$$

$$b_2 = 0.1144\ \text{cm} \quad \delta_2 = 0.0254\ \text{cm}$$

$$b_3 = 0.1144\ \text{cm} \quad \delta_3 = 0.0254\ \text{cm}$$

$$b_4 = 0.6350\ \text{cm} \quad \delta_4 = 0.0152\ \text{cm}$$

For fins 2 and 3,

$$m = \left(\frac{2h}{k\delta}\right)^{1/2} = \left[\frac{(2)(60)}{(150)(0.000254)}\right]^{1/2} = 56.1214\ \text{m}^{-1}$$

$$mb = (56.1214)(0.001144) = 0.0642$$

and

$$Y_o = (2hk\delta)^{1/2} L = [(2)(60)(150)(0.000254)]^{1/2}(1.00) = 2.1382\ \text{W/K}$$

Using eqs. (6.9) gives us

$$\tau_{11} = \cosh\ mb = \cosh\ 0.0642 = 1.0021$$

and

$$\tau_{21} = Y_o \sinh\ mb = (2.1382)(0.0642) = 0.1374\ \text{W/K}$$

and from eq. (7.6),

$$Y_{\text{in},2} = Y_{\text{in},3} = \frac{\tau_{21}}{\tau_{11}} = \frac{0.1374}{1.0021} = 0.1371\ \text{W/K}$$

[1]Data for this surface as well as many others can be found in Kays and London (1984). The designation 11.1 refers to the number of fins per *inch*.

For fins 1 and 4,

$$m = \left(\frac{2h}{k\delta}\right)^{1/2} = \left[\frac{(2)(60)}{(150)(0.000152)}\right]^{1/2} = 72.5476\ \text{m}^{-1}$$

$$mb = (72.5476)(0.00635) = 0.4607$$

and

$$Y_o = (2hk\delta)^{1/2}L = [(2)(60)(150)(0.000152)]^{1/2}(1.00) = 1.6541\ \text{W/K}$$

Using eqs. (6.9) gives

$$\gamma_{11} = \gamma_{22} = \cosh\ mb = \cosh\ 0.4607 = 1.1080$$

and

$$\gamma_{12} = -\frac{1}{Y_o}\ \sinh\ mb = \left(-\frac{1}{1.6541}\right)(0.4771) = -0.2885\ \text{K/W}$$

and

$$\gamma_{21} = -Y_o\ \sinh\ mb = (-1.6451)(0.4771) = -0.7892\ \text{W/K}$$

and from eq. (6.8),

$$\mathbf{\Gamma}_1 = \mathbf{\Gamma}_4 = \begin{bmatrix} 1.1080 & -0.2885 \\ -0.7892 & 1.1080 \end{bmatrix}$$

and

$$\mathbf{Y}_{\text{in}} = \begin{bmatrix} 1 & 0 \\ -(Y_{\text{in},2} + Y_{\text{in},3}) & 1 \end{bmatrix} = \begin{bmatrix} 1 & 0 \\ -0.2742 & 1 \end{bmatrix}$$

Then the product $\mathbf{\Gamma}_4 \mathbf{Y}_{\text{in}} \mathbf{\Gamma}_1$ is

$$\begin{bmatrix} 1.1080 & -0.2885 \\ -0.7892 & 1.1080 \end{bmatrix} \begin{bmatrix} 1 & 0 \\ -0.2742 & 1 \end{bmatrix} \begin{bmatrix} 1.1080 & -0.2885 \\ -0.7892 & 1.1080 \end{bmatrix}$$

$$= \begin{bmatrix} 1.1080 & -0.2885 \\ -0.7892 & 1.1080 \end{bmatrix} \begin{bmatrix} 1.1080 & -0.2885 \\ -1.0930 & 1.1871 \end{bmatrix}$$

$$= \begin{bmatrix} 1.5430 & -0.6621 \\ -2.0855 & 1.5430 \end{bmatrix}$$

and the sought-after relationship is

$$\begin{bmatrix} \theta_r \\ q_r \end{bmatrix} = \mathbf{\Gamma}_4 \mathbf{Y}_{\text{in}} \mathbf{\Gamma}_1 \begin{bmatrix} \theta_l \\ q_l \end{bmatrix} = \begin{bmatrix} 1.5430 & -0.6621 \\ -2.0855 & 1.5430 \end{bmatrix} \begin{bmatrix} \theta_l \\ q_l \end{bmatrix}$$

or in expanded form,

$$\theta_r = 1.5430\theta_l - 0.6621 q_l$$

and

$$q_r = -2.0855\theta_l + 1.5430q_l$$

Using the second of these with $q_r = -12$ W and $q_l = 20$ W,

$$\theta_l = \frac{-12 - (1.5430)(20)}{-2.0855} = 20.6 \text{ K}$$

and then from the first

$$\theta_r = (1.5430)(20.6) - (0.6621)(20) = 18.5 \text{ K}$$

8.3 NODE ANALYSIS OF FINNED ARRAYS

If the finned array possesses multiple heat inputs or contains loops in the graph-theoretical sense, such as the array shown in Fig. 8.2, it may be analyzed by the general array algorithm proposed by Snider and Kraus (1981) or by a method of node analysis proposed by Kraus et al. (1982). The general array algorithm is discussed in Section 8.4, and the node analysis method is described here in a matrix-oriented procedure.

8.3.1 Reciprocity

Recall that in Section 5.7 it was shown that the matrix $\mathbf{\Gamma}$ was observed to be the product of two matrices:

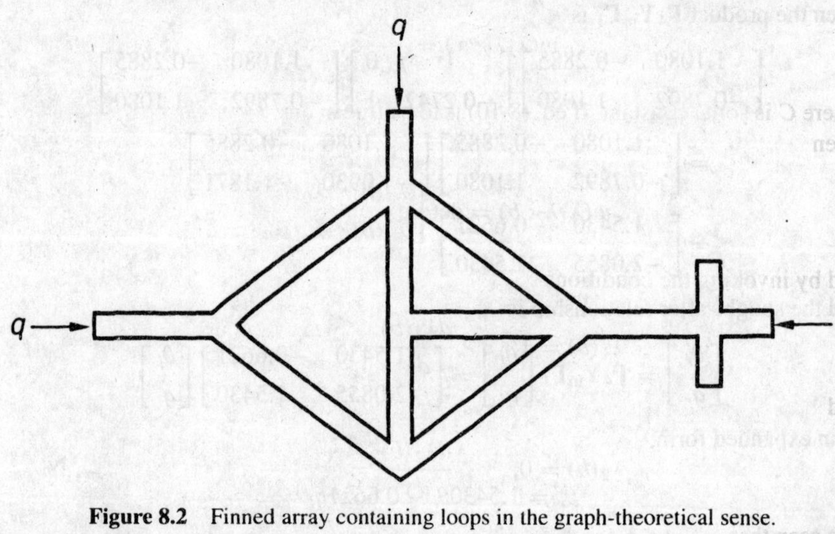

Figure 8.2 Finned array containing loops in the graph-theoretical sense.

$$\begin{bmatrix} \theta(x) \\ q(x) \end{bmatrix} = \begin{bmatrix} \lambda_1(x) & \lambda_2(x) \\ k(x)A(x)\lambda_1'(x) & k(x)A(x)\lambda_2'(x) \end{bmatrix} \begin{bmatrix} \theta_b \\ q_b \end{bmatrix} \tag{5.36}$$

With constant thermal conductivity so that $k(a) = k$, at the tip of the fin

$$\begin{bmatrix} \theta_a \\ q_a \end{bmatrix} = \begin{bmatrix} 1 & 0 \\ 0 & kA(a) \end{bmatrix} \begin{bmatrix} \lambda_1(a) & \lambda_2(a) \\ \lambda_1'(a) & \lambda_2'(a) \end{bmatrix} \begin{bmatrix} \theta_b \\ q_b \end{bmatrix} \tag{8.6}$$

Thus

$$\mathbf{\Gamma} = \begin{bmatrix} 1 & 0 \\ 0 & kA(a) \end{bmatrix} \begin{bmatrix} \lambda_1(a) & \lambda_2(a) \\ \lambda_1'(a) & \lambda_2'(a) \end{bmatrix}$$

and the determinant of Γ is

$$\det \mathbf{\Gamma} = \det \begin{bmatrix} 1 & 0 \\ 0 & kA(a) \end{bmatrix} \begin{bmatrix} \lambda_1(a) & \lambda_2(a) \\ \lambda_1'(a) & \lambda_2'(a) \end{bmatrix} \tag{8.7}$$

The second matrix in eq. (8.7) is the *Wronskian*, and eq. (8.7) may be written as

$$\det \mathbf{\Gamma} = kA(a)w(\lambda_1\lambda_2; a) \tag{8.8}$$

Abel's formula can be applied to the governing differential equation for the temperature excess in regular longitudinal fins given by a form of eq. (1.3):

$$\frac{d}{dx}\left[kA(x)\frac{d\theta}{dx}\right] - h\frac{dS(x)}{dx}\theta(x) = 0 \tag{8.9}$$

to evaluate the Wronskian. In terms of the parameters in eq. (8.9), Abel's formula states that

$$w(\lambda_1\lambda_2; x) = \frac{C}{k(x)A(x)} \tag{8.10}$$

where C is some constant. If eq. (8.10) is to be true at any x, it must be true at $x = b$. Then

$$w(\lambda_1\lambda_2; b) = \det \begin{bmatrix} \lambda_1(b) & \lambda_2(b) \\ \lambda_1'(b) & \lambda_2'(b) \end{bmatrix}$$

and by invoking the conditions

$$\lambda_1(b) = 1; \qquad \frac{d\lambda_1(b)}{dx} = 0 \tag{5.33a}$$

and

$$\lambda_2(b) = 0; \qquad \frac{d\lambda_2(b)}{dx} = \frac{1}{k(b)A(b)} \tag{5.33b}$$

it is seen that

$$w(\lambda_1\lambda_2; b) = \det \begin{bmatrix} 1 & 0 \\ 0 & 1/kA(b) \end{bmatrix} = \frac{1}{kA(b)} \tag{8.11}$$

A comparison of eq. (8.11) with eq. (8.10) with $x = b$ indicates that

$$w(\lambda_1\lambda_2; b) = \frac{C}{kA(b)} = \frac{1}{kA(b)}$$

which shows that $C = 1$ and that from eq. (8.8),

$$\det \mathbf{\Gamma} = kA(a)\frac{1}{kA(a)} = 1 \tag{8.12}$$

With $\det \mathbf{\Gamma} = 1$, Table 5.2 shows that the off-diagonal elements of $\mathbf{Y}$ are the negative of one another:

$$y_{12} = \frac{1}{\gamma_{12}}$$

and

$$y_{21} = -\frac{1}{\gamma_{12}}$$

Consider Fig. 8.3, which shows a hypothetical fin of arbitrary profile with a heat source, q, provided first at the the base (the left end) and then at the tip (the right end). Observe that in each case, a temperature excess (θ_a and θ_b) is also indicated. Moreover, a height coordinate, x, having origin at the fin tip and positive sense toward the fin base, appears in both cases.

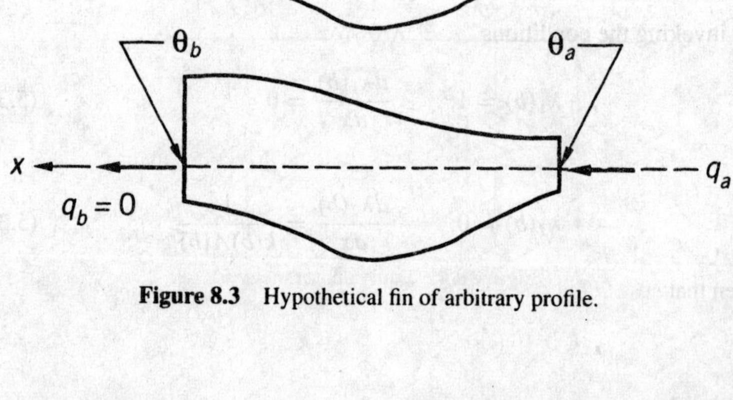

Figure 8.3 Hypothetical fin of arbitrary profile.

A fin will be termed *reciprocal* if an identical heat injection q, introduced at either end with the other end insulated, produces an equal temperature excess at the opposite end. The mathematical statement is: If $q_b = q$ produces θ_a, and if $q_a = -q$ (the heat is inserted in the opposite direction) produces θ_b, the fin is said to be reciprocal if $\theta_a = \theta_b$.

The heat flows and temperature exesses are related by the thermal admittance matrix

$$\begin{bmatrix} q_b \\ q_a \end{bmatrix} = \begin{bmatrix} y_{11} & y_{12} \\ y_{12} & y_{22} \end{bmatrix} \begin{bmatrix} \theta_b \\ \theta_a \end{bmatrix}$$

and θ_b and θ_a may be obtained by taking the inverse:

$$\begin{bmatrix} \theta_b \\ \theta_a \end{bmatrix} = \frac{1}{\det \mathbf{Y}} \begin{bmatrix} y_{22} & -y_{12} \\ -y_{21} & y_{11} \end{bmatrix} \begin{bmatrix} q_b \\ q_a \end{bmatrix} \tag{8.13}$$

In the first case with $q_b = q$ and the fin tip insulated so that $q_a = 0$,

$$\theta_a = \frac{-y_{21}}{\det \mathbf{Y}} q \tag{8.14a}$$

In the second case with $q_a = -q$ and the fin base insulated so that $q_b = 0$,

$$\theta_b = \frac{-y_{12}}{\det \mathbf{Y}}(-q) = \frac{y_{12}}{\det \mathbf{Y}} q \tag{8.14b}$$

Reciprocity requires that $\theta_a = \theta_b$. Thus it is seen that for a fin to be reciprocal,

$$y_{12} = -y_{21}$$

It is this reciprocity principle that permits a single regular fin to be represented as a π*-network* containing three thermal admittances, as shown in Fig. 8.4.

8.3.2 Conditions for Reciprocity

With $y_{12} = -y_{21}$, the conditions for reciprocity can be found from Table 5.2.

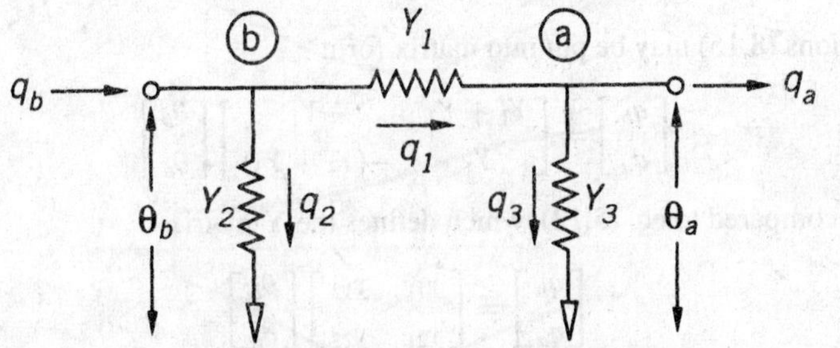

Figure 8.4 Equivalent π-network.

$$\det \mathbf{\Gamma} = 1 \qquad \det \mathbf{T} = 1$$

$$z_{12} = -z_{21} \qquad y_{12} = -y_{21}$$

$$h_{12} = h_{21} \qquad g_{12} = g_{21}$$

8.3.3 Equivalent π-Network

In Fig. 8.4, an application of continuity at points a and b, called *node points* or merely *nodes*, gives

$$q_b = q_1 + q_2$$

$$q_a = q_1 - q_3$$

However, each heat flow, q_1, q_2, and q_3 in Fig. 8.4, can be represented as a product of an admittance and a temperature excess (or a difference of temperature excesses)

$$q_1 = Y_1(\theta_b - \theta_a)$$

$$q_2 = Y_2\theta_b$$

$$q_3 = Y_3\theta_a$$

so that

$$q_b = Y_1(\theta_b - \theta_a) + Y_2\theta_b$$

$$q_a = Y_1(\theta_b - \theta_a) - Y_3\theta_a$$

or

$$q_b = (Y_1 + Y_2)\theta_b - Y_1\theta_a \tag{8.15a}$$

and

$$q_a = Y_1\theta_b - (Y_1 + Y_3)\theta_a \tag{8.15b}$$

Equations (8.15) may be put into matrix form

$$\begin{bmatrix} q_b \\ q_a \end{bmatrix} = \begin{bmatrix} Y_1 + Y_2 & -Y_1 \\ Y_1 & -(Y_1 + Y_3) \end{bmatrix} \begin{bmatrix} \theta_b \\ \theta_a \end{bmatrix}$$

and then compared to eq. (5.12), which defines the $\mathbf{Y}$-matrix

$$\begin{bmatrix} q_b \\ q_a \end{bmatrix} = \begin{bmatrix} y_{11} & y_{12} \\ y_{21} & y_{22} \end{bmatrix} \begin{bmatrix} \theta_b \\ \theta_a \end{bmatrix} \tag{5.12}$$

The comparison shows that

$$Y_1 + Y_2 = y_{11}$$
$$Y_1 = -y_{12} = y_{21}$$

and

$$Y_1 + Y_3 = -y_{22}$$

Then, it is just a matter of algebra to show that

$$Y_1 = -y_{12} = y_{21} \tag{8.16a}$$

$$Y_2 = y_{11} + y_{12} = y_{11} - y_{21} \tag{8.16b}$$

and

$$Y_3 = y_{12} - y_{22} = -y_{21} - y_{22} \tag{8.16c}$$

Example 8.2. Suppose that a single fin entity, whether an individual fin or an equivalent fin, possesses a y-parameter matrix

$$\mathbf{Y} = \begin{bmatrix} 1.6324 & -0.9143 \\ 0.9143 & -1.6324 \end{bmatrix}$$

Construct an equivalent π-network for this fin.

SOLUTION. First observe that $y_{12} = -y_{21}$ and that the matrix represents a reciprocal fin. Then, by eqs. (8.16),

$$Y_1 = -y_{12} = y_{21} = 0.9143 \text{ W/K}$$
$$Y_2 = y_{11} + y_{12} = 1.6324 + (-0.9143) = 0.7181 \text{ W/K}$$

and

$$Y_3 = y_{12} - y_{22} = -0.9143 - (-1.6324) = 0.7181 \text{ W/K}$$

The equivalent π-network of Fig. 8.4 is not some feeble attempt at constructing an electrical analog even though the assembly of such π-networks to model a finned array is best handled by techniques of network analysis. The π-network derives from the rigor inherent in reciprocity and provides a lumped-element representation of the fin. It is noted that as long as $y_{12} = -y_{21}$, any fin may be represented as an equivalent π. All regular fins are reciprocal and possess this important property. Hence arrays composed of regular fins may be treated as a combination of equivalent π-networks representing the individual fins in the array and connected as the array is connected. An example of a four-fin array with unequal heat inputs provided at opposite ends is displayed, along with its network representation, in Fig. 8.5.

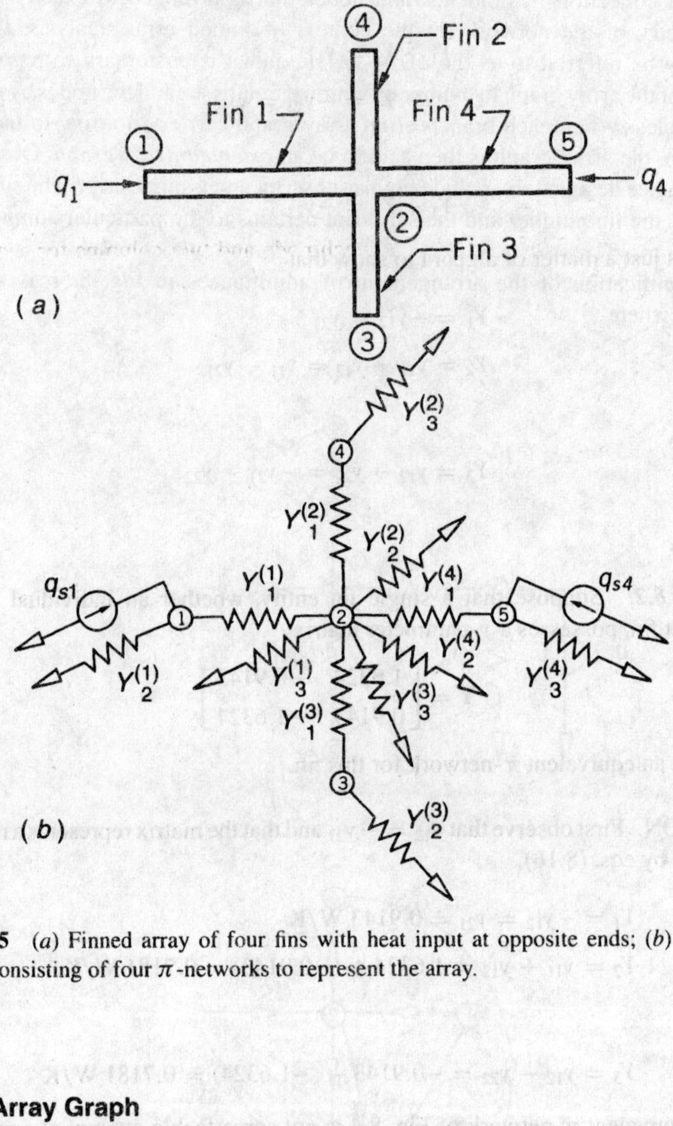

Figure 8.5 (*a*) Finned array of four fins with heat input at opposite ends; (*b*) equivalent network consisting of four π-networks to represent the array.

8.3.4 Array Graph

The several lumped admittances, arranged as in Fig. 8.5*b*, can be considered as a network of admittances, and reference may be made to Fig. 8.5*b* as a network. *Network topology* derives from a branch of mathematics that permits the determination of network properties on the basis of the structure, geometry, or interconnection of the network. It serves as a basis for the general method of node analysis developed in subsequent sections. The use of network topology does not depend on the types or the magnitude of the elements that constitute the network.

Every network element such as the lumped admittances in Fig. 8.5*b* may be replaced by a single line called a *branch*, which has two ends called *nodes*. A *linear* *graph* is a collection of branches and nodes that is arranged to exactly represent the geometry of the network. In the context of finned array analysis, the linear graph may be referred to as the *array graph*, and it is customary to represent the branches of the array graph by numerals within parentheses and the nodes by numerals within circles. When each branch of an array graph carries an arrow to indicate its orientation, the array graph is then said to be an *oriented array graph*. Observe that the admittance designators include both a subscript and a superscript. The superscript designates the fin number and the subscript pertains to the particular admittance for that fin.

A simplification of the arrangement of admittances in Fig. 8.5*b* is shown in Fig. 8.6*a*, where

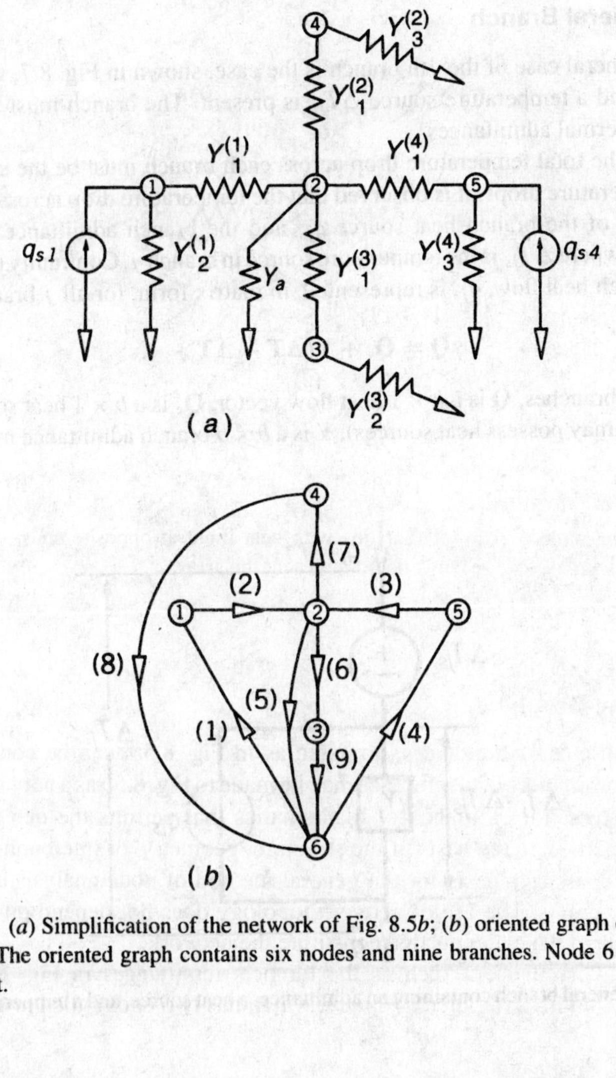

Figure 8.6 (*a*) Simplification of the network of Fig. 8.5*b*; (*b*) oriented graph of the network of part (*a*). The oriented graph contains six nodes and nine branches. Node 6 represents the environment.

$$Y_a = Y_3^{(1)} + Y_2^{(2)} + Y_3^{(3)} + Y_2^{(4)}$$

and Fig. 8.6*b* shows an oriented graph of the network in Fig. 8.6*a*. Again note the use of circles to designate nodes and numbers in parentheses to designate branches. Also note that branches containing heat sources must also contain an admittance. The heat sources are included in branches 1 and 4 and the branch orientations are discretionary, except for the branches containing the heat sources, where the positive orientation must correspond to the direction of the heat input. The objective is to obtain all node temperature excesses in a computationally efficient manner and then, knowing the temperature excesses and heat flows at nodes 1 and 5, to determine the input admittances at nodes 1 and 5.

8.3.5 General Branch

The most general case of the jth branch is the case, shown in Fig. 8.7, where a heat source q_{js} and a temperature source ΔT_{js} is present. The branch must also contain one of the thermal admittances.

Because the total temperature drop across each branch must be the sum of component temperature drops, it is observed that the temperature drop across the parallel combination of the branch heat source q_{js} and the branch admittance Y_j must be $\Delta T_j - \Delta T_{js}$, where ΔT_{js} is the temperature source in branch j. Continuity then dictates that the branch heat flow, q_j, is represented, in matrix form, for all j branches, by

$$\mathbf{Q} = \mathbf{Q}_s + \mathbf{Y}(\mathbf{\Delta T} - \mathbf{\Delta T}_s) \tag{8.17}$$

Here, with b branches, $\mathbf{Q}$ is a $b \times 1$ heat flow vector, $\mathbf{Q}_s$ is a $b \times 1$ heat source vector (all branches may possess heat sources), $\mathbf{Y}$ is a $b \times b$ branch admittance matrix which

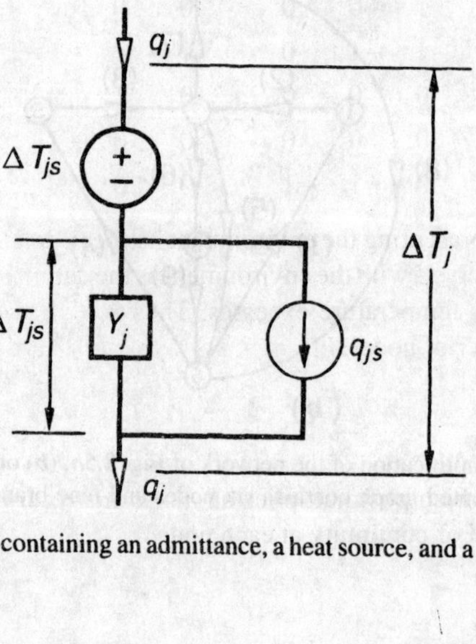

Figure 8.7 General branch containing an admittance, a heat source, and a temperature source.

is diagonal, $\mathbf{\Delta T}_s$ is a $b \times 1$ temperature source vector which may be (and most often is) null and $\mathbf{\Delta T}$ is a $b \times 1$ temperature vector.

8.3.6 Node Branch Incidence Matrix

For a graph containing n_t nodes and b branches, a matrix that exactly represents the graph may be formulated. This matrix, the *augmented node-branch incidence matrix*, will have n_t rows and b columns and will contain elements $a_{jk} = +1, -1$, or 0 in accordance with the scheme

$$a_{jk} = \begin{cases} +1 & \text{if branch } j \text{ leaves node } k \\ -1 & \text{if branch } j \text{ enters node } k \\ 0 & \text{if branch } j \text{ does not touch node } k \end{cases} \tag{8.18}$$

For example, the augmented node-branch incidence matrix for the oriented graph in Fig. 8.6b will be $n_t = 6 \times b$ or 6×9:

$$\mathbf{A}_a = \begin{bmatrix} -1 & 1 & 0 & 0 & 0 & 0 & 0 & 0 & 0 \\ 0 & -1 & -1 & 0 & 1 & 1 & 1 & 0 & 0 \\ 0 & 0 & 0 & 0 & 0 & -1 & 0 & 0 & 1 \\ 0 & 0 & 0 & 0 & 0 & 0 & -1 & 1 & 0 \\ 0 & 0 & 1 & -1 & 0 & 0 & 0 & 0 & 0 \\ 1 & 0 & 0 & 1 & -1 & 0 & 0 & -1 & -1 \end{bmatrix} \tag{8.19}$$

It can be noted that every column contains a single +1 and a single −1, and that a summation of all elements in each column yields a zero.

The node-branch incidence matrix contains $n_t - 1$ rows and b branches. It is obtained from $\mathbf{A}_a$ merely by deleting the row that represents the surrounding environment. In Fig. 8.6b, node 6 represents the environment and when it is deleted,

$$\mathbf{A} = \begin{bmatrix} -1 & 1 & 0 & 0 & 0 & 0 & 0 & 0 & 0 \\ 0 & -1 & -1 & 0 & 1 & 1 & 1 & 0 & 0 \\ 0 & 0 & 0 & 0 & 0 & -1 & 0 & 0 & 1 \\ 0 & 0 & 0 & 0 & 0 & 0 & -1 & 1 & 0 \\ 0 & 0 & 1 & -1 & 0 & 0 & 0 & 0 & 0 \end{bmatrix} \tag{8.20}$$

Because the row representing the environment has been deleted, the analysis becomes a node-to-datum analysis with the environment as the datum and all node temperatures to be considered as temperature excesses. Thus $n = n_t - 1$ may now be used to represent the number of nodes.

8.3.7 Continuity

The product $\mathbf{AQ}$, which postmultiplies an $n \times b$ matrix by a $b \times 1$ column vector, presents a statement of continuity at each node:

$$
\mathbf{AQ} = \begin{bmatrix} -1 & 1 & 0 & 0 & 0 & 0 & 0 & 0 & 0 \\ 0 & -1 & -1 & 0 & 1 & 1 & 1 & 0 & 0 \\ 0 & 0 & 0 & 0 & 0 & -1 & 0 & 0 & 1 \\ 0 & 0 & 0 & 0 & 0 & 0 & -1 & 1 & 0 \\ 0 & 0 & 1 & -1 & 0 & 0 & 0 & 0 & 0 \end{bmatrix} \begin{bmatrix} q_1 \\ q_2 \\ q_3 \\ q_4 \\ q_5 \\ q_6 \\ q_7 \\ q_8 \\ q_9 \end{bmatrix}
$$

$$
= \begin{bmatrix} q_2 - q_1 \\ q_5 + q_6 + q_7 - q_2 - q_3 \\ q_9 - q_6 \\ q_8 - q_7 \\ q_3 - q_4 \end{bmatrix} = \begin{bmatrix} 0 \\ 0 \\ 0 \\ 0 \\ 0 \end{bmatrix}
$$

and this can be confirmed by an inspection of the oriented graph in Fig. 8.6*b*. Thus the equation

$$
\mathbf{AQ} = \mathbf{0} \tag{8.21}
$$

is a matrix statement of the continuity at every node in the array.

A matrix that links the branch temperature drops can also be found. Define the elements c_{jk} of the matrix $\mathbf{C}$ in

$$
\mathbf{\Delta T} = \mathbf{C}\mathbf{\Theta} \tag{8.22}
$$

by

$$
c_{jk} = \begin{cases} +1 & \text{if branch } j \text{ leaves node } k \\ -1 & \text{if branch } j \text{ enters node } k \\ 0 & \text{if branch } j \text{ does not touch node } k \end{cases} \tag{8.23}
$$

The matrix $\mathbf{C}$ will contain b rows and n columns and the $b \times n$ matrix for the oriented graph of Fig. 8.6*b* (without node 6) will be

$$
\mathbf{C} = \begin{bmatrix} -1 & 0 & 0 & 0 & 0 \\ 1 & -1 & 0 & 0 & 0 \\ 0 & -1 & 0 & 0 & 1 \\ 0 & 0 & 0 & 0 & -1 \\ 0 & 1 & 0 & 0 & 0 \\ 0 & 1 & -1 & 0 & 0 \\ 0 & 1 & 0 & -1 & 0 \\ 0 & 0 & 0 & 1 & 0 \\ 0 & 0 & 1 & 0 & 0 \end{bmatrix} \tag{8.24}
$$

There are two kinds of branches. Those that touch two nodes, say nodes r and s, will possess a temperature drop, $\Delta T = \theta_r - \theta_s$. Those that touch a node and the datum node will have a temperature drop equal to the temperature excess for that node. In both cases, the branch temperature drops depend on the node temperature excesses. If the matrix of eq. (8.24) is put into eqs. (8.22), the result is

$$\mathbf{C\Theta} = \begin{bmatrix} -1 & 0 & 0 & 0 & 0 \\ 1 & -1 & 0 & 0 & 0 \\ 0 & -1 & 0 & 0 & 1 \\ 0 & 0 & 0 & 0 & -1 \\ 0 & 1 & 0 & 0 & 0 \\ 0 & 1 & -1 & 0 & 0 \\ 0 & 1 & 0 & -1 & 0 \\ 0 & 0 & 0 & 1 & 0 \\ 0 & 0 & 1 & 0 & 0 \end{bmatrix} \begin{bmatrix} \theta_1 \\ \theta_2 \\ \theta_3 \\ \theta_4 \\ \theta_5 \end{bmatrix} = \begin{bmatrix} -\theta_1 \\ \theta_1 - \theta_2 \\ \theta_5 - \theta_2 \\ -\theta_5 \\ \theta_2 \\ \theta_2 - \theta_3 \\ \theta_2 - \theta_4 \\ \theta_4 \\ \theta_3 \end{bmatrix}$$

which is entirely correct. Moreover, a comparison of eqs. (8.20) and (8.24) shows that $a_{jk} = c_{kj}$, which indicates that

$$\mathbf{C} = \mathbf{A}^{\mathrm{T}}$$

and that

$$\mathbf{\Delta T} = \mathbf{A}^{\mathrm{T}}\mathbf{\Theta} \tag{8.25}$$

8.3.8 Node-to-Datum Analysis of Finned Arrays

The method of node-to-datum analysis is based on the branch equations, which are a form of eqs. (8.17) rewritten here as

$$\mathbf{Q} = \mathbf{Q}_s - \mathbf{Y}\,\mathbf{\Delta T}_s + \mathbf{Y}\,\mathbf{\Delta T} \tag{8.26}$$

the expression of continuity

$$\mathbf{AQ} = \mathbf{0} \tag{8.21}$$

and the relationship between the branch temperature drops and the node temperature excesses,

$$\mathbf{\Delta T} = \mathbf{A}^{\mathrm{T}}\mathbf{\Theta} \tag{8.25}$$

If eqs. (8.26) are premultiplied by $\mathbf{A}$ and set equal to zero in accordance with eqs. (8.21),

$$\mathbf{AQ} = \mathbf{0} = \mathbf{AQ}_s - \mathbf{AY}\,\mathbf{\Delta T}_s + \mathbf{AY}\,\mathbf{\Delta T}$$

or

$$\mathbf{AY}\,\Delta\mathbf{T} = \mathbf{AY}\,\Delta\mathbf{T}_s - \mathbf{AQ}_s \tag{8.27}$$

Then define a node heat source vector

$$\tilde{\mathbf{Q}} = \mathbf{AY}\,\Delta\mathbf{T}_s - \mathbf{AQ}_s \tag{8.28}$$

so that

$$\mathbf{AY}\,\Delta\mathbf{T} = \tilde{\mathbf{Q}}$$

If eqs. (8.25) are inserted here,

$$\mathbf{AYA}^{\mathrm{T}}\boldsymbol{\Theta} = \tilde{\mathbf{Q}}$$

the node equations result:

$$\mathbf{Y}_n\boldsymbol{\Theta} = \tilde{\mathbf{Q}} \tag{8.29}$$

where the node admittance matrix $\mathbf{Y}_n$ is defined by

$$\mathbf{Y}_{\mathrm{n}} = \mathbf{AYA}^{\mathrm{T}} \tag{8.30}$$

Solution of eqs. (8.29),

$$\boldsymbol{\Theta} = \mathbf{Y}_n^{-1}\tilde{\mathbf{Q}} \tag{8.31}$$

yields the temperature excess at each node, and if desired, input admittances can then be evaluated at each point where heat is injected into the array.

Example 8.3: Node Analysis of a Finned Array. Consider the finned array shown in Fig. 8.8 and note that the three fins are identical. Suppose that the array is 16 cm long and that each of the three fins in the array is a longitudinal fin of rectangular profile with a thickness of 0.2000 cm and a height of 4.00 cm. All the fins are fabricated from a material with a thermal conductivity of $k = 160$ W/m · K and dissipation is to an environment at 25°C via a heat transfer coefficient of $h = 120$ W/m^2 · K. Heat is injected at the extremes of fins 1, 2, and 3 in the amount of 60, 50, and 40 W, respectively. Determine the tempeature excess vector.

SOLUTION. The temperatures at all the nodes can be obtained from a node-to-datum analysis. Each fin in the array possesses a **Y**-matrix whose elements can be obtained from eq. (6.11). With

$$Y_o = (2hk\delta)^{1/2}L = [(2)(120)(160)(0.0020)]^{1/2}(0.16) = 1.4022 \text{ W/K}$$

and

$$mb = \left(\frac{2h}{k\delta}\right)^{1/2} b = \left[\frac{(2)(120)}{(160)(0.0020)}\right]^{1/2} (0.0400) = 1.0955$$

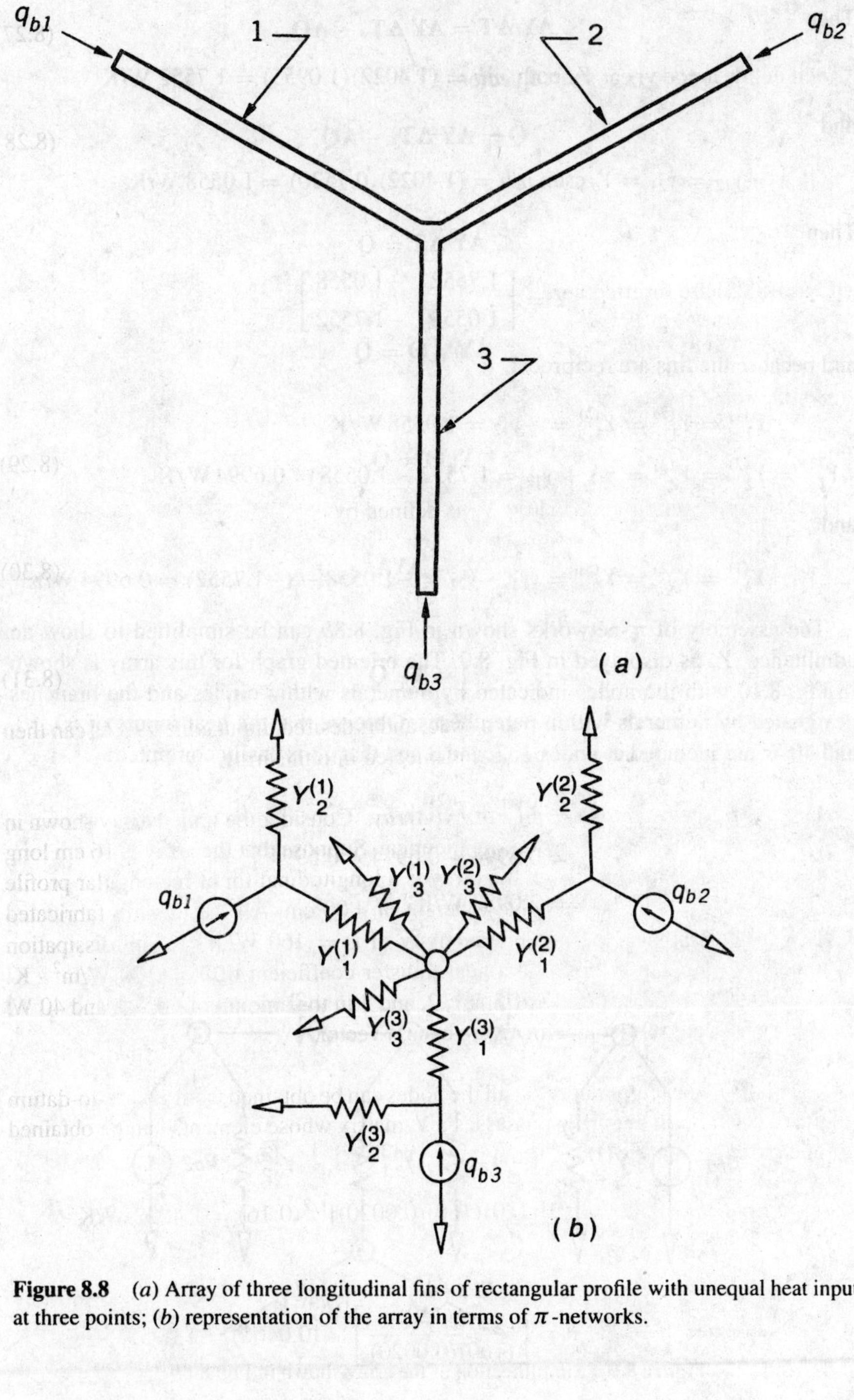

Figure 8.8 (*a*) Array of three longitudinal fins of rectangular profile with unequal heat input at three points; (*b*) representation of the array in terms of π-networks.

Then

$$y_{11} = -y_{22} = Y_o \coth mb = (1.4022)(1.0955) = 1.7552 \text{ W/K}$$

and

$$-y_{12} = y_{21} = Y_o \,\text{csch}\, mb = (1.4022)(0.7530) = 1.0558 \text{ W/K}$$

Then

$$\mathbf{Y} = \begin{bmatrix} 1.7552 & -1.0558 \\ 1.0558 & -1.7552 \end{bmatrix}$$

and because the fins are reciprocal,

$$Y_1^{(1)} = Y_1^{(2)} = Y_1^{(3)} = -y_{12} = 1.0558 \text{ W/K}$$

$$Y_2^{(1)} = Y_2^{(2)} = Y_2^{(3)} = y_{11} + y_{12} = 1.7552 - 1.0558 = 0.6994 \text{ W/K}$$

and

$$Y_3^{(1)} = Y_3^{(2)} = Y_3^{(3)} = y_{12} - y_{22} = -1.0558 - (-1.7552) = 0.6994 \text{ W/K}$$

The assembly of π-networks shown in Fig. 8.8*b* can be simplified to show an admittance, Y_a as displayed in Fig. 8.9. The oriented graph for this array is shown in Fig. 8.10 with the nodes indicated by numerals within circles and the branches designated by numerals within parentheses. Observe that the heat inputs of 60, 50, and 40 W are included at nodes 1, 2, and 3 and that Y_a is easily computed:

$$\begin{aligned} Y_a &= Y_3^{(1)} + Y_3^{(2)} + Y_3^{(3)} \\ &= 0.6994 + 0.6994 + 0.6994 \\ &= 2.0982 \text{ W/K} \end{aligned}$$

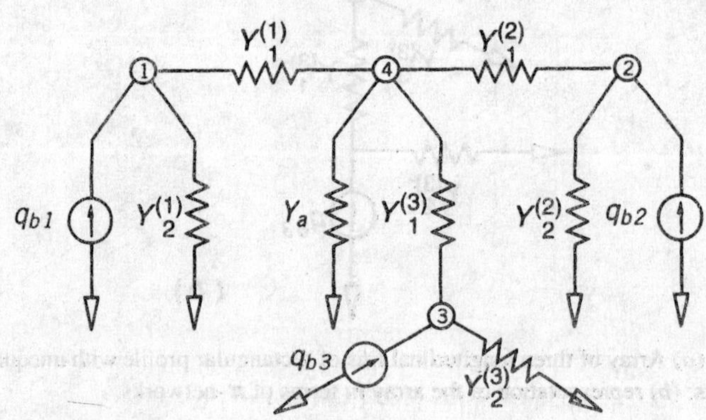

Figure 8.9 Simplification of the array shown in Fig. 8.8*b*.

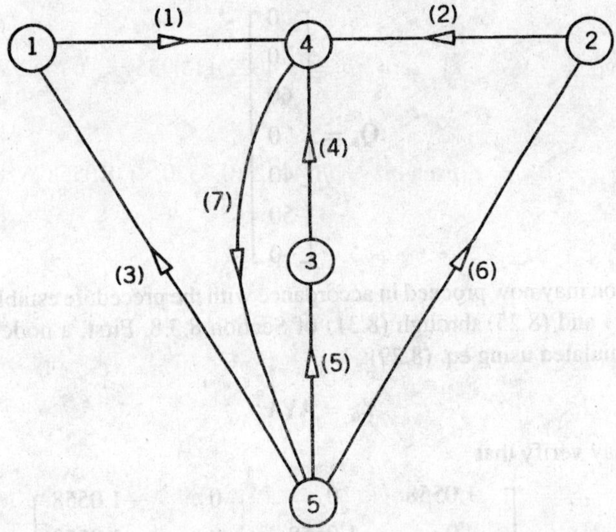

Figure 8.10 Oriented graph for the arrangement of admittances in Fig. 8.8.

The oriented graph is displayed in Fig. 8.10 and the node branch incidence matrix is 4×7

$$\mathbf{A} = \begin{bmatrix} 1 & 0 & -1 & 0 & 0 & 0 & 0 \\ 0 & 1 & 0 & 0 & 0 & -1 & 0 \\ 0 & 0 & 0 & 1 & -1 & 0 & 0 \\ -1 & -1 & 0 & -1 & 0 & 0 & 1 \end{bmatrix}$$

and with the help of Figs. 8.9 and 8.10, the node branch admittance matrix can be shown to be

$$\mathbf{Y} = \begin{bmatrix} Y_1^{(1)} & 0 & 0 & 0 & 0 & 0 & 0 \\ 0 & Y_1^{(2)} & 0 & 0 & 0 & 0 & 0 \\ 0 & 0 & Y_2^{(1)} & 0 & 0 & 0 & 0 \\ 0 & 0 & 0 & Y_1^{(3)} & 0 & 0 & 0 \\ 0 & 0 & 0 & 0 & Y_2^{(3)} & 0 & 0 \\ 0 & 0 & 0 & 0 & 0 & Y_2^{(2)} & 0 \\ 0 & 0 & 0 & 0 & 0 & 0 & Y_a \end{bmatrix}$$

where (all in W/K) in the entries bearing both a subscript and a superscript, the superscript identifies the branch and the subscript pertains to a particular admittance.

There will be no branch temperature source vector because there are no temperature sources. But because branches 3, 5, and 6 contain heat sources, the branch heat source vector will be

$$\mathbf{Q}_\mathrm{s} = \begin{bmatrix} 0 \\ 0 \\ 60 \\ 0 \\ 40 \\ 50 \\ 0 \end{bmatrix}$$

The solution may now proceed in accordance with the precedure established based on eqs. (8.21) and (8.25) through (8.31) of Section 8.3.8. First, a node admittance matrix is formulated using eq. (8.29):

$$\mathbf{Y}_\mathrm{n} = \mathbf{A}\mathbf{Y}\mathbf{A}^\mathrm{T} \tag{8.30}$$

The reader may verify that

$$\mathbf{Y}\mathbf{A}^T = \begin{bmatrix} 1.0558 & 0 & 0 & -1.0558 \\ 0 & 1.0558 & 0 & -1.0558 \\ -0.6944 & 0 & 0 & 0 \\ 0 & 0 & 1.0558 & -1.0558 \\ 0 & 0 & -0.6994 & 0 \\ 0 & -0.6994 & 0 & 0 \\ 0 & 0 & 0 & 2.0982 \end{bmatrix}$$

and

$$\mathbf{Y}_n = \begin{bmatrix} 1.7552 & 0 & 0 & -1.0558 \\ 0 & 1.7552 & 0 & -1.0558 \\ 0 & 0 & 1.7552 & -1.0558 \\ -1.0558 & -1.0558 & -1.0558 & 5.2656 \end{bmatrix}$$

The node heat source vector $\tilde{\mathbf{Q}}$ is computed from eq. (8.28):

$$\tilde{\mathbf{Q}} = \mathbf{A}\mathbf{Y}\,\Delta\mathbf{T}_s - \mathbf{A}\mathbf{Q}_s \tag{8.28}$$

but because the branch temperature excess vector is null, this becomes

$$\tilde{\mathbf{Q}} = -\mathbf{A}\mathbf{Q}_\mathrm{s}$$

$$= -\begin{bmatrix} 1 & 0 & -1 & 0 & 0 & 0 & 0 \\ 0 & 1 & 0 & 0 & 0 & -1 & 0 \\ 0 & 0 & 0 & 1 & -1 & 0 & 0 \\ -1 & -1 & 0 & -1 & 0 & 0 & 1 \end{bmatrix} \begin{bmatrix} 0 \\ 0 \\ 60 \\ 0 \\ 40 \\ 50 \\ 0 \end{bmatrix}$$

or

$$\tilde{\mathbf{Q}} = \begin{bmatrix} 60 \\ 50 \\ 40 \\ 0 \end{bmatrix} \text{W}$$

The formulation of the node equations is

$$\mathbf{Y}_n \mathbf{\Theta} = \tilde{\mathbf{Q}} \tag{8.29}$$

and the solution for the temperature excess vector is obtained from

$$\mathbf{\Theta} = \mathbf{Y}_n^{-1} \tilde{\mathbf{Q}}$$

or

$$\begin{bmatrix} \theta_1 \\ \theta_2 \\ \theta_3 \\ \theta_4 \end{bmatrix} = \begin{bmatrix} 0.6774 & 0.1077 & 0.1077 & 0.1790 \\ 0.1077 & 0.6774 & 0.1077 & 0.1790 \\ 0.1077 & 0.1077 & 0.6774 & 0.1790 \\ 0.1790 & 0.1790 & 0.1790 & 0.2976 \end{bmatrix} \begin{bmatrix} 60 \\ 50 \\ 40 \\ 0 \end{bmatrix}$$

so that

$$\tilde{\mathbf{Q}} = \begin{bmatrix} \theta_1 \\ \theta_2 \\ \theta_3 \\ \theta_4 \end{bmatrix} = \begin{bmatrix} 50.3 \\ 44.6 \\ 38.9 \\ 26.8 \end{bmatrix} \text{K}$$

8.4 GENERAL ANALYSIS METHOD

The general analysis method employs the **Y**-matrix, which represents a linear transformation from temperature excesses at both fin tip and fin base to heat flows at both tip and base:

$$\begin{bmatrix} q_b \\ q_a \end{bmatrix} = \mathbf{Y} \begin{bmatrix} \theta_b \\ \theta_a \end{bmatrix} = \begin{bmatrix} y_{11} & y_{12} \\ y_{12} & y_{22} \end{bmatrix} \begin{bmatrix} \theta_b \\ \theta_a \end{bmatrix}$$

For singular fins, the definition of the thermal transmission ratio

$$\mu = \frac{q_b}{\theta_b}$$

permits use of the matrix

$$\mathbf{Y} = \begin{bmatrix} \mu & 0 \\ 0 & 0 \end{bmatrix}$$

and it is noted that with this admittance formulation, the distinction between regular and singular fins is unnecessary.

8.4.1 General Array Algorithm

To analyze an arbitrary fin configuration by the general array algorithm, one first represents the array by a labeled, oriented graph with its branches corresponding to the individual fins and its nodes corresponding to the connection points. The orientation arrows always point from the base to the tip of the individual fins. The graph of the array shown in Fig. 8.11 is displayed in Fig. 8.12.

Consider the mathematical problem of solving for the heat flows and temperature excesses at the base and tip of every fin. For a rough count, disregard any singular fins and note that each regular fin carries four unknowns and that for e fins, there are $4e$

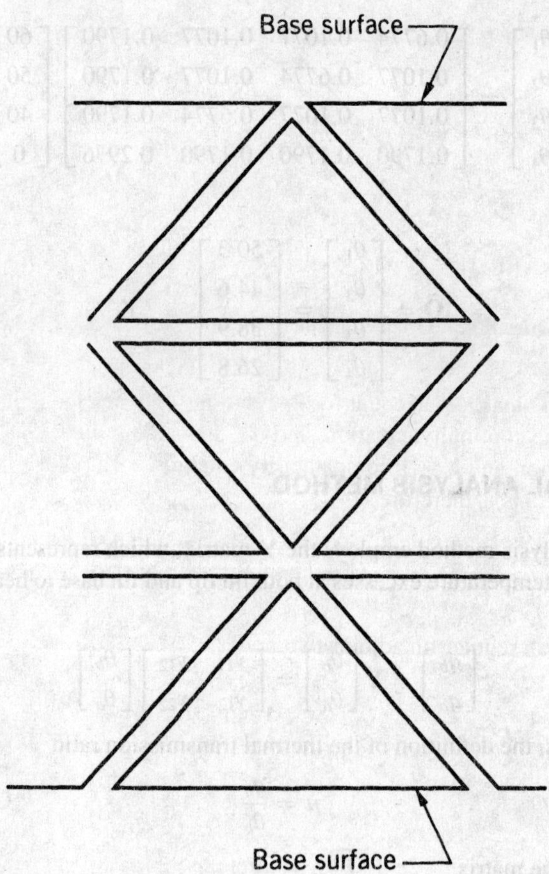

Figure 8.11 Repeating section of a triple stack configuration for a cross-flow heat exchanger application.

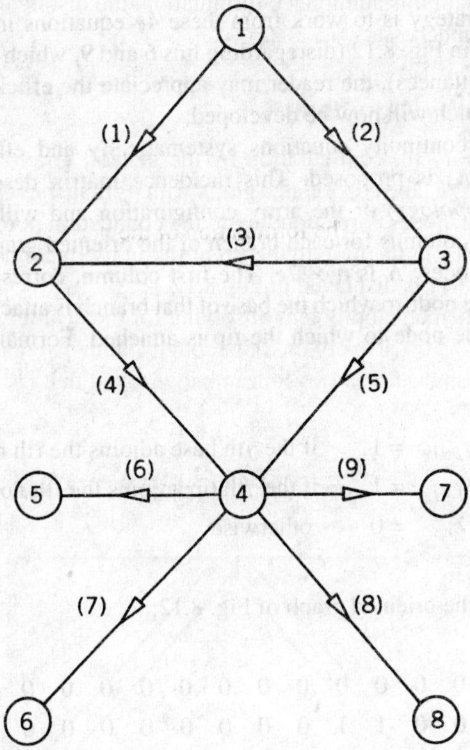

Figure 8.12 Oriented graph of the array shown in Fig. 8.11.

unknowns. The **Γ**-matrices account for $2e$ relations between the unknowns. At each node, there is one continuity equation, and if l_i fins adjoin the ith node, there is one continuity equation and $l_i - 1$ compatability equations. Thus for n nodes, the total number of equations is therefore

$$2e + \sum_{i=1}^{n} (1 + l_i - 1) = 2e + \sum_{i=1}^{n} l_i$$

But because every regular fin adjoins two nodes,

$$\sum_{i=1}^{n} l_i = 2e$$

so that

$$2e + \sum_{i=1}^{n} l_i = 2e + 2e = 4e$$

Thus there are $4e$ relationships available to determine the $4e$ unknowns.

The classical strategy is to work from these $4e$ equations in the $4e$ unknowns. Inasmuch as $e = 7$ in Fig. 8.12 (disregarding fins 6 and 9, which may be represented by their input admittances), the reader may appreciate the efficiency of the general array algorithm, which will now be developed.

To express the continuity equations systematically and efficiently, use of an *incidence matrix*, $\mathbf{\Lambda}$, is proposed. This incidence matrix describes the physical connections (the *topology*) of the array configuration and will have one row for each node and two columns for each branch of the oriented graph. If the graph has e-branches and n-nodes, $\mathbf{\Lambda}$ is $n \times 2e$. The first column, corresponding to a given branch, indicates the node to which the base of that branch is attached, and the second column indicates the node to which the tip is attached. Formally, the elements of $\mathbf{\Lambda}$ are

$$\begin{cases} \lambda_{i,2j-1} = 1 & \text{if the } j\text{th base adjoins the } i\text{th node} \\ \lambda_{i,2j} = 1 & \text{if the } j\text{th tip adjoins the } i\text{th node} \\ \lambda_{r,s} = 0 & \text{otherwise} \end{cases}$$

For example, in the oriented graph of Fig. 8.12:

$$\mathbf{\Lambda} = \begin{bmatrix} 1 & 0 & 1 & 0 & 0 & 0 & 0 & 0 & 0 & 0 & 0 & 0 & 0 & 0 & 0 & 0 & 0 & 0 \\ 0 & 1 & 0 & 0 & 0 & 1 & 1 & 0 & 0 & 0 & 0 & 0 & 0 & 0 & 0 & 0 & 0 & 0 \\ 0 & 0 & 0 & 1 & 1 & 0 & 0 & 0 & 1 & 0 & 0 & 0 & 0 & 0 & 0 & 0 & 0 & 0 \\ 0 & 0 & 0 & 0 & 0 & 0 & 0 & 1 & 0 & 1 & 1 & 0 & 1 & 0 & 1 & 0 & 1 & 0 \\ 0 & 0 & 0 & 0 & 0 & 0 & 0 & 0 & 0 & 0 & 0 & 1 & 0 & 0 & 0 & 0 & 0 & 0 \\ 0 & 0 & 0 & 0 & 0 & 0 & 0 & 0 & 0 & 0 & 0 & 0 & 0 & 1 & 0 & 0 & 0 & 0 \\ 0 & 0 & 0 & 0 & 0 & 0 & 0 & 0 & 0 & 0 & 0 & 0 & 0 & 0 & 0 & 0 & 0 & 1 \\ 0 & 0 & 0 & 0 & 0 & 0 & 0 & 0 & 0 & 0 & 0 & 0 & 0 & 0 & 0 & 1 & 0 & 0 \end{bmatrix} \tag{8.32}$$

Moreover, the continuity conditions state that at every node, the heat entering each adjacent fin base must be equal to the heat entering from each adjacent fin tip *plus* the heat input from heat sources, if there are any. Thus at the ith node

$$\sum q_b^{(j)} - \sum q_a^{(k)} = q_{s,i} \tag{8.33}$$

where the j sum extends over the fins with *base* incident to node i and the k sum extends over the fins with *tip* incident to node i. The heat input or the heat source at node i is represented by $q_{s,i}$.

The $+1$ entries in the ith row of $\mathbf{\Lambda}$ are precisely the terms that enter the sum of eq. (8.33). If the heat flows are arranged into a $2e \times 1$ column vector

$$\hat{\mathbf{Q}} = \begin{bmatrix} q_b^{(1)} \\ -q_a^{(1)} \\ q_b^{(2)} \\ -q_a^{(2)} \\ \cdots \\ q_b^{(e)} \\ -q_a^{(e)} \end{bmatrix} \tag{8.34}$$

the continuity conditions for heat flow can be expressed as

$$\mathbf{\Lambda}\hat{\mathbf{Q}} = \mathbf{Q}_s \tag{8.35}$$

where $\mathbf{Q}_s$ is an $n \times 1$ column vector of heat sources for the n nodes:

$$\mathbf{Q}_s = \begin{bmatrix} q_{s,1} \\ q_{s,2} \\ q_{s,3} \\ \cdots \\ q_{s,n} \end{bmatrix} \tag{8.36}$$

Compatability conditions for the temperature excesses simply state that all fins attached to a particular node must have the same temperature excess. The temperature excess vector is

$$\mathbf{\Theta} = \begin{bmatrix} \theta_b^{(1)} \\ \theta_a^{(1)} \\ \theta_b^{(2)} \\ \theta_a^{(2)} \\ \cdots \\ \theta_b^{(e)} \\ \theta_a^{(e)} \end{bmatrix} \tag{8.37}$$

and it is observed that the node to which the given base or tip is attached can be found by finding the +1 entry in the corresponding column of $\mathbf{\Lambda}$. Thus, if an $n \times 1$ column vector, $\hat{\mathbf{\Theta}}$, whose ith component is the common value of the temperature excess at node i is constructed:

$$\hat{\mathbf{\Theta}} = \begin{bmatrix} \theta_1 \\ \theta_2 \\ \theta_3 \\ \cdots \\ \theta_n \end{bmatrix} \tag{8.38}$$

then the compatability conditions for the temperature can be expressed as

$$\mathbf{\Theta} = \mathbf{\Lambda}^{\mathrm{T}}\hat{\mathbf{\Theta}} \tag{8.39}$$

Finally, all the thermal admittance equations, one for each fin, can be assembled into a single matrix equation:

$$\mathbf{Q} = \mathbf{Y}_g\mathbf{\Theta} \tag{8.40}$$

where $\mathbf{Y}_g$ is the general *fin-graph admittance matrix*:

$$\mathbf{Y}_g = \begin{bmatrix} y_{11}^{(1)} & y_{12}^{(1)} & 0 & 0 & \cdots & 0 & 0 \\ y_{21}^{(1)} & y_{22}^{(1)} & 0 & 0 & \cdots & 0 & 0 \\ 0 & 0 & y_{11}^{(2)} & y_{12}^{(2)} & \cdots & 0 & 0 \\ 0 & 0 & y_{21}^{(2)} & y_{22}^{(2)} & \cdots & 0 & 0 \\ \cdots & \cdots & \cdots & \cdots & \cdots & \cdots & \cdots \\ 0 & 0 & 0 & 0 & \cdots & y_{11}^{(n)} & y_{12}^{(n)} \\ 0 & 0 & 0 & 0 & \cdots & y_{21}^{(n)} & y_{22}^{(n)} \end{bmatrix} \tag{8.41}$$

Then, with a sign-correcting matrix,

$$\mathbf{S} = \begin{bmatrix} 1 & 0 & 0 & 0 & \cdots & 0 & 0 \\ 0 & -1 & 0 & 0 & \cdots & 0 & 0 \\ 0 & 0 & 1 & 0 & \cdots & 0 & 0 \\ 0 & 0 & 0 & -1 & \cdots & 0 & 0 \\ \cdots & \cdots & \cdots & \cdots & \cdots & \cdots & \\ 0 & 0 & 0 & 0 & \cdots & 1 & 0 \\ 0 & 0 & 0 & 0 & \cdots & 0 & -1 \end{bmatrix} \tag{8.42}$$

so that

$$\hat{\mathbf{Q}} = \mathbf{S}\mathbf{Q} \tag{8.43}$$

one can assemble eqs. (8.35), and (8.39) through (8.43) into a set of n node equations in n temperature excesses:

$$\mathbf{Q}_s = \mathbf{\Lambda}\mathbf{S}\mathbf{Y}_g\mathbf{\Lambda}^{\mathrm{T}}\hat{\mathbf{\Theta}} \tag{8.44}$$

The awkward appearance of $\mathbf{S}$ in eq. (8.44) could be eliminated by reversing the sign convention for q_a (Fig. 8.11), but this would cause considerable difficulty elsewhere as in the case of the cascade procedure.

8.4.2 Remarks

1. If the tip of a particular fin has no other fins or sources connected to it, as fins 6 and 7 in Fig. 8.12, the corresponding node equations can be reduced in number

by resorting to the fact that the fin can be represented by a **Y**-matrix containing the single element μ:

$$\mathbf{Y} = \begin{bmatrix} \mu & 0 \\ 0 & 0 \end{bmatrix}$$

Observe that the contribution of such a fin to the fin-graph admittance matrix $\mathbf{Y}_g$ produces a row of zeros. Thus the equation is redundant and the associated temperature excess does not appear in the system of node equations of eq. (8.44). Of course, there is no loss of information because the performance of such a fin does not depend on its tip temperature, as demonstrated by the definition of μ:

$$\mu = \frac{q_b}{\theta_b}$$

To take advantage of this, one deletes this fin from further consideration by replacing it in the fin graph by an equivalent negative heat source at its base node with a heat injection into the base node at a rate

$$q_i = -\mu_i \theta_{bi} \tag{8.45}$$

This substitution introduces no new unknowns and preserves the linearity of the node equations. It can be used whenever a subgraph of the fin graph contains no loops.

2. The recognition of the similarity of the node equations of eq. (8.44) to the node equations of electrical network theory is tempting but misleading. The temperature excesses θ are not *across variables* like voltage; that is, the operating characteristics of a fin depend on both the base and tip temperature excesses, θ_b and θ_a and not on their difference. For instance, one cannot treat any of the nodes as a datum and must always use the environment as the datum. Moreover, q is not a *through variable* such as current. Consequently, there is no thermal analogy to the mesh or loop currents in circuit theory. Yet a partial analogy with the electrical two-port theory does indeed exist and this analogy can be used in conjunction with the cascade and parallel algorithms of Chapter 7.
3. The matrix $\mathbf{\Lambda}$ contains precisely the same information as the $n \times e$ reduced branch incidence matrix, $\mathbf{A}$. This leads one to expect that the algorithm could be formulated in terms of $\mathbf{A}$, but there is some doubt as to the validity of this point. Certainly, $\mathbf{A}$ can be expressed in terms of $\mathbf{\Lambda}$ because the columns of the former are linear combinations of the columns of the latter. However, $\mathbf{\Lambda}$ cannot be obtained from $\mathbf{A}$ because the rank of $\mathbf{A}$ does not equal the rank of $\mathbf{\Lambda}$:

$$r(\mathbf{A}) = n - 1 \quad \text{and} \quad r(\mathbf{\Lambda}) = n$$

Example 8.4: The General Array Method. The temperature excess vector for the array in Example 8.2 is to be determined by the general array method.

SOLUTION. In Example 8.2, the admittance parameter matrix for each of the three fins in the array was found to be

$$\mathbf{Y} = \begin{bmatrix} 1.7552 & -1.0558 \\ 1.0558 & -1.7552 \end{bmatrix}$$

Using the oriented graph in Fig. 8.13, the incidence matrix is

$$\mathbf{\Lambda} = \begin{bmatrix} 1 & 0 & 0 & 0 & 0 & 0 \\ 0 & 0 & 1 & 0 & 0 & 0 \\ 0 & 0 & 0 & 0 & 1 & 0 \\ 0 & 1 & 0 & 1 & 0 & 1 \end{bmatrix}$$

The fin graph admittance matrix has the form

$$\mathbf{Y_g} = \begin{bmatrix} 1.7552 & -1.0558 & 0 & 0 & 0 & 0 \\ 1.0558 & -1.7552 & 0 & 0 & 0 & 0 \\ 0 & 0 & 1.7552 & -1.0558 & 0 & 0 \\ 0 & 0 & 1.0558 & -1.7552 & 0 & 0 \\ 0 & 0 & 0 & 0 & 1.7552 & -1.0558 \\ 0 & 0 & 0 & 0 & 1.0558 & -1.7552 \end{bmatrix}$$

and the reader may verify that the product $\mathbf{Y}_g \mathbf{\Lambda}^T$ is

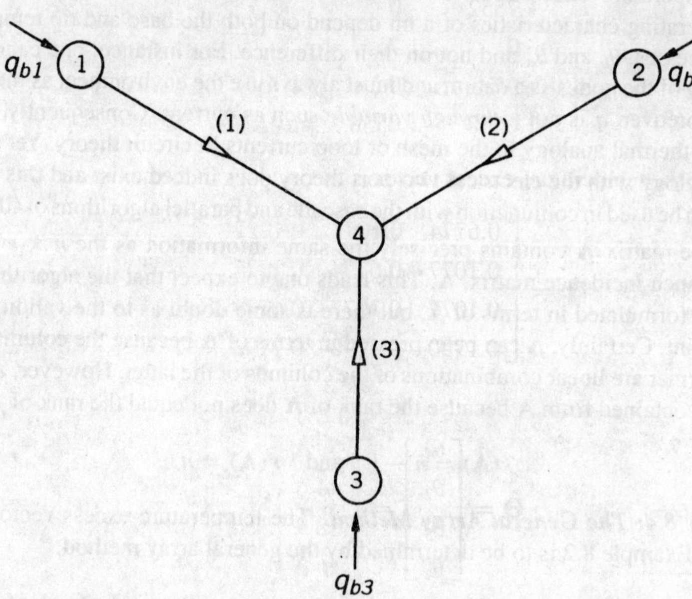

Figure 8.13 Oriented graph for the array shown in Fig. 8.8*a*.

$$\mathbf{Y}_g\mathbf{\Lambda}^{\mathrm{T}} = \begin{bmatrix} 1.7552 & 0 & 0 & -1.0558 \\ 1.0558 & 0 & 0 & -1.7552 \\ 0 & 1.7552 & 0 & -1.0558 \\ 0 & 1.0558 & 0 & -1.7552 \\ 0 & 0 & 1.7552 & -1.0558 \\ 0 & 0 & 1.0558 & -1.7552 \end{bmatrix}$$

With

$$\mathbf{\Lambda S} = \begin{bmatrix} 1 & 0 & 0 & 0 & 0 & 0 \\ 0 & 0 & 1 & 0 & 0 & 0 \\ 0 & 0 & 0 & 0 & 1 & 0 \\ 0 & -1 & 0 & -1 & 0 & -1 \end{bmatrix}$$

and the heat source vector

$$\mathbf{Q_s} = \begin{bmatrix} q_{s1} \\ q_{s2} \\ q_{s3} \\ q_{s4} \end{bmatrix} = \begin{bmatrix} 60 \\ 50 \\ 40 \\ 0 \end{bmatrix}$$

the matrix $\mathbf{K} = \mathbf{\Lambda S Y}_g\mathbf{\Lambda}^T$ can be shown to be

$$\mathbf{K} = \begin{bmatrix} 1.7552 & 0 & 0 & -1.0558 \\ 0 & 1.7552 & 0 & -1.0558 \\ 0 & 0 & 1.7552 & -1.0558 \\ -1.0558 & -1.0558 & -1.0558 & -5.2656 \end{bmatrix}$$

Finally, the temperature excess vector is

$$\hat{\mathbf{\Theta}} = \mathbf{K}^{-1}\mathbf{Q}_s = \begin{bmatrix} 0.6774 & 0.1077 & 0.1077 & 0.1790 \\ 0.1077 & 0.6774 & 0.1077 & 0.1790 \\ 0.1077 & 0.1077 & 0.6774 & 0.1790 \\ 0.1790 & 0.1790 & 0.1790 & 0.2976 \end{bmatrix} \begin{bmatrix} 60 \\ 50 \\ 40 \\ 0 \end{bmatrix}$$

or

$$\hat{\mathbf{\Theta}} = \begin{bmatrix} \theta_1 \\ \theta_2 \\ \theta_3 \\ \theta_4 \end{bmatrix} = \begin{bmatrix} 50.3 \\ 44.6 \\ 38.9 \\ 26.8 \end{bmatrix} \quad \mathrm{K}$$

which checks the result of Example 8.2.

8.5 OPTIMIZATION OF FINNED ARRAYS

Kern and Kraus (1972) provided a modest treatment of the optimization of longitudinal and radial fins. Later, Aziz (1992), in a comprehensive article, much of which has been summarized in Chapter 3, provided the optimum dimensions for several fin profiles. For the optimum longitudinal fin of rectangular profile, where the profile area A_p for a single fin is specified, Aziz gives

$$\delta_{\text{opt}} = \frac{A_p}{b_{\text{opt}}} = 0.791\left(\frac{2A_p^2 h}{k}\right)^{1/3} \tag{3.6}$$

and

$$b_{\text{opt}} = 1.262\left(\frac{A_p k}{2h}\right)^{1/3} \tag{3.7}$$

If the desired value of $q_b/\theta_b = Y_{\text{in}}$ is specified, Aziz provides

$$\delta_{\text{opt}} = \frac{0.632}{hk}\left(\frac{Y_{\text{in}}}{L}\right)^2 \tag{3.8}$$

and

$$b_{\text{opt}} = \frac{0.798}{h}\frac{Y_{\text{in}}}{L} \tag{3.9}$$

Notice that in eqs. (3.8) and (3.9), the optimum fin height varies inversely with the magnitude of the heat transfer coefficient and that in natural convection applications, the height of the individual optimized fin tends to be high.

8.5.1 Step-by-Step Optimization Method

In a complex finned array, each of the remote fins can be optimized through the use of eqs. (3.6) through (3.10). These equations provide optimum dimensions which are based on assumption that no heat flows from the tip of a remote fin.

Finned arrays also include fins in cascade and fins in cluster at the tips of nonremote connecting fins. In such fins, heat flows out of the tip and expressions for the optimum thickness and height need development. Aziz (1978) has addressed this problem. His expression for the heat dissipated is

$$q = (2hk\delta)^{1/2}\theta_b \frac{\sinh mb + H \cosh mb}{\cosh mb + H \sinh mb}$$

where $H = (\delta/2hk)^{1/2} h_a$ and where h_a is the tip heat transfer coefficient. For a fixed profile area, $A_p = \delta b$, the height b can be replaced by A_p/δ and the point where $dq/d\delta$ vanishes may be found. This procedure leads to a relationship between $\beta = mb$ and H (Aziz, 1978):

$$6\beta(1 - H^2) = (1 + H^2) \sinh 2\beta + 2h(1 + \cosh 2\beta)$$

and a trial-and-error procedure is necessary to fix the optimum fin dimensions, b_{opt} and δ_{opt}.

The foregoing discussion is based on the optimization of a single fin with tip heat loss. This optimization does not allow the user to consider the temperature excesses at the base or tip. Because of this, the optimization is difficult to apply to a finned array. Thus, attention turns to the *reflection relationship* for the longitudinal fin of rectangular profile

$$Y_{in} = \frac{Y_o \sinh mb + Y_a \cosh mb}{\cosh mb + Y_a/Y_o \sinh mb}$$

or with $\Upsilon = Y_a/Y_o$,

$$Y s_{in} = Y_o \frac{\sinh mb + \Upsilon \cosh mb}{\cosh mb + \Upsilon \sinh mb}$$

where the optimization will involve finding the point where $dY_{in}/d\delta$ vanishes.

The first step is to eliminate b in favor of δ via the profile area ($b = A_p/\delta$) so that

$$\beta = mb = \left(\frac{2h}{k\delta}\right)^{1/2} \frac{A_p}{\delta} = \left(\frac{2hA_p^2}{k}\right)^{1/2} \delta^{-3/2}$$

Then with

$$Y_o = (2hk\delta)^{1/2} L = [(2hk)^{1/2} L]\delta^{1/2}$$

Y_{in} becomes

$$Y_{in} = \frac{C_1\delta^{1/2} \sinh \beta + Y_a \cosh \beta}{\cosh \beta + C_2\delta^{-1/2} \sinh \beta}$$

where

$$C_1 = (2hk)^{1/2} L \quad \text{and} \quad C_2 = \frac{Y_a}{(2hk)^{1/2} L}$$

Let

$$u = Y_o \sinh mb + Y_a \cosh mb$$

and

$$v = \cosh mb + Y_a/Y_o \sinh mb$$

Then, because in

$$\frac{du}{dv} = \frac{v\,du - u\,dv}{v^2} = 0$$

the product, v^2, is not of any consequence,

$$\frac{dY_{\text{in}}}{d\delta} = A + B + C + D = 0$$

where

$$A = \frac{1}{2}(C_1\delta^{-1/2} + C_2Y_a\delta^{-3/2}) \sinh \beta \cosh \beta$$

$$B = \left[\frac{3}{2}\left(\frac{2hA_p^2}{k}\right)^{1/2} (C_1\delta^{-2} - C_2Y_a\delta^{-3}) \quad \sinh^2\beta\right]$$

$$C = C_1C_2\delta^{-1} \sinh^2\beta$$

and

$$D = \frac{3}{2}\left(\frac{2hA_p^2}{k}\right)^{1/2} (C_2Y_a\delta^{-3} - C_1\delta^{-2}) \cosh^2\beta$$

After multiplication by 2δ, recognition of the hyperbolic function identity,

$$\sinh^2\beta = 2 \sinh \beta \cosh \beta$$

and some algebraic simplification, the result, with $\Upsilon = Y_a/Y_o$, is

$$\sinh^2\beta + \frac{1+\Upsilon^2}{4\Upsilon} \sinh 2\beta = 3\left(\frac{1-\Upsilon^2}{2\Upsilon}\right)\beta \tag{8.46}$$

Observe that if eq. (8.46) is multiplied throughout by 4Υ, the result is

$$4\Upsilon \sinh^2\beta + (1+\Upsilon^2) \sinh 2\beta = 6(1-\Upsilon^2)\beta$$

and if $\Upsilon = 0$, that is, $Y_a/Y_o = 0$, this reduces to

$$\sinh 2\beta = 6\beta$$

which is the transcendental equation used by Kern and Kraus (1972) and Aziz (1992) in the development of the optimum dimensions of the longitudinal fin of rectangular profile when the profile area is specified and the tip is adiabatic.

Optimization Procedure. Equation (8.46) is the key to the optimization. But the value of the tip admittance is all that is specified, and a solution must be obtained by trial and error. Some preliminaries are necessary before the trial-and-error procedure to determine suitable values of b and δ.

1. Because the parameter β involves both the fin height b and the fin thickness δ, one of these must be selected to begin the procedure. However, because Y_o involves δ and not b, a value for the fin thickness is selected. This will force the value of Y_o:

$$Y_o = (2hk\delta)^{1/2}L$$

This value of fin thickness will be the optimum fin thickness, not only because it will help dictate the value of the optimum fin height b_{opt}, but it will form the fin profile area

$$A_p = \delta_{opt} b_{opt}$$

- Observe that Y_o must be greater than Y_a to avoid the heat flow choke considered in Section 7.2.4.
- A selection of δ that is too small will provide a value of Y_o that will yield, after the laborious trial-and-error procedure needed to solve eq. (8.46), a value of β that will cause the fin height b to be very large. A large value of b may cause the array to exceed the envelope permitted for the heat sink.
- On the other hand, too large a value of δ will yield a value of Y_o that will provide a value of β that will cause the value of the fin height to be quite low. A low value of b coupled with a high value of δ yields a low aspect ratio b/δ, and low aspect ratios mean that the fin height is comparable in magnitude to the fin thickness. This leads to a violation of one of the Murray–Gardner assumptions that the heat flow in the fin must be one-dimensional. Figure 8.14 shows the variation of β with Υ and may be used to make ballpark estimates for use in eq. (8.46). Observe that the value of Υ in the trial-and-error procedure should be restricted to values between 0.4 and 0.65. Values outside this range will yield values of β that provide unrealistic aspect ratios b/δ.

2. Find the value of β that will satisfy Eq. (8.46).
3. Then find the optimum fin height. If the selection of δ is called δ_{opt}, then

$$b_{opt} = \frac{\beta}{m} = \beta\left(\frac{k\delta_{opt}}{2h}\right)^{1/2}$$

4. A check of the aspect ratio should then be made. Aspect ratios should be at least 6:1 in order not to violate the Murray–Gardner assumption that the heat flow from base to tip is one-dimensional.

8.5.2 Forced-Convection Example

Finned arrays that are cooled by forced convection operate at higher values of the heat transfer coefficient. Because this mode provides more moderate values of the input admittance, it is deemed improper to attempt to cluster more than two or three remote fins at the tip of a supporting fin. A cluster admittance (equal to the tip admittance, $Y_a = q_a/\theta_a$) that possesses a high magnitude will require a high value of input

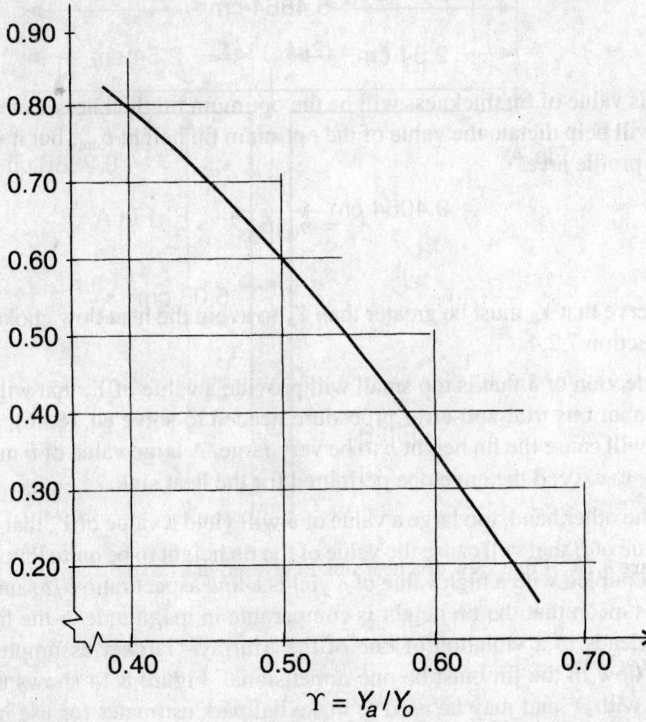

Figure 8.14 Optimization parameter β as a function of the ratio $\Upsilon = Y_a/Y_o$.

admittance to maintain Υ within the feasible range. All things being equal, a high value of the input admittance can only derive from a high value of fin thickness. This, in turn, may lead to a violation of the Murray–Gardner assumption that the fin height must be substantially higher than the fin thickness to assure that the heat flow is one-dimensional (from base to tip). Thus, when dealing with the forced convection mode, there are limitations to the number of fins that can be put into a cluster.

Example 8.5. Figure 8.15 shows a finned array that is to be employed in a forced-convection application where the heat transfer coefficient is 60 W/m^2 · K. The fins are fabricated from a material having a thermal conductivity of 180 W/m · K and the length of the array is 0.10 m. Pertinent computed values along with the length L, height b and thickness δ, for the fins are·

	Fins A	Fin B
δ(m)	2.286×10^{-3}	4.064×10^{-3}
b (m)	0.0254	0.0508
L (m)	0.100	0.100
AR	11.11	12.50

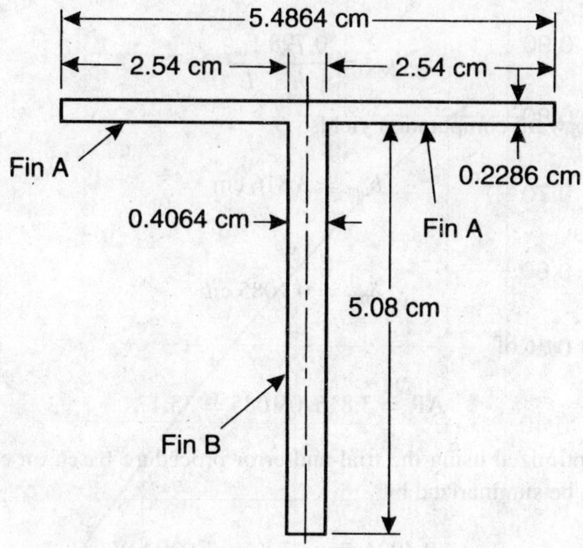

Figure 8.15 Edge view of a heat sink to be used in a forced-convection application.

Y_o (W/°C)	0.7027	0.9369
m (m^{-1})	17.0772	12.8079
mb	0.4338	0.6506

Fins A have negligible tip heat dissipation. Under this circumstance, $Y_a = q_a/\theta_a = 0$ and eq. (5.46) reduces to

$$Y_{in} = \frac{Y_o \sinh mb}{\cosh mb}$$

or

$$Y_{in} = Y_o \tanh mb$$

Use of this relationship with $Y_o = 0.7027$ W/°C and $mb = 0.4338$ gives $Y_{in} = 0.2870$ W/°C, and the admittance at the tip of fin B is $Y_a = 2Y_{in} = 2(0.2870) = 0.5740$ W/°C.

Equation (8.46) can now be employed using mb and Y_o for fin B with $\Upsilon = Y_a/Y_o = 0.5740/0.9369 = 0.6127$ W/°C. This gives $Y_{in} = 0.8220$ W/°C. An optimization can then be sought to maintain Y_{in} at this value.

Fins A can be optimized using eqs. (3.8) and (3.9):

$$\delta_{opt} = \frac{0.632}{hk}\left(\frac{Y_{in}}{L}\right)^2 \tag{3.8}$$

and

$$b_{\text{opt}} = \frac{0.798\ Y_{\text{in}}}{h \quad L} \tag{3.9}$$

This straightforward computation yields

$$b_{\text{opt}} = 3.816 \text{ cm}$$

and

$$\delta_{\text{opt}} = 0.1085 \text{ cm}$$

with an aspect ratio of

$$\text{AR} = 3.816/0.1085 = 35.17$$

Fin B is optimized using the trial-and-error procedure based on eq. (8.46). The procedure can be summarized by

$$\delta_{\text{opt}} = 0.4904 \text{ cm} \qquad Y_o = 1.0293 \text{ W/°C}$$

$$\Upsilon = Y_a/Y_o = 0.5740/1.0293 = 0.5577 \qquad \beta = 0.4655$$

$$m = 11.6595 \text{ m}^{-1} \qquad b_{\text{opt}} = 3.992 \text{ cm}$$

with an aspect ratio of

$$\text{AR} = 3.992/0.4904 = 8.14$$

The original design can be compared to the optimum design in a side-by-side listing, and this optimization is shown in Fig. 8.16.

Property	Original Design	Optimum Design
	Fin *A*	
δ (cm)	0.2286	0.1085
b (cm)	2.54	3.816
A_p (cm^2)	0.5806	0.4140
AR	11.11	35.17
	Fin *B*	
δ (cm)	0.4064	0.4904
b (cm)	5.08	3.992
A_p (cm^2)	2.0645	1.9557
AR	12.5	8.14
Total A_p	3.2258	2.7857

On a profile area basis, the optimum design uses only $2.7857/3.2258 = 0.864$ or 86.4% as much metal.

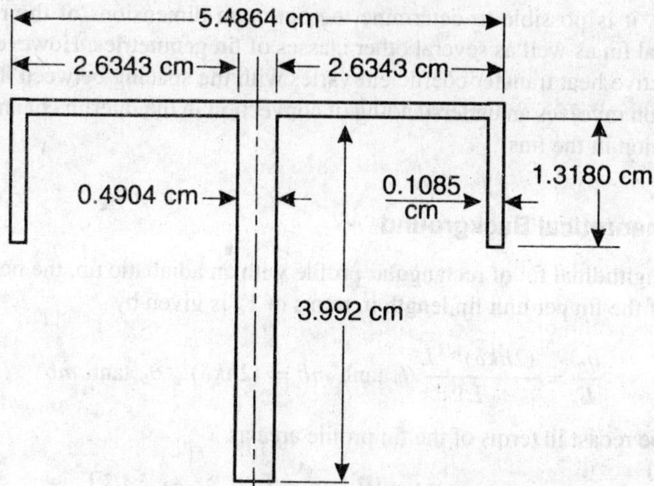

Figure 8.16 Optimized finned array. The optimization is an optimization of the array shown in Fig. 8.15 and is based on the optimization of fins A on the basis of Y_{in}.

What If? What if an attempt is made to design fins A for the specified profile area of 0.5806 cm^2? In this case, eqs. (3.7) and (3.8) are employed and the result is

$$b_{opt} = 5.80 \text{ cm} \quad \text{and} \quad \delta_{opt} = 0.100 \text{ cm}$$

These dimensions provide an aspect ratio of 58 and the fin is somewhat unwieldly. Moreover, compared with the original design, no metal is saved and the fin does not transfer as much heat as in the design where q_b or Y_{in} is specified.

Finned arrays that operate in the free or natural convection mode operate at lower values of the heat transfer coefficient. Because of this, attempts to optimize individual fins contained in a cluster at the tip of a supporting fin can meet with rather dismal results. However, because the input admittance for these individual fins will be rather low in a natural convection application, use of a cluster containing many fins is encouraged. In such cases, the supporting fin can be optimized because the value of $\Upsilon = Y_a/Y_o$, due to the low value of Y_a can still be made to fall in the range $0.40 \leq \Upsilon \leq 0.65$.

Attention now turns to a procedure for optimization of an array of fins proposed by Bar-Cohen and Jelinek (1985).

8.6 OPTIMUM (LEAST MATERIAL) ARRAYS OF FINS

The optimization of single fins, discussed in Chapter 3, provides the starting point for the optimization of the array of fins. For a specified convective heat transfer coefficient, it is possible to determine the optimum dimensions of the rectangular longitudinal fin as well as several other classes of fin geometries. However, because the convective heat transfer coefficient varies with the spacing between the fins, the optimization involves an understanding of convection in the interfin channels as well as conduction in the fins.

8.6.1 Theoretical Background

For the longitudinal fin of rectangular profile with an adiabatic tip, the heat entering the base of the fin per unit fin length in terms of Y_o is given by

$$\frac{q_b}{L} = \frac{(2hk\delta)^{1/2}L}{L}\theta_b \tanh mb = (2hk\delta)^{1/2}\theta_b \tanh mb$$

This can be recast in terms of the fin profile area as

$$\frac{q_b}{L} = \left(\frac{2hkA_p}{b}\right)^{1/2}\theta_b \tanh\left[b^{3/2}\left(\frac{2h}{kA_p}\right)^{1/2}\right] \tag{8.47}$$

Moreover, as shown in Chapter 3, the fin heat flow can be optimized by differentiating eq. (8.47) with respect to b and determining where this derivative vanishes. The result is given in terms of β:

$$\beta \equiv mb = b^{3/2}(2h/kA_p)^{1/2}$$

by the transcendental equation given in Section 3.3.1 by

$$3\beta \operatorname{sech}^2\beta = \tanh\beta$$

or

$$(\sinh 2\beta - 6\beta)_{s,\mathrm{opt}} = 0 \tag{8.48}$$

where the subscript s,opt indicates *single fin optimum*. The solution of eq. (8.48) can be expected to yield the rectangular fin geometry that will maximize the fin heat dissipation for a given profile area, volume, or mass. As indicated in Section 3.3.1, the single-fin optimum value of $\beta_{s,\mathrm{opt}}$ is given by Kern and Kraus (1972) and Aziz (1992) and is found to equal $\beta_{s,\mathrm{opt}} = 1.4192$.

With $\beta_{s,\mathrm{opt}} = 1.4192$, the aspect ratio δ/b of the single optimum fin can be derived from the expression for mb:

$$mb = \left(\frac{2h}{k\delta}\right)^{1/2} b = 1.4192$$

$$\left(\frac{2hb^2}{k\delta}\right)^{1/2} = 1.4192$$

and hence

$$\left(\frac{\delta}{b}\right)_{s,\text{opt}} = \frac{2}{(1.4192)^2}\frac{hb}{k} = 0.993\left(\frac{hb}{k}\right) \tag{8.49}$$

The profile area of the optimum fin can be found from eqs. (3.8) and (3.9):

$$\delta_{\text{opt}} = \frac{0.632}{hk}\left(\frac{Y_{\text{in}}}{L}\right)^2 \tag{3.8}$$

and

$$b_{\text{opt}} = \frac{0.798}{h}\frac{Y_{\text{in}}}{L} \tag{3.9}$$

to be

$$A_{p,\,\text{opt}} = b_{\text{opt}}\delta_{\text{opt}} = \frac{(0.632)(0.798)}{h^2 k}\left(\frac{Y_{\text{in}}}{L}\right)^3$$

Then, because $Y_{\text{in}} = q_b/\theta_b$,

$$\left(\frac{q_b}{\theta_b L}\right)_{s,\text{opt}} = \left(\frac{h^2 k A_p}{0.5043}\right)^{1/3}$$

and the optimum single fin heat dissipation per unit length is given by

$$\left(\frac{q_b}{L}\right)_{s,\text{opt}} = 1.258\theta_b(h^2 A_p k)^{1/3} \tag{8.50}$$

8.6.2 Optimum Multiple-Fin Array

Individually Optimum Fins. Studies concerning the optimization of arrays of fins go back to the mid-1960s. Appl and Hung (1964) studied the optimization of radial fins on circular tubes, and Sobel et al. (1966) experimentally studied natural-convection airflows in staggered and vertical channels with constant heat flux boundary conditions. Jones and Smith (1970) obtained expanded experimental average heat transfer coefficients for natural convection cooling of arrays of isothermal fins on horizontal surfaces, and Levy (1971) and Fitzroy (1971) studied the plate spacings that maximized the heat transfer from each plate in natural convection.

An optimization of radial fins on a horizontal tube was provided by Littlefield and Cox (1974), and Pnueli (1974) showed that heat dissipated from vertical fins in an array could be increased by inclining them with respect to some angle in the direction of the gravity force. Kovarik (1983, 1984) conducted studies on optimal heat transfer assemblies, and the 1984 work claimed to have found the single necessary condition of optimality for a finned heat exchanger in laminar flow, the ratio of the maximum heat transfer for the fins to the per unit cost of the finned duct. Topaldi and Kuehn (1984) studied a horizontal isothermal cylinder with infinitely large transverse nonisothermal plate fins.

Bar-Cohen and Jelinek (1985) have shown that the results of Section 8.6.1 can be applied to the optimization of an array of fins by assuming that the array is to be composed of n individually optimum fins. These fins are to possess an aspect ratio determined by eq. (8.49) and are capable of transferring heat according to eq. (8.50). In such an array, the required number of fins, n, is approximately equal to the desired or design point total heat dissipation divided by the heat transfer rate of each fin, Q/q. In examining eq. (8.50), it may be observed that the optimum fin profile area A_p is proportional to q^3/h^2. Consequently, the profile area of the entire array of fins nA_p is proportional to Qq^2/h^2.

As a consequence of this proportionality, minimization of the total array profile area and mass would appear to require either an increase in the heat transfer coefficient (an obvious approach) or a decrease in the heat dissipation of each fin by reducing the fin dimensions. Either approach, carried to its logical conclusion, would yield a thermally optimum array of a near-infinite number of vanishingly thin fins, which would together have a near-zero profile area and a near-zero mass!

Unfortunately, manufacturing constraints and available pumping power (or the buoyant driving force) generally limit the minimum fin thickness and minimum interfin spacing, respectively, and thus make realization of a near-zero added-mass optimum array impossible. Instead, the thermal optimization must seek to find the minimum viable array profile area (or mass) consistent with the appropriate system constraints. When the fin material, the prime mover, and the working fluid have been determined, the requisite analysis can often be cast in the form of minimizing the total fin profile area (or mass) for a specified number of fins and a given heat transfer coefficient.

In optimizing an array of longitudinal fins, as may be done in the design of a heat sink, it must be recognized that the base temperature, the local fluid temperature, and the surface heat transfer coefficient may all vary in the flow direction. Consequently, in applying eqs. (8.49) and (8.50) to a single fin in an array, the designer can elect to use average values for θ_b and h or to perform the calculations in discrete steps in the flow direction. Such discretization can, of course, follow naturally from the geometry of the array when an interrupted fin structure is used.

It must be noted that while the fin heat transfer rate is proportional to both θ_b and $h^{2/3}$, the optimum single-fin aspect ratio, as indicated by eq. (8.49), is independent of θ_b and, in fact, becomes invariant in the flow direction as the fully developed condition (h = constant) is attained. As a consequence, the use of average values can be expected to introduce only a minor error in the determination of the optimum aspect ratio for such a fin.

Optimum Array. The rate of heat transfer from a shrouded array of longitudinal rectangular fins, as shown in Fig. 8.17, can be found by summing the thermal contribution of the fins and the interfin area, according to

$$\frac{Q}{L} = n\left[\left(\frac{2hkA_p}{b}\right)^{1/2}\theta_b \tanh\beta + h_b\theta_b\left(\frac{W}{n} - \frac{A_p}{b}\right)\right] \tag{8.51}$$

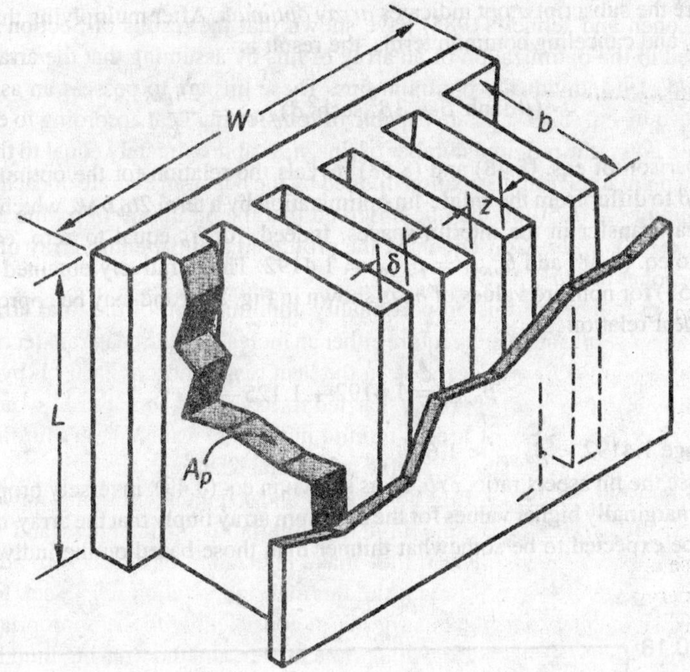

Figure 8.17 Geometric definition of an array of longitudinal fins.

where Q is the array dissipation, A_p the profile area of the single fin and equal to δb, W the array width, and h_b the heat transfer coefficient along the interfin surfaces (the prime or *base* surfaces). Although provision has been made in eq. (8.51) for differences between the heat transfer coefficient prevailing along the fin surfaces and in the interfin spaces, it has been assumed that the primary surface area (including both the fin base and interfin surface area) is isothermal. The validity of this common assumption can be expected to improve as the fin spacing decreases and the thermal conductivity, as well as thickness of the fin support plate, increases.

To find the maximum array dissipation at a fixed profile area nA_p or (with the number of fins predetermined by system considerations) for a fixed A_p, it is possible to differentiate $Q/Ln\theta_b$ with respect to b and to set the derivative equal to zero. Upon so doing, it is found that the array optimum is governed by

$$\left[\frac{1}{2}(2hkA_p)^{1/2} f(\beta, b) + \frac{h_b A_p}{b^2}\right]_{a,\text{opt}} = 0 \tag{8.52}$$

where

$$f(\beta, b) = 3\beta b^{-1/2} \operatorname{sech}^2 \beta - b^{-3/2} \tanh \beta$$

and where the subscript a,opt indicates *array optimum*. After multiplying through by $2b^3/kA_p$ and canceling common terms, the result is

$$(\beta \tanh \beta - 3\beta^2 \operatorname{sech}^2 \beta)_{a,\mathrm{opt}} = \frac{2h_b b}{k} \tag{8.53}$$

Comparison of eqs. (8.48) and (8.53) reveals the relation for the optimum array value of β to differ from the single-fin optimization by a term $2h_b b/k$, which pertains to the heat transfer in the interfin spaces. Indeed, for h_b equal to zero, eq. (8.53) reduces to eq. (8.48) and $\beta_{a,\mathrm{opt}} = \beta_{s,\mathrm{opt}} = 1.4192$. The iteratively obtained solution of eq. (8.53) for nonzero values of h_b is shown in Fig. 8.18 and may be approximated by the linear relation

$$\beta_{a,\mathrm{opt}} = 1.4192 + 1.125\frac{h_b b}{k} \tag{8.54}$$

in the range $1.4192 < \beta_{a,\mathrm{opt}} < 1.60$.

Because the fin aspect ratio, δ/b, is, as shown in eq. (8.49), inversely proportional to β, the marginally higher values for the optimum array imply that the array-optimum fins can be expected to be somewhat thinner than those based on the individual fin

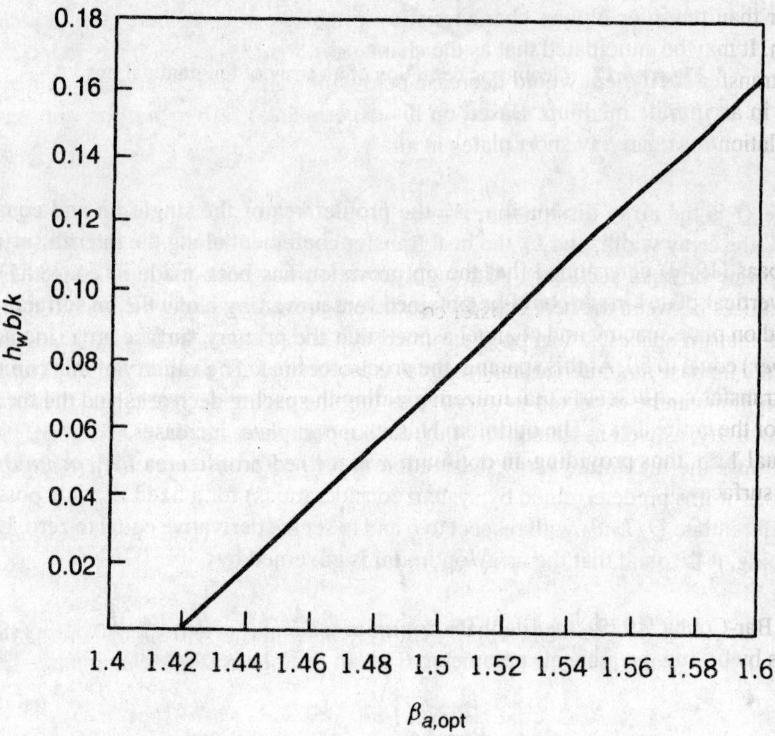

Figure 8.18 Optimizing value of β for an array of fins.

optimization. However, this difference may often be sufficiently small to justify the use, in commercial practice, of the individual-optimum fin values. The implications of this conclusion are explored in the next two sections, dealing with natural- and forced-convection optimum fin arrays in air, respectively.

As noted earlier, the validity of the foregoing analysis is limited to configurations in which the primary surface area is (or can be assumed) isothermal. Furthermore, this analysis is limited to situations in which the heat transfer coefficient along the finned surface can be assumed constant.

8.7 OPTIMUM NATURAL-CONVECTION HEAT SINKS

The design of natural convection fin arrays or heat sinks is complicated by the coupling between the flow and temperature fields. This is a characteristic of all natural convection heat transfer situations. Thermal optimization of such an array of fins requires selection of both the optimum fin spacing and optimum fin thickness or thickness, each determined to maximize total heat dissipation for a given added mass or profile area.

In natural convection, a balance between viscous dissipation and buoyant potential, rather than pump or blower characteristics, establishes the desired interfin channel width. It may be anticipated that as the channel width or plate spacing is reduced, the heat transfer coefficient would decrease below the value associated with an isolated plate in an infinite medium. Based on his experimental results, and his subsequent correlation for relatively short plates in air,

$$\mathrm{Nu} = \frac{hz}{k} = \frac{\mathrm{El}}{24}\left(1 - e^{-35/\mathrm{El}}\right)^{3/4} \tag{8.55}$$

Elenbaas (1942) determined that the optimum spacing between parallel isothermal thin vertical plates in air could be obtained by setting the channel Rayleigh number (based on plate spacing and channel aspect ratio and here referred to as the *Elenbaas number*) equal to 50. At this spacing, the product of the total plate surface area and the heat transfer coefficient is maximized, yielding the peak volumetric heat dissipation rate for the entire array. The optimum Nusselt number was found by Elenbaas (1942) to equal 1.25, thus providing an optimum average heat transfer coefficient along the plate surfaces of

$$h_{\mathrm{opt}} = 1.25\frac{k_g}{z_{\mathrm{opt}}} \tag{8.56}$$

In Bar-Cohen (1979), the Elenbaas optimum spacing was extended to optimum fin arrays by basing the plate air parameter P on an average surface temperature. Thus

$$P = \left(\frac{L\nu^2}{g\beta\bar{\theta}\mathrm{Pr}}\right)^{1/4}$$

and

$$z_{\text{opt}} = \left(\frac{50 L \nu^2}{\eta g \beta \bar{\theta} \text{Pr}} \right)^{1/4} = 2.66P \tag{8.57}$$

where the average temperature excess $\hat{\theta}$ can be expressed as the product of the fin efficiency η and the base temperature excess θ_b. The fin efficiency for the single optimum fin, for which $mb = 1.4192$, can be found to equal 0.627:

$$\eta = \frac{\tanh mb}{mb} = \frac{\tanh 1.4192}{1.4192} = \frac{0.8894}{1.4192} = 0.627$$

The extension of such an optimization approach to fins where the thickness is not negligible is facilitated by the use of the composite Nusselt number

$$\text{Nu}_o = \left[\left(\frac{1}{24} \text{El} \right)^{-n} + (0.59 \text{El}^{1/4})^{-n} \right]^{-1/n}$$

and is discussed by Bar-Cohen and Rohsenow (1983).

It is interesting to recall that the Elenbaas results were obtained with relatively tall fins where inflow from the open edges had a negligible impact on the heat transfer rate. However, studies on actual heat sinks, most recently by Bilitzky (1986), as well as of narrow vertical plates by Sparrow and Bahrami (1980), have shown that the Elenbaas correlation, eq. (8.55), underpredicts the heat transfer coefficients for such configurations at low Elenbaas numbers. However, for closely spaced narrow fins, the Nusselt numbers lie between the isolated plate limit and the Elenbaas correlation, thereby reflecting the contribution of inflow from the open edge. As the Elenbaas number increases, the Elenbaas correlation asymptotically approaches the isolated plate limit and in the range of $\text{El} = 50$, corresponding to the optimum fin spacing, the isolated plate relation yields Nusselt numbers that are just 25% higher than eq. (8.55). In this same intermediate range of Elenbaas numbers, the empirical heat transfer coefficients, for the heat sinks studied by Bilitzky (1986), are approximately 15% above the predicted values. Moreover, examination of eq. (8.49) reveals that the "least-material" optimization of a longitudinal fin of rectangular profile yields relatively tall fins, (e.g., a b of 10 cm for a 1-mm-thick fin with a thermal conductivity of 50 W/m · K).

Consideration of these twin factors, intermediate Elenbaas numbers and relatively tall fins suggests that the optimization of a multiple-fin array can proceed on the basis of the Elenbaas correlation for the natural convection heat transfer coefficient. Furthermore, because in Section 8.6 the optimizing parameter for the fin aspect ratio has been shown to be nearly independent of the fin array characteristics, it would now appear possible to use eqs. (8.50), (8.56), and (8.57) to determine the geometric parameters of an optimum heat sink using vertical natural-convection air-cooled fins.

8.7.1 Array Dissipation

To obtain the heat dissipation of an array of optimally spaced least-material fins, eq. (8.50) can be used with eq. (8.56) to quantify the heat flow through each fin, while eq. (8.57) sets the interfin spacing. Moreover, assuming that h_{opt} can also be used to characterize the heat transfer coefficient in the area between the fins and that the number of fins is equal to the number of fin spaces (as is nearly the case for a large number of fins), the total heat dissipation of the array is expressible as

$$\frac{Q}{L} = \frac{\left[(q_{\text{opt}}/L) + h_{\text{opt}} z_{\text{opt}} \theta_b\right] W}{z_{\text{opt}} + \delta_{\text{opt}}} \tag{8.58}$$

Following substitution for q_{opt}/L, h_{opt}, and z_{opt} and using the fin efficiency of 0.627 as is appropriate for optimum rectangular fins, eq. (8.58) can be modified to yield

$$\frac{Q}{LW\theta_b} = \frac{0.862(k_g k \delta_{\text{opt}}/P^{1/2}) + 1.25k_g}{2.66P + \delta_{\text{opt}}} \tag{8.59}$$

A graphical presentation of eq. (8.59) for aluminum, brass, and steel arrays of least-material fins, dissipating heat to ambient air, is shown in Figs. 8.19 through 8.21, respectively, for a typical parametric range. Array dissipation is seen to vary greatly with both the array parameter P and the thickness of the optimum fin, δ_{opt}. Furthermore, a maximizing value of δ_{opt} appears to exist for each value of P.

The locus of the array dissipation maxima, shown graphically in Figs. 8.19 through 8.21, can be obtained by differentiation of eq. (8.59) or, more simply, by differentiation of an approximate form of this relation, in which heat transfer from the area between the fins is neglected. With this additional assumption, clearly justified by the large surface area of the optimum natural convection fin, eq. (8.58) yields

$$\frac{Q}{LW\theta_b} \simeq 1.258 \frac{(hk\delta_{\text{opt}})^{1/2}}{z_{\text{opt}} + \delta_{\text{opt}}} \tag{8.60}$$

Following Bar-Cohen (1979), differentiating eq. (8.60) with respect to δ_{opt} and setting the derivative to zero, the maximizing value of δ_{opt} is found to equal z_{opt}.

This particular value of fin thickness defines an array geometry that maximizes heat dissipation from a prime area of given length L and specified temperature excess θ_b while using least-material fins of rectangular profile. Substitution of $\delta_{\text{opt}} = z_{\text{opt}}$ into eq. (8.59) yields the heat transfer rate for the maximum-dissipation optimum array as

$$\left(\frac{Q}{LW\theta_b}\right)_{\text{opt}} = 0.264\left(\frac{kk_g}{P^2}\right)^{1/2} \tag{8.61}$$

where k is the thermal conductivity of the fin material and k_g is the thermal conductivity of the fluid.

Example 8.6. Consider an example of an air-cooled natural-convection array of fins. A single side of an aluminum ($k = 175$ W/m · K) module dissipates 80 W and is to be cooled to a temperature of 65°C by natural convection in air. The module wall is 35.6 cm high and 30.5 cm wide and is exposed to ambient air at 35°C. Find the

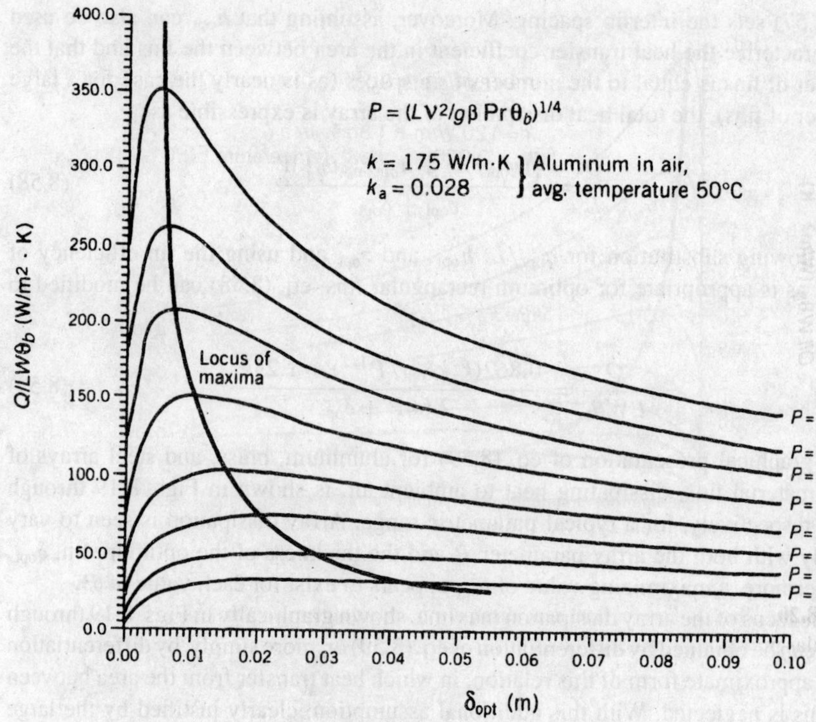

Figure 8.19 Natural-convection heat dissipation of an array of optimally spaced vertical rectangular-profile least-material aluminum fins.

maximum least-material heat transfer rate from this module wall and determine the added mass and geometry (fin height, thickness, and fin spacing) of the finned array.

SOLUTION. For the conditions stated,

$$\theta_b = 65°\text{C} - 35°\text{C} = 30°\text{C}$$

and

$$T_{\text{av}} = \frac{65°\text{C} + 35°\text{C}}{2} = 50°\text{C}$$

the thermal properties of air are

$$\text{Pr} = 0.701$$

$$\nu = 1.79 \times 10^{-5}\ \text{m}^2/\text{s}$$

$$k = 0.0274\ \text{W/m} \cdot \text{K}$$

and

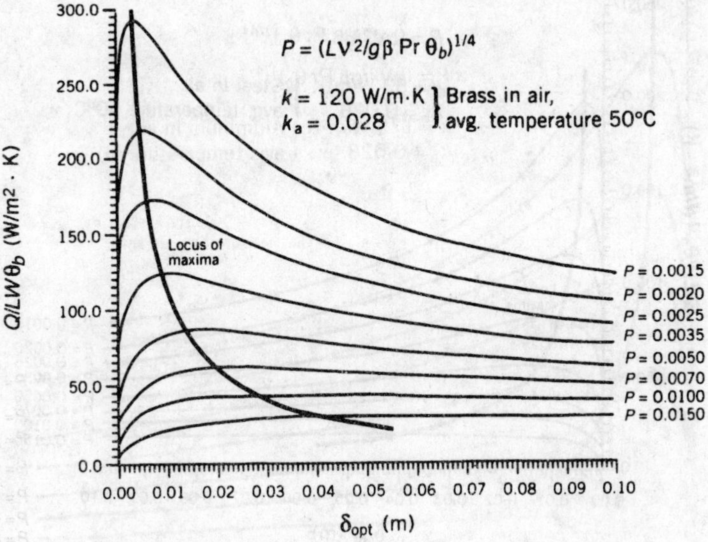

Figure 8.20 Natural-convection heat dissipation of an array of optimally spaced vertical rectangular-profile least-material brass fins.

$$\beta = \frac{1}{T_{av}} = \frac{1}{323} = 0.003096 \text{ K}^{-1}$$

The fin array parameter P using $\theta = \overline{\theta}$ is evaluated as

$$P = \left(\frac{L\nu^2}{g\beta\overline{\theta}\text{Pr}}\right)^{1/4} = \left(\frac{L\nu^2}{g\beta\eta\theta_b\text{Pr}}\right)^{1/4}$$

or

$$\begin{aligned} P &= \left[\frac{(0.356)(1.79 \times 10^{-5})^2}{(9.81)(0.003096)(0.627)(30)(0.701)}\right]^{1/4} \\ &= (2.853 \times 10^{-10})^{1/4} \\ &= 0.004110 \text{ m} \end{aligned}$$

The optimum fin spacing is

$$z_{opt} = 2.66P = (2.66)(0.004110) = 0.01093 \text{ m} \quad (10.93 \text{ mm})$$

and the optimum heat transfer coefficient is found using eq. (8.56).

$$h_{opt} = 1.25\frac{k_g}{z} = 1.25\left(\frac{0.0274}{0.01093}\right) = 3.13 \text{ W/m}^2 \cdot \text{K}$$

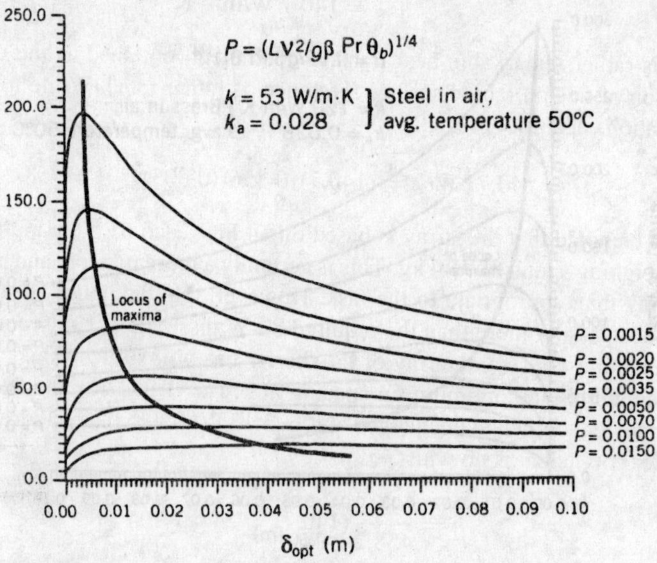

Figure 8.21 Natural-convection heat dissipation of an array of optimally spaced vertical rectangular-profile least-material steel fins.

In accordance with the findings of this section, the peak performance of such a natural-convection array is achieved by setting the fin thickness equal to the fin spacing (10.93 mm). Consequently, the fin thickness is set equal to $\delta = 0.01093$ m and the fin height may determined:

$$\left(\frac{\delta}{b}\right)_{s,\text{opt}} = \frac{2}{(1.4192)^2}\frac{hb}{k} = 0.993\left(\frac{hb}{k}\right) \tag{8.49}$$

or

$$b = \frac{k\delta}{0.993h} = \frac{(175)(0.01093)}{(0.993)(3.13)} = 0.615 \text{ m}$$

The dimensions of $\delta = 1.093$ cm and $b = 0.615$ m define an array that maximizes heat transfer, yielding an effective cooling heat transfer coefficient in accordance with eq. (8.61):

$$\begin{aligned}\left(\frac{Q}{LW\theta_b}\right)_{\text{opt}} &= 0.264\left(\frac{kk_g}{P^2}\right)^{1/2} \\ &= 0.264\left[\frac{(175)(0.0274)}{(0.004110)^2}\right]^{1/2} \\ &= (0.264)(2.839 \times 10^5)^{1/2}\end{aligned}$$

$$= 140.7 \text{ W/m}^2 \cdot \text{K}$$

This rather remarkable heat transfer coefficient, obtained as the combined contribution of fins on a base surface reference, is rather ideal. While it leads to a heat dissipation capability of

$$Q = 140.7LW\theta_b = (140.7)(0.356)(0.305)(30) = 458.2 \text{ W}$$

it is to be noted that this array is based on 14 fins, each 61.5 cm high, 1.1 cm thick, and weighing a total of 91.4 kg. This is certainly a more massive and more expensive fin array than appropriate to the task. However, the "optimum" heat dissipation is also substantially more than the required 80-W dissipation. Consequently, it is clear that a natural-convection array of far-shorter fins will suffice. This can be achieved by maintaining the optimum fin spacing at $z_{\text{opt}} = 1.093$ cm and keeping the local heat transfer coefficient unchanged at $h = 3.13 \text{ W/m}^2 \cdot \text{K}$. The required effective heat transfer coefficient is now just

$$h = \frac{80}{458.2}(140.7) = 24.6 \text{ W/m}^2 \cdot \text{K}$$

8.8 NOMENCLATURE

Roman Letter Symbols

- A cross-sectional or profile area, m^2; combination of terms, dimensions vary
- $\mathbf{A}$ node-branch incidence matrix, dimensionless
- a fin tip dimension, m; elements of node-branch incidence matrix, dimensionless
- B combination of terms, dimensions vary
- b fin height, m
- C constant, dimensionless; combination of terms, dimensions vary
- $\mathbf{C}$ matrix, dimensionless
- c elements of a matrix, dimensionless
- D combination of terms, dimensions vary
- d diameter, m; derivative or differential, dimensions vary
- El Elenbaas number, dimensionless
- e number of fins, dimensionless
- f function, dimensionless
- g acceleration of gravity, dimensionless
- H combination of terms, dimensionless
- h heat transfer coefficient, W/m$^2 \cdot$ K
- I modified Bessel function of the first kind
- i node designator, dimensionless
- k thermal conductivity, W/m $\cdot$ K
- L fin length, m; length of array, m
- l number of fins adjoining a node, dimensionless

- m fin performance parameter, m^{-1}
- Nu Nusselt number, dimensionless
- n number of fins, dimensionless
- P plate air parameter, m
- Pr Prandtl number, dimensionless
- $\mathbf{Q}$ heat flow vector
- $\hat{\mathbf{Q}}$ heat flow vector
- $\tilde{\mathbf{Q}}_s$ node heat source vector
- q heat flow, W
- R thermal resistance, K/W
- S surface area, m^2
- $\mathbf{S}$ sign correcting matrix, dimensionless
- T temperature, K
- $\mathbf{T}$ temperature vector, K
- u combination of terms, dimensionless
- v combination of terms, dimensionless
- W width of array, m
- x height coordinate, m
- Y thermal admittance, W/K
- $\mathbf{Y}$ thermal admittance matrix, W/K
- y elements of thermal admittance matrix, W/K
- Z thermal impedance, K/W
- z clear space between fins, dimensionless

Greek Letter Symbols

- β combination of terms, dimensionless; volumetric coefficient of expansion, K^{-1}
- $\mathbf{\Gamma}$ thermal transmission matrix, dimensions vary
- Γ gamma function, dimensionless
- γ elements of thermal transmission matrix, dimensions vary; parameter, dimensionless
- Δ change in, dimensionless
- δ fin thickness, m
- η fin efficiency, dimensionless
- $\mathbf{\Theta}$ temperature excess vector, K
- $\hat{\mathbf{\Theta}}$ node-temperature excess vector, K
- θ temperature excess, K; elements of temperature excess vector, K
- $\bar{\theta}$ average fin surface temperature excess, K
- $\mathbf{\Lambda}$ A matrix, dimensions vary; incidence matrix, dimensionless
- λ elements of a matrix, dimensions vary; elements of incidence matrix, dimensionless
- μ thermal transmission ratio, W/K
- ν kinematic viscosity, m^2/s
- ρ radius ratio, dimensionless
- $\boldsymbol{T}$ linear transmission matrix, dimensions vary

- τ elements of linear transmission matrix, dimensions vary
- υ ratio of thermal impedances, dimensionless

Roman Letter Subscripts

- a tip of fin; augmented matrix; array
- av average value
- b base of fin; interfin surface in an array of fins
- g gap quantity
- i a counter
- in input
- l left hand side
- o characteristic value; optimum or nominal value
- opt optimum value
- p profile area
- r right hand side
- s surroundings; source; single fin

Roman Letter Superscripts

- (i) counter
- (j) counter
- T transpose

CHAPTER 9

FINNED PASSAGES

9.1 INTRODUCTION

Interest in the efficiency of finned passages for the augmentation of convection heat transfer was originally stimulated by the *double-pipe heat exchanger*, with its low cost per unit heat transfer surfaces. This exchanger consists of two concentric pipes with longitudinal fins of rectangular profile bonded to the outside of the inner pipe. This produced a finned annular passage with heat flow through the inner pipe and the fins. It was apparent that if the fins were bonded to both the inner and outer pipes, the inside surface of the outer pipe could contribute as well to the gross heat transfer. The heat loss from the outer pipe to its surroundings could be made negligible by applying insulation to the outer surface of the outer pipe.

Kraus (1960) developed an expression for the efficiency of fins bonded to the inner and outer pipes and of the outer pipe, which acted as a fin. Kraus (1961) extended this to the forced-convection cooled electronic chassis or *cold plate heat exchanger*, which consisted of longitudinal fins between two exterior or *cover plates*. Kraus (1962) obtained an optimization for the fins in this rectangular finned passage, and the performance of such a passage with unequal inputs on opposite sides was studied by Smith (1966).

The growing availability of *compact heat exchanger cores*, as disclosed by Kays and London (1955), with later editions in 1964 and 1984, stimulated interest in the fin efficiency of *stacked cores*. Kays (1960) treated, as stack components, two- and three-layer cores, termed the *double* and *triple sandwich*, and developed efficiency relationships for equal heat input on both exterior plates. Kern and Kraus (1972) looked at *single*, *double*, and *triple stacks* (stacks used here rather than sandwiches) with heat input on one side and heat input on two sides.

In this chapter, methods for evaluation of the input admittance and fin efficiency of single and double stacks for three different conditions of heat loading and the triple and quadruple stacks for two conditions of heat loading are presented.

9.2 FINNED PASSAGE

A repeating section of a quadruple stack is shown in Fig. 9.1, where the vertical members are the fins and where the horizontal members or fins at the exteriors are called the *exterior* or *cover plates*. All of the horizontal fins that are not exterior or cover plates are categorized as *splitter* or *separation plates*.[1] Because the configuration is a repeating section of the overall passage, the tips of all horizontal fins may be considered adiabatic.

Heat may be applied at either cover plate or at both cover plates. If heat is applied evenly and equally at both cover plates, the passage is said to have *even loading*. In the event that heat is applied to one cover plate, the stack is considered to have *single-side loading*. The case of unequal heat input to the cover plates is known as *uneven loading*.

In Fig. 9.1, the fin heights b and the fin thicknesses δ are equal:

$$b = b_1 = b_2 = b_3 = b_4 \quad \text{and} \quad \delta_1 = \delta_2 = \delta_3 = \delta_4 = \delta$$

and because the number of fins per unit passage width is fixed for each stack, the cover and splitter plate heights s are equal:

$$s_1 = s_2 = s_3 = s_4 = s$$

However, the cover and separation plate thicknesses a_c and a_s may vary. All surfaces are presumed to be fabricated of the same material, having thermal conductivity k, and there is no bond resistance. The entire passage carries a fluid under a constant and uniform heat transfer coefficient h.

9.3 INPUT ADMITTANCE AND THE FIN EFFICIENCY

Evaluation of the input admittance and fin efficiency of the finned passages considered in this chapter involves a straightforward procedure that is based on:

- The matrix of the linear transformation (the **T**-matrix) given by eq. (5.16)
- The *reflection relationship*, given by eq. (7.6)
- The cluster algorithm, given by eq. (7.9)
- Conversion of the input admittance to the fin efficiency

[1]The horizontal members will be designated as cover and splitter plates even though a single cover or splitter plate may extend across and cover all of the repeating sections.

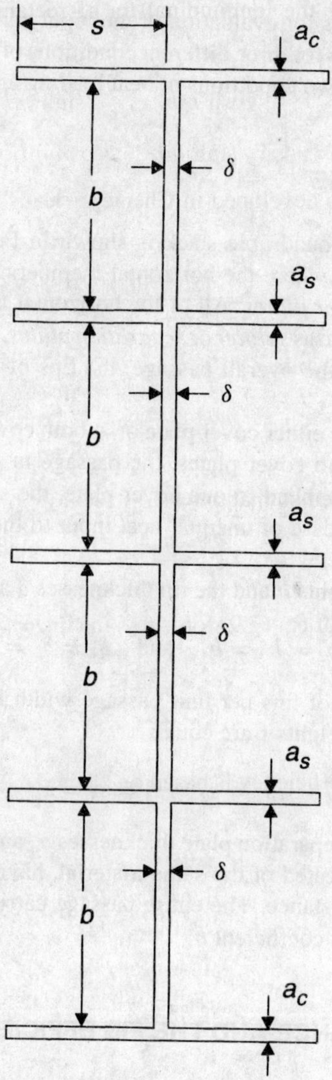

Figure 9.1 Configuration for the quadruple stack.

These relationships may be used to develop Y_{in} and η for cases of even loading, single-side loading, and unequal loading.

The matrix of the linear transformation is given by eq. (5.16) and is repeated here:

$$\begin{bmatrix} \theta_b \\ q_b \end{bmatrix} = \mathbf{T} \begin{bmatrix} \theta_a \\ q_a \end{bmatrix} = \begin{bmatrix} \tau_{11} & \tau_{12} \\ \tau_{21} & \tau_{22} \end{bmatrix} \begin{bmatrix} \theta_a \\ q_a \end{bmatrix} \tag{9.1}$$

where the parameters for the longitudinal fin of rectangular profile are given by eq. (6.9) and are repeated here:

$$\mathbf{T} = \begin{bmatrix} \cosh mb & \dfrac{1}{Y_o} \sinh mb \\ Y_o \sinh mb & \cosh mb \end{bmatrix} \tag{9.2}$$

The cascade algorithm developed in Chapter 7 leads to the following reflection relationship:

$$Y_{\text{in}} = \frac{\tau_{21} + (q_a/\theta_a)\tau_{22}}{\tau_{11} + (q_a/\theta_a)\tau_{12}} \tag{7.6}$$

If $Y_a = q_a/\theta_a$, eq. (7.6) becomes

$$Y_{\text{in}} = \frac{q_b}{\theta_b} = \frac{\tau_{21} + Y_a \tau_{22}}{\tau_{11} + Y_a \tau_{12}} \tag{9.3a}$$

and if the fin tip is insulated so that $q_a = 0$,

$$Y_{\text{in}} = \frac{q_b}{\theta_b} = \frac{\tau_{21}}{\tau_{11}} \tag{9.3b}$$

The cluster algorithm of eq. (7.9) for n-fins in cluster is

$$Y_C = \sum_{i=1}^{i=n} Y_{\text{in},i} \tag{9.4}$$

The conversion to fin efficiency is based on a comparison of the definitions of the fin efficiency for any fin:

$$\eta_f = \frac{q_b}{q_{\text{id}}} = \frac{q_b}{hS_f\theta_b}$$

and the input admittance

$$Y_{\text{in}} = \frac{q_b}{\theta_b}$$

or

$$q_b = Y_{\text{in}}\theta_b$$

Thus

$$\eta_f = \frac{q_b}{\theta_b} = \frac{Y_{\text{in}}\theta_b}{hS_f\theta_b}$$

or

$$\eta_f = \frac{Y_{\text{in}}}{hS_f} \tag{9.5}$$

9.4 OVERALL PASSAGE EFFICIENCY

Limitations of the fin efficiency were discussed in Section 5.2, where it was shown that one fin, at a lower fin efficiency, could dissipate more heat than another under identical operating conditions. It was also noted that the efficiency of a fin in a finned array depends on where the fin was mounted in the array. Although these points suggest that the fin efficiency might be abandoned, there is an application where the concept of fin efficiency is quite useful as a design parameter. This application is the overall passage efficiency used for finned passages.

Consider any finned passage where the total surface consists of a finned surface S_f and a base surfaces S_b,

$$S = S_f + S_b$$

Then presume that the entire surface operates at an overall passage efficiency, η_o, so that with the base surface operating with an efficiency of unity ($\eta_b = 1.0$) and the finned surface operating with an efficiency of η_f, the relationship for the $\eta_o S$ product is

$$\eta_o S = \eta_f S_f + (1) S_b$$

and with

$$S_b = S - S_f$$

then

$$\eta_o = \frac{\eta_f S_f + (1)(S - S_f)}{S}$$

or

$$\eta_o = 1 - \frac{S_f}{S}(1 - \eta_f) \tag{9.6}$$

Use of the overall passage efficiency permits the heat transfer equation to be written in terms of this efficiency, the heat transfer coefficient, the total surface, and the base temperature excess:

$$q = h \eta_o S \theta_b \tag{9.7}$$

9.5 SINGLE STACK

In the single stack, the vertical member is a fin designated by the subscript f and the horizontal surfaces are cover plates with thickness a_c. For cases of even or unequal loading, both cover plates are treated as base surfaces designated by the subscript b. For the case of single-side loading, one of the cover plates is designated by the subscript c and the other is the base surface with subscript b.

9.5.1 Even Loading

The evenly loaded single stack shown in Fig. 9.2*a* is a symmetrical configuration. In this case, half of the dissipation appears on each cover plate and the fin height is $b/2$. With

$$m_f = \left(\frac{2h}{k\delta}\right)^{1/2} \quad \text{and} \quad Y_{o,f} = (2hk\delta)^{1/2} L$$

and because the fin tip is adiabatic,

$$Y_{\text{in},f} = \frac{q_b}{\theta_b} = \frac{\tau_{22}}{\tau_{11}} = \frac{Y_{o,f} \cosh m_f b/2}{\sinh m_f b/2} = Y_{o,f} \tanh m_f b/2 \tag{9.8}$$

With the fin surface

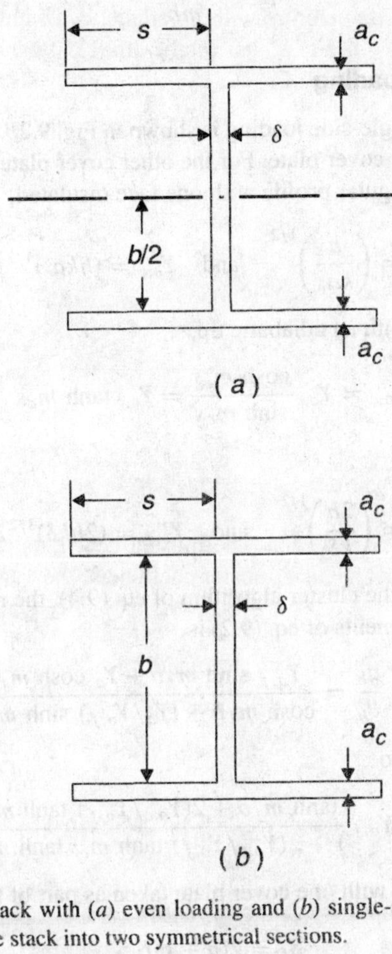

Figure 9.2 Single stack with (*a*) even loading and (*b*) single-side loading. In part (*a*), the dashed line divides the stack into two symmetrical sections.

$$S_f = 2\left(\frac{b}{2}\right)L = bL$$

eq. (9.5) is used to determine the fin efficency:

$$\eta_f = \frac{Y_{\text{in},f}}{hS_f} = \frac{Y_{o,f}}{bLh}\tanh m_f b/2 \tag{9.9}$$

Observe that if $Y_{o,f} = (2hk\delta)^{1/2}L$ is inserted into eq. (9.9), the result is

$$\eta_f = \frac{(2hk\delta)^{1/2}L \tanh m_f b/2}{2(b/2)Lh} = \frac{\tanh m_f b/2}{m_f b/2}$$

which is in the form

$$\eta = \frac{\tanh mb}{mb} \tag{1.11}$$

9.5.2 Single-Side Loading

The single stack with single-side loading is shown in Fig. 9.2*b*. In this case, all of the dissipation occurs at one cover plate. For the other cover plate, which is treated as a longitudinal fin of rectangular profile with one face insulated,

$$m_c = \left(\frac{h}{ka_c}\right)^{1/2} \quad \text{and} \quad Y_{o,c} = (hka_c)^{1/2}L$$

and for the cover plate with an adiabatic tip,

$$Y_{\text{in},c} = Y_{o,c}\frac{\cosh m_c s}{\sinh m_c s} = Y_{o,c}\tanh m_c s$$

For the fin

$$m_f = \left(\frac{2h}{k\delta}\right)^{1/2} \quad \text{and} \quad Y_{o,f} = (2hk\delta)^{1/2}L$$

and with $Y_a = 2Y_{\text{in},c}$ by the cluster algorithm of eq. (9.4), the reflection relationship of eq. (9.3*a*) with the elements of eq. (9.2) is

$$Y_{\text{in},f} = \frac{q_b}{\theta_b} = \frac{Y_{o,f}\sinh m_f b + Y_a \cosh m_f b}{\cosh m_f b + (Y_a/Y_{o,f})\sinh m_f b}$$

and this can be adjusted to

$$Y_{\text{in},f} = Y_{o,f}\frac{\tanh m_f b + 2(Y_{o,c}/Y_{o,f})\tanh m_c s}{1 + 2(Y_{o,c}/Y_{o,f})\tanh m_c s \tanh m_f b} \tag{9.10}$$

For this configuration, with one cover plate taken as part of the fin,

$$S_f = 2(b+s)L$$

so that by eq. (9.5),

$$\eta_f = \frac{Y_{o,f}}{2(b+s)Lh} \frac{\tanh mb + (2Y_{o,c}/Y_{o,f}) \tanh m_c s}{1 + (2Y_{o,c}/Y_{o,f}) \tanh m_c s \tanh mb} \tag{9.11}$$

Example 9.1. A 20-cm-long repeating section of a finned passage with 500 fins per meter is fabricated from an aluminum with thermal conductivity, $k = 200$ W/m · K. The fins are 0.0508 cm thick and the cover plate spacing (fin height) is 0.635 cm. If the cover plates are 0.0254 cm thick and the heat transfer coefficient is 50 W/m^2 · K, determine the cover plate temperature excess for **(a)** a heat dissipation of 2 W per section on each cover plate and **(b)** a heat dissipation of 4 W per section applied to one cover plate.

SOLUTION. **(a)** For the case of even loading, for the fin

$$m_f = \left(\frac{2h}{k\delta}\right)^{1/2} = \left[\frac{(2)(50)}{(200)(5.08 \times 10^{-4})}\right]^{1/2}$$

$$= (984.25)^{1/2} = 31.37 \text{ m}^{-1}$$

$$m_f b/2 = 31.37\left(\frac{0.00635}{2}\right) = 0.0996$$

and

$$Y_{o,f} = (2hk\delta)^{1/2} L$$

$$= [(2)(50)(200)(5.08 \times 10^{-4})]^{1/2}(0.20)$$

$$= (10.16)^{1/2}(0.20) = 0.6375 \text{ W/K}$$

Then

$$Y_{\text{in},f} = Y_{o,f} \tanh m_f b/2$$

$$= (0.6375) \tanh 0.0996$$

$$= (0.6375)(0.0993)$$

$$= 0.0633 \text{ W/K}$$

The surface area of each fin is

$$S_f = 2\left(\frac{b}{2}\right) L = bL = (6.35 \times 10^{-3})(0.20) = 1.270 \times 10^{-3} \text{ m}^2$$

so that the fin efficiency is

$$\eta_f = \frac{Y_{\text{in},f}}{hS_f} = \frac{0.0633}{(50)(1.270 \times 10^{-3})} = 0.997$$

The repeating section has the width

$$w = \frac{1}{500} = 0.002 \text{ m}$$

so that the cover plate height s is

$$s = \frac{w - \delta}{2} = \frac{0.002 - 5.08 \times 10^{-4}}{2} = 7.460 \times 10^{-4} \text{ m}$$

The base surface, which is the inside surface of a single cover plate, is

$$S_b = 2sL = 2(7.460 \times 10^{-4})(0.20) = 2.984 \times 10^{-4} \text{ m}^2$$

which makes the total surface

$$S = S_b + S_f = 2.984 \times 10^{-4} + 1.270 \times 10^{-3} = 1.568 \times 10^{-3} \text{ m}^2$$

With a heat flow of $q = 4/2 = 2$ W, the base temperature excess will be

$$\begin{aligned}
\theta_b &= \frac{q}{h(S_b + \eta S_f)} \\
&= \frac{2}{(50)[2.984 \times 10^{-4} + (0.997)(1.270 \times 10^{-3})]} \\
&= \frac{2}{(50)(1.565 \times 10^{-3})} = 25.6^\circ\text{C}
\end{aligned}$$

There are two alternative approaches that will provide the same result. The first is based on Y_{in} for the entire surface:

$$\begin{aligned}
Y_{\text{in,T}} &= Y_{\text{in},f} + Y_{\text{in},b} \\
&= Y_{\text{in},f} + hS_b \\
&= 0.0633 + (50)(2.984 \times 10^{-4}) \\
&= 0.0633 + 0.0149 = 0.0782 \text{ W/K}
\end{aligned}$$

and

$$\theta_b = \frac{q}{Y_{\text{in,T}}} = \frac{2}{0.0782} = 25.6^\circ\text{C}$$

The second alternative is based on the overall passage efficiency η_o. Equation (9.6) gives

$$\begin{aligned}
\eta_o &= 1 - \frac{S_f}{S}(1 - \eta_f) \\
&= 1 - \frac{1.270 \times 10^{-3}}{1.568 \times 10^{-3}}(1 - 0.997) \\
&= 1 - 0.002 \\
&= 0.998
\end{aligned}$$

Then, in accordance with eq. (9.5)

$$\theta_b = \frac{q}{h\eta_o S} = \frac{2}{(50)(0.998)(1.568 \times 10^{-3})} = 25.6°\text{C}$$

(**b**) For the case of single-side loading, the values of m_f and $Y_{o,f}$ remain the same as for the case of even loading.

$$m_f = 31.37 \text{ m}^{-1} \quad \text{and} \quad Y_{o,f} = 0.6375 \text{ W/°C}$$

but b is twice as large:

$$b = 6.35 \times 10^{-3} \text{ m}^{-1}$$

Thus

$$m_f b = (31.37)(6.35 \times 10^{-3}) = 0.1992$$

and

$$\tanh m_f b = 0.1966$$

For the cover plates,

$$m_c = \left(\frac{h}{ka_c}\right)^{1/2} = \left[\frac{50}{(200)(2.540 \times 10^{-4})}\right]^{1/2}$$

$$= (984.25)^{1/2} = 31.37 \text{ m}^{-1}$$

$$m_c s = (31.37)(7.460 \times 10^{-4}) = 0.0234$$

and

$$\tanh m_c s = \tanh 0.0234 = 0.0234$$

The characteristic admittance of the cover plates is

$$Y_{o,c} = (hka_c)^{1/2} L = [(50)(200)(2.540 \times 10^{-4})]^{1/2}(0.20)$$

$$= (2.540)^{1/2}(0.20) = 0.3187 \text{ W/°C}$$

so that

$$2\frac{Y_{o,c}}{Y_{o,f}} = 2\left(\frac{0.3187}{0.6375}\right) = 1.0000$$

The input admittance will be in the form of eq. (9.10):

$$Y_{\text{in},f} = Y_{o,f}\frac{\tanh m_f b + 2(Y_{o,c}/Y_{o,f}) \tanh m_c s}{1 + 2(Y_{o,c}/Y_{o,f}) \tanh m_c s \tanh m_f b}$$

$$= 0.6375 \left[\frac{0.1966 + (1.0000)(0.0234)}{1.0000 + (1.0000)(0.0234)(0.1966)} \right]$$
$$= 0.6375 \left(\frac{0.2200}{1.0046} \right)$$
$$= 0.1396 \text{ W/°C}$$

The finned surface is now composed of the fin and the inside surface of one cover plate:

$$S_f = 2bL + 2sL = 2(b + s)L$$
$$= (2)(6.35 \times 10^{-3} + 7.46 \times 10^{-4})(0.20)$$
$$= (2)(7.096 \times 10^{-3})(0.20)$$
$$= 2.838 \times 10^{-3} \text{ m}^2$$

The base surface is the surface of the other cover plate

$$S_b = 2sL = (2)(7.46 \times 10^{-4})(0.20) = 2.984 \times 10^{-4} \text{ m}^2$$

and the total surface is

$$S = S_f + S_b$$
$$= 2.838 \times 10^{-3} + 2.984 \times 10^{-4}$$
$$= 3.138 \times 10^{-3} \text{ m}^2$$

The fin efficiency [eq. (9.5)] is

$$\eta_f = \frac{Y_{\text{in}}}{hS_f} = \frac{0.1396}{(50)(2.838 \times 10^{-3})} = 0.984$$

and with $q = 4$ W (single-side loading), the base temperature excess will be

$$\theta_b = \frac{q}{h(S_b + \eta S_f)}$$
$$= \frac{4}{(50)[2.984 \times 10^{-4} + (0.984)(2.838 \times 10^{-3})]}$$
$$= \frac{4}{(50)(3.091 \times 10^{-3})} = 25.9\text{°C}$$

The approach based on the input admittance for the entire surface gives

$$Y_{\text{in,T}} = Y_{\text{in},f} + Y_{\text{in},b}$$
$$= Y_{\text{in},f} + hS_b$$

$$= 0.1396 + (50)(2.984 \times 10^{-4})$$
$$= 0.1396 + 0.0149 = 0.1545 \text{ W/K}$$

and

$$\theta_b = \frac{q}{Y_{\text{in}}} = \frac{4}{0.1545} = 25.9°\text{C}$$

The approach based on the overall passage efficiency η_o gives

$$\eta_o = 1 - \frac{S_f}{S}(1 - \eta_f)$$
$$= 1 - \frac{2.838 \times 10^{-3}}{3.138 \times 10^{-3}}(1 - 0.984)$$
$$= 1 - 0.014$$
$$= 0.986$$

Then, in accordance with eq. (9.7),

$$\theta_b = \frac{q}{h\eta_o S} = \frac{4}{(50)(0.986)(3.138 \times 10^{-3})} = 25.9°\text{C}$$

The solution using all of the available alternatives match for both cases.

9.5.3 Unequal Loading

Pieper and Kraus (1995a) have shown that to allow for unequal temperature excesses and heat flows at opposite ends of the fin, the single fin of total height b has been subdivided into two fins with fin heights b_1 and b_2 in Fig. 9.3*a*. The fin heights b_1 and b_2 have been chosen such that no heat crosses the interface between fins b_1 and b_2. Figure 9.3b shows this interface with the orientation of the height coordinates x_1 and x_2, which have a positive orientation from the fin tips at the interface toward the fin bases. In what follows, the subscript f will not be included.

Each of the two fins in Fig. 9.3*a* has its own **T**-matrix

$$\begin{bmatrix} \theta_{b1} \\ q_{b1} \end{bmatrix} = \begin{bmatrix} \cosh mb_1 & \dfrac{1}{Y_o} \sinh mb_1 \\ Y_o \sinh mb_1 & \cosh mb_1 \end{bmatrix} \begin{bmatrix} \theta_{a1} \\ q_{a1} \end{bmatrix} \tag{9.12a}$$

and

$$\begin{bmatrix} \theta_{b2} \\ q_{b2} \end{bmatrix} = \begin{bmatrix} \cosh mb_2 & \dfrac{1}{Y_o} \sinh mb_2 \\ Y_o \sinh mb_2 & \cosh mb_2 \end{bmatrix} \begin{bmatrix} \theta_{a2} \\ q_{a2} \end{bmatrix} \tag{9.12b}$$

and the fins are subject to the continuity and compatablity conditions shown in Fig. 9.3*b*:

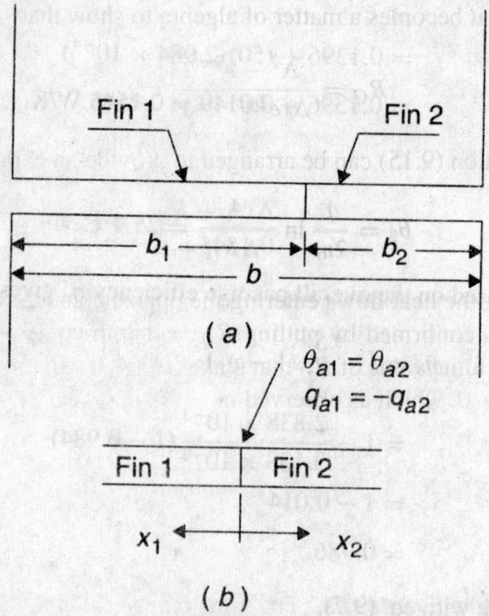

Figure 9.3 Single stack with unequal heat loading. The unequal heat loading causes an imbalance in the fin heights (*a*) the two fins with cover plates and (*b*) the interface showing the coordinate system and continuity and compatability conditions.

$$\begin{bmatrix} \theta_{a1} \\ q_{a1} \end{bmatrix} = \begin{bmatrix} \theta_{a2} \\ -q_{a2} \end{bmatrix} = \begin{bmatrix} \theta_a \\ 0 \end{bmatrix} \tag{9.13}$$

Observe that when either $q_{b1} \neq q_{b2}$ or $\theta_{b1} \neq \theta_{b2}$, symmetry will not apply and the fin heights b_1 and b_2 will not be equal ($b_1 \neq b_2$). Neither q_{b1} nor q_{b2} involve q_{a1} or q_{a2} because $q_{a1} = -q_{a2} = 0$. With $q_{a1} = q_{a2} = 0$, eqs. (9.12*a*) and (9.12*b*) can be expanded:

$$q_{b1} = [(Y_o \sinh mb_1)]\theta_a \tag{9.14a}$$

$$q_{b2} = [(Y_o \sinh mb_2)]\theta_a \tag{9.14b}$$

and

$$\theta_{b1} = (\cosh mb_1)\theta_a \tag{9.14c}$$

$$\theta_{b2} = (\cosh mb_2)\theta_a \tag{9.14d}$$

Unequal Heat Inputs. Equations (9.14*a*) and (9.14*b*) show that the heat flows are related through θ_a so that a heat flow ratio R_Q can be proposed:

$$R_Q = \frac{q_{b1}}{q_{b2}} = \frac{Y_o \sinh mb_1}{Y_o \sinh mb_2} = \frac{e^{mb_1} - e^{-mb_1}}{e^{mb_2} - e^{-mb_2}}$$

and with $b_2 = b - b_1$, it becomes a matter of algebra to show that

$$R_Q = \frac{\Lambda^2 - e^{2mb_2}}{\Lambda(e^{2mb_2} - 1)} \tag{9.15}$$

where $\Lambda \equiv e^{mb}$. Equation (9.15) can be arranged to provide an expression for b_2:

$$b_2 = \frac{1}{2m} \ln \frac{\Lambda(\Lambda + R_Q)}{\Lambda R_Q + 1} \tag{9.16}$$

Notice that if $R_Q = 1$, the heat flows entering the opposite ends of the fin are equal and $b_2 = b/2$. This is confirmed by putting $R_Q = 1$ into eq. (9.16). However, an effort to determine the finite value of R_Q that makes $b_2 \to 0$ will prove futile because this implies that $q_{b2} \to 0$, which as observed in

$$R_Q = \frac{q_{b1}}{q_{b2}} = \frac{Y_o \sinh mb_1}{Y_o \sinh mb_2}$$

causes $R_Q \to \infty$.

Unequal Temperature Excesses. The same type of thinking goes with the case where $\theta_{b1} \neq \theta_{b2}$. Here too, symmetry will not apply because the fin heights b_1 and b_2 will not be equal ($b_1 \neq b_2$). The strategy here is to form $R_\theta = \theta_{b1}/\theta_{b2}$ and to recognize that R_θ will not equal R_Q. This time, eqs. (9.14) provide

$$R_\theta = \frac{\theta_{b1}}{\theta_{b2}} = \frac{\cosh mb_1}{\cosh mb_2}$$

The hyperbolic cosines can be represented as exponentials

$$R_\theta = \frac{e^{mb_1} + e^{-mb_1}}{e^{mb_2} + e^{-mb_2}}$$

and with $b_2 = b - b_1$, it becomes a matter of algebra to show that

$$R_\theta = \frac{\Lambda^2 + e^{2mb_2}}{\Lambda(e^{2mb_2} + 1)} \tag{9.17}$$

The value of b_2 for a given R_θ can be found by a rearrangment of eq. (9.17) so that

$$b_2 = \frac{1}{2m} \ln \frac{\Lambda(\Lambda - R_\theta)}{\Lambda R_\theta - 1} \tag{9.18}$$

If $R_\theta = 1$ because $\theta_{b1} = \theta_{b2}$, then because of symmetry, b_2 must equal half the total fin height, This is confirmed by putting $R_\theta = 1$ into eq. (9.18). Moreover, the value of R_θ for $b_2 = 0$ can be found directly from eq. (9.17). For $b_2 = 0$,

$$R_\theta = \frac{\Lambda^2 + 1}{2\Lambda} \tag{9.19a}$$

which corresponds to $R_Q = \infty$, and for $b_2 = b$,

$$R_\theta = \frac{2\Lambda}{\Lambda^2 + 1} \tag{9.19b}$$

which corresponds to $R_Q = 0$.

Interdependence of the Heat Flow and Temperature Excess Ratios. It remains to show that R_Q and R_θ are not independent and can be related in a very general way via the linear transformation between the base of fin 1 and the base of fin 2 given by eq. (9.1). Once again, refer to Fig. 9.3*a* and note that q_{b2} flows *into* the fin. Thus

$$\begin{bmatrix} \theta_{b1} \\ q_{b1} \end{bmatrix} = \begin{bmatrix} \tau_{11} & \tau_{12} \\ \tau_{21} & \tau_{22} \end{bmatrix} \begin{bmatrix} \theta_{b2} \\ -q_{b2} \end{bmatrix}$$

Expansion gives

$$\theta_{b1} = \tau_{11}\theta_{b2} - \tau_{12}q_{b2}$$

or

$$R_\theta = \frac{\theta_{b1}}{\theta_{b2}} = \tau_{11} - \tau_{12}Y_{b2} \tag{a}$$

where

$$Y_{b2} = \frac{q_{b2}}{\theta_{b2}}$$

In similar fashion,

$$q_{b1} = \tau_{21}\theta_{b2} - \tau_{22}q_{b2}$$

or

$$R_Q = \frac{q_{b1}}{q_{b2}} = \frac{\tau_{21}}{Y_{b2}} - \tau_{22} \tag{b}$$

Equations (a) and (b) can be rewritten to provide expressions for Y_{b2}:

$$Y_{b2} = \frac{\tau_{11} - R_\theta}{\tau_{12}}$$

and

$$Y_{b2} = \frac{\tau_{21}}{R_Q + \tau_{22}}$$

and when these are equated, an expression for R_θ in terms of R_Q results:

$$R_\theta = \tau_{11} - \frac{\tau_{12}\tau_{21}}{R_Q + \tau_{22}}$$

With the τ-parameters given in eq. (9.2), this becomes

$$R_\theta = \cosh mb - \frac{\sinh^2 mb}{R_Q + \cosh mb} \tag{9.20}$$

Example 9.2. A finned passage with 400 fins per meter is used in a forced-convection application with a heat transfer coefficient of $h = 400\ \text{W/m}^2 \cdot \text{K}$ and a thermal conductivity of $k = 160\ \text{W/m} \cdot \text{K}$. The fins are 1.27 cm high and 0.2286 cm thick and the finned passage is 25 cm long. Determine the heat flow from each cover plate if they are held at temperature excesses of 28°C and 25°C, respectively.

SOLUTION. This example pertains only to fins and may be presented without the use of the subscript f. For the fins with $b = 0.0127$ m and $\delta = 2.286 \times 10^{-3}$ m,

$$m = \left(\frac{2h}{k\delta}\right)^{1/2} = \left[\frac{(2)(400)}{(160)(2.286 \times 10^{-3})}\right]^{1/2}$$

$$= (2187.23)^{1/2} = 46.77\ \text{m}^{-1}$$

$$mb = (46.77)(0.0127) = 0.5940$$

and

$$Y_o = (2hk\delta)^{1/2} L$$

$$= [(2)(400)(160)(2.286 \times 10^{-3})]^{1/2}(0.25)$$

$$= (292.61)^{1/2}(0.25)$$

$$= 4.2764\ \text{W/°C}$$

With fin 2 on the 25°C side,

$$R_\theta = \frac{28}{25} = 1.1200$$

and

$$\Lambda = e^{mb}$$

$$= e^{0.5940}$$

$$= 1.8111$$

Equation (9.18) can be used to find b_2. With $m = 46.77\ \text{m}^{-1}$,

$$b_2 = \frac{1}{2m} \ln \frac{\Lambda(\Lambda - R_\theta)}{\Lambda R_\theta - 1}$$

$$= \frac{1}{2(46.77)} \ln \frac{(1.8111)(1.8111 - 1.1200)}{(1.8111)(1.1200) - 1.0000}$$

$$= 0.0107 \ln \frac{1.2517}{1.0285}$$

$$= 0.0107 \ln 1.2171$$

$$= (0.0107)(0.1965)$$

$$= 2.100 \times 10^{-3} \text{ m}$$

Then

$$mb_2 = (46.77)(2.100 \times 10^{-3}) = 0.0982$$

and

$$\begin{aligned} Y_{\text{in},2} &= Y_o \tanh mb_2 \\ &= 4.2764 \tanh 0.0982 \\ &= (4.2764)(0.0979) \\ &= 0.4187 \text{ W/K} \end{aligned}$$

Neither the finned surface area nor the fin efficiency is needed.

$$\begin{aligned} q_{b2} &= Y_{\text{in},2}\theta_{b2} \\ &= (0.4187)(25) \\ &= 10.4679 \text{ W} \quad \text{(say, 10.47 W)} \end{aligned}$$

With

$$w = \frac{1}{400} = 2.500 \times 10^{-3} \text{ m}$$

the cover plate height s is

$$s = \frac{w - \delta}{2} = \frac{2.500 \times 10^{-3} - 2.286 \times 10^{-3}}{2} = 1.070 \times 10^{-4} \text{ m}$$

and the base surface, which is the inside surface of the cover plate, is

$$S_b = 2sL = (2)(1.070 \times 10^{-4})(0.25) = 5.350 \times 10^{-5} \text{ m}^2$$

The heat dissipation by the cover plate acting as the base surface is

$$\begin{aligned} q_{c2} &= hS_b\theta_{b2} \\ &= (400)(5.350 \times 10^{-5})(25) \\ &= 0.5350 \text{ W} \end{aligned}$$

The total heat dissipated by the side at 25°C is

$$q = q_{b2} + q_{c2}$$

$$= 10.4679 + 0.5350$$

$$= 11.0029 \text{ W} \quad (\text{say, } 11.00 \text{ W})$$

For b_1, which is on the 28°C side,

$$b_1 = b - b_2$$

$$= 1.27 \times 10^{-2} - 2.100 \times 10^{-3}$$

$$= 0.0106 \text{ m}$$

Then

$$mb_1 = (46.77)(0.0106) = 0.4958$$

and

$$Y_{\text{in},1} = Y_o \tanh mb_1$$

$$= 4.2764 \tanh 0.4958$$

$$= (4.2764)(0.4588)$$

$$= 1.9619 \text{ W/K}$$

The heat dissipated by the fin is

$$q_{b1} = Y_{\text{in},1}\theta_{b1}$$

$$= (1.9619)(28)$$

$$= 54.9337 \text{ W} \quad (\text{say, } 54.93 \text{ W})$$

and with the cover plate surface the same as on the 25°C side, the heat dissipation by the cover plate on the 28°C side will be

$$q_{c2} = hS_b\theta_{b1}$$

$$= (400)(5.350 \times 10^{-5})(28)$$

$$= 0.5992 \text{ W}$$

The total heat dissipated by the side at 25°C is

$$q = q_{b2} + q_{c2}$$

$$= 54.9337 + 0.5992 \text{ W}$$

$$= 55.5329 \text{ W} \quad (\text{say, } 55.53 \text{ W})$$

In summary,

$$q_{b1} = 55.53 \text{ W}$$

and

$$q_{b2} = 11.00 \text{ W}$$

Remarks:

1. It is observed that a temperature excess difference of only 28°C − 25°C = 3°C is almost sufficient for side 1 to overpower side 2 and push the heat from the cover plate at side 1 across the fin height and into side 2.
2. Note that

$$R_Q = \frac{q_{b1}}{q_{b2}} = \frac{54.9337}{10.4679} = 5.2478 \quad \text{(say, 5.25)}$$

This can be confirmed by eq. (9.20):

$$R_\theta = \cosh mb - \frac{\sinh^2 mb}{R_Q + \cosh mb}$$

and with $mb = 0.5940$ and $R_Q = 5.2478$,

$$= \cosh 0.5940 - \frac{\sinh^2 0.5940}{5.2478 + \cosh 0.5940}$$

$$= 1.1817 - \frac{(0.6296)^2}{5.2478 + 1.1817}$$

$$= 1.1817 - 0.0617$$

$$= 1.1200$$

which checks the previous result.

9.6 DOUBLE STACK

9.6.1 Even Loading

The configuration for the double stack with even loading is shown in Fig. 9.4*a*, which bears a striking resemblance to Fig. 9.2*b*, which pertains to the single stack with single-side loading. However, in Fig. 9.4*a*, the inner horizontal members are splitter plates with thickness a_s and are longitudinal fins of rectangular profile with one face insulated.

For the fins, m_f and $Y_{o,f}$ are the same as for the single stack:

$$m_f = \left(\frac{2h}{k\delta}\right)^{1/2} \quad \text{and} \quad Y_{o,f} = (2hk\delta)^{1/2} L$$

and for the splitter plates, longitudinal fins of rectangular profile with one face insulated and thickness $a_s/2$,

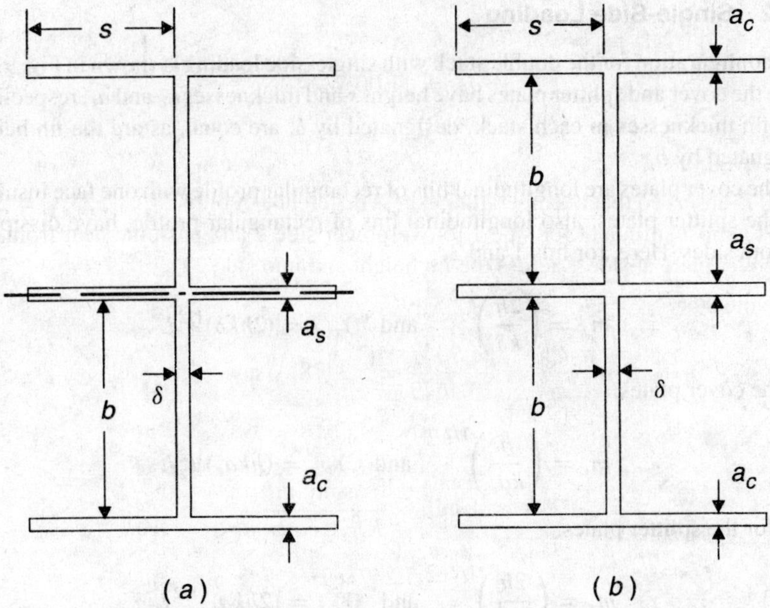

Figure 9.4 Double stack (*a*) with even loading and (*b*) with single-side loading. In part (*a*), the dashed line divides the stack into two symmetrical sections.

$$m_s = \left[\frac{h}{k(a_s/2)}\right]^{1/2} = \left(\frac{2h}{ka_s}\right)^{1/2} \quad \text{and} \quad Y_{o,s} = \left(\frac{hka_s}{2}\right)^{1/2} L$$

Then, because the tips of the splitter plates are insulated, $Y_{\text{in},s} = Y_{o,s} \tanh m_s s$, and with Y_a for the fin equal to

$$Y_a = 2Y_{o,s} \tanh m_s s$$

the input admittance for the fin will be in the form of eq. (9.10):

$$Y_{\text{in},f} = Y_{o,f} \frac{\tanh m_f b + 2(Y_{o,s}/Y_{o,f}) \tanh m_s s}{1 + 2(Y_{o,s}/Y_{o,f}) \tanh m_s s \tanh m_f b} \tag{9.21}$$

For this configuration, the surface area of both the fin and the cover plates is the finned surface

$$S_f = 2(b+s)L$$

so that by eq. (9.5), the fin efficiency is

$$\eta_f = \frac{Y_{o,f}}{2(b+s)Lh} \frac{\tanh m_f b + (2Y_{o,s}/Y_{o,f}) \tanh m_s s}{1 + (2Y_{o,s}/Y_{o,f}) \tanh m_s s \tanh m_f b} \tag{9.22}$$

9.6.2 Single-Side Loading

The configuration for the double stack with single-side loading is shown in Fig. 9.4[b]. Here the cover and splitter plates have height s and thicknesses a_c and a_s, respectively. The fin thicknesses in each stack, designated by δ, are equal, as are the fin heights, designated by b.

The cover plates are longitudinal fins of rectangular profile with one face insulated, but the splitter plates, also longitudinal fins of rectangular profile, have dissipation on both sides. Here, for fins 1 and 2,

$$m_f = \left(\frac{2h}{k\delta}\right)^{1/2} \quad \text{and} \quad Y_{o,f} = (2hk\delta)^{1/2}L$$

for the cover plates,

$$m_c = \left(\frac{h}{ka_c}\right)^{1/2} \quad \text{and} \quad Y_{o,c} = (hka_c)^{1/2}L$$

and for the splitter plates.

$$m_s = \left(\frac{2h}{ka_s}\right)^{1/2} \quad \text{and} \quad Y_{o,s} = (2hka_s)^{1/2}L$$

Begin with the cover plates with an insulated tip. In accordance with eq. (9.3b),

$$Y_{\text{in},c} = Y_{o,c}\frac{\cosh m_c s}{\sinh m_c s} = Y_{o,c} \tanh m_c s$$

Then, at the tip of fin 2,

$$Y_{a2} = 2Y_{o,c} \tanh m_c s$$

and for fin 2 in the form of eq. (9.10),

$$Y_{\text{in},f2} = Y_{o,f2}\frac{\tanh m_f b + 2(Y_{o,c}/Y_{o,f}) \tanh m_c s}{1 + 2(Y_{o,c}/Y_{o,f}) \tanh m_c s \tanh m_f b}$$

For the splitter plates, also with insulated tips, in accordance with eq. (9.3b),

$$Y_{\text{in},s} = Y_{o,s}\frac{\cosh m_s s}{\sinh m_s s} = Y_{o,s} \tanh m_s s$$

The cluster admittance at the tip of fin 1 will be

$$Y_{a1} = 2Y_{\text{in},s} + Y_{\text{in},f2} = 2Y_{o,s} \tanh m_s s + Y_{\text{in},f2}$$

and in the form of eq. (9.10),

$$Y_{\text{in},f1} = Y_{o,f}\frac{\tanh m_f b + \Psi_1}{1 + \Psi_1 \tanh m_f b} \tag{9.23}$$

where

$$\Psi_1 = \frac{Y_{a1}}{Y_{o,f}}$$

The entity of fins 1 and 2, the cover plates, and the splitter plates comprise the fin. The surface area is

$$S_f = 4bL + 4sL + 2sL = 2(2b + 3s)L$$

and the efficiency [eq. (9.5)] is

$$\eta_f = \frac{Y_{o,f}}{2(2b + 3s)Lh} \frac{\tanh m_f b + \Psi_1}{1 + \Psi_1 \tanh m_f b} \tag{9.24}$$

Example 9.3. Two passages, identical to the passage used in Example 9.1, are to be used to form a double stack. The pair of cover plates at the outside of the double stack are 0.0254 cm thick as in Example 9.1, but the splitter plates between the two passages are 0.0204 cm thick. The passage has a length of 20 cm, the thermal conductivity of all material is $k = 200$ W/m · K and the heat transfer coefficient is 50 W/m^2 · K. Determine the temperature excess of the cover plates for (**a**) a heat dissipation of 4 W per section on each pair of cover plates and (**b**) a heat dissipation of 8 W per section applied to one pair of cover plates.

SOLUTION. From Example 9.1 with $L = 0.20$ m:

Fin	*Cover Plate*
$m_f = 31.37\ \text{m}^{-1}$	$m_c = 31.37\ \text{m}^{-1}$
$b = 6.350 \times 10^{-3}$ m	$s_c = 7.460 \times 10^{-4}$ m
$m_f b = 0.1992$	$m_c s = 0.0234$
$Y_{o,f} = 0.6375$ W/K	$Y_{o,c} = 0.3187$ W/K

(**a**) For the case of even loading (Fig. 9.4*a*), $s = 7.460 \times 10^{-4}$ m and $a_s = 2.040 \times 10^{-4}$ m. With the splitter plates treated as longitudinal fins of rectangular profile with one face insulated,

$$m_s = \left[\frac{h}{k(a_s/2)}\right]^{1/2} = \left[\frac{(2)(50)}{(200)(2.04 \times 10^{-4})}\right]^{1/2}$$

$$= (2450.98)^{1/2} = 49.51\ \text{m}^{-1}$$

$$m_s s = (49.51)(7.460 \times 10^{-4}) = 0.0369$$

$$\tanh m_s s = \tanh 0.0369 = 0.0369$$

and

$$Y_{o,s} = \left(\frac{hka_s}{2}\right)^{1/2} L$$

$$= \left[\frac{(50)(200)(2.04 \times 10^{-4})}{2}\right]^{1/2} (0.20)$$

$$= (1.0200)^{1/2}(0.20) = 0.2020 \text{ W/K}$$

The surface of each fin (fin and inside surface of the splitter plates) is

$$S_f = 2(b+s)L$$
$$= (2)(6.35 \times 10^{-3} + 7.460 \times 10^{-4})(0.20)$$
$$= (2)(7.096 \times 10^{-3})(0.20) = 2.838 \times 10^{-3} \text{ m}^2$$

The fin efficiency will be, in accordance with eq. (9.22),

$$\eta_f = \frac{Y_{o,f}}{2(b+s)Lh} \frac{\tanh m_f b + (2Y_{o,s}/Y_{o,f}) \tanh m_s s}{1 + (2Y_{o,s}/Y_{o,f}) \tanh m_s s \tanh m_f b}$$

$$= \frac{0.6375}{(2.838 \times 10^{-3})(50)} \left[\frac{0.1966 + (0.6337)(0.0369)}{1.0000 + (0.6337)(0.0369)(0.1966)}\right]$$

$$= 4.4926 \left(\frac{0.2200}{1.0046}\right)$$

$$= 0.984$$

The base surface is the surface of the cover plates, which is equal to

$$S_b = 2sL = (2)(7.460 \times 10^{-4})(0.20) = 2.984 \times 10^{-4} \text{ m}^2$$

and with a 4-W dissipation,

$$\theta_b = \frac{q}{h(S_b + \eta S_f)}$$

$$= \frac{4}{(50)[2.984 \times 10^{-4} + (0.984)(2.838 \times 10^{-3})]}$$

$$= \frac{4}{(50)(3.091 \times 10^{-3})} = 25.9°\text{C}$$

(b) For the case of single-side loading (Fig. 9.4*b*),

$$m_f = 31.37 \text{ m}^{-1} \quad \text{and} \quad Y_{o,f} = 0.6375 \text{ W/K}$$

as in part (a) but for the cover plates with $a_c = 2.540 \times 10^{-4}$ m^{-1}:

$$m_c = \left(\frac{h}{ka_c}\right)^{1/2} = \left[\frac{50}{(200)(2.540 \times 10^{-4})}\right]^{1/2}$$

$$= (984.25)^{1/2} = 31.37 \text{ m}^{-1}$$

$$Y_{o,c} = (hka_c)^{1/2}L = [(50)(200)(2.540 \times 10^{-4})]^{1/2}(0.20)$$
$$= (2.540)^{1/2}(0.20) = 0.3187 \text{ W/K}$$

and for the splitter plates with $a_s = 2.040 \times 10^{-4}$,

$$m_s = \left(\frac{2h}{ka_s}\right)^{1/2} = \left[\frac{(2)(50)}{(200)(2.040 \times 10^{-4})}\right]^{1/2}$$
$$= (2450.98)^{1/2} = 49.51 \text{ m}^{-1}$$
$$Y_{o,s} = (2hka_c)^{1/2}L = [(2)(50)(200)(2.040 \times 10^{-4})]^{1/2}(0.20)$$
$$= (4.080)^{1/2}(0.20) = 0.4040 \text{ W/K}$$

With

$$2\frac{Y_{o,c}}{Y_{o,f}} = 2\left(\frac{0.3187}{0.6375}\right) = 1.0000$$

and

$$m_f b = (31.37)(6.350 \times 10^{-3}) = 0.1992$$
$$\tanh m_f b = \tanh 0.1992 = 0.1966$$
$$m_c s = (31.37)(7.460 \times 10^{-4}) = 0.0234$$
$$\tanh m_c s = \tanh 0.0234 = 0.0234$$

the input admittance of fin 2 is, in accordance with eq. (9.10),

$$Y_{\text{in},f2} = Y_{o,f}\frac{\tanh m_f b + 2(Y_{o,c}/Y_{o,f}) \tanh m_c s}{1 + 2(Y_{o,c}/Y_{o,f}) \tanh m_c s \tanh m_f b}$$
$$= 0.6375\left[\frac{0.1966 + (1.0000)(0.0234)}{1.0000 + (1.0000)(0.0234)(0.1996)}\right]$$
$$= (0.6375)(0.2190) = 0.1396 \text{ W/K}$$

For the splitter plates,

$$m_s s = (49.51)(7.460 \times 10^{-4}) = 0.0369$$

and

$$\tanh m_s s = \tanh 0.0369 = 0.0369$$

The input admittance of the splitter plates is in accordance with eq. (9.3*b*):

$$Y_{\text{in},s} = Y_{o,s} \tanh m_s s$$
$$= (0.4040)(0.0369)$$
$$= 0.0149 \text{ W/K}$$

Then

$$
\begin{aligned}
Y_{a1} &= 2Y_{\text{in},s} + Y_{\text{in},f2} \\
&= 2(0.0149) + 0.1396 \\
&= 0.1694 \text{ W/K}
\end{aligned}
$$

and

$$\Psi_1 = \frac{Y_{a1}}{Y_{o,f}} = \frac{0.1694}{0.6375} = 0.2658 \text{ W/K}$$

The total finned surface is the surface of both sides of two fins, two sides of the splitter plates and two sides of the cover plates:

$$
\begin{aligned}
S_f &= 4bL + 4sL + 2sL \\
&= 2(2b + 3s)L \\
&= 2[(2)(6.350 \times 10^{-3}) + (3)(7.460 \times 10^{-4})](0.20) \\
&= (2)(14.938 \times 10^{-3})(0.20) \\
&= 5.9752 \times 10^{-3} \text{ m}^2
\end{aligned}
$$

and the fin efficiency is obtained from eq. (9.24):

$$
\begin{aligned}
\eta_f &= \frac{Y_{o,f}}{2(2b + 3s)Lh} \frac{\tanh m_f b + \Psi_1}{1 + \Psi_1 \tanh m_f b} \\
&= \frac{0.6375}{(5.9752 \times 10^{-3})(50)} \left[\frac{0.1966 + 0.2658}{1.0000 + (0.2658)(0.1966)} \right] \\
&= (2.1338)(0.4394) = 0.938
\end{aligned}
$$

With the same base surface as in part (a)

$$S_b = 2.984 \times 10^{-4} \text{ m}^2$$

and with an 8-W dissipation,

$$
\begin{aligned}
\theta_b &= \frac{q}{h(S_b + \eta S_f)} \\
&= \frac{8}{(50)[2.984 \times 10^{-4} + (0.938)(5.9752 \times 10^{-3})]} \\
&= \frac{8}{(50)(5.903 \times 10^{-3})} = 27.1^\circ\text{C}
\end{aligned}
$$

9.6.3 Unequal Loading

Pieper and Kraus (1995b) have also investigated the performance of the double stack with unequal heat loading. A repeating section of the double stack is shown in

Fig. 9.5a, where the two fins with heat input at their bases are designated as fin l and fin r. While there may be conditions when an imbalance in heat flows or temperature excesses are such to preclude an adiabatic point in either fin l or fin r, in this section the cases where $q_{bl} > q_{br}$ and/or $\theta_{bl} > \theta_{br}$ does indeed produce an adiabatic point in fin r are considered.

The fin at the right, fin r, of total height, b, is subdivided into two fins with fin heights b_1 and b_2, to allow for $q_{bl} > q_{br}$ or $\theta_{bl} > \theta_{br}$. This point of subdivision is displayed in Fig. 9.5c, where it may be observed that the origins of the fin height coordinates, b_1 and b_2, are at the adiabatic point with positive orientation from fin tip to fin base. The temperature excess at this adiabatic point is designated as θ_i.

At the intersection point in Fig. 9.5a which is detailed in Fig. 9.5b, continuity and compatability dictate that

$$q_{al} = -q_{ar} + 2q_{bs}$$

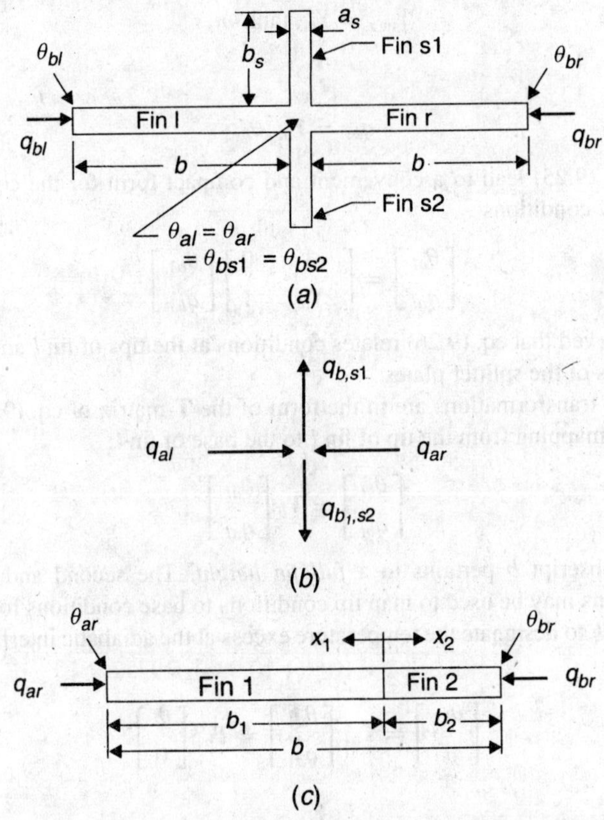

Figure 9.5 (a) Double stack with unequal heat loading; (b) detail of the fin intersection points; (c) right-hand fin, which has been decomposed into two subfins.

and

$$\theta_{al} = \theta_{ar} = \theta_{bs}$$

However, because $q_{ar} = -q_{b1}$ and $\theta_{ar} = \theta_{b1}$, these .conditions may be written as

$$q_{al} = q_{b1} + 2q_{bs} \tag{9.25a}$$

and

$$\theta_{al} = \theta_{b1} = \theta_{br} \tag{9.25b}$$

The splitter plates are longitudinal fins of rectangular profile so that

$$m_s = \left(\frac{2h}{ka_s}\right)^{1/2} \quad \text{and} \quad Y_{o,s} = (2hka_s)^{1/2}L$$

and in accordance with eq. (9.3*b*),

$$Y_{\text{in},s} = Y_{o,s} \tanh m_s s \tag{9.25c}$$

Moreover,

$$q_{bs} = Y_{\text{in},s}\theta_{bs} \tag{9.25d}$$

Equations (9.25) lead to a convenient and compact form for the continuity and compatability conditions

$$\begin{bmatrix} \theta_{al} \\ q_{al} \end{bmatrix} = \begin{bmatrix} 1 & 0 \\ 2Y_{\text{in},s} & 1 \end{bmatrix} \begin{bmatrix} \theta_{b1} \\ q_{b1} \end{bmatrix} \tag{9.26}$$

and it is observed that eq. (9.26) relates conditions at the tips of fin l and fin r at the opposite sides of the splitter plates.

The linear transformations are in the form of the **T**-matrix of eq. (9.1). The first concerns the mapping from the tip of fin l to the base of fin l:

$$\begin{bmatrix} \theta_{bl} \\ q_{bl} \end{bmatrix} = \mathbf{T}_b \begin{bmatrix} \theta_{al} \\ q_{al} \end{bmatrix} \tag{9.27a}$$

where the subscript b pertains to a *full fin height.* The second and third linear transformations may be used to map tip conditions to base conditions for both fins 1 and 2. Using θ_i to designate the temperature excess at the adiabatic interface between fins 1 and 2,

$$\begin{bmatrix} \theta_{b1} \\ q_{b1} \end{bmatrix} = \mathbf{T}_{b1} \begin{bmatrix} \theta_{a1} \\ q_{a1} \end{bmatrix} = \mathbf{T}_{b1} \begin{bmatrix} \theta_i \\ 0 \end{bmatrix} \tag{9.27b}$$

and

$$\begin{bmatrix} \theta_{b2} \\ q_{b2} \end{bmatrix} = \mathbf{T}_{b2} \begin{bmatrix} \theta_{a2} \\ q_{a2} \end{bmatrix} = \mathbf{T}_{b2} \begin{bmatrix} \theta_i \\ 0 \end{bmatrix} \tag{9.27c}$$

where $\mathbf{T}_{b1}$ and $\mathbf{T}_{b2}$ represent the T-matrix with b set equal to either b_1 or b_2. A combination of eqs. (9.26), (9.27*a*), and (9.27*b*) then yields

$$\begin{bmatrix} \theta_{bl} \\ q_{bl} \end{bmatrix} = \mathbf{T}_b \begin{bmatrix} 1 & 0 \\ 2Y_{\text{in},s} & 1 \end{bmatrix} \mathbf{T}_{bl} \begin{bmatrix} \theta_i \\ 0 \end{bmatrix} \tag{9.28}$$

The hyperbolic cosines and sines can be represented in terms of exponentials:

$$\cosh mb = \frac{e^{mb} + e^{-mb}}{2} = \frac{e^{2mb} + 1}{2e^{mb}} = \frac{\Lambda^2 + 1}{2\Lambda} \tag{9.29a}$$

and

$$\sinh mb = \frac{e^{mb} - e^{-mb}}{2} = \frac{e^{2mb} - 1}{2e^{mb}} = \frac{\Lambda^2 - 1}{2\Lambda} \tag{9.29b}$$

where $\Lambda = e^{mb}$.

Then, with the elements of the T-matrices taken from eq. (9.2), the product of the first two matrices on the right-hand side of eq. (9.28) will be

$$\mathbf{P} = \begin{bmatrix} \cosh mb & \dfrac{1}{Y_o} \sinh mb \\ Y_o \sinh mb & \cosh mb \end{bmatrix} \begin{bmatrix} 1 & 0 \\ 2Y_{\text{in},s} & 1 \end{bmatrix}$$

$$= \begin{bmatrix} \cosh mb + \dfrac{2Y_{\text{in},s}}{Y_o} \sinh mb & \dfrac{1}{Y_o} \sinh mb \\ Y_o \sinh mb + 2Y_{\text{in},s} \cosh mb & \cosh mb \end{bmatrix}$$

Use of eqs. (9.29) allows this product to be written as

$$\mathbf{P} = \begin{bmatrix} \dfrac{\Lambda^2 + 1}{2\Lambda} + \dfrac{2Z_o Y_{\text{in},s}(\Lambda^2 - 1)}{2\Lambda} & \dfrac{\Lambda^2 - 1}{2\Lambda Y_o} \\ Y_o \dfrac{\Lambda^2 - 1}{2\Lambda} + \dfrac{2Y_{\text{in},s}(\Lambda^2 + 1)}{2\Lambda} & \dfrac{\Lambda^2 + 1}{2\Lambda} \end{bmatrix}$$

or

$$\mathbf{P} = \frac{1}{2\Lambda} \begin{bmatrix} \Lambda^2 + 1 + 2Z_o Y_{\text{in},s}(\Lambda^2 - 1) & Z_o(\Lambda^2 - 1) \\ Y_o(\Lambda^2 - 1) + 2Y_{\text{in},s}(\Lambda^2 + 1) & (\Lambda^2 + 1) \end{bmatrix} \tag{9.30}$$

and eq. (9.28) to be written as

$$\begin{bmatrix} \theta_{bl} \\ q_{bl} \end{bmatrix} = \mathbf{P}\mathbf{T}_{b1} \begin{bmatrix} \theta_i \\ 0 \end{bmatrix}$$

With the elements of $\mathbf{T}_{b1}$ taken from eq. (9.2),

$$\mathbf{T}_{b1} \begin{bmatrix} \theta_i \\ 0 \end{bmatrix} = \begin{bmatrix} \cosh mb_1 & \dfrac{1}{Y_o} \sinh mb_1 \\ Y_o \sinh mb_1 & \cosh mb_1 \end{bmatrix} \begin{bmatrix} \theta_i \\ 0 \end{bmatrix}$$

$$= \begin{bmatrix} (\cosh mb_1)\theta_i \\ (Y_o \sinh mb_1)\theta_i \end{bmatrix}$$

and now eq. (9.28) may be written using the elements of **P** as

$$\begin{bmatrix} \theta_{bl} \\ q_{bl} \end{bmatrix} = \begin{bmatrix} p_{11} & p_{12} \\ p_{21} & p_{22} \end{bmatrix} \begin{bmatrix} (\cosh mb_1)\theta_i \\ (Y_o \sinh mb_1)\theta_i \end{bmatrix}$$

$$= \begin{bmatrix} (p_{11} \cosh mb_1 & p_{12} Y_o \sinh mb_1) \\ (p_{21} \cosh mb_1 & p_{22} Y_o \sinh mb_1) \end{bmatrix} \theta_i$$

Expansion gives

$$\theta_{bl} = \theta_i (p_{11} \cosh mb_1 + p_{12} Y_o \sinh mb_1)$$

and with the p's taken from eq. (9.30),

$$\theta_{bl} = \frac{\theta_i}{2\Lambda}[\Lambda^2 + 1 + 2Z_o Y_{\text{in},s}(\Lambda^2 - 1)] \cosh mb_1 + \frac{\theta_i}{2\Lambda}(\Lambda^2 - 1) \sinh mb_1 \tag{9.31a}$$

Another expansion gives

$$q_{bl} = \theta_i (p_{21} \cosh mb_1 + p_{22} Y_o \sinh mb_1)$$

and with the p's taken from eq. (9.30),

$$q_{bl} = \frac{\theta_i}{2\Lambda}[Y_o(\Lambda^2 - 1) + 2Y_{\text{in},s}(\Lambda^2 + 1)] \cosh mb_1 + \frac{\theta_i}{2\Lambda} Y_o(\Lambda^2 + 1) \sinh mb_1 \tag{9.31b}$$

The transformation from the tip of fin 2 to the base of fin 2 is

$$\begin{bmatrix} \theta_{b2} \\ q_{b2} \end{bmatrix} = \mathbf{T}_{b2} \begin{bmatrix} \theta_i \\ 0 \end{bmatrix}$$

or

$$\begin{bmatrix} \theta_{br} \\ q_{br} \end{bmatrix} = \begin{bmatrix} \cosh mb_2 & \dfrac{1}{Y_o} \sinh mb_2 \\ Y_o \sinh mb_2 & \cosh mb_2 \end{bmatrix} \begin{bmatrix} \theta_i \\ 0 \end{bmatrix}$$

and hence

$$\theta_{br} = (\cosh mb_2)\theta_i \tag{9.32a}$$

and

$$q_{br} = (Y_o \sinh mb_2)\theta_i \tag{9.32b}$$

Equations (9.31) and (9.32) provide the underpinning for the evaluation of the heat flow and temperature excess ratios, R_Q and R_θ.

Unequal Heat Loading. The heat flow ratio $R_Q = q_{bl}/q_{br}$ is obtained by dividing eq. (9.31*b*) by eq. (9.32*b*):

$$R_Q = \frac{q_{bl}}{q_{br}} = \frac{[Y_o(\Lambda^2 - 1) + 2Y_{\text{in},s}(\Lambda^2 + 1)]\cosh mb_1}{2\Lambda Y_o \sinh mb} + \frac{(\Lambda^2 + 1)\sinh mb_1}{2\Lambda \sinh mb_2} \tag{9.33}$$

It can be shown that the two hyperbolic function ratios can be written as

$$\frac{\cosh mb_1}{\sinh mb_2} = \frac{\Lambda^2 + z}{\Lambda(z - 1)}$$

and

$$\frac{\sinh mb_1}{\sinh mb_2} = \frac{\Lambda^2 - z}{\Lambda(z - 1)}$$

where $\Lambda = e^{mb}$ as before and $z = e^{2mb_2}$. Thus eq. (9.33) becomes

$$R_Q = \frac{1}{2\Lambda}\left[\frac{Y_o(\Lambda^2 - 1) + 2Y_{\text{in},s}(\Lambda^2 + 1)}{Y_o}\,\frac{\Lambda^2 + z}{\Lambda(z - 1)}\right. \left. + (\Lambda^2 + 1)\frac{\Lambda^2 - z}{\Lambda^2(z - 1)}\right]$$

or

$$R_Q = \frac{[\Lambda^2 - 1 + \alpha(\Lambda^2 + 1)](\Lambda^2 + z) + (\Lambda^2 + 1)(\Lambda^2 - z)}{2\Lambda^2(z - 1)} \tag{9.34}$$

where

$$\alpha = \frac{2Y_{\text{in},s}}{Y_o}$$

When eq. (9.34) is solved for b_2, the result is

$$b_2 = \frac{1}{2m}\ln\frac{\Lambda^2[2(R_Q + \Lambda^2) + \alpha(\Lambda^2 + 1)]}{2(1 + \Lambda^2 R_Q) - \alpha(\Lambda^2 + 1)} \tag{9.35}$$

The minimum value for R_Q that produces the adiabatic point *within fin r* can be obtained from eq. (9.34) after setting $b_2 = b$ or, equivalently, $z = \Lambda^2$. This procedure leads to

$$R_{Q_{\min}} = 1 + \alpha\left(\frac{\Lambda^2 + 1}{\Lambda^2 - 1}\right)$$

On the other hand, for $b_2 = 0$ so that $z = 1$, the heat flow ratio $R_Q \to \infty$ and therefore $R_{Q_{\max}} \to \infty$. It is stated with emphasis that eq. (9.34) is valid only for $R_Q \geq R_{Q_{\min}}$.

Unequal Temperature Excesses. The temperature excess ratio $R_\theta = \theta_{bl}/\theta_{br}$ is obtained by dividing eq. (9.31a) by eq. (9.32a):

$$R_\theta = \frac{\theta_{bl}}{\theta_{br}} = \frac{[\Lambda^2 + 1 + 2Z_o(\Lambda^2 - 1)Y_{\text{in},s}] \cosh mb_1}{2\Lambda \cosh mb_2} + \frac{(\Lambda^2 - 1) \sinh mb_1}{2\Lambda \cosh mb_2} \tag{9.36}$$

In this case, the two hyperbolic function ratios can be written as

$$\frac{\cosh mb_1}{\cosh mb_2} = \frac{\Lambda^2 + z}{\Lambda(1 + z)}$$

and

$$\frac{\sinh mb_1}{\cosh mb_2} = \frac{\Lambda^2 - z}{\Lambda(1 + z)}$$

where $z = e^{2mb_2}$, as in the case of unequal heat inputs. Use of these permits eq. (9.36) to be written as

$$R_\theta = \frac{[\Lambda^2 + 1 + \alpha(\Lambda^2 - 1)](\Lambda^2 + z) + (\Lambda^2 - 1)(\Lambda^2 - z)}{2\Lambda^2(1 + z)} \tag{9.37}$$

where $\alpha = 2Z_o Y_{\text{in},s} = 2Y_{\text{in},s}/Y_o$.

As in the case of the single stack, the minimum value of R_θ for which an adiabatic point exists in fin r (Fig. 9.5a) is obtained by setting $b_2 = b$ (or equivalently, $z = \Lambda^2$). It then follows that

$$(R_\theta)_{\text{min}} = 1 + \alpha \frac{\Lambda^2 - 1}{\Lambda^2 + 1} \tag{9.38}$$

which corresponds to the maximum R_Q. Moreover, a simple algebraic procedure leads to a rearrangement of eq. (9.36) and provides the location of the adiabatic point b_2 in terms of R_θ, Λ, and α as

$$b_2 = \frac{1}{2m} \ln \frac{\Lambda^2[2\Lambda^2 + \alpha(\Lambda^2 - 1) - 2R_\theta]}{2\Lambda^2 R_\theta - 2 - \alpha(\Lambda^2 - 1)} \tag{9.39}$$

Equation (9.39) is valid only for $R_\theta \geq (R_\theta)_{\text{min}}$. However, there is a finite maximum for R_θ beyond which eq. (9.39) is not valid. This maximum is obtained by setting $b_2 = 0$ (which means that $z = 1$) in eq. (9.37). This leads to

$$(R_\theta)_{\text{max}} = \frac{2(\Lambda^4 + 1) + \alpha(\Lambda^4 - 1)}{4\Lambda^2} \tag{9.40}$$

Of course,

$$(R_\theta)_{\text{min}} \leq R_\theta \leq (R_\theta)_{\text{max}} \tag{9.41}$$

Interdependence of the Heat Flow and Temperature Excess Ratios. Pieper and Kraus (1995b) have shown that there is a relationship that shows the interdependence of R_θ and R_Q:

$$R_\theta = \frac{AR_Q + 1}{R_Q + B} \tag{9.42}$$

where

$$A = \frac{(\Lambda^2 + 1)[\Lambda^2 + 1 + \alpha(\Lambda^2 - 1)] + (\Lambda^2 - 1)^2}{4\Lambda^2}$$

and

$$B = \frac{(\Lambda^2 - 1)[\Lambda^2 - 1 + \alpha(\Lambda^2 + 1)] + (\Lambda^2 + 1)^2}{4\Lambda^2}$$

Example 9.4. Two passages, identical to the passage used in Example 9.2, are assembled to form a double stack. The pair of cover plates at the outside of the double stack are 0.2032 cm thick and the single pair of splitter plates is 0.1524 cm thick. There are 400 fins per meter and the fins are 1.27 cm high, 0.2286 cm thick, and 0.25 cm long. The heat transfer coefficient is $h = 400\ \text{W/m}^2 \cdot \text{K}$ and the thermal conductivity is $k = 160\ \text{W/m} \cdot \text{K}$. Determine the heat flow from each of the cover plates if they are held at temperature excesses of 35.2°C and 32°C, respectively.

SOLUTION. Assume that an adiabatic point exists in fin r. The configuration will look like Fig. 9.5a. From Example 9.2 with $L = 0.25$ m, for the fins

$$m_f = 46.77\ \text{m}^{-1}$$

$$m_f b = 0.5940$$

and

$$Y_o = 4.2764\ \text{W/K}$$

With fin 2 on the 32°C side,

$$R_\theta = \frac{35.2}{32} = 1.1000$$

and

$$\Lambda = e^{m_f b} = e^{0.5940} = 1.8111$$

For the splitter plate with $a_s = 1.524 \times 10^{-3}$ and $L = 0.25$ m,

$$m_s = \left(\frac{2h}{ka_s}\right)^{1/2} = \left[\frac{(2)(400)}{(160)(1.524 \times 10^{-3})}\right]^{1/2}$$

$$(3280.84)^{1/2} = 57.28\ \text{m}^{-1}$$

and with $s = 1.070 \times 10^{-4}$ m taken from Example 9.2:

$$m_s s = (57.28)(1.070 \times 10^{-4}) = 6.129 \times 10^{-3}$$

Moreover,

$$Y_{o,s} = (2hka_s)^{1/2} = [(2)(400)(160)(1.524 \times 10^{-3})]^{1/2}(0.25)$$
$$(195.07)^{1/2}(0.25) = 3.4917 \text{ W/K}$$

so that

$$\begin{aligned} Y_{\text{in},s} &= Y_o \tanh mb_s \\ &= 3.4917 \tanh 6.129 \times 10^{-3} \\ &= (3.4917)(6.129 \times 10^{-3}) \\ &= 0.0214 \text{ W/K} \end{aligned}$$

Thus the parameter α is

$$\alpha = \frac{2Y_{\text{in},s}}{Y_{o,f}} = \frac{(2)(0.0214)}{4.2764} = 1.001 \times 10^{-2}$$

The minimum and maximum values of R_θ are given by eqs. (9.38) and (9.40):

$$\begin{aligned} (R_\theta)_{\min} &= 1 + \alpha \frac{\Lambda^2 - 1}{\Lambda^2 + 1} \\ &= 1 + (1.001 \times 10^{-2}) \left(\frac{2.2804}{4.2804} \right) \\ &= 1 + (1.001 \times 10^{-2})(0.5328) \\ &= 1.0053 \end{aligned}$$

and

$$\begin{aligned} (R_\theta)_{\max} &= \frac{2(\Lambda^4 + 1) + \alpha(\Lambda^4 - 1)}{4\Lambda^2} \\ &= \frac{(2)(11.7589) + (1.001 \times 10^{-2})(9.7589)}{(4)(3.2804)} \\ &= \frac{23.5179 + 0.0977}{13.1203} \\ &= 1.8000 \end{aligned}$$

Thus the criterion of eq. (9.41) is met:

$$(R_\theta)_{\min} \leq R_\theta \leq (R_\theta)_{\max}$$
$$1.0053 \leq 1.1000 \leq 1.8000$$

and an adiabatic point will indeed exist in fin r.

By eq. (9.39),

$$b_2 = \frac{1}{2m_f} \ln \frac{\Lambda^2[2\Lambda^2 + \alpha(\Lambda^2 - 1) - 2R_\theta]}{2\Lambda^2 R_\theta - 2 - \alpha(\Lambda^2 - 1)}$$

$$= \frac{1}{2(46.77)} \ln \frac{(3.2801)(6.5602 + 0.0228 - 2.2000)}{7.2162 - 2.0000 - 0.0228}$$

$$= 0.0107 \ln \frac{14.3766}{5.1934}$$

$$= 0.0107 \ln 2.7682$$

$$= (0.0107)(1.0182)$$

$$= 0.0109 \text{ m}$$

Then

$$m_f b_2 = (46.77)(0.0109) = 0.5098$$

$$\tanh m_f b_2 = 0.4698$$

and with $Y_{o,f} = 4.2764$ W/K,

$$Y_{\text{in},b2} = Y_o \tanh m_f b_2$$

$$= (4.2764)(0.4698)$$

$$= 2.0090 \text{ W/K}$$

The heat entering the fin is

$$q_{b2} = Y_{\text{in},2}\theta_{b2} = (2.0090)(32) = 64.29 \text{ W}$$

and for the cover plates with surface

$$S_c = S_b = 2sL = (2)(1.070 \times 10^{-4})(0.25) = 5.350 \times 10^{-5} \text{ m}^2$$

the heat total heat flow will be

$$q_r = q_{br} + hS_c\theta_{br}$$

$$= 64.29 + (400)(5.350 \times 10^{-5})(32)$$

$$= 64.29 + 0.68$$

$$= 64.97 \text{ W}$$

For fin 1, which is on the 35.2°C side, simple array analysis using eqs. (9.3) and (9.4) will yield the required heat flow. First,

$$b_1 = b - b_2$$

$$= 0.0127 - 0.0109$$
$$= 0.0018 \text{ m}$$

Then

$$m_f b_1 = (46.77)(0.0018) = 0.0842$$

and

$$\begin{aligned} Y_{\text{in},b1} &= Y_o \tanh m_f b_1 \\ &= 4.2764 \tanh 0.0842 \\ &= (4.2764)(0.0840) \\ &= 0.3592 \text{ W/K} \end{aligned}$$

For the splitter plates, $Y_{\text{in},s} = 2.140 \times 10^{-2}$ W/K · °C, and by the cluster algorithm of eq. (9.4), the admittance at the tip of the fin with height b is

$$\begin{aligned} Y_a &= 2Y_{\text{in},s} + Y_{\text{in},b1} \\ &= (2)(2.140 \times 10^{-2}) + 0.3592 \\ &= 0.0428 + 0.3592 \\ &= 0.4020 \text{ W/K} \end{aligned}$$

With the elements of the **T** matrix taken from eq. (9.2),

$$\tau_{11} = \cosh mb = \cosh 0.5940 = 1.1817$$
$$\tau_{12} = \frac{1}{Y_o} \sinh mb = \frac{1}{4.2764} \sinh 0.5940 = 0.1472$$
$$\tau_{21} = Y_o \sinh mb = 4.2764 \sinh 0.5940 = 2.6922$$

and

$$\tau_{22} = \cosh mb = \cosh 0.5940 = 1.1817$$

the reflection relationship of eq. (9.3*a*) can be employed to give

$$\begin{aligned} Y_{\text{in},b1} &= \frac{\tau_{21} + Y_a \tau_{22}}{\tau_{11} + Y_a \tau_{12}} \\ &= \frac{2.6922 + (0.4020)(1.1817)}{1.1817 + (0.4020)(0.1472)} \\ &= \frac{3.1672}{1.2409} \\ &= 2.5524 \text{ W/K} \end{aligned}$$

The heat entering the finned array is

$$q_{bl} = Y_{\text{in},1}\theta_{bl} = (2.5524)(35.2) = 89.86 \text{ W}$$

and with the surface of the cover plates the same as on the 32°C side, the heat dissipation on the 35.2°C side will be

$$\begin{aligned} q_l &= q_{bl} + hs_b\theta_{bl} \\ &= 89.86 + (400)(5.350 \times 10^{-5})(35.2) \\ &= 89.86 + 0.75 \\ &= 90.61 \text{ W} \end{aligned}$$

In summary,

$$q_{bl} = 89.86 \text{ W}$$

and

$$q_{br} = 64.29 \text{ W}$$

For this double-stack arrangement,

$$R_Q = \frac{q_{bl}}{q_{br}} = \frac{89.86}{64.29} = 1.3977$$

and this can be checked by eq. (9.42). With $\Lambda = 1.8111$, $\Lambda^2 = 3.2801$ and $\alpha = 1.001 \times 10^{-2}$:

$$\begin{aligned} A &= \frac{(\Lambda^2+1)[\Lambda^2+1+\alpha(\Lambda^2-1)] + (\Lambda^2-1)^2}{4\Lambda^2} \\ &= \frac{(3.2801+1)[3.2801+1+1.001\times 10^{-2}(3.2801-1)] + (3.2801-1)^2}{(4)(3.2801)} \\ &= \frac{(4.2801)(4.3029) + 5.1988}{13.1203} \\ &= \frac{18.4168 + 5.1988}{13.1203} \\ &= \frac{23.6156}{13.1203} \\ &= 1.8000 \end{aligned}$$

and

$$B = \frac{(\Lambda^2-1)[\Lambda^2-1+\alpha(\Lambda^2+1)] + (\Lambda^2+1)^2}{4\Lambda^2}$$

$$= \frac{(3.2801 - 1)[3.2801 - 1 + 1.001 \times 10^{-2}(3.2801 + 1)] + (3.2801 + 1)^2}{(4)(3.2801)}$$

$$= \frac{(2.2801)(2.3229) + 18.3191}{13.1203}$$

$$= \frac{5.2965 + 18.3191}{13.1203}$$

$$= \frac{23.6156}{13.1203} = 1.8000$$

so that

$$R_\theta = \frac{AR_Q + 1}{R_Q + B}$$

$$= \frac{(1.8000)(1.3977) + 1.0000}{1.3977 + 1.8000}$$

$$= \frac{3.5159}{3.1977}$$

$$= 1.1000$$

which checks the previous result.

9.7 *n*-STACK

This chapter concludes with the development of general relationships for a configuration with n-stacks with either even loading or single-side loading.

9.7.1 Even Loading

Even Number of Stacks. The configuration for the n-stack (n-even) with even loading on both cover plates is shown in Fig. 9.6. Let $p = n/2$ and observe in Fig. 9.6 that all horizontal members are splitter plates with thickness a (the subscript is not required) and that because of the stack symmetry, the outermost pair of splitter plates are longitudinal fins of rectangular profile with thickness $a/2$ and with one face insulated. Hence for these outermost plates,

$$Y_{o,s} = (hka)^{1/2} L$$

$$m_s s = \left[\frac{h}{k(a/2)}\right]^{1/2} s = \left(\frac{2h}{ka}\right)^{1/2} s$$

and because the configuration in Fig. 9.6 is a repeating section, the tips of these plates are insulated and, in accordance with eq. (9.3*b*),

$$Y_{\text{in},s} = Y_{o,s} \tanh m_s s$$

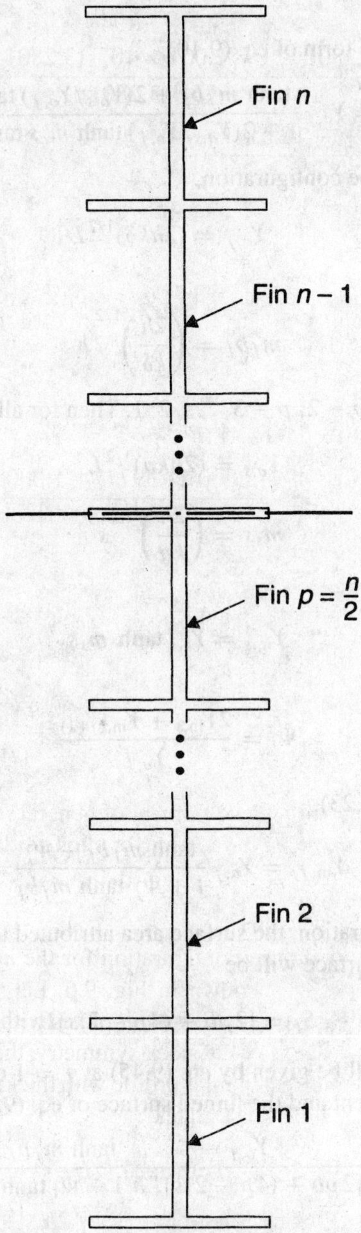

Figure 9.6 Stack with n even with even heat loading on both cover plates. The dashed line divides the stack into two symmetrical entities.

Then for fin p, in the form of eq. (9.10),

$$Y_{\text{in},fp} = Y_{o,f}\frac{\tanh m_f b_f + 2(Y_{o,s}/Y_{o,f})\tanh m_c s}{1 + 2(Y_{o,s}/Y_{o,f})\tanh m_c s \tanh m_f b_f} \tag{9.43}$$

where for every fin in the configuration,

$$Y_{o,f} = (2hk\delta)^{1/2}L$$

and

$$m_f b_f = \left(\frac{2h}{k\delta}\right)^{1/2} b$$

Now let $j = p-1, p-2, p-3, \ldots, 2, 1$. Then for all other splitter plates,

$$Y_{o,s} = (2hka)^{1/2}L$$

$$m_s s = \left(\frac{2h}{ka}\right)^{1/2} s$$

and

$$Y_{\text{in},s} = Y_{o,s}\tanh m_s s$$

so that

$$\Psi_j = \frac{2Y_{\text{in},s} + Y_{\text{in},(j+1)}}{Y_{o,f}} \tag{9.44}$$

and in the form of eq. (9.23),

$$Y_{\text{in},fj} = Y_{o,f}\frac{\tanh m_f b_f + \Psi_j}{1 + \Psi_j \tanh m_f b_f} \tag{9.45}$$

For the entire configuration, the surface area attributed to the fins with all splitter plates acting as finned surface will be

$$S_f = [2pb + (4p-2)s]L \tag{9.46}$$

and the fin efficiency will be given by eq. (9.45) at $j = 1$ divided by the product of the heat transfer coefficient and the finned surface of eq. (9.46):

$$\eta_f = \frac{Y_{o,f}}{[2pb + (4p-2)s]Lh}\frac{\tanh m_f b_f + \Psi_1}{1 + \Psi_1 \tanh m_f b_f} \tag{9.47}$$

Odd Number of Stacks. The configuration for the n-stack (n-odd) with even loading on both cover plates is shown in Fig. 9.7. This time, let $p = (n-1)/2$ and observe in Fig 9.7 that all horizontal members are splitter plates with thickness a, so that

$$Y_{o,s} = (2hka)^{1/2}L \qquad m_s s = \left(\frac{2h}{ka}\right)^{1/2} s$$

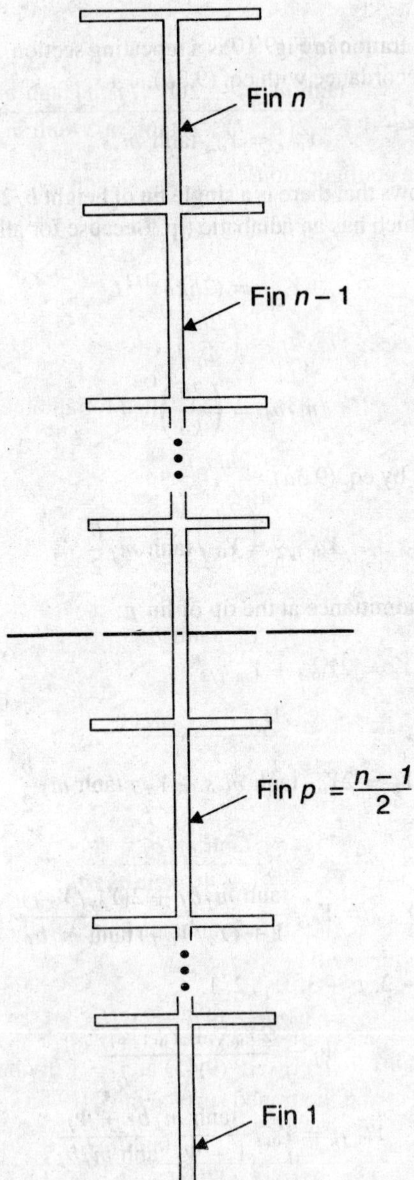

Figure 9.7 Stack with n odd with even heat loading on both cover plates. Note that the dashed line divides the stack into two symmetrical sections.

and because the configuration in Fig. 9.7 is a repeating section, the tips of these plates are insulated, and in accordance with eq. (9.3*b*),

$$Y_{\text{in},s} = Y_{o,s} \tanh\ m_s s$$

Here the symmetry shows that there is a single fin of height $b/2$ at the outer extremity of the configuration which has an adiabatic tip. Because for all fins

$$Y_{o,f} = (2hk\delta)^{1/2} L$$

and

$$m_f b_f = \left(\frac{2h}{k\delta}\right)^{1/2} b$$

and for the extreme fin by eq. (9.3*a*),

$$Y_{\text{in},f/2} = Y_{o,f} \tanh\ m_f \frac{b}{2}$$

there will be a cluster admittance at the tip of fin p:

$$Y_a = 2Y_{\text{in},s} + Y_{\text{in},f/2}$$

or

$$Y_a = 2Y_{o,s} \tanh\ m_s s + Y_{o,f} \tanh\ m_f \frac{b}{2} \tag{9.48}$$

Then for fin p,

$$Y_{\text{in},p} = Y_{o,f} \frac{\tanh\ m_f b_f + 2(Y_{ay}/Y_{o,f})}{1 + (Y_a/Y_{o,f}) \tanh\ m_f b_f} \tag{9.49}$$

With $j = p-1, p-2, p-3, \ldots, 2, 1$.

$$\Psi_j = \frac{2Y_{\text{in},s} + Y_{\text{in},(j+1)}}{Y_{o,f}} \tag{9.44}$$

$$Y_{\text{in},fj} = Y_{o,f} \frac{\tanh\ m_f b_f + \Psi_j}{1 + \Psi_j\ \tanh\ m_f b_f} \tag{9.45}$$

and the result for fin 1 will be the same as that for the even number of stacks:

$$Y_{\text{in},f1} = Y_{o,f} \frac{\tanh\ m_f b_f + \Psi_1}{1 + \Psi_1\ \tanh\ m_f b_f}$$

Here, however, the surface area attributed to the fins with all splitter plates acting as finned surface will be

$$S_f = [(2p+1)b + 4ps]L \tag{9.50}$$

and the fin efficiency will be given by eq. (9.47) at $j = 1$ divided by the product of the heat transfer coefficient and the finned surface of eq. (9.50):

$$\eta_f = \frac{Y_{o,f}}{[(2p+1)b + 4ps]Lh} \frac{\tanh m_f b_f + \Psi_1}{1 + \Psi_1 \tanh m_f b_f} \tag{9.51}$$

Example 9.5. A triple stack is to be formed from three 15.08 plain plate fin surfaces. Each 15.08 surface has 15.08 fins per inch, and each of the fins has a height of 0.418 in. and a thickness of 0.006 in. The stack is fabricated from material having a thermal conductivity $k = 160$ W/m · K and the heat transfer coefficient is $h = 120$ W/m^2 · K. The cover plates are 0.010 in. thick and the splitter plates are 0.008 in. thick. Determine the cover plate temperature if the stack dissipation is 36 W per meter per repeating section on each cover plate.

SOLUTION. Here for this arrangement with $n = 3$ (odd),

$$p = \frac{n-1}{2} = \frac{3-1}{2} = 1$$

and

$$b_f = (0.418)(0.0254) = 1.062 \times 10^{-2} \text{ m}$$

$$\delta_f = (0.006)(0.0254) = 1.524 \times 10^{-4} \text{ m}$$

$$s = \frac{0.3048}{(2)(15.08)(12)} = 8.422 \times 10^{-4} \text{ m}$$

and

$$a_s = (0.008)(0.0254) = 2.032 \times 10^{-4} \text{ m}$$

The configuration is shown in Fig. 9.8a and it is observed that the cover plate does not enter the problem except in its capacity as the base surface for the single pair of splitter plates with unit length

$$\begin{aligned} Y_{o,s} &= (2hka_s)^{1/2} L \\ &= [(2)(120)(160)(2.032 \times 10^{-4})]^{1/2}(1) \\ &= (7.8029)^{1/2} \\ &= 2.7934 \text{ W/K} \end{aligned}$$

$$\begin{aligned} m_s s &= \left(\frac{2h}{ka_s}\right)^{1/2} s \\ &= \left[\frac{(2)(120)}{(160)(2.032 \times 10^{-4})}\right]^{1/2} (8.422 \times 10^{-4}) \\ &= (85.9179)(8.422 \times 10^{-4}) \end{aligned}$$

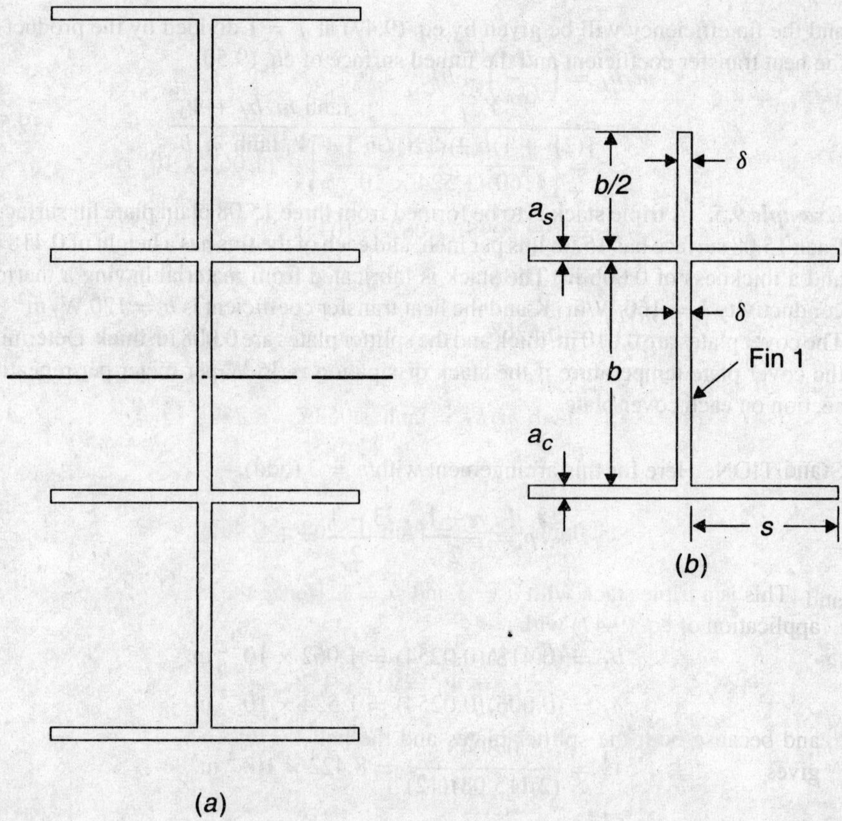

Figure 9.8 Configuration for Example 9.5. This is the triple stack with even loading on both sides: (*a*) the entire surface; (*b*) the symmetrical section.

$$= 0.0724$$

and

$$\tanh m_s s = \tanh 0.0724$$
$$= 0.0723$$

For the fins per unit length,

$$Y_{o,f} = (2hk\delta_f)^{1/2} L$$
$$= [(2)(120)(160)(1.524 \times 10^{-4})^{1/2}(1)$$
$$= (5.8522)^{1/2}$$
$$= 2.4191 \text{ W/K}$$

$$
\begin{aligned}
m_f b_f &= \left(\frac{2h}{k\delta_f}\right)^{1/2} b_f \\
&= \left[\frac{(2)(120)}{(160)(1.524\times 10^{-4})}\right]^{1/2} (1.062\times 10^{-2}) \\
&= (99.2095)(1.062\times 10^{-2}) \\
&= 1.0536 \\
\frac{m_f b_f}{2} &= 0.5268
\end{aligned}
$$

$$\tanh m_f b_f = \tanh 1.0536 = 0.7832$$

and

$$\tanh \frac{m_f b_f}{2} = \tanh 0.5268 = 0.4829$$

This is a triple stack with $n = 3$ and $p = 1$. Hence the problem requires a single application of eq. (9.47) with

$$Y_a = \Psi_1 = 2Y_{\text{in},s} + Y_{\text{in},f/2}$$

and because both the splitter plates and the half fin have adiabatic tips, eq. (9.3*a*) gives

$$
\begin{aligned}
Y_{\text{in},s} &= Y_{o,s} \tanh m_s s \\
&= (2.7934)(0.0723) \\
&= 0.2019 \text{ W/K}
\end{aligned}
$$

and

$$
\begin{aligned}
Y_{\text{in},f/2} &= Y_{o,f/2} \tanh \frac{m_f b_f}{2} \\
&= (2.4191)(0.4829) \\
&= 1.1682 \text{ W/K}
\end{aligned}
$$

Thus

$$
\begin{aligned}
Y_a = \Psi_1 &= 2Y_{\text{in},s} + Y_{\text{in},f/2} \\
&= (2)(0.2019) + 1.1682 \\
&= 1.5720 \text{ W/K}
\end{aligned}
$$

and eq. (9.45) gives

$$
\begin{aligned}
Y_{\text{in},f1} &= Y_{o,f}\,\frac{\tanh m_f b_f + \Psi_1}{1+\Psi_1 \tanh m_f b_f} \\
&= 2.4191\left[\frac{0.7832+1.5720}{1+(1.5720)(0.7832)}\right] \\
&= 2.4191\left(\frac{2.3552}{2.2312}\right) \\
&= 2.5536 \text{ W/K}
\end{aligned}
$$

The finned surface area is obtained from eq. (9.50) with $p = 1$,

$$S_f = \{[(2)(1)+1]b + (4)(1)s\}L$$

and for $L = 1$ m,

$$
\begin{aligned}
S_f &= (3)(1.062\times10^{-2}) + (4)(8.422\times10^{-4}) \\
&= 3.1860\times10^{-2} + 3.3688\times10^{-3} \\
&= 3.5229\times10^{-2}\ \text{m}^2
\end{aligned}
$$

Then

$$\eta_f = \frac{Y_{\text{in},f1}}{hS} = \frac{2.5536}{(120)(3.5229\times10^{-2})} = 0.604$$

The base surface is the surface of a single pair of cover plates,

$$
\begin{aligned}
S_b &= 2sL \\
&= (2)(8.422\times10^{-4})(1) \\
&= 1.6844\times10^{-3}\ \text{m}^2
\end{aligned}
$$

The temperature excess of the cover plate for an evenly distributed 36 W per meter per repeating section on each cover plate will be

$$
\begin{aligned}
\theta_b &= \frac{q}{h(S_b+\eta_f S_f)} \\
&= \frac{36}{(120)[1.6844\times10^{-3} + (0.604)(3.5229\times10^{-2})]} \\
&= \frac{36}{(120)(2.2964\times10^{-2})} \\
&= 13.1^\circ\text{C}
\end{aligned}
$$

9.7.2 Single-Side Loading

The analysis for single-side loading follows the analysis for the even number of stacks with even loading with a single exception. The configuration for the n-stack with single-side loading on a single cover plate is shown in Fig. 9.9, where it is observed that there is no symmetry and the plates at the tip of fin n are cover plates with thickness, a_c. These cover plates are longitudinal fins of rectangular profile with one face insulated and hence

$$Y_{o,c} = (hka_c)^{1/2} L$$

$$m_c s = \left(\frac{h}{ka_c}\right)^{1/2} s$$

and because the configuration in Fig. 9.9 is a repeating section, the tips of these plates

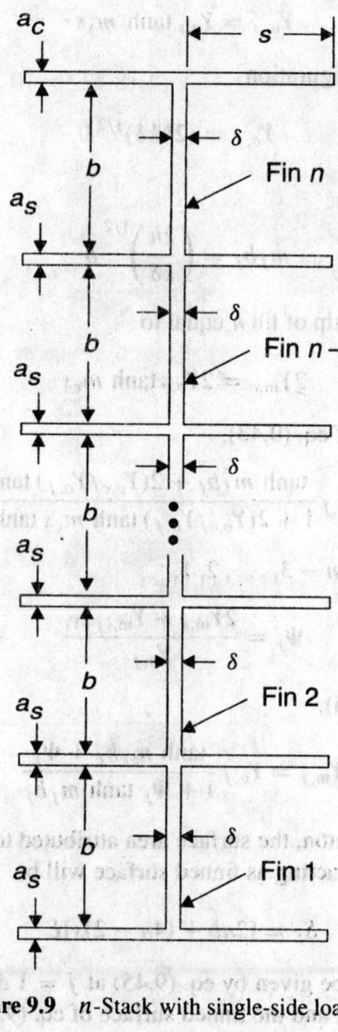

Figure 9.9 *n*-Stack with single-side loading.

are adiabatic and in accordance with eq. (9.3*b*),

$$Y_{\text{in},c} = Y_{o,c} \tanh m_c s$$

All other horizontal members are splitter plates with thickness a_s, so that

$$Y_{o,s} = (2hka_s)^{1/2} L$$

$$m_s s = \left(\frac{2h}{ka_s}\right)^{1/2} s$$

and here, too, because the tips of these plates are insulated, in accordance with eq. (9.3*b*),

$$Y_{\text{in},s} = Y_{o,s} \tanh m_s s$$

For every fin in the configuration

$$Y_{o,f} = (2hk\delta)^{1/2} L$$

and

$$m_f b_f = \left(\frac{2h}{k\delta}\right)^{1/2} b$$

There is a cluster at the tip of fin n equal to

$$2Y_{\text{in},c} = 2Y_{o,c} \tanh m_c s$$

and for fin n, in the form of eq. (9.43),

$$Y_{\text{in},n} = Y_{o,f} \frac{\tanh m_f b_f + 2(Y_{o,c}/Y_{o,f}) \tanh m_c s}{1 + 2(Y_{o,c}/Y_{o,f}) \tanh m_c s \tanh m_f b_f} \tag{9.52}$$

With $j = n-1, n-2, n-3, \ldots, 2, 1$,

$$\Psi_j = \frac{2Y_{\text{in},s} + Y_{\text{in},(j+1)}}{Y_{o,f}} \tag{9.44}$$

and in the form of eq. (9.45),

$$Y_{\text{in},j} = Y_{o,f} \frac{\tanh m_f b_f + \Psi_j}{1 + \Psi_j \tanh m_f b_f} \tag{9.45}$$

For the entire configuration, the surface area attributed to the fins with all splitter plates and the cover plates acting as finned surface will be

$$S_f = [2nb + (4n-2)s]L \tag{9.53}$$

and the fin efficiency will be given by eq. (9.45) at $j = 1$ divided by the product of the heat transfer coefficient and the finned surface of eq. (9.53):

$$\eta_f = \frac{Y_{o,f}}{[2nb + (4n-2)s]Lh} \frac{\tanh m_f b_f + \Psi_1}{1 + \Psi_1 \tanh m_f b_f} \tag{9.54}$$

Example 9.6. A quadruple stack is formed from four of the 15.08 plain plate fin surfaces of Example 9.5 and is shown in Fig. 9.10. If the dissipation is 96 W per meter per repeating section on one cover plate, determine the cover plate temperature on the side where the heat is dissipated.

SOLUTION. From Example 9.5,

$$b_f = 1.062 \times 10^{-2} \text{ m} \qquad \delta_f = 1.524 \times 10^{-4} \text{ m}$$

$$s = 8.422 \times 10^{-4} \text{ m} \qquad a_s = 2.032 \times 10^{-4} \text{ m}$$

and because the outer cover plate acts as a heat-dissipating surface,

$$a_c = (0.010)(0.0254) = 2.540 \times 10^{-4} \text{ m}$$

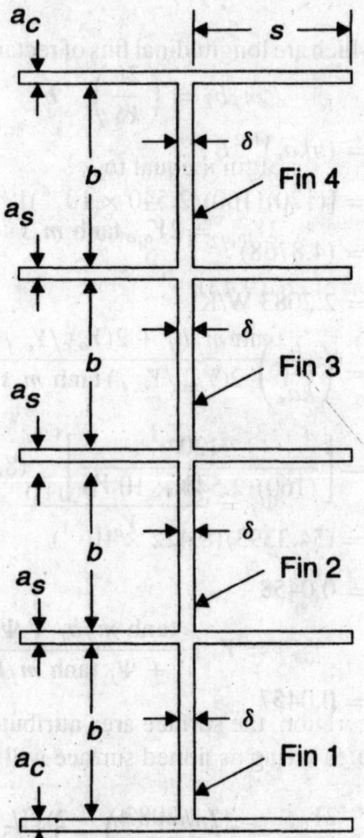

Figure 9.10 Quadruple stack with single-side loading.

From Example 9.5, for the splitter plates per unit length,

$$Y_{o,s} = 2.7934 \text{ W/K}$$

$$m_s s = 0.0724$$

$$\tanh m_s s = 0.0723$$

and

$$Y_{\text{in},s} = 0.2019 \text{ W/K}$$

and for the fins, per unit length,

$$Y_{o,f} = 2.4191 \text{ W/K}$$

$$m_f b_f = 1.0536$$

and

$$\tanh m_f b_f = 0.7832$$

For the cover plates, which are longitudinal fins of rectangular profile with adiabatic tips, per unit length,

$$\begin{aligned} Y_{o,c} &= (hka_c)^{1/2} L \\ &= [(120)(160)(2.540 \times 10^{-4})]^{1/2}(1) \\ &= (4.8768)^{1/2} \\ &= 2.2083 \text{ W/K} \end{aligned}$$

$$\begin{aligned} m_c s &= \left(\frac{h}{ka_s}\right)^{1/2} s \\ &= \left[\frac{120}{(160)(2.540 \times 10^{-4})}\right]^{1/2} (8.422 \times 10^{-4}) \\ &= (54.3393)(8.422 \times 10^{-4}) \\ &= 0.0458 \end{aligned}$$

and

$$\tanh m_c s = 0.0457$$

For $n = 4$ and with

$$\frac{2Y_{o,c}}{Y_{o,f}} = 2\left(\frac{2.2083}{2.4191}\right) = 1.8258$$

eq. (9.52) gives

$$Y_{\text{in},4} = Y_{o,f}\,\frac{\tanh m_f b_f + 2(Y_{o,c}/Y_{o,f})\tanh m_c s}{1 + 2(Y_{o,c}/Y_{o,f})\tanh m_c s \tanh m_f b_f}$$

$$= 2.4191\left[\frac{0.7832 + (1.8258)(0.0457)}{1 + (1.8258)(0.0457)(0.7832)}\right]$$

$$= 2.4191\left(\frac{0.8666}{1.0653}\right)$$

$$= 1.9679\ \text{W/K}$$

For $n = 3$ by eq. (9.44),

$$\Psi_3 = \frac{2Y_{\text{in},s} + Y_{\text{in},4}}{Y_{o,f}}$$

$$= \frac{(2)(0.2019) + 1.9679}{2.4191}$$

$$= \frac{2.3717}{2.4191}$$

$$= 0.9804\ \text{W/K}$$

and by eq. (9.45),

$$Y_{\text{in},3} = Y_{o,f}\,\frac{\tanh m_f b_f + \Psi_3}{1 + \Psi_3 \tanh m_f b_f}$$

$$= 2.4191\left[\frac{0.7832 + 0.9804}{1 + (0.9804)(0.7832)}\right]$$

$$= 2.4191\left(\frac{1.7636}{1.7678}\right)$$

$$= 2.4133\ \text{W/K}$$

For $n = 2$ by eq. (9.44),

$$\Psi_2 = \frac{2Y_{\text{in},s} + Y_{\text{in},3}}{Y_{o,f}}$$

$$= \frac{(2)(0.2019) + 2.4133}{2.4191}$$

$$= \frac{2.8171}{2.4191}$$

$$= 1.1645\ \text{W/K}$$

and by eq. (9.45),

$$Y_{\text{in},2} = 2.4191\left[\frac{0.7832 + 1.1645}{1 + (1.1645)(0.7832)}\right]$$

$$= 2.4191\left(\frac{1.9477}{1.9121}\right)$$

$$= 2.4642 \text{ W/K}$$

Finally, for $n = 1$ by eq. (9.44),

$$\Psi_2 = \frac{2Y_{\text{in},s} + Y_{\text{in},4}}{Y_{o,f}}$$

$$= \frac{(2)(0.2019) + 2.4642}{2.4191}$$

$$= \frac{2.8680}{2.4191}$$

$$= 1.1856 \text{ W/K}$$

and by eq. (9.45),

$$Y_{\text{in},1} = 2.4191\left[\frac{0.7832 + 1.1856}{1 + (1.1856)(0.7832)}\right]$$

$$= 2.4191\left(\frac{1.9688}{1.9285}\right)$$

$$= 2.4696 \text{ W/K}$$

The finned surface area is obtained from eq. (9.53) with $n = 4$:

$$S_f = [2nb + (4n - 2)s]L$$

$$= [(8)(1.062 \times 10^{-2}) + (14)(8.422 \times 10^{-4})](1.000)$$

$$= 8.496 \times 10^{-2} + 1.1791 \times 10^{-2}$$

$$= 9.6751 \times 10^{-2} \text{ m}^2$$

and the fin efficiency is given by

$$\eta_f = \frac{Y_{\text{in},1}}{hS_f} = \frac{2.4696}{(120)(9.6751 \times 10^{-2})} = 0.213$$

The base surface is

$$S_b = 2sL$$

$$= (2)(8.422 \times 10^{-4})(1.000)$$

$$= 1.6844 \times 10^{-3} \text{ m}^2$$

and the temperature excess at the base surface for this single-side-loaded quadruple stack for a 96 W/m repeating section will be

$$\theta_b = \frac{q}{h(S_b + \eta_f S_f)}$$

$$= \frac{96}{(120)[1.6844 \times 10^{-3} + (0.213)(9.6751 \times 10^{-2})]}$$

$$= \frac{96}{(120)(2.2292 \times 10^{-2})}$$

$$= 35.9°\text{C}$$

9.8 CLOSURE

With the ability to determine the efficiencies of finned passages in hand, attention can now turn to the discussion of the compact heat exchanger, which is the subject of Chapter 10.

9.9 NOMENCLATURE

Roman Letter Symbols

A	cross-sectional or profile area, m^2; combination of terms, dimensionless
a	cover plate thickness, m; separation plate or splitter plate thickness, m
B	combination of terms, dimensionless
b	fin height, m
h	heat transfer coefficient, $W/m^2 \cdot K$
I	modified Bessel function of the first kind
i	counter, dimensionless
k	thermal conductivity, $W/m \cdot K$
L	fin length, m
l	left side, dimensionless
m	fin performance parameter, m^{-1}
$\mathbf{P}$	matrix, dimensionless
p	elements of a matrix, dimensionless; particular fin number, dimensionless
q	heat flow, W
R	ratio, dimensionless
r	right side, dimensionless
S	surface area, m^2
s	separation plate or splitter plate height, m
T	temperature, K
w	fin pitch, m
x	height coordinate, m
Y	thermal admittance, W/K
z	combination of terms, dimensionless

Greek Letter Symbols

α combination of terms, dimensionless
δ fin thickness, m
η fin efficiency, dimensionless
θ temperature excess, K
Λ combination of terms, dimensionless
$\mathbf{T}$ linear transmission matrix, dimensions vary
τ elements of linear transmission matrix, dimensions vary
Ψ ratio of admittances, dimensionless

Roman Letter Subscripts

a tip of fin
b base of fin
c cover plate
f fin
i counter
id ideal
in input condition
j counter
l left-hand side
o characteristic value; overall value
p profile area
Q heat flow
s surroundings; separation plate or splitter plate

Greek Letter Subscript

θ temperature excess ratio

CHAPTER 10

COMPACT HEAT EXCHANGERS

10.1 INTRODUCTION

In spacecraft, aircraft, and missiles, space and weight are used sparingly. It is essential in these vehicles that on-board heat interchange duties be accomplished in equipment as compact and light as possible. In cryogenic systems operating with warmer and cooler fluids, both at low temperatures, as in the liquefaction of permanent gases, there is also a stringent need for the heat exchangers to be compact. Only in this way is it possible to minimize the containment surfaces through which heat might leak into the system. Where very low temperatures are involved, the removal of unwanted heat is both difficult and expensive. There are also many other applications for compact heat exchangers between these two extreme examples.

Common tubular exchangers employ a limited range of tube diameters with about 6 to 12 m^2 of surface per cubic meter of exchanger volume when employing tubes 1.5 to 2.5 cm in outside diameter. Compact heat exchangers, through the development of unusual stamping and brazing techniques, contain as much as 500 m^2 of surface per square meter of exchanger volume and are offered in a large variety of external shapes with an even larger geometrical variation of the internals whereby fluids are kept separate as they pass through the exchanger. A compact heat exchanger is an exchanger that contains at least 650 m^2 of surface per cubic meter of volume (approximately 200 ft^2 of surface per cubic foot). This means that a compact heat exchanger must contain some form of extended surface.

In their monumental and extremely useful work entitled *Compact Heat Exchangers*, Kays and London (1984) point out that until the mid-1940s, the only generally available basic heat transfer and flow friction data of sufficient accuracy for heat exchanger design was for flow through and over banks of tubes. This has led to the development of many heat transfer surfaces that are nuch more compact than can be achieved practically with circular tubes. However, for a long period of time, lack of basic heat transfer and flow friction design data and an understanding of the basic mechanisms involved restricted their use to heat exchangers that could be designed and developed by cut-and-try methods. It ultimately became apparent that rationally optimized heat exchanger design, the development of new surfaces of superior characteristics, and the development of methods of fabrication of compact surfaces for high-temperature service could take place only after the basic characteristics of already existing surfaces were well known and understood.

Recognizing the need for such data, the U.S. Navy Bureau of Ships initiated a test program at the Naval Engineering Experiment Station in Annapolis, Maryland, in 1945. In 1947, the Office of Naval Research, in cooperation with the Bureaus of Ships and Aeronautics, extended this work by establishing a similar program at Stanford University. Later, the Atomic Energy Commision joined in the support.

The primary objective of the Stanford program was to investigate the effects of geometry on convection heat transfer and flow friction performance with the hope that the geometric advantages would provide incentive for the development of high-temperature fabrication techniques and new superior surfaces. Since the first publications describing the results of this program, both kinds of developments have occurred.

The experimental apparatus used in the first program at Annapolis was described by London and Ferguson (1949). The experimental apparatus used in the second program at Stanford was described by Kays and London (1950) and Kays et al. (1951). Actually, the experimental systems were quite similar, and cross-checks using the same test cores yielded very consistent results. The American Society of Mechanical Engineers published the first results in a monograph by Kays et al. (1951).

Kays (1960) updated the Kays and London *Compact Heat Exchangers* (1955) by providing heat transfer and friction data for six more high-performance heat transfer surfaces. This work apparently contained the first attempt at the computation of the overall efficiency of a finned passage with unequal dissipation on its sides. Kraus (1960) independently developed an expression for the finned annular passage. This was inspired by a desire to obtain the overall efficiency of the annular region of a double-pipe heat exchanger with the fin tips connected to the outer pipe, which was insulated from the environment. Briggs and London (1961) provided heat transfer and flow friction characteristics for 11 more surfaces: five offset rectangular surfaces and six plain triangular plate fin surfaces.

Shah (1971a,b) tested three geometrically similar offset rectangular plate fin surfaces with area densities from 2393 to 6562 m^2 of surface per cubic meter of volume (900 to 2000 ft^2 of surface per cubic foot) to establish heat transfer and flow friction characteristics. The tests were conducted over a Reynolds number range from 60 to 3000. One surface was unbrazed, but because of a 2:1 difference in plate fin and offset dimensions, two of the surfaces had different degrees of "brazing roughness." The testing revealed that the brazing had little effect on the heat transfer characteristics but influenced the friction characteristics significantly.

Smith (1972) conducted computer analyses and large-scale model experiments to determine the air-side performance of louvered fin evaporator cores. He obtained heat transfer and pressure loss data for several configurations and compared them with calculated values. Effects of condensation, fins per inch, fin thickness, and core characteristics were determined.

Mondt and Siegla (1974) obtained performance data for several perforated fin heat exchanger surfaces, and Wong and Smith (1973) conducted experiments on a 5:1 scale model of a typical louvered fin heat exchanger. These experiments were conducted in a controlled turbulence wind tunnel in order to determine whether full-scale heat exchanger characteristics could be obtained from large-scale data. The air pressure drop and the heat transfer rate of the model core were measured and compared with the full-scale data in terms of drag coefficient, and Nusselt number in terms of Reynolds number based on the face velocity. These data showed a close correlation between the model and the full-scale cores.

Cox and Jallouk (1973) summerized the methods for evaluating the performance of compact heat exchanger surfaces, and Edwards et al. (1973b) used staggered fin tubes in their heat transfer pressure drop experiments. In these experiments, the heat transfer from the fins and the tube was measured separately. Empirical expressions that predicted the mean Stanton number for the fins and the pressure loss factor were produced.

Wieting (1975) developed empirical relationships for heat transfer and flow friction by correlating experimental data from 22 offset fin heat exchanger surfaces. These relationships predicted the performance within the data scatter and allowed for the effects of fin length, height, thickness, and spacing as well as hydraulic diameter to be assessed. The relationships also indicated that the flow passage aspect ratio is significant only in the laminar flow regime.

Mondt (1977) assessed the effect of nonuniform passages on deep-fold heat exchanger performance, and Mochizuki and Yagi (1977) showed that an optimum strip length exists for optimum performance in a strip fin heat exchanger core. This conclusion was based on careful experiments on seven aluminum fin test cores. Mandel et al. (1979) considered the optimal fin side design of compact tube-in-fin heat exchangers with rippled fins. Zhukauskas and Ulinskas (1979) gave a historical review of developments pertaining to compact heat exchangers in the USSR. They considered current (at that time) possibilities for heat transfer augmentation on rough and finned tubular surfaces. They also suggested experimental correlations for heat transfer and pressure drop in heat exchangers employing smooth tubes, and they discussed the applicability of bimetallic and plastic heat exchanger elements.

Abbott et al. (1979) traced the use of compact heat exchangers in the products of their employer, and Parker and Coombs (1979) suggested and illustrated new uses for compact heat exchangers for both aircraft and industrial applications.

Chen et al. (1981) extended a 1975 study by including the effect of the fluid velocity on the heat transfer rate to a boiling liquid in a compact plate fin heat exchanger. Confirmation of their method was obtained by a comparison of their predictions with data obtained from tests using refrigerant R-113 in an actual exchanger. Webb and Joshi (1982) provided a friction factor correlation for an offset strip fin matrix, and Shah (1982) provided a keynote address at an international heat transfer conference which dealt with advances in compact heat exchanger technology and design theory.

Davenport (1983b) gave correlations for heat transfer and flow friction in several louvered fin surfaces. Chen and Westwater (1983) showed the application of the local assumption in the design of compact heat exchangers for boiling heat transfer, and Robertson (1984) developed a model for predicting flow boiling coefficients for serrated fin packages.

With so many shapes from which to choose, it is not unusual that there may be some overlapping of nomenclature. To clarify terms as they are used in this chapter, one variation of the fundamental compact exchanger element, the core, is shown in Fig. 10.1. The core consists of a pair of parallel plates with connnecting metal members that are bonded to the plates. The arrangement of plates and bonded members provides both a fluid-flow channel and prime and extended surface. It is observed that if a plane were drawn midway between the two plates, each half of the connecting metal members could be considered as longitudinal fins.

Moreover, as shown in Chapter 9, two or more identical cores can be connected by *separation* or *splitter plates*, and this arrangement is called a *stack* or *sandwich*. Heat can enter a stack through either or both end plates. However, the heat is removed from the successive separating plates and fins by a fluid flowing in parallel through the entire network with a single average convection heat transfer coefficient. For this reason, the stack was treated in Chapter 9 as a finned passage rather than a fluid–fluid heat exchanger, and of course, due consideration must be given to the fact that as more and more fins are placed in a core, the hydraulic radius of the core is lowered while the pressure loss is increased significantly.

Next, consider a pair of cores arranged as components of a two-fluid exchanger in cross flow as shown in Fig. 10.2. Fluids enter alternate cores from separate headers

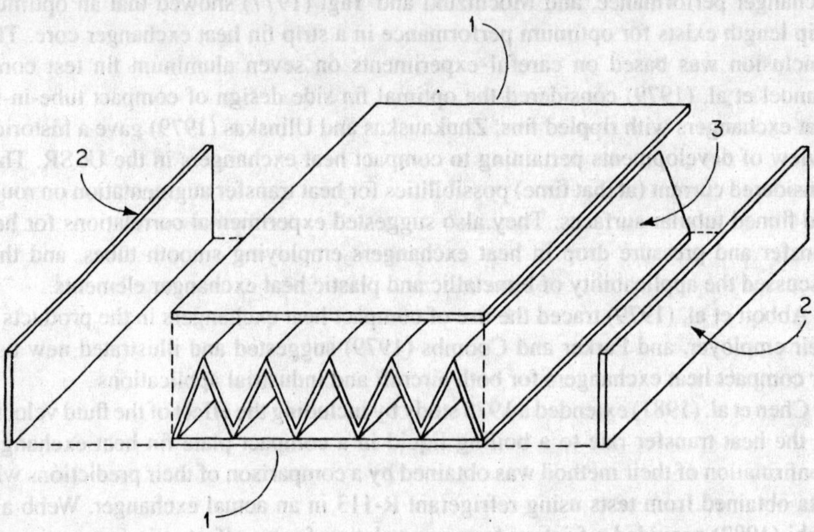

Figure 10.1 Exploded view of a compact heat exchanger core: 1, plates; 2, side bars; and 3, corrugated fins stamped from a continuous strip of metal. By spraying braze powder on the plates, the entire assembly of plates, fins, and bars can be thermally bonded in a single furnace operation.

at right angles to each other and leave through separate headers at opposite ends of the exchanger. The separation plate spacing need not be the same for both fluids, nor need the cores for both fluids contain the same numbers or kinds of fins. These are dictated by the allowable pressure drops for both fluids and the resulting heat transfer coefficients. While one coefficient is quite large compared with the other, it is entirely permissible to have no extended surface in the alternate cores through which the fluid with the higher coefficient travels. An exchanger built up with plates and fins as in Fig. 10.2 is a *plate fin* heat exchanger.

The discussion of plate fin exchangers has concentrated thus far on geometries involving two or more fluids which enter the body of the compact heat exchanger by means of headers. In many instances, one of the fluids may be merely air used as a cooling medium on a once-through basis. Typical examples include the air-fin cooler, which is treated in Chapter 11, and the *radiators* associated with various types of internal combustion engines. Similarly, there are examples in which the compact heat exchanger is a coil that is inserted into a duct as in air-conditioning applications. A small selection of compact heat exhanger elements available from various manufacturers is shown in Fig. 10.3.

10.1.1 Classification of Compact Heat Exchangers

Compact heat exchangers may be classified by the kinds of compact elements that they employ. The compact elements usually fall into five classes:

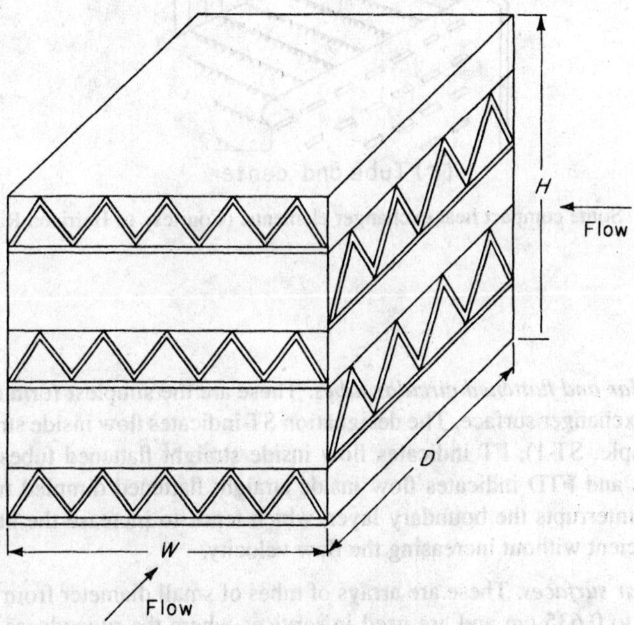

Figure 10.2 Two-fluid compact heat exchanger with headers removed.

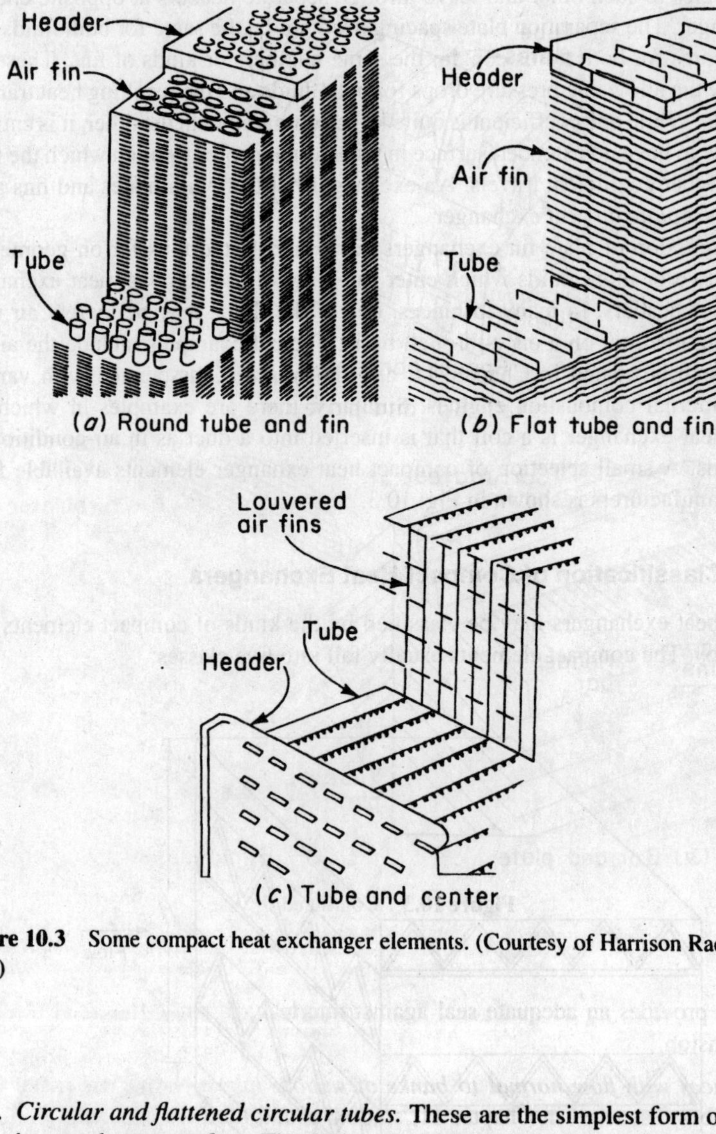

Figure 10.3 Some compact heat exchanger elements. (Courtesy of Harrison Radiator Division.)

1. *Circular and flattened circular tubes.* These are the simplest form of compact heat exchanger surface. The designation ST indicates flow inside straight tubes (example, ST-1), FT indicates flow inside straight flattened tubes (example, FT-1), and FTD indicates flow inside straight flattened dimpled tubes. Dimpling interrupts the boundary layer, which tends to increase the heat transfer coefficient without increasing the flow velocity.

2. *Tubular surfaces.* These are arrays of tubes of small diameter from 0.9525 cm down to 0.635 cm and are used in services where the ruggedness and cleanability of the conventional shell-and-tube exchanger are not required. Usually, tube sheets are comparatively thin, and soldering or brazing a tube to a tube

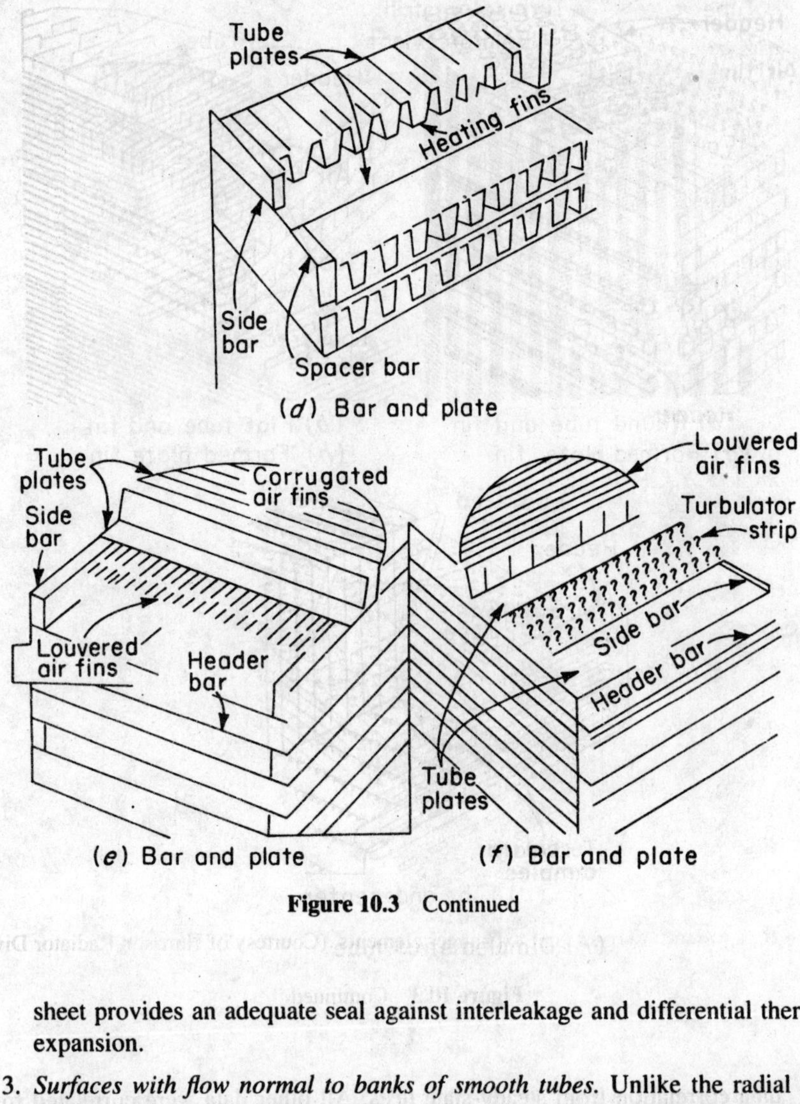

Figure 10.3 Continued

sheet provides an adequate seal against interleakage and differential thermal expansion.

3. *Surfaces with flow normal to banks of smooth tubes.* Unlike the radial low fin tubes discussed in Chapter 11, smooth round tube are expanded into fins that can accept a number of tube rows, as shown in Fig. 10.3*a*. Holes may be stamped in the fin with a drawn hub or foot to improve contact resistance or as a spacer between successive fins, as shown, or brazed directly to the fin with or without a hub. Other types reduce the flow resistance outside the tubes by using flattened tubes and brazing as indicated in Fig. 10.3*b* and *c*. Flat tubing is made from strips similar to the manufacture of welded circular tubing but is much thinner and is joined by soldering or brazing rather than welding.

The designation considers staggered (S) and in-line (I) arrangements of tubes and identifies transverse and longitudinal pitch ratios. The suffix (s) indicates

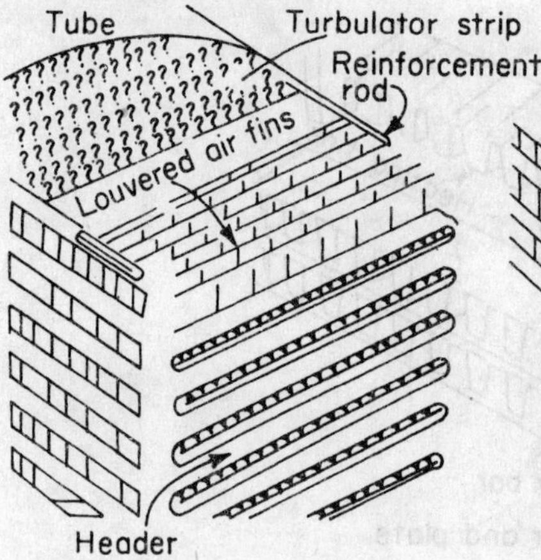

(*g*) Formed plate fin

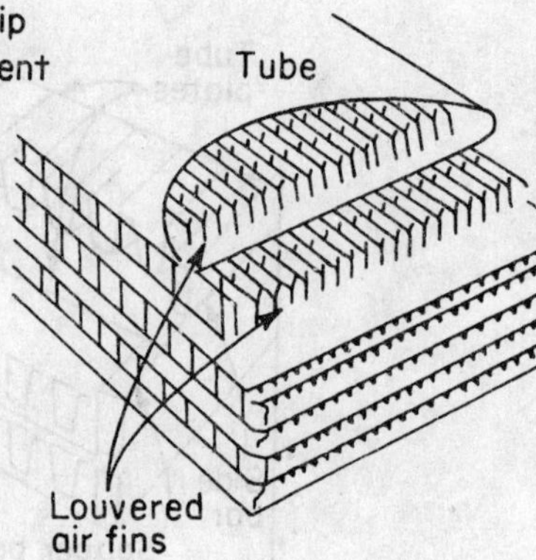

(*h*) Formed plate fin

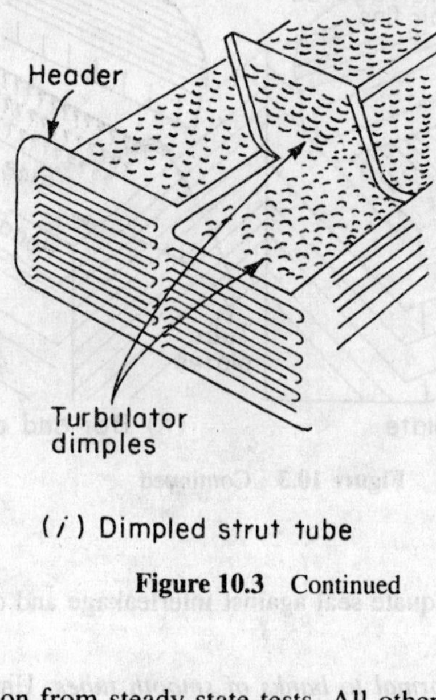

(*i*) Dimpled strut tube

Figure 10.3 Continued

data correlation from steady-state tests. All other data were correlated from a transient technique. Examples include the surface S1.50-1.25(s), which is a staggered arrangement with data obtained via steady-state tests with transverse pitch-to-diameter ratio of 1.50 and longitudinal pitch-to-diameter ratio of 1.25. The surface I1.25-1.25 has an in-line arrangement with data obtained from transient tests with both tranverse and longitudinal pitch-to-diameter ratios of 1.25.

4. *Plate fin surfaces.* These are shown in Fig. 10.3*d* through *i*.

a. The *plain fin* is characterized by long uninterrupted flow passages and is designated by a numeral that indicates the number of fins per inch. The suffix T is appended when the passages are of definite triangular shape. Examples are the surfaces 19.86, 15.08, and 46.45T.

b. The *louvered fin* is characterized by fins that are cut and bent into the flow stream at frequent intervals and is designated by a fraction that indicates the length of the fin in the flow direction (inches) followed by a numeral that indicates the number of fins per inch. For example, the designation

$$\tfrac{1}{2}\text{-}6.06$$

indicates 6.06 $\frac{1}{2}$-in.-long fins per inch.

c. The *strip fin* is designated in the same manner as the louvered fin. The suffixes (D) and (T) indicate double and triple stacks. The strip fins are frequently referred to as *offset fins* because they are offset at frequent intervals, and the exchanger is essentially a series of plate fins with alternate lengths offset.

d. The *wavy fin* is characterized by a continuous curvature. The change in flow direction introduced by the waves in the surface tends to interrupt the boundary layer as in the case of louvered and strip fins. Wavy fin designations are always followed by the letter W. For example, the

$$11.44\text{-}\tfrac{3}{8}\text{W}$$

is a wavy fin with 11.44 fins per inch and a $\frac{3}{8}$-in. wave.

e. The *pin fin* surface is constructed from small-diameter wires. This surface yields very high heat transfer coefficients because the effective flow length is very small. The designation of the pin fin surfaces is nondescriptive.

f. The *perforated fin* surface has holes cut in the fins to provide boundary layer interruption. These fins are designated by the number of fins per inch followed by the letter P.

5. *Finned-tube surfaces.* Circular tubes with spiral radial fins are designated by the letters CF followed by one or two numerals. The first numeral designates the number of fins per inch and the second, if one is used, refers to the nominal tube size. With circular tubes with continuous fins, no letter prefix is employed, and the two numerals have the same meaning as those used for circular tubes with spiral radial fins. For finned flat tubes no letter prefix is used; the first numeral indicates the fins per inch, and the second numeral indicates the largest tube dimension. When CF does not appear in the designation of the circular tube with spiral radial fins, the surface may be presumed to have continuous fins.

6. *Matrix surfaces.* These are surfaces that are used in rotating regenerative equipment such as combustion flue gas–air preheaters for conventional fossil furnaces. In this application, metal is deployed for its ability to absorb heat with minimal fluid friction while being exposed to hot flue gas and to give up this heat to incoming cold combustion air when it is rotated into the incoming cold airstream. No designation is employed.

Compact heat exchanger surfaces are described in the literature by geometric factors that have been standardized largely through the extensive work of Kays and London (1984). These factors and the relationships between them are essential for application of the basic heat transfer and flow friction data to a particular design problem. They are listed and defined in Table 10.1. Physical data for a number of compact heat exchanger surfaces are given in Table 10.2. The relationships between the geometric factors in Table 10.1 will now be established.

Consider an exchanger composed of n_1 layers of one type of plate fin surface and n_2 layers of a second type, as shown in Fig. 10.2. The separation plate thickness is established by the pressure differential to which it is exposed or through designer discretion. Retaining the subscripts, 1 and 2 for the respective types of surface, the overall exchanger height H is

$$H = n_1(b_1 + a) + n_2(b_2 + a) \tag{10.1}$$

where b_1 and b_2 are separation distances between the plates for the two kinds of surface. With the width W and depth D selected, the overall volume V is

$$V = WDH \tag{10.2}$$

In Fig. 10.2, the length L_1 is along the depth of the exchanger ($L_1 = D$) and the length L_2 is along the width ($L_2 = W$).

The frontal areas are also established. Again referring to Fig. 10.2,

$$A_{f1} = HW \tag{10.3a}$$

and

$$A_{f2} = HD \tag{10.3b}$$

If the entire exhanger consisted of a single exchanger surface, surface 1 or surface 2, the total surface area would be the product of the ratio of total surface to total volume β (m^2/m^3), and the total volume V. However, where there are two surfaces, it is necessary to employ the factor α, which is the ratio of the total surface on *one* side to the total surface on *both* sides of the exchanger. By taking simple proportions,

$$\alpha_1 = \frac{b_1}{b_1 + b_2 + 2a}\beta_1 \tag{10.4a}$$

and

$$\alpha_2 = \frac{b_2}{b_1 + b_2 + 2a}\beta_2 \tag{10.4b}$$

and the total surfaces will be

$$S_1 = \alpha_1 V \tag{10.5a}$$

and

$$S_2 = \alpha_2 V \tag{10.5b}$$

TABLE 10.1 Compact Heat Exchanger Geometric Factors

Factor and Symbol	Descriptive Comment
A	The free flow area on one side of the exchanger. To distinguish hot and cold sides, the free flow areas are frequently designated by A_h and A_c.
A_f	The frontal area on one side of the exchanger. This is merely the product of the overall exchanger width and height or depth and height.
a	The separation plate thickness. This applies to plate fin surfaces only and in Chapter 9 was designated by a_c or, if a splitter plate, by a_s.
b	The separation plate spacing. This dimension is an approximation of the fin height. Applies to plate fin surfaces only.
d_e	The equivalent diameter used to correlate flow friction and heat transfer and is four times the hydraulic radius r_h.
L	The flow length on one side of the exchanger. Note that this factor always concerns the flow length of a single side of the exchanger, although two sides may be present, and that the ambiguity is avoided with the overall exchanger dimensions which are designated *width*, *depth*, and *height*. It is therefore reasonable to have the overall exhanger depth be the length on one side of the exchanger and the overall width the length on the other side.
P	The perimeter of the passage.
p	The porosity, which is the ratio of the exchanger void volume to the total exhanger volume. Applies to matrix surfaces only.
r_h	The hydraulic radius, which is the ratio of the passage flow area to its wetted perimeter.
S	The heat transfer surface on one side of the exchanger. Subscripts are often appended to distinguish between hot- and cold-side surfaces.
S_f	The surface of the fins, only, on one side of the exchanger. Applies to finned surfaces only.
V	The total exchanger volume. This applies to both sides of the heat exchanger and is merely the product of the overall heat exchanger width, depth, and height.
α	The ratio of the total surface area on one side of the exchanger to the total volume on both sides of the exchanger. Applies to tubular, plate fin surfaces and crossed-rod matrices only.
β	The ratio of the total surface area to the total volume on one side of the exchanger. The surface alone is S. The total total volume includes the overall exchanger dimensions. Applies to plate fin surfaces only.
δ	The fin thickness.
η	The fin efficiency.
η_o	The overall passage efficiency.
σ	The ratio of the free flow area to the frontal area on one side of the exchanger.

The hydraulic radius is defined as the flow area divided by the wetted perimeter of the passage:

$$r_h = \frac{A}{P} = \frac{AL}{S} \tag{10.6}$$

TABLE 10.2 Surface Geometry of Some Plate Fin Surfaces

Plain Plate Fins	11.1	15.08	19.86	46.45T
b (10^{-3}m)	6.35	10.62	6.35	2.54
Fins per inch	11.1	15.08	19.86	46.45
d_e (10^{-3} m)	3.08	2.67	1.875	0.805
δ (10^{-3} m)	0.152	0.152	0.152	0.051
β (m^2/m^3)	1204	1358.3	1840.6	4371.7
S_f/S	0.756	0.870	0.849	0.837
Louvered Fins	$\frac{3}{8}$-6.06	$\frac{1}{2}$-6.06	$\frac{3}{16}$-11.1	$\frac{3}{4}$-11.1
b (10^{-3} m)	6.35	6.35	6.35	6.35
Fins per inch	6.06	6.06	11.1	11.1
d_e 10^{-3} m)	4.453	4.453	3.084	3.084
δ (10^{-3} m)	0.152	0.152	0.152	0.152
β (m^2/m^3)	840	840	1204	1204
S_f/S	0.640	0.640	0.756	0.756
Strip Fins	$\frac{1}{8}$-13.95	$\frac{1}{8}$-16.00D	$\frac{1}{8}$-19.82D	$\frac{1}{8}$-20.06
b (10^{-3} m)	9.54	6.48	5.21	5.11
Fins per inch	13.95	16.00	19.82	20.06
d_e (10^{-3} m)	2.68	1.86	1.54	1.49
δ (10^{-3} m)	0.254	0.152	0.102	0.102
β (m^2/m^3)	1250	1804	2231	2290
S_f/S	0.840	0.845	0.841	0.843
Wavy and Pin Fins	11.5-$\frac{3}{8}$W	17.8-$\frac{3}{8}$W	AP-1	PF-3
b (10^{-3} m)	9.53	10.49	6.10	19.1
Fins per inch or fin pattern	11.5	17.8	In-line	In-line
d_e (10^{-3} m)	3.02	2.12	4.40	1.636
δ or pin diameter (10^{-3} m)	0.254	0.152	1.02	0.79
β (m^2/m^3)	1138	1686	617	1112
S_f/S	0.822	0.892	0.512	0.834

and the ratio of the flow area to the frontal area is designated as σ:

$$\sigma = \frac{A}{A_f} \tag{10.7}$$

For all but matrix surfaces, because eq. (10.6) leads to $A = Sr_h/L$,

$$\sigma = \frac{Sr_h}{A_f L} = \frac{Sr_h}{V} = \alpha r_h \tag{10.8}$$

Thus the flow areas are given by

$$A_1 = \sigma_1 A_{f1} \tag{10.9a}$$

and

$$A_2 = \sigma_2 A_{f2} \tag{10.9b}$$

10.2 HEAT TRANSFER AND FLOW FRICTION DATA

10.2.1 Heat Transfer Data

Heat transfer data for compact heat exchangers are correlated on an individual surface basis using a Colburn type of representation. This representation plots the heat transfer factor j_h:

$$j_h = \text{St}\text{Pr}^{2/3} = \frac{h}{Gc_p}\left(\frac{c_p\mu}{k}\right)^{2/3} \tag{10.10}$$

as a function of the Reynolds number, which is obtained, as shown in Chapter 4, by employing the equivalent diameter $d_e = 4r_h$:

$$\text{Re} = \frac{4r_h G}{\mu} = \frac{d_e G}{\mu} \tag{10.11}$$

The Stanton number, St, is the ratio of the Nusselt number, Nu, to the product of the Reynolds and Prandtl numbers, RePr[1]:

$$\text{St} = \frac{\text{Nu}}{\text{RePr}} = \frac{hd_e/k}{(d_e G/\mu)(c\mu/k)} = \frac{h}{Gc} \tag{10.12}$$

The fluid properties in eqs. (10.10) and (10.11) are evaluated at the bulk temperature. Employing the nomenclature T_1 and T_2 for the hot fluid inlet and outlet temperatures, the hot-side bulk temperature is given by

$$T_b = \tfrac{1}{2}(T_1 + T_2) \tag{10.13a}$$

In similar fashion, with t_1 and t_2, for the cold fluid inlet and outlet temperatures, the cold-side bulk temperature is given by

$$t_b = \tfrac{1}{2}(t_1 + t_2) \tag{10.13b}$$

10.2.2 Flow Friction Data

Kays and London (1984) suggest that the pressure drop ΔP in a compact heat exchanger be computed from the equation

$$\frac{\Delta P}{P_1} = \frac{G^2 v_1}{2g_c P_1}[\Phi_1 + \Phi_2 + \Phi_3 - \Phi_4] \tag{10.14}$$

[1]Because compact heat exchangers find their greatest utility in gas-to-gas applications, in eqs. (10.10) and (10.12) the specific heat c is the value for constant pressure

where

$$\Phi_1 = 1 + K_c - \sigma^2$$

$$\Phi_2 = 2\left(\frac{v_2}{v_1} - 1\right)$$

$$\Phi_3 = f\frac{S}{A}\frac{v_m}{v_1}$$

$$\Phi_4 = (1 - \sigma^2 - K_e)\frac{v_2}{v_1}$$

Friction factors are correlated on an individual surface basis and are usually plotted as a function of the Reynolds number. The entrance and exit loss coefficients differ for the various types of passages and are plotted as functions of the parameter σ and the Reynolds number.

Four terms may be noted within the parentheses in eq. (10.14). These terms denote, respectively, the entrance or contraction loss as the fluid approaches the exchanger at line velocity and changes to the exchanger entrance velocity, acceleration loss or gain as the fluid expands or contracts during its passage through the exchanger, flow friction loss, and exit loss.

Kays and London (1984) have presented heat transfer and flow friction data for approximately 120 surfaces described by the foregoing. Some typical examples are shown in Figs. 10.4 through 10.7. Entrance and exit loss coefficients for plate fin cores and rectangular passages are plotted in Figs. 10.8 and 10.9.

10.2.3 Overall Heat Transfer Coefficient

The overall heat transfer coefficient may be based on the surface of the hot or cold side. In the absence of fouling, these relationships, based on the hot side are

$$U_h = \frac{1}{1/\eta_{oh}h_h + a/(S_w/S_h)k_w + 1/(S_c/S_h)\eta_{oc}h_c} \tag{10.15a}$$

where the subscript w refers to the wall between the fluids and where η_o refers to the overall surface efficiency discussed in Chapter 9.

Based on the cold side,

$$U_c = \frac{1}{1/\eta_{oc}h_c + a/(S_w/S_c)k_w + 1/(S_h/S_c)\eta_{oh}h_h} \tag{10.15b}$$

Metals have high values of thermal conductivity and with thin separation plates, the ratio a/k_w tends to be small, so that eqs. (10.15a) and (10.15b) can be written as

$$U_h = \frac{1}{1/\eta_{oh}h_h + 1/(S_c/S_h)\eta_{oc}h_c} \tag{10.15c}$$

$$U_c = \frac{1}{1/\eta_{oc}h_c + 1/(S_h/S_c)\eta_{oh}h_h} \tag{10.15d}$$

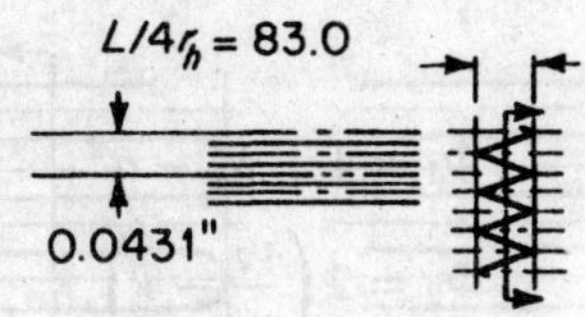

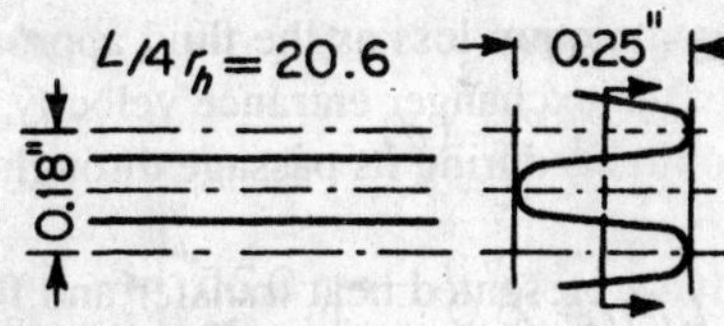

Figure 10.4 Heat transfer and flow friction characteristics of some plain plate fin compact heat exchanger surfaces. [From Kays and London (1984).]

10.3 ϵ–N_{tu} METHOD

The ϵ–N_{tu} method makes use of three dimensionless parameters: ϵ, R, and N_{tu}. The *exchanger effectiveness* or simply the effectiveness ϵ, is designated in the nomenclature of the ϵ–N_{tu} method by

$$\epsilon = \frac{C_h(T_1 - T_2)}{C_{\min}(T_1 - t_1)} = \frac{C_c(t_2 - t_1)}{C_{\min}(T_1 - t_1)} \tag{10.16}$$

where $C_h = \dot{m}_h c_h$ is the product of the mass flow $\dot{m}_h$ and the specific heat c_h of the hot fluid working between inlet temperature T_1 and outlet temperature T_2. In similar

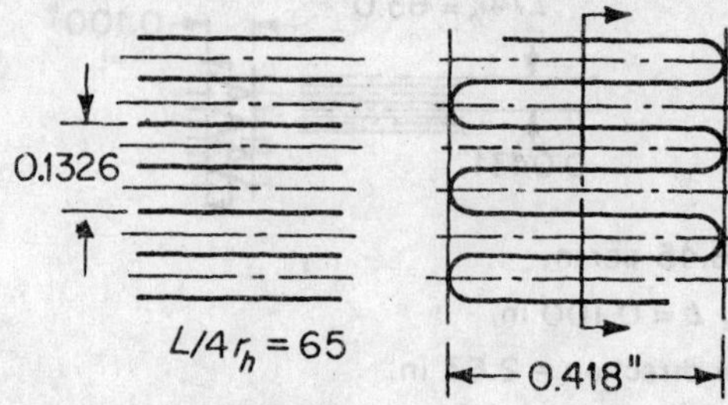

Fin pitch = 15.08 per in.

Plate spacing, $b = 0.418$ in.

Flow passage hydraulic diameter, $4r_h = 0.00876$ ft

Fin metal thickness = 0.006 in., aluminum

Total transfer area/volume between plates, $\beta = 414\ ft^2/ft^3$

Fin area/total area = 0.870

(*c*)

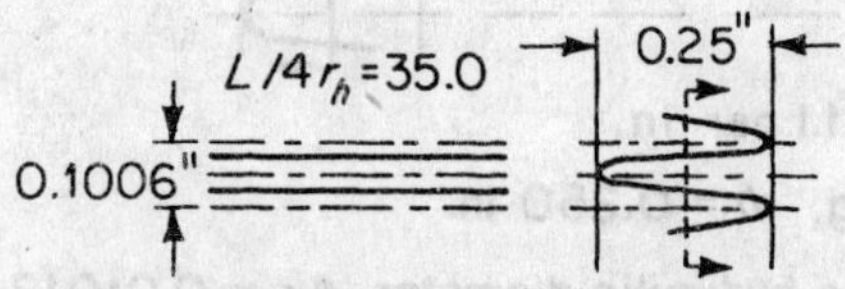

Fin pitch = 19.86 per in.

Plate spacing, $b = 0.250$ in.

Flow passage hydaulic diameter, $4r_h = 0.00615$ ft

Fin metal thickness = 0.006 in., aluminum

Total transfer area/volume between plates, $\beta = 561\ ft^2/ft^3$

Fin area/total area = 0.849

(*d*)

Figure 10.4 Continued

fashion, $C_c = \dot{m}_c c_c$ is the product of the mass flow $\dot{m}_c$ and the specific heat c_c of the cold fluid working between inlet temperature t_1 and outlet temperature t_2. Moreover, C_{min} is the smaller of the C_h or C_c values and it is noted that ϵ is a measure of the amount of heat transferred by the exchanger to the amount of heat that could be transferred if the exchanger had infinite surface.

The *number of transfer units* N_{tu} is defined by

$$N_{tu} = \frac{US}{C_{min}} = \frac{t_2 - t_1}{\text{LMTD}} \tag{10.17}$$

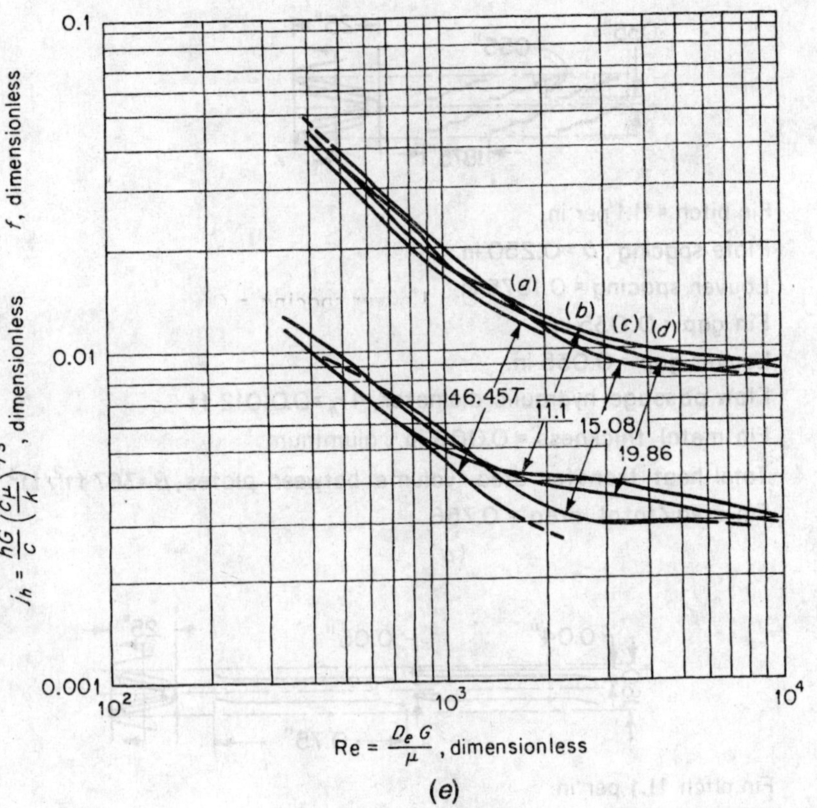

Figure 10.4 Continued

where U is the overall heat transfer coefficient, S the surface to which it refers, and LMTD the true temperature difference. Consider $t_2 - t_1$ as the temperature duty to be accomplished and LMTD as the potential for accomplishing it. Then N_{tu} has the significance of measuring the size of the heat transfer task, namely, the duty to be accomplished divided the temperature potential available for its accomplishment.

The third dimensionless parameter, the *capacity rate ratio*, is uniquely expressed by

$$R = \frac{C_{min}}{C_{max}} \tag{10.18}$$

The definitions of ϵ, R, and N_{tu} of eqs. (10.16), (10.17), and (10.18) may be applied to the heat balance:

$$q = US(\text{LMTD}) = C_h(T_1 - T_2) = C_c(t_2 - t_1)$$

to give when $C_c = C_{min}$:

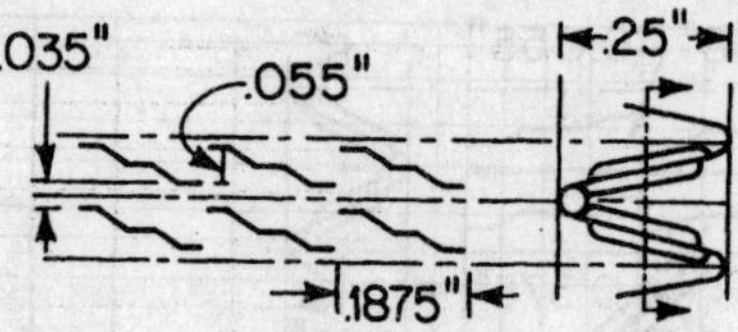

Fin pitch = 11.1 per in.
Plate spacing, b = 0.250 in.
Louver spacing = 0.1875 in.
Fin gap = 0.035 in.
Louver gap = 0.055 in.
Flow passage hydraulic diameter, $4r_h$ = 0.01012 ft
Fin metal thickness = 0.006 in., aluminum
Total heat transfer area / volume between plates, β = 367 ft^2/ft^3
Fin area / total area = 0.756

(a)

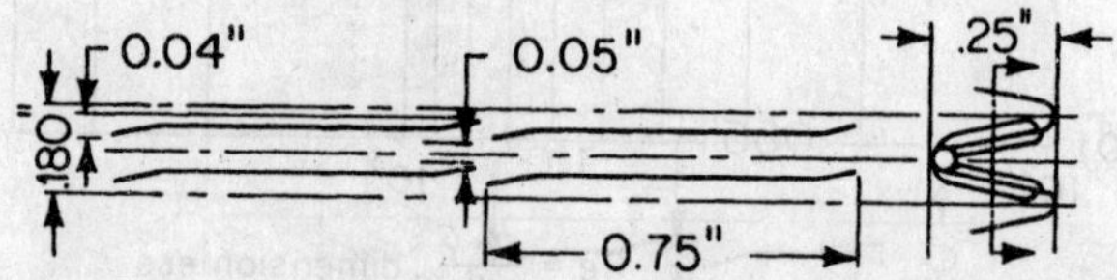

Fin pitch 11.1 per in.
Plate spacing, b = 0.250 in.
Louver spacing = 0.75 in.
Fin gap = 0.05 in.
Louver gap = 0.04 in.
Flow passage hydraulic diameter, $4r_h$ = 0.01012 ft.
Fin metal thickness = 0.006 in., aluminum
Total heat transfer area / volume between plates, β = 367 ft^2/ft^3
Fin area / total area = 0.756

(b)

Figure 10.5 Heat transfer and flow friction characteristics of some louvered fin compact heat exchanger surfaces. [From Kays and London (1984).]

$$q = C_c(t_2 - t_1) = \epsilon C_c \frac{C_{\min}}{C_c}(T_1 - t_1) = \epsilon C_{\min}(T_1 - t_1) \qquad (10.19a)$$

and when $C_h = C_{\min}$:

$$q = C_h(T_1 - T_2) = \epsilon C_h \frac{C_{\min}}{C_h}(T_1 - t_1) = \epsilon C_{\min}(T_1 - t_1) \qquad (10.19b)$$

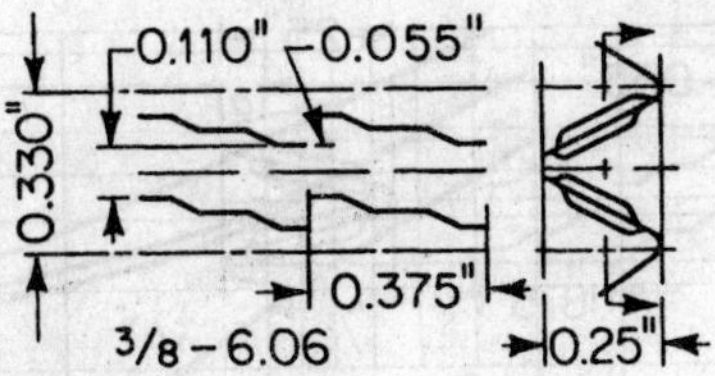

Fin pitch = 6.06 per in.

Plate spacing, b = 0.250 in.

Louver spacing = 0.375 in.

Fin gap = 0.110 in.

Louver gap = 0.055

Flow passage hydraulic diameter, $4r_h$ = 0.01460 ft

Fin metal thickness = 0.006 in., aluminum

Total transfer area/volume between plates, $\beta = 256\ ft^2/ft^3$

Fin area/total area = 0.640

(*c*)

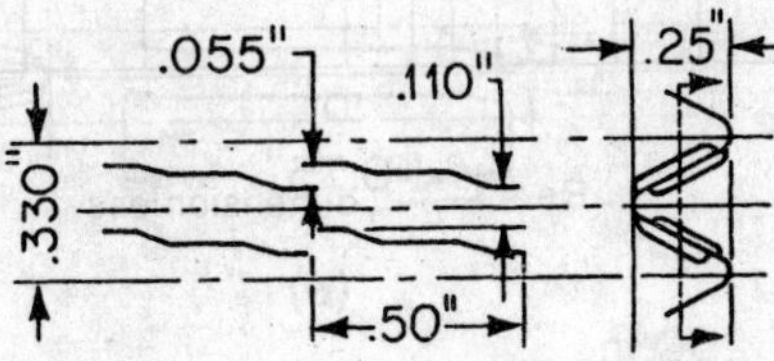

Fin pitch = 6.06 per in.

Plate spacing, b = 0.250 in.

Louver spacing = 0.50 in.

Fin gap = 0.110 in.

Louver gap = 0.055 in.

Flow passage hydraulic diameter, $4r_h$ = 0.01460 ft

Fin metal thickness = 0.006 in., aluminum

Total heat transfer area/volume between plates, $\beta = 256\ ft^2/ft^3$

Fin area/total area = 0.640

(*d*)

Figure 10.5 Continued

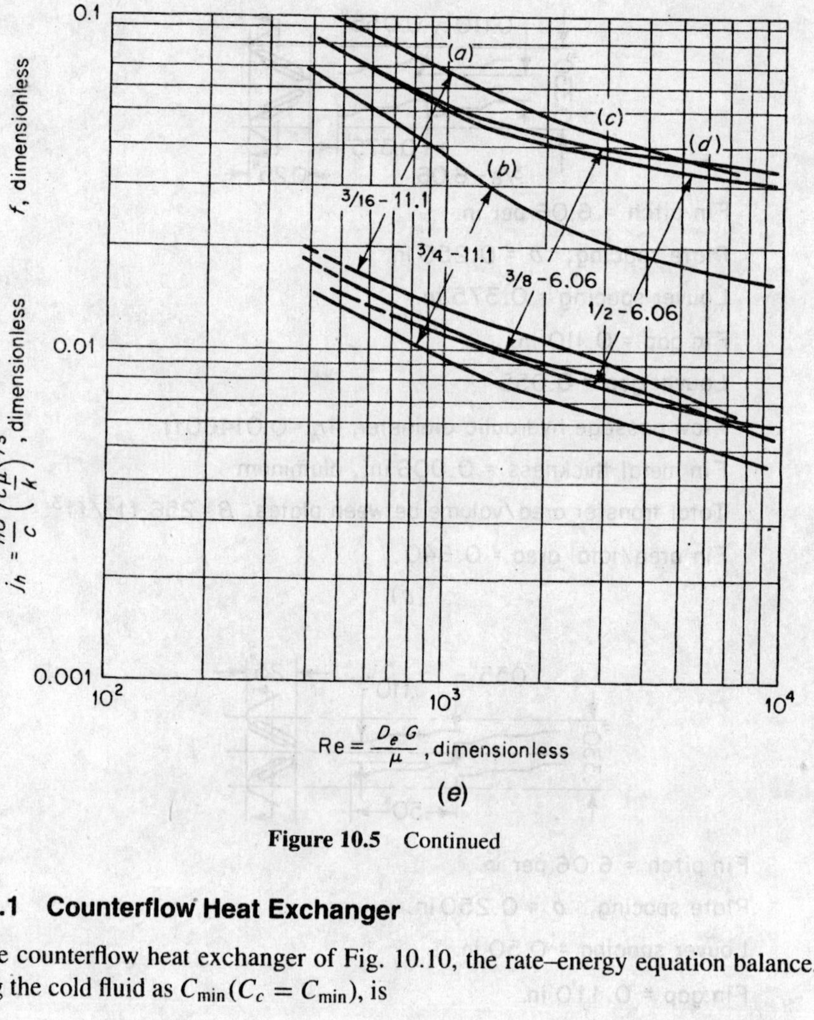

Figure 10.5 Continued

10.3.1 Counterflow Heat Exchanger

In the counterflow heat exchanger of Fig. 10.10, the rate–energy equation balance, using the cold fluid as $C_{min} (C_c = C_{min})$, is

$$q = US \frac{(T_1 - t_2) - (T_2 - t_1)}{\ln [(T_1 - t_2)/(T_2 - t_1)]} = C_c(t_2 - t_1)$$

Rearrangement gives

$$N_{tu} = \frac{US}{C_c} = \frac{t_2 - t_1}{(T_1 - T_2) - (t_2 - t_1)} \ln \frac{T_1 - t_2}{T_2 - t_1} \tag{10.20}$$

or, with further adjustment and substitution of eq. (10.18),

$$\frac{T_1 - t_2}{T_2 - t_1} = e^{N_{tu}(R-1)} \tag{10.21}$$

Adding and subtracting t_1 to the numerator and T_1 to the denominator of eq. (10.21) and then applying eq. (10.18) yields

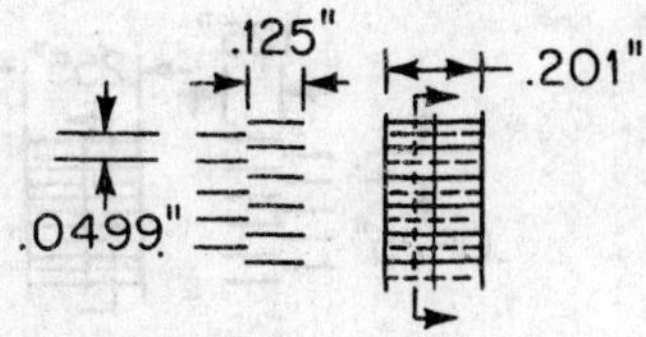

Fin pitch = 20.06 per in.
Plate spacing, b = 0.201 in.
Splitter symmetrically located
Fin length flow direction = 0.125 in.
Flow passage hydraulic diameter, $4r_h$ = 0.004892 ft.
Fin metal thickness = 0.004 in., aluminum
Splitter metal thickness = 0.006 in.
Total heat transfer area / volume plates, β = 698 ft^2/ft^3
Fin area (including splitter) / total area = 0.843

(*a*)

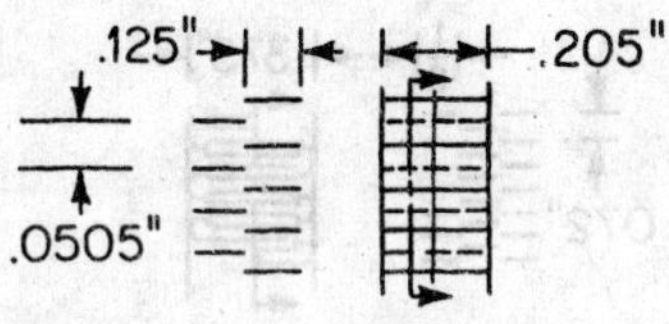

Fin pitch = 19.82 per in.
Plate spacing, b = 0.205 in.
Splitter symmetrically located
Fin length flow direction = 0.125 in.
Flow passage hydraulic diameter, $4r_h$ = 0.005049 ft
Fin metal thickness = 0.004 in., nickel
Splitter metal thickness = 0.006 in.
Total heat transfer area / volume between plates, β = 680 ft^2/ft^3
Fin area (including splitter) / total area = 0.841

(*b*)

Figure 10.6 Heat transfer and flow friction characteristics of some strip fin compact heat exchanger surfaces. [From Kays and London (1984).]

$$\frac{T_1 + t_1 - t_2 - t_1}{T_2 - t_1 + T_1 - T_1} = \frac{T_1 - t_1 - (t_2 - t_1)}{T_1 - t_1 - R(t_2 - t_1)} = e^{\mathrm{N_{tu}}(R-1)}$$

and with further expansion, factoring, and rearrangement employing eq. (10.16), the result is

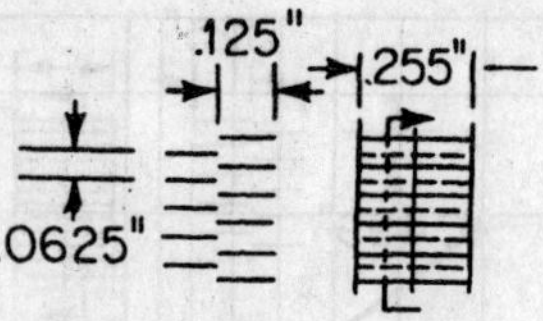

Fin pitch = 16.00 per in.

Plate spacing, b = 0.255 in.

Splitter symmetrically located

Fin length flow direction = 0.125 in.

Flow passage hydraulic diameter, $4r_h$ = 0.006112 ft.

Fin metal thickness = 0.006 in., aluminum

Splitter metal thickness = 0.006 in.

Total heat transfer area / volume between plates, β = 550 ft^2/ft^3

Fin area (including splitter) / total area = 0.845

(*c*)

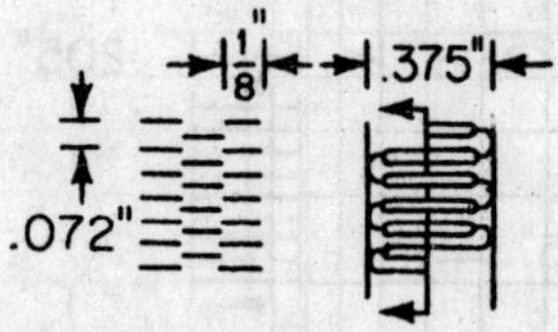

Fin pitch = 13.95 per in.

Plate spacing, b = 0.375 in.

Fin length = 0.125 in.

Flow passage hydraulic diameter, $4r_h$ = 0.00879 ft.

Fin metal thickness = 0.010 in., aluminum

Total heat transfer area / volume between plates, β = 381 ft^2/ft^3

Fin area / total area = 0.840

Note: The fin surface area on the leading and trailing edges of the fins have not been included in area computations.

(*d*)

Figure 10.6 Continued

$$\frac{1 - (t_2 - t_1)/(T_1 - t_1)}{1 - R[(t_2 - t_1)(T_1 - t_1)]} = e^{N_{tu}(R-1)} \tag{10.22}$$

Solving for ϵ yields

$$\epsilon = \frac{t_2 - t_1}{T_1 - t_1} = \frac{1 - e^{N_{tu}(R-1)}}{1 - Re^{N_{tu}(R-1)}}$$

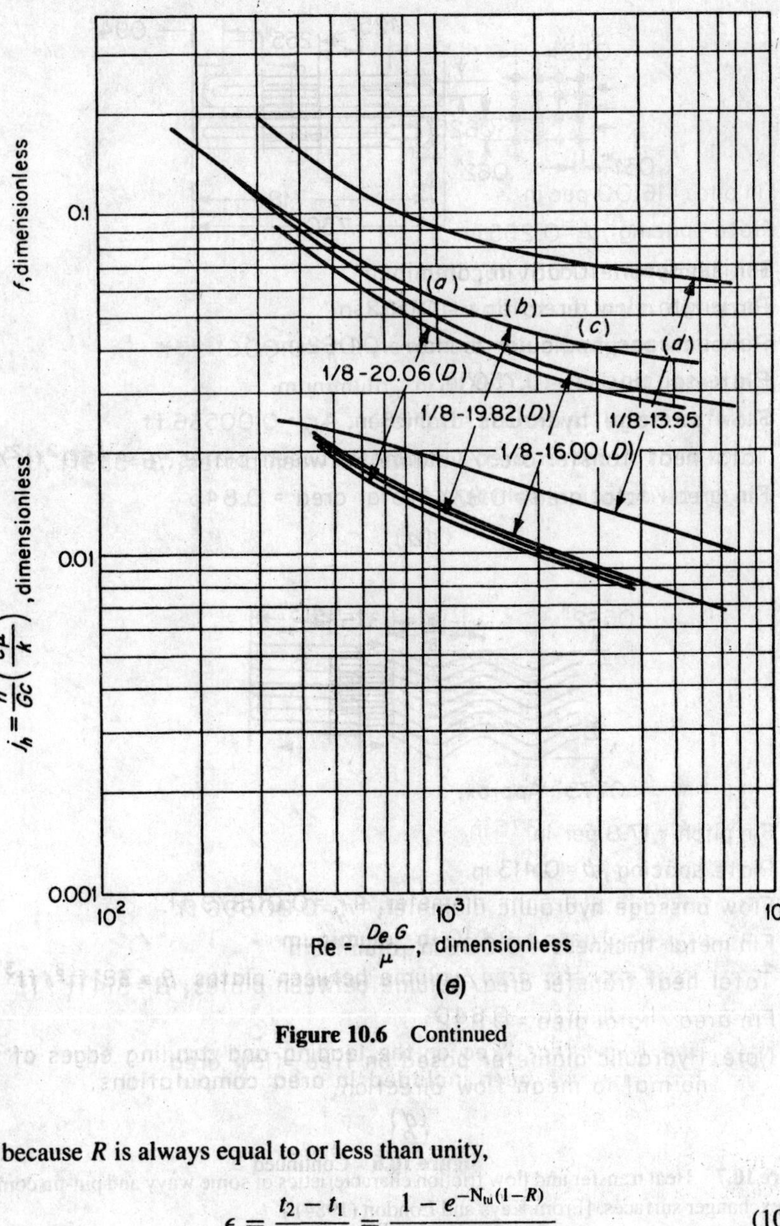

Figure 10.6 Continued

and because R is always equal to or less than unity,

$$\epsilon = \frac{t_2 - t_1}{T_1 - t_1} = \frac{1 - e^{-N_{tu}(1-R)}}{1 - Re^{-N_{tu}(1-R)}} \tag{10.23}$$

Using $c_h = C_{min}$ leads to an identical result but with ϵ having the hot fluid temperature range in the numerator:

$$\epsilon = \frac{T_1 - T_2}{T_1 - t_1} = \frac{1 - e^{-N_{tu}(1-R)}}{1 - Re^{-N_{tu}(1-R)}} \tag{10.24}$$

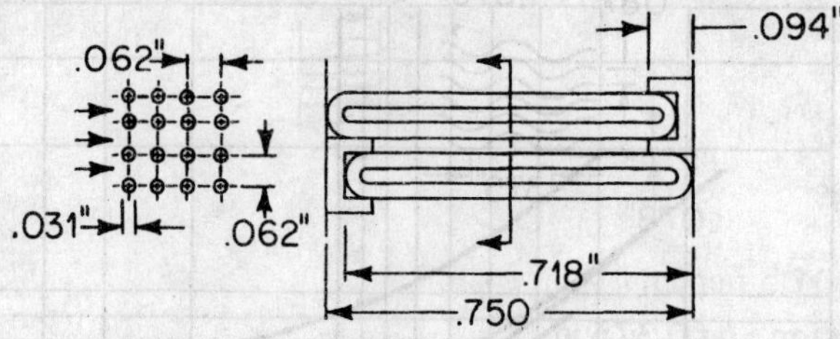

Pin diameter = 0.031 in., aluminum
Pin pitch parallel to flow = 0.062 in.
Pin pitch perpendicular to flow = 0.062 in.
Plate spacing, b = 0.750 in.
Flow passage hydraulic diameter, $4r_h$ = 0.00536 ft
Total heat transfer area / volume between plates, $\beta = 339\ \text{ft}^2/\text{ft}^3$
Fin area / total area = 0.834

(a)

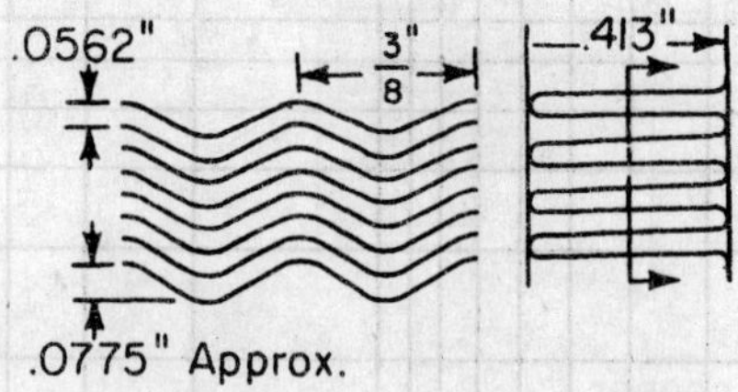

Fin pitch = 17.8 per in.
Plate spacing, b = 0.413 in.
Flow passage hydraulic diameter, $4r_h$ = 0.00696 ft
Fin metal thickness = 0.006 in., aluminum
Total heat transfer area / volume between plates, $\beta = 514\ \text{ft}^2/\text{ft}^3$
Fin area / total area = 0.892
Note: Hydraulic diameter based on free-flow area normal to mean flow direction.

(b)

Figure 10.7 Heat transfer and flow friction characteristics of some wavy and pin-fin compact heat exchanger surfaces. [From Kays and London (1984).]

Both eqs. (10.23) and (10.24) become indeterminate when $R = 1$. In this case the logarithmic mean temperature difference is LMTD = $T_1 - t_2 = T_2 - t_1$, and the heat balance may be written

$$US(T_1 - t_2) = C_c(t_2 - t_1) = C_h(T_1 - T_2)$$

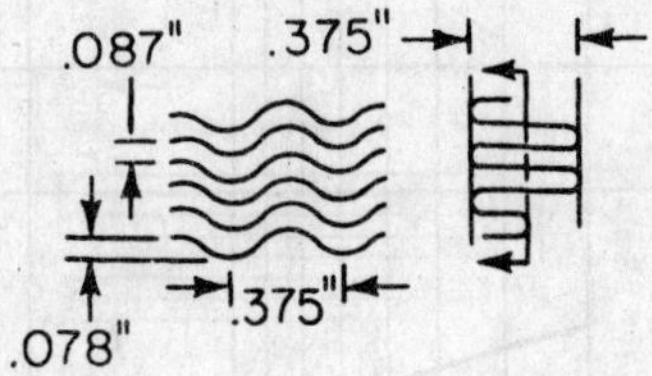

Fin pitch = 11.5 per in

Plate spacing, b = 0.375 in.

Flow passage hydraulic diameter, $4r_h$ = 0.00993 ft.

Fin metal thickness = 0.010 in., aluminum

Total heat transfer area / volume between plates, $\beta = 347\ ft^2/ft^3$

Fin area / total area = 0.822

Note: Hydraulic diameter based on free-flow area normal to mean flow direction.

(*c*)

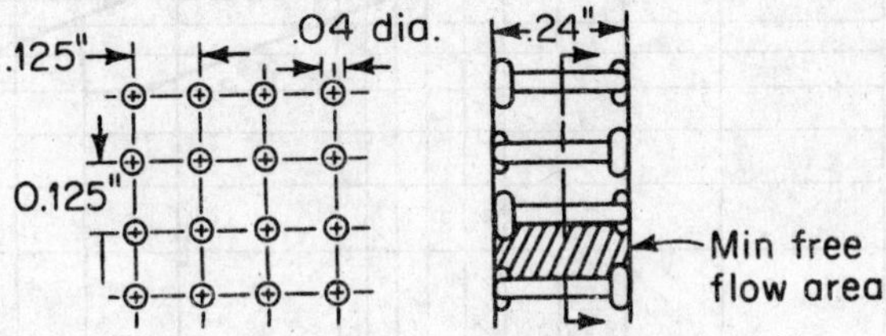

Pin diameter = 0.04 in., copper

Pin pitch parallel to flow = 0.125 in.

Pin pitch perpendicular to flow = 0.125 in.

Plate spacing, b = 0.24 in.

Flow passage hydraulic diameter, $4r_h$ = 0.01444 ft

Total heat transfer area / volume between plates, $\beta = 188\ ft^2/ft^3$

Fin area / total area = 0.512

(*d*)

Figure 10.7 Continued

or, using $C_c = C_{\min}$,

$$N_{tu} = \frac{US}{C_c} = \frac{t_2 - t_1}{T_1 - t_2} = \frac{t_2 - t_1}{T_1 - t_1 - (t_2 - t_1)}$$

Then, proceeding algebraically,

$$N_{tu} = \frac{(t_2 - t_1)/(T_1 - t_1)}{1 - (t_2 - t_1)/(T_1 - t_1)} = \frac{\epsilon}{1 - \epsilon}$$

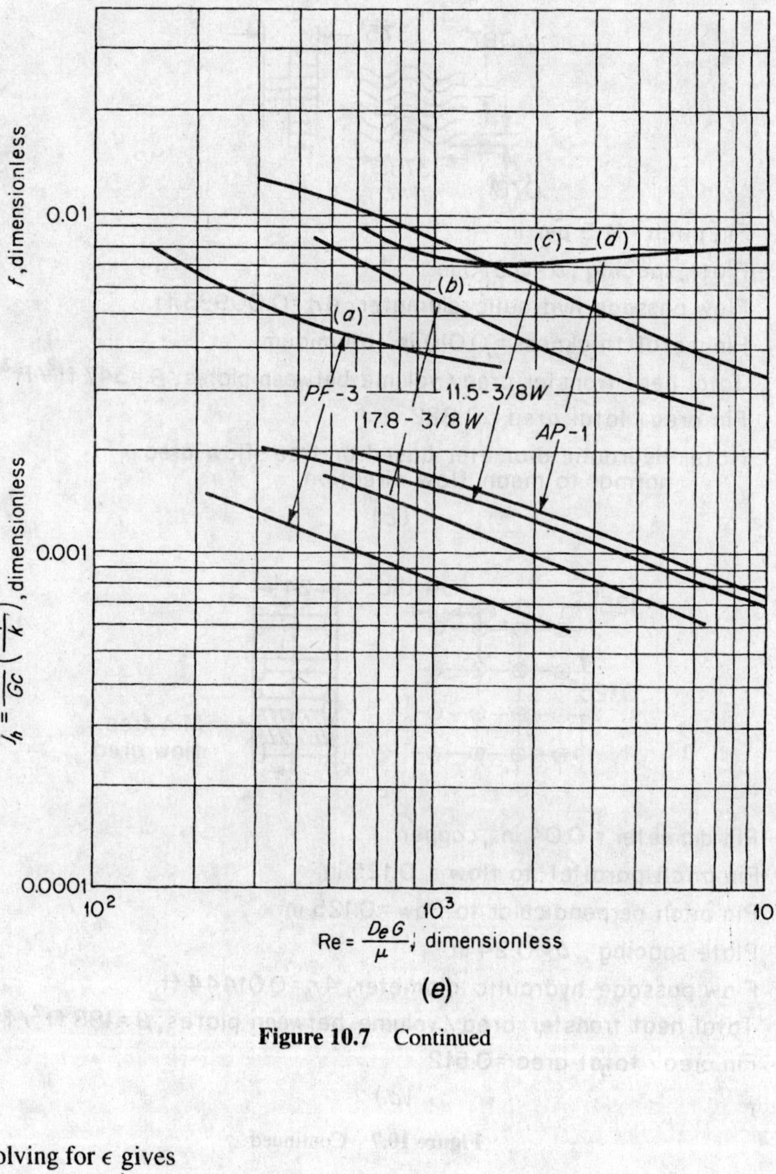

(e)

Figure 10.7 Continued

and solving for ϵ gives

$$\epsilon = \frac{t_2 - t_1}{T_1 - t_1} = \frac{N_{tu}}{N_{tu} + 1} \tag{10.25}$$

10.3.2 Other Flow Arrangements

Mueller (1985) and Baclic (1988) have provided catalogs of ϵ–N_{tu} relationships for many different heat exchanger configurations. Some of these are listed in this section,

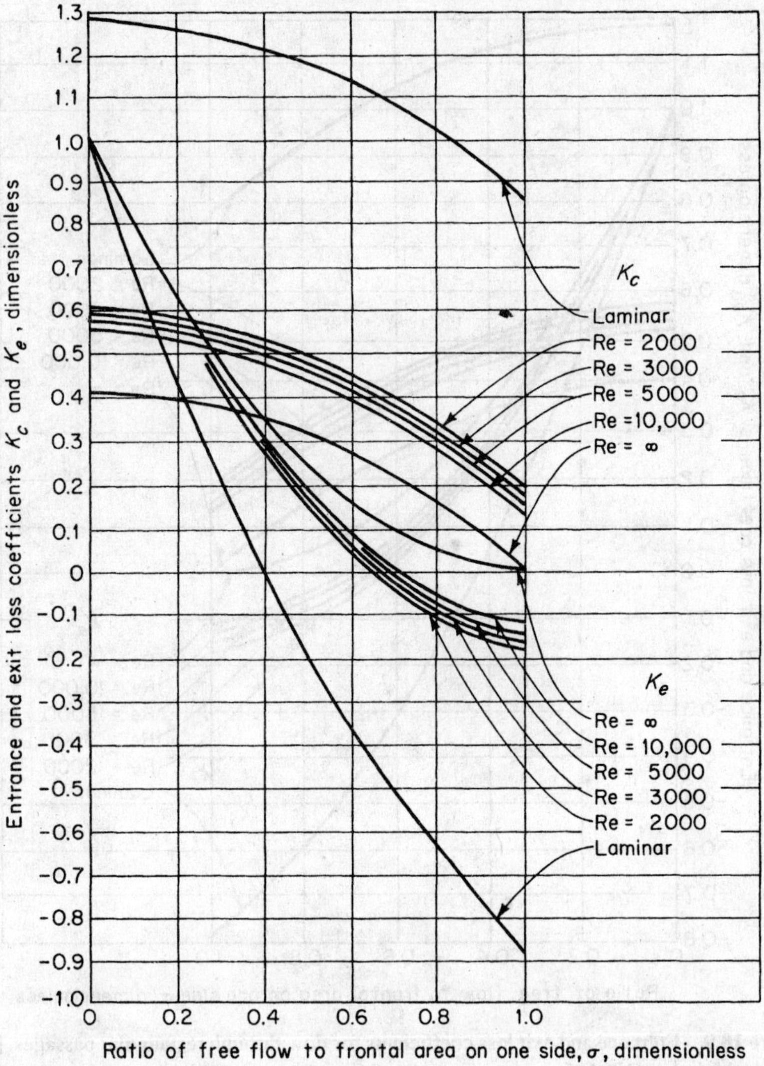

Figure 10.8 Entrance and exit loss coefficients for flow through plate fin exchanger cores. [From Kays and London (1984).]

and the listing is divided into two parts: Cross-flow configurations and parallel-counterflow configurations.

Cross Flow. For cross flow, with both fluids unmixed, effectiveness values were obtained by Nusselt (1911, 1930) whose work was extended by Smith (1934). Mason (1955) obtained a solution in the form of an infinite series by employing the Laplace transformation.

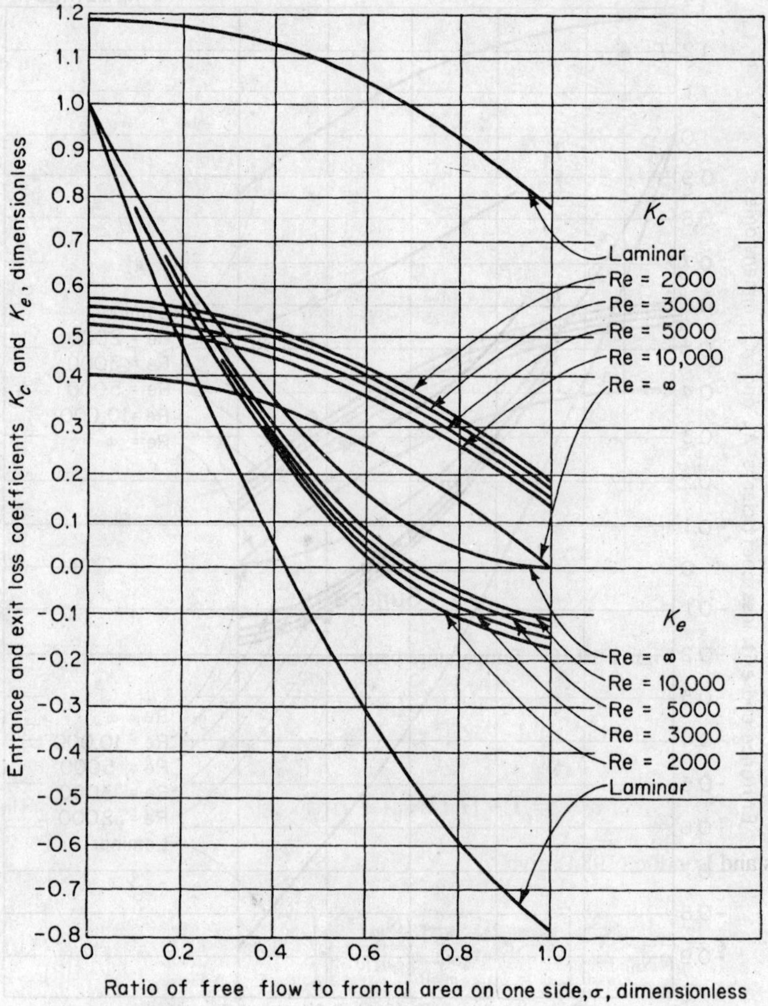

Figure 10.9 Entrance and exit loss coefficients for flow through rectangular passages. [From Kays and London (1984).]

Both Fluids Unmixed. For cross flow with both fluids unmixed, Mueller (1985) gives

$$\epsilon = 1 - e^{-\mathrm{N_{tu}}(R+1)}\left[I_0(2\mathrm{N_{tu}}R^{1/2}) + R^{1/2} I_1(2\mathrm{N_{tu}}R^{1/2}) - \frac{1-R}{R}\sum_{n=2}^{\infty} R^{n/2} I_n(2\mathrm{N_{tu}}R^{1/2}) \right] \quad (10.26a)$$

and if $R = 1$,

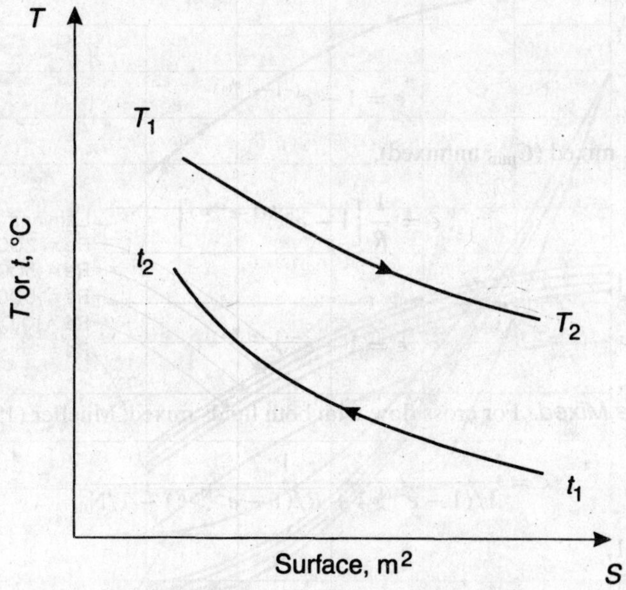

Figure 10.10 Temperature pattern in counterflow exchanger.

$$\epsilon = 1 - [I_0(2N_{tu}) + I_1(2N_{tu})]e^{-2N_{tu}} \tag{10.26b}$$

Kays and London (1984) give

$$\epsilon = \frac{1}{RN_{tu}} \sum_{n=0}^{\infty} \left[1 - e^{-N_{tu}} \sum_{m=0}^{n} \frac{(N_{tu})^m}{m!}\right]\left[1 - e^{-RN_{tu}} \sum_{m=0}^{n} \frac{(RN_{tu})^m}{m!}\right] \tag{10.26c}$$

which is based on the analysis of Mason (1955). Mills (1992) gives

$$\epsilon = 1 - e^{(N_{tu})^{0.22}\Lambda/R} \tag{10.26d}$$

where

$$\Lambda = e^{-R(N_{tu})^{0.78}} - 1$$

One Fluid Mixed and One Fluid Unmixed. The ϵ–N_{tu} cross-flow arrangement with one fluid mixed and one fluid unmixed depends on which fluid is mixed. With C_{min} mixed (C_{max} unmixed), Mueller (1985) gives

$$\epsilon = 1 - e^{-(1-e^{-RN_{tu}})/R} \tag{10.27a}$$

and if $R = 1$,

$$\epsilon = 1 - e^{(-1-e^{-N_{tu}})} \tag{10.27b}$$

For C_{max} mixed (C_{min} unmixed),

$$\epsilon = \frac{1}{R}\left[1 - e^{-R(1-e^{-N_{tu}})}\right] \tag{10.27c}$$

and if $R = 1$,

$$\epsilon = 1 - e^{-(1-e^{-N_{tu}})} \tag{10.27d}$$

Both Fluids Mixed. For cross flow with both fluids mixed, Mueller (1985) gives

$$\epsilon = \frac{1}{1/(1-e^{-N_{tu}}) + R/(1-e^{-N_{tu}R}) - 1/N_{tu}} \tag{10.28a}$$

and if $R = 1$,

$$\epsilon = \frac{1}{2/(1-e^{-N_{tu}}) - 1/N_{tu}} \tag{10.28b}$$

Parallel Flow

The parallel or cocurrent flow arrangement is but a specific case of a single parallel-flow channel with no counterflow channels. In this case the ϵ–N_{tu} relationship is easily derived. However, Mueller (1990) gives

$$\epsilon = \frac{1 - e^{-N_{tu}(R+1)}}{R+1} \tag{10.29a}$$

and when $R = 1$,

$$\epsilon = \tfrac{1}{2}\left(1 - e^{-2N_{tu}}\right) \tag{10.29b}$$

Shell-and-Tube Exchanger

For the 1–2 shell-and-tube exchanger with the shell fluid mixed and the tube fluid mixed between the two passes, Mueller (1990) gives

$$\epsilon = \frac{2}{(1+R) + (1+R^2)^{1/2}\coth\dfrac{N_{tu}(1+R^2)^{1/2}}{2}} \tag{10.30a}$$

and when $R = 1$,

$$\epsilon = \frac{2}{2 + 2^{1/2}\coth\dfrac{\sqrt{2}}{2}N_{tu}} \tag{10.30b}$$

10.4 DESIGN OF A COMPACT HEAT EXCHANGER

The steps in the design of or performance calculation for a compact heat exhanger with single hot and cold fluids follows a rather simple but possibly laborious procedure.

1. Establish the heat balance.
2. Assume an overall exchanger size (length, width, depth) and select the types of surfaces to be used. Using the geometric data for the assumed surfaces, compute the surface areas, free flow areas, and other physical parameters.
3. Obtain the thermal properties of the fluids at their bulk temperatures.
4. Compute the heat transfer coefficients.
5. Obtain the fin and overall passage efficiencies.
6. Determine the overall heat transfer coefficient.
7. Obtain R and N_{tu} and compute the actual exchanger surface effectiveness. Compare this value with the required effectiveness.
8. Calculate the pressure drops.

The sequence suggested by these steps is illustrated in the example that follows.

Example 10.1. An engine bleed air precooler is required for the air-conditioning system of a large commercial jet transport. Specifications for the unit entail the precooling of 0.575 kg/s of hot bleed air at 389.6 kPa from 356 to 177°C with a pressure drop not to exceed 1 kPa. Cool air at a flow rate of 0.545 kg/s is available at a temperature of 56°C and a pressure of 55.16 kPa with an allowable pressure drop of 2.20 kPa. The unit is to be installed in a space measuring 28 cm by 23 cm by 15.25 cm. Design a cross-flow compact heat exchanger for this application using aluminum throughout.

SOLUTION. (1) Heat balance. The hot airstream has a bulk temperature of

$$T_b = \tfrac{1}{2}(T_1 + T_2) = \tfrac{1}{2}(356 + 177) = \tfrac{1}{2}(533) = 266.5^\circ\text{C}$$

and at this temperature, Lienhard (1987) gives

$$c_{p,h} = 1.037 \text{ kJ/kg} \cdot \text{K}$$

Hence

$$C_h = \dot{m}_h c_{p,h} = (0.575)(1.037) = 0.596 \text{ kW/K}$$

and

$$\begin{aligned} q_h &= C_h(T_1 - T_2) \\ &= (0.596)(356 - 177) \end{aligned}$$

$$= (0.596)(179)$$

$$= 106.73 \text{ kW}$$

For the cold side, the outlet temperature is not known. Assume that it is 250°C, so that

$$t_b = \tfrac{1}{2}(t_2 + t_1) = \tfrac{1}{2}(250 + 56) = \tfrac{1}{2}(306) = 153°\text{C}$$

At this temperature, Lienhard (1987) gives

$$c_{p,c} = 1.016 \text{ kJ/kg} \cdot \text{K}$$

and

$$C_c = \dot{m}_c c_{p,c} = (0.545)(1.016) = 0.554 \text{ kW/K}$$

Hence with $q_h = q_c = 106.73$ kW,

$$t_2 = t_1 + \frac{q_c}{C_c}$$

$$= 56 + \frac{106.73}{0.554}$$

$$= 56 + 193$$

$$= 249°\text{C}$$

Notice that the assumed $t_2 = 250$°C closely approximates the calculated $t_2 = 249$°C. In summary, the heat balance is

Bleed air: $q_h = (0.596)(356 - 177) = 106.73$ kW
Cold air: $q_c = (0.554)(249 - 56) = 106.73$ kW

(2) *Exchanger physical data.* For the hot side, assume 21 single stacks of strip fin surface $\frac{1}{8}$-16.00(D), 21.60 cm long by 13.82 cm wide. As shown in Fig. 10.6, this surface is composed of a pair of surfaces, each 0.125 in. or 0.3175 cm high with a single splitter plate 0.005 in. or 0.0127 cm thick. The total height of this surface is 0.255 in. or 0.6477 cm.

For the cold side, assume 22 single stacks of strip fin surface $\frac{1}{8}$-19.82(D), 21.60 cm long by 13.82 cm wide. As shown in Fig. 10.6, this surface is composed of a pair of surfaces, each 0.100 in. or 0.2540 cm high with a single splitter plate 0.005 in. or 0.0127 cm thick. The total height of this surface is 0.205 in. or 0.5207 cm.

The separation plates are to be aluminum and are selected to be 0.0152 cm thick so that the exchanger height is in accordance with eq. (10.1):

$$H = n_1(b_1 + a) + n_2(b_2 + a)$$

$$= (21)(0.6477 + 0.0152) + (22)(0.5207 + 0.0152)$$

$$= (21)(0.6629) + (22)(0.5359)$$

$$= 13.9209 + 11.7898$$
$$= 25.7107 \text{ cm}$$

For the hot side, Table 10.2 gives for the $\frac{1}{8}$-16.00(D) surface,

$$\beta_h = 1804 \text{ m}^2/\text{m}^3$$
$$S_f/S = 0.845$$
$$d_{eh} = 1.86 \times 10^{-3} \text{ m}$$

and

$$\delta_h = 0.152 \times 10^{-3} \text{ m}$$

Then, from eq. (10.2),

$$V = WDH$$
$$= (21.60)(13.82)(25.71)$$
$$= 7674.7 \text{ cm}^3$$

or

$$= 7.675 \times 10^{-3} \text{ m}^3$$

from eq. (10.3a),

$$A_{fh} = HW$$
$$= (25.71)(13.82)$$
$$= 355.31 \text{ cm}^2$$

or

$$A_{fh} = 0.0355 \text{ m}^2$$

from eq. (10.4*a*),

$$\alpha_h = \frac{b_h}{b_h + b_c + a}\beta_h$$
$$= \left[\frac{0.6477}{0.6477 + 0.5207 + 2(0.0152)}\right](1804)$$
$$= (0.5472)(1804)$$
$$= 987.2 \text{ m}^2/\text{m}^3$$

from eq. (10.5*a*),

$$S_h = \alpha_h V$$

$$= (987.2)(7.675 \times 10^{-3})$$
$$= 7.577 \text{ m}^2$$

from eq. (10.8),

$$\sigma_h = \frac{\alpha_h d_{eh}}{4}$$
$$= \frac{(987.2)(1.86 \times 10^{-3})}{4}$$
$$= 0.459$$

and from eq. (10.9),

$$A_h = \sigma_h A_{fh}$$
$$= (0.459)(0.0355)$$
$$= 0.0163 \text{ m}^2$$

For the cold side, Table 10.2 gives for the $\frac{1}{8}$-19.82(D) surface

$$\beta_c = 2231 \text{ m}^2/\text{m}^3$$
$$S_f/S = 0.841$$
$$d_{ec} = 1.54 \times 10^{-3} \text{ m}$$

and

$$\delta_c = 0.102 \times 10^{-3} \text{ m}$$

Then, following the computations for the hot side with $V = 7.675 \times 10^{-3}$ m^3,

$$A_{fc} = HD$$
$$= (25.71)(21.60)$$
$$= 555.34 \text{ cm}^2$$

or

$$A_{fc} = 0.0555 \text{ m}^2$$

from eq. (10.4*a*),

$$\alpha_c = \frac{b_h}{b_h + b_c + a}\beta_h$$
$$= \left[\frac{0.5207}{0.6477 + 0.5207 + 2(0.0152)}\right](2231)$$

$$= (0.4399)(2231)$$
$$= 981.5 \text{ m}^2/\text{m}^3$$

$$S_c = \alpha_c V$$
$$= (981.5)(7.675 \times 10^{-3})$$
$$= 7.532 \text{ m}^2$$

$$\sigma_c = \frac{\alpha_c d_{ec}}{4}$$
$$= \frac{(981.5)(1.54 \times 10^{-3})}{4}$$
$$= 0.378$$

and

$$A_c = \sigma_c A_{fc}$$
$$= (0.378)(0.0555)$$
$$= 0.0210 \text{ m}^2$$

(3) *Fluid properties.* The fluid properties for the fluids at the already determined bulk temperatures can be obtained from Lienhard (1987):

Property	Bleed Air	Cold Air
Temperature (°C)	266.5	153
Specific heat c (kJ/kg · K)	1.037	1.016
Thermal conductivity k (W/m · K)	0.0419	0.0346
Dynamic viscosity μ (kg/m · s)	2.824×10^{-5}	2.394×10^{-3}
$\text{Pr} = c\mu/k$	0.698	0.702
$\text{Pr}^{2/3}$	0.787	0.790

(4) *Heat transfer coefficients.* For the hot side with $C_h = 0.596$ kJ/K,

$$G_h = \frac{\dot{m}_h}{A_h}$$
$$= \frac{0.575}{0.0163}$$
$$= 35.28 \text{ kg/m}^2 \cdot \text{s}$$

and

$$\text{Re}_h = \frac{d_{eh} G_h}{\mu}$$

$$= \frac{(1.86 \times 10^{-3})(35.28)}{2.824 \times 10^{-5}}$$

$$= 2323$$

From Fig. 10.6, read

$$j_h = \frac{h_h}{G_h c_h} \text{Pr}^{1/3} = 0.0105 \quad \text{and} \quad f_h = 0.0362$$

Then

$$h_h = \frac{j_h G_h c_h}{\text{Pr}^{2/3}}$$

$$= \frac{(0.0105)(35.28)(1037)}{0.787}$$

$$= 488.1 \text{ W/m}^2 \cdot \text{K}$$

The cold-side computations with $C_c = 0.554$ kJ/K,

$$G_c = \frac{\dot{m}_c}{A_c}$$

$$= \frac{0.545}{0.0210}$$

$$= 25.95 \text{ kg/m}^2 \cdot \text{s}$$

and

$$\text{Re}_c = \frac{d_{ec} G_c}{\mu}$$

$$= \frac{(1.54 \times 10^{-3})(25.95)}{2.394 \times 10^{-5}}$$

$$= 1669$$

From Fig. 10.6, read

$$j_h = \frac{h_c}{G_c c_c} \text{Pr}^{1/3} = 0.0111 \quad \text{and} \quad f_h = 0.0478$$

Then

$$h_c = \frac{j_c G_c c_c}{\text{Pr}^{2/3}}$$

$$= \frac{(0.0111)(25.95)(1016)}{0.790}$$

$$= 370.4 \text{ W/m}^2 \cdot \text{K}$$

(5) *Fin and overall passage efficiencies.* Both of the surfaces employed in this design may be treated as double stacks with equal heat dissipation on both sides. The configuration is shown in Fig. 9.4*a*, and eq. (9.22) will yield the fin efficiency. The thermal conductivity of aluminum (99% pure) is listed by Lienhard (1987) as $k = 211$ W/m · K.

For the hot side, pertinent data are

$$b_h = 3.175 \times 10^{-3} \text{ m}$$

$$s = \frac{1}{(2)(16)(12)}(0.3048)$$

$$= 7.938 \times 10^{-4} \text{ m}$$

$$\delta_{fh} = 1.52 \times 10^{-4} \text{ m}$$

$$a_s = 1.27 \times 10^{-4} \text{ m}$$

and

$$h_h = 488.1 \text{ W/m}^2 \cdot \text{K}$$

Then for the fin

$$m_f = \left(\frac{2h_h}{k\delta_f}\right)^{1/2} = \left[\frac{(2)(488.1)}{(211)(1.52 \times 10^{-4})}\right]^{1/2} = 174.46 \text{ m}^{-1}$$

$$m_f b = (174.46)(3.175 \times 10^{-3}) = 0.5539$$

$$\tanh m_f b = \tanh 0.5534 = 0.5035$$

and for unit fin length,

$$Y_{o,f} = (2h_h k\delta_{fh})^{1/2} = [(2)(488.1)(211)(1.52 \times 10^{-4})]^{1/2} = 5.5954 \text{ W/K}$$

For the splitter plate

$$m_s = \left(\frac{2h_h}{ka_s}\right)^{1/2} = \left[\frac{(2)(488.1)}{(211)(1.27 \times 10^{-4})}\right]^{1/2} = 190.86 \text{ m}^{-1}$$

$$m_s s = (190.86)(7.938 \times 10^{-4}) = 0.1515$$

$$\tanh m_s s = \tanh 0.1515 = 0.1504$$

and for unit fin length

$$Y_{o,s} = (h_c k a_s)^{1/2} = [(488.1)(211)(1.27 \times 10^{-4})]^{1/2} = 3.6166 \text{ W/K}$$

Then by eq. (9.22),

$$\eta_{fh} = \frac{Y_{o,f}}{2(b+s)Lh} \frac{\tanh m_f b + (2Y_{o,s}/Y_{o,f}) \tanh m_s s}{1 + (2Y_{o,s}/Y_{o,f}) \tanh m_s s \tanh m_f b}$$

With

$$2\frac{Y_{o,s}}{Y_{o,f}} = 2\left(\frac{3.6166}{5.5954}\right) = 1.2927$$

and the surface attributed to the fins,

$$\begin{aligned} S_f &= 2(b+s)L \\ &= (2)(3.175 \times 10^{-3} + 7.938 \times 10^{-4})(1) \\ &= 7.938 \times 10^{-3}\ \text{m}^2 \end{aligned}$$

then

$$\begin{aligned} \eta_{fh} &= \frac{5.5954}{(7.938 \times 10^{-3})(488.1)(1)}\left[\frac{0.5035 + (1.2927)(0.1504)}{1 + (1.2927)(0.5035)(0.1504)}\right] \\ &= 1.4442\left(\frac{0.6979}{1.0979}\right) \\ &= 0.918 \end{aligned}$$

and, in accordance with eq. (9.6),

$$\begin{aligned} \eta_{oh} &= 1 - \frac{S_f}{S}(1 - \eta_{fh}) \\ &= 1 - (0.845)(1 - 0.918) \\ &= 1 - (0.845)(0.082) \\ &= 1 - 0.069 \\ &= 0.931 \end{aligned}$$

For the cold side, the same procedure is followed. Pertinent data are

$$\begin{aligned} b_c &= 2.54 \times 10^{-3}\ \text{m} \\ s &= \frac{1}{(2)(19.82)(12)}(0.3048) \\ &= 6.408 \times 10^{-4}\ \text{m} \\ \delta_{fc} &= 1.02 \times 10^{-4}\ \text{m} \\ a_s &= 1.27 \times 10^{-4}\ \text{m} \end{aligned}$$

and

$$h_c = 370.4\ \text{W/m}^2 \cdot \text{K}$$

Then for the fin

$$m_f = \left(\frac{2h_c}{k\delta_f}\right)^{1/2} = \left[\frac{(2)(370.4)}{(211)(1.02 \times 10^{-4})}\right]^{1/2} = 185.53 \text{ m}^{-1}$$

$$m_f b = (185.53)(2.54 \times 10^{-3}) = 0.4712$$

$$\tanh m_f b = \tanh 0.4712 = 0.4392$$

and for unit fin length,

$$Y_{o,f} = (2h_c k\delta_{fc})^{1/2} = [(2)((370.4))(211)(1.02 \times 10^{-4})]^{1/2} = 3.9929 \text{ W/K}$$

For the splitter plate

$$m_s = \left(\frac{2h_c}{ka_s}\right)^{1/2} = \left[\frac{(2)(370.4)}{(211)(1.27 \times 10^{-4})}\right]^{1/2} = 166.27 \text{ m}^{-1}$$

$$m_s s = (166.27)(6.408 \times 10^{-4}) = 0.1065$$

$$\tanh m_s s = \tanh 0.1065 = 0.1061$$

and for unit fin length,

$$Y_{o,s} = (h_c k a_s)^{1/2} = [(370.4)(211)(1.27 \times 10^{-4})]^{1/2} = 3.1505 \text{ W/K}$$

Then by eq. (9.22),

$$\eta_{fc} = \frac{Y_{o,f}}{2(b+s)Lh} \frac{\tanh m_f b + (2Y_{o,s}/Y_{o,f}) \tanh m_s s}{1 + (2Y_{o,s}/Y_{o,f}) \tanh m_s s \tanh m_f b}$$

With

$$2\frac{Y_{o,s}}{Y_{o,f}} = 2\left(\frac{3.1505}{3.9929}\right) = 1.5780$$

and the surface attributed to the fins

$$\begin{aligned} S_f &= 2(b+s)L \\ &= (2)(2.54 \times 10^{-3} + 6.408 \times 10^{-4})(1) \\ &= 6.362 \times 10^{-3} \text{ m}^2 \end{aligned}$$

then

$$\begin{aligned} \eta_{fc} &= \frac{3.9929}{(6.362 \times 10^{-3})(370.4)(1)} \left[\frac{0.4392 + (1.5780)(0.1061)}{1 + (1.5780)(0.4392)(0.1061)}\right] \\ &= 1.6944\left(\frac{0.6066}{1.0735}\right) \\ &= 0.958 \end{aligned}$$

and in accordance with eq. (9.6),

$$\eta_{oc} = 1 - \frac{S_f}{S}(1 - \eta_{fh})$$
$$= 1 - (0.841)(1 - 0.958)$$
$$= 1 - (0.841)(0.042)$$
$$= 1 - 0.035$$
$$= 0.965$$

(6) *Overall heat transfer coefficient.* In the absence of fouling, the overall heat transfer coefficient is obtained from eq. (10.15*d*) because $C_c = C_{\min}$:

$$U_c = \frac{1}{1/\eta_{oc}h_c + S_c/S_h\eta_{oh}h_h}$$
$$= \frac{1}{1/(0.965)(370.4) + 7.532/(7.577)(0.931)(488.1)}$$
$$= \frac{1}{2.798 \times 10^{-3} + 2.188 \times 10^{-3}}$$
$$= \frac{1}{4.985 \times 10^{-3}}$$
$$= 200.6 \text{ W/m}^2 \cdot \text{K}$$

(7) *Effectiveness.* The required effectiveness with $C_c = C_{\min}$ is

$$\epsilon = \frac{t_2 - t_1}{T_1 - t_1} = \frac{249 - 56}{356 - 56} = \frac{193}{300} = 0.644$$

For this design

$$\text{N}_{\text{tu}} = \frac{U_c S_c}{C_{\min}} = \frac{(200.6)(7.532)}{554} = 2.727$$

and

$$R = \frac{C_{\min}}{C_{\max}} = \frac{554}{596} = 0.930$$

The actual effectiveness may be computed from eq. (10.26*d*). First,

$$\Lambda = e^{-R(\text{N}_{\text{tu}})^{0.78}} - 1$$
$$= e^{-(0.930)(2.727)^{0.78}} - 1$$
$$= e^{-2.034} - 1$$
$$= 0.1308 - 1$$
$$= -0.8692$$

Then

$$\begin{aligned}\epsilon &= 1 - e^{(N_{tu})^{0.22}\Lambda/R}\\ &= 1 - e^{(2.727)^{0.22}(-0.8692)/0.930}\\ &= 1 - e^{-1.1654}\\ &= 1 - 0.3118\\ &= 0.688\end{aligned}$$

This exceeds the required $\epsilon = 0.644$ and $\epsilon = 0.688$, by a factor of $0.688/0.644 = 1.068$ and is slightly conservative.

(8) *Pressure drops*. For the hot side,

$$\begin{aligned} Re_h &= 2323 & G_h &= 35.28\ \text{kg/m}^2\cdot\text{s}\\ f_h &= 0.0362 & S_h &= 7.577\ \text{m}^2\\ \sigma_h &= 0.459 & A_h &= 0.0163\ \text{m}^2\end{aligned}$$

Assuming that air is a perfect gas with $R = 0.287$ kPa · m³/kg · K at the inlet where $P_1 = 389.6$ kPa and $T_1 = 356 + 273 = 629$ K,

$$v_1 = \frac{RT_1}{P_1} = \frac{(0.287)(629)}{389.6} = 0.463\ \text{m}^3/\text{kg}$$

and at the outlet where $P_2 \approx 389.6$ kPa and $T_2 = 177 + 273 = 450$ K,

$$v_2 = \frac{RT_2}{P_2} = \frac{(0.287)(450)}{389.6} = 0.332\ \text{m}^3/\text{kg}$$

Then

$$v_m = \tfrac{1}{2}(v_1 + v_2) = \tfrac{1}{2}(0.463 + 0.332) = 0.398\ \text{m}^3/\text{kg}$$

$$\frac{v_m}{v_1} = \frac{0.398}{0.463} = 0.858\ \text{m}^3/\text{kg}$$

and

$$\frac{v_2}{v_1} = \frac{0.332}{0.463} = 0.717\ \text{m}^3/\text{kg}$$

From Fig. 10.8 at $\sigma_h = 0.460$ and Re = 2324,

$$K_c = 0.500 \quad \text{and} \quad K_e = 0.175$$

and the pressure drop is calculated from eq. (10.14):

$$\frac{\Delta P}{P_1} = \frac{G^2 v_1}{2g_c P_1}(\Phi_1 + \Phi_2 + \Phi_3 - \Phi_4)$$

where

$$\Phi_1 = 1 + K_c - \sigma_h^2$$
$$= 1 + 0.500 - (0.459)^2$$
$$= 1.289$$
$$\Phi_2 = 2\left(\frac{v_2}{v_1} - 1\right)$$
$$= (2)(0.717 - 1)$$
$$= -0.566$$
$$\Phi_3 = f_h \frac{S_h}{A_h}\frac{v_m}{v_1}$$
$$= (0.0362)\left(\frac{7.577}{0.0163}\right)(0.858)$$
$$= 14.438$$

and

$$\Phi_4 = [1 - \sigma_h^2 - K_e]\frac{v_2}{v_1}$$
$$= [1 - (0.459)^2 - 0.175](0.717)$$
$$= (0.6143)(0.717)$$
$$= 0.440$$

Then

$$\Delta P = \frac{G_h^2 v_1}{2g_c}(\Phi_1 + \Phi_2 + \Phi_3 - \Phi_4)$$
$$= \frac{(35.28)^2(0.463)}{(2)(9.81)}(1.289 - 0.566 + 14.438 - 0.440)$$
$$= (29.37)(14.721)$$
$$= 432.4\,\text{Pa} < 1000\,\text{Pa}$$

For the cold side,

$$\begin{aligned} Re_c &= 1669 & G_c &= 25.95\ \text{kg/m}^2\cdot\text{s} \\ f_c &= 0.0478 & S_c &= 7.532\ \text{m}^2 \\ \sigma_c &= 0.378 & A_c &= 0.0210\ \text{m}^2 \end{aligned}$$

With air at $P_1 = 55.16$ kPa and $t_1 = 56 + 273 = 329$ K,

$$v_1 = \frac{RT_1}{P_1} = \frac{(0.287)(329)}{55.16} = 1.712\ \text{m}^3/\text{kg}$$

and at $P_2 \approx 55.16$ kPa and $t_2 = 249 + 273 = 522$ K,

$$v_2 = \frac{RT_2}{P_2} = \frac{(0.287)(522)}{55.16} = 2.716 \text{ m}^3/\text{kg}$$

Then

$$v_m = \tfrac{1}{2}(v_1 + v_2) = \tfrac{1}{2}(1.712 + 2.716) = 2.214 \text{ m}^3/\text{kg}$$

$$\frac{v_m}{v_1} = \frac{2.214}{1.712} = 1.293 \text{ m}^3/\text{kg}$$

and

$$\frac{v_2}{v_1} = \frac{2.716}{1.712} = 1.587 \text{ m}^3/\text{kg}$$

From Fig. 10.8 at $\sigma_c = 0.378$ and Re $= 1669$,

$$K_c = 1.23 \quad \text{and} \quad K_e = 0.080$$

and the pressure drop is calculated from eq. (10.14):

$$\frac{\Delta P}{P_1} = \frac{G^2 v_1}{2g_c P_1}(\Phi_1 + \Phi_2 + \Phi_3 - \Phi_4)$$

where

$$\begin{aligned}
\Phi_1 &= 1 + K_c - \sigma_c^2 \\
&= 1 + 1.23 - (0.378)^2 \\
&= 2.087 \\
\Phi_2 &= 2\left(\frac{v_2}{v_1} - 1\right) \\
&= (2)(1.587 - 1) \\
&= 1.174 \\
\Phi_3 &= f_c \frac{S_c}{A_c}\frac{v_m}{v_1} \\
&= (0.0478)\left(\frac{7.532}{0.0210}\right)(1.293) \\
&= 22.168
\end{aligned}$$

and

$$\begin{aligned}
\Phi_4 &= [1 - \sigma_c^2 - K_e]\frac{v_2}{v_1} \\
&= [1 - (0.378)^2 - 0.080](1.587)
\end{aligned}$$

$$= (0.776)(1.587)$$

$$= 1.232$$

Then

$$\Delta P = \frac{G^2 v_1}{2g_c}(\Phi_1 + \Phi_2 + \Phi_3 - \Phi_4)$$

$$= \frac{(25.95)^2(1.712)}{(2)(9.81)}(2.087 + 1.174 + 22.168 - 1.232)$$

$$= (58.76)(24.197)$$

$$= 1422 \text{ Pa} < 2200 \text{ Pa}$$

The design is satisfactory from both a heat transfer and pressure loss standpoint. An example of such an aircraft precooler with air plenums is shown in Fig. 10.11, and the heat exchanger core is shown in Fig. 10.12.

10.5 NOMENCLATURE

Roman Letter Symbols

- A cross-sectional or frontal area, m^2
- a cover plate or separation plate thickness, m
- b fin height, m; passage height, m
- C capacity rate, W/K
- c specific heat, J/kg · K
- D exchanger core depth, m
- d diameter, m
- e Naperian base, dimensionless
- f friction factor, dimensionless
- H exchanger core height, m
- h heat transfer coefficient, $W/m^2 \cdot K$
- j heat transfer parameter, dimensionless
- K pressure loss coefficient, dimensionless
- k thermal conductivity, W/m · K
- L fin length, m; exchanger core length, m
- m fin performance parameter, m^{-1}
- $\dot{m}$ mass flow rate, kg/s
- Nu Nusselt number, dimensionless
- N_{tu} number of transfer units, dimensionless
- n number of exchanger layers, dimensionless
- P passage perimeter, m; pressure, N/m^2 (Pa)
- Pr Prandtl number, dimensionless
- q heat flow, W
- R capacity rate ratio, dimensionless

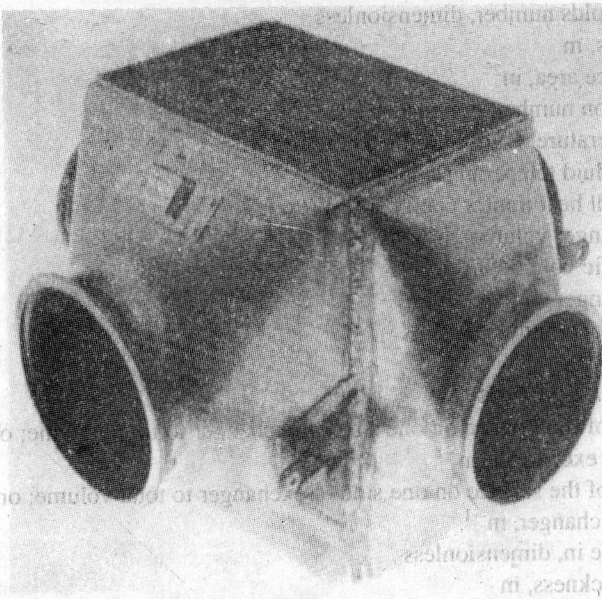

Figure 10.11 Aircraft precooler of Example 10.1. (Courtesy of Airesearch Division of Allied Signal Corporation.)

Figure 10.12 Heat exchanger core for the aircraft precooler of Example 10.1. (Courtesy of Airesearch Division of Allied Signal Corporation.)

Re Reynolds number, dimensionless
r radius, m
S surface area, m^2
St Stanton number, dimensionless
T temperature, K; hot fluid temperature, K
t cold fluid temperature, K
U overall heat tranfer coefficient, $W/m^2 \cdot K$
V exchanger volume, m^3
v specific volume, m^3/kg
W exchanger core width, m
Y thermal admittance, W/K

Greek Letter Symbols

α ratio of the surface on one side of exchanger to total volume; on both sides of the exchanger, m^{-1}
β ratio of the surface on one side of exchanger to total volume; on one side of the exchanger, m^{-1}
Δ change in, dimensionless
δ fin thickness, m
ϵ exchanger effectiveness, dimensionless
Λ combination of terms, dimensionless
μ dynamic viscosity, $kg/m \cdot s$
Σ summation, dimensionless
σ ratio of exchanger flow area to frontal area, dimensionless
Φ term in the flow friction equation, dimensionless

Roman Letter Subscripts

b bulk quantity
c entrance or contraction loss coefficient; cold fluid; gravitational constant
e equivalent; exit or expansion loss coefficient
f frontal area
h hot fluid; hydraulic radius
m mean value
max maximum value
min minimum value
o average value; overall value
p constant pressure condition
tu transfer units
w wall

CHAPTER 11

LONGITUDINAL FIN DOUBLE-PIPE EXCHANGERS

11.1 INTRODUCTION

The double-pipe exchanger consists of a pair of concentric tubes or pipes. One process stream flows through the inner pipe and the other flows, either in counter or cocurrent (parallel) flow, in the annular region between the two pipes. The inner pipe may be bare or it may contain as many as 48 longitudinal fins equally spaced around its periphery.

The study of double-pipe heat exchangers begins with a discussion of the unfinned or bare inner tube double-pipe heat exchanger, In this discussion, which is provided prior to the study of the longitudinal fin double-pipe heat exchanger, attention is focused on the overall heat transfer coefficient with no fouling, a discussion of fouling itself including the mechanism of fouling and then the use of the fouling factor or fouling resistance. The discussion concludes with a discussion of the overall heat transfer coefficient with fouling.

The thrust, however, is in the finned double-pipe exchanger, and the treatment of this type of exchanger includes a development of the physical parameters and the overall heat transfer coefficient, which is complicated by the presence of the fins and the need for the computation of the fin efficiency, with account taken for the annulus heat transfer coefficient. A catalog of heat transfer and friction loss data is given and examples are provided.

11.2 PLAIN DOUBLE-PIPE EXCHANGER

Consider the plain double-pipe exchanger as shown in Fig. 11.1. It usually consists of *two pairs* of concentric pipes with a return bend and a return head made leaktight by packing glands. The packing glands and bends returning outside rather than inside the return head are used only where the annulus has a low fluid pressure. If there is no problem with differential thermal expansion, the glands may be omitted and the outer and inner pipes may be welded together to provide a leaktight construction.

Two pairs of concentric pipes are used to form a *hairpin* because of the convenience the hairpin affords for manifolding streams and the natural loop it can provide for differential thermal expansion between the inner and outer pipes. The hairpin brings all inlets and outlets close together at one end, which is particularly important when multiple hairpins are connected in batteries. Moreover, the hairpins need not have the same length. An additional merit of the double-pipe heat exchanger is the ease in which it usually can be disassembled for inspection and cleaning or reused in another service whenever a process becomes obsolete.

11.2.1 The Overall Heat Transfer Coefficient with No Fouling

The heat transfer coefficent inside the inner pipe of diameter d_i is designated as h_i. It is computed using a Reynolds number formed from d_i, which is shown in Chapter 4 to be

$$\text{Re} = \frac{\rho d_i V_i}{\mu} = \frac{d_i G_i}{\mu} = \frac{d_i V_i}{\nu}$$

where V_i (m/s) is the velocity of the fluid inside the tubes. Use is often made of the mass velocity

$$G_i = \frac{\dot{m}_i}{A_i} \qquad (\text{kg/m}^2 \cdot \text{s})$$

where $\dot{m}_i$ (kg/s) is the flow rate and A_i is the cross-sectional area of the inside of the tube.

The heat transfer coefficient in the annular region between the inner pipe of outer diameter d_o and outer pipe of inner diameter D_i is designated as h_o. It is computed using a Reynolds number formed from the equivalent diameter d_e, which is shown in Chapter 4 to be

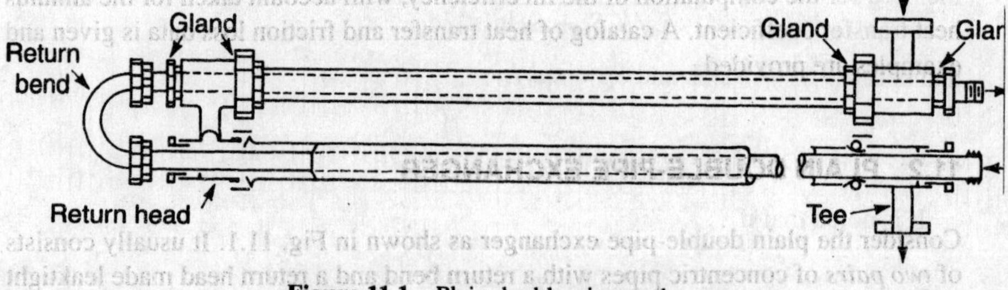

Figure 11.1 Plain double-pipe exchanger.

$$\text{Re} = \frac{\rho d_e V_o}{\mu} = \frac{d_e G_o}{\mu} = \frac{d_e V_o}{\nu}$$

where V_o is the velocity of the fluid inside the annulus (m/s) and G_o is the mass velocity:

$$G_o = \frac{\dot{m}_o}{A_o} \qquad \text{kg/m}^2 \cdot \text{s}$$

Here the equivalent diameter is obtained from

$$d_e = \frac{4A_o}{P_W} = 4\frac{(\pi/4)(D_i^2 - d_o^2)}{\pi(D_i + d_o)} = D_i - d_o$$

The heat flow by conduction through the inner pipe is calculated from the simple relationship for the heat flow through a curved surface:

$$q = \frac{2\pi k_m L}{\ln(d_o/d_i)}(T_i - T_o)$$

where k_m is the thermal conductivity of the metal and L is the length of the pipe. The heat transfer through the configuration between the inlet and outlet temperatures, T_i and T_o, where $(T_i > T_o)$ will be

$$q = \frac{T_i - T_o}{1/h_i S_i + \ln(d_o/d_i)/2\pi k_m L + 1/h_o S_o}$$

The *overall heat transfer coefficient* is defined, in general, by the relation

$$q = US(\text{LMTD})$$

and it observed that it depends whether the surface S_i or S_o is used for the computation of U. The two US products will be $U_i S_i$ and $U_o S_o$:

$$U_i S_i = U_o S_o = \frac{1}{1/h_i S_i + \ln(d_o/d_i)/2\pi k_m L + 1/h_o S_o}$$

The surface of reference is usually taken as the outside surface of the inner pipe S_o, and in this case, the overall heat transfer coefficient is

$$U_o = \frac{1}{S_o/h_i S_i + S_o \ln(d_o/d_i)/2\pi k_m L + 1/h_o}$$

and because $S_o = \pi d_o L$ and $S_i = \pi d_i L$,

$$U_o = \frac{1}{(d_o/d_i)(1/h_i) + d_o \ln(d_o/d_i)/2k_m + 1/h_o} \tag{11.1}$$

It is noted that if

$$h_{io} = h_i \frac{d_i}{d_o}$$

is the inside heat transfer coefficient referred to the outside of the inner pipe, and if the pipe wall is thin and has a fairly high thermal conductivity,

$$U_o = \frac{h_{io} h_o}{h_{io} + h_o}$$

11.2.2 Fouling

Fouling Mechanisms. Somerscales and Knudsen (1981) have identified six categories of fouling:

1. *Particulate fouling.* In this type of fouling, solid particles suspended in the process stream accumulate on the heat transfer surfaces. Typical examples include dust deposition, particles carried in condenser cooling water, and unburned fuel or fly ash. If the solid deposition is due to gravity, the process is referred to as *sedimentation fouling*.

2. *Precipitation fouling.* This fouling is characterized by the precipition of dissolved substances carried in the process stream on the heat transfer surfaces. Examples include carbonates, sulfates, and carbonates. *Scaling* occurs when precipitation occurs on heated rather than cooled surfaces.

3. *Chemical reaction fouling.* In certain cases, deposits on the heat transfer surfaces which are not themselves reactants are formed by chemical reactions. Typical examples of this type of fouling are, cracking and coking of hydrocarbons, and polymerization.

4. *Corrosion fouling.* In this type of fouling, the heat transfer surface reacts at certain pH levels to produce products that adhere to the heat transfer surfaces, and in turn, this may promote the attachment of additional fouling materials. Sulfur in fuel oil and sulfur products in the flue gas, such as sulfur dioxide, can lead to sulfuric acid. This has caused, for example, significant damage to heat exchange surfaces in air heaters in the power industry.

5. *Biological fouling.* Materials such as algae, bacteria, molds, seaweed, and barnacles carried in the process stream cause biological fouling of the heat transfer surfaces. A prime example of biological fouling is in marine power plant condensers.

6. *Freezing fouling.* In this type of fouling, a liquid or some of its higher-melting-point components will deposit on a subcooled heat transfer surface.

Fouling Factors. When a heat exchanger is placed in service, the heat transfer surfaces are, presumably, clean. With time, in some services in the power and process industries, the apparatus may undergo a decline in its ability to transfer heat. This is due to the accumulation of heat-insulating substances on either or both of the heat transfer surfaces. The Tubular Exchanger Manufacturers Association (TEMA) undertook the establishment of standards defining design practices not covered by the ASME Code for Unfired Pressure Vessels. Because the ASME Code is concerned primarily with safe pressure containment and the means for inspecting for it during construction, the contribution of TEMA to sound mechanical construction has been substantial.

In addition, TEMA published a table of *fouling factors* to assist the designer in preventing the fouling of a single item in a process, including several items of heat transfer equipment. Resistances were tabulated that were to be added to the film resistances ($1/S_i h_i$ and $1/h_o S_o$) of specific process streams so that the operating period of each would be similar and assure some desired period of continuous operation. The tables of fouling factors were intended as a crude guide toward the equalization of cumulative fouling in all fouling streams in the assembly.

The fouling factors published by TEMA became entrenched in industrial heat exchanger design. By the TEMA definition, fouling factors, are time dependent. They are not present when the apparatus is placed on stream, yet at some definite time in the future, when the apparatus has lost some of its heat transfer capabilities, the fouling factor is deemed to have arrived. TEMA does not delineate the in-between fouling process, and the fouling factor has shed little light on the nature of fouling. Significant is the fact that an item of equipment, which failed to comply with the TEMA notion of a desired period of continuous operation, became a *fouling problem*. Within the scope of the definition of a fouling factor, the only means for ameliorating fouling was to employ larger fouling factors for repetitive services.

The entire concept of the fouling factor is somewhat indefinite. It is an unsteady-state effect that is added indiscriminately to steady-state heat transfer resistances. The difference between a clean and a fouled exchanger is that an intolerable portion of the available temperature difference between fluids must be used to overcome fouling. Thus, if the outside surface S_o of a pipe or tube is the reference and r_{do} is the fouling or *dirt* factor, performance is defined by

$$q = \frac{1}{r_{do}} S \, \Delta T_{do}$$

where ΔT_{do} is the temperature drop across the fouling factor. After solving for the temperature drop,

$$\Delta T_{do} = r_{do} \frac{q}{S}$$

it is observed that the fouling factor only partially controls fouling. The heat flux q/S is of equal importance. If a fouling factor must be used, it would be more appropriate to specify a fouling factor for a particular process stream along with some range of appropriate temperature differences. TEMA has made some attempt to anticipate temperature differences by classifying fouling factors according to various process services. The result is inexact, as there is much latitude in establishing a process service. In any event, fouling factors which are, in reality, *fouling resistances* are given by TEMA and may be specified by other standards.

Overall Coefficient with Fouling. When the inner tube outside surface is the reference surface, the inside fouling resistance, designated by r_{di}, must be referred to the outside of the tube. With the outside fouling resistance designated as r_{do}, the overall heat transfer coefficient based on the outside surface of the inner tube becomes

$$U_o = \frac{1}{(d_o/d_i)(1/h_i) + (d_o/d_i)r_i + d_o \ln(d_o/d_i)/2k_m + r_{do} + 1/h_o} \tag{11.2}$$

11.3 LONGITUDINAL FIN DOUBLE-PIPE EXCHANGER

The longitudinal fin double-pipe exchanger is used advantageously where an appreciable inequity appears in the composite thermal resistance of a pair of fluids in a plain double-pipe exchanger. Because heat transfer equipment is usually purchased on the basis of its performance in the fouled condition, the composite thermal resistance is the sum of the convective film resistances and the fouling resistances. The advantage of the finned annulus lies in its ablility to offset the effects of poorer heat transfer in one fluid by exposing more surface to it than the other. Indeed, even if the composite resistances of both fluid are low, as discussed subsequently, there may still be an advantage in the use of the finned inner pipe.

Fins are usually 0.0889 cm thick (0.035 in. and 20 BWG). A steel fin with a thermal conductivity of 50 W/m·K and a height of 1.27 cm ($\frac{1}{2}$ in.) on exposure to a composite resistance of 0.004 $m^2 \cdot$ K/W (corresponding to a film coefficient of 250 W/$m^2 \cdot$ K) has a fin efficiency of about 0.650. Exposed to a composite resistance of 0.002 $m^2 \cdot$ K/W, the efficiency drops to about 0.500. Hence the high fin has its limitations, although metals of higher thermal conductivity extend the range of application. Fin surface is inexpensive compared with prime surface, but its usefulness diminishes significantly below a composite resistance of 0.002 $m^2 \cdot$ K/W. For the case where both composite resistances are very large, any improvement in the surface exposed to the higher resistance may save considerable linear meters of exchanger. Moreover, inner pipes are available with fins on the inside as well as the outside of the pipe, and the inner pipes are also available with continuous twisted longitudinal fins which cause some mixing in the annulus. As a class, however, these show a small increase in heat transfer coefficient for a large expenditure of pressure loss, and for viscous fluids, the mixing and its effects decay rapidly.

The disposition of the fins about the pipe are shown clearly in Fig. 11.2. They form a radial array of channels, with each channel composed of two fins. Channels may be attached by continuously spot-welding them to the outside of the inner pipe or by other brazing or welding procedures. It should be noted that the contact between the channels and the other pipe should be continuous over the entire channel length but need not be very wide. In another method of attaching longitudinal fins, grooves are plowed in the outside diameter of the inner pipe. Metal ribbon is then inserted into the grooves as fins, and the plowed-up metal is peened back to form a tight bond between fins and the inner pipe. In the laminar or transition flow regimes, fins are sometimes offset every 30 to 100 cm.

The common double-pipe exchanger units available are summarized in Table 11.1.

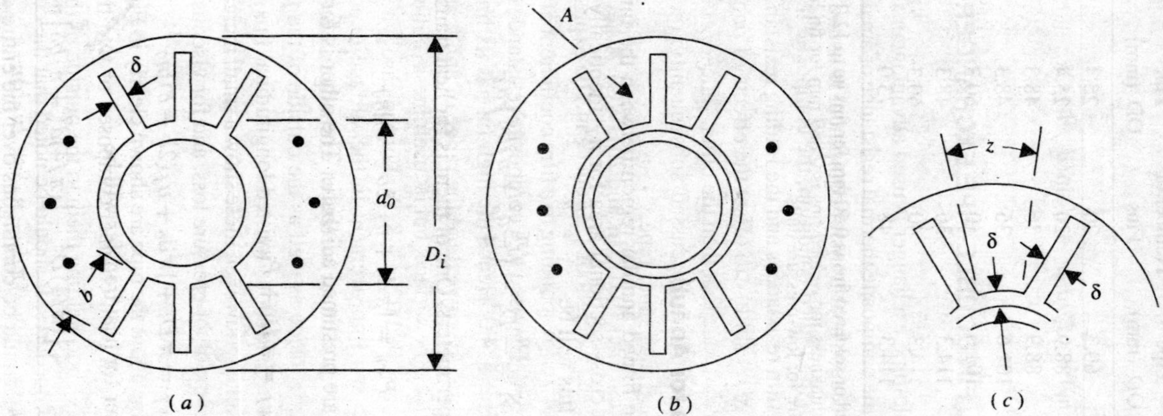

Figure 11.2 Configurations to be used for the determination of the flow area, wetted perimeter, and surfaces for the annular region in the double-pipe heat exchanger: (*a*) extruded fins; (*b*) welded U-fins; (*c*) detail of two welded U-fins.

TABLE 11.1 Dimensions of Double-Pipe Exchangers (One Outer Tube–One Inner Tube: Standard Units)[a]

Nominal Diameter (in.)	Outer Pipe Thickness (mm)	Outer Pipe OD (mm)	Maximum Number of Fins	Inner Pipe OD (mm)	Inner Pipe Thickness (mm)	Fin Height (mm)
2	3.91	60.3	20	25.4	2.77	11.1
3	5.49	88.9	20	25.4	2.77	23.8
	5.49	88.9	36	48.3	3.68	12.7
$3\frac{1}{2}$	5.74	101.6	36	48.3	3.68	19.05
	5.74	101.6	40	60.3	3.91	12.7
4	6.02	114.3	36	48.3	3.68	25.4
	6.02	114.3	40	60.3	3.91	10.05
	6.02	114.3	48	73.0	5.16	12.7

[a] The fin thickness for extruded or soldered fins is 0.50 mm for fins up to 12.7 mm high and 0.80 mm high for greater heights. Fin thickness for welded fins is 0.889 mm for fin heights up to 25.4 mm. The dimensions shown here are for low-pressure units.

11.3.1 Physical Data for Annuli

Extruded Fins. For the finned annular region between the inner and outer pipes shown in Fig. 11.2*a*, the cross-sectional area for n_T identically finned inner pipes each having n_f extruded fins will be

$$A = \frac{\pi}{4} D_i^2 - \left(\frac{\pi}{4} d_o^2 + n_f b\delta\right) n_T \tag{11.3}$$

There are two wetted perimeters. One of them is for heat transfer:

$$P_{Wh} = [\pi d_o + n_f(2b - \delta)]n_T \tag{11.4}$$

where the tips of the fins are presumed adiabatic. The other is for pressure loss,

$$P_{Wf} = \pi D_i + P_{Wh}$$

or

$$P_{Wf} = \pi D_i + [\pi d_o + n_f(2b - \delta)]n_T \tag{11.5}$$

The equivalent diameter for heat transfer will be

$$d_e = \frac{4A}{P_{Wh}} = \frac{4\left\{(\pi/4) D_i^2 - \left[(\pi/4) d_o^2 + n_f b\delta\right] n_T\right\}}{(\pi d_o - n_f\delta + 2n_f b)n_T} \tag{11.6}$$

and the equivalent diameter for pressure drop will be

$$d_e = \frac{4A}{P_{Wf}} = \frac{4\left\{(\pi/4) D_i^2 - \left[(\pi/4) d_o^2 + n_f b\delta\right] n_T\right\}}{\pi D_i + (\pi d_o - n_f\delta + 2n_f b)n_T} \tag{11.7}$$

The surface area per unit length per tube will be

$$S' = S = S_b + S_f$$

where S_b is the unfinned surface of the inner tube per unit length:

$$S_b = \pi d_o - n_f \delta \tag{11.8a}$$

Then, per unit length, with

$$S_f = 2bn_f \tag{11.8b}$$

the surface area on the annulus side of the inner pipe per unit length is

$$S' = S = \pi d_o + (2b - \delta)n_f \tag{11.8c}$$

Welded U-Fins. The configuration for the annular region that accommodates welded U-fins is shown in Fig. 11.2*b* and a detail for a pair of the fins is shown in Fig. 11.2*c*. Observe that z is the fin root width *and* thickness, which is usually taken as 2δ. The free area for flow for n_f fins and n_T inner tubes with $\delta = z/2$ is

$$A = \frac{\pi}{4} D_i^2 - \left[\frac{\pi}{4} d_o^2 + n_f \delta \left(b + \frac{z}{2}\right)\right] n_T \tag{11.9}$$

If $d_o >> z$, the wetted perimeter for heat flow will be the circumference of the inner tube less the thicknesses of the n_f fins plus twice the heights of the n_f fins:

$$P_{Wh} = [\pi d_o + n_f(2b - \delta)]n_T \tag{11.10}$$

Here, too, the tips are presumed to be adiabatic. The wetted perimeter for pressure loss is

$$P_{Wf} = \pi D_i + P_{Wh}$$

or

$$P_{Wf} = \pi D_i + [\pi d_o + n_f(2b - \delta)]n_T \tag{11.11}$$

Then the two equivalent diameters are: for heat transfer,

$$d_e = \frac{4A}{P_{Wh}} = \frac{4\left\{(\pi/4)D_i^2 - \left[(\pi/4)d_o^2 + n_f\delta\,(b + z/2)\right]n_T\right\}}{[\pi d_o + n_f(2b - \delta)]n_T} \tag{11.12}$$

and for pressure drop,

$$d_e = \frac{4A}{P_{Wf}} = \frac{4\left\{(\pi/4)D_i^2 - \left[(\pi/4)d_o^2 + n_f\delta\,(b + z2)\right]n_T\right\}}{\pi D_i + [\pi d_o + n_f(2b - \delta)]n_T} \tag{11.13}$$

The surface areas S_b, S_f, and S and the surface area per unit length S' will be the same as those for the extruded fin configuration and are given by eqs. (11.8).

11.3.2 Overall Heat Transfer Coefficient

The equation for an overall heat transfer coefficient is a complicated expression because of the annulus fouling and the fin efficiency. It can be developed from a series summation of several thermal resistances that are identified in Fig. 11.3. These resistances are in $m^2 \cdot K/W$.

After both inside and outside heat transfer coefficients, h_i and h_o, have been determined and after both fouling resistances, r_{di} and r_{do}, have been specified,[1] the steps can be arranged in a systematic order.

1. With $h_{io} = h_i(d_i/d_o)$, form the inside *film resistance*:

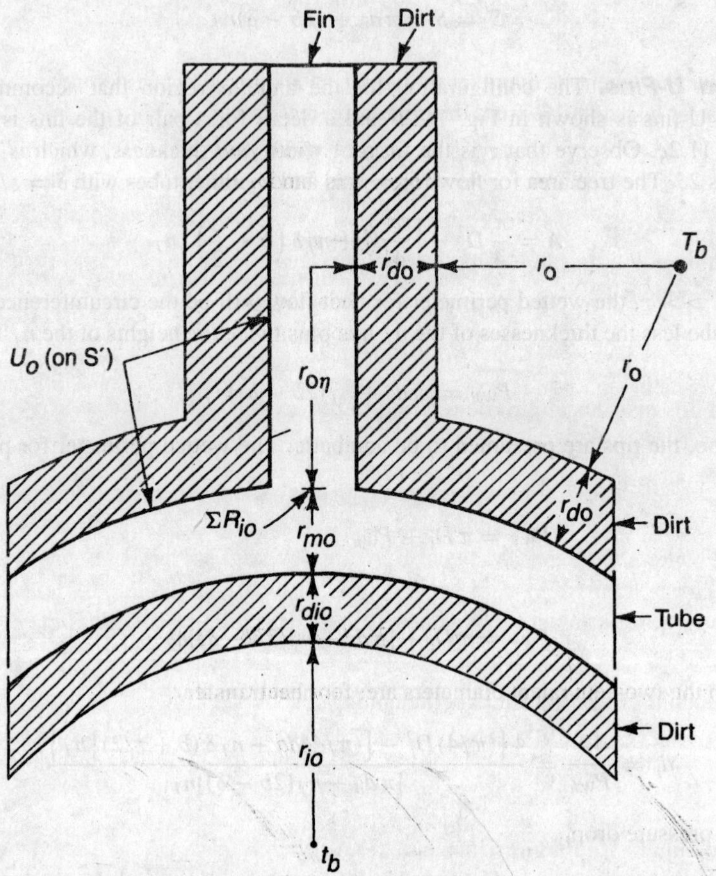

Figure 11.3 Location of thermal resistances for a fouled longitudinal fin double-pipe exchanger. The thermal resistances are based on gross fin and outer-pipe surface and the tip of the fin is considered adiabatic.

[1]Either one or both can be zero.

$$r_{io} = \frac{1}{h_{io}} \tag{11.14}$$

2. The inner-pipe fouling resistance is r_{di} and it must be referred to the outer tube surface. Hence

$$r_{dio} = \frac{d_o}{d_i} r_{di} \tag{11.15}$$

3. The pipe wall resistance referred to the outside of the inner pipe is

$$r_{mo} = \frac{d_o \ln(d_o/d_i)}{2k_m}$$

However, when the diameter ratio $d_i/d_o \geq 0.75$, r_{mo} can be computed with an error of less than 1% from the arithmetic mean diameter:

$$r_{mo} = \frac{d_o - d_i}{2k_m} \frac{2\pi d_o}{\pi(d_i + d_i)} = \frac{d_o(d_o - d_i)}{k_m(d_i + d_i)} \tag{11.16}$$

4. At this point, the sum of the internal resistances referred to the outside of the inner pipe is

$$\sum R_{io} = r_{io} + r_{dio} + r_{mo}$$

and this resistance must be referred to the gross outside surface of each inner pipe per meter:

$$S' = \pi d_o + (2b - \delta)n_f \tag{11.8c}$$

Thus

$$\sum R_{is} = \sum R_{io} \frac{S'}{\pi d_o} \tag{11.17}$$

5. The annulus heat transfer coefficient is h_o, so that

$$r_o = \frac{1}{h_o} \tag{11.18}$$

6. The annulus fouling resistance, r_{do}, must be combined with h_o to obtain the value of the annulus coefficient working on the fin and prime outer surface. Let this resistance be designated as r'_o, so that

$$r'_o = \frac{1}{h_o} + r_{do} = r_o + r_{do}$$

and

$$h'_o = \frac{1}{r_o + r_{do}}$$

The fin efficiency will be given be eq. (1.11):

$$\eta = \frac{\tanh mb}{mb} \tag{1.11}$$

where

$$m = \left(\frac{2h'_o}{k_m \delta}\right)^{1/2}$$

Then, with the weighted fin efficiency defined by eq. (9.6):

$$\eta_o = 1 - \frac{S_f}{S}(1 - \eta) \tag{9.6}$$

or

$$\eta_o = \frac{\eta S_f + S_b}{S_f + S_b}$$

the value of the heat transfer coefficient to the finned and prime surface corrected for the weighted fin efficiency and based on the outside surface of the inner pipe will be

$$h'_{o\eta} = h'_o \eta_o$$

so that the resistance is

$$r'_{o\eta} = \frac{1}{h'_{o\eta}} \tag{11.19}$$

7. The overall resistance is the sum of eqs. (11.17) and (11.19). Thus

$$\frac{1}{U_o} = \sum R_{is} + r'_{o\eta}$$

or

$$U_o = \frac{1}{\sum R_{is} + r'_{o\eta}} \tag{11.20}$$

The overall heat transfer coefficient given by eq. (11.20) is the coefficient to be used in the *rate equation*:

$$q = U_o S_o (\text{LMTD}) \tag{11.21}$$

Example 11.1. Determine the overall heat transfer coefficient for a double-pipe heat exchanger with an inner pipe having $d_o = 25.4$ mm and a thickness of 2.77 mm. The tube accommodates 20 fins with height $b = 11.1$ mm and thickness $\delta = 0.50$ mm. The tube-side heat transfer coefficient is $h_i = 250\ \text{W/m}^2 \cdot \text{K}$, and the annulus heat transfer coefficient is $h_o = 125\ \text{W/m}^2 \cdot \text{K}$. The thermal conductivity of the tube material is 46 W/m·K, and fouling resistances of 0.0001 m^2·K/W should be provided for each stream.

SOLUTION. In this configuration,

$$d_o = 25.4\ \text{mm} \qquad (0.02540\ \text{m})$$

$$\begin{aligned} d_i &= 25.4 - (2)(2.77) \\ &= 25.4 - 5.54 \\ &= 19.86\ \text{mm} \qquad (0.01986\ \text{m}) \end{aligned}$$

and

$$\frac{d_i}{d_o} = \frac{0.01986}{0.02540} = 0.7819$$

With a single inner pipe and 20 fins, $n_T = 1$ and $n_f = 20$ and with unit length ($L =$ 1.00 m),

$$\begin{aligned} S_f &= 2bn_f \\ &= (2)(0.0111)(20) \\ &= 0.4440\ \text{m}^2 \\ S_b &= [\pi d_o - n_f \delta] L n_T \\ &= [\pi(0.0254) - (20)(0.0005)](1.00)(1.00) \\ &= 0.0698\ \text{m}^2 \end{aligned}$$

and

$$\begin{aligned} S = S' &= S_b + S_f \\ &= 0.0698 + 0.4440 \\ &= 0.5138\ \text{m}^2 \end{aligned}$$

(1) First obtain h_{io}:

$$\begin{aligned} h_{io} &= h_i \frac{d_i}{d_o} \\ &= (250)(0.7819) \\ &= 195.47\ \text{W/m}^2\text{·K} \end{aligned}$$

and from eq. (11.14),

$$r_{io} = \frac{1}{h_{io}} = \frac{1}{195.47} = 5.1158 \times 10^{-3}\ \text{m}^2\cdot\text{K/W}$$

(2) With $r_{di} = 0.0001\ \text{m}^2\cdot\text{K/W}$, eq. (11.15) yields

$$r_{dio} = \frac{d_o}{d_i} r_{di} = \frac{0.0001}{0.7819} = 1.2790 \times 10^{-4}\ \text{m}^2\cdot\text{K/W}$$

(3) Noting that $d_i/d_o = 0.7819 > 0.75$, eq. (11.16) is used to obtain r_{mo}:

$$r_{mo} = \frac{d_i(d_o - d_i)}{k_m(d_o + d_i)}$$

$$= \frac{(0.01986)(0.02540 - 0.01986)}{(46)(0.02540 + 0.01986)}$$

$$= 5.2847 \times 10^{-5}\ \text{m}^2\cdot\text{K/W}$$

(4) Then

$$\sum R_{io} = r_{io} + r_{dio} + r_{mo}$$

$$= 5.1158 \times 10^{-3} + 1.2790 \times 10^{-4} + 5.2857 \times 10^{-5}$$

$$= 5.2965 \times 10^{-3}\ \text{m}^2\cdot\text{K/W}$$

and by eq. (11.17),

$$\sum R_{is} = \sum R_{io} \frac{S'}{\pi d_o}$$

$$= (5.2965 \times 10^{-3}) \left[\frac{0.5138}{\pi(0.0254)}\right]$$

$$= (5.2965 \times 10^{-3})(6.4839)$$

$$= 0.0341\ \text{m}^2\cdot\text{K/W}$$

(5) For the annulus r_o is given by eq. (11.18):

$$r_o = \frac{1}{h_o} = \frac{1}{125} = 0.0080\ \text{m}^2\cdot\text{K/W}$$

(6) Then

$$h'_o = \frac{1}{r_o + r_{do}}$$

$$= \frac{1}{0.0080 + 0.0001}$$

$$= \frac{1}{0.0081}$$

$$= 123.45\ \text{W/m}^2\cdot\text{K}$$

and with

$$m=\left(\frac{2h'_o}{k_m\delta}\right)^{1/2}=\left[\frac{(2)(123.45)}{(46)(0.0005)}\right]^{1/2}=103.6116\ \text{m}^{-1}$$

then

$$mb=(103.6116)(0.0111)=1.1501$$

The fin efficiency is

$$\eta=\frac{\tanh mb}{mb}=\frac{\tanh 1.1501}{1.1501}=0.711$$

and the weighted fin efficiency is given by eq. (9.6):

$$\begin{aligned}\eta_o&=1-\frac{S_f}{S}(1-\eta)\\&=1-\left(\frac{0.4440}{0.5138}\right)(1-0.711)\\&=1-(0.8641)(0.289)\\&=1-0.2497\\&=0.750\end{aligned}$$

Then

$$h'_{o\eta}=h'_o\eta_o=(123.45)(0.750)=92.6245$$

and

$$r'_{o\eta}=\frac{1}{h'_o\eta_o}=\frac{1}{92.6245}=0.0108\ \text{W/m}^2\cdot\text{K}$$

(7) The overall heat transfer coefficient is

$$\begin{aligned}U_o&=\frac{1}{\sum R_{is}+r'_{o\eta}}\\&=\frac{1}{0.0341+0.0108}\\&=\frac{1}{0.0449}\\&=22.27\ \text{W/m}^2\cdot\text{K}\end{aligned}$$

11.4 HEAT TRANSFER COEFFICIENTS IN PIPES AND ANNULI

Investigations that pertain to heat transfer and friction data within tubes and the annular passage created by longitudinal fins have been reported by Pohlhausen (1921),

DeLorenzo and Anderson (1945), Deissler (1951), McAdams (1954), Hausen (1959, 1974), Stefan (1959), Barnes and Jackson (1961), Yang (1962), Dalle Donne and Bowditch (1963), Petukhov and Popov (1963), Perkins and Worsœ-Schmidt (1965), Worsœ-Schmidt (1966), Test (1968), Webb (1971), Kusznetsova (1972), Oskay and Kakac (1973), Gnielinski (1976), Shah and London (1978), Rogers (1980), Kakac et al. (1985, 1987), Shah and Bhatti (1987), Kays and Crawford (1993), and Kakac and Yener (1994).

Some heat transfer correlations depend on a viscosity correction,

$$\phi^n = \left(\frac{\mu}{\mu_w}\right)^n \tag{11.22}$$

where μ and μ_w are the dynamic viscosities at the bulk and wall temperature, respectively, and where n is an exponent depending on whether the process is one of heating or one of cooling.

The heat transfer correlations that follow are subdivided into three listings: for laminar flow, for transition flow, and for turbulent flow, and in all of these, unless otherwise indicated, all fluid properties are based on the bulk temperature

$$T_b = \frac{T_o + T_i}{2}$$

11.4.1 Laminar Flow, Re ≤ 2300

For situations in which the thermal and velocity profiles are fully developed, the Nusselt number depends only on the thermal boundary conditions. For circular tubes with $\text{Pr} \geq 0.60$ and $\text{RePr}(L/d) > 0.05$, the Nusselt numbers have been shown to be

$$\text{Nu} = 3.657 \tag{11.23}$$

for constant-temperature conditions and

$$\text{Nu} = 4.364 \tag{11.24}$$

for constant-heat-flux conditions. Here Re_d is the Reynolds number based on the tube diameter, x is the distance along the tube, and L is the tube length.

In many cases, the Graetz number, which is the product of the Reynolds and Prandtl number and the diameter-to-length ratio

$$\text{Gz} \equiv \text{RePr}\frac{d}{L}$$

is employed.

At the entrance of a tube, the Nusselt number is infinite and decreases asymptotically to the value for fully developed flow as the flow progresses along the length of the tube. The Sieder–Tate (1936) equation gives good correlation for both liquids and gases in the region where both the thermal and velocity profiles are developing:

$$\text{Nu} = \frac{hd}{k} = 1.86\left(\text{RePr}\frac{d}{L}\right)^{1/3}\phi^{0.14} \tag{11.25}$$

where

$$\phi = \frac{\mu}{\mu_w}$$

Equation (11.25) applies when T_w is constant, $0.48 \le \text{Pr} \le 16{,}700$, $0.0044 \le \phi \le 9.75$, and $[\text{RePr}(d/L)]^{1/3}(\mu/\mu_w)^{0.14} \ge 2$. The limitations should be observed carefully, as the Sieder–Tate equation yields a zero heat transfer coefficient for extremely long tubes.

The correlation of Hausen (1943) is good for both liquids and gases at constant wall temperature:

$$\text{Nu} = \frac{hd}{k} = 3.66 + \frac{0.668\text{RePr}(d/L)}{1 + 0.40\,[\text{RePr}(d/L)]^{2/3}} \tag{11.26}$$

The heat transfer coefficient obtained from this correlation is the average value for the entire length of the tube, and it may be observed that when the tube is sufficiently long, the Nusselt number approaches the constant value of 3.657.

11.4.2 Transition, 2300 ≤ Re ≤ 10,000

For transition flow for both liquids and gases, the Hausen (1943) correlation for both liquids and gases may be employed:

$$\text{Nu} = \frac{hd}{k} = 0.116(\text{Re}^{2/3} - 125)\text{Pr}^{1/3}\phi^{0.14}\left[1 + \left(\frac{d}{L}\right)^{2/3}\right] \tag{11.27}$$

11.4.3 Turbulent Flow, Re ≥ 10,000

For both liquids and gases, Dittus and Boelter (1930) recommend

$$\text{Nu} = \frac{hd}{k} = 0.023\text{Re}^{0.80}\text{Pr}^n \tag{11.28}$$

where $n = 0.30$ for cooling and $n = 0.40$ for heating.

Sieder and Tate (1936) removed the dependency on heating and cooling by setting the exponent on Pr to $\frac{1}{3}$ and adding a viscosity correction

$$\text{Nu} = \frac{hd}{k} = 0.023\text{Re}^{0.80}\text{Pr}^{1/3}\phi^{0.14} \tag{11.29}$$

This correlation is valid for liquids and gases for $L/d > 60$, $\text{Pr} > 0.60$, and moderate $T_w - T_b$.

The correlation of Petukhov (1970) is cited in Chapter 4 as eq. (4.32) and is repeated here:

$$\mathrm{Nu} = \frac{hd}{k} = \frac{(f/8)\mathrm{RePr}}{1.07 + 12.7(f/8)^{1/2}(\mathrm{Pr}^{2/3} - 1)} \tag{11.30}$$

where

$$f = \frac{1}{(1.82 \log_{10} \mathrm{Re} - 1.64)^2} \tag{11.31}$$

is valid for $1.0 \le \mathrm{Pr} \le 2000$ and $10^4 \le \mathrm{Re} \le 5 \times 10^5$.

Bejan (1995) has suggested that the most accurate correlation is that of Gnielinski (1976), who provided a modification of the Petukhov (1970) correlation of eq. (11.30) with f given by eq. (11.31):

$$\mathrm{Nu} = \frac{hd}{k} = \frac{(f/8)(\mathrm{Re} - 1000)\mathrm{Pr}}{1.00 + 12.7(f/8)^{1/2}(\mathrm{Pr}^{2/3} - 1)} \tag{11.32}$$

in order to extend the range to $1.0 < \mathrm{Pr} < 10^6$ and $2300 < \mathrm{Re} < 5 \times 10^6$.

Two simpler alternatives to eq. (11.32) have been suggested by Gnielinski (1976):

$$\mathrm{Nu} = \frac{hd}{k} = 0.0214(\mathrm{Re}^{0.80} - 100)\mathrm{Pr}^{0.40} \tag{11.33}$$

for $0.50 < \mathrm{Pr} < 1.50$ and $10^4 < \mathrm{Re} < 5 \times 10^6$, and

$$\mathrm{Nu} = \frac{hd}{k} = 0.012(\mathrm{Re}^{0.87} - 280)\mathrm{Pr}^{0.40} \tag{11.34}$$

for $0.50 < \mathrm{Pr} < 500$ and $3 \times 10^3 < \mathrm{Re} < 10^6$.

Equation (11.30) can be modified to account for variable properties:

$$\mathrm{Nu} = \frac{hd}{k} = \frac{(f/8)\mathrm{RePr}}{1.07 + 12.7(f/8)^{1/2}(\mathrm{Pr}^{2/3} - 1)}\phi^n \tag{11.35}$$

where $n = 0.11$ for heating and $n = 0.25$ for cooling, and where f is given by eq. (11.31). In addition to the restrictions on L/d and Re cited with eq. (11.31), $1.0 \le \phi \le 40$ and $0.50 \le \mathrm{Pr} \le 140$.

Sleicher and Rouse (1975) give

$$\mathrm{Nu} = \frac{hd}{k} = 5.0 + 0.015\mathrm{Re}^m\mathrm{Pr}^n \tag{11.36}$$

where

$$m = 0.88 - \frac{0.24}{\mathrm{Pr} + 4.00}$$

and

$$n = \frac{1}{3} + 0.50e^{-0.60\mathrm{Pr}}$$

for $1.00 \le \mathrm{Pr} \le 1000$ and $10^4 \le \mathrm{Re} \le 10^6$.

11.5 PRESSURE LOSS IN PIPES AND ANNULI

11.5.1 Friction Loss

The friction loss inside tubes of circular cross section is given by a modification of eq. (4.6):

$$\Delta P_f = \frac{4 f \rho v^2}{2} \frac{L}{d} \quad \text{Pa} \tag{11.37a}$$

or

$$\Delta P_f = \frac{4 f G^2}{2\rho} \frac{L}{d} \quad \text{Pa} \tag{11.37b}$$

where f is the friction factor. For the annulus, eqs. (11.37) apply but with the tube diameter replaced with the equivalent diameter for pressure loss:

$$\Delta P_f = \frac{4 f \rho v^2}{2} \frac{L}{d_e} \quad \text{Pa} \tag{11.38a}$$

or

$$\Delta P_f = \frac{4 f G^2}{2\rho} \frac{L}{d_e} \quad \text{Pa} \tag{11.38b}$$

The total friction loss in the inner pipe of a double-pipe exchanger containing n_{hp} hairpins is

$$\Delta P_f = \frac{(2n_{\text{hp}})4 f \rho v^2}{2} \frac{L}{d} = 4 n_{\text{hp}} f \rho v^2 \frac{L}{d_i} \tag{11.39}$$

For the annulus, the equivalent diameter for pressure loss is substituted for d_i in eq. (11.39).

11.5.2 Turn Losses

The fluid will undergo an additional pressure loss due to contractions and expansions that occur during fluid turnaround between tube passes. Kern (1950) and Kern and Kraus (1972) have proposed that this loss be given by one velocity head per turn. For both inner tube and annulus,

$$\Delta P_t = \frac{4 \rho v^2}{2} \quad \text{Pa} \tag{11.40}$$

In an exchanger with n_{hp} hairpins, there will be $2n_{\text{hp}} - 1$ turns. Hence the turn loss will be

$$\Delta P_t = 2(2n_{\text{hp}} - 1)\rho v^2 \tag{11.41}$$

11.5.3 Friction Factors

Friction factors may be obtained from Fig. 11.4, which plots friction factor as a function of Reynolds number (for either inner pipe or annulus) and relative roughness, ϵ/d or ϵ/d_e. Figure 11.4 represents the work of Moody (1944) and is in the same form as Fig. 4.4. Observe that in laminar flow

$$f = \frac{16}{\text{Re}} \tag{11.42}$$

Many investigators have developed friction factor relationships as functions of Reynolds number for smooth tubes. The use of f given by eq. (11.31) in the Petukhov and Gnielinski correlations of eqs. (11.30) and (11.32) has been noted. Other functions can be fitted to the smooth curve in Fig. 11.4. Two of them, for smooth tubes, are

$$f = \frac{0.046}{\text{Re}^{0.20}} \tag{11.43}$$

for $3 \times 10^4 \leq \text{Re} \leq 10^6$ and

$$f = \frac{0.079}{\text{Re}^{0.25}} \tag{11.44}$$

for $4 \times 10^3 \leq \text{Re} \leq 10^5$.

11.6 COMPLETE DESIGN

Example 11.2: Finned Double-Pipe Heat Exchanger. In a marine application, 1.530 kg/s of engine oil is to be cooled from 117°C to 77°C using seawater entering at 17°C and leaving at 37°C. Fouling resistances of 0.00018 m^2·K/W for the engine oil and 0.0001 m^2·K/W for the water are to be used. The allowable pressure drops are 65 kPa (≈10 psig) and 80 kPa (≈ 15 psig) for the water and oil, respectively.

Determine the arrangement of longitudinal fin double-pipe exchangers for this service if both the inner pipe and the annulus are connected in series and if the length of each hairpin is 3 m. The thermal conductivity of the exchanger material is 52 W/m·K.

SOLUTION.

(1) *Thermal properties:*

$$T_b = \frac{T_1 + T_2}{2} = \frac{117 + 77}{2} = \frac{194}{2} = 97^\circ\text{C} \quad (370\,\text{K})$$

$$t_b = \frac{t_2 + t_1}{2} = \frac{37 + 17}{2} = \frac{54}{2} = 27^\circ\text{C} \quad (300\,\text{K})$$

At these bulk temperatures:

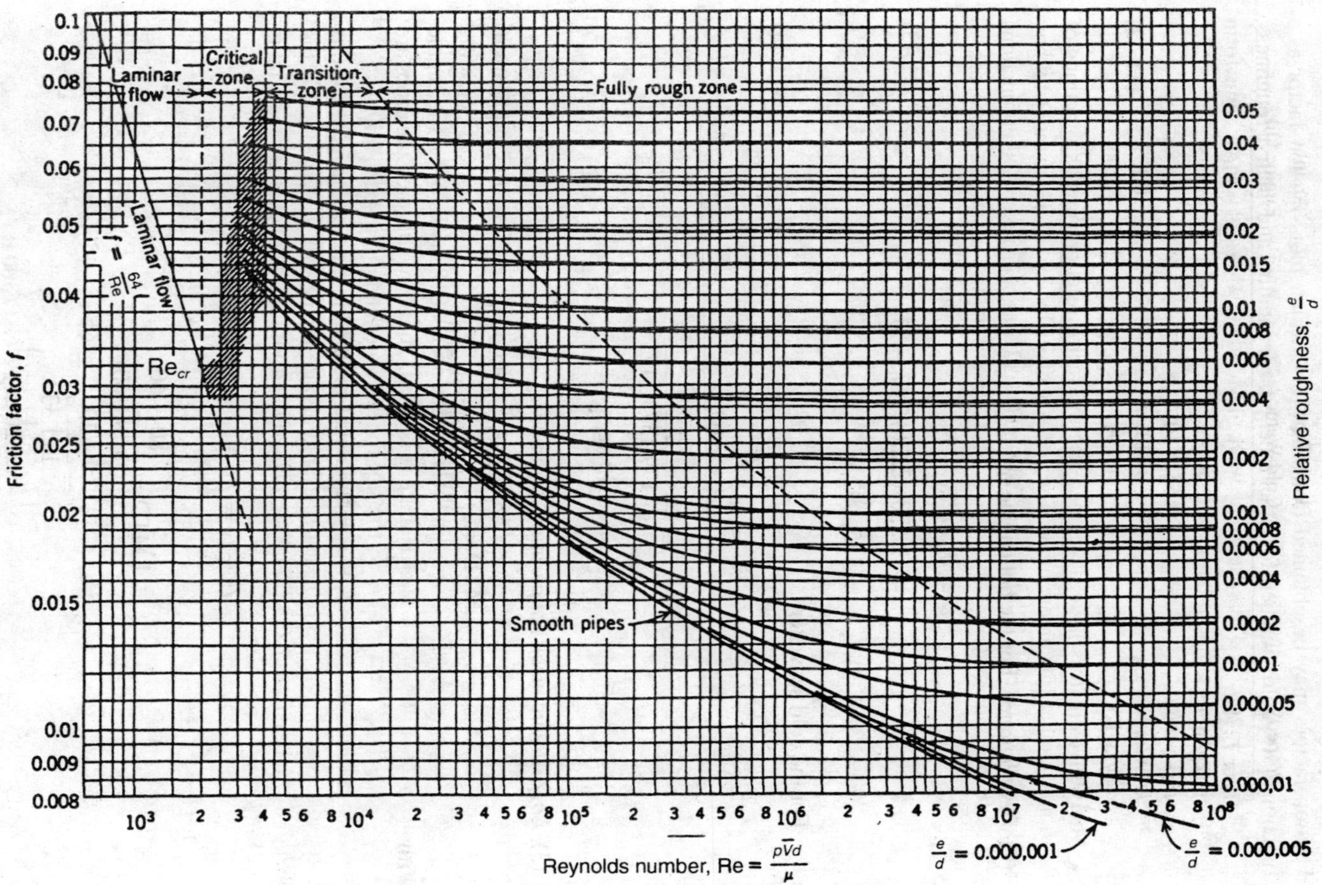

Figure 11.4 Friction factors for flow in tubes and annuli, (Adapted from *Trans. ASME*, 66, 671–684.)

Property	Inner Pipe: Cold Water	Annulus: Engine Oil
t_b or T_b (K)	300	370
ρ (kg/m^3)	997.0	841.8
μ (kg-m/s)	8.55×10^{-4}	0.0186
k (W/m·K)	0.613	0.137
c (J/kg·K)	4179	2206
Pr	5.83	300

(2) *Heat balance.* For the hot engine oil:

$$q = \dot{m}_h c_h (T_1 - T_2)$$
$$= (1.530)(2206)(117 - 77)$$
$$= 135,007 \text{ W}$$

For the cold seawater:

$$\dot{m}_c = \frac{q}{c_c(t_2 - t_1)}$$
$$= \frac{135,007}{(4179)(37 - 17)}$$
$$= 1.615 \text{ kg/s}$$

(3) *Logarithmic mean temperature difference:*

$$\text{LMTD} = \frac{(T_1 - t_2) - (T_2 - t_1)}{\ln(T_1 - t_2)(T_2 - t_1)}$$

With

$$T_1 - t_2 = 117 - 37 = 80°\text{C}$$

and

$$T_2 - t_1 = 77 - 17 = 60°\text{C}$$

$$\text{LMTD} = \frac{80 - 60}{\ln(80/60)}$$
$$= \frac{20}{\ln 1.333}$$
$$= \frac{20}{0.2877}$$
$$= 69.52°\text{C}$$

(4) *Exchanger data.* Select a $3\frac{1}{2}$ in. nominal heat exchanger with dimensions given in line 4 of Table 11.1.

$$d_o = 48.3\ \text{mm} \qquad (0.0483\ \text{m})$$

$$d_i = 48.3 - (2)(3.68)$$
$$= 48.3 - 7.36$$
$$= 40.94\ \text{mm} \qquad (0.04094\ \text{m})$$

$$\frac{d_i}{d_o} = \frac{0.04094}{0.0483}$$
$$= 0.8476$$

$$D_i = 101.6 - (2)(5.74)$$
$$= 101.6 - 11.48$$
$$= 90.12\ \text{mm} \qquad (0.09012\ \text{m})$$

$$n_T = 1$$

$$n_f = 36$$

$$b = 19.05\ \text{mm} \qquad (0.01905\ \text{m})$$

$$\delta = 0.80\ \text{mm} \qquad (0.0008\ \text{m})$$

The pertinent parameters may be computed. For extruded fins:

$$A_i = \frac{\pi}{4} d_i^2$$
$$= \frac{\pi}{4}(0.04094)^2$$
$$= 1.3164 \times 10^{-3}\ \text{m}^2$$

$$A_o = \frac{\pi}{4} D_i^2 - \left(\frac{\pi}{4} d_o^2 + n_f b \delta\right) n_T$$
$$= \frac{\pi}{4}(90.12)^2 - \left[\frac{\pi}{4}(48.3)^2 + (36)(19.05)(0.8)\right](1.0)$$
$$= 6.3787 \times 10^3 - (1.8322 \times 10^3 + 5.4864 \times 10^2)$$
$$= 6.3787 \times 10^3 - 2.3809 \times 10^3$$
$$= 3.9978 \times 10^3\ \text{mm}^2 \qquad (3.9978 \times 10^{-3}\ \text{m}^2)$$

$$P_{Wh} = [\pi d_o + n_f(2b - \delta)] n_T$$
$$= \{\pi(48.3) + (36)[2(19.05) - 0.8]\}(1.0)$$
$$= 1.5174 \times 10^2 + 1.3428 \times 10^3$$
$$= 1.4945 \times 10^3\ \text{mm} \qquad (1.4945\ \text{m})$$

$$P_{Wf} = \pi D_i + P_{Wh}$$
$$= \pi(0.01902) + 1.4945$$
$$= 0.0598 + 1.4945$$
$$= 1.5543 \text{ m}$$

Then

$$d_e = \frac{4A_o}{P_{Wh}} = \frac{(4)(3.9978 \times 10^{-3})}{1.4945} = 0.0107 \text{ m}$$

and

$$d_{ef} = \frac{4A}{P_{Wf}} = \frac{(4)(3.9978 \times 10^{-3})}{1.5543} = 0.0103 \text{ m}$$

Using eqs. (11.8) with $L = 1$ m yields

$$S_b = \pi d_o - n_f \delta$$
$$= \pi(0.0483) - (36)(0.0008)$$
$$= 1.5174 \times 10^{-1} - 2.8800 \times 10^{-2}$$
$$= 0.1229 \text{ m}^2$$
$$S_f = 2bn_f$$
$$= (2)(0.01905)(36)$$
$$= 1.3716 \text{ m}^2$$
$$S = S_b + S_f$$
$$= 0.1229 + 1.3716$$
$$= 1.4945 \text{ m}^2$$

and the surface per inner tube per meter is

$$S' = 1.4945 \text{ m}^2$$

(5) *Heat transfer coefficients.* For the inner pipe with cold water:

$$G_c = \frac{\dot{m}_c}{A_i} = \frac{1.615}{1.3164 \times 10^{-3}} = 1226.8 \text{ kg/m}^2\cdot\text{s}$$
$$v_c = \frac{G_c}{\rho} = \frac{1226.8}{997} = 1.2305 \text{ m/s}$$

and

$$\text{Re} = \frac{d_i G_c}{\mu} = \frac{(0.04094)(1226.8)}{8.55 \times 10^{-4}} = 58,755$$

Use Gnielinski's correlation given by eq. (11.32), which can be applied if $1.00 < \Pr < 10^5$ and if $2300 < \mathrm{Re} < 5 \times 10^6$. First, by eq. (11.31),

$$f_c = \frac{1}{(1.82 \log_{10} \mathrm{Re} - 1.64)^2}$$

$$= \frac{1}{(1.82 \log_{10} 58,755 - 1.64)^2}$$

$$= \frac{1}{[(1.82)(4.7690) - 1.64]^2}$$

$$= \frac{1}{(7.0397)^2}$$

$$= 2.0179 \times 10^{-2}$$

Then, with $f/8 = 2.0179 \times 10^{-2}/8 = 2.5223 \times 10^{-3}$,

$$\mathrm{Nu} = \frac{(f/8)(\mathrm{Re} - 1000)\Pr}{1.00 + 12.7(f/8)^{1/2}(\Pr^{2/3} - 1)}$$

$$= \frac{(2.5223 \times 10^{-3})(58,755 - 1000)(5.83)}{1.00 + (12.7)(2.5223 \times 10^{-3})^{1/2}[(5.83)^{2/3} - 1]}$$

$$= \frac{849.24}{1.00 + (12.7)(0.0502)(3.2393 - 1)}$$

$$= \frac{849.24}{1.00 + 1.4283}$$

$$= 349.73$$

and

$$h_i = \frac{k}{d_i}\mathrm{Nu} = \frac{0.613}{0.04094}(349.88) = 5236.5 \ \mathrm{W/m^2{\cdot}K}$$

For the annulus with engine oil:

$$G_h = \frac{\dot{m}_h}{A_o} = \frac{1.530}{3.9978 \times 10^{-3}} = 382.7 \ \mathrm{kg/m^2{\cdot}s}$$

$$v_h = \frac{G_h}{\rho} = \frac{382.7}{841.8} = 0.455 \ \mathrm{m/s}$$

and

$$\mathrm{Re} = \frac{d_e G_h}{\mu} = \frac{(0.0107)(382.7)}{0.0186} = 220.1$$

The engine oil in the annulus is in laminar flow. To use the Sieder–Tate correlation of eq. (11.26), a wall temperature must be assumed. Assume that the wall temperature is one-fourth of the temperature difference, $T_b - t_b$, above t_b. Thus, subject to later verification, assume that

$$T_w = t_b + (T_b - t_b) = 27 + (0.25)(70) = 44.5°\text{C} \qquad (317.5\ \text{K})$$

At T_w = 317.5 K, the engine oil has a viscosity of 0.1693, so that

$$\phi = \frac{\mu}{\mu_w} = \frac{0.0186}{0.1693} = 0.1099$$

Observe that the Sieder–Tate correlation requires that

$$\left(\text{RePr}\frac{d_e}{L}\right)^{1/3} \phi^{0.14} \geq 2$$

Here

$$\left[\text{RePr}\left(\frac{d_e}{L}\right)\right]^{1/3} \phi^{0.14} = \left[(220.1)(300)\left(\frac{0.0107}{3}\right)\right](0.1099)^{0.14}$$

$$= (235.5)^{1/3}(0.7340)$$

$$= 4.533 > 2$$

Thus

$$\text{Nu} = 1.86\left(\text{RePr}\frac{d_e}{L}\right)^{1/3} \phi^{0.14}$$

$$= (1.86)(4.533)$$

$$= 8.4314$$

and

$$h = \frac{k}{d_e}\text{Nu}$$

$$= \frac{0.137}{0.0107}(8.4314)$$

$$= 108.0\ \text{W/m}^2\cdot\text{K}$$

(6) *Overall coefficient of heat transfer, U_o*. First obtain h_{io}:

$$h_{io} = h_i\frac{d_i}{d_o}$$

$$= (5236.5)(0.8476)$$

$$= 4438.5\ \text{W/m}^2\cdot\text{K}$$

and from eq. (11.14),

$$r_{io} = \frac{1}{h_{io}} = \frac{1}{4438.5} = 2.2530 \times 10^{-4} \text{ m}^2\cdot\text{K/W}$$

With $r_{di} = 0.0001 \text{ m}^2\cdot\text{K/W}$, eq. (11.15) yields

$$r_{dio} = \frac{d_o}{d_i} r_{di} = \frac{0.0001}{0.8476} = 1.1798 \times 10^{-4} \text{ m}^2\cdot\text{K/W}$$

Noting that $d_i/d_o = 0.8476 > 0.75$, eq. (11.16) is used to obtain r_{mo}:

$$\begin{aligned} r_{mo} &= \frac{d_i(d_o - d_i)}{k_m(d_o + d_i)} \\ &= \frac{(0.04094)(0.0483 - 0.04094)}{52(0.0483 + 0.04094)} \\ &= 6.4933 \times 10^{-5} \text{ m}^2\cdot\text{K/W} \end{aligned}$$

Then

$$\begin{aligned} \sum R_{io} &= r_{io} + r_{dio} + r_{mo} \\ &= (2.2530 + 1.1798 + 0.6493) \times 10^{-4} \text{ m}^2\cdot\text{K/W} \\ &= 4.0821 \times 10^{-4} \text{ m}^2\cdot\text{K/W} \end{aligned}$$

and by eq. (11.17),

$$\begin{aligned} \sum R_{is} &= \sum R_{io} \left(\frac{S'}{\pi d_o} \right) \\ &= (4.0821 \times 10^{-4}) \left[\frac{1.4945}{\pi(0.0483)} \right] \\ &= 4.0821 \times 10^{-4}(9.8492) \\ &= 4.0205 \times 10^{-3} \text{ m}^2\cdot\text{K/W} \end{aligned}$$

For the annulus, r_o is given by eq. (11.18):

$$r_o = \frac{1}{h_o} = \frac{1}{108.0} = 9.2593 \times 10^{-3} \text{ m}^2\cdot\text{K/W}$$

Then

$$\begin{aligned} h_o' &= \frac{1}{r_o + r_{do}} \\ &= \frac{1}{9.2593 \times 10^{-3} + 1.8000 \times 10^{-4}} \\ &= \frac{1}{9.4393 \times 10^{-3}} \\ &= 105.94 \text{ W/m}^2\cdot\text{K} \end{aligned}$$

and with

$$m = \left(\frac{2h'_o}{k_m\delta}\right)^{1/2} = \left[\frac{(2)(105.94)}{(52)(0.0008)}\right]^{1/2} = 71.3671 \text{ m}^{-1}$$

then

$$mb = (71.3671)(0.01905) = 1.3595$$

The fin efficiency is

$$\eta = \frac{\tanh mb}{mb} = \frac{\tanh 1.3931}{1.3931} = 0.645$$

and the weighted fin efficiency is given by eq. (9.6):

$$\begin{aligned}\eta_o &= 1 - \frac{S_f}{S}(1-\eta)\\ &= 1 - \left(\frac{1.3716}{1.4945}\right)(1 - 0.645)\\ &= 1 - (0.9177)(0.355)\\ &= 1 - 0.326\\ &= 0.674\end{aligned}$$

Then

$$h'_{o\eta} = h'_o\eta_o = (105.94)(0.674) = 71.40$$

and

$$r'_{o\eta} = \frac{1}{h'_o\eta_o} = \frac{1}{71.40} = 1.4005 \times 10^{-2} \text{ W/m}^2\cdot\text{K}$$

The overall heat transfer coefficient is

$$\begin{aligned}U_o &= \frac{1}{\sum R_{is} + r'_{o\eta}}\\ &= \frac{1}{4.0205 \times 10^{-3} + 1.4005 \times 10^{-2}}\\ &= \frac{1}{1.8025 \times 10^{-2}}\\ &= 55.48 \text{ W/m}^2\cdot\text{K}\end{aligned}$$

(7) *Surface required.* The total surface required is

$$S_T = \frac{q}{U_o(\text{LMTD})} = \frac{135{,}007}{(55.48)(69.52)} = 35.00 \text{ m}^2$$

and with the surface of one hairpin equal to

$$S_1 = (2)(4.4836) = 8.9672 \text{ m}^2$$

the number of hairpins will be

$$n_{\text{hp}} = \frac{S_T}{S_1} = \frac{35.00}{8.9672} = 3.903 \quad (\text{say, } 4)$$

Use four hairpins.

(8) *Pressure losses.* The pressure losses are obtained from eqs. (11.39) and (11.41). For the inner pipe:

$$\Delta P_f = 4n_{\text{hp}} f_c \rho v_c^2 \frac{L}{d_i}$$

$$= (4)(4)(2.0179 \times 10^{-2})(997)(1.2305)^2 \left(\frac{3}{0.04094}\right)$$

$$= 35{,}715 \text{ Pa}$$

and

$$\Delta P_t = 2(2n_{\text{hp}} - 1)\rho v_c^2$$

$$= 2[(2)(4) - 1](997)(1.2305)^2$$

$$= 21{,}134 \text{ Pa}$$

and

$$\Delta P = \Delta P_f + \Delta P_t$$

$$= 35{,}715 + 21{,}134$$

$$= 56{,}849 \text{ Pa} < 65{,}000 \text{ Pa}$$

For the annulus, using the equivalent diameter for pressure loss,

$$\text{Re} = \frac{D_{ef} G_h}{\mu} = \frac{(0.0103)(382.7)}{0.0186} = 211.9$$

and

$$f = \frac{16}{\text{Re}} = \frac{16}{211.9} = 0.0755$$

Then

$$\Delta P_f = 4n_{\text{hp}} f_h \rho v_h^2 \frac{L}{d_{ef}}$$

$$= (4)(4)(0.0755)(841.8)(0.455)^2 \left(\frac{3}{0.0103}\right)$$

$$= 61{,}317 \text{ Pa}$$

and

$$\begin{aligned}\Delta P_t &= 2(2n_{\text{hp}} - 1)\rho v_h^2 \\ &= (2)[(2)(4) - 1](841.8)(0.455)^2 \\ &= 2440\ \text{Pa}\end{aligned}$$

so that

$$\begin{aligned}\Delta P &= \Delta P_f + \Delta P_t \\ &= 61{,}317 + 2440 \\ &= 63{,}757\ \text{Pa} < 80{,}000\ \text{Pa}\end{aligned}$$

It remains to confirm that $T_w = 44.5°\text{C}$ (317.5 K). This calculation is presented in the next section.

11.6.1 Wall Temperature and Further Remarks

Wall Temperature. It may be noted that the wall resistance presents the lowest resistance to the flow of heat between the hot and cold fluids. Hence an excellent approximation to the wall temperature may be obtained via the computation of the product of $\sum R_{is}$ and the heat flux. Then if the hot fluid is carried within the inner tube, the wall temperature will be

$$T_w = T_b - \sum R_{is} \frac{q}{S} \tag{11.45}$$

where

$$\sum R_{is} = \sum R_{io} \frac{S'}{\pi d_o} \tag{11.17}$$

and

$$\sum R_{io} = r_{io} + r_{dio} + r_{mo}$$

In the event that the cold fluid is carried in the inner tube, the wall temperature will be

$$T_w = t_b + \sum R_{is} \frac{q}{S} \tag{11.46}$$

Example 11.3: Verification of the Wall Temperature in Example 11.2. Determine the wall temperature for the double-pipe exchanger in Example 11.2.

SOLUTION. In Example 11.2 it was found that

$$T_b = 97°\text{C}$$

$$t_b = 27°\text{C}$$

and

$$\sum R_{is} = 4.0205 \times 10^{-3}\ \text{m}^2\cdot\text{K/W}$$

With the cold water inside the inner tube, eq. (11.46) may be used:

$$\begin{aligned} T_w &= t_b + \sum R_{is}\frac{q}{S} \\ &= 27 + (4.0205 \times 10^{-3}\ \text{m}^2\cdot\text{K/W})\left(\frac{135{,}007}{35.00}\right) \\ &= 27 + 15.5 \\ &= 42.5°\text{C} \end{aligned}$$

This is close enough to the assumed value of $T_w = 44.5°\text{C}$ to merit acceptance of this design.

What If? An attempt to reverse the streams by putting the engine oil within the inner pipe and the seawater in the annulus will result in the exchanger not meeting the heat transfer specification and an excessive pressure drop on the inner-pipe side. The fluid that tends to give the lower heat transfer coefficient should always be placed in the annulus for two reasons. The first of these concerns the advantage of a lower thermal resistance in the inner tube. This resistance must be transferred from the inner tube to the gross outside surface via a high multiplier as shown in the computation for $\sum R_{is}$ from $\sum R_{io}$. This also shows that fluids with low heat transfer coefficients should never be put into the inner pipe.

The second reason concerns the fin efficiency in the annulus, which tends to be low at higher heat transfer coefficients. There is a significant resistance that involves the product of h'_o with the overall fin efficiency, η_o. A higher value of η_o tends to compensate somewhat for the lower value of h'_o.

Finally, if the exchanger is to operate with the same liquid in the inner pipe and in the annulus consistent with the pressure loss specification, the hot fluid should be placed in the inner pipe. This takes advantage of the lower viscosity and higher Reynolds number to make the heat transfer coefficient on the inside as high as possible.

11.7 SERIES–PARALLEL ARRANGEMENTS

When two streams are arranged for counterflow, the LMTD represents the maximum thermal potential for heat transfer that can be obtained. Often, on the industrial scale, a single process service may entail the use of more than a single long hairpin. It then follows that it is desirable to connect the hairpins in series on both the annulus and inner-pipe sides, as in Fig. 11.5, and Example 11.2 illustrated the use of four hairpins connected in this manner. In this configuration, the temperature potential remains the LMTD for counterflow.

In some services there may be a large quantity of one fluid undergoing a small temperature change and a small quantity of another fluid undergoing a large temperature change. It may not be possible to circulate the large volume of fluid through the required number of hairpins with the available pressure drop. Under these circumstances, the larger volume of fluid may be manifolded in the series–parallel arrangement shown in Fig. 11.6. The inner-pipe fluid has been split between the exchangers designated I and II. Both of these exchangers are in counterflow relative to each other but not in the same sense as Fig. 11.5. In Fig. 11.6, the T's refer to the *series* stream and the t's refer to the *parallel* streams.

Departures from true counterflow and true cocurrent (parallel) flow can be handled by the *logarithmic mean temperature difference correction factor* F, which is defined as the ratio of the true temperature difference to the logarithmic mean temperature difference:

$$F = \frac{\text{true temperature difference}}{\text{logarithmic mean temperature difference}} \tag{11.47}$$

Charts of the logarithmic mean temperature difference correction factor are available in all heat transfer textbooks, heat transfer and engineering handbooks, and specialized books on process heat transfer. Kern (1950) and Kern and Kraus (1972) present a derivation, an equation, and charts for F based on the number of parallel streams.

Kern (1950) presents a derivation for the factor γ to be used in a modification of the heat transfer rate equation

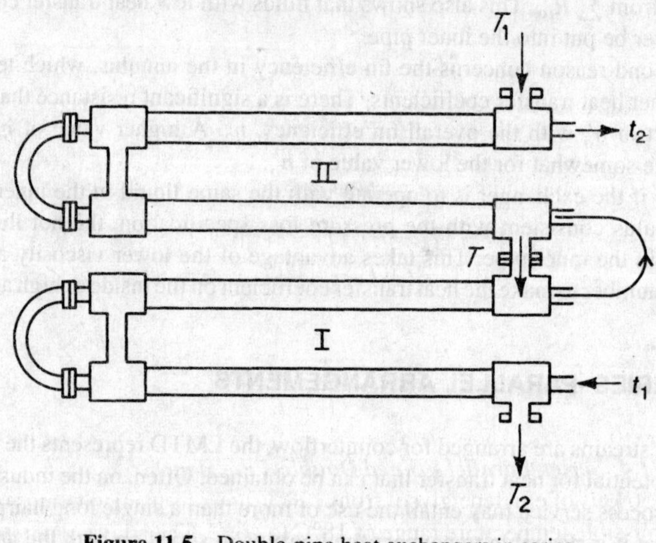

Figure 11.5 Double-pipe heat exchangers in series.

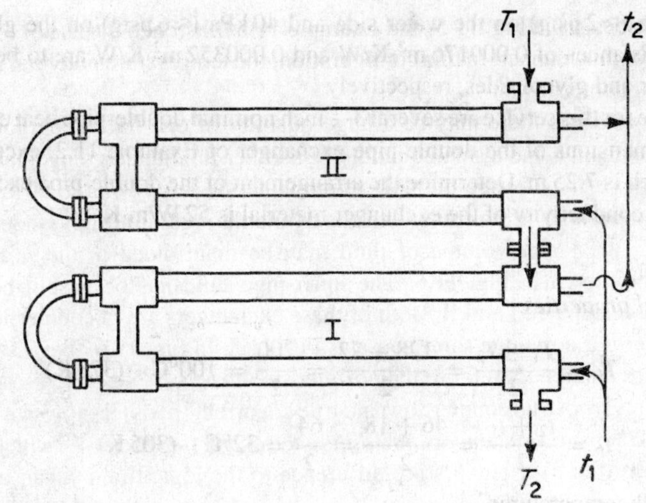

Figure 11.6 Series–parallel arrangement of double-pipe heat exchangers.

$$q = U_o S \gamma (T_1 - t_1) \tag{11.48}$$

where $T_1 - t_t$ represents the *total temperature potential*, the difference in the fluid stream inlet temperatures, in the exchanger configuration. After a laborious and detailed derivation, Kern (1950) gives for one series hot fluid and n parallel cold fluid streams,

$$\gamma = \frac{[1 - (T_2 - t_1)/(T_1 - t_1)][(Z - 1)/nZ]}{\ln\{[(Z - 1)/Z][(T_1 - t_1)/(T_2 - t_1)]^{1/n} + 1/Z\}} \tag{11.49}$$

where

$$Z = \frac{T_1 - T_2}{n(t_2 - t_1)}$$

For one series cold-fluid and n parallel hot-fluid streams, Kern (1950) gives

$$\gamma = \frac{[1 - (T_1 - t_2)/(T_1 - t_1)][(1 - Z)/nZ]}{\ln\{(1 - Z)(T_1 - t_1)/(T_1 - t_2)^{1/n} + Z\}} \tag{11.50}$$

where

$$Z = \frac{n(T_1 - T_2)}{t_2 - t_1}$$

Example 11.4: Series-Parallel Finned Double-Pipe Heat Exchanger. It is required to cool 2.556 kg/s of ethylene glycol from 128°C to 72°C using cooling-tower water with a permissible temperature range of 18°C to 46°C. The allowable pressure drops are 13.5 kPa (≈2 psig) on the water side and 40 kPa (≈6 psig) on the glycol side. Fouling resistances of 0.000176 m²·K/W and 0.000352 m²·K/W are to be provided on the water and glycol sides, respectively.

Available for this service are several 3-$\frac{1}{2}$ inch nominal double-pipe heat exchangers with the dimensions of the double-pipe exchanger of Example 11.2, except for the length, which is 7.25 m. Determine the arrangement of the double-pipe exchangers if the thermal conductivity of the exchanger material is 52 W/m·K.

SOLUTION.

(1) *Thermal properties:*

$$T_b = \frac{T_1 + T_2}{2} = \frac{128 + 72}{2} = \frac{200}{2} = 100°\text{C} \quad (373\,\text{K})$$

$$t_b = \frac{t_2 + t_1}{2} = \frac{46 + 18}{2} = \frac{64}{2} = 32°\text{C} \quad (305\,\text{K})$$

At these bulk temperatures

Property	Inner Pipe: Cooling-Tower Water	Annulus: Ethylene Glycol
t_b or T_b (K)	305	373
ρ (kg/m³)	995.0	1058.5
μ (kg-m/s)	7.69×10^{-4}	2.15×10^{-3}
k (W/m-K)	0.620	0.263
c (J/kg-K)	4178	2742
Pr	5.20	22.4

(2) *Heat balance.* For the hot ethylene glycol:

$$\begin{aligned} q &= \dot{m}_h c_h (T_1 - T_2) \\ &= 2.556(2742)(128 - 72) \\ &= 392,478\ \text{W} \end{aligned}$$

For the cold cooling tower water

$$\begin{aligned} \dot{m}_c &= \frac{q}{c_c(t_2 - t_1)} \\ &= \frac{392,478}{4179(46 - 18)} \\ &= 3.354\ \text{kg/s} \end{aligned}$$

(3) *Logarithmic mean temperature difference:*

$$\text{LMTD} = \frac{(T_1 - t_2) - (T_2 - t_1)}{\ln[(T_1 - t_2)/(T_2 - t_1)]}$$

With

$$T_1 - t_2 = 128 - 46 = 82°\text{C}$$

and

$$T_2 - t_1 = 72 - 18 = 54°\text{C}$$

$$\begin{aligned}
\text{LMTD} &= \frac{82 - 54}{\ln(82/54)} \\
&= \frac{28}{\ln 1.5185} \\
&= \frac{28}{0.4177} \\
&= 67.03°\text{C}
\end{aligned}$$

(4) *Exchanger data from Example 11.2:*

$d_o = 0.0483$ m	$A_i = 1.3164 \times 10^{-3}$ m^2
$d_i = 0.04094$ m	$A_o = 3.9978 \times 10^{-3}$ m^2
$d_o/d_i = 0.8476$	$P_{Wh} = 1.4945$ m
$D_i = 0.09012$ m	$P_{Wf} = 1.7777$ m
$n_T = 1$	$d_e = 0.0107$ m
$n_f = 36$	$d_{ef} = 0.0103$ m
$b = 0.01905$ m	$S' = 1.4945$ m^2/m
$\delta = 8 \times 10^{-4}$ m	$S_f/S = 0.9177$

(5) *Heat transfer coefficients:* Assume that $T_w = 67°\text{C}$ (340 K) to be verified. For the inner pipe with cooling-tower water:

$$G_c = \frac{\dot{m}_c}{A_i} = \frac{3.354}{1.3164 \times 10^{-3}} = 2547.9 \text{ kg/m}^2\text{-s}$$

and

$$v_c = \frac{G_c}{\rho} = \frac{2547.9}{995} = 2.561 \text{ m/s}$$

This value of v_c will lead to an excessive pressure loss. Suppose that three inner pipes are employed making this a three parallel pass (inner tube) and one series pass (annulus) exchanger. Then

$$G_c = \frac{2547.9}{3} = 849.3 \text{ kg/m}^2 \cdot \text{s}$$

$$v_c = \frac{2.561}{3} = 0.854 \text{ m/s}$$

and

$$\text{Re} = \frac{d_i G_c}{\mu} = \frac{(0.04094)(849.3)}{7.69 \times 10^{-4}} = 45{,}215$$

At $T_w = 67°\text{C}$ (340 K), $\mu = 4.20 \times 10^{-4}$ kg-m/s and

$$\phi = \frac{\mu}{\mu_w} = \frac{6.31 \times 10^{-4}}{4.20 \times 10^{-4}} = 1.5024$$

and this is not insignificant. Use Petukhov's correlation given by eq. (11.35), which can be applied if $1.0 < \phi < 40$ and if $1.00 < \text{Pr} < 140$. First, by eq. (11.31),

$$\begin{aligned}
f_c &= \frac{1}{(1.82 \log_{10} \text{Re} - 1.64)^2} \\
&= \frac{1}{(1.82 \log_{10} 45{,}215 - 1.64)^2} \\
&= \frac{1}{[1.82(4.6552) - 1.64]^2} \\
&= \frac{1}{(6.8326)^2} \\
&= 2.1420 \times 10^{-2}
\end{aligned}$$

Then, with $f/8 = 2.1420 \times 10^{-2}/8 = 2.6775 \times 10^{-3}$ and $n = 0.11$ for heating

$$\begin{aligned}
\text{Nu} &= \frac{(f/8)\text{RePr}\phi^{0.11}}{1.07 + 12.7(f/8)^{1/2}(\text{Pr}^{2/3} - 1)} \\
&= \frac{(2.6775 \times 10^{-3})(45{,}215)(5.20)(1.5024)^{0.11}}{1.07 + 12.7(2.6775 \times 10^{-3})^{1/2}[(5.20)^{2/3} - 1]} \\
&= \frac{624.25}{1.07 + 12.7(0.0517)(3.0015 - 1)} \\
&= \frac{624.25}{2.3853} \\
&= 261.70
\end{aligned}$$

and

$$h_i = \frac{k}{d_i}\text{Nu} = \frac{0.620}{0.04094}(261.70) = 3963 \text{ W/m}^2 \cdot \text{K}$$

For the annulus with ethylene glycol:

$$G_h = \frac{\dot{m}_h}{A_o} = \frac{2.556}{3.9978 \times 10^{-3}} = 639.35 \text{ kg/m}^2 \cdot \text{s}$$

$$v_h = \frac{G_h}{\rho} = \frac{639.35}{1058.5} = 0.604 \text{ m/s}$$

and

$$\text{Re} = \frac{d_e G_h}{\mu} = \frac{(0.0107)(639.35)}{2.15 \times 10^{-3}} = 3181.9$$

The ethylene glycol is in the annulus is in transition from laminar to turbulent flow and the correlation of Hausen (1943) given by eq. (11.27) should be employed. At $T_w = 340$ K, the ethylene glycol has a viscosity of 4.31×10^{-3} kg-m/s, so that

$$\phi = \frac{\mu}{\mu_w} = \frac{2.15 \times 10^{-3}}{4.31 \times 10^{-3}} = 0.4988$$

The Hausen (1943) correlation gives

$$\text{Nu} = 0.116[(\text{Re})^{2/3} - 125]\text{Pr}^{1/3}\phi^{0.14}\left[1 + \left(\frac{d_e}{L}\right)^{2/3}\right]$$

and with

$$1 + \left(\frac{d_e}{L}\right)^{2/3} = 1 + \left(\frac{0.0107}{7.25}\right)^{2/3} = 1.0130$$

$$\begin{aligned}\text{Nu} &= 0.116[(3181.9)^{2/3} - 125](22.4)^{1/3}(0.4988)^{0.14}(1.0130)\\ &= 0.116[216.3 - 125](2.8189)(0.9072)(1.0130)\\ &= 27.43\end{aligned}$$

Then

$$\begin{aligned}h &= \frac{k}{d_e}\text{Nu}\\ &= \frac{0.263}{0.0107}(27.43)\\ &= 674.4\ \text{W/m}^2 \cdot \text{K}\end{aligned}$$

(6) *Overall coefficient of heat transfer, U_o:* First obtain h_{io}

$$\begin{aligned}h_{io} &= h_i\left(\frac{d_i}{d_o}\right)\\ &= 3963(0.8476)\\ &= 3359\ \text{W/m}^2 \cdot \text{K}\end{aligned}$$

and from eq. (11.14)

$$r_{io} = \frac{1}{h_{io}} = \frac{1}{3359} = 2.9771 \times 10^{-4}\ \text{m}^2 \cdot \text{K/W}$$

With $r_{di} = 0.000176\ \text{m}^2 \cdot \text{K/W}$, eq. (11.15) yields

$$r_{dio} = \frac{d_o}{d_i} r_{di} = \frac{0.000176}{0.8476} = 2.0765 \times 10^{-4}\,\text{m}^2 \cdot \text{K/W}$$

The value of r_{mo} will be the same as in Example 11.2

$$r_{mo} = 6.4933 \times 10^{-5}\,\text{m}^2 \cdot \text{K/W}$$

Then

$$\sum R_{io} = r_{io} + r_{dio} + r_{mo}$$

$$= (2.9771 + 2.0765 + 0.6493) \times 10^{-4}\,\text{m}^2 \cdot \text{K/W}$$

$$= 5.7329 \times 10^{-4}\,\text{m}^2 \cdot \text{K/W}$$

and by eq. (11.17),

$$\sum R_{is} = \sum R_{io} \left(\frac{S'}{\pi d_o} \right)$$

$$= (5.7329 \times 10^{-4}) \left[\frac{1.4945}{\pi(0.0483)} \right]$$

$$= 5.7329 \times 10^{-4}(9.8492)$$

$$= 5.6464 \times 10^{-3}\,\text{m}^2 \cdot \text{K/W}$$

For the annulus, r_o is given by eq. (11.18)

$$r_o = \frac{1}{h_o} = \frac{1}{674.4} = 1.4828 \times 10^{-3}\,\text{m}^2 \cdot \text{K/W}$$

Then

$$h'_o = \frac{1}{r_o + r_{do}}$$

$$= \frac{1}{1.4828 \times 10^{-3} + 3.5200 \times 10^{-4}}$$

$$= \frac{1}{1.8348 \times 10^{-3}}$$

$$= 545.02\,\text{W/m}^2 \cdot \text{K}$$

and with

$$m = \left(\frac{2h'_o}{k_m \delta} \right)^{1/2} = \left[\frac{2(545.02)}{52(0.0008)} \right]^{1/2} = 161.8727\,\text{m}^{-1}$$

then

$$mb = 161.8727(0.01905) = 3.0837$$

The fin efficiency is

$$\eta = \frac{\tanh mb}{mb} = \frac{\tanh 3.0837}{3.0837} = 0.3229$$

and the weighted fin efficiency is given by eq. (9.6)

$$\begin{aligned}\eta_o &= 1 - \frac{S_f}{S}(1-\eta) \\ &= 1 - \frac{1.3716}{1.4945}(1-0.3229) \\ &= 1 - 0.9177(0.6771) \\ &= 1 - 0.6213 \\ &= 0.3787\end{aligned}$$

Then

$$h'_{o\eta} = h'_o \eta_o = 545.02(0.3787) = 206.37$$

and

$$r'_{o\eta} = \frac{1}{h'_o \eta_o} = \frac{1}{206.37} = 4.8457 \times 10^{-3}\ \mathrm{W/m^2 \cdot K}$$

The overall heat transfer coefficient is

$$\begin{aligned}U_o &= \frac{1}{\sum R_{is} + r'_{o\eta}} \\ &= \frac{1}{5.6464 \times 10^{-3} + 4.8457 \times 10^{-2}} \\ &= \frac{1}{1.0492 \times 10^{-2}} \\ &= 95.31\ \mathrm{W/m^2 \cdot K}\end{aligned}$$

(7) *Surface required:* The logarithmic mean temperature difference will not be needed. Instead, use eq. (11.48):

$$q = U_o S \gamma (T_1 - t_1) \tag{11.48}$$

where γ is given by eq. (11.49). Let

$$P = \frac{T_2 - t_1}{T_1 - t_1}$$

so that eq. (11.49) becomes

$$\gamma = \frac{(1-P)[(Z-1)/nZ]}{\ln\{[(Z-1)/nZ](1/P)^{1/n} + 1/Z\}}$$

where

$$Z = \frac{T_1 - T_2}{n(t_2 - t_1)}$$

Here with $n = 3$ (three cold-side parallel passes),

$$Z = \frac{T_1 - T_2}{n(t_2 - t_1)} = \frac{128 - 72}{3(46 - 18)} = \frac{56}{3(28)} = \frac{2}{3}$$

and

$$P = \frac{T_2 - t_1}{T_1 - t_1} = \frac{72 - 18}{128 - 18} = \frac{54}{110} = 0.4909$$

Thus

$$\begin{aligned}
\gamma &= \frac{(1 - 0.4909)\{[(2/3) - 1]/(3)(2/3)\}}{\ln\{[(2/3) - 1]/(2/3)\}[(1/0.4909)^{1/3} + 3/2]} \\
&= \left(-\frac{1}{6}\right)\frac{1 - 0.4909}{\ln[(-1/2)(2.0370)^{1/3} + 1.50]} \\
&= -\frac{0.0849}{\ln[-0.6338 + 1.50]} \\
&= -\frac{0.0849}{\ln[0.8662]} \\
&= \frac{-0.0849}{-0.1437} \\
&= 0.5906
\end{aligned}$$

The total surface required is obtained from eq. (11.48)

$$\begin{aligned}
S_T &= \frac{q}{U_o \gamma (T_1 - t_1)} \\
&= \frac{392,478}{(95.31)(0.5906)(128 - 18)} \\
&= 63.3856\,\mathrm{m}^2
\end{aligned}$$

The surface per hairpin is

$$S_1 = 2(7.25)(1.4945) = 21.6703\,\mathrm{m}^2$$

and the number of hairpins will be n_{hp}

$$n_{hp} = \frac{S_T}{S_1} = \frac{63.3856}{21.6703} = 2.926 \quad (\text{say, } 3)$$

Use three (3) hairpins arranged with one ethylene glycol pass and three parallel water passes.

(8) *Pressure losses:* The pressure losses are obtained from eqs. (11.39) and (11.41). For the inner pipe:

$$\Delta P_f = 4n_{\text{hp}} f_c \rho v_c^2 \left(\frac{L}{d_i}\right)$$

$$= 4(1)(2.1420 \times 10^{-2})(995)(0.854)^2 \left(\frac{7.25}{0.04094}\right)$$

$$= 11,101 \text{ Pa}$$

and

$$\Delta P_t = 2(2n_{\text{hp}} - 1)\rho v_c^2$$

$$= 2[2(1) - 1](995)(0.854)^2$$

$$= 1451 \text{ Pa}$$

and

$$\Delta P = \Delta P_f + \Delta P_t$$

$$= 11,101 + 1451$$

$$= 12,462 \text{ Pa} < 13,500 \text{ Pa}$$

For the annulus, using the equivalent diameter for pressure loss and eq. (11.43) for f_h

$$f_h = \frac{0.046}{\text{Re}^{0.20}} = \frac{0.046}{(3181.9)^{0.20}} = 9.1669 \times 10^{-3}$$

Then

$$\Delta P_f = 4n_{\text{hp}} f_h \rho v_h^2 \left(\frac{L}{d_{ef}}\right)$$

$$= 4(3)(9.1669 \times 10^{-3})(1058.5)(0.604)^2 \left(\frac{7.25}{0.0103}\right)$$

$$= 29,900 \text{ Pa}$$

and

$$\Delta P_t = 2(2n_{\text{hp}} - 1)\rho v_h^2$$

$$= 2[2(3) - 1](1058.5)(0.604)^2$$

$$= 3863 \text{ Pa}$$

and

$$\Delta P = \Delta P_f + \Delta P_t$$

$$= 29,900 + 3863$$

$$= 33,763 \text{ Pa} < 40,000 \text{ Pa}$$

(9) *Verification of assumed wall temperature*. The wall temperature will be

$$T_w = t_b + \sum R_{is} \left(\frac{q}{S}\right)$$

$$= 32 + (5.6364 \times 10^{-3})\left(\frac{392,478}{63.3856}\right)$$

$$= 32 + 35 = 67°\text{C} \quad (340\text{K}) \ \checkmark$$

11.8 MULTIPLE-FINNED DOUBLE-PIPE EXCHANGERS

There are numerous applications for longitudinal fin pipes and tubes. Closest to the double-pipe exchanger is the hairpin with multiple longitudinal fin pipes. A variety of pipes and tubes are available with longitudinal fins whose numbers, heights, thicknesses, and materials differ. Data for some of these configurations are shown in Table 11.2.

The procedure for the design and analysis of the multiple-tube exchanger differs little for that used for the single-tube exchanger.

Example 11.5: Multiple-Tube Finned Double-Pipe Heat Exchanger. 0.80 kg/s of oxygen entering at 250 kPa and 25°C is to be heated to 129°C using condensing ($h_i = 8750$ W/m^2·K) steam at 350 kPa. The maximum allowable pressure drop for the oxygen is 5 kPa, and although both fluids are clean, token fouling resistances of 0.0001 m^2·K/W should be provided for each stream.

TABLE 11.2 Dimensions of Multitube Double-Pipe Exchangers[a]

Nominal Diameter (in.)	Pipe Thickness (mm)	Pipe OD (mm)	Number of Tubes	Number of Fins	Tube OD (mm)	Tube Thickness (mm)	Fin Height (mm)
4	6.02	114.3	7	16	19.02	2.11	5.33
4	6.02	114.3	7	20	22.2	2.11	5.33
6	7.11	168.3	19	16	19.02	2.11	5.33
6	7.11	168.3	14	16	19.02	2.11	5.33
6	7.11	168.3	7	20	20.04	2.77	12.7
8	8.18	219.1	19	16	19.02	2.11	8.64
8	8.18	219.1	19	20	22.2	2.11	7.11
8	8.18	219.1	19	20	25.4	2.77	5.33
8	8.18	219.1	19	16	19.02	2.11	7.11
8	8.18	219.1	19	20	22.2	2.11	5.33

[a]Fin thicknesses are identical to those listed in Table 11.1. The dimensions shown here are for low-pressure units.

Available for this service is a single 8-in. nominal multiple-tube double-pipe heat exchanger containing 19 tubes, which have an outside diameter of 25.4 m and are 2 m long. The tubes have 20 welded fins which are 12.7 mm high and 0.889 mm thick. The thermal conductivity of the exchanger material is 50 W/m·K.

Can this exchanger fulfill this service?

SOLUTION

(1) *Thermal properties.* The hot fluid is steam at 350 kPa, where $T_b = 138.88°C$ and $h_{fg} = 2148.1$ kJ/kg. The flow of steam will be very small, and the pressure drop within the tubes will be negligible.

$$t_b = \frac{t_2 + t_1}{2} = \frac{129 + 25}{2} = \frac{154}{2} = 77°C \quad (350\,K)$$

For oxygen at 350 K:

Property	Annulus Oxygen
t_b (K)	350
μ (kg·m/s)	2.34×10^{-5}
k (W/m·K)	0.0306
c (J/kg·K)	929
Pr	0.710
R (kPa·m^2/kg·K)	0.2598

With oxygen as a perfect gas and assuming that the pressure drop is 5 kPa,

$$P = 250 - \frac{5}{2} = 247.5\,\text{kPa}$$

and the density of the oxygen at this pressure and 350 K will be

$$\rho = \frac{P}{RT} = \frac{247.5}{(0.2598)(350)} = 2.7217\,\text{kg/m}^3$$

(2) *Heat balance.* For the cold fluid—oxygen:

$$\begin{aligned} q &= \dot{m}_c c_c (t_2 - t_1) \\ &= (0.80)(929)(129 - 25) \\ &= 77{,}294\,\text{W} \end{aligned}$$

(3) *Logarithmic mean temperature difference:*

$$\text{LMTD} = \frac{(T_1 - t_2) - (T_2 - t_1)}{\ln(T_1 - t_2)/(T_2 - t_1)}$$

and with $T_1 - T_2 = T = 138.88°\text{C}$, then

$$\text{LMTD} = \frac{(t_2 - t_1)}{\ln(T - t_2)/(T - t_1)}$$

Then

$$T - t_2 = 138.88 - 129 = 9.88°\text{C}$$

and

$$T - t_1 = 138.88 - 25 = 113.88°\text{C}$$

$$\begin{aligned} \text{LMTD} &= \frac{129 - 25}{\ln(113.88/9.88)} \\ &= \frac{104}{\ln 11.5263} \\ &= \frac{104}{2.4446} \\ &= 42.54°\text{C} \end{aligned}$$

(4) *Exchanger data.* The exchanger physical properties are summarized in line 8 of Table 11.2. The inner-pipe properties are the same as in Example 11.1.

$$d_o = 0.0254 \text{ m} \qquad A_i = 3.0978 \times 10^{-4} \text{ m}^2$$
$$d_i = 0.01986 \text{ m} \qquad d_i/d_o = 0.7819$$

Then

$$\begin{aligned} D_i &= 219.1 - (2)(8.18) \\ &= 202.74 \text{ mm} \quad (0.20274 \text{ m}) \\ n_T &= 19 \\ n_f &= 20 \\ b &= 5.33 \text{ mm} \\ \delta &= 0.889 \text{ mm} \end{aligned}$$

For the annulus with $z = 2\delta = (2)(0.889) = 1.778$, by eq. (11.9)

$$A_o = \frac{\pi}{4} D_i^2 - \left[\frac{\pi}{4} d_o^2 + n_f \delta \left(b + \frac{z}{2}\right)\right] n_T$$

$$= \frac{\pi}{4}(202.74)^2 - \left[\frac{\pi}{4}(25.4)^2 + (20)(0.889)\left(5.33 + \frac{1.778}{2}\right)\right](19)$$

$$= 3.2283 \times 10^4 - (5.0671 \times 10^2 + 1.1057 \times 10^2)(19)$$

$$= 3.2283 \times 10^4 - 1.1728 \times 10^4$$

$$= 2.0554 \times 10^4 \text{ mm}^4 \qquad (2.0554 \times 10^{-2} \text{ m}^2)$$

From eq. (11.10),

$$P_{Wh} = [\pi d_o + n_f(2b - \delta)]n_T$$

$$= \{\pi(25.4) + (20)[(2)(5.33) - 0.889]\}(19)$$

$$= (79.7965 + 195.4200)(19)$$

$$= 5.2291 \times 10^3 \text{ mm} \qquad (5.2291 \text{ m})$$

and from eq. (11.11),

$$P_{Wf} = \pi D_i + P_{Wh}$$

$$= \pi(0.20274) + 5.2291$$

$$= 0.6369 + 5.2291$$

$$= 5.8660 \text{ m}$$

Then from eq. (11.12),

$$d_e = \frac{4A}{P_{Wh}} = \frac{(4)(2.0554 \times 10^{-2})}{5.2291} = 0.0157 \text{ m}$$

and from eq. (11.13),

$$d_{ef} = \frac{4A}{P_{Wf}} = \frac{(4)(2.0554 \times 10^{-2})}{5.8660} = 0.0140 \text{ m}$$

For unit length ($L = 1$ m) per tube, using eqs. (11.8),

$$S_b = [\pi d_o - n_f\delta]$$

$$= [\pi(0.0254) - (20)(8.89 \times 10^{-4})]$$

$$= 7.9796 \times 10^{-2} - 1.7780 \times 10^{-2}$$

$$= 0.0620 \text{ m}^2/\text{m}$$

$$S_f = 2bn_f$$

$$= (2)(0.00533)(20)$$

$$= 0.2132 \text{ m}^2/\text{m}$$

$$\begin{aligned} S &= S_b + S_f \\ &= 0.0620 + 0.2132 \\ &= 0.2752\ \text{m}^2/\text{m} \end{aligned}$$

and this also gives the gross outside surface of the inner pipe per meter and per tube:

$$S' = 0.2752\ \text{m}^2/\text{m}$$

(5) *Heat transfer coefficients.* For the, inner pipe with condensing steam the heat transfer coefficient for the condensing steam is given as

$$h_i = 8750\ \text{W/m}^2 \cdot \text{K}$$

For the annulus with oxygen, because the viscosity of oxygen (and other gases) does not vary significantly with temperature, let $\phi = 1.00$.

$$G_c = \frac{\dot{m}_c}{A_o} = \frac{0.80}{2.0554 \times 10^{-2}} = 38.92\ \text{kg/m}^2\cdot\text{s}$$

$$v_c = \frac{G_c}{\rho} = \frac{38.92}{2.722} = 14.30\ \text{m/s}$$

and

$$\text{Re} = \frac{d_e G_c}{\mu} = \frac{(0.0157)(38.92)}{2.34 \times 10^{-5}} = 26,114$$

The flow in the annulus is turbulent and the correlation of Sieder and Tate given by eq. (11.29) with ϕ taken as unity can be employed. The Sieder and Tate correlation gives

$$\begin{aligned} \text{Nu} &= 0.023\text{Re}^{0.80}\text{Pr}^{1/3} \\ &= (0.023)(26,114)^{0.80}(0.710)^{1/3} \\ &= (0.023)(3415.8)(0.892) \\ &= 70.09 \end{aligned}$$

and

$$h_o = \frac{k}{d_e}\text{Nu} = \left(\frac{0.0306}{0.0157}\right)(70.09) = 136.61\ \text{W/m}^2\cdot\text{K}$$

(6) *Overall coefficient of heat transfer, U_o.*
First obtain h_{io}:

$$h_{io} = h_i \frac{d_i}{d_o}$$

$$= (8750)(0.7819)$$
$$= 6841.5 \text{ W/m}^2\cdot\text{K}$$

and from eq. (11.14),

$$r_{io} = \frac{1}{h_{io}} = \frac{1}{6841.5} = 1.4617 \times 10^{-4} \text{ m}^2\cdot\text{K/W}$$

With $r_{di} = 0.0001 \text{ m}^2\cdot\text{K/W}$, eq. (11.15) yields

$$r_{dio} = \frac{d_o}{d_i} r_{di} = \frac{0.0001}{0.7819} = 1.2789 \times 10^{-4} \text{ m}^2\cdot\text{K/W}$$

The value of r_{mo} will be thc same as in Example 11.1:

$$r_{mo} = 5.2847 \times 10^{-5} \text{ m}^2\cdot\text{K/W}$$

Then

$$\sum R_{io} = r_{io} + r_{dio} + r_{mo}$$
$$= (1.4617 + 1.2789 + 0.5285) \times 10^{-4} \text{ m}^2\cdot\text{K/W}$$
$$= 3.2691 \times 10^{-4} \text{ m}^2\cdot\text{K/W}$$

and by eq. (11.17),

$$\sum R_{is} = \sum R_{io} \frac{S'}{\pi d_o}$$
$$= (3.2691 \times 10^{-4}) \left[\frac{0.2752}{\pi(0.0254)} \right]$$
$$= (3.2691 \times 10^{-4})(3.4488)$$
$$= 1.1274 \times 10^{-3} \text{ m}^2\cdot\text{K/W}$$

For the annulus r_o is given by eq. (11.18):

$$r_o = \frac{1}{h_o} = \frac{1}{136.61} = 7.3201 \times 10^{-3} \text{ m}^2\cdot\text{K/W}$$

Then

$$h'_o = \frac{1}{r_o + r_{do}}$$
$$= \frac{1}{7.3201 \times 10^{-3} + 1.0000 \times 10^{-4}}$$
$$= \frac{1}{7.4201 \times 10^{-3}}$$
$$= 134.77 \text{ W/m}^2\cdot\text{K}$$

and with

$$m = \left(\frac{2h'_o}{k_m\delta}\right)^{1/2} = \left[\frac{(2)(134.77)}{(50)(8.89 \times 10^{-4})}\right]^{1/2} = 77.8710\ \text{m}^{-1}$$

then

$$mb = (77.8710)(0.00533) = 0.4151$$

The fin efficiency is

$$\eta = \frac{\tanh mb}{mb} = \frac{\tanh 0.4151}{0.4151} = 0.946$$

and the weighted fin efficiency is given by eq. (9.6):

$$\eta_o = 1 - \frac{S_f}{S}(1-\eta)$$

$$= 1 - \left(\frac{0.2132}{0.2752}\right)(1 - 0.946)$$

$$= 1 - (0.7747)(0.054)$$

$$= 1 - 0.0042$$

$$= 0.958$$

Then

$$h'_{o\eta} = h'_o\eta_o = (134.77)(0.958) = 129.11$$

and

$$r'_{o\eta} = \frac{1}{h'_o\eta_o} = \frac{1}{129.11} = 7.7454 \times 10^{-3}\ \text{W/m}^2\cdot\text{K}$$

The overall heat transfer coefficient is

$$U_o = \frac{1}{\sum R_{is} + r'_{o\eta}}$$

$$= \frac{1}{1.1274 \times 10^{-3} + 7.7454 \times 10^{-3}}$$

$$= \frac{1}{8.8728 \times 10^{-3}}$$

$$= 112.70\ \text{W/m}^2\cdot\text{K}$$

(7) *Surface required.* The total surface required is obtained from eq. (11.48):

$$S_T = \frac{q}{U_o(\text{LMTD})}$$

$$= \frac{77,294}{(112.70)(42.54)}$$

$$= 16.1222 \text{ m}^2$$

The surface per hairpin is

$$S_1 = 2Ln_T S' = (2)(2)(19)(0.2752) = 20.9152 \text{ m}^2$$

and the number of hairpins will be

$$n_{hp} = \frac{S_T}{S_1} = \frac{16.1222}{20.9152} = 0.771 \quad (\text{say, } 1)$$

The exchanger will meet the heat transfer requirements.

(8) *Pressure losses.* The pressure losses are obtained from eqs. (11.39) and (11.41). The pressure loss in the inner pipe is negligible. For the, annulus, using the equivalent diameter for pressure loss and eq. (11.44) for f_c yields

$$\text{Re} = \frac{d_{ef} G_c}{\mu} = \frac{(0.0140)(38.92)}{2.34 \times 10^{-5}} = 23,287$$

and

$$f_c = \frac{0.079}{\text{Re}^{0.25}} = \frac{0.079}{(23,287)^{0.25}} = 6.3957 \times 10^{-3}$$

Then

$$\Delta P_f = 4n_{hp} f_h \rho v_h^2 \frac{L}{d_{ef}}$$

$$= (4)(1)(6.3957 \times 10^{-3})(2.7217)(14.30)^2 \left(\frac{2}{0.0140}\right)$$

$$= 2034 \text{ Pa}$$

and

$$\Delta P_t = 2(2n_{hp} - 1)\rho v_c^2$$

$$= 2[(2)(1) - 1](2.7217)(14.30)^2$$

$$= 1113 \text{ Pa}$$

and

$$\Delta P = \Delta P_f + \Delta P_t$$

$$= 2034 + 1113$$

$$= 3147 \text{ Pa} < 5000 \text{ Pa}$$

The exchanger meets the pressure loss requirements.

11.9 CLOSURE

Radial fins are also employed in heat exchangers. A study of radial fin heat exchangers is the subject of Chapter 12.

11.10 NOMENCLATURE

Roman Letter Symbols

A	cross-sectional area, m^2
b	fin height, m
c	specific heat, J/kg·K
D	outer pipe diameter, m
d	inner pipe diameter, m
F	logarithmic mean temperature difference correction factor, dimensionless
f	friction factor, dimensionless
G	mass velocity, $kg/m^2 \cdot s$
Gz	Graetz number, dimensionless
h	heat transfer coefficient, $W/m^2 \cdot K$
h'_o	annulus heat transfer coefficient, $W/m^2 \cdot K$
$h'_{o\eta}$	annulus heat transfer coefficient corrected for fin efficiency, $W/m^2 \cdot K$
j	heat transfer parameter, dimensionless
k	thermal conductivity, W/m·K
L	fin length, m; exchanger length, m
LMTD	logarithmic mean temperature difference, K
m	fin performance parameter, m^{-1}
$\dot{m}$	mass flow rate, kg/s
Nu	Nusselt number, dimensionless
N_{tu}	number of transfer units, dimensionless
n	number of cold-side passes, dimensionless; number annuli or hairpins, dimensionless
P	pressure, N/m^2 (Pa); temperature ratio, dimensionless
Pr	Prandtl number, dimensionless
q	heat flow, W
R	thermal resistance, $K \cdot m^2/W$
Re	Reynolds number, dimensionless
r	thermal resistance, $K \cdot m^2/W$
S	surface area, m^2
S'	surface area per unit length of annulus, m^2
s'	surface area per unit length of tube, m^2
T	temperature, K; hot fluid temperature, K

t	cold fluid temperature, K
V	velocity, m/s

Greek Letter Symbols

γ	correction factor, dimensionless
Δ	change in, dimensionless
δ	fin thickness, m
ε	exchanger effectiveness, dimensionless
η	fin efficiency, dimensionless
μ	dynamic viscosity, kg/m·s
ν	kinematic viscocity, m^2/s
ρ	density, kg/m^3
Σ	summation, dimensionless
ϕ	viscosity ratio, dimensionless

Roman Letter Subscripts

b	bulk quantity; bare tube
c	cold fluid
di	inside fouling factor
dio	inside fouling factor referred to outside
do	outside fouling factor
e	equivalent
f	fin or fins
h	hot fluid
hp	number of hairpins
i	inside pipe or tube
io	inside referred to outside
is	sum of internal resistances referred to outer surface
m	mean value; metal resistance
mo	metal resistance referred to outside surface
n	number of fins or tubes
o	outside
t	total number

Roman Letter Superscripts

m	exponent in convection analysis
n	exponent in convection analysis

CHAPTER 12

TRANSVERSE HIGH-FIN EXCHANGERS

12.1 INTRODUCTION

Pipes, tubes, and cast tubular sections with external transverse high fins have been used extensively for heating, cooling, and dehumidifying air and other gases. The fins are preferably called *transverse* rather than radial because they need not be circular, as the latter term implies, and are often helical. The air-fin cooler is a device in which hot-process fluids, usually liquids, flow inside extended surface tubes and atmospheric air is circulated outside the tubes by forced or induced draft over the extended surface. Figures 12.1 and 12.2 show such devices.

Unlike liquids, gases are compressible, and it usually necessary to allocate very small pressure drops for their circulation through industrial equipment, or the cost of

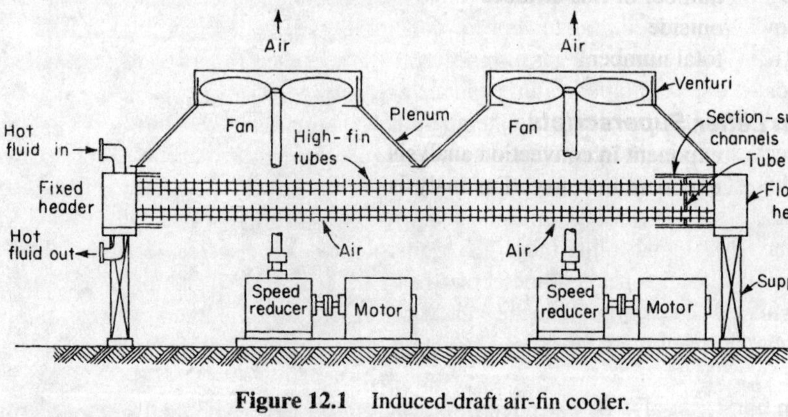

Figure 12.1 Induced-draft air-fin cooler.

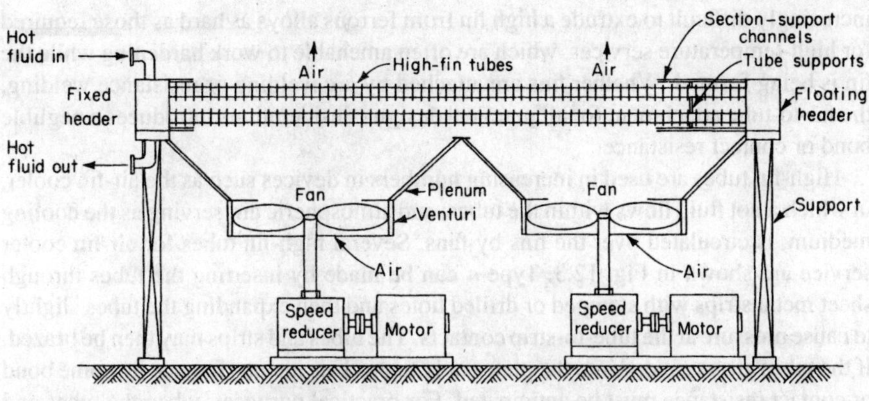

Figure 12.2 Forced-draft air-fin cooler.

the compression work may entail a substantial operating charge. Except for hydrogen and helium, which have relatively high thermal conductivities, the low thermal conductivities of gases coupled with small allowable pressure drops tend toward low external convection heat transfer coefficients.

In the discussion of longitudinal high-fin tubes in Section 11.3, it was noted that a steel fin 1.27 cm high and 0.0889 cm thick could be used advantageously with a fluid producing a heat transfer coefficient as high as 250 $W/m^2 \cdot K$. Aluminum and copper have thermal conductivities much higher than steel, 200 and 380 versus 45 W/m·K. It would appear that thin high fins made of aluminum or copper would have excellent fin efficiencies when exposed to various heating and cooling applications of air and other gases at or near atmospheric pressure.

In air-fin cooler services, the allowable pressure drop is measured in centimeters or inches of water and air can be circulated over a few rows of high-fin tubes with large transverse fin surfaces and at the same time require a very small pressure drop. Transverse high-fin tubular elements are found in such diverse places as economizers of steam power boilers, cooling towers, air-conditioning coils, indirect-fired heaters, waste-heat recovery systems for gas turbines and catalytic reactors, gas-cooled nuclear reactors, convectors for home heating, and air-fin coolers.

In the services cited involving high temperatures, hot gases flow over the fins and water or steam flows inside the tubes. The extended surface element usually consists of a chromium steel tube whose chromium content is increased with higher anticipated service temperatures. A ribbon, similar in composition to the tube, is helically wound and continuously welded to the tube. The higher and thicker the fins, the fewer the maximum number of fins per centimeter of tube which can be arc-welded because the fin spacing must also accommodate the welding electrode. High-temperature high-fin tubes on a closer spacing are fabricated by electrical resistance welding of the fins to the tube.

High-fin tubes can also be extruded directly from the tube wall metal, as in the case of the integral low-fin tubing discussed in Chapter 11. However, it becomes increaingly difficult to extrude a high fin from ferrous alloys as hard as those required for high-temperature services, which are often amenable to work hardening while the fin is being formed. Whether fins are attached by arc welding or resistance welding, the fin-to-tube attachment for all practical design considerations introduces a neglible bond or contact resistance.

High-fin tubes are used in increasing numbers in devices such as the air-fin cooler, in which a hot fluid flows within the tubes, and atmospheric air, serving as the cooling medium, is circulated over the fins by fans. Several high-fin tubes for air-fin cooler service are shown in Fig. 12.3. Type *a* can be made by inserting the tubes through sheet metal strips with stamped or drilled holes and then expanding the tubes slightly to cause pressure at the tube-to-strip contacts. The tubes and strips may then be brazed. If the tubes are expanded into the plates only to produce an interference fit, some bond or contact resistance must be anticipated. For practical purposes, when the tubes and strips are brazed together, the joint may be considered a metallurgical bond, and the bond resistance can be neglected.

In Fig. 12.3, tubes *b* through *e* are made by winding a metal ribbon in tension around the tube. These types are not metallurgically bonded and rely entirely on the tension in the ribbon to provide good contact. Type *f* combines tension winding with brazing and for the combination of a steel tube and an aluminum fin, the common tin–lead solder is not compatible and a zinc solder is used. Type *g* employs a tube as a liner and high fins are extruded from aluminum which, like copper, is a metal that can be manipulated to a considerable fin height. Types *d*, *e*, and *f* employ aluminum for the fins and are also arranged to protect the tube from the weather because air-fin coolers are installed outdoors. Type *g*, sometimes called a *muff-type* high-fin tube, has its contact resistance between the inside of the integral finned tube and the liner or plain tube. Type *h* has a mechanical bond that can closely match a metallurgical bond for contact resistance. Type *i*, an elliptical tube with rectangular fins, may employ steel fins that are galvanized. When tube ends are circular, they are rolled into headers.

Consider a typical air-fin cooler application with a hot fluid inside the tubes. In many instances, carbon steel meets the corrosion-resistance requirements of the tube side fluid. From the standpoint of high thermal conductivity and cost, aluminum ribbon is very suitable for the tension-wound fins. However, aluminum has twice the thermal coefficient of expansion of steel, and the higher the operating temperature of the fluid inside the tubes, the greater the tendency of the fins to elongate away from their room-temperature tension-wound contact with the tube, and the greater is the bond or contact resistance. In one variation of type *d* in Fig. 12.3, the ribbon is wound with *J* rather than *L* feet, with the *J*'s pressing against each other and the tube at room temperature. As the feet become heated during operation, they expand against each other.

12.2 BOND OR CONTACT RESISTANCE OF HIGH-FIN TUBES

The bond resistance of several types of interference-fit high-fin tubes shown in Fig. 12.3 has been studied by Gardner and Carnavos (1960), Shlykov and Ganin

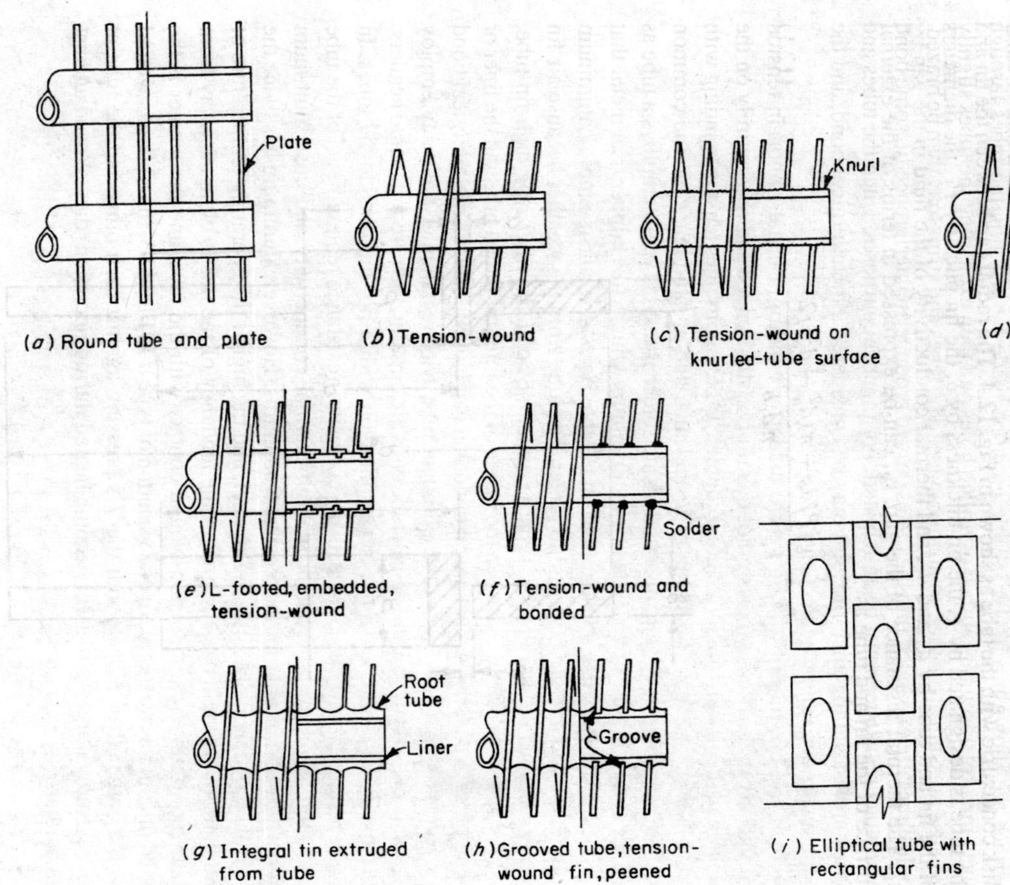

Figure 12.3 Various types of high-fin tubing.

(1964), and Yovanovich (1981). Gardner and Carnavos pointed out that in its most general sense, the term *interference fit* implies the absence of a metallurgical bond as opposed to the extrusion of a fin from a tube wall, and the welding, soldering, or brazing of a fin to the tube. The interference fit is produced by mechanically developing contact pressure through elastic deformation either by winding a ribbon under tension about a tube, as in types *b* through *e* in Fig. 12.3, or by expanding a tube against the fins as in type *a*, or a combination of pressing the root tube against the liner or the liner against the root tube, as in type *g*.

To simplify the analysis, consider the edge-wound ribbon fin to be a thin circular disk concentric with the tube as shown in Fig. 12.4. The radial gap between the fin and the tube is designated by g, the fin thickness by δ, the fin pitch by P_f, the diameters of the fin and tube by d_a and d_o, and thermal conductivity of the fluid in the gap by k_v. The thermal resistance of the gap, r_g, can be expressed in terms of the external surface of the fin and tube as

$$r_g = \frac{g}{k_v} \frac{P_f d_o + \pi (d_a^2 - d_o^2)/2}{\pi d_o \delta} \tag{12.1}$$

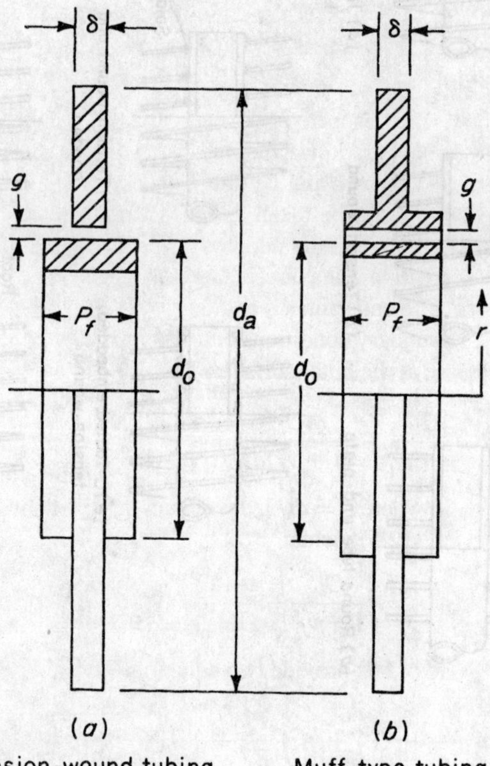

Figure 12.4 Nomenclature for the analysis of bond or contact resistance with high-fin tubing.

or

$$r_g = \frac{g}{k_v}\left(\frac{d_a^2 - d_o^2}{2d_o\delta} + \frac{P_f}{\pi\delta}\right) \tag{12.2}$$

The quantity $P_f/\pi\delta$ is often negligible with respect to $(d_a^2 - d_o^2)/2d_o\delta$, so that an approximate expression for r_g is given by

$$r_g = \frac{g(d_a^2 - d_o^2)}{2k_v d_o\delta} \tag{12.3}$$

To determine when there will be a radial gap between the tension-wound fin and a tube, and its magnitude, the thermal stresses in the fin and the tube must be considered. Timoshenko and Goodier (1951) developed relationships for the radial displacement of the fin, u_f, and the radial stress, σ_f, as functions of the temperature T and the radius r. For a thin circular disk, these relationships are

$$u_f = \frac{(1+\nu_f)\alpha_f}{r}\int_{d_o/2}^{r}(T - T_o)r\,dr + C_1 r + \frac{C_2}{r} \tag{12.4}$$

and

$$\sigma_f = \frac{\alpha_f E_f}{r^2}\int_{d_o/2}^{r}(T - T_o)r\,dr - E_f\left[\frac{C_1}{1-\nu_f} + \frac{C_2}{(1+\nu_f)r^2}\right] \tag{12.5}$$

where E_f, α_f, and ν_f are, respectively, the modulus of elasticity, coefficient of thermal expansion, and Poisson's ratio for the fin, T_o is the temperature at which there are no residual stresses in the fin, and C_1 and C_2 are arbitrary constants of integration. For this discussion, T_o can be taken as the temperature at which the finned tube was fabricated. The conventions adopted for eqs. (12.4) and (12.5) are that the displacement u_f is positive when directed radially outward and the stress σ_f is positive when compressive. The constants C_1 and C_2 may be determined from the problem at hand, namely, the boundary conditions on the stress. The radial stress is equal to the atmospheric pressure at the outer surface where $r = d_a/2$. Using eq. (12.5) yields

$$\frac{C_1}{1-\nu_f} + \frac{4C_2}{(1+\nu_f)d_a^2} = \frac{4\alpha_f}{d_a^2}\int_{d_o/2}^{d_a/2}(T - T_o)r\,dr \tag{12.6}$$

At the base of the fin where $r = d_o/2$, the radial stress is equal to the contact pressure P_c. For this case, eq. (12.5) yields

$$\frac{C_1}{1-\nu_f} + \frac{C_2}{(1+\nu_f)d_o^2} = \frac{P_c}{E_f} \tag{12.7}$$

Equations (12.6) and (12.7) provide two equations for the evaluation of C_1 and C_2:

$$C_1 = \left(\frac{d_o^2 - \nu_f d_a^2}{d_a^2 - d_o^2} + \nu_f\right)\frac{P_c}{E_f} + \frac{4\alpha_f(1-\nu_f)}{d_a^2 - d_o^2}\int_{d_o/2}^{d_a/2}(T - T_o)r\,dr \tag{12.8}$$

and

$$C_2 = \frac{(1-\nu_f)d_o^2}{d_a^2 - d_o^2}\frac{P_c}{E_f} + \frac{\alpha_f(1+\nu_f)d_o^2}{d_a^2 - d_o^2}\int_{d_o/2}^{d_a/2}(T - T_o)r\,dr \qquad (12.9)$$

To obtain the radial displacement at the fin base, the constants C_1 and C_2 must be substituted into eq. (12.4). After the substitutions, the equation is evaluated at the fin base where $r = d_o/2$. Then the displacement at the fin base, u_{fb}, is given by

$$u_{fb} = \frac{d_o}{2}\left[\left(\frac{d_a^2 + d_o^2}{d_a^2 - d_o^2} + \nu_f\right)\frac{P_c}{E_f} + \frac{8\alpha_f}{d_a^2 - d_o^2}\int_{d_o/2}^{d_a/2}(T - T_o)r\,dr\right] \qquad (12.10)$$

It is noted that at $T = T_o$, there is still a displacement. If the contact pressure between the fin and the tube at fabrication conditions is designated as P_{co}, the initial displacement is given by

$$u_{fo} = \frac{d_o}{2}\left(\frac{d_a^2 + d_o^2}{d_a^2 - d_o^2} + \nu_f\right)\frac{P_{co}}{E_f} \qquad (12.11)$$

The change in radial displacement contributes to the gap between the fin and the tube. When the change in the gap is designated as Δu_f, it can be expressed by

$$\Delta u_f = u_{fb} - u_{fo} \qquad (12.12)$$

or, by employing eqs. (12.10) and (12.11),

$$\Delta u_f = \frac{d_o}{2}\left[\left(\frac{d_a^2 + d_o^2}{d_a^2 - d_o^2} + \nu_f\right)\frac{P_c - P_{co}}{E_f} + \frac{8\alpha_f}{d_a^2 - d_o^2}\int_{d_o/2}^{d_a/2}(T - T_o)r\,dr\right] \qquad (12.13)$$

The integral in eq. (12.13) may be expressed in familiar terms by introducing the fin efficiency η:

$$\eta = \frac{8}{(d_a^2 - d_o^2)(T_b - T_s)}\int_{d_o/2}^{d_a/2}(T - T_s)r\,dr \qquad (12.14)$$

where T_b is the fin base temperature and T_s is the surrounding bulk fluid temperature. Writing $T - T_o$ as $T - T_o = (T - T_s) - (T_s - T_o)$, eq. (12.14) may be put into eq. (12.13), with the result

$$\Delta u_f = \frac{d_o}{2}\left[\left(\frac{d_a^2 + d_o^2}{d_a^2 - d_o^2} + \nu_f\right)\frac{P_c - P_{co}}{E_f} + \alpha_f(T_s - T_o) + \alpha_f\eta(T_b - T_s)\right] \qquad (12.15)$$

The tube itself undergoes expansion when its temperature is raised and the radial displacement of the tube must also be taken into account. Similar considerations may be applied to the tube, assuming that the contact from the fin is uniformly distributed over the tube wall and neglecting the pressure inside the tube. The expression defining the change in radial displacement of the tube Δu_t is

$$\Delta u_t = \frac{d_o}{2}\left\{\alpha_t(T_w - T_o) - \frac{P_c - P_{co}}{E_t}\frac{\delta}{P_f}\left[\frac{d_o^2 + (d_o - 2\delta_w)^2}{d_o^2 - (d_o - 2\delta_w)^2} - \nu_t\right]\right\} \qquad (12.16)$$

where α_t, E_t, and ν_t pertain to the tube, δ_w is the tube wall thickness, and T_w is the wall temperature. The assumption of a uniform distribution of the contact pressure over the tube wall is valid only if the fin spacing is sufficiently close so that

$$P_f < 0.43[\delta_w(d_o - \delta_w)]^{1/2}$$

The fin base temperature T_b and the wall temperature T_w are not generally known and must be computed from the known fluid temperatures and the thermal resistances. The overall resistance $\sum R$ can be expressed by

$$\sum R = \sum R^* + r_g \tag{12.17}$$

where $\sum R^*$ is the radial resistance in the absence of a radial gap:

$$\sum R^* = \frac{r_o'}{\eta} + r_{co}' + r_{mo} + r_{io}' \tag{12.18}$$

where all resistances are referred to the gross outside surface. In eq. (12.18), r_{io}' and r_o' are the inside and outside fluid film resistances (the reciprocal of the heat transfer coefficients), respectively, where both include the appropriate fouling resistances. The tube wall metal resistance is r_{mo}, η is the fin efficiency, and r_{co}' is the contact resistance referred to the gross outside surface. One expression for r_{co}' is given by Yovanovich (1981):

$$r_{co}' = \frac{d_a^2 - d_o^2}{2d_o P_f\left\{(k_v/\gamma_v) + 4.2\times 10^4(P_c/\sigma_t b_p)[k_f k_t/(k_f + k_t)]\right\}} \tag{12.19}$$

where γ_v is the arithmetic average of the surface roughnesses for the two solids in contact,

$$\gamma_v = \frac{\gamma_f + \gamma_t}{2}$$

k_v is the thermal conductivity of the void material (air), and b_p is a function of the cold workability of the metal, taken as 3 for aluminum and 5 for copper.

The heat flow through each portion of the finned tube is constant and for resistances in series between the hot fluid at T_h and the surrounding fluid at T_s is obtained by summing the individual resistances in series:

$$\frac{T_h - T_s}{\sum R^* + r_g} = \frac{T_h - T_w}{r_i'} = \frac{\eta(T_b - T_s)}{r_o'} \tag{12.20}$$

where T_h is the hot fluid temperature at the inside of the tube. When the first two terms of eq. (12.20) are equated, rearranged, and T_o subtracted from both sides, the result is

$$T_w - T_o = (T_h - T_o) - \frac{r_i'}{\sum R^* + r_g}(T_h - T_s) \tag{12.21}$$

This determines $(T_w - T_o)$ in terms of the known temperatures and resistances. Equating the first and third terms of eq. (12.20) yields

$$\eta(T_b - T_s) = \frac{r'_o}{\sum R^* + r_g}(T_h - T_s) \tag{12.22}$$

which relates $(T_b - T_s)$ to the known fluid temperatures and resistances. Equations (12.21) and (12.22) can, in turn, be substituted into eqs. (12.13) and (12.16) to obtain the changes in the radial displacement of the fin and tube:

$$\Delta u_f = \frac{d_o}{2}\left\{\left(\frac{d_a^2 + d_o^2}{d_a^2 - d_o^2} + \nu_f\right)\frac{P_c - P_{co}}{E_f} + \alpha_f\left[(T_h - T_o) - \left(1 - \frac{r'_o}{\sum R^* + r_g}\right)(T_h - T_s)\right]\right\} \tag{12.23}$$

and

$$\Delta u_t = \frac{d_o}{2}\left\{\alpha_t\left[(T_h - T_s) - \left(\frac{r'_i}{\sum R^* + r_g}\right)(T_h - T_s)\right] - \frac{P_c - P_{co}}{E_t}\frac{\delta}{P_f}\left[\frac{d_o^2 + (d_o - 2\delta_w)^2}{d_o^2 - (d_o - 2\delta_w)^2} - \nu_t\right]\right\} \tag{12.24}$$

The difference between Δu_f and Δu_t establishes the radial gap. The subtraction gives

$$g = \frac{d_o}{2}\left\{(\alpha_f - \alpha_t)(T_h - T_s) + \mu(P_c - P_{co}) - \left[\alpha_f\left(1 - \frac{r'_o}{\sum R^* + r_g}\right) - \alpha_t\frac{r'_i}{\sum R^* + r_g}\right](T_h - T_s)\right\} \tag{12.25}$$

where μ is a constant that depends only on the properties of the fin and tube:

$$\mu = \frac{1}{E_f}\left(\frac{d_a^2 + d_o^2}{d_a^2 - d_o^2} + \nu_f\right) + \frac{1}{E_t}\frac{\delta}{P_f}\left[\frac{d_o^2 + (d_o - 2\delta_w)^2}{d_o^2 - (d_o - 2\delta_w)^2} - \nu_t\right] \tag{12.26}$$

Naturally, negative values of g are not physically realistic, and g is effectively described by eq. (12.25) as the radial gap only when positive.

According to eq. (12.25) for the interference-fit fin tube, when g is positive, the contact pressure P_c is zero and the gap must be calculated by substituting the expression for g into eq. (12.3):

$$r_g = \varrho\left\{(\alpha_f - \alpha_t)(T_h - T_o) + \mu P_c - \left[\alpha_f\left(1 - \frac{r'_o}{\sum R^* + r_g}\right) - \alpha_t\frac{r'_i}{\sum R^* + r_g}\right](T_h - T_s)\right\} \tag{12.27}$$

where

$$\varrho = \frac{d_a^2 - d_o^2}{4k_v\delta} \tag{12.28}$$

It is anticipated that r_g will be much smaller than $\sum R^*$, and it is possible to approximate $1/(\sum R^* + r_g)$ by

$$\frac{1}{\sum R^* + r_g} = \frac{1}{\sum R^*}\left(1 - \frac{r_g}{\sum R^*}\right) \tag{12.29}$$

With

$$\Upsilon_1 = 1 - \frac{r_o'}{\sum R^*} \quad \text{and} \quad \Upsilon_2 = \frac{r_i'}{\sum R^*}$$

the approximation of eq. (12.29) may be substituted into eq. (12.27) to give an approximate result for r_g:

$$r_g = \frac{\varrho\{(\alpha_f - \alpha_t)(T_h - T_o) - \mu P_{co} - [\alpha_f \Upsilon_1 - \alpha_t \Upsilon_2](T_h - T_s)\}}{1 + \varrho(1/\sum R^*)^2(\alpha_f r_o' + \alpha_t r_i')(T_h - T_s)} \tag{12.30}$$

Equation (12.27) may also be used to determine the conditions when the gap begins to form, that is, when $g = 0$ and $P_c = 0$. This situation correpsonds to $r_g = 0$, or

$$(\alpha_f - \alpha_t)(T_h - T_o) = \mu P_{co} + (\alpha_f \Upsilon_1 - \alpha_t \Upsilon_2)(T_h - T_s) \tag{12.31}$$

Equation (12.31) may be considered as a constraint on the operating conditions of a finned tube. It takes into account the temperature of the fluid in the tube T_h at which a gap begins to form in terms of T_o, P_{co}, T_s, and the thermal resistances. It can also provide the temperature of the surrounding fluid outside the fins in terms of T_o, P_{co}, T_s, and the thermal resistances. Written as an inequality

$$(\alpha_f - \alpha_t)(T_h - T_o) > \mu P_{co} + (\alpha_f \Upsilon_1 - \alpha_t \Upsilon_2)(T_h - T_s) \tag{12.32}$$

it gives the conditions under which a radial gap exists.

Manufacturers of the muff-type tube offer test data up to a liner temperature of 200°C, stating that the *bond resistance* between tube and liner will not exceed a maximum value. The bond resistances of other fin types may be calculated from the foregoing equations of Gardner and Carnavos (1963).

12.3 FIN EFFICIENCY APPROXIMATION

The fin efficiency of the radial fin of rectangular profile was given in Chapter 1:

$$\eta = \frac{2r_b}{m(r_a^2 - r_b^2)} \frac{I_1(mr_a)K_1(mr_b) - K_1(mr_a)I_1(mr_b)}{I_0(mr_b)K_1(mr_a) + I_1(mr_a)I_o(mr_b)} \tag{1.35}$$

or alternatively, by

$$\eta = \frac{2\rho}{\phi(1+\rho)} \frac{I_1(\phi R_a)K_1(\phi R_b) - K_1(\phi R_a)I_1(\phi R_b)}{I_0(\phi R_b)K_1(\phi R_a) + I_1(\phi R_a)K_o(\phi R_b)} \tag{1.42}$$

where

$$m = \left(\frac{2h}{k\delta}\right)^{1/2}, \qquad \rho = \frac{r_b}{r_a}, \qquad \phi = (r_a - r_b)^{3/2}\left(\frac{2h}{kA_p}\right)^{1/2}$$

$$R_a = \frac{1}{1-\rho} \quad \text{and} \quad R_b = \frac{\rho}{1-\rho}$$

The modified Bessel functions in eqs. (1.35) and (1.42) are obtained from tables or from software and their employment to obtain the efficiency involves a somewhat laborious procedure. An alternative has been provided by McQuiston and Tree (1972), who suggest the approximation

$$\eta = \frac{\tanh m\psi}{m\psi} \tag{12.33}$$

where

$$m = \left(\frac{2h}{k\delta}\right)^{1/2}$$

and

$$\psi = r_b \frac{1-\rho}{\rho}\left(1 + 0.35 \ln \frac{1}{\rho}\right) \tag{12.34}$$

where ρ is the radius ratio defined after eq. (1.42).

Example 12.1. A radial fin of rectangular profile is exposed to a heat transfer coefficient of 40 W/m^2 · K and is fabricated from steel having a thermal conductivity of 40 W/m·K. Its outer and inner diameters are 25 cm and 10 cm, respectively, and the fin has a thickness of 0.25 cm. Use eq. (12.34) to determine its efficiency.

SOLUTION. This is a rework of a portion of Example 1.3, where it was found that the efficiency of this fin is $\eta = 0.343$. Here

$$\rho = \frac{d_b}{d_a} = \frac{0.10}{0.25} = 0.40$$

$$m = \left(\frac{2h}{k\delta}\right)^{1/2} = \left[\frac{(2)(40)}{(40)(0.0025)}\right]^{1/2} = 28.2843 \text{ m}^{-1}$$

$$\psi = r_b \frac{1-\rho}{\rho}\left(1 + 0.35 \ln \frac{1}{\rho}\right)$$

$$= 0.05\left(\frac{1-0.40}{0.40}\right)\left(1 + 0.35 \ln \frac{1}{0.40}\right)$$

$$= (0.05)(1.50)(1 + 0.35 \ln 2.50)$$

$$= (0.075)[1 + 0.35(0.9163)]$$

$$= 0.0991$$

and

$$m\psi = (28.2843)(0.0991)$$

$$= 2.8016$$

The fin efficiency is

$$\eta = \frac{\tanh m\psi}{m\psi} = \frac{\tanh 2.8016}{2.8016} = 0.354$$

This approximates the value of the efficiency found in Example 1.3.

12.4 AIR-FIN COOLERS

The air-fin cooler consists of one or more horizontal rows of tubes constituting a *section* through which air is circulated upward by mechanical draft. The fan that moves the air may be above the section, providing an *induced draft*, as in Fig. 12.1, or may be below the section, providing a *forced draft*, as in Fig. 12.2.

In the induced-draft air-fin cooler of Fig. 12.1, the heated air is thrown upward to a good height by its high exit velocity. A relatively small amount of the heated air is sucked back to reenter the air intake below the section and thereby cuts down the temperature difference available between the ambient air and the process fluid. In a forced-draft unit, the air leaves at a low velocity at a point not far from the high entrance velocity of the air to the fan below the section. Hot air is more apt to be sucked back into the fan intake, causing *recirculation*. Following a trend in cooling towers that started some years ago, induced-draft units now appear to be preferred. Usually, the section has cross bracing and baffles, as shown in Fig. 12.5, to increase rigidity and reduce vibration.

The design and analysis of air-fin coolers differs only in a few respects from the longitudinal fin exchangers in Chapter 11. The principal difference is on the air side, where air competes with other fluids as a coolant. Because air is incompressible and liquids are not, only a small pressure drop can be expended for air circulation across the finned tubes lest the cost of air-compression work becomes prohibitive. In most applications, the allowable air-side pressure drop is only about 1.25 cm ($\frac{1}{2}$ in.) of water. The air passes over the finned tubing in cross flow and this merely requires use of the proper heat transfer and flow friction data. The temperature excursion of the air usually cannot be computed at the start of the calculations because the air volume, and hence the air temperature rise, depend on the air pressure drop and flow area of the cooler.

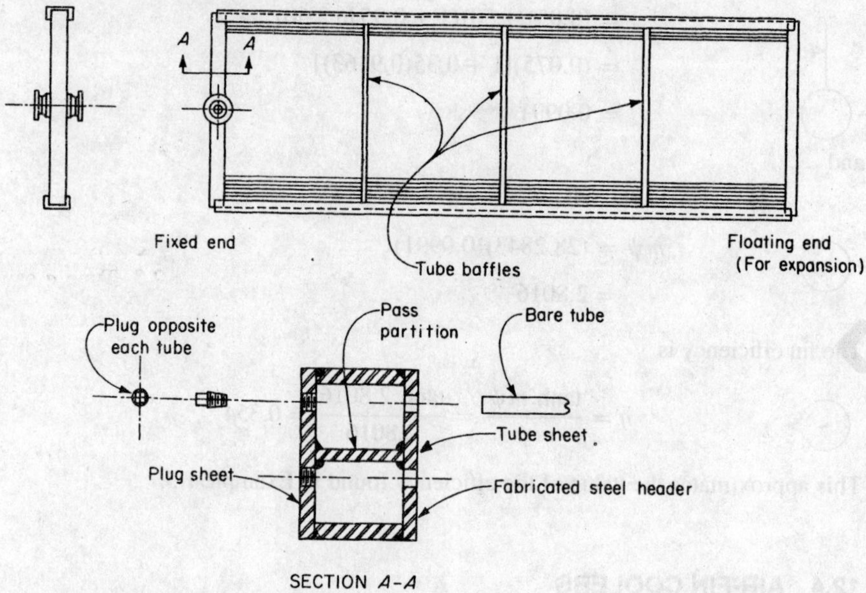

Figure 12.5 Tube section.

Most widely used are the integral-fin muff-type tube (Fig. 12.3 *g*), the L-footed tension wound tube (Fig. 12.3 *g*) and the grooved and peened tension-wound tube (Fig. 12.3 *h*). These tubes usually employ 9 or 11 fins per inch. Numerous other tubes are manufactured in accordance with the types shown in Fig. 12.3 *b*, *c*, and *f*. Other tubes, with serrated or discontinuous fins, are fabricated by manufacturers of air-fin coolers to their own standards.

12.4.1 Physical Data

As indicated in Fig. 12.6, tubes may be arranged in either triangular or in-line arrangements. Observe that the pitch in these arrangements is designated by P_t, P_ℓ, or P_d, where P_t is the transverse pitch, (m), P_ℓ is the longitudinal pitch (m), and P_d is the diagonal pitch (m). The diagonal pitch is related to the transverse and longitudinal pitch by

$$P_d = \left[\left(\frac{P_t}{2}\right)^2 + P_\ell^2\right]^{1/2} \tag{12.35}$$

and in the case of an equilateral triangular arrangement,

$$P_d = P_t$$

With n tubes in a row and n_r rows, the total number of tubes will be

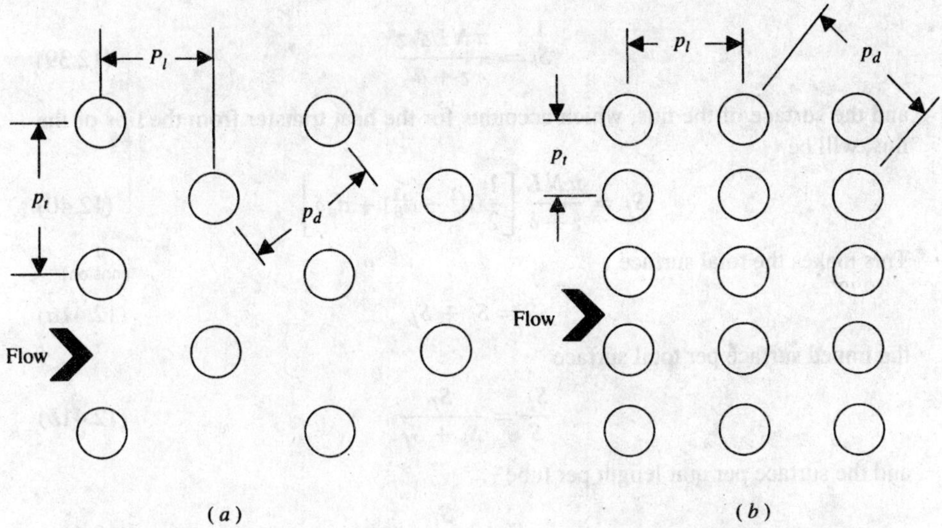

Figure 12.6 Tube arrangements: (*a*) triangular; (*b*) in-line.

$$N = nn_r \tag{12.36}$$

Let z be the clear space between the tubes, which are L meters long. The fins are b meters high:

$$b = \frac{d_a - d_b}{2}$$

where d_a and d_b are, respectively, the outer and inner diameters of the fin. The fins are δ thick and the minimum flow area $A = A_{\min}$ will depend on the transverse pitch P_t.

For

$$P_t > 2P_d - d_b - \frac{2z\delta}{z+\delta}$$

$$A = A_{\min} = nL\left(P_t - d_b - \frac{2z\delta}{z+\delta}\right) \tag{12.37}$$

and for

$$P_t < 2P_d - d_b - \frac{2z\delta}{z+\delta}$$

$$A = A_{\min} = 2nL\left(P_d - d_b - \frac{2z\delta}{z+\delta}\right) \tag{12.38}$$

The surface area of the tube (between the fins) will be

$$S_b = \frac{\pi N L d_b z}{z + \delta} \tag{12.39}$$

and the surface of the fins, which accounts for the heat transfer from the tips of the fins, will be

$$S_f = \frac{\pi N L}{z + \delta}\left[\frac{1}{2}(d_a^2 - d_b^2) + d_a \delta\right] \tag{12.40}$$

This makes the total surface

$$S = S_b + S_f \tag{12.41a}$$

the finned surface per total surface

$$\frac{S_f}{S} = \frac{S_f}{S_b + S_f} \tag{12.41b}$$

and the surface per unit length per tube

$$S' = \frac{S}{NL} \tag{12.41c}$$

12.4.2 Heat Transfer Correlations

Early investigations that pertain to heat transfer and friction data in tube bundles containing high-fin tubes have been reported by Jameson (1945) and Schmidt (1963). The correlation of Briggs and Young (1963) is based on a wide range of data. Their general equation for tube banks containing six rows of tubes on equilateral triangular pitch is

$$\mathrm{Nu} = \frac{h d_b}{k} = 0.134 \mathrm{Re}^{0.681} \mathrm{Pr}^{1/3} \left[\frac{2(P_f - \delta)}{d_a - d_b}\right]^{0.20} \left(\frac{P_f - \delta}{\delta}\right)^{0.1134} \tag{12.42}$$

where

$$\mathrm{Re} = \frac{d_b G}{\mu}$$

and where the range of parameters is

$1000 < \mathrm{Re} < 18{,}000$	$0.33\ \mathrm{mm} < \delta < 2.02\ \mathrm{mm}$
$11.13\ \mathrm{mm} < d_b < 40.89\ \mathrm{mm}$	$1.30\ \mathrm{mm} < P_f < 4.06\ \mathrm{mm}$
$1.42\ \mathrm{mm} < b = (d_a - d_b)/2 < 16.57\ \mathrm{mm}$	$24.99\ \mathrm{mm} < P_t < 111\ \mathrm{mm}$

Vampola (1966) proposed a correlation based on extensive study of different finned tubes. For more than three tube rows,

$$\mathrm{Nu} = \frac{h d_e}{k} = 0.251 \mathrm{Re}^{0.67} \left(\frac{P_t - d_b}{d_b}\right)^{-0.20} \times \left(\frac{P_t - d_b}{P_f - \delta} + 1\right)^{-0.20} \left(\frac{P_t - d_b}{P_d - d_b}\right)^{0.40} \tag{12.43}$$

where

$$\mathrm{Re} = \frac{d_e G}{\mu}$$

and where

$$d_e = \frac{S_b d_b + S_f (S_f P_f / 2NL)^{1/2}}{S}$$

with the diagonal pitch given by eq. (12.35). In eq. (12.43), the range of parameters is

$1000 < \mathrm{Re} < 10{,}000$	$24.78\ \mathrm{mm} < P_t < 49.55\ \mathrm{mm}$
$10.67\ \mathrm{mm} < d_b < 26.01\ \mathrm{mm}$	$16.20\ \mathrm{mm} < d_e < 34.00\ \mathrm{mm}$
$5.20\ \mathrm{mm} < b = (d_a - d_b)/2 < 9.70\ \mathrm{mm}$	$0.48\ \mathrm{mm} < (P_t - d_b)/d_b < 1.64$
$0.25\ \mathrm{mm} < \delta < 0.70\ \mathrm{mm}$	$4.34 < (P_t - d_b)/(P_f - \delta) + 1 < 25.2$
$2.28\ \mathrm{mm} < P_f < 5.92\ \mathrm{mm}$	$0.45 < (P_t - d_b)/(P_d - d_r) < 2.50$
$20.32\ \mathrm{mm} < P_\ell < 52.40\ \mathrm{mm}$	

Ganguli et al. (1985) proposed the following correlation for three or more rows of finned tubes:

$$\mathrm{Nu} = \frac{h d_b}{k} = 0.38 \mathrm{Re}^{0.6} \mathrm{Pr}^{1/3} \left(\frac{S_b}{S} \right)^{0.15} \tag{12.44}$$

where

$$\mathrm{Re} = \frac{d_b G}{\mu}$$

The correlation of eq. (12.44) is valid for

$1800 < \mathrm{Re} < 100{,}000$	$2.30\ \mathrm{mm} < P_f < 3.629\ \mathrm{mm}$
$11.176\ \mathrm{mm} < d_b < 19.05\ \mathrm{mm}$	$27.432\ \mathrm{mm} < P_t < 98.552\ \mathrm{mm}$
$5.842\ \mathrm{mm} < b = (d_a - d_b)/2 < 19.05\ \mathrm{mm}$	$1 < S/S_b < 50$
$0.254\ \mathrm{mm} < \delta < 0.559\ \mathrm{mm}$	

Other correlations include those of Brauer (1964), Schulenberg (1965), Kuntysh and Iokhvedov (1971), and Mirkovic (1974). More recent correlations include those of Zhukauskas (1974), Weierman (1976), Hofmann (1976), Elmahdy and Biggs (1979), Biery (1981), Gianolio and Cuti (1981), Brandt and Wehle (1983), and Nir (1991).

12.4.3 Pressure Loss Correlations for Staggered Tubes

Some of the earlier correlations for the static pressure drop through bundles of circular finned tubes are those of Jameson (1945), Gunter and Shaw (1945), and Ward and

Young (1959). A frequently used correlation is that of Robinson and Briggs (1966) for staggered tubes:

$$\Delta P = 18.03\frac{G^2}{\rho}n_r \mathrm{Re}^{-0.316}\left(\frac{P_t}{d_b}\right)^{-0.927}\left(\frac{P_t}{P_d}\right)^{0.515} \tag{12.45}$$

for n_r rows, where

$$\mathrm{Re} = \frac{d_b G}{\mu}$$

and where P_d is given by eq. (12.35). Equation (12.45) is valid for

$2000 < \mathrm{Re} < 50{,}000$	$2.31\text{ mm} < P_f < 2.82\text{ mm}$
$18.64\text{ mm} < d_b < 40.89\text{ mm}$	$42.85\text{ mm} < P_t < 114.3\text{ mm}$
$39.68\text{ mm} < d_a < 69.85\text{ mm}$	$37.11\text{ mm} < P_\ell < 98.89\text{ mm}$
$10.52\text{ mm} < (d_a - d_b)/2 < 14.48\text{ mm}$	$1.8 < P_t/d_b < 4.6$

Vampola (1966) proposed the correlation

$$\Delta P = 0.7315\frac{G^2}{\rho}n_r \mathrm{Re}^{-0.245}\left(\frac{P_t - d_b}{d_b}\right)^{-0.90} \times\left(\frac{P_t - d_b}{P_f - \delta}+1\right)^{0.70}\left(\frac{d_e}{d_b}\right)^{0.90} \tag{12.46}$$

where the Reynolds number, the equivalent diameter, and the limits of applicability are identical to those following eq. (12.43).

12.4.4 Overall Heat Transfer Coefficient

Because the air- and tube-side heat transfer coefficients, the bond and tube metal resistances, and the tube-side fouling factor all apply at very dissimilar surfaces, it is important that all of these resistances be properly corrected and summed. No provision need be made for air-side fouling because the air-side heat transfer coefficient is low and becomes the controlling resistance. Usually, with the muff-type tube, the resistances are first referred to a hypothetical bare tube having outside diameter, d_b. With diameter designations in Fig. 12.7, there are five "inside" resistances:

1. The inside film resistance

$$r_{io} = \frac{1}{h_i}\frac{d_b}{d_i} \tag{12.47}$$

2. The inside fouling resistance

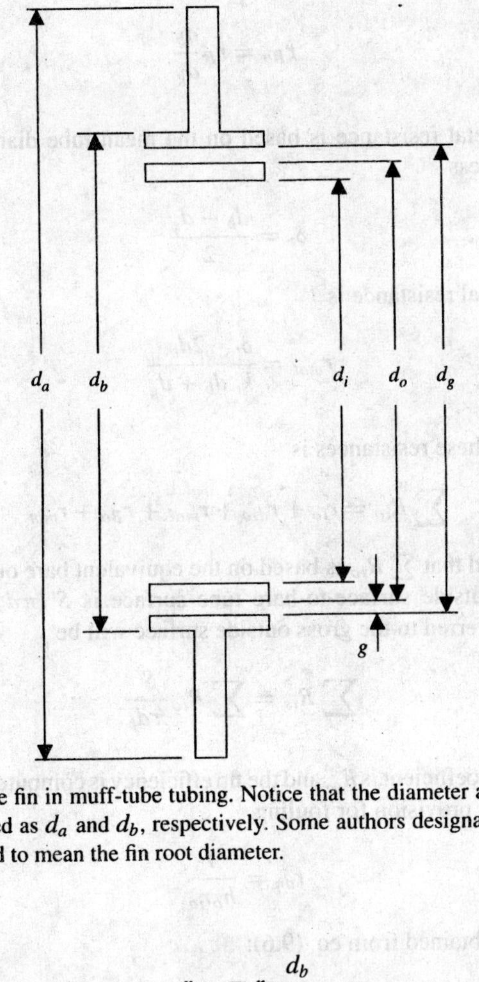

Figure 12.7 Single fin in muff-tube tubing. Notice that the diameter at the tips and base of the fin are designated as d_a and d_b, respectively. Some authors designate these as d_f and d_r where d_r is intended to mean the fin root diameter.

$$r_{dio} = r_{di} \frac{d_b}{d_i} \tag{12.48}$$

3. The tube metal resistance is based on the mean liner diameter, and with the metal thickness

$$\delta_l = \frac{d_o - d_i}{2}$$

the liner metal resistance is

$$r_{mol} = \frac{\delta_l}{k_l} \frac{2d_b}{d_o + d_i} \tag{12.49}$$

4. The bond resistance given by the tube manufacturer or calculated from the procedure of Section 12.2 is appropriately transferred via

$$r_{Bo} = r_B \frac{d_b}{d_g} \tag{12.50}$$

5. The tube metal resistance is based on the mean tube diameter, and with the metal thickness

$$\delta_t = \frac{d_b - d_g}{2}$$

the liner metal resistance is

$$r_{mot} = \frac{\delta_t}{k_t} \frac{2d_b}{d_b + d_g} \tag{12.51}$$

6. The sum of these resistances is

$$\sum R_{io} = r_{io} + r_{dio} + r_{mo\ell} + r_{Bo} + r_{mot}$$

and it is noted that $\sum R_{io}$ is based on the equivalent bare outside tube surface. The gross outside surface-to-bare tube surface is $S'/\pi d_b$, so that the total resistance referred to the gross outside surface will be

$$\sum R_{is} = \sum R_{io} \frac{S'}{\pi d_b} \tag{12.52}$$

7. The air-side coefficient is h_o, and the fin efficiency is computed from eq. (12.33). Then with no provision for fouling,

$$r_{o\eta} = \frac{1}{h_o \eta_o} \tag{12.53}$$

where η_o is obtained from eq. (9.6):

$$\eta_o = 1 - \frac{S_f}{S}(1 - \eta)$$

8. The overall heat transfer coefficient is then given by

$$U_o = \frac{1}{\sum R_{is} + r_{o\eta}} \tag{12.54}$$

12.4.5 ϵ–N_{tu} Relationships

Assume that baffles indicated by the vertical lines in Fig. 12.8 are placed across the tubes in the direction of the airflow in order to support the tube bundle and/or to prevent tube vibration. These baffles provide channels in the airflow direction, and the air flowing across the tubes is unmixed. The liquid flowing within the tubes is certainly unmixed, and the surface required can be determined from the logarithmic

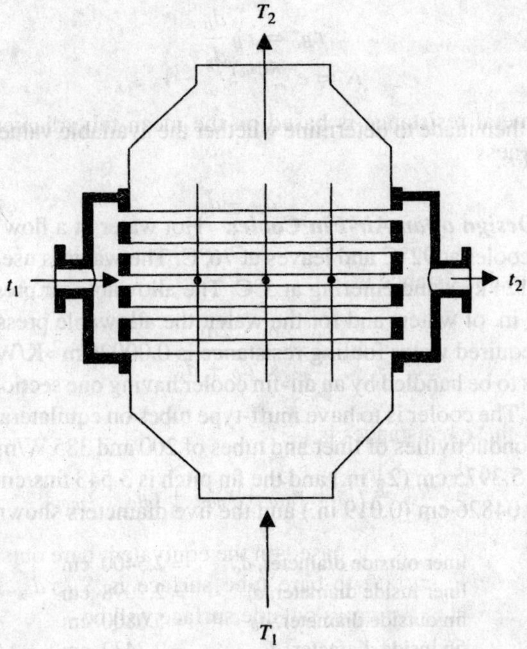

Figure 12.8 Cross-flow temperature arrangement with both fluids unmixed.

mean temperature difference and the correction factor for cross flow with both fluids unmixed:

$$S = \frac{q}{U_o \eta_o F(\text{LMTD})} \tag{12.55}$$

An alternative procedure is to compute the *required effectiveness* from the process temperatures,

$$\epsilon = \frac{T_1 - T_2}{T_1 - t_1}$$

if the hot fluid is C_{min}, or

$$\epsilon = \frac{t_2 - t_1}{T_1 - t_1}$$

if the cold fluid is C_{min}. Then, after the surface is calculated,

$$\text{N}_{\text{tu}} = \frac{U_o S}{C C_{\text{min}}}$$

is formed and the *available effectiveness* is found from eq. (10.26d)

$$\epsilon = 1 - e^{(\text{N}_{\text{tu}})^{0.22} \Lambda / R} \tag{10.26d}$$

where

$$\Lambda = e^{-R(N_{tu})^{0.78}} - 1$$

A comparison is then made to determine whether the available value of ϵ exceeds the required value of ϵ.

Example 12.2: Design of an Air-Fin Cooler. Hot water at a flow rate of 46.5 kg/s enters an air-fin cooler at 92°C and leaves at 76°C. The water is used to heat cold air at a flow rate of 124 kg/s and entering at 5°C. The allowable air pressure loss for the air is 250 Pa (≈1 in. of water) and for the water, the allowable pressure loss is 7 kPa (≈1 psig). The required water fouling resistance is 0.00026 m^2·K/W.

This service is to be handled by an air-fin cooler having one section that is 3 m high and 5.50 m wide. The cooler is to have muff-type tubes on equilateral triangular pitch having thermal conductivities of liner and tubes of 200 and 385 W/m·K, respectively. The tube pitch is 5.3975 cm ($2\frac{1}{8}$ in.) and the fin pitch is 3.543 fins/cm (9 fins/in.). The fin thickness is 0.04826 cm (0.019 in.) and the five diameters shown in Fig. 12.7 are

liner outside diameter, d_o	= 2.5400 cm
liner inside diameter, d_i	= 2.2098 cm
fin outside diameter, d_a	= 5.0800 cm
fin inside diameter, d_b	= 2.7432 cm
gap outside diameter, d_g	= 2.6467 cm

Design the air-fin cooler if the tube manufacturer specifies that the bond resistance up to 200°C will not exceed 0.000118 m^2·K/W.

SOLUTION.

(1) *Thermal Properties.* Assume that the air outlet temperature is 30°C, which is to be verified:

$$T_b = \frac{T_1 + T_2}{2} = \frac{92 + 76}{2} = \frac{168}{2} = 84°\text{C} \quad (357\,\text{K})$$

$$t_b = \frac{t_2 + t_1}{2} = \frac{30 + 5}{2} = \frac{35}{2} = 17.5°\text{C} \quad (290.5\,\text{K})$$

At these bulk temperatures:

Property	Inside Tubes: Hot Water	Outside Tubes: Cold Air
T_b or t_b (K)	357	290.5
μ (kg·m/s)	3.35×10^{-4}	1.798×10^{-5}
k (W/m·K)	0.672	0.0255
c or c_p, (kJ/kg-K)	4.206	1.007
ρ (kg/m^3)	969.4	See below
Pr	2.19	0.710

With air as a perfect gas ($R = 0.2870\,\text{kJ·m}^2/\text{kg·K}$) at atmospheric pressure and 288 K:

$$\rho = \frac{P}{RT} = \frac{101.33}{(0.2870)(290.5)} = 1.2154\,\text{kg/m}^3$$

(2) *Heat balance.* For hot water:

$$\begin{aligned} q &= \dot{m}_h c_h (T_1 - T_2) \\ &= (46.5)(4.206)(92 - 76) \\ &= 3129.3\text{ kW} \end{aligned}$$

For air:

$$\begin{aligned} t_2 &= t_1 + \frac{q}{\dot{m}_c c_{pc}} \\ &= 5 + \frac{3129.3}{(124)(1.007)} \\ &= 5 + 25.06 \\ &= 30.06°\text{C} \quad (\text{say, } 30°\text{C}) \quad \checkmark \end{aligned}$$

(3) *Effectiveness and capacity rate ratio:*

$$C_c = \dot{m}_c c_{pc} = (124)(1.007) = 124.868\text{ kW/K}$$

$$C_h = \dot{m}_h c_h = (46.5)(4.206) = 195.579\text{ kW/K}$$

C_c is $C_{\min}$ and

$$R = \frac{C_{\min}}{C_{\max}} = \frac{124.868}{195.579} = 0.6385$$

The required effectiveness is

$$\epsilon = \frac{t_2 - t_1}{T_1 - t_1} = \frac{30 - 5}{92 - 5} = \frac{25}{87} = 0.2874$$

(4) *Exchanger data.* The number of tubes in each row will be the section height divided by the tube pitch:

$$n = \frac{H}{P_t} = \frac{3.000}{5.3975 \times 10^{-2}} = 55.58 \qquad (\text{say, } 55\text{ tubes})$$

Select $n_r = 6$ rows deep so that

$$N = nn_r = (55)(6) = 330\text{ tubes}$$

The hot-side cross-sectional area is

$$A_h = A_i = \frac{\pi}{4} N d_i^2 = \frac{\pi}{4}(330)(2.2098 \times 10^{-2})^2 = 0.1266\text{ m}^2$$

With

$$P_f = z + \delta = \frac{1}{3.543} = 0.2822\,\text{cm} \quad \text{or} \quad 2.822 \times 10^{-3}\ \text{m}$$

then

$$z = \frac{1}{3.543} - \delta = 2.822 \times 10^{-3} - 4.826 \times 10^{-4} = 2.340 \times 10^{-3}\ \text{m}$$

For equilateral triangular pitch,

$$P_t = P_d = 5.3975\ \text{cm}$$

and working with centimeters,

$$\begin{aligned} 2P_d - d_b - \frac{2z\delta}{z+\delta} &= (2)(5.3975) - 2.7432 - \frac{(2)(0.2340)(4.826 \times 10^{-2})}{0.2822} \\ &= 10.7950 - 2.7432 - 0.0800 \\ &= 7.9718\,\text{cm} \quad \text{or} \quad 7.9718 \times 10^{-2}\ \text{m} \end{aligned}$$

and

$$P_t = 5.3975\,\text{cm} < 2P_d - d_b - \frac{2z\delta}{z+\delta} = 7.9718\ \text{cm}$$

Thus the minimum flow area must be obtained from eq. (12.38):

$$\begin{aligned} A_o = A_{\min} &= 2nL\left(P_d - d_b - \frac{2z\delta}{z+\delta}\right) \\ &= (2)(55)(5.50)[(5.3975 - 2.7432 - 0.0800) \times 10^{-2}] \\ &= (605)(2.5743 \times 10^{-2}) \\ &= 15.5745\ \text{m}^2 \end{aligned}$$

From eq. (12.39),

$$\begin{aligned} S_b &= \frac{2NLd_bz}{z+\delta} \\ &= \frac{(2)(330)(5.50)(2.7432 \times 10^{-2})(2.340 \times 10^{-3})}{2.822 \times 10^{-3}} \\ &= 82.57\ \text{m}^2 \end{aligned}$$

From eq. (12.40),

$$S_f = \frac{\pi NL}{z+\delta}\left[\frac{1}{2}(d_a^2 - d_b^2) + d_a\delta\right]$$

$$= \frac{\pi(330)(550)}{0.2822}\left\{\frac{1}{2}[(5.0800)^2 - (2.7432)^2] + 5.0800(4.826 \times 10^{-2})\right\}$$

$$= 2.0205 \times 10^6 \left[\frac{1}{2}(18.2813) + 0.2452\right]$$

$$= 2.0205 \times 10^6 (9.3858)$$

$$= 1.8964 \times 10^7 \text{ cm}^2 \quad \text{or} \quad 1896.44 \text{ m}^2$$

Then, by eqs. (12.41),

$$S = 82.57 + 1896.44$$

$$= 1979.01 \text{ m}^2$$

$$\frac{S_f}{S} = \frac{1896.44}{1979.01}$$

$$= 0.9583$$

and

$$S' = \frac{S}{NL}$$

$$= \frac{1979.01}{(330)(5.50)}$$

$$= 1.0904$$

(5) *Heat transfer coefficients.* For the inner pipe hot water:

$$G_h = \frac{\dot{m}_h}{A_i} = \frac{46.5}{0.1266} = 367.30 \text{ kg/m}^2\cdot\text{s}$$

$$v_h = \frac{G_h}{\rho} = \frac{367.30}{969.4} = 0.379 \text{ m/s}$$

and

$$\text{Re} = \frac{d_i G_h}{\mu} = \frac{(2.2098 \times 10^{-2})(367.30)}{3.35 \times 10^{-4}} = 24,228$$

Use Gnielinski's correlation given by eq. (11.32), which can be applied if $1.00 < \text{Pr} < 10^5$ and $2300 < \text{Re} < 5 \times 10^6$. First, by eq. (11.31),

$$f_c = \frac{1}{(1.82 \log_{10} \text{Re} - 1.64)^2}$$

$$= \frac{1}{(1.82 \log_{10} 24,228 - 1.64)^2}$$

$$= \frac{1}{[1.82(4.3843) - 1.64]^2}$$

$$= \frac{1}{(6.3395)^2}$$

$$= 2.4883 \times 10^{-2}$$

Then, with $f/8 = 2.4883 \times 10^{-2}/8 = 3.1103 \times 10^{-3}$,

$$\text{Nu} = \frac{(f/8)(\text{Re} - 1000)\text{Pr}}{1.00 + 12.7(f/8)^{1/2}(\text{Pr}^{2/3} - 1)}$$

$$= \frac{(3.1103 \times 10^{-3})(24,228 - 1000)(2.19)}{1.00 + (12.7)(3.1103 \times 10^{-3})^{1/2}[(2.19)^{2/3} - 1]}$$

$$= \frac{158.22}{1.00 + (12.7)(0.0558)(1.6864 - 1)}$$

$$= \frac{158.22}{1.00 + 0.4864}$$

$$= 106.44$$

and

$$h_i = \frac{k}{d_i}\text{Nu} = \frac{0.672}{2.2098 \times 10^{-2}}(106.44) = 3236.8 \text{ W/m}^2\cdot\text{K}$$

For the air side:

$$G_c = \frac{\dot{m}_c}{A_o} = \frac{124}{15.5745} = 7.961 \text{ kg/m}^2\cdot\text{s}$$

$$v_c = \frac{G_c}{\rho} = \frac{7.961}{1.2154} = 6.551 \text{ m/s}$$

and

$$\text{Re} = \frac{d_b G_c}{\mu} = \frac{(2.7432 \times 10^{-2})(7.961)}{1.798 \times 10^{-5}} = 12,146$$

Use the Briggs and Young correlation given by eq. (12.42) after first observing that the range of parameters are all fulfilled:

$$\text{Nu} = \frac{h d_b}{k} = 0.134\text{Re}^{0.681}\text{Pr}^{1/3}\left[\frac{2(P_f - \delta)}{d_a - d_b}\right]^{0.20}\left(\frac{P_f - \delta}{\delta}\right)^{0.1134}$$

Here

$$\frac{2(P_f - \delta)}{d_a - d_b} = \frac{(2)(0.2822 - 0.0483)}{5.0800 - 2.7432} = 0.2002$$

and

$$\frac{P_f - \delta}{\delta} = \frac{0.2822 - 0.0483}{0.0483} = 4.8427$$

and eq. (12.42) gives

$$\begin{aligned}\mathrm{Nu} &= (0.134)(12,146)^{0.681}(0.710)^{1/3}(0.2002)^{0.20}(4.8427)^{0.1134}\\ &= (0.134)(604.64)(0.8921)(0.7249)(1.1959)\\ &= 62.66\end{aligned}$$

and

$$h_o = \frac{k}{d_b}\mathrm{Nu} = \frac{0.0255}{2.7432\times10^{-2}}(62.66) = 58.25\ \mathrm{W/m^2\cdot K}$$

(6) *Overall coefficient of heat transfer, U_o.* From eq. (12.47),

$$r_{io} = \frac{1}{h_i}\frac{d_b}{d_i} = \frac{1}{3236.8}\left(\frac{2.7432}{2.2098}\right) = 3.8352\times10^{-4}\ \mathrm{m^2\cdot K/W}$$

From eq. (12.48),

$$r_{dio} = r_{di}\frac{d_b}{d_i} = (2.6\times10^{-4})\left(\frac{2.7432}{2.2098}\right) = 3.2276\times10^{-4}\ \mathrm{m^2\cdot K/W}$$

With $\delta_l = (d_o - d_i)/2 = (2.5400 - 2.2098)/2 = 0.1651$ cm and $k_l = 200$ W/m·K, eq. (12.49) gives

$$\begin{aligned}r_{mol} &= \frac{\delta_\ell}{k_\ell}\frac{2d_b}{d_o + d_i}\\ &= \frac{1.651\times10^{-3}}{200}\left[\frac{(2)(2.7432)}{2.5400 + 2.2098}\right]\\ &= 9.5352\times10^{-6}\ \mathrm{m^2\cdot K/W}\end{aligned}$$

Using eq. (12.50) yields

$$r_{Bo} = r_B\frac{d_b}{d_g} = (1.18\times10^{-4})\left(\frac{2.7432}{2.6467}\right) = 1.2230\times10^{-4}\ \mathrm{m^2\cdot K/W}$$

With $\delta_t = (d_b - d_g)/2 = (2.7432 - 2.6467)/2 = 0.04825$ cm and $k_t = 385$ W/m·K, eq. (12.51) gives

$$\begin{aligned}r_{mot} &= \frac{\delta_t}{k_t}\frac{2d_b}{d_b + d_g}\\ &= \frac{1.4825\times10^{-4}}{385}\left[\frac{(2)(2.7432)}{2.7432 + 2.6467}\right]\\ &= 1.2757\times10^{-6}\ \mathrm{m^2\cdot K/W}\end{aligned}$$

Then

$$\sum R_{io} = r_{io} + r_{dio} + r_{mol} + r_{Bo} + r_{mot}$$

and the summation is

$$
\begin{aligned}
r_{io} &= 3.8352 \times 10^{-4}\ \text{m}^2\cdot\text{K/W} \\
r_{dio} &= 3.2276 \times 10^{-4}\ \text{m}^2\cdot\text{K/W} \\
r_{mol} &= 9.5352 \times 10^{-6}\ \text{m}^2\cdot\text{K/W} \\
r_{Bo} &= 1.2230 \times 10^{-4}\ \text{m}^2\cdot\text{K/W} \\
r_{mot} &= \underline{1.2757 \times 10^{-6}\ \text{m}^2\cdot\text{K/W}} \\
\sum R_{io} &= 8.3939 \times 10^{-4}\ \text{m}^2\cdot\text{K/W}
\end{aligned}
$$

and

$$
\begin{aligned}
\sum R_{is} &= \sum R_{io}\frac{S'}{\pi d_b} \\
&= (8.3939 \times 10^{-4})\left[\frac{1.0904}{\pi(2.7432 \times 10^{-2})}\right] \\
&= (8.3939 \times 10^{-4})(12.6521) \\
&= 1.0620 \times 10^{-2}\ \text{m}^2 \cdot \text{K/W}
\end{aligned}
$$

Because there is no fouling on the air side,

$$
m = \left(\frac{2h_o}{k_t\delta}\right)^{1/2} = \left[\frac{(2)(58.25)}{(200)(4.826 \times 10^{-4})}\right]^{1/2} = 34.7420\ \text{m}^{-1}
$$

and with

$$
\rho = \frac{d_b}{d_a} = \frac{2.7432}{5.0800} = 0.5400
$$

then

$$
\begin{aligned}
\psi &= \frac{d_b}{2}\frac{1-\rho}{\rho}\left(1 + 0.35\ln\frac{1}{\rho}\right) \\
&= \frac{2.7432 \times 10^{-2}}{2}\left(\frac{1-0.54}{0.54}\right)\left(1 + 0.35\ln\frac{1}{0.54}\right) \\
&= (1.3716 \times 10^{-2})(0.8519)(1 + 0.35\ln 1.8519) \\
&= 1.1684 \times 10^{-2}[1 + (0.35)(0.6162)] \\
&= 1.1684 \times 10^{-2}(1.2157) \\
&= 1.4204 \times 10^{-2}
\end{aligned}
$$

and

$$
m\psi = (34.7420)(1.4204 \times 10^{-2}) = 0.4935
$$

This makes the fin efficiency

$$\eta = \frac{\tanh m\psi}{m\psi} = \frac{\tanh 0.4935}{0.4935} = 0.9260$$

and the weighted fin efficiency

$$\begin{aligned}\eta_o &= 1 - \frac{S_f}{S}(1-\eta)\\ &= 1 - (0.9583)(1 - 0.9260)\\ &= 1 - (0.9583)(0.0740)\\ &= 1 - 0.0709\\ &= 0.9291\end{aligned}$$

Hence

$$r_{o\eta} = \frac{1}{(58.25)(0.9291)} = \frac{1}{54.12} = 0.0185$$

The overall heat transfer coefficient is

$$\begin{aligned}U_o &= \frac{1}{\sum R_{is} + r_{o\eta}}\\ &= \frac{1}{1.0620 \times 10^{-2} + 0.0185}\\ &= \frac{1}{0.0291}\\ &= 34.37\,\text{W/m}^2\cdot\text{K}\end{aligned}$$

(7) *Actual effectiveness.* Here

$$\text{N}_{\text{tu}} = \frac{U_o S}{C_{\min}} = \frac{(34.37)(1979.01)}{124,868} = 0.5447$$

and with $R = 0.6385$ for this case of both fluids unmixed, use eq. (10.26*d*):

$$\epsilon = 1 - e^{(\text{N}_{\text{tu}})^{0.22}\Lambda/R}$$

where

$$\begin{aligned}\Lambda &= e^{-R(\text{N}_{\text{tu}})^{0.78}} - 1\\ &= e^{-(0.6385)(0.5447)^{0.78}} - 1\\ &= e^{-(0.6385)(0.6226)} - 1\\ &= e^{-0.3975} - 1\\ &= 0.6720 - 1\\ &= -0.3280\end{aligned}$$

and the effectiveness is

$$\begin{aligned}\epsilon &= 1 - e^{(N_{tu})^{0.22}\Lambda/R} \\ &= 1 - e^{(0.5447)^{0.22}(-0.3280)/0.6385} \\ &= 1 - e^{-0.4995} \\ &= 1 - 0.6380 \\ &= 0.3620\end{aligned}$$

The actual effectiveness exceeds the required effectiveness:

$$0.3620 > 0.2874$$

The heat transfer design is acceptible (about 20% conservative) and the pressure losses are the final items for consideration.

(8) *Pressure losses.* The pressure losses are obtained from eqs. (11.37) and (12.45). For the inside of tubes, using eq. (11.37*b*) with f_h, G, and ρ already obtained:

$$f_h = 0.0248, \quad G_h = 367.30\,\text{kg/m}^2\cdot\text{s} \quad \text{and} \quad \rho = 969.4\,\text{kg/m}^3$$

$$\begin{aligned}\Delta P_f &= \frac{4fG_h^2}{2\rho}\frac{L}{d_i} \\ &= \frac{(4)(0.0248)(367.30)^2}{(2)(969.4)}\,\frac{5.50}{2.2098\times 10^{-2}} \\ &= 1718\,\text{Pa}\end{aligned}$$

There is no turn loss, so that

$$\Delta P = \Delta P_f = 1718\,\text{Pa} < 7000\,\text{Pa}$$

For the outside of tubes, using the correlation of Robinson and Briggs (1966) given by eq. (12.45) gives

$$\Delta P = 18.03\frac{G^2}{\rho}n_r\text{Re}^{-0.316}\left(\frac{P_t}{d_b}\right)^{-0.927}\left(\frac{P_t}{P_d}\right)^{0.515} \qquad (12.45)$$

Here $P_t = P_d$, so that $P_t/P_t = 1.00$, $(P_t/P_d)^{0.515} = 1.00$, and

$$\frac{P_t}{d_b} = \frac{5.3975}{2.7432} = 1.9676$$

and with

$$G_c = 7.961\,\text{kg/m}^2\cdot\text{s}, \quad \rho = 1.2154\,\text{kg/m}^3, \quad \text{and} \quad \text{Re} = 12,146$$

$$\Delta P = (18.03)\left[\frac{(7.961)^2}{1.2154}\right](6)(12,146)^{-0.316}(1.9676)^{-0.927}$$

$$= (18.03)(52.1454)(6)(0.0512)(0.5340)$$

$$= 154.2\,\text{Pa} < 250\,\text{Pa}$$

The design is satisfactory for both heat transfer and pressure drop.

12.5 NOMENCLATURE

Roman Letter Symbols

A	cross-sectional area, m^2
b	fin height, m
b_p	function of cold workability
C	arbitrary constant, dimensionless; capacity rate, W/K
c	specific heat, J/kg·K
d	diameter, m; derivative or differential, dimensionless
E	modulus of elasticity, N/m^2
F	logarithmic mean temperature difference correction factor, dimensionless
G	mass velocity, $kg/m^2 \cdot s$
g	radial gap, m
H	section height, m
h	heat transfer coefficient, $W/m^2 \cdot K$
I	modified Bessel function of first kind, dimensionless
K	modified Bessel function of second kind, dimensionless
k	thermal conductivity, W/m·K
L	fin length, m; exchanger length, m
LMTD	logarithmic mean temperature difference, K
m	fin performance parameter, m^{-1}
$\dot{m}$	mass flow rate, kg/s
N	total number of tubes, dimensionless
Nu	Nusselt number, dimensionless
N_{tu}	number of transfer units, dimensionless
n	number of tubes in a row
n_r	number of rows deep
P	pitch, m; pressure, N/m^2 (Pa)
Pr	Prandtl number, dimensionless
q	heat flow, W
R	thermal resistance, K/W; capacity rate ratio, dimensionless; gas constant, $kPa \cdot m^3/kg \cdot K$; combination of terms, dimensionless
Re	Reynolds number, dimensionless
r	radius, m; radial coordinate, m; thermal resistance, K/W

S	surface area, m^2; surface area per unit length, m^2
T	temperature, K; hot fluid temperature, K
t	cold fluid temperature, K
U	overall heat transfer coefficient, $W/m^2 \cdot K$
u	radial displacement, m
v	velocity, m/s
z	clear space between tubes, m

Greek Letter Symbols

α	coefficient of thermal expansion, K^{-1}
γ	surface roughness, m
Δ	change in, dimensionless
δ	fin thickness, m; thickness, m
ϵ	exchanger effectiveness, dimensionless
η	fin efficiency, dimensionless
Λ	combination of terms, dimensionless
μ	dynamic viscosity, kg/m·s
ν	Poisson ratio, dimensionless
ρ	radius ratio, dimensionless; density, kg/m^3
ϱ	combination of terms, $m^2 \cdot K/W$
σ	radial stress, N/m^2
υ	combination of terms, dimensionless
ϕ	combination of terms, dimensionless
ψ	combination of terms

Roman Letter Subscripts

a	tip of fin; fin outer diameter
B	bond resistance
Bo	bond resistance referred to outside of tube
b	base of fin; tube outer diameter; bare surface; bulk
c	contact
co	initial contact value; contact resistance referred to outside surface; cold fluid
d	diagonal pitch
di	dirt or fouling resistance
dio	dirt or fouling resistance referred to outside of tube
f	fin
fb	fin base
fo	initial displacement
h	hot fluid; hydraulic radius
i	inside
io	inside referred to outside
l	longitudinal pitch; liner
m	metal
mo	metal resistance referred to outside surface

max	maximum value
min	minimum value
mol	liner metal resistance referred to outside surface
mot	tube metal resistance referred to outside surface
o	outside of tube; diameter of tube; initial value; condition of no residual stress; overall; air side
$o\eta$	outside condition corrected for fin efficiency
p	profile; constant pressure condition
r	row
s	surroundings
t	tube; transverse pitch
v	gap fluid; arithmetic average
w	tube wall

Superscripts

′	referred to outside surface
*	radial

CHAPTER 13

FINS WITH RADIATION

13.1 INTRODUCTION

One of the first papers to treat the radiation mode as the sole means of heat dissipation from the faces of a fin was that of Callinan and Berggren (1959). This paper considered flat and convex tubes with radiation from one side and fin-and-tube double-surface radiators. The radiation interchange between fin and tube was approximated except that interreflections were not considered for the gray body case, and no account was taken of the incident radiation from the fin on the tube. Even at this early stage in the technology pertaining to the investigation of radiative dissipation from fins, an attempt was made to maximize heat rejection on a per unit mass basis.

Chambers and Somers (1959) analyzed the radiation from one side of a circular disk to an absolute zero heat sink, and in this and the next chapter, radiation to an absolute zero heat sink will be referred to as *radiation to free space*. In a NASA Technical Note, Lieblein (1959) unknowingly covered much of the ground covered by Callinan and Berggren (1959), but he provided the basis for the consideration of an equivalent heat sink temperature which would enable later investigators to take account of a variety of environmental conditions. Leiblein also considered the mass minimization problem and provided radiation fin efficiency curves for both finite- and infinite-length plates for various source-to-sink temperature ratios.

Bartas and Sellars (1960) provided efficiency curves for minimum mass fins, and Nilson and Curry (1960) gave a numerical solution to obtain the mass minimization of a straight (longitudinal) fin of rectangular profile from fin surfaces radiating with an emmissivity of 0.50 to free space. Mackay (1960) published a text that provided a wealth of interesting data that would enable a designer to obtain solutions to both fin analysis and fin synthesis problems involving rectangular profile and optimized profile fins radiating to free space and to nearby surfaces.

Eckert et al. (1960) provided a brief note on the formulation of models for the radiation from fins with mutual irradiation. However, in their study, the radiant energy incident on the fin from the base or prime surface was assumed to be negligible. An analytical solution to the problem of mass minimization was given by Liu (1960). His solution involved determination of the optimum dimensions of the longitudinal fin of rectangular profile radiating to free space with no energy absorption from the environment.

Mackay (1960), taking the lead from Schmidt (1926), developed fin shapes to yield linear temperature gradients when dissipation from the fin faces is purely by radiation. Although he made no claim that the linear temperature gradient fins were optimum, he did show that these fins were somewhat lighter than their companion rectangular profile fins designed for identical conditions. However, it was Wilkins (1960a,b) who showed that the radiating longitudinal fin of least material is not the one that exhibits a linear temperature gradient. He stated that the problem is one of finding three functions, one of heat flow, one of temperature profile, and one relating to a variable fin thickness that will match the boundary conditions at fin tip and fin base and make the profile area a minimum. Through the use of some clever transformations with regard to temperature and heat flow, he was able to determine the profile required. It is interesting to note that Haley and Westwater (1966) employed Wilkins's procedure to optimize the shape of a spine dissipating heat to a boiling liquid.

Granet and McIlroy (1961) described a procedure that enabled the optimization of any fin whose profile could be expressed as a sum of exponentials. Liu (1961) extended his previous work by giving an analytical solution for the optimum rectangular cooling fin in terms of the beta function. It was Heaslet and Lomax (1961) who appeared to be among the first to analyze radiating fins with variable thermal conductivity and variable emissivity on the fin faces. His treatment pertained to a family of very thin longitudinal fins of rectangular profile. These fins extended symmetrically from a common edge and mutual diffuse radiation was taken into account. At about the same time, Sparrow et al. (1961a) published an analysis that pertained to two longitudinal fins of rectangular profile with a common edge at a general angle with respect to each other. Diffuse radiant interchange between the fins was considered, but no attempt to include the radiant energy incident on the fins from the base (prime) surface was made.

In 1962, Kreith (1962) published his book entitled *Radiation in Space*, which contained a section on radiating fins, and Wilkins (1962a) addressed the problem of optimization with a constant and linear temperature gradient. Wilkins (1962b) also considered a minimum mass radiating fin with internal heat generation.

Sparrow and Eckert (1962) provided a more comprehensive treatment of radiation between fin and base surface in the case of fin-and-tube construction. For blackbody conditions, they included the effect of radiation from the tubes falling on the fins and observed that under practical operating conditions, the radiation from the base surface could cause a significant reduction in the heat dissipation from the fin. For gray body conditions, they set up equations that would account for diffuse reflections, but they did not solve them.

Liu (1962) proposed a general differential equation and solution that applied to longitudinal fins of arbitrary profile and which considered a possible variation of surface emissivity as a function of fin height.

Sparrow et al. (1961b) developed a model for annular (radial) fins of rectangular profile with blackbody surfaces and included the fin and base radiant interaction. It was shown that the radiant interchange between fin and tube was quite significant, leading to appreciable error if this interaction is not considered. Equations for gray body radiation with constant surface emissivity were presented but were not solved.

Reynolds (1963) pointed out that a typical fin-and-tube space radiator such as the one shown in Fig. 13.1 would have manifold tubes with manifold lengths dependent on the length of the fins. He also considered that the mass of the system, consisting of the tube, the manifolds, the fluid that the tubes and manifolds contain, and any protective armor, may be such that shorter and thicker fins may be more desirable than fins whose design depended solely on individual fin optimization. Reynolds's (1963) work provided a mass optimization for longitudinal fins of rectangular, trapezoidal, and triangular profile spaced (Fig. 13.1) 180° apart on the periphery of a circular tube, which, in turn, was attached to a manifold at each end. The mass optimization included the supporting manifolds.

Stockman and Kramer (1963) considered the variation of thermal conductivity and emissivity as linear functions of temperature in a fin-and-tube configuration assuming

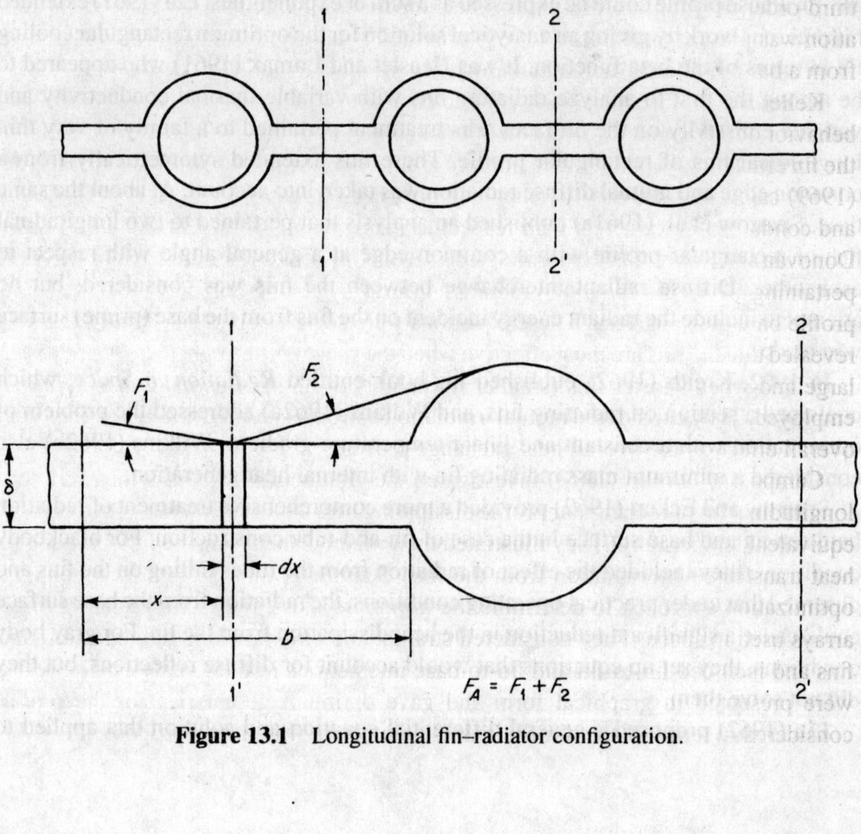

Figure 13.1 Longitudinal fin–radiator configuration.

one-dimensional heat flow and with radiation to an equivalent heat sink temperature. Stockman and Bittner (1965) provided what was one of the first treatments of two-dimensional heat flow in radiating fins. The study was based on a fin-and-tube configuration with stainless steel cladding on a copper fin and radiation to free space. Kotan and Arnas (1965) studied the parabolic profile longitudinal radiating fin and provided an optimization.

Okamoto (1966a,b) also made a two-dimensional study for the rectangular profile fin. He concluded that the one-dimensional model was accurate down to height-to-thickness ratios as low as 3, although he also ignored the base and fin radiation interaction. Hering (1966) considered the specular radiation interaction in the angular space between rectangular plates of rectangular cross section having a common edge. This work was significant because it showed that the total heat loss from the plates that were acting as fins was greater for specular radiation than for the diffuse radiation considered earlier.

Sarabia and Hitchcock (1966) extended the work of Sparrow and Eckert (1962) by solving the problem of gray body interchange between fin and base for a configuration of infinite length. Stockman et al. (1966) enhanced the work of Stockman and Bittner (1965) by comparing one- and two-dimensional solutions and including the radiation interchange between fins and tube (base surface). The study showed that the one-dimensional solutions were in good agreement with the two-dimensional solutions.

Tien (1967) commented on the work of Hering (1966) and provided an approximate third-order polynomial solution for the temperature distribution when specular radiation was present, and Cohen (1969) provided a numerical solution for heat transfer from a bar of variable thermal conductivity by radiation.

Keller and Holdredge (1969) conducted a numerical solution for the steady-state behavior of the annular (radial) fin of trapezoidal profile and provided charts relating the fin efficiency to other dimensionless groups defined in their analysis. Koshelyaev (1969) calculated the fin efficiency and obtained optimum combinations of thermal and conduction parameters with respect to mass for straight radiating fins on tubes. Donovan and Rohrer (1971) formulated a set of nonlinear integrodifferential equations pertaining to heat dissipation by radiation in an array of longitudinal fins of rectangular profile on a plane surface. These equations were solved numerically, and the results revealed that the fins are most effective when the spacings between them are relatively large and when shorter fins (smaller fin height) of higher thermal conductivity are employed. They considered mutual irradiation and observed that this had an important overall effect on the overall heat exchange process.

Campo and Wolko (1973) investigated the conduction–radiation interplay for a longitudinal fin of rectangular profile dissipating heat to the surroundings at a constant equivalent temperature. They illustrated their mathematical scheme for obtaining the heat transferred by radiation from the fins. Schnurr et al. (1976) used a nonlinear optimization approach to determine the minimum mass design for radiating finned arrays used in space. They considered straight and circular (longitudinal and radial) fins and included fin-to-fin and fin-to-base interactions in their analysis. The results were presented in graphical form and gave optimum geometries for the profiles considered in terms of the dimensionless parameters which they proposed.

Chiou and Na (1977) developed an initial value method for the solution of nonlinear two-point boundary value problems that pertain to the analysis of radiating fins. This method is noniterative, computationally efficient, and gives good agreement with the solutions of identical problems solved by more conventional methods.

Mehta (1978) obtained minimum mass designs for radiating finned arrays which he called *heat sinks*, using a direct search procedure using pseudorandom numbers. This analysis included fin-to-fin and fin-to-base interactions. Crawford (1978) compared three methods of calculating the heat transfer by radiation from fins of arbitrary shape, and Karam and Eby (1978) showed that the differential equation for the temperature profile when radiation and convection are both present could be simplified considerably if the temperature to the fourth power in the radiation term is replaced by a linear expansion about a term known as the *mean temperature*. Solution of the linearized steady-state equation was provided and a method was indicated in which the mean temperature was optimized as a function of the fin properties in order to minimize the errors introduced by the process of linearization.

Truong and Mancuso (1980) treated the problem of radiation from an annular (radial) fin whose surfaces had different emissivities. The study included various profile shapes and the results were obtained by the shooting method in conjunction with the Runge–Kutta–Verner fifth- and sixth-order integration method. The results were plotted as a function of dimensionless parameters proposed by the authors.

Delfour et al. (1983) used a finite element method as a first step toward the solution of a minimum mass radiating fin in a satellite application. Colle (1983) provided a general solution for heat transfer in arrays of radiating fins of arbitrary shape, and Chang (1985) obtained an analytical solution for a radiating annular (radial) fin by linearizing the radiating term in the differential equation for the temperature profile. The errors resulting from the linearization process were claimed to have been minimized. Chung and Nguyen (1986) provided a general relationship for the optimized dimensions of longitudinal fins of rectangular, trapezoidal, triangular, and parabolic profile radiating to free space, and Smith (1992) presented a single equation for the profile area of longitudinal fins of rectangular, trapezoidal, and triangular profile as a function of the taper ratio.

In Chapter 1 it was convenient to fix the environments in which extended surfaces were studied merely by assigning a temperature to the environment and a convection coefficient through which heat was transported from the fin faces to the environment. Heat rejection systems for space vehicles usually employ some form of extended surface which must perform in surroundings in which radiation is the only form of heat communication. Similarly, the generation of steam power from fossil fuel has progressively utilized the advantages of radiation over convection in furnace designs.

Although radiation and convection frequently coexist, a study of systems involving radiation alone provides a tidier approach to systems depending wholly or in part on heat transfer by radiation. The mathematic analysis may be based on the modifications of two of the Murray–Gardner assumptions cataloged in Chapter 1. Assumption 3 requires that the fin be exposed to a uniform heat transfer coefficient, and it is apparent that in space vehicles, portions of a fin may be turned toward or away from a radiation heat sink. With this assumption removed, consideration must be given to the purely radiative heat transport that exists between various points on fin surfaces and their surroundings. The removal of assumption 3 nullifies assumption 10, which requires that the heat transferred be proportional to the temperature excess, $\theta = T - T_s$, whereas radiation implies a fourth-power temperature dependency.

13.2 LONGITUDINAL RADIATING FIN OF RECTANGULAR PROFILE

13.2.1 Radiation to Free Space

Bartas and Sellars (1960) investigated radiation from rectangular fins joining circular tubes as shown in Fig. 13.1. The governing differential equation was formulated using a nonzero surroundings temperature, referred to here as *non-free space*, and considered the radiation or shape factors that exist between fins and tubes. These interchange factors, designated here as F_A,[1] account for the effects of mutual radiation between a fin and other surfaces and have been studied by Sparrow and Eckert (1962), and Kreith (1962), and many others.

Mackay and Leventhal (1960) derived basic relationships for parameters affecting heat transfer from a plate uniformly heated at one edge. This was the case for radiation to free space where the environmental temperature was taken as absolute zero. They also considered heat input to the fin from sources such as the sun, earth, and other bodies in space.

The terminology and coordinate system for the longitudinal fin of rectangular profile radiating to deep space at $T_s = 0\,\text{K}$ are shown in Fig. 13.2. The fin receives no heat input on its faces from other bodies in space or from other fins, solar panels, or conduits in the space vehicle and space radiator configuration. Heat enters uniformly at the fin base at $x = 0$ and passes from the fin faces by radiation.

Referring to the differential element, dx, in Fig. 13.2, the difference between the heat entering and leaving by conduction is

$$dq = k\delta L \frac{d^2T}{dx^2}\,dx \tag{13.1}$$

where T is the absolute temperature in kelvin and k is the thermal conductivity. In the steady state, this heat must be equal to the heat dissipated by radiation from the faces of the element dx:

$$dq = 2\sigma\epsilon L T^4\,dx \tag{13.2}$$

where σ is the Stefan–Boltzmann constant,

$$\sigma = 5.669 \times 10^{-8}\ \text{W/m}^2 \cdot \text{K}^4$$

An energy balance over the differential element dx requires that eqs. (13.1) and (13.2) be equated and after some simplification

[1] F_A for *arrangement factor.*

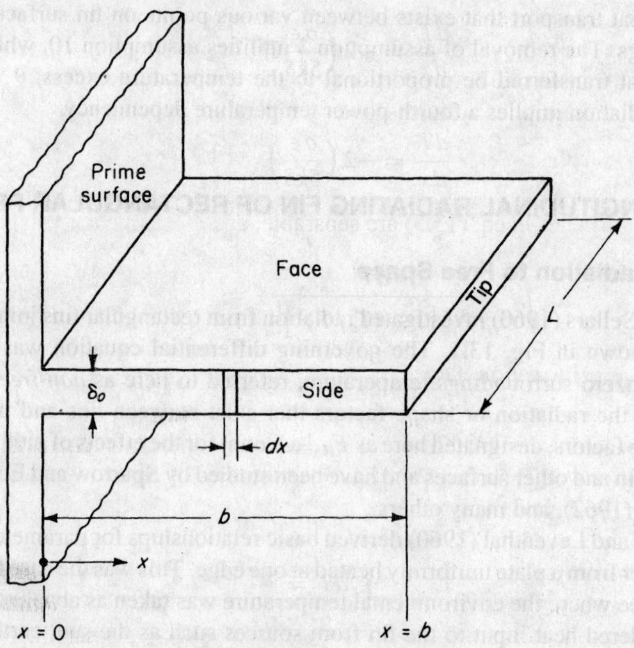

Figure 13.2 Terminology and coordinate system for a radiating longitudinal fin of rectangular profile.

$$\frac{d^2T}{dx^2} = \frac{2\sigma\epsilon}{k\delta}T^4 \tag{13.3}$$

which is a second-order nonlinear differential equation. Equation (13.3) governs the temperature profile, and its solution can be obtained by successive integration.

Let $p = dT/dx$, so that

$$\frac{d^2T}{dx^2} = \frac{dp}{dx} = \frac{dp}{dT}\frac{dT}{dx} = p\frac{dp}{dT}$$

Substituting into eq. (13.3) gives

$$p\frac{dp}{dt} = \frac{2\sigma\epsilon}{k\delta}T^4$$

and a separation of variables and integration yields

$$p = \frac{d\theta}{dx} = -2\left(\frac{\sigma\epsilon}{5k\delta}T^5 + C\right)^{1/2} \tag{13.4}$$

where C is the arbitrary constant of integration. The minus sign in eq. (13.4) is required because the temperature gradient is negative everywhere. The arbitrary constant is evaluated at the fin tip where $x = b$ and where $dT/dx = 0$ and $T = T_a$. Then eq. (13.4) becomes

$$\frac{dT}{dx} = -2\left(\frac{\sigma\epsilon}{5k\delta}T^5 - \frac{\sigma\epsilon}{5k\delta}T_a^5\right)^{1/2}$$

or

$$\frac{dT}{dx} = -2\left(\frac{\sigma\epsilon}{5k\delta}\right)^{1/2}(T^5 - T_a^5)^{1/2} \tag{13.5}$$

The variables in eq. (13.5) are separable, and after algebraic manipulation,

$$\int_{T_b}^{T_a} \frac{dT}{T^{5/2}\left[1 - (T_a/T)^5\right]^{1/2}} = -2\left(\frac{\sigma\epsilon}{5k\delta}\right)^{1/2}\int_0^b dx \tag{13.6}$$

Let the transformation

$$u \equiv \left(\frac{T_a}{T}\right)^5 \tag{13.7}$$

be made so that

$$dT = -\frac{T_a\,du}{5u^{6/5}} \tag{13.8}$$

and

$$T^{5/2} = \left(\frac{T_a^5}{u}\right)^{1/2} \tag{13.9}$$

Moreover, let

$$Z \equiv \frac{T_b}{T_a}$$

so that at $x = 0$, $T = T_b$ and $u = Z^{-5}$, and at $x = b$, $T = T_a$ and $u = 1$. These terms and eqs. (13.7) through (13.9) may be put into eq. (13.6) to give

$$\int_{u=Z^{-5}}^{u=1} \frac{u^{-0.7}(1-u)^{-0.5}\,du}{5T_a^{3/2}} = 2\left(\frac{\sigma\epsilon}{5k\delta}\right)^{1/2}\int_0^b dx$$

or

$$\int_{u=Z^{-5}}^{u=1} u^{-0.7}(1-u)^{-0.5}\,du = \left(\frac{20\sigma\epsilon T_a^3}{k\delta}\right)^{1/2}\int_0^b dx \tag{13.10}$$

The integral on the left of eq. (13.10) may be evaluated in terms of the *beta functions*,

$$\int_{u=Z^{-5}}^{u=1} u^{-0.7}(1-u)^{-0.5}\,du = B(a,b) - B_u(a,b) \tag{13.11}$$

where $B(a, b)$ is the *complete beta function*,

$$B(a,b) = \int_{u=0}^{u=1} u^{a-1}(1-u)^{b-1}\,du = \frac{\Gamma(a)\Gamma(b)}{\Gamma(a+b)} \tag{13.12a}$$

and $B_u(a,b)$ is the *incomplete beta function*,

$$B_u(a,b) = \int_{u=0}^{u=Z^{-5}} u^{a-1}(1-u)^{b-1}\,du \tag{13.12b}$$

In eqs. (13.12), the constants a and b are numerically equal to $a = 0.3$ and $b = 0.5$. Thus the temperature profile is described completely by

$$B(0.3,0.5) - B_u(0.3,0.5) = b\left(\frac{20\sigma\epsilon T_a^3}{k\delta}\right)^{1/2} \tag{13.13}$$

The heat dissipated by the fin is obtained from

$$q = -kL\delta \left.\frac{dT}{dx}\right|_{x=0}$$

and using eq. (13.5) for the temperature gradient at $x = 0$ where $T = T_b$ gives

$$q_b = 2k\delta L\left(\frac{\sigma\epsilon}{5k\delta}\right)^{1/2}(T_b^5 - T_a^5)^{1/2} \tag{13.14}$$

As in Chapter 1, the fin efficiency may be defined as the ratio of the actual heat dissipation to the ideal heat dissipation if the entire fin were to operate at the base temperature. With the ideal heat dissipation, $q_{\text{id}} = 2\sigma\epsilon bLT_b^4$,

$$\eta = \frac{2k\delta L\,(\sigma\epsilon/5k\delta)^{1/2}\,(T_b^5 - T_a^5)^{1/2}}{2\sigma\epsilon bLT^4}$$

or by using $Z = T_b/T_a$, after algebraic adjustment,

$$\eta = \frac{2\left(1/Z^3 - 1/Z^8\right)^{1/2}}{b\left(20\sigma\epsilon T_a^3/k\delta\right)^{1/2}} \tag{13.15}$$

The ratio Z relates the temperatures at the base and tip of the fin. As shown by eq. (13.13), there is a relationship between the parameter Z and the parameter

$$\psi \equiv b\left(\frac{20\sigma\epsilon T_a^3}{k\delta}\right)^{1/2}$$

This relationship, evaluated by using the hypergeometric function (Abramowitz and Stegun, 1964) which, in turn, is related to the incomplete beta function, is given in Fig. 13.3 with ψ plotted against Z in accordance with eq. (13.13). The fin efficiency is plotted in Fig. 13.4 using the values obtained from Fig. 13.3 in eq. (13.15). Unlike the convective case, the fin efficiency is a function of the dimensions and thermal properties of the fin (the parameter ψ) as well as the temperatures at the extremes

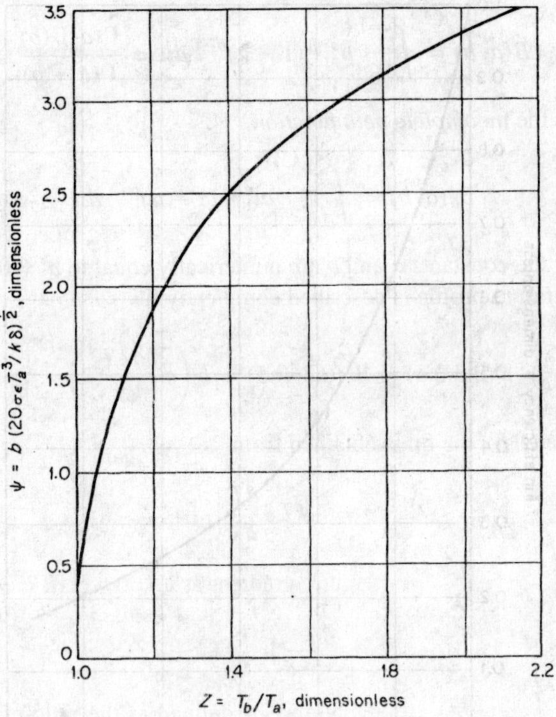

Figure 13.3 Fin parameter ψ as a function of Z for a longitudinal fin of rectangular profile radiating to free space.

of the fir (the parameter Z). For this reason, an analysis of the performance of the radiating fin requires a trial-and-error solution.

Example 13.1: Radiation to Free Space (Fin Analysis). A longitudinal fin of rectangular profile is 4 m long, 50 cm high, and 0.635 cm thick and is fabricated of magnesium ($k = 152$ W/m·K). Its surface has been treated so that its emissivity is 0.85. For a base temperature of 350 K (77°C) and radiation to free space, determine (*a*) the tip temperature, (*b*) the efficiency, and (*c*) the heat dissipation.

SOLUTION. **(a)** For the tip temperature, observe that everything in the parameter ψ is known except T_a:

$$\psi \equiv b\left(\frac{20\sigma\epsilon T_a^3}{k\delta}\right)^{1/2} = 0.50\left[\frac{20(5.669\times10^{-8})(0.85)}{152(0.00635)}\right]^{1/2} T_a^{3/2}$$

or

$$\psi = 4.996\times10^{-4}T_a^{3/2}$$

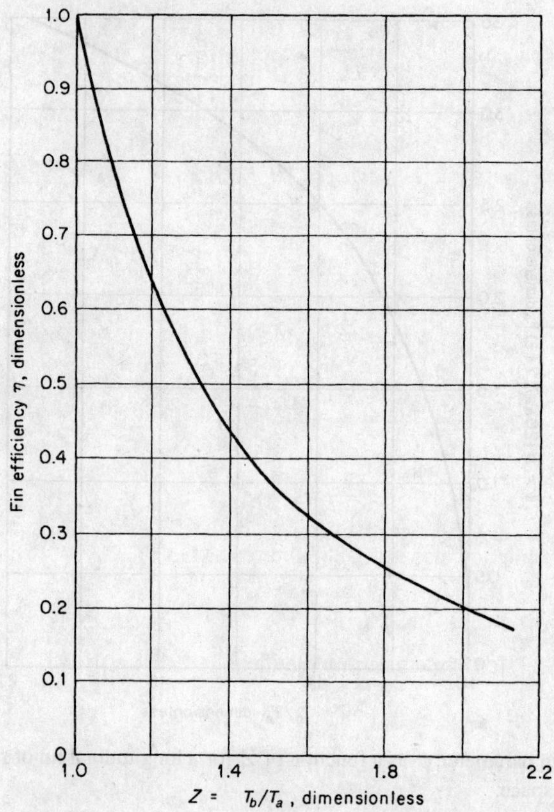

Figure 13.4 Radiation fin efficiency for a longitudinal fin of rectangular profile radiating to free space.

Trial I. Assume that $T_a = 300$ K (27°C). Then

$$\psi = 4.996 \times 10^{-4}(300)^{3/2} = 2.596$$

and

$$Z = \frac{T_b}{T_a} = \frac{350}{300} = 1.167$$

From Fig. 13.3 read $Z = 1.448$ at $\psi = 2.596$ and note that the Z value of 1.167 does not match the Z value of 1.448.

Trial II. Reduce T_a to $T_a = 273$ K (0°C). Then

$$\psi = (4.996 \times 10^{-4})(273)^{3/2} = 2.254$$

and

$$Z = \frac{T_b}{T_a} = \frac{350}{273} = 1.282$$

From Fig. 13.3, read Z = 1.302 at $\psi = 2.254$ and note that the Z value of 1.282 does not match the Z value of 1.302. However, the agreement is close.

Trial III. Assume that $T_a = 270$ K (−3°C). Then

$$\psi = (4.996 \times 10^{-4})(270)^{3/2} = 2.217$$

and

$$Z = \frac{T_b}{T_a} = \frac{350}{270} = 1.296$$

From Fig. 13.3, read Z = 1.292 at $\psi = 2.217$. This is close enough to be considered as the solution and

$$T_a = 270 \text{ K}$$

(b) For the fin efficiency, use Fig. 13.4 and read at Z = 1.296,

$$\eta = f(Z) = 0.524$$

(c) For the heat dissipation determine the ideal heat dissipation first:

$$\begin{aligned} q_{id} &= 2\sigma\epsilon bLT_b^4 \\ &= (2)(5.669 \times 10^{-8})(0.85)(4)(0.50)(350)^4 \\ &= 2892.4 \text{ W} \end{aligned}$$

and this makes the heat dissipation

$$\begin{aligned} q_b &= \eta q_{id} \\ &= (0.524)(2892.4) \\ &= 1515.6 \text{ W} \end{aligned}$$

Example 13.2: Radiation to Free Space (Fin Synthesis). A longitudinal fin of rectangular profile is 5 m long and 0.635 cm thick and is fabricated of aluminum ($k = 202$ W/m·K). Its surface has been treated so that its emissivity is 0.90. For a base temperature of 300 K (27°C), determine the fin height required for a 1350-W dissipation.

SOLUTION. Work on a 1-m-length basis. With $T_b = 300$ K and $q_b = 1350/5 = 270$ W,

$$\begin{aligned} q_{id} &= 2\sigma\epsilon bLT_b^4 \\ &= (2)(5.669 \times 10^{-8})(0.90)(1.00)(300)^4 b \end{aligned}$$

or

$$q_{\text{id}} = 826.5b \text{ W}$$

Moreover,

$$\eta = \frac{q_b}{q_{\text{id}}} = \frac{270}{826.5b} = \frac{0.3267}{b}$$

and

$$\psi = b\left(\frac{20\sigma\epsilon T_a^3}{k\delta}\right)^{1/2}$$

$$= \left[\frac{(20)(5.669 \times 10^{-8})(0.90)}{(202)(0.00635)}\right]^{1/2} bT_a^{3/2}$$

or

$$\psi = 8.9192 \times 10^{-4} bT_a^{3/2}$$

Trial I. Assume that $b = 0.40$ m and obtain

$$\eta = \frac{0.3267}{0.40} = 0.817$$

Read from Fig. 13.4 at $\eta = 0.817$ a value of $Z = 1.092$. Then

$$T_a = \frac{T_b}{Z} = \frac{300}{1.092} = 274.7 \text{ K}$$

Now calculate ψ:

$$\psi = (8.9192 \times 10^{-4})(0.40)(274.7)^{3/2} = 1.624$$

From Fig. 13.3, read $Z = 1.153$ at $\psi = 1.624$, and note that the Z value of 1.092 does not match the Z value of 1.153.

Trial II. Assume that $b = 0.60$ m and obtain

$$\eta = \frac{0.3267}{0.60} = 0.545$$

Then read from Fig. 13.4 at $\eta = 0.545$ a value of $Z = 1.283$ and calculate T_a:

$$T_a = \frac{T_b}{Z} = \frac{300}{1.283} = 233.8 \text{ K}$$

Now calculate ψ:

$$\psi = (8.9192 \times 10^{-4})(0.60)(233.8)^{3/2} = 1.913$$

From Fig. 13.3, read $Z = 1.218$ at $\psi = 1.913$ and note that the Z value of 1.283 does not match the Z value of 1.218.

Trial III. Assume that $b = 0.50$ m and obtain

$$\eta = \frac{0.3267}{0.50} = 0.653$$

Then read from Fig. 13.4 at $\eta = 0.653$ a value of $Z = 1.186$ and calculate T_a:

$$T_a = \frac{T_b}{Z} = \frac{300}{1.186} = 252.9 \text{ K}$$

Now calculate ψ:

$$\psi = (8.9192 \times 10^{-4})(0.50)(252.9)^{3/2} = 1.795$$

From Fig. 13.3, read $Z = 1.186$ at $\psi = 1.795$ and note that a match has been obtained. The fin height for a dissipation of 270 W/m is 0.50 m or 50 cm.

13.2.2 Radiation to Non-Free Space

The longitudinal fin of rectangular profile radiating to a sink that is not at or near absolute zero and with radiant interchange to and from other bodies has been treated by Lieblein (1959), Bartas and Sellars (1960), and Mackay and Leventhal (1960). Both the Lieblein and Bartas and Sellars works considered these environmental effects as part of a fictitious sink temperature. Mackay and Leventhal lumped these effects into what eventually was incorporated into an environmental parameter. The Mackay and Leventhal procedure is considered here.

Refer to the longitudinal fin of rectangular profile in Fig. 13.2 and consider the radiant heat exchange between the differential element of fin surface, $L\,dx$, and the surroundings. This radiant heat exchange will be composed of two terms,

$$K_1 T^4 L\,dx$$

where the constant K_1 embraces all factors that modify the fin temperature as multipliers and where both sides of the fin dissipate, and

$$K_2 L\,dx$$

where the constant K_2 consists of all terms that do not multiply the fin temperature and may include solar and/or terrestrial radiation, radiant interchange factors between the fin, and other elements in the configuration and the temperature of the surroundings.

The total radiant heat dissipated by the differential fin element will be

$$dq = (K_1 T^4 - K_2) L\,dx \tag{13.16}$$

In accordance with the steady-state heat balance, this heat can be equated to the difference in heat entering and leaving the element dx by conduction. Thus

$$k\delta L \frac{d^2T}{dx^2} dx = (K_1 T^4 - K_2) L\,dx$$

or

$$\frac{d^2T}{dx^2} = \frac{K_1}{k\delta}T^4 - \frac{K_2}{k\delta} \qquad (13.17)$$

Equation (13.17) is the differential equation governing the temperature profile on the fin. By a procedure similar to that employed for radiation to free space, a single integration gives

$$\frac{dT}{dx} = -\left(\frac{2K_1}{5k\delta}T^5 - \frac{2K_2}{k\delta}T + C\right)^{1/2} \qquad (13.18)$$

where C is the constant of integration and where the minus sign assures a temperature gradient that is everywhere negative. The arbitrary constant can be evaluated at $x = b$, where dT/dx is set equal to zero ($dT/dx = 0$) and $T = T_a$:

$$C = -\frac{2K_1}{5k\delta}T_a^5 + \frac{2K_2}{k\delta}T_a$$

Upon substitution of this value of C into eq. (13.18), one obtains

$$\frac{dT}{dx} = -\left[\frac{2K_1}{5k\delta}(T^5 - T_a^5) - \frac{2K_2}{k\delta}(T - T_a)\right]^{1/2}$$

or after algebraic adjustment,

$$\frac{dT}{dx} = -\left(\frac{2K_1T_a^5}{5k\delta}\right)^{1/2}\left(\frac{T}{T_a} - 1\right)^{1/2}\Upsilon \qquad (13.19)$$

where

$$\Upsilon = \left[\left(\frac{T}{T_a}\right)^4 + \left(\frac{T}{T_a}\right)^3 + \left(\frac{T}{T_a}\right)^2 + \frac{T}{T_a} + 1 - \frac{5K_2}{K_1T_a^4}\right]^{1/2}$$

Now make the transformation

$$v \equiv \frac{T}{T_a} - 1 \qquad (13.20)$$

so that

$$\frac{T}{T_a} = 1 + v^2 \qquad (13.21)$$

and

$$dT = 2T_a v\, dv \qquad (13.22)$$

Equations (13.20) through (13.22) may be substituted into eq. (13.19), and after algebraic manipulation to separate the variables, the result is

$$\int_{v_b}^{v_a} \frac{dv}{[(1+v^2)^4+(1+v^2)^3+(1+v^2)^2+(1+v^2)+1-5K_2/K_1T_a^4]^{1/2}}$$

$$= -\int_0^b \left(\frac{K_1T_a^3}{10k\delta}\right)^{1/2} dx = -b\left(\frac{K_1T_a^3}{10k\delta}\right)^{1/2} \tag{13.23}$$

where the limits are

$$v(x=0) = v_b = (Z-1)^{1/2}$$

where $Z = T_b/T_a$ and where

$$v(x=b) = v_a = 0$$

When the limits of integration on the left of eq. (13.23) are reversed, the minus sign on the right may be removed. The integral on the left may be evaluated graphically by plotting the reciprocal of the denominator against the variable v, as shown in Fig. 13.5. The area under each curve will be equal numerically to the parameter on the right of eq. (13.23). Note that the limits of integration are $v_b = (Z-1)^{1/2}$ and $v_a = 0$. Associated with the limit, v_b is a value of Z. Hence one may actually plot

$$b\left(\frac{K_1T_a^3}{10k\delta}\right)^{1/2} = f(Z)$$

This has been carried out in the preparation of Fig. 13.6.

The use of Fig. 13.6 is limited because the ordinate and family of parameters are functions of the fin tip temperature T_a. The ordinate may be adjusted by multiplication by $(5Z^3)^{1/2}$:

$$\zeta = b\left(\frac{K_1T_b^3}{2k\delta}\right)^{1/2} = b(5Z^3)^{1/2}\left(\frac{K_1T_a^3}{10k\delta}\right)^{1/2}$$

which may be referred to as the *profile number* for the rectangular fin.

The family parameter may be adjusted by multiplying by $1/5Z^4$ so that the new value becomes

$$\frac{K_2}{K_1T_b^4} = \frac{1}{5Z^4}\frac{5K_2}{K_1T_a^4}$$

Thus Fig. 13.6 may be adjusted by multiplying values read from the curves by factors containing the parameter Z. Such a procedure leads to the development of Fig. 13.7, which facilitates design and performance calculations.

The heat transferred from the fin faces will be equal to the quantity of heat entering the fin at its base:

$$q_b = -k\delta L \left.\frac{dT}{dx}\right|_{x=0}$$

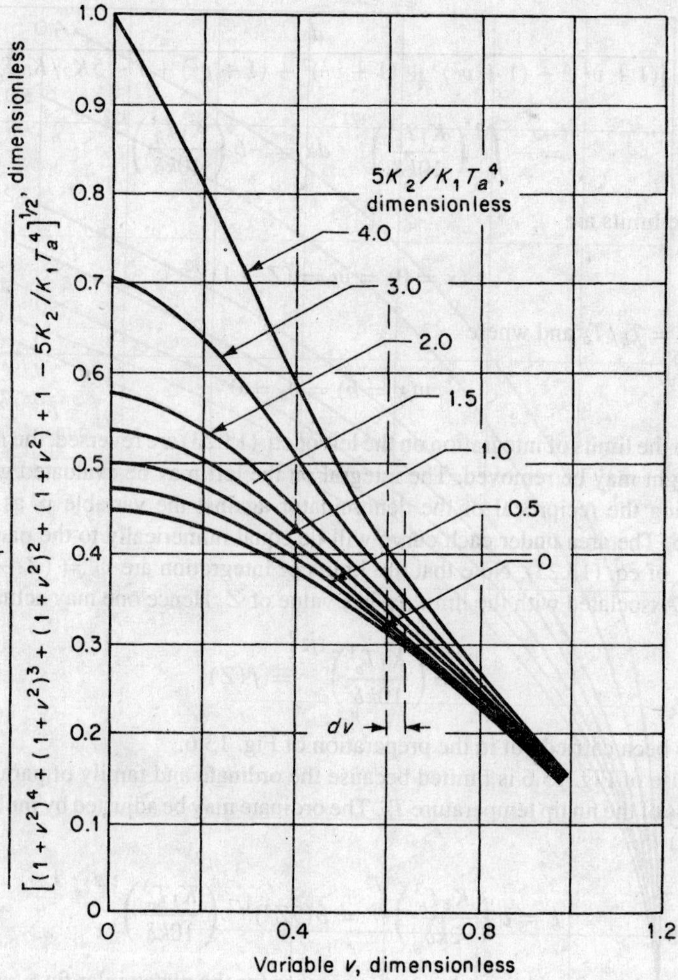

Figure 13.5 Curves for the graphical evaluation of eq. (13.23) for a longitudinal fin of rectangular profile radiating to non-free space.

The derivative is given by eq. (13.19) and is evaluated at $x = 0$ where $T = T_b$. The heat transferred becomes

$$q_b = k\delta L \left[\frac{2K_1 T_a^5}{5k\delta}(Z-1)\left(\Phi - \frac{5K_2}{K_1 T_a^4}\right)\right]^{1/2} \tag{13.24}$$

where

$$\Phi = Z^4 + Z^3 + Z^2 + Z + 1$$

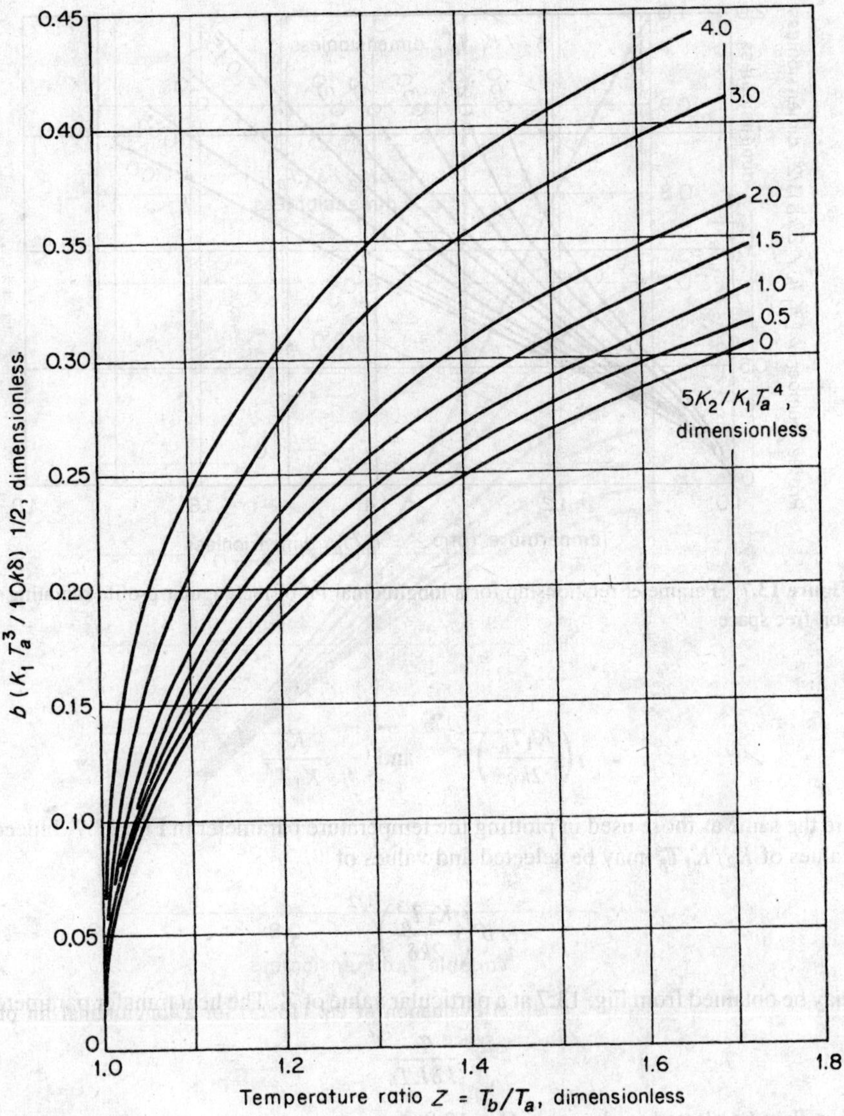

Figure 13.6 Parameter relationship for a longitudinal fin of rectangular profile radiating to non-free space.

Rearrangement of terms in eq. (13.24) gives

$$\frac{q_b}{k\delta L T_b} = b\left\{\frac{K_1 T_b^3}{2k\delta}\left[\frac{4(Z-1)}{5Z^5}\right]^{1/2}\left(\Phi - 5Z^4\frac{K_2}{K_1 T_b^4}\right)\right\}^{1/2} \tag{13.25}$$

Observe that the parameters

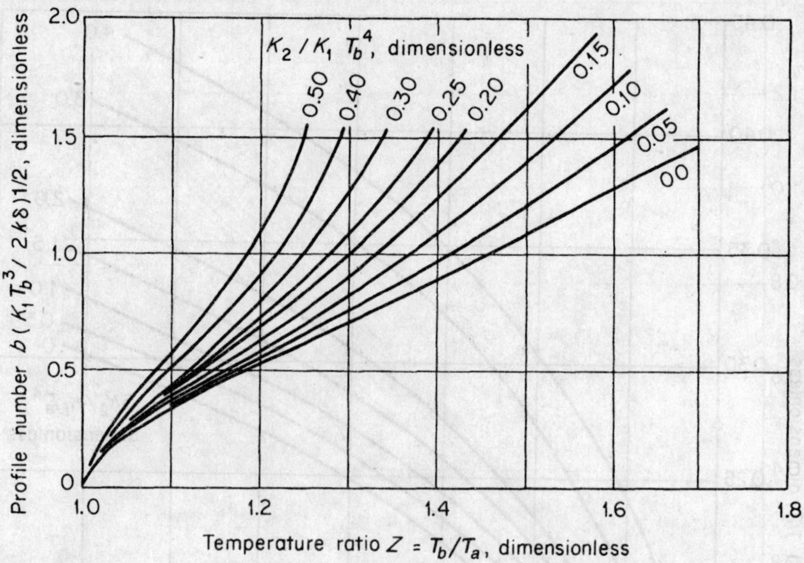

Figure 13.7 Parameter relationship for a longitudinal fin of rectangular profile radiating to non-free space.

$$\left(\frac{K_1 T_b^3}{2k\delta}\right)^{1/2} \quad \text{and} \quad \frac{K_2}{K_1 T_b^4}$$

are the same as those used in plotting the temperature parameter in Fig. 13.7. Indeed, values of $K_2/K_1T_b^4$ may be selected and values of

$$b\left(\frac{K_1 T_b^3}{2k\delta}\right)^{1/2}$$

may be obtained from Fig. 13.7 at a particular value of Z. The heat transfer parameter

$$\frac{q_b}{k\delta L T_b}$$

may then be plotted as shown in Fig. 13.8.

The fin efficiency is the ratio of the actual heat dissipated to the ideal heat dissipated if the entire fin were to operate at the base temperature T_b with no heat received from the environment ($K_2 = 0$). The actual heat dissipation is given by eq. (13.25). With $K_1 = 2\sigma\epsilon$,

$$\eta = \frac{q_b}{2\sigma\epsilon b L T_b^4}$$

This may be modified to accommodate the ordinate and abscissa values shown in Fig. 13.8. Multiplying numerator and denominator by $b/k\delta L T_b$ yields

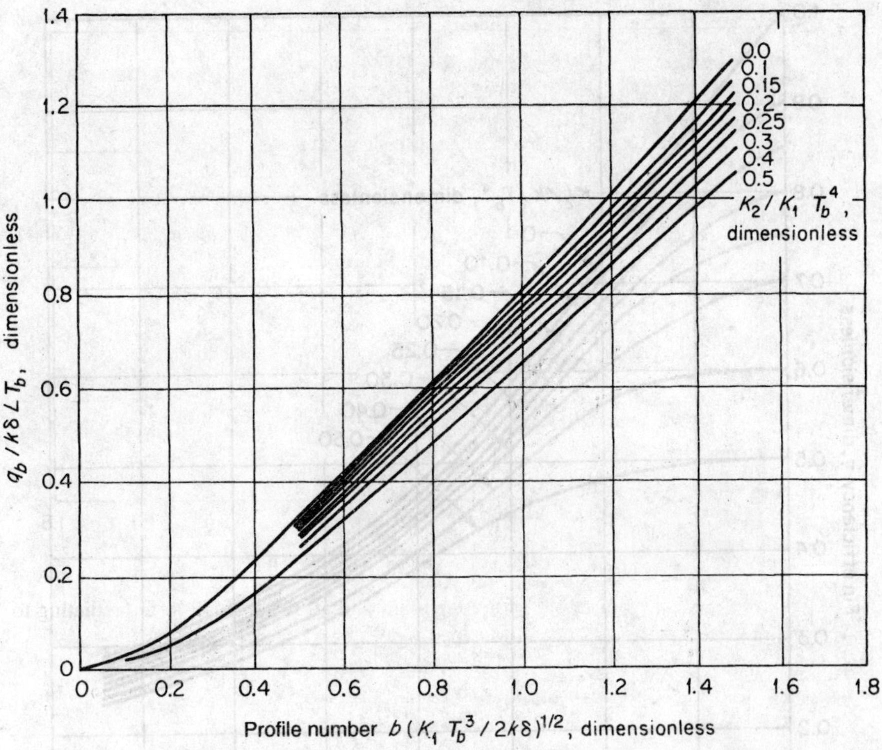

Figure 13.8 Heat flow parameter relationship for a longitudinal fin of rectangular profile radiating to non-free space.

$$\eta = \frac{q_b b / k\delta L T_b}{2\sigma\epsilon b^2 T_b^3 / k\delta} = \frac{q_b b / k\delta L T_b}{\left[b\left(2\sigma\epsilon T_b^3 / k\delta\right)^{1/2}\right]^2} \tag{13.26}$$

It is seen that the numerator is the ordinate and the denominator is twice the square of the abscissa of Fig. 13.8. The fin efficiency is plotted in Fig. 13.9.

Example 13.3: Radiation to Non-Free Space (Fin Analysis). It is the intent of this example to show that Figs. 13.7, 13.8, and 13.9 can be used for the free-space case. A good basis for comparison is Example 13.1. All that is necessary is to set $K_2 = 0$ so that $K_2/K_1 T_b^4 = 0$.

A longitudinal fin of rectangular profile, 4 m long, 50 cm high, and 0.635 cm thick is fabricated of magnesium ($k = 152$ W/m·K). Its surface has been treated so that its emissivity is 0.85. For a base temperature of 350 K (77°C), determine **(a)** the tip temperature, **(b)** the efficiency, and **(c)** the heat dissipation.

SOLUTION. **(a)** For the tip temperature with $K_1 = 2\sigma\epsilon$, form the profile number

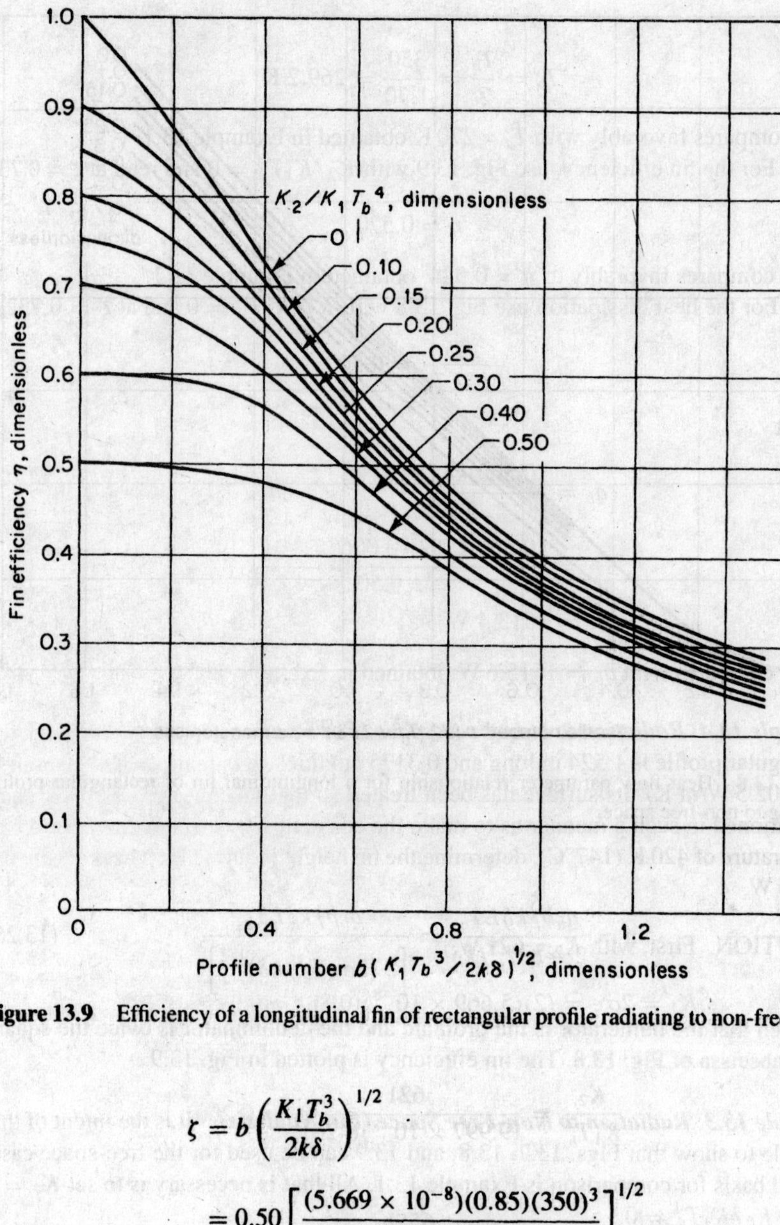

Figure 13.9 Efficiency of a longitudinal fin of rectangular profile radiating to non-free space.

and

$$\zeta = b\left(\frac{K_1 T_b^3}{2k\delta}\right)^{1/2}$$

Then

$$= 0.50\left[\frac{(5.669 \times 10^{-8})(0.85)(350)^3}{(152)(0.00635)}\right]^{1/2}$$

$$= 0.732$$

From Fig. 13.7, at $\zeta = 0.732$ and $K_2/K_1T_b^4 = 0$, read

$$Z = \frac{T_b}{T_a} = 1.30$$

and

$$T_a = \frac{T_b}{Z} = \frac{350}{1.30} = 269.2\ \text{K}$$

This compares favorably with $T_a = 270$ K obtained in Example 13.1.

(b) For the fin efficiency, use Fig. 13.9 with $K_2/K_1T_b^4 = 0$ and read at $\zeta = 0.732$,

$$\eta = 0.520$$

which compares favorably to $\eta = 0.524$ obtained in Example 13.1.

(c) For the heat dissipation use Fig. 13.8 with $K_2/K_1T_b^4 = 0$ and at $\zeta = 0.732$,

$$\frac{q_b}{k\delta L T_b} = 0.560$$

so that

$$q_b = \frac{(0.560)k\delta L T_b}{b}$$

$$= \frac{(0.560)(152)(0.00635)(4)(350)}{0.50}$$

$$= 1513.4\ \text{W}$$

which compares with $q_b = 1515.6$ W obtained in Example 13.1.

Example 13.4: Radiation to Non-Free Space (Fin Synthesis). A longitudinal fin of rectangular profile is 1.524 m long and 0.3175 cm thick. It is fabricated of aluminum ($k = 202.5$ W/m·K). Its surface has been treated so that its emissivity is 0.88, and it is positioned in such a manner as to make the constant $K_2 = 621\ \text{W/m}^2$. For a base temperature of 420 K (147°C), determine the fin height required for a heat dissipation of 655 W.

SOLUTION. First, with $K_2 = 621$ W,

$$K_1 = 2\sigma\epsilon = (2)(5.669 \times 10^{-8})(0.88) = 9.997 \times 10^{-8}$$

and

$$\frac{K_2}{K_1T_b^4} = \frac{621}{(9.997 \times 10^{-8})(420)^4} = 0.200$$

Then

$$\frac{q_b b}{k\delta L T_b} = \frac{655b}{(202.5)(0.003175)(1.524)(420)} = 1.592b$$

and

$$\zeta = b\left(\frac{K_1T_b^3}{2k\delta}\right)^{1/2} = b\left[\frac{(9.997 \times 10^{-8})(420)^3}{(2)(202.5)(0.003175)}\right]^{1/2} = 2.398b$$

Trial I. Assume that $b = 0.50$ m and obtain

$$\frac{q_b b}{k\delta L T_b} = (1.592)(0.50) = 0.796$$

and at this value, Fig. 13.8 gives $\zeta = 1.02$. But

$$b = \frac{\zeta}{2.398} = \frac{1.02}{2.398} = 0.425$$

and the two values of b do not match.

Trial II. Assume that $b = 0.30$ m and obtain

$$\frac{q_b b}{k\delta L T_b} = (1.592)(0.30) = 0.478$$

and Fig. 13.8 gives $\zeta = 0.68$. But

$$b = \frac{\zeta}{2.398} = \frac{0.68}{2.398} = 0.284$$

and the two values of b almost match. One more try is in order.

Trial III. Assume that $b = 0.25$ m and obtain

$$\frac{q_b b}{k\delta L T_b} = (1.592)(0.25) = 0.398$$

and Fig. 13.8 gives $\zeta = 0.59$. But

$$b = \frac{\zeta}{2.398} = \frac{0.59}{2.398} = 0.247$$

and the two values of b match.

The fin height should be $b = 0.875$ m.

13.3 LONGITUDINAL RADIATING FINS OF TRAPEZOIDAL AND TRIANGULAR PROFILE

The terminology and coordinate system for the longitudinal fin of trapezoial profile is displayed in Fig. 13.10. The triangular profile is an extension of the trapezoidal profile when the tip thickness $\delta_a = 0$. Note that the coordinate system has its origin at the fin tip and the height coordinate is taken positive in a direction toward the fin base.

Mackay and Bacha (1961) were among the first to treat the radiating longitudinal fin of trapezoidal profile. As in previous cases, the differential equation for the temperature distribution is obtained from a heat balance on the differential height element dx. The difference in heat entering and leaving by conduction is

$$dq = kL\frac{d}{dx}\left[\delta(x)\frac{dT}{dx}\right]dx \tag{13.27}$$

This is the same as the heat leaving the element by radiation:

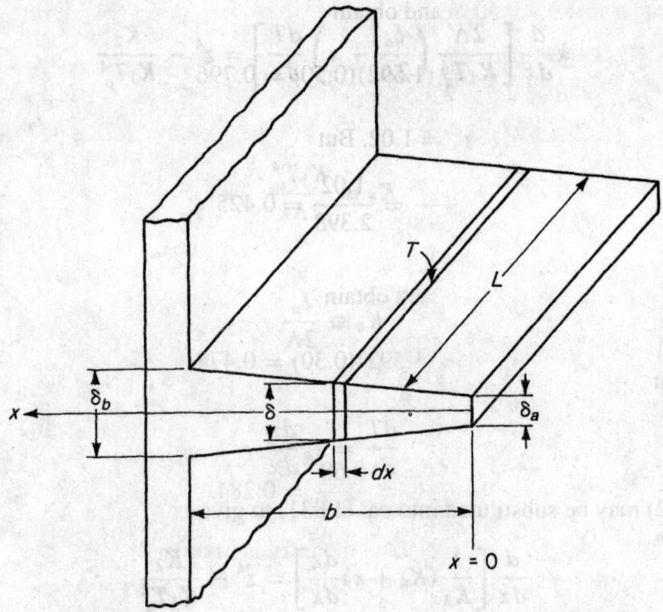

Figure 13.10 Coordinate system for a longitudinal fin of trapezoidal profile. The triangular profile is a special case when $\delta_a = 0$.

$$dq = (K_1 T^4 - K_2) L\,dx \tag{13.28}$$

The resulting nonlinear differential equation is

$$k \frac{d}{dx}\left[\delta(x) \frac{dT}{dx}\right] = K_1 T^4 - K_2 \tag{13.29}$$

From the geometry in Fig. 13.10 it is seen that

$$\delta(x) = \delta_a + (\delta_b - \delta_a)\frac{x}{b} = 2\Lambda\left(\frac{\delta_a}{2\Lambda} + x\right) \tag{13.30}$$

where the *fin taper* is

$$\Lambda \equiv \frac{\delta_b - \delta_a}{2b}$$

Let

$$z \equiv \frac{T}{T_a}$$

and substitute eq. (13.30) into eq. (13.29). After dividing by $K_1 T_a^4$ and algebraic adjustment, one obtains

$$k\frac{d}{dx}\left[\frac{2\Lambda}{K_1T_a^4}\left(\frac{\delta_a}{2\Lambda}+x\right)\frac{dT}{dx}\right]=z^4-\frac{K_2}{K_1T_a^4} \tag{13.31}$$

Then let

$$K_3 \equiv \frac{K_1T_a^4}{2\Lambda k} \tag{13.32a}$$

and

$$K_4 \equiv \frac{\delta_a}{2\Lambda} \tag{13.32b}$$

Noting that

$$\frac{dT}{dx}=T_a\frac{dz}{dx}$$

eqs. (13.32) may be substituted into eq. (13.31) to give

$$\frac{d}{dx}\left[\frac{1}{K_3}(K_4+x)\frac{dz}{dx}\right]=z^4-\frac{K_2}{K_1T_a^4} \tag{13.33}$$

Now introduce the variable

$$w \equiv K_3(K_4+x) \tag{13.34}$$

Its derivative is $dw/dx = K_3$, and dz/dx can be represented in terms of dz/dw as

$$\frac{dz}{dx}=\frac{dz}{dw}\frac{dw}{dx}=K_3\frac{dz}{dw}$$

With this in eq. (13.33), the result is

$$\frac{d}{dw}\left(w\frac{dz}{dw}\right)-z^4+\frac{K_2}{K_1T_a^4}=0$$

Because z is a function of w, this may be written, with primes denoting derivatives, as

$$wf''(w)-f'(w)-[f(w)]^4+\frac{K_2}{K_1T_a^4}=0 \tag{13.35}$$

Equation (13.35) is the governing differential equation for the temperature profile of the fin. Its solution may be obtained via a numerical integration procedure that has starting points as shown in Table 13.1. The numerical integration of eq. (13.35) will yield a set of values and curves of $z = f(w)$ as a function of w. Typical plots as functions of $K_2/K_1T_a^4$ and $w_a = K_3K_4$ (or zero for the triangular profile) are shown in Fig. 13.11. At the same time, the computer may be used to generate values of $f'(w)$, $f''(w)$, and $\int_{w_a}^{w} dw$, all as functions of w. These will be seen to be most useful.

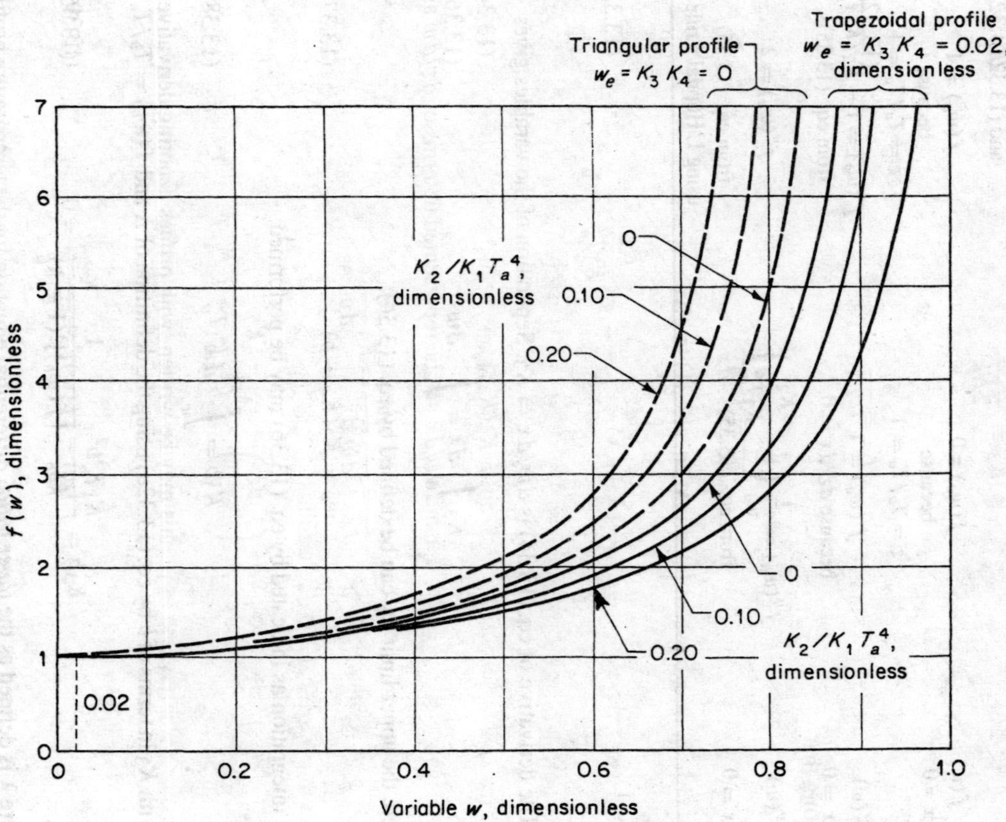

Figure 13.11 Relationship of $f(w)$ to w for longitudinal fins of trapezoidal and triangular profiles.

TABLE 13.1 Starting Points for Numerical Integration of Eq. (13.35).

Item	Trapezoidal Profile: δ_a Is Finite	Triangular Profile: $\delta_a = 0$
w at $x = 0$	$w = w_a = K_3 K_4$ from eq. (13.34)	$w = w_a = 0$ from eqs. (13.34) and (13.32b)
$z = f(w)$ at $x = 0$	$f(w_a) = 0$ because $z = T_a/T_a = 1$	$f(w_a) = 1$ because $z = T_a/T_a = 1$
$f'(w)$ at $x = 0$	$f'(w_a) = 1$ because $dz/dx = 0$	$f'(w_a) = 1 - K_2/K_1 T_a^4$ from eq. (13.35)
$f''(w)$ at $x = 0$	$f''(w_a) = \dfrac{1}{w_a}\left(1 - \dfrac{K_2}{K_1 T_a^4}\right)$ from eq. (13.35)	$f''(w_a) = 2$ from eq. (13.35) using L'Hôpital's rule

The derivative of eq. (13.34) is $dw/dx = K_3$. Separation of the varables gives

$$K_3 \int_0^b dx = \int_{w_a}^{w_b} dw \tag{13.36}$$

where the upper limit w_b can be defined by eq. (13.34):

$$w_b = K_3(K_4 + b) \tag{13.37}$$

The integration as indicated by eq. (13.36) may be performed:

$$K_3 b = \int_{w_a}^{w_b} dw \tag{13.38}$$

or with K_3 in terms of b by eq. (13.32a) using the definition of Λ and $f(w_b) = T_b/T_a$:

$$K_3 b = \frac{K_1 T_b^3 b^2}{k\delta_b} \frac{1}{[f(w_b)]^3(1 - \lambda)} \tag{13.39}$$

where λ is defined as the *taper ratio*[2]

$$\lambda = \frac{\delta_a}{\delta_b} \tag{13.40}$$

The first term on the right-hand side in eq. (13.39) may be designated as the *profile number* for the trapezoidal or triangular fin:

[2]Not the fin taper, Λ,

$$\zeta \equiv \frac{K_1 T_b^3 b^2}{k\delta_b} \tag{13.41}$$

which is dimensionless. Then a combination of eqs. (13.38), (13.39), and (13.41) gives

$$\zeta = [f(w_b)]^3(1-\lambda)\int_{w_a}^{w_b} dw \tag{13.42}$$

The combining of eqs. (13.30), (13.32*a*), (13.37) (13.38), and (13.40) gives

$$\frac{1}{\lambda} = \frac{\delta_b}{\delta_a} = 1 + \frac{1}{w_a}\int_{w_a}^{w_b} dw \tag{13.43}$$

Equation (13.43) may be substituted into eq. (13.42) to give a complete representation of the profile number in terms of parameters generated by the computer:

$$\zeta = \frac{[f(w_b)]^3\left(\int_{w_a}^{w_b} dw\right)^2}{w_a + \int_{w_a}^{w_b} dw} \tag{13.44}$$

The heat flowing into the fin may be obtained from·

$$q_b = k\delta_b L \left.\frac{dT}{dx}\right|_{x=b}$$

or in terms of the transformed variable w,

$$q_b = k\delta_b L T_a K_3[f'(w_b)] \tag{13.45}$$

The fin efficiency is

$$\eta = \frac{q_b}{q_{\text{id}}} = \frac{k\delta_b T_a K_3[f'(w_b)]}{K_1 b T_b^4} \tag{13.46}$$

and a combination of eqs. (13.30), (13.38), (13.44), and (13.46) provides an expression for the fin efficiency in terms of the parameters generated by the computer:

$$\eta = \frac{f'(w_b)\left(w_a + \int_{w_a}^{w_b} dw\right)}{[f(w_b)]^4\int_{w_a}^{w_b} dw} \tag{13.47}$$

The numerical solution to eq. (13.35) is the key to the development of design charts for the fin efficiency as a function of the profile number. The reasoning is:

1. The value of $w_a = K_3K_4$, which is used as a starting point, basically fixes the fin taper.

2. The numerical solutions, which are based on a fixed value of $K_2/K_1T_a^4$, yield at a value of $Z = T_b/T_a = f(w_b)$ a parameter that is more useful:

$$\frac{K_2}{K_1T_b^4} = \frac{1}{Z^4}\frac{K_2}{K_1T_a^4}$$

This is the environmental parameter used in Figs. 13.7 to 13.9 for the rectangular profile.

3. For any value of $Z = f(w_b)$ and $w_a = K_3K_4$, the environmental parameter $K_2/K_1T_b^4$ is fixed, and the profile number and fin efficiency may be calculated from eqs. (13.44) and (13.47), respectively.
4. Design plots with the fin efficiency as the ordinate as a function of the profile number as the abscissa may be obtained for several values as the environmental parameter $K_2/K_1T_b^4$. Each set of plots is for a particular value of $\lambda = \delta_a/\delta_b$.

Figures 13.12 through 13.15 are the result of this procedure. These figures are for the trapezoidal fin with $\lambda = 0.25, 0.50$, and 0.75. Because of computational difficulties, the triangular profile (Fig. 13.15) with a theoretical $\lambda = 0.00$ was obtained using $\lambda = 0.01$. The curves designated *theoretical optimum* have been inserted to show the ideal for a design. These optima occur when the slope of each of the curves is $-\frac{1}{3}$, and this is derived in Section 14.2.2.

Example 13.5: Radiation to Free and Non-Free Space. A longitudinal fin of trapezoidal profile is fabricated of a lampblack-coated steel fin (k = 30 W/m·K, $\epsilon = 0.95$) is 15.24 cm high, 0.9525 cm thick at the base, 0.4763 cm thick at the tip, and 1.3716 m long, and the base temperature is 445 K. How much heat is dissipated by the fin (**a**) to free space, and (**b**) if the fin receives 845 W from the surroundings?

SOLUTION. Here, in both cases,

$$\lambda = \frac{\delta_a}{\delta_b} = \frac{0.4763}{0.9525} = 0.50$$

$$K_1 = 2\sigma\epsilon = (2)(5.669 \times 10^{-8})(0.95) = 1.077 \times 10^{-7}$$

and

$$\zeta = \frac{K_1T_b^3b^2}{k\delta_b}$$

$$= \frac{(1.077 \times 10^{-7})(445)^3(0.1524)^2}{(30)(0.009525)}$$

$$= 0.771$$

(**a**) For free space, use Fig. 13.13 and read at $\zeta = 0.771$ and $K_2/K_1/T_b^4 = 0$

$$\eta = 0.548$$

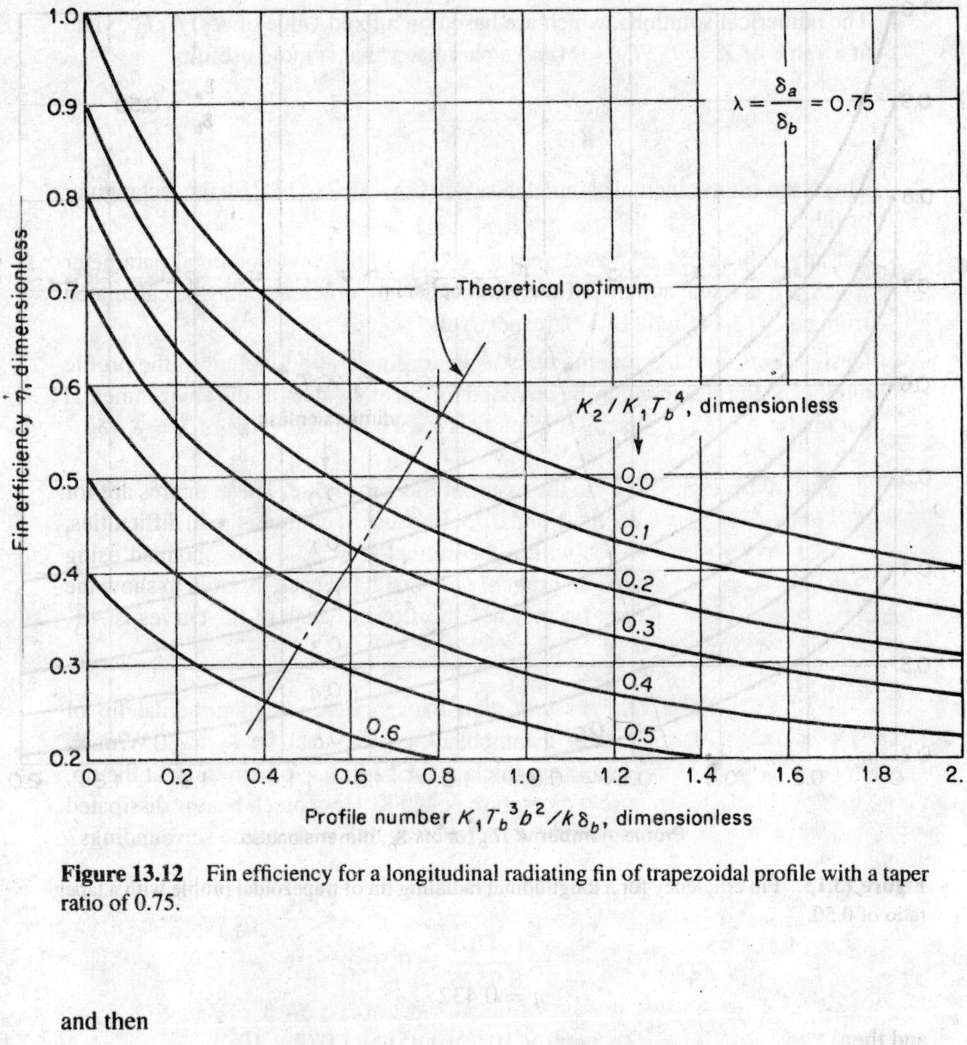

Figure 13.12 Fin efficiency for a longitudinal radiating fin of trapezoidal profile with a taper ratio of 0.75.

and then

$$
\begin{aligned}
q_b &= K_1 L b T_b^4 \eta \\
&= (1.077 \times 10^{-7})(1.3716)(0.1524)(445)^4(0.548) \\
&= 483.8 \text{ W}
\end{aligned}
$$

(b) If the fin receives 845 W from the surroundings, the environmental parameter will be

$$
\frac{K_2}{K_1 T_b^4} = \frac{845}{(1.077 \times 10^{-7})(445)^4} = 0.20
$$

and from Fig. 13.13, read at $\zeta = 0.771$ and $K_2/K_1/T_b^4 = 0.20$,

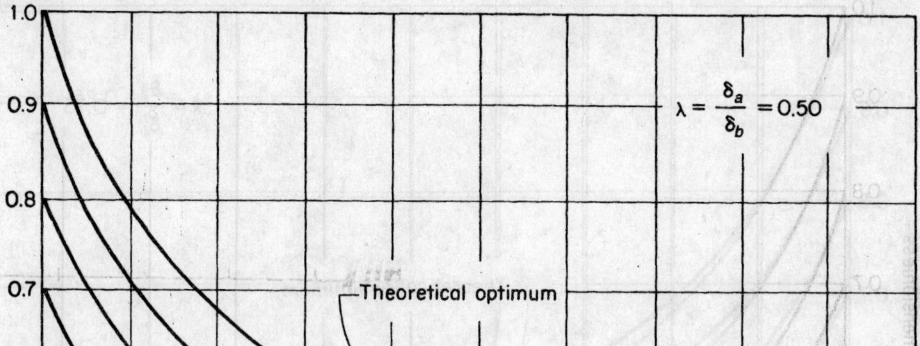

Figure 13.13 Fin efficiency for a longitudinal radiating fin of trapezoidal profile with a taper ratio of 0.50.

$$\eta = 0.432$$

and then

$$\begin{aligned} q_b &= K_1 L b T_b^4 \eta \\ &= (1.077 \times 10^{-7})(1.3716)(0.1524)(445)^4(0.432) \\ &= 381.4 \text{ W} \end{aligned}$$

13.4 USE OF THE CASCADE ALGORITHM

Smith (1992) has used the cascade algorithm

$$Y_{\text{in}} = \frac{\tau_{22} + \tau_{21}(q_a/\theta_a)}{\tau_{11} + \tau_{12}(q_a/\theta_a)} \tag{7.6}$$

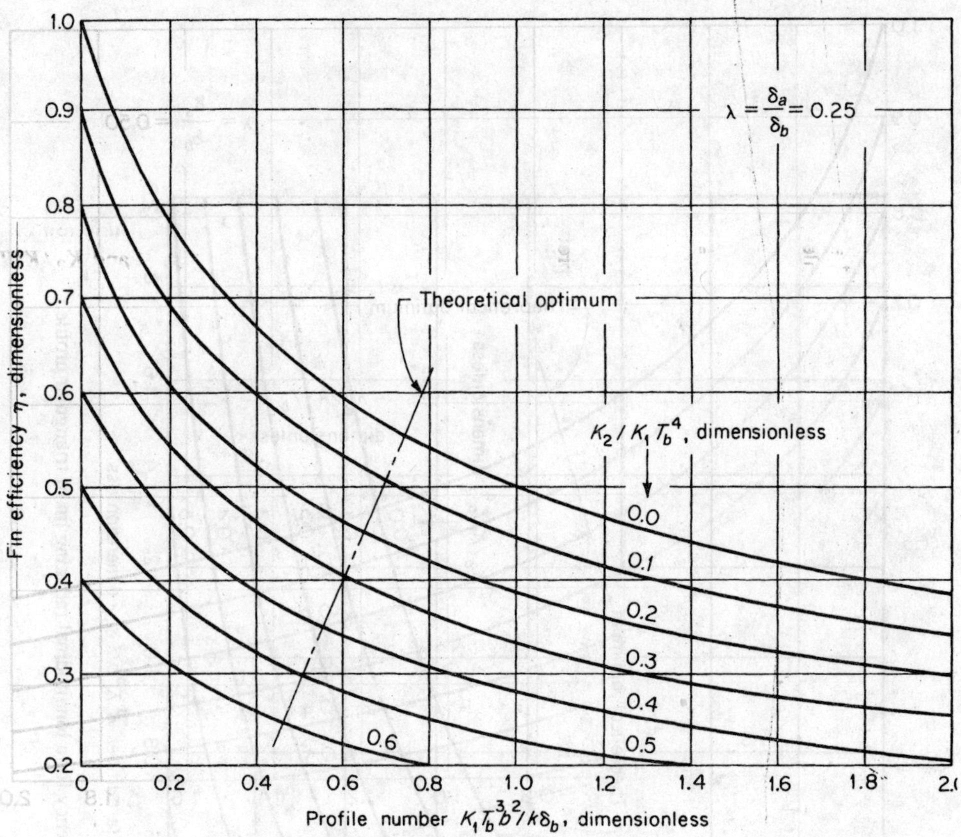

Figure 13.14 Fin efficiency for a longitudinal radiating fin of trapezoidal profile with a taper ratio of 0.25.

to predict the performance of longitudinal radiating fins of rectangular, trapezoidal, and triangular profile. His analysis method divides the fin under consideration into 100 subfins and uses eq. (6.6) to effectively cascade the 100 subfins into an overall single entity.

The heat transfer coefficient employed derives from the radiative dissipation from an element of fin with surface area $2L\,dx$ (two-sided dissipation)

$$dq_r = \sigma F_a F_\epsilon (2L\,dx)(T_{av}^4 - T_s^4)$$

where T_{av} is the average temperature of the fin surface element. For a small body in a large enclosure (the fin in outer space), the shape factor $F_A = 1$, and the emissivity factor $F_\epsilon = \epsilon$. Thus with T_s effectively at 0 K,

$$dq_r = 2\sigma \epsilon L T_{av}^4\,dx$$

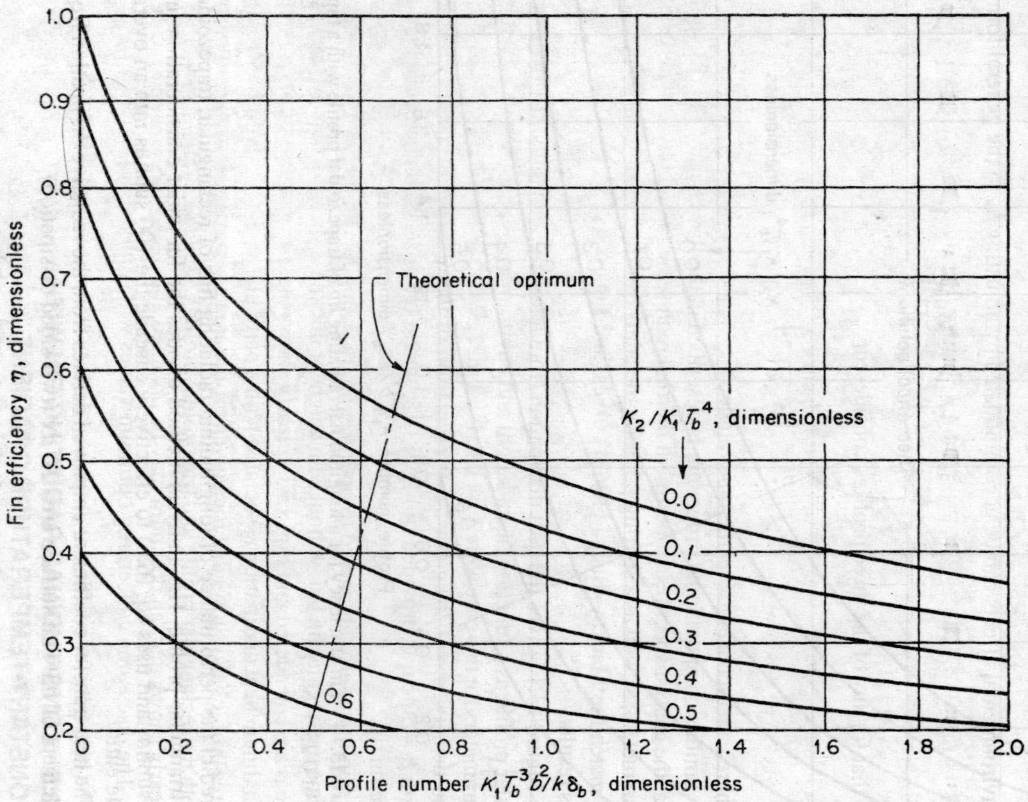

Figure 13.15 Fin efficiency for a longitudinal radiating fin of triangular profile.

and a radiative heat transfer coefficient may be proposed to fit

$$dq_r = h_r dST_{av}^3$$

where $dS = 2L\,dx$ and where

$$h_r = \sigma\epsilon T_{av}^3$$

Environmental effects can be accommodated by adjusting h_r to the general form

$$h_r = \sigma\epsilon\left(T_{av}^3 - \frac{K_2}{T_{av}}\right)$$

so that for any of the three profiles considered,

$$m = \left(\frac{h\sin\kappa}{k}\right)^{1/2}$$

where $\kappa = \arctan \delta_b/2b$ is the taper angle.

Smith's computational procedure is described in the flowchart of Fig. 13.16. Note that the analyst specifies all pertinent thermal requirements and physical and thermal parameters, as well as a starting point for the analysis. Moreover, Smith has been able to match the Mackay–Bacha (1962) curves, and Figs. 13.17 to 13.20 show some of his results.

Figure 13.17 gives the heat dissipation from a longitudinal radiating fin of rectangular profile as a function of fin height and environmental parameter, $K_2/K_1T_a^4$. Fin parameters for this case were $\delta = 1$ cm, $L = 1$ m, $T_b = 400$ K, $\epsilon = 0.85$, $k = 209.4$ W/m·K, and density, $\rho = 0.26$ kg/cm^3. A presentation of the temperature profile for this fin with $b = 50$ cm is provided in Fig. 13.18, where the abscissa reckons distance from the fin tip.

Smith has also attempted to optimize the radiating fin. Figure 13.19 considers the same rectangular fin but without a fin height or thickness specification. Several profile areas are considered and Fig. 13.19 clearly shows that for each profile area, there is a maximum heat dissipation associated with a particular width. Figure 13.20 compares three profiles and shows that for a given profile area, the triangular profile outperforms both the rectangular profile and trapezoidal profiles, with taper ratio $\lambda = 0.50$. It also clearly shows that, an optimum may be obtained at a particular fin width.

The discussion concerning optimum values will continue in Chapter 14.

13.5 LONGITUDINAL RADIATING FIN WITH CONSTANT-TEMPERATURE GRADIENT

Mackay (1960) took the idea from the convective case that the longitudinal radiating fin of least material would be a fin with a constant-temperature gradient and studied a radiating fin with a linear temperature profile

$$T = T_a + \frac{x}{b}(T_b - T_a) \tag{13.48}$$

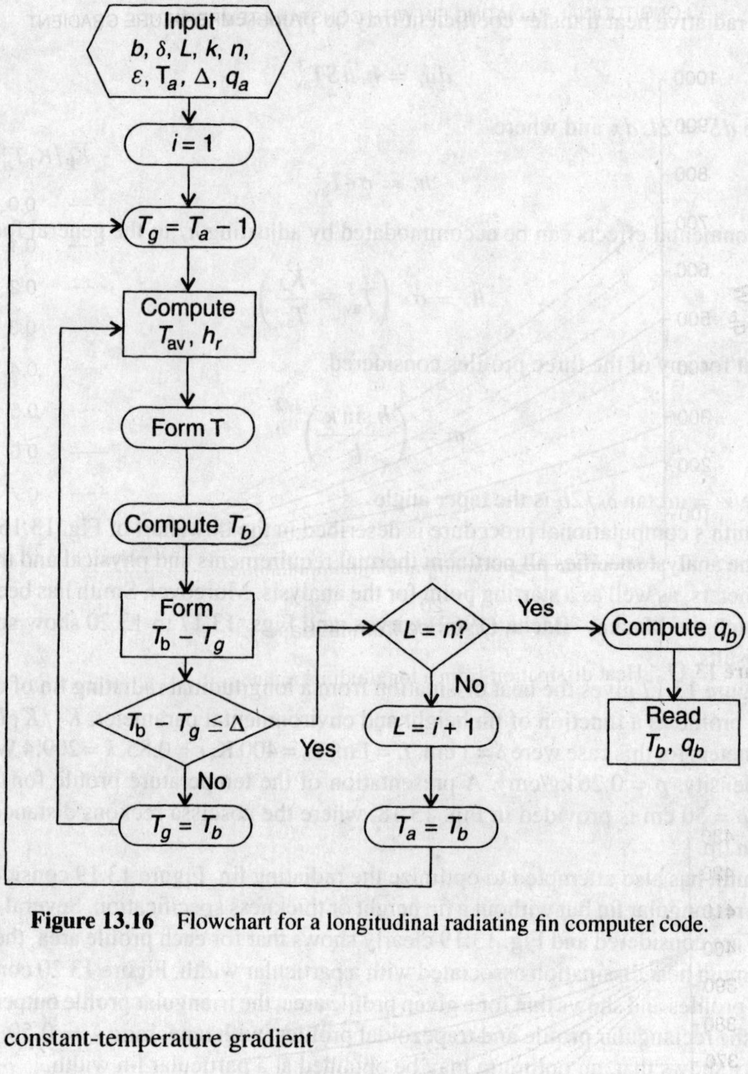

Figure 13.16 Flowchart for a longitudinal radiating fin computer code.

with a constant-temperature gradient

$$\frac{dT}{dx} = \frac{T_b - T_a}{b} \tag{13.49}$$

This fin does not represent the contour of least material, but it entails only a slight increase in weight over the profile described by Wilkins (1960*a*). However, it does not have a requirement that the tip temperature be zero.

Suppose that the fin with constant-temperature gradient possesses the arbitrary profile shown in Fig. 13.21. The origin of the height coordinate x is at the fin tip and the heat conducted past any plane located at x will be

$$q_x = 2kf_2(x)L\frac{dT}{dx}$$

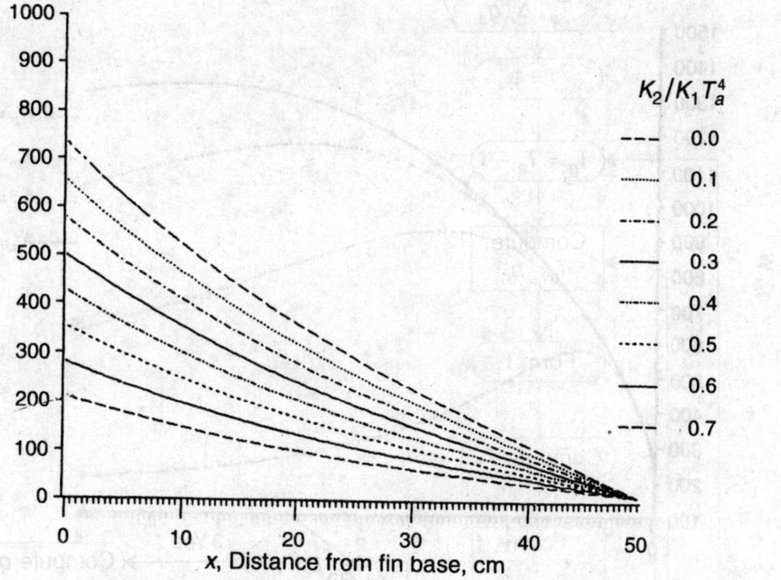

Figure 13.17 Heat dissipation from a longitudinal radiating fin of rectangular profile.

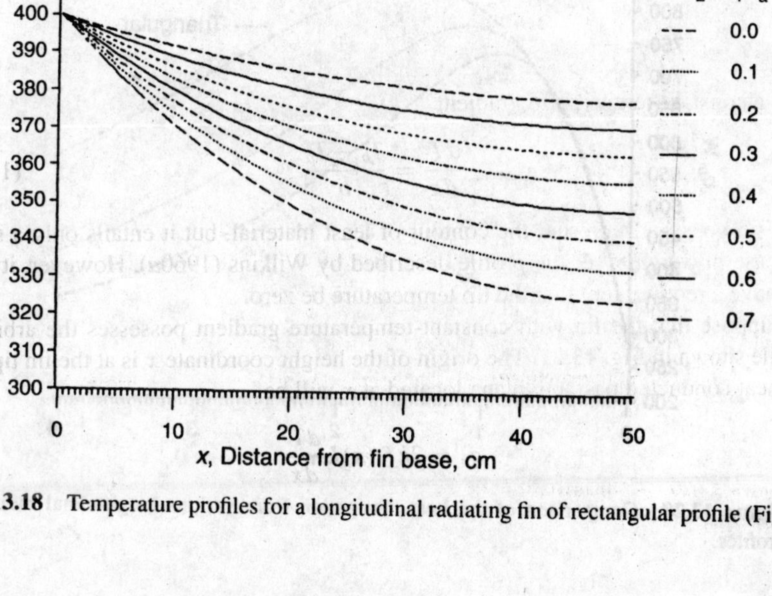

Figure 13.18 Temperature profiles for a longitudinal radiating fin of rectangular profile (Fig. 13.17).

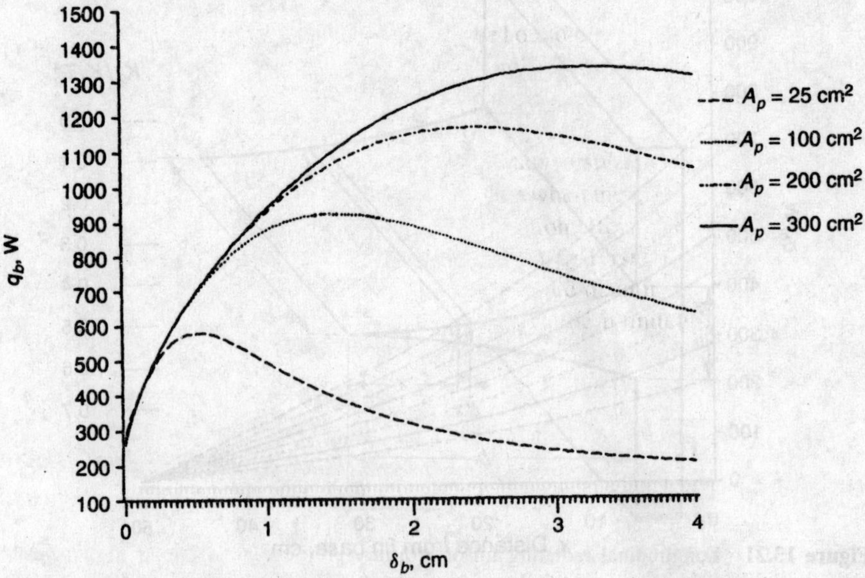

Figure 13.19 Heat dissipation from a longitudinal radiating fin of rectangular profile as a function of fin width for several profile areas.

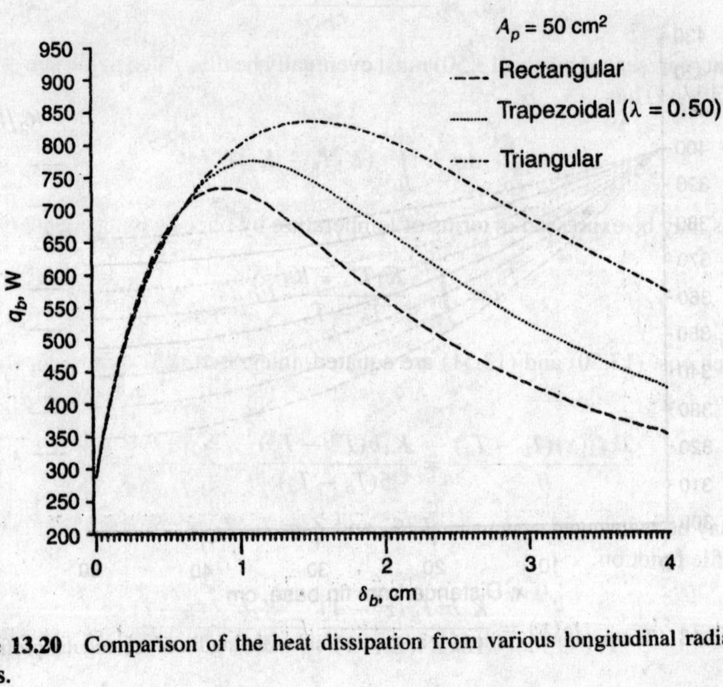

Figure 13.20 Comparison of the heat dissipation from various longitudinal radiating fin profiles.

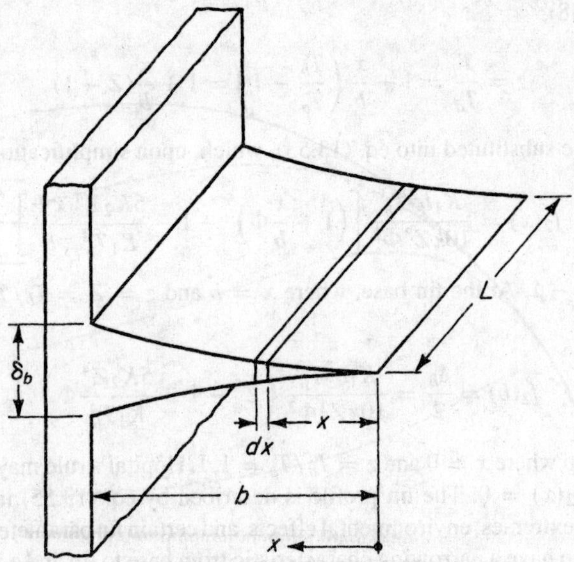

Figure 13.21 Longitudinal radiating fin of arbitrary profile for analysis of the constant-temperature gradient.

If this is substituted into eq. (13.49), the result is

$$q_x = \frac{2kf_2(x)L(T_b - T_a)}{b} \tag{13.50}$$

The heat represented by eq. (13.50) must eventually be dissipated to the surroundings by radiation. Thus

$$\int_0^q dq = \int_0^x (K_1T_4 - K_2)L\,dx$$

and this may be expressed in terms of temperature by once again using eq. (13.49):

$$q = \int_{T_a}^{T} \frac{K_1T^4 - K_2}{T_b - T_a} Lb\,dT \tag{13.51}$$

When eqs. (13.50) and (13.51) are equated, integrated, and simplified, the result is

$$\frac{2kf_2(x)(T_b - T_a)}{b} = \frac{K_1b(T^5 - T_a^5)}{5(T_b - T_a)} - \frac{K_2b(T - T_a)}{T_b - T_a} \tag{13.52}$$

This may be rearranged using $z = T/T_a$ and $Z = T_b/T_a$ to provide a solution for the fin profile function:

$$f_2(x) = \frac{K_1b^2T_b^3(z^5 - 1)}{10kZ^3(Z - 1)^2} - \frac{K_2b^2Z(z - 1)}{2kT_b(Z - 1)^2} \tag{13.53}$$

From eq. (13.48),

$$z = \frac{T}{T_a} = 1 + \frac{x}{b}\left(\frac{T_b}{T_a} - 1\right) = 1 + \frac{x}{b}(Z - 1) \tag{13.54}$$

and this may be substituted into eq. (13.53), which, upon simplification, gives

$$f_2(x) = \frac{K_1 b^2 T_b^3}{10kZ^3\Phi^2}\left[\left(1 + \frac{x}{b}\Phi\right)^5 - 1 - \frac{5K_2 Z^4}{K_1 T_b^4}\frac{x\Phi}{b}\right] \tag{13.55}$$

where $\Phi = Z - 1$. At the fin base, where $x = b$ and $z = Z = T_b/T_a$, eq. (13.55) reduces to

$$f_2(b) = \frac{\delta_b}{2} = \frac{K_1 b^2 T_b^3}{10kZ^3\Phi^2}\left(Z^5 - 1 - \frac{5K_2 Z^4}{K_1 T_b^4}\Phi\right) \tag{13.56}$$

and at the fin tip where $x = 0$ and $z = T_a/T_a = 1$, L'Hôpital's rule may be employed to show that $f_2(x) = 0$. The fin profile is described by eq. (13.55) in terms of the environmental extremes, environmental effects, and certain fin parameters. The profile is constrained to have a narrowing characteristic from base to tip and a zero thickness at the tip. The base-to-tip temperature ratio Z must be chosen discriminately. Its limit may be determined by focusing on the portion of the fin close to the tip, where x/b is small and by working with eq. (13.55). The fin profile must always be positive: hence the term in brackets must always be positive. This requires that

$$\left(1 + \frac{x}{b}\Phi\right)^5 > 1 + \frac{5K_2 Z^4}{K_1 T_b^4}\left(\frac{x}{b}\Phi\right)$$

and the term on the left may be subjected to a binomial expansion. Because x/b is small, a good approximation is obtained by considering only the first two terms of the expansion, and thus

$$1 + 5\frac{x}{b}\Phi + \cdots > 1 + \frac{5K_2 Z^4}{K_1 T_b^4}\left(\frac{x}{b}\Phi\right)$$

Cancellation of the common terms gives the inequality

$$\frac{K_2 Z^4}{K_1 T_b^4} < 1$$

from which it is observed that

$$Z_{\max} = \left(\frac{K_1 T_b^4}{K_2}\right)^{1/4} \tag{13.57}$$

and

$$T_{a,\min} = \left(\frac{K_2}{K_1}\right)^{1/4} \tag{13.58}$$

The approximate limits on the tip temperature are given by eqs. (13.57) and (13.58). Observe that for free space where $K_2 = 0$, $T_a = 0$, which is the approximate temperature of free space. The heat radiated by the fin with constant-temperature gradient is given by eq. (13.51). Integration between T_a and T_b and simplification with $Z = T_b/T_a$ yields

$$q_b = \frac{K_1 L b T_a^4 (Z^5 - 1)}{5\Phi} - K_2 L b \tag{13.59}$$

The ideal heat dissipation is $q_{id} = K_1 L b T_b^4$, and the fin efficiency is

$$\eta = \frac{q_b}{q_{id}} = \frac{Z^5 - 1}{5Z^4\Phi} - \frac{K_2}{K_1 T_b^4} \tag{13.60}$$

The profile number is useful in design and performance calculations and may be obtained from eq. (13.56), which may be written as

$$\delta_b = \frac{K_1 b^2 T_b^3}{k} \frac{Z^5 - 1 - (5K_2 Z^4/K_1 T_b^4)\Phi}{5Z^3\Phi^2}$$

so that

$$\zeta = \frac{K_1 b^2 T_b^3}{k\delta_b} = \frac{5Z^3\Phi^2}{Z^5 - 1 - (5K_2 Z^4/K_1 T_b^4)\Phi} \tag{13.61}$$

Figure 13.22 shows the efficiency of the fin with constant-temperature gradient as a function of the profile number. The dashed line represents the approximate limit given by eq. (13.57). Figure 13.23 is a plot of the profile number as a function of the base-to-tip temperature-excess ratio Z, as given by eq. (13.61). The several curves in these figures are due to different values of the environmental parameter $K_2/K_1 T_b^4$. Figures 13.22 and 13.23 contain all the data necessary for design and performance calculations for the fin with a constant-temperature gradient.

13.6 PARABOLIC RADIATING PROFILES

Kotan and Arnas (1965) investigated radiation from longitudinal fins of concave and convex parabolic profiles whose cross-sectional area functions for unit fin length are given, respectively, by

$$A(x) = A_b \left(1 - \frac{x}{b}\right)^2 \tag{13.62a}$$

and

$$A(x) = A_b \left(1 - \frac{x}{b}\right)^{1/2} \tag{13.62b}$$

where $A_b = \delta_b L$. The differential equation for the temperature ratio $y = T/T_b$ is obtained by considering a differential length element dx. The difference between the

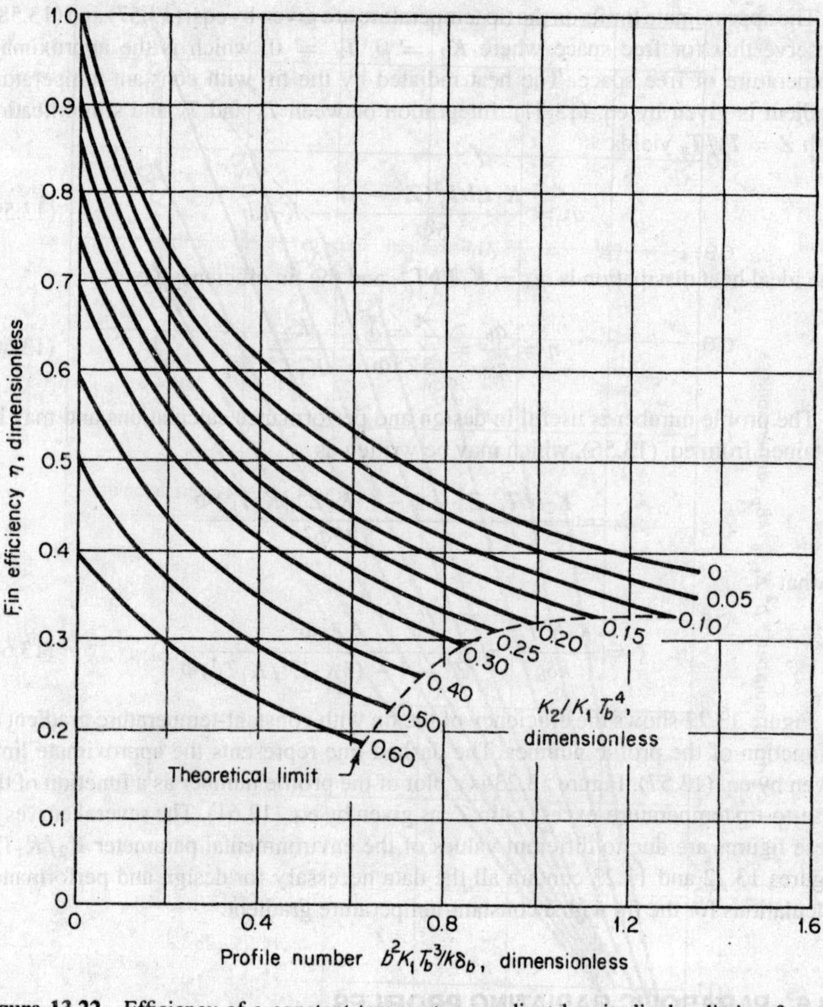

Figure 13.22 Efficiency of a constant-temperature-gradient longitudinal radiating fin as a function of profile number.

heat entering and leaving the element by conduction is equated to the heat dissipated from the faces of the element by radiation. In dimensionless form with $X = x/b$ and ζ, K_1 and K_2 as defined previously, the differential equation for the concave parabolic fin is

$$(1-X)^2\frac{d^2y}{dx^2} + 2(1-X)\frac{dy}{dX} - \zeta\left(y^4 - \frac{K_2}{K_1T_b^4}\right) = 0 \qquad (13.63a)$$

and for the convex parabolic fin

$$(1-X)^{1/2}\frac{d^2y}{dx^2} + \frac{1}{2(1-X)^{1/2}}\frac{dy}{dX} - \zeta\left(y^4 - \frac{K_2}{K_1T_b^4}\right) = 0 \qquad (13.63b)$$

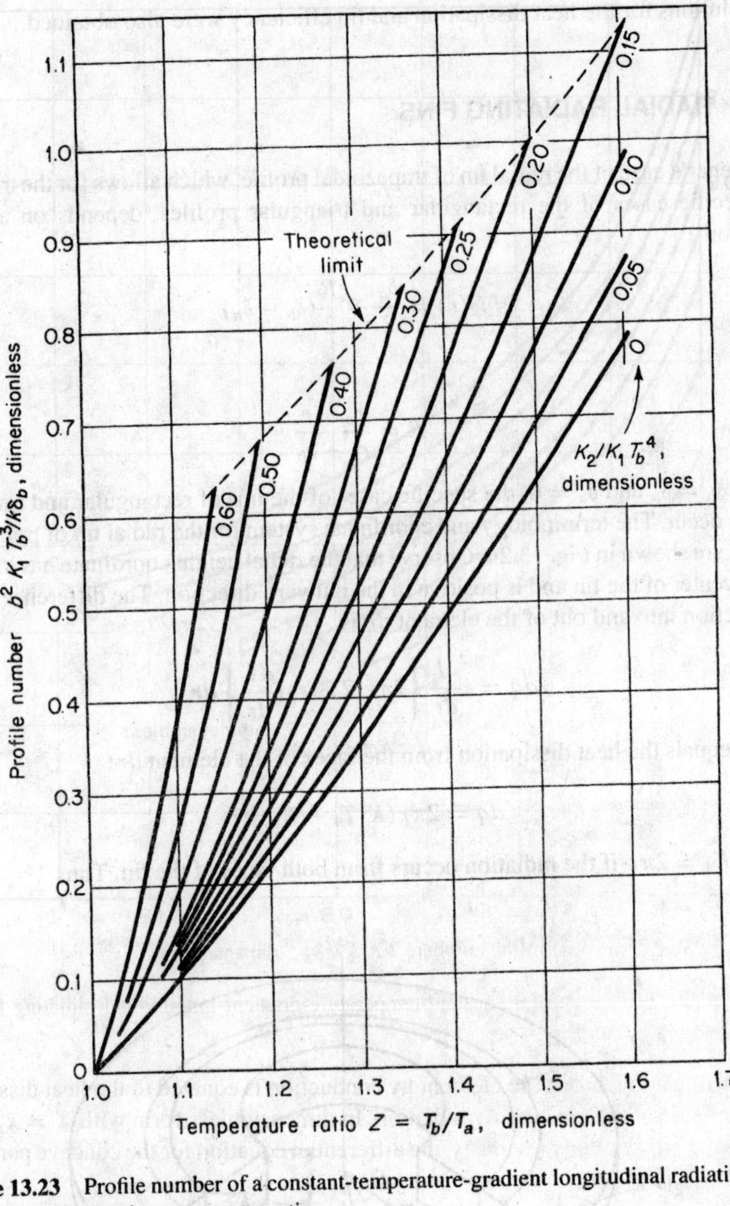

Figure 13.23 Profile number of a constant-temperature-gradient longitudinal radiating fin as function of base-to-tip temperature ratio.

Kotan and Arnas (1965) obtained computer solutions for eqs. (13.63) with the boundary conditions

$$y(X = 0) = 1 \quad \text{and} \quad \left.\frac{dy}{dX}\right|_{X=1} = 0$$

Solutions for the heat dissipation and fin efficiency were also obtained.

13.7 RADIAL RADIATING FINS

The general case of the radial fin of trapezoidal profile, which allows for the treatment of specific cases of the rectangular and triangular profiles, depends on a profile function

$$f_2(r) = \frac{\delta_a}{2} + \frac{R}{2}(\delta_b - \delta_a) \tag{13.64}$$

where

$$R = \frac{r_a - r}{r_a - r_b}$$

When $\delta_b = \delta_a$ and $\delta_a = 0$, the specific cases of the fins of rectangular and triangular profile occur. The terminology and coordinate system for the radial fin of trapezoidal profile are shown in Fig. 13.24. Observe that the radial height coordinate has its origin at the center of the fin and is positive in the outward direction. The difference in heat conduction into and out of the element dr is

$$dq = k\frac{d}{dr}\left\{2\pi r[2f_2(r)]\frac{dT}{dr}\right\}dr$$

which equals the heat dissipation from the faces of the element dr:

$$dq = 2\pi r(K_1 T_4 - K_2)\,dr$$

where $K_1 = 2\pi\epsilon$ if the radiation occurs from both faces of the fin. Thus

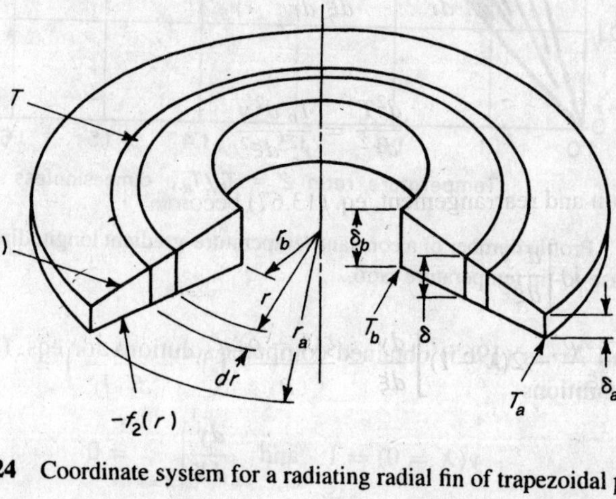

Figure 13.24 Coordinate system for a radiating radial fin of trapezoidal profile.

$$k\frac{d}{dr}\left\{r[2f_2(r)]\frac{dT}{dr}\right\} - K_1 r T_4 + K_2 r = 0 \qquad (13.65)$$

The taper and radius ratios have been defined previously:

$$\lambda = \frac{\delta_a}{\delta_b} \qquad (13.40)$$

and

$$\rho = \frac{r_b}{r_a} \qquad (1.36)$$

Equation (13.64) can be written in terms of these parameters:

$$\delta(r) = 2f_2(r) = \frac{\delta_b \rho}{1-\rho}\left[\frac{1}{\rho} - \lambda + \frac{r}{r_b}(\lambda - 1)\right] \qquad (13.66)$$

Then

$$2f_2'(r) = \frac{\delta_b \rho(\lambda - 1)}{(1-\rho)r_b}$$

and differentiation of eq. (13.65) gives

$$\frac{r\delta_b\rho}{1-\rho}\left[\frac{1}{\rho} - \lambda + \frac{r}{r_b}(\lambda - 1)\right]\frac{d^2T}{dr^2} + \left\{\frac{r\delta_b\rho(\lambda-1)}{(1-\rho)r_b} + \frac{\delta_b\rho}{1-\rho}\left[\frac{1}{\rho} - \lambda + \frac{r}{r_b}(\lambda-1)\right]\right\}\frac{dT}{dr} - \frac{K_1}{k}T^4 r + \frac{K_2}{k} r = 0 \qquad (13.67)$$

With $y = T/T_b$ and $\xi = r/r_b$,

$$\frac{dT}{dr} = T_b\frac{dy}{d\xi}\frac{d\xi}{dr} = \frac{T_b}{r_b}\frac{dy}{d\xi}$$

and

$$\frac{d^2T}{dr^2} = \frac{T_b}{r_b^2}\frac{d^2y}{d\xi^2}$$

With substitution and rearrangement, eq. (13.67) becomes

$$\left[\frac{1}{\rho} - \lambda + \xi(\lambda - 1)\right]\frac{d^2y}{d\xi^2} + \left[\frac{1-\lambda\rho}{\rho\xi} - 2(\lambda-1)\right]\frac{dy}{d\xi} - \frac{\zeta(1-\rho)}{\rho}\left(y^4 - \frac{K_2}{K_1 T_b^4}\right) = 0 \qquad (13.68)$$

where

$$\zeta = \frac{K_1 T_b^3 r_b^2}{k\delta}$$

is the profile number for the radial radiating fin.

The governing differential equation for the temperature profile is eq. (13.68). It is nonlinear and must be solved numerically for the dimensionless temperature y as a function of the dimensionless ratio ξ for assumed values of the parameters ζ, ρ, λ, and $K_2/K_1T_b^4$. The solution is subject to the condition that $dy/d\xi = dT/dr$ is equal to zero at $\xi = r_a/r_b = 1/\rho$.

The heat transferred from the fin may be evaluated by applying Fourier's law at the fin base,

$$q_b = -2\pi k\delta_b r_b \left.\frac{dT}{dr}\right|_{r=r_b}$$

or in terms of the dimensionless parameters,

$$q_b = -2\pi k\delta_b T_b \left.\frac{dy}{d\xi}\right|_{\xi=1} \tag{13.69}$$

The fin efficiency will be this actual heat flow divided by the ideal heat flow with no heat gained from the surroundings:

$$\eta = \frac{q_b}{q_{\text{id}}} = \frac{-2\pi k\delta_b T_b \left.\dfrac{dy}{d\xi}\right|_{\xi=1}}{\pi K_1 T_b^4 (r_a^2 - r_b^2)} = -\frac{2\rho^2 \left.\dfrac{dy}{d\xi}\right|_{\xi=1}}{\zeta(1-\rho^2)} \tag{13.70}$$

The numerical solution of eq. (13.68) yields the values of y and $dy/d\xi$ as a function of ξ. Typical plots of the numerical data for a fin 10.16 cm by 20.32 cm by 0.635 cm thick at the base, fabricated of a metal with a thermal conductivity of 106.4 W/m·K and an emissivity of 0.90 radiating from *one face*, are given in Fig. 13.25. Note that the curves are for several taper ratios λ and for several values of K_2/K_1Tb^4. For these dimensions, the profile number is $\zeta = 0.0147$ and $\rho = 0.50$.

The value of the derivative $dy/d\xi$ at $\xi = 1.0$ is always negative, and the fin efficiency can be computed using eq. (13.70) with particular values of ρ, λ, ζ, and the environmental parameter. Unfortunately, the large number of variables requires a multiplicity of curves for the presentation of the fin efficiency. Table 13.2 lists and summarizes the curves included here.

Example 13.6: Radiation from Radial Fins (Fin Analysis). A radial fin has a base temperature of 445 K and is fabricated of a material having thermal conductivity $k = 82.1$ W/m·K. The fin is 10.16 cm inside diameter by 25.4 cm outside diameter by 0.3125 cm thick at the base and radiates from one face, which has an emissivity of 0.809. Solar energy is received by the dissipating face at a rate of 1303.7 W/m^2 and the fin material has an absorbtivity of 0.276 at the wavelength of solar radiation. Determine the heat dissipation from radial fins of (**a**) rectangular profile, (**b**) trapezoidal profile with a tip width of 0.1588 cm, and (**c**) triangular profile.

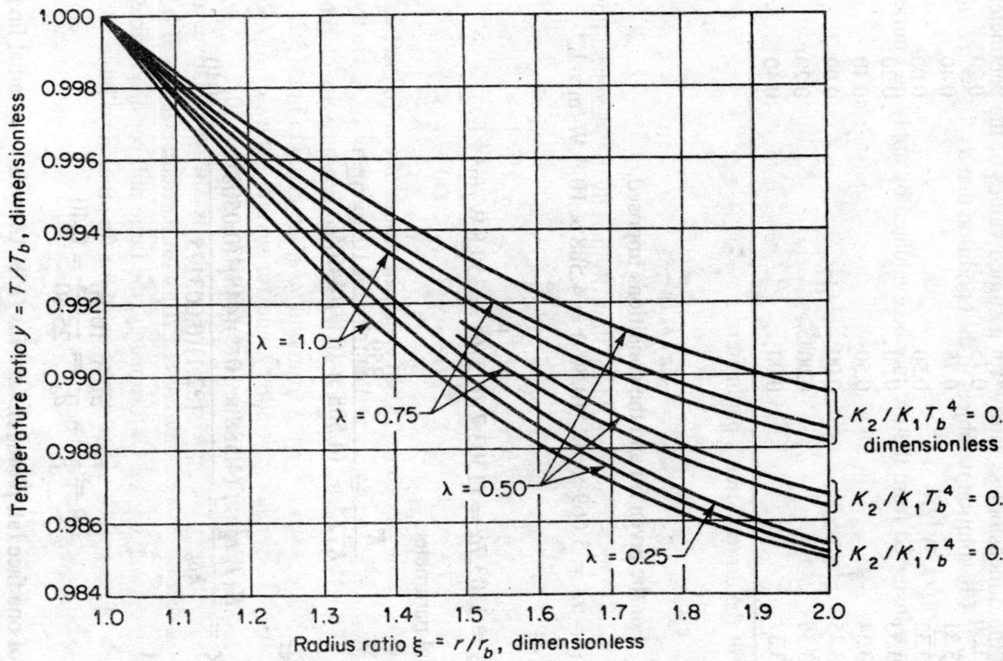

Figure 13.25 Some temperature profiles obtained from a computer solution of eq. (13.70). For these profiles, $\zeta = 0.0147$ and $\rho = 0.5$.

TABLE 13.2 Summary of Radial Fin Efficiency Data

Figure	λ	$K_2/K_1T_b^4$
13.26	1.00	0.00
13.27	1.00	0.20
13.28	1.00	0.40
13.29	0.75	0.00
13.30	0.75	0.20
13.31	0.75	0.40
13.32	0.50	0.00
13.33	0.50	0.20
13.34	0.50	0.40
13.35	0.00	0.00
13.36	0.00	0.20
13.37	0.00	0.40

SOLUTION. For the given fin and the conditions imposed,

$$K_1 = \sigma\epsilon = 5.669 \times 10^{-8}(0.809) = 4.588 \times 10^{-8}\ \text{W/m}^2 \cdot \text{K}^4$$

and

$$K_2 = 1303.7\alpha = (1303.7)(0.276) = 359.8\ \text{W/m}^2$$

Environmental parameter:

$$\frac{K_2}{K_1 T_b^4} = \frac{359.8}{(4.588 \times 10^{-8})(445)^4} = 0.20$$

Profile number:

$$\zeta = \frac{K_1 T_b^3 r_b^2}{k\delta_b} = \frac{(4.588 \times 10^{-8})(445)^3(0.0508)^2}{(82.1)(0.003175)} = 0.040$$

Radius ratio:

$$\rho = \frac{r_b}{r_a} = \frac{d_b}{d_a} = \frac{10.16}{25.40} = 0.40$$

Surface area of one face (all profiles):

$$S = \pi(r_a^2 - r_b^2) = \pi[(0.1270)^2 - (0.0508)^2] = 0.0426\ \text{m}^2$$

Ideal heat dissipation (all profiles):

$$q_{\text{id}} = K_1 S T_b^4 = (4.588 \times 10^{-8})(0.0426)(445)^4 = 76.6\ \text{W}$$

(a) *Rectangular profile*, $\lambda = 1$. From Fig. 13.27 at $\rho = 0.40$ and $\zeta = 0.04$, read

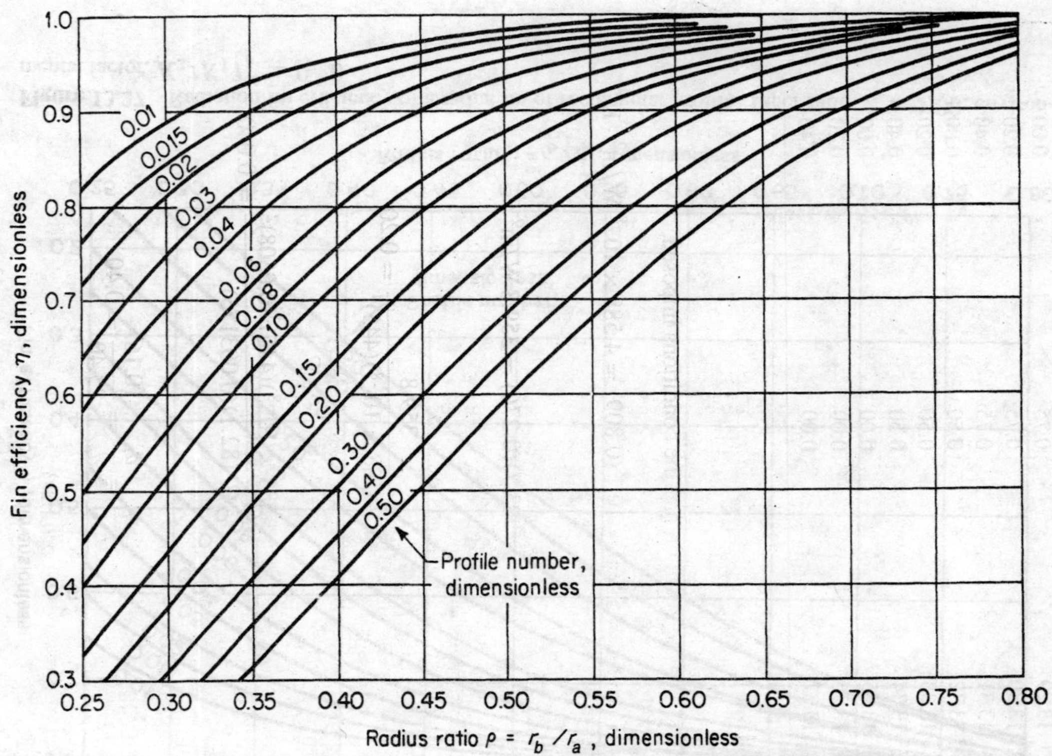

Figure 13.26 Radiation fin efficiency of a radial fin of rectangular profile. Taper ratio, $\lambda = 1.00$; environmental factor, $K_2/K_1 T_b^4 = 0.00$.

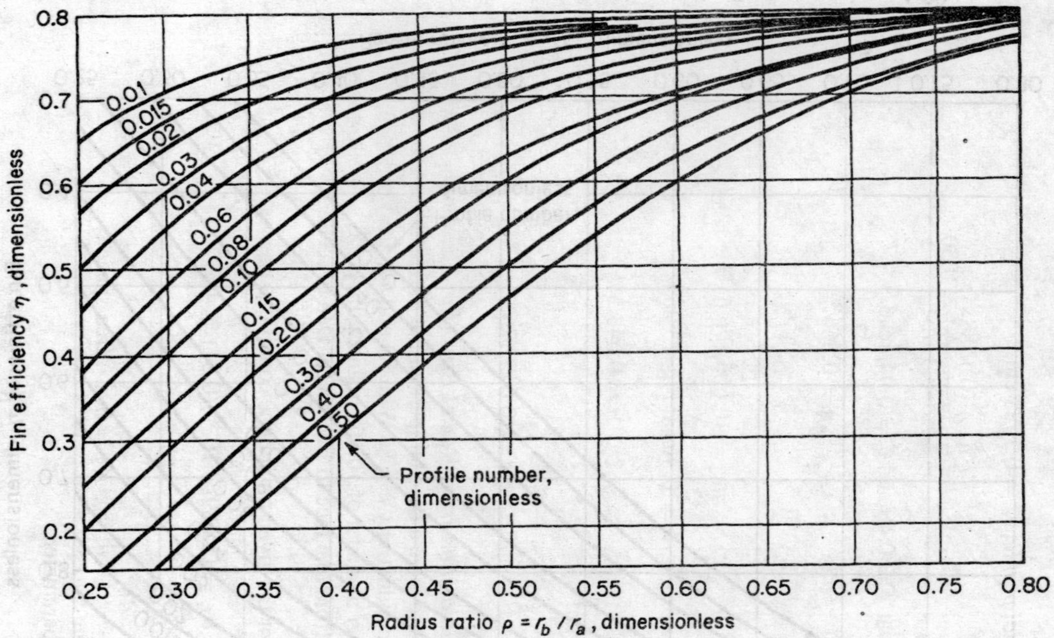

Figure 13.27 Radiation fin efficiency of a radial fin of rectangular profile. Taper ratio, $\lambda = 1.00$; environmental factor, $K_2/K_1T_b^4 = 0.20$.

$$\eta = 0.676$$

and then

$$q_b = \eta q_{\text{id}} = (0.676)(76.6) = 51.8 \text{ W}$$

(b) *Trapezoidal profile*, $\delta_a = 0.1588$ cm:

$$\lambda = \frac{\delta_a}{\delta_b} = \frac{0.1588}{0.3175} = 0.50$$

From Fig. 13.33 at $\rho = 0.40$ and $\zeta = 0.04$, read

$$\eta = 0.666$$

and then

$$q_b = \eta q_{\text{id}} = (0.666)(76.6) = 51.0 \text{ W}$$

(c) *Triangular profile*, $\lambda = 0$. From Fig. 13.36 at $\rho = 0.40$ and $\zeta = 0.04$, read

$$\eta = 0.643$$

and then

$$q_b = \eta q_{\text{id}} = (0.643)(76.6) = 49.3 \text{ W}$$

Example 13.7: Radiation from Radial Fins (Fin Synthesis). A radial fin of rectangular profile is to have a base temperature of 450 K and is fabricated of a material having thermal conductivity $k = 86$ W/m·K) and an emissivity of $\epsilon = 0.81$. The fin is to have an inner diameter of 10.16 cm and a thickness of 0.635 cm and is required to radiate 76 W from one face to free space. Determine the outer diameter required.

SOLUTION. For the given fin and the conditions imposed,

$$K_1 = \sigma\epsilon = (5.669 \times 10^{-8})(0.81) = 4.592 \times 10^{-8} \text{ W/m}^2 \cdot \text{K}^4$$

and

$$K_2 = 0 \qquad \text{free space}$$

Environmental parameter:

$$\frac{K_2}{K_1 T_b^4} = 0$$

Profile number:

$$\zeta = \frac{K_1 T_b^3 r_b^2}{k\delta_b} = \frac{(4.592 \times 10^{-8})(450)^3(0.0508)^2}{86(0.00615)} = 0.020$$

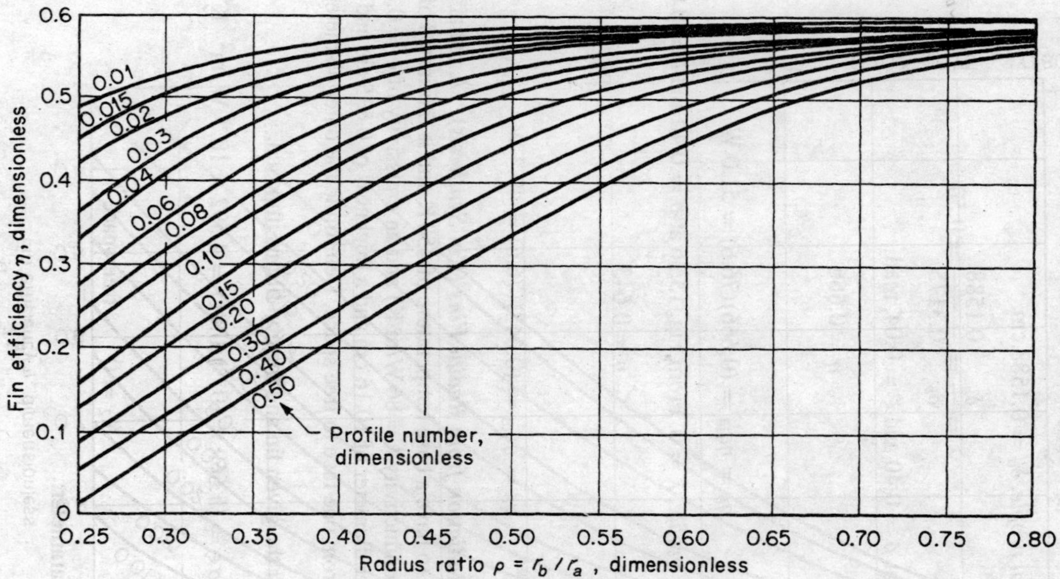

Figure 13.28 Radiation fin efficiency of a radial fin of rectangular profile. Taper ratio, $\lambda = 1.00$; environmental factor, $K_2/K_1 T_b^4 = 0.40$.

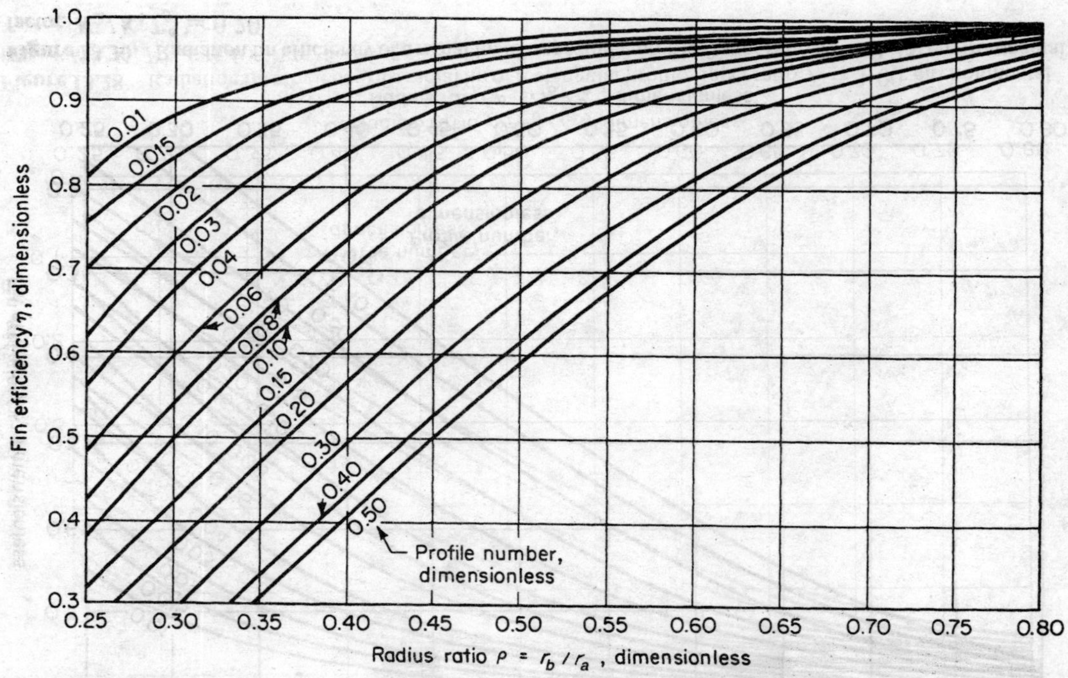

Figure 13.29 Radiation fin efficiency of a radial fin of trapezoidal profile. Taper ratio, $\lambda = 0.75$; environmental factor, $K_2/K_1T_b^4 = 0.00$.

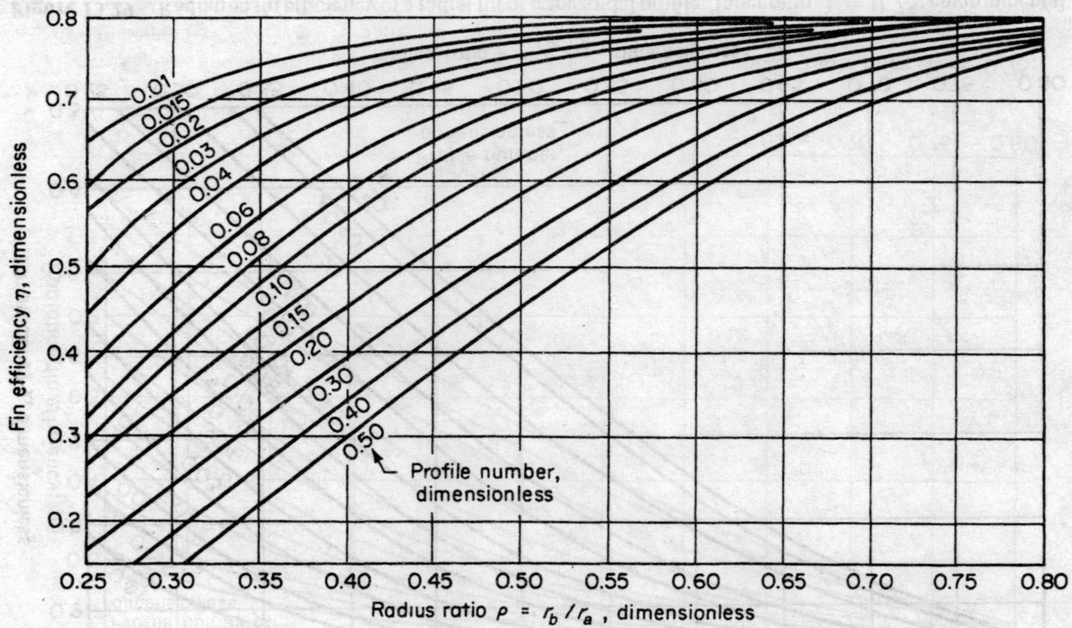

Figure 13.30 Radiation fin efficiency of a radial fin of trapezoidal profile. Taper ratio, $\lambda = 0.75$; environmental factor, $K_2/K_1 T_b^4 = 0.20$.

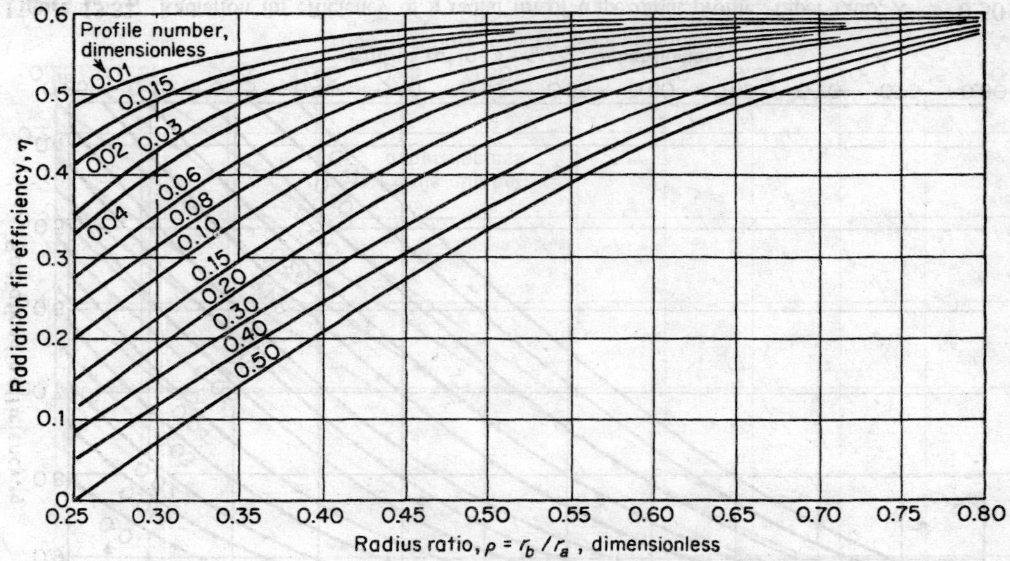

Figure 13.31 Radiation fin efficiency of a radial fin of trapezoidal profile. Taper ratio, $\lambda = 0.75$; environmental factor, $K_2/K_1 T_b^4 = 0.40$.

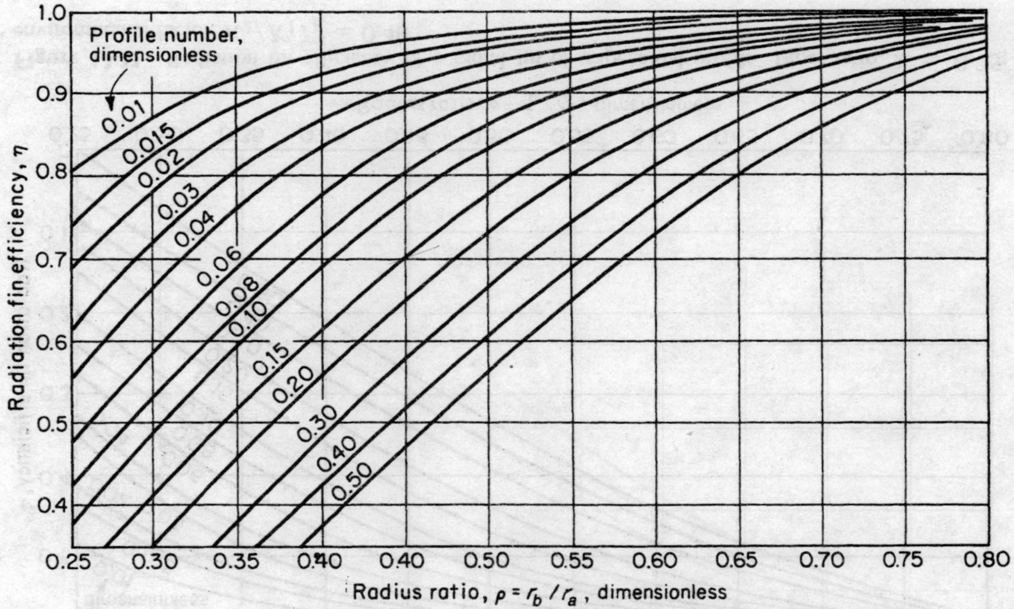

Figure 13.32 Radiation fin efficiency of a radial fin of trapezoidal profile. Taper ratio, $\lambda = 0.50$; environmental factor, $K_2/K_1T_b^4 = 0.00$.

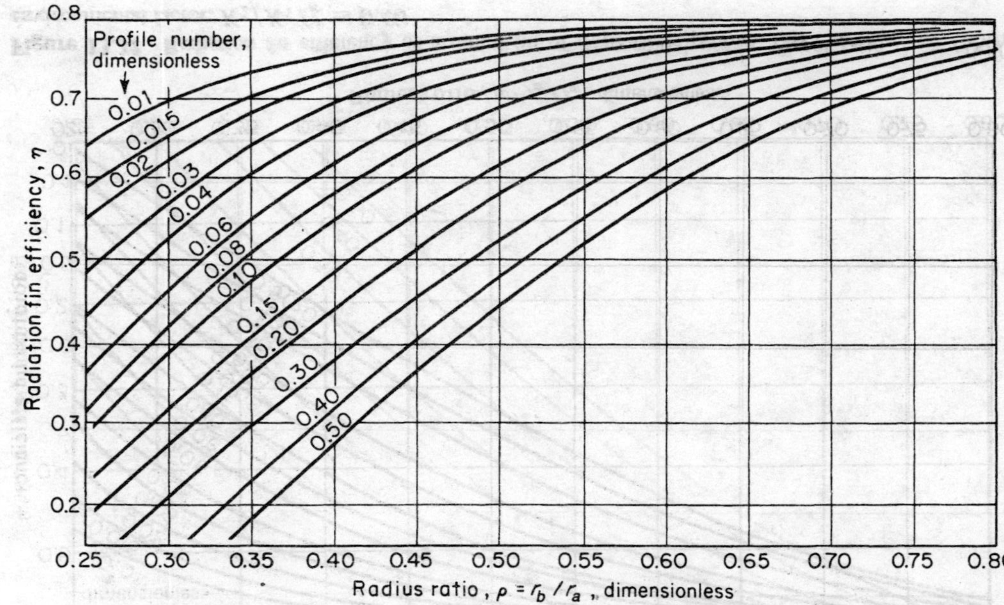

Figure 13.33 Radiation fin efficiency of a radial fin of trapezoidal profile. Taper ratio, $\lambda = 0.50$; environmental factor, $K_2/K_1T_b^4 = 0.20$.

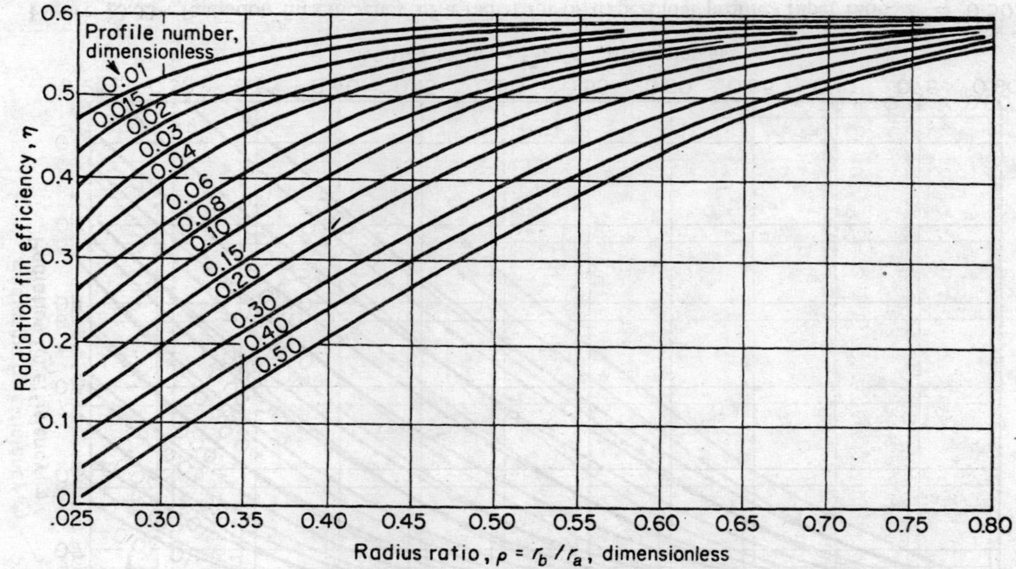

Figure 13.34 Radiation fin efficiency of a radial fin of trapezoidal profile. Taper ratio, $\lambda = 0.50$; environmental factor, $K_2/K_1T_b^4 = 0.40$.

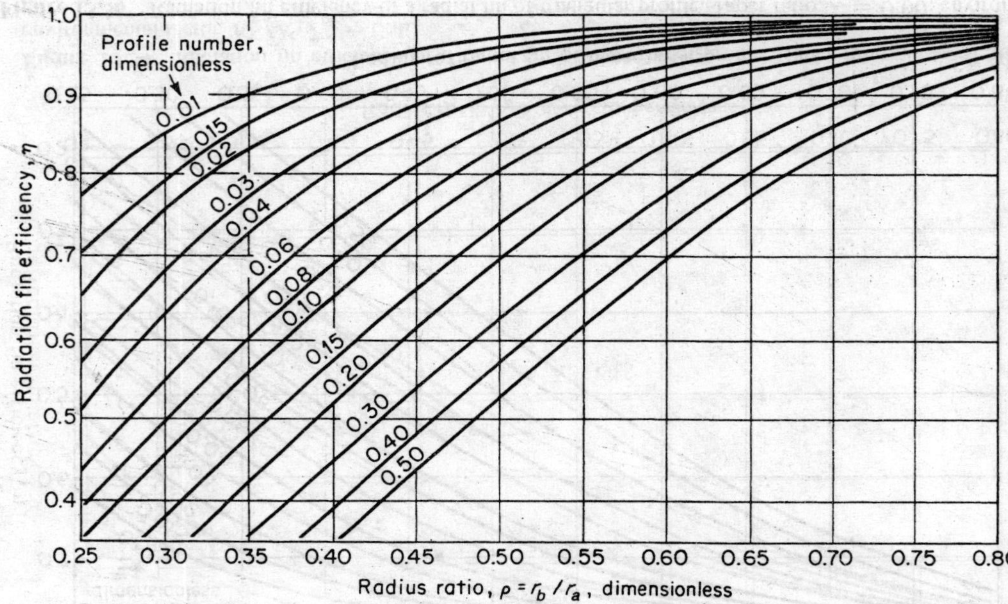

Figure 13.35 Radiation fin efficiency of a radial fin of triangular profile. Taper ratio, $\lambda = 0.00$; environmental factor, $K_2/K_1T_b^4 = 0.00$.

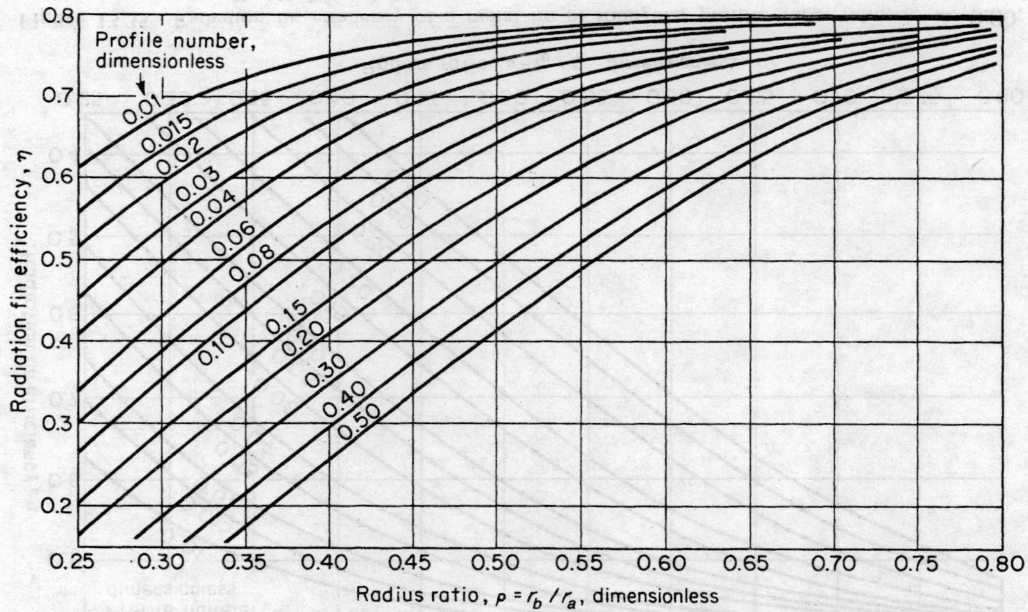

Figure 13.36 Radiation fin efficiency of a radial fin of triangular profile. Taper ratio, $\lambda = 0.00$; environmental factor, $K_2/K_1T_b^4 = 0.20$.

Trial I. Assume an outer diameter $d_a = 25.4$ cm. The radius ratio is

$$\rho = \frac{r_b}{r_a} = \frac{d_b}{d_a} = \frac{10.16}{25.40} = 0.40$$

From Fig. 13.26 at $\rho = 0.40$ and $\zeta = 0.020$, read

$$\eta = 0.916$$

With

$$S = \pi(r_a^2 - r_b^2) = \pi[(0.1270)^2 - (0.0508)^2] = 0.0426 \text{ m}^2$$

and with the ideal heat dissipation

$$q_{\text{id}} = K_1 S T_b^4 = (4.592 \times 10^{-8})(0.0426)(450)^4 = 80.1 \text{ W}$$

the required vale of η will be

$$\eta = \frac{q_b}{q_{\text{id}}} = \frac{76}{80.2} = 0.948 > 0.916 \quad \text{(no check)}$$

Trial II. Assume an outer diameter $d_a = 26.5$ cm. The radius ratio is

$$\rho = \frac{r_b}{r_a} = \frac{d_b}{d_a} = \frac{10.16}{26.50} = 0.383$$

From Fig. 13.26 at $\rho = 0.383$ and $\zeta = 0.020$, read

$$\eta = 0.905$$

With

$$S = \pi(r_a^2 - r_b^2) = \pi[(0.1325)^2 - (0.0508)^2] = 0.0470 \text{ m}^2$$

and with the ideal heat dissipation

$$q_{\text{id}} = K_1 S T_b^4 = (4.592 \times 10^{-8})(0.0470)(450)^4 = 88.6 \text{ W}$$

the required value of η will be

$$\eta = \frac{q_b}{q_{\text{id}}} = \frac{76}{88.6} = 0.858 < 0.905 \quad \text{(no check)}$$

Trial III. Assume an outer diameter $d_a = 26.0$ cm. The radius ratio is

$$\rho = \frac{r_b}{r_a} = \frac{d_b}{d_a} = \frac{10.16}{26.00} = 0.391$$

From Fig. 13.26 at $\rho = 0.391$ and $\zeta = 0.020$, read

$$\eta = 0.902$$

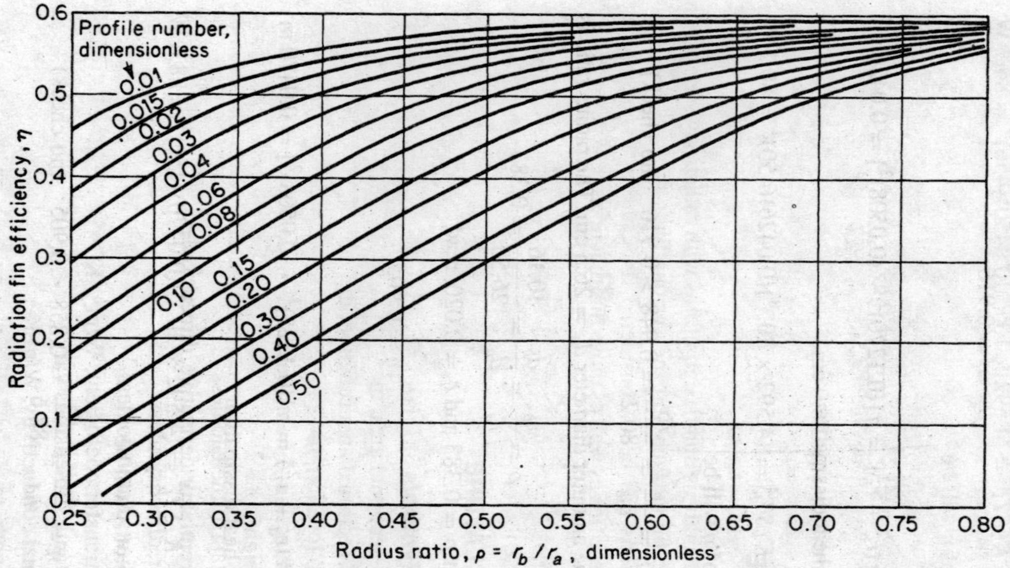

Figure 13.37 Radiation fin efficiency of a radial fin of triangular profile. Taper ratio, $\lambda = 0.00$; environmental factor, $K_2/K_1T_b^4 = 0.40$.

With

$$S = \pi(r_a^2 - r_b^2) = \pi[(0.1300)^2 - (0.0508)^2] = 0.0450 \text{ m}^2$$

and with the ideal heat dissipation

$$q_{id} = K_1 S T_b^4 = (4.592 \times 10^{-8})(0.0450)(450)^4 = 84.7 \text{ W}$$

the required vale of η will be

$$\eta = \frac{q_b}{q_{id}} = \frac{76}{84.7} = 0.897 \approx 0.902 \quad \checkmark$$

The required outer diameter is 26.0 cm.

13.8 CLOSURE

In this chapter, the analysis and design of fins with radiation as the sole heat transfer mode between the fin surface and the surroundings for both free space and non-free space conditions has been considered. In Chapter 14, attention turns to the optimization of these fins as well as those fins that operate in a combined convection–radiation mode.

13.9 NOMENCLATURE

Roman Letter Symbols

A	cross-sectional area, m^2
$A(x)$	cross-sectional area function, m^2
a	fin tip location, m
$B(a, b)$	beta function, dimensionless
b	fin height, m
C	constant, dimensionless
d	differential or derivative, dimensionless
F	factor, dimensionless
f	function, dimensionless
h	heat transfer coefficient, W/m^2 · K
K	constants, dimensions vary
k	thermal conductivity, W/m·K
L	fin length, m
m	fin performance parameter, m^{-1}
p	derivative of temperature with respect to height, K/W
q	heat flow, W
R	radius ratio, dimensionless
r	radius, m; radial coordinate, m
S	surface area, m^2
T	temperature, K

u	temperature ratio, dimensionless
v	transformed temperature variable, dimensionless
w	transformed variable, K
X	height, dimensionless
x	height coordinate, m
Y	thermal admittance, W/K
y	temperature ratio, dimensionless
Z	temperature ratio, dimensionless
z	temperature ratio, dimensionless

Greek Letter Symbols

Γ	gamma function, dimensionless
Δ	change in, dimensionless
δ	fin thickness, m
$\delta(x)$	thickness function, m
$\delta(r)$	thickness function, m
ϵ	emissivity, dimensionless
ζ	profile function, dimensionless
η	fin efficiency, dimensionless
κ	taper angle, rad
Λ	fin taper, dimensionless
λ	thickness ratio, dimensionless
ξ	radius function, dimensionless
ρ	radius ratio, dimensionless
σ	Stefan–Boltzmann constant, $W/m^2 \cdot K^4$
Υ	combination of terms
υ	combination of terms, dimensionless
Φ	combination of terms, dimensionless
ψ	combination of terms, dimensionless

Roman Letter Subscripts

a	arrangement; fin tip
av	average
b	fin base
id	ideal
in	input condition
max	maximum value
min	minimum value
r	radiation
s	surroundings
u	incomplete beta function
x	x-coordinate direction

Greek Letter Superscript

ϵ	emissivity

Symbolic Superscripts

$'$	first derivative
$''$	second derivative

CHAPTER 14

OPTIMUM DESIGN OF RADIATING AND CONVECTING–RADIATING FINS

14.1 INTRODUCTION

In Chapter 13 fins with radiation as the sole heat transfer mode between fin surface and surroundings, with no consideration given to their optimization were discussed. This chapter considers to their optimization, and the works of Wilkins (1960a,b), Bartas and Sellars (1960), Chung and Nguyen (1985, 1986), Smith (1992), Nilson and Curry (1960), and Sparrow and Eckert (1962) are discussed.

This chapter also includes radiation from fins with temperature-dependent thermal conductivity and emissivity (Campo and Wolko, 1973), mutual irradiation between adjacent fins (Sparrow et al., 1961a), the combined effects of environmental radiation (Karlekar and Chao, 1963), the effects of associated structure (Reynolds, 1963), and fin-to-fin and fin-to-base radiation–radiation interchange (Schnurr, 1975; Schnurr et al., 1976).

The optimum dimensions of longitudinal convecting–radiating fins are discussed using the work of Sparrow and Niewerth (1968), and studies by Chambers and Somers (1959), Keller and Holdridge (1969), Truong and Mancuso (1980), Sparrow et al. (1962), and Schnurr and Cothran (1974) pertaining to the optimum dimensions of radial radiating fins are then explored.

14.2 OPTIMUM DIMENSIONS OF LONGITUDINAL RADIATING FINS

14.2.1 Fin of Least Material

As shown by Wilkins (1960a), the radiating fin of least material treated in Section 13.5 does not possess a linear temperature profile. Of course, this is contrary to Schmidt's

(1926) contention pertaining to convecting fins. Consider a fin of perfectly arbitrary profile radiating to free space at $T_s = 0$ with no radiant energy received from the surroundings. Let the origin of the coordinate system be the fin tip. The fin has a height b and the profile is defined by the function $f_2(x)$. The heat passing through the fin by conduction at any point x is

$$q = 2kf_2(x)\frac{dT}{dx} \tag{14.1}$$

and the heat dissipated from a differential element by radiation is

$$dq = 2\sigma\epsilon T^4\, dx \tag{14.2}$$

The boundary conditions associated with these equations are

$$q(x = 0) = 0, \quad q(x = b) = q_b, \quad \text{and} \quad T(x = b) = T_b$$

and the problem may be stated simply as one of finding three functions, $q(x)$, $T(x)$, and $f_2(x)$, that satisfy the foregoing equations and boundary conditions for a given value of b for which the profile area

$$A_p = 2\int_0^b f_2(x)\, dx$$

is a minimum. Let

$$u \equiv \left(\frac{T}{T_b}\right)^9 \tag{14.3}$$

and

$$v \equiv \left(\frac{q}{q_b}\right)^{3/2} \tag{14.4}$$

Then

$$T^4 = T_b^4 u^{4/9} \tag{14.5a}$$

$$dT = \frac{1}{9}T_b u^{-8/9}\, du \tag{14.5b}$$

$$q = q_b v^{2/3} \tag{14.5c}$$

and

$$dq = \frac{2}{3}q_b v^{-1/3}\, dv \tag{14.5d}$$

Solving eq. (14.2) for dx and integrating between limits using eqs. (14.5) gives

$$x = \int_0^q \frac{dq}{2\sigma\epsilon T^4} = \frac{q_b}{3\sigma\epsilon T_b^4}\int_{u_a}^u u^{-4/9}v^{-1/3}\frac{dv}{du}\,du \tag{14.6}$$

where the limit $u_a = (T_a/T_b)^9$. Then eq. (14.1) may be solved for $f_2(x)$ using eqs. (14.2) and (14.5):

$$f_2(x) = \frac{3q_b^2 u^{4/9} v^{1/3}}{2\sigma\epsilon k T_b^5} \frac{dv}{du} \tag{14.7}$$

Hence

$$A_p = 2\int_0^b f_2(x)\,dx = \frac{q_b^3}{k\sigma^2\epsilon^2 T_b^9} \int_{u_a}^{1} \left(\frac{dv}{du}\right)^2 du \tag{14.8}$$

where the upper limit $u_b = 1$, because at $x = b$, $u = u_b = (T_b/T_b)^9 = 1$.

Equation (14.8) may be adjusted to

$$A_p = \frac{q_b^3}{k\sigma^2\epsilon^2 T_b^9} \int_{u_a}^{1} \left[2\frac{dv}{du} - 1 + \left(\frac{dv}{du} - 1\right)^2\right] du \tag{14.9}$$

and because at $x = 0$ where $u = u_a$ and $v = v_a = 0$ and at $x = b$ where $u = u_b = 1$ and $v = v_b = (q_b/q_b)^3 = 1$, eq. (14.9) can be rearranged to an inequality:

$$\frac{k\sigma^2\epsilon^2 T_b^9 A_p}{q_b^3} = 1 + u_a + \int_{u_a}^{1} \left(\frac{dv}{du} - 1\right)^2 du \geq 1 \tag{14.10}$$

because $0 \leq u_a \leq 1$.

The inequality of eq. (14.10) can hold only if $u_a = 0$ and $u = v$. Then, from eq. (14.7) using eq. (14.3),

$$f_2(x) = \frac{3q_b^2 u^{4/9} v^{1/3}}{2\sigma\epsilon k T_b^5} = \frac{3q_b^2 u^{7/9}}{2\sigma\epsilon k T_b^5} = \frac{3q_b^2 T^7}{2\sigma\epsilon k T_b^{12}} \tag{14.11}$$

and from eq. (14.6) using eq. (14.4),

$$x = \frac{q_b}{3\sigma\epsilon T_b^4} \int_0^u u^{-4/9} u^{-1/3}\,du = \frac{q_b}{3\sigma\epsilon T_b^4} \int_0^u u^{-7/9}\,du = \frac{3q_b T^2}{2\sigma\epsilon T_b^6} \tag{14.12}$$

Equations (14.3), (14.4), (14.10), (14.11), and (14.12) may now be used to establish the pertinent relationships. With q_b and T_b specified, eq. (14.10) gives the profile area

$$A_p = \frac{q_b^3}{k\sigma^2\epsilon^2 T_b^9} \tag{14.13}$$

The fin height is determined from eq. (14.12) at $x = b$ where $T = T_b$:

$$b = \frac{3q_b}{2\sigma\epsilon T_b^4} \tag{14.14}$$

From a combination of eqs. (14.3) and (14.4), because $u = v$,

$$\frac{q}{q_b} = \left(\frac{T}{T_b}\right)^6 \tag{14.15}$$

From eqs. (14.12) and (14.14),

$$x = \frac{3q_b}{2\sigma\epsilon T_b^4}\left(\frac{T}{T_b}\right)^2 = b\left(\frac{T}{T_b}\right)^2 \tag{14.16a}$$

or

$$\frac{T}{T_b} = \left(\frac{x}{b}\right)^{1/2} \tag{14.16b}$$

and with eqs. (14.15) and (14.16) combined,

$$\frac{q}{q_b} = \left(\frac{x}{b}\right)^3 \tag{14.17}$$

Finally, using eqs. (14.11) and (14.16) yields

$$f_2(x) = \frac{3q_b^2}{2\sigma\epsilon k T_b^5}\left(\frac{T}{T_b}\right)^7 = \frac{3q_b^2}{2\sigma\epsilon k T_b^5}\left(\frac{x}{b}\right)^{7/2} \tag{14.18}$$

Observe that eq. (14.16*b*) indicates that the tip temperature must equal zero. This is purely a theoretical consideration and is based on the assumption that the fin radiates to surroundings at absolute zero and that the fin receives no radiation from the surroundings. Moreover, it would be extremely difficult to put a fin with zero tip thickness into a space environment.

Example 14.1: Design of a Longitudinal Radiating Fin of Least Material. An aluminum fin ($k = 202.5$ W/m·K) has been coated with a paint so that its emissivity is 0.88. The fin is 1.8288 m long and is required to radiate 1000 W at a base temperature of 560 K (287°C). Design a theoretically optimum (Wilkins) fin.

SOLUTION. Work on a 1-m basis,

$$q_b = \frac{1000}{1.8288} = 546.8 \text{ W/m}$$

The profile area is obtained from eq. (14.13):

$$A_p = \frac{q_b^3}{k\sigma^2\epsilon^2 T_b^9}$$

$$= \frac{(546.8)^3}{(202.5)(5.669 \times 10^{-8})^2(0.88)^2(560)^9}$$

$$= 5.9895 \times 10^{-5} \text{ m}^2$$

The fin height by eq. (14.14) is

$$b = \frac{3q_b}{2\sigma\epsilon T_b^4}$$

$$= \frac{3(546.8)}{(2)(5.669 \times 10^{-8})(0.88)(560)^4}$$
$$= 0.1672 \text{ m}$$

or 16.72 cm. The fin base thickness at $x = b$ by eq. (14.18) is

$$2f_2(x) = \delta_b = \frac{3q_b^2}{\sigma \epsilon k T_b^5} \left(\frac{x}{b} \right)^{7/2}$$
$$= \frac{(3)(546.8)^2}{(5.669 \times 10^{-8})(0.88)(202.5)(560)^5}$$
$$= 1.612 \times 10^{-3} \text{ m}$$

or 0.1612 cm.

14.2.2 Specific Profiles

In this section the optimization of particular longitudinal fins that dissipate heat solely by radiation is considered. The problem to be solved may be posed as follows: Given a fin geometry (shape or profile), find the dimensions of the fin so that the heat dissipation for a given volume of material is at a maximum. This is equivalent to the statement that given a fin geometry and a desired heat dissipation, find the dimensions of the fin so that the volume of material used is at a minimum. The solution to this optimization problem will be presented for five longitudinal fin profile geometries: rectangular, triangular, trapezoidal, concave parabolic, and convex parabolic. The solution strategy involves either a purely mathematical approach or a combined numerical–mathematical approach. For each geometry (profile), two sets of expressions are given, one set when the profile area is specified and the other when the heat dissipation is specified. The section concludes with a comparison of the material used by different optimum fins. The analysis is based on the Murray–Gardner assumptions listed in Chapter 1.

Longitudinal Fin of Rectangular Profile. Consider a longitudinal fin of rectangular profile with length L, height b, thickness δ, thermal conductivity k, and emissivity ϵ, losing heat, solely by radiation, from its exposed surfaces to free space at absolute zero temperature, as shown in Fig. 14.1a. For the boundary conditions of constant base temperature T_b, and an insulated tip, the heat dissipation per unit length ($L = 1$ m) can be obtained from a slight modification of eq. (13.14):

$$q = 2k\delta \left[\frac{\sigma \epsilon \left(T_b^5 - T_a^5 \right)}{5k\delta} \right]^{1/2} \tag{14.19}$$

Here T_a, the fin tip temperature, is related to the temperature distribution in the fin, which can be expressed in terms of the complete beta function $B(a, b)$ and the incomplete beta function $B_u(a, b)$ as

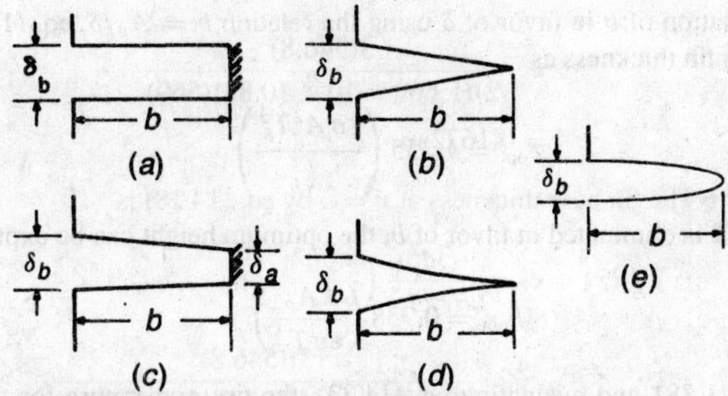

Figure 14.1 Longitudinal fin geometries: (*a*) rectangular profile with insulated tip; (*b*) triangular profile; (*c*) trapezoidal profile with insulated tip; (*d*) concave parabolic profile; (*e*) convex parabolic profile. All fins possess a thermal conductivity k, have a surface emissivity ϵ, and operate at a base temperature T_b.

$$f(\delta, T_a) = B(0.3, 0.5) - B_u(0.3, 0.5) - b\left(\frac{20\epsilon\sigma T_a^3}{k\delta}\right)^{1/2} = 0 \qquad (14.20)$$

The optimization problem is to find δ and T_a such that q is maximized while the profile area $A_p = \delta b$ remains fixed. Liu (1960, 1961) has solved this problem using the method of Lagrange multipliers. The equations defining the optimum q are

$$\frac{\partial q}{\partial \delta} + \lambda \frac{\partial f(\delta, T_a)}{\partial \delta} = 0 \qquad (14.21)$$

and

$$\frac{\partial q}{\partial T_a} + \lambda \frac{\partial f(\delta, T_a)}{\partial T_a} = 0 \qquad (14.22)$$

where λ is the unknown Lagrange multiplier. Equations (14.20) through (14.22) constitute three equations for the three unknowns, δ, T_a, and λ. By eliminating λ from eqs. (14.21) and (14.22), it can be shown that T_a is given by

$$T_a = 0.425 T_b \left[12 - G + G^{1/2}(G + 120)^{1/2}\right]^{1/5} \qquad (14.23)$$

where

$$G = \frac{5k\delta}{9\epsilon\sigma T_b^3 b^2}$$

Substitution of eq. (14.23) into eq. (14.20) yields a transcendental equation in G that can be solved to give the solution $G = 1.381$. Then, by using the definition of G, it is observed that for the optimum condition, the relationship between δ and b must be

$$\frac{\delta}{b^2} = 2.486 \frac{\epsilon\sigma T_b^3}{k} \qquad (14.24)$$

Upon elimination of b in favor of δ using the relation $b = A_p/\delta$, eq. (14.24) gives the optimum fin thickness as

$$\delta_{\text{opt}} = 1.355 \left(\frac{\epsilon\sigma A_p^2 T_b^3}{k} \right)^{1/3} \tag{14.25}$$

Similarly, if δ is eliminated in favor of b, the optimum height can be expressed as

$$b_{\text{opt}} = 0.738 \left(\frac{k A_p}{\epsilon\sigma T_b^3} \right)^{1/3} \tag{14.26}$$

Using $G = 1.381$ and evaluating eq. (14.23), the tip temperature for the fin with optimum dimensions is determined as

$$T_{a,\text{opt}} = 0.799 T_b \tag{14.27}$$

If eqs. (14.25) and (14.27) are used in eq. (14.19), the heat dissipation from the optimum fin is obtained as

$$q_{\text{opt}} = 0.855 \left[k(\sigma\epsilon)^2 A_p T_b^9 \right]^{1/3} \tag{14.28}$$

and because the fin efficiency for these optimum conditions η_{opt} is given by

$$\frac{q_{\text{opt}}}{2b_{\text{opt}}\epsilon\sigma T_b^4}$$

one can use eqs. (14.26) and (14.28) to obtain

$$\eta_{\text{opt}} = 0.579 \tag{14.29}$$

This problem has also been considered by Wilkins (1960a) and by Bartas and Sellers (1960). Wilkins used a calculus of variations approach, and Bartas and Sellers adopted a numerical method. Wilkins's results are convenient to use when the heat dissipation q is specified rather than the profile area A_p. For example, Wilkins has shown that the optimum dimensions in terms of the heat dissipation q are given by

$$b_{\text{opt}} = 0.8844 \left(\frac{q}{\epsilon\sigma T_b^4} \right) \tag{14.30}$$

$$\delta_{\text{opt}} = 1.8484 \left(\frac{q^2}{k\epsilon\sigma T_b^5} \right) \tag{14.31}$$

and

$$A_{\text{opt}} = 1.6347 \left(\frac{q^3}{k\epsilon^2\sigma^2 T_b^9} \right) \tag{14.32}$$

On the other hand, Bartas and Sellers provide a graph of fin efficiency,

$$\eta = \frac{q}{2b\epsilon\sigma T_b^4}$$

as a function of the *profile number*,

$$\zeta = \frac{2\sigma\epsilon b^2 T_b^3}{k\delta}$$

for the optimum conditions. It appears that the results of Wilkins are more convenient to use than those of Bartas and Sellers. More recently, Chung and Nguyen (1986) have used a numerical approach to determine the characteristics of the optimum fin. Their strategy was to solve the fin equation numerically and obtain the fin efficiency as a function of the same profile number:

$$\zeta = \frac{2\sigma\epsilon b^2 T_b^3}{k\delta}$$

To find the optimum point, they expressed the heat dissipation in terms of the fin efficiency as

$$q = \left[4k(\sigma\epsilon)^2 A_p T_b^9\right]^{1/3} \eta\zeta^{1/3} \tag{14.33}$$

which shows that for fixed A_p, q is a function of η, ζ, and a constant C (Mackay and Bacha, 1962; Kern and Kraus, 1972):

$$q^3 = C\zeta\eta^3$$

This may be written in differential form as

$$3q^2\,dq = C(3\zeta\eta^2\,d\eta + \eta^3\,d\zeta)$$

For q to be a maximum, $dq = 0$ and hence

$$\frac{d\eta}{d\zeta} = -\frac{1}{3}\frac{\eta}{\zeta} \tag{14.34}$$

or

$$\frac{d(\ln\eta)}{d(\ln\zeta)} = -\frac{1}{3} \tag{14.35}$$

Equation (14.35) shows that at the optimum point, the slope of the $\ln\eta$ versus $\ln\zeta$ curve should have a slope of $-\frac{1}{3}$. Thus by plotting the data (η and ζ) on log–log coordinates and identifying the point where the slope is $-\frac{1}{3}$, the optimum values of η and ζ can be established. When substituted into eq. (14.33), these values give the heat dissipation from the optimum fin. The same information can also be used to obtain δ_{opt} and b_{opt}. Moreover, this indicates that the curves of the fin efficiency η as a function of the profile number ζ plotted on log–log paper will reach optimum mass when the slope of the curves is $-\frac{1}{3}$. The loci of these points is plotted in Figs. 13.12 through 13.15[1] to show a line marked "theoretical optimum." The final results reported by Chung and Nguyen vary slightly from those quoted here. For example, they give the

[1]The same reasoning and development apply to longitudinal fins of trapezoidal and triangular profile.

numerical constants for eqs. (14.25) through (14.29) as 1.332, 0.751, 0.798, 0.849, and 0.565, respectively.

In a 1992 thesis, Smith (1992) used the cascade algorithm developed by Kraus et al. (1978) to optimize fins radiating to free space. His basic idea was division of the fin into a number of subfins, with subsequent generation of a thermal transmission matrix for each subfin by analogy with a segment of the electrical transmission line. By multiplying the individual transmission matrices for each subfin, an overall thermal transmission matrix for the entire fin was derived. The results of an optimization study using the cascade algorithm are close to the values quoted here. For example, Smith gives the numerical constants for eqs. (14.30) through (14.32) as 0.8675, 1.8648, and 1.6178, respectively.

Example 14.2. A longitudinal fin of rectangular profile is radiating to free space. The fin is to be made of a steel with $k = 58.3$ W/m·K and $\epsilon = 0.85$, and its mass should not exceed 1 kg. The base temperature is expected to be 875 K. Assume the fin length to be 1 m, and take the density, as $\rho = 7832$ kg/m^3 for steel and determine the optimum dimensions. Then determine the heat dissipated by the fin and the tip temperature.

SOLUTION. The volume of the fin for a mass of 1 kg is

$$V = \frac{1}{\rho} = \frac{1}{7832} = 1.2768 \times 10^{-4}\ \text{m}^3$$

and for a length of 1 m, the profile area is

$$A_p = 1.2768 \times 10^{-4}\ \text{m}^2$$

Then use of eq. (14.25) gives

$$\delta_{\text{opt}} = 1.355 \left(\frac{\epsilon\sigma A_p^2 T_b^3}{k} \right)^{1/3}$$

$$= 1.355 \left[\frac{(0.85)(5.669 \times 10^{-8})(1.2768 \times 10^{-4})^2 (875)^3}{58.3} \right]^{1/3}$$

$$= 2.82 \times 10^{-3}\ \text{m} \quad \text{or} \quad 2.82\ \text{mm}$$

and

$$b_{\text{opt}} = \frac{A_p}{\delta_{\text{opt}}}$$

$$= \frac{1.2768 \times 10^{-4}}{2.82 \times 10^{-3}}$$

$$= 0.0453\ \text{m} \quad \text{or} \quad 45.3\ \text{mm}$$

To calculate the tip temperature, use eq. (14.27):

$$T_a = 0.799T_b = (0.799)(875) = 699 \text{ K}$$

and then, using eq. (14.28)

$$q = (0.855)\left[k(\sigma\epsilon)^2 A_p T_b^9\right]^{1/3}$$

$$= (0.855)\{(58.3)[(5.669 \times 10^{-8})(0.85)]^2(1.2748 \times 10^{-4})(875)^9\}^{1/3}$$

$$= 1481 \text{ W/m}$$

Example 14.3. Determine the optimum dimensions of a rectangular fin capable of radiating 1481 W/m of heat to free space at 0 K. The fin is fabricated of steel with $k = 58.3$ W/m·K and $\epsilon = 0.85$. Assume the fin base temperature to be at 875 K.

SOLUTION. In this example it is more convenient to use eqs. (14.30) and (14.31). Thus

$$\delta_{\text{opt}} = 1.8484\left(\frac{q^2}{k\epsilon\sigma T_b^5}\right)$$

$$= 1.8484\left[\frac{(1481)^2}{(58.3)(0.85)(5.669 \times 10^{-8})(875)^5}\right]$$

$$= 2.81 \times 10^{-3} = 2.81 \text{ mm}$$

and

$$b_{\text{opt}} = 0.8844\left(\frac{q}{\epsilon\sigma T_b^4}\right)$$

$$= 0.8844\left[\frac{1481}{(0.85)(5.669 \times 10^{-8})(875)^4}\right]$$

$$= 0.0464 \text{ m} \quad \text{or} \quad 46.4 \text{ mm}$$

It is noted that the dimensions obtained here from a consideration of the heat dissipation are very close to those obtained in Example 14.2 based on considerations of minimum metal. This demonstrates the consistency between the Liu (1961) and Wilkins (1960a) analyses. Furthermore, the results of Examples 14.2 and 14.3 also check with the graphical results of Bartas and Sellers (1960) and the numerical results of Chung and Nguyen (1986).

Longitudinal Fin of Triangular Profile. The optimization problem for the longitudinal fin of triangular profile shown in Fig. 14.1*b* has been the subject of several studies. For example, Wilkins (1960b) formulated a novel similarity transformation and used it to replace the nonlinear fin differential equation with an expression for the profile A_p in terms of an integral. The minimum of this integral establishes the optimum dimensions of the fin. Kern and Kraus (1972) tackled the same problem using a numerical procedure to solve for the efficiency η of the fin as a function of the profile number $\zeta = 2\sigma\epsilon b^2 T_b^3 / k\delta_b$, and identified the optimum using eq. (14.35), which also holds for the triangular geometry. Essentially the same procedure has been used by Chung and Nguyen (1986) and Nilson and Curry (1960) to establish the optimum dimensions of a triangular fin. Using the cascade algorithm referred to in Section 13.4, Smith (1992) also reported on the results for the optimum triangular fin. The results of all of these studies can be expressed in the forms

$$\delta_{b,\text{opt}} = C_1 \frac{q^2}{\sigma\epsilon k T_b^5} \tag{14.36}$$

and

$$b_{\text{opt}} = C_2 \frac{q}{\sigma\epsilon T_b^4} \tag{14.37}$$

where the values of the constants C_1 and C_2, taken from the different studies appear in Table 14.1. The results of Kern and Kraus (1972) and Smith (1992) provide somewhat of a departure from those of Wilkins (1960b), Chung and Nguyen (1986), and Nilson and Curry (1960), which are very close.

If δ_{opt} and b_{opt} are desired in terms of the profile area A_p, the expressions are

$$\delta_{b,\text{opt}} = C_3 \left(\frac{\sigma\epsilon A_p^2 T_b^3}{k} \right)^{1/3} \tag{14.38}$$

and

$$b_{\text{opt}} = C_4 \left(\frac{k A_p}{\sigma\epsilon T_b^3} \right)^{1/3} \tag{14.39}$$

where C_3 and C_4 are also provided in Table 14.1. The table also contains the results for the efficiency η and the temperature ratio T_a/T_b for the optimum fin.

Example 14.4. Design a longitudinal fin of triangular profile with optimum dimensions for the data of Example 14.3.

SOLUTION. Use of eqs. (14.36) and (14.37) with C_1 and C_2 from Wilkins's work provides

TABLE 14.1 Constants C_1, C_2, C_3, and C_4 for Eqs. (14.36)–(14.39)

	C_1	C_2	C_3	C_4	η_{opt}	$(T_a/T_b)_{\text{opt}}$
Wilkins (1960b)	2.2986	0.9598	—	—	—	—
Kern and Kraus (1972)	2.4255	0.9091	2.274	0.880	0.550	—
Chung and Nguyen (1986)	2.3104	0.9485	2.174	0.920	0.527	0.714
Nilson and Curry (1960)	2.3027	0.9546	2.162	0.925	0.524	0.711
Smith (1992)	2.4852	0.9040	—	—	—	—

$$\delta_{b,\mathrm{opt}} = 2.2986\left(\frac{q^2}{\sigma\epsilon k T_b^5}\right)$$

$$= 2.2986\left[\frac{(1481)^2}{(5.669\times10^{-8})(0.85)(58.3)(875)^5}\right]$$

$$= 3.499\times10^{-3}\ \mathrm{m} \quad \text{or} \quad 3.499\ \mathrm{mm}$$

and

$$b_{\mathrm{opt}} = 0.9598\left(\frac{q}{\sigma\epsilon T_b^4}\right)$$

$$= 0.9598\left[\frac{1481}{(5.669\times10^{-8})(0.85)(875)^4}\right]$$

$$= 0.0503\ \mathrm{m} \quad \text{or} \quad 50.3\ \mathrm{mm}$$

A comparison of the results of Examples 14.3 and 14.4 shows that for the same heat dissipation, the optimum triangular fin is longer and thicker at the base than the optimum rectangular fin.

Longitudinal Fin of Trapezoidal Profile. The optimization of the longitudinal fin of trapezoidal profile shown in Fig. 14.1*c* was first considered by Mackay and Bacha (1961). This work has been discussed extensively by Kern and Kraus (1972) as well as in Section 13.3. The approach used by Mackay and Bacha was to solve the governing differential equation numerically and then obtain the fin efficiency η as a function of the profile number $\zeta = \sigma\epsilon T_b^3 b^2/k\delta_b$ and the taper ratio $\lambda = \delta_a/\delta_b$. The optimum was then located by applying the criterion given by

$$\frac{d(\ln\eta)}{d(\ln\zeta)} = \frac{1}{3} \tag{14.35}$$

which is also valid for the trapezoidal profile. The results for η_{opt} and $(\sigma\epsilon T_b^3 b^2/k\delta_b)_{\mathrm{opt}}$ for three values of the taper ratio are given in Table 14.2.

Example 14.5. Design a longitudinal fin of trapezoidal profile with optimum dimensions for the data of Example 14.3. The taper ratio is to be 0.25.

SOLUTION. From Table 14.2, $\eta_{\mathrm{opt}} = 0.5588$ and $(\sigma\epsilon T_b^3 b^2/k\delta_b)_{\mathrm{opt}} = 0.3747$. Because η is known, the ideal heat flow, q_{id} can be computed:

TABLE 14.2 Optimum Parameters for a Trapezoidal Fin

δ_a/δ_b	η_{opt}	$(\sigma\epsilon T_b^3 b^2/k\delta_b)_{\mathrm{opt}}$
0.25	0.5588	0.3747
0.50	0.5684	0.3737
0.75	0.5711	0.4053

$$q_{id} = 2\sigma\epsilon b_{opt} T_b^4 = \frac{q}{\eta} = \frac{1481}{0.5588} = 2650 \text{ W/m}$$

Solving for b_{opt} yields

$$\begin{aligned} b_{opt} &= \frac{q_{id}}{2\sigma\epsilon T_b^4} \\ &= \frac{2650}{(2)(5.669 \times 10^{-8})(0.85)(875)^4} \\ &= 0.0469 \text{ m} \quad \text{or} \quad 46.9 \text{ mm} \end{aligned}$$

Now from Table 14.2 at $\delta_a/\delta_b = 0.25$,

$$\left.\frac{\sigma\epsilon T_b^3 b^2}{k\delta_b}\right|_{opt} = 0.3747$$

and $\delta_{b,opt}$ is then determined:

$$\begin{aligned} \delta_{b,opt} &= \frac{\sigma\epsilon T_b^3 b_{opt}^2}{0.3747k} \\ &= \frac{(5.669 \times 10^{-8})(0.85)(875)^3(0.0469)^2}{(0.3747)(58.3)} \\ &= 3.25 \times 10^{-3} \text{ m} \quad \text{or} \quad 3.25 \text{ mm} \end{aligned}$$

and then

$$\begin{aligned} \delta_{a,opt} &= 0.25\delta_{b,opt} \\ &= (0.25)(3.25 \times 10^{-3}) \\ &= 8.125 \times 10^{-4} \text{ m} \quad \text{or} \quad 0.81 \text{ mm} \end{aligned}$$

Longitudinal Fin of Concave Parabolic Profile. Studies pertaining to the optimization of the longitudinal fin of concave parabolic profile shown in Fig. 14.1*d* have apparently been sparse. One study, by Chung and Nguyen (1986), used the approach that was described in some detail earlier. The various relationships for the optimum fin are

$$\delta_{b,opt} = 2.621\left(\frac{\sigma\epsilon A_p^2 T_b^3}{k}\right)^{1/3} \tag{14.40}$$

$$b_{opt} = 1.145\left(\frac{kA_p}{\sigma\epsilon T_b^3}\right)^{1/3} \tag{14.41}$$

$$T_{a,opt} = 0 \tag{14.42}$$

$$q_{opt} = 0.998\left[k(\sigma\epsilon)^2 A_p T_b^9\right]^{1/3} \tag{14.43}$$

and

$$\eta_{opt} = 0.436 \tag{14.44}$$

If q is specified instead of A_p, one can use eq. (14.43) to calculate A_p and then use eqs. (14.40) and (14.41) to find $\delta_{b,opt}$ and b_{opt}. Observe that the tip temperature for the optimum fin is equal to the effective sink temperature.

Example 14.6. Design a longitudinal fin of concave parabolic profile with optimum dimensions for the data of Example 14.3.

SOLUTION. Rewriting eq. (14.43) and then solving for A_p gives

$$A_p = 1.006\left[\frac{q^3}{k(\sigma\epsilon)^2 T_b^9}\right]$$

$$= 1.006\left[\frac{(1481)^3}{(58.3)(5.669\times10^{-8})^2(0.85)^2(875)^9}\right]$$

$$= 8.029\times10^{-5}\ \text{m}^2$$

and then use of eq. (14.40) yields $\delta_{b,opt}$:

$$\delta_{b,opt} = 2.621\left(\frac{\sigma\epsilon A_p^2 T_b^3}{k}\right)^{1/3}$$

$$= 2.621\left[\frac{(5.669\times10^{-8})(0.85)(8.029\times10^{-5})^2(875)^3}{58.3}\right]^{1/3}$$

$$= 4.01\times10^{-3}\ \text{m} \quad \text{or} \quad 4.01\ \text{mm}$$

Equation (14.41) then provides

$$b_{opt} = 1.145\left(\frac{kA_p}{\sigma\epsilon T_b^3}\right)^{1/3}$$

$$= 1.145\left[\frac{58.3(8.026\times10^{-5})}{(5.669\times10^{-8})(0.85)(875)^3}\right]^{1/3}$$

$$= 0.060\ \text{m} \quad \text{or} \quad 60\ \text{mm}$$

Longitudinal Fin of Convex Parabolic Profile. Figure 14.1*e* shows a longitudinal fin of convex parabolic profile. Apparently, the only study that relates to the optimization of this geometry is that of Chung and Nguyen (1986), who found the optimum relationships to be

$$\delta_{b,opt} = 1.778\left(\frac{\sigma\epsilon A_p^2 T_b^3}{k}\right)^{1/3} \tag{14.45}$$

$$b_{\text{opt}} = 0.843 \left(\frac{kA_p}{\sigma\epsilon T_b^3} \right)^{1/3} \tag{14.46}$$

$$T_{a,\text{opt}} = 0.768T_b \tag{14.47}$$

$$q_{\text{opt}} = 0.927[k(\sigma\epsilon)^2 A_p T_b^9]^{1/3} \tag{14.48}$$

and

$$\eta_{\text{opt}} = 0.550 \tag{14.49}$$

Example 14.7. Design a longitudinal fin of convex parabolic profile with optimum dimensions for the data of Example 14.3.

SOLUTION. Equation (14.48) can be modified to give the profile area A_p:

$$\begin{aligned} A_p &= 1.2553 \left[\frac{q^3}{k(\sigma\epsilon)^2 T_b^9} \right] \\ &= 1.2553 \left[\frac{(1481)^3}{(58.3)(5.669 \times 10^{-8})^2(0.85)^2(875)^9} \right] \\ &= 1.0015 \times 10^{-4}\ \text{m}^2 \end{aligned}$$

Then eq. (14.45) is used to determine $\delta_{b,\text{opt}}$:

$$\begin{aligned} \delta_{b,\text{opt}} &= 1.778 \left(\frac{\sigma\epsilon A_p^2 T_b^3}{k} \right)^{1/3} \\ &= 1.778 \left[\frac{(5.669 \times 10^{-8})(0.85)(1.0015 \times 10^{-4})^2(875)^3}{58.3} \right]^{1/3} \\ &= 3.15 \times 10^{-3}\ \text{m} \quad \text{or} \quad 3.15\ \text{mm} \end{aligned}$$

Finally, use of eq. (14.46) gives b_{opt}:

$$\begin{aligned} b_{\text{opt}} &= 0.843 \left(\frac{kA_p}{\sigma\epsilon T_b^3} \right)^{1/3} \\ &= 0.843 \left[\frac{(58.3)(1.0015 \times 10^{-4})}{(5.669 \times 10^{-8}(0.85)(875)^3} \right]^{1/3} \\ &= 0.0477\ \text{m} \quad \text{or} \quad 47.7\ \text{mm} \end{aligned}$$

Comparison of Optimum Fins. It is interesting to compare the profile areas for the optimum fins of rectangular, triangular, concave parabolic, and convex parabolic profiles. Expressing $A_{p,\text{opt}}$ in terms of q gives

$$\text{Rectangular :} \quad A_{p,\text{opt}} = 1.6347 \left(\frac{q^3}{k\epsilon^2\sigma^2 T_b^9} \right) \tag{14.50}$$

$$\text{Triangular :} \quad A_{p,\text{opt}} = 1.10131 \left(\frac{q^3}{k\epsilon^2\sigma^2 T_b^9} \right) \tag{14.51}$$

$$\text{Concave parabolic :} \quad A_{p,\text{opt}} = 1.006 \left(\frac{q^3}{k\epsilon^2\sigma^2 T_b^9} \right) \tag{14.52}$$

and

$$\text{Convex parabolic :} \quad A_{p,\text{opt}} = 1.2553 \left(\frac{q^3}{k\epsilon^2\sigma^2 T_b^9} \right) \tag{14.53}$$

For the same heat dissipation and operational parameters, eqs. (14.50) through (14.53) show that the concave parabolic fin has the least profile area and consequently, is the lightest of the four shapes. It uses only 61.5% as much material as the rectangular fin. However, the triangular fin, which uses only 9% more material than the concave parabolic fin, may be preferable for ease of fabrication.

As in the convective case, because A_p in each case is proportional to q^3, the increase in A_p is eightfold if q is to be doubled. Thus if a single fin is used to accommodate twice the heat dissipation, the fin becomes very bulky. In this case it is better to employ two identical fins instead of one. Indeed, the use of a larger number of shorter and lighter fins as opposed to fewer longer and heavier fins results in a better design.

When selecting the fin material, a variety of choices are available. Because the weight of the fin is proportional to $\gamma/k\epsilon^2$, the material with the lowest value of $\gamma/k\epsilon^2$ gives the lightest fin for a given geometry. Notice that this is similar to, but not exactly the same as, in the convective case, where the weight of the fin was proportional to γ/k.

14.3 OPTIMUM DIMENSIONS OF RADIATING SPINES

Unlike the large number of papers on the optimum design of longitudinal fins, only two papers dealing with the optimum design of radiating spines have been identified: Wilkins (1960a) deals with the optimum dimensions of a cylindrical spine, and Chung and Nguyen (1985) consider a family of spines. The analyses in both papers pertain to a single fin and are based on the Murray–Gardner assumptions. The presentation here follows the pattern of Section 14.2 and the optimum dimensions, volume, and efficiency are developed for cylindrical, conical, and parabolic spines.

14.3.1 Specific Profiles

Cylindrical Spine. Consider a cylindrical spine of radius r and height b, as shown in Fig. 14.2*a*. For a given spine volume $V = \pi r^2 b$, the problem is to find r and b so

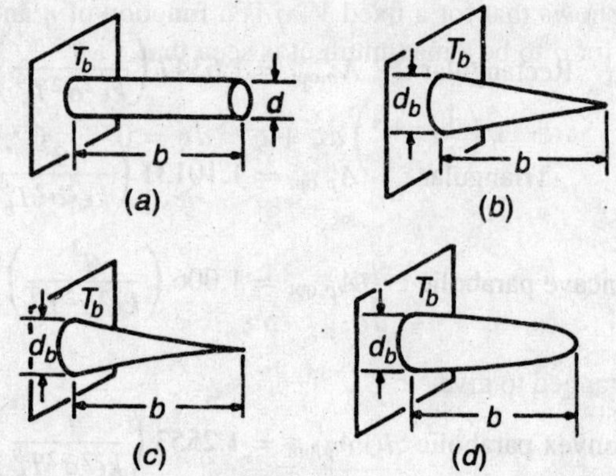

Figure 14.2 Spine geometries: (*a*) cylindrical; (*b*) conical; (*c*) concave parabolic; (*d*) convex parabolic. All spines possess a thermal conductivity k, have a surface emissivity ϵ, and operate at a base temperature T_b.

that q is maximized. By utilizing the similarity transformation mentioned in Section 14.2.1, Wilkins (1960a) successfully reduced the problem to the minimization of an integral representing the volume of the spine. The optimum characteristics were then obtained by finding the minimum of the integral employing the calculus of variations. Wilkins's results for r_{opt}, b_{opt}, V_{opt}, and η_{opt} are

$$r_{\text{opt}} = 0.6870 \left(\frac{q^2}{\sigma \epsilon k T_b^5} \right)^{1/3} \tag{14.54}$$

$$b_{\text{opt}} = 0.2999 \left[\frac{qk}{(\sigma\epsilon)^2 T_b^7} \right]^{1/3} \tag{14.55}$$

$$V_{\text{opt}} = 0.4447 \left[\frac{q^5}{(\sigma\epsilon)^4 k T_b^{17}} \right]^{1/3} \tag{14.56}$$

and

$$\eta_{\text{opt}} = 0.7725 \tag{14.57}$$

Chung and Nguyen's (1985) approach was to solve the spine equation numerically and obtain η as a function of the profile number $\zeta = \sigma\epsilon T_b^3 b^2 / kr$. To locate the optimum, they expressed the heat dissipation q as

$$q = \eta q_{\text{id}} = 2\eta\pi r b \sigma \epsilon T_b^4$$

which can be rearranged using the volume V and the definition of the profile number ζ to give

$$q = [32\pi^2 k (\sigma\epsilon)^4 V^3 T_b^{17}]^{1/5} \eta \zeta^{1/5} \tag{14.58}$$

Equation (14.58) shows that for a fixed V, q is a function of η and ζ. Invoking the condition $dq = 0$ for q to be a maximum, it is seen that

$$\eta\left(\frac{1}{5}\zeta^{-4/5}\right)d\zeta + \zeta^{1/5}\,d\eta = 0$$

or

$$\frac{d\eta}{d\zeta} = -\frac{1}{5}\frac{\eta}{\zeta}$$

which can be rearranged to give

$$\frac{d(\ln\eta)}{d(\ln\zeta)} = -\frac{1}{5} \tag{14.59}$$

Thus the optimum point can be determined by plotting η versus ζ on a log–log scale and finding the point where the slope is $-\frac{1}{5}$. The final results for the optimum parameters in terms of V are

$$r_{\text{opt}} = 0.9455\left(\frac{\sigma\epsilon V^2 T_b^3}{k}\right)^{1/5} \tag{14.60}$$

$$b_{\text{opt}} = 0.356\left[\frac{k^2 V}{(\sigma\epsilon)^2 T_b^6}\right]^{1/5} \tag{14.61}$$

$$q_{\text{opt}} = 1.626[k(\sigma\epsilon)^4 V^3 T_b^{17}]^{1/5} \tag{14.62}$$

$$T_{a,\text{opt}} = 0.904 T_b \tag{14.63}$$

and

$$\eta_{\text{opt}} = 0.769 \tag{14.64}$$

Conical Spine. For the conical spine shown in Fig. 14.2*b*, the optimum parameters are

$$r_{b,\text{opt}} = 1.214\left(\frac{\sigma\epsilon V^2 T_b^3}{k}\right)^{1/5} \tag{14.65}$$

$$b_{\text{opt}} = 0.648\left[\frac{k^2 V}{(\sigma\epsilon)^2 T_b^6}\right]^{1/5} \tag{14.66}$$

$$q_{\text{opt}} = 1.846[k(\sigma\epsilon)^4 V^3 T_b^{17}]^{1/5} \tag{14.67}$$

$$T_{a,\text{opt}} = 0.806 T_b \tag{14.68}$$

and

$$\eta_{\text{opt}} = 0.747 \tag{14.69}$$

Concave Parabolic Spine. The optimum relationships for the concave parabolic spine shown in Fig. 14.2*c* are

$$r_{b,\text{opt}} = 1.2935\left(\frac{\sigma\epsilon V^2 T_b^3}{k}\right)^{1/5} \tag{14.70}$$

$$b_{\text{opt}} = 0.951\left[\frac{k^2 V}{(\sigma\epsilon)^2 T_b^6}\right]^{1/5} \tag{14.71}$$

$$q_{\text{opt}} = 1.865[k(\sigma\epsilon)^4 V^3 T_b^{17}]^{1/5} \tag{14.72}$$

$$T_{a,\text{opt}} = 0 \tag{14.73}$$

and

$$\eta_{\text{opt}} = 0.723 \tag{14.74}$$

Convex Parabolic Spine. For the convex parabolic spine shown in Fig. 14.2*d*, the optimum relationships are

$$r_{b,\text{opt}} = 1.124\left(\frac{\sigma\epsilon V^2 T_b^3}{k}\right)^{1/5} \tag{14.75}$$

$$b_{\text{opt}} = 0.504\left[\frac{k^2 V}{(\sigma\epsilon)^2 T_b^6}\right]^{1/5} \tag{14.76}$$

$$q_{\text{opt}} = 1.800[k(\sigma\epsilon)^4 V^3 T_b^{17}]^{1/5} \tag{14.77}$$

$$T_{a,\text{opt}} = 0.868 T_b \tag{14.78}$$

and

$$\eta_{\text{opt}} = 0.759 \tag{14.79}$$

Example 14.8. It is desired to dissipate 100 W from a primary surface at 875 K by radiation to free space at 0 K by attaching a spine. Determine the optimum base radius and optimum height for the (**a**) cylindrical, (**b**) conical, (**c**) concave parabolic, and (**d**) convex parabolic shapes. The spines are to be made of a material having a thermal conductivity of 65 W/m·K and an emissivity of 0.85. Compare the heat dissipation per unit volume for each shape.

SOLUTION. (**a**) Using eq. (14.54) gives

$$r_{\text{opt}} = 0.6870\left(\frac{q^2}{\sigma\epsilon k T_b^5}\right)^{1/3}$$

$$= 0.6870\left[\frac{(100)^2}{(5.669\times10^{-8})(0.85)(65)(875)^5}\right]^{1/3}$$

$$= 0.0126 \text{ m} \quad \text{or} \quad 12.6 \text{ mm}$$

Using eq. (14.55) yields

$$b_{\text{opt}} = 0.2999 \left[\frac{qk}{(\sigma\epsilon)^2 T_b^7} \right]^{1/3}$$

$$= 0.2999 \left[\frac{100(65)}{(5.669 \times 10^{-8})^2 (0.85)^2 (875)^7} \right]^{1/3}$$

$$= 0.0577 \text{ m} \quad \text{or} \quad 57.7 \text{ mm}$$

(**b**) Rewriting eq. (14.67) gives

$$V = 0.3600 \left[\frac{q^5}{k(\sigma\epsilon)^4 T_b^{17}} \right]^{1/3}$$

$$= 0.3600 \left[\frac{(100)^5}{(65)(5.669 \times 10^{-8})^4 (0.85)^4 (875)^{17}} \right]^{1/3}$$

$$= 2.346 \times 10^{-5} \text{ m}^3$$

Using eq. (14.65) gives

$$r_{b,\text{opt}} = 1.214 \left(\frac{\sigma\epsilon V^2 T_b^3}{k} \right)^{1/5}$$

$$= 1.214 \left[\frac{(5.669 \times 10^{-8})(0.85)(2.346 \times 10^{-5})^2 (875)^3}{65} \right]^{1/5}$$

$$= 0.0148 \text{ m} \quad \text{or} \quad 14.8 \text{ mm}$$

Then because $V = \pi r_b^2 b/3$,

$$b_{\text{opt}} = \frac{3V}{\pi r_{b,\text{opt}}^2}$$

$$= \frac{(3)(2.346 \times 10^{-5})}{\pi (0.0148)^2}$$

$$= 0.1023 \text{ m} \quad \text{or} \quad 102.3 \text{ mm}$$

(**c**) Rewriting eq. (14.72) yields V

$$V = 0.3539 \left[\frac{q^5}{k(\sigma\epsilon)^4 T_b^{17}} \right]^{1/3}$$

$$= 0.3539 \left[\frac{(100)^5}{(65)(5.669 \times 10^{-8})^4 (0.85)^4 (875)^{17}} \right]^{1/3}$$

$$= 2.306 \times 10^{-5} \text{ m}^3$$

Using eq. (14.70) gives

$$r_{b,\mathrm{opt}} = 1.2935\left(\frac{\sigma\epsilon V^2 T_b^3}{k}\right)^{1/5}$$

$$= 1.2935\left[\frac{(5.669\times 10^{-8})(0.85)(2.306\times 10^{-5})^2(875)^3}{65}\right]^{1/5}$$

$$= 0.0157\ \mathrm{m} \quad \text{or} \quad 15.7\ \mathrm{mm}$$

Because $V = \pi r_b^2 b/5$,

$$b_{\mathrm{opt}} = \frac{5V}{\pi r_{b,\mathrm{opt}}^2}$$

$$= \frac{(5)(2.306\times 10^{-5})}{\pi(0.0157)^2}$$

$$= 0.1489\ \mathrm{m} \quad \text{or} \quad 148.9\ \mathrm{mm}$$

(d) Rewriting eq. (14.77) yields V

$$V = 0.3755\left[\frac{q^5}{k(\sigma\epsilon)^4 T_b^{17}}\right]^{1/3},$$

$$= 0.3755\left[\frac{(100)^5}{(65)(5.669\times 10^{-8})^4(0.85)^4(875)^{17}}\right]^{1/3}$$

$$= 2.447\times 10^{-5}\ \mathrm{m}^3$$

Using eq. (14.75) yields

$$r_{b,\mathrm{opt}} = 1.124\left(\frac{\sigma\epsilon V^2 T_b^3}{k}\right)^{1/5}$$

$$= 1.124\left[\frac{(5.669\times 10^{-8})(0.85)(2.447\times 10^{-5})^2(875)^3}{65}\right]^{1/5}$$

$$= 0.0140\ \mathrm{m} \quad \text{or} \quad 14\ \mathrm{mm}$$

Because $V = \pi r_b^2 b/2$,

$$b_{\mathrm{opt}} = \frac{2V}{\pi r_{b,\mathrm{opt}}^2}$$

$$= \frac{(2)(2.447\times 10^{-5})}{\pi(0.0140)^2}$$

$$= 0.0795\ \mathrm{m} \quad \text{or} \quad 79.5\ \mathrm{mm}$$

The heat dissipation per unit volume q/V for the cylindrical, conical, concave parabolic, and convex parabolic shapes are 3475, 4263, 4337, and 4087 kW/m^3, respectively. The concave parabolic shape has the highest heat dissipation per unit volume, and the cylindrical spine has the lowest heat dissipation per unit volume.

14.4 IMPROVED OPTIMIZATION ANALYSES FOR RADIATING FINS

The optimization procedures discussed in Section 14.2 have been based on a number of assumptions. These procedures have been improved to include the effects of:

- A nonzero sink temperature
- Radiative interaction between the fin and the primary surface
- Temperature-dependent thermal conductivity and emissivity
- Mutual irradiation between adjacent fins
- Combined effect of mutual irradiation and environmental radiation
- Radiation interaction with the associated structure

This section is devoted to a discussion of these effects.

14.4.1 Nonzero Sink Temperature

Consider the the longitudinal fin of rectangular profile shown in Fig. 14.1*a*. Let the effective sink temperature for radiation be T_s. In this case, an analytical solution for q is not possible. The fin equation must therefore be solved numerically. Among others, Bartas and Sellers (1960) have solved this problem for the boundary conditions of constant base temperature and an insulated tip. They obtained the fin efficiency η as a function of the profile number $\zeta = 2\sigma\epsilon b^2 T_b^3/k\delta$. Next, using eq. (14.35), the optimum fin efficiency η_{opt} and the optimum profile number ζ_{opt} were determined. Figure 14.3 shows the variation of η_{opt} and ζ_{opt} as a function of the ratio $\theta_s = T_s/T_b$. As θ_s increases, η_{opt} increases while ζ_{opt} decreases.

Example 14.9. A longitudinal fin of rectangular profile is to be considered using the data of Example 14.3. Recall that the fin base temperature was 875 K and the sink temperature was taken as 0 K. Here the effect of nonzero sink temperatures is examined and the sink temperature values to be considered are 525, 612.5, 700, and 787.5 K, which give the values $\theta_s = 0.6, 0.7, 0.8$, and 0.9, respectively. Values for b_{opt} and δ_{opt} will be calculated for each case and the effect on the ratio q/V due to changes in the heat sink temperature will be studied.

SOLUTION. Use Fig. 14.3 to find η_{opt} and ζ_{opt} for $\theta_s = 0.6, 0.7, 0.8$, and 0.9 as follows:

	θ_s			
	0.6	0.7	0.8	0.9
η_{opt}	0.577	0.589	0.602	0.616
ζ_{opt}	0.770	0.700	0.635	0.565

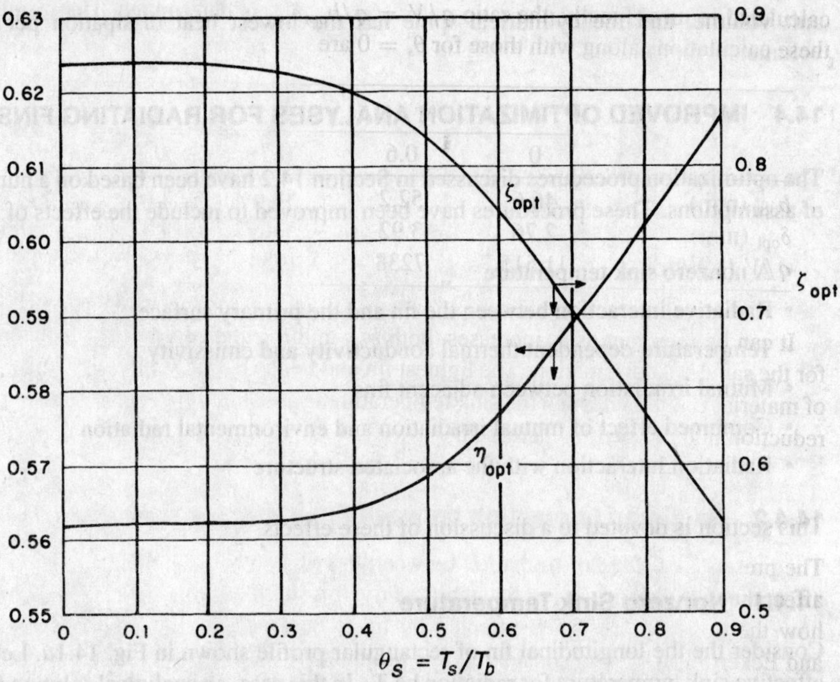

Figure 14.3 Efficiency and profile number for a longitudinal radiating fin of rectangular profile. [Adapted from Bartas and Sellars (1960).]

From the definition of η_{opt},

$$b_{\text{opt}} = \frac{q}{2\epsilon\sigma T_b^4(1-\theta_s^4)\eta_{\text{opt}}}$$

For all cases of θ_s, $q = 1481$ W/m, $\sigma = 5.669 \times 10^{-8}$ W/m^2·K^4, $\epsilon = 0.85$, and $T_b = 875$ K. Hence

$$b_{\text{opt}} = \frac{0.02621}{(1-\theta_s^4)\eta_{\text{opt}}}$$

Because $\zeta_{\text{opt}} = 2\epsilon\sigma T_b^3 b_{\text{opt}}^2/k\delta_{\text{opt}}$, δ_{opt} can be found:

$$\delta_{\text{opt}} = \frac{2\sigma\epsilon T_b^3}{k}\frac{b_{\text{opt}}^2}{\zeta_{\text{opt}}}$$

$$= \frac{(2)(5.669 \times 10^{-8})(0.85)(875)^3}{58.3}\frac{b_{\text{opt}}^2}{\zeta_{\text{opt}}}$$

$$= 1.1076\frac{b_{\text{opt}}^2}{\zeta_{\text{opt}}}$$

Using the prescribed values of θ_s, η_{opt}, and ζ_{opt} in each case, b_{opt} and then δ_{opt} are calculated in turn. Finally, the ratio $q/V = q/b_{opt}\delta_{opt}$ is determined. The results of these calculations along with those for $\theta_s = 0$ are

	θ_s				
	0	0.6	0.7	0.8	0.9
b_{opt} (mm)	46.6	52.2	58.6	73.7	123.7
δ_{opt} (mm)	2.76	3.92	5.43	9.47	30.0
q/V (kW/m^3)	11,515	7238	4654	2122	399

It can be seen that as θ_s increases, both b_{opt} and δ_{opt} increase, giving heavier fins for the same heat dissipation. The lightest fin and hence the most effective utilization of material is attained when the sink temperature is absolute zero. There is a dramatic reduction in q/V as θ_s is increased.

14.4.2 Radiation Interaction between Fin and Base Surfaces

The presence of mutual irradiation between the fin and its base surface is known to affect the performance of the fin significantly. It is therefore interesting to investigate how the optimum design is affected by fin-to-base radiative interaction. Sparrow and Eckert (1962) have considered this problem by studying the tube sheet radiator shown in Fig. 14.4*a*, where all surfaces are assumed to be black. This configuration has frequently been proposed for heat rejection applications in space. Because of thermal symmetry, results are presented for the quarter segment of the system (Fig. 14.4*b*). Allowing for the radiative exchange between fin and tube, Sparrow and Eckert determined the heat dissipation from the tube as well as the heat dissipation from the fin. Both these dissipations were found to depend on two parameters,

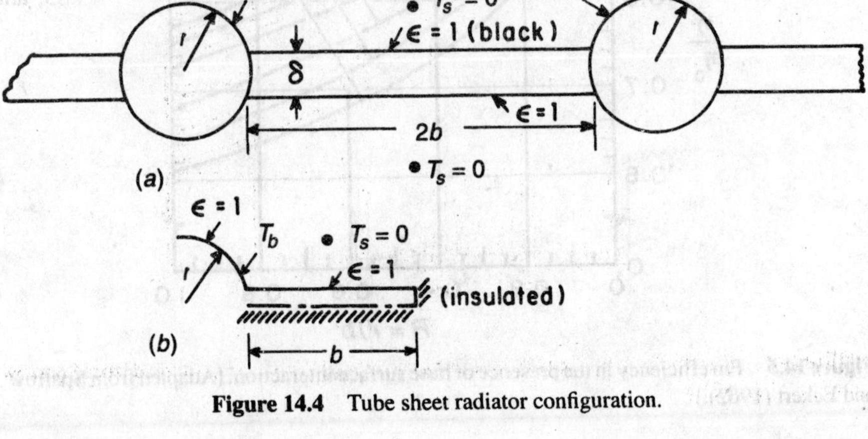

Figure 14.4 Tube sheet radiator configuration.

$$\xi_1 = \frac{2\sigma b^2 T_b^3}{k\delta}$$

and

$$R = \frac{r}{b}$$

The results are shown in Figs. 14.5 through 14.8, where η is the fin efficiency with fin–tube interaction, η_0 is the fin efficiency without fin–tube interaction, and q/b is the heat dissipation from the tube per unit length (into the plane of the page) of the radiator.

The optimum fin dimensions are found by maximizing the total heat dissipation q, which can be expressed in dimensionless form as

$$\frac{q}{\sigma T_b^4 b} = \frac{\pi}{2}\frac{r}{b}\left[\frac{2q_t}{\pi r \sigma T_b^4}\right] + \eta \tag{14.80}$$

This process resulted in a relationship between ξ_1 and

$$\xi_2 = r\left(\frac{2\sigma T_b^3}{k\delta b}\right)^{1/3}$$

which is shown in Fig. 14.8.

Example 14.10. A tube sheet radiator is to be designed to dissipate 2400 W per tube

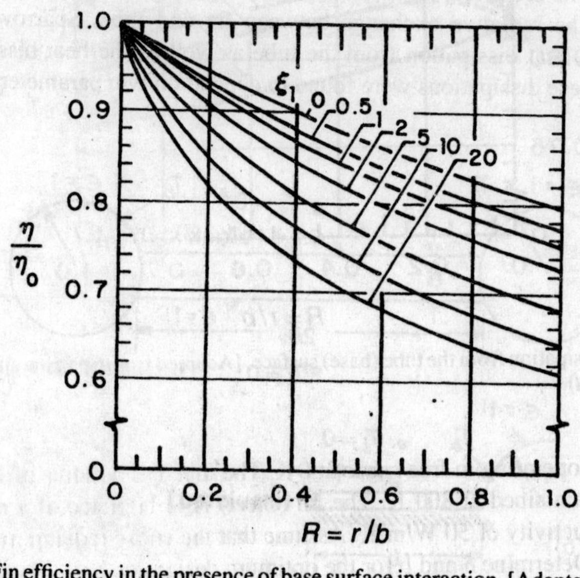

Figure 14.5 Fin efficiency in the presence of base surface interaction. [Adapted from Sparrow and Eckert (1962).]

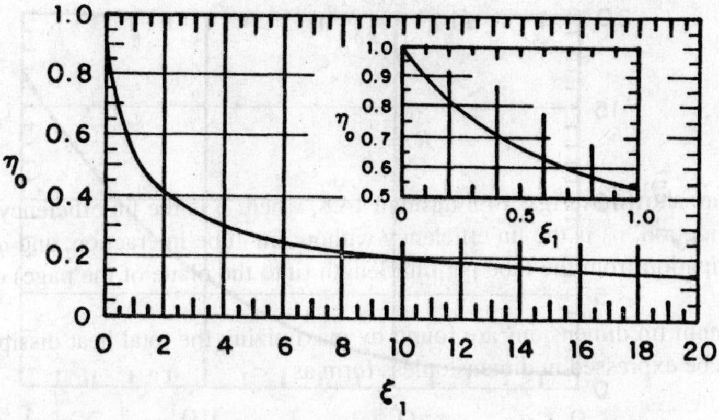

Figure 14.6 Fin efficiency in the absence of base surface interaction. [Adapted from Sparrow and Eckert (1962).]

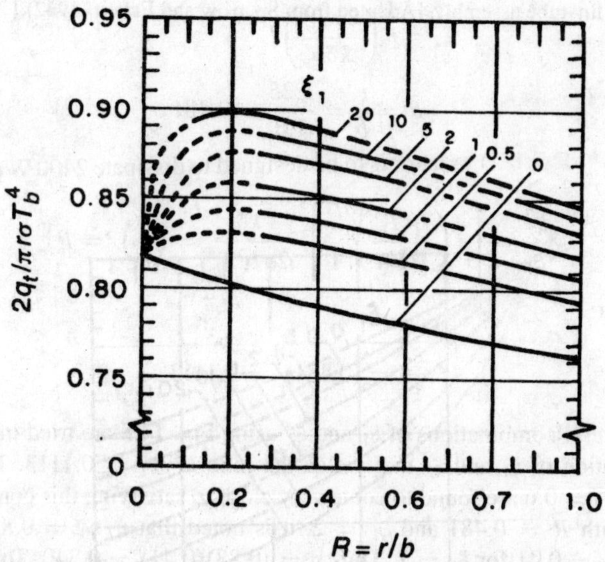

Figure 14.7 Heat dissipation from the tube (base) surface. [Adapted from Sparrow and Eckert (1962).]

when the radiator is operating in free space at 0 K. The tube has a radius of 1.25 cm and its surface is maintained at 800 K. The fin (sheet) is to be made of a material with a thermal conductivity of 50 W/m·K. Assume that the entire radiator assembly is black ($\epsilon = 1.00$). Determine δ and b for the optimum design.

SOLUTION. Because both ξ_1 and ξ_2 involve δ and b, a trial-and-error solution is necessary. Assume that $\delta = 0.5$ mm and $b = 26$ mm. Then

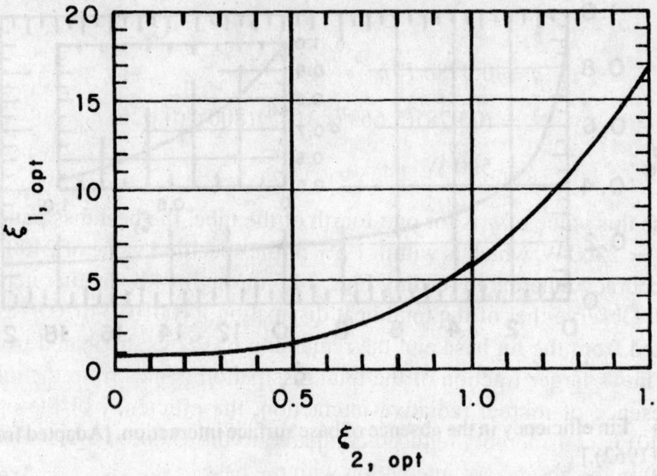

Figure 14.8 Optimum relationship between profile numbers for maximum total heat dissipation from a fin–tube assembly. [Adapted from Sparrow and Eckert (1962).]

$$R = \frac{r}{b} = \frac{1.25}{2.6} = 0.481$$

Noting that

$$\frac{\xi_2^3}{\xi_1} = \left[r\left(\frac{2\sigma T_b^3}{k\delta b}\right)^{1/3}\right]^3 \frac{k\delta}{2\sigma b^2 T_b^3} = \left(\frac{r}{b}\right)^3 = R^3$$

it is seen that

$$\frac{\xi_2^3}{\xi_1} = (0.481)^3 = 0.1113$$

Next, several combinations of ξ_1 and ξ_2 using Fig. 14.8 are tried in order to find the combination of ξ_1 and ξ_2 that yields the ratio $\xi_2^3/\xi_1 = 0.1113$. The values of $\xi_1 = 2$ and $\xi_2 = 0.6$ are found to come very close to satisfying this condition. Using Fig. 14.5 with $R = 0.481$ and $\xi_1 = 2$, it is noted that $\eta/\eta_0 = 0.83$. Next from Fig. 14.6, $\eta_0 = 0.41$ for $\xi_1 = 2$. Thus $\eta = (0.83)(0.41) = 0.340$. To calculate the heat dissipation from the tube, use Fig. 14.7 and find that for $\xi_1 = 2$ and $r/b = 0.481$,

$$\frac{q_t}{(\pi/2)r\sigma T_b^4} = 0.845$$

Finally, these results are put into eq. (14.80) to give

$$\frac{\pi}{2}\frac{r}{b}\frac{2q_t}{\pi r\sigma T_b^4 b} + \eta = \frac{\pi}{2}(0.481)(0.845) + 0.340$$

$$= 0.638 + 0.340$$

$$= 0.978$$

and

$$q = 0.978\sigma T_b^4 b$$
$$= (0.978)(5.669 \times 10^{-8})(800)^4(0.026)$$
$$= 590 \text{ W}$$

Because this value of q is for one-fourth of the tube, the heat dissipation per tube is $4 \times 590 = 2360$ W, which is within 1.6% of the specified value of 2400 W. In view of the inaccuracy inherent in reading Figs. 14.5 through 14.8, another iteration is not warranted. Observe that of the total heat dissipation, $(100)(0.340/0.978) = 34.8\%$ is dissipated from the fin base and the remaining 65.2% is dissipated from the tube surface. Thus a larger fraction of the total dissipation comes from the tube surface. In the presence of mutual radiative interaction, the efficiency of the optimum fin ($\eta = 0.340$) is considerably reduced compared to the value of 0.579 obtained from eq. (14.29) in the absence of interaction with the base.

In addition to the results for black fin-and-tube surfaces, Sparrow and Eckert (1962) also developed a generalized analytical formulation which allowed surfaces to be selectively gray and included the effect of a coordinate dependent distribution of incident energy due to an external source. The actual numerical results for the simpler case of nonselective gray surfaces without external radiation are given by Sarabia and Hitchcock (1966). However, no optimization study was included.

14.4.3 Temperature-Dependent Thermal Conductivity and Emissivity

Campo and Wolko (1973) extended the optimization study of Liu (1961) to include power law variations of thermal conductivity and emissivity with temperature. They considered a rectangular fin (Fig. 14.1a) radiating to free space at 0 K with k and ϵ varying with temperature according to

$$\epsilon = \epsilon_0 T^m \tag{14.81}$$

and

$$k = k_0 T^n \tag{14.82}$$

The fin was assumed to be operating with constant base temperature and an insulated tip. By using an algebraic transformation, the governing equations were reduced to the form obtained by Liu (1961). Following Liu's analysis, the heat dissipation per unit depth was found to be

$$q = \left[\frac{4\epsilon_0 \sigma k_o \delta}{(m+n+5)^{1/2}}\right]^{1/2} \left(T_b^{m+n+5} - T_a^{m+n+5}\right)^{1/2} \tag{14.83}$$

which reduces to eq. (14.19) if $m = n = 0$ (constant k and ϵ). The tip temperature T_a is given by a relationship involving the complete beta function B and the incomplete beta function B_u:

$$B\left(\nu, \frac{1}{2}\right) - B_u\left(\nu, \frac{1}{2}\right) - M\delta^{-3/2}T_a^{(m-n+3)/2} = 0 \tag{14.84}$$

where

$$\nu = \frac{m-n+3}{2(m+n+5)}$$

$$u = \left(\frac{T_a}{T_b}\right)^{m+n+5}$$

and

$$M = \left[\frac{4(m+n+5)\epsilon_0\sigma A_p^2}{k_0}\right]^{1/2}$$

The optimization problem is to maximize $q = q(\delta, T_a)$, as given by eq. (14.83), subject to the constraint of eq. (14.84). The solution requires a simultaneous solution of eq. (14.84) and

$$\delta^{1/2} = -R\frac{\left[T_b^{m+n+5} - (S+1)T_a^{m+n+5}\right]^{1/3}}{\left(T_b^{m+n+5} - T_a^{m+n+5}\right)^{1/6}} \tag{14.85}$$

to yield δ_{opt} and $T_{a,\text{opt}}$. In eq. (14.85),

$$R = \left[\frac{\epsilon_0\sigma A_p^2(m+n+3)^2}{k_0(m+n+5)T_b^{2(n+1)}}\right]^{1/6}$$

and

$$S = 3\left(\frac{m+n+5}{m+n+3}\right)$$

14.4.4 Mutual Irradiation between Adjacent Fins

In previous sections the focus was on single radiating fins that exchange heat with space but do not interact with neighboring fins. When two fins are attached to a tube as in Fig. 14.4*a*, no radiative interaction between the fins can occur. However, when more than two fins are attached to the tube, the fins "see" each other in a radiative sense and such interaction can significantly affect the performance of the ensemble. Sparrow et al. (1961a) were apparently the first to analyze the effect of mutual radiative interaction between adjacent fins. The specific configuration considered appears in Fig. 14.9, which shows two longitudinal fins of rectangular profile sharing a common edge at their base. This configuration approximates two adjacent fins on a longitudinally finned cylinder of small radius. Note, in particular, that the fins are assumed to radiate to space at zero absolute temperature.

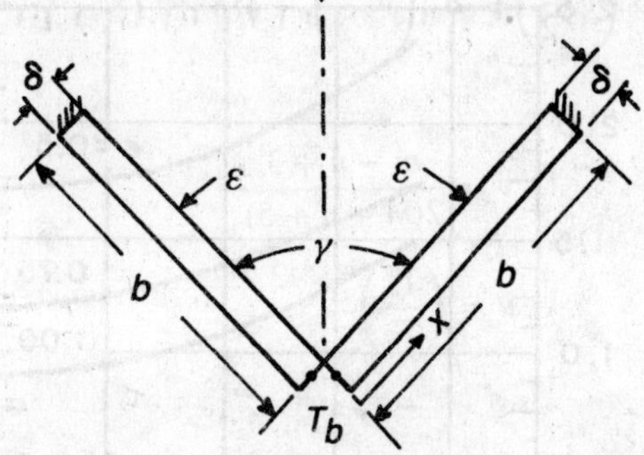

Figure 14.9 Radiative interaction between two adjacent fins.

An energy balance between conduction and the radiation flux in a differential element of fin height with unit length (into the plane of the paper) shows that the mathematical model consists of an algebraic equation, a differential equation, and an integral, all of which are coupled. The three unknowns are the temperature, radiosity, and irradiation at any location on the fin. An iterative solution of the system of equations gives the heat dissipation from each fin and hence the efficiency η of the fin. The ideal heat dissipation for calculating η is based on the radiation from a black surface at a uniform temperature T_b, stretched tightly between the tips of the fins. Thus, for one face of a fin,

$$q_{id} = \sigma T_b^4 \left(b \sin \frac{\gamma}{2} \right) \tag{14.86}$$

The efficiency η depends on three parameters: the emissivity ϵ, the included angle between the fins γ, and the profile number, which in reality is a conduction parameter, $\zeta = 2\epsilon\sigma T_b^3 b^2 / k\delta$. The fin of optimum dimensions is realized when the condition given by eq. (14.35),

$$\frac{d(\ln \eta)}{d(\ln \zeta)} = -\frac{1}{3} \tag{14.35}$$

is met.

The optimum ζ as a function of ϵ and γ is shown in Fig. 14.10. From this figure it is observed that as ϵ or γ increase, ζ_{opt} decreases. This indicates that for a given fin thickness, the optimum fin becomes shorter. Alternately, if ϵ, γ, and δ are all fixed, b_{opt} increases with increasing thermal conductivity k.

If δ is specified, Fig. 14.10 provides the information necessary to establish b_{opt} for given values of ϵ and γ. However, if q is known, the curves of η as a function ϵ, γ, and ζ must be available to design the fin. A sample set of curves for $\epsilon = 0.75$ is given in Fig. 14.11. Graphs for $\epsilon = 0.50$ and 0.10 are available (Sparrow et al., 1961a). The use of Figs. 14.10 and 14.11 to design an optimum radiator is now illustrated with an example.

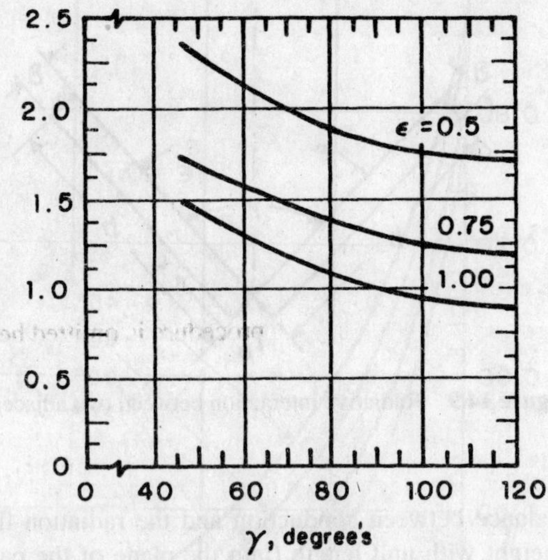

Figure 14.10 Optimum profile number as a function of angle and emissivity for the configuration of Fig. 14.9. [Adapted from Sparrow et al. (1961a).]

Example 14.11. A small-radius tube is to be equipped with eight longitudinal fins of rectangular profile so that the assembly dissipates 2400 W/m when operating with a base temperature of 500 K in a zero-absolute-temperature environment. The fin material has a thermal conductivity of 173 W/m·K and the surfaces are coated to give an emissivity of 0.75. Taking into account fin-to-fin radiative interaction, calculate the fin thickness and length for an optimum design.

SOLUTION. Assume the fins to be uniformly spaced around the tube. Then

$$\gamma = 360/8 = 45^\circ$$

For both sides of each fin to be active,

$$q_{\text{fin}} = 2400/8 = 300 \text{ W/m}$$

Read Fig. 14.10 for $\epsilon = 0.75$ and $\gamma = 45^\circ$ to obtain $\zeta_{\text{opt}} = 1.75$. Next refer to Fig. 14.11 and find that at $\zeta_{\text{opt}} = 1.75$ and $\gamma = 45$ deg, $\eta_{\text{opt}} = 0.50$.

Assume a trial value of $\delta = 2$ mm. Then

$$b = \left(\frac{1.75k\delta}{2\sigma\epsilon T_b^3} \right)^{1/2}$$

$$= \left[\frac{1.75(173)(0.002)}{(2)(5.669 \times 10^{-8})(0.75)(500)^3} \right]^{1/2}$$

$$= 0.238 \text{ m}$$

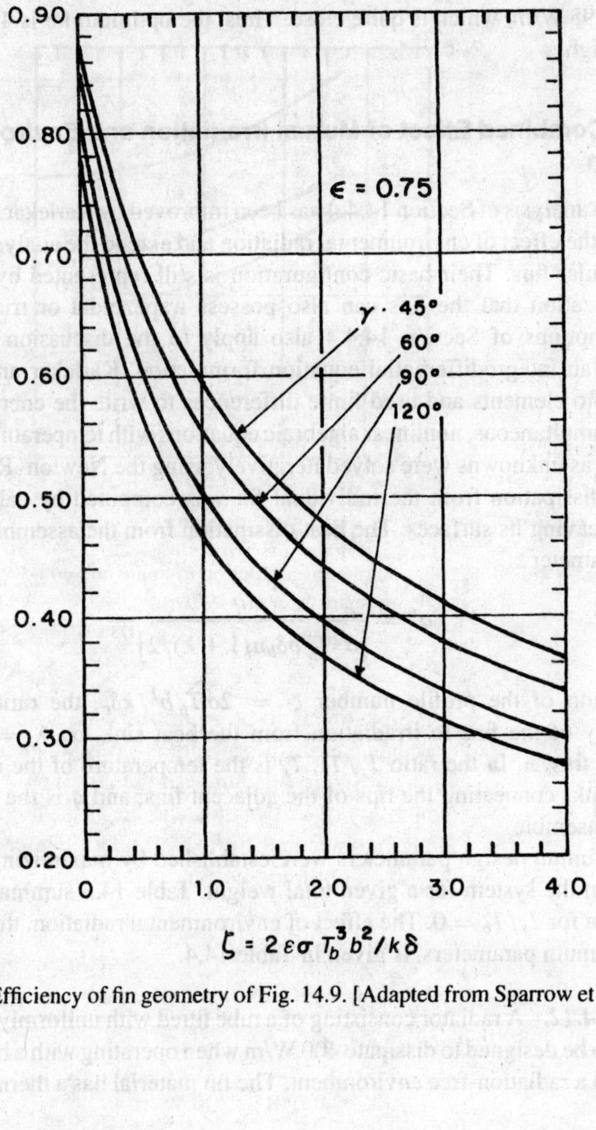

Figure 14.11 Efficiency of fin geometry of Fig. 14.9. [Adapted from Sparrow et al. (1961a).]

and

$$
\begin{aligned}
q &= \eta_{\text{opt}} q_{\text{id}} \\
&= \eta_{\text{opt}} \sigma T_b^4 b \sin \frac{\gamma}{2} \\
&= (0.5)(5.669 \times 10^{-8})(500)^4 (0.238) \sin(45/2) \\
&= 161.4 \text{ W/m}
\end{aligned}
$$

The value of 161.4 W/m falls short of the desired value of 300 W/m. Increasing values of δ are then employed and it is found that $\delta = 4.9$ mm gives $b = 0.439$ m and $q = 298$ W/m, which is quite close. Thus, the optimum fin is 4.9 mm thick and 439 mm high.

14.4.5 Combined Effect of Mutual Irradiation and Environmental Radiation

The design analysis of Section 14.4.4 has been improved by Karlekar and Chao (1963) to include the effect of environmental radiation and extend the analysis to trapezoidal and triangular fins. Their basic configuration is still represented by Fig. 14.9, with the modification that the fins can also possess trapezoidal or triangular profiles. The assumptions of Section 14.4.4 also apply to the discussion here. However, instead of an integrodifferential equation formulation, Karlekar and Chao divided each fin into elements and used finite differences to write the energy balance. The resulting simultaneous, nonlinear algebraic equations with temperature, radiosity, and irradiation as unknowns were solved iteratively using the Newton–Raphson method. The heat dissipation from the individual fin was computed by calculating the net radiation leaving its surfaces. The heat dissipation from the assembly, characterized by the parameter

$$q^* = \frac{q}{\left[\sigma^2 T_b^9 b \delta_b n(1+\lambda)/2\right]^{1/3}}$$

is a function of the profile number $\zeta = 2\sigma T_b^3 b^2/k\delta_b$, the ratio T_s/T_b, ϵ, the absorptivity of the fins to irradiation from the heat sink, α_s, $\lambda = \delta_a/\delta_b$, and the number of fins, n. In the ratio T_s/T_b, T_s is the temperature of the imaginary black surface (sink) connecting the tips of the adjacent fins, and q is the heat dissipation from the ensemble.

The optimum design parameters were established by maximizing the heat dissipation from the system for a given total weight. Table 14.3 summarizes the design information for $T_s/T_b = 0$. The effect of environmental radiation, that is, $T_s/T_b \neq 0$ on the optimum parameters, is given in Table 14.4.

Example 14.12. A radiator consisting of a tube fitted with uniformly spaced triangular fins is to be designed to dissipate 400 W/m when operating with a base temperature of 500 K in a radiation-free environment. The fin material has a thermal conductivity

TABLE 14.3 Optimum Design Parameters: T_s/T_b, $\alpha_s = 0.20$

	$\lambda = 1.00$ (Rectangular Fin)			$\lambda = 0.01$ (Very Nearly Triangular Fin)			$\lambda = 0.25$ (Trapezoidal Fin)		
ϵ	q^*_{opt}	n_{opt}	ζ_{opt}	q^*_{opt}	η_{opt}	ζ_{opt}	q^*_{opt}	n_{opt}	ζ_{opt}
0.50	1.2850	8	2.35	1.4350	7	2.00	—	—	—
0.75	1.4690	6	1.58	1.6390	6	1.31	—	—	—
0.90	1.5620	5	1.28	1.7563	5	1.15	1.6890	5	1.17
1.00	1.6245	5	1.15	1.8310	4	0.95			

TABLE 14.4 Optimum Design Parameters: $\epsilon = 0.9, \alpha_s = 0.2$

	$\lambda = 1$ (Rectangular)			$\lambda = 0.01$ (Nearly Triangular)		
T_s/T_b	q^*_{opt}	n_{opt}	ζ_{opt}	q^*_{opt}	n_{opt}	ζ_{opt}
0.25	1.5585	5	1.27	1.7525	5	1.13
0.50	1.5040	5	1.25	1.6930	4	1.00

of 173 W/m·K, and the fin surfaces are coated to give an emisssivity of 0.90. Using the optimum design information provided by Karlekar and Chao, calculate the number of fins and the dimensions of each if the design is to result in minimum weight. If this radiator were to operate in an environment with $T_s = 250$ K, how much heat would it dissipate?

SOLUTION. From Table 14.3, for $\epsilon = 0.90$, read

$$n_{opt} = 5, \quad \zeta_{opt} = 1.15, \quad \text{and} \quad q^* = 1.7563$$

Use the definition of ζ,

$$b = \left(\frac{\zeta k \delta_b}{2\sigma T_b^3}\right)^{1/2}$$

and assume that $\delta_b = 2$ mm. Then

$$b = \left[\frac{(1.15)(173)(0.002)}{(2)(5.669 \times 10^{-8})(500)^3}\right]^{1/2}$$

$$= 0.168 \text{ m}$$

Using the definition of q^* with $\lambda = 0$ yields

$$q = q^* \left(\frac{\sigma^2 T_b^9 b \delta_b n}{2}\right)^{1/3}$$

$$= 1.7563 \left[\frac{(5.669 \times 10^{-8})^2 (500)^9 (0.168)(0.002)(5)(1)}{2}\right]^{1/3}$$

$$= 305.6 \text{ W/m}$$

which falls short of the desired value of 400 W/m. With an assumption of $\delta_b = 4$ mm, then

$$b = 0.168 \left(\frac{0.004}{0.002}\right)^{1/2}$$

$$= 0.238 \text{ m}$$

and

$$q = 305.6 \left[\frac{(0.238)(0.004)}{(0.168)(0.002)} \right]^{1/3}$$

$$= 432.4 \text{ W/m}$$

which exceeds the desired value of 400 W/m. Finally, a trial of $\delta_b = 3.5$ mm provides

$$b = 0.168 \left(\frac{0.0035}{0.002} \right)^{1/2}$$

$$= 0.222 \text{ m}$$

and

$$q = 305.6 \left[\frac{(0.222)(0.0035)}{(0.168)(0.002)} \right]^{1/3}$$

$$= 404 \text{ W/m} \quad \text{(close and sufficient)}$$

To evaluate the performance of this design in an environment with $T_s = 250$ K, a graph of q^* as a function of ζ for $T_s/T_b = 250/500 = 0.5$, $\epsilon = 0.9$, and $n = 5$ is required. Figure 14.12 provides the necessary information. For $\zeta = 1.15$ and $n = 5$, Fig. 14.12 gives $q^* = 1.69$. Thus

$$q = q^* \left(\frac{\sigma^2 T_b^9 b \delta_b n}{2} \right)^{1/3}$$

$$= 1.69 \left[\frac{(5.669 \times 10^{-8})^2 (500)^9 (0.222)(0.0035)(5)(1)}{2} \right]^{1/3}$$

$$= 389 \text{ W/m}$$

The heat dissipation is reduced by about 4%. This slight reduction is due to the fact that the fin absorptivity to external radiation is low ($\alpha_s = 0.2$).

14.4.6 Combined Effect of Environmental Radiation and Associated Structure

The optimization studies presented thus far considered only the weight of the fins, with no consideration given to the associated structure. Reynolds (1963) pointed out that a more realistic optimization must include the weight of the tubes, the manifolds, the fluid contained therein, and their protective armor. He considered the configuration shown in Fig. 14.13 with fins of rectangular, trapezoidal, and triangular profiles. To take into account the environmental radiation, the fin surface heat dissipation was assumed to be of the form $K_1 T^4 - K_2$, which is identical to the form used in Chapter 13 except that K_1 incorporates a fin view factor F_f, so that

$$K_1 = 2\epsilon\sigma F_f$$

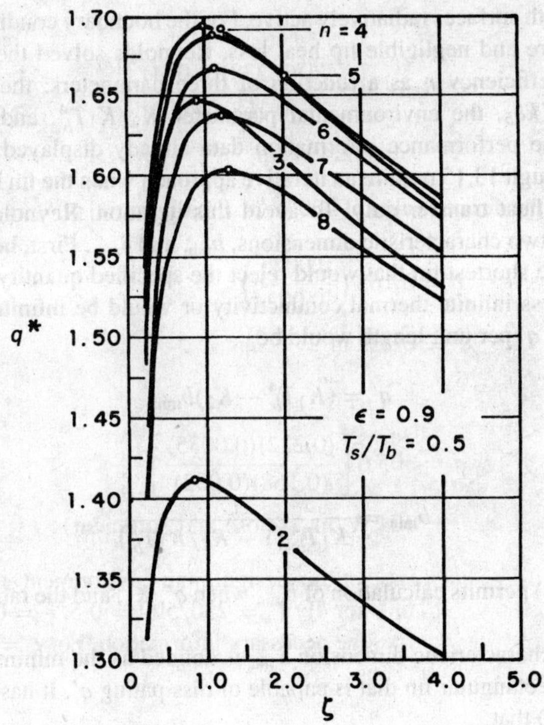

Figure 14.12 Heat dissipation rate for a radiator consisting of a tube with uniformly spaced triangular fins. [Adapted from Karlekar and Chao (1963).]

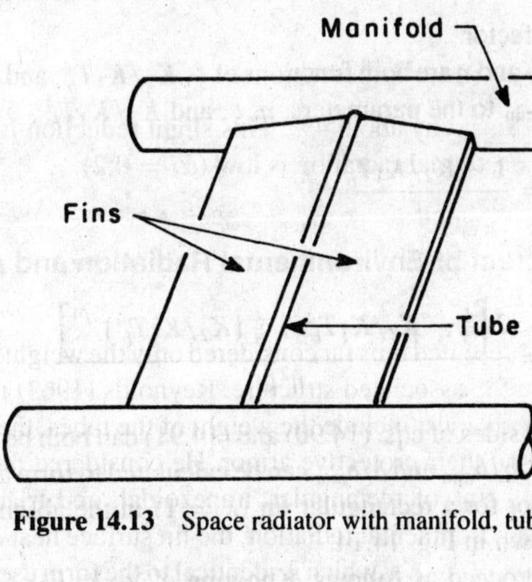

Figure 14.13 Space radiator with manifold, tube, and fins.

for a fin with both surfaces radiatively active. For the boundary conditions of constant base temperature and negligible tip heat loss, Reynolds solved the fin equation to obtain the fin efficiency η as a function of three parameters: the profile number $\zeta = K_1 T_b^3 b^2 / k\delta_b$, the environmental parameter $K_2/K_1 T_b^4$, and the taper ratio $\lambda = \delta_a/\delta_b$. The performance information data already displayed in this form in Figs. 13.12 through 13.15 require an iterative approach when the fin is to be designed for a specified heat transfer duty. To avoid this iteration, Reynolds proposed the introduction of two characteristic dimensions, b_{min} and δ_{min}. First, he defined b_{min} as the length of the shortest fin that would reject the specified quantity of heat. Such a fin would possess infinite thermal conductivity or would be infinitely wide and its heat dissipation q' per unit length would be

$$q' = (K_1 T_b^4 - K_2) b_{min} \tag{14.87}$$

This makes

$$b_{min} = \frac{q'}{K_1 T_b^4 \left(1 - K_2/K_1 T_b^4\right)} \tag{14.88}$$

Equation (14.88) permits calculation of b_{min} when q', K_1, and the ratio $K_2/K_1 T_b^4$ are known.

The second characteristic dimension δ_{min} is defined as the minimum width of an infinitely long rectangular fin that is capable of dissipating q'. It has been shown by Reynolds (1963) that

$$\delta_{min} = \frac{(q')^2}{4k\sigma\epsilon F_f T_b^5 \left[\frac{1}{5} - K_2/K_1 T_b^4 + \frac{4}{5}\left(K_2/K_1 T_b^4\right)^{5/4}\right]} \tag{14.89}$$

where F_f is the fin view factor.

Because $q' = \eta K_1 T_b^4 b$ and η are both functions of ζ, $K_2/K_1 T_b^4$, and λ, it is possible to relate b/b_{min} and δ_b/δ_{min} to the parameters, η, ζ, and $K_2/K_1 T_b^4$:

$$\frac{b}{b_{min}} = \frac{1 - K_2/K_1 T_b^4}{\eta} \tag{14.90}$$

and

$$\frac{\delta_b}{\delta_{min}} = \frac{2\left[\frac{1}{5} - K_2/K_1 T_b^4 + \frac{4}{5}\left(K_2/K_1 T_b^4\right)^{5/4}\right]}{\eta^2 \zeta} \tag{14.91}$$

Because the right-hand sides of eqs. (14.90) and (14.91) can both be obtained from Figs. 13.12 through 13.15, b/b_{min} and δ/δ_{min} can be calculated to form a plot of b/b_{min} versus δ_b/δ_{min}. Such a plot for a rectangular fin ($\lambda = 1$) in the absence of external radiation ($K_2 = 0$) is shown in Fig. 14.14.

The design can now proceed as follows. Knowing q', ϵ, k, F_f, T_b, and K_2, the parameters b_{min} and δ_{min} can be calculated from eqs. (14.88) and (14.89), respectively. One can now select a fin height b that gives b/b_{min}. Figure 14.14 may then be used

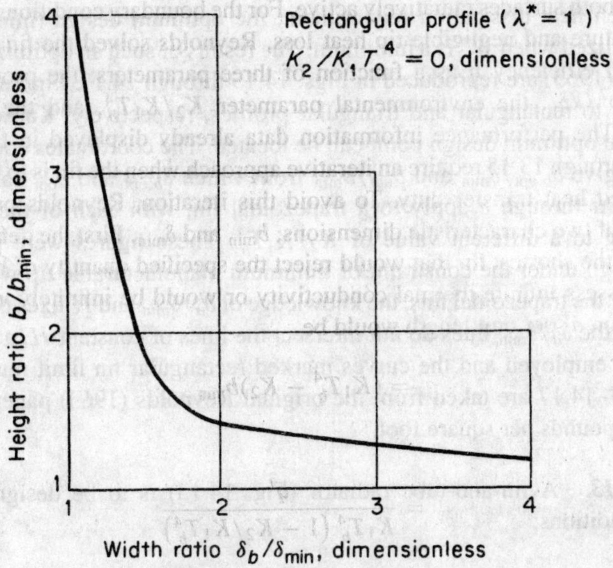

Figure 14.14 Relationship between height and width for a rectangular fin capable of a specified heat dissipation in the absence of environmental radiation. [Adapted from Reynolds (1963).]

to fix the value of δ_b/δ_{min} and hence δ_b. However, this procedure does not result in an optimum design.

Turning to the problem of optimum design of minimum weight, let W_s be the weight of the associated structure per unit of fin plan area. Because the entire analysis is based on a unit length, the fin plan area is simply b. Thus the weight of the fin and the weight of the structure chargeable to the per unit length is

$$W = W_s b + \tfrac{1}{2} b(\delta_b + \delta_a)\gamma \tag{14.92}$$

where γ is the specific weight of the fin material. Defining the dimensionless quantities W^* and H as

$$W^* = \frac{W}{b_{min}\delta_{min}\gamma} \tag{14.93}$$

and

$$H = \frac{W_s}{\gamma \delta_{min}} \tag{14.94}$$

it follows that

$$W^* = H\frac{b}{b_{min}} + \frac{1}{2}\frac{b}{b_{min}}\left(\frac{\delta_b}{\delta_{min}} + \frac{\delta_a}{\delta_{min}}\right) \tag{14.95}$$

For a given heat dissipation requirement, the optimum design (minimum total weight) is realized when W^* is a minimum. The results of such an optimization given by Reynolds (1963) are reproduced in Figs. 14.15 through 14.17. Figures 14.15 and 14.16 pertain to rectangular and triangular profiles, respectively. Knowing H and $K_2/K_1T_b^4$, the optimum design point can be located. The coordinates corresponding to this point give $\delta_{b,\mathrm{opt}}/\delta_{\min}$ and $b_{\mathrm{opt}}/b_{\min}$ from which $\delta_{b,\mathrm{opt}}$ and b_{opt} can be found. Figures 14.17*a* through *e* apply to a trapezoidal fin, with each of the subfigures corresponding to a different value of $K_2/K_1T_b^4$. These figures were prepared to facilitate design under the constraint of minimum manufacturable tip width, $\delta_{a,\min}$. Therefore, for the trapezoidal fins, the knowledge of $\delta_a/\delta_{\min}$ and H fixes the optimum point. Where the $\delta_a/\delta_{\min}$ lines do not intersect the lines of constant H, a trapezoidal fin cannot be employed and the curves marked rectangular fin limit must be used. Figures 14.15–14.17 are taken from the original Reynolds (1963) paper, where he showed H in pounds per square foot.

Example 14.13. A fin-and-tube radiator (Fig. 14.13) is to be designed for the following conditions:

1. Tube length (fin length) = 3 m
2. Tube outside diameter = 2 cm
3. Heat dissipation (one tube and two fins) = 5000 W
4. External radiation (both fin faces) = 1200 W/m^2
5. Fin base temperature = 340°C = 613 K

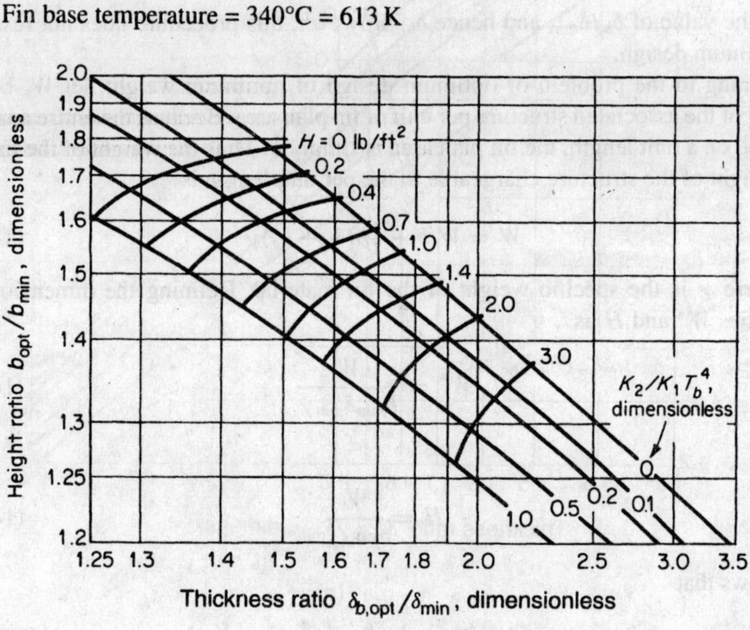

Figure 14.15 Optimum design chart for a rectangular profile fin with associated structure. [Adapted from Reynolds (1963).]

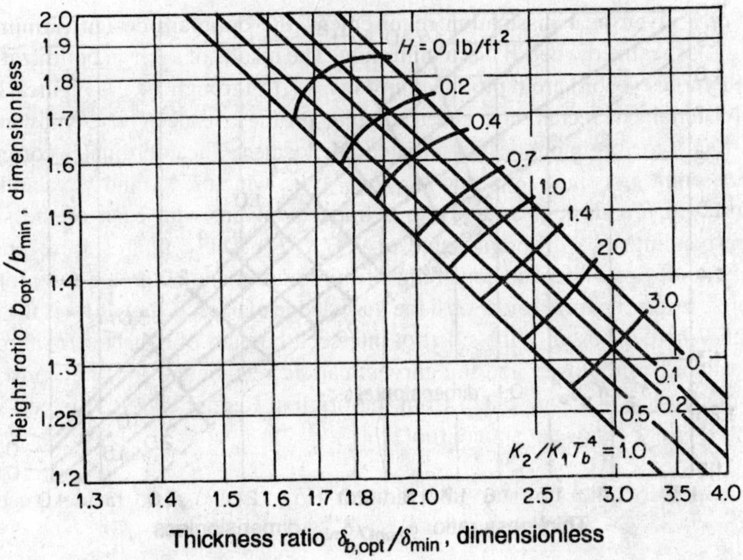

Figure 14.16 Optimum design chart for a triangular profile fin with associated structure. [Adapted from Reynolds (1963).]

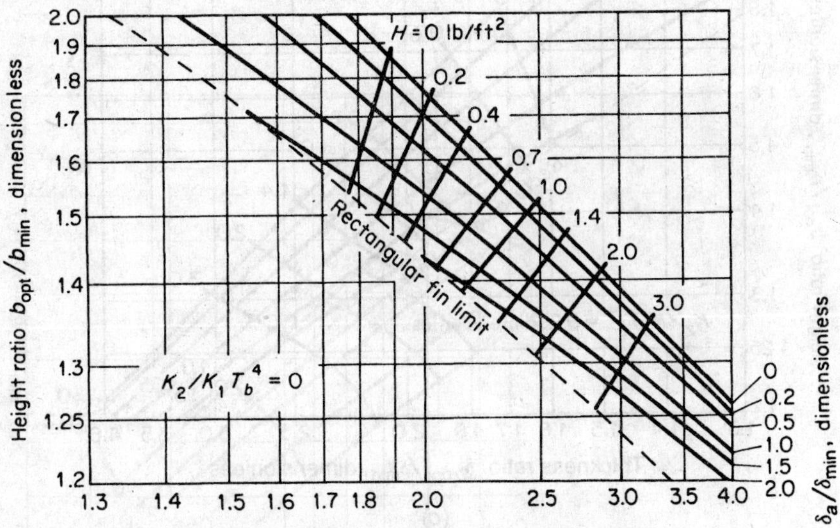

(*a*)

Figure 14.17 Optimum design chart for trapezoidal profile fins with associated structure. [Adapted from Reynolds (1963).]

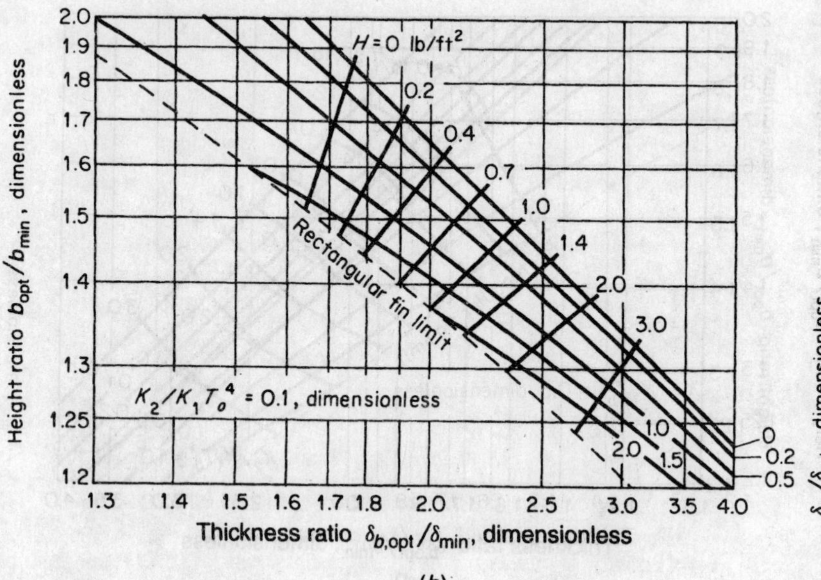

(*b*)

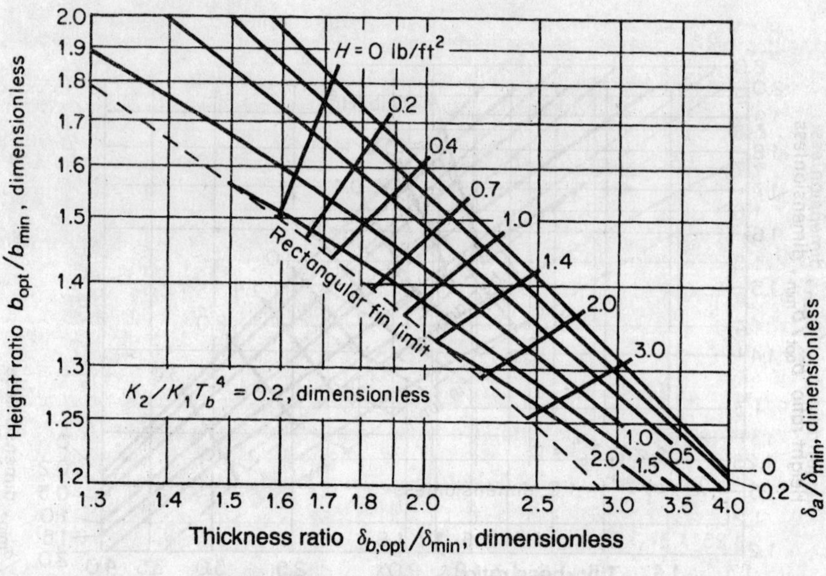

(*c*)

Figure 14.17 continued

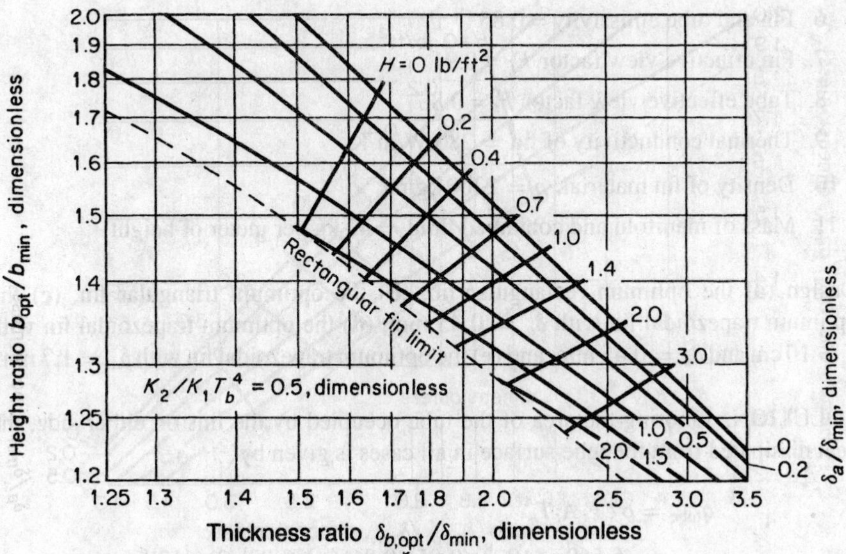

(*d*)

Figure 14.17 continued

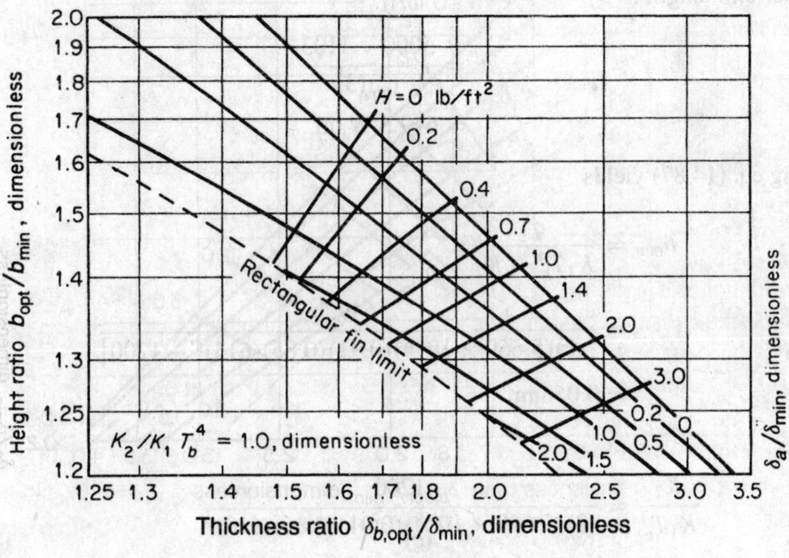

(*e*)

Figure 14.17 continued

6. Fin and tube emissivity = 0.85
7. Fin effective view factor $F_f = 0.91$
8. Tube effective view factor $F_t = 0.86$
9. Thermal conductivity of fin = 280 W/m·K
10. Density of fin material, $\rho = 2200\,\text{kg/m}^3$
11. Mass of manifold and contained fluid = 43 kg per meter of height

Design (**a**) the optimum rectangular fin, (**b**) the optimum triangular fin, (**c**) the optimum trapezoidal fin with $\delta_a = 0.45$ mm, (**d**) the optimum trapezoidal fin with $b = 10$ cm and $\delta_a = 0.45$ mm, and (**e**) the optimum trapezoidal fin with $\delta_a = 1.7$ mm.

SOLUTION. Ignoring the area of the tube occupied by the fins on either side, the heat dissipated from the tube surface in all cases is given by

$$
\begin{aligned}
q_{\text{tube}} &= \sigma \epsilon F_t A_t T_b^4 \\
&= (5.669 \times 10^{-8})(0.85)(0.86)(\pi)(0.02)(3)(613)^4 \\
&= 1103 \text{ W}
\end{aligned}
$$

Because there are two fins on each side of the tube, the heat dissipation required per fin per unit length is

$$
\begin{aligned}
q' &= \frac{5000 - 1103}{(2)(3)} \\
&= 649.5 \text{ W/m}
\end{aligned}
$$

Using eq. (14.87) yields

$$
\begin{aligned}
b_{\min} &= \frac{q'}{K_1 T_b^4 - K_2} \\
&= \frac{649.5}{\left[(2)(5.669 \times 10^{-8})(0.91)(0.85)(613)^4 - 1200\right]} \\
&= 0.058 \text{ m}
\end{aligned}
$$

Then

$$
\begin{aligned}
\frac{K_2}{K_1 T_b^4} &= \frac{1200}{(2)(5.669 \times 10^{-8})(0.91)(0.85)(613)^4} \\
&= 0.097 \simeq 0.10
\end{aligned}
$$

Using eq. (14.89),

$$
\delta_{\min} = \frac{(q')^2}{4k\sigma\epsilon F_f T_b^5 \left[\frac{1}{5} - K_2/K_1 T_b^4 + \frac{4}{5}\left(K_2/K_1 T_b^4\right)^{5/4}\right]}
$$

$$= \frac{(649.5)^2}{(4)(280)(5.669 \times 10^{-8})(0.85)(0.91)(613)^5[0.20 - 0.10 + 0.8(0.10)^{1.25}]}$$

$$= 0.00068 \text{ m} \quad \text{or} \quad 0.68 \text{ mm}$$

Then, using eq. (14.94),

$$H = \frac{W_s}{\rho \delta_{\min}}$$

$$= \frac{(43)(9.81)(0.0209)}{(2200)(9.81)(0.00068)(3)}$$

$$= 0.20$$

(**a**) For the optimum rectangular fin, use Fig. 14.15. For $H = 0.20$ and $K_2/K_1T_b^4 = 0.10$,

$$b_{\text{opt}}/b_{\min} = 1.67 \quad \text{and} \quad \delta_{b,\text{opt}}/\delta_{\min} = 1.95$$

Thus

$$b = b_{\text{opt}} = (1.67)(0.058) = 0.0969 \text{ m} \quad \text{or} \quad 96.9 \text{ mm}$$

$$\delta = \delta_{b,\text{opt}} = (1.95)(0.00068) = 0.0010 \text{ m} \quad \text{or} \quad 1.0 \text{ mm}$$

Using eq. (14.92) with $\delta_a = \delta_b$ gives

$$W = W_s b + \tfrac{1}{2}b(\delta_b + \delta_a)\gamma$$

$$= \frac{43(9.81)(0.0969)}{3} + \frac{1}{2}(0.0969)(0.0010 + 0.0010)(2200)(9.81)$$

$$= 13.63 + 2.09 = 15.72 \text{ N/m}$$

(**b**) For the optimum triangular fin, use Fig. 14.16. For $H = 0.20$ and $K_2/K_1T_b^4 = 0.1$,

$$b_{\text{opt}}/b_{\min} = 1.70 \quad \text{and} \quad \delta_b/\delta_{\min} = 1.91$$

Thus

$$b = b_{\text{opt}} = (1.70)(0.058) = 0.0986 \text{ m} \quad \text{or} \quad 98.6 \text{ mm}$$

$$\delta = \delta_{b,\text{opt}} = (1.91)(0.00068) = 0.0013 \text{ m} \quad \text{or} \quad 1.3 \text{ mm}$$

Using eq. (14.92) gives

$$W = W_s b + \tfrac{1}{2}b(\delta_b + \delta_a)\gamma$$

$$= \frac{(43)(9.81)(0.0986)}{3} + \frac{1}{2}(0.0986)(0.0013)(2200)(9.81)$$

$$= 13.86 + 1.38$$

$$= 15.24 \text{ N/m}$$

(c) For the optimum trapezoidal fin with $K_2/K_1T_b^4 = 0.1$, use Fig. 14.17b. For $H = 0.20$ and $\delta_a/\delta_{\min} = 0.45/0.68 = 0.66$, interpolation gives

$$b_{\text{opt}}/b_{\min} = 1.59 \quad \text{and} \quad \delta_{b,\text{opt}}/\delta_{\min} = 1.81$$

Thus

$$b_{\text{opt}} = (1.59)(0.058) = 0.092 \text{ m} \quad \text{or} \quad 92 \text{ mm}$$

$$\delta_{b,\text{opt}} = (1.81)(0.00068) = 0.0012 \text{ m} \quad \text{or} \quad 1.2 \text{ mm}$$

Using eq. (14.92) gives

$$\begin{aligned} W &= W_s b + \tfrac{1}{2}b(\delta_b + \delta_a)\gamma \\ &= \frac{(43)(9.81)(0.092)}{3} + \frac{1}{2}(0.092)(0.0012 + 0.00045)(2200)(9.81) \\ &= 12.94 + 1.64 \\ &= 14.58 \text{ N/m} \end{aligned}$$

(d) For the optimum trapezoidal fin with $b = 10$ cm or 0.10 m and $\delta_a = 0.45$ mm or 0.00045 m,

$$\frac{b}{b_{\min}} = \frac{0.1}{0.058} = 1.72$$

$$\frac{\delta_a}{\delta_{\min}} = \frac{0.00045}{0.00068} = 0.66$$

Find the intersection of $b/b_{\min} = 1.72$ and $\delta_a/\delta_{\min} = 0.66$ on Fig. 14.17b, and read $\delta_{b,\text{opt}}/\delta_{\min} = 1.65$. Thus

$$\delta_{b,\text{opt}} = (1.65)(0.00068) = 0.0011 \text{ m} \quad \text{or} \quad 1.1 \text{ mm}$$

Using eq. (14.92) gives

$$\begin{aligned} W &= W_s b + \tfrac{1}{2}b(\delta_b + \delta_a)\gamma \\ &= \frac{43(9.81)(0.1)}{3} + \frac{1}{2}(0.1)(0.0011 + 0.00045)(2200)(9.81) \\ &= 14.06 + 1.67 = 15.73 \text{ N/m} \end{aligned}$$

(e) For the fin with $\delta_a = 1.7$ mm,

$$\frac{\delta_a}{\delta_{\min}} = \frac{1.7}{0.68} = 2.5$$

There is no curve for $\delta_a/\delta_{\min} = 2.5$ in Fig. 14.17b. If such a curve is drawn using extrapolation, it will not intersect the $H = 0.20$ curve, but it will intersect the rectangular fin limit at $\delta_b/\delta_{\min} \simeq 2.5$. Because the weight of the fin decreases as the point moves to the rectangular fin limit, the appropriate choice is a rectangular fin that is 1.7 mm wide. The corresponding value of $b/b_{\min} = 1.246$, giving

$$b_{\text{opt}} = (1.246)(0.058) = 0.0723 \text{ m} \quad \text{or} \quad 72.3 \text{ mm}$$

Using eq. (14.92) yields

$$\begin{aligned} W &= W_s b + \tfrac{1}{2} b(\delta_b + \delta_a)\gamma \\ &= \frac{(43)(9.81)(0.0723)}{3} + \frac{1}{2}(0.0723)(0.0017 + 0.0017)(2200)(9.81) \\ &= 10.17 + 2.69 = 12.86 \text{ N/m} \end{aligned}$$

14.5 FIN-TO-FIN AND FIN-TO-BASE RADIATION INTERCHANGE

The effect of radiative exchange between adjacent fins was discussed in Sections 14.4.4 and 14.4.5. There it was assumed that the interaction between the fins and the base was negligible. Such an assumption can be justified if the fins are mounted on a small-radius cylindrical surface or if the fins are high and closely spaced. However, when this is not the case, the analysis must take into account both fin-to-fin and fin-to-base radiative exchanges. Schnurr (1975) conducted such an analysis for longitudinal fins of triangular profile arranged uniformly around the tube surface as shown in Fig. 14.18. Considering a nonisothermal gray enclosure formed by two neighboring fins, the segment of tube surface between them and free space (no external radiation),

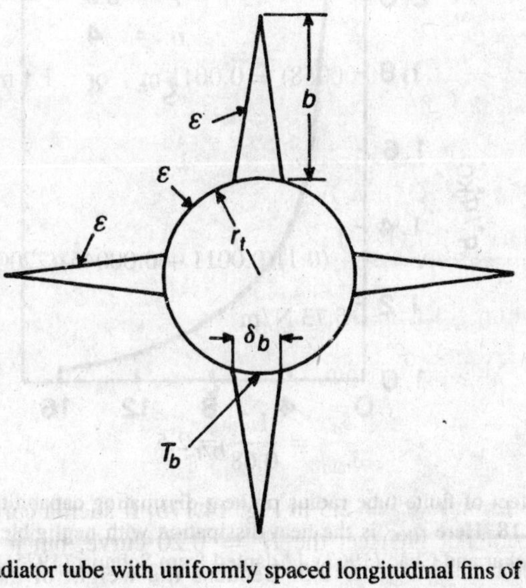

Figure 14.18 Radiator tube with uniformly spaced longitudinal fins of triangular profile.

he solved the coupled conductive–radiative transport equations numerically. The heat dissipation from the system

$$q^* = \frac{q}{2\pi r_t \sigma T_b^4}$$

was found as a function of four parameters: the number of fins n, emissivity of the fins and tube ϵ, profile number $\zeta = 2\sigma\epsilon r_t^2 T_b^3 / k\delta_b$, and geometrical ratio b/r_t.

To study the effect of tube radius r_t on the heat dissipation capability of the system, Schnurr calculated q^* for $n = 4$, $\epsilon = 0.9$, $\zeta = 4$, and values of b/r_t ranging from 2 to 16. These results were then compared with the q^*_{KC} (KC for Karlekar and Chao) values predicted by Karlekar and Chao (see Section 14.4.5) for a small-radius tube with negligible base interaction. This comparison, which is shown in Fig. 14.19, clearly demonstrates that q^* is affected significantly by the fin-to-base interaction. For example, the heat dissipation at $b/r_t = 2$ is twice as much as that when $b/r_t = 16$. Note that the effect of base interaction becomes small (within 5%) when $b/r_t > 12$.

Figures 14.20 and 14.21 show sample results for q^* for black and gray surfaces, respectively. These figures can be used to design a minimum-weight fin array. Knowing the tube radius r_t, base temperature T_b, emissivity ϵ, and heat dissipation required q, the value of $q^* = q/2\pi r_t \sigma T_b^4$ can be calculated.

Figures 14.20 and 14.21 may then be used to determine ζ for any selected value of n (4 or 6) and b/r_t. The profile number ζ may be expressed in the form

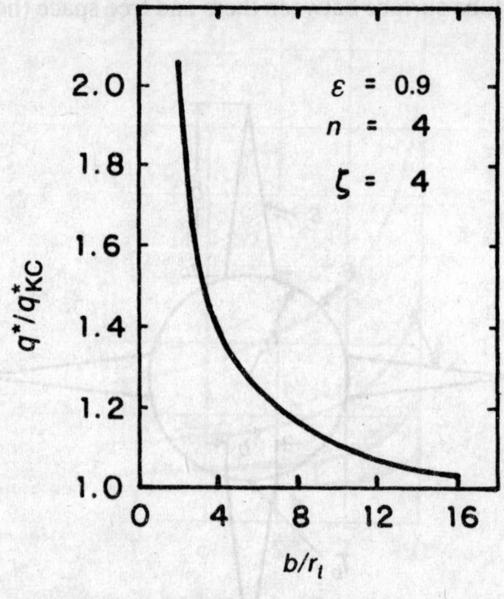

Figure 14.19 Effect of finite tube radius on heat-dissipating capability of the radiator geometry of Fig. 14.18. Here q_{KC} is the heat dissipation with negligible tube–fin interaction developed by Karlekar and Chao (1963). [Adapted from Schnurr (1975).]

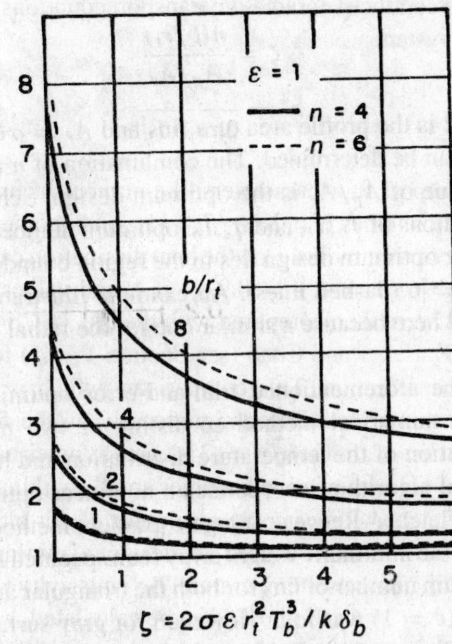

Figure 14.20 Effect of tube (base) surface interaction radius on heat dissipation from four- and six-fin radiator arrangements with black surfaces. [Adapted from Schnurr (1975).]

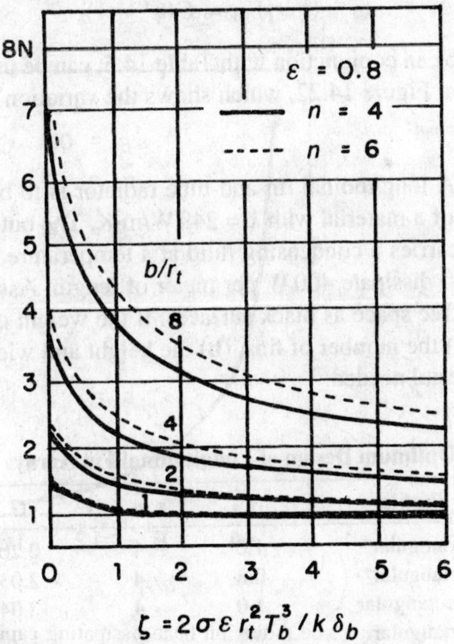

Figure 14.21 Effect of tube (base) surface interaction radius on heat dissipation from four- and six-fin radiator arrangements with gray surfaces. [Adapted from Schnurr (1975).]

$$\zeta = \frac{n(b/r_t)}{A_p/A_r}$$

where $A_p = \delta_b bn/2$ is the profile area of n fins and $A_r = \sigma\epsilon r_t^3 T_b^3/k$ is a reference area. Thus A_p/A_r can be determined. The combination of n and b/r_t which results in the minimum value of A_p/A_r is the optimum design. Schnurr has reported that for various combinations of T_b, r_t, and q, the optimum number of fins lies between 4 and 6. Therefore, the optimum design lies in the region bounded by curves for $n = 4$ (solid lines) and $n = 6$ (dashed lines). An example illustrating this trial-and-error procedure is omitted here because a similar one for the radial geometry is given in a subsequent section.

To circumvent the aforementioned trial-and-error optimization, Schnurr et al. (1976) developed a numerical method consisting of two major algorithms. The first involves calculation of the temperature distribution and heat dissipation for the fin array. The second algorithm incorporates a nonlinear optimization technique (a combination of the Fletcher–Reeves conjugate gradient method and a Hooke–Jeeves direct search) to find the minimum-weight array for a specified heat dissipation. It was found that the optimum number of fins for both the triangular and rectangular profiles with black surfaces ($\epsilon = 1$) was four. However, for gray surfaces with $\epsilon = 0.8$, the optimum number of fins was five. Table 14.5 summarizes the results.

The quantity C appearing in Table 14.5 relates to b/r_t and q^* via

$$(b/r_t)_{\text{opt}} = C(q^* - 1) \tag{14.96}$$

Equation (14.96), in conjunction with Table 14.5, can be used to obtain b_{opt} when q^* an r_t are known. Figure 14.22, which shows the variation of ζ_{opt} with q^*, can be used to establish $\delta_{b,\text{opt}}$.

Example 14.14. A longitudinal fin-and-tube radiator is to be designed using rectangular fins made of a material with $k = 249$ W/m·K. The outer radius of the tube is 1.5 cm. The tube carries a condensing fluid at a temperature of 140°C. The system is to be designed to dissipate 400 W per meter of length. Assume that the tube and the fins radiate to free space as black surfaces. If the weight of the fin array is to be minimized, find **(a)** the number of fins, **(b)** the height and width of each fin, and **(c)** the volume of material needed.

TABLE 14.5 Optimum Design of Longitudinal Fin Arrays

Case	Profile	ϵ	n	C	Optimum
1	Triangular	1.0	4	2.20	Yes
2	Triangular	0.8	4	2.03	No
3	Rectangular	1.0	4	2.04	Yes
4	Triangular	0.8	5	1.90	Yes
5	Rectangular	0.8	4	1.89	No
6	Rectangular	0.8	5	1.76	Yes

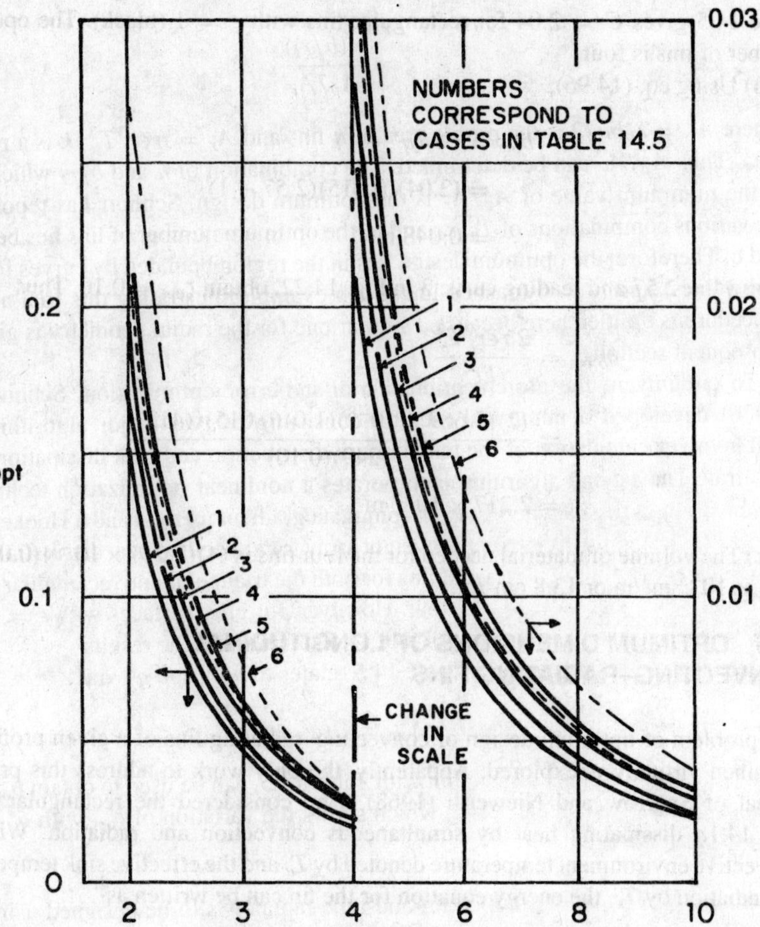

Figure 14.22 Optimum relationship between profile number and heat dissipation for an array of longitudinal fins of rectangular and triangular profiles. [Adapted from Schnurr et al. (1976).]

SOLUTION. **(a)** Assuming the inside convection resistance and tube conduction resistance to be negligible, the base temperature of the fins is equal to the fluid temperature. Thus $T_b = 140°C$ or 413 K and

$$q^* = \frac{q}{2\pi r_t \sigma T_b^4}$$

$$= \frac{400}{2(\pi)(0.015)(5.669 \times 10^{-8})(413)^4}$$

$$= 2.57$$

Table 14.5 gives $C = 2.04$ for rectangular fins with $\epsilon = 1$ (black). The optimum number of fins is four.

(**b**) Using eq. (14.96),

$$b_{\text{opt}} = Cr_t(q^* - 1)$$
$$= (2.04)(0.015)(2.57 - 1)$$
$$= 0.048 \text{ m}$$

Using $q^* = 2.57$ and reading curve 3 in Fig. 14.22, obtain $\zeta_{\text{opt}} = 0.10$. Thus

$$\delta_{b,\text{opt}} = \frac{2\sigma\epsilon r_t^2 T_b^3}{k\zeta}$$
$$= \frac{(2)(5.669 \times 10^{-8})(1.0)(0.015)^2(413)^3}{(249)(0.10)}$$
$$= 7.217 \times 10^{-5} \text{ m}$$

(**c**) The volume of material needed for the four fins is $(4)(7.217 \times 10^{-5})(0.048) = 1.38 \times 10^{-5}$ m³/m or 13.8 cm³/m.

14.6 OPTIMUM DIMENSIONS OF LONGITUDINAL CONVECTING–RADIATING FINS

The problem of optimum design of convecting–radiating fins of a given profile has remained virtually unexplored. Apparently, the only work to address this problem is that of Sparrow and Niewerth (1968), who considered the rectangular fin of Fig. 14.1*a* dissipating heat by simultaneous convection and radiation. With the convective environment temperature denoted by T_a and the effective sink temperature for radiation by T_s, the energy equation for the fin can be written as

$$\frac{d^2\theta}{dX^2} = \zeta_1(\theta^4 - \theta_s^4) + \zeta_2(\theta - \theta_a) \tag{14.97}$$

with boundary conditions

$$X = 0, \quad \theta = 1 \tag{14.98a}$$

and

$$X = 1, \quad \frac{d\theta}{dX} = 0 \tag{14.98b}$$

where

$$\theta = \frac{T}{T_b} \qquad X = \frac{x}{b}$$

$$\zeta_1 = \frac{2\epsilon\sigma T_b^3 b^2}{k\delta} \qquad \zeta_2 = \frac{2hb^2}{k\delta}$$

$$\theta_s = \frac{T_s}{T_b} \quad \text{and} \quad \theta_a = \frac{T_a}{T_b}$$

Equations (14.98) imply the boundary conditions of constant base temperature and insulated tip.

As usual, the optimum design was sought such that for a given profile area $A_p = \delta b$, the heat dissipation q was maximized. To facilitate the search for the optimum, Sparrow and Niewerth (1968) linearized the radiation term by letting $\theta^4 = \theta\overline{\theta}^3$, where $\overline{\theta}$ is taken to be the mean of the base and the tip temperature. With the help of this linearized model, they were able to determine the relationship between ζ_1 and ζ_2 for the optimum condition. Figure 14.23 shows their results. The main figure pertains to cases for which $\theta_a = \theta_s = 0, 0.4, 0.7$, and 0.9. The inset shows two cases for which $\theta_a \neq \theta_s$. Observe that the ordinate gives ζ_1 values for purely radiating fins, while the abscissa gives ζ_2 values for purely convecting fins.

Figure 14.24 gives the efficiency of the fin for the case of $\theta_s = \theta_a = 0.7$. This figure is based on the direct numerical solutions of eqs. (14.97) and (14.98) and is used in the design example that follows.

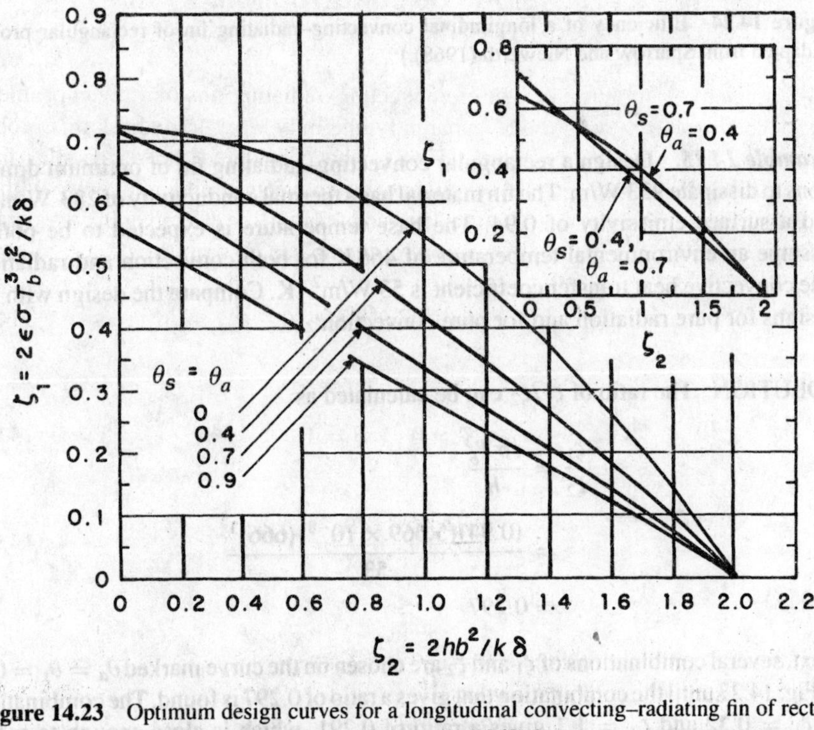

Figure 14.23 Optimum design curves for a longitudinal convecting–radiating fin of rectangular profile. [Adapted from Sparrow and Niewerth (1968).]

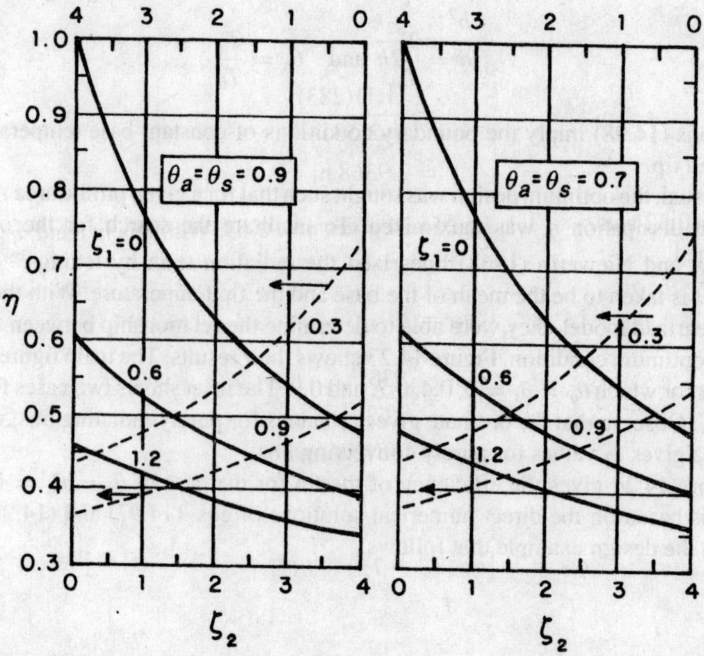

Figure 14.24 Efficiency of a longitudinal convecting–radiating fin of rectangular profile. [Adapted from Sparrow and Niewerth (1968).]

Example 14.15. Design a rectangular convecting–radiating fin of optimum dimensions to dissipate 933 W/m. The fin material has a thermal conductivity of 283 W/m·K, and a surface emissivity of 0.94. The base temperature is expected to be 666 K. Assume an environmental temperature of 466 K for both convection and radiation. The convective heat transfer coefficient is 53 W/m^2 · K. Compare the design with the designs for pure radiation and for pure convection.

SOLUTION. The ratio of ζ_1/ζ_2 can be calculated as

$$\frac{\zeta_1}{\zeta_2} = \frac{\epsilon\sigma T_b^3}{h}$$

$$= \frac{(0.94)(5.669 \times 10^{-8})(666)^3}{53}$$

$$= 0.297$$

Next, several combinations of ζ_1 and ζ_2 are chosen on the curve marked $\theta_a = \theta_s = 0.7$ in Fig. 14.23 until the combination that gives a ratio of 0.297 is found. The combination of $\zeta_1 = 0.32$ and $\zeta_2 = 1.1$ gives a ratio of 0.291, which is close enough to merit acceptance. Then, using the definition of ζ_2

$$\frac{b^2}{\delta} = \frac{\zeta_2 k}{2h} = \frac{(1.1)(283)}{2(53)} = 2.9368 \text{ m}$$

For $\zeta_1 = 0.32$ and $\zeta_2 = 1.1$, Fig. 14.24 gives $\eta = 0.63$. With this in hand, q_{id} can be determined:

$$q_{id} = \frac{q}{\eta} = \frac{933}{0.63} = 1481 \text{ W/m}$$

Noting that

$$q_{id} = 2b\left[\epsilon\sigma(T_b^4 - T_s^4) + h(T_b - T_s)\right]$$

then

$$b = \frac{q_{id}}{2\left[\epsilon\sigma(T_b^4 - T_s^4) + h(T_b - T_s)\right]} = \frac{1481}{2\left[(0.94)(5.669 \times 10^{-8})(666^4 - 466^4) + 53(666 - 466)\right]} = 0.0399 \text{ m}$$

and

$$\delta = \frac{b^2}{2.9368} = \frac{(0.0399)^2}{2.9368} = 0.00054 \text{ m}$$

For pure radiation, $\eta_{opt} = 0.589$ and $\zeta_{opt} = 0.70$ for $\theta_s = 0.7$ (see Example 14.9). Then

$$b_{opt} = \frac{q}{2\epsilon\sigma T_b^4(1 - \theta_s^4)\eta_{opt}} = \frac{933}{(2)(0.94)(5.669 \times 10^{-8})(666)^4(1 - 0.7^4)(0.589)} = 0.0994 \text{ m}$$

$$\begin{aligned}\delta &= \frac{2\epsilon\sigma T_b^3 b^2}{k\zeta} \\ &= \frac{(2)(0.94)(5.669 \times 10^{-8})(666)^3(0.2)^2}{(283)(0.70)} \\ &= 0.0064 \text{ m}\end{aligned}$$

For pure convection, Aziz (1992) gives in eqs. (3.8) and (3.9)

$$\begin{aligned}b_{\text{opt}} &= \frac{0.798q}{h(T_b - T_a)} \\ &= \frac{(0.798)(933)}{(53)(200)} \\ &= 0.0702 \text{ m}\end{aligned}$$

$$\begin{aligned}\delta_{\text{opt}} &= \frac{0.632}{hk}\left(\frac{q}{T_b - T_a}\right)^2 \\ &= \frac{0.632}{(53)(283)}\left(\frac{933}{200}\right)^2 \\ &= 0.0009 \text{ m}\end{aligned}$$

Comparison of the dimensions shows that the fin is highest and widest for pure radiation and is shortest and thinnest for simultaneous convection and radiation. The dimensions for the pure convection fin fall in between. Calculating the heat dissipation per unit volume, the values are 5.87, 14.77, and 87.40 W/cm^3 for pure radiation, pure convection, and simultaneous convection–radiation.

14.7 OPTIMUM DIMENSIONS OF RADIAL RADIATING FINS

The radial fin-and-tube radiator shown in Fig. 14.25 has been suggested as an alternative to the tube sheet radiator of Fig. 14.4*a*. The thermal performance of such an arrangement has been studied by several investigators, but the information on an optimum design appears to be lacking in the literature. Here, the literature concerned with the performance of radial radiating fins is reviewed briefly and illustrated with two examples of how an optimum design may be achieved.

The performance of a radial fin of rectangular profile was first reported by Chambers and Somers (1959), who gave a graph of fin efficiency η as a function of $(\epsilon\sigma T_b^3/k\delta)^{1/2}$ for parametric values of r_a/r_b. The fin was assumed to operate with boundary conditions of constant base temperature and no tip heat loss. The analysis ignored any external radiation. Using these same assumptions, Keller and Holdredge (1969) analyzed radial fins of trapezoidal and triangular profiles and gave numerical results for η. Truong and Mancuso (1980) introduced the effect of environmental radiation and calculated the efficiency of radial fins of three different profile shapes.

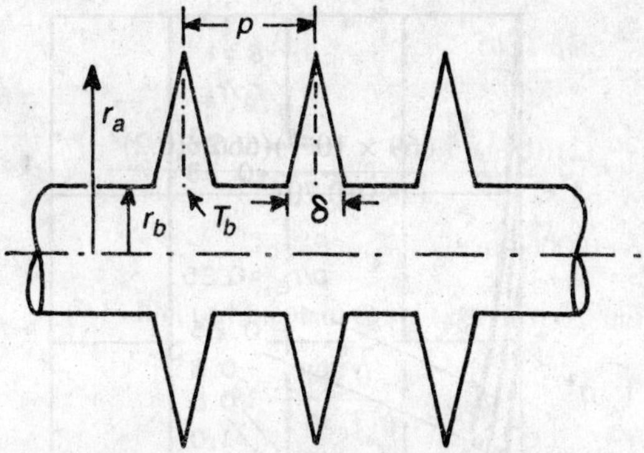

Figure 14.25 Radial fin-and-tube radiator.

The effects of fin-to-fin and fin-to-base radiative exchange was included in the analysis of radial fins of rectangular profile by Sparrow et al. (1962) using an integrodifferential equation formulation. Their results can be used to compute the heat dissipation from the fin and the tube separately. A finite difference approach was adopted by Schnurr and Cothran (1974) to study radial fins of trapezoidal and triangular profiles. They also included radiative exchange between neighboring fins and between the fins and the tube. Sample results for heat dissipation from a black radiator is shown in Fig. 14.26, where

$$q^* = \frac{q}{2\pi r_b p \sigma T_b^4}, \quad \rho = \frac{r_b}{r_a}, \quad \text{and} \quad \xi = \frac{2\sigma\epsilon T_b^4 r_b^2}{k\delta_b}$$

and p is the fin spacing. Observe that q^* is the ratio of the heat dissipations from finned and unfinned tubes (Fig. 14.25).

The following example can be used to optimize a radial fin-and-tube radiator. The example is patterned after the one given by Schnurr and Cothran (1974).

Example 14.16. A radial fin-and-tube radiator is to be designed using triangular fins made of aluminum. The tube is 0.61 m long and has a radius of 2.54 cm. The tube surface is maintained at 340 K and radiates to an environment at 0 K. The finned tube is to enhance the heat dissipation by a factor of 2 compared to the unfinned tube. Assume all surfaces to be black and investigate a range of values of r_a, b, and δ_b. Compute the volume of material used in each case. Which design gives the minimum volume? If copper or steel fins of these optimum r_b and b values were used, what would be the volumes and base widths?

SOLUTION. Assume that $\rho = r_b/r_a = 0.25$ and $p/r_a = 0.25$. From Fig. 14.26, the value of ζ for $q^* = 2$ is 0.12. Using the definition of ζ, solve for δ_b:

$$\delta_b = \frac{2\sigma\epsilon T_b^3 r_b^2}{k\zeta}$$

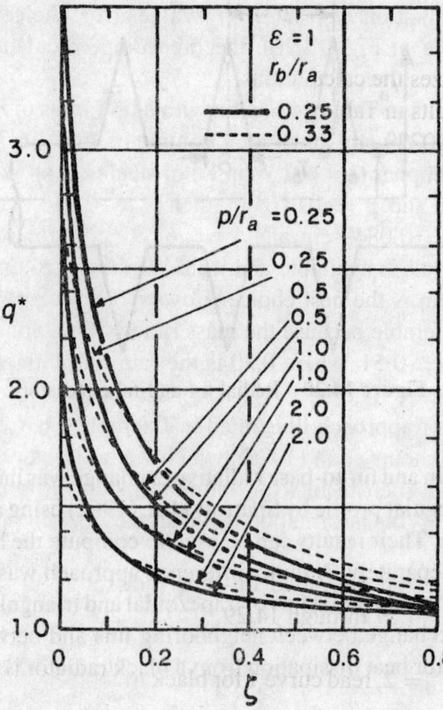

Figure 14.26 Heat dissipated from a radial fin (triangular)-and-tube radiator of Fig. 14.25 with black surfaces. [Adapted from Schnurr and Cothran (1974).]

$$= \frac{(2)(5.669 \times 10^{-8})(1)(340)^3(0.0254)^2}{(249)(0.12)}$$

$$= 9.62 \times 10^{-5} \text{ m}$$

Now

$$r_a = \frac{r_b}{\rho} = \frac{2.54}{0.25} = 10.16 \text{ cm}$$

$$p = 0.25 r_a = (0.25)(10.16) = 2.54 \text{ cm}$$

$$V = 2\pi \delta_b \left[\tfrac{1}{2} r_a (r_b + r_a) - \tfrac{1}{3}(r_a^2 + r_b r_a + r_b^2) \right]$$

$$= 2\pi (0.0962) \left\{ \tfrac{1}{2}(101.6)(25.4 + 101.6) - \tfrac{1}{3}\left[(101.6)^2 + (25.4)(101.6) + (25.4)^2\right] \right\}$$

$$= 1170 \text{ mm}^3 \quad \text{or} \quad 1.170 \text{ cm}^3$$

For $p = 2.54$ cm on a 61-cm-long tube, there would be $61/2.54 = 24$ fins. Thus the total volume of fins is $(24)(1.170) = 28.08 \text{ cm}^3$.

Selecting a combination of ρ and p/r_a values (the choice being limited by the information available in Fig. 14.26), the preceding calculations can be repeated. Table 14.6 summarizes the calculations.

Based on the results in Table 14.6, a fin with a tip radius of $r_a = 10.16$ cm, a base thickness of $\delta_b = 0.0229$ cm, and a total volume of $V = 16.70$ cm^3 would give an optimum design. If copper ($k = 401$ W/m·K) or steel ($k = 64$ W/m·K) fins are chosen with $r_a = 10.16$ cm and $p = 10.16$ cm, then $\delta_b = 0.0135$ cm and $V = 9.88$ cm^3 for copper and $\delta_b = 0.0849$ cm and $V = 61.97$ cm^3 for steel. Thus if a choice of these three materials is available and if the minimum volume is the optimization criterion, a copper fin is the best choice. However, if weight is a consideration, an aluminum fin is preferable because the mass ratio of an aluminum to a copper fin is $(16.70/9.88)(0.30) = 0.51$, where 0.30 is the ratio of aluminum to copper specific weights.

The trial-and-error approach illustrated in Example 14.16 can be avoided if one uses the results of Schnurr et al. (1976), which were discussed in Section 14.4.6 for the longitudinal fins. Their results for optimum radial fins are given in Figs. 14.27 through 14.29. Use of this information is illustrated in a rework of Example 14.16.

Example 14.17. Design the radial fin–tube radiator of Example 14.16 using the information in Figs. 14.27 through 14.29.

SOLUTION. For $q^* = 2$, read curve 1 for black triangular fins in Fig. 14.27 to obtain

$$(r_a/r_b)_{\text{opt}} = 3.6$$

or

$$r_{a,\text{opt}} = 3.6r_b = (3.6)(2.54) = 9.14 \text{ cm}$$

Next, read curve 1 in Fig. 14.28 to obtain

$$\frac{p_{\text{opt}}}{r_b} = 1.85$$

or

$$p_{\text{opt}} = 1.85r_b = (1.85)(2.54) = 4.70 \text{ cm}$$

TABLE 14.6 Results of Calculations of Example 14.16

ρ	r_a (cm)	p/r_a	p (cm)	ζ	δ_b (mm)	V (cm^3)
0.25	10.16	0.25	2.540	0.12	0.095	27.72
0.33	7.62	0.25	1.905	0.11	0.104	22.47
0.33	7.62	0.50	3.810	0.06	0.191	20.61
0.25	10.16	0.50	5.080	0.08	0.143	20.88
0.25	10.16	1.00	10.160	0.05	0.229	16.70
0.33	7.62	1.00	7.62	0.02	0.572	30.89

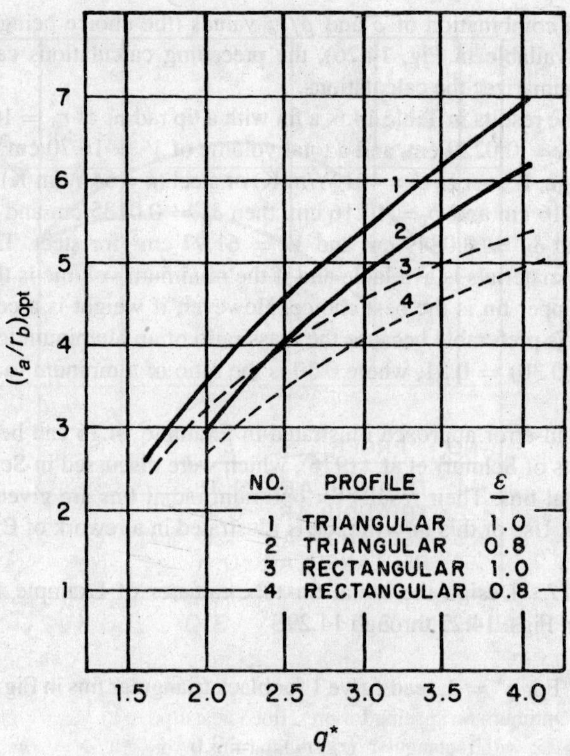

Figure 14.27 Optimum tip-to-base radius for given heat dissipation from a radial fin-and-tube radiator (Fig. 14.25) with rectangular or triangular fins. [Adapted from Schnurr et al. (1976).]

Finally, reading curve 1 in Fig. 14.29 provides

$$\zeta_{\text{opt}} = \frac{2\epsilon\sigma T_b^3 r_b^2}{k\delta_{b,\text{opt}}} = 0.07$$

or

$$\delta_{b,\text{opt}} = \frac{2\epsilon\sigma T_b^3 r_b^2}{k\zeta_{\text{opt}}}$$

$$= \frac{(2)(1)(5.669 \times 10^{-8})(340)^3(0.0254)^2}{(249)(0.07)}$$

$$= 1.65 \times 10^{-4} \text{ m}$$

Although the values for the radius r_a are within 10% of the values obtained in Example 14.16, the spacing p and the base width δ_b are significantly different. It is believed that the results presented here are more accurate than the ones in Example 14.16, which were based on a trial-and-error approach.

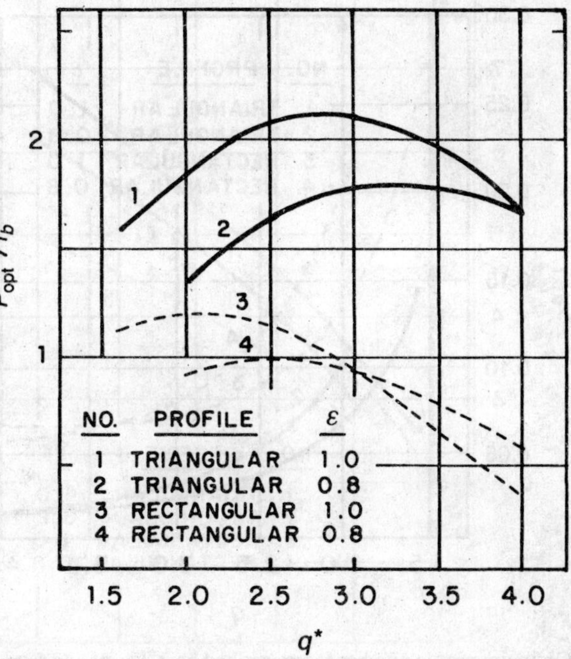

Figure 14.28 Optimum fin spacing for prescribed heat dissipation from a radial fin-and-tube radiator (Fig. 14.25) with rectangular or triangular fins. [Adapted from Schnurr et al. (1976)].

14.8 MISCELLANEOUS STUDIES

Okamoto (1964) looked at the longitudinal fin of rectangular profile, and treated the simultaneous modes of convection and radiation from the fin to the surrounding gas in a one-dimensional model numerically. In this study, h, k, ϵ, and the temperature of the surrounding gas were assumed constant. Collicott et al. (1965) did a free convection–radiation study that was somewhat less comprehensive than the one by Cobble (1964), and Shouman (1965) showed that a general solution can be found for one-dimensional configurations.

Eno (1967) considered the longitudinal fin of rectangular profile, and his linearized approximate solution was apparently less accurate than the numerically based solution of Cobble (1964) for the cylindrical spine. Shouman (1968) provided an exact solution for the one-dimensional problem that was formulated in his previous (Shouman, 1965) work.

Dent (1968) developed an electrical analog method for the determination of the temperature distribution in the radial fin of rectangular profile in a one-dimensional framework together with blackbody radiation and natural convection, and Frost and Eraslan (1968) provided an analysis of the longitudinal fin of rectangular profile that included gray body diffuse radiation interchange between fin and base.

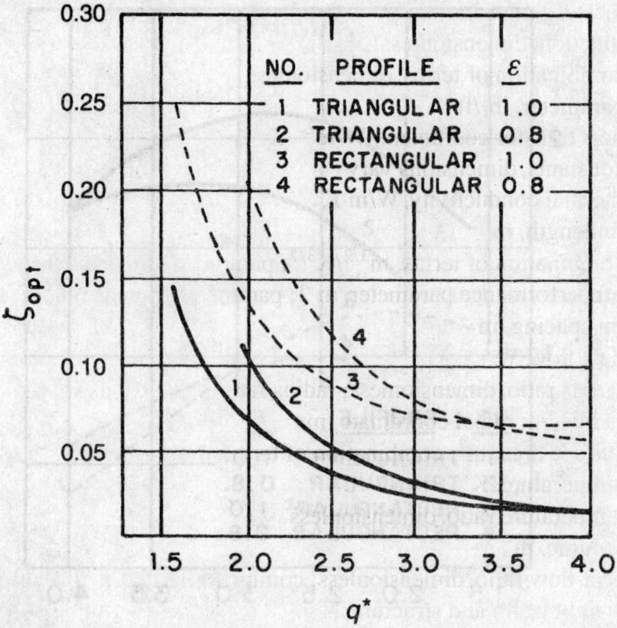

Figure 14.29 Optimum profile number for prescribed heat dissipation from a radial fin-and-tube radiator (Fig. 14.25) with rectangular or triangular fins. [Adapted from Schnurr et al. (1976).]

Kumar et al. (1993) considered the optimization of an array of rectangular fins standing on a nonisothermal horizontal surface and losing heat to the environment by simultaneous convection and radiation. This analysis, which includes fin-to-fin and fin-to-base radiative interactions, leads to correlations for the optimum fin effectiveness and optimum number of fins as functions of geometrical and thermal parameters. In application-oriented studies, Baker and Lund (1992) and Lund and Baker (1993) have considered two-dimensional conduction in an anisotropic fin brazed to a heat pipe and radiating to space. These studies optimized the total weight of the fin, heat pipe, and header section, and present simple algebraic expressions for calculating the fin dimensions (height and width), optimum efficiency, and optimum weight-to-heat transfer rate ratio.

14.9 NOMENCLATURE

Roman Letter Symbols

A	cross-sectional or profile area, m^2
$B(a, b)$	beta function, dimensionless
b	fin height, m
C	constant, dimensionless
d	diameter, dimensionless; differential or derivative, dimensionless

F	factor, dimensionless
f	function, dimensionless
G	combination of terms, dimensionless
H	parameter, lb_f/ft^2
h	heat transfer coefficient, $W/m^2 \cdot K$
K	constants, dimensions vary
k	thermal conductivity, W/m·K
L	fin length, m
M	combination of terms, $m^{3/2}/K^{3/2}$; parameter, dimensionless
m	fin performance parameter, m^{-1}; parameter, dimensionless
p	fin spacing, m
q	heat flow, W
R	radius ratio, dimensionless; radius ratio, dimensions vary
r	radius, m; radial coordinate, m
S	surface area, m^2; combination of terms, dimensionless
T	temperature, K
u	temperature ratio, dimensionless
V	volume, m^{-3}
υ	heat flow ratio, dimensionless; combination of terms, dimensionless
W	weight of fin and structure, N
W_s	mass of structure, W/m
w	transformed variable, K
X	height, dimensionless
x	height coordinate, m

Greek Letter Symbols

Γ	gamma function, dimensionless
γ	specific weight, N/m^3; angle between adjacent fins, dimensionless
Δ	change in, dimensionless
δ	fin thickness, m
∂	partial derivative, dimensionless
ϵ	emissivity, dimensionless
ζ	profile number, dimensionless
η	fin efficiency, dimensionless
θ	temperature ratio, dimensionless
θ	mean temperature, K (°R)
λ	Lagrange multiplier, W; taper ratio, dimensionless
ν	combination of terms, dimensionless
ξ	parameter, dimensionless; combination of terms, dimensionless
ρ	radius ratio, dimensionless; density, kg/m^3
σ	Stefan–Boltzmann constant, $W/m^2 \cdot K^4$

Roman Letter Subscripts

a	fin tip
b	fin base
e	emissivity

f	fin
id	ideal
max	maximum value
min	minimum value
o	nominal value
opt	optimum
p	profile
s	surroundings; height of associated structure per unit area
t	tube
u	incomplete beta function

Roman Letter Superscripts

m	exponent
n	exponent

Symbolic Superscript

	heat dissipation per unit length; heat flow parameter; dimensionless parameter

CHAPTER 15

MULTIDIMENSIONAL HEAT TRANSFER IN FINS AND FIN ASSEMBLIES

15.1 INTRODUCTION

A study of the references cited thus far shows that most extended surface analyses employ a one-dimensional heat conduction model. This model has proved adequate for the design of fins for a large variety of applications. Indeed, Razelos and Georgiou (1992) have argued that if fins are designed such that the transverse Biot number and the aspect ratio are on the order of 0.10 and 1.00, respectively, the one-dimensional model does not introduce appreciable error. Even these design criteria, while applicable to many engineering situations, are not applicable in all cases.

For example, as observed by Look (1988), if a horizontal fin is operating in the natural convection mode, the heat transfer coefficients for the top and bottom surfaces can differ by a factor of 2. If this difference is to be accommodated, a two-dimensional analysis becomes inevitable. In another situation, if the fin supporting structure is included in the analysis of the fin, it is found that the presence of the fin induces two-dimensional effects within the supporting structure and causes the base temperature of the fin to be nonuniform across the fin thickness. As Manzoor et al. (1983b) have shown, this nonuniformity produces two-dimensional effects within the fin. In yet another situation, a fin made of a strongly anisotropic composite material such as the material attached to a heat pipe radiator has been proposed by Lund and Baker (1993). Here a two-dimensional model also becomes necessary. Finally, Huang and Shah (1992) cite the example of a plate fin compact heat exchanger where the variation of surrounding temperature over the fin surface necessitates a two-dimensional analysis for the prediction of the temperature distribution in the fin.

This chapter reviews the literature on multidimensional heat conduction in extended surfaces, provides a comprehensive collection of important results, and identifies the practical situations where these results are applicable. The chapter begins with a study of the longitudinal fin of rectangular profile with convection. The situations discussed include constant base temperature, nonuniform ambient temperature, asymmetrical thermal conditions, nonconstant base temperature, anisotropic fin material, variable heat transfer coefficients, and internal heat generation. Next, the analyses and results for the longitudinal fin of triangular profile are given, and following this, two-dimensional effects in radial fins are considered. The final sections of the chapter deal with composite fins, convecting fin assembles, radiating–convecting fins, and fin arrays.

15.2 LONGITUDINAL FIN OF RECTANGULAR PROFILE WITH CONVECTION

Because the longitudinal fin of rectangular profile is most common in engineering applications, this geometry appears to have attracted the most attention in the literature. The two-dimensional performance of this fin has been studied under a myriad of conditions.

15.2.1 Constant Base Temperature

Consider the longitudinal fin of rectangular profile shown in Fig. 15.1. Observe that the origin of the coordinate system is taken at the fin base with positive orientation toward the fin tip, the fin height is b, and the half thickness in the y-coordinate direction is $\delta/2$. The fin has a thermal conductivity k, and the heat transfer coefficient on the faces of the fin is h. The fin has a base temperature T_b and a surrounding temperature T_s. Because of thermal symmetry, only one half of the fin shown in Fig. 15.1 need be considered.

The partial differential equation that governs the two-dimensional temperature distribution in the fin will be

$$\frac{\partial^2 T}{\partial x^2} + \frac{\partial^2 T}{\partial y^2} = 0 \tag{15.1}$$

with boundary conditions

$$T(0, y) = T_b \tag{15.2a}$$

$$\frac{\partial T}{\partial y}(x, 0) = 0 \qquad \text{(thermal symmetry)} \tag{15.2b}$$

$$\frac{\partial T}{\partial x}(b, y) = -\frac{h}{k}[T(b, y) - T_s] \tag{15.2c}$$

and

$$\frac{\partial T}{\partial y}(x, \delta/2) = -\frac{h}{k}[T(x, \delta/2) - T_s] \tag{15.2d}$$

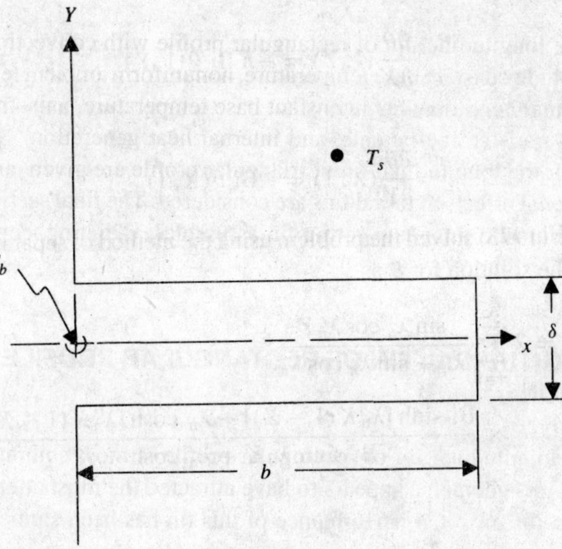

Figure 15.1 Two-dimensional longitudinal fin of rectangular profile.

After introducing the dimensionless variables

$$\theta = \frac{T - T_s}{T_b - T_s} \tag{15.3a}$$

$$X = \frac{x}{b} \tag{15.3b}$$

$$Y = \frac{2y}{\delta} \tag{15.3c}$$

$$K = \frac{2b}{\delta} \tag{15.3d}$$

and

$$\mathrm{Bi} = \frac{h\delta}{2k} \tag{15.3e}$$

eqs. (15.1) and (15.2) take the form

$$\frac{\partial^2 \theta}{\partial X^2} + K^2 \frac{\partial^2 \theta}{\partial Y^2} = 0 \tag{15.4}$$

and

$$\theta(0, Y) = 1 \tag{15.5a}$$

$$\frac{\partial \theta}{\partial Y}(X, 0) = 0 \tag{15.5b}$$

$$\frac{\partial \theta}{\partial X}(1, Y) = -K\text{Bi}\ \theta(1, Y) \tag{15.5c}$$

and

$$\frac{\partial \theta}{\partial Y}(X, 1) = -\text{Bi}\ \theta(X, 1) \tag{15.5d}$$

Lau and Tan (1973) solved the problem using the method of separation of variables and obtained the solution for θ as

$$\theta = 2\sum_{n=1}^{\infty} \frac{\sin \lambda_n \cos \lambda_n Y}{\lambda_n + \sin \lambda_n \cos \lambda_n} \times \frac{\text{Bi} \sinh [\lambda_n K(1-X)] + \lambda_n \cosh [\lambda_n K(1-X)]}{\text{Bi} \sinh \lambda_n K + \lambda_n \cosh \lambda_n K} \tag{15.6}$$

where

$$\lambda_n \tan \lambda_n = \text{Bi} \tag{15.7}$$

Using eq. (15.6), the dimensionless heat transfer rate from the base of the fin per unit fin length, $q = q_b/L$, can be defined as

$$Q = \frac{q}{k(T_b - T_s)}$$

and then obtained as

$$Q = 4\sum_{n=1}^{\infty} \frac{\sin^2 \lambda_n}{\lambda_n + \sin \lambda_n \cos \lambda_n} \frac{\text{Bi} \cosh \lambda_n K + \lambda_n \sinh \lambda_n K}{\text{Bi} \sinh \lambda_n K + \lambda_n \cosh \lambda_n K} \tag{15.8}$$

Figure 15.2 shows the dimensionless heat dissipation Q as a function of the aspect ratio K for Bi = 0.01, 0.05, and 0.10. As expected, the results for Q show a strong dependence on Bi. For a given Bi, Q increases as K increases but appears to level off beyond $K = 100$.

If $Q_{\text{1-D}}$ denotes the one-dimensional heat dissipation, it can be shown that

$$Q_{\text{1-D}} = \frac{q_{\text{1-D}}}{k(T_b - T_s)} = 2\text{Bi}^{1/2} \frac{\sinh K\text{Bi}^{1/2} + \text{Bi} \cosh K\text{Bi}^{1/2}}{\cosh K\text{Bi}^{1/2} + \text{Bi} \sinh K\text{Bi}^{1/2}} \tag{15.9}$$

The percent error

$$100\left(\frac{Q_{\text{1-D}} - Q}{Q}\right)$$

is plotted in Fig. 15.3 as a function of K for Bi = 0.01, 0.10, 1.0, and 10. It can be noted that that for Bi $\leq$ 0.10, the maximum error is about 1%. As Bi increases, the error increases and is as high as 60% for Bi = 10 and $K > 1$. These results indicate that the one-dimensional assumption is valid if Bi is much less than unity.

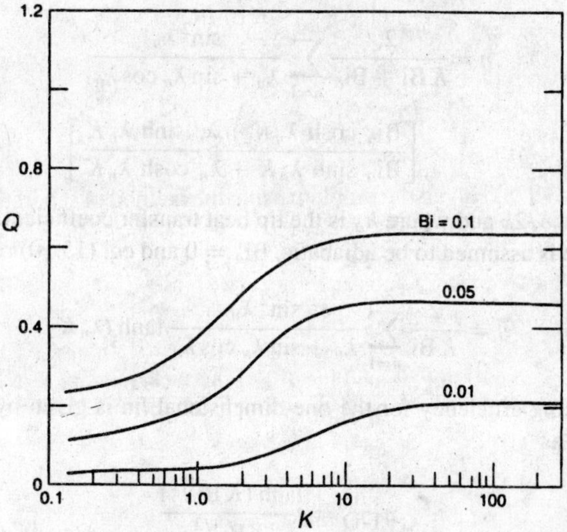

Figure 15.2 Two-dimensional heat transfer in a longitudinal fin of rectangular profile.

Levitsky (1972) obtained a double series solution for three-dimensional conduction in an infinitely high rectangular cross-section fin and proved that the three-dimensional solution reduces to a one-dimensional solution if Bi $<<$ 1.

This problem was revisited by Huang and Shah (1992), who allowed the tip heat transfer coefficient to differ from that of the fin faces. They derived an expression for the fin efficiency

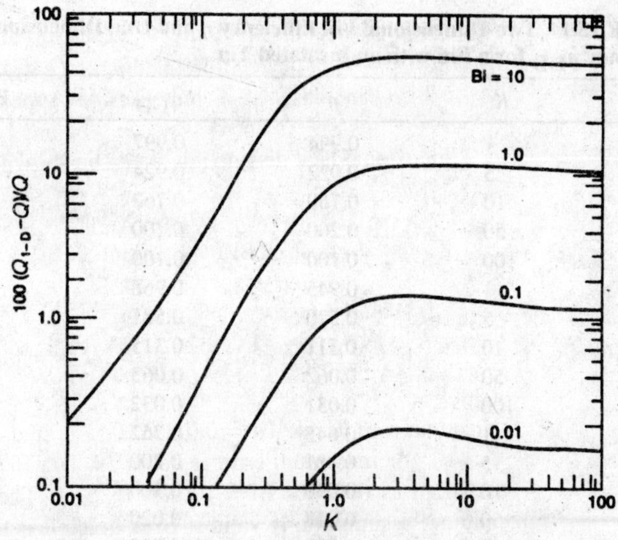

Figure 15.3 Error due to one-dimensional model for a longitudinal fin of rectangular profile.

$$\eta = \frac{2}{K\text{Bi} + \text{Bi}_a} \sum_{n=1}^{\infty} \frac{\sin^2 \lambda_n}{\lambda_n + \sin \lambda_n \cos \lambda_n} \times \left[\frac{\text{Bi}_a \cosh \lambda_n K + \lambda_n \sinh \lambda_n K}{\text{Bi}_a \sinh \lambda_n K + \lambda_n \cosh \lambda_n K} \right] \tag{15.10}$$

where $\text{Bi}_a = h_a \delta / 2k$ and where h_a is the tip heat transfer coefficient.

If the fin tip is assumed to be adiabatic, $\text{Bi}_a = 0$ and eq. (15.10) reduces to

$$\eta = \frac{2}{K\text{Bi}} \sum_{n=1}^{\infty} \frac{\sin^2 \lambda_n}{\lambda_n + \sin \lambda_n \cos \lambda_n} \tanh(\lambda_n K) \tag{15.11}$$

The corrsponding efficiency for the one-dimensional fin is given by eq. (1.11) and can be written as

$$\eta_{1\text{-D}} = \frac{\tanh(K\text{Bi}^{1/2})}{K\text{Bi}^{1/2}} \tag{15.12}$$

Huang and Shah (1992) have computed the one- and two-dimensional efficiencies using eqs. (15.11) and (15.12), respectively. Their results, which are given in Table 15.1 show once again that the two-dimensional effects are negligible for Bi << 1. Moreover, the two-dimensional efficiency is lower than the one-dimensional efficiency. In Table 15.1, the error is given as a percentage:

$$\text{error} = 100 \left(\frac{\eta_{1\text{-D}} - \eta}{\eta} \right)$$

TABLE 15.1 Two-Dimensional Fin Efficiency η and One-Dimensional Fin Efficiency $\eta_{1\text{-D}}$ for a Fin with an Insulated Tip

Bi	K	η	$\eta_{1\text{-D}}$	Error %
0.01	1	0.994	0.997	0.255
	5	0.922	0.924	0.292
	10	0.760	0.762	0.249
	50	0.200	0.200	0.159
	100	0.100	0.100	0.159
0.10	1	0.945	0.968	2.454
	5	0.570	0.581	1.876
	10	0.311	0.315	1.480
	50	0.062	0.063	1.441
	100	0.031	0.032	1.441
1.00	1	0.645	0.762	18.042
	5	0.181	0.200	10.792
	10	0.090	0.100	10.764
	50	0.018	0.020	10.764
	100	0.009	0.010	10.764

The present problem of two-dimensional conduction in the longitudinal fin of rectangular profile was first solved by Avrami and Little (1942). Unfortunately, they did not use the Biot number explicitly as the independent parameter and reached conclusions that are strictly applicable for small Biot number values.

The nature of the infinite series in the two-dimensional solutions make them cumbersome to use for quick calculations. To circumvent this difficulty, Sfeir (1976) and Aparecido and Cotta (1990) developed approximate solutions that resemble the classical one-dimensional solutions but account for the two-dimensional heat conduction. Using a second-order polynomial for the temperature variation in the y-direction, Sfeir was able to reduce eq. (15.1), the governing partial differential equation, to an ordinary differential equation similar to the classical one-dimensional fin equation. With Q defined as

$$Q = \frac{q}{k(T_b - T_s)}$$

Sfeir obtained

$$Q = 2(\mathrm{Bi}^+)^{1/2} \frac{(\mathrm{Bi}^+)^{1/2} \sinh[K(\mathrm{Bi}^+)^{1/2}] + \mathrm{Bi}_a \cosh[K(\mathrm{Bi}^+)^{1/2}]}{(\mathrm{Bi}^+)^{1/2} \cosh[K(\mathrm{Bi}^+)^{1/2}] + \mathrm{Bi}_a \sinh[K(\mathrm{Bi}^+)^{1/2}]} \tag{15.13}$$

where

$$\mathrm{Bi}^+ = \frac{\mathrm{Bi}}{1 + \mathrm{Bi}/3}$$

While Sfeir (1976) used a heat balance integral approach, Aparecido and Cotta (1990) employed the coupled integral equation to derive the same equation as eq. (15.13) except that their Bi^+ has the $\mathrm{Bi}/3$ in the denominator replaced by $\mathrm{Bi}/4$. Regardless of the definition of Bi^+, when used for the fin with an insulated tip, eq. (15.13) gives results within 1% of the exact solution for $\mathrm{Bi} \leq 1$ and $1 \leq K \leq 100$. Under these restrictions, eq. (15.13) provides a quick and accurate means for computing two-dimensional heat transfer.

15.2.2 Asymmetric Thermal Boundary Conditions

In many situations, the fin extends between two surfaces at different temperatures and thus operates under asymmetric boundary conditions. Aziz and Tesarik (1992) have considered the asymmetrically heated rectangular fin of Fig. 15.4. For this physical situation, the two-dimensional steady conduction in the fin is still governed by eqs. (15.4), (15.5a), and (15.5b) except that

$$\theta = \frac{T - T_s}{T_1 - T_s} \tag{15.14}$$

and the boundary condition of eq. (15.5c) becomes

$$\theta(1, Y) = \theta^* = \frac{T_2 - T_s}{T_1 - T_s} \tag{15.15}$$

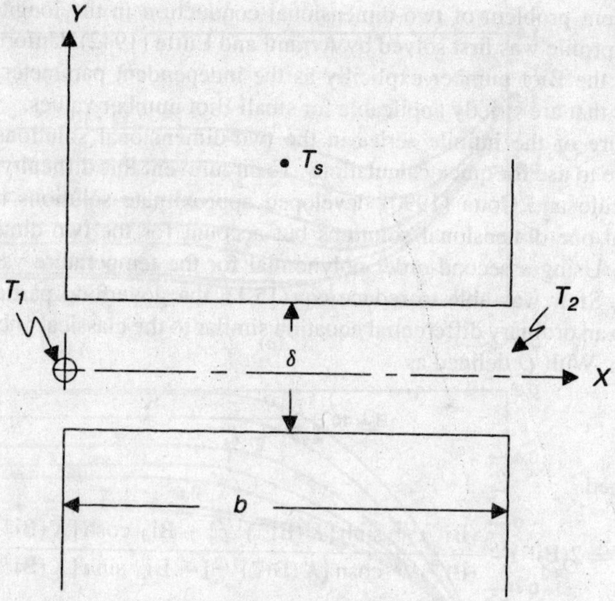

Figure 15.4 Asymmetrically heated rectangular profile fin. Here $T_1 > T_2$.

Here θ^* characterizes the temperature asymmetry.

Use of the method of separation of variables in conjunction with superposition gives the solution for θ:

$$\theta = 2\sum_{n=1}^{\infty} \frac{\sin^2 \lambda_n \cos \lambda_n Y}{\lambda_n + \sin \lambda_n \cos \lambda_n} \frac{\theta^* \sinh \lambda_n K X - \sinh [\lambda_n K(1 - X)]}{\sinh \lambda_n K} \tag{15.16}$$

where λ_n is obtained from eq. (15.7):

$$\lambda_n \tan \lambda_n = \mathrm{Bi} \tag{15.7}$$

The heat transfer rates per unit length, Q_1 and Q_2, from the two ends of the fin are given by

$$Q_1 = \frac{q_1}{4k(T_1 - T_s)} = \sum_{n=1}^{\infty} \frac{\sin^2 \lambda_n}{\lambda_n + \sin \lambda_n \cos \lambda_n} \frac{\cosh \lambda_n K - \theta^*}{\sinh \lambda_n K} \tag{15.17}$$

and

$$Q_2 = \frac{q_2}{4k(T_1 - T_s)} = \sum_{n=1}^{\infty} \frac{\sin^2 \lambda_n}{\lambda_n + \sin \lambda_n \cos \lambda_n} \frac{\theta^* \cosh \lambda_n K - 1}{\sinh \lambda_n K} \tag{15.18}$$

Figure 15.5 shows the results for Q_1 and Q_2 for the case of $\theta^* = 0.75$. It is seen that the heat transfer from the high-temperature end, Q_1, decreases as K increases and

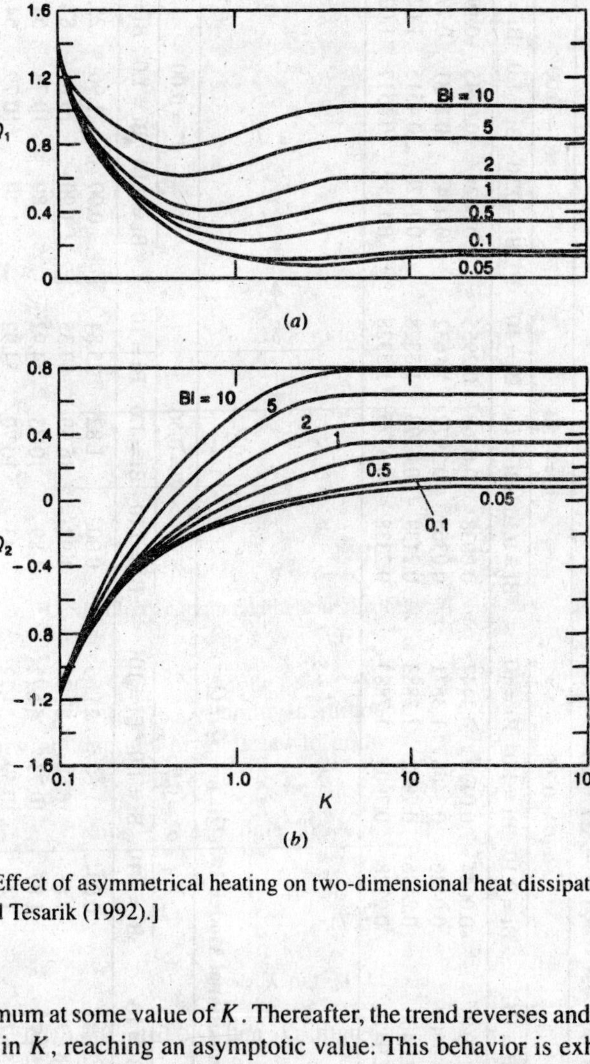

Figure 15.5 Effect of asymmetrical heating on two-dimensional heat dissipation, $\theta^* = 0.75$. [From Aziz and Tesarik (1992).]

attains a minimum at some value of K. Thereafter, the trend reverses and Q_1 increases with increase in K, reaching an asymptotic value. This behavior is exhibited for all values of Bi considered, but the pattern is most distinct at Bi = 10. The lower portion of Fig. 15.5 shows that Q_2 is negative at low values of K, indicating that the low-temperature end receives heat through the fin. This effect is especially pronounced for small aspect ratios K when the walls are relatively close together. Moreover, for each Bi, there is a value of $K = K_o$ for which $Q_2 = 0$ and K_o decreases as Bi increases. Thus, if the fin is to be designed so that the low-temperature end rejects heat, K must exceed K_o. Aziz and Tesarik (1972) provide results for Q_1 and Q_2 for other values of θ.*

The total convective heat loss from the fin, $Q = Q_1 + Q_2$, is given in Table 15.2 for $\theta^* = 1.00, 0.75, 0.50$, and 0.00. These results show that Q is a maximum when the fin is symmetrically heated ($\theta^* = 1$) and decreases as θ^* decreases. The heat loss is at

TABLE 15.2 Total Convective Heat Loss When $Q = Q_1 + Q_2$

	$\theta^* = 1.00$			$\theta^* = 0.75$			$\theta^* = 0.50$			$\theta^* = 0.00$		
K	Bi = 0.10	Bi = 1.0	Bi = 10	Bi = 0.10	Bi = 1.0	Bi = 10	Bi = 0.10	Bi = 1.0	Bi = 10	Bi = 0.10	Bi = 1.0	Bi = 10
0.10	0.0050	0.0491	0.3940	0.0044	0.0430	0.3447	0.0038	0.0368	0.2955	0.0025	0.0245	0.1970
1.00	0.0487	0.3963	1.5509	0.0426	0.3468	1.3571	0.0365	0.2972	1.1632	0.0243	0.1981	0.7755
10	0.2852	0.9031	2.0438	0.2496	0.7902	1.7883	0.2139	0.6773	1.5328	0.1426	0.4515	1.0219
100	0.3118	0.9034	2.0438	0.2728	0.7905	1.7883	0.2338	0.6775	1.5328	0.1559	0.4517	1.0219

TABLE 15.3 Errors in One-Dimensional Model, $100(Q_{1-D} - Q)/Q$

	$\theta^* = 1.00$			$\theta^* = 0.75$			$\theta^* = 0.50$			$\theta^* = 0.00$		
K	Bi = 0.10	Bi = 1.0	Bi = 10	Bi = 0.10	Bi = 1.0	Bi = 10	Bi = 0.10	Bi = 1.0	Bi = 10	Bi = 0.10	Bi = 1.0	Bi = 10
0.10	0.00	1.75	25.84	0.00	1.66	25.86	0.00	1.82	25.84	0.00	1.96	25.84
1.00	1.84	16.65	87.35	1.88	16.62	87.34	1.92	16.99	87.35	1.90	16.65	87.34
10	1.89	10.73	54.72	1.87	10.73	54.72	1.89	10.73	54.92	1.89	10.74	54.72
100	1.38	10.69	54.72	1.42	10.69	54.72	1.43	10.70	54.92	1.40	10.79	54.72

a minimum when $\theta^* = 0$, which corresponds to the case when the low-temperature end of the fin at T_2 equals the surrounding temperature at T_s.

It is useful to assess the accuracy of the one-dimensional solution in light of the two-dimensional results of Table 15.2. The total heat dissipated by the one-dimensional fin can be shown to be

$$Q_{1\text{-D}} = \frac{q_{1\text{-D}}}{4k(T_1 - T_s)} = \frac{1}{2}(1 + \theta^*)\text{Bi}^{1/2} \tanh \frac{K\text{Bi}^{1/2}}{2} \tag{15.19}$$

The error between the one-dimensional heat loss $Q_{1\text{-D}}$ and the two-dimensional heat loss Q, expressed as

$$\text{error} = 100\left(\frac{Q_{1\text{-D}} - Q}{Q}\right)$$

was evaluated for the parametric values used in Table 15.2. The results are given in Table 15.3 and show that the percent error is affected only slightly by the temperature asymmetry. However, the error is a strong function of K and Bi. At Bi = 0.1, the maximum error is under 2%, while for Bi = 1, the corresponding figure is about 17%. For Bi = 10, the maximum error is about 87%. For every Bi, the maximum error occurs in the neighborhood of $K = 1$. With regard to Table 15.3, it should be noted that all combinations of K and Bi may not be encountered in practice.

15.2.3 Unequal Convection Coefficients on Opposite Fin Faces

Consider the horizontal fin shown in Fig. 15.6, which operates under natural convection conditions. As shown in Chapter 4, the convection heat transfer coefficients on the top and bottom surfaces differ markedly. For the top, which is a heated horizontal surface facing upward,

$$\text{Nu}_L = 0.54\text{Ra}_L^{1/4} \qquad (10^4 < \text{Ra}_L < 10^7) \tag{4.43}$$

and for the bottom, which is a heated horizontal surface facing downward,

$$\text{Nu}_L = 0.27\text{Ra}_L^{1/4} \qquad (10^5 < \text{Ra}_L < 10^{10}) \tag{4.45}$$

At corresponding Rayleigh numbers in the range 10^5 to 10^7, the average heat transfer coefficient for the top surface is twice that for the bottom surface, that is, $h_1 = 2h_2$. For this situation, Look (1988) obtained an exact analytical solution, and Sfeir (1976) developed an approximate solution based on the heat balance integral approach.

The mathematical model of Look (1988) requires that the boundary conditions of eqs. (15.2*b*) through (15.2*d*) be changed to

$$\frac{\partial T}{\partial y}(x, \delta/2) = -\frac{h_1}{k}[T(x, \delta/2) - T_s] \tag{15.20a}$$

$$\frac{\partial T}{\partial y}(x, -\delta/2) = \frac{h_2}{k}[T(x, -\delta/2) - T_s] \tag{15.20b}$$

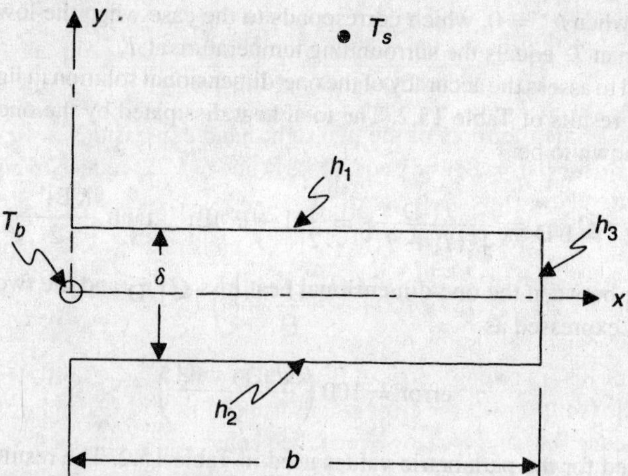

Figure 15.6 Longitudinal fin of rectangular profile with different surface convection coefficients.

$$\frac{\partial T}{\partial x}(b, y) = -\frac{h_3}{k}[T(b, y) - T_s] \tag{15.20c}$$

Equations (15.1) and (15.2*a*) still apply. Using the method of separation of the variables, Look (1988) obtained solutions for θ,

$$\theta = \sum_{n=1}^{\infty} \frac{(\cos \lambda_n Y + C_n \sin \lambda_n Y) \sin \lambda_n}{\lambda_n + \sin \lambda_n \cos \lambda_n + C_n^2(\lambda_n - \sin \lambda_n \cos \lambda_n)} \times \frac{\text{Bi}_3 \sinh[\lambda_n K(1 - X)] + \lambda_n \cosh[\lambda_n K(1 - X)]}{\text{Bi}_3 \sinh \lambda_n K + \lambda_n \cosh \lambda_n K} \tag{15.21}$$

and Q, defined by

$$Q = \frac{q}{k(T_b - T_s)}$$

as

$$Q = \sum_{n=1}^{\infty} \frac{\sin^2 \lambda_n}{\lambda_n + \sin \lambda_n \cos \lambda_n + C_n^2(\lambda_n - \sin \lambda_n \cos \lambda_n)} \times \frac{\text{Bi}_3 \cosh \lambda_n K + \lambda_n \sinh \lambda_n K}{\text{Bi}_3 \sinh \lambda_n K + \lambda_n \cosh \lambda_n K} \tag{15.22}$$

In eqs. (15.21) and (15.22),

$$\text{Bi}_1 = \frac{h_1 \delta}{2k}, \quad \text{Bi}_2 = \frac{h_2 \delta}{2k}, \quad \text{and} \quad \text{Bi}_3 = \frac{h_3 \delta}{2k}$$

and

$$C_n = \frac{\lambda_n \tan\lambda_n - \mathrm{Bi}_1}{\lambda_n + \mathrm{Bi}_1 \tan\lambda_n} = \frac{\mathrm{Bi}_2 - \lambda_n \tan\lambda_n}{\lambda_n + \mathrm{Bi}_2 \tan\lambda_n} \tag{15.23}$$

The values of λ_n are the roots of the two right-hand expressions in eq. (15.23). For $C_n = 0$, and $\mathrm{Bi}_1 = \mathrm{Bi}_2 = \mathrm{Bi}_3 = 0$, eqs. (15.21) and (15.22) reduce to eqs. (15.6) and (15.8), respectively.

Figure 15.7 shows the percent error

$$100\left(\frac{Q_{1\text{-D}} - Q}{Q}\right)$$

as a function of K for various values of Bi_1 and Bi_3 with $\mathrm{Bi}_2 = 0.5\mathrm{Bi}_1$. For the case of an insulated tip (solid lines), the error ranges from 15 to 50% depending on the values of Bi_1 (and hence Bi_2) and K. However, when convection from the fin tip is allowed ($\mathrm{Bi}_3 = 0.25$), the error is considerably reduced, particularly at low values of K. The case of $\mathrm{Bi}_3 = \infty$ corresponds to the condition when the fin-tip and environmental temperatures are the same.

Look (1988) also made a comparison of the heat dissipation $Q^{=}$ by the fin when $\mathrm{Bi}_1 = \mathrm{Bi}_2$ and the heat dissipation $Q^{\neq}$ by the fin when $\mathrm{Bi}_1 > \mathrm{Bi}_2$. As anticipated, the superscripts, = and ≠ denote whether the heat transfer coefficients are equal or

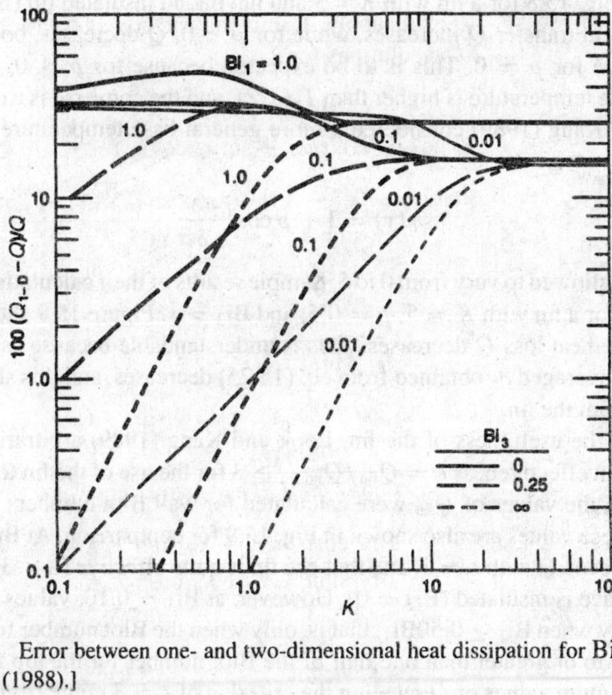

Figure 15.7 Error between one- and two-dimensional heat dissipation for $\mathrm{Bi}_2 = 0.50\mathrm{Bi}_1$. [From Look (1988).]

unequal, respectively. Table 15.4, which is based on an additional study by Look (1989), gives the ratio $Q^{=}/Q^{\neq}$ for $K = 5$ and various combinations of the Biot numbers. For a given Bi_1, as $\mathrm{Bi}_2/\mathrm{Bi}_1$ decreases, that is, as Bi_2 decreases, $Q^{\neq}$ is expected to decrease because of the weaker convection at the bottom surface. The decrease in $Q^{\neq}$ is reflected in the higher values of $Q^{=}/Q^{\neq}$. This behavior is exhibited for all three tip conditions.

15.2.4 Nonuniform Base Temperature

Studies of simultaneous conduction in the fin and the wall to which the fin is attached have shown that the fin base temperature varies slightly with y, being lowest at the center ($y = 0$) and highest at the edge ($y = \delta/2$). To account for this nonuniformity, Look (1989) extended the analysis of Section 15.2.3 by modeling the base temperature variation as

$$\theta_b(y) = \frac{T_b(y) - T_s}{T_b(\delta/2) - T_s} = 1 + p\cos\frac{\pi y}{\delta} \tag{15.24}$$

where $T_b(\delta/2)$ is equal to the temperature of the unfinned wall. The effect of the parameter p, which is the maximum base temperature depression or elevation, on

$$Q = \frac{q}{k[T_b(\delta/2) - T_s]}$$

is shown in Fig. 15.8 for a fin with $K = 5$ and having an insulated tip ($\mathrm{Bi}_3 = 0$). For $p > 0$, the heat transfer Q increases, while for $p < 0$, Q decreases, both compared with the value for $p = 0$. This is to be expected because for $p > 0$, the spatially averaged base temperature is higher than $T_b(\delta/2)$, and the converse is true for $p < 0$.

Look and Kang (1989) considered a more general base temperature variation of the form

$$\theta_b(y) = 1 + p\cos^n\frac{\pi y}{\delta} \tag{15.25}$$

where n was allowed to vary from 0 to 5. Sample results of their calculations are given in Fig. 15.9 for a fin with $K = 5$, $p = 0.5$, and $\mathrm{Bi}_3 = 1$. Figure 15.9 shows that as n increases, the heat loss Q decreases. This is understandable because as n increases, the spatially averaged θ_b obtained from eq. (15.25) decreases, and this should lead to a lower Q from the fin.

To assess the usefulness of the fin, Look and Kang (1989) arbitrarily chose the criterion of fin effectiveness $\epsilon = Q_{\text{fin}}/Q_{\text{no fin}} \geq 3$ for the use of the fin to be justified. Using $\epsilon = 3$, the values of Q_{fin} were calculated for wall Biot numbers, $\mathrm{Bi}_w = 0.01$ and 0.10. These values are also shown in Fig. 15.9 for comparison. At $\mathrm{Bi}_1 = 0.01$, the actual Q exceeds Q with $\epsilon = 3$, and thus the fin is quite effective ($\epsilon > 3$), even when the bottom face is insulated ($\mathrm{Bi}_2 = 0$). However, at $\mathrm{Bi}_1 = 0.10$, values of $\epsilon > 3$ are achieved only when $\mathrm{Bi}_2 \geq 0.50\mathrm{Bi}_1$, that is, only when the Biot number for the bottom face is equal to or greater than one-half of the Biot number for the top face. For $\mathrm{Bi}_1 = 1$, the minimum values of Q meeting the criterion of $\epsilon = 3$ range from 9 for $n = 0$

TABLE 15.4 $Q^{=}/Q^{\neq}$ **for a Two-Dimensional Rectangular Fin with** $K = 5$

	$Bi_1 = 0.01$			$Bi_1 = 0.10$			$Bi_1 = 1.0$		
Bi_2/Bi_1	$Bi_3 = 0$	$Bi_3 = 0.25$	$Bi_3 = 1.00$	$Bi_3 = 0$	$Bi_3 = 0.25$	$Bi_3 = 1.00$	$Bi_3 = 0$	$Bi_3 = 0.25$	$Bi_3 = 1.00$
1.00	1.0000	1.0000	1.0000	1.0000	1.0000	1.0000	1.0000	1.0000	1.0000
0.75	1.1320	1.0236	1.0094	1.0911	1.0684	1.0412	1.0720	1.0719	1.0716
0.50	1.3085	1.0487	1.0191	1.2106	1.1506	1.1039	1.1493	1.1489	1.1478
0.25	1.5567	1.0754	1.0290	...	...	...	1.2777	1.2763	1.2728
0.10	1.7607	1.0916	1.0351	1.5242	1.3339	1.2125	1.4018	1.3983	1.3902
0.00	1.9314	1.1038	1.0392	1.6478	1.3955	1.2454	1.5249	1.5174	1.5019

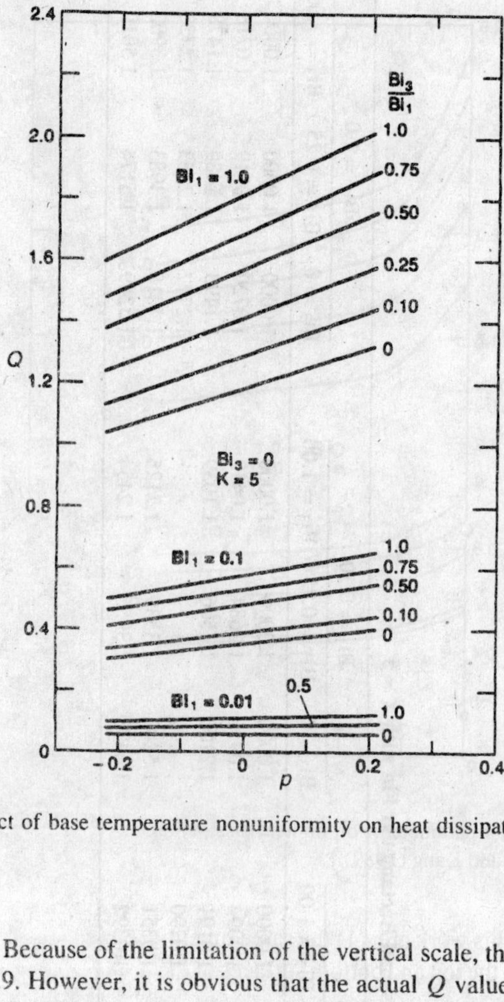

Figure 15.8 Effect of base temperature nonuniformity on heat dissipation. [Adapted from Look (1989).]

to 7.02 for $n = 5$. Because of the limitation of the vertical scale, these values are not shown in Fig. 15.9. However, it is obvious that the actual Q values for $Bi_1 = 1$ are much lower than the values needed to achieve an effectiveness value of 3. Thus the fin fails to meet the criterion of $\epsilon \geq 3$ at $Bi_1 = 1$.

An optimization study for a two-dimensional rectangular fin with a nonuniform base temperature of the form of eq. (15.25) and unequal heat transfer coefficients for top, bottom, and tip surfaces has also been published by Look and Kang (1992).

15.2.5 Coordinate-Dependent Heat Transfer Coefficient

Sections 15.2.3 and 15.2.4 have dealt with a fin having different heat transfer coefficients on the top and bottom surfaces, but the analyses did not allow for the heat transfer coefficient to vary with x. To overcome this, Ma et al. (1991) adopted a Fourier series approach and solved this problem for three types of tip conditions: a convecting tip, an insulated tip, and a tip in thermal equilibrium with the surroundings.

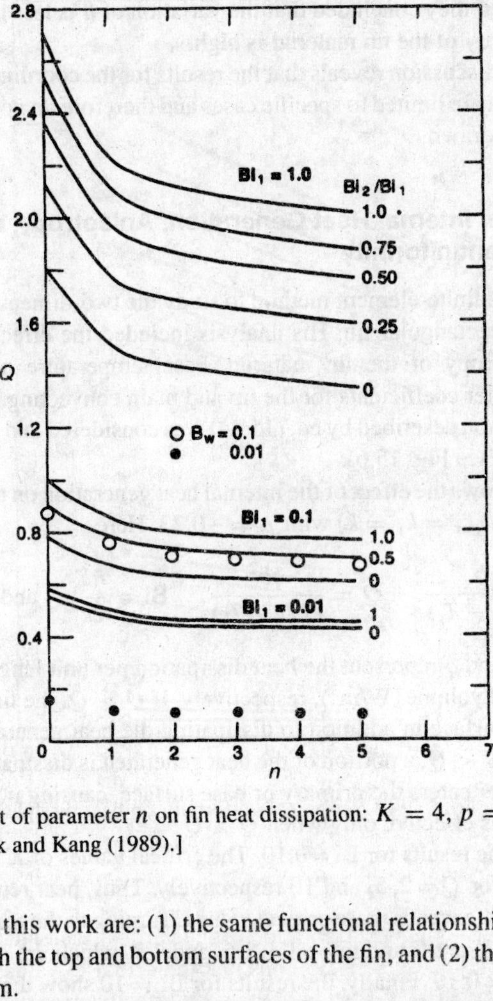

Figure 15.9 Effect of parameter n on fin heat dissipation: $K = 4$, $p = 0.5$, and $Bi_3 = 1$. [Adapted from Look and Kang (1989).]

The limitations of this work are: (1) the same functional relationship between h and x is assumed for both the top and bottom surfaces of the fin, and (2) the base temperature is taken as uniform.

The numerical results presented by the authors are limited to one example in which they consider a rectangular fin with $b = 4$ cm, $\delta = 1$ cm, and a thermal conductivity of 100 W/m·K. The heat transfer coefficient for the first 1 cm of height was taken to be 50 W/m^2 · K while for the remaining 3 cm, the heat transfer coefficient was assumed to be 100 W/m^2·K. The tip of the fin was regarded as adiabatic, and the value of $q/(T_b - T_s)$ was found to be 0.1259 W/K. The corresponding one-dimensional figure was calculated as 0.1150 W/K, indicating that the one-dimensional solution underestimates q by 8.7%. This result is in contrast to the results of Fig. 15.3 for constant h, which show that the one-dimensional $Q_{1\text{-D}}$ overestimates the two-dimensional q.

The effect of variable h on the two-dimensional rectangular fin has also been considered by Barrow et al. (1986). Using a finite difference approach and a linearly increasing h with x, they reported a 39% reduction in heat transfer for a specific situation. However, they concluded that the variation of h is less important when the thermal conductivity of the fin material is high.

The foregoing discussion reveals that the results for the coordinate-dependent heat transfer coefficient are limited to specific cases and therefore do not allow any general conclusions to be drawn.

15.2.6 Effect of Internal Heat Generation, Anisotropy, and Base Temperature Nonuniformity

Aziz (1993) used a finite-element method to study the two-dimensional performance of a longitudinal rectangular fin. His analysis included the effects of uniform heat generation, anisotropy of the fin material, base temperature nonuniformity, and unequal heat transfer coefficients for the tip and main convecting surfaces. The base temperature variation described by eq. (15.24) was considered and Aziz also assumed that $h_1 = h_2 = h$ (see Fig. 15.6).

Figure 15.10 shows the effect of the internal heat generation on the heat dissipation for an isotropic fin ($k_x = k_y = k$) with $p = -0.23$. Here

$$Q = \frac{qb}{k\delta(T_b - T_s)}, \quad \dot{Q} = \frac{\dot{q}b^2}{k(T_b - T_s)}, \quad \text{Bi} = \frac{h\delta}{2k}, \quad \text{and} \quad K = \frac{2b}{\delta}$$

The quantities $\dot{Q}$ and $\dot{q}$ represent the heat dissipation per unit length (W/m) and heat generation per unit volume (W/m^3), respectively. If $Q > \dot{Q}$, the fin is removing heat from its primary surface in addition to dissipating the heat generated internally. On the other hand, if $Q < \dot{Q}$, a portion of the heat generated is dissipated by convection, while the remainder enters the primary or base surface, causing it to heat rather than cool. Thus the fin is effective only when $Q > \dot{Q}$.

First consider the results for Bi = 0.10. The critical values of K for which $Q = \dot{Q}$ are 4.8, 7, and 10 for $\dot{Q} = 2, 5$, and 10 respectively. Thus, heat removal from the fin occurs only when K exceeds these critical values. Turning to the results for Bi = 1, the critical values for K are found to be 1.1, 1.6, and 1.9, which are significantly lower than those for Bi = 0.10. Finally, the results for Bi = 10 show that the critical value of K for $\dot{Q} = 10$ is 0.6. For $\dot{Q} = 5$, this value lies below $K = 0.1$.

The effect of anisotropy on the performance of the fin is illustrated in Fig. 15.11 for $\dot{Q} = 2$, $p = 0.23$, $K = 2$, and $h_1 = h_2 = h_3$. Here

$$Q = \frac{qb}{k_x\delta(T_b - T_s)}, \quad \dot{Q} = \frac{\dot{q}b^2}{k_x(T_b - T_s)},$$

$$\text{Bi}_1 = \frac{h_1\delta}{2k_y}, \quad \text{Bi}_2 = \frac{h_2\delta}{2k_y}, \quad \text{and} \quad \text{Bi}_3 = \frac{h_3 b}{k_x}$$

For a given Bi, Q is seen to decrease as the ratio k_x/k_y increases. This is plausible because as the longitudinal thermal conductivity k_x increases or as the transverse conductivity k_y decreases, the longitudinal conduction is increased while the transverse conduction is reduced. The reduction in transverse conduction means

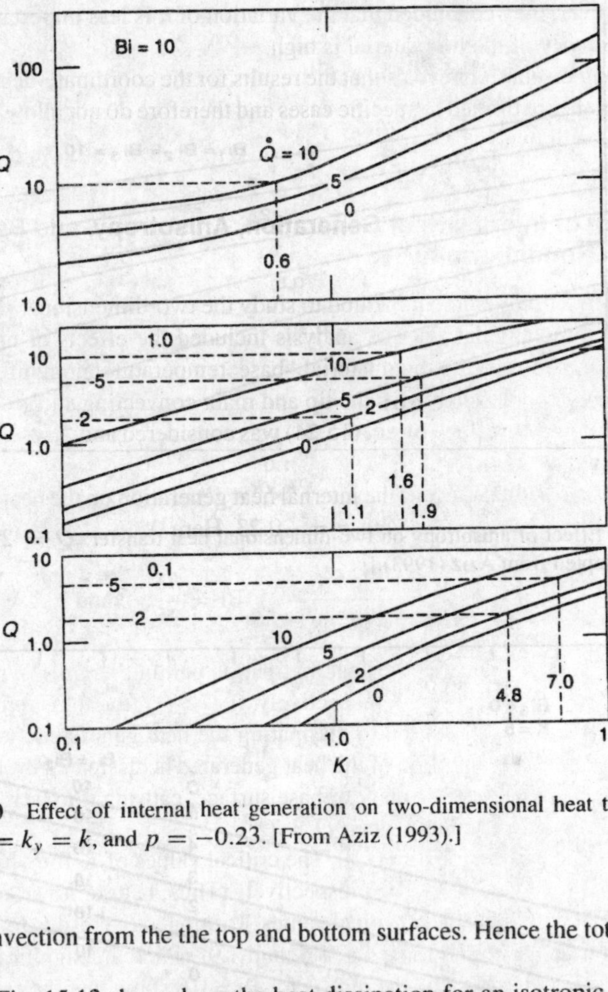

Figure 15.10 Effect of internal heat generation on two-dimensional heat transfer: $h_1 = h_2 = h$, $k_x = k_y = k$, and $p = -0.23$. [From Aziz (1993).]

reduced convection from the the top and bottom surfaces. Hence the total dissipation is reduced.

Finally, Fig. 15.12 shows how the heat dissipation for an isotropic fin ($k_x = k_y$) is affected by the base temperature nonuniformity. For this figure, $\text{Bi}_3 = 0$ (insulated tip) and $K = 5$. The quantities, Q, $\dot{Q}$, Bi_1, and Bi_2 have the same definitions as those given in the foregoing paragraph. For $p > 0$, Q is higher than that for $p = 0$, while the reverse is true for $p < 0$. This observation is consistent with that of Look (1989) discussed in Section 15.2.4.

15.3 CONVECTING LONGITUDINAL FIN OF TRIANGULAR PROFILE

Compared with rectangular geometry, two-dimensional studies of triangular fins appear to be rather limited. Sfeir (1976) was perhaps the first to analyze a two-dimensional triangular fin. He considered the fin shown in Fig. 15.13 with height b and base thickness δ_b convecting heat to the environment at T_s via a heat transfer

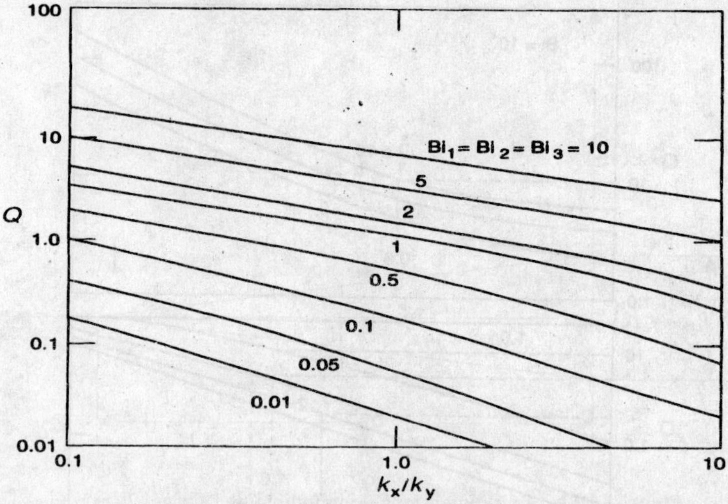

Figure 15.11 Effect of anisotropy on two-dimensional heat transfer: $\dot{Q} = 2$, $p = 0.23$, and $K = 2$. [Adapted from Aziz (1993).]

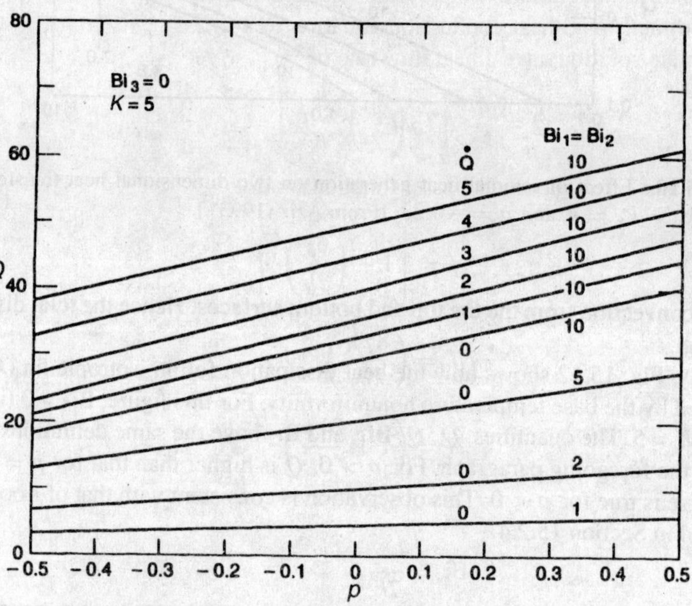

Figure 15.12 Effect of base temperature depression or elevation on two-dimensional heat transfer. [Adapted from Aziz (1993).]

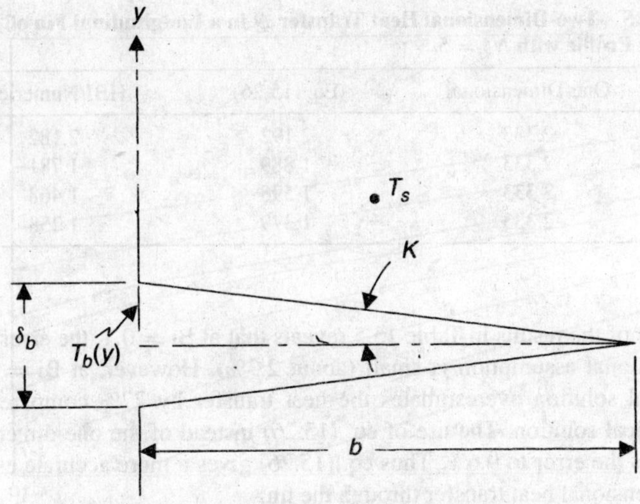

Figure 15.13 Two-dimensional longitudinal fin of triangular profile.

coefficient, h. Using a second-order polynomial for the temperature variation in the y-direction, he was able to employ the heat balance integral approach and reduce the two-dimensional heat conduction equation to a one-dimensional equation. The approximate solution gave a heat flow rate of

$$Q = \frac{qb}{k\delta_b}\left[T_b\left(\frac{\delta_b}{2}\right) - T_s\right] = \frac{1}{2}N\frac{I_1(N)}{I_0(N)} \tag{15.26}$$

where I_0 and I_1 are modified Bessel functions and

$$f = \left[1 + \left(\frac{\delta_b}{2b}\right)^2\right]^{1/2} \tag{15.27a}$$

$$N^* = \left(\frac{8fh}{k\delta_b}\right)^{1/2} b \tag{15.27b}$$

$$N = \left(1 + \frac{\text{Bi}\,f}{3}\right)^{-1/3} N^* \tag{15.27c}$$

and

$$\text{Bi} = \frac{h\delta_b}{2k} \tag{15.27d}$$

Note that the base temperature variation along y is dictated by the second-order polynomial used to account for the transverse temperature variations. Sfeir (1976) computed the Q values for $N^* = 5$ and Bi $= 0.1, 1, 2,$ and 3 using a heat balance integral (HBI)/numerical approach, the approximate analytical solution of eq. (15.26) and the standard one-dimensional solution. Table 15.5 summarizes the results.

TABLE 15.5 Two-Dimensional Heat Transfer Q in a Longitudinal Fin of Triangular Profile with $N^* = 5$

Bi	One-Dimensional	Eq. (15.26)	HBI/Numerical
0.1	2.333	2.192	2.182
1.0	2.333	1.889	1.781
2.0	2.333	1.598	1.468
3.0	2.333	1.379	1.258

Examination of the results in Table 15.5 reveals that at Bi $= 0.1$, the error due to the one-dimensional assumption is small (about 2.3%). However, at Bi $= 3$, the one-dimensional solution overestimates the heat transfer by 77% compared with the HBI/numerical solution. The use of eq. (15.26) instead of the one-dimensional solution reduces the error to 9.6%. Thus eq. (15.26) gives a more accurate estimate of the two-dimensional heat transfer through the fin.

Following Sfeir's (1976) work, Burmeister (1979) showed that the ordinary differential equation resulting from application of the heat balance integral method (which Sfeir solved numerically) actually admits solution in terms of the hypergeometric function. Using this solution, Burmeister developed a graph showing the fin efficiency as a function of the parameter $hfb/k \tan \kappa$, where $\kappa = \arctan(\delta_b/2b)$, as shown in Fig. 15.13. Burmeister's results showed that inclusion of two-dimensional effects lowers the efficiency and hence the heat transfer from the fin.

Look (1992) and Aziz and Nguyen (1992) also examined the problem of two-dimensional conduction in a triangular fin. Look (1992) explored the possibility of a brute-force application of the separation of variables method. He considered a fin with identical convection conditions on top and bottom faces and a constant base temperature. To account for convection from the sloping surfaces, Look wrote an energy balance in which the energy conducted into the fin at the base matched the energy convected out of the sloping surfaces. This energy balance provided a relationship between the temperature gradient and the integral of the temperature distribution on the sloping surfaces. When this energy-match condition was introduced into the separation of variables solution, it gave rise to an unusual eigenvalue problem in which the orthogonal functions (sine and cosine) and nonorthogonal functions (hyperbolic cosine and tangent) appeared simultaneously.

The eigenvalues in this case depend on two variables, K and Bi, and not on Bi alone. Moreover, it was found that the first two eigenvalues failed to satisfy the orthogonality conditions and, consequently, introduce error into the solution. However, the error is very small for $K \leq 1$ and Bi, not exceeding unity. As K and Bi increase, the error increases rapidly. Thus the separation of variables approach, even if forced, enjoys only limited success in predicting the two-dimensional performance of a triangular fin.

Realizing that adequate information on two-dimensional effects in a triangular fin was not available, Aziz and Nguyen (1992) decided to solve the problem using a finite-element approach. Their numerical results for the heat transfer rate $Q = q/k(T_b - T_s)$

as a function of K and Bi are shown in Fig. 15.14. The trend exhibited in Fig. 15.14 for Bi = 0.01 and 0.10 is similar to that of Fig. 15.2 for a two-dimensional rectangular fin. A closer comparison of Figs. 15.2 and 15.14 reveals that for a given K and Bi, the triangular fin dissipates more heat than its rectangular counterpart.

Figure 15.15 shows the percent error

$$100\left(\frac{Q_{1\text{-D}} - Q}{Q}\right)$$

between the one- and two-dimensional heat transfer rates. The error ranges between −55 and 110%. depending on the values of K and Bi. It can be seen that the Biot number alone cannot be used to establish the validity of a one-dimensional model for a triangular fin, It is the combination of Bi and K that dictates whether or not the one-dimensional assumption is valid.

15.4 CONVECTING CYLINDRICAL SPINE OR PIN FIN

Two-dimensional analysis of a cylindrical spine or pin fin was first presented by Irey (1968). He considered a cylindrical spine of diameter d, height b, and thermal conductivity k, attached to a wall at T_b as shown in Fig. 15.16. The convection process from the cylindrical surface is characterized by a constant heat transfer coefficient h and an environmental temperature T_s. Allowing for conduction in both z (axial

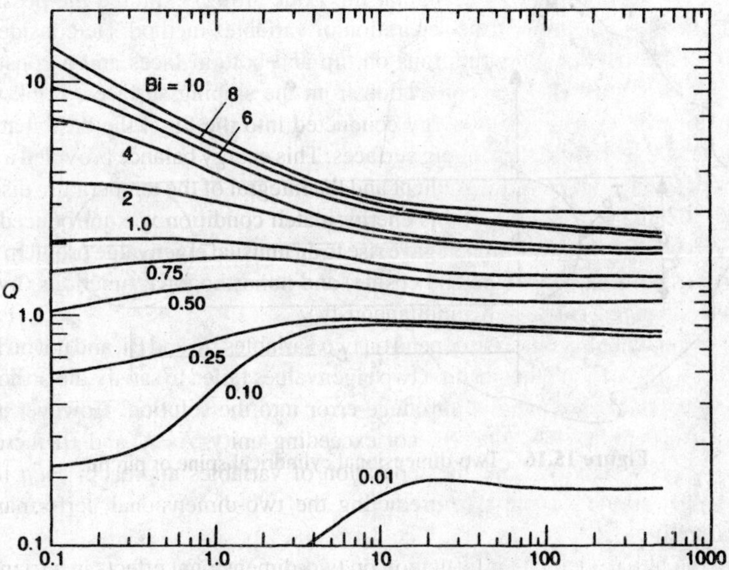

Figure 15.14 Two-dimensional heat transfer in a longitudinal fin of triangular profile.

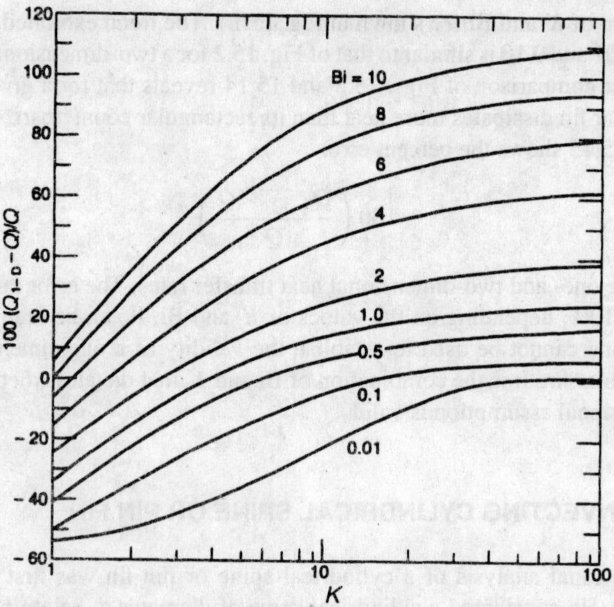

Figure 15.15 Error due to one-dimensional assumption for a longitudinal fin of triangular profile. [From Aziz and Nguyen (1992).]

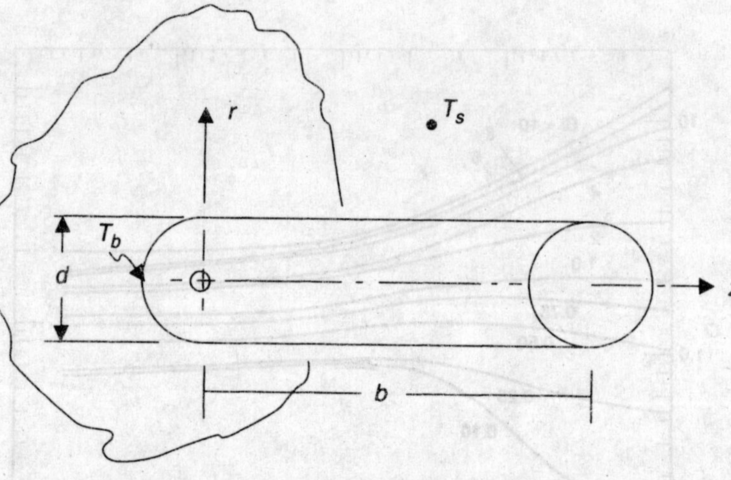

Figure 15.16 Two-dimensional cylindrical spine or pin fin.

direction) and r (radial direction), the mathematical model for this fin can be written in dimensionless form as

$$\frac{\partial^2\theta}{\partial R^2}+\frac{1}{R}\frac{\partial\theta}{\partial R}+\frac{\partial^2\theta}{\partial Z^2}=0 \tag{15.28}$$

with boundary conditions

$$\theta(R,0)=1 \tag{15.29a}$$

$$\frac{\partial\theta}{\partial R}(0,Z)=0 \tag{15.29b}$$

$$\frac{\partial\theta}{\partial R}(1,Z)=-\mathrm{Bi}\,\theta(1,Z) \tag{15.29c}$$

and

$$\frac{\partial\theta}{\partial R}(R,1)=-\mathrm{Bi}\,\gamma\theta(R,1) \tag{15.29d}$$

where the dimensionless variables are

$$\theta=\frac{T-T_s}{T_b-T_s} \tag{15.30a}$$

$$R=\frac{2r}{d} \tag{15.30b}$$

$$Z=\frac{z}{b} \tag{15.30c}$$

$$\gamma=\frac{2b}{d} \tag{15.30d}$$

and

$$\mathrm{Bi}=\frac{hd}{2k} \tag{15.30e}$$

The solution for θ obtained using the method of separation of variables is given by the infinite series

$$\theta=\sum_{n=1}^{\infty}\frac{2\lambda_n J_1(\lambda_n)J_0(\lambda_n R)}{(\lambda_n^2+\mathrm{Bi}^2)[J_0(\lambda_n)]^2}\times\left(\cosh\lambda_n\gamma Z-\frac{\lambda_n\sinh\lambda_n\gamma+\mathrm{Bi}\cosh\lambda_n\gamma}{\lambda_n\cosh\lambda_n\gamma+\mathrm{Bi}\sinh\lambda_n\gamma}\sinh\lambda_n\gamma Z\right) \tag{15.31}$$

where J_0 and J_1 are Bessel functions of the first kind and where λ_n (the eigenvalues) are given by

$$\lambda_n J_1(\lambda_n)=\mathrm{Bi}\,J_0(\lambda_n) \tag{15.32}$$

The heat transfer from the fin

$$Q = \frac{2q}{dk(T_b - T_s)}$$

follows from eq. (15.31) as

$$Q = 4\pi \sum_{n=1}^{\infty} \frac{\lambda_n [J_1(\lambda_n)]^2}{(\lambda_n^2 + \text{Bi}^2)[J_0(\lambda_n)]^2} \frac{\lambda_n \sinh \lambda_n \gamma + \text{Bi} \cosh \lambda_n \gamma}{\lambda_n \cosh \lambda_n \gamma + \text{Bi} \sinh \lambda_n \gamma} \tag{15.33}$$

A solution similar to the one presented here was also derived by Levitsky (1972) for the case of tip heat transfer coefficient different from that for the cylindrical surface. Many heat conduction texts, such as those of Ozisik (1993), Poulikakos (1994), and Gebhart (1993), also give the two-dimensional solution for a pin fin with $T_b = f(r)$ but with uniform heat transfer coefficient h for the entire surface, including the tip.

The heat flow of eq. (15.33) can be compared with the corresponding one-dimensional solution for the heat transfer rate. With

$$Q_{\text{1-D}} = \frac{2q_{\text{1-D}}}{kd(T_b - T_s)}$$

then

$$Q_{\text{1-D}} = \sqrt{2\text{Bi}}\,\pi \frac{\sinh \gamma\sqrt{2\text{Bi}} + (\text{Bi}/2)^{1/2} \cosh \gamma\sqrt{2\text{Bi}}}{\cosh \gamma\sqrt{2\text{Bi}} + (\text{Bi}/2)^{1/2} \sinh \gamma\sqrt{2\text{Bi}}} \tag{15.34}$$

The comparison is made in terms of the percent error

$$100\left(\frac{Q_{\text{1-D}} - Q}{Q}\right)$$

Figure 15.17 shows the percent error as a function of $\gamma = 2b/d$ for Bi ranging from 0.04 to 10. Once again, for $\text{Bi} \le 0.10$, the error is about 1%, but as Bi increases, the error increases rapidly and is as much as 60% for $\text{Bi} = 10$ and $\gamma > 1$.

Because the infinite series solution, such as eq. (15.33), is cumbersome to use, Aparecido and Cotta (1990) developed an approximate solution using the coupled integral equation approach. They showed that eq. (15.34) can be used for the approximate two-dimensional heat transfer Q if Bi is replaced by Bi^-, where

$$\text{Bi}^- = \frac{6\text{Bi}}{\text{Bi} + 6}$$

The approximate two-dimensional solution provides an accurate alternative to eq. (15.33).

To obtain some insight unto the accuracy of these solutions, consider the case of $\gamma = 5$ amd $\text{Bi.} = 1$. The error between the exact solution of eq. (15.33) and the one-dimensional solution of eq. (15.34) is about 8%. With the use of the approximate two-dimensional solution, the error is reduced to -0.1%. The corresponding errors for $\gamma = 5$ and $\text{Bi} = 10$ are 41 and 14%. This demonstrates the superiority of approximate two-dimensional solution over the one-dimensional solution of eq. (15.34).

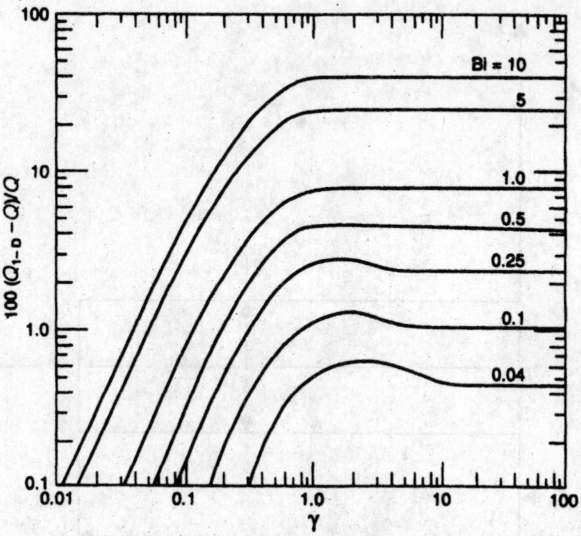

Figure 15.17 Error due to one-dimensional assumption for a cylindrical spine. Irey (1968).

15.5 RADIAL RECTANGULAR PROFILE CONVECTING FIN

Keller and Somers (1959) considered the two-dimensional analysis of a radial fin of inner radius r_b, outer radius r_a, and thickness δ, as shown in Fig. 15.18. The two faces and the tip were assumed to convect to the surrounding environment at T_s through a heat transfer coefficient h. For the case of a constant base temperature T_b, they derived expressons for the two-dimensional temperature distribution and heat flow. However, the transverse Biot number $\mathrm{Bi} = h\delta/2k$ did not appear explicitly in their calculations. In fact, their analysis implicitly assumed that Bi was much smaller than unity.

Later, Lau and Tan (1973) reexamined the fin configuration of Fig. 15.18 and wrote the dimensionless equation for the temperature distribution in the fin:

$$\frac{\partial^2\theta}{\partial R^2} + \frac{1}{R}\frac{\partial\theta}{\partial R} + \gamma^2\frac{\partial^2\theta}{\partial Y^2} = 0 \tag{15.35}$$

with dimensionless boundary conditions

$$\theta(R_b, Y) = 1 \tag{15.36a}$$

$$\frac{\partial\theta}{\partial Y}(R, 0) = 0 \tag{15.36b}$$

$$\frac{\partial\theta}{\partial Y}(R, 1) = -\mathrm{Bi}\,\theta(R, 1) \tag{15.36c}$$

$$\frac{\partial\theta}{\partial R}(1, Y) = -\mathrm{Bi}\,\gamma\theta(1, y) \tag{15.36d}$$

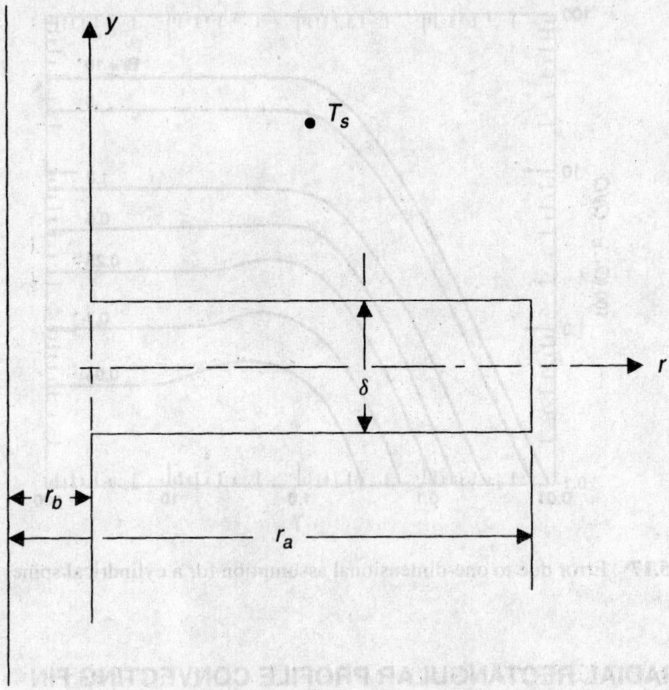

Figure 15.18 Two-dimensional radial fin of rectangular profile.

where the dimensionless quantities are

$$\theta = \frac{T - T_s}{T_b - T_s} \tag{15.37a}$$

$$R = \frac{r}{r_a} \tag{15.37b}$$

$$R_b = \frac{r_b}{r_a} \tag{15.37c}$$

$$Y = \frac{2y}{\delta} \tag{15.37d}$$

$$\gamma = \frac{2r_a}{\delta} \tag{15.37e}$$

and

$$\mathrm{Bi} = \frac{h\delta}{2k} \tag{15.37f}$$

Using the method of separation of variables, the solution for θ was derived as

$$\theta = 2\sum_{n=1}^{\infty} \frac{\sin\lambda_n \cos\lambda_n Y}{\lambda_n + \sin\lambda_n \cos\lambda_n} \times \frac{K_0(\lambda_n\gamma R)[\mathrm{Bi}\, I_0(\lambda_n\gamma) + \lambda_n I_1(\lambda_n\gamma)] - I_0(\lambda_n\gamma R)[\mathrm{Bi}\, K_0(\lambda_n\gamma) - \lambda_n K_1(\lambda_n\gamma)]}{K_0(\lambda_n\gamma R_b)[\mathrm{Bi}\, I_0(\lambda_n\gamma) + \lambda_n I_1(\lambda_n\gamma)] - I_0(\lambda_n\gamma R_b)[\mathrm{Bi}\, K_0(\lambda_n\gamma) - \lambda_n K_1(\lambda_n\gamma)]} \tag{15.38}$$

Then, using eq. (15.38) to calculate the conduction heat transfer through the base of the fin, the heat lost from the fin q can be obtained on the basis of Q,

$$Q = \frac{q}{2\pi r_b k(T_b - T_s)}$$

as

$$Q = 4\sum_{n=1}^{\infty} \frac{\sin^2\lambda_n}{\lambda_n + \sin\lambda_n \cos\lambda_n} \times \frac{K_1(\lambda_n\gamma R_b)[\mathrm{Bi}\, I_0(\lambda_n\gamma) + \lambda_n I_1(\lambda_n\gamma)] + I_1(\lambda_n\gamma R_b)[\mathrm{Bi}\, K_0(\lambda_n\gamma) - \lambda_n K_1(\lambda_n\gamma)]}{K_0(\lambda_n\gamma R_b)[\mathrm{Bi}\, I_0(\lambda_n\gamma) + \lambda_n I_1(\lambda_n\gamma)] - I_0(\lambda_n\gamma R_b)[\mathrm{Bi}\, K_0(\lambda_n\gamma) - \lambda_n K_1(\lambda_n\gamma)]} \tag{15.39}$$

In eqs. (15.38) and (15.39), I_0, I_1, K_0, and K_1 are modified Bessel functions, and the eigenvalues λ_n are obtained from the transcendental equation

$$\lambda_n \tan\lambda_n = \mathrm{Bi} \tag{15.40}$$

The two-dimensional heat flow predicted by eq. (15.39) is shown in Fig. 15.19. The heat flow Q depends on three parameters: R_b, γ, and Bi. For a given R_b and Bi, Q increases with increasing γ, reaches a maximum, and then decreases to an asymptotic value. The peak in Q is more distinct for $R_b = 0.30$ than for $R_b = 0.70$. For $\gamma(1 - R_b) > 100$, Q is virtually insensitive to $\gamma(1 - R_b)$. Insofar as the effect of Bi is concerned, Q increases as Bi increases (stronger convection).

The one-dimensional analysis of the fin in Fig. 15.18 was considered in Chapter 2. The heat flow expressed in the present nomenclature is $Q_{1\text{-D}}$, defined as

$$Q_{1\text{-D}} = \frac{q_{1\text{-D}}}{2\pi r_b k(T_b - T_s)}$$

and equal to

$$Q_{1\text{-D}} = (2\mathrm{Bi})^{1/2} \frac{I_1(\gamma \mathrm{Bi}^{1/2})K_1(\gamma R_b \mathrm{Bi}^{1/2}) - K_1(\gamma \mathrm{Bi}^{1/2})I_1(\gamma R_b \mathrm{Bi}^{1/2})}{I_0(\gamma R_b \mathrm{Bi}^{1/2})K_1(\gamma \mathrm{Bi}^{1/2}) + I_1(\gamma \mathrm{Bi}^{1/2})K_0(\gamma R_b \mathrm{Bi}^{1/2})} \tag{15.41}$$

To assess the validity of the one-dimensional model, the percent error

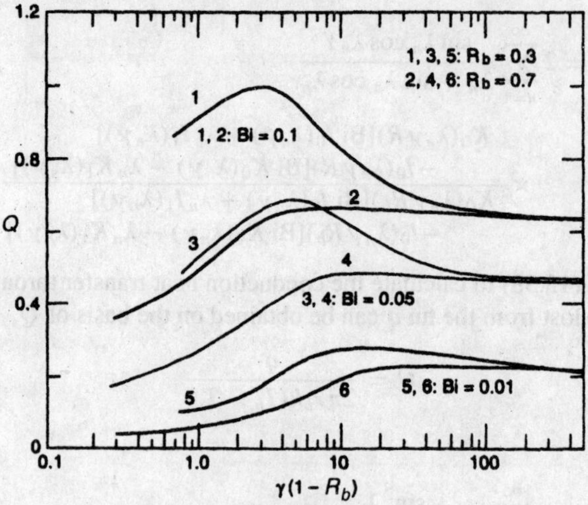

Figure 15.19 Two-dimensional heat transfer in a radial fin of rectangular profile. [From Lau and Tan (1973).]

$$100\left(\frac{Q_{\text{1-D}} - Q}{Q}\right)$$

is plotted in Fig. 15.20 for a range of values of $\gamma(1 - R_b)$ and Bi $= 0.01, 0.10, 1.0$, and 10. For Bi < 0.1, the error is under 1% indicating that the one-dimensional model is valid. As Bi increases, the error increases and is about 50% at Bi $= 10$ and for $\gamma(1 - R_b) > 1$.

The analytical solution of Lau and Tan (1973) discussed in the foregoing paragraphs has been generalized by Yovanovich et al. (1988) by replacing the constant base temperature condition of eq. (15.36a) by a convective condition, thereby accounting for the contact resistance at the base. The generalized solution also allowed for the heat transfer coefficients for the fin faces and the fin tip to be different. In terms of the nomenclature used in this chapter, the solutions for θ and Q are

$$\theta = \frac{T - T_s}{T_f - T_s} = \sum_{n=1}^{\infty} \frac{A_n}{C_n}[I_0(\lambda_n \gamma R) + B_n K_0(\lambda_n \gamma R)] \cos \lambda_n Y \tag{15.42}$$

and

$$Q = \frac{q}{2\pi r_b k(T_f - T_s)} = 2\sum_{n=1}^{\infty} \frac{D_n}{C_n}[C_n K_1(\lambda_n \gamma R_b) - I_1(\lambda_n \gamma R_b)] \tag{15.43}$$

with

$$A_n = \frac{2 \sin \lambda_n}{\lambda_n + \sin \lambda_n \cos \lambda_n} \tag{15.44a}$$

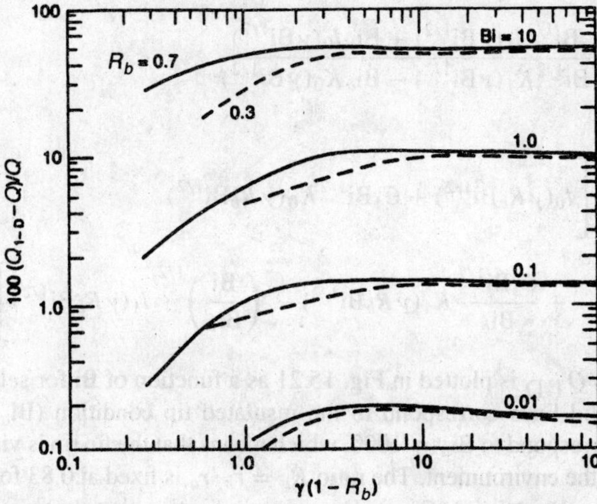

Figure 15.20 Error due to one-dimensional model for a radial fin of rectangular profile. [From Lau and Tan (1973).]

$$B_n = \frac{\lambda_n I_1(\lambda_n \gamma) + \mathrm{Bi}_a I_0(\lambda_n \gamma)}{\lambda_n K_1(\lambda_n \gamma) - \mathrm{Bi}_a K_0(\lambda_n \gamma)} \tag{15.44b}$$

$$C_n = I_0(\lambda_n \gamma R_b) + B_n K_0(\lambda_n R_b) + \frac{B_n \lambda_n}{\mathrm{Bi}_b} K_1(\lambda_n \gamma R_b) - \frac{\lambda_n}{\mathrm{Bi}_b} I_1(\lambda_n \gamma R_b) \tag{15.44c}$$

$$D_n = A_n \sin \lambda_n \tag{15.44d}$$

where the eigenvalues λ_n are given by eq. (15.40) and where

$$\mathrm{Bi}_b = \frac{h_b \delta}{2k} \quad \text{and} \quad \mathrm{Bi}_a = \frac{h_a \delta}{2k}$$

are the Biot numbers characterizing convection at the base and the tip respectively. The temperature T_f is the temperature of the fluid at the base of the fin. For $\mathrm{Bi}_b \to \infty$ and $\mathrm{Bi} = \mathrm{Bi}_b$, eqs. (15.42) and (15.43) reduce to eqs. (15.38) and (15.39), respectively, and the hot fluid temperature equals the base temperature ($T_f = T_b$).

To compare the one- and two-dimensional heat flows, Yovanovich et al. (1988) also derived a solution for one-dimensional heat flow,

$$Q_{1\text{-D}} = \frac{q_{1\text{-D}}}{2\pi k r_b (T_f - T_a)}$$

or

$$Q_{1\text{-D}} = 2\mathrm{Bi}^{1/2} C_2 [I_1(\gamma R_b \mathrm{Bi}^{1/2}) - C_1 K_1(\gamma R_b \mathrm{Bi}^{1/2})] \tag{15.45}$$

where

$$C_1 = \frac{\mathrm{Bi}^{1/2} I_1(\gamma \mathrm{Bi}^{1/2}) + \mathrm{Bi}_a I_0(\gamma \mathrm{Bi}^{1/2})}{\mathrm{Bi}^{1/2} K_1(\gamma \mathrm{Bi}^{1/2}) - \mathrm{Bi}_a K_0(\gamma \mathrm{Bi}^{1/2})} \tag{15.46a}$$

and

$$C_2 = \left[I_0(\gamma R_b \mathrm{Bi}^{1/2}) + C_1 \mathrm{Bi}^{1/2} K_0(\gamma R_b \mathrm{Bi}^{1/2}) + \frac{C_1 \mathrm{Bi}^{1/2}}{\mathrm{Bi}_b} K_1(\gamma R_b \mathrm{Bi}^{1/2}) - \left(\frac{\mathrm{Bi}}{\mathrm{Bi}_b}\right)^{1/2} I_1(\gamma R_b \mathrm{Bi}^{1/2}) \right]^{-1} \tag{15.46b}$$

The ratio $Q/Q_{1\text{-D}}$ is plotted in Fig. 15.21 as a function of Bi for selected values of Bi_b. The solid lines correspond to the insulated tip condition ($\mathrm{Bi}_a = 0$). The dashed lines correspond to $\mathrm{Bi}_a = 1000$, which means that the fin tip is virtually at the temperature of the environment. The ratio $R_b = r_b/r_a$ is fixed at 0.83 for Fig. 15.21. As Bi increases, the ratio $Q/Q_{1\text{-D}}$ decreases, indicating that the one-dimensional calculation increasingly overestimates the heat flow. The difference between the one- and two-dimensional heat flows is more pronounced as convective heating at the base becomes stronger, that is, as Bi_b increases.

Figure 15.22 illustrates the effect of the geometrical parameter R_b on the ratio of the heat flows for two tip conditions, $\mathrm{Bi}_a = 0$ (insulated tip) and $\mathrm{Bi}_a = 1000$ (strongly convecting tip). Consider the curves for $\mathrm{Bi}_a = 0$. For a fixed base radius r_b, the results show that as the fin height or tip radius is increased, that is, as R_b decreases, the ratio of the heat flows increases which means that the error between the two- and one-dimensional heat flow decreases. However, at the other extreme,

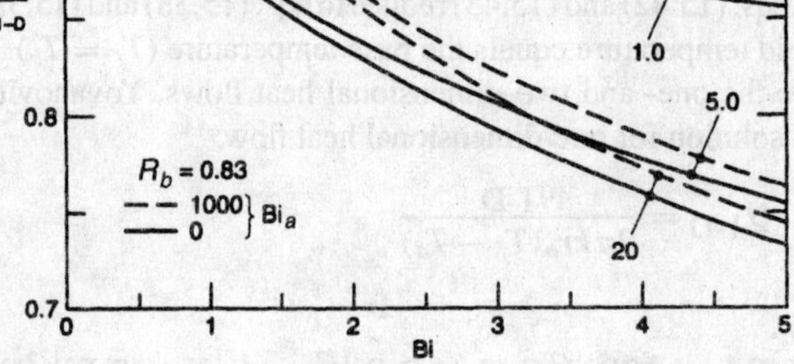

Figure 15.21 Ratio of two- to one-dimensional heat transfer in radial fin of rectangular profile with contact resistance. [From Yovanovich et al. (1988).]

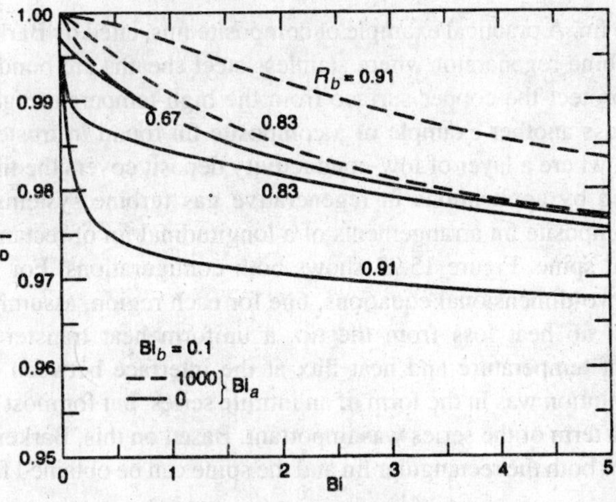

Figure 15.22 Effect of geometry on $Q_{1\text{-D}}/Q$ for a radial fin of rectangular profile with contact resistance. [From Yovanovich et al. (1988).]

when Bi = 1000, the effect of increasing the tip radius (decreasing R_b) for a fixed base radius is to reduce the ratio of heat flows or increase the error between two- and one-dimensional heat flows.

To facilitate the rapid computation of two-dimensional heat flow in a radial fin, Sfeir (1976) used a heat balance integral method (described briefly in Section 15.2.1). Aparecido and Cotta (1988) adopted a numerical approach to obtain an approximate solution. Considering the boundary conditions of constant base temperature and identical heat transfer coefficients for the fin faces and fin tip, they showed that the one-dimensional solution, which in this case is given by eq. (15.41), can be used for obtaining the two-dimensional heat flow, provided that Bi is replaced by

$$\frac{4\text{Bi}}{4+\text{Bi}}$$

For Bi < 1 and $1 \leq \gamma \leq 5$, this modified solution has a maximum error of about 0.5%. For Bi = 10 and $\gamma = 5$, the actual one-dimensional solution overestimates the heat flow by 48% whereas the modified one-dimensional solution underestimates the heat flow by 15%.

15.6 COMPOSITE FINS

When a finned heat exchanger is designed for operation in a high temperature and/or corrosive environment, it becomes necessary to protect the high-thermal-conductivity expensive surface from the damaging effects of the environment. This can be achieved by covering the surface with less expensive material with relatively low thermal conductivity that can withstand the harsh environment. The resulting arrangement is a composite fin. A practical example of composite fins, cited by Barker (1958), is that of a gas turbine regenerator where stainless steel sheaths are bonded to the copper surface to protect the copper surface from the high-temperature gas. Barrow et al. (1986) discuss another example of a composite fin found in frosted or fouled heat exchangers, where a layer of low-conductivity deposit covers the finned surface.

Motivated by applications in regenerative gas turbine systems, Barker (1958) analyzed composite fin arrangements of a longitudinal fin of rectangular profile and a cylindrical spine. Figure 15.23 shows both configurations. For both fins Barker solved the two-dimensional equations, one for each region, assuming constant base temperature, no heat loss from the tip, a uniform heat transfer coefficient, and continuity of temperature and heat flux at the interface between the two regions. The exact solution was in the form of an infinite series, but for most cases of interest, only the first term of the series was important. Based on this, Barker showed that the efficiency for both the rectangular fin and the spine can be obtained from the equation

$$\eta = \frac{\tanh mb}{mb} \tag{1.11}$$

but with

$$m = \left(\frac{hP}{\bar{k}A}\right)^{1/2}$$

where $\bar{k}$ is the volumetric average thermal conductivity of the fin.

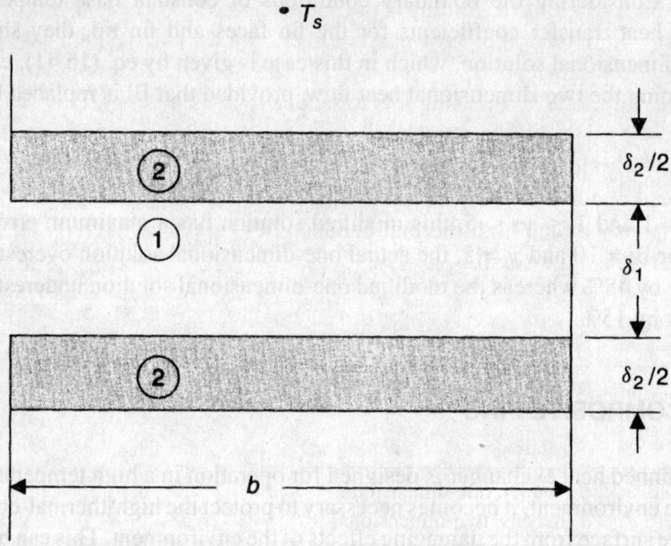

Figure 15.23 Composite longitudinal fin of rectangular profile and cylindrical spine. For the cylindrical spine, δ is the spine diameter.

The rectangular profile composite fin has also been studied by Barrow et al. (1986). This work was performed to quantify the effect of frosting or fouling on the heat dissipation capability of finned coils. Besides developing a general separation of variables type of solution, they also solved the specific problem arising in a heat pump application using finite differences and finite elements. The specific problem considered has the following specifications:

- Copper fin, $k_1 = 238$ W/m·K, $b = 6.2$ mm, and $\delta_1 = 0.10$ mm
- Frost deposit, $k_2 = 0.5$ W/m·K, $b = 6.2$ mm and $\delta_2 = 0.60$ mm
- $h = 79.8$ W/m^2·K, $T_b - T_s = 100$°C, and an insulated tip

The heat flow results obtained with different solution methods are summarized in Table 15.6. The parallel resistance model result is based on the method proposed by Epstein and Sandhu (1978). The exact two-dimensional result is calculated from the separation of variables solution. This exact solution assumes that the interface temperature decreases linearly with x. The last entry in Table 15.6 is the prediction of the standard exact solution for a single fin with no fouling deposit.

Comparison of the parallel resistance model with the exact solution (no deposit) for one dimension shows that the former result is slightly higher than the latter (92.35 W/m versus 92.31 W/m). Predictions of the finite difference, finite element, and exact two-dimensional solution are close, and all three indicate a reduction in heat flow of about 8% compared with the parallel resistance model. This shows that the heat transfer is not affected significantly even when the thickness of the deposit is many times that of the fin proper. Thus deterioration of the heat transfer performance of a closely spaced finned heat exchanger due to fouling is not so much due to the insulating effect of the deposit as it is to reduction of flow passage area between the fins that are caused by the buildup of surface deposit.

The effect of linearly increasing h is to reduce two-dimensional heat transfer, but reduction compared with constant h for any of the specific consideratons does not exceed 5%. This clearly shows that heat transfer in a composite fin consisting of a high-thermal-conductivity core covered with a low-thermal-conductivity deposit is controlled largely by the high-thermal-conductivity core. The effects of the deposit and/or of variable h are small.

TABLE 15.6 Heat Transfer q(W/m) for a Fouled Copper Fin

Method	Constant, h	h, Linearly Increasing with x
Parallel resistance model, one-dimensional	92.35	—
Finite difference analysis, two-dimensional	84.80	82.98
Finite element analysis, two-dimensional	85.06	80.96
Exact solution, two dimensional	85.15	
Exact solution, one-dimensional, no deposit	92.31	—

15.7 CONVECTING FIN ASSEMBLIES

To this point, the analyses have dealt almost exclusively with a single fin in isolation from the primary or supporting surface to which it was attached. This decoupling of the fin from its primary surface simplified the analyses considerably and permitted analytical solutions to be obtained for many cases. In this section in some detail studies pertaining to convecting fin assemblies are described.

15.7.1 Base Temperature Depression or Elevation

Consider a fin with height b and thickness δ attached to a wall and transferring heat by convection from its faces to the environment. If the fin is designed properly, heat flow through its base must greatly exceed heat flow through an identical unfinned area. Thus, compared with the unfinned surface, attachment of the fin must introduce spatial nonuniformities in the wall and lower the temperature at the base of the fin. This effect, termed *base temperature depression*, reduces the heat transfer capability of the fin and is not accounted for in classical fin theory.

Sparrow and Hennecke (1970) were first to examine the phenomenon of base temperature depression in detail. Considering a single rectangular fin attached to a thick wall, they obtained a solution for the coupled two-dimensional conduction in the fin and in the wall. Figure 15.24 shows the effect of this coupling on the temperature distribution in the fin. Each set of curves corresponds to a certain ratio K of fin height to half thickness ($K = 2b/\delta$). It can be observed that the base temperature at

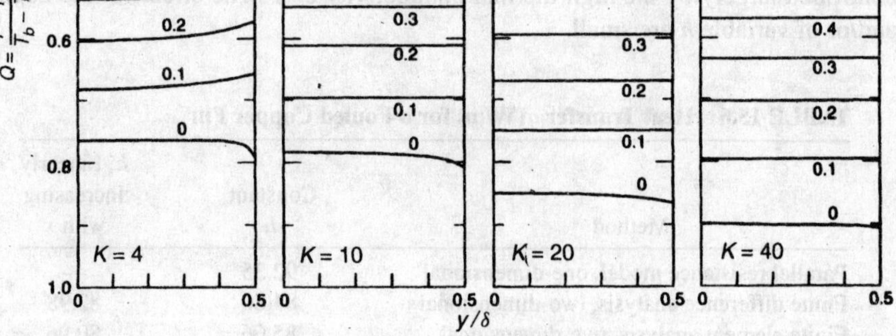

Figure 15.24 Base temperature depression in a longitudinal fin of rectangular profile. [Adapted from Sparrow and Hennecke (1970).]

$x/b = 0$ varies slightly with y, being lowest at the fin centerline, where $y/\delta = 0$, and highest at the fin face, where $y/\delta = 0.5$. The magnitude of the base temperature depression depends on the parameter K. For $K = 40$ (rather slender fins), the ratio $\theta = (T - T_s)/T_b - T_s)$ at the base of the fin where $x/b = 0$ is about 0.90, indicating a base temperature depression of 10%. As K decreases, the base temperature becomes progressively more depressed and at $K = 4$, the depression is about 23%. Thus the classical assumption that the fin base is at the same temperature as the unfinned primary can significantly overestimate heat dissipation of the fin.

The original work of Sparrow and Hennicke (1970) was extended by Klett and McCulloch (1972) by allowing thermal conductivity of the primary surface, k_w, to be different from the thermal conductivity of the fin, k_f. But as pointed out by Huang and Shah (1992), the parameter k_w/k_f in the graphs given by Klett and McCulloch should be replaced by its reciprocal, k_f/k_w. Moreover, Huang and Shah have performed their calculations considering the primary or base surface to be of finite thickness w. Table 15.7 is a summary of their results. It shows that for a given K and $h\delta/k_w$, the temperature ratio θ at $x/b = 0$ decreases as k_f/k_w increases. Thus the higher the thermal conductivity of the fin compared with that of the primary surface, the greater the base temperature depression.

The observation of base temperature depression in a single fin–primary surface model prompted Sparrow and Lee (1975) to investigate this phenomenon in a more realistic arrangement. They considered a circular tube fitted externally with longitudinal fins of rectangular profile and included in their analysis the presence of adjacent fins as well as the effect of the convection process on the inside of the tube. The number of fins on the tube was found to have a marked effect on the base temperature depression. As the number of fins increased, the base temperature depression effect

TABLE 15.7 Base Temperature Depression in a Longitudinal Fin of Rectangular Profile

$\dfrac{k_f}{k_w}$	$\dfrac{h\delta}{k_w}$	K	$\theta = \dfrac{T - T_s}{T_b - T_s}$ at $x/b = 0$
1	0.25	2	0.770
1	0.0625	4	0.780
1	0.0156	8	0.850
4	0.25	2	0.630
4	0.0625	4	0.700
4	0.0156	8	0.800
10	0.25	2	0.590
10	0.0625	4	0.680
10	0.0156	8	0.760
20	0.25	2	0.570
20	0.0625	4	0.665
20	0.0156	8	0.745

became progressively less important and was negligible when 12 or more fins were present. However, the effect on temperature depression of convective heating on the inside of the tube could not be clearly identified.

The topic of base temperature nonuniformity continues to attract attention. Jucá and Prata (1993) have considered a multifin array of rectangular fins attached to a wall of finite thickness as shown in Fig. 15.25*a*. A representative half-section of one of the fins (Fig. 15.25*b*) was analyzed numerically. Note that the temperature on the unfinned side of the primary surface is assumed to have a constant value T_b. The two-dimensional temperature field is controlled by four parameters:

$$\text{Bi} = \frac{h\delta}{2k}, \qquad K = \frac{2b}{\delta}, \qquad \frac{z}{\delta}, \qquad \text{and} \qquad \frac{2w}{\delta}$$

Figure 15.26 shows typical base ($x = 0$) temperatures for various values of Bi. At $\text{Bi} = 0.001$, the base temperature is virtually uniform but higher than the temperature of the unfinned portion, where $y/\delta > 0.50$ and $y/\delta < -0.50$. This exhibits a base temperature elevation and not a depression. The effect of base temperature elevation becomes more pronounced at higher Biot numbers. It is only at $\text{Bi} = 0.10$ that some base temperature depression is noticeable. The presence of base temperature elevation implies that the fins are impeding the flow of heat, and consequently, do not serve any useful purpose. Contrary to the case of base temperature depression, the highest base temperature occurs at the center, where $y/\delta = 0$, and the lowest at the fin edges, where $y/\delta = \pm 0.50$.

The nonuniformity of base temperature was taken into account in analyzing heat transfer from a single fin in Section 15.2.4. In the following section attention is focused on two-dimensional heat flow in a fin assembly in which the base temperature variation along the fin thickness as well as the base temperature depression or elevation are taken into account simultaneously.

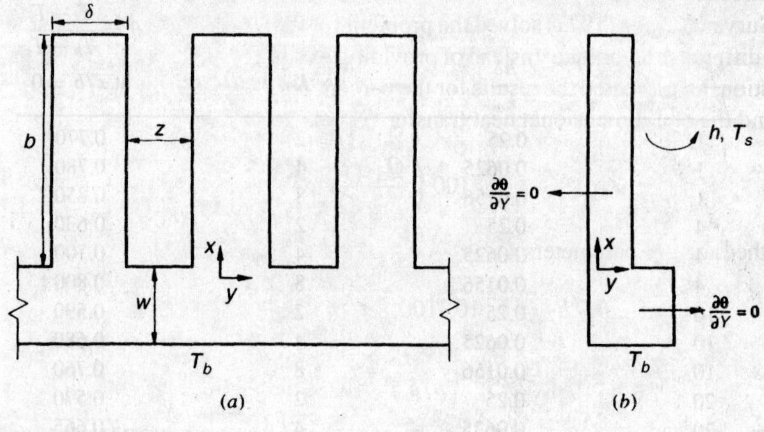

Figure 15.25 Array of vertical longitudinal fins of rectangular profile: (a) section of the array; (b) detail of half of one fin.

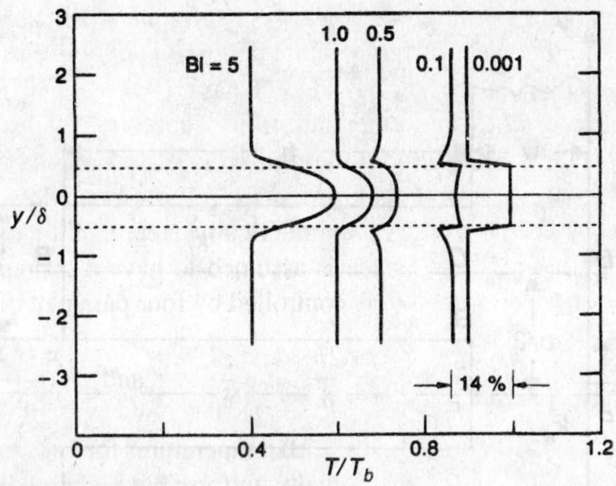

Figure 15.26 Effect of Bi on base temperature depression for a longitudinal fin of rectangular profile. [Adapted from Juca and Prata (1993).]

15.7.2 Fin Assembly Heat Transfer

Most fin assembly heat transfer studies have focused on an array of rectangular fins attached to a plane wall as shown in Fig. 15.27. The wall has thickness w and thermal conductivity k_w, while the fin has height b, thickness δ, and thermal conductivity k_f. The fins are spaced such that the fin spacing is p. The fin assembly is heated on the unfinned side of the wall by a fluid at temperature T_1, and the convective heat transfer coefficient is h_1. The entire surface of the fin, including the fin tip and the unfinned surface between the fins, is exposed to the environment at T_2 via a heat transfer coefficient h_2. Because of thermal symmetry, only the section $ABCDEFGA$ requires analysis.

Suryanarayana (1977) solved the problem for the case of an insulated tip using a finite difference technique. Instead of providing explicit results for the two-dimensional solution, he presented the results for the percent error between the actual heat transfer Q and the one-dimensional heat transfer $Q_{1\text{-D}}$,

$$100\left(\frac{Q_{1\text{-D}}-Q}{Q}\right)$$

for the range of parameters

$$h_1/h_2 = 1,\ 10,\ 100, \qquad 2b/\delta = 1,\ 2.5,\ 5$$

$$2w/\delta = 0.50,\ 1,\ 2,\ 5, \quad p/\delta = 1.5,\ 2,\ 3, \quad \text{and} \quad h_2\delta/2k_f = 0.1$$

The first model assumed that the fin base temperature along CG in Fig. 15.27 is the same as the temperature of the primary surface in the absence of fins. The second

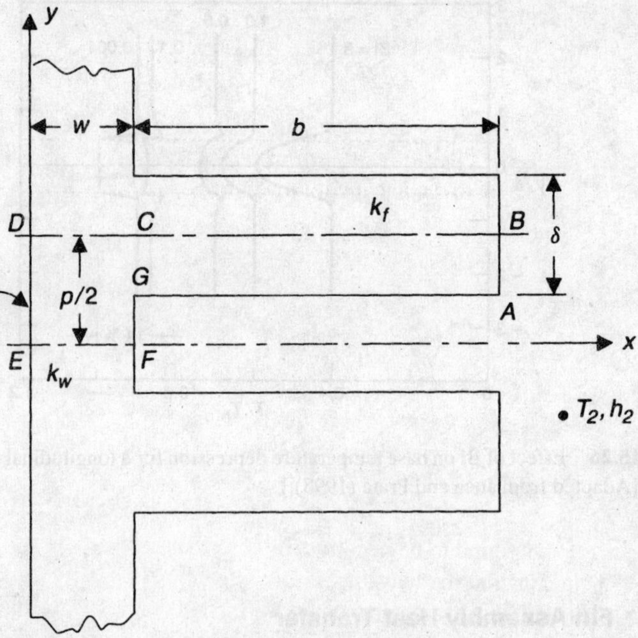

Figure 15.27 Array of horizontal longitudinal fins of rectangular profile.

model ignored the thermal resistance of the wall in area $CDEF$ and assumed that the base along CG was exposed directly to the hot convective environment at T_1 with h_1. The third model included the thermal resistance of the wall with convective heating along DE. The study concluded that the first and second one-dimensional models overestimate the heat transfer (the error is positive), while the third model underestimates the heat transfer (the error is negative). The maximum error of about 80% was noted for the case of $h_1/h_2 = 1$, $2w/\delta = 5$, $p/\delta = 1.5$, and $h_2\delta/2k = 0.10$. In addition, it was found that the error increases as the wall thickness w and fin height b increase and as the the fin spacing p and ratio h_1/h_2 decrease. Some of these conclusions, however, seem to contradict those of related studies. For example, Sparrow and Lee (1975) conclude that decreasing the spacing between longitudinal rectangular fins on a circular tube decreases rather than increases the error.

Heggs and Stones (1980) also obtained a finite difference solution for the arrangement of Fig. 15.27. They calculated a heat loss Q from the assembly for the range of parameters

$$b/\delta = 1 \text{ to } 160, \quad w/\delta = 1 \text{ to } 100, \quad p/\delta = 1 \text{ to } 51,$$

$$k_f/k_w = 1 \text{ to } 25, \quad h_1/h_2 = 2 \text{ to } 20{,}000, \quad \text{and} \quad h_1\delta/k_w = 2$$

These results were presented as percent error

$$100\left(\frac{Q_{1\text{-D}}-Q}{Q}\right)$$

where $Q_{1\text{-D}}$ was derived by constructing a thermal network consisting of a hot-side convection resistance, a wall resistance in series with a parallel combination of a fin resistance, and convection resistance of the unfinned portion. The resulting expression for $Q_{1\text{-D}}$ is

$$Q_{1\text{-D}}=\frac{T_1-T_2}{\dfrac{1}{h_1 p}+\dfrac{w}{k_w p}+h_2(p-\delta)+k_f m\delta\dfrac{(1+H)\sinh mb}{\cosh mb+H\sinh mb}} \tag{15.47}$$

where

$$m=\left(\frac{2h_2}{k_f\delta}\right)^{1/2} \quad \text{and} \quad H=\frac{h_2}{k_f\delta}$$

The error was found to be negative for all combinations of the controlling parameters. This means that the two-dimensional heat flow always overestimates the one-dimensional heat flow evaluated from eq. (15.47). Thus the use of eq. (15.47) to design a fin asembly should be on the conservative side. Heggs and Stones (1980) noted a maximum error of – 20% for $h_1/h_2=2$, $p/\delta=6$, $w/\delta=10$, $b/\delta=5$, $k_f/k_w=25$, and $h_1\delta/k_w=2$. Furthermore, they concluded that the overestimation is more pronounced for greater values of k_f/k_w, shorter fins, greater fin spacing, and thinner tube walls. It must be emphasized that a rigorous comparison between the works of Heggs and Stones (1980), Sparrow and Lee (1975), and Suryanarayana (1977) is not possible because both Suryanarayana and Sparrow and Lee calculated the heat transfer from the fin separately and not as a finned assembly.

In a subsequent investigation, Heggs et al. (1982) proposed and implemented a semianalytical solution of the problem. Their strategy consisted of extending the separation of variables technique to derive a truncated series solution. The coefficients associated with the series was obtained by solving a system of linear algebraic equations. The authors claimed that the series truncation method (ST) was not only computationally more efficient, but yielded accurate solutions even for problems for which finite difference and finite element methods fail to produce satisfactory results.

In a text devoted to heat flow through extended surface exchangers, Manzoor (1984) applied the boundary integral equation method (BIE) to the same problem. The BIE formulation was found to reduce the computational storage and time requirement, an advantage that was also a feature of the ST method. The BIE method is also flexible enough to handle curved fins, variable heat transfer coefficients, and different thermal conductivities of fin and wall materials.

An interesting comparison of finite difference (FD), finite element (FE), boundary integral equation (BIE), and series truncation (ST) methods is given by Ingham et al. (1981). For comparison purposes, the configuration in Fig. 15.27 was solved for the following set of system parameters:

Problem	$Bi_1 = h_1 p/2k_w$	$Bi_2 = h_2 p/2k_w$	k_f/k_w	$2b/p$	δ/p	$2w/p$
A	0.20	0.001	10.0	2	0.20	0.50
B	1.00	0.010	1.00	5	0.25	1.00
C	5.00	0.200	20.0	4	0.20	2.00

The results for an augmentation factor, defined as the ratio of the heat loss from the fin assembly to that from the unfinned wall operating under the same conditions is shown in Table 15.8.

Heggs and Stones (1980) also presented heat transfer results for a fin assembly consisting of radial fins mounted on a circular tube. Figure 15.27 can also be reviewed as a radial fin assembly, with DE representing the inner surface of the tube. For this radial fin arrangement, Heggs and Stones reported the maximum error between one- and two-dimensional heat transfer as a percentage of the one-dimensional heat transfer as +26% for $h_1/h_2 = 2$, $p/\delta = 6$, $w/\delta = 10$, $b/\delta = 60$, and $k_f/k_w = 5$. Unlike the longitudinal geometry, for which the error was negative, the error for the radial fin arrangement was found to be positive or negative, depending on the values of the parameters.

15.7.3 Optimum Design

Buccini and Soliman (1986) considered the optimum design of an array of radial fins of rectangular profile attached to a tube of inner radius r_i in contact with a hot fluid at temperature, T_1, transferring heat via a heat transfer coefficient h_1. The fin has base radius r_b, tip radius r_a, and thickness δ. The heat transfer coefficient on the fin side is h_2, and the environmental temperature is T_2. Both the wall and the fins are assumed to have the same thermal conductivity ($k_w = k_f = k$). The fin half-distance between two adjacent fins is p.

Allowing for conduction in the radial and transverse (thickness) directions, Buccini and Soliman (1986) computed the fin loss Q using a finite difference approach. The heat loss Q was found to depend on six parameters

$$\frac{r_i}{r_b}, \quad \frac{h_1}{h_2}, \quad \frac{r_b}{\delta}, \quad \frac{p}{r_b}$$

the dimensionless volume

TABLE 15.8 Augmentation Factor Comparison of Different Solutions

Method	Problem A	Problem B	Problem C
FD	4.278	2.963	1.901
FE	4.274	2.963	1.900
BIE	4.279	2.963	1.902
ST	4.227	2.963	1.899

$$V = \frac{\delta(r_a^2 - r_b^2)}{r_b^3}$$

and the Biot number

$$\text{Bi} = \frac{2h_2 r_b}{k}$$

For a fixed value of $r_i/r_b = 0.90$, optimization of Q with respect to the parameter r_b/δ showed that $(r_b/\delta)_{opt}$ was controlled essentially by the parameters V and Bi with a weak influence of h_1/h_2 and p/r_b.

Figure 15.28 shows $\delta^*_{opt}/\delta_{opt}$ as a function of V for various parametric values of $\sqrt{\text{Bi}}$. Here δ^*_{opt} is the optimum thickness based on the one-dimensional model and can be obtained from the work of Brown (1965) discussed in Chapter 3. The curves in Fig. 15.28 were plotted using the data for $h_1/h_2 = 100$ and $p/r_b = 0.70$. As noted earlier, the optimum is not affected significantly by the variations in h_1/h_2 and p/r_b. The results show that for $\sqrt{\text{Bi}} = 0.125, 0.25, 0.50$, and 1.0, the two-dimensional

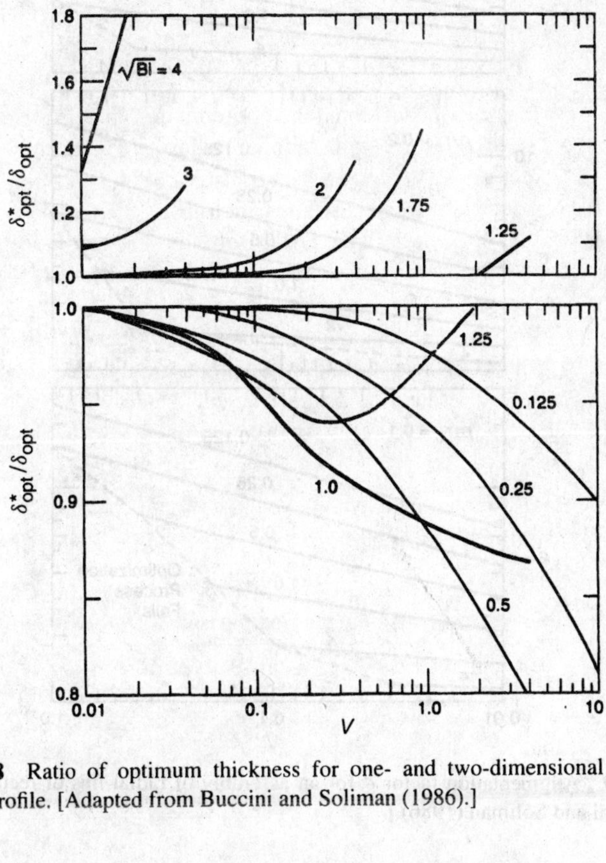

Figure 15.28 Ratio of optimum thickness for one- and two-dimensional radial fins of rectangular profile. [Adapted from Buccini and Soliman (1986).]

optimum fin is thicker than its one-dimensional counterpart; that is, $\delta^*_{opt}/\delta_{opt} < 1$ for all values of V investigated.

The behavior of the curve for $\sqrt{\text{Bi}} = 1.25$ shows that the thickness for the two-dimensional fin (compared with the one-dimensional fin) increases as V increases, reaches a maximum value at about $V = 0.20$, and then decreases again. At $V = 2$, the two- and one-dimensional optimum fins have identical thicknesses. For $V > 2$ and $\sqrt{\text{Bi}} = 1.75, 2, 3$, and 4, the two-dimensional optimum fin is thinner than the one-dimensional optimum fin for all values of V.

Although δ_{opt} is quite insensitive to variations in h_1/h_2 and p/r_b, the optimum heat loss Q_{opt} is strongly influenced by these parameters. Figure 15.29 shows the results for the enhancement or augmentation factor

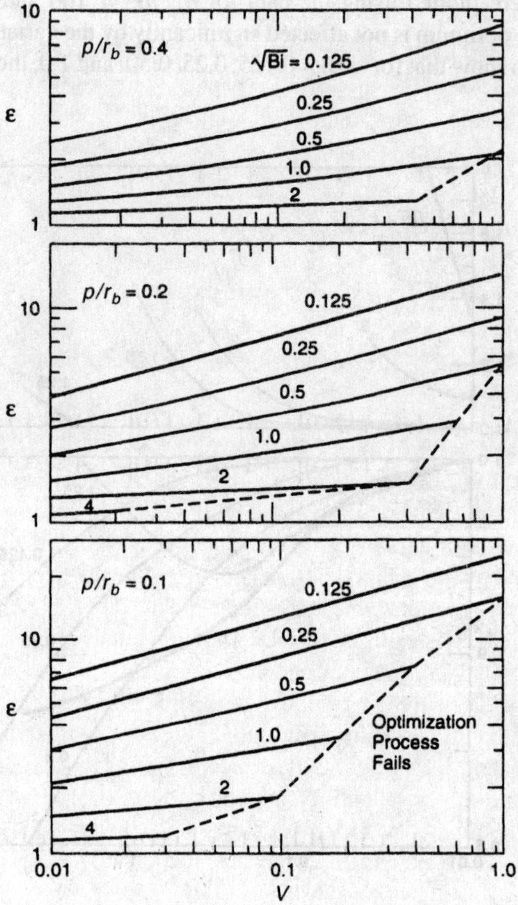

Figure 15.29 Augmentation factor ϵ for an assembly of radial fins of rectangular profile. [From Buccini and Soliman (1986).]

$$\epsilon = \frac{Q_{\text{opt}}}{Q_{\text{unfinned}}}$$

where Q_{unfinned} is given by

$$Q_{\text{unfinned}} = \frac{2(h_1h_2)(r_i/r_b)(\text{Bi})}{1 + (h_1h_2)(r_i/r_b)(\text{Bi})[2 - \text{Bi}\ln(r_1/r_b)]} \tag{15.48}$$

An examination of Fig. 15.29 reveals that the fins become more effective as Bi decreases. For a fixed Bi, the augmentation factor ϵ, increases with an increase in V and a decrease in p/r_b. The dashed lines indicate the region in which an optimum does not exist.

15.8 RADIATING AND CONVECTING–RADIATING FINS AND FIN ASSEMBLIES

15.8.1 Radiating Fins

The earliest work on the two-dimensional performance of radiating fins appears to have been conducted by Sparrow et al. (1963). They considered a fin–tube radiator and allowed for conduction along the length of both the fin and the tube. The model, which also allowed for fin-to-base surface interaction, was solved using a finite difference approach. A similar study was reported by Stockman et al. (1966), in which they provided a comparison of one- and two-dimensional heat transfer in fin and tube radiators.

Another application, in which two-dimensional conduction in a rectangular plate or fin becomes necessary, is the design of the *thermal doubler*, which is basically a lightweight high-thermal-conductivity metal sheet upon which heat-generating electronic components are mounted. In space applications, the device serves to dissipate heat by radiating it to free space. Figure 15.30 shows a typical design. The area occupied by the components, called the *footprint*, is represented by a region of uniform heat flux q''. The unheated area on the inside face radiates to the ambient at T_s. The outside face is exposed to solar radiation but also radiates to free space at 0 K. Analysis is necessary to determine the two-dimensional (x, y) temperature field in the plate, particularly in the footprint area, to assure that the maximum temperature does not exceed the safe (design) operating limit of the components.

This problem and its variations have been studied by several researchers. For example, Iqbal and Aggarwala (1968) adopted a variational approach to solve for the two-dimensional temperature distribution in the plate when one face was exposed to a uniform radiant heat flux while the other face radiated to free space at 0 K. The same problem was later solved by Moszynski and Champanèria (1971) using a semiiterative procedure. Bobco and Starkovs (1970) linearized the radiation terms and obtained the solution using the method of separation of variables. Narayana and Kumari (1988) avoided the linearization by using a strongly implicit numerical scheme to generate

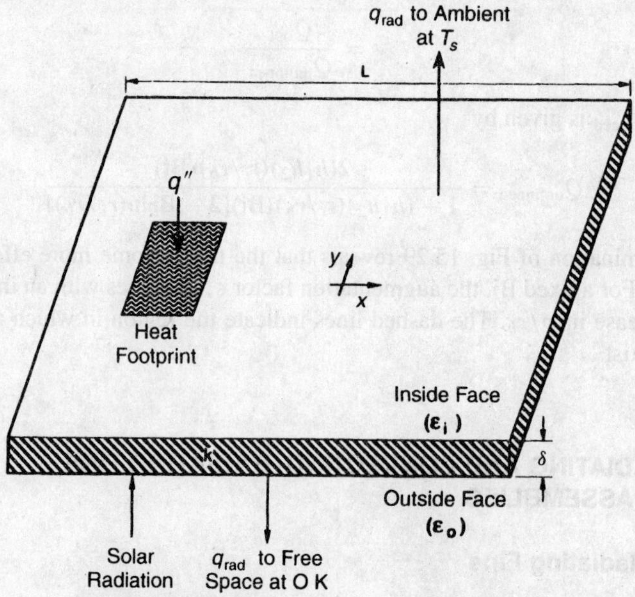

Figure 15.30 Typical thermal doubler design.

the temperature results. The results of Narayana and Kumari are discussed here in some detail.

Figure 15.31 shows the effect of the heat flux parameter, $\phi = q''x_o^2/k\delta T_o$ on the dimensionless temperatures, T/T_c and T_c/T_o for a square plate with a centrally located square footprint. The quantity T_c is the temperature at the origin where $X = x/x_o = 0$ and $Y = y/y_o = 0$ and where x_o and T_o are the reference distance and reference temperature, respectively. The figure corresponds to a situation where the emissivity ratio $W = \epsilon_i/\epsilon_o = 1$, the radiation conduction parameter $\xi = \sigma\epsilon_o x_o^2 T_o^3/k\delta = 0.50$, and the environmental parameter $\psi = C/\sigma\epsilon_o T_o^4 = 0.2$. The constant C in the definition of ψ is a measure of the solar radiation absorbed by the plate. As ϕ increases, T_c increases, as expected. However, the temperature ratio T/T_c at any location $X = x/x_o$ decreases as ϕ increases. Although T increases with ϕ, the corresponding increase in T_c causes the ratio T/T_c to decrease.

Figure 15.32 illustrates the effect of varying the emissivity ratio W with ϕ, ξ, and ψ values fixed. As anticipated, the highest temperatures occur for $W = 0$, that is, when the inside surface is radiatively inactive ($\epsilon_i = 0$). As the emissivity ϵ_i of the inside surface increases, causing W to increase, the temperature levels fall off. Thus for effective thermal management, both the inside and outside faces must be radiatively active.

The effect of the environmental parameter ψ on the temperature ratios T/T_c and T_c/T_o is shown in Fig. 15.33. Although the temperature T_c at the origin is significantly affected by ψ, the ratio T/T_c is not very sensitive to the variation of ψ. The temperature level throughout the plate is the lowest when the solar radiation is

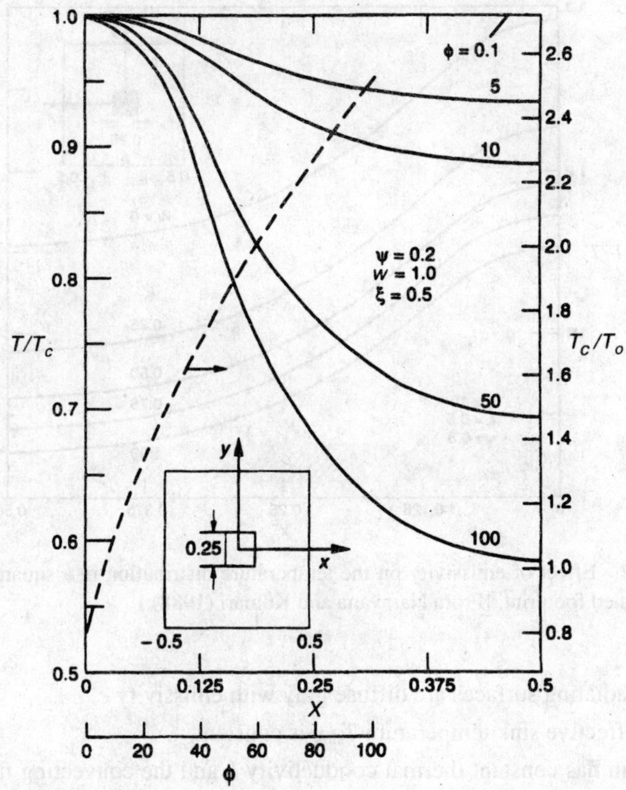

Figure 15.31 Effect of the heat flux parameter on the temperature distribution in a square plate with a centrally located footprint. [From Narayana and Kumari (1988).]

zero ($\psi = 0$). Narayana and Kumari (1988), provide results for the isotherms and show how the isotherm pattern is affected by the location of the heat footprint and provides an example of the design of radiatiing rectangular plates.

15.8.2 Convecting–Radiating Fins

The two-dimensional heat transfer chracteristics of individual longitudinal convecting–radiating fins have been reported by Aziz and Nguyen (1993). They considered fins of rectangular, trapezoidal, triangular, and concave parabolic profiles, shown in Fig. 15.34. The analysis is based on the following considerations:

1. The fin base is maintained at a constant and uniform temperature T_b.
2. All exposed surfaces, including the tip, lose heat by simultaneous convection and radiation.
3. The convective heat transfer coefficient h and the environmental temperature T_s are constant.

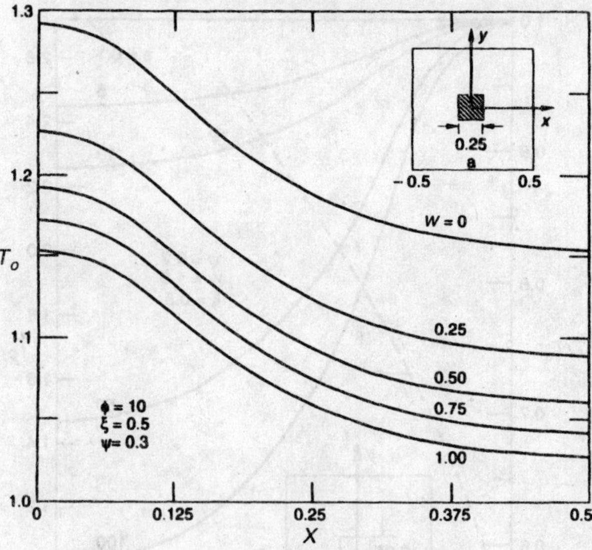

Figure 15.32 Effect of emissivity on the temperature distribution in a square plate with a centrally located footprint. [From Narayana and Kumari (1988).]

4. The radiating surfaces are diffuse gray with emissivity ϵ.
5. The effective sink temperature T_{sink} is constant.
6. The fin has constant thermal conductivity k and the convecting fluid is transparent to radiation.

Using a finite element approach, Aziz and Nguyen (1993) computed the two-dimensional heat transfer rate per unit length, expressing it as $Q = \sigma / kT_b$ for a range of parameters. A sample of their results is given in Fig. 15.35, where Q is plotted against $K = 2b/\delta$ for the four profile shapes. The top portion of the figure is for the radiation–conduction parameter $N_r = \epsilon\sigma\delta T_b^3/2k = 5$ and for $\mathrm{Bi} = h\delta/2k = 1$; the lower portion corresponds to $N_r = 1$ and $\mathrm{Bi} = 1$. For all values of K, the highest heat dissipation is achieved with the concave parabolic fin and the lowest with the rectangular fin. The performance curves for the trapezoidal and triangular fins fall in between. It is interesting to note that the heat dissipation is hardly affected by K beyond $K = 16$.

Table 15.9 presents data for the maximum error for each of the four geometries, the error being calculated as

$$\text{error} = 100\left(\frac{Q_1 - Q}{Q}\right)$$

with $Q_1 = q_1/kT_b$, where q_1 is the one-dimensional heat transfer per unit depth. The maximum positive error of 144 % occurs for a rectangular fin with $K = 1$, and the maximum negative error of 47 % occurs for a triangular fin with $K = 20$.

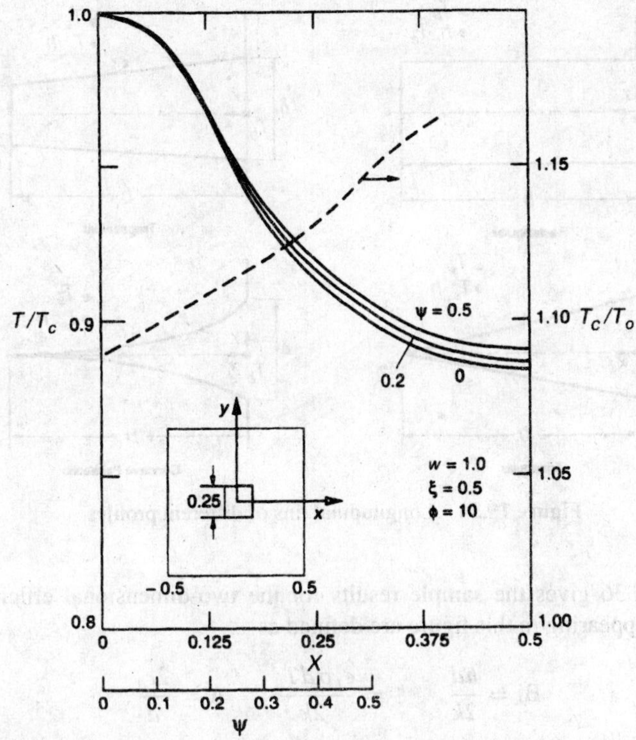

Figure 15.33 Effect of the environmental parameter on the temperature distribution in a square plate with a centrally located square footprint. [From Narayana and Kumari (1988).]

TABLE 15.9 Maximum Percent Error: Bi = 10, $N_r = 5$, $T_s/T_b = T_{sink}/T_b = 0.70$

K	Rectangular	Trapezoidal	Triangular	Conc. Parabolic
1	144	52	33	12
5	67	40	36	23
10	10	−4	−5	−10
20	−38	−44	−47	−46

Attention now turns to the work of Sikka and Iqbal (1970) dealing with the two-dimensional performance of a cylindrical spine or pin fin losing heat by simultaneous convection and radiation. The analysis considered a pin fin attached to a circular base at uniform temperature. Using the same assumptions used by Aziz and Nguyen (1993), but allowing for fin-to-base radiative interaction and conduction in the radial and axial directions, Sikka and Iqbal (1970) obtained a series solution for the temperature distribution in the fin. The coefficients were determined from least-squares fitting at a finite number of points on the surface of the fin.

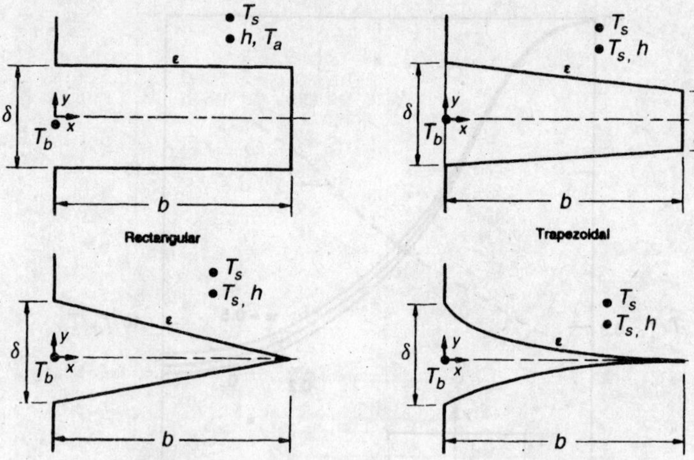

Figure 15.34 Longitudinal fins of different profiles.

Figure 15.36 gives the sample results for the two-dimensional efficiency. The parameters appearing in this figure are defined as

$$\mathrm{Bi} = \frac{hd}{2k}, \qquad \xi = \frac{\epsilon_f \sigma d T_b^3}{2k}, \qquad \Lambda = \frac{2r_b}{d},$$

$$\gamma = \frac{2b}{d}, \qquad \theta_a = \frac{T_a}{T_b} \quad \text{and} \quad \theta_{\text{sink}} = \frac{T_{\text{sink}}}{T_b}$$

where d is the fin diameter, r_b the radius of the circular base, h the convective heat transfer coefficient, k the thermal conductivity of the fin, b the fin height, T_s and T_{sink} the temperatures of the convective environment and the radiation sink, T_b the fin base temperature, ϵ_f and ϵ_b the emissivities of the fin and base, and σ the Stefan–Boltzmann constant. The curve for $\xi = 10^{-7}$ corresponds to negligible radiative heat transfer and is therefore a very close representation of a purely convective fin. In this case, the fin has the highest efficiency. As radiation becomes stronger, that is, as ξ increases, the efficiency drops sharply, paricularly at low values of Bi. This behavior is consistent with that observed by Sparrow and Nieworth (1968) for a one-dimensional convecting–radiating longitudinal fin of rectangular profile. Sikka and Iqbal (1970) noted that the efficiency of a two-dimensional convecting–radiating fin was lower than that of its one-dimensional counterpart.

15.8.3 Convecting–Radiating Fin Assembly

Manzoor et al. (1983a) considered an array of longitudinal fins of rectangular profile attached to a plane wall as shown in Fig. 15.27. The analysis considered pure convection heating on the unfinned side of the wall but allowed for combined convection

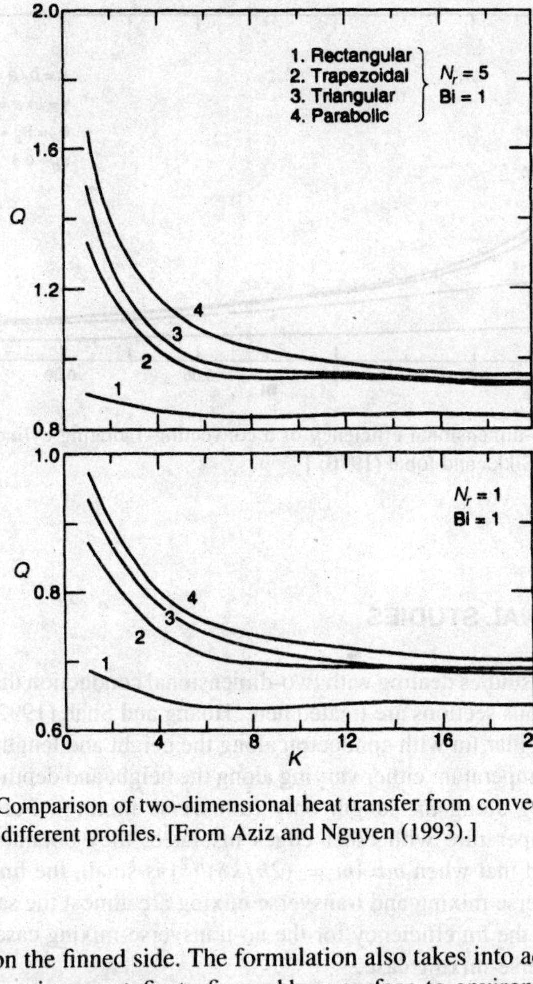

Figure 15.35 Comparison of two-dimensional heat transfer from convecting–radiating longitudinal fins of different profiles. [From Aziz and Nguyen (1993).]

and radiation on the finned side. The formulation also takes into account fin-to-base surface, fin-to-environment, fin-to-fin, and base surface-to-environment interactions. In addition, provision was made for the thermal conductivity and emissivity of the fin to be different from those of the base surface. The rather complex model was solved using a nonlinear boundary integral equation technique.

Three important conclusions emerged from this study. First, the addition of the fin acts to depress the temperature at the base of the fin. For realistic values of the system parameters, the reduction in base temperature can be as high as 50%. Second, the assumption of uniform base temperature can, in some cases, overpredict the heat dissipation of the fin assembly by over 150%. Third, the error between the one- and two-dimensional analyses depends on the system parameters. In some cases the one-dimensional model overestimates the heat flow by over 30%. Additional thoughts on the two-dimensional performance of convecting–radiating assemblies may be found in Manzoor (1984).

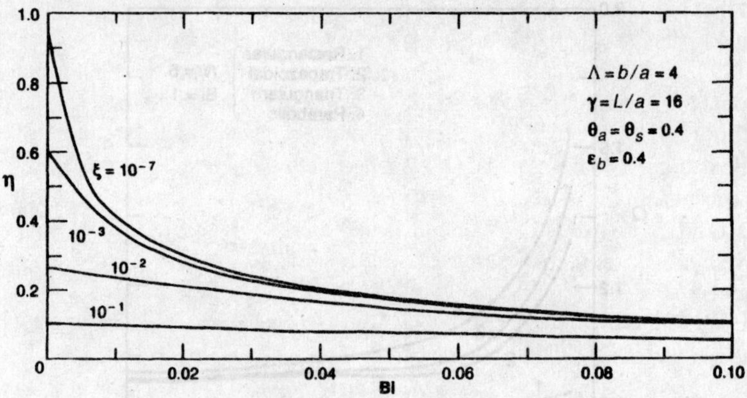

Figure 15.36 Two-dimensional efficiency of a convecting–radiating cylindrical spine (pin fin). [Adapted from Sikka and Iqbal (1970).]

15.9 ADDITIONAL STUDIES

Several additional studies dealing with two-dimensional conduction that do not fit the scope of the previous sections are treated here. Huang and Shah (1992) considered a convecting rectangular fin with conducton along the height and length of the fin and the surrounding temperature either varying along the height and depth (no transverse mixing) or varying along the length only (transverse mixing). For conditions of constant base temperature with other edges insulated, they obtained a numerical solution and found that when mb [$m = (2h/k\delta)^{1/2}$] is small, the fin efficiency for cases of no transverse mixing and transverse mixing are almost the same. However, when mb is large, the fin efficiency for the no-transverse-mixing case is lower than that for the transverse-mixing case.

The analysis of Aziz and Tesarik (1992), discussed in Section 15.2.2, has been extended by Lam and Hardt (1994) to include the effect of nonuniform base temperatures. Their results indicate that the error in Q due to the assumption of constant T_1 and T_2 can be as high as 30%.

Nakajima (1992) considered two-dimensional conduction in a sandwich panel made of two rectangular plates. The top surface of the upper plate was assumed to carry a rectangular heat footprint, while the lower face of the lower plate was subjected to a uniform solar heat flux. Both of the exposed surfaces were allowed to radiate to their respective environments. The surfaces of the plates facing each other were regarded as being convectively coupled. By linearizing the radiation terms, Nakajima (1992) was able to solve the coupled equations using an iterative procedure.

The two-dimensional performance of fins with simultaneous heat and mass transfer has been investigated by Chen (1991) and by Kazeminejad et al. (1993). The analysis of Chen (1991) deals with a rectangular fin in contact with a moist stream of air. Chen found that the effect of two-dimensional conduction was to reduce the efficiency of the fin compared with that of the one-dimensional model. The work of Kazeminejad et al. (1993) considers an eccentric fin of rectangular profile operating in a moist air environment. Using the bipolar coordinate system, Kazeminejad et al. were able to transform the eccentric radial fin, with conduction in the radial and circumferential directions, into a rectangular strip with subsequent numerical solution. The efficiency of an eccentric fin was found to be lower than that of a concentric fin with the same radius ratio. Moreover, it was found that the efficiency of the fin deceases as the relative humidity of the air increases.

Newhouse (1964) considered the solution of the radial fin of rectangular profile with constant thermal conductivity and constant heat transfer coefficient on its faces for the special case of a fin base variation of the form

$$T = \hat{T} - F\cos\theta$$

where θ is the circumferential coordinate, $\hat{T}$ is the mean base temperature, and F represents the peak amplitude of the circumferential variation.

Heggs and Stones (1980) considered the effect of temperature depression at the base of a fin and showed that in multifin arrays, heat fluxes can be significantly in error (80%) compared to those obtained in one-dimensional analyses. Their studies compared one- and two-dimensional solutions for the heat flow in both radial and longitudinal fin assemblies for a wide range of system parameters, such as fin height, interfin spacing, and fin materials. This was followed by an additional study by Heggs and Stones (1983). Indeed, much of the work considered in this chapter has been summarized in the article by Aziz and Lunardini (1995).

15.10 NOMENCLATURE

Roman Letter Symbols

A	combination of terms, dimensionless; constant, dimensionless
B	combination of terms, dimensionless; constant, dimensionless
Bi	Biot number, dimensionless
b	fin or spine height, m
C	constant, dimensionless; combination of terms, dimensionless
D	combination of terms, dimensionless; constant, dimensionless
d	diameter, m; differential or derivative, dimensionless
F	peak temperature amplitude, K
f	combination of terms, dimensionless
H	combination of terms, dimensionless
h	heat transfer coefficient, $W/m^2 \cdot K$
I	modified Bessel function of first kind, dimensionless
J	Bessel function of first kind, dimensionless
K	modified Bessel function of second kind, dimensionless; thickness, dimensionless

k	thermal conductivity, W/m·K
L	fin length, m
m	fin performance parameter, m^{-1}
N	combination of terms, dimensionless; radiation parameter, dimensionless
N^*	combination of terms, dimensionless
P	fin perimeter, m
p	fin spacing, m; maximum base temperature depression, K
Q	heat flow per unit length, dimensionless; heat flow per unit length, W/m
$\dot{Q}$	dimensionless heat generation
q	heat flow, W; heat flow per unit length, W/m
$\dot{q}$	heat generated per unit volume, W/m^3
R	radius, dimensionless
r	radius, m; radial coordinate, m
S	surface area, m^2
T	temperature, K
$\hat{T}$	mean base temperature, K
V	volume, dimensionless
W	emissivity radius, dimensionless
X	height, dimensionless
x	height coordinate, m
Y	Bessel function of second kind, dimensionless; thickness, dimensionless
y	thickness coordinate, m
Z	height, dimensionless
z	height coordinate, m; fin spacing, m

Greek Letter Symbols

Γ	aspect ratio, dimensionless
Δ	change in, dimensionless
δ	fin thickness, m
∂	partial derivative, dimensionless
ϵ	emissivity, dimensionless
ζ	profile function, dimensionless
η	fin efficiency, dimensionless
θ	circumferential coordinate, rad; temperature ratio, dimensionless
Λ	thickness ratio, dimensionless
λ	eigenvalue, dimensionless;
ξ	radiation conduction parameter, dimensionless; combination of terms, dimensionless
σ	Stefan–Boltzmann constant, $W/m^2 \cdot K^4$
ϕ	heat flux parameter, dimensionless
ψ	environmental parameter, dimensionless

Roman Letter Subscripts

a	ambient; fin tip
b	fin base
c	condition at origin
i	inner surface
id	ideal
L	fin length
n	nth eigenvalue
o	referenced quantity
opt	optimum
r	radiation
s	surroundings
sink	radiation sink
unfinned	unfinned condition
1−D	one-dimensional

Symbolic Superscripts

$+$	modified Biot modulus
$-$	modified Biot modulus; condition of temperature asymmetry
$''$	indicates heat flux
$\cdot$	indicates volumetric heat generation
$=$	indicates "equal" condition
$\neq$	indicates "unequal" condition

CHAPTER 16

TRANSIENT HEAT TRANSFER IN EXTENDED SURFACES

16.1 INTRODUCTION

Although designs of systems employing extended surfaces are based on steady-state analyses which are adequate for most applications, there are situations in which a knowledge of the transient response is necessary. This is true for fins used in high-speed aircraft, intermittently operating heat exchangers, electronic components, automatic control equipment, and solar energy systems.

This chapter is organized following a rather prescribed format. First, the transient response of the longitudinal fin of rectangular profile is presented. Four types of time-dependent conditions are considered: a step change in base temperature, a step change in base heat flux, a step change in the temperature of the fluid on the unfinned side of the primary surface, and a step change in environmental temperature. The solution techniques involve separation of variables and the Laplace transformation. Next, the analysis for radial fins of different profiles that experience a step change in base temperature is presented. Following this, the response of a longitudinal fin of rectangular profile with a power law type of surface heat dissipation is discussed, including solutions obtained by coordinate perturbation expansions. Later discussion focuses on the transient responses of a two-dimensional cylindrical spine, a two-dimensional composite rectangular fin, and one-dimensional longitudinal and radial fins with stochastic base temperature. The chapter concludes with a discussion of a conjugate study in which conduction, natural convection, and radiation are coupled.

16.2 LONGITUDINAL FIN OF RECTANGULAR PROFILE

The transient response of a longitudinal fin of rectangular profile operating in the convection mode has been studied for four types of time-dependent boundary conditions: (1) step change in base temperature, (2) step change in base heat flow, (3) step change in the temperature of the hot fluid on the unfinned side of the primary surface, and (4) step change in the temperature of the environment. The analytical tools employed are the method of separation of variables, Laplace transformation, and the Kantorovich method. For cases 1 and 2, the separation of variables solutions presented here were developed by Donaldson and Shouman (1972) and extended here to case 3. Laplace transform solutions have been reported by Suryanarayana (1975, 1976) and Papadopoulos et al. (1990). The Kantorovich method has been used by Kim (1976) to study the case of step change in base temperature.

16.2.1 Step Change in Base Temperature

A physical system consisting of a rectangular fin of length L, thickness δ, height b, and thermal conductivity k is shown in Fig. 16.1. Initially, the fin is at a uniform temperature T_i, interacting with a convecting environment at temperature T_s, and with a heat transfer coefficient h. At time $t = 0$, the temperature at the base is suddenly changed from T_i to T_b and the problem is to establish the transient temperature distribution in the fin for all $t \geq 0$. It will be assumed that the heat loss from the fin tip is negligible at all times and that all of the Murray–Gardner assumptions listed in Section 1.2.1 are in force except for the assumption that steady state prevails.

The partial differential equation governing the transient response of the fin can be written as

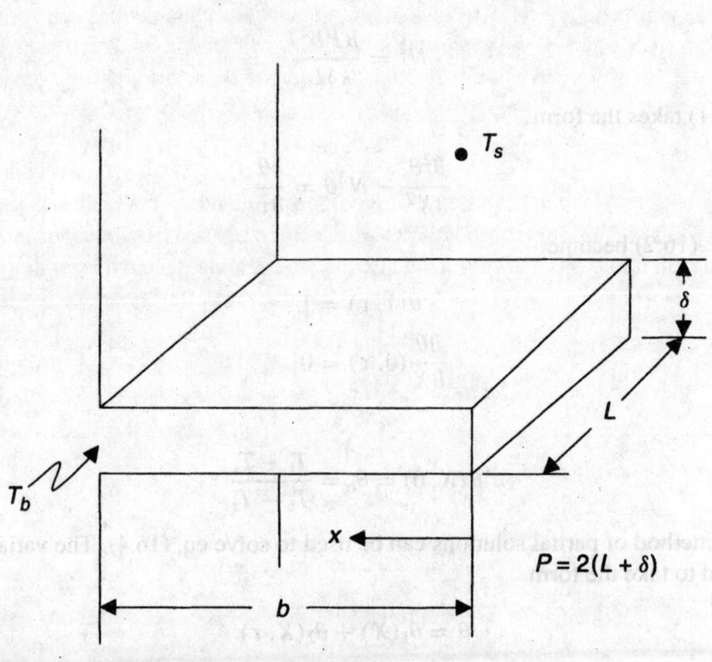

Figure 16.1 Longitudinal fin of rectangular profile.

$$\frac{\partial^2 T}{\partial x^2} - \frac{hP}{k\delta L}(T - T_s) = \frac{1}{\alpha}\frac{\partial T}{\partial t} \tag{16.1}$$

and this equation is to be solved subject to the boundary conditions

$$T(b, t) = T_b \tag{16.2a}$$

and

$$\frac{\partial T}{\partial x}(0, t) = 0 \tag{16.2b}$$

and the initial condition

$$T(x, 0) = T_i \tag{16.2c}$$

After introducing the dimensionless variables

$$\theta = \frac{T - T_s}{T_b - T_s} \tag{16.3a}$$

$$X = \frac{x}{b} \tag{16.3b}$$

$$\tau = \frac{\alpha t}{b^2} \tag{16.3c}$$

and

$$N^2 = \frac{hPb^2}{k\delta L} \tag{16.3d}$$

eq. (16.1) takes the form

$$\frac{\partial^2 \theta}{\partial X^2} - N^2\theta = \frac{\partial \theta}{\partial \tau} \tag{16.4}$$

and eqs. (16.2) become

$$\theta(1, \tau) = 1 \tag{16.5a}$$

$$\frac{\partial \theta}{\partial X}(0, \tau) = 0 \tag{16.5b}$$

and

$$\theta(X, 0) = \theta_i = \frac{T_i - T_s}{T_b - T_s} \tag{16.5c}$$

The method of partial solutions can be used to solve eq. (16.4). The variable θ is assumed to take the form

$$\theta = \theta_1(X) + \theta_2(X, \tau) \tag{16.6}$$

which then leads to an equation for θ_1 as

$$\frac{d^2\theta_1}{dX^2} - N^2\theta_1 = 0 \tag{16.7}$$

with

$$\frac{d\theta_1}{dX}(0) = 0 \tag{16.8a}$$

$$\theta_1(1) = 1 \tag{16.8b}$$

and for θ_2,

$$\frac{\partial^2\theta_2}{\partial X^2} - N^2\theta_2 = \frac{\partial\theta_2}{\partial\tau} \tag{16.9}$$

with

$$\frac{\partial\theta_2}{\partial X}(0, \tau) = 0 \tag{16.10a}$$

$$\theta_2(1, \tau) = 0 \tag{16.10b}$$

and

$$\theta_2(X, 0) = \theta_i - \theta_1(X) \tag{16.10c}$$

The general solution of eq. (16.7) is

$$\theta_1 = C_1 e^{NX} + C_2 e^{-NX}$$

and with eqs. (16.8) used to evaluate C_1 and C_2,

$$\theta_1 = \frac{\cosh NX}{\cosh N} \tag{16.11}$$

The solution for θ_2 can be obtained using the method of separation of variables:

$$\theta_2 = 2\sum_{n=1}^{\infty}(-1)^{n+1}\left(\frac{\theta_i}{\lambda_n} - \frac{\lambda_n}{N^2+\lambda_n^2}\right)\cos\lambda_n X e^{-(N^2+\lambda_n^2)\tau} \tag{16.12}$$

where

$$\lambda_n = \frac{2n-1}{2}\pi, \qquad n = 1, 2, 3, \cdots$$

The instantaneous heat flow rate q from the base of the fin can then be obtained from

$$q_b = k\delta L\frac{\partial T}{\partial X}(b, t)$$

or in dimensionless form,

$$\frac{q_b b}{k\delta L(T_b - T_s)} = \frac{\partial\theta}{\partial X}(1, \tau) \tag{16.13}$$

Equations (16.11) and (16.12) are then added and the derivative $\partial\theta/\partial X$ is taken. Then use of eq. (16.13) yields

$$\frac{q_b b}{k\delta L(T_b - T_s)} = N \tanh N - 2\sum_{n=1}^{\infty}(-1)^{n+1}\left(\theta_i - \frac{\lambda_n^2}{N^2+\lambda_n^2}\right)\sin\lambda_n e^{-(N^2+\lambda_n^2)\tau} \tag{16.14}$$

Observe that as $\tau \to \infty$, the transient term in eq. (16.14) decays to zero and that the solution approaches the well-known steady-state solution.

Figure 16.2 shows how the tip of the fin responds at different values of N. As N increases, the response time decreases. For a fin of given material and fixed dimensions, it implies that the higher the convective heat transfer coefficient, the faster the response of the fin to the temperature disturbance at the base. This behavior is due to the presence of N^2 in the exponentially decaying factor associated with eq. (16.12). Figure 16.3 shows the instantaneous base heat flow rate q_b as a function of time for different values of N. These results also substantiate the conclusion from Fig. 16.2 that as N increases, the response time decreases.

Because series solutions such as those given by eqs. (16.12) and (16.14) converge slowly for small values of time, it is advantageous to use the Laplace transformation as a solution method because for small time, the technique allows rapidly convergent approximate solutions to be obtained. Suryanarayana (1975) has reported the Laplace transform solution of eq. (16.4) for the case of $\theta_i = 0$, which means that the fin is initially in thermal equilibrium with its surroundings. Following this approach, eq. (16.4) can be written in the Laplace domain (the s-domain) as

$$\frac{d^2\overline{\theta}(X,s)}{dX^2} - (N^2+s)\overline{\theta}(X,s) = 0 \tag{16.15}$$

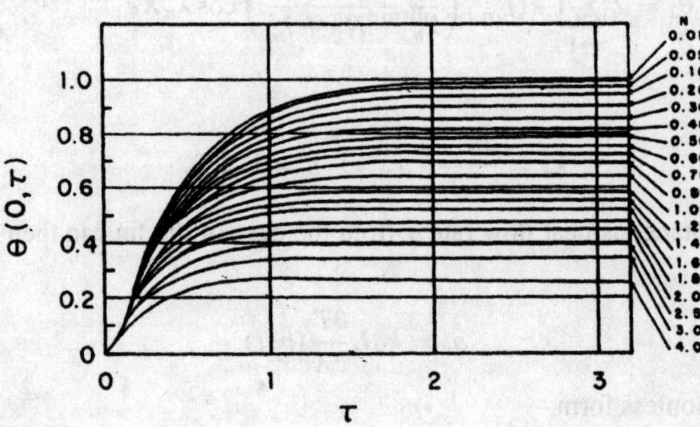

Figure 16.2 Response of fin-tip temperature due to step change in base temperature. [From Donaldson and Shouman (1972).]

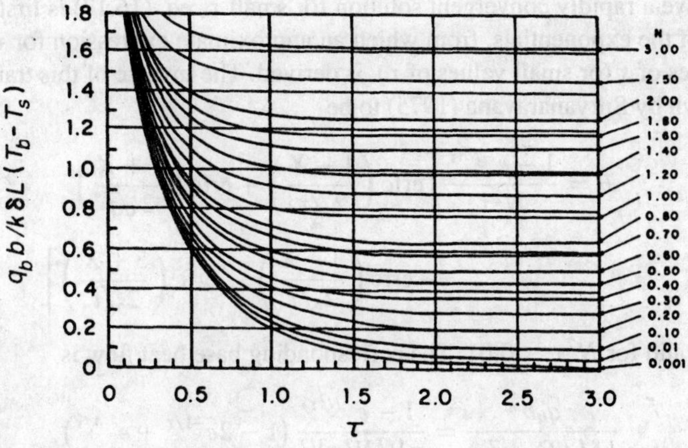

Figure 16.3 Instantaneous base heat flow due to step change in base temperature. [From Donaldson and Shouman (1972).]

where the Laplace transform of $\theta(X, \tau)$ is

$$\overline{\theta}(X, s) = \mathcal{L}[\theta(X, \tau)] \tag{16.16a}$$

and where

$$\frac{d\overline{\theta}(s, 0)}{dX} = 0 \tag{16.16b}$$

and

$$\overline{\theta}(s, 1) = \frac{1}{s} \tag{16.16c}$$

The solution for $\overline{\theta}(X, s)$ can be obtained as

$$\overline{\theta}(X, s) = \frac{1}{s} \frac{\cosh (N^2 + s)^{1/2} X}{\cosh (N^2 + s)^{1/2}} \tag{16.17}$$

and then using the complex inversion integral, the solution for $\theta(X, \tau)$ can be expressed as an integral:

$$\theta(X, \tau) = \frac{1}{2\pi i} \int_{\gamma - i\infty}^{\gamma + i\infty} e^{\tau s} \left[\frac{\cosh (N^2 + s)^{1/2} X}{s \cosh (N^2 + s)^{1/2}} \right] ds \tag{16.18}$$

Here, the integration is to be performed in the complex s-plane along the infinite line $X = \gamma$, with γ so chosen that all singularities of $\overline{\theta}(X, s)$ lie to the left of the line $X = \gamma$. When the integral in eq. (16.18) is evaluated, it leads exactly to the sum of θ_1 and θ_2 as represented by eqs. (16.11) and (16.12) with $\theta_i = 0$. Thus the two techniques give the same end result, as they should.

To derive a rapidly convergent solution for small τ, eq. (16.17) is first expressed in terms of the exponentials, from which an approximate expression for τ, valid for large values of s (or small values of τ), is derived. The inverse of this transform has been shown by Suryanarayana (1975) to be

$$\theta = \frac{1 - e^{-N^2\tau}}{N^2\tau}\left[\operatorname{erfc}\left(\frac{1-X}{2\sqrt{\tau}}\right) + \operatorname{erfc}\left(\frac{1+X}{2\sqrt{\tau}}\right) - \operatorname{erfc}\left(\frac{3-X}{2\sqrt{\tau}}\right) - \operatorname{erfc}\left(\frac{3+X}{2\sqrt{\tau}}\right)\right] \tag{16.19}$$

which is valid for $N^2\tau \le 0.01$. The corresponding base heat flow is

$$\frac{q_b b}{k\delta L(T_b - T_s)} = \frac{1 - e^{-N^2\tau}}{\pi^{1/2}N^2\tau^{3/2}}\left(1 - 2e^{-1/\tau} + e^{-4/\tau}\right) \tag{16.20}$$

Using 10 terms in eqs. (16.12) and (16.19), Suryanarayana (1975) plotted the transient temperature distributions in the fin for different values of N and τ. The result is shown in Fig. 16.4, which has been adapted from his paper. For $\tau \le 0.2$, the temperature distributions are virtually independent of the value of N. This is why a single curve is shown for $\tau = 10^{-4}$, 10^{-3}, and 10^{-2}. For $\tau = 0.1$ and 1, three curves are shown, each corresponding to a different N. At small values of τ, the bulk of the thermal energy entering at the base remains stored in the fin with only a small fraction available for dissipation through surface convection. Thus the heat transfer coefficient or N has very little impact on the temperature profiles. As τ increases, the role of

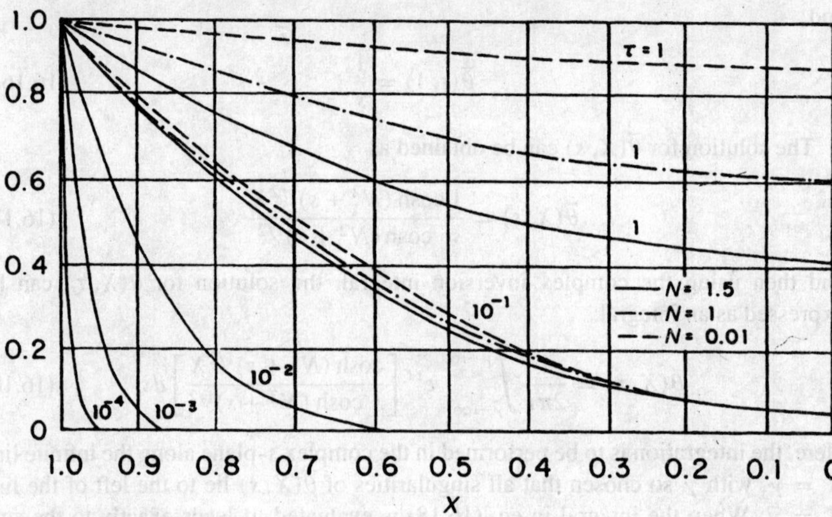

Figure 16.4 Transient temperature distributions in a longitudinal fin of rectangular profile: Step change in base temperature. [From Suryanarayana (1975).]

surface convection (and hence N) becomes progressively significant. At $\tau = 0.1$, the effect of N on the transient temperature distribution becomes noticeable. The curves for $\tau = 1$ clearly show that the temperature distribution is affected significantly by N.

Based on his calculations, Suryanarayana (1975) recommends the use of eq. (16.19) for $\tau \leq 0.01$ and $0.01 \leq N \leq 10$. For $\tau > 0.01$, he recommends that 10 terms of eq. (16.12) be used. Suryanarayana (1975) also found that the time required for the temperature to reach within 1% of its steady-state value is given by

$$\tau_s = \frac{1}{N^2 + \pi/4} \ln \frac{100\pi \cosh N}{N^2 + \pi/4} \tag{16.21}$$

Values of τ_s computed from eq. (16.21) are provided in Table 16.1, and the values shown confirm the earlier conclusion that the transient effects decay more rapidly as N increases.

In Fig. 16.3, the congestion of curves makes it very difficult to read values of q_b at small values of τ. To overcome this difficulty, numerical data for $q_b b/k\delta L(T_b - T_s)$ are displayed in Table 16.2.

As an alternative to the separation of variables and the Laplace transform approaches, Kim (1976) has solved this problem with $\theta_i = 0$, using the variational method of Kantorovich described by Kantorovich and Krylov (1964) and Arpaci (1966). He gives both the first- and second-order approximate solutions. The second-order solution, which is more accurate, can be couched in the nomenclature of this chapter as

$$\theta = 1 + \frac{21N^2(X^2 - 1)}{2(N^4 + 28N^2 + 63)}(\Lambda_1 + \Lambda_2) \tag{16.22}$$

TABLE 16.1 Values of τ_s as a Function of the Parameter N

N	0.01	0.1	1.0	2.0	5.0	10
τ_s	1.96	1.95	1.42	0.805	0.245	0.101

TABLE 16.2 Instantaneous Dimensionless Base Heat Flow for Different Values of N and τ

	$\tau =$						
N	10^{-4}	10^{-3}	0.01	0.1	1	10	100
0.01	56.32	17.8	11.28	3.57	0.339	0.999×10^{-4}	0.999×10^{-4}
0.1	56.32	17.8	11.27	3.56	0.334	0.997×10^{-2}	0.997×10^{-2}
0.6	56.4	17.8	11.0	3.37	0.529	0.322	0.322
0.8	56.4	17.8	11.82	3.26	0.673	0.534	0.531
1.0	56.4	17.8	10.6	3.16	0.850	0.762	0.762
1.5	56.4	17.8	10.18	2.98	1.376	1.358	1.358
2.0	56.4	17.8	9.8	2.98	1.930	1.928	1.928
5.0	56.3	17.6	8.99	5.04	4.995	4.999	4.999
10.0	56.15	16.98	11.01	10.00	10.00	10.00	10.00

where

$$\Lambda_1 = \left(\frac{N^2}{12} + 3\right)(1 + c_1 e^{-\kappa_1 \tau} + c_2 e^{-\kappa_2 \tau}) \tag{16.23a}$$

and

$$\Lambda_2 = \frac{N^2 X^2}{4}\left(1 + c_3 e^{-\kappa_1 \tau} + c_4 e^{-\kappa_2 \tau}\right) \tag{16.23b}$$

where in Λ_1 and Λ_2

$$\kappa_1 = N^2 + 25.53256 \tag{16.23c}$$

$$\kappa_2 = N^2 + 2.46744 \tag{16.23d}$$

In eqs. (16.23), the c's are given by

$$c_1 = \frac{0.45106023N^4 - 7.67296673N^2 - 21.67328855}{1.75N^4 + 63N^2} \tag{16.24}$$

$$c_2 = \frac{-0.20016623N^4 + 0.67296673N^2 + 147.67328855}{1.75N^4 + 63N^2} \tag{16.25}$$

$$c_3 = \frac{-0.04418669N^4 + 56.14855739N^2 + 158.5988741}{5.25N^4} \tag{16.26}$$

and

$$c_4 = \frac{0.04418669N^4 - 0.14855739N^2 - 32.5999741}{5.25N^4} \tag{16.27}$$

Kim has not assessed the accuracy of his solution against the exact solution discussed earlier. However, some spot checks have shown his solution to be accurate within a few percent. Thus the use of eqs. (16.22) through (16.27) is recommended for rapid calculations.

Example 16.1. A longitudinal fin of rectangular profile is 4.75 mm thick, 51 mm high, and 1 m long. The fin is initially in thermal equilibrium with its environment maintained at 10°C. The thermal conductivity and the thermal diffusivity of the fin material are 31 W/m·K and 8.5×10^{-6} m²/s, respectively. Assume that the convective heat dissipation from the fin surfaces is characterized by a convective heat transfer coefficient of 114 W/m²·K. If the base temperature is suddenly elevated to 116°C, calculate (**a**) the base heat flow rate after a lapse of 30 s, (**b**) the time elapsed before the response attains a value within 1% of its new steady state, and (**c**) the new steady-state base heat flow

SOLUTION. (**a**) Here, with $L = 1$ m, $\delta = 0.00475$ m, and $b = 0.051$ m, the fin parameter N can be calculated as

$$N = \left(\frac{hPb^2}{k\delta L}\right)^{1/2}$$

$$= \left\{ \frac{(114)[2(1 + 4.75 \times 10^{-3})](0.051)^2}{(31)(4.75 \times 10^{-3})(1)} \right\}^{1/2}$$

$$= 2.01$$

The dimensionless time τ corresponding to $t = 30$ s is

$$\tau = \frac{\alpha t}{b^2}$$

$$= \frac{(8.5 \times 10^{-6})(30)}{(0.051)^2}$$

$$= 0.098$$

For $N = 2.01$ and $\tau = 0.098$, in Table 16.2 read the entry at $N = 2$ and $\tau = 0.1$, so that

$$\frac{q_b b}{k\delta L(T_b - T_s)} = 2.98$$

Thus

$$q_b = \frac{2.98 k\delta L(T_b - T_s)}{b}$$

$$= \frac{(2.98)(31)(4.75 \times 10^{-3})(1)(106)}{0.051}$$

$$= 912 \text{ W}$$

(b) From Table 16.1, $\tau_s = 0.805$ for $N = 2$. Thus

$$t = \frac{0.805 b^2}{\alpha}$$

$$= \frac{(0.805)(0.051)^2}{8.5 \times 10^{-6}}$$

$$= 246 \text{ s} \quad \text{or} \quad 4.1 \text{ min}$$

(c) The steady-state base heat flow, which also equals the surface convection heat dissipation, is obtained from eq. (16.14) and is given by

$$q_b = \frac{k\delta L(T_b - T_s)}{b} N \tanh N$$

$$= \left[\frac{(31)(4.75 \times 10^{-3})(1)(106)}{0.051} \right] 2.01 \tanh 2.01$$

$$= 593 \text{ W}$$

16.2.2 Step Change in Base Heat Flow

The case of a step change in base heat flow is described by

$$\frac{\partial^2 T}{\partial x^2} - \frac{hP}{k\delta L}(T - T_s) = \frac{1}{\alpha}\frac{\partial T}{\partial t} \tag{16.1}$$

except for the boundary condition at the base, which now becomes

$$\frac{\partial T}{\partial x}(b, t) = \frac{q_b}{k\delta L} \tag{16.28}$$

where q_b is the step change in base heat flow. With the dimensionless temperature θ defined as

$$\theta = \frac{k\delta L(T - T_s)}{q_b b} \tag{16.29}$$

the governing equation in dimensionless form becomes

$$\frac{\partial^2 \theta}{\partial X^2} - N^2\theta = \frac{\partial \theta}{\partial \tau} \tag{16.30}$$

with boundary conditions

$$\frac{\partial \theta}{\partial X}(1, \tau) = 1 \tag{16.31a}$$

and

$$\frac{\partial \theta}{\partial X}(0, \tau) = 0 \tag{16.31b}$$

and initial condition

$$\theta(X, 0) = \theta_i = \frac{k\delta L(T_i - T_s)}{q_b b} \tag{16.31c}$$

where the remaining dimensionless variables retain their definitions given in eqs. (16.3).

Using eq. (16.6) as the solution form for θ, the equations for θ_1 and θ_2 can be obtained as

$$\frac{d^2\theta_1}{dX^2} - N^2\theta_1 = 0 \tag{16.32}$$

with

$$\frac{d\theta_1}{dX}(1) = 1 \tag{16.33a}$$

and

$$\frac{d\theta_1}{dX}(0) = 0 \tag{16.33b}$$

for θ_1 and

$$\frac{\partial^2 \theta_2}{\partial X^2} - N^2 \theta_2 = \frac{\partial \theta_2}{\partial \tau} \tag{16.34}$$

with

$$\frac{\partial \theta_2}{\partial X}(1, \tau) = 0 \tag{16.35a}$$

$$\frac{\partial \theta_2}{\partial X}(0, \tau) = 0 \tag{16.35b}$$

and

$$\theta_2(X, 0) = \theta_i - \theta_1(X) \tag{16.35c}$$

for θ_2.

The solution of eq. (16.32) is

$$\theta_1 = \frac{\cosh NX}{N \sinh N} \tag{16.36}$$

and by employing the method of separation of variables, the solution for θ_2 can be obtained as

$$\theta_2 = \left(\theta_i - \frac{1}{N^2} \right) e^{-N^2 \tau} + 2 \sum_{n=1}^{\infty} \frac{(-1)^{n+1}}{N^2 + n^2 \pi^2} \cos n\pi X e^{-(N^2 + n^2 \pi^2)\tau} \tag{16.37}$$

As an alternative to the separation of variables approach, one can use the Laplace transformation to recast eq. (16.30) into the form

$$\frac{d^2 \overline{\theta}(X, s)}{dX^2} - (N^2 + s)\overline{\theta}(X, s) = 0 \tag{16.38}$$

with

$$\frac{d\overline{\theta}}{dX}(s, 1) = \frac{1}{s} \tag{16.39a}$$

and

$$\frac{d\overline{\theta}}{dX}(s, 0) = 0 \tag{16.39b}$$

where θ_i is taken to be zero; that is, the fin is initially at the same temperature as the environment. The solution for $\overline{\theta}(X, s)$ is

$$\overline{\theta}(X, s) = \frac{\cosh (N^2 + s)^{1/2} X}{s(N^2 + s)^{1/2} \sinh (N^2 + s)^{1/2}} \tag{16.40}$$

When the inverse transform is taken, the same result as obtained by the separation of variables with $\theta_i = 0$ is obtained. Suryanarayana (1975) has shown that for $\tau << 1$ and $N^2 \tau << 1$, a more rapidly converging solution for small values of τ is

$$\theta = \frac{1 - e^{-N^2\tau}}{N^2\tau}\left[2\left(\frac{\tau}{N}\right)^{1/2}\Lambda_1 - \Lambda_2 - \Lambda_3 - \Lambda_4 + \Lambda_5\right] \tag{16.41}$$

where

$$\Lambda_1 = e^{-(1-X)^2/4\tau} + e^{-(1+X)^2/4\tau} + e^{-(3-X)^2/4\tau} + e^{-(3+X)^2/2\tau}$$

$$\Lambda_2 = (1 - X)\,\text{erfc}\left(\frac{1 - X}{2\sqrt{\tau}}\right)$$

$$\Lambda_3 = (1 + X)\,\text{erfc}\left(\frac{1 + X}{2\sqrt{\tau}}\right)$$

$$\Lambda_4 = (3 - X)\,\text{erfc}\left(\frac{3 - X}{2\sqrt{\tau}}\right)$$

and

$$\Lambda_5 = (3 + X)\,\text{erfc}\left(\frac{3 + X}{2\sqrt{\tau}}\right)$$

The transient temperature distributions obtained from eqs. (16.6), (16.36), (16.37), and (16.41) are shown in Fig. 16.5 for $N = 1.0$ and $N = 1.5$. At low values of τ, the effect of N on the transient temperature profiles is very small. In fact, for $\tau \leq 0.1$, the curves for $N = 1.0$ and 1.5 are graphically indistinguishable. However, as τ increases, the effect of N becomes more pronounced. The curves for $\tau = 1$ and 10 clearly show

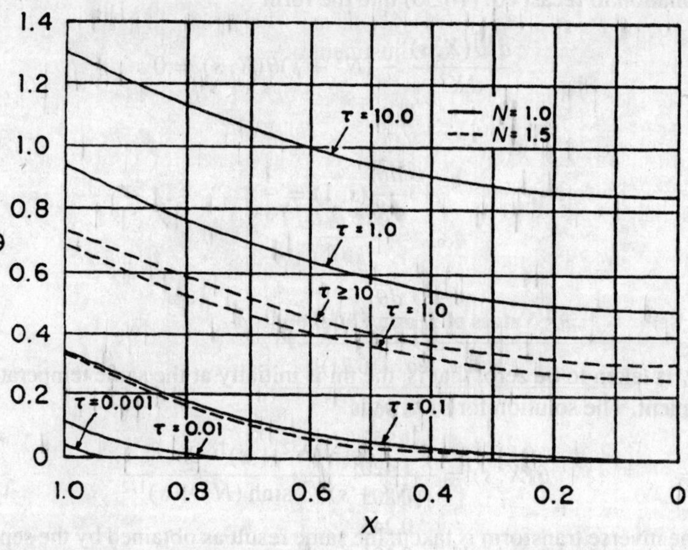

Figure 16.5 Transient temperature distributions in a longitudinal fin of rectangular profile for a step change in base heat flow. [From Suryanarayana (1975).]

the marked influence of N. This behavior can be explained using the same reasoning as that given in Section 16.2.1.

It has been shown by Suryanarayana (1975) that the time τ_s needed for the temperature to reach within 1% of its steady-state values is given by

$$\tau_s = \frac{1}{N^2} \ln \frac{100 \sinh N}{N} \tag{16.42}$$

Comparisons of the values of τ_s obtained from eq. (16.21) for a step change in base temperature and eq. (16.42) for a step change in base heat flow are given in Table 16.3. Comparison of the τ_s values in Tables 16.1 and 16.3 shows that for a given N, the transient lasts longer for a step change in base heat flow than for a step change in base temperature. The response times in the two cases are vastly different at low values of N.

Example 16.2. A longitudinal fin of rectangular profile is 4 mm thick, 40 mm high, and has a length of 1 m. Initially, the fin is in thermal equilibrium with its environment, which is maintained at 15°C. The thermal conductivity and thermal diffusivity of the fin are 50 W/m·K and 5.33 × 10^{-6} m^2/s, respectively. Assume a convective heat transfer coefficient of 62 W/m^2·K, and if the base heat flow suddenly changes to 100 W, **(a)** determine the instantaneous convective heat loss and the rate of energy storage at time $t = 30$ s; and **(b)** show how these quantities compare with the corresponding values at $t = 5$ min.

SOLUTION. **(a)** The heat loss can be found from the instantaneous temperature distribution via an integration

$$q_{\text{conv}} = hP \int_0^b (T - T_s)\, dx$$

and if $T - T_s$ is replaced after an adjustment of eq. (16.29) by

$$T - T_s = \frac{\theta q_b b}{k \delta L}$$

The result is

$$q_{\text{conv}} = \frac{hPq_b b}{k\delta L} \int_0^b \theta\, dx$$

TABLE 16.3 Values of τ_s as a Function of the Parameter N

N	τ_s [eq. (16.21)]	τ_s [eq. (16.42)]
0.01	1.960	46052
0.10	1.950	460.7
1.00	1.420	4.76
2.00	0.805	1.30
5.00	0.245	0.292
10.0	0.101	—

With $x = bX$ so that $dx = b\,dX$,

$$x = 0 \quad \text{and} \quad X = 0$$

$$x = b \quad \text{and} \quad X = 1$$

Thus

$$q_{\text{conv}} = \frac{hPq_b b^2}{k\delta L}\int_0^1 \theta\,dX = N^2 q_b \int_0^1 \theta\,dX$$

The dimensionless quantities N and τ are

$$N = \left(\frac{hPb^2}{k\delta L}\right)^{1/2}$$

$$= \left[\frac{(62)(2)(1 + 4.0 \times 10^{-3})(0.04)^2}{(50)(4.0 \times 10^{-3})(1)}\right]^{1/2}$$

$$= 1$$

and

$$\tau = \frac{\alpha t}{b^2} = \frac{(5.33 \times 10^{-6})(30)}{(0.04)^2} = 0.10$$

The integral $\int_0^1 \theta\,dX$ represents the area under one of the curves in Fig. 16.5 and can be evaluated numerically. Using the trapezoidal rule, this area for $\tau = 0.1$ and $N = 1.0$ is found to be 0.112. Thus

$$q_{\text{conv}} = N^2 q_b \int_0^1 \theta\,dX$$

$$= (1)^2(100)(0.112) = 11.2\,\text{W}$$

The energy storage rate after a time interval of 30 s is the difference between the base heat flow and the convective heat loss. Hence

$$q_{\text{stored}} = q_b - q_{\text{conv}} = 100 - 11.2 = 88.8\,\text{W}$$

(b) For $t = 5$ min or 300 s, $\tau = 1$. The integral $\int_0^1 \theta\,dX$ for $N = 1$ and $\tau = 1$ is obtained from Fig. 16.5 as 0.659. Thus

$$q_{\text{conv}} = N^2 q \int_0^1 \theta\,dX$$

$$= (1)^2(100)(0.659) = 65.9\,\text{W}$$

and

$$q_{\text{stored}} = q_b - q_{\text{conv}} = 100 - 65.9 = 34.1\,\text{W}$$

These calculations reveal that during the early part of the transient, the bulk of the energy entering through the base is stored in the fin. At $t = 30$ s, 88.8% of the energy entering the base is stored and only 11.2% is dissipated from the fin faces. However, at $t = 300$ s, only 34.1% is stored and 65.9% is dissipated to the environment.

16.2.3 Step Change in Base Fluid Temperature

In this case it is assumed that the base fluid, the fin, and the environment are all initially at the same temperature, T_s. At $t \geq 0$, the temperature of the fluid flowing inside the base structure is suddenly elevated to T_f, and the fluid heats the base by convection under heat transfer coefficient h_f. The equation

$$\frac{\partial^2 T}{\partial x^2} - \frac{hP}{k\delta L}(T - T_s) = \frac{1}{\alpha}\frac{\partial T}{\partial t} \tag{16.1}$$

remains valid, but one of the boundary conditions and the initial condition change. Equation (16.1) is to be solved subject to

$$\frac{\partial T}{\partial x}(0, t) = 0 \tag{16.43a}$$

$$k\delta L\frac{\partial T}{\partial x}(b, t) = h_f\delta L[T_f - T(b, t)] \tag{16.43b}$$

and

$$T(x, 0) = T_s \tag{16.43c}$$

The Biot number is defined as

$$\text{Bi} \equiv \frac{h_f b}{k} \tag{16.44}$$

and with the rest of the dimensionless variables as defined by eqs. (16.3), the mathematical model will be

$$\frac{\partial^2 \theta}{\partial X^2} - N^2\theta = \frac{\partial \theta}{\partial \tau} \tag{16.45}$$

and the solution will be subject to the boundary conditions

$$\frac{\partial \theta}{\partial X}(1, \tau) = \text{Bi}(1 - \theta) \tag{16.46a}$$

$$\frac{\partial \theta}{\partial X}(0, \tau) = 0 \tag{16.46b}$$

and the initial condition

$$\theta(X, 0) = 0 \tag{16.46c}$$

Assuming a solution for θ in the form of eq. (16.6), the equation governing θ_1 can be obtained as

$$\frac{d^2\theta_1}{dX^2} - N^2\theta_1 = 0 \tag{16.47}$$

with

$$\frac{d\theta_1}{dX}(1) = \text{Bi}[1 - \theta_1(1)] \tag{16.48a}$$

and

$$\frac{d\theta_1}{dX}(0) = 0 \tag{16.48b}$$

and for θ_2,

$$\frac{\partial^2\theta_2}{\partial X^2} - N^2\theta_2 = \frac{\partial\theta_2}{\partial\tau} \tag{16.49}$$

with

$$\frac{\partial\theta_2}{\partial X}(1,\tau) = -\text{Bi}\theta_2(1,\tau) \tag{16.50a}$$

$$\frac{\partial\theta_2}{\partial X}(0,\tau) = 0 \tag{16.50b}$$

and

$$\theta_2(X,0) = -\theta_1(X) \tag{16.50c}$$

The solution for θ_1 is

$$\theta_1 = \frac{\text{Bi}\cosh NX}{N\sinh N + \text{Bi}\cosh N} \tag{16.51}$$

and in using the method of separation of variables, the solution for θ_2 emerges as

$$\theta_2 = -2\text{Bi}\sum_{n=1}^{\infty} \frac{\lambda_n \cos\lambda_n e^{-(N^2+\lambda_n^2)\tau}\cos\lambda_n X}{(\lambda_n + \sin\lambda_n\cos\lambda_n)(N^2+\lambda_n^2)} \tag{16.52}$$

where the λ_n are given by

$$\lambda_n \tan\lambda_n = \text{Bi}$$

Using the form of eq. (16.13), the base heat flow q_b can be obtained as

$$\frac{q_b b}{k\delta L(T_f - T_s)} = \left.\frac{\partial\theta}{\partial X}\right|_{X=1}$$

or

$$\frac{q_b b}{k\delta L(T_f - T_s)} = \mathrm{Bi}\frac{N \sinh N}{N \sinh N + \mathrm{Bi} \cosh N}$$

$$+ 2\mathrm{Bi} \sum_{n=1}^{\infty} \frac{\lambda_n^2 \sin \lambda_n \cos \lambda_n e^{-(N^2+\lambda_n^2)\tau}}{(\lambda_n + \sin \lambda_n \cos \lambda_n)(N^2 + \lambda_n^2)} \tag{16.53}$$

As shown by Suryanarayana (1976), this problem can also be solved using the Laplace transformation, which gives exactly the same result. However, with the Laplace transform approach, one can derive solutions that are rapidly convergent for small τ. Suryanarayana (1976) reports such solutions as

$$\theta = \frac{1 - e^{-N^2\tau}}{N^2\tau} \mathrm{Bi} \sum_{n=1}^{4} \left[2\left(\frac{\tau}{\pi}\right)^{1/2} e^{-f_n^2(X)4\tau} + f_n(X)\,\mathrm{erfc}\frac{f_n(X)}{2\sqrt{\tau}} \right]$$

$$- \frac{\mathrm{Bi}^2}{N^4\tau}\left[1 - (1 + N^2\tau)e^{-N^2\tau}\right] \sum_{n=1}^{6} a_n\, \mathrm{erfc}\frac{f_n(X)}{2\sqrt{\tau}} \tag{16.54}$$

and

$$\frac{q_b b}{k\delta L(T_f - T_s)} = \frac{1 - e^{-N^2\tau}}{N^2\tau}\mathrm{Bi}\left(1 - \mathrm{erfc}\frac{2}{\sqrt{\tau}}\right)$$

$$- \frac{\mathrm{Bi}^2}{N^4\tau\sqrt{\pi\tau}}[1 - (1 + N^2\tau)e^{-N^2\tau}](1 + 2e^{-1/\tau} - e^{-4/\tau} - 2e^{-9/\tau}) \tag{16.55}$$

where in eq. (16.54),

$$f_n(X) = (1 - X), \quad (1 + X), \quad (3 - X), \quad (3 + X), \quad (5 - X) \quad \text{and} \quad (5 + X)$$

and for $n = 1, 2, 3, 4, 5$, and 6 in that order,

$$a_n = 1, \quad 1, \quad 3, \quad 3, \quad 2 \quad \text{and} \quad 2$$

Figure 16.6 shows the base heat flow as a function of N and Bi. These results were obtained from eqs. (16.53) and (16.55). It can be observed that for $\tau < 0.001$, the dimensionless base heat flow is equal to the Biot number. This can be seen from the convective boundary condition of eq. (16.46*a*). For $\tau << 1, \theta$ is small and $(\partial\theta/\partial X)(1, \tau)$ (which equals the dimensionless heat flow) nearly equals Bi. As in the preceding two cases, the early part of the transient is hardly affected by N because the bulk of the thermal energy entering the base is stored in the fin and little is dissipated from the fin faces.

In this case, the time τ_s required for the temperature at any point to be within 1% of its steady-state value is given by

$$\tau_s = \frac{1}{\Psi} \ln \frac{200\lambda_1 \cos \lambda_1 (N \sinh N + \mathrm{Bi} \cosh N)}{\Psi(\lambda_1 + \sin \lambda_1 \cos \lambda_1)} \tag{16.56}$$

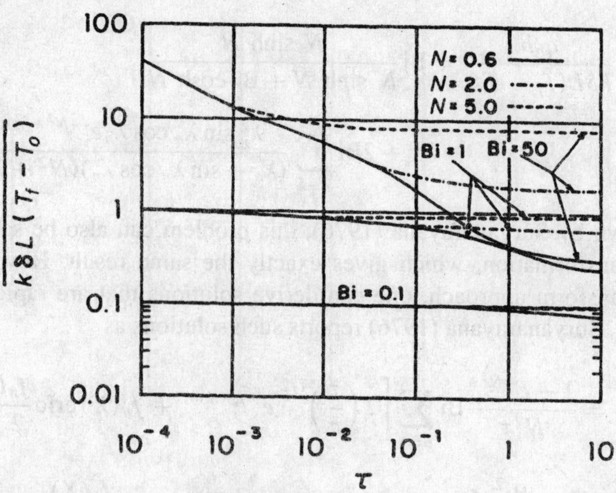

Figure 16.6 Base heat flow in a longitudinal fin of rectangular profile: step change in base fluid temperature. [From Suryanarayana (1976).]

where λ_1 is the first root of $\lambda_n \tan \lambda_n = \text{Bi}$ and $\Psi = N^2 + \lambda_1^2$. Figure 16.7 shows τ_s as a function of Bi and N. For a given N, τ_s decreases as Bi increases, indicating that as base convection becomes stronger, the response time becomes shorter. As $\text{Bi} \longrightarrow \infty$, the condition of a step change in base temperature is approached and τ_s values approach those given in Table 16.1. In fact, for $\text{Bi}/N > 50$, the response of the fin to a step change in base fluid temperature is essentially the same as that due to a step change in base temperature.

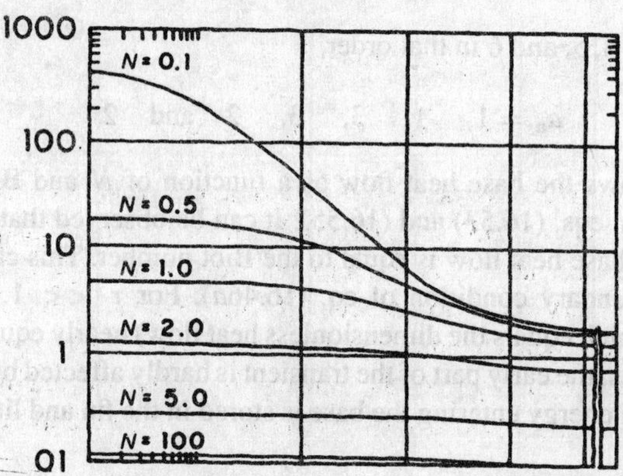

Figure 16.7 Time to attain steady state for longitudinal fin of rectangular profile: step change in base fluid temperature. [From Suryanarayana (1976).]

Example 16.3. Consider the fin in Example 16.1. The prime surface and the base of the fin are heated by convection with a heat transfer coefficient of 61 W/m²·K. If the base fluid temperature is suddenly elevated to 116°C, **(a)** determine the base heat flow rate after a lapse of 30 s, **(b)** the time elapsed before the response attains a value within 1% of its new steady state, and **(c)** the new steady-state base heat flow

SOLUTION. **(a)** From the solution of Example 16.1,

$$N = 2 \quad \text{and} \quad \tau = 0.1$$

The Biot number is

$$\text{Bi} = \frac{h_f b}{k} = \frac{(61)(0.051)}{31} = 0.10$$

For $N = 2$, $\tau = 0.1$ and Bi $= 0.1$, Fig. 16.6 gives

$$\frac{q_b b}{k\delta L(T_f - T_s)} = 0.10$$

or

$$q_b = \frac{0.10 k\delta L(T_f - T_s)}{b}$$

$$= \frac{(0.1)(31)(4.75 \times 10^{-3})(1)(116 - 10)}{0.051}$$

$$= 30.6 \text{ W}$$

(b) For $N = 2$ and Bi $= 0.10$, Fig. 16.7 gives

$$\tau_s = \frac{\alpha t}{b^2} = 1.3$$

or

$$t = \frac{1.3 b^2}{\alpha}$$

$$= \frac{(1.3)(0.051)^2}{8.6 \times 10^{-6}}$$

$$= 393 \text{ s} \quad \text{or} \quad 6.55 \text{ min}$$

(c) The steady-state base heat flow is given by the first term of eq. (16.53):

$$q_b = \frac{k\delta L(T_f - T_s)\text{Bi}N \sinh N}{b(N \sinh N + \text{Bi} \cosh N)}$$

$$= \frac{(31)(4.75 \times 10^{-3})(1)(116 - 10)(0.10)(2) \sinh 2}{(0.051)(2 \sinh 2 + 0.10 \cosh 2)}$$

$$= 29.1 \text{ W}$$

A comparison of the instantaneous base flow q_b at $t = 30$ s with the corresponding value of Example 16.1 shows that q_b is considerably reduced (30.6 versus 912). This, of course, is due to the high convection resistance at the base due to a low value of Bi. One also notes that with convective heating at the base, the response time is increased by about 60% [(393 − 246)100/246] compared to the value for a step change in base temperature.

16.2.4 Step Change in Environmental Temperature

Suppose that the fin and its convective environment is initially at a temperature T_b. At time $t \geq 0$, the environmental temperature is suddenly decreased from T_b to T_s while the temperature at the base of the fin is maintained at the initial temperature T_b. The equation to be solved is

$$\frac{\partial^2 \theta}{\partial X^2} - N^2 \theta = \frac{\partial \theta}{\partial \tau} \tag{16.4}$$

and with the boundary and initial conditions of eqs. (16.5),

$$\theta(1, \tau) = 1 \tag{16.5a}$$

$$\frac{\partial \theta}{\partial X}(0, \tau) = 0 \tag{16.5b}$$

$$\theta(X, 0) = \theta_i = \frac{T_i - T_s}{T_b - T_s} = 1 \tag{16.5c}$$

Thus the solution for θ is given by eqs. (16.6), (16.11), and (16.12), except that in eq. (16.12), $\theta_i = 1$. Similarly, eq. (16.14) with $\theta_i = 1$ gives the base heat flow.

The analyses in Sections 16.2.1, 16.2.2, and 16.2.4 have been extended by Mao and Rook (1993) to include the effects of tip heat loss. This work used the Laplace transformation to derive the solutions and it was concluded that errors due to the neglect of the tip heat loss become significant only when $h\delta/2k$ and b/δ are small.

16.3 RADIAL CONVECTING FIN

In this section, the transient response of a radial fin of rectangular profile, which is also known as an annular, circular, or circumferential fin is considered. The analyses presented assume that convection is the only mode of surface heat dissipation and that the fin experiences a step change in its base temperature or a step change in the temperature of the environment.

16.3.1 Step Change in Base Temperature

An exact analytical solution for this case was presented by Chapman (1959), who considered a radial fin of thickness δ, base radius r_b, and tip radius r_a, as shown in Fig. 16.8. The fin is initially assumed to be at the same temperature, T_s, as its environment. At time $t = 0$, the base temperature is suddenly raised to T_b ($T_b > T_s$) and remains at T_b for all $t \geq 0$. The fin dissipates heat by convection to the environment through a heat transfer coefficient h and during the entire transient, the tip of the fin is assumed to be insulated.

The differential equation governing the transient response is

$$\frac{\partial^2 T}{\partial r^2} + \frac{1}{r}\frac{\partial T}{\partial r} - \frac{2h}{k\delta}(T - T_s) = \frac{1}{\alpha}\frac{\partial T}{\partial t} \tag{16.57}$$

and this equation is to be solved subject to the boundary conditions

$$T(r_b, t) = T_b \tag{16.58a}$$

and

$$\frac{\partial T}{\partial r}(r_a, t) = 0 \tag{16.58b}$$

and the initial condition

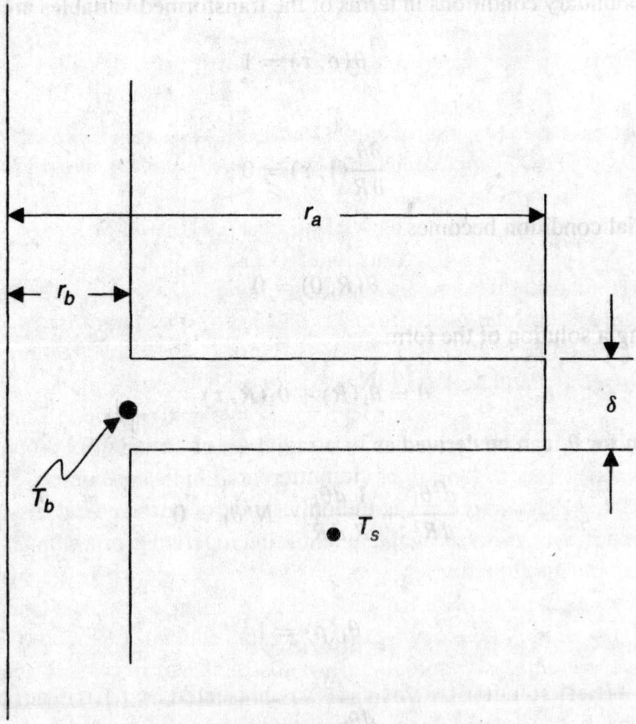

Figure 16.8 Radial fin of rectangular profile.

$$T(r, 0) = T_s \tag{16.58c}$$

Use of the dimensionless variables

$$\theta = \frac{T - T_s}{T_b - T_s} \tag{16.59a}$$

$$R = \frac{r}{r_a} \tag{16.59b}$$

$$\rho = \frac{r_b}{r_a} \tag{16.59c}$$

$$\tau = \frac{\alpha t}{r_a^2} \tag{16.59d}$$

$$N^2 = \frac{2hr_a^2}{k\delta} \tag{16.59e}$$

puts the equation into the form

$$\frac{\partial^2 \theta}{\partial R^2} + \frac{1}{R}\frac{\partial \theta}{\partial R} - N^2\theta = \frac{\partial \theta}{\partial \tau} \tag{16.60}$$

where the boundary conditions in terms of the transformed variables are

$$\theta(\rho, \tau) = 1 \tag{16.61a}$$

and

$$\frac{\partial \theta}{\partial R}(1, \tau) = 0 \tag{16.61b}$$

and the initial condition becomes

$$\theta(R, 0) = 0 \tag{16.61c}$$

Assuming a solution of the form

$$\theta = \theta_1(R) + \theta_2(R, \tau) \tag{16.62}$$

the equation for θ_1 can be derived as

$$\frac{d^2\theta_1}{dR^2} + \frac{1}{R}\frac{d\theta_1}{dR} - N^2\theta_1 = 0 \tag{16.63}$$

with

$$\theta_1(\rho) = 1 \tag{16.64a}$$

and

$$\frac{d\theta_1}{dR}(1) = 0 \tag{16.64b}$$

and for θ_2

$$\frac{\partial^2\theta_2}{\partial R^2} + \frac{1}{R}\frac{\partial\theta_2}{\partial R} - N^2\theta_2 = \frac{\partial\theta_2}{\partial\tau} \tag{16.65}$$

with

$$\theta_2(\rho, \tau) = 0 \tag{16.66a}$$

$$\frac{\partial\theta_2}{\partial R}(1, \tau) = 0 \tag{16.66b}$$

and

$$\theta_2(R, 0) = -\theta_1(R) \tag{16.66c}$$

The solution for θ_1 is

$$\theta_1 = \frac{K_1(N)I_0(NR) + I_1(N)K_0(NR)}{K_1(N)I_0(N\rho) + I_1(N)K_0(N\rho)} \tag{16.67}$$

where the I's and K's are modified Bessel functions.

The solution for θ_2 can be obtained by the method of separation of variables as

$$\theta_2 = \sum_{n=1}^{\infty} \frac{\pi\lambda_n^2[J_0(\lambda_n R)Y_0(\lambda_n\rho) - Y_0(\lambda_n R)J_0(\lambda_n\rho)]}{(\lambda_n^2 + N^2)\left\{[J_0(\lambda_n\rho)/J_1(\lambda_n\rho)]^2 - 1\right\}} e^{-(N^2+\lambda_n^2)\tau} \tag{16.68}$$

where the eigenvalues λ_n are determined from

$$J_1(\lambda_n)Y_0(\lambda_n\rho) - Y_1(\lambda_n)J_0(\lambda_n\rho) = 0 \tag{16.69}$$

and where the J's and Y's are Bessel functions.

The instantaneous heat flow q through the base of the fin is given by

$$q_b = -2\pi r_b k\delta \left.\frac{\partial T}{\partial r}\right|_{r=r_b} \tag{16.70}$$

or in dimensionless form as

$$\frac{q_b}{2\pi k\delta(T_b - T_s)} = -\rho \left.\frac{\partial\theta}{\partial R}\right|_{R=\rho} \tag{16.71}$$

The solution for θ is used to evaluate the derivative $\partial\theta/\partial R|_{R=\rho}$, so that eq. (16.71) becomes

$$\frac{q_b}{2\pi k\delta(T_b - T_s)} = \gamma \frac{I_1(N)K_1(\gamma) - K_1(N)I_1(\gamma)}{I_0(\gamma)K_1(N) + I_1(N)K_0(\gamma)} + \sum_{n=1}^{\infty} \frac{\pi\lambda_n^2[J_0(\zeta)Y_0(\zeta) - Y_0(\zeta)J_0(\zeta)]}{\phi\left\{[J_0(\zeta)/J_1(\zeta)]^2 - 1\right\}} e^{-\phi\tau} \tag{16.72}$$

where

$$\gamma = \rho N, \quad \zeta = \lambda_n \rho, \quad \text{and} \quad \phi = N^2 + \lambda_n^2$$

Expressions for the tip temperature T_a, the cumulative heat flow from the base, $Q = \int_0^t q_b\, dt$, and the cumulative dissipation to the environment, Q_s can be derived using the results for θ_1 given by eq. (16.67) plus θ_2 given by eq. (16.68) and q_b obtained from eq. (16.72). These relationships have been graphed and are presented in charts taken from Chapman (1959) in Figs. 16.9 through 16.14. The charts employ six parameters—F_1, F_2, F_3, F_4, G_1, and G_2—and are displayed for $\rho = 0.5$. The relationships for T_a, q_b, Q, and Q_s are

$$\frac{T_a - T_s}{T_b - T_s} = G_1 - F_1 \tag{16.73}$$

$$\frac{q_b}{2\pi k\delta(T_b - T_s)} = N^2 G_2 + F_3 \tag{16.74}$$

$$\frac{Q\alpha}{2\pi k\delta r_a^2(T_b - T_s)} = N^2 G_2 \tau + F_4 \tag{16.75}$$

$$\frac{Q_s\alpha}{2\pi k\delta r_a^2(T_b - T_s)} = (N^2\tau - 1)G_2 + F_2 + F_4 \tag{16.76}$$

Example 16.4. A radial fin of uniform thickness (rectangular profile) has a base radius of 76 mm and a tip radius of 152 mm. The fin is 5 mm thick and is made of a material with $k = 65$ W/m·K and $\alpha = 6.45 \times 10^{-6}$ m²/s. The fin is initially in thermal equilibrium with its environment, which is at 20°C. The base of the fin is suddenly elevated to a temperature of 120°C, while the environment temperature

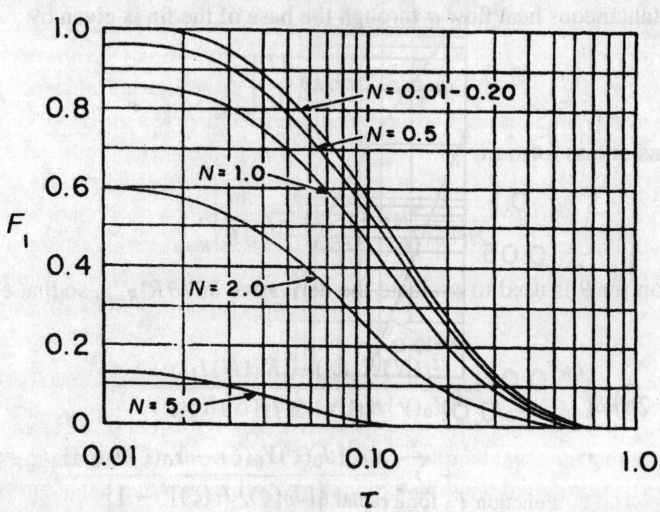

Figure 16.9 Function F_1 for a radial fin with $\rho = 0.5$. [From Chapman (1959).

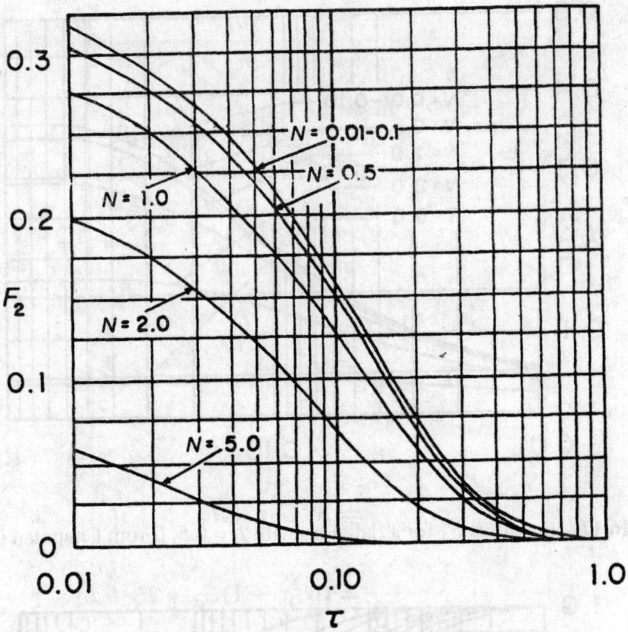

Figure 16.10 Function F_2 for a radial fin with $\rho = 0.5$. [From Chapman (1959).]

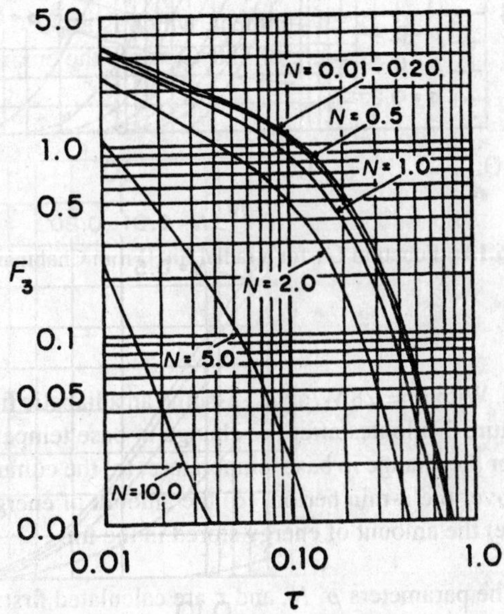

Figure 16.11 Function F_3 for a radial fin with $\rho = 0.5$. [From Chapman (1959).]

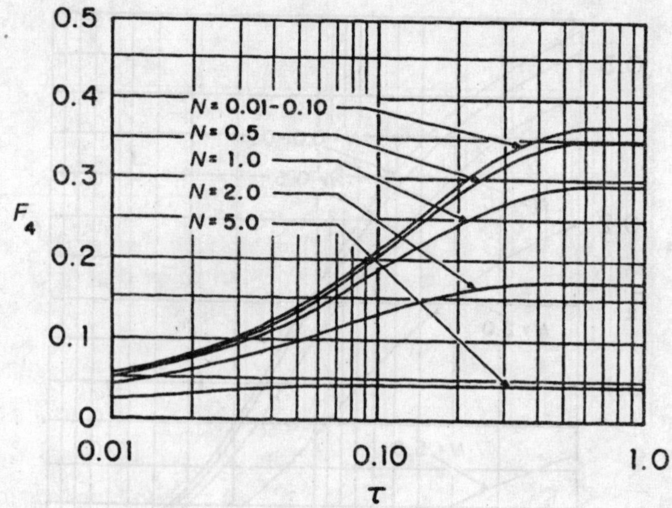

Figure 16.12 Function F_4 for a radial fin with $\rho = 0.5$. [From Chapman (1959).]

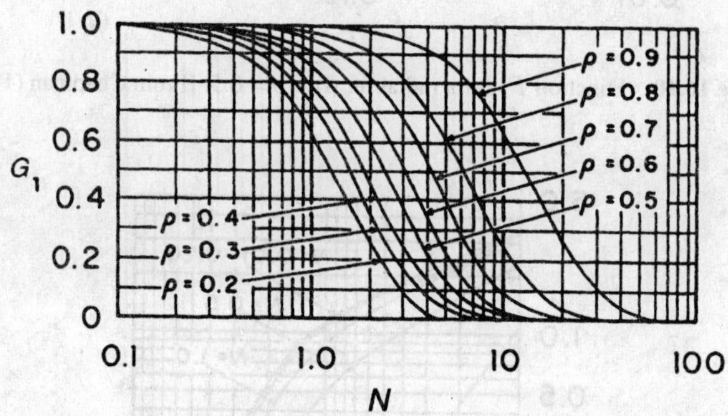

Figure 16.13 Function G_1 for a radial fin. [From Chapman (1959).]

remains unchanged. With $h = 28$ W/m^2·K, assume an adiabatic fin tip and determine **(a)** the tip temperature 3 minutes after the change in base temperature, **(b)** the base heat flow 3 min after the change in base temperature, **(c)** the cumulative heat through the base of the fin over the 3-min period, **(d)** the amount of energy dissipated to the environment, and **(e)** the amount of energy stored in the fin.

SOLUTION. **(a)** The parameters ρ, N, and τ are calculated first:

$$\rho = \frac{r_b}{r_a} = \frac{76}{152} = 0.50$$

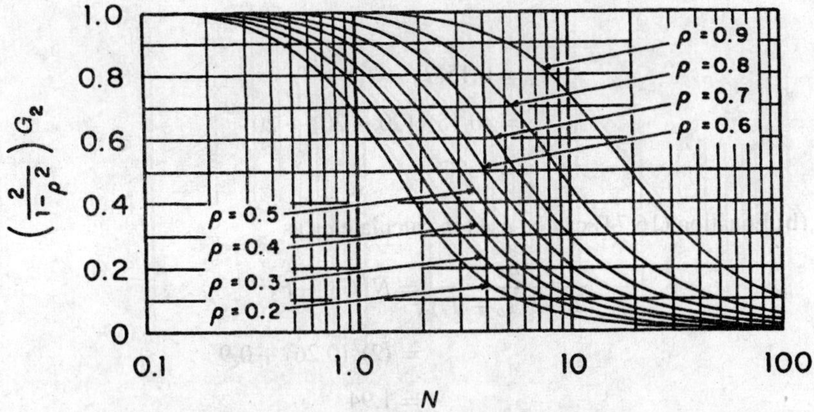

Figure 16.14 Function $2/(1-\rho^2)G_2$ for a radial fin. [From Chapman (1959).]

$$N = \left(\frac{2hr_a^2}{k\delta}\right)^{1/2} = \left[\frac{(2)(28)(0.152)^2}{(65)(5\times10^{-3})}\right]^{1/2} = 2.00$$

and

$$\tau = \frac{\alpha t}{r_a^2} = \frac{(6.45\times10^{-6})(180)}{(0.152)^2} = 0.05$$

Read Figs. 16.9 through 16.13 to obtain

$$F_1 = 0.45 \qquad F_2 = 0.125$$

$$F_3 = 0.90 \qquad F_4 = 0.10$$

$$G_1 = 0.60$$

and then obtain G_2 from Fig. 16.14,

$$\frac{2}{1-\rho^2}G_2 = 0.69$$

or

$$G_2 = \frac{(0.69)[1-(0.5)^2]}{2} = 0.26$$

Use the values of F_1 and G_1 in eq. (16.73) to calculate the tip temperature:

$$\frac{T_a - T_s}{T_b - T_s} = G_1 - F_1$$

$$= 0.6 - 0.45$$

$$= 0.15$$

or

$$T_a = 0.15(T_b - T_s) + T_s$$
$$= (0.15)(120 - 20) + 20$$
$$= 35°\text{C}$$

(b) Equation (16.74) can be used to calculate q_b as

$$\frac{q_b}{2\pi k\delta(T_b - T_s)} = N^2 G_2 + F_3$$
$$= (2)^2(0.26) + 0.9$$
$$= 1.94$$

so that

$$q_b = (1.94)2\pi k\delta(T_b - T_s)$$
$$= (1.94)(2\pi)(65)(5 \times 10^{-3})(120 - 20)$$
$$= 396 \text{ W}$$

(c) Using eq. (16.75) gives

$$\frac{Q\alpha}{2\pi k\delta r_a^2(T_b - T_s)} = N^2 G_2 \tau + F_4$$
$$= (2)^2(0.26)(0.05) + 0.10$$
$$= 0.152$$

and then

$$Q = \frac{(0.152)2\pi k\delta r_a^2(T_b - T_s)}{\alpha}$$
$$= \frac{(0.152)(2\pi)(65)(5 \times 10^{-3})(0.152)^2(120 - 20)}{6.45 \times 10^{-6}}$$
$$= 111,182 \text{ J} \quad \text{or} \quad 111.2 \text{ kJ}$$

(d) From eq. (16.76),

$$\frac{Q_s\alpha}{2\pi k\delta r_a^2(T_b - T_s)} = (N^2\tau - 1)G_2 + F_2 + F_4$$
$$= [(2)^2(0.05) - 1](0.26) + 0.125 + 0.10$$
$$= 0.017$$

or

$$Q_s = \frac{(0.017)2\pi k\delta r_a^2(T_b - T_s)}{\alpha}$$

$$= \frac{(0.017)(2)(\pi)(65)(5 \times 10^{-3})(0.152)^2(120 - 20)}{6.45 \times 10^{-6}}$$

$$= 12,435 \text{ J} \quad \text{or} \quad 12.4 \text{ kJ}$$

(e)

$$Q_{\text{stored}} = Q - Q_s$$

$$= 111,182 - 12,435$$

$$= 98,747 \text{ J} \quad \text{or} \quad 98.7 \text{ kJ}$$

Thus during the first 3 min of the transient, about 11% of the energy is dissipated to the environment and the remaining 89% is stored in the fin.

16.3.2 Step Change in Environment Temperature

This case has been studied by Papadopoulos et al. (1990), who considered a radial fin, initially operating in a steady-state mode, with a base temperature of T_b and an environment temperature of T_{s1}. At time $t = 0$, the temperature of the environment was allowed to change suddenly from T_{s1} to T_{s2} while the base temperature T_b remained unchanged. Using the Laplace transform approach, an analytical solution similar to the one discussed in Section 16.3.1 was developed. Figure 16.15 provides the limited results for $N = 2$ and $\rho = 0.5$ and is a plot of $(T - T_b)/(T_{s2} - T_{s1})$ as a function of $R = r/r_b$.

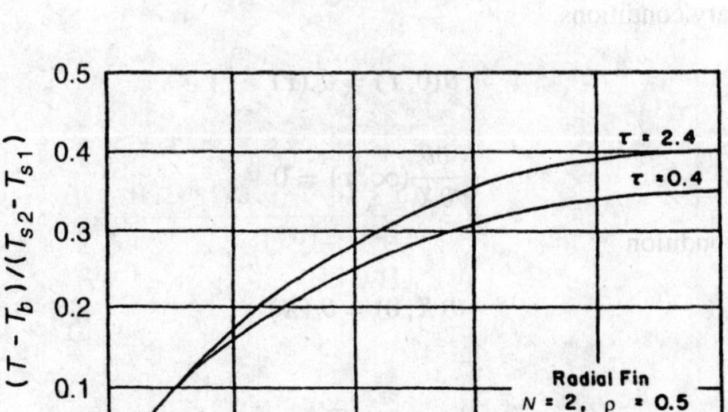

Figure 16.15 Transient temperature profiles in a radial fin: step change in environmental temperature. [Adapted from Papadopoulos et al. (1990).]

16.4 LONGITUDINAL FINS WITH POWER LAW DISSIPATION

In the previous sections, the focus has been on fins with pure convection as the surface heat dissipation mechanism. The partial differential equations were linear and therefore amenable to solution techniques such as the separation of variables and the Laplace transformation. In many applications, the surface cooling is proportional to a power of the temperature difference, $\Delta T = T - T_s$, other than unity, that is, $q = K(T - T_s)^n$, where $n \neq 1$. For example, the exponent n takes on values of 0.75, 1.25, 1.33, 3, and 4 when the cooling is due to film boiling, laminar natural convection, turbulent natural convection, nucleate boiling, and radiation to free space, respectively (Aziz and Benzies, 1976). The presence of this nonlinear cooling term precludes an exact analytical solution, and consequently, the solutions are generated using either approximate analytical, numerical, or hybrid procedures. These solutions are considered in this section.

16.4.1 Similarity Solution

A similarity analysis has been used by Russell and Chapman (1967) to study the transient response of an infinitely long purely radiating fin. They considered a longitudinal fin of rectangular profile, initially at steady state and having a temperature distribution $T_i(x)$. At $t = 0$, the temperature of the base, T_b, is varied in some prescribed manner while the tip at $x = \infty$ remains insulated. The fin is assumed to radiate energy to the environment at a temperature of absolute zero (free space or a zero effective sink temperature) from both faces. For this situation, the transient is governed by the dimensionless differential equation

$$\frac{\partial^2\theta}{\partial X^2} - N^2\theta^4 = \frac{\partial\theta}{\partial\tau} \tag{16.77}$$

with boundary conditions

$$\theta(0, \tau) = \theta_b(\tau) \tag{16.78a}$$

$$\frac{\partial\theta}{\partial X}(\infty, \tau) = 0 \tag{16.78b}$$

and initial condition

$$\theta(X, 0) = \theta_i(X) \tag{16.78c}$$

where

$$\theta = \frac{T}{T_{bi}}$$

with T_{bi}, the initial temperature at the base of the fin,

$$X = \frac{x}{x_r}$$

where x_r is a convenient reference length

$$\tau = \frac{\alpha t}{x_r^2}$$

and

$$N^2 = \frac{2\epsilon_s \sigma T_{bi}^3 x_r^2}{\delta L} \tag{16.78d}$$

where ϵ_s is the surface emissivity and σ is the Stefan–Boltzmann constant (σ = 5.669×10^{-8} W/m^2·K^4) and

$$\theta_b(0) = \theta_i(0) = 1$$

is the initial condition at the base of the fin.

The main goal of the analysis is to identify the base temperature–time variations, that is, $\theta_b(\tau)$, so that the partial differential equation of eq. (16.77) can be reduced to an ordinary differential equation in a similarity variable η. Using the free-parameter method of Hansen (1964), Russell and Chapman (1967) found that if $\theta_b(\tau)$ and η are of the form

$$\theta_b(\tau) = \left(\frac{1}{1 - C_1 N^2 \tau} \right)^{1/3} \tag{16.79a}$$

and

$$\eta = \frac{XN}{(1 + C_1 \tau N^2)^{1/2}} \tag{16.79b}$$

where C_1 is an arbitrary constant, the governing equation then reduces to

$$\phi'' + \tfrac{1}{2} C_1 \eta \phi' - \tfrac{1}{3} C_1 \phi - \phi^4 = 0 \tag{16.80}$$

with

$$\phi(0) = 1 \tag{16.81a}$$

and

$$\phi'(\infty) = 0 \tag{16.81b}$$

where

$$\phi = \frac{\theta}{\theta_b(\tau)} \tag{16.81c}$$

The initial steady-state temperature distribution $\theta(X, 0) = \theta_i(X)$ in terms of ϕ is

$$\theta(X, 0) = \theta_i(X) = \theta_b(0)\phi(\eta)\Big|_{\tau=0} = \phi(\eta)\Big|_{\tau=0} \tag{16.82}$$

and once eq. (16.80) has been solved numerically, the initial temperature distribution can be obtained from eq. (16.82). It is interesting to note that the solution for ϕ also represents the initial temperature distribution if η is interpreted as NX because for $\tau = 0$, $\eta = NX$ [see eq. (16.79*b*)].

Figure 16.16*a* shows the solutions of eq. (16.80) for $C_1 = 0$, 0.1, 1, and 10. The corresponding base temperature variations described by eq. (16.79*a*) are shown in

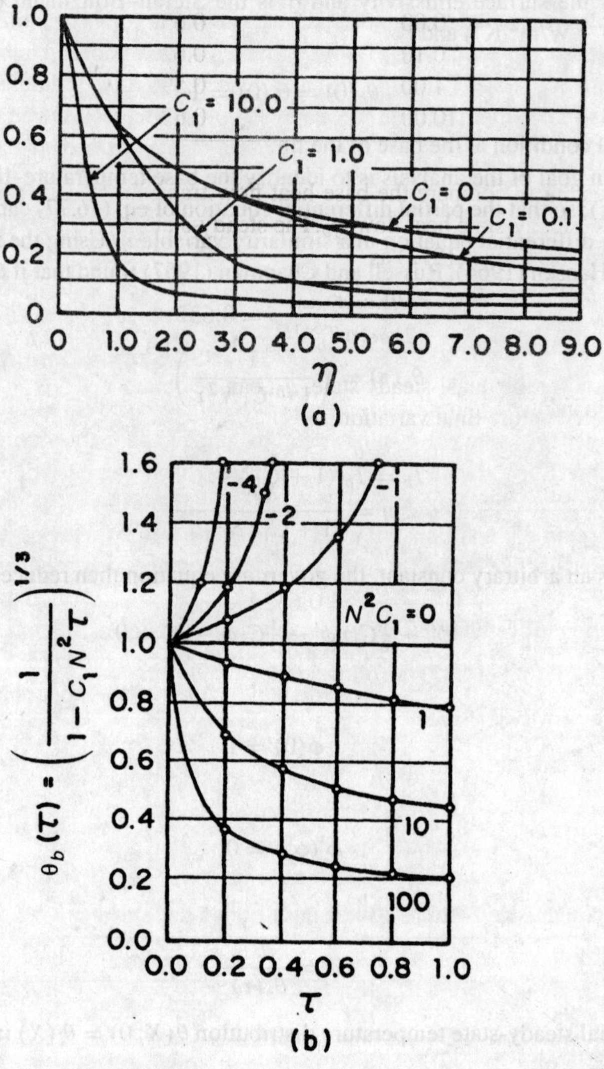

Figure 16.16 (*a*) Similarity solutions for infinitely long longitudinal radiating fin; (*b*) admissible base temperature variations. [From Russell and Chapman (1967).]

Fig. 16.16*b*. For negative values of C_1, Russell and Chapman (1967) could not obtain the numerical solutions of eqs. (16.80) and (16.81).

The base heat flux per unit depth q_b can be found in terms of $\phi'(0)$ as

$$\frac{q_{b,\text{transient}}}{(2k\delta\epsilon_s\sigma T_{bi}^5)^{1/2}} = -\frac{\phi'(0)}{(1 + C_1 N^2\tau)^{5/6}} \tag{16.83}$$

where the values of $\phi'(0)$ for different values of C_1 are

C_1	$-\phi'(0)$
0.00	0.632
0.10	0.620
1.00	0.590
10.00	0.630

It is interesting to compare the base heat flow transient q_b given by eq. (16.83) with the quasi-steady-state heat flow, q_b. The steady-state base heat flux for the fin is given by Kern and Kraus (1972) as

$$\frac{q_{b,\text{steady state}}}{2k\delta\epsilon_s\sigma T_b^5)^{1/2}} = 0.632$$

The expression for the quasi-steady state, q_b, can be obtained by replacing T_b by the actual base temperature time variation,

$$T_b = T_{bi}(1 + C_1 N^2\tau)^{1/3} \tag{16.84}$$

giving

$$\frac{q_{b,\text{quasi-steady state}}}{(2k\delta\epsilon_s\sigma T_{bi}^5)^{1/2}} = 0.632\left(\frac{1}{1 + C_1 N^2\tau}\right)^{5/6} \tag{16.85}$$

From eqs. (16.83) and (16.85), it follows that

$$\frac{q_{b,\text{transient}}}{q_{b,\text{quasi-steady state}}} = -\frac{\phi'(0)}{0.632} \tag{16.86}$$

An examination of the values of $\phi'(0)$ shows that the ratio of the heat fluxes is nearly unity. Thus for all practical purposes, the quasi-steady-state solution of eq. (16.85) can be employed to obtain the transient base heat flux for situations that meet the requirements of the similarity solutions.

16.4.2 Coordinate Perturbation Solution

The coordinate perturbation expansion method has been used by Aziz and Na (1984) to analyze the transient response of an infinitely long fin with a power law type of surface heat dissipation. They considered a longitudinal fin of uniform thickness δ

initially in equilibrium with the environment at temperature T_s. At $t = 0$, the base experiences a step increase in temperature from T_s to T_b ($T_b > T_s$). The surface heat dissipation is assumed to be proportional to $(T - T_s)^n$. The equations governing the transient appear in dimensionless form as

$$\frac{\partial^2 \theta}{\partial X^2} - N^2 \theta^n = \frac{\partial \theta}{\partial \tau} \tag{16.87}$$

where the boundary conditions are

$$\theta(0, \tau) = 1 \tag{16.88a}$$

$$\theta(\infty, \tau) = 0 \tag{16.88b}$$

and the initial condition is

$$\theta(X, 0) = 0 \tag{16.88c}$$

where $X = x/x_r$, $\tau = \alpha t/x_r^2$, and θ and N are defined appropriately in accordance with the surface heat transfer mechanism. For example, with a convecting fin ($n = 1$),

$$\theta = \frac{T - T_s}{T_b - T_s}$$

and for unit fin length ($L = 1$)

$$N^2 = 2hx_r^2/k\delta$$

whereas for a radiating fin operating in a zero-sink temperature environment,

$$\theta = \frac{T}{T_b}$$

and

$$N^2 = 2\epsilon_s \sigma T_b^3 x_r^2 / k\delta$$

A perturbation series in ϵ,

$$\theta = \theta_0(\eta) + \epsilon\theta_1(\eta) + \epsilon^2\theta_2(\eta) + \cdots \tag{16.89}$$

with

$$\eta = \frac{X}{2\sqrt{\tau}} \quad \text{and} \quad \epsilon = 4N^2\tau$$

may be assumed and substitution of η and ϵ into eqs. (16.87) and (16.88) shows that the equations governing θ_0, θ_1, and θ_2 are

$$\epsilon^0: \quad \theta_0'' + 2\eta\theta_0' = 0 \tag{16.90}$$

$$\eta = 0, \ \theta_0 = 1; \quad \eta = \infty, \ \theta_0 = 0 \tag{16.91}$$

$$\epsilon^1: \quad \theta_1'' + 2\eta\theta_1' - 4\theta_1 = \theta_0^n \tag{16.92}$$

$$\eta = 0, \quad \theta_1 = 0; \quad \eta = \infty, \quad \theta_1 = 0 \tag{16.93}$$

$$\epsilon^2: \quad \theta_2'' + 2\eta\theta_2' - 8\theta_2 = n\theta_0^{n-1}\theta_1 \tag{16.94}$$

$$\eta = 0, \quad \theta_2 = 0; \quad \eta = \infty, \quad \theta_2 = 0 \tag{16.95}$$

For $n = 1$ (pure convection), eqs. (16.90) through (16.95) can be solved analytically. The solution is

$$\theta = \text{erfc}\,\eta + \epsilon\left(I^2\,\text{erfc}\,\eta - \tfrac{1}{4}\text{erfc}\,\eta\right) + \epsilon^2\left(I^4\,\text{erfc}\,\eta - \tfrac{1}{4}I^2\text{erfc}\,\eta + \tfrac{1}{32}\text{erfc}\,\eta\right) \tag{16.96}$$

where $I^n\text{erfc}\ n^{\text{th}}$ is the repeated integral of the error function. The base heat flux q_b can be derived from eq. (16.96) as

$$Q_b = \frac{q_b x_r}{k\delta(T_b - T_s)} = -\frac{1}{2\sqrt{\tau}}\theta'(0) = \frac{1}{\sqrt{\pi\tau}}\left(1 + \frac{1}{4}\epsilon - \frac{1}{96}\epsilon^2\right) \tag{16.97}$$

It is interesting to note that as Aziz and Na (1984) have shown, the perturbation solution given by eqs. (16.96) and (16.97) is identical to the exact solution of the problem, which can be obtained using the Laplace transformation.

For $n \neq 1$, the solutions for θ_1 and θ_2 must be obtained numerically. However, because the equations governing θ_1 and θ_2 are linear, the method of superposition can be used. The numerical results for θ_1 and θ_2 appear in Aziz and Na (1984). Only the results for Q_b for $n = 2$, $n = 3$, and $n = 4$ are given here.

$$n = 2: \quad Q_b\sqrt{\pi\tau} = 1 + 0.183617\epsilon - 0.010246\epsilon^2 \tag{16.98}$$

$$n = 3: \quad Q_b\sqrt{\pi\tau} = 1 + 0.147253\epsilon - 0.010076\epsilon^2 \tag{16.99}$$

$$n = 4: \quad Q_b\sqrt{\pi\tau} = 1 + 0.123757\epsilon - 0.007081\epsilon^2 \tag{16.100}$$

Equations (16.97) through (16.100) clearly show that the coefficients of the terms diminish very rapidly, and therefore the solution can be expected to be accurate for $\epsilon > 1$ despite the basic assumption of the perturbation analysis that $\epsilon << 1$. For example, for $n = 1$, when compared with the exact solution, eq. (16.97) possesses negligible error for values of ϵ up to $\epsilon = 4$. Even for $\epsilon = 10$, the error does not exceed 13%. This figure can be reduced further if one applies the Shanks transformation cited by Aziz and Na (1984) to the three-term series of eq. (16.97). The transformed result is

$$Q_b = \frac{1}{\sqrt{\pi\tau}}\,\frac{1 + 0.291667\epsilon}{1 + 0.041760\epsilon} \tag{16.101}$$

Equation (16.101) agrees to within 3% with the exact solution up to $\epsilon = 10$ and within 10% up to $\epsilon = 30$. This is a significant improvement over eq. (16.97). Application of the Shanks transformation to eqs. (16.98) through (16.100) gives

$$n = 2: \quad Q_b = \frac{1}{\sqrt{\pi\tau}} \frac{1 + 0.239418\epsilon}{1 + 0.055801\epsilon} \tag{16.102}$$

$$n = 3: \quad Q_b = \frac{1}{\sqrt{\pi\tau}} \frac{1 + 0.215679\epsilon}{1 + 0.06842\epsilon} \tag{16.103}$$

$$n = 4: \quad Q_b = \frac{1}{\sqrt{\pi\tau}} \frac{1 + 0.180976\epsilon}{1 + 0.057216\epsilon} \tag{16.104}$$

The results for Q_b from eqs. (16.101) through (16.104) are shown in Fig. 16.17 and compared with the exact solution for ($n = 1$) and the numerical solutions for $n = 2, 3$, and 4. During the early part of the transient, the exponent n does not have much effect. As mentioned earlier, this is due to the fact that initially, the energy entering the base of the fin contributes mainly to an increase in the internal energy of the fin, and the surface heat dissipation mechanism has very little influence. As

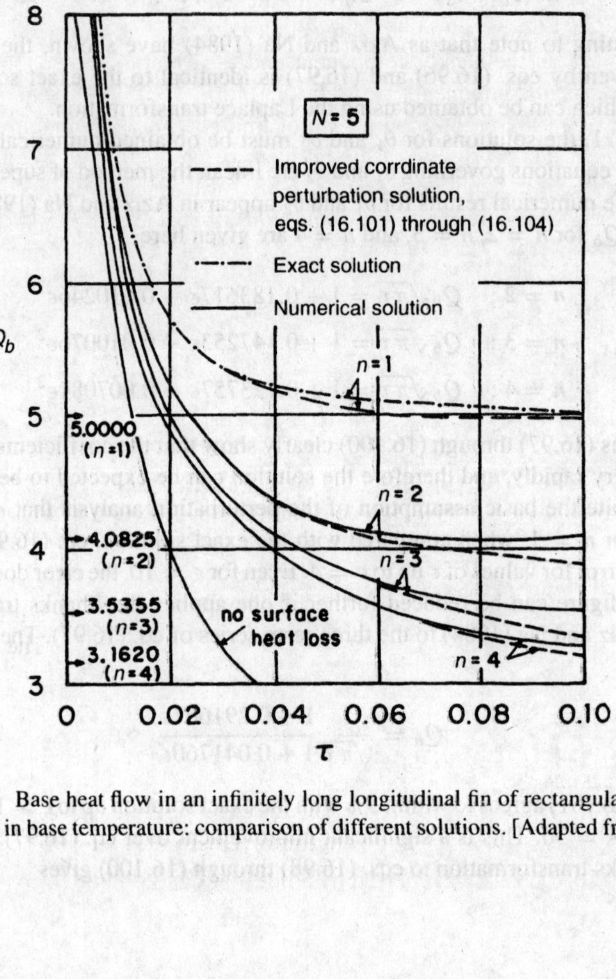

Figure 16.17 Base heat flow in an infinitely long longitudinal fin of rectangular profile due to step change in base temperature: comparison of different solutions. [Adapted from Aziz and Na (1980).]

time progresses, the influence of n becomes quite significant. In Fig. 16.17, $Q_{b,ss}$ represents the steady-state value.

Because for $N > 3$ the performance of a finite-length fin rapidly approaches that of an infinitely long fin, the restriction of infinite length may not be too severe for some applications. For example, when the fin is being employed for laminar condensation, a good fin design requires N to be of the order of 10, as shown by Lienhard and Dhir (1974). Thus the transient analysis presented here is applicable to a fin supporting laminar condensation.

16.4.3 Solution Using Optimal Linearization and Variational Embedding

The coordinate perturbation solution of the preceding section has two shortcomings. First, it is valid only for the early portion of the transient, although the application of Shanks transformation extends its accuracy to longer time frames. However, in the limit as $\tau \to \infty$, the solutions fail to approach the correct steady-state values. The second shortcoming is associated with the assumption of an infinitely long fin. To overcome these shortcomings, Chang et al. (1982b) used an optimal linearization approach to obtain the steady component and a variational embedding method to derive the transient component.

Chang et al. (1982b) considered a fin of finite height b having an adiabatic tip, with the remaining description of the problem being the same as before. Equation (16.87) with the initial condition of eq. (16.88*c*) still applies. However, because the fin tip is not at infinity, the boundary conditions of eqs. (16.88*a*) and (16.88*b*) must be adjusted to

$$\theta(1, \tau) = 1 \tag{16.105a}$$

and

$$\frac{\partial \theta}{\partial X}(0, \tau) = 0 \tag{16.105b}$$

where X is measured from the tip of the fin.

The first step in the solution is to consider the steady-state governing equation

$$\frac{d^2\theta}{dX^2} - N^2\theta^n = 0 \tag{16.106}$$

with

$$\frac{d\theta}{dX}(0) = 0 \tag{16.107a}$$

and

$$\theta(1) = 1 \tag{16.107b}$$

Equation (16.106) is next replaced by a linear equation of the form

$$\frac{d^2\theta}{dX^2} - \lambda_n\theta = 0 \tag{16.108}$$

where the parameter λ_n is the optimal value of λ, which minimizes the integral

$$I(\lambda) = \int_0^1 (\lambda\theta - N^2\theta^n)^2\, dX$$

Imposing the condition $dI/d\lambda = 0$ leads to the optimum value of $\lambda = \lambda_n$ as

$$\lambda_n = \frac{N^2 \int_0^1 \theta^{n+1}\, dX}{\int_0^1 \theta^2\, dX} \tag{16.109}$$

Of course, a solution for θ is needed to evaluate λ_n from eq. (16.109). The solution for the pure convection case, which is

$$\theta = \frac{\cosh NX}{\cosh N} \tag{16.110}$$

can be used for this purpose. Use of eq. (16.110) in eq. (16.109) leads to the value of λ_n:

$$\lambda_n = \frac{B}{A} \tag{16.111}$$

where

$$A = \frac{1}{\cosh^2 N}\left(\frac{1}{2} + \frac{\sinh 2N}{4N}\right)$$

and

$$\begin{aligned}
B &= \frac{N^2}{\cosh^2 N}\left(\frac{1}{2} + \frac{\sinh 2N}{4N}\right) && \text{for } n = 1 \\
&= \frac{N^2}{\cosh^3 N}\left(\frac{\sinh^3 N}{3N} + \frac{\sinh N}{N}\right) && \text{for } n = 2 \\
&= \frac{N^2}{\cosh^4 N}\left(\frac{\sinh 4N}{32N} + \frac{\sinh 2N}{4N} + \frac{3}{8}\right) && \text{for } n = 3 \\
&= \frac{N^2}{\cosh^5 N}\left[\frac{1}{5N}\sinh N \cosh^4 N + \frac{4}{5N}\left(\frac{\sinh^3 N}{3} + \sinh N\right)\right] && \text{for } n = 4
\end{aligned}$$

Once λ_n has been determined, the solution of eq. (16.108) subject to eqs. (16.107) is readily obtained as

$$\theta = \frac{\cosh \lambda_n^{1/2} X}{\cosh \lambda_n^{1/2}} \tag{16.112}$$

To obtain the solution of the transient equation, eq. (16.87), it is first linearized by replacing the nonlinear term $N^2\theta^n$ with $\lambda_n\theta$, giving

$$\frac{\partial^2\theta}{\partial X^2} - \lambda_n\theta = \frac{\partial\theta}{\partial\tau} \tag{16.113}$$

Next, the Lagrangian functional, $J[\theta, A]$, embedding eq. (16.113), is created to give

$$J[\theta, A] = \int_0^\gamma \int_0^1 A(X,\tau)\left(\frac{\partial^2\theta}{\partial X^2} - \frac{\partial\theta}{\partial\tau} - \lambda_n\theta\right) dX d\tau \tag{16.114}$$

where γ is an arbitrary upper limit of τ. The trial functions for $\theta(X,\tau)$ and $A(X,\tau)$ are selected such that they satisfy the boundary conditions of the problem. Assuming trial functions of the form

$$\theta(X,\tau) = \frac{\cosh\lambda_n^{1/2}X}{\cosh\lambda_n^{1/2}} + \sum_{i=1}^{\infty} G_i(\tau)\cos\left(i - \frac{1}{2}\right)\pi X \tag{16.115}$$

where $i = 1, 2, 3, \ldots$ and

$$A(X,\tau) = \sum_{j=0}^{\infty} H_j(\tau)\cos\left(j - \frac{1}{2}\right)\pi X \qquad j = 1, 2, 3, \ldots \tag{16.116}$$

where $j = 1, 2, 3, \ldots$, integration of eq. (16.114) with respect to X yields the Euler–Lagrange equation

$$-\frac{dG_j}{d\tau} + \left[\left(j - \frac{1}{2}\right)^2 \pi^2 + \lambda_n\right] G_j = 0 \tag{16.117}$$

which has a solution

$$G_j(\tau) = C_j e^{-[(j-1/2)^2\pi^2+\lambda_n]\tau} \tag{16.118}$$

Substitution of eq. (16.118) into eq. (16.115) gives

$$\theta(X,\tau) = \frac{\cosh\lambda_n^{1/2}X}{\cosh\lambda_n^{1/2}} + \sum_{i=1}^{\infty} C_i e^{-[(j-1/2))^2\pi^2+\lambda_n]\tau}\cos\left(i - \frac{1}{2}\right)\pi X \tag{16.119}$$

The constant C_i can be found by an application of the initial condition, $\theta(X, 0) = 0$ and then utilizing the orthogonality property of $\cos(i - \frac{1}{2})\pi X$:

$$C_i = 2(-1)^i \frac{(i - \frac{1}{2})\pi}{(i - \frac{1}{2})^2\pi^2 + \lambda_n} \tag{16.120}$$

The instantaneous base heat flow rate Q_b can be obtained from eq. (16.119) as

$$Q_b = \frac{q_b b}{k\delta(T_b - T_s)} = \lambda_n^{1/2}\tanh\lambda_n^{1/2}$$

$$+2\sum_{i=1}^{\infty}\frac{\left(i-\frac{1}{2}\right)\pi}{\left(i-\frac{1}{2}\right)^2\pi^2+\lambda_n}e^{-[(i-1/2)^2\pi^2+\lambda_n]\tau} \tag{16.121}$$

Equation (16.121) is plotted in Fig. 16.18 for $N = 5$ and $n = 1, 2, 3$, and 4. Shown for comparison are the results of the coordinate perturbation solution (Fig. 16.17) and the numerical solution. Unlike the perturbation solution, eq. (16.121) covers the entire transient in quite a satisfactory manner and does approach the correct steady-state values as $\tau \to \infty$. Furthermore, the largest error between eq. (16.121) and the numerical solution is about 35%.

16.4.4 Hybrid Numerical/Analytical Solution

Cotta (1990) adopted a hybrid numerical/analytical approach to solve for the transient response of a radiating fin dissipating heat to an environment with a nonzero sink temperature. He considered a longitudinal fin of rectangular profile initially at a uniform temperature and subjected to a step change in base temperature. The tip of the fin was assumed to be insulated.

As a first step, Cotta (1990) applied the generalized integral transform technique to reduce the partial differential equation into a denumerable system of ordinary

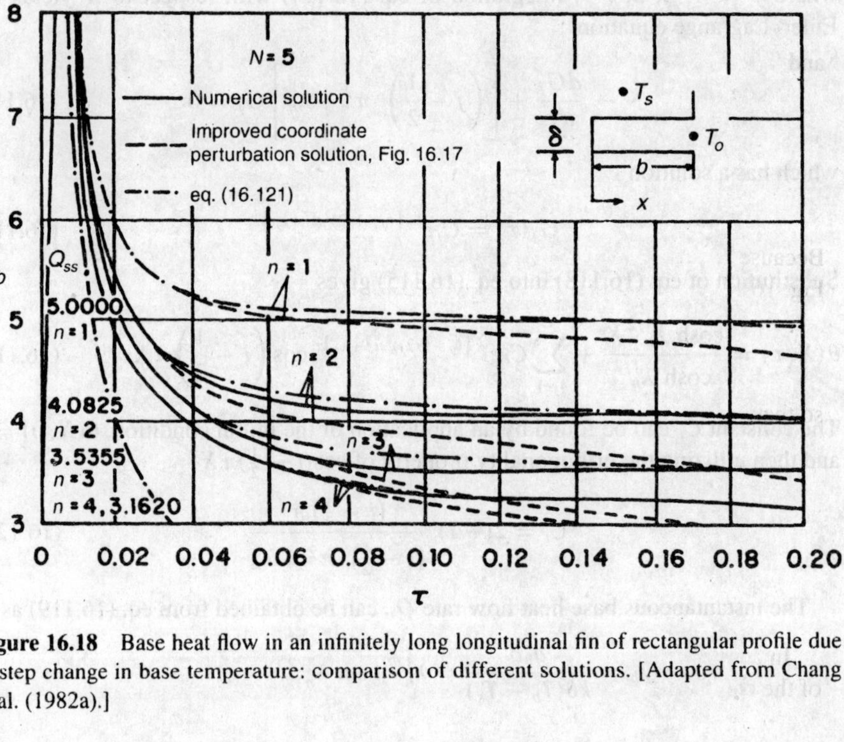

Figure 16.18 Base heat flow in an infinitely long longitudinal fin of rectangular profile due to step change in base temperature: comparison of different solutions. [Adapted from Chang et al. (1982a).]

differential equations. In the second step, these equations were truncated at the nth row and column and then solved using a standard finite difference scheme. The final step consisted of invoking the explicit inversion formula to recover the temperature distribution. The accuracy of the solution depends on the choice of the number of equations, n, with the larger the number of equations, the better the accuracy. However, it was found that $n = 15$ was sufficient to produce a converged solution.

Example 16.5. A longitudinal radiating fin of rectangular profile is 260 mm high and 2.5 mm thick. The fin is made of a material whose thermal conductivity and thermal diffusivity are 60 W/m·K and 8.5×10^{-6} m^2/s, respectively. The fin is initially in equilibrium with its environment at 0 K (free space). Assuming that faces of the fin possess an emissivity of $\epsilon_s = 0.95$ and that both faces of the fin are dissipating heat by radiation: **(a)** calculate the base heat flow 5 min after the base temperature experiences a step change to 800 K, **(b)** estimate the approximate duration of the transient, and **(c)** determine the steady-state base heat flow after the transient has elapsed.

SOLUTION. **(a)** First, compute the values of N and τ:

$$N = \left(\frac{2\sigma\epsilon_s T_b^3 b^2}{k\delta}\right)^{1/2}$$

$$= \left[\frac{(2)(5.669 \times 10^{-8})(0.95)(800)^3(0.260)^2}{(60)(0.0025)}\right]^{1/2}$$

$$= 4.99 \approx 5.00$$

and

$$\tau = \alpha t/b^2$$

$$= (8.5 \times 10^{-6})(60)(5)/(0.26)^2$$

$$= 0.038$$

Because $N \simeq 5$, Fig. 16.18 is applicable. With $n = 4$ (a radiating fin) and $\tau = 0.038$, Fig. 16.18 gives

$$Q_b = \frac{q_b b}{k\delta(T_b - T_s)} = 4$$

so that

$$q_b = \frac{4k\delta(T_b - T_s)}{b}$$

$$= \frac{(4)(60)(0.0025)(800 - 0)}{0.26}$$

$$= 1846 \text{ W/m}$$

In addition, because $N \simeq 5$, the approximation of infinite length permits the use of the coordinate perturbation solution of eq. (16.104). This gives

$$Q_b = \frac{1}{\sqrt{\pi\tau}} \frac{1 + 0.180976\epsilon}{1 + 0.057216\epsilon}$$

Here $\epsilon = 4N^2\tau = (4)(5)^2(0.038) = 3.8$. Thus

$$Q_b = \frac{1}{[\pi(0.038)]^{1/2}} \left[\frac{1 + (0.180976)(3.8)}{1 + (0.057216)(3.8)} \right]$$
$$= 4.01$$

which is very close to the value read from Fig. 16.18.

(b) From Fig. 16.18, the dimensionless time τ for the attainment of steady state is estimated as 0.12, which gives

$$t = \frac{b^2\tau}{\alpha}$$
$$= \frac{(0.26)^2(0.12)}{8.5 \times 10^{-6}}$$
$$= 954 \text{ s} \quad \text{or} \quad 15.9 \text{ min}$$

(c) From fig. 16.18, the steady-state Q_b for $n = 4$ is 3.1620, giving

$$q_b = \frac{3.1620k\delta(T_b - T_s)}{b}$$
$$= \frac{(3.1620)(60)(0.0025)(800)}{0.26}$$
$$= 1459 \text{ W/m}$$

16.5 LONGITUDINAL CONVECTING–RADIATING FINS

The transient response of fins with simultaneous convection and radiation has been studied by Campo (1976), who considered the longitudinal fin of rectangular profile of Fig. 16.1 and assumed that the fins are initially at steady state with a temperature profile of the form

$$\theta = \frac{T}{T_b} = 1 - a\left[1 - \left(\frac{x}{b}\right)^r\right] \tag{16.122}$$

where the parameters a and r are to be determined. The initial sink temperatures for convection and radiation are $T_{sc,i}$ and $T_{sr,i}$, respectively. At $t = 0$, the two sink temperatures experience a change with time which persists for all $t \geq 0$ in accordance with

$$T_{sc} = T_{sc,i} + (T_{cf} - T_{sc,i})(1 - e^{-c_o\tau}) \tag{16.123}$$

and

$$T_{sr} = T_{sr,i} + (T_{rf} - T_{sr,i})(1 - e^{-c_r\tau}) \tag{16.124}$$

where the subscript f denotes the values for $\tau \to \infty$, and c_c and c_r are constants. The unsteady-state fin governing equation for simultaneous convection and radiation is

$$\frac{\partial^2\theta}{\partial X^2} - N_c(\theta - \theta_c) - N_r(\theta^4 - \theta_r^4) = \frac{\partial\theta}{\partial\tau} \tag{16.125}$$

where

$$\theta = T/T_b$$

$$\theta_c = T_c/T_b$$

$$\theta_r = T_r/T_b$$

$$X = x/b$$

$$N_c = 2h^2b^2/k\delta$$

$$N_r = 2\epsilon_s\sigma T_b^3 b^2/k\delta$$

and

$$\tau = \alpha t/b^2$$

The boundary conditions for eq. (16.125) are

$$\frac{\partial\theta}{\partial X}(0, \tau) = 0 \tag{16.126a}$$

and

$$\theta(1, \tau) = 1 \tag{16.126b}$$

and the initial condition is

$$\theta(X, 0) = 1 - a(1 - X^r) \tag{16.126c}$$

where the tip heat loss has been neglected. It should be noted that θ_c and θ_r in eq. (16.125) are functions of time in accordance with eqs. (16.123) and (16.124), which can be recast in dimensionless form as

$$\theta_c = \theta_{ci} + (\theta_{cf} - \theta_{ci})(1 - e^{-c_c\tau}) \tag{16.127}$$

and

$$\theta_r = \theta_{ri} + (\theta_{rf} - \theta_{ri})(1 - e^{-c_r\tau}) \tag{16.128}$$

Using the Kantorovich method described by Kantorovich and Krylov (1964) and utilizing a transient temperature distribution which is in the form

$$\theta = 1 - (1 - X^r)\phi(\tau) \tag{16.129}$$

an ordinary differential equation for $\phi(\tau)$ can be derived as shown by Campo (1976). By comparing the steady-state solution based on the Kantorovich method with the corresponding finite difference solution, Campo (1976) found that $r = 2.6$ gave the best agreement between the two solutions. Sample results of integrating the ϕ equation appear numerically in Fig. 16.19. These results are for the case of $\theta_{ci} = \theta_{ri} = 0.1$, $\theta_{cf} = \theta_{rf} = 0.9$, and $c_c = c_r = 0.02$. Table 16.4 gives the duration of the transient τ_s up to 1% of the steady-state value for some combinations of N_c and N_r.

Table 16.4 shows that at $N_r = 0.3$, which describes a situation with weak radiation, the duration of the transient increases only slightly (from $\tau_s = 78$ to $\tau_s = 82$) as N_c is

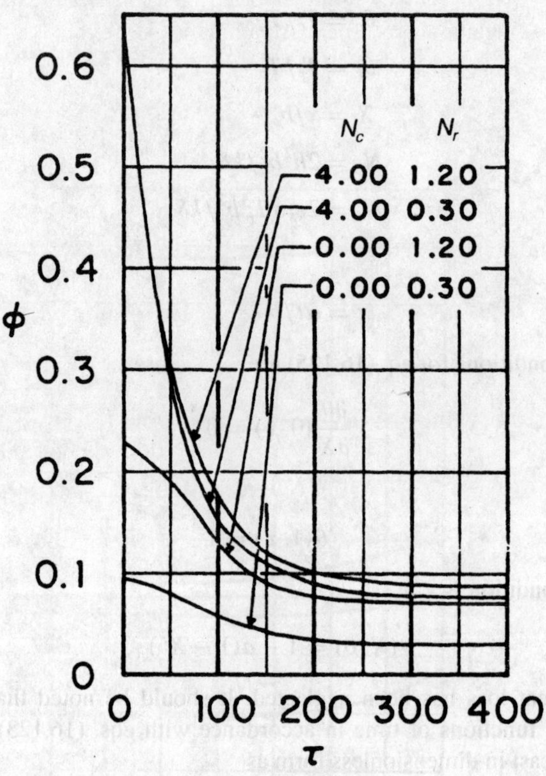

Figure 16.19 Solutions for $\Phi(\tau)$ for longitudinal convecting-radiationg fin of rectangular profile. [From Campo (1976).]

TABLE 16.4 Values of τ_s: $\theta_{ci} = \theta_{ri} = 0.1, \theta_{cf} = \theta_{rf} = 0.9, c_c = c_r = 0.02$

$N_c = 0$		$N_c = 4$	
$N_r = 0.3$	$N_r = 1.2$	$N_r = 0.3$	$N_r = 1.2$
$\tau_s = 78$	$\tau_s = 157$	$\tau_s = 82$	$\tau_s = 134$

increased from 0 (no convection) to 4 (strong convection). However, the same change in N_c at $N_r = 1.2$ (moderately strong radiation) produces a significant decrease in the duration of the transient ($\tau_s = 157$ to $\tau_s = 124$). The unsteady-state fin efficiency for the case of Fig. 16.19 is shown in Fig. 16.20.

16.6 TWO-DIMENSIONAL EFFECTS AND TRANSIENT HEAT FLOW

It is well known that the one-dimensional conduction model for fins is valid provided that the Biot number based on the half thickness is small, that is, $\mathrm{Bi} = h\delta/2k << 1$. When Bi is large, the error due to the one-dimensional assumption can be quite significant. For example, it has been observed by Lau and Tan (1973) that for a longitudinal fin of rectangular profile dissipating heat by convection, the error in heat transfer rate at Bi = 10 can be as high as 80%. With this in mind, several investigators

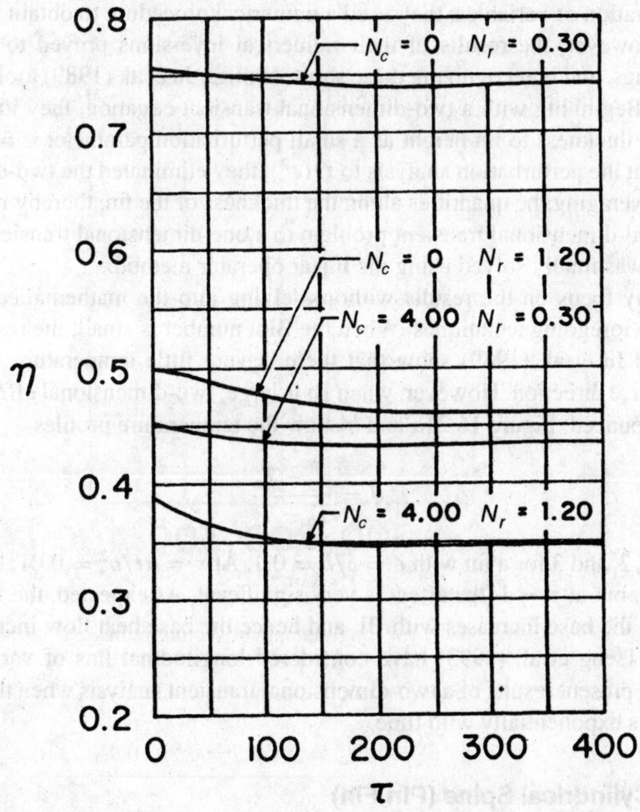

Figure 16.20 Transient efficiency of a longitudinal convecting-radiating fin of rectangular profile. [From Campo (1976).]

have studied the transient response of fins allowing for two-dimensional conduction. In this section the important results obtained from these studies are highlighted and reviewed.

16.6.1 Longitudinal Fin of Rectangular Profile

The two-dimensional model of the the longitudinal fin of rectangular profile shown in Fig. 16.1 has been considered by Chu (1977), Chu et al. (1982), and Ju et al. (1989). Each of these studies assumes that the fin is initially in thermal equilibrium with its environment, and that at $t = 0$, the base of the fin experiences a step change in temperature and remains at this temperature for all $t \geq 0$. The fin loses heat to the environment by pure convection. Chu's (1977) approach was to eliminate the time variable by applying the Laplace transformation. The resulting partial differential equation in x and y was then solved using the method of separation of variables, giving rise to a series type of solution. As noted earlier for the one-dimensional case, the convergence of such a series solution is painfully slow for small time values.

In a later paper, Chu et al. (1982) examined the same problem, but instead of using separation of variables, they used a numerical procedure to obtain the Laplace inverse. However, the results of their numerical inversions proved to have some shortcomings, and in recognizing these shortcomings, Ju et al. (1989) took a different approach. Beginning with a two-dimensional transient equation, they identified the ratio of fin thickness to fin height as a small perturbation parameter $\epsilon = \delta/b$. After carrying out the perturbation analysis to $O(\epsilon^2)$, they eliminated the two-dimensional effect by averaging the quantities along the thickness of the fin, thereby reducing the original two-dimensional transient problem to a one-dimensional transient problem. The latter was finally solved using the linear operator method.

One may focus on the results without delving into the mathematical details of each of the foregoing techniques. When the Biot number is small, the results of Chu (1977) and Ju et al. (1989) show that there is very little temperature variation in the transverse direction. However, when Bi is large, two-dimensional effects become more pronounced. Figure 16.21*a* and *b* show the temperature profiles

$$\bar{\theta} = \frac{\bar{T} - T_s}{T_b - T_s}$$

at Bi = 0.1, 2, and 3 for a fin with $\epsilon = \delta/b = 0.1$. At $\tau = \alpha t/b^2 = 0.01$, the effect of Bi is small, but at $\tau = 1$, the effect is very significant. As expected, the temperature gradient at the base increases with Bi, and hence the base heat flow increases as Bi increases. Tseng et al. (1993) have considered longitudinal fins of various profile shapes and present results of a two-dimensional transient analysis when the base heat flow decays exponentially with time.

16.6.2 Cylindrical Spine (Pin Fin)

The two-dimensional transient analysis for a cylindrical spine or pin fin has been reported by Chu et al. (1983a). They considered the cases of a step change in base

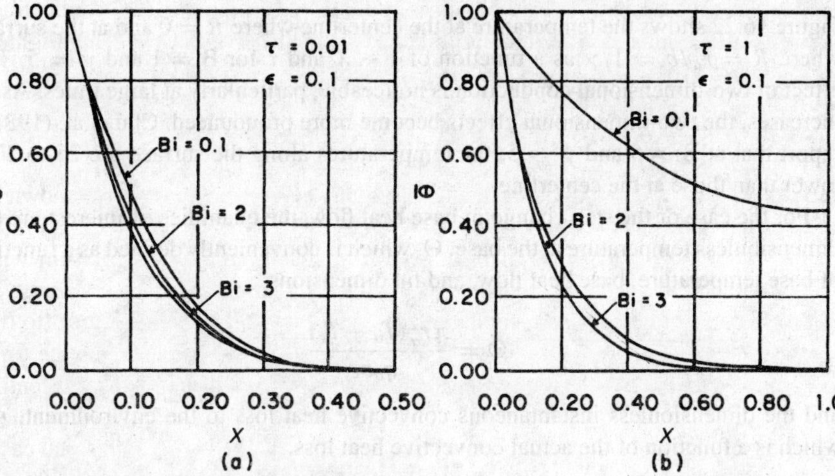

Figure 16.21 Effect of the Biot number on the transient temperature distribution in a two-dimensional longitudinal fin of rectangular profile. (*a*) $\tau = 0.01$; (*b*) $\tau = 1.0$. [From Ju et al. (1989).]

temperature and a step change in base heat flow. In each case, the Laplace transform technique was used to eliminate the time variable, and then the method of separation of variables was used to solve the transformed equation. Finally, the inverse was determined.

First, consider the case of a step change in base temperature for a spine with outer radius r_a, radial coordinate r, height coordinate x, with origin taken at the spine base and height b. The spine temperature

$$\theta = \frac{T - T_s}{T_b - T_s}$$

is a function of five parameters:

$$R = \frac{r}{b}$$

$$X = \frac{x}{b}$$

$$\gamma = \frac{b}{r_a}$$

$$\text{Bi} = \frac{hr_a}{k}$$

and

$$\iota = \frac{\alpha t}{b^2}$$

Figure 16.22 shows the temperature at the centerline where $R = 0$ and at the surface where $R = r_a/b = 1/\gamma$ as a function of $1 - X$ and τ for Bi = 1 and $\gamma = 5$. The effect of two-dimensional conduction is noticeable, particularly at large times. As Bi increases, the two-dimensional effects become more pronounced. Chu et al. (1983a) report that at Bi = 5 and $\gamma = 5$, the temperatures along the surface are 25 to 30% lower than those at the centerline.

For the case of the step change in base heat flow, the quantities of interest are the dimensionless temperature at the base, Θ, which is conveniently defined as a function of base temperature, base heat flow, and fin dimensions.

$$\Theta = \frac{\pi r_a^2 (T_b - T_s)}{q_b b}$$

and the dimensionless instantaneous convective heat loss to the environment, Q_s, which is a function of the actual convective heat loss:

$$Q_s = \frac{q_s b}{\pi r_a^2 k (T_b - T_s)}$$

A typical set of results for θ appears in Fig. 16.23 for the case of Bi = 1 and $\gamma = 5$. The closeness of the curves for the surface and the center temperatures shows that the temperature distribution is virtually one-dimensional. Figure 16.24 shows the ratio Q_s/Q_b, where

$$Q_b = \frac{q_b b}{\pi r_a^2 k (T_b - T_s)}$$

as a function of τ. As the transient dies out ($\tau \to \infty$), q_s approaches q_b and $Q_s/Q_b \to 1$.

Example 16.6. A cylindrical spine, fabricated from a material with $k = 15$ W/m·K

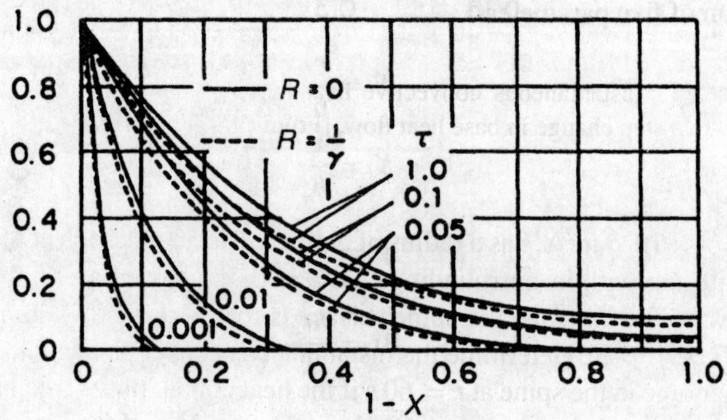

Figure 16.22 Transient temperature distributions in a two-dimensional cylindrical spine for Bi = 1 and $\gamma = b/r_a = 5$ and a step change in base temperature. [Adapted from Chu et al. (1983a).]

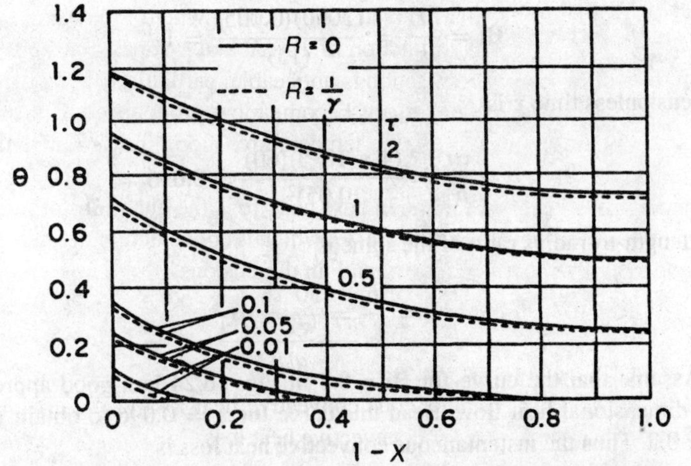

Figure 16.23 Transient temperature distributions in a two-dimensional cylindrical spine for Bi = 1 and $\gamma = b/r_a = 5$ and a step change in base heat flow. [Adapted from Chu et al. (1983a).]

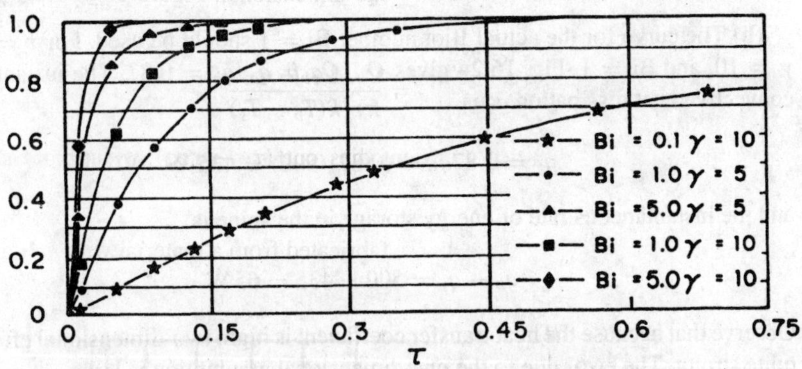

Figure 16.24 Instantaneous convective heat loss in a two-dimensional cylindrical spine subjected to a step change in base heat flow. [From Chu et al. (1983a).]

and $\alpha = 4 \times 10^{-6}$ m^2/s, has a radius of 5 mm and is 50 mm high. The spine is initially in equilibrium with its surroundings. At $t = 0$, the spine base is subjected to a step heat flow of 500 W while the spine surface is forced cooled by a liquid coolant with $h = 3000$ W/m^2·K. Determine the instantaneous convective heat loss and the rate of energy storage in the spine at $t = 60$ s if the heat conduction is assumed to be **(a)** one- and **(b)** two-dimensional.

SOLUTION. The Biot number is

$$\text{Bi} = \frac{hr_a}{k} = \frac{(3000)(0.005)}{(15)} = 1$$

the dimensionless time τ is

$$\tau = \frac{\alpha t}{b^2} = \frac{(4 \times 10^{-6})(60)}{(0.05)^2} = 0.096$$

and the length-to-radius ratio of the spine is

$$\gamma = \frac{b}{r_a} = \frac{50}{5} = 10$$

(a) Assume that the curve for Bi = 0.1 in Fig. 16.24 is a good approximation for one-dimensional heat flow. Read this curve for $\tau = 0.096$ to obtain $Q_s/Q_b = q_s/q_b = 0.2$. Thus the instantaneous convective heat loss is

$$q_s = 0.2q_b = (0.2)(500) = 100 \text{ W}$$

and the instantaneous rate of storage of energy in the spine is

$$q_b - q_s = 500 - 100 = 400 \text{ W}$$

(b) The curve for the actual Biot number Bi = 1 should be used. For $\tau = 0.096$, $\gamma = 10$, and Bi = 1, Fig. 16.24 gives $Q_a/Q_b = q_a/q_b = 0.87$. The instantaneous convective heat dissipation q_s is

$$q_s = 0.87q_b = (0.87)(500) = 435 \text{ W}$$

and the instantaneous rate of energy storage in the spine is

$$q_b - q_s = 500 - 435 = 65 \text{ W}$$

Observe that because the heat transfer coefficient is high, two-dimensional effects are quite strong. The error due to the one-dimensional assumption is large.

16.6.3 Radial Fins

The effect of two-dimensional conduction on the transient response of radial fins has been studied by Huang (1985). His analysis covered four profile shapes: the radial fins of rectangular, triangular, convex parabolic, and concave parabolic profiles. For each shape, the fin was assumed to be initially at a uniform temperature and allowed to experience a step change in base temperature while the tip of the fin remained adiabatic. The three-step solution strategy employed was first to use the Laplace transform to eliminate the time variable, then to utilize the integral method to solve the transformed equation, and finally, to employ Fourier series to obtain the inverse Laplace transform.

16.7 COMPOSITE FINS

When a finned heat exchanger is designed for operation in a high-temperature and/or corrosive environment, it becomes necessary to protect the high-thermal-conductivity expensive surface from the damaging effects of the environment. This can be achieved by covering the surface with a relatively low-thermal-conductivity and less expensive material that can withstand the harsh environment. The resulting arrangement is a composite fin. A practical example of composite fins, cited by Baker (1958), is that of a gas turbine regenerator where stainless steel sheaths are bonded to the copper surface to protect the copper surface from the high-temperature gas.

A composite fin of rectangular profile and consisting of two materials is shown in Fig. 16.25. The transient response of this arrangement has been analyzed by Chu et al. (1983b). They considered a situation in which the fin was initially in equilibrium with its environment at temperature T_s and suddenly experienced a step change in its base temperature from $T_i = T_s$ to T_b. The outer surfaces of the covering was allowed to dissipate heat by convection, the process being characterized by a constant heat transfer coefficient h.

With the x and y coordinates reckoned from the top interface point at the base (see Fig. 16.25), two-dimensional transient heat conduction equations were set up for each material. The interface boundary conditions were based on a perfect contact

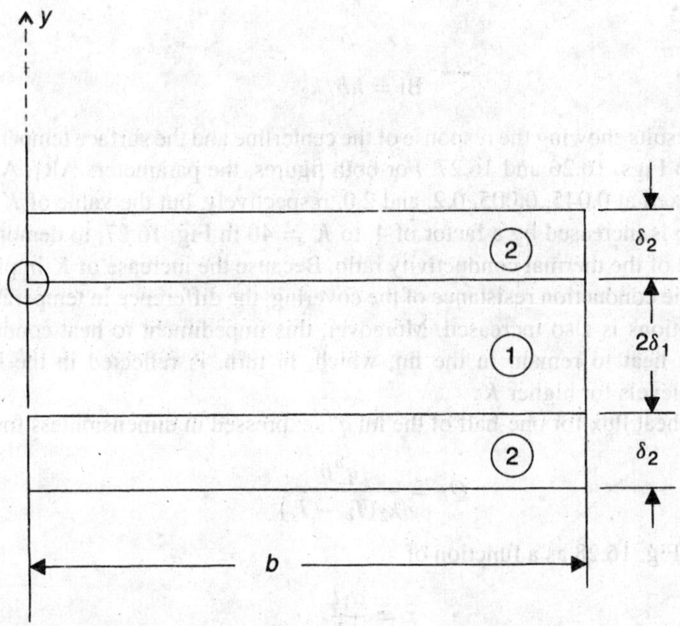

Figure 16.25 Composite longitufinal fin of rectangular profile. [Adapted from Chu et al. (1983a).]

assumption and the tip of the fin was assumed to be insulated. The system of equations was solved using the Laplace transform and the eigenfunction expansion methods. The final step for inversion was carried out with the help of a Fourier series.

In addition to the dimensionless spatial and time variables

$$X = \frac{x}{b}$$

$$Y = \frac{y}{b}$$

and

$$\tau = \frac{\alpha_1 t}{b^2}$$

the transient response in this case is characterized by five parameters:

$$\text{AR}_1 = \frac{\delta_1}{b}$$

$$\text{AR}_2 = \frac{\delta_2}{b}$$

$$\alpha = \frac{\alpha_1}{\alpha_2}$$

$$K = \frac{k_1}{k_2}$$

and

$$\text{Bi} = hb/k_2$$

Typical results showing the response of the centerline and the surface temperatures are shown in Figs. 16.26 and 16.27. For both figures, the parameters AR_1, AR_2, α, and Bi are fixed at 0.045, 0.005, 0.2, and 2.0, respectively, but the value of $K = 10$ in Fig. 16.26 is increased by a factor of 4, to $K = 40$ in Fig. 16.27, to demonstrate the influence of the thermal conductivity ratio. Because the increase of K implies an increase in the conduction resistance of the covering, the difference in temperature at the two locations is also increased. Moreover, this impediment to heat conduction causes more heat to remain in the fin, which, in turn, is reflected in the higher temperature levels for higher K.

The base heat flux for one-half of the fin q'', expressed in dimensionless form,

$$Q_b = \frac{q''b}{k_2(T_b - T_s)}$$

is plotted in Fig. 16.28 as a function of

$$\tau = \frac{\alpha_1 t}{b^2}$$

for a range of values of K. Two features of Fig. 16.28 are noteworthy. First, as K increases, Q_b decreases. This is a consequence of the increased conduction resistance

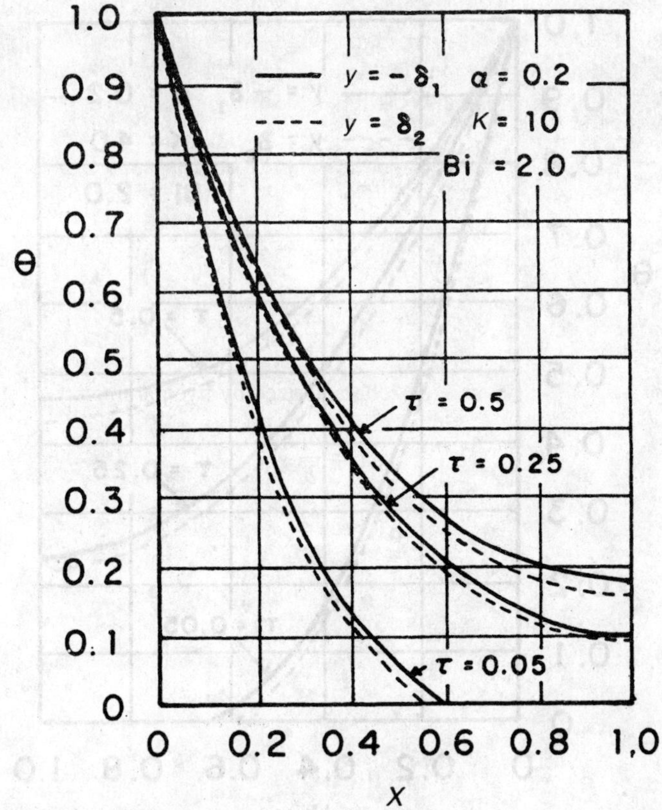

Figure 16.26 Transient responses of the center and outer surfaces of a composite fin with $\alpha = 0.20$ and $K = 40$ and with $\mathrm{Bi} = 2$. [From Chu et al. (1983b).]

of the covering, which, as noted, enhances the temperature levels in the fin and thus suppresses the axial temperature gradient. Second, the greater the value of K, the shorter the duration of the transient.

Besides the results for the step change in base temperature discussed here, Chu et al. (1983b) also give results for the condition of base temperature decaying exponentially with time.

The accuracy of the Chu et al. (1983b) results has been checked by Goshdastidar and Mukbopadhyay (1989) against a direct numerical solution based on an alternating direction implicit scheme. The numerical results were found to be in good agreement with the analytical predictions except for small time and large axial distance values. However, this discrepancy does not affect the base heat flow predictions because the analytical and numerical results match very closely near the base.

In a related study, Ju et al. (1991) considered the composite fin of Fig. 16.25, but their analysis allowed the top and bottom coverings to be of different materials and having different thicknesses. Their approach was to apply the continuum mixture

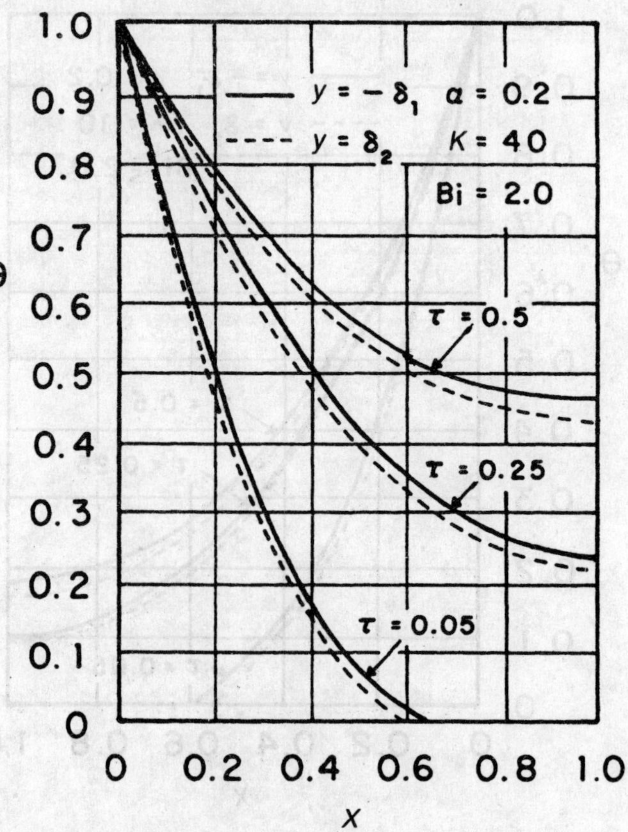

Figure 16.27 Transient responses of the center and outer surfaces of a composite fin with $\alpha = 0.2$, $K = 10$ and $\mathrm{Bi} = 2$. [From Chu et al. (1983b).]

theory in conjunction with the linear operator method. These authors also noticed some discrepancy between their results and those of Chu et al. (1983a) at small times values. Figure 16.29 shows the base heat flux

$$Q = \frac{q''b}{k(T_b - T_a)}$$

for the four cases reported by Ju et al. (1991). Curve 1 is for a bare copper fin. As expected, this fin gives the highest base heat flow. Curve 2 is for an asymmetrically coated fin in which copper is sandwiched between stainless steel on the top and gold on the bottom. The fractional thicknesses of copper, stainless steel, and gold are 0.9, 0.09, and 0.01, respectively. Curve 3 is also for an asymmetrical fin except that both coverings are of stainless steel. Curve 4 is for a symmetrical fin with stainless steel coverings of identical thicknesses on the top and bottom. These curves show that the best design for a composite fin is an asymmetrically coated fin (curves 2 and 3).

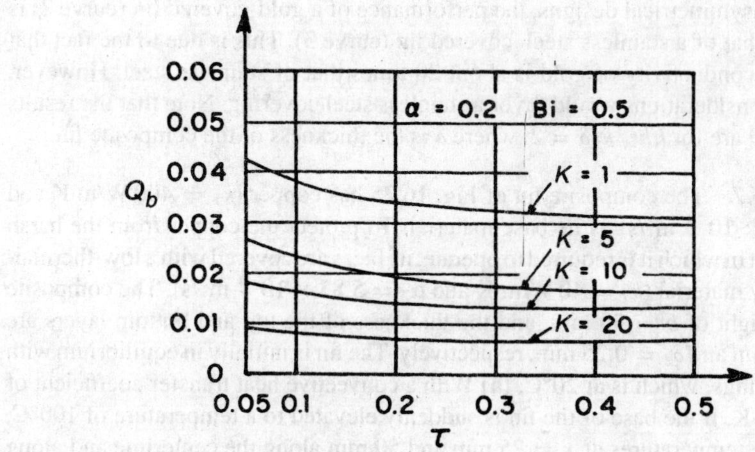

Figure 16.28 Effect of conductivity ratio on transient heat flow in a composite fin. $\alpha = 0.2$ and Bi = 2. [From Chu et al. (1983b).]

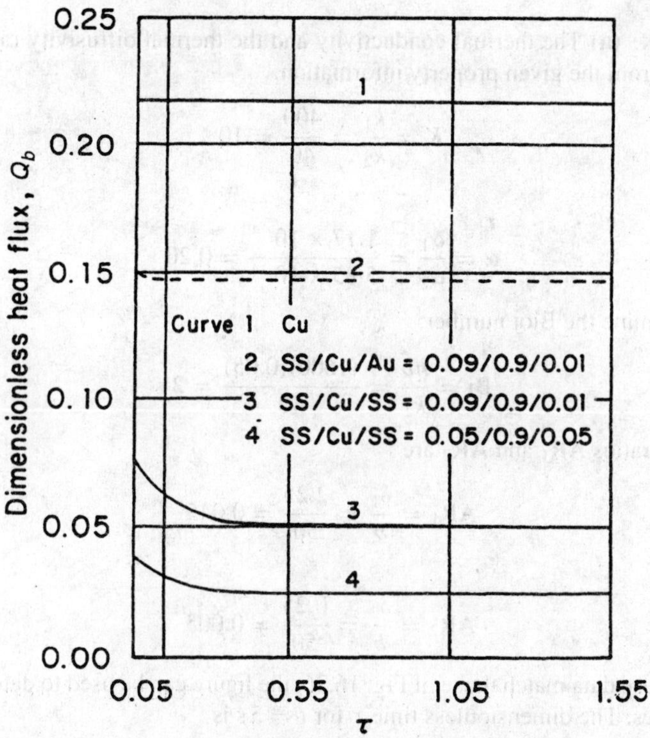

Figure 16.29 Base heat flow in tri-laminated fins of different material combinations. [From Ju et al. (1991).]

Of the two asymmetrical designs, the performance of a gold-covered fin (curve 2) is superior to that of a stainless steel–covered fin (curve 3). This is due to the fact that the thermal conductivity of gold is about 20 times that of stainless steel. However, economic considerations would favor a stainless steel covering. Note that the results in Fig. 16.29 are for $hb^2/k_b\delta = 2$, where δ is the thickness of the composite fin.

Example 16.7. The composite fin of Fig. 16.25 has copper ($k_1 = 400$ W/m·K and $\alpha_1 = 1.17 \times 10^{-4}$ m^2/s) as its base material. To protect the copper from the harsh environment in which it is required to operate, its faces are covered with a low-thermal-conductivity material ($k_2 = 40$ W/m·K and $\alpha_2 = 5.85 \times 10^{-4}$ m^2/s). The composite fin has a height of $b = 50$ mm, and the thickness of the top and bottom layers are $\delta_1 = 2.25$ mm and $\delta_2 = 0.25$ mm, respectively. The fin is initially in equilibrium with its surroundings, which is at 20°C. **(a)** With a convective heat transfer coefficient of 1600 W/m^2·K, if the base of the fin is suddenly elevated to a temperature of 100°C, compute the temperatures at $x = 25$ mm and 50 mm along the centerline and along the exposed surface of the top covering at time $t = 5$ s, and **(b)** determine how these temperatures are affected if the heat transfer coefficient is reduced to 400 W/m^2·K and the covering material is changed such that $k_2 = 10$ W/m·K while α_2 remains unchanged.

SOLUTION. **(a)** The thermal conductivity and the thermal diffusivity ratio can be calculated from the given property information:

$$K = \frac{k_1}{k_2} = \frac{400}{40} = 10$$

and

$$\alpha = \frac{\alpha_1}{\alpha_2} = \frac{1.17 \times 10^{-4}}{5.85 \times 10^{-4}} = 0.20$$

Next, determine the Biot number:

$$\text{Bi} = \frac{hb}{k_2} = \frac{(1600)(0.05)}{40} = 2$$

The aspect ratios AR_1 and AR_2 are

$$\text{AR}_1 = \frac{\delta_1}{b} = \frac{2.25}{50} = 0.045$$

and

$$\text{AR}_2 = \frac{\delta_2}{b} = \frac{0.25}{50} = 0.005$$

Because these data match those of Fig. 16.26, the figure can be used to determine the temperatures. The dimensionless time τ for $t = 5$ s is

$$\tau = \frac{\alpha_1 t}{b^2} = \frac{(1.17 \times 10^{-4})(5)}{(0.05)^2} = 0.234$$

For the centerline, read the solid lines in Fig. 16.26:

$$x = 25\text{ mm} \quad \text{or} \quad X = 0.5, \qquad \theta = 0.27$$

$$x = 50\text{ mm} \quad \text{or} \quad X = 1, \qquad \theta = 0.09$$

Thus the temperatures at $x = 25$ mm and $x = 50$ mm are

$$(0.27)(100 - 20) + 20 = 41.6°\text{C}$$

and

$$(0.09)(100 - 20) + 20 = 27.2°\text{C}$$

respectively. For the exposed surface of the covering, the dashed lines in Fig. 16.26 provide

$$x = 25\text{ mm} \quad \text{or} \quad X = 0.5, \qquad \theta = 0.25$$

$$x = 50\text{ mm} \quad \text{or} \quad X = 1, \qquad \theta = 0.08$$

This gives temperature values of 40 and 26.4°C, respectively.

(b) When the heat transfer coefficient is reduced to 400 W/m^2·K and the covering material is changed such that $k_2 = 10$ W/m·K, α_2 does not change and, hence, $\alpha = 0.2$. The Biot number is now

$$\text{Bi} = \frac{hb}{k_2} = \frac{(400)(0.05)}{10} = 2$$

Reading Fig. 16.27 for $\tau = 0.25$ and $X = 0.5$ and 1.0 gives, for the centerline:

$$x = 25\text{ mm} \quad \text{or} \quad X = 0.5, \qquad \theta = 0.43 \quad \text{or} \quad T = 54.4°\text{C}$$

$$x = 50\text{ mm} \quad \text{or} \quad X = 1, \qquad \theta = 0.22 \quad \text{or} \quad T = 37.6°\text{C}$$

and for the top surface:

$$x = 25\text{ mm} \quad \text{or} \quad X = 0.5, \qquad \theta = 0.4 \quad \text{or} \quad T = 52°\text{C}$$

$$x = 50\text{ mm} \quad \text{or} \quad X = 1, \qquad \theta = 0.21 \quad \text{or} \quad T = 36.8°\text{C}$$

A comparison of the temperatures for the two cases indicates that the temperatures are higher in the second case. This is because of substantially lower thermal conductivity (10 as opposed to 40) and a much lower heat transfer coefficient (1600 as opposed to 400) in the second case.

Example 16.8. A longitudinal fin of rectangular profile is fabricated of copper ($k = 400$ W/m·K and $\alpha = 1.17 \times 10^{-4}$ m^2/s). The fin is 9 mm thick and 50 mm high. Because the fin is to operate in a harsh environment, protective coverings are to be provided on either side. Three design possibilities are considered:

Design	Top Covering	Bottom Covering
1	0.9-mm stainless steel	0.1-mm gold
2	0.9-mm stainless steel	0.1-mm stainless steel
3	0.5-mm stainless steel	0.5-mm stainless steel

For each design, the composite fin is initially at 20°C. At time $t > 0$, the base temperature is stepped up to 100°C and remains there for all $t \geq 0$. Assume that the convective heat transfer coefficient is 3200 W/m^2·K and **(a)** calculate the base heat flow for each design after 1 s, **(b)** discuss how these values compare with the corresponding base heat flow for the bare fin, and **(c)** compare these values with the steady-state base heat flows.

SOLUTION. **(a)** The width or thickness δ of the composite fin for each design is 10 mm. The dimensionless time τ is

$$\tau = \frac{\alpha_1 t}{b^2} = \frac{(1.17 \times 10^{-4})(1)}{(0.05)^2} = 0.047 \approx 0.05$$

For design 1, read curve 2 of Fig. 16.29 and obtain

$$Q = \frac{q''b}{k}(T_b - T_s) = 0.15$$

or

$$q'' = \frac{0.15k(T_b - T_s)}{b}$$

$$= \frac{(0.15)(400)(100 - 20)}{0.05}$$

$$= 96{,}000 \text{ W/m}^2$$

For a unit depth ($L = 1$ m)

$$q_1 = (9600)(0.01)(1) = 960 \text{ W}$$

Read curves 3 and 4 for the second and third designs and obtain $Q = 0.08$ and 0.04, respectively. From these values, $q_2 = 512$ W and $q_3 = 256$ W.

For the bare fin, read curve 1 and find $Q = 0.217$. Thus

$$q''_{\text{bare}} = 138{,}880 \text{ W/m}^2$$

or

$$q_{\text{bare}} = (138{,}880)(0.009)(1) = 1249.9 \text{ W}$$

(b) It is seen that the heat flow is highest (1249.9 W) for the bare fin. Of the three composite fin designs, the asymmetrical design 1 with 0.9 mm of stainless steel on one side and 0.1 mm of gold on the other side performs best with a base heat flow of 960 W. The next-best design is the asymmetrical fin, with 0.9 mm of stainless steel on one side and 0.1 mm of stainless steel on the other, giving a base heat flow of 512 W. Design 3, which is a symmetrically coated fin with 0.5 mm of stainless steel on either side, is the worst and gives a base heat flow of 256 W.

(c) Figure 16.29 shows that fins approach steady state at $\tau = 1.55$. Reading the corresponding values of Q and converting them into q values provides

$$q_1(\text{steady state}) = 947 \text{ W}$$

$$q_2(\text{steady state}) = 320 \text{ W}$$

$$q_3(\text{steady state}) = 160 \text{ W}$$

$$q_{\text{bare}}(\text{steady state}) = 1249.9 \text{ W}$$

16.8 VARIABLE HEAT TRANSFER COEFFICIENT

The assumption common to the cases discussed thus far has been that the convective heat transfer coefficient is constant throughout the transient process. This assumption facilitates both analysis and numerical simulations, but the predicted results may not be very precise because the heat transfer coefficient does change as the transient progresses. However, the constant h analysis does provide results that can be used for preliminary analysis and design.

The variability of the heat transfer coefficient is not difficult to envision. For example, consider the situation when the fin is initially in equilibrium with the environment. If the base temperature is suddenly elevated, the process will begin as pure transient conduction with no surface dissipation. As the fin responds to the disturbance, its temperature begins to increase and the surface dissipation comes into play. If the fin is cooled by natural convection, then as the temperature difference between fin and environment, ΔT, increases, the heat transfer coefficient increases, being proportional to a power of $\frac{1}{4}$ or $\frac{1}{3}$, depending on whether the natural convection is laminar or turbulent.

The rejection of heat to the environment would increase the environmental temperature, particularly if the natural convection is weak. Indeed, the experimental work of Mutlu and Al-Shemmeri (1993) on forced-convection-cooled shrouded arrays of longitudinal fins shows that both the heat transfer coefficient and environmental temperature change with time as the transient progresses. Mutlu and Al-Shemmeri (1993) found that the heat transfer coefficient initially increases with time, reaches a peak value, and then decays rapidly to a constant value. The period of increase is short compared with the period over which the decay occurs.

Cotta (1990) has applied the finite integral transform technique to study the transient response of a truncated conical pin fin with a time-dependent heat transfer coefficient h. The geometry considered is shown in Fig. 16.30. The analysis assumes one-dimensional conduction and allows h to vary with dimensionless time $\tau = \alpha t/x_a^2$ in accordance with

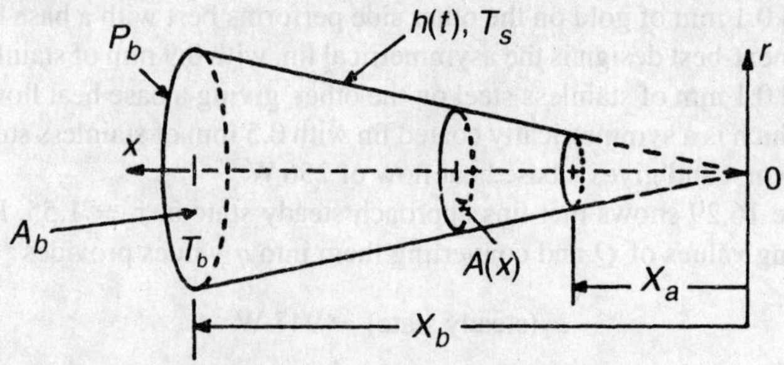

Figure 16.30 Truncated conical spine.

$$h(\tau) = \frac{h(0)}{1 + B\tau} \tag{16.130}$$

where $h(0)$ is the heat transfer coefficient at t or $\tau = 0$ and B is a dimensionless parameter. Using the approach described in Section 16.4.4, Cotta solved the problem for $N = [h(0)P_b b^2/kA_b]^{1/2} = 1$, $x_a/x_b = 0.5$, $B = 1$ and 10, and using different levels of truncation, that is, $n = 5$, 10, 20, and 50. He found that $n = 10$ was sufficient to obtain fully converged solutions. Table 16.5 gives the sample results for the temperature distribution obtained with $n = 10$.

Equation (16.130) shows that as B increases, the heat transfer coefficient decreases more sharply with time. This means that the time-averaged h decreases as B increases. With this in mind, one can anticipate higher temperatures in the fin as B increases. This anticipation is confirmed by the results of Table 16.5.

16.9 MISCELLANEOUS STUDIES

Thus far, the discussion of transient effects in an extended surface has focused on deterministic models. Recognizing the fact that random conditions can occur during the operation of a fin, some studies have considered the performance of fins under stochastic conditions. For example, Hung (1969) presented an analysis of the

TABLE 16.5 Transient Temperature Distributions in a Truncated Conical Pin Fin

	$B = 1$			$B = 10$		
X	$\tau = 0.01$	0.1	1.0	0.01	0.1	1.0
0.500	0.88541	0.88981	0.93819	0.88612	0.91762	0.98891
0.625	0.89686	0.90114	0.94446	0.89748	0.92614	0.98922
0.750	0.92372	0.92750	0.95910	0.92423	0.94613	0.99203
0.875	0.95970	0.96203	0.97837	0.96005	0.97207	0.99576
1.000	0.10000	1.00000	1.00000	1.00000	1.00000	1.00000

temperature distribution in longitudinal and radial fins when the base temperature is assumed to be stationary and Gaussian with a zero mean. Hung gives results for the root-mean-square (*rms*) temperature and power spectral density for both purely stochastic and Markovian base temperature models. A similar study for stochastic surrounding temperature has been reported by Yoshimura and Campo (1981).

Besides the papers on radiating and convecting–radiating fins discussed in previous sections, there have been a few studies that are closely related to these topics but more application specific. In a study that was apparently inspired by a desire to determine when a coolant in a space radiator would begin to freeze, Palmquist et al. (1970) conducted a study of the transient cooling of a fin–tube system when the steady-state operation is suddenly halted. In another application-oriented study, Okamoto and Kameoka (1974) calculated the transient performance of radial fins attached to fuel rods in gas-cooled nuclear reactors numerically. Their model allowed for simultaneous convection and radiation heat dissipation with the fin base either experiencing a step change in temperature or heat flux. With space power plant application in mind, French (1964) and Okamoto and Negishi (1969) studied the transient response of finned tube sheet radiators. A finite element analysis for the transient response of an array of fins with coupled radiation has been reported by Singh and Stasa (1982). A transient study by Assis and Kalman (1993) reported on numerical solutions for the transient response of two types of fins (longitudinal and radial), each with three possible profiles (rectangular, triangular, and parabolic). The cylindrical, conical, and parabolic spines were also considered.

Although the problem of conjugate heat transfer in extended surfaces under steady conditions has been studied extensively in recent years, the transient counterpart of the problem has virtually remained unexplored. Apparently, there is only the paper of Hsu and Tsai (1992), which analyzes the transient response of a vertical plate fin when the top is subjected to a sudden heat flux with the bottom insulated. The fin is losing heat by simultaneous natural convection and radiation. The formulation couples the one-dimensional transient heat conduction equation for the fin with the natural convection boundary layer equations incorporating a radiation heat flux term in the energy equation. The important parameters governing the problem are the width-to-height ratio δ/b, the conduction–convection number, the conduction–radiation number, and the base heat flux. Hsu and Tsai (1992) present numerical results showing how the base temperature–time variation is influenced by the problem parameters.

Reference can also be made to three experimental studies. The first was reported by Donaldson and Shouman (1972) as part of a predominantly theoretical work. Their experimental arrangement consisted of a copper rod instrumented with copper–constantan thermocouples for temperature measurements. The rod was heated with a heating tape wrapped around it and the setup enabled the authors to simulate the boundary condition of a step change in base heat flow in an approximate manner. Another experimental study is that of Mutlu and Al-Shemmeri (1993). This work reports the results of experiments conducted on a forced-convection-cooled shrouded array of longitudinal fins. Experimentally determined temperature data were used to deduce the instantaneous and time-average fin efficiencies as a function of power input and air mass flow rate. The authors conclude that both the heat transfer coefficient and air bulk temperature vary with time, and consequently, it is not appropriate to assume a constant heat transfer coefficient or a constant surrounding temperature when the fin is analyzed for transient effects. Finally, it is to be noted that French (1964) experimentally verified the transient response of a space radiator.

16.10 CLOSURE

The topic of transient heat flow in extended surface appears to have been well explored. Nonetheless, several ideas for future research activity can be identified. For example, there exist no transient studies in the open literature that include effects such as internal heat generation, contact resistance between the wall and the fin, and thermal coupling between the wall and the fin. Similarly, the information on transient heat transfer in arrays of fins is limited and needs to be expanded. There is also a need for transient studies that take the variation of heat transfer coefficient and the temperature dependence of thermal properties into account. Conjugate heat transfer, in which the transient heat conduction in the fin is coupled to the flow equations, is another topic for future research endeavors.

16.11 NOMENCLATURE

Roman Letter Symbols

A	combination of terms, dimensionless; function, dimensionless
AR	aspect ratio, dimensionless
a	coefficient, dimensionless
B	combination of terms, dimensionless; parameter, dimensionless
Bi	Biot number, dimensionless
b	fin or spine height, m
C	constant, dimensionless
c	combination of terms, dimensionless
d	diameter, m; differential or derivative, dimensionless
F	parameter, dimensionless
f	function, dimensionless
G	parameter, dimensionless; function, dimensionless
h	heat transfer coefficient, $W/m^2 \cdot K$
I	modified Bessel function of first kind, dimensionless
i	counter, dimensionless
J	Bessel function of first kind, dimensionless; Lagrangian functional, dimensionless
j	counter, dimensionless
K	modified Bessel function of second kind, dimensionless; thermal conductivity ratio, dimensionless
k	thermal conductivity, $W/m \cdot K$

L	fin length, m
$\mathcal{L}$	Laplace transform, dimensionless
m	fin performance parameter, m^{-1}
N	combination of terms, dimensionless
n	number of equation or terms
P	fin perimeter, m
Q	base heat flux, dimensionless
q	heat flow, W
q_b	base heat flow per unit length, W/m
$\dot{q}$	heat generated per unit volume, W/m^3
R	radius, dimensionless
r	radius, m; radial coordinate, m
s	Laplace transform variable, dimensionless
T	temperature, K
t	time, s
X	height, dimensionless
x	height coordinate, m
Y	Bessel function of second kind, dimensionless
Z	distance, dimensionless
z	distance, m

Greek Letter Symbols

α	thermal diffusivity, m^2/s; thermal diffusivity ratio, dimensionless
γ	combination of terms, dimensionless; height-to-radius ratio, dimensionless; upper limit of integration, dimensionless
Δ	change in, dimensionless
δ	fin thickness, m
∂	partial derivative, dimensionless
ϵ	emissivity, dimensionless; perturbation parameter, dimensionless
ζ	combination of terms, dimensionless
η	fin efficiency, dimensionless; combination of terms, dimensionless; similarity variable, dimensionless
Θ	base temperature, dimensionless
θ	term in perturbation series, dimensionless; temperature ratio, dimensionless
$\bar{\theta}$	Laplace transform of temperature, dimensionless
κ	combination of terms, dimensionless
Λ	combination of terms, dimensionless
λ	eigenvalue, dimensionless; optimum value, dimensionless
ρ	radius ratio, dimensionless
σ	Stefan-Boltzmann constant, $W/m^2 \cdot K^4$
τ	time, dimensionless
ϕ	combination of terms, dimensionless
ψ	combination of terms, dimensionless

Roman Letter Subscripts

a	fin tip
b	fin base
bi	initial condition at fin base
conv	convection
f	fluid final
i	initial value
n	nth eigenvalue
r	radiation; reference
s	surroundings
sc	sink, convection
sr	sink, radiation
ss	steady state
stored	stored
transient	transient

Roman Letter Superscripts

n	number of terms
r	power law exponent

Symbolic Superscripts

$'$	first derivative
$''$	second derivative; heat flux

CHAPTER 17

PERIODIC HEAT FLOW IN FINS

17.1 INTRODUCTION

As discussed in Chapter 16, it is essential to understand the performance of fins in transient or unsteady-state thermal conditions as found in many cooling applications, such as in electronic equipment, cylinders of air-cooled aircraft or jet engines, automatic control mechanisms, and energy storage devices. These diverse applications have spawned many investigations, and these, to single out a few, have been attributed to Suryanarayana (1975), Russell and Chapman (1967), Campo (1977), Aziz and Na (1980), Chu et al. (1983a,b), and Papadopoulos et al. (1990).

When fins operate under thermal conditions that are periodic in nature, the temperature response consists of two components, an initial transient component which dies away after a certain number of cycles and a steady periodic component which prevails indefinitely. The latter has been the subject of many studies and the problems involving steady periodic heat transfer in fins have been solved using a variety of analytical, numerical, and combined analytical–numerical approaches. These include the method of Frobenius, perturbation expansions, finite differences, finite elements, and straight-line and boundary element methods.

17.2 LONGITUDINAL FIN OF RECTANGULAR PROFILE

The method of complex combination has proved to be an effective tool for analyzing several periodic heat transfer conditions. Myers (1971) has described the method in detail and has used it to analyze steady periodic heat conduction in a semi-infinite medium. The method can be described briefly as follows:

1. Let $\theta_1 = \theta_1(x, t)$ describe the unknown steady periodic temperature distribution, where x and t are the spatial and time variables, respectively. Then introduce a new function, $\theta_2 = \theta_2(x, t)$, such that the equations for θ_2 are the same as those for θ_1 except that the periodic term, whether appearing in the equation or in the boundary conditions, is shifted by an angle $\pi/2$.
2. A complex temperature ψ whose real part is θ_1 and whose imaginary part is θ_2 is then created:

$$\psi = \theta_1 + \imath\theta_2$$

3. The equations for θ_1 and θ_2 are then combined to generate the equations governing ψ.
4. A solution for ψ of the form

$$\psi = F(X)e^{\pm \imath B\tau}$$

is assumed where $F(X)$, to be determined, is a function of a dimensionless variable X alone, B is a dimensionless frequency related to the frequency of the periodic term in the problem, and τ is a dimensionless time.
5. The assumed solution, ψ, is substituted into the governing equations for ψ. This results in an ordinary differential equation for $F(X)$ but with a complex coefficient. Using standard techniques, the equation is solved (often analytically) to obtain $F(X)$ and then ψ.
6. The final step is to extract the real part of ψ to obtain θ_1 because, by definition,

$$\theta_1 = \text{Re}[\psi]$$

The method of complex combination has been employed to study periodic heat transfer in a longitudinal fin of rectangular profile with various types of periodic conditions listed along with the sections in which they are considered in detail: oscillating base temperature (Section 17.2.1), oscillating base heat flux (Section 17.2.2), oscillating base fluid temperature (Section 17.2.3), and oscillating environmental temperature (Section 17.2.4).

17.2.1 Oscillating Base Temperature

The case of oscillating temperature has been studied using both one- and two-dimensional models.

One-Dimensional Model. Figure 17.1*a* shows a longitudinal fin of rectangular profile with thickness δ and height b. The fin, having thermal conductivity k and thermal diffusivity α, is exposed on both faces to a convective environment at temperature T_s via a heat transfer coefficient h. The fin tip is assumed to be adiabatic, while the base temperatute T_b is allowed to vary periodically around a mean temperature T_{bm} with frequency ω, as indicated in Fig. 17.1*b*.

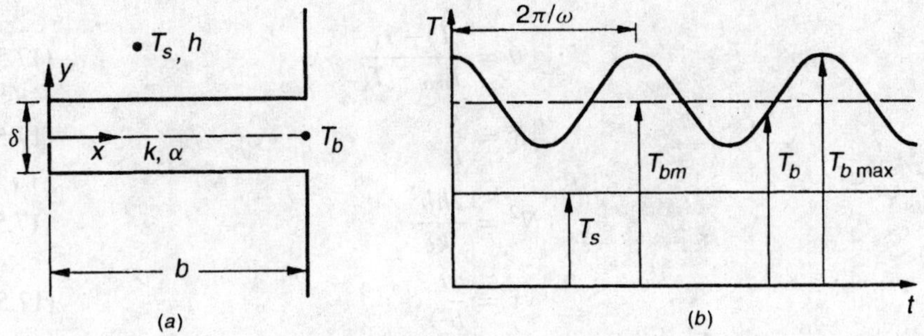

Figure 17.1 Longitudinal fin of rectangular profile: (*a*) fin geometry; (*b*) base temperature oscillation.

Assuming one-dimensional conduction, unit length, and constant thermal properties, the energy equation and boundary conditions can be written as

$$\frac{\partial^2 T}{\partial x^2} - \frac{2h}{k\delta}(T - T_s) = \frac{1}{\alpha}\frac{\partial T}{\partial t} \tag{17.1}$$

with boundary conditions

$$\frac{\partial T(0,t)}{\partial x} = 0 \tag{17.2a}$$

and

$$T(b,t) = T_b = T_{\mathrm{bm}} + (T_{\mathrm{bm}} - T_s)A\cos\omega t \tag{17.2b}$$

where

$$A = \frac{T_{b,\max} - T_{\mathrm{bm}}}{T_{\mathrm{bm}} - T_s} < 1$$

is the dimensionless amplitude of oscillation. Because the steady periodic solution is the only solution sought, there is no need to specify an initial condition.

In dimensionless form, eqs. (17.1) and (17.2) become

$$\frac{\partial^2 \theta}{\partial X^2} - N^2\theta = \frac{\partial \theta}{\partial \tau} \tag{17.3}$$

with boundary conditions

$$\frac{\partial \theta(0,\tau)}{\partial X} = 0 \tag{17.4a}$$

and

$$\theta(1,\tau) = 1 + A\cos B\tau \tag{17.4b}$$

where the dimensionless quantities are defined by

$$\theta = \frac{T - T_s}{T_{\text{bm}} - T_s} \tag{17.5a}$$

$$X = \frac{x}{L} \tag{17.5b}$$

$$N^2 = \frac{2hb^2}{k\delta} \tag{17.5c}$$

$$\tau = \frac{\alpha t}{b^2} \tag{17.5d}$$

and

$$B = \frac{\omega b^2}{\alpha} \tag{17.5e}$$

In preparation for the method of complex combination, first write

$$\theta = \theta_1(X) + \theta_2(X, \tau) \tag{17.6}$$

Then, substitution of this into eqs. (17.3) and (17.4) gives equations for θ_1 and θ_2. For θ_1,

$$\frac{d^2\theta_1}{dX^2} - N^2\theta_1 = 0 \tag{17.7}$$

with

$$\frac{d\theta_1(0)}{dX} = 0 \tag{17.8a}$$

and

$$\theta_1(1) = 1 \tag{17.8b}$$

and for θ_2,

$$\frac{\partial^2\theta_2}{\partial X^2} - N^2\theta_2 = \frac{\partial\theta_2}{\partial\tau} \tag{17.9}$$

with

$$\frac{\partial\theta_2(0, \tau)}{\partial X} = 0 \tag{17.10a}$$

and

$$\theta_2(1, \tau) = A \cos B\tau \tag{17.10b}$$

The solution for θ_1 is readily obtained as

$$\theta_1 = \frac{\cosh NX}{\cosh N} \tag{17.11}$$

Following the method of complex combination for the solution of θ_2, define a θ_3 problem such that

$$\frac{\partial^2 \theta_3}{\partial X^2} - N^2 \theta_3 = \frac{\partial \theta_3}{\partial \tau} \tag{17.12}$$

with

$$\frac{\partial \theta_3(0, \tau)}{\partial X} = 0 \tag{17.13a}$$

and

$$\theta_3(1, \tau) = A \sin B\tau \tag{17.13b}$$

Observe that the equations for θ_3 are the same as those for θ_2 except that the periodic condition has been shifted by $\pi/2$; that is, $\cos B\tau$ has been replaced by $\sin B\tau$. Next, a complex temperature ψ is defined such that its real part is θ_2 and its imaginary part is θ_3.

$$\psi = \theta_2 + \imath \theta_3 \tag{17.14}$$

When eqs. (17.12) and (17.13) are multiplied by $\imath$ and added to eqs. (17.9) and (17.10), the equations governing ψ become

$$\frac{\partial^2 \psi}{\partial X^2} - N^2 \psi = \frac{\partial \psi}{\partial \tau} \tag{17.15}$$

with

$$\frac{\partial \psi(0, \tau)}{\partial X} = 0 \tag{17.16a}$$

and

$$\psi(1, \tau) = Ae^{\imath B\tau} \tag{17.16b}$$

At this stage, ψ is assumed to be of the form

$$\psi = AF(X)e^{\imath B\tau} \tag{17.17}$$

where $F(X)$ is the function of X that is sought. If eq. (17.17) is substituted into eqs. (17.15) and (17.16), the equations for $F(X)$ appear as

$$\frac{d^2 F(X)}{dX^2} - (N^2 + \imath B)F(X) = 0 \tag{17.18}$$

with

$$\left.\frac{dF(X)}{dX}\right|_{X=0} = 0 \tag{17.19a}$$

and

$$F(X = 1) = 1 \tag{17.19b}$$

Because eqs. (17.18) and (17.19) are in the same form as eqs. (17.7) and (17.8), the solution for $F(X)$ follows as

$$F(X) = \frac{\cosh\,(N^2 + \imath B)^{1/2} X}{\cosh\,(N^2 + \imath B)^{1/2}} \tag{17.20}$$

Thus

$$\psi = A \frac{\cosh\,(N^2 + \imath B)^{1/2} X}{\cosh\,(N^2 + \imath B)^{1/2}} e^{\imath B \tau} \tag{17.21}$$

According to eq. (17.14), θ_2 is the real part of ψ. Defining

$$u = \left[\frac{N^2 + (N^4 + B^2)^{1/2}}{2} \right]^{1/2} \tag{17.22a}$$

and

$$v = \left[\frac{-N^2 + (N^4 + B^2)^{1/2}}{2} \right]^{1/2} \tag{17.22b}$$

the solution for ψ can be written as

$$\psi = A \frac{\cosh\,[(u + \imath v) X]}{\cosh\,(u + \imath v)} e^{\imath B \tau} \tag{17.23}$$

Expansion of the hyperbolic and exponential functions permits the real part of ψ to be expressed as

$$\mathrm{Re}[\psi] = \theta_2 = \frac{A}{C} [(\Lambda_1 + \Lambda_2) \cos B\tau + (\Lambda_3 - \Lambda_4) \sin B\tau] \tag{17.24}$$

where

$$\begin{aligned} \Lambda_1 &= \cosh u \cos v \cosh uX \cos vX \\ \Lambda_2 &= \sinh u \sin v \sinh uX \sin vX \\ \Lambda_3 &= \sinh u \sin v \cosh uX \cos vX \\ \Lambda_1 &= \cosh u \cos v \sinh uX \sin vX \end{aligned} \tag{17.25}$$

and

$$C = (\cosh u \cos v)^2 + (\sinh u \sin v)^2 \tag{17.26}$$

Thus the solution for θ is

$$\theta = \frac{\cosh NX}{\cosh N} + \frac{A}{C} [(\Lambda_1 + \Lambda_2) \cos B\tau + (\Lambda_3 - \Lambda_4) \sin B\tau] \tag{17.27}$$

The foregoing solution is developed from the work of Yang (1972), and the same solution was also reported by Houghton et al. (1990).

The heat transfer through the base of the fin can be evaluated in terms of the temperature gradient at the base, that is,

$$q_b = k\delta \frac{\partial T}{\partial x}\bigg|_{x=b}$$

or in dimensionless form,

$$\frac{q_b b}{k\delta(T_{\mathrm{bm}} - T_s)} = \frac{\partial \theta(1,\tau)}{\partial X} \tag{17.28}$$

Use of eq. (17.27) to evaluate the gradient gives

$$\frac{q_b b}{k\delta(T_{\mathrm{bm}} - T_s)} = N \tanh N + \frac{A}{2C}(\Upsilon_1 \cos B\tau - \Upsilon_2 \sin B\tau) \tag{17.29}$$

where

$$\begin{aligned} \Upsilon_1 &= u \sinh 2u - v \sin 2v \\ \Upsilon_2 &= v \sinh 2u + u \sin 2v \end{aligned} \tag{17.30}$$

As pointed out by Houghton et al. (1990, 1991), unlike the steady-state case, the base heat transfer rate q_b does not equal the fin surface heat transfer rate. The latter must be calculated by integrating the convective heat loss, that is,

$$q = 2\int_0^b h(T - T_s)dx \tag{17.31}$$

or in dimensionless form,

$$\begin{aligned} &\frac{q}{2hb(T_{\mathrm{bm}} - T_s)} = \\ &\int_0^1 \theta\, dX = \frac{\tanh N}{N} + \frac{A}{C}[(\Upsilon_3 + \Upsilon_4)\cos B\tau + (\Upsilon_5 - \Upsilon_6)\sin B\tau] \end{aligned} \tag{17.32}$$

where

$$\begin{aligned} \Upsilon_3 &= \frac{\cosh u \cos v(v \cosh u \sin v + u \sinh u \cos v)}{u^2 + v^2} \\ \Upsilon_4 &= \frac{\sinh u \sin v(u \cosh u \sin v - v \sinh u \cos v)}{u^2 + v^2} \\ \Upsilon_5 &= \frac{\sinh u \sin v(v \cosh u \sin v + u \sinh u \cos v)}{u^2 + v^2} \\ \Upsilon_6 &= \frac{\cosh u \cos v(u \cosh u \sin v - v \sinh u \cos v)}{u^2 + v^2} \end{aligned} \tag{17.33}$$

The average fin surface heat transfer rate over a cycle is given by

$$\bar{q} = \frac{\omega}{2\pi} \int_0^{2\pi/\omega} q\, dt = \frac{k\delta(T_{\text{bm}} - T_s)}{b} N \tanh N \tag{17.34}$$

or in dimensionless form,

$$\frac{\bar{q}b}{k\delta(T_{\text{bm}} - T_s)} = N \tanh N \tag{17.35}$$

which means that the average heat transfer rate equals the steady-state base flow, which is given by the first term of eq. (17.32).

The usual definition of the fin efficiency gives

$$\eta = \frac{q}{2hb(T_b - T_s)} \tag{17.36}$$

Because q and T_b are functions of time, η must be considered as an *intantaneous efficiency*. Use of eqs. (17.2) and (17.32) gives η as

$$\eta = \frac{1}{1 + A\cos B\tau}\left\{\frac{\tanh N}{N} + \frac{A}{C}[(\Upsilon_3 + \Upsilon_4)\cos B\tau + (\Upsilon_5 - \Upsilon_6)\sin B\tau]\right\} \tag{17.37}$$

where Υ_3 through Υ_6 are given by eq. (17.33).

In this situation, the performance of the fin is better indicated by the average fin efficiency over a cycle rather than the instantaneous efficiency η. Using eq. (17.37), $\bar{\eta}$ is given by

$$\bar{\eta} = \frac{\omega}{2\pi}\int_0^{2\pi/\omega} \eta\, dt = \frac{\tanh N}{N(1 - A^2)^{1/2}} + \frac{A}{C}\left[(\Upsilon_3 I_1 - \Upsilon_6 I_2) + (\Upsilon_4 I_1 + \Upsilon_5 I_2)\right] \tag{17.38}$$

where the integrals I_1 and I_2 are given by

$$I_1 = \int_0^{2\pi} \frac{\cos(B\tau)}{1 + A\cos(B\tau)}\, d(B\tau) \tag{17.39a}$$

and

$$I_2 = \int_0^{2\pi} \frac{\sin(B\tau)}{1 + A\cos(B\tau)}\, d(B\tau) \tag{17.39b}$$

The analysis is now complete and it is instructive to study the results. Examination of eq. (17.27) reveals that the temperature distribution is a function of three parameters: the amplitude parameter A, the frequency parameter B, and the fin parameter N. This equation shows that the oscillatory component is directly proportional to A. Figure 17.2 shows the effect of parameters B and N on temperature oscillation for $A = 0.1$. Temperature variations at the midpoint ($X = 0.5$) and the tip ($X = 0$) are shown together with the imposed temperature at the base ($X = 1$). At any axial location, both the amplitude and the phase angle are affected by changes in N and B.

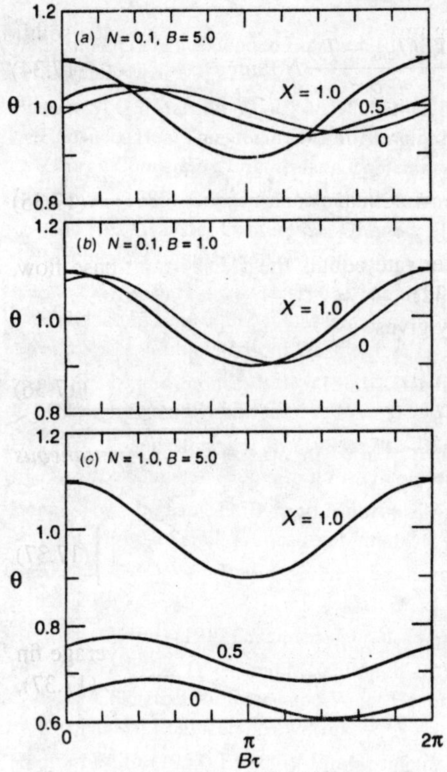

Figure 17.2 Effect of the parameters B and N on the temperature oscillation for $A = 0.1$. [Adapted from Yang (1972).]

As B increases, the amplitude decreases but the phase angle increases. For the set of parameters used, the parameter N has little effect on temperature oscillations.

Figure 17.3 shows the effect of N on the axial temperature distribution with $A = 0.1$ and $B = 1.0$. The dashed line corresponds to the steady-state distribution ($A = 0$). It is interesting to note that for $N = 0.2$, the temperature gradient at the base ($X = 1$) becomes negative for $B\tau = \pi/2$. This means that the heat is flowing back into the fin base at that instant. This phenomenon can be attributed to the thermal inertia of the fin. The effect of parameter B on the axial temperature distribution is shown in Fig. 17.4, with $A = 0.1$ and $N = 0.1$. Again, note that the back heat flow occurs at $B\tau = \pi/2$ and π for $B = 1$ and $B = 5$.

According to eq. (17.29), the base heat flux oscillates around the steady-state value. The oscillating component is directly proportional to A. Figure 17.5 demonstrates the effect of parameter N with A and B fixed at 0.1 and 1.0, respectively. The amplitude of oscillation is affected little as N changes from 0.2 to 1.0, but the phase angle changes significantly. For $N = 0.2$, the minimum heat flux ccurs at $\tau = 0.6\pi/\omega$, while for $N = 1.0$, it occurs at $\tau = 0.8\pi/\omega$. Comparing these with the base temperature variation where the minimum occurs at $\tau = \pi/\omega$, it is seen that the heat flux leads the base temperature. The curve for $N = 0.2$ indicates negative heat flow over a portion

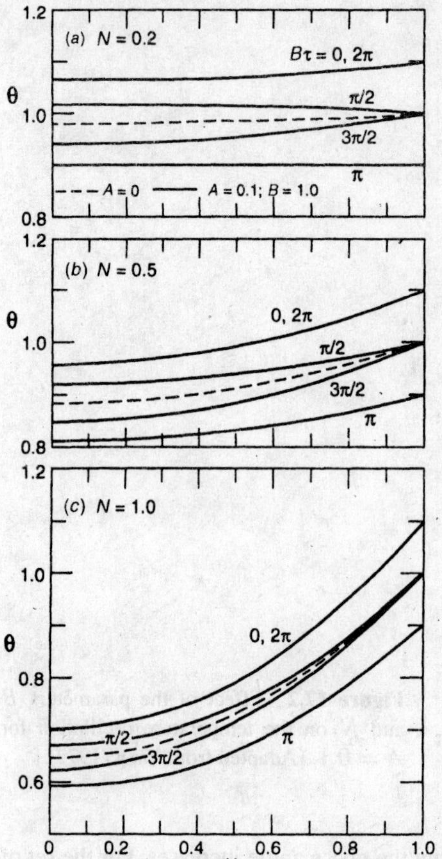

Figure 17.3 Effect of the parameter N on the axial temperature distribution for $A = 0.1$ and $B = 1.0$. [Adapted from Yang (1972).]

of the cycle, which substantiates the earlier observation made from the corresponding temperature distribution of Fig. 17.3*a*. The effect of the parameter B is illustrated in Figs. 17.6 and 17.7. In both cases the amplitude of oscillation increases significantly as B increases. Once again, negative base heat flows are observed for $B = 5$. Note that the effect is more pronounced at $N = 0.1$ (Fig. 17.7). In physical terms, this means that if the fin parameter is small, negative heat flow can occur if the base temperature oscillation is high and/or the fin material has a low thermal diffusivity.

Finally, results based on the average fin efficiency $\bar{\eta}$ can be considered. In Fig. 17.8, $\bar{\eta}$ is plotted as a function of N for $B = 1, 5$, and 10. Two sets of curves are shown, one for $A = 0.1$ and the other for $A = 0.05$. As B increases, $\bar{\eta}$ decreases. The reduction in efficiency compared to steady-state operation ($A = 0$) becomes noticeable for $N < 1$ and becomes quite substantial at low values of N.

The problem considered in this section has also been solved by Suryanarayana (1975) using the Laplace transformation and employing the residue theorem for evaluating the inverse. The algebra is more tedious, but the method has the advantage

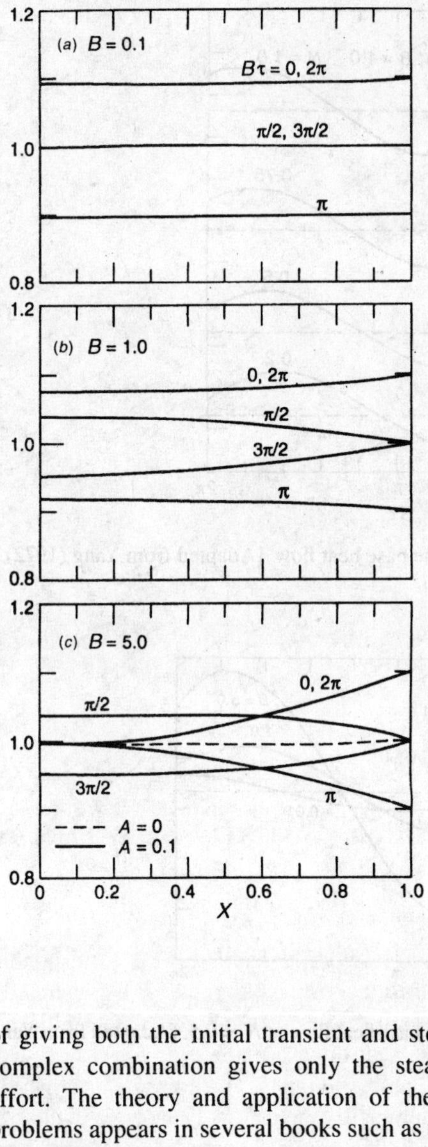

Figure 17.4 Effect of the parameter B on the axial temperature distribution for $N = 0.1$. [Adapted from Yang (1972).]

of giving both the initial transient and steady periodic responses. The method of complex combination gives only the steady periodic response but with reduced effort. The theory and application of the Laplace transform to heat conduction problems appears in several books such as those of Myers (1971) and Ozisik (1993). Approximate closed-form solutions for unsteady temperatures in fins are given by Beck et al. (1992) using Green's functions.

Two-Dimensional Model. For two-dimensional heat conduction, the energy equation and the boundary conditions take the form

$$\frac{\partial^2\theta}{\partial X^2} + \frac{\partial^2\theta}{\partial Y^2} = \frac{\partial\theta}{\partial\tau} \tag{17.40}$$

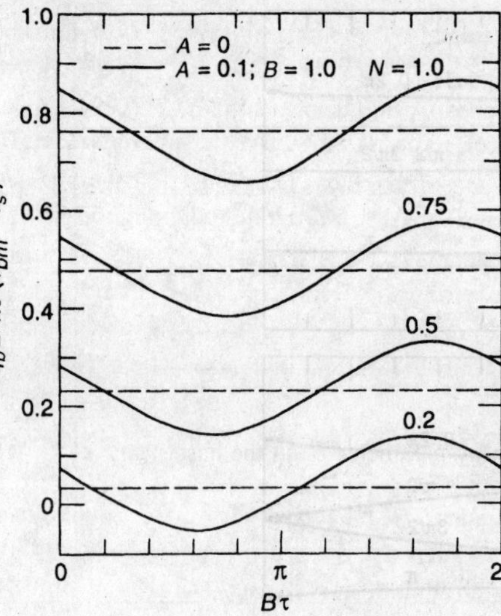

Figure 17.5 Effect of the parameter N on the base heat flow. [Adapted from Yang (1972).]

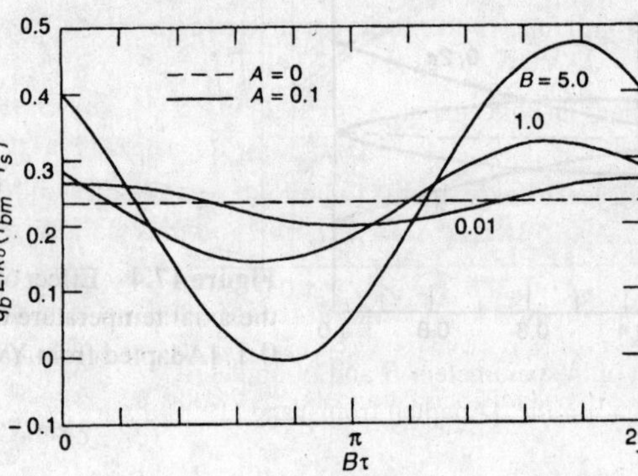

Figure 17.6 Effect of the parameter B on the base heat flow at $N = 0.5$. [Adapted from Yang (1972).]

$$\left.\frac{\partial \theta}{\partial X}\right|_{(X=0)} = 0 \tag{17.41a}$$

$$\theta(X = 1) = 1 + A \cos B\tau \tag{17.41b}$$

$$\left.\frac{\partial \theta}{\partial Y}\right|_{(Y=0)} = 0 \tag{17.41c}$$

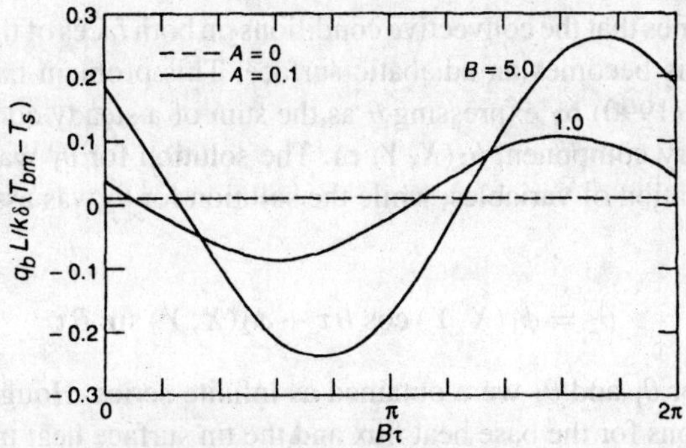

Figure 17.7 Effect of the parameter B on the base heat flow at $N = 0.1$. [Adapted from Yang (1972).]

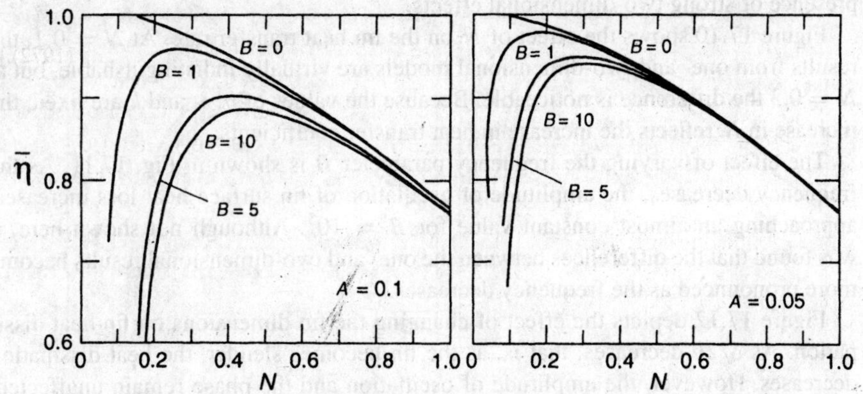

Figure 17.8 Effect of the parameters B and N on the time-averaged efficiency of a longitudinal fin of rectangular profile. [Adapted from Yang (1972).]

and

$$\left.\frac{\partial \theta}{\partial Y}\right|_{(Y=\delta/2b)} = -\text{Bi}\,\theta \tag{17.41d}$$

where the dimensionless quantities are defined in eqs. (17.5) except for

$$Y = \frac{y}{L} \tag{17.42a}$$

and

$$\text{Bi} = \frac{hb}{k} \tag{17.42b}$$

The model assumes that the convective conditions on both faces of the fin are identical, so that the x-axis becomes an adiabatic surface. This problem has been solved by Houghton et al. (1990) by expressing θ as the sum of a steady component $\theta_1(X, Y)$ and an oscillatory component $\theta_2(X, Y, \tau)$. The solution for θ_1 was obtained by the method of separation of variables, while the solution for θ_2 was assumed as being of the form

$$\theta_2 = \phi_1(X, Y) \cos B\tau + \phi_2(X, Y) \sin B\tau$$

The solutions for θ_1 and θ_2 were obtained as infinite series. Houghton et al. (1990) also gave solutions for the base heat flux and the fin surface heat transfer rate.

The results to be presented were generated using a material with $k = 204$ W/m·K and $\alpha = 8 \times 10^{-5}$ m²/s. A typical two-dimensional temperature distribution in such a fin with $\delta/2b = 0.5$ (a short, stubby fin) is shown in Fig. 17.9. The figure shows contours for $\theta = 1.0, 0.985, 0.970, 0.963, 0.959$, and 0.956 and clearly indicates the presence of strong two-dimensional effects.

Figure 17.10 shows the effect of N on the fin heat transfer rate. At $N = 0.1$, the results from one- and two-dimensional models are virtually indistinguishable, but at $N = 0.3$ the difference is noticeable. Because the values of b, δ, and k are fixed, the increase in N reflects the increase in heat transfer coefficients.

The effect of varying the frequency parameter B is shown in Fig. 17.11. As the frequency decreases, the amplitude of oscillation of fin surface heat loss increases, approaching an almost constant value for $B = 10^5$. Although not shown here, it was found that the differences between the one- and two-dimensional results become more pronounced as the frequency decreases.

Figure 17.12 depicts the effect of changing the fin dimensions on fin heat dissipation. As $\delta/2b$ decreases, that is, as the fin becomes slender, the heat dissipation decreases. However, the amplitude of oscillation and the phase remain unaffected.

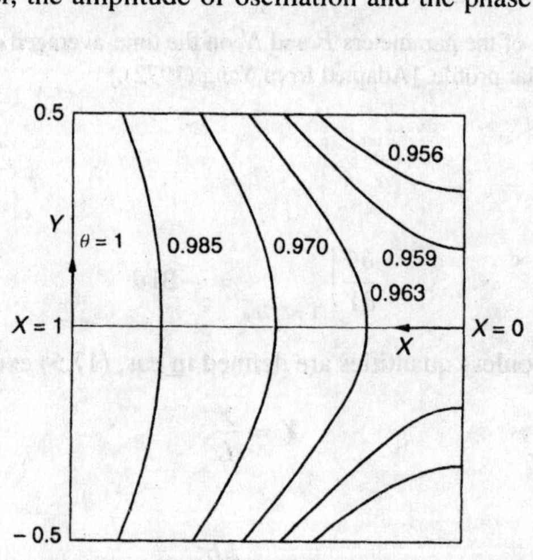

Figure 17.9 Typical two-dimensional temperature contours in a fin with $\delta/2b = 0.5$. [Adapted from Houghton et al. (1990).]

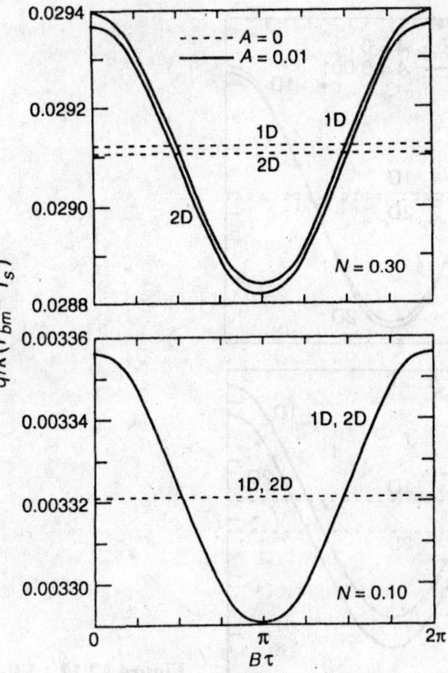

Figure 17.10 Effect of the heat transfer coefficient (the parameter N) on the fin heat transfer rate. [Adapted from Houghton et al. (1990).]

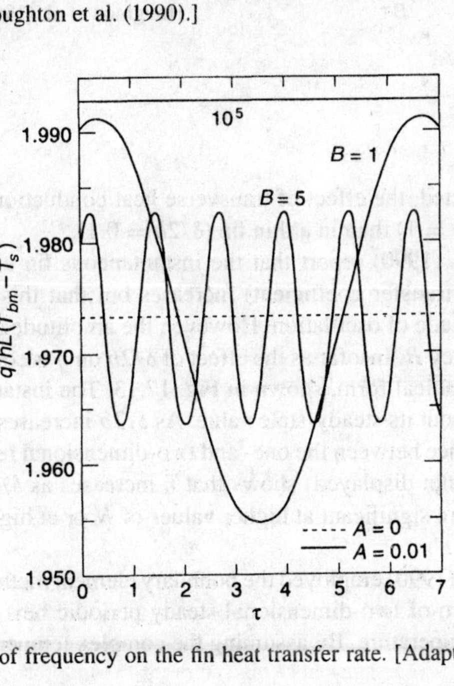

Figure 17.11 Effect of frequency on the fin heat transfer rate. [Adapted from Houghton et al. (1990).]

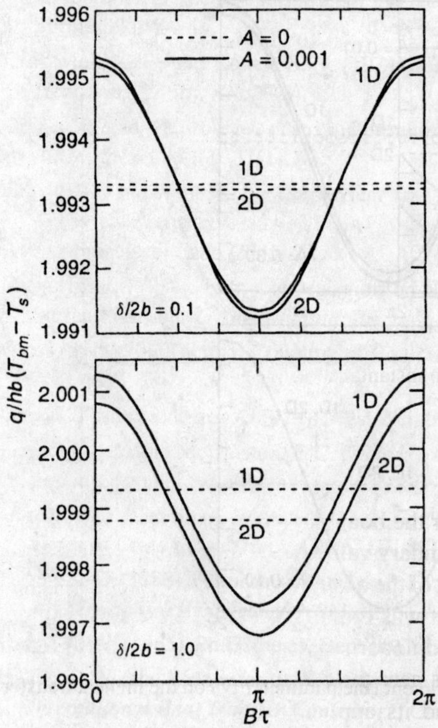

Figure 17.12 Effect of fin dimensions on the fin heat transfer rate. [Adapted from Houghton et al. (1990).]

As might be expected, the effect of transverse heat conduction is more pronounced in a thick fin ($\delta/2b = 1$) than in a thin fin ($\delta/2b = 0.1$).

Houghton et al. (1990) report that the instantaneous fin efficiency, η decreases as N (or the heat transfer coefficient) increases but that this has no effect on the phase or the amplitude of oscillation. However, the amplitude does increase with the increase in frequency B. Insofar as the effect of $\delta/2b$ on η is concerned, they give the information in graphical form, shown in Fig. 17.13. The instantaneous efficiency is seen to oscillate about its steady-state value. As $\delta/2b$ increases, η also increases and so does the difference between the one- and two-dimensional results. The average fin efficiency results (not displayed) shows that $\bar{\eta}$ increases as $\delta/2b$ increases and that this increase is more significant at higher values of N or at higher values of the heat transfer coefficient.

Houghton et al. (1990) employed the boundary element method (BEM) for solving the present problem of two-dimensional steady periodic heat transfer in a fin with oscillating base temperature. By assuming the complex temperature in the form

$$T = \phi(X, Y)e^{B\tau}$$

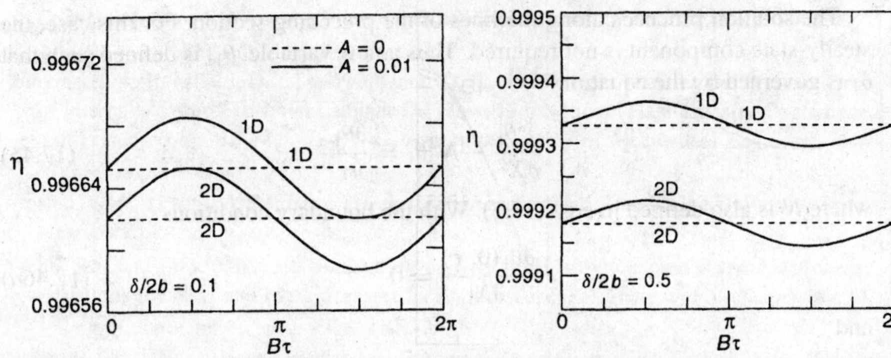

Figure 17.13 Effect of fin dimensions on instantaneous fin efficiency for one- and two-dimensional fins.

they were able to pose the problem for the boundary element method as two steady-state problems, a Laplacian mixed boundary value problem for the steady component and a Helmholtz equation for the ϕ part of the oscillatory component. In implementing the constant elements BEM method, it was found that the usual problems of corner singularities and slow convergence did not arise. The analytical and BEM solutions agreed closely with slight inaccuracies near the corners.

The boundary element method and its application to engineering problems in general has been described in detail by Brebbia et al. (1984). The book by Manzoor (1984) focuses exclusively on application of the technique to steady-state extended surface problems.

17.2.2 Oscillating Base Heat Flux

Unlike the case of oscillating base temperature, this case has been studied only with a one-dimensional heat conduction model. The problem is described by eqs. (17.1) and (17.2), but the boundary condition at the base now changes to

$$k\frac{\partial T(L,t)}{\partial x} = q'' \cos \omega t \tag{17.43}$$

In dimensionless form,

$$k\frac{\partial \theta(1,\tau)}{\partial X} = \cos B\tau \tag{17.44}$$

where q'' is the amplitude of the base heat flux and

$$\theta = \frac{k(T - T_s)}{q''b}$$

where X, B, and τ are defined by eqs. (17.5).

The solution proceeds along the lines of the preceding section. For this case, the steady-state component is not required. Thus a new variable, θ_1, is defined such that θ_1 is governed by the equation

$$\frac{\partial^2 \theta_1}{\partial X^2} - N^2 \theta_1 = \frac{\partial \theta_1}{\partial \tau} \tag{17.45}$$

where N is also defined in eqs. (17.5). With the boundary conditions

$$\frac{\partial \theta_1(0, \tau)}{\partial X} = 0 \tag{17.46a}$$

and

$$\frac{\partial \theta_1(1, \tau)}{\partial X} = \sin B\tau \tag{17.46b}$$

a complex temperature can be proposed:

$$\psi = \theta + \iota \theta_1 \tag{17.47}$$

The resulting equations for ψ are exactly the same as eqs. (17.15) and (17.16) with $A = 1$. Assuming that

$$\psi = F(X) e^{\iota B\tau}$$

the equations for $F(X)$ become

$$\frac{d^2 F(X)}{dX^2} - (N^2 + \iota B) F(X) = 0 \tag{17.48}$$

with

$$\frac{dF(0)}{dX} = 0 \tag{17.49a}$$

and

$$\frac{dF(1)}{dX} = 1 \tag{17.49b}$$

Their solution is

$$F(X) = \frac{\cosh (N^2 + \iota B)^{1/2} X}{(N^2 + \iota B)^{1/2} \sinh (N^2 + \iota B)^{1/2}} \tag{17.50}$$

Consequently,

$$\psi = \frac{\cosh (N^2 + \iota B)^{1/2} X}{(N^2 + \iota B)^{1/2} \sinh (N^2 + \iota B)^{1/2}} e^{\iota B\tau} \tag{17.51}$$

Extraction of the real part of ψ gives

$$\theta = \frac{f_1 \cos B\tau + f_2 \sin B\tau}{(u^2 + v^2)[(\cosh u \sin v)^2 + (\sinh u \cos v)^2]} \tag{17.52}$$

where u and v are defined by eq. (17.22) and

$$f_1 = \cosh uX \cos vX(u \sinh u \cos v - v \cosh u \sin v) + \sinh uX \sin vX(u \cosh u \sin v - v \sinh u \cos v) \qquad (17.53)$$

and

$$f_2 = \cosh uX \cos vX(u \cosh u \sin v - v \sinh u \cos v) + \sinh uX \sin vX(u \sinh u \cos v - v \cosh u \sin v) \qquad (17.54)$$

Now define

$$\beta = \arctan \frac{f_2}{f_1}$$

so that

$$\cos\beta = \frac{f_1}{(f_1^2 + f_2^2)^{1/2}} \quad \text{and} \quad \sin\beta = \frac{f_2}{(f_1^2 + f_2^2)^{1/2}}$$

then

$$\theta = \frac{(f_1^2 + f_2^2)^{1/2} \cos(B\tau - \beta)}{(u^2 + v^2)(\cosh^2 u \sin^2 v + \sinh^2 u \cos^2 v)} \qquad (17.55)$$

The amplitude

$$\frac{(f_1^2 + f_2^2)^{1/2}}{(u^2 + v^2)(\cosh^2 u \sin^2 v + \sinh^2 u \cos^2 v)}$$

and the phase angle β can be calculated for any given value of X, N, and B. Figure 17.14 shows the amplitude and phase angle plotted against X for $N = 1$ and $B = 0.01, 0.1, 1$, and 10. It can be observed that the amplitude decreases and the phase angle increases as the frequency B increases. Calculations for $N = 2$ reveal that amplitude and phase-angle values are smaller than those for $N = 1$.

The problem of oscillating base heat flux has also been solved by Suryanarayana (1975) using the Laplace transform approach. His conclusions regarding the steady periodic response of the fin are consistent with those reported here.

17.2.3 Oscillating Base Fluid Temperature

If the fin is heated at its base by convection through a fluid whose temperature T_f oscillates around its mean value T_{fm} in accordance with

$$T_f = T_{fm} + (T_{fm} - T_s)A \sin \omega t \qquad (17.56)$$

eq. (17.1) applies but the boundary condition at $x = b$ now takes the form

$$k\frac{\partial T(b,t)}{\partial x} = h_f[T_f - T(b,t)] \qquad (17.57)$$

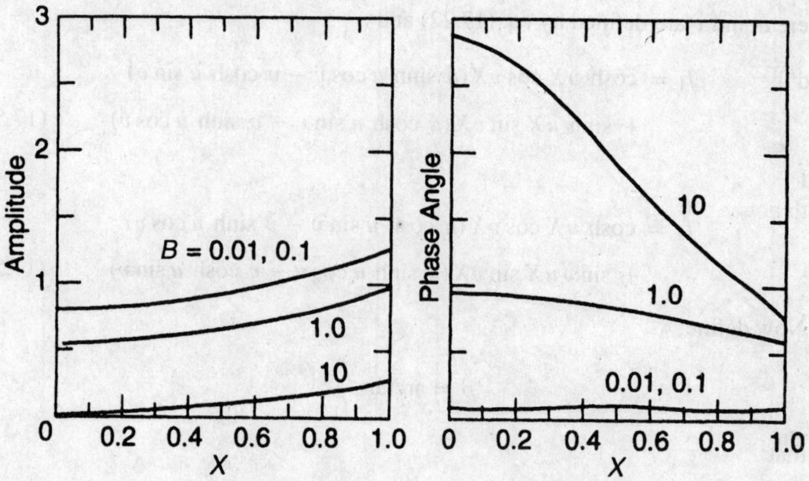

Figure 17.14 Effect of the parameter B on the amplitude and phase angle for $N = 1$ and with oscillating base heat flux.

where h_f is the convective heat transfer coefficient characterizing the heating at the base. In dimensionless form, eq. (17.57) becomes

$$\frac{\partial \theta(1, \tau)}{\partial X} + \mathrm{Bi}\,\theta(1, \tau) = \mathrm{Bi}(1 + A \sin B\tau) \tag{17.58}$$

where $\mathrm{Bi} = h_f b/k$ and

$$\theta = \frac{T - T_s}{T_{\mathrm{fm}} - T_s}$$

The remaining dimensionless variables have their definitions of eq. (17.5).

This problem was solved by Suryanarayana (1976) with the aid of the Laplace transformation. As remarked earlier, the technique gives both the initial transient as well as the steady periodic solution. The method of complex combination can also be employed to find the steady periodic solution. Suryanarayana (1976) does not present any results for the temperature distribution but does conclude that the temperature amplitude is attenuated as the frequency is increased. The amplitude of the base heat flux for a given Bi increases as N increases, but this effect appears only in the range $0.01 \le \mathrm{Bi} \le 10$. For $\mathrm{Bi} \ge 10$, the amplitude is not influenced by changes in the fin parameter N.

17.2.4 Oscillating Environmental Temperature

This case has been studied by Aziz and Sofrata (1981) using the method of complex combination. They assumed that the environmental temperature T_s oscillates around its mean value T_{fm} in accordance with

$$T_s = T_{sm} + (T_b - T_{sm}) A \cos \omega t \tag{17.59}$$

and they derived the governing partial differential equation in dimensionless form

$$\frac{\partial^2 \theta}{\partial X^2} - N^2(\theta - A \cos B\tau) = \frac{\partial \theta}{\partial \tau} \tag{17.60}$$

with boundary conditions

$$\frac{\partial \theta(0, \tau)}{\partial X} = 0 \tag{17.61a}$$

and

$$\theta(1, \tau) = 0 \tag{17.61b}$$

where

$$\theta = \frac{T - T_{sm}}{T_b - T_{sm}}$$

and where the remaining dimensionless variables have their definitions of eq. (17.5). Note that the base temperature T_b is fixed.

The conclusions of Aziz and Sofrata (1981) can be compared to the case of oscillating base temperature given in Section 17.2.1. The effect of the frequency parameter B in the two cases is similar. However, the parameter N has very little impact on the temperature oscillation when the environmental temperature is oscillating. This is in contrast with the oscillating base temperature case, where N significantly affects the temperature oscillations. Furthermore, because of different imposed boundary conditions, the directions of temperature amplitude attenuation along the fin are reversed. Whereas for the case of oscillating base temperature, the attenuation occurs from the base to the tip, the reverse is true for the case of oscillating environmental temperature.

With regard to the time-averaged fin efficiency, Aziz and Sofrata (1981) have shown that for $N < 2$,

$$\bar{\eta} = \frac{\tanh N}{N(1 - A^2)^{1/2}}$$

Thus $\bar{\eta}$ increases as A increases. For example, with $A = 0.5$, the increase is about 15%. This is in contrast to the case of oscillating base temperature, where $\bar{\eta}$ was found to decrease as A was increased (see Fig. 17.8). This problem has also been solved using the Laplace transformation by Papadopoulos et al. (1990).

17.3 RADIAL FIN OF RECTANGULAR PROFILE

The steady-state periodic heat transfer in radial convecting fins has been studied solely for the radial fin of rectangular profile. The first study in this area was conducted by Aziz (1975) when he extended the analysis of Section 17.2.1 to a radial fin. The solution for $F(X)$ given by eq. (17.17) appears in this case in terms of modified Bessel functions of complex argument. Using the series expansions for these functions, Aziz extracted the real and imaginary parts and computed the oscillating components. The numerical results for the temperature distribution, base heat flux, and time-average fin efficiency exhibit essentially the same behavior as results for the longitudinal fin.

Because the heat capacity in a radial fin increases with radial distance from the fin base, the damping of the temperature oscillation is greater than in the longitudinal fin, where the heat capacity does not change in moving from the fin base to the fin tip. Indeed, at high frequencies, the fin tip ceases to respond to the base temperature variation, and the temperature at the fin tip remains practically at the steady-state value.

Figure 17.15 shows a set of results for $\bar{\eta}$ for a radial fin with

$$R_b = \frac{r_b}{r_a - r_b} = 1 \quad \text{and} \quad R_a = \frac{r_a}{r_a - r_b} = 2$$

The parameters N and B are defined as

$$N^2 = \frac{2h(r_a - r_b)^2}{k\delta} \quad \text{and} \quad B = \frac{\omega(r_a - r_b)^2}{\alpha}$$

The definition of A is the same as in Section 17.2.1.

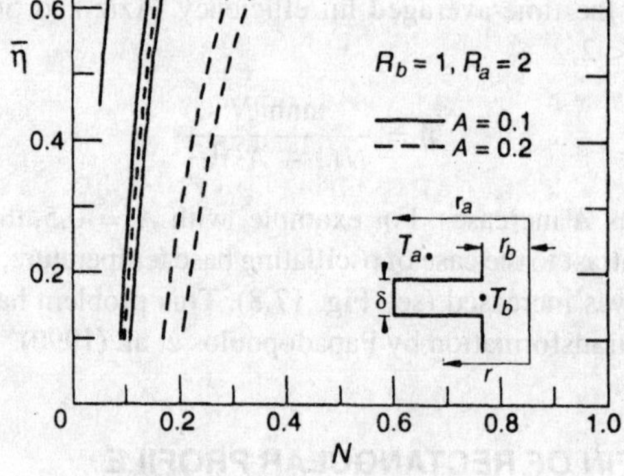

Figure 17.15 Effect of the parameters B and N on the time-averaged efficiency of a radial fin. [Adapted from Aziz (1975).]

Al Mujahid (1989) has studied the radial fin with periodic base heat flux and shows, in detail, how the modified Bessel functions of complex argument arising in the solution can be split into their real and imaginary parts, each part being in the form of a series. This paper is of both mathematical and engineering interest.

The case of oscillating environmental temperature with a fixed base temperature has been reported by Papadopoulos et al. (1990). The solution is derived using the Laplace transformation. Based on the numerical results generated, it was concluded that for periodic changes in the environmental temperature, the heat removed over a period of time is the same as that in steady-state operation at the mean environmental temperature.

17.4 ADDITIONAL STUDIES

17.4.1 Longitudinal Fins

Aziz and Na (1981b) considered the problem of oscillating base temperature (Section 17.2.1) with the added condition of heat loss to the environment by radiation. The analysis of Section 17.2.1 was extended to a trapezoidal fin by Crawford (1982) and a generalized analysis for the longitudinal fin of arbitrary profile was considered by Aziz and Na (1980), who considered the problem of oscillating base temperature and oscillating enviromental temperature separately.

Aziz and Na (1981a,b) analyzed the problem of oscillating base temperature (Section 17.2.1) with two additional effects. In the 1981a paper, they allowed the heat transfer coefficient to vary with respect to the fin height coordinate and the thermal conductivity to vary with temperature. The 1981b paper added the condition of heat loss to the environment by radiation. Using the method of complex combination, Al Mujahid (1987), developed an analytical solution for the longitudinal fin of rectangular profile with both base heat flux and environmental temperature oscillating simultaneously.

17.4.2 Arrays of Longitudinal Fins

Periodic heat transfer in fins and arrays of fins with simultaneous convection and radiation has been studied by Eslinger and Chung (1979) and Houghton et al. (1992, 1993a,b) have investigated the phenomenon of steady periodic heat transfer in a convecting fin assembly. The 1993b paper extended the problem to include two-dimensional heat conduction in the fin while allowing the fluid temperature on the unfinned side of the wall to experience an arbitrary variation rather than the cosine variation studied previously.

17.4.3 Radial Fins

The radial fin of rectangular profile under steady-state periodic conditions but with radiation has been been studied by Campo (1977).

17.5 NOMENCLATURE

Roman Letter Symbols

A	amplitude of oscillation
B	frequency, dimensionless
Bi	Biot number, dimensionless
b	fin or spine height, m
C	constant, dimensionless
F	function, dimensionless
f	combination of terms, dimensionless
h	heat transfer coefficient, $W/m^2{\cdot}K$
I	Integral, dimensionless
ι	complex number $\sqrt{-1}$, dimensionless
k	thermal conductivity, $W/m{\cdot}K$
N	combination of terms, dimensionless
q	heat flow, W
$\bar{q}$	average heat flow, W
q''	amplitude of base heat flux, W
R	radius, dimensionless
Re	real part, dimensionless
r	radius, m; radial coordinate, m
T	temperature, K
t	time, s
u	combination of terms, dimensionless
v	combination of terms, dimensionless
X	height, dimensionless
x	height coordinate, m
Y	thickness, dimensionless
y	thickness coordinate, m

Greek Letter Symbols

α	thermal diffusivity, m^2/s
β	angle, rad
Δ	change in, dimensionless
δ	fin thickness, m
∂	partial derivative, dimensionless
η	fin efficiency, dimensionless
$\bar{\eta}$	average fin efficiency, dimensionless
θ	temperature function, dimensionless
λ	eigenvalue, dimensionless
τ	time, dimensionless
γ	combination of terms, dimensionless
υ	combination of terms, dimensionless
Υ	combination of terms, dimensionless
ϕ	function, K

ψ	complex temperature, dimensionless
ω	angular frequency, s^{-1}

Roman Letter Subscripts

a	fin tip
b	fin base
bm	mean base temperature
f	fluid
fm	mean fluid temperature
s	surroundings

Symbolic Superscript

$''$	heat flux

CHAPTER 18

BOILING FROM FINNED SURFACES

18.1 INTRODUCTION

There is a modest literature concerning boiling off of extended surfaces. Haley and Westwater (1966) employed the procedures suggested by Wilkins (1960a) to optimize a spine dissipating heat to a boiling liquid. The Haley–Westwater adaption was necessary because in dealing with a spine, the functions of heat flow, temperature profile, and spine radius that were to satisfy conditions at the spine base and the spine tip were selected for a minimum spine volume rather than a minimum fin profile area. Although their general development could admit both a variation in exterior heat transfer coefficient and fin material thermal conductivity, their final expressions for fin height and spine radius (as functions of the height coordinate) employed a constant thermal conductivity. These relationships also clearly showed that the linear temperature gradient and parabolic spine shape suggested by Schmidt (1926) do not apply to the optimum spine in boiling because the value of the heat transfer coefficient is not constant over the spine surface.

In proposing a heat transfer coefficient

$$h(T) = K_n(T - T_s)^n$$

it is seen that if several forms of boiling exist simultaneously on the spine, for example, nucleate and film boiling, n and K will assume different values at different points on the spine. Haley and Westwater used the boiling curves for isothermal surfaces and water, isopropyl alcohol, and refrigerent R-113 to obtain the product $h(T - T_s)$, which yields the appropriate value for $h(T)$ for use in the equation for the spine height and the spine radius developed using Wilkins's procedures. The optimum spine was then determined from a numerical solution of these equations using Simpson's rule and a

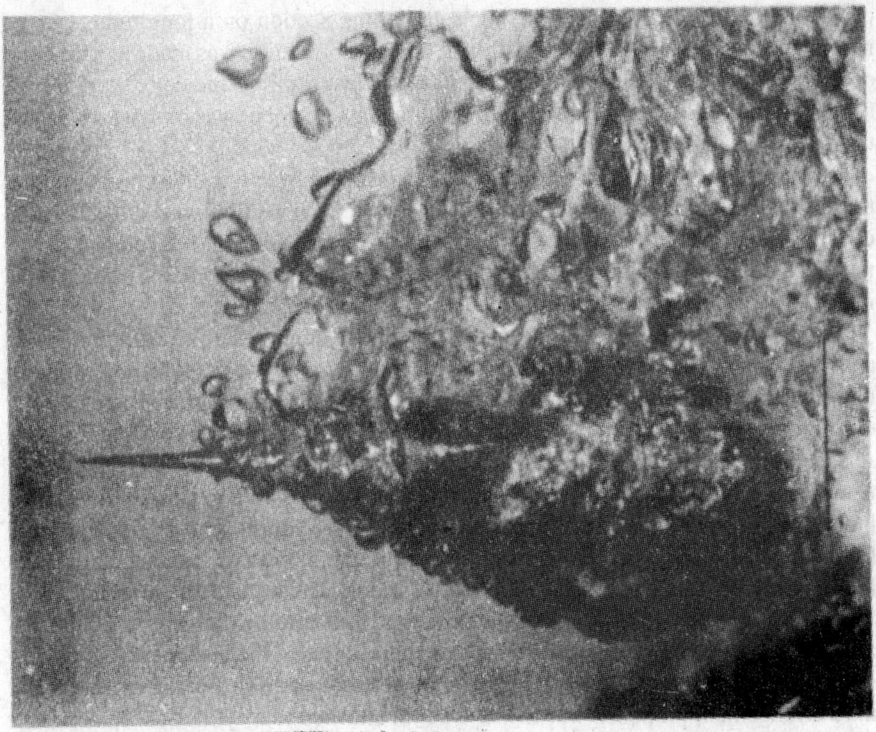

Figure 18.1 Haley–Westwater optimum spine shown operating at peak duty in R-113 at 1 atm with a base temperature excess of 91.7°C.

fourth-order Runge–Kutta approach. The optimum spine, so determined, possesses a spadelike shape and is shown operating at peak duty in R-113 in Fig. 18.1.

Here the usefulness of the spine with heat dissipation to a boiling liquid is clearly demonstrated. This unusual spine profile was proved logical and appears to be more clearly so when the distribution of the surface heat flux of the spine is considered. It is obviously desirable to minimimize the surface in the natural convection and film boiling regimes and to maximize the surface in the nucleate and transition boiling regimes. The surface in film boiling is minimized by using a very small cross section at the spine base so that the temperature drop in the metal required for passing through the film boiling region is accomplished over a very short distance. At the point on the transition region where the heat transfer coefficient begins to increase, the spine diameter is rapidly increased. This increase in diameter reduces the temperature gradient in the section so that the highly effective transition and nucleate boiling regimes are spread over a relatively large surface. Finally, as the heat transfer coefficient decreases at lower temperatures, the spine cross section tapers to a point. Thus the spine effectively utilizes both sides of the peak on the boiling curve.

The Haley–Westwater paper stimulated a great deal of activity in the field of boiling from extended surfaces. Lai and Hsu (1967) proposed a simple model for the determination of the length of the nucleate boiling section on a longitudinal fin of rectangular profile and the base heat flux of the fin. Correlation was made as a function of the ratios of the heat transfer coefficients of nucleate boiling and convective modes as well as the characteristics of the fin. Qualitative agreement was observed between the analytical and experimental results.

The analysis conducted by Lai and Hsu (1967) and by Hsu (1968) proposed that the fin be treated as the total of several sections. Each section was subjected to a different heat transfer coefficient and the key to the mathematical development was the fact that that the heat transferred through the fin at the exit from one section must match, by continuity, the heat entering the fin at the adjacent section.

Siman-Tov (1970) pointed out that the spadelike spine resulting from the Haley and Westwater (1966) study may be difficult and expensive to manufacture, difficult to weld, and that the very thin spine base may make the spine-base surface structurally unsound. He proposed three fin shapes that blend the shape concept calling for more surface in the nucleate boiling portion of the spine with a practicability for ease of manufacture and structural strength. The three shapes proposed by Siman-Tov (1970) include a crossed longitudinal fin which consists of two additional rectangular profile longitudinal fins attached in the long-dimension direction to the rectangular profile longitudinal fin that is attached to the base surface. The second shape proposed by Siman-Tov was a ring that is attached to a series of longitudinal fins protruding from a cylinder. In a view looking toward the end of the cylinder, the fin appears as an outer shell that appears as a finned annular passage between the ring and the cylinder. The third arrangement proposed was a disk-type fin attached to a cylindrical spine, a sort of "thumbtack" arrangement.

Siman-Tov used the computer to solve the governing equations that evolved for each of his arrangements. However, a constant heat transfer coefficient was assumed. He did show, however, that the peak heat flux could be moved along the fin as a function of the fin base temperature and he claimed a 93-fold increase in performance over a bare surface.

Cash et al. (1971) recognized that the spadelike spine, often referred to as the "turnip," would indeed be difficult to manufacture. They used a Runge–Kutta fourth-order integration to show that two cones attached to a small cylindrical neck is a good approximation to the optimum turnip-shaped spine. The two-cone arrangement, as expected, gave higher heat duties per unit volume than those calculated for a cylindrical spine. Experiments not only confirmed this result but showed the design by computer to be conservative and that the actual fins gave higher performance than that computed.

Thomas and Young (1970) used thin longitudinal fins spaced 0.318 cm apart on the inside of a vertical tube to markedly enhance the evaporative heat transfer coefficient of thin water films. For a given heat flux, there was an optimum feed rate for maximum heat transfer coefficient. At the optimum feed rate, the relative increase in heat transfer coefficient was directly proportional to the cube of the number of fins and inversely proportional to the heat flux. Jones (1970) described a tube with axial capillary grooves on the inside that was operable with qualities as high as 98% without water film dryout. The critical quality for a constant critical heat flux was found to be 15% higher than for an unprepared plain tube and required only 20% more pumping power. When the grooves were stuffed with a spiral 0.2362-cm-diameter wire, critical qualities as high as 99% were acheived but at the expense of 11 times the pumping power.

Klein and Westwater (1971) investigated the effect of fin (spine) spacing on heat duty using horizontal cylindrical spines in water and in R-113 at atmospheric pressure. Up to five parallel spines were used in a single horizontal row and up to three were used in a single vertical column. Multiple rows and columns were also tested with a maximum of 10 fins arranged in three horizontal rows of three, four, and three fins each. They found that the spines could be brought very close together before any effect of spacing on the heat duty was detected. A horizontal spacing of about 0.7938 cm was wide enough to allow all spines to act independently of one another, even for an array of nine or 10 spines. They pointed out that caution should be used in applying their results to arrays of hundreds of spines because new effects might occur when huge amounts of vapor are produced.

Bondurant and Westwater (1971) employed a fourth-order Runge–Kutta approach to solve the heat balance equations for the radial fin of rectangular profile with both the fin thermal conductivity and the heat transfer coefficient from its faces functions of temperature. Data for the heat transfer coefficient were taken from the Haley and Westwater (1966) paper. An entire boiling curve of total heat duty as a function of base temperature difference was generated. They concluded that radial fins could be put as close together as about a 0.1588-cm clearance.

Lienhard and Funk (1971) pointed out that the boiling heat flux remained within 90% of the maximum heat flux over a substantial range between two temperatures T_R and T_L where the subscripts for "right" and "left" are obvious and where T is the temperature above the saturated temperature. In a sense, they proposed a "bandwidth" from T_R to T_L and suggested that a fin or spine should be operated over the bandwidth with its base at T_R and its tip at T_L. They showed that such a fin would not only operate at a point near the point of maximum heat flux but would exhibit an effectiveness higher than unity.

Petukhov et al. (1972) conducted experiments using R-113 and showed that the stable occurrence of nucleate, transition, and film boiling can occur simultaneously on a single fin. Jones (1970) used a computer analysis and large-scale-model experiments to determine the air-side performance of louvered-fin evaporator cores. Heat transfer and pressure drops were determined for several configurations and compared with calculated values. Effects of condensation, fin spacing, fin thickness, and the core characteristics were determined. The computer model gave valid results over a rather wide range of Reynolds numbers, but it was stated that the model did not have the ability to obtain the liquid-side heat transfer coefficient, the effect of louver angle or louver spacing or the effect of water droplet formation on the airflow and pressure drop.

Gogonin and Svorkova (1973) made an experimental determination of heat flow densities for the boiling of R-21 on smooth tubes and tubes with trapezoidal profile fins of various aspect ratios. Takeyama et al. (1974) approximated the boiling curve as a combination of straight lines on log-log coordinates and their experimental study of a fin covered entirely with a fluid in film boiling showed that the local heat transfer coefficient depends on the temperature distribution and other factors. Shih and Westwater (1974) provided the results of experiments conducted to obtain the optimum dimensions of spheres, hemispheres, and disks as high-performance fins using two metals and two boiling liquids.

Edwards et al. (1974) demonstrated that a power law solution exists for the capillary flow supplying liquid to the tip of an evaporator fin. This solution indicated a maximum heat transfer coefficient limited by fluid properties but differing for different fin surfaces. They gave an estimate of the heat transfer coefficient obtainable using water on titanium fins. Ouchi and Takeyama (1978) conducted an experimental study of some extended surfaces in boiling water at atmospheric pressure. Their purpose was to emphasize the differences in heat transfer coefficients between surfaces in general and surfaces equipped with extended surfaces.

Kubanek and Miletti (1979) took heat transfer and pressure drop measurements on three integral spiraled inner-fin tubes with two-phase flow of R-22 under evaporating conditions. The data were compared with the performance of smooth tubes with and without a star-shaped insert. Enhancements in heat transfer coefficient for the internally finned tubes over those for smooth tubes were claimed to range from 30 to 760% and they typically increased with mass velocity. Tighter fin spacing was also shown to increase the heat transfer coefficient. For the star-shaped inserts, enhancements from 40 to 370% were claimed, but these enhancements decreased with mass velocity. However, these heat transfer enhancements were seen to come at the expense of pressure drop which increased from 10 to 290% for the internally finned tubes and from 300 to over 2000% for the star-shaped inserts.

Chen et al. (1981) improved on a method proposed in 1975 for predicting the heat transfer rates to a boiling liquid in a compact plate–fin heat exchanger. The improvement consisted of an extension to include the effect of velocity on the local heat transfer coefficient. Confirmation was obtained by a comparison of predicted and actual data obtained using R-113 in an actual heat exchanger. Petukhov et al. (1981) experimented with boiling R-113 in the narrow horizontal slot which modeled the space between two adjoining longitudinal fins. They found that boiling heat transfer in such a constrained space differs in a number of significant respects from boiling under nonconstrained (pool boiling) conditions. They pointed out that the most important of these differences is in an increase in heat flux in transition boiling with an increase in the rate of liquid circulation through the spaces.

Robertson and Lovegrove (1983) considered the boiling of R-11 in brazed aluminum pin fin exchangers, and Chen and Westwater (1983) made an application of the local assumption (the local heat transfer coefficient during flow boiling is uniquely fixed by the metal-to-liquid ΔT and the local fluid velocity) for the design of compact heat exchangers for boiling applications. Robertson (1984) developed a model for predicting convective boiling for what he referred to as the "peculiar" two-phase-flow condition within serrated fin passages where the liquid film flowing on the fin is totally entrained at the trailing edge of each serration. The liquid fin is subsequently deposited from the entrained liquid onto the following fins. The model made use of a droplet deposition mechanism, and by fitting the predictions of the model to the results from convective boiling heat transfer experiments it was claimed that it was possible to obtain the best values of nondimensional deposition coefficients from previously reported tests with liquid nitrogen and R-11 in upflow over a range of pressures, mass flows, and qualities.

Biyikli (1985) developed design charts for the optimum use of longitudinal fins of rectangular profile in boiling liquids to be used in engineering design applications. The boiling liquids to which the charts pertain are water, isopropyl alcohol, and R-113.

18.2 EVAPORATION

18.2.1 Thermodynamic Basis

When a liquid contained in a vessel is uniformly heated at constant pressure and at a slow rate, the resulting changes in temperature and volume are related in a characteristic way as shown in Fig. 18.2. Starting at point A, the addition of heat results in a temperature rise and a modest rise in volume until point 1 is reached. Along this all-liquid locus, the amount of heat required to raise the temperature of a unit mass 1 degree is nearly constant and is referred to as the liquid *constant-pressure specific heat*, c_p. In the region between points 1 and 2, vapor and liquid are present as coexisting phases in equilibrium. The addition of heat increases the vapor fraction, and hence the volume, without altering the temperature of the two-phase mixture. The amount of heat needed to convert a unit mass of liquid into vapor is termed the *latent heat of vaporization*, h_{fg}. At point 2, all the liquid has vaporized, and further heating results in a temperature rise of the vapor along a locus appropriate to the *vapor* constant-pressure specific heat. The terminaton of the all-liquid state, point 1, and the beginning of the all-vapor state, point 2, are generally referred to as as the liquid and vapor *saturation points*, respectively.

This description pertains to addition of heat to a quantity of liquid at a single constant pressure. If the process is repeated for many different pressures, the *saturation curve* shown in Fig. 18.3 is obtained. The *saturated liquid line* shown in the figure

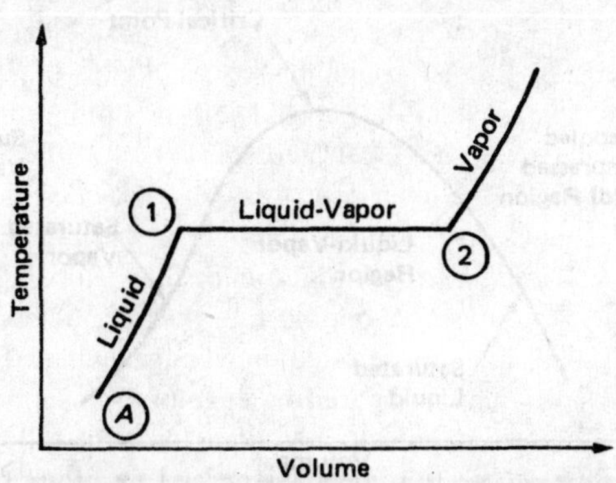

Figure 18.2 Constant pressure heat addition.

represents the locus of all points 1, and the *saturated vapor line* represents the locus of all points 2. It is of interest to note that a *critical* temperature and pressure, at which the liquid and vapor molecules are energetically identical, and at which no difference exists between the specific volume of the liquid and vapor, can be defined for each fluid. This point corresponds to the top of the *dome* in Fig. 18.3.

For convenience, the relation between the saturation temperature (line 1 to line 2 in Fig. 18.2) and pressure implicit in Fig. 18.3 is often plotted directly, to reflect the dependence of vapor pressure on the liquid temperature. The appropriate curves for several of the fluorocarbon fluids are shown in Fig. 18.4.

18.2.2 Thermal Transport Mechanism

Because of the higher internal energy of a vapor molecule of the same species and at the same temperature, evaporation of a fluid, under considerations far removed from the critical point, generally results in a very substantial heat flux. The migration or diffusion away from a free liquid–vapor interface is governed by the difference in vapor pressure or vapor concentration between the interface and the ambient. Along the interface, it is generally assumed that the vapor and liquid are in thermodynamic equilibrium, and therefore that the vapor at the interface corresponds to the saturation value at the local temperature. Because, as seen in Fig. 18.4, the saturation pressure increases with temperature, the rate of evaporation from a free surface in an environement of fixed vapor pressure increases as the liquid temperature approaches the boiling point. At a given total ambient pressure, that is, vapor pressure plus pressure of other gas components, the liquid temperature cannot exceed the boiling point, and consequently, the maximum interface vapor pressure and evaporation rate are attained at this point.

It must be noted, however, that when the heat of vaporization needed for this process is transferred to the liquid by boiling along surfaces submerged in the liquid, the bursting of vapor bubbles at the liquid vapor interface may enhance the evaporative

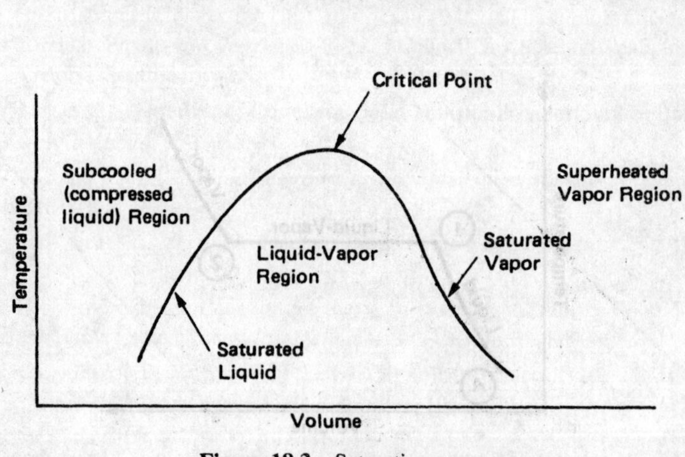

Figure 18.3 Saturation curve.

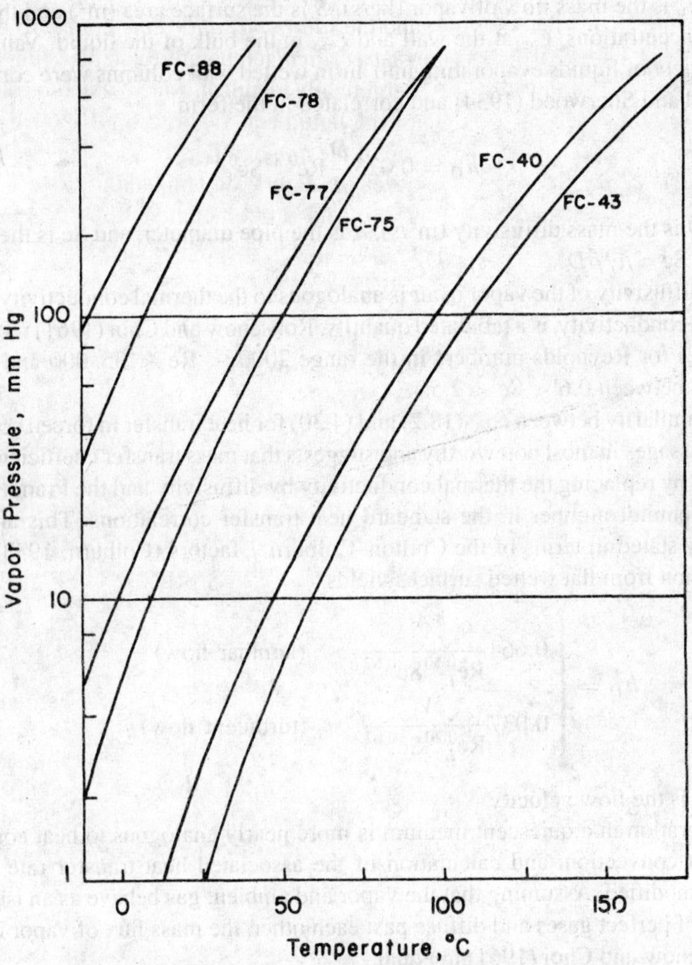

Figure 18.4 Vapor pressure of several fluorocarbon liquids.

process and result in heat transfer rates much above those associated with diffusive vapor transport from the interface.

18.2.3 Predicting Evaporation Rates

By analogy to convection heat transfer as discussed in Chapter 4, the convection of vapor from wetted surfaces can be related to a mass transfer coefficient h_D and can be described by an equation of the form (Rohsenow and Choi, 1961).

$$\dot{m}_v = h_D S(C_w - C_\infty) \tag{18.1}$$

where $\dot{m}_v$ is the mass flow of vapor (kg/s), S is the surface area (m^2) and the C's are mass concentrations, C_w at the wall and C_∞ in the bulk of the liquid. Values of h_D for a variety of liquids evaporating into air in wetted wall columns were compiled by Gilliland and Sherwood (1934) and correlated in the form

$$h_D = 0.023\frac{D}{d}\text{Re}^{0.83}\text{Sc}^{0.44} \tag{18.2}$$

where D is the mass diffusivity (m^2/s), d is the pipe diameter, and Sc is the Schmidt number, $\text{Sc} = \mu/\rho D$.

The diffusivity of the vapor in air is analogous to the thermal conductivity of solids and, like conductivity, is a tabulated quantity. Rohsenow and Choi (1961) recommend eq. (18.2) for Reynolds numbers in the range $2000 < \text{Re} < 35,000$ and Schmidt numbers between $0.6 < \text{Sc} < 2.5$.

The similarity between eqs. (18.2) and (4.30) for heat transfer in forced convection inside passages in most noteworthy and suggests that mass transfer coefficients can be obtained by replacing the thermal conductivity by diffusivity and the Prandtl number by the Schmidt number in the standard heat transfer correlations. This analogy is generally stated in terms of the Chilton–Colburn j-factors (Colburn, 1933) and for evaporation from flat wetted surfaces yields

$$h_D = \begin{cases} 0.664\dfrac{\hat{V}}{\text{Re}_L^{0.50}\text{Sc}^{0.67}} & \text{(laminar flow)} \quad (18.3) \\ 0.037\dfrac{\hat{V}}{\text{Re}_L^{0.20}\text{Sc}^{0.67}} & \text{(turbulent flow)} \quad (18.4) \end{cases}$$

where $\hat{V}$ is the flow velocity.

Evaporation in a quiescent medium is more nearly analogous to heat conduction than heat convection, and calculation of the associated heat transfer rate must be suitably modified. Assuming that the vapor and ambient gas behave as an isothermal mixture of perfect gases and diffuse past each other, the mass flux of vapor is shown by Rohsenow and Choi (1961) to equal

$$\frac{\dot{m}_v}{S} = \frac{M_v D p}{RTL} \ln \frac{p - p_\ell}{p - p_w} \tag{18.5}$$

where M_v is the molecular weight of the vapor, L the diffusion path, p the total pressure, R the universal gas constant, p_ℓ the partial pressure of the liquid, and p_v the partial pressure of the vapor. Multiplication of the mass flux of the vapor $\dot{m}_v/S$ by the latent heat of vaporization h_{fg} yields the desired surface heat flux.

18.3 BOILING

Although boiling, like evaporation, involves a change from the liquid to the vapor phase, in contrast to evaporation, boiling generally refers to a process that occurs along surfaces submerged in a fluid and is characterized by the dominant influence of vapor bubbles. Boiling typically begins when the surface temperature exceeds the liquid saturation temperature (the *boiling point*) by 3 to 10°C, and the vapor bubbles are then found to grow and issue from minute cavities in the surface.

In the analysis and correlation of boiling phenomena, it is convenient to distinguish between *pool boiling*, referring to the boiling of an initially quiescent liquid, and *flow boiling*, referring to boiling in the presence of a strong velocity field, as may occur, for example, in pipe flow. Furthermore, because the boiling regime depends primarily on the temperature of the heated surface, both *subcooled boiling*, during which the bulk fluid is below the saturation temperature, and *saturated boiling*, during which the bulk fluid is at the saturation temperature, may be observed.

In boiling, the variation of heat flux along a heated element or surface results in a characteristic temperature response at the surface, reflecting a progression through particular regimes of ebullient heat transfer. This behavior is generally represented by a log–log plot of the heat flux q'' versus the surface superheat $T_w - T_s$. A typical boiling curve for the saturated pool boiling of R-113 at atmospheric pressure is shown in Fig. 18.5.

The initial, preboiling part of the curve, labeled I, represents thermal transport by natural convection, resulting from the temperature difference between the heated surface and the liquid. At point *a*, sufficient superheat is available to initiate the growth of vapor bubbles at nucleation sites on the surface and boiling begins. In region II, modest increases in surface superheat result in the activation of many more

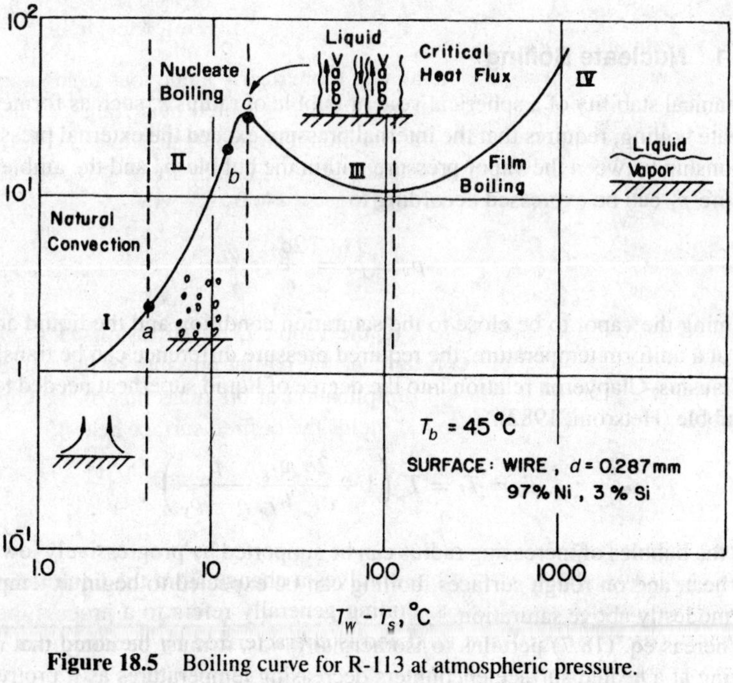

Figure 18.5 Boiling curve for R-113 at atmospheric pressure.

nucleation sites and a rapid increase in the frequency of bubble departure at each site. The violent disturbance of the hot boundary layer along the heated surface and gross fluid circulation induced by the motion of the vapor bubbles leads to a very steep rise in the heat flux from the heater in this *nucleate boiling regime*.

At point b in Fig. 18.5, bubble departure frequency at each nucleation site is so high that the trailing bubble merges with the leading bubble and vapor columns rooted at particular nucleation sites appear. Beyond this inflection point, bubble interference constrains the incremental increase in heat flux resulting in higher wall superheat, and at point c, bubble interference is so severe that the flow of liquid to the surface is inhibited and *burnout* occurs. This peak or *critical heat flux* represents a local maximum in the boiling curve, and higher heat fluxes can be obtained only in the *film boiling regime*, labeled IV, at much higher wall superheats than are encountered in nucleate boiling. In region IV a vapor layer blankets the heater, and relatively ineffective thermal conduction through the vapor (augmented by radiation) must be relied on to transfer the heat from the heated surface to the liquid. Region III on the boiling curve corresponds to unstable film boiling and is characterized by rapid local oscillations between nucleate and film boiling.

Because of the complex interactions of the distinct heat transfer mechanisms active in ebullient thermal transport, it is not yet possible to present the entire boiling curve by a single analytical function. Rather, it is common practice to provide separate analytical and empirical relationships for each inflection point and boiling regime. Predictive expressions for the boiling curve and salient features of the individual boiling regimes are discussed in succeeding sections.

18.3.1 Nucleate Boiling

Mechanical stability of a spherical void or bubble of radius r, such as formed during nucleate boiling, requires that the internal pressure exceed the external pressure. The relationship between the vapor pressure within the bubble p_v and the ambient liquid pressure p_ℓ can be expressed according to

$$p_v - p_\ell = \frac{2\sigma}{r} \tag{18.6}$$

Assuming the vapor to be close to the saturation condition and the liquid and vapor to be at a uniform temperature, the required pressure difference can be translated via the Clausius–Clapyeron relation into the degree of liquid superheat needed to sustain the bubble (Hetsroni, 1982):

$$T_v = T_\ell = T_s\left(1 + \frac{2\sigma}{r}\frac{v_{fg}}{h_{fg}}\frac{\rho}{\rho - \rho_v}\right) \tag{18.7}$$

Thus the bubbles of increasing radius can be supported by progressively lower liquid superheat, and on rough surfaces, boiling can be expected to begin at temperatures only modestly above saturation.

Whereas eq. (18.7) pertains to isothermal fluids, it must be noted that a bubble growing at a heated surface encounters decreasing temperatures as it protrudes into the liquid boundary layer. Under such circumstances it is the liquid superheat at the bubble "roof" that must follow eq. (18.7), and the wall temperature must exceed this value. Following Bergles and Rohsenow (1964), the wall superheat required for formation of the first bubble or boiling incipience at a heated surface can be extimated by

$$(T_w - T_s)_{\text{incip}} = \left(\frac{8q''T\sigma}{\rho_v h_{fg} k_f}\right)^{1/2} \tag{18.8}$$

where the subscript f indicates a film condition. The preboiling or incipience heat flux q'' is itself a function of the wall-to-liquid temperature difference and a convection heat transfer coefficient h such that $q'' = h(T_w - T_s)_{\text{incip}}$. Consequently, in a saturated liquid, the estimated incipience superheat can be expressed as

$$(T_w - T_s)_{\text{incip}} = \frac{8hT\sigma}{\rho_v h_{fg} k_f} \tag{18.9}$$

It must be noted that eqs. (18.8) and (18.9) presume that cavities of all radii exist on the surface and that the liquid temperature decreases linearly from the wall. As a result, when applied to real surfaces with low heat fluxes or heat transfer coefficients, these equations are found to underpredict the superheat required for boiling incipience.

Bubble Departure Diameter. A vapor bubble growing in a surface cavity will remain attached to the surface until the appropriate combination of buoyancy, fluid drag, and inertia forces overcome the retarding force of surface tension. For the commonly encountered slow-growing bubbles, prime attention can be focused on buoyancy and surface tension forces and the bubble departure diameter for saturated pools (Cole and Rohsenow, 1969):

$$\text{Eu}^{1/2} = (1.5 \times 10^{-4})\text{Ja}^{*5/4} \qquad \text{for water} \tag{18.10a}$$

and

$$\text{Eu}^{1/2} = (4.65 \times 10^{-4})\text{Ja}^{*5/4} \qquad \text{for other liquids} \tag{18.10b}$$

where $\text{Eu} = g(\rho - \rho_v)d_b^2/g_o\sigma$ is the *Eustis number* and $\text{Ja}^* = \rho c T_s/\rho_v h_{fg}$ is the *modified Jakob number*.

Bubble Departure Frequency. The frequency with which bubbles depart a specific nucleation site is dependent on the heat flux and thermofluid properties and has been found to be related as well to the bubble departure diameter. Much of the data are inconsistent, but for slowly growing bubbles (thermally controlled), the frequency can be estimated by (Hetsroni, 1982).

$$f = \frac{9\pi\alpha(\text{Ja})^2}{16d_b^2} \tag{18.11}$$

where

$$\text{Ja} = \frac{\rho c(T_w - T_s)}{h_{fg}\rho_v}$$

is the *Jakob number*. The departure frequency of dynamically controlled rapidly growing bubbles is given by Cole (1960):

$$f = \left[\frac{4g(\rho - \rho_v)}{3C_D\rho d_b}\right]^{1/2} \tag{18.12}$$

where C_D is the drag coefficient.

The foregoing bubble frequency expressions can be used to determine the bubble flux or total number of bubbles departing a heated boiling surface when the density and size of bubble nucleation sites are known. This information, however, is rarely available. Alternatively, the bubble flux can be determined from the vapor flux and bubble departure diameter. Unfortunately, the vapor flux Q_v is itself difficult to predict, and although Q_v can be expressed as

$$Q_v = \eta \frac{q''}{h_{fg}\rho_v} \tag{18.13}$$

the value of η varies in complex fashion, from 0.01 to 0.02 at low heat fluxes to $\eta > 0.50$ at 20% of the critical heat flux (Graham and Hendricks, 1967) and to unity as the critical heat flux is approached.

Nucleate Boiling Curve. Because increased heat flux in the nucleate boiling regime is generally associated with the appearance of more vapor bubbles, and according to Graham and Hendricks (1967), the content of the vapor in the bubbles typically accounts for only a small percentage of the total heat transfer, it appears possible to relate the observed heat flux to "bubble pumping" of hot liquid away from the heated surface. Based on this postulate and the definition of a bubble Reynolds number, Rohsenow (1952) was able to obtain one of the more successful correlations of nucleate pool boiling:

$$\frac{c_\ell(T_w - T_s)}{h_{fg}} = C_{sf}\left[\frac{q/A}{\mu h_{fg}}\sqrt{\frac{g_o\sigma}{g(\rho - \rho_v)}}\right]^r \text{Pr}^s \tag{18.14}$$

The C_{sf} coefficient is interpreted as reflecting the influence of cavity size distribution of the surface on vapor bubble generation, and hence ebullient heat transfer, and has been tabulated (Rohsenow and Choi, 1961; Hetsroni, 1982) for various fluid–surface combinations. The surface quality can also affect the exponents r and s in eq. (18.14), but nevertheless, r is generally taken as $\frac{1}{3}$ and $s = 1.0$ for water, but 1.70 for all other fluids. With these exponents, Hetsroni (1982) points out that a C_{sf} value of 0.013 correlates a wide spectrum of data to within $\pm 20\%$.

To obtain a precise relation between heat flux and wall superheat in the nucleate boiling regime, it is generally advisable to perform a limited number of experiments with the surface–fluid combination of interest and to use the results to establish C_{sf}, r, and s empirically.

The wall superheat encountered during boiling in flowing liquids is influenced not only by the heat flux, fluid properties, and surface characteristics, but also reflects the liquid flow rate and the vapor fraction of the liquid. Accurate prediction of flow-boiling heat transfer rates is thus extremely difficult and requires, in addition to precise property values, detailed knowledge of the geometry and orientation of the heated element and the flow regime and thermal history of the working fluid.

In the absence of such detail, it is possible to obtain a first estimate of flow boiling behavior by superimposing the single-phase convective heat transfer rate on the *bubble pumping effect* (Rohsenow and Choi, 1961) by

$$\left.\frac{q}{A}\right|_{fb} = \left.\frac{q}{A}\right|_{c} + \left.\frac{q}{A}\right|_{b} \tag{18.15}$$

In using eq. (18.15), $q/A|_c$ can be calculated from the standard natural and forced convection heat transfer coefficient correlations presented in Chapter 4 and the surface-to-liquid (actual, not saturation) temperature difference. The *pure* boiling heat flux can be obtained from the Rohsenow (1952) correlation given by eq. (18.14). Under most circumstances this same relation can also be used to account approximately for the effect of significant liquid subcooling on the pool boiling curve.

Parametric Effects. Because of the bubble pumping nature of nucleate pool boiling, the heat transfer rates and wall superheats attained in this regime are sensitive to both bubble and thermal conduction parameters. Some of the important dependencies are embodied in eq. (18.14), and this relation can generally be used to obtain at least a first estimate of the effects of variations in the fluid properties. Thus it is found that with increasing pressure, the boiling curve shifts toward the left or toward lower wall superheats, as shown, for example, in Fig. 18.6.

As noted earlier, the character of the surface can influence the location and shape of the boiling curve and is one of the primary variables in ebullient behavior. Rougher surfaces generally contain more and larger cavities than polished surfaces of the same material, and as a consequence, nucleate boiling from rough surfaces begins and continues at lower wall superheats than from smooth surfaces (Hetsroni, 1982) and is characterized by lower values of C_{sf}.

Although not shown explicity in the previous equations, the presence of noncondensible gases dissolved in the working fluid can have a profound influence on boiling behavior and is especially significant in dielectric fluids that possess a very high solubility for air. In such fluids, bubble formation and growth are aided by the presence of gas in the cavities along the heated surface, which reduces the vapor partial pressure and wall superheat needed for nucleation and shifts the nucleate boiling curve to lower wall superheats.

In this boiling regime, vapor and possibly noncondensible gas occupying minute surface cavities serve as nuclei for vapor bubbles. Evaporation at the bubble–liquid interface provides the vapor needed for the bubble growth and the available superheat determines the size of three activated cavities. The available boiling data suggest that for most working fluid–surface combinations, the wetting angle, shown in Fig. 18.7, is

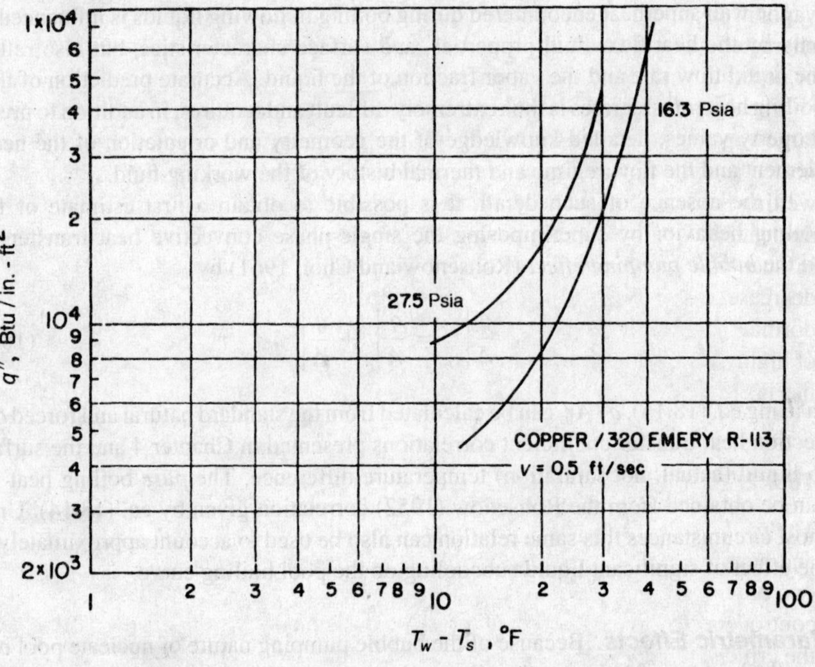

Figure 18.6 Effect of pressure on nucleate boiling of R-113. Adapted from Bergles et al. (1968).]

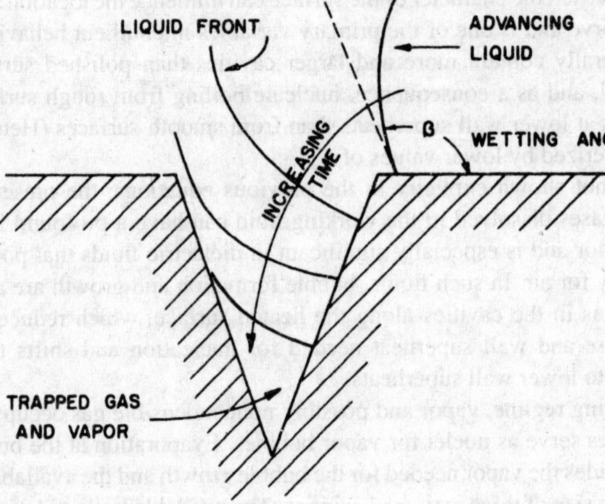

Figure 18.7 Vapor trapping in surface cavities by an advancing liquid front (time increasing).

sufficiently large that the inflow of liquid in the wake of a departing bubble traps some vapor in the cavity and provides a nucleus for a subsequent bubble. The proper wetting characteristics are thus central to the nucleate boiling process, and extremely wetting or nonwetting surface–fluid combinations can result in an anamolous behavior.

The boiling of nonwetting liquids is accompanied by the generation of very large vapor bubbles, which, according to Hetsroni (1982), blanket wide areas of the heated surface and result in rather substantial wall superheats. Alternatively, highly wetting fluids penetrate and even fill small cavities and thus eliminate many potential nucleation sites. Boiling incipience, via the small-diameter cavities remaining active in such fluids, may necessitate very high wall superheats followed by a sudden decrease in surface temperature as the growing bubble activates the previously dormant larger-diameter cavities. Such behavior may be encountered in the boiling of highly wetting dielectric fluids and may help explain the thermal hysteresis, shown in Fig. 18.8, observed in the boiling of R-113.

18.3.2 Critical Heat Flux

The *peak*, *critical*, or *burnout* heat flux marks the end of the nucleate boiling regime, and heat fluxes in excess of this value result in wall superheats that are often one to two orders of magnitude above those prevailing in nucleate boiling. Although some controversy still exists in the literature over the physical mechanism responsible for the critical heat flux, it appears possible to explain much of the available data within the framework of Zuber's hydrodynamic instability model (Zuber, 1958; Bergles, 1975). This model asserts that the critical heat flux results when sufficient liquid can no longer flow past the vapor columns to the heated surface as a result of the breakdown of the liquid–vapor interface. For boiling on large, horizontal surfaces, Hetsroni (1982) points out that the critical heat flux q''_{crit} is found to equal

$$q''_{\text{crit}} = \Phi = 0.18\rho_v h_{fg}\left[\frac{\sigma g g_o(\rho - \rho_v)}{\rho_v^2}\right]^{1/4}\left(\frac{\rho + \rho_v}{\rho}\right)^{-1/2} \tag{18.16}$$

where the constant, 0.18, is determined empirically. Experimental results for saturated pool boiling of many fluids are closer to the values of q''_{crit} calculated via eq. (18.16), but heater size, fluid velocity, and liquid subcooling exert a very profound influence on the critical heat flux. In contrast with nucleate boiling, surface characteristics are generally found not to affect the magnitude of the critical heat flux.

Heater Configuration. A series of investigations by Sun and Lienhard (1970) and Ded and Liehhard (1972) produced a unified approach to geometric effects on pool boiling critical heat flux and showed that the variation between experimental and predicted values could be related to a characteristic length

$$L' = L\sqrt{\frac{g(\rho - \rho_v)}{\sigma g_o}}$$

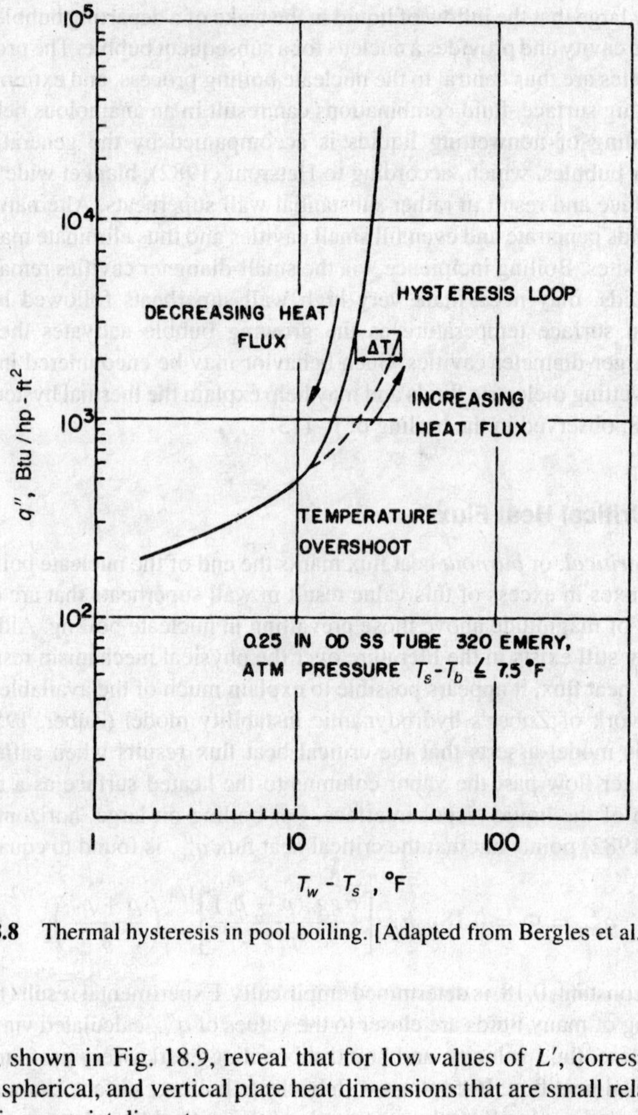

Figure 18.8 Thermal hysteresis in pool boiling. [Adapted from Bergles et al. (1968).]

The results, shown in Fig. 18.9, reveal that for low values of L', corresponding to cylindrical, spherical, and vertical plate heat dimensions that are small relative to the characteristic vapor jet diameter,

$$d_j \approx \frac{\sigma g_o}{g(\rho - \rho_v)}$$

the critical heat flux is considerably larger than that predicted by eq. (18.16), but the values of q''_{crit} approach the *Zuber relation* (Hetsroni, 1982) asymptotically as L' increases toward 10.

It is interesting to note that at very small values of L', the first bubble nucleated may blanket the heater and lead to an abrupt transtion from natural convection to film boiling. Under these circumstances, as pointed out by Bakhru and Lienhard (1972), it

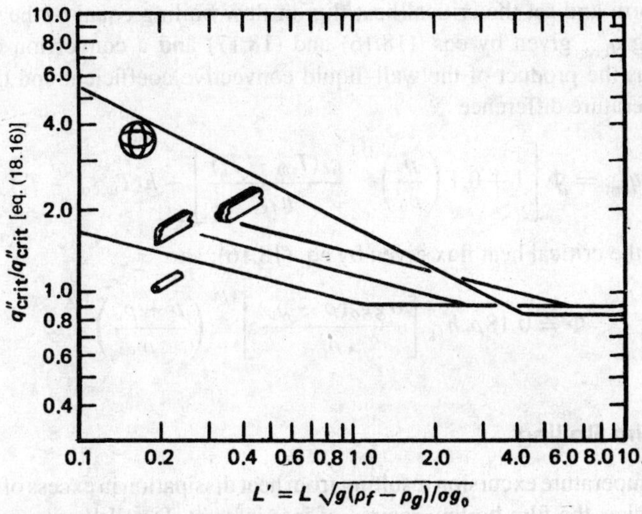

Figure 18.9 Predictions of critical heat flux for various heater configurations. [Adapted from Bergles (1975).]

is difficult to define the critical heat flux, and extrapolation of the curves of Fig. 18.9 below values of $L' = 0.10$ must therefore be done with caution.

Liquid Subcooling and Velocity. Experimental results suggest that the degree of subcooling in the bulk fluid can markedly affect the critical heat flux. This parametric variation has been correlated by several investigators, including Ivey and Morris (1962), who found the simple expression

$$\frac{q''_{\text{crit,sub}}}{q''_{\text{crit,sat}}} = 1 + 0.1\left(\frac{\rho}{\rho_v}\right)^{3/4} \frac{c_\ell(T_s - T_\ell)}{h_{fg}} \tag{18.17}$$

The flow of liquid toward and along the heated element provides an alternative path for liquid to reach the boiling surface and eases the removal or collapse of vapor columns generated at the surface. These two mechansims can be expected to increase the critical heat flux, and according to Lienhard and Keeling (1970), even a modest natural circulation velocity in a nominally quiescent liquid can increase the critical heat flux substantially.

In the presence of forced convection, the critical heat flux is often as much as an order of magnitude higher than in pool boiling, but precise prediction is made difficult by the dependence of q''_{crit} on the two-phase flow regime, system dynamics, and thermal boundary conditions. Distinct correlations, many of which are summarized by Hetsroni (1982), are available for particular parametric combinations but are generally limited to a high-pressure water flow, of interest to the nuclear industry.

For other fluids and as a general tool for obtaining a first estimate of q'' in subcooled flow boiling, it may be convenient to follow Gambill's (1962) recommendation in modified form and set the critical heat flux in flow boiling equal to the sum of the pool boiling q''_{crit} given by eqs. (18.16) and (18.17) and a convection component calculated as the product of the wall–liquid convective coefficient and the wall-to-liquid temperature difference

$$q''_{\text{crit}} = \Phi\left[1 + 0.1\left(\frac{\rho}{\rho_v}\right)^{3/4}\frac{c_\ell(T_{\text{sat}} - T_\ell)}{h_{fg}}\right] + h(T_{w,\text{crit}} - T_\ell) \tag{18.18}$$

where Φ is the critical heat flux given by eq. (18.16):

$$\Phi = 0.18\rho_v h_{fg}\left[\frac{\sigma g g_o(\rho - \rho_v)}{\rho_v^2}\right]^{1/4}\left(\frac{\rho + \rho_v}{\rho}\right)^{1/2} \tag{18.16}$$

18.3.3 Film Boiling

The large temperature excursion resulting from heat dissipation in excess of the critical heat flux makes the film boiling regime of less interest. Detailed experimental data are generally available for cryogenic fluids and low-molecular-weight fuels.

The minimum film boiling heat flux has been shown by Zuber and Tribus (1958) to be derivable from a stability analysis of the liquid–vapor boundary that yields

$$q''_{\text{min}} = 0.09\rho_{v,f} h_{fg}\left[\frac{g(\rho - \rho_v)}{\rho + \rho_v}\right]^{1/2}\left[\frac{g_o\sigma}{g(\rho - \rho_v)}\right]^{1/4} \tag{18.19}$$

where $\rho_{v,f}$ is the vapor density evaluated at the average film temperature and all other properties are at the saturation temperature.

The minimum wall superheat to sustain film boiling was found by Berenson (1960) to be expressible as

$$\Delta T_{\text{min}} = 0.127\frac{\rho_{v,f} h_{fg}}{k_{v,f}}\left[\frac{g(\rho - \rho_v)}{\rho + \rho_v}\right]^{2/3}\left[\frac{g_o\sigma}{g(\rho\rho_v)}\right]^{1/2}\left[\frac{\mu_f}{g_o\sigma(\rho - \rho_v)}\right]^{1/3} \tag{18.20}$$

The effective heat transfer coefficient associated with thermal conduction through the vapor film blanketing the heated surface in stable film boiling can be determined by analogy to the Nusselt analysis commonly employed in film condensation. A balance between buoyant and frictional forces and consideration of evaporation at the liquid–vapor interface yields the film thickness and subsequently the heat transfer coefficient. For horizontal tubes in pool boiling, Bromley et al. (1953) found

$$h_c = 0.62\left\{\frac{k_v^3\rho_v(\rho - \rho_v)g[h_{fg} + 0.4c_{p,v}(T_w - T_s)]}{d\mu_v(T_w - T_s)}\right\}^{1/4} \tag{18.21}$$

where d is the tube diameter. After the addition of thermal radiation, the total heat transfer coefficient is expressed as

$$h = h_c\left(\frac{h_c}{h}\right)^{1/3} + h_r \tag{18.22}$$

where h_r is calculated for radiation between parallel plates. Following Berenson (1960), h_c in film boiling on horizontal surfaces can be obtained by replacing the coefficient 0.62 in eq. (18.21) by 0.425 and the pipe diameter d by

$$\sqrt{\frac{g_o \sigma}{g(\rho - \rho_v)}}$$

It may be observed that the diameter of vapor bubbles released during film boiling is proportional to this factor and ranges, according to Bromley et al. (1953), as

$$3.14\sqrt{\frac{g_o \sigma}{g(\rho - \rho_v)}} \leq d_b \leq 5.45\sqrt{\frac{g_o \sigma}{g(\rho - \rho_v)}} \tag{18.23}$$

For flow film boiling along horizontal cylinders with significant velocity, that is, for $V_\infty > 2\sqrt{gd_o}$, Bromley et al. (1953) suggest

$$h_c = 2.7\sqrt{\frac{\hat{V}_\infty k_v \rho_v [h_{fg} + 0.4c_p(T_w - T_s)]}{d(T_w - T_s)}} \tag{18.24}$$

and

$$h = h_c + \frac{7h_r}{8}$$

18.4 PERFORMANCE OF A SINGLE CYLINDRICAL SPINE

Consider a horizontal cylindrical spine of radius r and height b, which, as shown in Fig. 18.10, is immersed in a saturated liquid at T_{sat}. If $h(x)$ is the value of the heat transfer coefficient at any location x, and k is the thermal conductivity of the spine, the temperature distribution in the spine is governed by the differential equation

$$\frac{dT}{dx} - \frac{2h(x)}{kr}(T - T_{\text{sat}}) = 0 \tag{18.25}$$

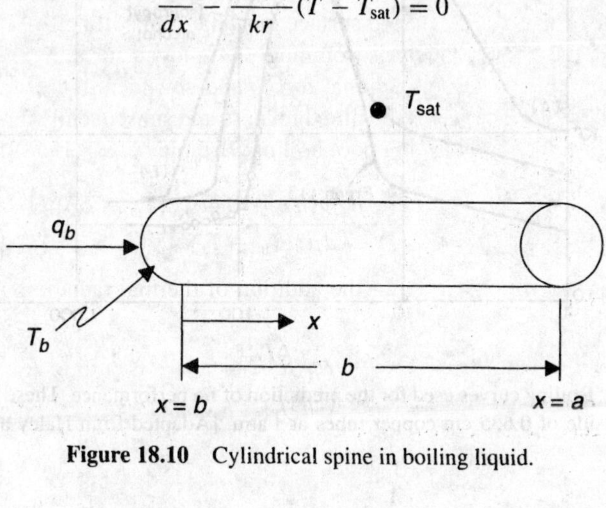

Figure 18.10 Cylindrical spine in boiling liquid.

and the solution of this differential equation is subject to boundary conditions at the fin tip (at $x = 0$) and the fin base at $x = b$. For the fin tip

$$\left.\frac{dT}{dx}\right|_{x=0} = h(a)(T_a - T_{\text{sat}}) \qquad (18.26a)$$

where T_a is the tip temperature and $h(a)$ is the heat transfer coefficient at the tip. For the fin base

$$\left.\frac{dT}{dx}\right|_{x=b} = \frac{q_b}{k\pi r^2} \qquad (18.26b)$$

where q_b is a specified quantity.

To solve eq. (18.25), the functional relationship between $h(x)$ and $T - T_{\text{sat}}$ must be specified. This, of course, depends on the boiling regime that is established over the surface of the spine and the boiling fluid. To seek some insight, refer to the experimental evidence provided by Haley and Westwater (1966), which is displayed in Fig. 18.11. The curves in this figure are based on the data of Haley (1965) for R-113, Breen and Westwater (1962) for isopropyl alcohol, and Braunlich (1941) for water. The fluids shown in Fig. 18.11 are all in contact with a copper spine 0.635 cm in diameter at 1 atmosphere.

Because the curves in Fig. 18.11 show that the natural convection, nucleate boiling,

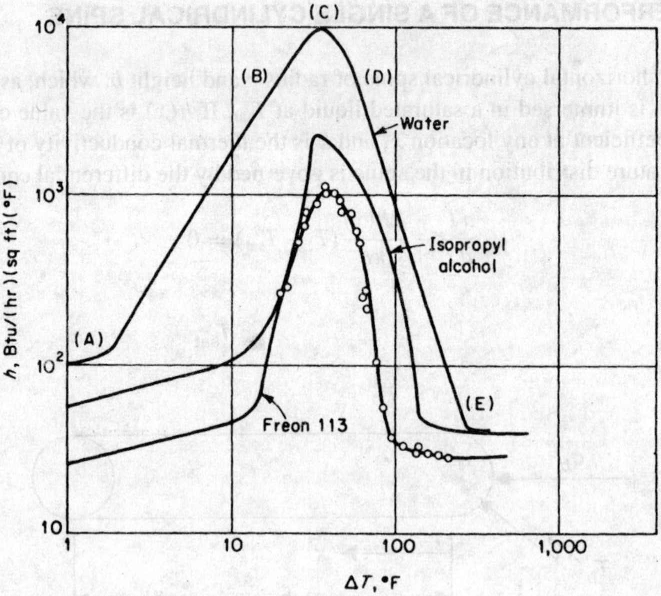

Figure 18.11 Boiling curves used for the prediction of fin performance. These were obtained for liquids outside of 0.635 cm copper tubes at 1 atm. [Adapted from Haley and Westwater (1966).]

and film boiling regimes all coexist, the possibility ofobtaining a single relationship of the form

$$h(T) = K_n \, \Delta T^n$$

is precluded. Instead, the actual data in Fig. 18.11 must be employed.

There is a conceptual difficulty in using the data of Fig. 18.11 with eqs. (18.25) and (18.26). The data are for an isothermal surface, but the actual spine surface will be anything but isothermal. The difficulty, however, can be overcome by using the *local assumption*, which says that the heat transfer coefficient h at any location on the spine is the same as would exist if the entire fin were at that temperature. In other words, the local temperature $T - T_{sat}$ can be interpreted as ΔT when reading the values of h from the boiling curve.

The numerical integration of eq. (18.25) with the boundary conditions of eqs. (18.26) begins with a knowledge, or assumption, of k, b, r, q_b, T_{sat}, and the experimental curve of h as a function of ΔT, such as any of the curves in Fig. 18.11. The value of the tip temperature may be estimated and the integration can be performed along the x-direction from tip to base to see if the boundary condition at $x = b$ is satisfied. If not, the process must be repeated with a different value of the tip temperature until a match is obtained.

Such a numerical solution, obtained by Haley and Westwater (1966) for a 0.635-cm copper spine of different heights and operating in water at 1 atmosphere is shown in Fig. 18.12. The figure gives

$$\frac{q_b}{k\pi r^2} = \left.\frac{dT}{dx}\right|_{x=b}$$

as a function of $T_b - T_{sat}$ for $b = 0.25, 0.50, 0.75, 1.0$, and 1.5 in.[1] Similar curves for isopropyl alcohol and Freon-113 are displayed in Figs. 18.13 and 18.14, respectively.

The information in Figs. 18.12, 18.13 and 18.14 is quite enlightening. Consider, for example, the curve for $b = 1.5$ in in Fig. 18.13. For $0 \leq \Delta T_b \leq 35$°F, the base temperature gradient increases rapidly with ΔT_b, which means that as T_b increases, the nucleate boiling becomes more pronounced. The curve exhibits an inflection point at $\Delta T_b = 35$°F, which coincides with the peak value of h for isopropyl alcohol in Fig. 18.11. As ΔT_b increases further, nucleate boiling is gradually replaced by transition boiling, causing the slope of the curve to diminish until at $\Delta T_b = 140$°F, the slope is a minimum. Any further increase in ΔT_b gives way to film boiling at the base, which is indicated by a more or less constant slope of the curve.

While film boiling is being established at the base, nucleate and transition regions occur farther along the fin. The curve reaches its peak value of 3900 °F/ft at $\Delta T_b = 400$°F. At this point, the nucleate boiling region has been stretched out to the fin tip.

For the dashed portion of the curve, most of the fin surface is experiencing film boiling; the tip is in transition boiling. The base temperature gradient is reduced drastically under these conditions. The remaining solid portion represents the condition

[1]The Haley–Westwater data employ English engineering units, and these data have not been transcribed here to the SI system.

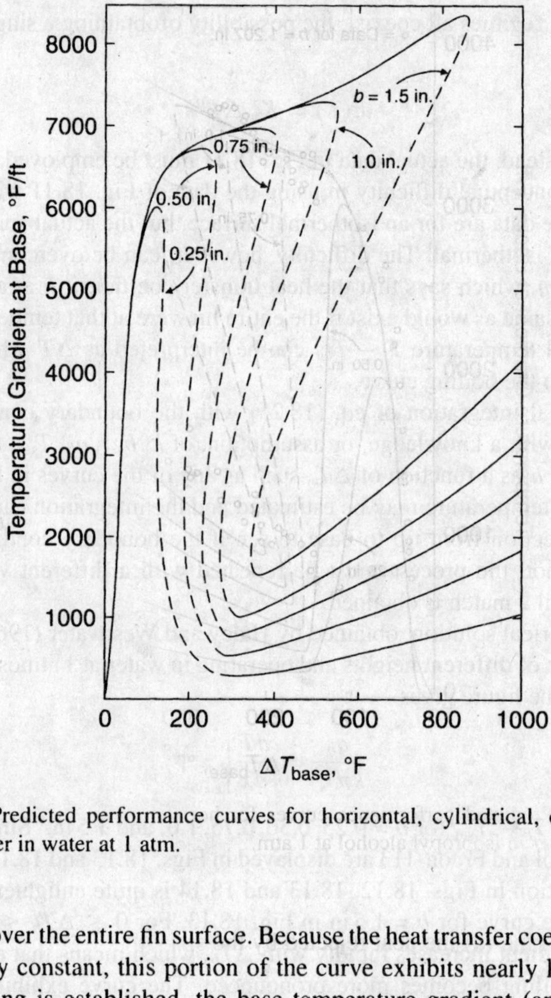

Figure 18.12 Predicted performance curves for horizontal, cylindrical, copper spines of 0.635 cm diameter in water at 1 atm.

of film boiling over the entire fin surface. Because the heat transfer coefficient for film boiling is nearly constant, this portion of the curve exhibits nearly linear behavior. Once film boiling is established, the base temperature gradient (and hence q_b) is limited by how far ΔT_b can be increased without damage to the fin material.

It is also interesting to note that for a given base temperature gradient or, equivalently, a given base heat dissipation, there are, in general, three values of ΔT_b but only two of them (on the solid line) may be practically realizable. Moreover, for a given ΔT_b, there are three values of heat dissipation, but again, only the two that correspond to the solid line may be achievable in practice. The experimental data in Figs. 18.12, 18.13, and 18.14 confirm, in general, the validity of these semitheoretical predictions, although some discrepancies exist at high heat dissipations. As pointed out by Haley and Westwater (1966), the higher heat dissipations could be attributed to the disturbing effect on film boiling, due to the more vigorous neighboring nucleate boiling.

Example 18.1. A copper spine of 0.25 in. diameter and 1.0 in. height is used to enhance boiling of R-113 at atmospheric pressure. If the base temperature difference

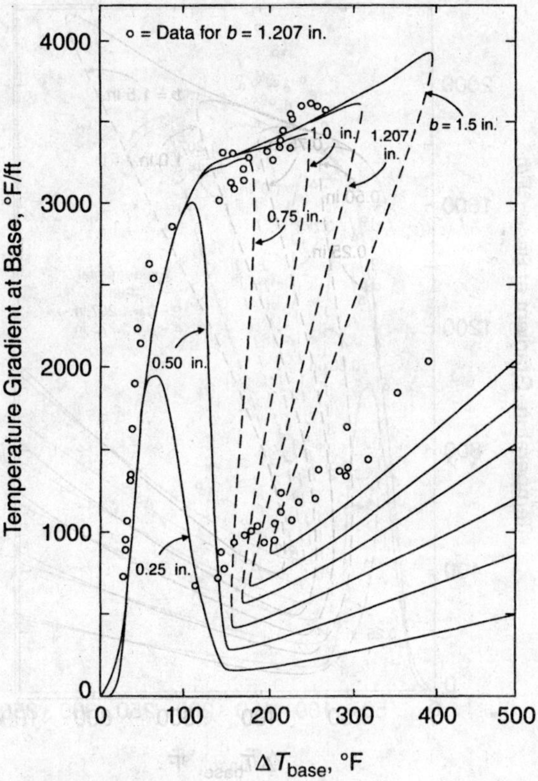

Figure 18.13 Predicted performance curves for horizontal, cylindrical, copper spines of 0.635 cm diameter in isopropyl alcohol at 1 atm.

is 100°F, determine **(a)** the heat rejected by the spine and **(b)** the heat rejection rate without the spine.

SOLUTION. **(a)** For $\Delta T_b = 100°\text{F}$ and $b = 1.0\,\text{in.}$, Fig. 18.14 gives a base temperature gradient of 1780°F/ft. Taking the thermal conductivity of copper to be 223 Btu/ft-hr-°F, q_b can be calculated as

$$q_b = k\pi r^2 \left.\frac{dT}{dx}\right|_{x=b}$$

$$= (223)\pi\left(\frac{0.125}{12}\right)^2 (1780)$$

$$= 135.3\,\text{Btu/hr}$$

(b) Referring to Fig. 18.11, for $\Delta T_b = 100°\text{F}$, $h = 44\,\text{Btu/ft}^2\text{-hr-°F}$. Thus, without the fin,

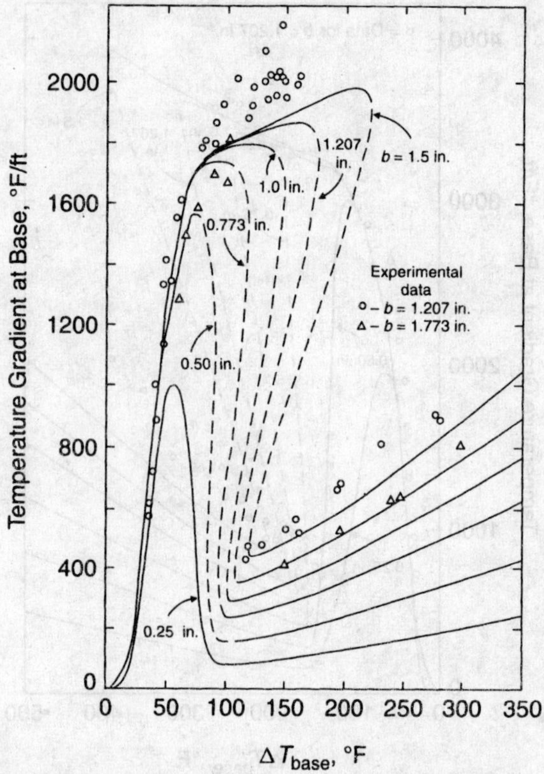

Figure 18.14 Predicted performance curves for horizontal, cylindrical, copper spines of 0.635 cm diameter in R-113 at 1 atm.

$$q_b = h\pi r^2 \Delta T_b$$

$$= (44)\pi \left(\frac{0.125}{12}\right)^2 (100)$$

$$= 1.5 \text{ Btu/hr}$$

The presence of the spine improves the heat transfer by the enhancement ratio of $135.3/1.5 \approx 90$. This is due to the fact that with the spine, all four boiling regimes occur simultaneously, while without the spine, the surface is experiences only film boiling.

18.5 OPTIMUM DIMENSIONS FOR THE CYLINDRICAL SPINE

From performance curves such as those shown in Figs. 18.12, 18.13, and 18.14, the optimum dimensions of a cylindrical spine for a given required heat dissipation can be deduced. To illustrate this, consider again the curve for $b = 1.5$ in. in Fig. 18.13. As noted in Section 18.4, the peak value of 3900°F/ft occurs at $\Delta T_b = 400$°F. A different interpretation says that for a heat dissipation of $q_b/k\pi r^2 = 3900$°F/ft, the optimum height of a spine with 0.25 in. diameter for boiling isopropyl alcohol at 1 atmosphere has a height of 1.5 in. In fact, each peak point defines a heat dissipation for which that particular value of b is optimum. If a set of curves such as Fig. 18.13 can be constructed for different diameters, this peak point information can be employed to construct a design chart for the optimum dimensioned cylindrical spine. Such a chart for R-113 is shown in Fig. 18.15. The solid lines form a grid of constant-volume lines and constant-b/d lines. The dashed lines pertain to the minimum mass spine, which is discussed in the next section.

Example 18.2.[2] Use Fig. 18.15 to design a cylindrical copper fin for boiling Freon-113 that will dissipate 100 Btu/hr at a base temperature difference of 275°F.

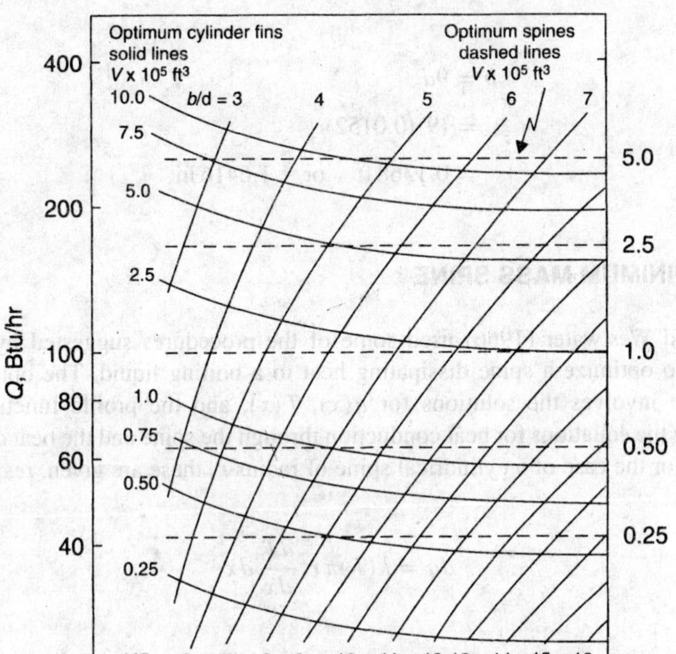

Figure 18.15 Optimum dimensioned cylindrical spines and minimum mass spines (both copper) for boiling R-113 at atmospheric pressure.

[2]This example also employs English units.

SOLUTION. From Fig. 18.15 for $\Delta T_b = 275°\text{F}$ and $q_b = 100$ Btu/hr, read

$$V = 2.5 \times 10^{-5}\ \text{ft}^3 \quad \text{and} \quad b/d = 9$$

Then

$$V = \frac{\pi}{4}d^2 b$$
$$= \frac{\pi}{4}d^2(9d)$$
$$= 2.5 \times 10^{-3}$$

and

$$d = \left[\frac{4}{9\pi}(2.5 \times \times 10^{-5})\right]^{1/3} = 0.0152\ \text{ft} \quad \text{or} \quad 0.1828\ \text{in.}$$

and

$$b = 9d$$
$$= (9)(0.0152)$$
$$= 0.1368\ \text{ft} \quad \text{or} \quad 1.6416\ \text{in}$$

18.6 MINIMUM MASS SPINE

Haley and Westwater (1966) used some of the procedures suggested by Wilkins (1960a) to optimize a spine dissipating heat to a boiling liquid. The optimization procedure involves the solutions for $q(x)$, $T(x)$, and the profile function $f_2(x)$ satisfying the equations for heat conduction through the spine and the heat dissipated from it. For the case of a cylindrical spine of radius r, these are given, respectively, by

$$dq = k(T)\pi r^2 \frac{dT}{dx}\,dx \qquad (18.27)$$

and

$$dq = 2\pi r h(T)\,dx \qquad (18.28)$$

The dependence of the thermal conductivity on temperature is to be noted in eq. (18.27). In eq. (18.28), $h(T)$ is any known heat transfer function of the form

$$h(T) = K_n(T - T_{\text{sat}})^n \qquad (18.29)$$

where K_n is a constant and n can take on any values except $-\frac{1}{4}$ and $-\frac{3}{2}$. These are excluded for mathematical rather than physical reasons.

The functions $q(x)$, $T(x)$, and $f_2(x)$ must satisfy the boundary conditions for the case where $x = 0$ at the spine tip and $x = b$ at the spine base:

$$q(x = 0) = 0$$

$$q(x = b) = q_b$$

and

$$T(x = b) = T_b$$

yielding a minimum spine volume

$$V = \int_0^b \pi r^2 \, dx \tag{18.30}$$

Haley and Westwater (1966) used the transformations of Wilkins (1960a):

$$u = \frac{1}{U_b} \int_{T_{\text{sat}}}^{T} k(T)[h(T)]^4 \, dT \tag{18.31}$$

where U_b is defined by

$$U_b = \int_{T_{\text{sat}}}^{T_b} k(T)[h(T)]^4 \, dT \tag{18.32}$$

to obtain the dimensions of the optimum spine:

$$x = \left(\frac{4q_b}{25\pi U_b^2} \right)^{1/3} \int_{T_{\text{sat}}}^{T} k(T)[h(T)]^2 u^{-2/5} \, dT \tag{18.33}$$

and

$$r = \left(\frac{2q_b^2}{5\pi^2 U_b} \right)^{1/3} u^{1/5} h(T) \tag{18.34}$$

with

$$q = q_b u^{4/5} \tag{18.35}$$

Equations (18.33) through (18.35) give the spine shape and local heat flux for any set of physical constants q_b, T_b, T_{sat}, and the functions $k(T)$ and $h(T)$. With constant thermal conductivity and use of eq. (18.29) with eqs. (18.33) and (18.34), the dimensions for the optimum spine are

$$b = \left[\frac{20(4n+1)^2 k q_b}{(2n+3)^3 \pi K_n^2 (T_b - T_{\text{sat}})^{(2n-1)}} \right]^{1/3} \tag{18.36}$$

$$\frac{T - T_{\text{sat}}}{T_b - T_{\text{sat}}} = \left(\frac{x}{b} \right)^{5/(2n+3)} \tag{18.37}$$

and

$$r = \left[\frac{2(4n+1)q_b^2}{5\pi^2 k K_n (T_b - T_{\text{sat}})^{n+1}}\right]^{1/3} \left(\frac{x}{b}\right)^{(9n+1)/(2n+3)} \tag{18.38}$$

from which r_b follows as

$$r_b = \left[\frac{2(4n+1)q_b^2}{5\pi^2 k K_n (T_b - T_{\text{sat}})^{n+1}}\right]^{1/3} \tag{18.39}$$

and again, it is clear that the linear profile and the parabolic shapes suggested by Schmidt (1926) are correct only for the case of a uniform heat transfer coefficient.

18.7 EXTENDED SURFACES IN MOIST AIR

The phenomenon of heat transfer with extended surfaces in moist air has been treated, in considerable detail, by Abramzon and Shmurak (1998). Applications where these conditions exist include building air-conditioning and aircraft environmental control systems. These applications embrace evaporation, which is the subject of this chapter, and condensation, which is considered in Chapter 19.

When a fluid adjacent to a fin (such as air) contains a second phase (such as water vapor) so that condensation or evaporation may occur on the fin surface, the heat transfer becomes coupled with mass transfer, and the entire process becomes nonlinear. Under such conditions, the fin temperature varies in a complex fashion and the fin efficiency is not a constant.

Abramzon and Shmurak (1998) considered the two typical fin geometries shown in Fig. 18.16. These are the longitudinal fin of rectangular profile and the radial fin of rectangular profile.

18.7.1 Problem Formulation

In the event that heat and mass are being transferred simultaneously, as with evaporation from a fin surface, the energy and vapor mass fluxes are coupled. The two fluxes can be expressed as

$$q'' = -k_g \frac{dT}{dy} + J[h_{fg} + C_p(T - T_s)] \tag{18.40}$$

and

$$J = -\rho D \frac{dZ}{dy} + \hat{V} \rho Z \tag{18.41}$$

In each of eqs. (18.40) and (18.41), the first term on the right-hand side describes molecular diffusion, and the second is associated with convective transport. Moreover, the heat flux and both the temperature and vapor mass fraction in the air–water vapor mixture are functions of y, the distance from the faces of the fin.

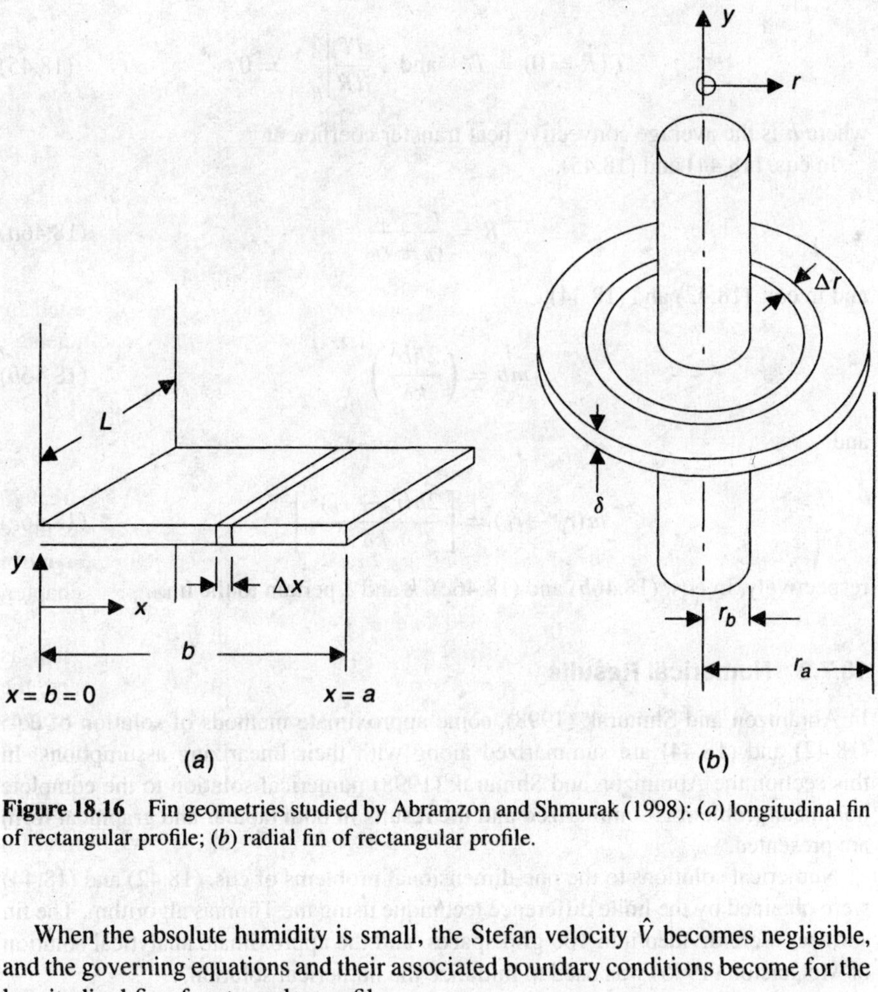

Figure 18.16 Fin geometries studied by Abramzon and Shmurak (1998): (*a*) longitudinal fin of rectangular profile; (*b*) radial fin of rectangular profile.

When the absolute humidity is small, the Stefan velocity $\hat{V}$ becomes negligible, and the governing equations and their associated boundary conditions become for the longitudinal fin of rectangular profile,

$$\frac{d^2T}{dx^2} = (mb)^2 \left[(T_b - T_s) + \frac{h_{fg}}{h} J \right] \tag{18.42}$$

subject to

$$T(x = 0) = T_b \quad \text{and} \quad \left. \frac{dT}{dx} \right|_{x=b} = 0 \tag{18.43}$$

and for the radial fin of rectangular profile,

$$\frac{d^2T}{dR^2} + \frac{1}{r}\frac{dT}{dR} = [m(r_a - r_b)]^2 \left[(T_b - T_s) + \frac{h_{fg}}{h} J \right] \tag{18.44}$$

subject to

$$T(R = 0) = T_b \quad \text{and} \quad \left.\frac{dT}{dR}\right|_{R=1} = 0 \tag{18.45}$$

where h is the average convective heat transfer coefficient.

In eqs. (18.44) and (18.45),

$$R = \frac{r - r_b}{r_a - r_b} \tag{18.46a}$$

and in eqs. (18.42) and (18.44),

$$mb = \left(\frac{2hb^2}{k\delta}\right)^{1/2} \tag{18.46b}$$

and

$$m(r_a - r_b) = \left[\frac{2h(r_a - r_b)^2}{k\delta}\right]^{1/2} \tag{18.46c}$$

respectively. In eqs. (18.46*b*) and (18.46*c*), k and δ pertain to the fin.

18.7.2 Numerical Results

In Abramzon and Shmurak (1998), some approximate methods of solution of eqs. (18.42) and (18.44) are summarized along with their linearizing assumptions. In this section the Abramzon and Shmurak (1998) numerical solution to the complete nonlinear problem is summarized and the results in both tabular and graphical form are presented.

Numerical solutions to the one-dimensional problems of eqs. (18.42) and (18.44) were obtained by the finite difference technique using the Thomas algorithm. The fin surfaces were divided into 100 grid spaces, and the approximate analytical solution of McQuiston (1975) was used to initialize the numerical solution.

Figure 18.17 displays the finite difference results for the temperature $T(x)$ or $T(R)$ and vapor flux $J(x)$ distributions along the longitudinal fin of rectangular profile and radial fin of rectangular profile for both evaporating ($J > 0$) and condensing ($J < 0$) conditions. These results are specifically for base and environmental temperatures of $T_b = 50°C$ and $T_s = 10°C$ (the evaporating case), respectively. The figure is for a fin with $mb = 1.00$ and a relative humidity of $\phi = 1.0$. It is clear that the influence of moisture in the ambient air is significant, as the fin temperature profile changes much more rapidly from T_b toward the environmental temperature T_s when moisture is present.

Table 18.1 lists values for the heat transferred and the fin efficiency for several sets of boundary conditions for several values of the environmental relative humidity ϕ for the longitudinal fin of rectangular profile and the radial fin of rectangular profile. Results for cases 1 and 3 suggest that the heat transferred and the fin efficiency are significantly different for different values of the environmental relative humidity.

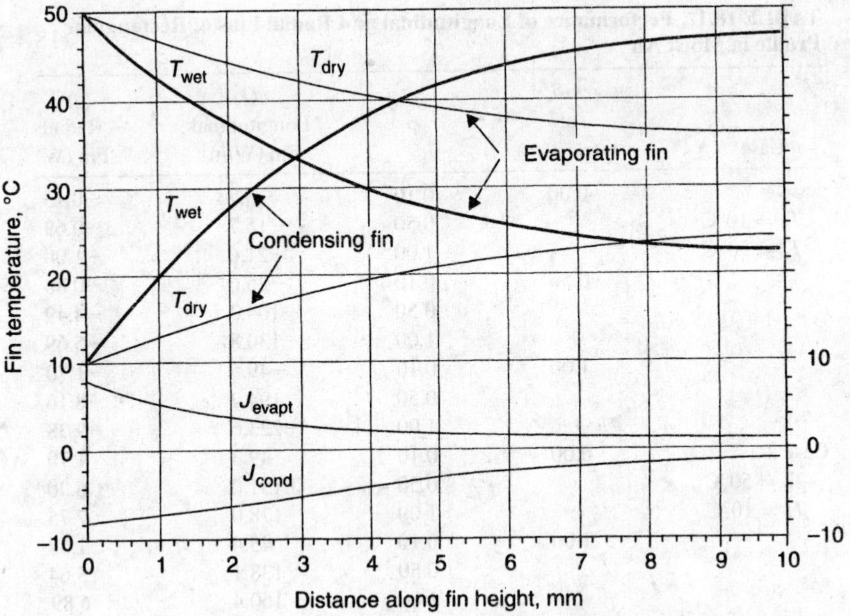

Figure 18.17 Temperature and mass flux as a function of distance along the fin for a longitudinal fin of rectangular profile and a radial fin of rectangular profile.

Example 18.3. For a longitudinal fin of rectangular profile and a radial fin of rectangular profile of equal surface areas, compare the heat dissipation rates and determine the tip temperature of the longitudinal fin of rectangular profile if the air at 10°C is considered dry. The conditions are:

T_b	$= 50°\text{C}$	δ	$= 0.10\,\text{mm}$
T_s	$= 10°\text{C}$	b	$= 1.0\,\text{cm}$
k	$= 200\,\text{W/m·K}$	r_b	$= 0.5\,\text{cm}$
h	$= 100\,\text{W/m}^2\text{·K}$	$r_a - r_b$	$= 1.0\,\text{cm}$
ϕ	$= 1.00$		

SOLUTION. The outer radius of the radial fin of rectangular profile is

$$r_a = r_b + 0.5\,\text{cm} = 1.5\,\text{cm}$$

so that its surface area is

$$\begin{aligned} S &= 2\pi(r_a^2 - r_b^2) \\ &= 2\pi[(1.5)^2 - (0.50)^2] \\ &= 12.57\,\text{cm}^2 \end{aligned}$$

TABLE 18.1 Performance of Longitudinal and Radial Fins of Rectangular Profile in Moist Air

Case	mb or $m(r_a - r_b)$	ϕ	Q Longitudinal Fin (W/m)	q Radial Fin (W)
Case 1	0.00	0.10	−3.73	−0.13
$T_b = 10°C$		0.50	−15.7	−0.69
$T_s = 50°C$		1.00	−21.6	−1.00
	0.50	0.10	−25.6	−0.88
		0.50	−104.3	−4.49
		1.00	−130.8	−5.69
	1.00	0.10	−49.5	−1.70
		0.50	−194.2	−8.16
		1.00	−223.6	−9.38
Case 2	0.00	0.10	49.3	1.70
$T_b = 50°C$		0.50	151.0	6.20
$T_s = 10°C$		1.00	178.0	7.75
	1.00	0.10	45.4	1.57
		0.50	138.1	5.64
		1.00	160.4	6.89
Case 3	0.00	0.10	−9.60	−0.33
$T_b = 80°C$		0.50	−40.2	−1.76
$T_s = 10°C$		1.00	−53.8	−2.47
	1.00	0.10	−256.7	−8.85
		0.50	−726.7	−27.1
		1.00	−727.7	−27.2
Case 4	0.00	0.10	182.3	6.27
$T_b = 10°C$		0.50	370.3	14.4
$T_s = 80°C$		1.00	394.0	15.9
	1.00	0.10	179.3	6.17
		0.50	361.6	14.4
		1.00	382.2	15.3

A longitudinal fin of rectangular profile with $S = 12.57\ \text{cm}^2$ will have a fin length of

$$L = \frac{S}{2b} = \frac{12.57\ \text{cm}^2}{(2)(1\ \text{cm})} = 6.28\ \text{cm}$$

The parameters mb and $m(r_a - r_b)$ may be determined from eqs. (18.46*b*) and (18.46*c*):

$$mb = \left(\frac{2hb^2}{k\delta}\right)^{1/2} = \left[\frac{(2)(100)(0.01)^2}{(200)(0.0001)}\right]^{1/2} = 1.00$$

and

$$m(r_a - r_b) = \left[\frac{2h(r_a - r_b)^2}{k\delta}\right]^{1/2} = \left[\frac{(2)(100)(0.01)^2}{(200)(0.0001)}\right]^{1/2} = 1.00$$

With these values and the data specified, it is seen that there is consistency with the values provided in case 2 of Table 18.1.

For the longitudinal fin of rectangular profile,

$$Q = 160.4\ \text{W/m}$$

and for a fin length of 0.0628 m (6.28 cm),

$$q = (160.4)(0.0628) = 10.07\ \text{W}$$

For the radial fin of rectangular profile,

$$q = 6.89\ \text{W}$$

The tip temperatures can be read from Fig. 18.17. For this evaporative situation ($T_b > T_s$) at 10 mm (1 cm), read

$$T_{a,\text{wet}} = 22°\text{C}$$

and

$$T_{a,\text{dry}} = 36°\text{C}$$

This example was contrived to fit the operating conditions used to generate Table 18.1 and Fig. 18.17. For other operating conditions, the problems formulated by eqs. (18.42) through (18.45) would require a numerical solution using the stipulated operating values.

18.8 NOMENCLATURE

Roman Letter Symbols

A	area, m^2
b	fin or spine height, m
C	coefficient, dimensionless
c	specific heat, J/kg·K
D	mass diffusivity, m^2/s; binary diffusion constant of water in air, m^2/s; bubble diameter, m; mass concentration, kg/m^3
d	spine diameter, m; bubble departure diameter, m; differential or derivative, dimensionless; vapor jet diameter, m
Eu	Eustis number, dimensionless
f	frequency, s^{-1}
g	acceleration, m/s^2

h	heat transfer coefficient, W/m^2·K; function, dimensionless; mass transfer coefficient, m/s; specific enthalpy, J/kg
h_{fg}	latent heat of vaporization, J/kg
J	mass flux, kg/m^2
Ja	Jakob number, dimensionless
K	constant, dimensionless
k	thermal conductivity, W/m·K
k_g	average thermal conductivity of air-water vapor mixture, W/m·K
L	length, m
M	molecular weight, kg-mol
m	fin performance parameter, m^{-1}
$\dot{m}$	mass flow, kg/s
Q	vapor flux, kg/m^2; heat flow per meter, W/m
q	heat flow, W
q''	heat flux, W/m^2
R	radius, dimensionless; universal gas constant, J/kg mol·K
Re	Reynolds number, dimensionless
r	radius, m; radial coordinate, m
S	surface area, m^2
Sc	Schmidt number, dimensionless
T	temperature, K
U	transformed variable, W^5/m^9·K^4
u	transformed variable, dimensionless
V	volume, m^3
$\hat{V}$	flow velocity, m/s; Stefan velocity of vapor diffusion, m/s
v	transformed variable, dimensionless; specific volume, m^3/kg
x	height coordinate, m
Z	vapor mass fraction, dimensionless

Greek Letter Symbols

δ	fin thickness, m
∂	partial derivative, dimensionless
η	fraction of heat flux resulting in; net vapor generation, dimensionless
μ	dynamic viscosity, kg/s·m
ρ	density, kg/m^3
σ	surface tension, N/m
Φ	critical heat flux, W/m^2
ϕ	relative humidity, dimensionless

Roman Letter Subscripts

a	fin tip
b	fin base; bubble
c	convection; collapse
crit	critical
dry	dry condition

f	fluid; flow condition
fb	flow boiling
fg	latent heat of vaporization
g	vapor (gas)
incip	incipient
j	characteristic
ℓ	liquid
min	minimum
o	gravitiational acceleration
p	pressure, N/m^2; constant pressure
r	radiation
s	surroundings; surface
sat	saturated
sf	solid–fluid
sub	subcooled
v	vapor
wet	wet condition

Symbolic Subscript

∞	bulk of fluid

Roman Letter Superscript

n	exponent on temperature difference

Symbolic Superscripts

$'$	characteristic
$''$	flux; modified value

CHAPTER 19

CONDENSATION ON FINNED SURFACES

19.1 INTRODUCTION

Extended surfaces have traditionally been employed to reduce the convective resistance associated with low values of the heat transfer coefficient, h. In the case of condensation, the typical values of h are high, and the need for and effectiveness of extended surface for augmentation of the condensation process may not be immediately apparent. However, in the last 50 years or so, several engineering situations have been identified where the augmentation of condensation can be beneficial.

For example, consider the condensation of organic vapors where poor thermophysical properties result in low values of h and, consequently, leave room for enhancement. Even with fluids having favorable thermophysical properties, the condensing side resistance may be significant and warrant reduction, expecially if the cold side is augmented in some manner. The widespread use of integral-finned tubes in surface condensers for refrigeration and process industries clearly demonstrates the usefulness of extended surface for enhancing condensation.

Condensation of a vapor on a surface occurs if the temperature of the surface is kept below the vapor saturation temperature. Although four basic condensation mechanisms occur—homogeneous condensation, direct contact condensation, dropwise condensation, and film condensation—most condensers are designed to operate under the film condensation mode. The process is characterized by the formation of a thin film of liquid which drains under the influence of gravity and/or surface tension. The presence of a film creates a barrier between the vapor and the cooled surface and thus retards the condensation process. If condensation is to be enhanced, the film thickness must be reduced, and this reduction may be achieved by, among other methods, using finned surfaces instead of plain surfaces.

The use of extended surface to enhance film condensation involves low-profile integral fins. Among the early investigators of this technique were Beatty and Katz (1948), who measured condensing film coefficients for several refrigerants (low-surface-tension liquids) on a horizontal integal-fin tube containing 630 fins per meter.

In the work of Webb (1981a), Cooper and Rose (1981), and Nakayama (1982), it is shown that the fins were introduced initially to provide additional surface area. However, fin geometries, dimensions, and spacings are limited, due to manufacturing techniques, and condensate flow was assumed to be governed only by viscous and gravitational forces.

For many years, finned tubes were believed to be inappropriate for high-surface-tension liquids because of the possibility of condensate flooding between the fins. While surface tension forces appeared to be neglected until the appearance of the work of Gregorig (1954) and Thomas (1967), experimental work by Yau et al. (1985) and Wanniarachchi et al. (1986) on water and Masuda and Rose (1985) on ethylene glycol has shown that sizable enhancements can exist provided that the number of fins is chosen judiciously.

Work in the 1970s was not confined to the low-profile integral fin on the exterior of horizontal tubes. Williams et al. (1971) carried out experiments to determine the effect of centrifugal force on the the condensation rate. Using methanol and isopropanol on a vertical rotating tube, they showed that the condensation heat transfer coefficient can be increased significantly through the use of centrifugal force. A unique double-grooved surface was developed by Markowitz et al. (1972) for horizontal plate facing down condensers. Condensing coefficients based on a nominal surface area for R-113 were improved by almost 100%. This configuration was developed for a submerged condenser, where direct condensation occurred in the bulk of the subcooled liquid.

Reisbig (1974) conducted an experimental study of condensing R-12 as it flowed through the inside of both finned and smooth-bore aluminum tubes. Oshima and Iuchi (1974) provided a calculation method in cooler condensers for the effective heat transfer area of plate-fin tubes and showed that the results obtained utilizing this calculation method were in good agreement with experimental results. Smirnov (1976) provided data on the condensation of binary mixtures on finned surfaces.

Patankar and Sparrow (1979) made a numerical analysis of film condensation on a vertical fin attached to a cooled vertical plate or cylinder. They showed that in this three-dimensional process, the condensation on the surface is tightly coupled to the conduction within the fin. A physically meaningful similarity problem was formulated and evaluated numerically. The results obtained from the similarity solution show that the heat transfer can be represented by some simple algebraic relationships that appear to be applicable to the operating range encountered in practice. It was found that the fin heat transfer calculated was markedly less than that which would be predicted by an isothermal fin model. This indicates that the use of the isothermal fin model substantially overestimates the performance of the configuration.

Hazeler (1980) conducted experiments on condensing nitrogen in downflow in a plate heat exchanger. He obtained stream and wall temperature profiles in both the wet-wall desuperheating and condensing regions, and his heat transfer coefficients were presented as a function of quality. Mil'man and Shklover (1980) looked at various techniques for enhancing the heat transfer coefficient for steam on tubes such as finned tubes, wire-type turbulence promoters, and hydrophobic coatings and additives.

Kharku and Borovkov (1971), Carnavos (1980), Zozulya et al. (1977), Edwards et al. (1973a), and Joos (1984) all showed the importance of surface tension forces in the condensate film, and Rifert (1982) provided a comprehensive review of steam condensation on extended surface. He showed that the heat transfer enhancement with finned surfaces can vary substantially, depending on the fin geometry and spacing. He also noted that flooding of the interfin surfaces by the condensate, which increases as the fin spacing decreases, can alter the performance substantially. Prior to this, Rifert (1980) analyzed condensation on horizontal finned tubes, including the effect of surface tension forces. In this endeavor he divided the tube into flooded and unflooded zones and solved a two-dimensional form of the energy equation for each zone. Solutions of these equations revealed that in most cases, the fin temperature was markedly nonuniform.

Hirasawa et al. (1980) showed that on a finned surface, there exists a region of high heat transfer coefficients in the trough between the crests of the fins. This analytical study predicted a local thinness of the condensate film produced by the capillary action in the trough. This surface tension phenomenon was confirmed by optical measurements.

The trend toward higher fin density has directed more attention to the lower portion of the condenser. Rudy and Webb (1981) made static measurements of condensate retention angles on finned tubes with different fin densities using water, R-111, and *n*-pentane. Honda et al. (1983) took data for both static and condensing conditions using ethanol and R-113 and found the results to be essentially the same. Yau et al. (1985) took 13 tubes with varying fin pitch but with the same fin thickness and fin height and measured the heat transfer enhancement over a plain tube having the same inside and outside diameters. All measurements were made at near atmospheric pressure with the vapor flowing downward with velocities between 0.5 and 1.1 m/s. It was shown that the heat transfer enhancement obtained with finned tubes significantly exceeded that which was to be expected on the basis of increased surface area. Plots of enhancement against fin density were repeatable and indicated local maxima and minima. It was also shown that the dependence of enhancement on fin density was not appreciably influenced by vapor velocity or condensation flow rate, at least for the ranges employed.

Rudy and Webb (1981) provided a theoretical model for condensation on integral-fin tubes and developed (Rudy and Webb, 1985) a condensate retention model based on the experimental observation that the vertical rise height of the condensate between the fins must be the same for vertical tubes as it is for a vertical plate. This reasoning showed that for a fin of arbitrary cross section, the summation of surface tension and body forces acting on the retained liquid between the fins will eventually lead to an expression for the retention angle. A similar expression was also derived by Honda et al. (1983) and it shows that high-surface-tension fluids such as water can completely flood the configuration, whereas for low-surface-tension fluids such as the Freons, the fin density can be increased before flooding occurs.

The results of several systematic experimental programs have provided more insight into the condensation of various fluids on integral-fin tubes. Works by Rudy (1982), Wanniarachchi et al. (1985), and Marto et al. (1986), as well as Wanniarachchi et al. (1986), provide a wide variety of data for film condensation of steam. Over 60 finned tubes with the same nominal fin root diameter were manufactured and tested. The effect of fin spacing, fin thickness, fin height, geometry, and material were tested at atmospheric pressure. It was found that the optimum fin spacing occurs somewhere near 1.5 mm, which translates to about 400 fins per meter or 10 fins per inch. This shows that at the optimum spacing, enhancements of 3.5 or 4 to 1 are possible. Moreover, the Wanniarachchi et al. (1985) work shows that steam-side enhancement increased with fin height. However, the rate of this performance enhancement is not very bold, as higher rates of enhancement were seen to be obtained with a similar increase in prime surface area. This, of course, demonstrates that there is a fin efficiency effect, and indeed, the fin efficiency has been investigated by Lienhard and Dhir (1974), Nader (1978), and Burmeister (1982).

Marto et al. (1986) showed that the fin shape can alter the performance, and Mitrou (1986) showed that the wall material can have a marked effect on the results. Mitrou's work confirmed that as the thermal conductivity of the fin decreases, the result is an increase in the thermal resistance of the fin. This translates into a decrease in fin efficiency, leading to a lower heat transfer capability. This has been confirmed by Mills et al. (1975) and Shklover et al. (1985).

Rudy and Webb (1981) also proposed a modification to the Beatty and Katz (1948) model, which neglects surface tension forces on the fins and neglects heat transfer through the flooded portion of the tube. The work of Webb et al. (1985) assumed condensate flow radially inward along the fin flanks due to the surface tension forces in the unflooded portion of the tube. Here, gravity drainage was neglected. In formulating their model, they used the heat transfer coefficient model proposed by Adamek (1981) for heat transfer to the fin faces. The Adamek work was based on a theoretical analysis for the surface-tension-driven condensation on a family of condensate surface profiles.

Kilic and Onat (1981) pointed out that there may be condensation on the finned surfaces of air-conditioning systems, and their study considered both heat and mass transfer from the saturated air layer to the liquid water film with the latent heat of vaporization being transferred to the fins. A study of Toner et al. (1983) compared rectangular and triangular profile fins, and optimum dimensions of the fins as a function of the Biot number were claimed. Additional guidance on the condensation of superheated vapor on finned surfaces can be found in the work of Chen et al. (1983) and Takazawa et al. (1983). Wang and Chen (1984) considered condensation on a circumferential fin, and Chen et al. (1984) have looked at laminar film condensation on a downward-facing vertical fin. Rudy et al. (1984) considered the use of integral-type condenser tubes for process applications.

After this brief literature review, attention should rightfully turn to a more detailed discussion of condensation on single fins, which now follows. Nusselt-type models using the Nusselt theory are considered first.

19.2 CONDENSATION ON SINGLE FINS

19.2.1 Nusselt Models

The main difficulty in the analysis of film condensation on single fins or spines is the variability of the heat transfer coefficient h. Unlike classical extended surface analysis, which by virtue of the Murray (1938) and Gardner (1945) assumptions assumes h to be constant and uniform over the lateral surface of fin or spine, h for laminar condensation is a function of the difference between the local fin or spine temperature and the saturation temperature of the condensing vapor. Moreover, the surface of the fin or spine is not isothermal, whereas the simple Nussult theory applies to an isothermal surface. Despite these difficulties, it will be shown in the following sections that localized application of the Nusselt theory can be used for at least a preliminary analysis or design.

Nusselt Theory. Nusselt himself described the physical nature of laminar film condensation in 1916 (Nusselt, 1916). The results of his analysis for vertical and inclined plates and horizontal cylinders are in good agreement with experiment. A flat plate of length L and unit width into the plane of the paper and inclined at an angle κ with the horizontal is shown in Fig. 19.1a. Note that the length coordinates are x, having an origin at the top of the plate with positive direction downward, and y, having an origin at the surface of the plate with positive direction out from the surface of the plate. The liquid film that flows along and parallel to the plate is at velocity v. The differential element of condensate film dx is located at a distance x along the plate and this element has film thickness Y. At the plate surface where

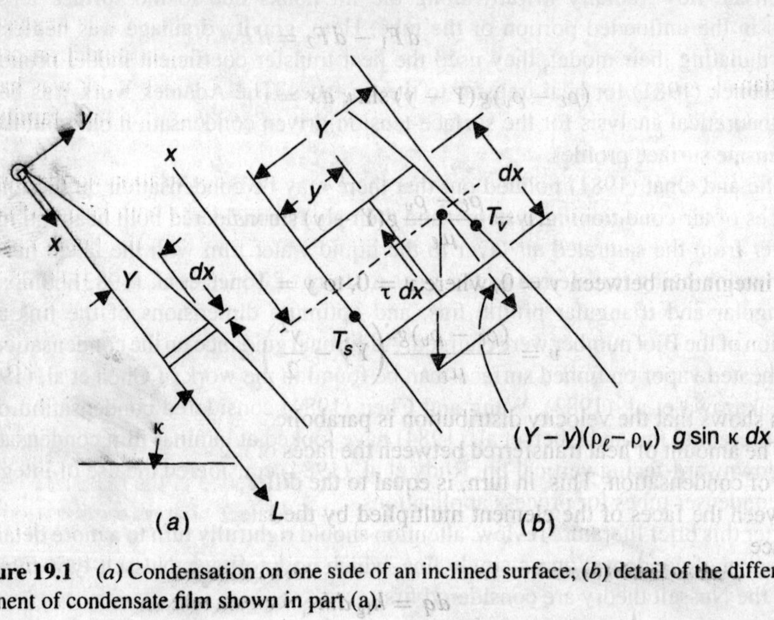

Figure 19.1 (a) Condensation on one side of an inclined surface; (b) detail of the differential element of condensate film shown in part (a).

$y = 0$, the temperature of the plate is T_s, and at the liquid–vapor interface, where $y = Y$, the temperature is T_v.

A detail of the differential element dx is provided in Fig. 19.1b, and a balance of forces in the section of the element from y to Y is to be considered. Because vapor condenses at the exposed surface at $y = Y$, there is conduction of heat through the film from $y = Y$ to $y = 0$. If a linear temperature gradient is assumed, the local heat transfer coefficient (film coefficient) will be $h(x) = k/Y$.

For the differential element between y and Y, three forces parallel to the plate are apparent:

1. The component of the gravity force parallel to the plate on a mass of $\rho(Y-y)\,dx$:

$$dF_1 = \rho_\ell g(Y - y) \sin\kappa \, dx$$

2. A balancing force due to the difference in pressure between the faces of the differential element due to the hydrostatic effect in the surrounding vapor:

$$dF_2 = \rho_v g(Y - y) \sin\kappa \, dx$$

3. The shear force due to the velocity gradient dv/dy at a distance y from the surface of the plate:

$$dF_3 = \tau \, dx = \mu_\ell \frac{dv}{dy} dx$$

where, as discussed in Chapter 4, τ is the viscous shear stress.

The sum of the forces parallel to the plate must be zero. Hence

$$dF_1 - dF_2 = dF_3$$

$$(\rho_\ell - \rho_v)g(Y - y) \sin\kappa \, dx = \mu_\ell \frac{dv}{dy} dx$$

or

$$dv = \frac{\rho_\ell - \rho_v}{\mu_\ell} g(Y - y) \sin\kappa \, dy$$

and integration between $y = 0$, where $v = 0$, to $y = Y$, where $v = v$, gives

$$v = \frac{(\rho_\ell - \rho_v)g}{\mu_\ell} \left(Yy - \frac{y^2}{2} \right) \sin\kappa \qquad (19.1)$$

This shows that the velocity distribution is parabolic.

The amount of heat transferred between the faces of the element dx is equal to the rate of condensation. This, in turn, is equal to the difference in the mass rate of flow between the faces of the element multiplied by the latent heat of vaporization, h_{fg}. Hence

$$dq = h_{fg} \, d\dot{m}$$

This heat transferred is also equal to

$$dq = h(T_v - T_s)\,dx = \frac{k_\ell}{Y}(T_v - T_s)\,dx$$

so that

$$d\dot{m} = \frac{k_\ell(T_v - T_s)}{Y h_{fg}}\,dx \tag{19.2}$$

and because the velocity distribution is given by

$$v = \frac{1}{\rho}\frac{d\dot{m}}{dy}$$

the change in $\dot{m}$ can be related to the change in y via

$$\dot{m} = \int_0^Y \rho v\,dy = \rho \int_0^Y v\,dy = \rho Y v_{av}$$

Thus

$$v_{av} = \frac{1}{Y}\int_0^Y v\,dy$$

$$= \frac{(\rho_\ell - \rho_v) g \sin\kappa}{\mu_\ell Y}\int_0^Y \left(Yy - \frac{y^2}{2}\right) dy$$

$$= \frac{(\rho_\ell - \rho_v) g \sin\kappa\, Y^2}{3\mu_\ell}$$

and therefore

$$\dot{m} = \frac{\rho_\ell(\rho_\ell - \rho_v) g\ \sin\kappa}{3\mu_\ell} Y^3 \tag{19.3}$$

and

$$d\dot{m} = \frac{\rho_\ell(\rho_\ell - \rho_v) g\ \sin\kappa}{\mu_\ell} Y^2\,dY \tag{19.4}$$

Elimination of $d\dot{m}$ between eqs. (19.2) and (19.4) yields

$$Y^3 dY = \frac{k_\ell \mu_\ell (T_v - T_s)}{\rho_\ell(\rho_\ell - \rho_v) g h_{fg} \sin\kappa}\,dx$$

and an integration between $x = 0$ where $Y = 0$ and some point (x, Y) gives

$$\frac{Y^4}{4} = \frac{k\mu_l(T_v - T_s)x}{\rho(\rho - \rho_v) g h_{fg}\ \sin\kappa}$$

and

$$Y = \left[\frac{4k_\ell \mu_\ell (T_v - T_s) x}{\rho_\ell(\rho_\ell - \rho_v) g h_{fg} \sin\kappa}\right]^{1/4}$$

With a linear variation of temperature across the film from T_v to T_s, the heat transfer coefficient will be

$$h = \frac{k_l}{Y} = \left[\frac{\rho_\ell(\rho_\ell - \rho_v) k^3 g h_{fg} \sin\kappa}{4\mu_\ell (T_v - T_s) x}\right]^{1/4}$$

and if this is written in terms of the local Nusselt number Nu_x,

$$\mathrm{Nu}_x = \frac{hx}{k} = \left[\frac{(\rho_\ell - \rho_v) g h_{fg} \sin\kappa\, x^3}{4\nu_\ell k_\ell (T_v - T_s)}\right]^{1/4}$$

The total heat passing through the condensate layer from $x = 0$ to $x = x$ is q_x:

$$q_x = \int_0^x h_x (T_v - T_s)\, dx$$

$$= \int_0^x \left[\frac{\rho_\ell(\rho_\ell - \rho_v) k_\ell^3 g h_{fg} \sin\kappa}{4\mu_\ell (T_v - T_s)}\right]^{1/4} \frac{T_v - T_s}{x^{1/4}} dx$$

$$= \frac{4}{3}\left[\frac{\rho_\ell(\rho_\ell - \rho_v) k_\ell^3 g h_{fg} \sin\kappa}{4\mu_\ell (T_v - T_s)}\right]^{1/4} (T_v - T_s) x^{3/4}$$

and with the average heat transfer coefficient between points $x = 0$ and $x = L$ designated as

$$\bar{h} = \frac{q_x|_{x=L}}{(T_v - T_s) L}$$

the average heat transfer coefficient becomes

$$\bar{h} = \frac{4^{3/4}}{3}\left[\frac{\rho_\ell(\rho_\ell - \rho_v) k_\ell^3 g h_{fg} \sin\kappa}{\mu_\ell (T_v - T_s)}\right]^{1/4} \frac{(T_v - T_s) L^{3/4}}{(T_v - T_s) L}$$

or

$$\bar{h} = 0.943\left[\frac{\rho_\ell(\rho_\ell - \rho_v) k_l^3 g h_{fg} \sin\kappa}{\mu_\ell L (T_v - T_s)}\right]^{1/4} \tag{19.5}$$

where the fluid properties are evaluated at the film temperature

$$T_f = \frac{T_v + T_s}{2}$$

With $\rho >> \rho_v$, eq. (19.5) becomes

$$\bar{h} = 0.943\left[\frac{\rho_\ell^2 k_\ell^3 g h_{fg} \sin\kappa}{\mu_\ell L (T_v - T_s)}\right]^{1/4} \tag{19.6}$$

and the Nusselt number can be written as

$$\mathrm{Nu} = \frac{hL}{k_\ell} = 0.943\left[\frac{\rho_\ell g L^3 h_{fg} \sin\kappa}{\nu_\ell k_\ell (T_v - T_s)}\right]^{1/4} \tag{19.7}$$

In the event that the plate is vertical, $\kappa = 90°$, so that $\sin\kappa = 1.00$ and eqs. (19.6) and (19.7) reduce to

$$h = 0.943\left[\frac{\rho_\ell^2 k_\ell^3 g h_{fg}}{\mu_\ell L (T_v - T_s)}\right]^{1/4} \tag{19.8}$$

and

$$\mathrm{Nu} = \frac{hL}{k_l} = 0.943\left[\frac{\rho_\ell g L^3 h_{fg}}{\nu_\ell k_\ell (T_v - T_s)}\right]^{1/4} \tag{19.9}$$

For the film condensation on a horizontal cylinder, the analysis is somewhat more complicated. In this case, the results for the heat transfer coefficient and Nusselt number are

$$h = 0.725\left[\frac{\rho_\ell^2 k_\ell^3 g h_{fg}}{\mu_\ell d (T_v - T_s)}\right]^{1/4} \tag{19.10}$$

and

$$\mathrm{Nu} = \frac{hd}{k_\ell} = 0.725\left[\frac{\rho_\ell g d^3 h_{fg}}{\nu_\ell k_\ell (T_v - T_s)}\right]^{1/4} \tag{19.11}$$

Horizontal Cylindrical Spine (Pin Fin). Consider a horizontal cylindrical spine or pin fin of diameter d, height b, and thermal conductivity k, as shown in Fig. 19.2. The spine is in contact with a pure saturated quiescent vapor at T_{sat} and is attached to a cooled surface at temperature $T_b < T_{sat}$. Thus the spine provides a cooled surface to accommodate the condensation process, and under steady-state conditions, the latent heat is extracted from the vapor and conducted toward the colder fin base. The condensate film formed on the surface of the spine drips down under the action of gravity.

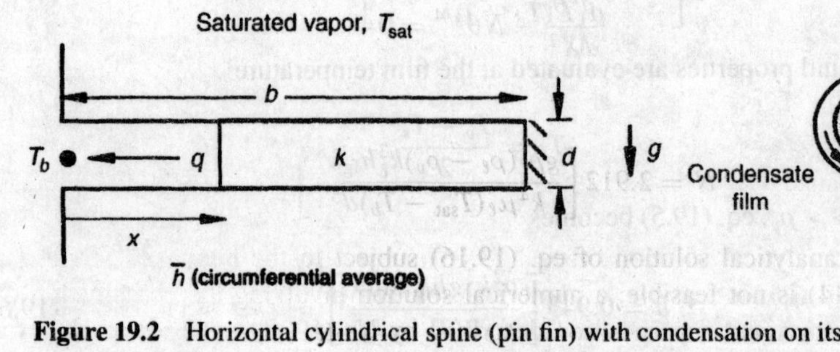

Figure 19.2 Horizontal cylindrical spine (pin fin) with condensation on its surface.

Let $\bar{h}$ be the circumferentially averaged heat transfer coefficient at any axial location on the spine. Because the values of $\bar{h}$ are usually large, high values of the Biot number $\mathrm{Bi} = hd/2k$ occur, which, in turn, induce two-dimensional effects into the spine. In what follows, this fact is ignored to avoid further complications and axial conduction from spine tip to spine base is the dominant mode of heat transfer through the spine. With

$$\theta = \frac{T_{\text{sat}} - T}{T_{\text{sat}} - T_b} \tag{19.12a}$$

and

$$X = \frac{x}{b} \tag{19.12b}$$

the governing differential equation for the temperature distribution in the spine can be written as

$$\frac{d^2\theta}{dX^2} - \frac{4\bar{h}b^2}{kd}\theta = 0 \tag{19.13}$$

and this may be solved using the boundary conditions

$$\theta(X = 0) = 1 \tag{19.14a}$$

and

$$\left.\frac{d\theta}{dX}\right|_{X=1} = 0 \tag{19.14b}$$

Note that eq. (19.14*b*) implies an insulated spine tip. The value of $\bar{h}$ to be used in eq. (19.13) derives from the Nusselt theory for laminar condensation on a single horizontal tube with zero interfacial shear on the condensate film as given by Webb (1994):

$$\bar{h} = 0.728\left[\frac{g\rho_\ell(\rho_\ell - \rho_v)k_\ell^3 h_{\text{fg}}}{\mu_\ell(T_{\text{sat}} - T_b)\theta d}\right]^{1/4} \tag{19.15}$$

With this value of $\bar{h}$ substituted into eq. (19.13), the result is

$$\frac{d^2\theta}{dX^2} - N\theta^{3/4} = 0 \tag{19.16}$$

where

$$N = 2.912\left[\frac{g\rho_\ell(\rho_\ell - \rho_v)k_\ell^3 h_{\text{fg}} b^8}{k^4 \mu_\ell(T_{\text{sat}} - T_b)d^5}\right]^{1/4} \tag{19.17}$$

Because an analytical solution of eq. (19.16) subject to the boundary conditions of eqs. (19.14) is not feasible, a numerical solution involving the combination of quasilinearization and superposition described by Na (1979) was used to obtain a solution. Figure 19.3 displays the numerical results for $N = 5$, 10, 50, and 100. These results match the results of Lienhard and Dhir (1974), who used a shooting technique to generate their numerical solution.

Figure 19.3 provides some insight into fin design. For $N = 50$ and 100, the remote half of the fin does not sustain any condensation because the temperature difference $T_{sat} - T$ is virtually zero. Such a design is wasteful of fin material. At low values of N, say at $N = 5$, the entire fin surface is effective in supporting condensation, but in this event, the fin is too stubby for substantial augmentation of the condensation. It is therefore reasonable to conclude that for good design, N should be on the order of $N = 10$.

The efficiency of the fin η can be found by taking the heat conducted into the base of the fin and dividing it by the ideal dissipation of the fin:

$$q_{id} = \pi\, db(T_{sat} - T_b)[\bar{h}(x = 0)]$$

This gives

$$\eta = -\frac{1}{N}\frac{d\theta}{dX}\bigg|_{X=0} \tag{19.18}$$

The efficiency values calculated from the numerical solution (Fig. 19.3) and eq. (19.18) are plotted in Fig. 19.4. These values also apply to the vertical spine (pin fin). If $N = 10$ is the value selected for the design of the spine or fin, the corresponding value of η read from Fig. 19.4 is 0.34 (34%). This low value is inevitable if a significant increase of condensation is to be realized.

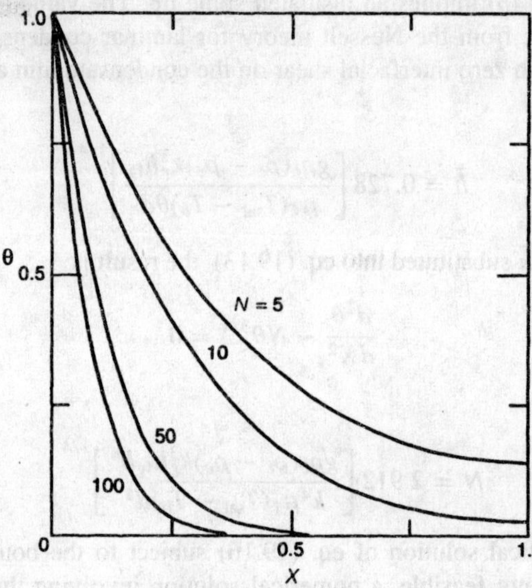

Figure 19.3 Temperature distributions in a horizontal spine with condensation.

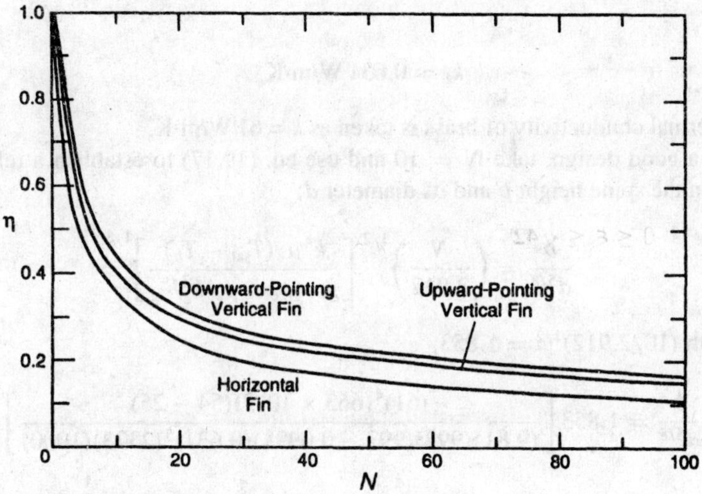

Figure 19.4 Efficiencies of horizontal and vertical spines with condensation.

Example 19.1. Saturated steam at 0.15 bar condenses on a surface at 25°C. It is desired to enhance the rate of condensation by attaching a cylindrical fin of brass to the primary surface. Take the thermal conductivity of brass as 61 W/m·K and suggest some satisfactory design options.

SOLUTION. For saturated steam at 0.15 bar (0.015 MPa), the steam tables give

$$T_{\text{sat}} = 53.97°\text{C} \ (\approx 54°\text{C})$$

$$v_v = 10.022 \text{ m}^3/\text{kg}$$

and

$$h_{\text{fg}} = 2373 \text{ kJ/kg}$$

The properties of the condensate are evaluated at a mean film temperature of

$$T_f = \frac{25°\text{C} + 54°\text{C}}{2} = 39.5°\text{C}$$

and with

$$\rho_v = \frac{1}{v_v} = \frac{1}{10.022 \text{ m}^3/\text{kg}} = 0.0998 \text{ kg/m}^3$$

three liquid properties are obtained:

$$\rho_\ell = 992 \text{ kg/m}^3$$

$$\mu_\ell = 663 \times 10^{-6} \text{ N·s/m}^2$$

and

$$k_\ell = 0.631 \text{ W/m·K}$$

The thermal conductivity of brass is taken as $k = 61$ W/m·K.

For a good design, take $N = 10$ and use eq. (19.17) to establish a relationship between the spine height b and its diameter d:

$$\frac{b}{d^{5/8}} = \left(\frac{N}{2.912}\right)^{1/2} \left[\frac{k^4 \mu_\ell (T_{sat} - T_b)}{g\rho_\ell(\rho_\ell - \rho_v)k_\ell^3 h_{fg}}\right]^{1/8}$$

and with $(10/2.912)^{1/2} = 1.853$,

$$\frac{b}{d^{5/8}} = 1.853 \left[\frac{(61)^4(663 \times 10^{-6})(54 - 25)}{(9.81)(992)(992 - 0.0998)(0.631)^3(2373)(1000)}\right]^{1/8}$$

or

$$\frac{b}{d^{5/8}} = 0.2244 \text{ m}^{3/8}$$

Five possibilities may be obtained by assuming values for d and obtaining the corresponding values of b:

d (mm)	b mm
5	8.18
7.5	10.54
10	12.62
12.5	14.51
15	16.26

Vertical Cylindrical Spine (Pin Fin). Figure 19.5 shows a cylindrical spine in two vertical orientations. For the upward-pointing spine, the distance x is measured from the spine tip, while for the downward-pointing fin, the distance x is measured from the spine base. The essential difference between horizontal and vertical spines is that for the horizontal spine, the surface is isothermal in the circumferential direction of condensate flow, while for the vertical spines, in the direction of condensate flow, the temperature decreases for the upward-pointing spine and increases for the downward-pointing spine. Thus, the original Nusselt theory, which assumes isothermal conditions in the condensate flow direction, is not directly applicable for the vertical fins shown in Fig. 19.5. However, Lienhard and Dhir (1974) have shown that if appropriately modified to account for a nonisothermal surface, Nusselt's theory gives results that are close to the predictions obtained from the full boundary layer equations. For a nonisothermal flat vertical surface, the modified Nusselt theory gives for the value of $\bar{h}$ (the circumferential average) at any location, x from the leading edge of the spine

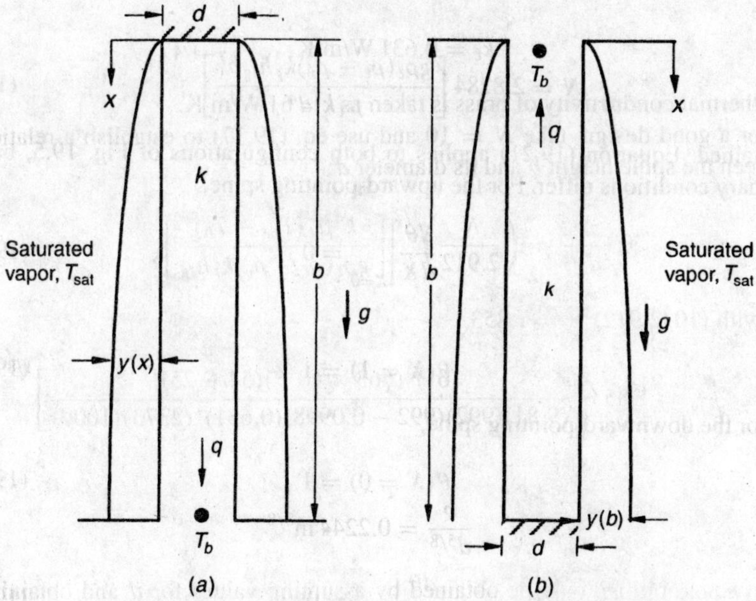

Figure 19.5 Vertical cylindrical spine (pin fin) in (*a*) upward-pointing and (*b*) downward-pointing orientations.

$$\bar{h} = 0.7071 \left[\frac{g\rho_\ell(\rho_\ell - \rho_v)k_\ell^3 h_{fg}}{\mu_\ell \int_0^x (T_{sat} - T)\, dx} \right]^{1/4} \tag{19.19}$$

Equation (19.19) applies to a cyclindrical surface if the curvature effects are small, that is, if

$$y(b) << \frac{d}{2}$$

Use of the definitions of θ and X given by eqs. (19.12) allows eq. (19.19) to be recast as

$$\bar{h} = 0.7071 \left[\frac{g\rho_\ell(\rho_\ell - \rho_v)k_\ell^3 h_{fg}}{\mu_\ell b(T_{sat} - T_b) \int_0^X \theta\, dx} \right]^{1/4} \tag{19.20}$$

If this is substituted into eq. (19.13), an integrodifferential equation is obtained:

$$\frac{d^2\theta}{dX^2} - N\theta \left[\int_0^X \theta\, dx \right]^{-1/4} = 0 \tag{19.21}$$

where

$$N = 2.8284 \left[\frac{g\rho_\ell(\rho_\ell - \rho_v)k_\ell^3 h_{fg} b^7}{\mu_\ell k^4 d^4} \right]^{1/4} \tag{19.22}$$

is obtained. Equation (19.21) applies to both configurations of Fig. 19.5, but the boundary conditions differ. For the upward-pointing spine,

$$\left. \frac{d\theta}{dX} \right|_{X=0} = 0 \tag{19.23a}$$

and

$$\theta(X = 1) = 1 \tag{19.23b}$$

and for the downward-pointing spine,

$$\theta(X = 0) = 1 \tag{19.24a}$$

and

$$\left. \frac{d\theta}{dX} \right|_{X=1} = 0 \tag{19.24b}$$

Numerical solutions of eq. (19.21) subject to the boundary conditions of either eqs. (19.23) or (19.24) have been reported by Lienhard and Dhir (1974). Figure 19.6 is an adaption of their results. A close examination of the left and right portions of

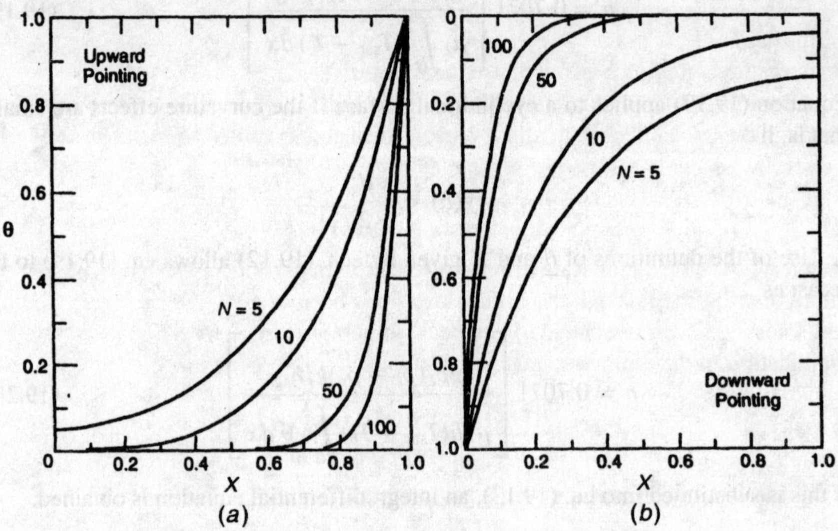

Figure 19.6 Temperature distributions in vertical cylindrical spines: (*a*) upward pointing; (*b*) downward pointing.

Fig. 19.6 reveals that the temperature profiles for the two orientations are slightly different. As in the case of the horizontal spine, a good design value is on the order of 10.

Equation (19.18) for the fin efficiency also applies to the vertical cylindrical spines. As pointed out, the efficiencies for the vertical spines are plotted in Fig. 19.4 to allow a comparison with the horizontal spines. Figure 19.4 shows that vertical spines are more efficient than horizontal spines and of the two vertical arrangements in Fig. 19.5, the downward-pointing spine has a higher efficiency.

Vertical Rectangular Fin. The results of the foregoing discussion pertaining to cylindrical spines are also applicable to a vertical longitudinal fin of rectangular profile as long as the definition of N is modified appropriately to represent the rectangular geometry.

19.2.2 Conjugate Models

In the conjugate models, the heat conduction equation for the fin and the condensate boundary layer equations are solved simultaneously. In the simple model that has been used by Nader (1978), Burmeister (1982), and Acharya et al. (1986), both the fin temperature T and the film thickness $y(x)$ are allowed to vary solely along the condensate flow direction, that is, a one-dimensional fin model is used with a two-dimensional condensate film. The improved model, proposed by Patankar and Sparrow (1979), considers a two-dimensional fin with a three-dimensional condensate layer.

Simple Conjugate Model. Consider a vertical longitudinal fin of rectangular profile as shown in Fig. 19.7. The fin has height b, thickness δ, and thermal conductivity k. Both faces of the fin are exposed to a saturated vapor at $T_{\text{sat}} > T_b$. As will be seen, the boundary conditions for the fin are those of a constant and uniform base temperature and no heat flow through the fin tip.

Consider a slice of fin and two films of height dx. Equating the heat conducted through the two condenstate films to the net heat conducted through the fin slice gives

$$\frac{d^2T}{dx^2} = -\frac{2k_\ell(T_{\text{sat}} - T)}{k\delta y(x)} \tag{19.25}$$

where $y(x)$ is the condensate film thickness at x. In deriving eq. (19.25), the temperature distribution through the fin thickness has been assumed to be linear. The film thickness $y(x)$ can be related to the local temperature difference $T_{\text{sat}} - T$ through an application of the momentum equation giving

$$\frac{d[y(x)]^4}{dx} = \frac{4\mu_\ell k_\ell(T_{\text{sat}} - T)}{g\rho_\ell(\rho_\ell - \rho_v)h_{fg}} \tag{19.26}$$

It is convenient to introduce the dimensionless quantities

$$\theta = \frac{T_{\text{sat}} - T}{T_{\text{sat}} - T_b} \tag{19.27a}$$

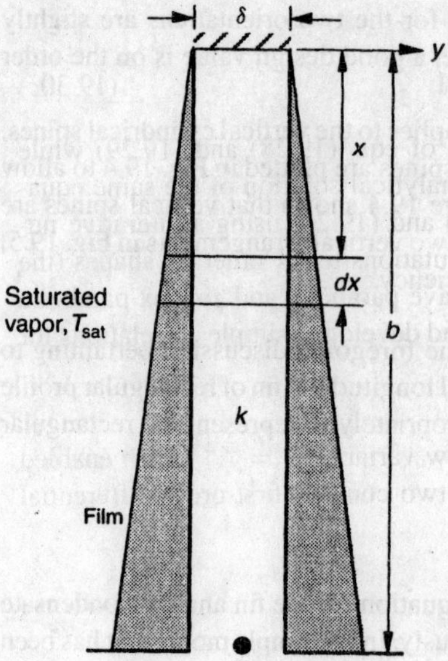

Figure 19.7 Vertical longitudinal fin of rectangular profile.

$$X = \frac{x}{b} \tag{19.27b}$$

$$\Lambda = \frac{y(x)}{b} \tag{19.27c}$$

$$F_1 = \frac{g\rho_\ell(\rho_\ell - \rho_v)h_{fg}b^3}{\mu_\ell k_\ell(T_{sat} - T_b)} \tag{19.27d}$$

and

$$F_2 = \frac{k\delta}{2k_\ell b} \tag{19.27e}$$

into eqs. (19.25) and (19.26) to obtain

$$\frac{d^2\theta}{dX^2} = \frac{\theta}{\Lambda F_2} \tag{19.28}$$

$$\frac{d(\Lambda)^4}{dX} = \frac{4\theta}{F_1} \tag{19.29}$$

The boundary conditions for eqs. (19.28) and (19.29) are

$$\Lambda(X = 0) = 0 \tag{19.30a}$$

$$\left.\frac{d\theta}{dX}\right|_{X=0} = 0 \tag{19.30b}$$

and

$$\theta(X = 1) = 1 \tag{19.30c}$$

Nader (1978) obtained a numerical solution of eqs. (19.28) and (19.29) while Burmeister (1982) developed an approximate analytical solution of the same equations. Acharya et al. (1986) solved eqs. (19.25) and (19.26) using an iterative numerical scheme. They also extended their computations to six other fin shapes (the longitudinal fins of triangular, trapezoidal, concave parabolic and convex parabolic profiles and the cylindrical and conical spines) and developed simple correlations for the fin efficiency.

Nader Solution. Nader (1978) introduced a new variable $\psi = \Lambda^4$ which enabled him to transform eqs. (19.28) and (19.29) into two coupled first-order differential equations:

$$\frac{d\psi}{dX} = \frac{4\theta}{F_1} \tag{19.31}$$

$$\frac{d\theta}{dX} = \frac{F_1}{3F_2}\psi^{3/4} \tag{19.32}$$

subject to the boundary conditions

$$\psi(X = 0) = 0 \tag{19.33a}$$

and

$$\theta(X = 1) = 1 \tag{19.33b}$$

In solving the foregoing system of equations numerically, Nader obtained values of the fin tip temperature $\theta(0)$ for a range of values of F_1 and F_2. These values are tabulated in Table 19.1.

The rate of heat flow into the base of the fin, q_b can be obtained as

$$q_b = 2k_\ell(T_{\text{sat}} - T_b)F_2 \left.\frac{d\theta}{dX}\right|_{X=1} \tag{19.34}$$

and the ideal heat heat transfer q_{id} occurs when the entire fin is at temperature T_b,

$$q_{\text{id}} = 2\bar{h}b(T_{\text{sat}} - T_b) \tag{19.35}$$

Here $\bar{h}$ is the average heat transfer coefficient for condensation on an isothermal vertical surface. It is based on Nusselt's theory,

$$\bar{h} = 0.943 \left[\frac{g\rho_\ell(\rho_\ell - \rho_v)k_\ell^3 h_{\text{fg}}}{\mu_\ell(T_{\text{sat}} - T_b)b}\right]^{1/4} \tag{19.36}$$

which in terms of F_1 becomes

TABLE 19.1 Tip Temperatures for a Vertical Longitudinal Fin of Rectangular Profile

	F_2			
F_1	10^4	10^3	10^2	10
10^7	0.9970	0.9703	0.7460	0.0875
10^8	0.9947	0.9479	0.6011	0.0169
10^9	0.9905	0.9096	0.4182	0.0007
10^{10}	0.9832	0.8461	0.2307	—
10^{11}	0.9703	0.7460	0.0875	—
10^{12}	0.9479	0.6011	0.0169	—
10^{13}	0.9096	0.4183	0.0007	—
10^{14}	0.8461	0.2307	—	—
10^{15}	0.7460	0.0875	—	—
10^{16}	0.6011	0.0169	—	—
10^{17}	0.4183	—	—	—

$$\bar{h} = 0.943\frac{k_\ell}{b}F_1^{1/4} \tag{19.37}$$

The efficiency can be obtained from eqs. (19.34) and eq. (19.35):

$$\eta = \frac{q_b}{q_{\text{id}}} = 1.0604\frac{F_2}{F_1^{1/4}}\left.\frac{d\theta}{dX}\right|_{X=1} \tag{19.38}$$

and the efficiency values obtained from this equation appear in Table 19.2 as a function of F_1 and F_2.

TABLE 19.2 Condensation Fin Efficiency for Condensation on a Vertical Longitudinal Fin of Rectangular Profile

	F_2			
F_1	10^4	10^3	10^2	10
10^7	0.9988	0.9857	0.8745	0.4603
10^8	0.9979	0.9747	0.7981	0.3612
10^9	0.9955	0.9563	0.6969	0.2823
10^{10}	0.9919	0.9254	0.5776	—
10^{11}	0.9857	0.8745	0.4603	—
10^{12}	0.9747	0.7989	0.3612	—
10^{13}	0.9562	0.6969	0.2823	—
10^{14}	0.9254	0.5776	—	—
10^{15}	0.8745	0.4603	—	—
10^{16}	0.7989	0.3612	—	—
10^{17}	0.6969	—	—	—

The film thickness, the condensate flow rate rate, and the heat conducted into the base of the fin, all at $x = b$, can be expressed in terms of F_1 and F_2:

$$y(b) = 1.412b\eta^{1/3}F_1^{-1/4} \tag{19.39}$$

$$\dot{m}(b) = \frac{1.8856k_\ell(T_{\rm sat} - T_b)\eta F_1^{1/4}}{h_{\rm fg}} \tag{19.40}$$

and

$$q = 1.8856k_\ell(T_{\rm sat} - T_b)\eta F_1^{1/4} \tag{19.41}$$

Example 19.2. Pure saturated steam at 60°C condenses on the outside surface of a 50-mm-diameter horizontal tube maintained at a uniform temperature of 34°C. To enhance the rate of condensation, the tube is fitted with a vertical longitudinal fin of rectangular profile. The fin is 2 mm thick, 7.50 mm high, and has a thermal conductivity of 48 W/m·K. **(a)** Calculate the rate at which heat is transferred to the bare tube, **(b)** for the bare tube, determine the condensate flow rate, and **(c)** evaluate the heat transfer and the condensate flow rate when the fin is installed.

SOLUTION. For saturated steam at 60°C,

$$v_v = 7.671\ \text{m}^3/\text{kg} \quad \text{and} \quad h_{\rm fg} = 2358\ \text{kJ/kg}$$

With

$$\rho_v = \frac{1}{v_v} = \frac{1}{7.671} = 0.130\ \text{kg/m}^3$$

and for the condensate at a mean temperature of

$$T_m = \frac{34 + 60}{2} = 47°\text{C}$$

the pertinent properties are

$$\rho_\ell = 989.1\ \text{kg/m}^3$$

$$\mu_\ell = 577 \times 10^{-6}\ \text{N·s/m}^2$$

and

$$k_\ell = 0.640\ \text{W/m·K}$$

The average heat transfer coefficient for laminar film condensation on an isothermal horizontal tube is given by eq. (19.15) with $\theta = 1.00$ for the isothermal condition

$$\bar{h} = 0.728\left[\frac{g\rho_\ell(\rho_\ell - \rho_v)k_\ell^3 h_{\rm fg}}{\mu_\ell(T_{\rm sat} - T_b)\theta d}\right]^{1/4}$$

$$= 0.728\left[\frac{(9.81)(989.1)(989.1 - 0.130)(0.640)^3(2358)(1000)}{(577 \times 10^{-6})(60 - 34)(1.00)(0.05)}\right]^{1/4}$$

or

$$\bar{h} = 6865\ \mathrm{W/m^2{\cdot}K}$$

(a) The rate at which the heat is transferred to the bare tube is

$$q = \bar{h}\pi d(T_{\mathrm{sat}} - T_b) = (6865)\pi(0.05)(60 - 34) = 28,037\ \mathrm{W/m}$$

(b) The condensate flow rate for the bare tube is given by

$$\dot{m} = \frac{q}{h_{\mathrm{fg}}} = \frac{28,037}{2358 \times 10^3} = 0.0119\ \mathrm{kg/m{\cdot}s}$$

(c) To evaluate the heat transferred and the condensate flow rate when the fin is installed, first determine F_1 and F_2:

$$F_1 = \frac{g\rho_\ell(\rho_\ell - \rho_v)h_{\mathrm{fg}}b^3}{\mu_\ell k_\ell(T_{\mathrm{sat}} - T_b)}$$

$$= \frac{(9.81)(989.1)(989.1 - 0.130)(2358)(1000)(0.0075)^3}{(577 \times 10^{-6})(0.640)(60 - 34)}$$

$$= 9.942 \times 10^8 \approx 10^9$$

and

$$F_2 = \frac{k\delta}{2k_\ell b}$$

$$= \frac{(48)(0.002)}{(2)(0.640)(0.0075)}$$

$$= 10$$

With $F_1 = 10^9$ and $F_2 = 10$, read Table 19.2 to obtain

$$\eta = 0.2823$$

The heat transferred to the fin is given by eq. (19.41):

$$q = 1.8856k_\ell(T_{\mathrm{sat}} - T_b)\eta F_1^{1/4}$$

$$= (1.8856)(0.640)(60 - 34)(0.2823)(9.94 \times 10^8)^{1/4}$$

$$= 1572.8\ \mathrm{W/m}$$

The condensate flow rate $\dot{m}(b)$ is obtained from eq. (19.40):

$$\dot{m}(b) = \frac{1.8856k_\ell(T_{\mathrm{sat}} - T_b)\eta F_1^{1/4}}{h_{\mathrm{fg}}}$$

$$= \frac{(1.8856)(0.640)(60 - 34)(0.2823)(9.94 \times 10^8)^{1/4}}{(2358)(1000)}$$

$$= 6.670 \times 10^{-4}\ \mathrm{kg/m{\cdot}s}$$

Burmeister Solution. Burmeister (1982) combined eqs. (19.28) and (19.29) into a single equation and obtained an approximate analytical solution for the resulting equation. His solutions for the tip temperature $\theta(0)$, heat transfer rate q, and fin efficiency η are

$$\theta(0) = \frac{1}{\cosh F} \tag{19.42}$$

$$q = 1.8263 k_\ell (T_{\text{sat}} - T_b)\left(F_1 F_2^3\right)^{1/7} (\tanh F)^{6/7} \tag{19.43}$$

and

$$\eta = \left(\frac{\tanh F}{F}\right)^{6/7} \tag{19.44}$$

where

$$F = 1.038(F_1 F_2^4)^{1/8} \tag{19.45}$$

The efficiency values predicted by eq. (19.44) are in close agreement with the values given in Table 19.2.

Acharya et al. Solution. Acharya et al. (1986) considered vertical longitudinal fins of several profile shapes (rectangular, triangular, concave parabolic, trapezoidal, and convex parabolic) and two spines (cylindrical and conical). For each geometry they solved eqs. (19.25) and (19.26) numerically and obtained values of η. For all seven shapes, it was found that the efficiency could be represented by a simple relationship of the form

$$\eta = \eta_c^{0.855} \tag{19.46}$$

where η_c is the fin efficiency calculated from the classical fin theory for one dimension, assuming the heat transfer coefficient to be constant for all the shapes and given by eq. (19.36):

$$\bar{h} = 0.943\left[\frac{g\rho_\ell(\rho_\ell - \rho_v)k_\ell^3 h_{fg}}{\mu_\ell(T_{\text{sat}} - T_b)b}\right]^{1/4} \tag{19.36}$$

Tabulations for η_c for the various fin shapes have been provided by Kern and Kraus (1972).

19.2.3 Improved Conjugate Model

Kazeminejad (1993) improved the simple conjugate model described by eqs. (19.28) through (19.30) by including the effect of vapor velocity. As expected, the effect of vapor shear was to reduce the condensate film thickness and hence increase the surface temperature. Moreover, the heat transfer to the fin and its capability to support condensation are enhanced considerably.

Patankar and Sparrow (1979) considered film condensation on a vertical longitudinal fin of rectangular profile that is attached to a cooled vertical base at temperature T_b. As shown in Fig. 19.8, the fin has height b, length L, thickness δ, and is made of a material with thermal conductivity k. The fin is immersed in a pure saturated vapor at temperature $T_{sat} > T_b$, and the analysis assumes the fin temperature T and film thickness $y(x)$ to be functions of x and z, that is, $T(x, z)$ and $y(x, z)$. These assumptions are justified by considerations that now follow.

As the condensate film flows downward along the fin, more condensate is added and its thickness increases along z to accommodate the increased flow. Along the x-direction, the temperature difference, $T_{sat} - T$, decreases from the base of the fin at $x = 0$ to its tip at $x = b$. Consequently, $y(x)$ also decreases along the x-direction. Thus the physics of the process dictates that $y(x)$ is a function of x and z. The effect of the growth of $y(x)$ with z is to increase the thermal resistance of the film, thereby reducing the heat conduction into the fin. The decreased heat flow implies that the fin temperature at a given x must decrease along the z-direction. Thus the temperature distribution in the fin is two-dimensional.

An energy balance over a fin element of dimensions dx, dy, and δ gives

$$\frac{\partial^2 T}{\partial x^2} = \frac{2k_\ell (T - T_{sat})}{k\delta y(x)} \tag{19.47}$$

This equation assumes that in the fin, conduction in the x-direction is dominant,

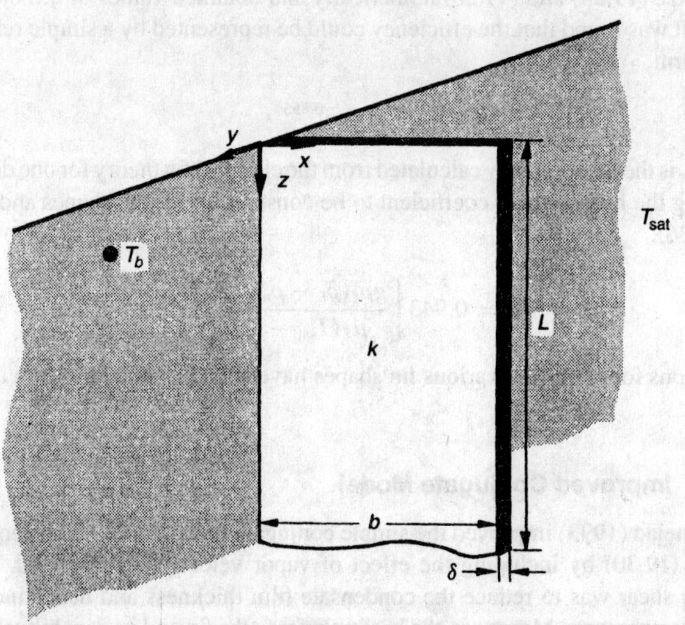

Figure 19.8 Two-dimensional vertical longitudinal fin of rectangular profile.

whereas in the film, conduction in the y-direction is dominant. The momentum equation for the z-direction takes the form

$$\frac{\partial}{\partial z}[y(x)]^4 = \frac{4\mu_\ell k_\ell (T_{\text{sat}} - T)}{g\rho_\ell(\rho_\ell - \rho_v)h_{\text{fg}}} \tag{19.48}$$

Equations (19.47) and (19.48) constitute two coupled partial differential equations for $T(x, z)$ and $y(x, z)$.

The following dimensionless quantities may be introduced into eqs. (19.47) and (19.48):

$$\theta = \frac{T_{\text{sat}} - T}{T_{\text{sat}} - T_b} \tag{19.49a}$$

$$X = \frac{x}{b} \tag{19.49b}$$

$$\Lambda = \frac{k\delta}{2k_\ell b^2} y(x) \tag{19.49c}$$

and

$$Z = \frac{\mu_\ell k_\ell (T_{\text{sat}} - T_b)}{4g\rho_\ell(\rho_\ell - \rho_v)h_{\text{fg}}} \left(\frac{k\delta}{k_\ell b^2}\right)^4 z \tag{19.49d}$$

which then become

$$\frac{\partial^2\theta}{\partial X^2} = \frac{\theta}{\Lambda} \tag{19.50}$$

$$\frac{\partial(\Lambda^4)}{\partial Z} = \theta \tag{19.51}$$

where the boundary condtions on θ and Λ are

$$\theta(X = 0) = 1 \tag{19.52a}$$

$$\left.\frac{\partial\theta}{\partial X}\right|_{X=1} = 0 \tag{19.52b}$$

and

$$\Lambda(Z = 0) = 0 \tag{19.52c}$$

Patankar and Sparrow (1979) sought a similarity solution of eqs. (19.50) to (19.52) by arguing that Nusselt's theory on a vertical isothermal surface shows that the local heat transfer coefficient h_Z is proportional to $Z^{-1/4}$, resulting in high values for h_Z at small values of Z. The high values of h_z cause the fin temperature to increase rapidly from T_b at $x = 0$ to T_{sat} significantly before $x = b$. Thus the behavior of the fin closely approximates that of an infinitely high fin, permitting the condition of eq. (19.52b) to be replaced by

$$\theta(X = \infty) = 0 \tag{19.53}$$

In the limit $Z \to 0$, h_z becomes infinite and the temperature distribution in the fin takes the form of a step increase from T_b to T_{sat}. Mathematically, this means that

$$\theta(Z = 0, X > 0) = 0 \tag{19.54}$$

In examining the behavior of $y(x)$, it is observed that at small values of Z, $y(x)$ must diminish quite rapidly with X to reflect the rapid decrease of $T_{sat} - T$ with X. This permits writing

$$\Lambda(X = \infty) = 0 \tag{19.55}$$

Thus the behavior of θ and Λ at $Z = 0$ and $X = \infty$ indicates the possibility of a similarity solution.

Similarity Solutions. The introduction of a similarity variable ξ as

$$\xi = \frac{X}{Z^{1/8}} \tag{19.56}$$

with dependent variables

$$\theta^*(\xi) = \theta \quad \text{and} \quad \Lambda^*(\xi) = \frac{\Lambda^4}{Z} \tag{19.57}$$

reduces the partial differential equations of eqs. (19.50) and (19.51) to the ordinary differential equations

$$\frac{d^2\theta^*}{d\xi^2} - \theta^*(\Lambda^*)^{-1/4} = 0 \tag{19.58}$$

and

$$\frac{d\Lambda^*}{d\xi} - 8(\theta^* - \Lambda^*)\xi^{-1} = 0 \tag{19.59}$$

with the boundary conditions

$$\theta^*(\xi = 0) = 1 \tag{19.60a}$$

$$\theta^*(\xi = \infty) = 0 \tag{19.60b}$$

and

$$\Lambda^*(\xi = \infty) = 0 \tag{19.60c}$$

Patankar and Sparrow (1979) observed that the analytical solution of eqs. (19.58) and (19.59) with the boundary conditions of eqs. (19.60) was out of reach, but Wilkins (1980) showed that an analytical solution does exist and can be written as

$$\theta = \theta^* = \begin{cases} \left(1 - \dfrac{\xi}{\sqrt{42}}\right)^7 & 0 \le \xi \le \sqrt{42} \\ 0 & \xi > \sqrt{42} \end{cases} \tag{19.61}$$

and

$$\Lambda^* = \begin{cases} \left(1 - \dfrac{\xi}{\sqrt{42}}\right)^8 & 0 \le \xi \le \sqrt{42} \\ 0 & \xi > \sqrt{42} \end{cases} \tag{19.62}$$

The temperature gradient $d\theta/dX$ evaluated at $X = 0$ can be found from eq. (19.61). An integration of $d\theta/dX|_{X=0}$ from $Z = 0$ to $Z = Z$ allows the heat conducted into the base of the fin over a distance Z to be evaluated. The final result is

$$q(Z) = 4.9371 \frac{g\rho_\ell(\rho_\ell - \rho_v)k_\ell^3 h_{fg} b^7}{\mu_\ell k^3 \delta^3} Z^{7/8} \tag{19.63}$$

The ideal heat transfer $q_{id}(Z)$ can be found by assuming the entire fin to be isothermal at temperature T_b

$$q_{id}(Z) = 5.333 \frac{g\rho_\ell(\rho_\ell - \rho_v)k_\ell h_{fg} b^7}{\mu_\ell k^3 \delta^3} Z^{3/4} \tag{19.64}$$

The fin efficiency η is the ratio of $q(Z)$ to $q_{id}(Z)$:

$$\eta = 0.9257 Z^{1/8} \tag{19.65}$$

The similarity solutions for θ, Λ^*, and η are shown in Figs. 19.9 through 19.11. These figures also show the nonsimilarity solutions, which are discussed next.

Nonsimilarity Solutions. The applicability of the similarity solutions is limited to those z locations for which the boundary condition of eq. (19.53) is justified, that is, at locations where the tip temperature is nearly equal to the vapor saturation temperature ($\theta = 0$). For z locations where this condition is not met, eqs. (19.50) through (19.52) were solved numerically by Patankar and Sparrow (1979). These results appear in Figs. 19.9 through 19.11.

Figure 19.9 shows the temperature profiles at various Z locations. For low values of Z, the temperature distribution in the fin is quite steep, as envisioned earlier. As Z increases, the profiles become less and less steep, because larger values of θ mean lower values for T. It may therefore be concluded from Fig. 19.9 that the general level of fin temperature decreases as Z increases. This confirms the earlier hypothesis about the dependence of T on Z. Because the nonsimilarity temperature profiles must terminate at $X = 1$, the terminal point on the $\xi = X/Z^{1/8}$ scale occurs at lower and lower values of ξ as Z increases. The similarity solution, on the other hand, extends up to $\xi = \sqrt{42} = 6.48$.

The condensate film thickness results are shown in Fig. 19.10. For a fixed Z location, the film thickness decreases as x or ξ increases. This is consistent with

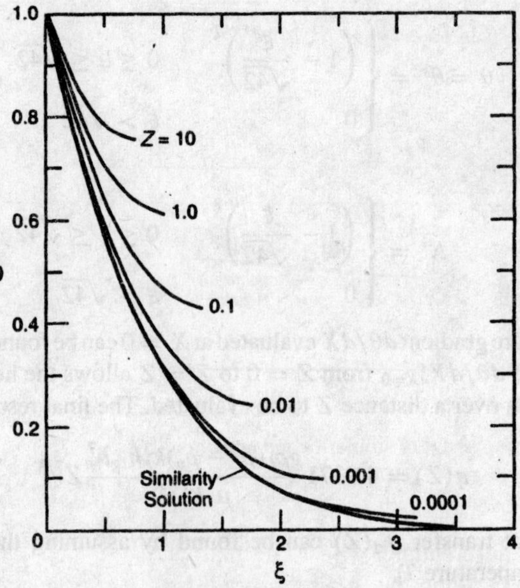

Figure 19.9 Nonsimilar temperature profiles in a vertical longitudinal fin of rectangular profile. [Adapted from Patankar and Sparrow (1979).]

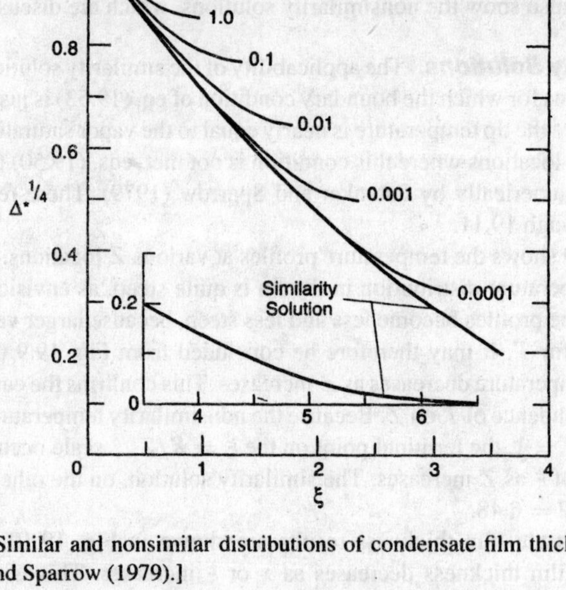

Figure 19.10 Similar and nonsimilar distributions of condensate film thickness. [Adapted from Patankar and Sparrow (1979).]

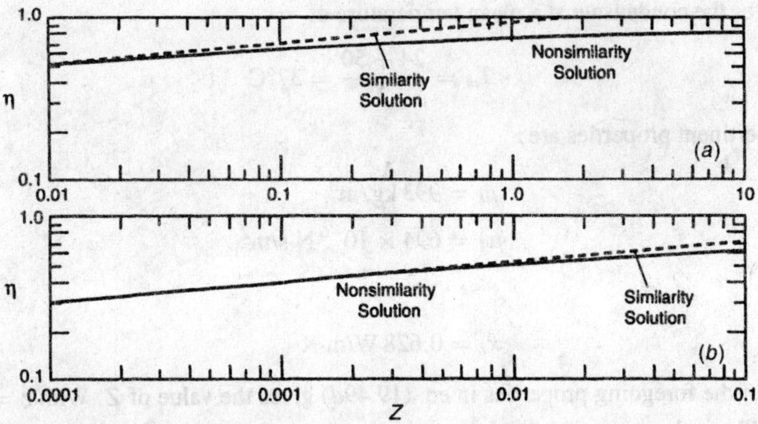

Figure 19.11 Comparison of similarity and nonsimilarity results for fin efficiency. [Adapted from Patankar and Sparrow (1979).]

the temperature differential $T_{sat} - T$, decreasing along X. For small Z the film is highly nonuniform along the X-direction but becomes more and more uniform as Z increases. This clearly shows that the assumption that $y(x) = f(Z)$ employed in previous sections is not strictly valid.

Finally, Fig. 19.11 shows the fin efficiency as a function of Z. The lower two curves cover the range $0.0001 \leq Z \leq 0.1$, while the upper two curves cover the range $0.01 \leq Z \leq 10$. It is interesting to note that the similarity and nonsimilarity solutions for η are virtually identical up to $Z = 0.01$. Thus for $Z \leq 0.01$, eq. (19.65) for η and hence eq. (19.63) for $q(Z)$ give accurate predictions. However, this is not true for the similarity results for θ and Λ^*. For example, Figs. 19.9 and 19.10 show that for $Z = 0.01$, there is a significant difference between the similarity and nonsimilarity solutions.

Example 19.3. A vertical longitudinal fin of rectangular profile with thermal conductivity $k = 400$ W/m·K is attached to a vertical surface as shown in Fig. 19.8. The fin dimensions are $b = 1.5$ cm, $\delta = 1.5$ mm, and $L = 25$ cm. The fins are attached to a cold surface at 24°C, and the environment surrounding the fin is saturated steam at 50°C. Calculate **(a)** the rate at which heat is removed by the cooled surface, **(b)** the condensation rate supported by the fin, and **(c)** the fin temperature and film thickness at $L = 1.5$ cm and $y = 25$ cm.

SOLUTION. For saturated steam at 50°C,

$$v_v = 12.032 \text{ m}^3/\text{kg} \quad \text{and} \quad h_{fg} = 2383 \text{ kJ/kg}$$

Thus

$$\rho_v = \frac{1}{v_v} = \frac{1}{12.032} = 0.083 \text{ kg/m}^3$$

and for the condensate at a mean temperature of

$$T_m = \frac{24 + 50}{2} = 37°\text{C}$$

the pertinent properties are

$$\rho_\ell = 993\,\text{kg/m}^3$$
$$\mu_\ell = 694 \times 10^{-6}\,\text{N·s/m}^2$$

and

$$k_\ell = 0.628\,\text{W/m·K}$$

Use of the foregoing properties in eq. (19.49*d*) gives the value of Z. With $z = L = 0.25$ m,

$$Z = \frac{\mu_\ell k_\ell (T_{\text{sat}} - T_b)}{4g\rho_\ell(\rho_\ell - \rho_v)h_{\text{fg}}} \left(\frac{k\delta}{k_\ell b^2}\right)^4 z$$

and with

$$\frac{k\delta}{k_\ell b^2} = \frac{(400)(0.0015)}{(0.628)(0.015)^2} = 4246.3\,\text{m}^{-1}$$

With $\rho_\ell - \rho_v = 993 - 0.083 = 992.9$,

$$Z = \frac{(694 \times 10^{-6})(0.628)(50 - 24)}{(4)(9.81)(993)(992.9)(2383)(1000)}(4246.3)^4(0.25)$$

or

$$Z = 0.01$$

(a) Because Z is within the range of applicability of the similarity solution, eq. (19.63) can be used to compute the rate at which heat is removed by the cooled surface:

$$q(Z) = 4.9371 \left[\frac{g\rho_\ell(\rho_\ell - \rho_v)k_\ell^3 h_{\text{fg}} b^7}{\mu_\ell k^3 \delta^3}\right] Z^{7/8}$$

With

$$\left[\frac{(9.81)(993)(992.9)(0.628)^3(2383)(1000)(0.015)^7}{(694 \times 10^{-6})(400)^3(0.0015)^3}\right] = 6500$$

$$q = 4.9371(6500)(0.01)^{7/8} = 570.7\,\text{W}$$

(b) The condensate rate supported by the fin is given by

$$\dot{m} = \frac{q}{h_{\text{fg}}} = \frac{570.7}{(2383)(1000)} = 2.395 \times 10^{-4}\,\text{kg/s} = 0.862\,\text{kg/h}$$

(c) Use the terminal value of the curve for $Z = 0.01$ in Fig. 19.9 to obtain the terminal tip temperature of $\theta_a = 0.25$. Thus

$$\theta_a = \frac{T_a - T_{sat}}{T_b - T_{sat}} = 0.25$$

or

$$\begin{aligned} T_a &= 0.25(T_b - T_{sat}) + T_{sat} \\ &= (0.25)(24 - 50) + 50 \\ &= 43.5°\text{C} \end{aligned}$$

In similar fashion, the value of $\Lambda^{*1/4}$ is read at the terminal value for $Z = 0.01$ from Fig. 19.10 as $\Lambda^{*1/4} = 0.65$. Then using eq. (19.57),

$$\Lambda = \Lambda^{*1/4}(Z)^{1/4} = (0.65)(0.01)^{1/4} = 0.2055$$

Then from the definition of Λ in eq. (19.49*c*), the film thickness may be found:

$$\begin{aligned} y(L = 1.5\,\text{cm}) &= \frac{2k_\ell b^2 \Lambda}{k\delta} \\ &= \frac{(2)(0.628)(0.015)^2(0.2055)}{(400)(0.0015)} \\ &= 9.681 \times 10^{-5}\,\text{m} \quad \text{or} \quad 0.0968\,\text{mm} \end{aligned}$$

19.3 DEHUMIDIFICATION OF AIR ON FINS

In air-conditioning applications, finned cooling coils are often employed to cool and dehumidify air. The thermal performance of these coils is not only affected by geometry, materials, and psychrometric conditions, but also by the efficiency of the fin. If the fin temperature is lower than the dew point of the air passing over the coil, the moisture will be condensed on the fin surface, and this affects the fin efficiency.

In this section, the performance of fins operating in moist air streams with moisture condensation occurring on their surface is considered. Section 19.3.1 discusses simple models in which the classical fin theory for dry fins is modified to take into account the effect of mass transfer. Section 19.3.2 describes two conjugate models for simultaneous heat and mass transfer to a cooling and dehumidifying vertical longitudinal fin of rectangular profile. Experimental studies of dehumidification in finned coiled heat exchangers is covered in Section 19.3.3, and the design of optimum-dimensional longitudinal fins of rectangular and triangular profile is covered in Section 19.3.4.

19.3.1 Simple Models

Longitudinal Fins. McQuiston (1975) considered moisture condensation on a longitudinal fin of rectangular profile having height b, thickness δ, thermal conductivity k, and with the average heat transfer coefficient for dry operating conditions, designated as h_d. For the moisture condensation situation, neglecting the thermal resistance of the condensate, McQuiston, postulated that the local driving potential for simultaneous heat and mass transfer was the difference between the enthalpy of air adjacent to the fin and that of saturated air at the local fin temperature. By approximating the saturation curve on the psychrometric chart by a straight line over a small range of temperatures, he expressed the slope a as

$$a = \frac{\omega_{s,2} - \omega_{s,1}}{T_2 - T_1} \tag{19.66}$$

where ω_s is the specific humidity of the saturated air. The heat transfer coefficient h_w for wet conditions was expressed in terms of h_d and a by

$$h_w = h_d \left(1 + \frac{a h_{fg}}{c_p}\right) \tag{19.67}$$

where h_{fg} is the latent heat of condensation and c_p is the specific heat of moist air at constant pressure. Another expression for h_w is

$$h_w = \frac{h_d b}{c_p} \tag{19.68}$$

where b is the slope of the enthalpy–temperature curve for saturated air.

With h_w specified by either of eqs. (19.67) or (19.68), conventional fin theory can be employed to obtain the efficiency of a wet fin. For boundary conditions of constant fin base temperature and insulated fin tip, the fin efficiency can be expressed as

$$\eta_w = \frac{\tanh N^*}{N^*} \tag{19.69}$$

where

$$N^* = \left(\frac{2h_w}{k\delta}\right)^{1/2}$$

Radial Fins. Elmahdy and Biggs (1983) considered a radial fin of rectangular profile of base radius r_b and tip radius r_a, thickness δ, and thermal conductivity k, exposed to a stream of moist air at temperature T_w and with specific humidity ω_a. With the average heat and mass transfer coefficients taken as h and h_m, respectively, the differential equation governing the temperature distribution in the fin can be written as

$$\frac{d^2T}{dr^2} + \frac{1}{r}\frac{dT}{dr} - \frac{2h}{k\delta}(T_{\text{air}} - T) - \frac{2h_m}{k\delta}(\omega_a - \omega_{T,s})h_{fg} = 0 \tag{19.70}$$

where $\omega_{T,s}$ is the saturated specific humidity of air corresponding to the local fin temperature T with specific humidity ω_a. Assuming a constant fin base temperature T_b and an insulated fin tip, the boundary conditions for eq. (19.70) can be written as

$$T(r = r_b) = T_b \tag{19.71a}$$

and

$$\left.\frac{dT}{dr}\right|_{r=r_a} = 0 \tag{19.71b}$$

Next, $\omega_{T,s}$ is assumed to be a linear function of the temperature T, that is,

$$\omega_{T,s} = c + aT \tag{19.72}$$

where the constants, a, given by eq. (19.66), and c are to be determined from the psychrometric data for the range of temperatures considered. After substitution of $\omega_{T,s}$ from eq. (19.72) into eq. (19.70), the differential equation for T becomes

$$\frac{d^2T}{dr^2} + \frac{1}{r}\frac{dT}{dr} - \frac{2h}{k\delta}(T_{\text{air}} - T) - \frac{2h_m}{k\delta}(\omega_a - c - aT)h_{\text{fg}} = 0 \tag{19.73}$$

A sampling of numerical solutions of eq. (19.73) subject to the boundary conditions of eqs. (19.71) is shown in Fig. 19.12. In this figure the dimensionless temperature

$$\theta = \frac{T_{\text{air}} - T}{T_{\text{air}} - T_b}$$

is plotted against the dimensionless radius,

$$R = \frac{r - r_b}{r_a - r_b}$$

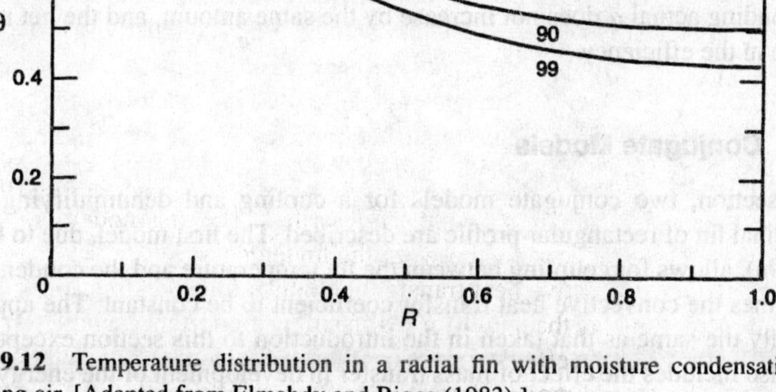

Figure 19.12 Temperature distribution in a radial fin with moisture condensation from surrounding air. [Adapted from Elmahdy and Biggs (1983).]

for dry as well as wet operating conditions. The results are based on the data

$$T_{\text{air}} = 16°\text{C}, \quad T_b = 7°\text{C}, \quad h = 57\,\text{W/m}^2\cdot\text{K}$$

and

$$N = (r_a - r_b)\left(\frac{2h}{k\delta}\right)^{1/2} = 0.82$$

It can be seen that the temperature profiles for the wet fin lie below those of a dry fin. As the relative humidity of the air increases, the driving potential for mass transfer increases. This leads to a higher latent heat transfer and a higher fin temperature. Observe that the lower values of θ mean higher fin temperatures.

Consider a typical fin surface area element $ds = 2\pi r\,dr$, and note that the heat transfer to this element can be expressed as

$$dq = [h(T_{\text{air}} - T) + h_m(\omega_a - \omega_{T,x})h_{\text{fg}}](2\pi r\,dr) \tag{19.74}$$

Then, allowing for heat transfer to the two surfaces of the fin, integration yields the total heat transfer as

$$q = \int_{r_b}^{r_a} 4\pi[h(T_{\text{air}} - T) + h_m(\omega_a - \omega_{T,x})h_{\text{fg}}]r\,dr \tag{19.75}$$

The maximum or ideal heat transfer to the fin occurs if the entire fin surface is maintained at temperature T_b:

$$q_{\text{id}} = 2\pi(r_a^2 - r_b^2)[h(T_{\text{air}} - T_b) + h_m(\omega_a - \omega_{T_b,s})] \tag{19.76}$$

The ratio q/q_{id} gives the efficiency of the fin. Figure 19.13 shows the efficiency as a function of N for both wet and dry fins. This figure is based on the same data as Fig. 19.12. It can be seen that the efficiency of the wet fin is lower than that of a dry fin and decreases as the relative humidity increases. The explanation for this involves the recognition that as the relative humidity increases, the driving potential for mass transfer also increases. This, in turn, causes q_{id} to increase. However, the corresponding actual q does not increase by the same amount, and the net result is a decrease in the efficiency.

19.3.2 Conjugate Models

In this section, two conjugate models for a cooling and dehumidifying vertical longitudinal fin of rectangular profile are described. The first model, due to Coney et al. (1989a), allows for coupling between the fin temperature and the condensate film but assumes the convective heat transfer coefficient to be constant. The approach is essentially the same as that taken in the introduction to this section except that the model also includes the effect of mass transfer in development of the energy balance for the fin. The second model, described by Kazeminejad et al. (1993), neglects the

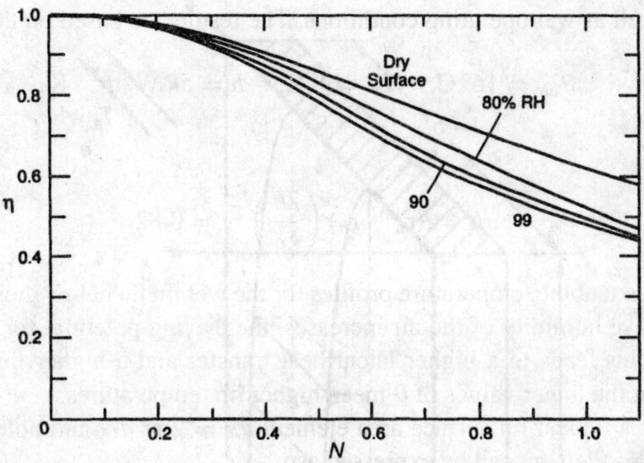

Figure 19.13 Efficiencies of wet and dry radial fins. [Adapted from Elmahdy and Biggs (1983).]

thermal resistance of the condensate film but allows for the variation of h through the solution of the boundary layer equations. Both methods are discussed below.

Coney et al. (1989a) Model. The model considers a vertical longitudinal fin of rectangular profile as shown in Fig. 19.14. Taking a slice of fin of volume $b\delta\,dz$ and equating the net energy conducted through the slice to the energy convected to the surface of area $2(b+\delta)dz$ by simultaneous heat and mass transfer gives

$$\frac{d^2T}{dz^2} = -\frac{2(b+\delta)q_t''}{k\delta b} \tag{19.77}$$

where q_t'' is the total heat flux through the condensate film. Assuming a linear temperature profile for the condensate film, q_t'' can be expressed as

$$q_t'' = \frac{k_\ell(T_i - T)}{y(x)} \tag{19.78}$$

where T_i is the condensate–air interface temperature.

The presence of the condensate film can enhance the heat and mass transfer at the condensate–air interface due to increased turbulence and interface roughness. This can be taken into account by multiplying the single-phase heat transfer coefficient h by an interface enhancement factor C_f, which depends on geometry and surface conditions and which must be determined experimentally. Because the minimum value of C_f must be unity (for a smooth surface at low vapor velocity), the use of $C_f = 1$ would be conservative. The effect of mass transfer on the temperature profile is taken into account by the *Ackerman correction factor* C_a. Thus the sensible heat flux q_s'' between the air and condensate film can be expressed as

$$q_s'' = C_f C_a h(T_{\text{air}} - T_i) \tag{19.79}$$

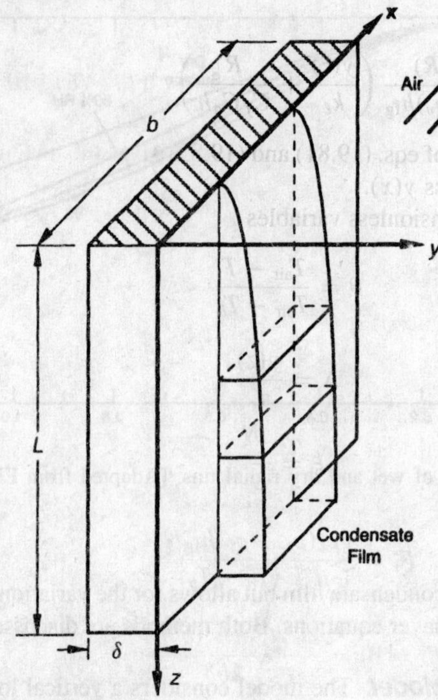

Figure 19.14 Dehumidification of air on a vertical longitudinal fin of rectangular profile.

Elimination of T_i between eqs. (19.78) and (19.79) and denoting the ratio of the heat fluxes by R,

$$R = \frac{q_s''}{q_t''}$$

yields

$$q_t'' = \frac{T_{\text{air}} - T}{y(z)/k_\ell + R/C_f C_a h} \tag{19.80}$$

Then a substitution for q_t'' from eq. (19.80) into eq. (19.77) provides the differential equation

$$\frac{d^2T}{dz^2} + \frac{2(L+\delta)}{k\delta L}\left(\frac{y(z)}{k_\ell} + \frac{R}{C_f C_a h}\right)^{-1}(T_a - T) = 0 \tag{19.81}$$

Equation (19.26), the momentum equation for the condensate film, can be modified for the present analysis:

$$[y(x)]^2 \frac{dy(z)}{dz} = \frac{\mu_\ell (1-R) q_t''}{g\rho_\ell(\rho_\ell - \rho_v) h_{\text{fg}}} \tag{19.82}$$

and after elimination of q_i'' between eqs. (19.80) and (19.82), the governing differential equation can be written as

$$[y(z)]^2 \frac{dy(z)}{dz} - \frac{\mu_\ell(1-R)}{g\rho_\ell(\rho_\ell - \rho_v)h_{fg}} \left(\frac{y(z)}{k_\ell} + \frac{R}{C_f C_a h} \right)^{-1} (T_a - T) = 0 \tag{19.83}$$

The simultaneous solution of eqs. (19.81) and (19.83) gives the fin temperature $T(z)$ and the condensate thickness $y(x)$.

Substitution of the dimensionless variables

$$\theta = \frac{T_{\text{air}} - T}{T_{\text{air}} - T_b} \tag{19.84a}$$

$$\Lambda = \frac{y(z)}{L} \tag{19.84b}$$

$$\xi = \frac{z}{L} \tag{19.84c}$$

$$F_1 = \frac{g\rho_\ell(\rho_\ell - \rho_v)h_{fg}L^3}{\mu_\ell k_\ell (T_a - T_b)} \tag{19.84d}$$

$$F_2 = \text{Bi} = \frac{2C_f C_a h L^2 (b + \delta)}{kb\delta} \tag{19.84e}$$

$$F_3 = \frac{k_\ell}{C_f C_a h L} \tag{19.84f}$$

and

$$F_4 = \frac{1}{F_2 F_3} = \frac{kb\delta}{2k_\ell L(b + \delta)} \tag{19.84g}$$

into eqs. (19.81) and (19.83) gives

$$\frac{d^2\theta}{d\xi^2} = \frac{\theta}{F_4(\Lambda + F_3 R)} \tag{19.85a}$$

and

$$\Lambda^2 \frac{d\Lambda}{d\xi} = \frac{(1-R)\theta}{F_4(\Lambda + F_3 R)} \tag{19.85b}$$

The case of $R = 0$ represents the condensation of pure vapor on a fin, and eqs. (19.85) reduce to eqs. (19.28) and (19.29) of Section 19.2. Observe that for a thin fin, $\delta/L << 1$ and $F_4 = k\delta/2k_\ell$, which is equal to F_2 defined by eq. (19.27e). It is also interesting to note that the case of purely convecting fins with no condensation is represented by $R = 1$ and $\Lambda = 0$. The range $0 < R < 1$ represents the case of simultaneous heat and mass transfer.

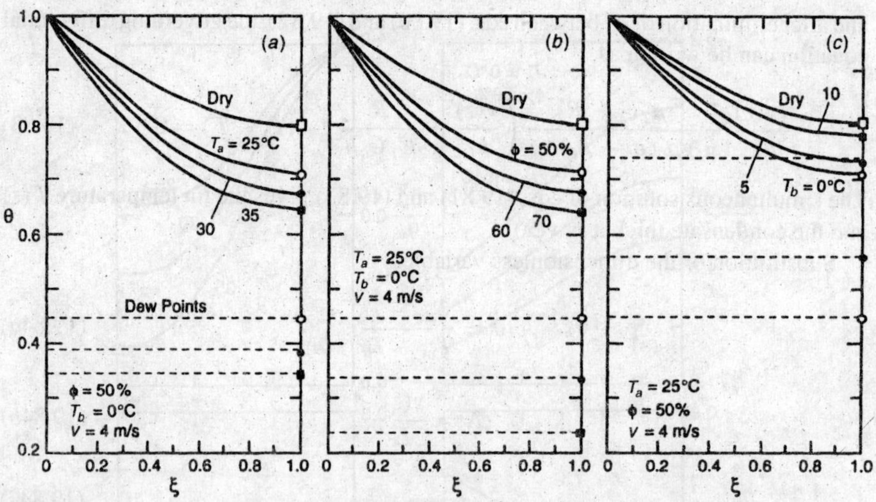

Figure 19.15 Effects of (*a*) dry bulb temperature, (*b*) relative humidity, and (*c*) fin base temperature distributions in a vertical longitudinal fin of rectangular profile. [Adapted from Coney et al. (1989a).]

Figure 19.15 shows typical results for θ obtained from a numerical solution of eqs. (19.85) subject to the boundary conditions of constant base temperature and insulated fin tip:

$$\theta(\xi = 0) = 1 \quad \text{and} \quad \left.\frac{d\theta}{d\xi}\right|_{\xi=1} = 0$$

The parameters F_1 through F_4 were computed assuming moisture condensation on a copper fin having $k = 380$ W/m·K and having dimensions of $L = 240$ mm, $b = 220$ mm, and $\delta = 20$ mm. The value of h was calculated assuming a free-stream velocity of $v = 4$ m/s, which is typical of air-conditioning systems. Figure 19.15*a* illustrates the effect of dry bulb temperature, T_{db}, with $T_b = 0°\text{C}$ and $\phi = 50\%$. As T_a increases, θ decreases, indicating an increase in fin temperature T. The increase in T reflects higher sensible and latent heat transfer to the fin surface. The effect of relative humidity ϕ, shown in Fig. 19.15*b*, is similar to that for a radial fin (Fig. 19.12). Finally, Fig. 19.15*c* shows that as the fin base temperature T_b is reduced, the driving potential for heat and mass transfer is increased, causing the fin temperature to increase or θ to decrease.

A comparison of dry and wet fin heat transfer is presented in Fig. 19.16. Figure 19.16*a* shows the ratio of total heat transfer for wet conditions, q_{wt}, to that for dry conditions, q_d. This ratio decreases with increase in Biot number. The ratio of sensible heat transfer for wet conditions, q_{ws}, to q_d is plotted in Fig. 19.16*b*, where it can be noted that the sensible heat transfer during condensation is appreciably reduced as Bi increases. Figure 19.16*c* and *d* show how the sensible and latent heat transfer, as a fraction of the total heat transfer for wet conditions, q_{wt}, are affected by Bi and T_{db}.

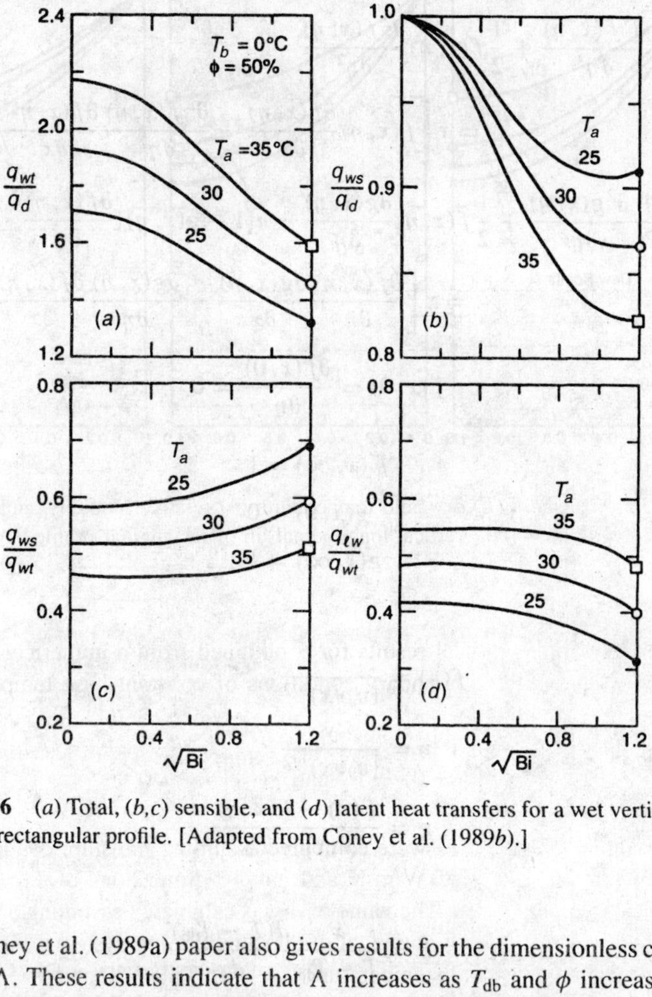

Figure 19.16 (*a*) Total, (*b,c*) sensible, and (*d*) latent heat transfers for a wet vertical longitudinal fin of rectangular profile. [Adapted from Coney et al. (1989*b*).]

The Coney et al. (1989a) paper also gives results for the dimensionless condensate thickness Λ. These results indicate that Λ increases as T_{db} and ϕ increase or as T_b decreases. However, the study concluded that the effect of film thermal resistance can be neglected without introducing significant error for normal conditions encountered in practice. On the other hand, if dropwise condensation occurs, the increased surface roughness and the resultant high turbulence intensity of the fin surface can slightly enhance the heat and mass transfer to the fin.

Kazeminejad et al. (1993) Model. Kazeminejad et al. (1993) considered the downward-pointing longitudinal fin of rectangular profile of Fig. 19.5*b* with moist air at temperature T_{db} and relative humidity ϕ flowing upward with a uniform velocity v. They neglected the condensate film thickness Λ in eq. (19.85*a*) but allowed h appearing in the definition for F_3 in eq. (19.84*f*) to be a function of x. To obtain $h(x)$, where x is measured from the fin tip, the nonsimilar boundary layer equations for upward flow over a nonisothermal vertical surface were written as

$$\frac{\partial^3 f(x,\eta)}{\partial\eta^3} + \frac{1}{2} f(x,\eta)\frac{\partial^2 f(x,\eta)}{\partial\eta^2} = x\left[f(x,\eta)\frac{\partial f(x,\eta)}{\partial x} - \frac{\partial^2 f(x,\eta)}{\partial\eta^2}\frac{\partial f(x,\eta)}{\partial x}\right] \tag{19.86}$$

$$\frac{1}{\mathrm{Pr}}\frac{\partial^2 g(x,\eta)}{\partial\eta^2} + \frac{1}{2} f(x,\eta)\frac{\partial g(x,\eta)}{\partial\eta} + n[1-g(x,\eta)]\frac{\partial f(x,\eta)}{\partial\eta} = x\left[\frac{\partial f(x,\eta)}{\partial\eta}\frac{\partial g(x,\eta)}{\partial x} - \frac{\partial g(x,\eta)}{\partial\eta}\frac{\partial f(x,\eta)}{\partial x}\right] \tag{19.87}$$

$$f(x,0) = \frac{\partial f(x,0)}{\partial\eta} = 0 \tag{19.88a}$$

$$f'(x,\infty) = 1 \tag{19.88b}$$

$$g(x,0) = 0 \tag{19.88c}$$

$$g(x,\infty) = 1 \tag{19.88d}$$

where

$$f(\eta) = \frac{\psi}{(v_v \nu x)^{1/2}} \tag{19.89a}$$

$$\eta = \frac{y}{(v_v \nu x)^{1/2}} \tag{19.89b}$$

$$g = \frac{T(x) - T(x,\eta)}{T(x) - T_{\mathrm{db}}} \tag{19.89c}$$

and

$$h = \frac{x}{T - T_{\mathrm{db}}}\frac{d(T - T_{\mathrm{db}})}{dx} \tag{19.89d}$$

The local heat transfer coefficient $h(x)$ relates to $\partial g(x,0)/\partial\eta$ by

$$h(x) = k_{\mathrm{air}}\left(\frac{v_\infty}{\nu_{\mathrm{air}} x}\right)^{1/2}\frac{\partial g(x,0)}{\partial\eta} \tag{19.90}$$

The coupled problem consisting of eq. (19.85*a*) with $\Lambda = 0$ and the boundary layer equations [eqs. (19.86) through (19.88)] was solved numerically by Kazeminejad et al. (1993) to obtain the fin temperature distribution, total heat transfer to the fin, and the fin efficiency. These results show that the conjugate conduction–boundary layer analysis gives higher fin temperatures and higher fin efficiencies than those predicted by the Coney et al. (1989*a*) model. This conclusion applies to both dry and wet fins.

Example 19.4. Air at 25°C and 4 m/s flows over a vertical longitudinal fin of rectangular profile as shown in Fig. 19.17*a*. The fin with thermal conductivity

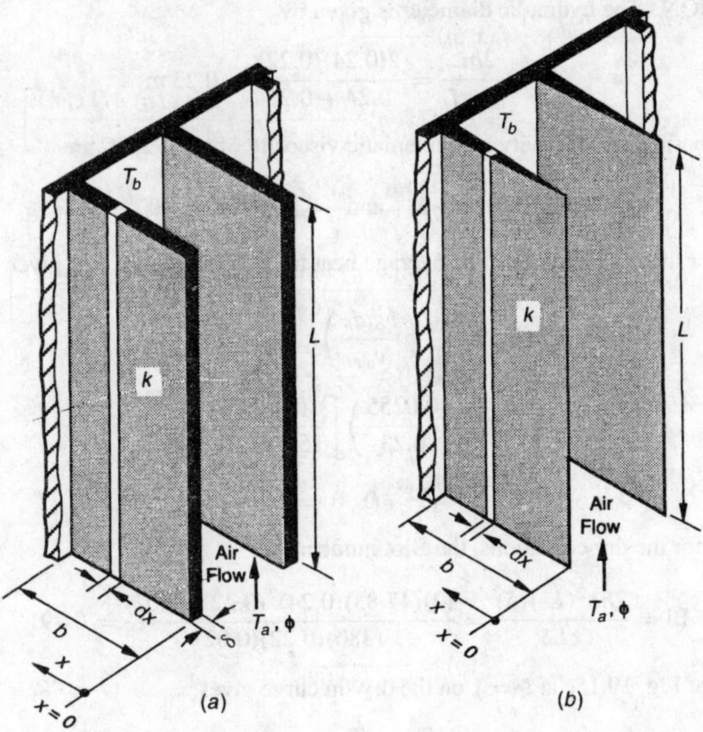

Figure 19.17 Coordinate systems for longitudinal fins of (*a*) rectangular and (*b*) triangular profile.

$k = 380$ W/m·K has dimensions $b = 240$ mm, $\delta = 20$ mm, and $L = 220$ mm. The base of the fin is cooled and maintained at a temperature of 0°C. Assuming that the results of Figs. 19.14 and 19.15 apply to this fin, calculate the tip temperature and the total heat transferred for **(a)** perfectly dry air and **(b)** moist air with a relative humidity of 50%. In addition, for the moist air, calculate **(c)** the sensible and latent heat contributions to the heat transfer. Take $C_f = C_a = 1.00$ and use the following correlations for calculating the average convective heat transfer coefficient h:

$$\frac{hd_h}{k_{\text{air}}} = \begin{cases} 0.590 \left(\dfrac{v_\infty d_h}{\nu_{\text{air}}} \right)^{0.60} & \text{for dry conditions} \\ 0.231 \left(\dfrac{v_\infty d_h}{\nu_{\text{air}}} \right)^{0.69} & \text{for wet conditions} \end{cases}$$

where d_h is the hydraulic diameter,

$$d_h = \frac{2Lb}{L+b}$$

These correlations are discussed in the next section.

SOLUTION. The hydraulic diameter is given by

$$d_h = \frac{2bL}{b+L} = \frac{2(0.24)(0.22)}{0.24+0.22} = 0.23 \text{ m}$$

The thermal conductivity and kinematic viscosity of air at 25°C are

$$k_{\text{air}} = 0.0255 \text{ W/m}\cdot\text{K} \quad \text{and} \quad \nu_{\text{air}} = 15.5 \times 10^{-6} \text{ m}^2/\text{s}$$

(a) For the dry conditions, the average heat transfer coefficient h is given by

$$\begin{aligned} h &= 0.590 \frac{k_{\text{air}}}{d_h}\left(\frac{v_\infty d_h}{\nu_{\text{air}}}\right)^{0.60} \\ &= 0.590\left(\frac{0.0255}{0.23}\right)\left[\frac{(4)(0.23)}{15.5\times 10^{-6}}\right]^{0.60} \\ &= 47.83 \text{ W/m}^2\cdot\text{K} \end{aligned}$$

Then for the dry conditions, the Biot number is

$$\text{Bi} = \frac{2hb^2(L+\delta)}{kL\delta} = \frac{(2)(47.83)(0.24)^2(0.22+0.02)}{(380)(0.22)(0.02)} = 0.791$$

and use of Fig. 19.15c at $\xi = 1$ on the dry fin curve gives

$$\theta_a = \frac{T_{\text{air}} - T_{\text{db}}}{T_{\text{air}} - T_b} = 0.80$$

or

$$\begin{aligned} T_{\text{db}} &= T_{\text{air}} - 0.80(T_{\text{air}} - T_b) \\ &= 25 - (0.80)(25 - 0) \\ &= 5°\text{C} \end{aligned}$$

The efficiency of a dry fin is given by

$$\eta_d = \frac{\tanh\sqrt{\text{Bi}}}{\sqrt{\text{Bi}}} = \frac{\tanh\sqrt{0.791}}{\sqrt{0.791}} = 0.800$$

The ideal heat transfer is

$$\begin{aligned} q_{\text{id}} &= 2h(b+\delta)L(T_{\text{air}} - T_b) \\ &= (2)(47.83)(0.24+0.02)(0.22)(25-0) \\ &= 136.8 \text{ W} \end{aligned}$$

and the actual heat transfer is given by

$$q_d = \eta_d q_{id}$$

$$= (0.800)(136.8)$$

$$= 109.4 \text{ W}$$

(b) For wet conditions, the average heat transfer coefficient h is given by

$$h = 0.231 \frac{k_{air}}{d_h} \left(\frac{v_\infty d_h}{\nu_{air}} \right)^{0.69}$$

$$= 0.231 \left(\frac{0.0255}{0.23} \right) \left[\frac{(4)(0.23)}{15.5 \times 10^{-6}} \right]^{0.69}$$

$$= 50.36 \text{ W/m}^2 \cdot \text{K}$$

The Biot number for wet conditions is related to that for dry conditions.

$$\text{Bi} = 0.791 \frac{50.36}{47.83} = 0.833$$

Then use of Fig. 19.15c at $\xi = 1$ (the fin tip) on the curve, for $T_b = 0°\text{C}$ gives

$$\theta_a = \frac{T_{air} - T_{db}}{T_{air} - T_b} = 0.72$$

or

$$T_{db} = T_{air} - (0.72)(T_{air} - T_b)$$

$$= 25 - 0.72(25 - 0)$$

$$= 7°\text{C}$$

For $\sqrt{\text{Bi}} = \sqrt{0.833} = 0.913$, Fig. 19.16$a$ for $T_{air} = 25°\text{C}$ gives

$$\frac{q_{wt}}{q_d} = 1.45$$

or

$$q_{wt} = 1.45 q_d = (1.45)(109.4) = 158.6 \text{ W}$$

Then reading the curve for $T_{air} = 25°\text{C}$ in Fig. 19.16c, the ratio q_{ws}/q_{wt} for $\sqrt{\text{Bi}} = 0.913$ is

$$\frac{q_{ws}}{q_{wt}} = 0.63$$

or

$$q_{ws} = (0.63)(158.6) = 99.9 \text{ W}$$

(c) The latent heat transfer is $158.6 - 99.9 = 58.7$ W.

19.3.3 Experimental Studies

Experimental studies of finned coil heat exchangers have been carried by several workers, including Bryan (1962), Bettanini (1970), Yoshi et al. (1971), and Guillory and McQuiston (1973). These studies have confirmed that the performance of finned coils is reduced significantly when dehumidification occurs. This reduction is the consequence of lower fin efficiency for wet conditions.

Kazeminejad et al. (1993) and Coney et al. (1989a,b) conducted experimental studies to investigate the performance of a vertical longitudinal fin of rectangular profile (Fig. 19.17*a*) when moist air in turbulent flow dehumidifies on the surface of the fin. The studies revealed that the wet fin surface temperature increases with increase in free-stream velocity, relative humidity, and dry bulb temperature. The increase in wet fin surface temperature also occurs when the fin base temperature is decreased. The studies also noted that smooth and clean surface copper fins promote dropwise rather than film condensation. Although the effect of mass transfer on the heat transfer coefficient was small, the fin efficiency was reduced markedly under wet conditions. Another interesting conclusion was that the shape of the leading edge of the fin affected the heat and fluid flow. The heat transfer to the fin was higher when the leading edge of the fin was blunt, with attendant flow separation and reattachment, than when the leading edge of the fin was elliptical.

Based on their experimental work, Kazeminejad et al. (1993) and Coney et al. (1989b) presented the following correlations for the heat transfer coefficient for vertical longitudinal fins of rectangular profile: For a blunt-edged dry fin,

$$\frac{hd_h}{k_{\text{air}}} = 0.590\left(\frac{v_\infty d_h}{\nu_{\text{air}}}\right)^{0.60} \tag{19.91}$$

for a blunt-edged wet fin,

$$\frac{hd_h}{k_{\text{air}}} = 0.231\left(\frac{v_\infty d_h}{\nu_{\text{air}}}\right)^{0.69} \tag{19.92}$$

for an elliptical-edged dry fin,

$$\frac{hd_h}{k_{\text{air}}} = 0.420\left(\frac{v_\infty d_h}{\nu_{\text{air}}}\right)^{0.60} \tag{19.93}$$

and for an elliptical-edged wet fin,

$$\frac{hd_h}{k_{\text{air}}} = 0.146\left(\frac{v_\infty d_h}{\nu_{\text{air}}}\right)^{0.69} \tag{19.94}$$

19.3.4 Optimum Fin Design

In this section we consider the design of optimum dimensioned fins for use in a moist airstream. The discussion is based on the works of Kilic and Onat (1981) and Toner et al. (1983) and covers longitudinal fins of rectangular and triangular profile.

Rectangular Fins. Kilic and Onat (1981) considered a vertical longitudinal fin of rectangular profile as displayed in Fig. 19.17a and modified the classical convecting fin equation to allow for simultaneous heat and mass transfer. The modified equation can be expressed as

$$\frac{d^2T}{dx^2} = m^2\left[(T - T_{\text{air}}) - \frac{h_m h_{\text{fg}}}{h R_v T_{\text{air}}}\left(P_{v,\text{air}} - P e^{A-B/T}\right)\right] \tag{19.95}$$

where

$$m = \left(\frac{2h}{k\delta}\right)^{1/2}$$

h_m the mass transfer coefficient, R_v the gas constant for the water vapor, $P_{v,\text{air}}$ the partial pressure of the water vapor at the air temperature T_{air}, P the total pressure, and A and B are constants.

A model similar to eq. (19.95) has also been used by Karniven et al. (1990) to study moisture on fins. This model allows for radiation heat transfer in addition to convective heat and mass transfer. Furthermore, the model considers a fin that is partially rather than fully wet and determines the line of demarcation separating the wet and dry regions.

Assuming a constant base temperature and an insulated fin tip, an approximate analytical solution of eq. (19.95) was derived using the quasilinearization technique. For a fixed profile area $b\delta$, the expression for the heat transfer rate was maximized to yield relationships for δ and b. A reference temperature is established:

$$T_r = T_{\text{air}} - (T_{\text{air}} - T_b)\frac{\tanh mb}{mb} \tag{19.96}$$

and then with $P_{s,r}$ taken as the saturation pressure at the reference temperature T_r,

$$m_1 = 1 + \frac{B h_m h_{\text{fg}} P_{s,r}}{h R_v T_{\text{air}} T_r^2} \tag{19.97}$$

Then the optimum fin height and aspect ratio are given by

$$b = \left(\frac{\delta b k}{m_1 h}\right)^{1/3} \tag{19.98}$$

and

$$\frac{\delta}{b} = m_1 \frac{hb}{k} \tag{19.99}$$

A sample solution of eqs. (19.96) through (19.99) for atmospheric air with a relative humidity of 50% is shown in Fig. 19.18. The details for the evaluation of h_m/h and B can be found in Kilic and Onat (1981). Figure 19.18 also shows that the optimum value for N_{opt} for a wet fin is lower than that for a dry fin where N_{opt} = 1.4192. Moreover, N_{opt} for a wet fin decreases as the fin base temperature T_b and free-stream temperature T_{air} increase.

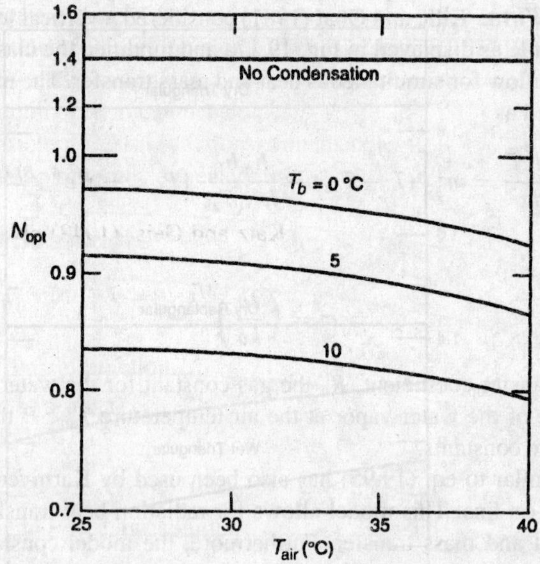

Figure 19.18 Optimum fin parameter N_{opt} for a vertical longitudinal fin of rectangular profile with moisture condensation. [Adapted from Kilic and Onat (1981).]

Triangular Fins. Toner et al. (1983) extended the analysis of the foregoing section to the vertical longitudinal fin of triangular profile shown in Fig. 19.17*b*. Figure 19.19 provides their final results showing how the optimum value of N is affected by the variations in the ambient temperature T_{air} and the fin base temperature T_b. Shown for comparison are the results for the optimum rectangular fins. It can be seen that for a given T_{air} and T_b, the optimum value for N for a wet triangular fin is lower than that for a dry triangular fin and that N_{opt} decreases as T_{air} and T_b increase. As noted earlier, the wet rectangular fin exhibits the same behavior. It is also interesting to note that values for N_{opt} for triangular fins, both dry and wet, are higher than for those of the rectangular fins.

19.4 HORIZONTAL INTEGRAL-FIN TUBES

As shown in Fig. 19.20, the use of horizontal integral-fin tubes, to enhance condensation is quite common in the design of surface condensers in the refrigeration and process industries. The tubes have therefore been studied extensively. An excellent review of the pertinent literature provided by Marto (1988), and Webb (1994) also discusses at length the analysis and design of integral-fin tubes.

Commercial integral-fin tubes are made of various materials (aluminum, stainless steel, and copper and its alloys) and are available in densities ranging from 433 to 1675 fins per meter (11 to 40 fins per inch). Although the standard integral-fin design

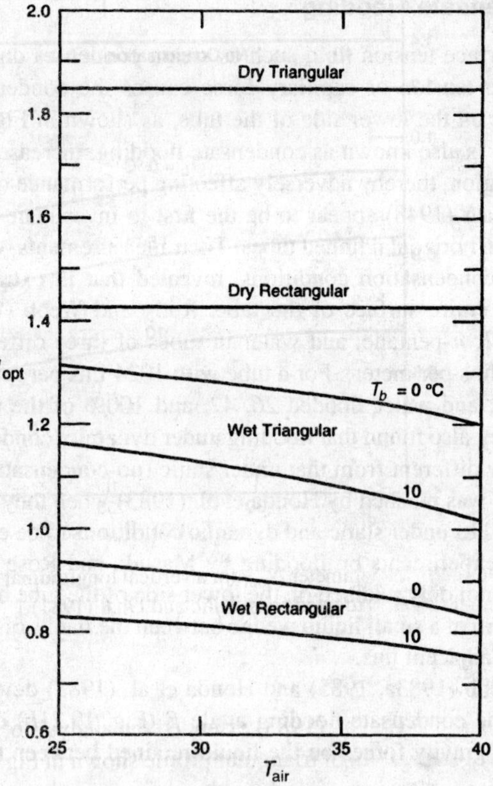

Figure 19.19 Comparison of N_{opt} for vertical longitudinal fins of rectangular and triangular profile under dry and wet conditions. [Adapted from Toner et al. (1983).]

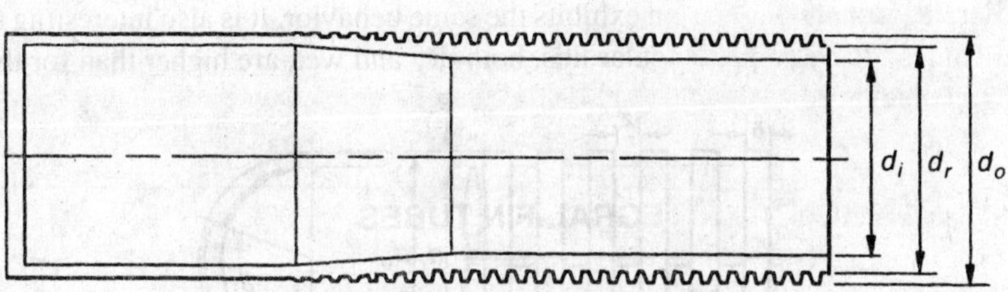

Figure 19.20 Horizontal integral-fin tube.

provides a significant enhancement in condensation over plain tubes, more advanced designs, such as the Hitachi-Thermorxcel-C, the Wieland GEWA-SC, the Wolverine Turbo-C, and the Sumitomo Tred-26D improve the performance still further. These designs use a sawtooth fin shape.

19.4.1 Condensate Flooding

When a high-surface-tension fluid such as steam condenses on a horizontal finned tube, the surface tension or capillary force causes the condensate to be retained between the fins on the lower side of the tube, as shown in Fig. 19.21*a*. This phenomenon, which is also known as condensate flooding, increases thermal resistance in the flooded region, thereby adversely affecting performance of the tube.

Beatty and Katz (1948) appear to be the first to investigate the phenomenon of liquid retention in horizontal finned tubes. Their measurements, which were obtained under static no-condensation conditions, revealed that in extreme cases, flooding could cover the entire surface of the tube. Rudy and Webb (1981) measured the retention of R-11, *n*-pentane, and water in tubes of three different densities (748, 1024, and 1378 fins per meter). For a tube with 1024 fins per meter, they found that R-11, *n*-pentane, and water flooded 26, 42, and 100% of the tube circumference, respectively, They also found that flooding under dynamic (condensation) conditions was only slightly different from that under static (no-condensation) conditions. The same conclusion was reached by Honda et al. (1983) when they found their data for methanol and R-113 under static and dynamic conditions to be essentially the same.

More recent experiments on flooding by Masuda and Rose (1987) have shown that the liquid is not only retained on the lower side of the tube but also on the upper portion in the form of a small liquid wedge between the flanks of the fins and the tube surface between adjacent fins.

Rudy and Webb (1983a, 1985) and Honda et al. (1983) developed a theoretical expression for the condensate flooding angle β (Fig. 19.21*b*) by equating the surface tension and gravity forces on the liquid retained between the fins (rectangular greometry):

$$\beta = \arccos\left(1 - \frac{4\sigma}{d_o \rho_\ell g z}\right) \tag{19.100}$$

where σ is the surface tension, d_o the tube outside diameter, ρ_ℓ the condensate density,

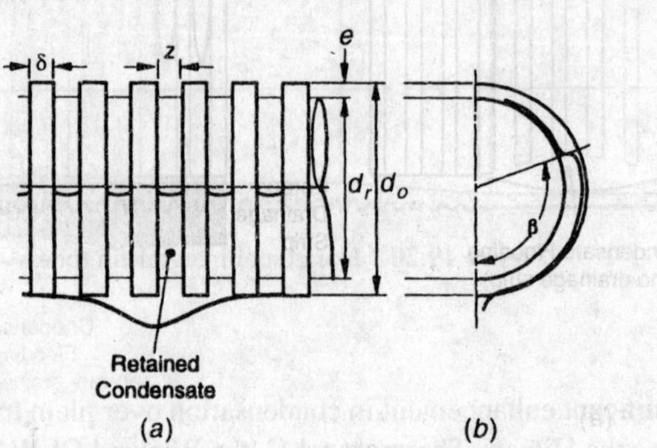

Figure 19.21 Condensate flooding on a horizontal integral-fin tube: (*a*) side and (*b*) end views.

g the acceleration of gravity, and z the spacing between the fins. Equation (19.100) shows that the flooding angle β increases with increasing surface tension σ and with decreasing fin spacing z.

It is possible to reduce the amount of condensate retained between the fins by attaching a rectangular porous plate to the bottom of the tube. This attachment, known as a *drainage strip*, increases the downward capillary force by drawing the condensate into its pores. Figure 19.22 illustrates how reduction of condensate flooding is achieved with a porous drainage strip. A solid plate can also serve as a drainage strip, but such a plate is not as effective as the porous plate. For example, Honda and Nozu (1987a) measured the effect of different drainage strips for R-113 condensing on an 18.9-mm-diameter (d_o) tube having a fin density of 2000 fins per meter and a fin height of 1.13 mm. With no drainage strip, they found that 62% of the tube circumference was flooded. When a polyvinyl chloride (PVC) solid strip of height 12.6 mm was employed, the percent of the circumference flooded was reduced to 57, a reduction of about 8%. However, when a porous nickel strip of the same height (12.6 mm) was attached, only 32% of tube circumference was flooded. Clearly, the porous strip is much more effective than a solid strip. The measured heat transfer coefficients for no strip, a PVC strip, and a porous nickel strip were 6200, 6900, and 10,200 W/m$^2\cdot$K respectively. The effectiveness of draining strips was also demonstrated by others, including Yau et al. (1986) and Marto et al. (1988). Although the drainage strips have proved effective under laboratory conditions, their use in an actual condenser tube bundle may not be feasible.

Example 19.5. Horizontal integral-fin tubes are to be designed to condense steam, ethylene glycol, and R-113 at atmospheric conditions. **(a)** If the maximum circumference flooded is to be 50%, calculate the number of fins per meter if the outer tube diameter is 19 mm and radial fins of rectangular profile of 0.25 mm thickness are used. **(b)** If the outside diameter and fin thickness are changed to 21.05 and 0.50 mm,

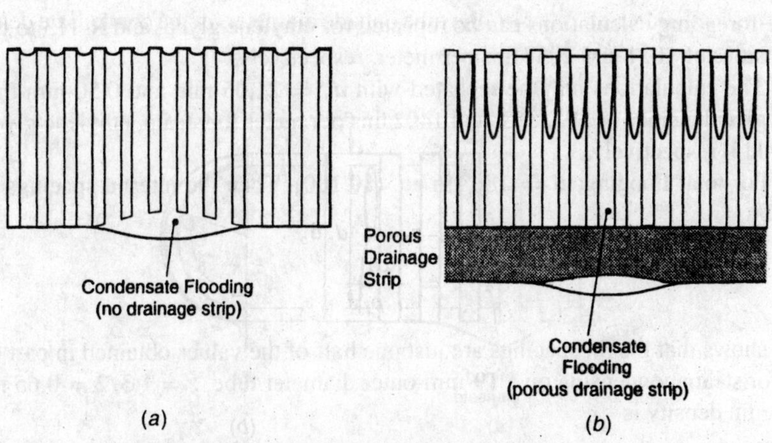

Figure 19.22 Effect of a porous drainage strip on condensate flooding (a) with no strip and (b) with strip. [Adapted from Marto (1988).]

respectively, find the new fin density for each of the three fluids. **(c)** Calculate the maximum fin density for steam, ethylene glycol, and R-113 condensing on 19-mm-outer diameter tubes fitted with fins of 0.25 mm thickness.

SOLUTION. **(a)** For 50% flooding, $\beta = 90°$ in eq. (19.100). The interfin spacing is then given by

$$z = \frac{4\sigma}{d_o \rho_\ell g}$$

The values of the ratio σ/ρ_ℓ at atmospheric pressure for the three fluids have been quoted by Marto (1988) as follows:

Steam	$61 \times 10^{-6}\ m^3/s^2$
Ethylene glycol	$34 \times 10^{-6}\ m^3/s^2$
R-113	$11 \times 10^{-6}\ m^3/s^2$

Thus, for steam, the spacing is

$$z = \frac{4\sigma}{d_o \rho_\ell g}$$

$$= \frac{(4)(61 \times 10^{-6})}{(0.019)(9.81)}$$

$$= 1.3 \times 10^{-3}\ \text{m} \quad \text{or} \quad 1.3\ \text{mm}$$

and the fin density is

$$\frac{1}{(0.25 + 1.30) \times 10^{-3}} = 645 \text{ fins per meter}$$

The foregoing calculations can be repeated for ethylene glycol and R-113 to give fin densities of 1021 and 2057 fins per meter, respectively.

(b) The calculations may be repeated with $d_o = 21.05$ mm and 0.50-mm-thick fins to give densities of 595, 863, and 1402 fins per meter for steam, ethylene glycol, and R-113, respectively.

(c) For total flooding, $\beta = 180°$ in eq. (19.100). Then the interfin spacings are given by

$$z = \frac{2\sigma}{d_o \rho_\ell g}$$

which shows that the fin spacings are just one half of the values obtained in part (a). Thus for steam condensing on a 19-mm-outer diameter tube, $z = 1.3/2 = 0.65$ mm and the fin density is

$$\frac{1}{(0.25 + 0.65) \times 10^{-3}} = 1111 \text{ fins per meter}$$

The corresponding figures for ethelyne glycol and R-113 work out to be 1626 and 2717 fins per meter. The value of 1111 fins per meter confirms Webb's (1994) statement that condensation of steam on a 19-mm-diameter tube with a 0.25-mm fin thickness would result in total flooding if the fin density were greater than 1000 fins per meter. Indeed, the experimental work of Jaber and Webb (1994) on enhanced tubes for steam condensers shows that the integral-fin tubes for steam condensation should use no more than 630 fins per meter (16 fins per inch).

19.4.2 Theoretical Models for the Heat Transfer Coefficient

Perhaps the first theoretical model for predicting the condensation heat transfer for integral-fin tubes was proposed by Beatty and Katz (1948). They assumed condensate flow to be purely gravity driven and applied a Nusselt type of analysis for both the fin surface and the tube surface between the fins. The average heat transfer coefficient $\bar{h}$ for the configuration of Fig. 19.21 was expressed as

$$\bar{h}\eta_o = h_h \frac{S_r}{S} + h_f \eta_f \frac{S_f}{S} \tag{19.101}$$

where S is the total heat transfer surface area,

$$S = S_r + S_f$$

S_b the surface area of the tube at the base of the fins, S_f the surface area of the fin, η_f the fin efficiency, η_o the overall surface efficiency, h_h the heat transfer coefficient for the horizontal tube surface, and h_f the heat transfer coefficient for the fin surface. The expressions for h_h and h_f and the relationship between η_o and η_f are given by

$$h_h = 0.728 \left[\frac{g\rho_\ell(\rho_\ell - \rho_v)k_\ell^3 h_{fg}}{\mu_\ell(T_{sat} - T_b)d_r} \right]^{1/4} \tag{19.102}$$

$$h_f = 0.943 \left[\frac{g\rho_\ell(\rho_\ell - \rho_v)k_\ell^3 h_{fg}}{\mu_\ell(T_{sat} - T_b)b_{av}} \right]^{1/4} \tag{19.103}$$

and

$$\eta_o = 1 - (1 - \eta_f)\frac{S_f}{S} \tag{19.104}$$

where b_{av} is the average fin height over the diameter d_o:

$$b_{av} = \frac{\pi}{4} \frac{d_o^2 - d_r^2}{d_o}$$

Beatty and Katz (1948) used eq. (19.101) to predict their test data for six low-surface-tension fluids condensing on finned tubes ranging in fin density from 433 to 630 fins per meter. They found that the coefficient in eq. (19.102) was, in reality, not 0.728 but 0.689. Equation (19.101) predicted their data within 10%. Despite the fact that the Beatty and Katz model does not account for condensate retention and surface tension effects, the model has enjoyed success in the refrigeration industry for many years.

The first model to include the effect of surface tension in draining condensate from horizontal finned tubes was presented by Karkhu and Borovkov (1971). They postulated that condensate drainage from the fin surface was purely surface-tension driven. From experiments in which R-12 was condensed on the face of a cylinder with machined longitudinal fins of rectangular profile, Webb et al. (1982) concluded that surface tension was indeed the force controlling the condensate drainage from the fin surface. Using a linear-surface-tension model, Rudy and Webb (1983b) developed an expression for the heat transfer coefficient on the fin surface:

$$h_f = 0.943 \left[\frac{\rho_\ell k_\ell^3 h_{\mathrm{fg}}}{\mu_\ell (T_{\mathrm{sat}} - T_b)} \right]^{1/4} \left[\frac{8\sigma}{(d_o - d_r)^2} \left(\frac{1}{\delta} + \frac{1}{z} \right) \right]^{1/4} \tag{19.105}$$

Pursuing the surface-tension-drainage approach, Adamek (1981) considered a family of convex liquid–vapor interfaces that promote surface tension drainage. The radius of curvature of the family of profiles is described by

$$\frac{1}{r} = \frac{\phi_m}{s_m} \frac{\xi + 1}{\xi} \left[1 - \left(\frac{s}{s_m} \right)^{\xi} \right] \tag{19.106}$$

where r the local radius of interface, s the distance along the curved interface profile, s_m the maximum length of the curved surface, ϕ_m the rotation angle from fin tip to fin base, and ξ an interface shape parameter, $-1 \le \xi \le \infty$. These quantities are illustrated for a typical convex liquid–vapor interface in Fig. 19.23. The relationship between s_m, ϕ_m, ξ, and the tip radius r_a is given by

$$\frac{s_m}{r_a \phi_m} = \frac{\xi + 1}{\xi} \tag{19.107}$$

For the condensate profiles described by eq. (19.107), Adamek (1981) obtained an expression for the heat transfer coefficient:

$$h_f = 2.149 \frac{k_\ell}{s_m} \left[\frac{\rho_\ell \sigma h_{\mathrm{fg}} \phi_m s_m}{\mu_\ell k_\ell (T_b - T_{\mathrm{sat}})} \frac{\xi + 1}{(\xi + 2)^3} \right]^{1/4} \tag{19.108}$$

A new family of practical fin profiles for surface-drained condensation has been described by Kedzierski and Webb (1990), and more precise models for surface-tension-controlled condensation have been developed by Honda and Nozu (1987b), Honda et al. (1987a,b) and Adamek and Webb (1990).

To account for condensate flooding, Webb et al. (1985) suggested that the heat transfer coefficients for the flooded and unflooded parts be computed separately. They recommended that eq. (19.101) should be modified to

$$\bar{h}\eta_o = \left(1 - \frac{\beta}{\pi} \right) \left(h_h \frac{S_r}{S} + h_f \eta_f \frac{S_f}{S} \right) + \frac{\beta}{\pi} h_b \tag{19.109}$$

where

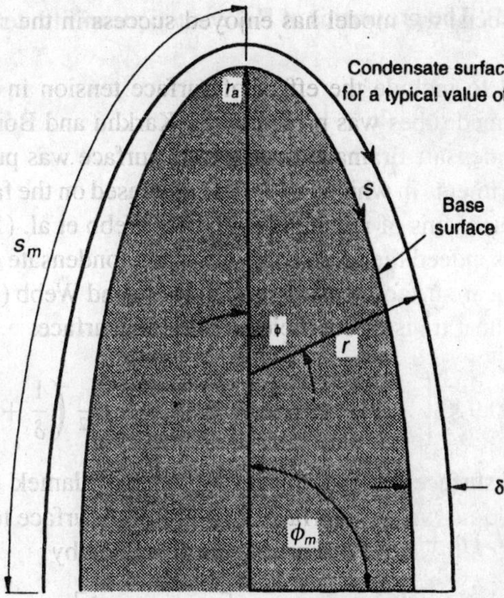

Figure 19.23 Typical fin profile for condensation on a fin with small tip radius, with increasing radius along the arc length.

$$1 - \frac{\beta}{\pi} \quad \text{and} \quad \frac{\beta}{\pi}$$

represent the fractions of the circumference that are unflooded and flooded, respectively, and h_b is the heat tansfer coefficient for the flooded region. Webb et al. (1985) found that for steam condensing on a 19-mm-diameter tube having 203 fins per meter, heat transfer through the flooded region was only 1.6% of the total. Thus the second term in eq. (19.109) can be neglected for most practical purposes.

Rose (1994) used some simplifying assumptions together with dimensional analysis to develop an equation for calculating the ratio of coefficients for the finned tube to the plain or bare tube for condensation on horizontal trapezoidal integral-fin tubes. In a contemporary paper, Briggs and Rose (1994) modified the equation given by Rose (1994) to include the effect of fin efficiency.

Example 19.6. Saturated refrigerant R-12 at 32°C condenses on a horizontal integral-fin tube of outer diameter $d_o = 19.1$ mm and a root diameter of $d_r = 15.88$ mm. The tube has 748 fins per meter and the temperature at the root of the fins is maintained at 22°C. The fins are trapezoidal in shape, 0.38 mm thick at the base and 0.23 mm thick at the tip. Assuming that the fin efficiency is $\eta_f = 1.00$, calculate the condensation heat transfer coefficient and the enhancement ratio using **(a)** the Beatty and Katz (1948) gravity-drained model, and **(b)** the Webb et al. (1985) surface-tension-drained model. For the surface-tension-drained model, assume that the condensate film shape is descibed by the Adamek profile parameters, $\xi = -0.857$, $s_m = 1.5936$ mm, and

$\phi_m = 85°$ (1.4835 rad). The properties of R-12 at the mean film temperature of 27°C (300 K) are

$$\rho_\ell = 1305.8\ \text{kg/m}^3$$

$$\rho_v = 40\ \text{kg/m}^3$$

$$\mu_\ell = 2.07 \times 10^{-4}\ \text{m}^2/\text{s}$$

$$k_\ell = 0.072\ \text{W/m·K}$$

$$\sigma = 0.0158\ \text{N/m}$$

and

$$h_{fg} = 133.79\ \text{kJ/kg}$$

SOLUTION. The fin surface per meter of tube length is

$$\begin{aligned} S_f &= 2\frac{\pi}{4}\left(d_o^2 - d_r^2\right) n + \pi d_o \delta_a n \\ &= 2\frac{\pi}{4}\left[(0.0191)^2 - (0.01588)^2\right](748) + \pi(0.0191)(0.23 \times 10^{-3})(748) \\ &= 0.1427\ \text{m}^2/\text{m} \end{aligned}$$

The unfinned surface area of the tube per meter is

$$\begin{aligned} S_b &= \pi d_r - n\pi d_r \delta_b \\ &= \pi(0.01588) - (748)\pi(0.01588)(0.38 \times 10^{-3}) \\ &= 0.0499 - 0.0142 \\ &= 0.0357\ \text{m}^2/\text{m} \end{aligned}$$

Thus the total surface of the finned tube per meter is

$$\begin{aligned} S &= S_f + S_b \\ &= 0.1427 + 0.0357 \\ &= 0.1784\ \text{m}^2/\text{m} \end{aligned}$$

The heat transfer coefficient for the horizontal (unfinned) surface of the tube is given by eq. (19.102):

$$h_h = 0.728\left[\frac{g\rho_\ell(\rho_\ell - \rho_v)k_\ell^3 h_{fg}}{\mu_\ell(T_{sat} - T_b)d_r}\right]^{1/4}$$

and with $\rho_\ell - \rho_v = 1305.8 - 40 = 1265.8$,

$$\begin{aligned} h_h &= 0.728\left[\frac{(9.81)(1305.8)(1265.8)(0.072)^3(133.79)(1000)}{(2.07 \times 10^{-4})(32 - 22)(0.01588)}\right]^{1/4} \\ &= 1622\ \text{W/m}^2 \cdot \text{K} \end{aligned}$$

(a) The average fin height over the diameter d_o is

$$b_{av} = \frac{\pi(d_o^2 - d_r^2)}{4d_o}$$

$$= \frac{\pi[(0.0191)^2 - (0.01588)^2]}{(4)(0.0191)}$$

$$= 0.0046 \text{ m}$$

and the heat transfer coefficient is given by eq. (19.103):

$$h_f = 0.943\left[\frac{g\rho_\ell(\rho_\ell - \rho_v)k_\ell^3 h_{fg}}{\mu_\ell(T_{sat} - T_b)b_{av}}\right]^{1/4}$$

$$= 0.943\left[\frac{(9.81)(1305.8)(1265.8)(0.072)^3(133.79)(1000)}{(2.07 \times 10^{-4})(32 - 22)(0.0046)}\right]^{1/4}$$

$$= 2864 \text{ W/m}^2 \cdot \text{K}$$

Equation (19.104) shows that if $\eta_f = 1.00$, then $\eta_o = 1.00$, and by eq. (19.101),

$$\bar{h} = h_h\frac{S_r}{S} + h_f\frac{S_f}{S}$$

$$= 1622\left(\frac{0.0357}{0.1784}\right) + 2864\left(\frac{0.1427}{0.1784}\right)$$

$$= 324.6 + 2290.9$$

$$= 2615 \text{ W/m}^2 \cdot \text{K}$$

The enhancement ratio is

$$\epsilon = \frac{\bar{h}}{h_h} = \frac{2615}{1622} = 1.612$$

and based on the envelope area over the fins, $\pi d_o L$, the average heat transfer coefficient is

$$\bar{h} = \frac{(2615)(0.1784)}{\pi(0.0191)(1.00)} = 7775 \text{ W/m}^2 \cdot \text{K}$$

(b) For the Adamek condensate film profile, the Webb et al. model uses eq. (19.108) for h_f. Thus

$$h_f = 2.149\frac{k_\ell}{s_m}\left[\frac{\rho_\ell\sigma h_{fg}\phi_m s_m}{\mu_\ell k_\ell(T_b - T_{sat})}\frac{\xi + 1}{(\xi + 2)^3}\right]^{1/4}$$

and with

$$\left[\frac{\xi + 1}{(\xi + 2)^3}\right]^{1/4} = \left[\frac{1 - 0.857}{(2 - 0.857)^3}\right]^{1/4}$$

$$= 0.5563$$

and

$$\left[\frac{\rho_\ell \sigma h_{fg} \phi_m s_m}{\mu_\ell k_\ell (T_b - T_{sat})}\right]^{1/4} = \left[\frac{(1305.8)(0.0158)(133{,}790)(1.4835)(0.0015936)}{(2.07 \times 10^{-4})(0.072)(10)}\right]^{1/4}$$

$$= 81.34$$

Then

$$h_f = 2.419 \frac{0.072}{(1.5963 \times 10^{-3})}(81.34)(0.5563)$$

$$= 4937 \text{ W/m}^2\cdot\text{K}$$

The average interfin spacing z is calculated by assuming the fin to have an average thicknes of (0.38 + 0.23)/2 = 0.305 mm. Thus

$$z = \frac{1000 - (748)(0.305)}{748} = 1.032 \text{ mm}$$

The flooding angle β can then be obtained from eq. (19.100):

$$\beta = \arccos\left(1 - \frac{4\sigma}{d_o \rho_\ell g z}\right)$$

$$= \arccos\left[1 - \frac{(4)(0.0158)}{(0.0191)(1305.8)(9.81)(1.032 \times 10^{-3})}\right]$$

$$= 41.44^\circ \quad \text{or} \quad 0.723 \text{ rad}$$

Finally, using eq. (19.109) with the last term neglected and with $\eta_f = \eta_o = 1.00$ yields

$$\bar{h} = \left(1 - \frac{\beta}{\pi}\right)\left(h_h \frac{S_b}{S} + h_f \frac{S_f}{S}\right)$$

$$= \left(1 - \frac{0.723}{\pi}\right)\left[\frac{(1622)(0.0357)}{0.1784} + \frac{(4937)(0.1427)}{0.1784}\right]$$

$$= 3290 \text{ W/m}^2 \cdot \text{K}$$

The enhancement ratio is

$$\frac{\bar{h}}{h_h} = \frac{3290}{1622} = 2.028$$

and, based on the envelope area over the fins, the average heat transfer coefficient is

$$\bar{h} = \frac{(3290)(0.1784)}{\pi(0.0191)(1.00)} = 9781 \text{ W/m}^2 \cdot \text{K}$$

The Beatty–Katz model predicts a value of h lower than that predicted by the Webb et al. model. The value of 9781 W/m^2·K is about the same as the value measured by Webb et al. (1985) for condensation of R-11 on a tube with 748 fins per meter.

19.4.3 Experimental Heat Transfer Coefficients

Experimental investigations concerning film condensation on horizontal integral-fins entail many difficulties. For example, the presence of noncondensable gases, partial dropwise condensation, or a substantial vapor velocity near the tube can affect the data significantly but are rarely discussed in the published literature. In many cases the uncertainties in the measurements remain obscure and the technique used to measure the average condensation heat transfer coefficient can itself introduce a 10 to 15% discrepancy between different data. The lack of consistency in the choice of the surface area on which $\bar{h}$ is based often confuses the end user, and the interpretation and use of experimental data must therefore be handled with great care. The difficulties and uncertainties associated with the measurement of condensation on horizontal integral-fin tubes are discussed in detail by Marto (1992).

A vast amount of data for condensation of various refrigerants and steam on tubes having different fin geometries and spacing has been generated, and only a representative sample is discussed here. Figure 19.24 shows the enhancement ratio

$$\epsilon = \frac{h_{\text{finned}}}{h_{\text{plain}}}$$

from the data of Wanniarachchi et al. (1985) for steam condensing on a family of cooled copper tubes (with a coolant temperature T_c and a coolant velocity v_c, with a root diameter of 19.05 mm, and having longitudinal fins of rectangular profile 1 mm thick and 1 mm high. The figure also shows the predictions of the theoretical models of Beatty and Katz (1948), Honda and Nozu (1987b), and Honda et al. (1987b). The Beatty–Katz model overpredicts the experimental data at small fin spacings but underpredicts at large fin spacings. The Honda–Nozu model underpredicts the data, and the discrepancy gets worse as the fin spacing increases. The predictions of Honda et al. (1987b) appear to fit the data best except at full fin spacing, where the flooding is very nearly complete. Although not shown here, the model of Adamek and Webb (1990) predicts the steam data within ±10 to ±15% over the complete range of fin spacings.

The same family of copper tubes as used by Wanniarachchi et al. (1985) were tested by Marto et al. (1988) for R-113. These results, together with the theoretical predictions of Beatty and Katz (1948), Honda et al. (1987a), and Webb at al. (1985), are shown in Fig. 19.25. The Beatty–Katz data vastly underpredict the data. The Webb et al. (1985) theory provides a much better prediction at $\xi = -0.85$ than at $\xi = -0.95$. However, the closest agreement with the data is achieved by the Honda et al. (1987b) model.

From the brief discussion here and the more extensive coverage provided by Marto (1988), it appears that virtually all available experimental data pertain to steam and currently popular refrigerants such as R-11, R-12, R-22, and R-113.

19.4.4 Effect of Interfacial Shear

It is well known that the interfacial shear stress of the liquid–vapor interface increases the heat transfer coefficient if the vapor and liquid (condensate) flow in the same

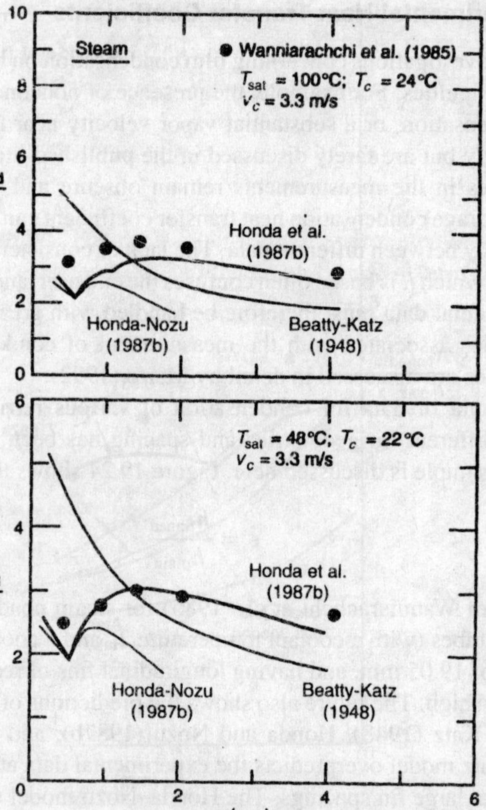

Figure 19.24 Effect of fin spacing on the enhacement ratio for steam condensing on horizontal integral-fin tubes. [Adapted from Honda and Nozu (1987c).]

direction and decreases it if the two constituents flow in opposite directions. For condensation on plain (smooth) tubes, the effect of vapor velocity has been studied extensively by Honda et al. (1986) and Fujii (1991). However, the corresponding information for integral-fin tubes is very limited. Webb (1984) and Gogonin and Dorokhov (1981) report data for R-12 condensing on two different 800-fin per meter tubes, under a range of operating pressure and heat fluxes and a maximum vapor veloocity of 8 m/s. Their data indicate that the effect of velocity for finned tubes is very small compared with the effect for smooth tubes. This observation is in contradiction with the results of Yau et al. (1986), who measured the condensation coefficient for steam condensing on a family of finned tubes and found that the heat transfer enhancement due to assisting vapor shear stress was essentially the same for finned and plain tubes. This latter result has been confirmed in a study by Bella et al. (1993), who condensed R-11 and R-113 on a finned horizontal tube with vapor velocities ranging from 2 to 30 m/s. The work of Bella et al. (1993) concluded that

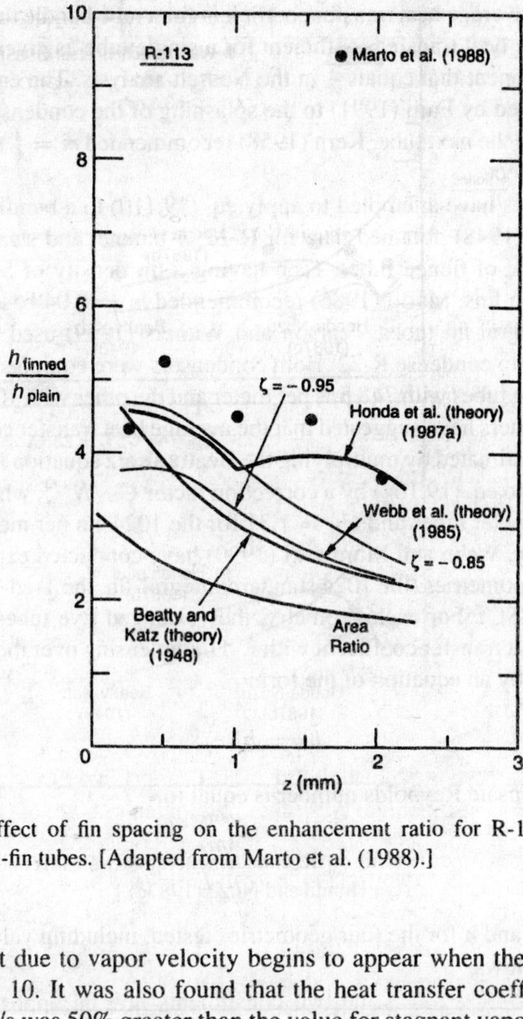

Figure 19.25 Effect of fin spacing on the enhancement ratio for R-113 condensing on horizontal integral-fin tubes. [Adapted from Marto et al. (1988).]

the enhancement due to vapor velocity begins to appear when the vapor Reynolds number exceeds 10. It was also found that the heat transfer coefficient at a vapor velocity of 30 m/s was 50% greater than the value for stagnant vapor. Studies by Lee and Rose (1984) and Michael et al. (1989) also underscore the beneficial effect of vapor shear in finned tubes.

19.4.5 Effects of Tube Bundle Geometry

When condensation occurs over a horizontal bundle of tubes, the condensate drains from tube to tube, causing the film thickness to increase on the the lower tubes. The net effect is to lower the overall heat transfer coefficient compared with a single tube. The classical Nusselt analysis takes this condensate inundation effect into account by multiplying the single-tube heat transfer coefficient by a factor N^{-m}, giving

$$\bar{h}_N = \bar{h}_1 N^{-m} \tag{19.110}$$

where $\bar{h}_N$ is the average heat transfer coefficient for a tube bundle that is N rows high, $\bar{h}_1$ is the average heat transfer coefficient for a single tube as given by eq. (19.102), and m is an exponent that equals $\frac{1}{4}$ in the Nusselt analysis. The enhancement in $\bar{h}_N$ has been attributed by Fujii (1991) to the splashing of the condensate from one tube as it impinges on the next tube. Kern (1958) recommended $m = \frac{1}{6}$ to account for this enhancement effect.

Several studies have attempted to apply eq. (19.110) to a bundle of finned tubes. Katz and Geist (1948) obtained data for R-12, n-butane, and steam consensing on a six-tube bundle of finned tubes, each having a fin density of 590 fins per meter and 1.6-mm-high fins. Marto (1986) recommended $m = 0.04$ based on his data for a bundle of integral-fin tubes. Pearson and Withers (1969) used two identical 60-tube condensers to condense R-22. Both condensers were equipped with integral-fin tubes, one having tubes with 748 fins per meter and the other with 1024 fins per meter. Pearson and Withers have suggested that the average heat transfer coefficient for tube bundles can be estimated by multiplying the Beatty–Katz equation for $\bar{h}$ as developed in eqs. (19.101) to eq. (19.104) by a correction factor $C_N/N^{1/4}$, where $C_N = 1.34$ for the 748-fin per meter tubes and $C_N = 1.31$ for the 1024-fin per meter tubes.

More recently, Webb and Murawski (1990) have conducted experiments on four enhanced tube geometries (the 1024 standard integral-fin, the Tred-26D, the Turbo-C and the GEWA-SC). For each geometry, they arranged five tubes in a bundle and measured the heat transfer coefficent with R-11 condensing over the bundle. The data were correlated by an equation of the form

$$\bar{h}_N = a\text{Re}_\ell^{-n} \tag{19.111}$$

where the condensate Reynolds number is equal to

$$\text{Re}_\ell = \frac{4\dot{m}_\ell}{\mu_\ell L}$$

The constants a and n for the four geometries tested, including values of $\bar{h}_N$ for $\text{Re}_\ell = 100$, are as follows:

Tube Type	a (W/m$^2\cdot$K)	n	$\bar{h}_N$ (W/m$^2\cdot$K)
1024 integral-fin	12.90×10^3	0	12,900
Tred-26D	269.90×10^3	0.576	18,956
Turbo-C	257.80×10^3	0.507	24,885
GEWA-SC	54.14×10^3	0.220	19,657

The 1024 standard integral-fin exhibits a zero row effect ($n = 0$) for the same tubing using R-112. The results for $\bar{h}_N$ show that the Turbo-C give the best performance and that the Tred-26D and GEWA-SC perform comparably. The lowest performance is exhibited by the 1024 standard integral-fin tubes.

The combined effects of interfacial shear and tube bundle geometry with integral-fin tubes have been studied for downflow by Smirnov and Lukanov (1972). Their results show that the effect of tube bundle geometry is more pronounced in finned tubes than in plain tubes. Webb (1994) reviewed the work done on the effects of interfacial shear and tube bundle geometry for both plain and integral-fin tubes and recommended that designers of shell-side refrigeration condensers should maintain a certain minimum vapor velocity in all regions of the tube bundle.

19.4.6 Effect of Tube Thermal Conductivity

The bulk of the experiments described in Section 19.4.3 pertained to condensation on copper tubes. Because copper has a high thermal conductivity, any "fin efficiency" effects due to the temperature drop within the fin are small. However, this effect can become more pronounced if tubes of lower thermal conductivity are employed. To determine the effect of thermal conductivity on the performance of horizontal integral-fin tubes, Huang et al. (1994) condensed steam and CFC113 on tubes made of copper, brass, and bronze. Each tube had a root diameter of 12.7 mm with longitudinal fins of rectangular profile 1 mm thick and spaced 0.5 mm apart. Four tubes of each material were tested with fin heights of 0.5, 0.90, 1.30, and 1.69 mm. For comparison, plain tubes (with an outside diameter of 12.7 mm) of each material were also tested.

The measured data expressed as the enhancement ratio are given in Table 19.3. As one may have expected, the effect of lowering the thermal conductivity is to reduce the enhancement ratio. The effect is stronger for steam than it is for CFC113. For example, in the condensation of steam on copper and bronze tubes with 1.60-mm-high fins, the enhancement ratio of 1.39 for a bronze tube is 42% less than the enhancement ratio of 2.40 for a copper tube. For CFC113, the corresponding figure is less than 5%. The enhancement ratio increases as the fin height increases, except for steam condensing on bronze tubes, where the trend is a little erratic. It is also worth noting that the enhancement ratios for CFC113 are much higher than those for steam.

19.5 INTERNALLY FINNED TUBES

When saturated vapor flows into a cooled tube, vapor condenses on the tube wall, forming a condensate film on the wall. A variety of flow patterns can arise, depending

TABLE 19.3 Effect of Tube Material on Enhancement Ratios $k = 315$, 112, and 78 W/m·K for Copper, Brass, and Bronze, Respectively

Fin Height (mm)	Steam			CFC113		
	Copper	Brass	Bronze	Copper	Brass	Bronze
0.50	1.74	1.50	1.50	3.16	3.15	2.96
0.90	1.90	1.63	1.43	4.24	4.35	3.90
1.30	2.05	1.68	1.37	4.60	4.72	4.28
1.60	2.40	1.77	1.39	5.16	5.09	4.91

on the mass velocity. For example, when the condensate layer is symmetric and the liquid–vapor interface is sharply defined, the flow is classified as annular flow. Other flow possibilities include bubbly, dispersed, wavy, stratified, plug, and slug flow. Even for a smooth tube or pipe, the flow is quite complex and difficult to analyze. The presence of internal fins complicates the situation further and makes it extremely difficult, if not impossible, to develop an accurate, physics-based model describing the heat and fluid flow processes. Consequently, the discussion in this section, including the pressure drop and heat transfer correlations, are based largely on experimental studies.

Reisbig (1974) and Vrable et al. (1974) both reported on the condensation of R-12 in internally finned tubes. Reisbig found the condensation heat transfer coefficient to be 20 to 40% greater than the smooth tube value. However, Reisbig did not propose any correlation. On the other hand, Vrable et al. conducted 26 experiments with two different internally finned tubes, varying the inlet reduced pressure p/p_{cr} from 0.18 to 0.46 and the mass flux from 86.7 to 853 kg/m^2·s. The more effective tube was found to give a maximum enhancement of 300%. The correlation

$$\bar{h} = 0.01\frac{k_\ell}{d_h}\left(\frac{2d_h G_\ell}{\mu_\ell}\right)^{0.80} \mathrm{Pr}_\ell^{1/3}\left(\frac{p}{p_{cr}}\right)^{-0.65} \tag{19.112}$$

where

$$G_\ell = \left[\left(\frac{\rho_\ell}{\rho_v}\right)^{1/2} x + (1-x)\right] G \tag{19.113}$$

represented their data within $\pm$30%. In this correlation, G is the mass flux, d_h the tube diameter, x the vapor quality, p the pressure, and p_{cr} the critical pressure.

Royal and Bergles (1978) condensed low-pressure steam in four internally finned copper tubes, three of which had spiral fins. The geometric data for these fins are given in Table 19.4, and Figs. 19.26 and 19.27 represent the heat transfer and pressure drop, respectively. The heat transfer data for finned tubes indicate a significant improvement over the smooth or plain tube. The highest enhancement ratio of 2.3 is achieved with a 15.9-mm-outside diameter tube containing 16 spiraled fins of 1.45 mm height and a helix angle of 2.33° (tube 5). It is also interesting to note that the ratio of the surface area S to the surface area of the plain tube, S_p, for tubes 3 and 4 is the same, but the heat transfer of tube 3 is superior to that of tube 4. Clearly, the area ratio is not the controlling factor. The pressure drop characteristics of Fig. 19.27 show that the highest pressure drop occurs in tube 3 and that this pressure drop is higher than that in tube 4.

Based on their experimental data, Royal and Bergles (1978) proposed a correlation that represented 95% of their data to within $\pm$ 30%:

$$\bar{h} = 0.0265\frac{k_\ell}{d_h}\left(\frac{d_h G_\ell}{\mu_\ell}\right)^{0.80} \mathrm{Pr}_\ell^{1/3}\left[160\left(\frac{b^2}{\bar{z}d_h}\right)^{1.91} + 1\right] \tag{19.114}$$

where $\bar{z}$ is the average fin spacing. It is important to note that this correlation was derived for steam at 1 bar and may not be accurate for other fluids or pressures.

TABLE 19.4 Geometric Data for Internally Finned Tubes

Tube	d_o (mm)	e (mm)	n	α (deg)	S/S_p
1	15.9	—	—	—	1.00
2	15.9	0.60	32	2.95	1.70
3	12.8	1.74	6	5.25	1.44
4	12.8	1.63	6	0.00	1.44
5	15.9	1.45	16	3.22	1.73

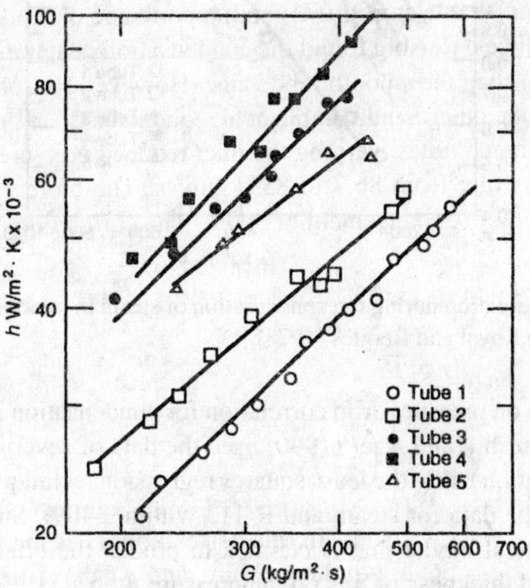

Figure 19.26 Condensation heat transfer coefficient for condensation of steam in internally finned horizontal tubes. [Adapted from Royal and Bergles (1978).]

Furthermore, the correlation takes no account of surface tension effects or temperature gradients in the fin.

Luu and Bergles (1979a) tested the tubes in Table 19.4 with R-113 as the condensing fluid and found that tube 2, with the shortest fins and the highest number of fins, exhibited the best heat transfer performance at low flow rates. At high mass flow rates, the measured heat transfer coefficient ranged between 70 and 120% of the smooth tube value. A subsequent investigation by Luu and Bergles (1979b) tested eq. (19.114) against their R-113 data and found it to be unsatisfactory. A new correlation was developed to predict condensation of R-113 in internally finned tubes.

Kaushik and Azer (1988) appled regression analysis to the steam data of Royal and Bergles (1978), the R-113 data of Luu and Bergles (1980) and Said and Azer (1983), and the R-11 data of Venkatesh (1984) to develop a general heat transfer correlation. This correlation was claimed to predict 71% of the data with ±30%. An analytical model to predict condensation heat transfer in internally finned tubes was proposed by Kaushik and Azer (1989), but the model remains unvalidated against experimental data and, as such, can only be recommended with caution.

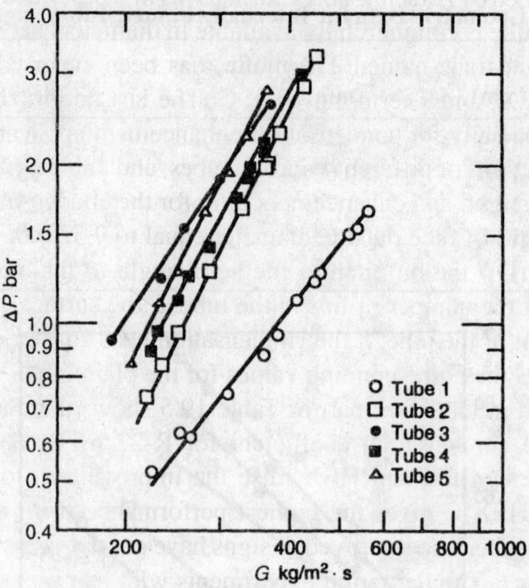

Figure 19.27 Pressure drop during the condensation of steam in internally finned horizontal tubes. [Adapted from Royal and Bergles (1978).]

The information on pressure drop correlation for condensation in internally finned tubes is limited. Kaushik and Azer (1990) used the data of several investigators and developed a correlation using the least-squares regression technique. This correlation predicted 68% of the data for steam and R-113 within ±40%. Sur and Azer (1991) proposed an analytical model that is claimed to predict the effect of changing the number, height, and thickness of fins on the pressure drop.

Thus far, the discussion has focused on the condensation of pure vapor. When internally finned tubes are employed in the condensers of vapor compression refrigeration systems, the refrigerant vapor is often mixed with small quantities of compressor lubricating oil. This situation has been studied by Schlager et al. (1989a,b) in a two-part paper. In the first part (1989a) the heat transfer and pressure drop correlations for single- and two-phase flows inside smooth and internally finned tubes are reviewed. It was found that there was no published correlation for refrigerant–oil mixtures in finned tubes. This conclusion prompted the authors to embark on an experimental program to study the evaporation and condensation of refrigerant–oil mixtures in smooth and internally finned tubes. The results of this study revealed that the presence of oil lowers the condensation heat transfer coefficient for both smooth and finned tubes. This degradation of heat transfer perfomance increases as the oil concentration increases. The pressure drop was also found to increase with an increase in oil concentration.

19.6 MICROFIN TUBES

The microfin tube is one of the most propular enhancement devices for residential air-cooled air conditioners, whether installed in a window or as a central unit. This geometry was first developed by Hitachi Cable Ltd. and was first described in a patent by Fujie et al. (1977). The tube is commercially available in diameters between 4 and 16 mm. The original design, trade-named Thermofin, has been superseded by Thermofin-EX, Thermofin-HEX, and Thermofin-HEX-C. The last design, Thermofin-HEX-C, was developed especially for condensation-enhancement applications. Figure 19.28 shows the cross sections of different microfin tubes, and Table 19.5 gives information on the geometry, weight, and enhancement ratio for the condensation of R-22 for the four microfin designs of tube outside diameter equal to 9.52 mm. In Table 19.5, b is the fin height (mm), p the fin pitch, α the helix angle, β the included angle of the fin cross section, n the number of fins in the tube, a the surface area of the microfin tube, W_t the weight of the tube, h the condensation heat transfer coefficient, and the subscript p denotes the corresponding values for the plain tube.

The last column in the lower part of Table 19.5 shows that the original microfin tube improves the condensation coefficient for R-22 by 80% over a plain tube. The subsequent design EX and HEX raise the improvement to 140% and 150%, respectively. The HEX-C gives the highest performance, with an improvement of 210% over a plain tube. The improved designs have greater fin heights b and smaller fin included angles β. Condensation experiments with microfin tubes conducted by Shinohara and Tobe (1985), Shinohara et al. (1987), and Schlager et al. (1990a) have shown that the heat transfer coefficient for R-22 gradually increases with helix angle α, from 7° up to 30°. Although these results do not clearly identify the optimum helix angle, it does appear that $\alpha = 30°$ results in the best condensation performance.

Figure 19.29 shows the R-22 condensation heat transfer coefficient as a function of mass velocity for the tube geometries in Table 19.5 and demonstrates the superiority of microfin tubes over plain fin tubes markedly. The HEX-c design stands out as distinctly superior to the HEX and EX designs for R-22 condensation, and the tube

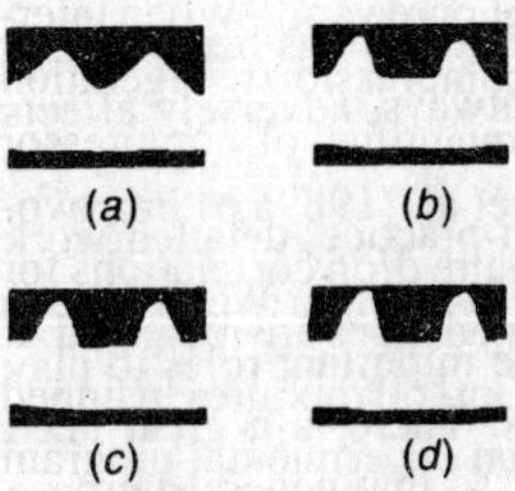

Figure 19.28 Cross sections of Hitachi Thermofin tubes: (*a*) Thermofin; (*b*) Thermofin-EX; (*c*) Thermofin-HEX; (*d*) Thermofin-HEX-C. [Adapted from Yasuda et al. (1990).]

TABLE 19.5 Characteristics of Microfin Tube Designs

Design	b (mm)	p/b	α (deg)	β (deg)	n	S/S_p	W_t/W_{tp}	h/h_p
Original	0.15	2.14	25	90	65	1.28	1.22	1.8
EX	1.90	2.32	18	53	60	1.51	1.19	2.4
HEX	0.20	2.32	18	40	60	1.60	1.19	2.5
HEX-C	0.25	2.32	30	40	60	1.73	1.20	3.1

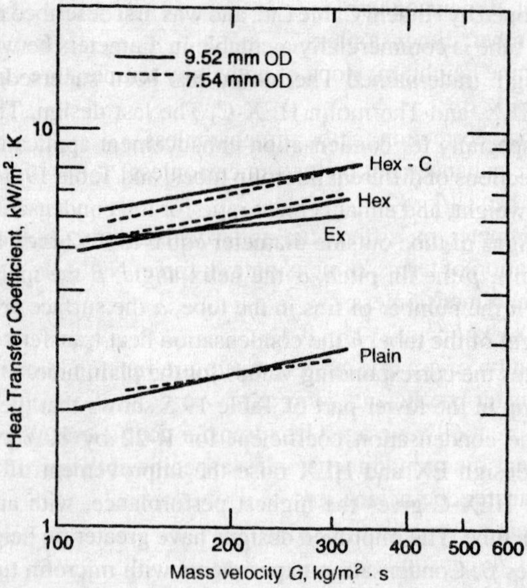

Figure 19.29 Heat transfer coefficients for R-22 condensing in micro-fin tubes.

outside diameter seems to have little effect on the heat transfer performance.

Pressure drop measurements for R-22 condensing in 12.7-mm-diameter microfin tubes have been reported by Schlager et al. (1990b). Their results show that the pressure drop in microfin tubes is somewhat higher than that of the plain tube. Because of the uncertainties in the pressure drop measurements, the relative differences between the diffierent microfin tubes could not be clearly established. The effect of small concentrations of oil in R-22 condensation in microfin tubes has been studied by Schlager (1989a,b). The presence of oil often, but not always, adversely affects the heat transfer and pressure drop in microfin tubes.

Despite the proven effectiveness of the microfin tube in practice, detailed work on the understanding of enhancement mechanism involved is lacking. Webb (1994) has proposed that both vapor shear and surface tension have important roles to play in augmenting the heat transfer process in microfin tubes. There is a great need for experimental work on fluids other than the refrigerants, so that general-purpose correlations for heat transfer and pressure drop in microfin tubes can be developed.

19.7 NOMENCLATURE

Roman Letter Symbols

A constant, dimensionless

a area of microfin tube, m^2; constant, $kg/m^3 \cdot K$; slope of saturation curve, $kg/m^3 \cdot K$

B	constant, dimensionless
Bi	Biot number, dimensionless
b	fin or spine height, m; slope of enthalphy temperature curve, kJ/K or J/K
C	constant, kg/m^3
C_a	Ackerman correction factor, dimensionless
C_f	interface enhancement factor, dimensionless
c	specific heat, J/kg·K
d	spine diameter, m; differential or derivative, dimensionless
F	combination of terms, dimensionless; force, N
f	function, dimensionless
G	mass flux, $kg/m^2 \cdot s$
g	acceleration of gravity, m/s^2; temperature functon, dimensionless
h	heat transfer coefficient, $W/m^2 \cdot K$
$\bar{h}$	circumferentially averaged heat transfer coefficient, $W/m^2 \cdot K$; heat transfer coefficient function, dimensionless; specific enthalpy, J/kg
h_m	mass transfer coefficient, m/s
h_{fg}	latent heat of vaporization, J/kg
K	constant, dimensionless
k	thermal conductivity, W/m·K
L	length, m
m	fin performance parameter, m^{-1}; combination of terms, dimensionless
$\dot{m}$	mass flow, kg/s
N	number of tube rows, dimensionless; number of tubes, dimensionless; combination of terms, dimensionless; combination of terms, s/m;
N^*	combination of terms, m^{-1}
Nu	Nusselt number, dimensionless
P	pressure, N/m^2 (Pa)
Pr	Prandtl number, dimensionless
p	fin pitch, m
q	heat flow, W/m^2
q_s''	sensible heat flux, W/m^2
q_t''	heat flux through condensate film, W/m^2
R	heat flux ratio, dimensionless; radius, dimensionless; universal gas constant, $kPa \cdot m^3/kg \cdot K$
Re	Reynolds number, dimensionless
Re_ℓ	condensate Reynolds number, dimensionless
r	radius, m; radial coordinate, m
S	surface area, m^2
s	distance along curved surface, m
T	temperature, K
u	airstream velocity, m/s
v	flow velocity, m/s
W	weight, N
X	height, dimensionless
x	height coordinate, m; length coordinate in Nusselt analysis (Section 19.2),

m; vapor quality, dimensionless

Y film thickness, dimensionless; thickness coordinate, m

y length coordinate, m; film thickness coordinate in Nusselt analysis (Section 10.2), m

Z combination of terms, dimensionless

z length coordinate, m; spacing between fins, m

$\bar{z}$ average fin spacing

Greek Letter Symbols

α helix angle, rad

β condensate flooding angle, rad; included angle of fin cross section, rad

Δ change in, dimensionless

δ fin thickness, m

∂ partial derivative, dimensionless

ϵ enhancement ratio, dimensionless

η similarity variable, dimensionless

θ function, dimensionless; temperature ratio, dimensionless

κ inclination angle, rad

Λ function, dimensionless; ratio of condensate film thickness to fin height, dimensionless

μ dynamic viscosity, kg/s·m

ν kinematic viscosity, m^2/s

ξ interface shape parameter, dimensionless; similarity variable, dimensionless; length ratio, dimensionless

ρ density, kg/m^3

σ surface tension, N/m

τ shear stress, N/m^2

ϕ rotation angle, rad; relative humidity, dimensionless

ψ combination of terms, dimensionless

ω specific humidity, kg/kg dry air

Roman Letter Subscripts

a fin tip

air air

av average

b fin base

c classical fin theory

cr critical

db dry bulb

dry dry condition

f film; fin

fg latent heat of vaporization

h hydraulic; horizontal tube

i condensate–air interface

id ideal

ℓ	liquid
m	maximum
n	tube rows high
o	overall; gravitational acceleration; outside
p	constant pressure; plain tube
r	fin root
s	surroundings
s, r	saturation at reference temperature
sat	saturated
T, s	saturated specific humidity
v	vapor
w	wet
ws	sensible heat for wet fin
wt	total heat for wet fin
x	x-coordinate direction

Roman Letter Superscripts

m	exponent
n	exponent

Symbolic Superscript

$''$	heat flux

CHAPTER 20

AUGMENTATION AND ADDITIONAL STUDIES

20.1 AUGMENTATION

Economic considerations, which most certainly include the desire to reduce energy consumption, have led to expanded research and development efforts to produce more efficient heat exchange equipment. The goal may be to reduce the size of a given heat exchanger or, alternatively, to upgrade the performance to obtain higher heat fluxes at lower temperature differences in the equipment.

The heat transfer augmentation literature was reviewed by Bergles (1973) and again by Bergles et al. (1978). The latter review listed over 2000 references arranged in various categories, some of which deal directly or obliquely with extended surfaces. Previously cited references appropriate to four of these categories are listed here with no further amplification.

Forced Convection

- Transverse ribs (annular geometry)
  - Dalle Donne and Meerwald (1970)
  - Zappa and Geiger (1971)
- Extended surfaces (internally finned tubes)
  - Carnavos (1980)
  - Hilding and Coogan (1964)
  - Hu and Chang (1973)
  - Masliyah and Nandakumar (1976)
  - Royal and Bergles (1978)

- Watkinson et al. (1975)
- Extended surfaces (finned tube heat exchangers)
  - Ackerman and Brunsvold (1969)
  - Briggs and Young (1963)
  - Goldstein and Sparrow (1976)
  - Rabas and Eckels (1975)
  - Rich (1973)
  - Saboya and Sparrow (1974a,b, 1976)
  - Zozulya et al. (1970a)
- Extended surfaces (compact heat exchangers)
  - Briggs and London (1961)
  - Cox and Jallouk (1973)
  - Kays and London (1984)
  - Mondt and Siegla (1974)
  - Sparrow et al. (1978)
  - Wieting (1975)

Pool Boiling

- Shih and Westwater (1974)

Flow Boiling (Internal Fins)

- Kubanek and Miletti (1979)

Condensation

- Beatty and Katz (1948)

Webb (1994) has provided additional data in the areas of compact heat exchanger surfaces, externally finned tubes, fouling, boiling, and condensation. Germaine to the study of extended surface heat transfer are some details in the area of compact heat exchanger surfaces and externally finned tubes which provide the subject matter of the discussion follows.

Chapter 10 provided a modest study of the compact heat exchanger. There it was shown that the exchanger surfaces range from about 80 to about 1825 fins per meter. Because of the very small equivalent or hydraulic diameters and the low density of the gases employed, the compact heat exchanger surface is usually operated between $500 \leq Re \leq 1500$, where Re is based on the equivalent diameter.

Webb (1994) points out that because roughened surfaces do not provide appreciable enhancement at these low Reynolds numbers, two additional concepts have been extensively employed to provide enhancement:

1. Special channel shapes, such as the wavy channel, which provide mixing due to secondary flows or boundary layer separation within the channel.
2. Repeated growth and wake destruction of the boundary layer. This concept is used in the offset strip fin, or merely the strip fin, in the louvered fin, and to some extent, in the perforated fin.

20.1.1 Offset or Strip Fin

The strip fin surfaces described by Kays and London (1984) have fin spacings ranging from 437 to 1065 fins per meter. The enhancement principle of the strip fin is based on the development of a laminar boundary layer on the short strip length followed by its dissipation in the wake region between the strips. This results in about a twofold increase in both the heat transfer parameter j and the friction factor f. Webb (1994) gives an example where the strip fin surface requires only about 45% of the surface area of the plain plate fin surface area to obtain the same heat transfer coefficient–surface area product. This requires about a 10% increase in free flow area to maintain the same fluid pumping power.

Prediction of j and f Factors. While Kays and London (1984) proposed a simple, accurate model to predict j and f versus the Reynolds number, multiple regression power law correlations were developed by Wieting (1975) and Joshi and Webb (1987) based on 21 strip fin geometries. Both present separable correlations for the laminar and nonlaminar wake regions, where transition from the laminar region occurs when the heat transfer parameter j departs from its log–log linear characteristic.

Joshi and Webb (1987) determined the *transition Reynolds number* $\mathrm{Re_{tr}}$ for 21 core geometries:

$$\mathrm{Re_{tr}} = 257 \left(\frac{L_p}{z}\right)^{1.23} \left(\frac{\delta}{L_p}\right)^{0.58} \frac{d_e}{\delta_{\mathrm{mom}}} \tag{20.1}$$

$$\delta_{\mathrm{mom}} = \delta + 1.328 \frac{L_p}{\mathrm{Re_L^{0.50}}}$$

where G is the mass velocity and

$$\mathrm{Re_L} = \frac{GL_p}{\mu}$$

In the foregoing equations, L_p is the strip length, δ the fin thickness, z the spacing between adjacent fins, and d_e the equivalent (or hydraulic) diameter.

Manglik and Bergles (1990) used the asymptotic correlating method developed by Churchill and Usagi (1972). Their correlation was based on 18 of the 21 surface geometries in the Joshi and Webb (1987) data set, and eq. (20.1) was used to define the upper limit of the laminar correlation. With b taken as the plate spacing, for the laminar region, $\mathrm{Re} \leq \mathrm{Re_{tr}}$,

$$j = 0.652\text{Re}^{-0.540}\left(\frac{z}{b}\right)^{-0.154}\left(\frac{\delta}{L_p}\right)^{0.150}\left(\frac{\delta}{z}\right)^{-0.068} \tag{20.2}$$

and

$$f = 9.624\text{Re}^{-0.742}\left(\frac{z}{b}\right)^{-0.186}\left(\frac{\delta}{L_p}\right)^{0.305}\left(\frac{\delta}{z}\right)^{-0.266} \tag{20.3}$$

For the nonlaminar region, $\text{Re} \geq \text{Re}_{tr} + 1000$,

$$j = 0.244\text{Re}^{-0.406}\left(\frac{z}{b}\right)^{-0.104}\left(\frac{\delta}{L_p}\right)^{0.196}\left(\frac{\delta}{z}\right)^{-0.173} \tag{20.4}$$

and

$$f = 1.870\text{Re}^{-0.299}\left(\frac{z}{b}\right)^{-0.094}\left(\frac{\delta}{L_p}\right)^{0.682}\left(\frac{\delta}{z}\right)^{-0.242} \tag{20.5}$$

The Churchill and Usagi (1972) method can be used to develop a general friction factor by taking eqs. (20.3) and (20.5) and finding the asymptotes as $\text{Re} \to 0$ and $\text{Re} \to \infty$. This provides

$$f = 9.624\text{Re}^{-0.742}\left(\frac{z}{b}\right)^{-0.186}\left(\frac{\delta}{L_p}\right)^{0.305}\left(\frac{\delta}{z}\right)^{-0.266}\Lambda_1 \tag{20.6a}$$

where

$$\Lambda_1 = \left[1.000 + 7.669 \times 10^{-8}\text{Re}^{4.43}\left(\frac{z}{b}\right)^{0.92}\left(\frac{\delta}{L_p}\right)^{3.77}\left(\frac{\delta}{z}\right)^{0.236}\right]^{0.10} \tag{20.6b}$$

A similar procedure can be used to obtain

$$j = 0.652\text{Re}^{-0.540}\left(\frac{z}{b}\right)^{-0.154}\left(\frac{\delta}{L_p}\right)^{0.150}\left(\frac{\delta}{z}\right)^{-0.068}\Lambda_2 \tag{20.7a}$$

where

$$\Lambda_2 = \left[1.000 + 5.269 \times 10^{-5}\text{Re}^{1.34}\left(\frac{z}{b}\right)^{0.50}\left(\frac{\delta}{L_p}\right)^{0.46}\left(\frac{\delta}{z}\right)^{-1.06}\right]^{0.10} \tag{20.7b}$$

In eqs. (20.1) through (20.7), the equivalent diameter is defined as

$$d_e = \frac{4zbL_p}{2(zL_p + bL_p + \delta b) + \delta z} \tag{20.8}$$

20.1.2 Louvered Fin

The louvered fin can provide heat transfer coefficients that are comparable to the offset or strip fin for the same strip width (Webb, 1994). Shah and Webb (1982) have provided descriptions of modified louvered-fin geometries, and Davenport (1983a,b) has provided data on 32 one-row cases. Other studies include those of Sunden and Svantesson (1990), Aoki et al. (1989), Fujikake et al. (1983), Kajino and Hiramatsu (1987), Achaichia and Cowell (1988), and Suga and Aoki (1991).

The results of Davenport (1983b) for the heat transfer paramater j and the friction factor f are in the form of the correlations

$$j = 0.259\mathrm{Re}_{\mathrm{L}}^{-0.42} L_h^{0.33} H^{0.26} \left(\frac{L_L}{H}\right)^{1.10} \qquad (300 < \mathrm{Re} < 4000) \qquad (20.9)$$

where Re is based on the equivalent diameter, L_L the louver length, and H the louver fin height. For the friction factor between $70 < \mathrm{Re} < 1000$,

$$f = 5.47\mathrm{Re}_L^{-0.72} L_h^{0.37} L_p^{0.20} \left(\frac{L_L}{H}\right)^{0.89} \qquad (20.10a)$$

where L_p is the louver pitch, and for $1000 < \mathrm{Re} < 4000$,

$$f = 0.494\mathrm{Re}_L^{-0.39} H^{0.46} \left(\frac{L_h}{L_p}\right)^{0.33} \left(\frac{L_L}{H}\right)^{1.10} \qquad (20.10b)$$

20.1.3 Plain Plate Fins on Round Tubes

Several tubes, either in an in-line or staggered arrangement, can be put into sheets of metal that form continuous fins. Webb (1994) asserts that the in-line arrangement is seldom used because it provides substantially lower performance than the staggered arrangement. For the staggered arrangement, correlations to provide the heat transfer parameter j and friction factor h have been provided by McQuiston (1978) and Gray and Webb (1986). The Gray and Webb heat transfer correlation is for four rows,

$$j_4 = 0.14\mathrm{Re}_d^{-0.328} \left(\frac{S_t}{S_\ell}\right)^{-0.502} \left(\frac{z}{d}\right)^{0.031} \qquad (20.11)$$

where Re_d is the Reynolds number based on the tube diameter, z the fin spacing, and S_t and S_ℓ the transverse and longitudinal fin pitches, respectively.

Equation (20.11) assumes that the heat transfer coefficient is stabilized by the fourth row. For less than four tube rows, an adjustment may be made for the number of tube rows N:

$$j_n = 0.991 j_4 \left[2.24\mathrm{Re}_d^{-0.092} \left(\frac{N}{4}\right)^{-0.031}\right]^{0.607(4-N)} \qquad (20.12)$$

According to Gray and Webb (1986), the friction factor is composed of a term that accounts for the drag force of the fins, f_f, and a term that accounts for the drag force on the tubes f_t:

$$f = f_f \frac{S_f}{S} + f_t \left(1 - \frac{S_f}{S}\right)\left(1 - \frac{\delta}{p_f}\right) \qquad (20.13)$$

where S_f is the fin surface, S the total surface, and p_f the fin pitch. The friction factor assiociated with the fin, f_f, is given by

$$f_f = 0.508\mathrm{Re}_d^{-0.521}\left(\frac{S_t}{d}\right)^{1.318} \tag{20.14}$$

The friction factor associated with the tubes is obtained from a correlation for flow normal to banks of staggered banks of plain tubes. One such correlation is that of Zhukauskas (1972), which is described in almost all of the heat transfer texts.[1]

20.1.4 Plain Individually Finned Tube Banks

The correlation of Briggs and Young (1963) for the heat transfer parameter j and Robinson and Briggs (1966) for the friction factor f served for many years. Rabas et al. (1981) developed more accurate correlations for lower fin heights and smaller fin spacings. The Briggs and Young (1963) and Robinson and Briggs (1966) correlations are, respectively,

$$j = 0.134\mathrm{Re}_d^{-0.319}\left(\frac{z}{b}\right)^{0.20}\left(\frac{z}{\delta}\right)^{0.11} \tag{20.15}$$

where z is the fin spacing and

$$f = 9.47\mathrm{Re}_d^{-0.316}\left(\frac{S_t}{d}\right)^{-0.927}\left(\frac{S_t}{S_d}\right)^{0.515} \tag{20.16}$$

where S_d is the diagonal pitch.

The range of application of eq. (20.15) is

$$\begin{gathered} 1100 \le \mathrm{Re}_d \le 18,000 \\ 0.13 \le z/b \le 0.63 \\ 0.01 \le z/\delta \le 6.6 \\ 0.09 \le b/d \le 0.69 \\ 0.01 \le \delta/d \le 0.15 \\ 1.5 \le S_t/d \le 8.2 \end{gathered}$$

The Rabas et al. (1981) correlations are for heat transfer

$$j = 0.292\mathrm{Re}_d^n\left(\frac{z}{d}\right)^{1.12}\left(\frac{z}{b}\right)^{0.26}\left(\frac{\delta}{z}\right)^{0.67}\left(\frac{d_a}{d}\right)^{0.47}\left(\frac{d_a}{\delta}\right)^{0.27} \tag{20.17}$$

where d_a is the fin tip diameter and where

$$n = -0.415 + 0.0346\left(\frac{d_a}{z}\right)$$

[1]Incropera and Dewitt (1990), for example.

and for the friction factor

$$j = 3.805\mathrm{Re}_d^{-0.234}\left(\frac{z}{d_a}\right)^{0.25}\left(\frac{b}{z}\right)^{0.76}\left(\frac{d}{d_a}\right)^{0.73}\left(\frac{d}{S_t}\right)^{0.71}\left(\frac{S_t}{S_\ell}\right)^{0.38} \tag{20.18}$$

The Rabas et al. (1981) are valid for staggered tubes with

$$N > 6$$
$$5000 \le \mathrm{Re}_d \le 25,000$$
$$1.3 \le z/b \le 1.50$$
$$0.01 \le z/\delta \le 0.06$$
$$b/d \le 0.10$$
$$0.01 \le \delta/d \le 0.02$$
$$1.3 \le S_t/d \le 1.5$$

20.1.5 Louvered Plate Fins on Flat Tubes

Achaichia and Cowell (1988) provided correlations for the automobile radiator configuration that has louvered plate fins on flat tubes. They tested 16 core geometries, and for the heat transfer parameter j they proposed

$$j = 1.234\gamma\mathrm{Re}_L^{-0.59}\left(\frac{S_t}{L_p}\right)^{-0.09}\left(\frac{p_f}{L_p}\right)^{-0.04} \tag{20.19}$$

where

$$\gamma = \frac{1}{\alpha}\left(0.936 - \frac{243}{\mathrm{Re}_L} - 1.76\frac{p_f}{L_p} + 0.995\alpha\right) \tag{20.20}$$

where

$$\alpha = \frac{z}{a}$$

is the aspect ratio for a rectangular duct, Re_L the Reynolds number based the louver interruption length, and H the louver fin height. Moreover, the range of application of eq. (20.19) is

$$1.70 \le p_f \le 3.44\ \mathrm{mm}$$
$$0.81 \le L_p \le 1.40\ \mathrm{mm}$$
$$8 \le S_t \le 14\ \mathrm{mm}$$
$$22 \le \theta \le 30^\circ$$

For the friction factor

$$f = 533 p_f^{-0.22} L_p^{0.25} S_t^{0.26} H^{0.33}\left(\mathrm{Re}_L^{0.318\log_{10}\mathrm{Re}_L - 2.25}\right)^{0.17} \tag{20.21}$$

with all dimensions in millimeters and where H is the louver fin height, p_f the fin pitch, L_p the louver pitch, and θ the louver angle.

20.1.6 Closure

The subject of augmentation or enhancement through the use of extended surface is vast. The discussion here has touched on only a few areas. Noticeable among the subject areas not covered are the use of wavy fins and perforated fins as augmentation devices as well as an extended surface inserts within tubes, internally finned tubes, and annuli; the effect of extended surface on fouling on both the inside and outside of tubes; and the effect of extended surfaces on boiling on condensation. The field of augmentation or enhancement using extended surfaces is still burgeoning and the use of a treatise on enhancement such as the one by Webb (1994) is recommended.

20.2 HEAT TRANSFER IN ELECTRONIC EQUIPMENT

20.2.1 From 1954 to 1975

Fins and fin structures have been used effectively in thermal control of electrical and electronic equipment. Motor casings that contain finned passages, switching transformers in power stations and substations (and in front yards and on street corners), and transistor heat sinks are but few of the many examples of the employment of extended surfaces to carry away the dissipated heat in electrical and electronic equipment.

London (1954) may well have been the first to point out that it is essential to address heat transfer design considerations in parallel with the electrical design so that high-power vacuum tubes would eventually operate at their "electronic" rather than their "heat transfer" rating. He presented factors relating to the heat transfer design of an annular air cooler of the axial flow type in which the extended surface consisted of a large number of longitudinal fins mounted radially in the annular region. He derived performance equations from basic heat transfer and fluid flow theory and constructed a finned heat exchanger that was capable of cooling a 25-kW tube using air at a moderate flow rate. His design was verified in a testing program.

Kraus (1958) set up two differential equations for temperature excess in the top and wraparound in a transistor cap. These were solved simultaneously and efficiency curves were provided for the top and side. In the late 1950s, it was not recognized that this was a classic case of two dissipating entities in cascade which could later be analyzed in the computationally efficient procedure provided by Kraus et al. (1978). Indeed, the techniques presented in the 1978 work (and discussed here in Chapters 5 and 7) could replace the two fin efficiencies (top and side) that were developed in the former with a single input admittance that incorporates the effects of both top and side.

Kraus (1961, 1962) provided guidance on the efficiency and optimization of a cold plate heat exchanger. Dyer and Schaedle (1970) discussed the performance of the Apollo pin fin cold plate and presented an analytical approach for calculating the forced-convection heat transfer coefficient and pressure drop. They also discussed flow control and basic baffle design for cold plates.

Mochalov (1970) gave the heat sinks that were applied to semiconductor triodes the name *finned radiators* and provided graphs for determination of the performance of these devices under natural-convection conditions. Zhadan et al. (1970) provided an empirical relationship for calculation of the mean heat transfer coefficient for motors with finned enclosures with operation under natural-convection conditions for an orientation with shafts in a horizontal position. Finlay (1971) formulated equations that would describe the process of heat and mass transfer occurring within a pin-finned cold plate cooled by an air–water spray. These equations were used for the thermal design of spray-cooled cold plates and to study the way in which a number of the more important design variables influence the heat transfer and thermodyamic performance. Grakovich and Konev (1974) showed how inclined fins made for a more efficient heat sink, confirming, to some extent, what was contended by Pnueli (1974).

Kraus and Morales (1983) showed that magnesium fins could outperform aluminum fins over a broad range of heat transfer coefficients. This study was based on a weight optimization trade-off and produced magnesium fins with a slightly higher fin height than that of their aluminum counterparts at the design conditions. The efficiency, of course, was lower for the magnesium fin, but because its surface area was greater (the fin height was greater), the magnesium fin transferred more heat. Moreover, once the slightly greater fin height was established for the magnesium fin at a markedly lower weight, reduction of the heat transfer coefficient from the design value indicated that the magnesium fin clearly outperformed the aluminum fin over the entire range of heat transfer coefficients from zero to the design value.

Yovanovich (1983) showed a noniterative control volume approach to the problem of one-dimensional steady conduction with convection, and Kishimoto et al. (1984) looked at gas cooling enhancement technology for integrated-circuit chips. They described new approaches for increasing the capability of forced-gas-convection cooling for chips using an enhanced heat transfer technique and a higher gas flow velocity in a closed cycle flow.

Schenone and Tanda (1990) reported on heat transfer data obtained for five staggered fin arrays over a range of Reynolds numbers from 2000 to 20,000. The configurations investigated included fin heights of 20 and 35 mm and spacings between fins of 6.3, 7.8, and 10 mm. The measured data were found to be substantially in agreement with standard correlations available in the literature for offset strip fin heat exchangers. Lee et al. (1990) investigated the effects of airflow bypass on the heat transfer performance of straight fin heat sinks. The geometric parameters considered included clearance between the fin tips and the adjacent printed circuit board, lateral spacing between the heat sinks, fin height, and spacing between fins. The results demonstrated that the effect of airflow bypass due to lateral spacing or tip clearance can be significant, with more than a 20% reduction in heat transfer performance. Cengel and Ngai (1991) conducted experiments to determine the natural-convection heat transfer characteristics of vertically oriented shrouded heat sinks. Commercially available heat sinks specifically designed to cool electronic components were used for the experiments. It was reported that shrouding, in general, improves heat transfer from heat sinks, significantly, with as much as a 20% reduction in the maximum base temperature obtained without shrouding at a given power level.

Kraus and Bar-Cohen (1995) have published a book on the design and analysis of heat sinks. The book includes a description of much of the work of Snider and Kraus (and Kraus and Snider) and Aziz (1992), heretofore cited, and details of methods for obtaining coefficients in natural convection in parallel plate channels proposed by Welling and Wooldridge (1965), Van de Pol and Tierney (1973), Bilitsky (1986), and Yovanovich and his co-workers (Yovanovich, 1987a,b; Lee et al. 1991) cited in Chapter 4.

20.2.2 Microchannel Heat Sink

There are a wide variety of applications where there is a contradictory requirement for high rates of heat dissipation with a low allowable temperature rise. A typical example is a high-speed high-heat-density microelectronic device where heating rates on the order of 10 W/cm^2 are produced with an allowable temperature rise of 80°C above a 20°C environment. As indicated in Fig. 20.1, this level of heat flux is comparable to that from a nuclear blast, and it must be accommodated at a much lower temperature.

The microchannel heat sink, shown in Fig. 20.2, is an innovative structure with numerous small channels and fins arranged parallel to one another very close to the heated surface. The dissipated heat passes through the heat sink by conduction and to the coolant by forced convection. The coolant is usually a liquid with single- rather than two-phase flow. The distinguishing feature is the size of the channels, and by the early 1980s, researchers at Stanford University (Tuckerman and Pease, 1981a,b,c; Tuckerman, 1984) showed that microfabrication techniques used for microelectronic devices could also be employed to fabricate heat sinks. With these techniques they were able to produce large-aspect-ratio channels with channel widths z on the order of 50 μm and channel heights b on the order of several hundred micrometers. These channels were typically separated by fins with thickness δ on the order of 50 μm. The

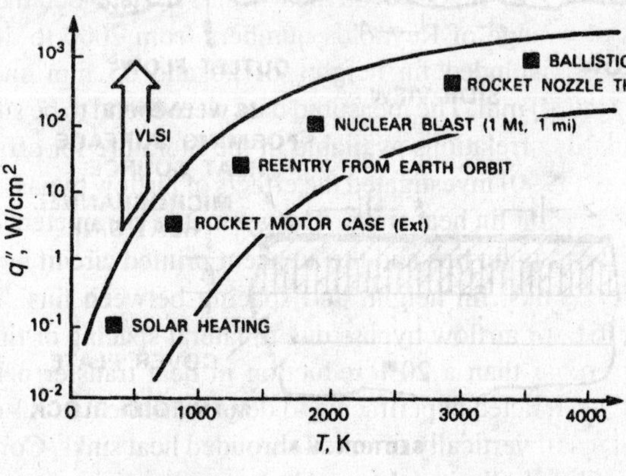

Figure 20.1 Comparison of heat fluxes.

thickness of the substrate δ_p between the channel bottoms and the heated surface was on the order of 100 μm.

An idea of the magnitude of the heat transfer coefficient can be derived from a consideration of a channel with an aspect ratio of $\alpha = b/z = 4.0$. If it is assumed that the coolant flow through each coolant channel is fully developed and laminar, the Nusselt number for a uniformly heated channel will be approximately 5.35. For $z = 50\ \mu$m, the equivalent or hydraulic diameter will be

$$d_e = \frac{4bz}{2(b+z)} = \frac{2\alpha z}{1+z} = \frac{(2)(4)(50 \times 10^{-6})}{1+4} = 80 \times 10^{-6}\ \text{m}$$

If the coolant is water with a thermal conductivity of 0.598 W/m·K, with Nu = 5.35, the heat transfer coefficient in the channel will be

$$h = \frac{k}{d_e}\text{Nu} = \frac{0.598}{8 \times 10^{-6}}(5.35) \approx 400{,}000\ \text{W/m}^2 \cdot \text{K}$$

The magnitude of h for the microchannel is comparable to the heat transfer coefficients obtained in flow boiling heat transfer. Moreover, the surface area for heat transfer is much larger than a surface without fins, and heat dissipation rates in excess of 1000 W/cm^2 with a thermal resistance of less than 0.1°C/W·cm^2 have been demonstrated with operation at room temperature.

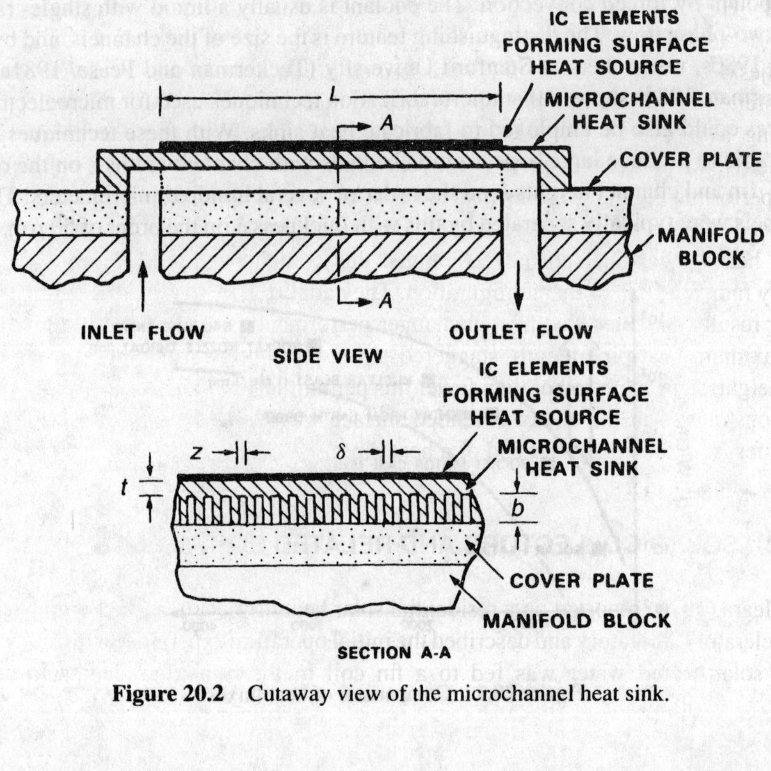

Figure 20.2 Cutaway view of the microchannel heat sink.

Tuckerman and Pease (1981a,b) were able to show theoretically that there is an optimum fully developed laminar flow heat sink design. In particular, their concern was with water-cooled silicon heat sinks 1.0 cm by 1.0 cm square, and they found that for a 20.7-kPa (30-psi) water pressure drop, there was an optimal design channel width of $z = \delta = 57 \mu m$ with a corresponding aspect ratio of $\alpha = 6.4$. Heat dissipation as high as 790 W/cm^2 and thermal resistance as low as 0.90 °C/W·cm^2 were obtained.

Since its introduction in 1981, several other researchers have published theoretical and/or experimental works on the microchannel heat sink. Goldberg (1984) studied thermal and fluid performance of laminar-flow air-cooled microchannel heat sinks fabricated in copper. Sixteen 0.508 cm by 0.508 cm chips on 1.27-cm centers were each provided with a copper heat sink containing 0.635-cm-long channels. The channel widths studied were $z = 127, 254$, and $635 \mu m$, and the fin width was equal to the channel width in all cases. Thermal resistances as low as 3.4 °C/W were measured. Although this thermal resistance is higher than that obtained by Tuckerman and Pease (1981a,b), it was based on air cooling and with an air pressure drop limited to a few inches of water.

Additional studies include those of Keyes (1984), Acosta et al. (1985), Mahalingham (1985), Swift et al. (1985), Koh and Colony (1986), Reichl (1986), Kishimoto and Ohsaki (1986), Sasaki and Kishimoto (1986), Kishimoto and Sasaki (1987), Mahalingham and Andrews (1987), Phillips (1987, 1988a,b), Hwang et al. (1987), and Nayak et al. (1987).

20.3 HEAT PIPES

Madjeski and Mikielewicz (1971) described a liquid fin as a new device for heat transfer equipment, and Ruch (1976) showed that counterflow heat exchangers comprising a bundle of heat pipes with secondary finned surfaces attached offered attractive advantages as energy recovery units. Feldman and Lu (1976) optimized the performance and showed that the cost of heat pipe heat exchangers could be reduced. They developed a computer code describing a fin–tube puddle-artery heat pipe heat exchanger, and they employed this computer code to generate data for many ramifications of the configuration. They analyzed two types of heat pipes, and their results indicated that heat exchanger performance could be inproved by using a maximum heat pipe length, staggered pipes, pipes of larger diameter, reasonable fin heights, and more rather than fewer fins per unit length. Marto and Weigel (1981) demonstrated that the use of extended surface was well conceived in treating heat transfer from the condenser end of a heat pipe.

20.4 SOLAR COLLECTORS AND RELATED EQUIPMENT

O'Meara (1976) reported on a residential solar heating system at the Fermi National Accelerator Laboratory and described the initial operating experience with the system. The solar-heated water was fed to a fin coil in the return air duct, and design considerations were given for the sizing of this coil. Hewitt et al. (1978) discussed the design of high-efficiency flat plate solar collectors for space and water heating, and Smith and Lee (1979) considered the problem of radiant heat exchange in a fin-and-tube solar collector.

Christopher and Pearson (1980) conducted a parametric analysis of air-cooled solar collectors. Their analysis of a family of collectors having inclined louvered absorber surfaces included conduction, forced convection, and thermal radiation exchange using a two-band model to describe the surface and material properties. The parametric study involved the determination of the effects of the collector on louver geometry, solar angle, and operational parameters. El-Refaie et al. (1983) incorporated the fin effect in predicting the performance of cascaded solar stills, and Bansal and Garg (1985) reported on an experimental study performed on two nonporous solar air heaters with and without fins.

20.5 FINNED REGENERATORS

O'Reilly (1978) presented a high-effectiveness regenerator design that was an adaptation of criteria employed in the aerospace industry. The reasoning leading to the design of this stainless steel plate-and-fin regenerator was presented, and material, fabrication, and test data were included. Fucinari (1977) presented basic heat transfer and pressure drop characteristics of the then existing state-of-the-art matrix fin configurations in what he termed a "shuttle rig." He discussed the effects of fin geometry and manufacturing processes on ceramic regenerator performance and presented a simplified analysis for an estimation of the effects of alterations in package size and/or fin parameters.

Cuffe at al. (1978) described a "new" regenerator intended for high-cyclic and/or high-temperature operation. They discussed how this was proven effective in improving the cyclic ability of the regenerator and in reducing the field maintenance required.

20.6 NUMERICAL ANALYSES

In this section, some of the numerical analyses that have appeared in the literature during the course of the development of extended surface technology are discussed. Some of them have already been cited elsewhere in the book.

Apparently, the first extended numerical analyses of extended surface heat transfer were the two works of Cumo et al. (1965a,b). Hung and Appl (1967) published a numerical method for solving the one-dimensional differential equation arising from the combined convection–radiation dissipation of a longitudinal fin of arbitrary profile with variable fin metal conductivity, heat transfer coefficient, and surface emissivity.

Donovan and Rohrer (1971) analyzed the heat transfer by radiation from a finned array consisting of longitudinal rectangular fins and a plane surface. The study included radiant heat exchange between a fin and its neighboring elements and resulted in a set of simultaneous nonlinear integrodifferential equations that were solved by a numerical integration method. Kern and Kraus (1972) considered 21 different but contrived cases of steady-state and five cases of unsteady-state analysis problems which were solved numerically using a finite difference computer code.

Sane and Sukhatme (1974) conducted a numerical analysis of natural convection heat transfer from a short horizontal finned array in which a single chimney flow pattern occurred. Results were obtained for a wide range of governing parameters for the case of air as the surrounding environmental medium. Good agreement was obtained with the experimental data available.

Chiou and Na (1977) developed an initial value method for solution of the linear two-point boundary value problem resulting from the analysis of radiation heat transfer from fins. Mehta (1978) employed a direct search procedure utilizing pseudorandom numbers over a region to obtain the minimum weight design of radiating fin arrays, and Truong and Mancuso (1980) obtained one-dimensional steady-state solutions for radiation from the front and rear surfaces of a radial fin having different front and rear emissivities. The Truong and Mancuso work considered different profile shapes, and the results were obtained by the shooting method, along with the use of the Runge–Kutta–Verner fifth- and sixth-order integration methods.

A numerical conjugate conduction–convection analysis was performed by Sparrow and Acharya (1981) for a vertical plate fin exchanging heat with the environment by natural convection. The analysis was based on a first-principles approach whereby the heat conduction equation for the fin was solved simultaneously with the conservation equation for mass, momentum, and energy in the fluid boundary layer adjacent to the fin. Nakamura et al. (1981) used a numerical method to develp a triangular coordinate system in order to solve the two-dimensional problem in a radial fin of triangular profile. The actual triangular shape, in general, was arbitrary, but for the specific case of an isosceles triangle, they discussed the effect of the Biot number on fin efficiency and compared the solutions graphically for one- and two-dimensional analyses.

Khader (1981) used a panel numerical scheme to investigate the heat conduction in extended surfaces, and Prakash and Patankar (1981) performed an analysis of the fully developed laminar flow and heat transfer in vertical tubes with internal radial fins. The governing equations for velocity and temperature were solved by a finite difference technique that incorporated a special scheme for treating the two coupled variables. Results were presented for a range of Rayleigh numbers and for various values of fin height and the number of fins.

Sunden (1983a,b) looked at the problem of coupled conduction and mixed convection for rectangular fins. The thickness of the fins was small and it was assumed that boundary layer flow existed around the fin surfaces. The buoyancy force was taken into account by applying Boussinesq's approximation. An iterative finite difference technique with an overall iteration loop and several subiteration loops was then developed to solve the problem. Details of the numerical method and an assessment of its accuracy were also presented. Sunden (1984) also used a numerical technique to investigate the heat transfer from rectangular fins by laminar and/or turbulent forced convection. The results were compared with conventional fin theory, and it was shown that a satisfactory comparison occurred if most of the boundary layer is either laminar or turbulent. It was observed that if the boundary layer contained laminar and turbulent parts of almost equal spatial extent, serious discrepancies between the prediction methods would occur.

Murthy and Patankar (1983) conducted a numerical study of heat transfer from a rotating cylinder with external longitudinal fins, and Kelkar and Patankar (1984) made a prediction of flow and heat transfer in a parallel-plate channel with staggered fins using a numerical method. Zhang and Patankar (1984) showed that heat transfer from the fins in a shrouded fin array is dependent on the distribution of flow in the duct cross section. It was shown that the amount of clearance between the fin tip and the shroud has a significant effect on the heat transfer. A numerical analysis of laminar flow and heat transfer with twisted fins was reported by MacArthur and Patankar (1985). A coordinate system that twisted with the fins was employed so that a fully developed flow could be defined with reference to the coordinate system. The thermal boundary condition chosen was that of axially uniform heat input and peripherally uniform wall temperature for the inner pipe wall. The fins were assumed to be highly conducting, so that they acquired the temperature of the inner pipe wall. The coupled axial flow and secondary velocity field were solved numerically and the corresponding temperature field was also determined. Results were presented in terms of overall friction factor–Reynolds number product and in terms of the overall Reynolds number. Circumferential, radial, and axial velocity contours were also provided.

Sunden (1985) treated the problem of conjugate heat transfer from rectangular fins numerically. He provided results for three different fluids numerically and comparisons with conventional theory showed that while the conventional theory yields acceptable results with regard to fin efficiency, the total heat transferred, heat flux distribution, and temperature variation are not accurately predicted.

Acharya and Patankar (1985) performed an analytical study to investigate the effect of buoyancy on laminar free convection in a shrouded fin array. Two heating conditions were considered: In one, the fins and base temperature were hotter than the fluid, and in the other, they were colder. Results were obtained numerically for a wide range of the governing buoyancy parameter. It was found that with a hot fin and base, the secondary flow pattern is made up primarily by a single eddy. The inflence of buoyancy appeared to be significant and leads to Nusselt numbers and friction factors that are much higher than for pure forced convection. Choudhury and Patankar (1985) made an analysis of the laminar fluid flow and heat transfer in the entrance region of an isothermal tube with radial internal fins. The governing equations of mass, momentum, and energy were solved numerically by means of of a three-dimensional parabolic procedure. Detailed results were presented for the developing velocity profiles and friction factors for a range of geometries.

20.7 MATHEMATICAL TECHNIQUES

One often loses sight of the elegance provided by the mathematics that is involved in the analysis and evaluation of extended surfaces. As a case in point, consider the five papers of Wilkins (1960a,b, 1961, 1962a,b) that pertain to radiating fins, which were cited in Chapters 13 and 14. The idea of finding three functions, one involving heat flow, one involving the temperature distribution, and one involving the fin profile that simultaneously minimized the profile area of the fin was contained in the Wilkins (1960b) paper. As discussed in Chapter 18, this approach was followed by Haley and Westwater (1966) in designing an optimum fin for a boiling application.

As discussed in Chapter 2, Smith and Sucec (1969) presented an exact solution for the efficiency of the radial fin of triangular profile which had apparently eluded prior investigators, such as Gardner (1945) and Harper and Brown (1922), who provided correction factors for this profile. The solution was obtained through the use of the infinite series method (the method of Frobenius), which in itself is quite fundamental. The trick, however, was to make a transformation to a new independent variable, and then the solution followed. Smith and Sucec (1969) gave an example of three different fins of this shape, of various sizes and thermal parameters of thermal conductivity and heat transfer coefficient, and showed a marked difference between the efficiencies computed by the Harper and Brown (1922) correction factor and those computed from Gardner's (1945) relationship for the radial fin of constant cross section.

Bilenas and Jiji (1970) presented a regular perturbation solution for a particular fin problem. The configuration considered was a fin with constant base temperature and a dissipating tip and the study included the interaction of conduction with convection and radiation. They showed that the perturbation solution becomes increasingly accurate as the perturbation parameter decreases because at these smaller values, the nonlinear radiation effects become small relative to the conduction effect. The accuracy of the solution was seen to improve for increasing values of the Biot number.

Bhargava and Duffin (1972) proposed a network in which the conduction in certain branches was variable. They used a dual variational principle which led to a duality inequality and gave sharper upper and lower estimates of the maximum joint conductance. It was claimed that such a network serves as a discrete model for a cooling fin subjected to a weight limitation.

Marsh and Costello (1973) pointed out that in the design of convective fins, the stochastic variations in the fin dimensions were traditionally handled by the use of safety factors. They showed that this process often resulted in a multiplication of safety factors and thus yielded an overly expensive design. The paper presented a probabilistic approach that not only analyzed for the probability of system failure, but allowed for synthesis of the optimal design. They described four methods of varying accuracy and difficulty and showed that the method based on the RMS approximation for the variances appeared to be the most useful for design purposes.

Date (1974) formulated the problem of fully developed laminar and turbulent uniform-property flow in a tube containing a twisted tape in terms of the partial differential equations of momentum and heat transfer. These equations were then solved by adapting an existing numerical procedure for two-dimensional elliptic equations to predict the heat transfer and friction characteristics of the flow. Chumakov (1974) presented an analytical solution for the one-dimensional problem of a fin with with nonlinearly varying radiation from its surfaces. In the method of perturbation used, the zero approximation was based on the solution of the linearized problem with the nonlinear boundary condition satisfied exactly. The error in the linear approximation of the nonlinear function in the energy equation was then treated as the perturbing function.

Aziz and Na (1981a,b) investigated the problem of periodic heat transfer in a convecting fin with temperature-dependent thermal conductivity and coordinate-dependent heat transfer coefficient using a perturbation analysis. The zero-order problem, which corresponds to steady-state behavior, was solved by quasilinearization. A method of complex combination was used in conjunction with a noniterative numerical scheme. Sparrow and Chyu (1982) conducted a conjugate forced convection–convection analysis of the heat transfer in a vertical fin.

In a work cited previously, Bobco and Starkovs (1970) developed an infinite series solution for the steady temperature field in a thin rectangular region exposed to a piecewise continuous heat flux and losing energy from the exposed surfaces by either linearized radiation or true convection. These environmental conditions were shown to approxmate a spacecraft in which a finlike plate (a *thermal doubler*) is used to enhance the radiating area of energy-dissipating electronic equipment mounted in an equipment bay. A numerical example based on typical spacecraft power and environmental conditions was employed to show how the closed-form solution could be used to investigate geometrical influences. Both direct (specified fin thickness and unknown temperature) and design (specified temperature and unknown fin thickness) problems were examined to show how the temperature and fin thickness depended on the heated area (the "footprint"), the shape of the fin, and the location of the heat-dissipating area within the fin boundaries.

In a work cited in Chapter 4, Netrakanti and Huang (1985) showed how the heat dissipated in a fin of given volume could be optimized by the invariant embedding technique, in which the inherent instability in a numerical integration can be avoided.

20.8 TURBINE BLADES

Konopolev and Bychkovskii (1970) calculated three-dimensional temperature fields in turbine water wall tubes by what they termed a temperature balance method, and they provided nomograms for the calculation of fin and wall temperatures. Martin et al. (1978) reported on heat transfer measurements around a PVD rotor blade using a transient method. Measured transition lengths on the suction surface, over which the heat transferred appeared to be nearly trebled, were observed to be rather short in comparison to other measurements.

Bogomolov and Orlova (1984) suggested a method for calculating the temperature distribution in a perforated blade under conditions of nonisothermal injection with allowance for the effect exerted by the coolant jets on the heat transfer from the main stream. The method was derived by correlation of experimental data and by approximate solution of the problem of heat conduction in a perforated wall coated by a thin insulating layer.

20.9 DESIGN METHODS

Wilkie (1971) gave criteria for the choice of the surface form in gas-cooled reactors, and Shcherbakov and Bosyy (1974) presented fin selection criteria for various types and profiles of fins. Mineur and Dunstan (1979) presented a rational approach to the design of gas-fired finned-tube exchangers. Novel equations were developed to facilitate design procedures, and comparisons were made between the theoretical performance of a finned-tube boiler and those determined experimentally.

Shvets et al. (1985) presented a technique for designing variable-thickness fins in longitudinal flow under conjugate boundary conditions with heat sources distributed in arbitrary fashion and with tip heat loss. In a work already cited, Biyikli (1985) developed design charts for longitudinal fins of rectangular profile in boiling water.

20.10 FREEZING OR MELTING ON THE FIN FACES

In an experimental endeavor, Bathelt and Viskanta (1981) studied the effectiveness of extended surfaces on a horizontal cylindrical heat source–sink during solid–liquid phase-change heat transfer on the fin faces. Melting and freezing experiments were conducted in a test cell suitable for photographic and shadowgraphic observation using a circular cylinder and three rectangular profile fins parallel to the axis and evenly distributed around the circumference of the heat exchanger. Results were reported for *n*-heptadecane as the phase-change material.

Sparrow et al. (1981) conducted experiments to study freezing on a finned vertical tube when either conduction in the solid or natural convection in the liquid controlled the heat transfer. Auxiliary experiments were also performed with an unfinned tube to obtain comparison data. It was observed that for conduction control, the enhancement of freezing due to finning is less than the area ratio of the finned to the unfinned tubes, whereas for natural-convection control, the enhancement is very nearly equal to the area ratio.

Sheffield (1982) showed the effects of double-connected finned surfaces on heat transfer during solidification, and Smith (1983) conducted an analytical and experimental study of freezing adjacent to a pin fin. Chepernoi et al. (1985) studied the process of frost formation in finned air coolers and reported on the effect of fin spacing on frost formation.

20.11 HEAT AND MASS TRANSFER

McQuiston (1975) presented an analysis of a fin of uniform cross section, obtaining the fin efficiency in the presence of mass transfer. He proposed a new parameter that takes surface temperature and moist air conditions into account. El-Ridi et al. (1976) conducted experimental studies of combined heat and mass transfer on horizontal finned tubes cooled by natural convection in air and claimed to have obtained expressions for the optimum geometry.

The analysis of Kreid et al. (1978) showed that the equations describing the combined latent and sensible heat transfer on a wet surface may be transformed to a single, approximate equation involving the product of a heat transfer coefficient and a driving potential, in a direct analog to the "usual" heat transfer equation. For the wet model, the appropriate driving potential was shown to be the moist air enthalpy difference evaluated between the surface and the coolant stream. Estimates were given for the accuracy of the approximations inherent in the model, and examples were given to illustrate the potential benefits of deluge for a typical plate fin heat exchanger. Ehlmady and Biggs (1983) gave data on the efficiency of extended surfaces with simultaneous heat and mass transfer.

20.12 MISCELLANEOUS CITATIONS

In this, the concluding section of the text, several references that were to difficult to categorize are cited.

The interesting problem of temperature measurement using a thermocouple in a well has been studied by West and Westwater (1953). The well is a tube that is sealed at the free end and secured to the confining surface at the other end. This may be thought of as a capped, hollow cylindrical spine having a base temperature equal to the confining surface temperature. The thermocouple in the well is in contact with the cap but not with the fluid whose temperature is to be measured. Ths yields a thermometry error that, ideally, should be small and predictable. The West and Westwater (1953) work considered the case for which radiation must be included and derived an equation to yield the fluid temperature.

Hilding (1953) treated the problem of the effect of fin geometry on the performance of heat exchanger surfaces, and Kraemer and Westwater (1954) looked further into the radiation effect in a thermocouple well and using a numerical solution showed that the West and Westwater (1953) result would be attained to within 1% if a variation of the radiation as a function of position within the well was ignored, with the total effect taken as a constant.

Dusinberre (1956) pointed out that for the longitudinal fin of rectangular profile with fin efficiencies higher than 0.75, an accurate representation could be obtained by truncating the infinite series for tanh mb. He also fitted functions to Gardner's (1945) curves and came up with approximations for a longitudinal fin of triangular profile, a radial fin of rectangular profile, and a conical spine.

Beauvais and Nickol (1964) provided a rational approach to improved fin performance, and Chapman (1984) presented a grapical method to determine the temperature excess and heat flow in a radial fin of rectangular profile. This was deemed necessary at the time because of the presence of eight modified Bessel functions (and the repeated evaluation of them) in the equations for the temperature excess and heat flow. Yudin and Tokhtarova (1967) reported on the heat dissipation and flow resistance in what they called checkerwork stacks, which are, in reality, crossed tubes. Their studies were comparative because they took measurements on plain tubes, tubes with round and flat pin fins, tubes with disk-shaped fins, and wire-fin systems.

Evenko and Shishkov (1969) gave results of an investigation of a surface finned with a corrugated strip welded to the tubes by a condenser welder. They contended that the configuration possessed a high finning ratio and that certain manufacturing economies would prevail. Shmerkovich (1969) cited a "new" air-cooled system for cooling a medium with a small temperature drop for operation under vacuum or under high pressure.

Wimberley and Costello (1970) developed an analytical model to predict the effects of circumferential wall temperature gradients in spacecraft radiator tubes. They derived exact solutions for the fin effectiveness of the tube wall in terms of an infinite system of linear equations. Gross (1970) showed that interleaving fins can significantly increase the rate of heat transfer in a space application where thermal control of internal components is desired. His concept involved the use of "hot" fins radiating to "colder" fins, with the bases of the fins directed in opposite directions. It was claimed that the interleaving fins were better thermally and lighter and more rigid than what could be obtained with a conduction approach.

Coleman and Lemmon (1973) measured turbulent heating distributions in the interference region on and around several small highly swept fin configurations mounted on a cone, and the results were presented by Lemmon and Coleman (1973). Lubny-Gerczyk et al. (1973) presented the results of calculations of temperature fields and stresses in the wall of a heated fin tube and a fin under thermal and mechanical loads. The investigation was claimed to have a direct bearing on the conditions of operation of heated waterwalls with unilateral heating in steam boiler furnaces with supercritical parameters. Their calculations showed that the highest thermal stresses in the tube are the axial stresses. They also showed that the hot waterwalls of transversely finned tubes could be used in furnaces of gas- and oil-fired boilers. Dolgii et al. (1973) looked at what were at that time new possibilities for finning tubes. They discussed the advantages and disadvantages of tubes with internal longitudinal fins, all welded platens of finned tubes, finned and membrane convective heating surfaces, and heating surfaces composed of tubes with transverse finning.

Sherwin (1973) discussed the performance of circular finned tubes in relation to both construction and fin formation. He discussed the effect of fin spacing, fin thickness, and height-to-tube diameter ratio and showed the particular importance of the height-to-tube diameter ratio.

Hunn (1974) presented a systematic procedure for assessing the effect of external fouling on both heat transfer and friction power performance in finned-tube heat exchangers in dry cooling-tower service. His procedure provided a basis for the development of cost–benefit relationships for heat exchanger cleaning. Tolubinsky and Zozulya (1974) considered the application of finned surfaces in high-temperature heat exchangers, and heat transfer between finned tubes and air in the presence of a corona discharge was investigated by Reynolds and Holmes (1975). They used wires stretched parallel to finned tubes and applied a high voltage. They showed that the heat transfer coefficients were a function of the power input to the corona for various wire locations.

Soland et al. (1978) made an attempt at ranking the heat exchanger surfaces given in a prior edition (1964) of the book by Kays and London (1984). Their ranking was based on comparisons of the same exchanger shape and volume, the same volume and pumping power, the same pumping power and heat transfer capability represented by the number of transfer units N_{tu}, and the same volume and N_{tu}. O'Brien and Ehrlich (1977) showed that the nonuniform temperature on the periphery of a straight noncircular duct had a significant effect on the local heat conductance and overall heat transfer. They showed that the physics of a thermally active fin required a matching of a heat conduction solution within the fin material to heat convection in the duct.

Russell (1974) measured the heat transfer performance of three corroded fin tube samples and in all cases found increases in both heat transfer and air-side pressure loss compared to new tubes. Eriksen (1979) gave a basic discussion of the application of waste heat boilers to recover energy from the exhaust of gas turbine engines. He presented thermal design parameters for the various heat exchangers encountered in this application, along with mechanical considerations.

Hauser et al. (1981) designed and built an apparatus to measure the transient cooling effect on a gas inside a closed space in a Stirling engine with interleaving fins. Karvinen (1981) performed an analysis of heat transfer from a plate fin cooled by forced and natural convection. He used approximate solutions to relate convective heat flux and temperature in the cases of laminar and turbulent boundary layers in forced flow and in the case of natural convection on a vertical fin. He concluded his study by giving a simple procedure for the solution of the conjugate problem composed of convection within the fin.

Mikk (1982) provided a nomograph for the determination of the fin efficiency of radial fins of trapezoidal profile, and Kuehn et al. (1983) contended that fins with large height fabricated of materials of moderate to low thermal conductivity will not remain isothermal and required solution of a conjugate problem. The conjugate problem of conduction within the fin and natural convection in the fluid was solved, and the results were presented for a uniform conductivity plate fin as a function of the fluid Prandtl number, which was claimed to be the only independent governing parameter.

Bott and Bemrose (1983) considered particulate fouling on the gas side of finned-tube heat exchangers, and Egorov et al. (1984) reported on experiments involving the double-walled finned pipe as a structural element for finned boiler economizers. Migai et al. (1984) presented the results of their research on heat transfer in vertical-lengthwise finned tubes and membrane surfaces within fluidized beds.

Kwon et al. (1984) studied the influence of fin height on steady conjugate natural-convection heat transfer from an isothermal cylinder with one vertical-longitudinal conducting plate fin. They compared their results with the limiting case of an isothermal free cylinder and a cylinder with an infinitely high conducting fin. They observed that short fins exhibited large local coefficients near the tip, with a rapid drop to zero at the base, and that high fins follow the distribution obtained for the infinitely high fin except near the tip.

Small et al. (1984) reported on an investigation of asymptotically large-area fins, and Aziz (1985b) described a proposed graduate course in extended surface heat transfer to complement the more traditional graduate courses in heat transfer. Huang and Chen (1984, 1985) studied analytically the heat transfer characteristics of laminar mixed convection over a vertical pin by conjugate convection–conduction theory. Numerical results were obtained for the local heat transfer coefficient, local heat flux, temperature distribution, and efficiency of the fin as influenced by the surface curvature and buoyancy effect.

Finally, Pop et al. (1985) conducted analyses of steady-state conjugate free convection about a vertical fin embedded in a porous medium at high Rayleigh numbers, and Hollworth and Gero (1985) measured convective heat transfer for a heated axisymmetric air jet impinging on a flat surface.

20.13 NOMENCLATURE

Roman Letter Symbols

b	fin or spine height, m
d	spine diameter, m
f	friction factor, dimensionless
G	mass velocity, $kg/m^2 \cdot h$
H	louver fin height, dimensionless
h	heat transfer coefficient, $W/m^2 \cdot K$
j	heat transfer parameter, dimensionless
k	thermal conductivity, $W/m \cdot K$
L	length, m
L_p	louver pitch, m
N	number of tube rows in flow direction, dimensionless
N_{tu}	number of transfer units, dimensionless
n	an exponent, dimensionless
Nu	Nusselt number, dimensionless
p	pitch, m
Re	Reynolds number, dimensionless
s	fin pitch, m

Greek Letter Symbols

α	aspect ratio, dimensionless
γ	combination of terms, dimensionless
γ_p	substrate thickness, m
δ	fin thickness, m
θ	louver angle, rad
Λ	combination of terms, dimensionless
μ	dynamic viscosity, $kg/s \cdot m$

Roman Letter Subscripts

a	fin tip
d	diameter; bubble

e	equivalent
f	fins
L	length
ℓ	longitudinal
mom	momentum
p	strip length; pitch; substrate
t	transverse; tubes
tr	transition

APPENDIX A

GAMMA AND BESSEL FUNCTIONS

A.1 GAMMA FUNCTION

A.1.1 Development

The gamma function is defined by

$$\Gamma(n) = \int_0^\infty x^{n-1} e^{-x}\, dx \tag{A.1}$$

where *n* may be any positive number although not necessarily an integer. This function is important because it is a generalization of the *factorial*.

The gamma function $\Gamma(n+1)$ can be evaluated from the definition of eq. (A.1) through an integration by parts:

$$\Gamma(n+1) = \int_0^\infty x^n e^{-x}\, dx$$

Here

$$u = x^n \quad \text{and} \quad du = nx^{n-1}$$

and

$$dv = e^{-x}\, dx \quad \text{and} \quad v = -e^{-x}$$

Therefore,

$$\int u\, dv = uv - \int v\, du$$

or

$$\Gamma(n+1) = \int_0^\infty x^n e^{-x}\,dx = -x^n e^{-x}\Big|_0^\infty - \int_0^\infty -nx^{n-1}e^{-x}\,dx$$

However, either limit will render the first term zero; the second term is

$$-\int_0^\infty -nx^{n-1}e^{-x}\,dx = n\int_0^\infty x^{n-1}e^{-x}\,dx = n\Gamma(n)$$

Therefore, the fundamental relation

$$\Gamma(n+1) = n\Gamma(n) \tag{A.2}$$

is established.

The value of $\Gamma(1)$ can be evaluated from the definition of the gamma function [eq. (A.1)] as follows:

$$\Gamma(1) = \int_0^\infty x^{n-1}e^{-x}\,dx = \int_0^\infty x^0 e^{-x}\,dx = -e^{-x}\Big|_0^\infty = 1$$

Then according to eq. (A.2),

$$\begin{aligned}
\Gamma(2) &= 1\Gamma(1) = 1\cdot 1 = 1 = 1! \\
\Gamma(3) &= 2\Gamma(2) = 2\cdot 1\Gamma(1) = 2\cdot 1\cdot 1 = 2\cdot 1 = 2! \\
\Gamma(4) &= 3\Gamma(3) = 3\cdot 2\Gamma(2) = 3\cdot 2\cdot 1\Gamma(1) = 3\cdot 2\cdot 1 = 3! \\
\Gamma(5) &= 4\Gamma(4) = 4\cdot 3\Gamma(3) = 4\cdot 3\cdot 2\Gamma(2) = 4\cdot 3\cdot 2\cdot 1 = 4!
\end{aligned}$$

and the general relationship can be deduced:

$$\Gamma(n+1) = n! \tag{A.3}$$

This is an important relationship between the gamma function and the factorial $n!$.

The Taylor series expansion of a function $f(x)$ about a point $x = b$, using primes to denote derivatives, is known to be

$$f(x) = f(b) + \frac{(x-b)f'(b)}{1} + \frac{(x-b)^2 f''(b)}{2} + \frac{(x-b)^3 f'''(b)}{6} + \cdots + \frac{(x-b)^n f^n(b)}{n!}$$

In a strictly mathematical sense, and to make the expression completely symmetrical, the first term of the Taylor series expansion should be set down as

$$\frac{f(b)}{0!}$$

which has no meaning in terms of the ordinary definition of the factorial:

$$n! = n(n-1)(n-2)(n-3)\cdots(3)(2)(1)$$

On the other hand, 0! can be defined using the gamma function and specifically through the use of eq. (A.3) with $n = 0$:

$$0! = \Gamma(0+1) = \Gamma(1) = 1$$

The gamma function can be defined for noninteger values of the argument n. Indeed, it can be seen that there is no need to suggest integral values of the argument n. The gamma function may be written

$$\Gamma(x) = \int_0^\infty t^{x-1} e^{-t}\, dt$$

where the variable x, for the present, can take on any value whatever. However, because there is a connection between the gamma function and the ordinary factorial, one must exercise care. Referring again to eq. (A.2) for any value of x,

$$\Gamma(x+1) = x\Gamma(x)$$

from which it can be seen that

$$\Gamma(x) = \frac{\Gamma(x+1)}{x}$$

Therefore, as x approaches zero, the value of $\Gamma(x)$ becomes undefined. Furthermore, the gamma function is undefined for $x = -1, -2, -3\ldots$ because

$$\Gamma(-1) = \frac{\Gamma(0)}{-1} = \frac{\infty}{-1}, \quad \text{undefined}$$

$$\Gamma(-2) = \frac{\Gamma(-1)}{-2} = \frac{\infty}{-2}, \quad \text{undefined}$$

$$\Gamma(-3) = \frac{\Gamma(-2)}{-3} = \frac{\infty}{-3}, \quad \text{undefined}$$

A.1.2 Polynomial Approximations

Abramowitz and Stegun (1964) provide two polynomial approximations for the gamma function. For $0 \leq x \leq 1$,

$$\Gamma(x+1) = x! = 1 + a_1 x + a_2 x^2 + a_3 x^3 + a_4 x^4 + a_5 x^5 + e(x)$$

where

$$a_1 = -0.5748646$$

$$a_2 = 0.9512363$$

$$a_3 = -0.6998588$$

$$a_4 = 0.4245549$$

$$a_5 = -0.1010678$$

where the absolute value of the error is

$$|e(x)| \leq 5 \times 10^{-5}$$

and for $0 \leq x \leq 1$

$$\Gamma(x+1) = x! = 1 + b_1 x + b_2 x^2 + \cdots + b_7 x^7 + b_8 x^8 + e(x)$$

where

$$b_1 = -0.577191652$$

$$b_2 = 0.988205891$$

$$b_3 = -0.897056937$$

$$b_4 = 0.918206857$$

$$b_5 = -0.756704078$$

$$b_6 = 0.482199394$$

$$b_7 = -0.193527818$$

$$b_8 = 0.035868343$$

where the the absolute value of the error is

$$|e(x)| \leq 3 \times 10^{-7}$$

A.2 BESSEL FUNCTIONS

A.2.1 Bessel Equations and the Bessel Functions

The exposition of the functions attributed to F. W. Bessel (1784–1846) appeared in *Berliner Abhandlungen* (1826) and the comprehensive modern summary is that of G. N. Watson (1966).

Consider the linear differential equation

$$x^2 \frac{d^2 y}{dx^2} + x \frac{dy}{dx} + (x^2 - n^2) y = 0 \tag{A.4}$$

This equation with variable coefficients, known as the *Bessel differential equation* is said to be of order n. Because it is a linear differential equation of second order, it must have two linearly independent solutions. Its general solution *when n is not an integer* is

$$y = C_1 J_n(x) + C_2 J_{-n}(x) \tag{A.5a}$$

and *when n is an integer,*

$$y = C_1 J_n(x) + C_2 Y_n(x) \tag{A.5b}$$

In eqs. (A.5), C_1 and C_2 are arbitrary constants, $J_n(x)$ is the *Bessel function of the first kind, of order n and argument x*, and $Y_n(x)$ is the *Bessel function of the second kind, of order n and argument x*. $J_n(x)$ and $J_{-n}(x)$ are infinite series of the forms

$$J_n(x) = \sum_{m=0}^{\infty} \frac{(-1)^m x^{2m+n}}{2^{2m+n} m!\,\Gamma(m+n+1)} \tag{A.6a}$$

and

$$J_{-n}(x) = \sum_{m=0}^{\infty} \frac{(-1)^m x^{2m-n}}{2^{2m-n} m!\,\Gamma(m-n+1)} \tag{A.6b}$$

where $\Gamma(m+n+1)$ and $\Gamma(m-n+1)$ are the gamma functions of $m+n+1$ and $m-n+1$, respectively. $Y_n(x)$ is also an infinite series, which may be written, in general terms, as

$$Y_{-n}(x) = A J_n(x) \int \frac{dx}{x[J_n(x)]^2} + B J_n(x) \tag{A.7a}$$

Various forms for $Y_n(x)$ may be obtained, depending on the values assigned to the constants A and B. One such form is

$$Y_n(x) = \frac{J_n(x)\cos n\pi - J_{-n}(x)}{\sin n\pi} \tag{A.7b}$$

Known as *Weber's form*, it is commonly used and is valid as long as n is not an integer. When n is an integer, $Y_n(x)$ is the limit of eq. (A.7*b*) because n approaches the value of the integer; hence $Y_n(x)$ may be expressed as a power series.

An equation that closely resembles the Bessel equation is

$$x^2 \frac{d^2 y}{dx^2} + x \frac{dy}{dx} - (x^2 + n^2) y = 0 \tag{A.8a}$$

Equation (A.8*a*) differs from eq. (A.4) only in the last term, which is negative. Equation (A.8*a*) is called the *modified Bessel equation*. It may be adjusted to yield

$$x^2 \frac{d^2 y}{dx^2} + \frac{dy}{dx} + (\imath^2 x^2 - n^2) y = 0 \tag{A.8b}$$

Equations (A.8*a*) and (A.8*b*) are alternate forms. The formal solution of these equations is, *when n is not an integer*,

$$y = C_1 J_n(\imath x) + C_2 J_{-n}(\imath x) \tag{A.9a}$$

and *when n is an integer*,

$$y = C_1 J_n(\imath x) + C_2 Y_n(\imath x) \tag{A.9b}$$

The actual solutions of eqs. (A.9) are real functions in that they do not contain the imaginary quantity $\imath = \sqrt{-1}$. Accordingly, the *modified Bessel function of the first kind, of order n and argument x*, is defined by

$$I_n(x) = (\imath)^{-n} J_n(\imath x) = \sum_{m=0}^{\infty} \frac{x^{2m+n}}{2^{2m+n} m!\,\Gamma(m+n+1)} \tag{A.10a}$$

and

$$I_{-n}(x) = \imath^n J_{-n}(\imath x) = \sum_{m=0}^{\infty} \frac{x^{2m-n}}{2^{2m-n} m!\,\Gamma(m-n+1)} \tag{A.10b}$$

The second independent solution of the modified Bessel equation is the *modified Bessel function of the second kind, of order n and argument x*. This solution, denoted as $K_n(x)$, is similar in form to $Y_n(x)$ and may be written *when n is not an integer* as

$$K_n(x) = \frac{\pi/2}{\sin n\pi}[I_{-n}(x) - I_n(x)] \tag{A.11a}$$

and *when n is an integer*, $K_n(x)$ is the limit approached by eq. (A.11*a*) as n approaches the value of the integer:

$$K_n(x) = \frac{2}{\cos n\pi}\left[\frac{\partial I_{-n}(x)}{\partial n} - \frac{\partial I_n(x)}{\partial n}\right] \tag{A.11b}$$

The solution of eq. (A.8*a*) may therefore be written, *when n is not an integer*,

$$y = C_1 I_n(x) + C_2 I_{-n}(x) \tag{A.12a}$$

and *when n is an integer*,

$$y = C_1 I_n(x) + C_2 K_n(x) \tag{A.12b}$$

A.2.2 Generalized Form of the Bessel Equation

Bessel functions are of particular interest because they provide solutions for many of the equations that arise continuously in engineering applications, especially in the analysis of extended surfaces. Often, equations having Bessel functions as their solutions are not readily recognizable. Consequently, it is desirable to investigate given linear differential equations with variable coefficients to determine if they are reducible to the Bessel equation.

For example, take the general differential equation

$$\frac{d}{dx}\left(x^p \frac{dy}{dx}\right) + (ax^j + bx^k)y = 0 \qquad j > k \tag{A.13}$$

an attempt to reduce it to a form of the Bessel equation

$$\frac{d^2y}{dx} + \frac{1}{x}\frac{dy}{dx} + \left(m^2 - \frac{n^2}{x^2}\right)y = 0 \tag{A.14}$$

To accomplish this reduction, it is necessary to make two transformations

$$x = u^{\alpha} \tag{A.15a}$$

and

$$y = u^{\beta}v \tag{A.15b}$$

where α and β are arbitrary parameters yet to be determined. Then

$$\begin{aligned}\frac{dx}{du} &= \alpha u^{\alpha-1}\\ \frac{du}{dx} &= \frac{1}{\alpha u^{\alpha-1}}\\ \frac{dy}{du} &= v\beta u^{\beta-1} + u^{\beta}\frac{dv}{du}\end{aligned} \tag{A.16}$$

Equations (A.16) can be utilized to yield

$$\frac{dy}{dx} = \frac{dy}{du}\frac{du}{dx} = \left(v\beta u^{\beta-1} + u^{\beta}\frac{dv}{du}\right)\frac{1}{\alpha u^{\alpha-1}}$$

and

$$x^p\frac{dy}{dx} = \frac{1}{\alpha}\left(u^{\alpha p+\beta-\alpha+1}\frac{dv}{du} + \beta v u^{\alpha p+\beta-\alpha}\right) \tag{A.17}$$

Now the derivative with respect to x of eq. (A.17), together with eqs. (A.15), can be substituted into eq. (A.13) to give, upon simplification,

$$\begin{aligned}0 = {}& \frac{d^2v}{du^2} + \frac{\alpha p + 2\beta - \alpha + 1}{u}\frac{dv}{du}\\ &+ \left[\frac{\beta(\alpha p + \beta - \alpha)}{u^2} + \alpha^2 a u^{\alpha j+2\alpha-\alpha p-2} + \alpha^2 b u^{\alpha k-\alpha p+2\alpha-2}\right]v\end{aligned} \tag{A.18a}$$

Equation (A.18a) is in the same form as Bessel's equation,

$$\frac{d^2v}{du^2} + \frac{1}{u}\frac{dv}{du} + \left(\omega^2 - \frac{n^2}{u^2}\right)v = 0 \tag{A.18b}$$

and it is seen that for complete identity between eqs. (A.18a) and (A.18b), the following must occur with regard to eq. (A.18b):

1. The coefficient of $(1/u)(dv/du)$ must equal unity, or

$$\alpha p + 2\beta - \alpha + 1 = 1 \tag{A.19}$$

2. The exponent of u in the second term within the brackets in eq. (A.18a) must be zero, or

$$\alpha j + 2\alpha - \alpha p - 2 = 0 \tag{A.20}$$

3. The exponent of u in the third term within the brackets must be -2, so that this term may be combined with the first term within the brackets:

$$\alpha k - \alpha p + 2\alpha - 2 = -2 \tag{A.21}$$

Equations (A.19) through (A.21) permit the evaluation of α, β, and k. First, from eq. (A.20),

$$\alpha = \frac{2}{2 - p + j} \tag{A.22}$$

Then, by substitution into eq. (A.19),

$$\beta = \frac{\alpha(1-p)}{2} = \frac{1-p}{2-p+j} \tag{A.23}$$

and finally, from eq. (A.21),

$$k = p - 2 \tag{A.24}$$

Observe that if $b = 0$ in eq. (A.13), the condition that $k = p - 2$ is irrelevant. But when $b \neq 0$ and $k = p - 2$, the original equation can be reduced to the form of the Bessel equation where α and β are given by eqs. (A.22) and (A.23):

$$\frac{d^2 v}{du^2} + \frac{1}{u}\frac{dv}{du} + \left[a\alpha^2 + \frac{\beta(\alpha p + \beta - \alpha) + \alpha^2 b}{u^2}\right] v = 0 \tag{A.25}$$

From eq. (A.18b)

$$-n^2 = \beta(\alpha p + \beta - \alpha) + \alpha^2 b$$

or, inserting α and β from eqs. (A.22) and (A.23),

$$n = \frac{\sqrt{(1-p)^2 - 4b}}{2 - p + j} \tag{A.26}$$

and finally,

$$\omega^2 = a\alpha^2$$

or

$$\omega = \alpha\sqrt{a} \tag{A.27}$$

The general solution for v in eq. (A.18*b*) can now be written, *when n is not an integer*,

$$v = C_1 J_n(\omega u) + C_2 J_{-n}(\omega u) \tag{A.28a}$$

and *when n is an integer*,

$$v = C_1 J_n(\omega u) + C_2 Y_n(\omega u) \tag{A.28b}$$

Next, the solution of eq. (A.18*b*) in terms of y and x is obtained. From eqs. (A.15)

$$u = x^{1/\alpha}$$

and

$$y = u^{\beta} v = x^{\beta/\alpha} v$$

When n is not an integer,

$$y = x^{\beta/\alpha}[C_1 J_n(\omega x^{1/\alpha}) + C_2 J_{-n}(\omega x^{1/\alpha})] \tag{A.29a}$$

and *when n is an integer*,

$$y = x^{\beta/\alpha}[C_1 J_n(\omega x^{1/\alpha}) + C_2 Y_n(\omega x^{1/\alpha})] \tag{A.29b}$$

Example A.1: Solution of the Bessel Equation. Determine the general solution of

$$\frac{d^2y}{dx^2} + \left(3x - \frac{6}{x^2}\right)y = 0$$

SOLUTION. Here

$$p = 0 \qquad b = -6$$

$$a = 3 \qquad k = -2$$

$$j = 1$$

First, does $k = p - 2$?

$$k = -2 = p - 2 = 0 - 2 = -2 \qquad \text{yes}$$

Hence, k does indeed equal $p - 2$, and the equation is reducible to Bessel's equation

$$\alpha = \frac{2}{2 - p + j} = \frac{2}{2 - 0 + 1} = \frac{2}{3}$$

$$\beta = \frac{1 - p}{2 - p + j} = \frac{1 - 0}{2 - 0 + 1} = \frac{1}{3}$$

$$n = \frac{\sqrt{(1-p)^2 - 4b}}{2 - p + j} = \frac{\sqrt{(1-0)^2 - 4(-6)}}{2 - 0 + 1} = \frac{\sqrt{25}}{3} = \frac{5}{3}$$

$$\omega = \alpha\sqrt{a} = \frac{2}{3}\sqrt{3} = \frac{2\sqrt{3}}{3}$$

$$\frac{\beta}{\alpha} = \frac{1/3}{2/3} = \frac{1}{2}$$

The solution is (n is not an integer)

$$y = x^{\beta/\alpha} C_1 J_n(\omega x^{1/\alpha}) + C_2 J_{-n}(\omega x^{1/\alpha})]$$

or

$$y = \sqrt{x}\left[C_1 J_{5/3}\left(\frac{2\sqrt{3}}{3}x^{3/2}\right) + C_2 J_{-5/3}\left(\frac{2\sqrt{3}}{3}x^{3/2}\right)\right]$$

A.3 FORMAL SOLUTION OF THE BESSEL EQUATION

A.3.1 Bessel Function of the First Kind

Consider the Bessel equation

$$x^2\frac{d^2y}{dx^2} + x\frac{dy}{dx} + (x^2 - n^2)y = 0 \tag{A.4}$$

known as the *Bessel equation of order n*.

To find the solution of eq. (A.4), the method of Frobenius is used by assuming that the dependent variable y is in the form of a power series in the independent variable x. Let this representation be

$$y = x^p(a_0 + a_1x + a_2x^2 + a_3x^3 + \cdots) \tag{A.30}$$

The first and second derivatives of y with respect to x are therefore

$$\frac{dy}{dx} = px^{p-1}(a_0 + a_1x + a_2x^2 + \cdots) + x^p(a_1 + 2a_2x + 3a_3x^2 + \cdots)$$

and

$$\begin{aligned}\frac{d^2y}{dx^2} &= p(p-1)x^{p-2}(a_0 + a_1x + a_2x^2 + a_3x^3 + \cdots)\\ &\quad + px^{p-1}(a_1 + 2a_2x + 3a_3x^2 + \cdots)\\ &\quad + px^{p-1}(a_1 + 2a_2x + 3a_3x^2 + \cdots) + x^p(2a_2 + 6a_3x + \cdots)\end{aligned}$$

The power series representation of y and its derivatives may be substituted into the original eq. (A.4). This substitution causes the second derivative to be multiplied by x^2, the first derivative to be multiplied by x, and the function itself to be multiplied by $(x^2 - n^2)$. If the substitutions are made, the expression for the original differential equation in terms of a power series in x will resolve itself into the ordered array shown in Table A.1.

The coefficient of each power of x may then be determined.

$$\begin{aligned} 0 = &\left[p(p-1) + p - n^2\right] a_0 x^p \\ &+ \left[p(p-1) + 3p + 1 - n^2\right] a_1 x^{p+1} \\ &+ \left\{\left[p(p-1) + 5p + 4 - n^2\right] a_2 + a_0\right\} x^{p+2} \\ &+ \left\{\left[p(p-1) + 7p + 9 - n^2\right] a_3 + a_1\right\} x^{p+3} \\ &+ \left\{\left[p(p-1) + 9p + 16 - n^2\right] a_4 + a_2\right\} x^{p+4} \\ &+ \cdots \end{aligned} \tag{A.31}$$

Equation (A.31) will be true when the coefficient of each power of x is equal to zero. Take the coefficient of the x^p term. This is the *indicial equation*

$$\left[p(p-1) + p - n^2\right] a_0 = 0$$

Assuming that $a_0 \neq 0$,

$$p^2 - p + p - n^2 = p^2 - n^2 = 0$$

from which

$$p = \pm n$$

Let $p = +n$. Then the coefficient of the x^{p+1} term may be found.

$$\begin{aligned} \left[p(p-1) + 3p + 1 - n^2\right] a_1 &= 0 \\ \left[n(n-1) + 3n + 1 - n^2\right] a_1 &= 0 \\ (1 + 2n) a_1 &= 0 \\ a_1 &= 0 \end{aligned}$$

TABLE A.1 Ordered Array of Coefficients

	x^p	x^{p+1}	x^{p+2}	x^{p+3}	x^{p+4}
x^2	$p(p-1)a_0$	$p(p-1)a_1$ $+2pa_1$	$p(p-1)a_2$ $+4pa_2$ $+2a_2$	$p(p-1)a_3$ $+6pa_3$ $+6a_3$	$p(p-1)a_4$ $+8pa_4$ $+12a_4$
$x\dfrac{dy}{dx}$	pa_0	$pa_1 + a_1$	$pa_2 + 2a_2$	$pa_3 + 3a_3$	$pa_4 + 4a_4$
$x^2 y$	—	—	a_0	a_1	a_2
$n^2 y$	$-n^2 a_0$	$-n^2 a_1$	$-n^2 a_2$	$-n^2 a_3$	$-n^2 a_4$

Next, determine the coefficient of the x^{p+2} term again with $p = +n$.

$$\left[p(p-1) + 5p + 4 - n^2\right] a_2 + a_0 = 0$$
$$\left[n(n-1) + 5n + 4 - n^2\right] a_2 + a_0 = 0$$
$$(4n+4)a_2 + a_0 = 0$$
$$a_2 = \frac{-a_0}{4(n+1)}$$

The coefficients of the x^{p+3} and the x^{p+4} terms may be used in similar fashion to determine that

$$a_3 = 0$$

and

$$a_4 = \frac{-a_2}{8n+16} = \frac{-a_2}{4(2n+4)}$$

This suggests the general *recurrence relationship* for the coefficients, namely,

$$a_k = \frac{-a_{k-2}}{k(2n+k)} \tag{A.32}$$

which shows that every coefficient a_k is proportional to the second coefficient preceding it, a_{k-2}. Because it has been proved that a_1 is equal to zero, it is seen that every coefficient with an odd subscript must disappear. This is true even in the specific case for $n = -\frac{1}{2}$ because the generality is not altered if a_1 is still assumed to be equal to zero.

All of the coefficients can be determined through use of the recurrence relationship of eq. (A.32), by selecting the first coefficient and determining all of the others in terms of the first. Let $a_0 = a_0$. Then

$$a_2 = \frac{-a_0}{2(2n+2)} = \frac{-a_0}{4(n+1)} = \frac{-a_0}{2^2 \cdot 1!\,(n+1)}$$

$$a_4 = \frac{-a_2}{4(2n+4)} = \frac{-a_2}{8(n+2)} = \frac{a_0}{8 \cdot 2^2 \cdot 1!\,(n+2)(n+1)}$$
$$= \frac{a_0}{2^4 \cdot 2!(n+2)(n+1)}$$

$$a_6 = \frac{-a_4}{6(2n+6)} = \frac{-a_4}{12(n+3)} = \frac{-a_0}{2^6 \cdot 3!\,(n+3)(n+2)(n+1)}$$

Then

$$a_k = \frac{(-1)^{k+1} a_0}{2^k (k/2)!\,(n+k/2)\cdots(n+2)(n+1)}$$

or if $k = 2m$,

$$a_{2m} = \frac{(-1)^m a_0}{2^{2m} m!\,(n+m)\cdots(n+2)(n+1)} \tag{A.33}$$

with $m = 1, 2, 3, 4, \ldots$.

Both the numerator and denominator of eq. (A.33) can be multiplied by 2^n, for a reason that will be apparent presently. This multiplication results in the following:

$$a_{2m} = \frac{(-1)^m 2^n a_0}{2^{2m} 2^n m!(n+m)\cdots(n+2)(n+1)} \tag{A.34}$$

In addition, the factors $(n+m)\cdot(n+m-1)\cdots(n+3)\cdot(n+2)\cdot(n+1)$ suggest a factorial if the numerator and denominator of eq. (A.34) were to be multiplied by $n!$. However, n is not necessarily an integer, and therefore the generalization of $n!$, namely, the gamma function, is used. The numerator and denominator are therefore multiplied by $\Gamma(n+1)$, to give

$$a_{2m} = \frac{(-1)^m 2^n \Gamma(n+1) a_0}{2^{2m+n} m!(n+m)\cdots(n+2)(n+1)\Gamma(n+1)} \tag{A.35}$$

Substitution of eq. (A.2) into eq. (A.35) gives

$$a_{2m} = \frac{(-1)^m 2^n \Gamma(n+1) a_0}{2^{2m+n} m!\,\Gamma(m+n+1)}$$

A selection may be made for a_0:

$$a_0 = \frac{1}{2^n \Gamma(n+1)}$$

a relationship that is quite arbitrary but which need cause no concern because particular solutions of the Bessel equation will eventually be sought.

Using these equations, the value of a_{2m} is established.

$$a_{2m} = \frac{(-1)^m}{2^{2m+n} m!\,\Gamma(n+m+1)} \tag{A.36}$$

Thus one solution of the differential equation is

$$y = x^n \left[\frac{1}{2^n \Gamma(n+1)} - \frac{x^2}{2^{n+2}\Gamma(n+2)} + \frac{x^4}{2^{n+4} 2!\,\Gamma(n+3)} \cdots \right]$$

or

$$y = \sum_{m=0}^{\infty} \frac{(-1)^m x^{2m+n}}{2^{2m+n} m!\,\Gamma(n+m+1)} = J_n(x) \tag{A.37}$$

a function already defined as the Bessel function of the first kind of order n.

If the function $J_n(x)$ defined by eq. (A.37) is convergent for all values of x, it is indeed a unique solution of the Bessel equation. To determine this region of convergence, the Cauchy ratio test is employed by taking the ratio of the $(m+1)$th term to the mth term. Hence

$$R = \left|\frac{m+1}{m}\right| = \frac{(-1)^{m+1}x^{2m+n+2}/2^{2m+n+2}(m+1)!\,\Gamma(n+m+2)}{(-1)^m x^{2m+n}/2^{2m+n}m!\,\Gamma(n+m+1)}$$

or

$$R = \frac{-x^2}{2^2(m+1)(n+m+1)}$$

In the limit, as m approaches infinity, R approaches zero and the infinite series of eq. (A.37) is convergent for all values of x.

A.3.2 Bessel Function of the Second Kind

Equation (A.4) is a second-order differential equation with two independent solutions. In seeking the second solution, $J_n(x)$ being one solution, it can reasonably be assumed that $p = -n$ and an attempt can be made to determine whether $J_{-n}(x)$ is the second solution. Substituting $p = -n$ into eq. (A.37), it is found that all series expansions for the Bessel function of the first kind become

$$J_{-n}(x) = \sum_{m=0}^{\infty} \frac{(-1)^m x^{2m-n}}{2^{2m-n}m!\,\Gamma(m-n+1)} \tag{A.38}$$

This clearly shows that $J_{-n}(x)$ is a solution of the differential equation if n is not an integer. If n is an integer, $\Gamma(m+n+1)$ is undefined, and eq. (A.38) does not represent the second solution. Moreover, because $J_{-n}(x)$ is seen to contain some negative powers of x, whereas $J_n(x)$ does not, it is obvious that $J_{-n}(x)$ and $J_n(x)$ cannot be proportional. Therefore, as long as n is not an integer, the solution will be

$$y = C_1 J_n(x) + C_2 J_{-n}(x)$$

where C_1 and C_2 are arbitrary constants.

However, when n is an integer, a second independent solution can be found by assuming that

$$y = f(x)J_n(x) = y(x)$$

and then determining the value of $f(x)$.

Begin by differentiating $y(x)$ twice, again denoting derivatives by primes:

$$y'(x) = f(x)J_n'(x) + f'(x)J_n(x)$$

$$y''(x) = f(x)J_n''(x) + 2f'(x)J_n'(x) + f''(x)J_n(x)$$

These derivatives and the function itself are substituted into the original equation, which is eq. (A.4):

$$x^2[f(x)J_n''(x) + 2f'(x)J_n'(x) + f''(x)J_n(x)]$$
$$+\,x[f(x)J_n'(x) + f'(x)J_n(x)] + (x^2 - n^2)[f(x)J_n(x)] = 0$$

Rearranging yields

$$\left[x^2 J_n(x)\right] f''(x) + \left[2x^2 J_n'(x) + x J_n(x)\right] f'(x)$$
$$+ \left[x^2 J_n''(x) + x J_n'(x) + (x^2 - n^2) J_n(x)\right] f(x) = 0 \qquad \text{(A.39)}$$

The coefficient of the term $f(x)$ reduces to zero because $J_n(x)$ is itself a solution of the Bessel equation of order n. In other words, if $J_n(x)$ is one solution of the equation

$$x^2 \frac{d^2 y}{dx^2} + x \frac{dy}{dx} + (x^2 - n^2) y = 0$$

then

$$x^2 J_n''(x) + x J_n'(x) + (x^2 - n^2) J_n(x) = 0$$

Only the first two terms of eq. (A.39) remain:

$$[x^2 J_n(x)] f''(x) + [2x^2 J_n'(x) + x J_n(x)] f'(x) = 0 \qquad \text{(A.40)}$$

which can be solved by first separating the variables and then integrating.

First, rewrite eq. (A.40),

$$x^2 J_n(x) \frac{d}{dx}[f'(x)] + [2x^2 J_n'(x) + x J_n(x)] f'(x) = 0$$

and then separate the variables,

$$\frac{df'(x)}{f'(x)} + \left[\frac{2J_n'(x)}{J_n(x)} + \frac{1}{x}\right] dx = 0$$

Then, by integration

$$\ln f'(x) + 2 \ln J_n(x) + \ln x = \ln A$$

where A is an arbitrary constant. Therefore,

$$f'(x) = \frac{A}{x[J_n(x)]^2}$$

and

$$f(x) = A \int \frac{dx}{x[J_n(x)]^2} + B$$

where B is another arbitrary constant.

The value of the second independent solution can now be determined as

$$y(x) = f(x) J_n(x)$$

or

$$y = AJ_n(x)\int \frac{dx}{x[J_n(x)]^2} + BJ_n(x) \tag{A.41}$$

which is, of course, eq. (A.7*a*).

Refer to the original expansion for $J_n(x)$, as given by eq. (A.6*a*), where note can be taken of the leading term. When $m = 0$, the leading term contains x^n, which when squared and multiplied by x, yields x^{2n+1}. The denominator of the integrand in eq. (A.41) therefore contains x^{2n+1}, which, when integrated, yields $(-1/2n)x^{2n}$, or $\ln x$, dependent on the value of n, that is, whether n does or does not equal zero. In either case, at $x = 0$, the leading term of eq. (A.41) is infinite. Thus the second independent solution, as established by eq. (A.41), is not proportional to, and is independent of, the first. This second independent solution, as stated previously, is the Bessel function of the second kind.

A.3.3 Modified Bessel Functions

Throughout the development of the Bessel functions of the first and second kind, the solutions of

$$x^2\frac{d^2y}{dx^2} + x\frac{dy}{dx} + (x^2 - n^2)y = \tag{A.4}$$

have been considered. However, the Bessel equation may take the form

$$x^2\frac{d^2y}{dx^2} + x\frac{dy}{dx} - (x^2 + n^2)y = 0 \tag{A.8a}$$

Recognizing that $\imath = \sqrt{-1}$, eq. (A.8*a*) can be adjusted to

$$x^2\frac{d^2y}{dx^2} + \frac{dy}{dx} + (\imath^2x^2 - n^2)y = 0 \tag{A.8b}$$

This equation must have as solutions, *when n is not an integer*,

$$y = C_1J_n(\imath x) + C_2J_{-n}(\imath x)$$

and *when n is an integer*,

$$y = C_1J_n(\imath x) + C_2Y_n(\imath x)$$

$J_n(\imath x)$ is a function of an imaginary variable and is not equal to $J_n(x)$, a function of a real variable. The solution of eqs. (A.8) is not the same as the solution of eq. (A.4). A representation of $J_n(\imath x)$ must be found to make it a function of the real variable x. For this purpose, the infinite series representing $J_n(x)$ as given by eq. (A.6*a*) is written in terms of the imaginary variable $\imath x$:

$$J_n(\imath x) = \sum_{m=0}^{\infty} \frac{(-1)^m(\imath x)^{2m+n}}{2^{2m+n}m!\,\Gamma(n+m+1)}$$

$$= \sum_{m=0}^{\infty} \frac{(\imath^2)^m (\imath)^{2m+n} x^{2m+n}}{2^{2m+n} m!\, \Gamma(n+m+1)}$$

$$= \sum_{m=0}^{\infty} \frac{\imath^{4m} \imath^n x^{2m+n}}{2^{2m+n}!\, \Gamma(n+m+1)}$$

Because m can take on integer values only and merely indicates a particular term in the series expansion, i^{4m} must always equal unity. Furthermore, $\imath^n$ can be removed from within the summation. Thus

$$J_n(\imath x) = \imath^n \sum_{m=0}^{\infty} \frac{x^{2m+n}}{2^{2m+n} m!\, \Gamma(n+m+1)}$$

Making a small algebraic adjustment, a series of expansion for the *modified Bessel function of the first kind of order n*, $I_n(x)$ is obtained:

$$(\imath)^{-n} J_n(\imath x) = I_n(x) = \sum_{m=0}^{\infty} \frac{x^{2m+n}}{2^{2m+n} m!\, \Gamma(n+m+1)} \qquad \text{(A.42)}$$

Recognizing $J_n(\imath x)$ to be a solution of eq. (A.8*b*), and seeking only particular solutions of the Bessel equation, it is certain that a solution multiplied by a constant is still a solution. The solution to eq. (A.8*b*) may therefore also be written if n is not an integer:

$$y = C_1 I_n(x) + C_2 I_{-n}(x)$$

When n is an integer, it is seen that eq. (A.42) shows $J_{-n}(x)$ to be undefined and that the denominators of the first n terms of the series expansion are infinite. Then these terms are not present, and the expansion For $J_{-n}(x)$ really begins at the point where $m = n$. Hence

$$J_{-n}(x) = \sum_{m=n}^{\infty} \frac{(-1)^m x^{2m-n}}{2^{2m-n} m!\, \Gamma(m-n+1)}$$

which is eq. (A.38). Let h' be a substitution term so that $h' = m - n$. Then $m = h' + n$ and

$$J_{-n}(x) = \sum_{h'=0}^{\infty} \frac{(-1)^{h'+n} x^{2(h'+n)-n}}{2^{2(h'+n)-n} (h'+n)!\, \Gamma(h'+n-n+1)}$$

$$J_{-n}(x) = (-1)^n \sum_{h'=0}^{\infty} \frac{(-1)^{h'} x^{2h'+n}}{2^{2h'+n} h'!\, \Gamma(h'+n+1)}$$

and

$$J_{-n}(x) = (-1)^n I_n(x) \qquad \text{(A.43)}$$

Equation (A.43) is further proof that when n is an integer, the second solution of eq. (A.8b) is proportional to the first. Furthermore, when n is an integer,

$$J_{-n}(\imath x) = (-1)^n J_n(\imath x)$$

$$\imath^{-n} J_{-n}(\imath x) = (-1)^n [\imath^{-n} J_n(\imath x)]$$

$$\imath^{-2n}[\imath^n J_{-n}(\imath x)] = (-1)^n [\imath^{-n} J_n(\imath x)]$$

$$\imath^{-2n} I_{-n}(x) = (-1)^n J_n(x)$$

$$I_{-n}(x) = I_n(x) \tag{A.44}$$

so that a second independent solution of eq. (A.8b) must now be found. The second solution of eq. (A.8b) is deduced from $I_n(x)$ in the same manner as $Y_n(x)$ was deduced from $J_n(x)$. Again, one assumes a relationship of the form

$$y = f(x) I_n(x) = y(x)$$

and establishes

$$y(x) = A I_n(x) \int \frac{dx}{x[I_n(x)]^2} + B I_n(x) \tag{A.45}$$

where A and B are again arbitrary constants. Equation (A.45) defines *the modified Bessel function of the second kind of order n*, which is usually indicated in the literature by $K_n(x)$, as stated previously.

A.4 RECURRENCE RELATIONSHIPS

Many basic identities and recurrence relationships are associated with the Bessel functions. Several of these are given without proof in Table A.2.

Example A.2: Solution Involving Recurrence Relationships. If $J_0(2.0) = 0.2239$ and $J_1(2.0) = 0.5767$, determine $J_4(2.0)$

SOLUTION. Refer to Table A.2 and use relationship 6 with $n = 3$.

$$J_{n+1}(x) = \frac{2n}{x} J_n(x) - J_{n-1}(x)$$

$$J_4(x) = \frac{6}{x} J_3(x) - J_2(x)$$

Now, with $n = 2$ and then $n = 1$,

$$J_4(x) = \frac{6}{x}[J_{2+1}(x)] - J_{1+1}(x)$$

$$= \frac{6}{x}\left[\frac{4}{x} J_2(x) - J_1(x)\right] - \left[\frac{2}{x} J_1(x) - J_0(x)\right]$$

$$= \frac{24}{x^2} J_2(x) - \frac{6}{x} J_1(x) - \frac{2}{x} J_1(x) + J_0(x)$$

$$= \frac{48}{x^3} J_1(x) - \frac{24}{x^2} J_0(x) - \frac{8}{x} J_1(x) + J_0(x)$$

$$= \left(\frac{48}{x^3} - \frac{8}{x}\right) J_1(x) + \left(1 - \frac{24}{x^2}\right) J_0(x)$$

With $x = 2.0$ and $J_0(x)$ and $J_1(x)$ as given,

$$J_4(2.0) = \left(\frac{48}{2^3} - \frac{8}{2}\right)(0.5767) + \left(1 - \frac{24}{2^2}\right)(0.2239) = 0.0339$$

TABLE A.2 Recurrence Relationships

1. $\frac{d}{dx}[x^{n+1} J_{n+1}(x)] = x^{n+1} J_n(x)$	14. $x\frac{d}{dx}[Y_n(x)] = -nY_n(x) + xY_{n-1}(x)$
2. $\frac{d}{dx}[x^{-n} J_n(x)] = -x^{-n} J_{n+1}(x)$	15. $\frac{d}{dx}[Y_n(x)] = \frac{1}{2}[Y_{n-1}(x) - Y_{n+1}(x)]$
3. $\frac{d}{dx}[J_n(x)] = J_{n-1}(x) - \frac{n}{x} J_n(x)$	16. $\frac{2n}{x} Y_n(x) = Y_{n+1}(x) + Y_{n-1}(x)$
4. $\frac{d}{dx}[J_n(x)] = \frac{n}{x} J_n(x) - J_{n+1}(x)$	17. $\frac{d}{dx}[J_0(x)] = -J_1(x)$
5. $\frac{d}{dx}[J_n(x)] = \frac{1}{2}[J_{n-1}(x) - J_{n+1}(x)]$	18. $\frac{d}{dx}[Y_0(x)] = -Y_1(x)$
6. $J_{n+1}(x) = \frac{2n}{x} J_n(x) - J_{n-1}(x)$	19. $\int x^{-n} Y_{n+1}(x)\, dx = -x^{-n} Y_n(x) + C$
7. $J_{n-1}(x) = \frac{2n}{x} J_n(x) - J_{n+1}(x)$	20. $\int x^n Y_{n-1}(x)\, dx = -x^n Y_n(x) + C$
8. $J_n(x) = \frac{x}{2n}[J_{n+1}(x) + J_{n-1}(x)]$	21. $\frac{d}{dx}[I_n(x)] = I_{n-1}(x) - \frac{n}{x} I_n(x)$
9. $\int x^{n+1} J_n(x)\, dx = x^{n+1} J_{n+1}(x) + C$	22. $\frac{d}{dx}[I_n(x)] = \frac{1}{2}[I_{n-1}(x) + I_{n+1}(x)]$
10. $\int x^{-n} J_{n+1}(x)\, dx = -x^{-n} J_n x) + C$	23. $\frac{d}{dx}[I_n(x)] = \frac{n}{x}[I_n(x) + I_{n+1}(x)]$
11. $x\frac{d}{dx}[J_n(x)] = nJ_n(x) - xJ_{n+1}(x)$	24. $\frac{2n}{x} I_n(x) = I_{n-1}(x) - I_{n+1}(x)$
12. $x\frac{d}{dx}[Y_n(x)] = nY_n(x) - xY_{n+1}(x)$	25. $\frac{d}{dx}[I_0(x)] = I_1(x)$
13. $x\frac{d}{dx}[J_n(x)] = -nJ_n(x) + xJ_{n-1}(x)$	26. $\frac{d}{dx}[K_0(x)] = -K_1(x)$

A.5 POLYNOMIAL APPROXIMATIONS FOR THE BESSEL FUNCTIONS

Abramowitz and Stegun (1964) also provide polynomial approximations for $J_0(x)$, $J_1(x)$, $Y_0(x)$, $Y_1(x)$, $I_0(x)$, $I_1(x)$, $K_0(x)$, and $K_1(x)$. These are presented here in a manner that differs from their source, and they depend on the use of the following forms:

Form A

$$u_a(x) = \alpha_0 + \alpha_1 t^2 + \alpha_2 t^4 + \alpha_3 t^6 + \alpha_4 t^8 + \alpha_5 t^{10} + \alpha_6 t^{12} + \epsilon \tag{A.46}$$

Form B

$$u_b(x) = \frac{2}{\pi} \ln\left(\frac{1}{2}x\right) J_0(x) + u_a(x) \tag{A.47}$$

Form C

$$u_c(x) = x^{-1/2} \beta(t) \cos\theta(t) \tag{A.48}$$

with

$$\beta(t) = \beta_0 + \beta_1 t + \beta_2 t^2 + \beta_3 t^3 + \beta_4 t^4 + \beta_5 t^5 + \beta_6 t^6 + \epsilon \tag{A.49}$$

and

$$\theta(t) = x + \theta_0 + \theta_1 t + \theta_2 t^2 + \theta_3 t^3 + \theta_4 t^4 + \theta_5 t^5 + \theta_6 t^6 + \epsilon \tag{A.50}$$

Form D

$$u_d(x) = x^{-1/2} \beta(t) \sin\theta(t) \tag{A.51}$$

where $\beta(t)$ and $\theta(t)$ are given, respectively, by eqs. (A.49) and (A.50).

Form E

$$u_e(x) = \gamma_0 + \gamma_1 t + \gamma_2 t^2 + \gamma_3 t^3 + \gamma_4 t^4 + \gamma_5 t^5 + \gamma_6 t^6 + \gamma_7 t^7 + \gamma_8 t^8 \tag{A.52}$$

Form F

$$u_f(x) = \frac{e^{-x}}{\sqrt{x}} \beta(t) \tag{A.53}$$

where $\beta(t)$ is given by eq. (A.49).

A.5.1 $J_0(x)$

In the range $-3 \le x \le 3$, use form A with

$$t = \frac{x}{3}$$

and

$$\alpha_0 = 1.0000000$$
$$\alpha_1 = -2.2499997$$
$$\alpha_2 = 1.2656208$$
$$\alpha_3 = -0.3163866$$
$$\alpha_4 = 0.0444479$$
$$\alpha_5 = -0.0039444$$
$$\alpha_6 = 0.0002100$$

with thc absolute value of the error

$$\epsilon < 5 \times 10^{-8}$$

In the range $3 \leq x < \infty$, use form C with

$$t = \frac{3}{x}$$

For $\beta(t)$, use

$$\beta_0 = 0.79788456$$
$$\beta_1 = -0.00000077$$
$$\beta_2 = -0.00552740$$
$$\beta_3 = -0.00009512$$
$$\beta_4 = 0.00137237$$
$$\beta_5 = -0.00072805$$
$$\beta_6 = 0.00014476$$

with the absolute value of the error

$$\epsilon < 1.6 \times 10^{-8}$$

and for $\theta(t)$, use

$$\theta_0 = -0.78539816$$
$$\theta_1 = -0.04166397$$
$$\theta_2 = -0.00003954$$
$$\theta_3 = 0.00262573$$
$$\theta_4 = -0.00054125$$

$$\theta_5 = -0.00029333$$

$$\theta_6 = 0.00013558$$

with the absolute value of the error

$$\epsilon < 7 \times 10^{-8}$$

A.5.2 $J_1(x)$

In the range $-3 \le x \le 3$, first obtain $u_a(x)$ from form A with

$$t = \frac{x}{3}$$

and

$$\alpha_0 = 1.00000000$$

$$\alpha_1 = -0.56249985$$

$$\alpha_2 = 0.21093573$$

$$\alpha_3 = -0.03954289$$

$$\alpha_4 = 0.00443319$$

$$\alpha_5 = -0.00031761$$

$$\alpha_6 = 0.00001109$$

with the absolute value of the error

$$\epsilon < 1.3 \times 10^{-8}$$

Then determine $J_1(x)$ from

$$J_1(x) = \left[u_a(x) - \frac{1}{2} \right] x$$

In the range $3 \le x < \infty$, use form C with

$$t = \frac{3}{x}$$

For $\beta(t)$, use

$$\beta_0 = 0.79788456$$

$$\beta_1 = 0.00000156$$

$$\beta_2 = 0.01659667$$

$$\beta_3 = 0.00017105$$

$$\beta_4 = -0.00249511$$

$$\beta_5 = 0.00113653$$

$$\beta_6 = -0.00020033$$

with the absolute value of the error

$$\epsilon < 4 \times 10^{-8}$$

For $\theta(t)$, use

$$\theta_0 = -2.35619449$$

$$\theta_1 = 0.12499612$$

$$\theta_2 = 0.00005650$$

$$\theta_3 = -0.00637879$$

$$\theta_4 = 0.00074348$$

$$\theta_5 = 0.00079824$$

$$\theta_6 = -0.00029166$$

with the absolute value of the error

$$\epsilon < 9 \times 10^{-8}$$

A.5.3 $Y_0(x)$

In the range $0 < x < 3$, determine $J_0(x)$ and then use

$$Y_0(x) = u_b(x) = \frac{2}{\pi} \ln\left(\frac{1}{2}x\right) J_0(x) + u_a(x)$$

where in form A,

$$t = \frac{x}{3}$$

and

$$\alpha_0 = 0.36746691$$

$$\alpha_1 = 0.60559366$$

$$\alpha_2 = -0.74350384$$

$$\alpha_3 = 0.25300117$$

$$\alpha_4 = -0.04261214$$

$$\alpha_5 = 0.00427916$$

$$\alpha_6 = -0.00024846$$

with the absolute value of the error

$$\epsilon < 1.4 \times 10^{-8}$$

In the range $3 \le x < \infty$, use form D with

$$t = \frac{3}{x}$$

The values of ϵ, β, and θ are identical to those used for $J_0(x)$ in Section A.5.1.

A.5.4 $Y_1(x)$

In the range $0 \le x \le 3$, determine $J_1(x)$ and then use

$$Y_1(x) = \frac{1}{x}\left[\frac{2}{\pi}x \ln\left(\frac{1}{2}x\right) J_1(x) - u_a(x)\right]$$

where in form A,

$$t = \frac{x}{3}$$

and

$$\alpha_0 = -0.6366198$$

$$\alpha_1 = 0.2212091$$

$$\alpha_2 = 2.1682709$$

$$\alpha_3 = -1.3164827$$

$$\alpha_4 = 0.3123951$$

$$\alpha_5 = -0.0400976$$

$$\alpha_6 = 0.0027873$$

with the absolute value of the error

$$\epsilon < 1.1 \times 10^{-7}$$

In the range $3 \le x < \infty$, use form D with

$$t = \frac{3}{x}$$

The values of ϵ, β, and θ are identical to those used for $J_1(x)$ in Section A.5.2.

A.5.5 $I_0(x)$

In the range $-3.75 \le t \le 3.75$, use form A with

$$t = \frac{t}{3.75}$$

and

$$\alpha_1 = 1.0000000$$

$$\alpha_2 = 3.5156229$$

$$\alpha_3 = 3.0899424$$

$$\alpha_4 = 1.2067492$$

$$\alpha_5 = 0.2659732$$

$$\alpha_6 = 0.0360768$$

$$\alpha_7 = 0.0045813$$

with the absolute value of the error

$$\epsilon < 1.6 \times 10^{-7}$$

In the range $3.75 \le x < \infty$, use form E in

$$I_0(x) = \frac{e^x}{\sqrt{x}} u_e(x)$$

with

$$t = \frac{3.75}{x}$$

and

$$\gamma_0 = 0.39894228$$

$$\gamma_1 = 0.01328592$$

$$\gamma_2 = 0.00225319$$

$$\gamma_3 = -0.00157565$$

$$\gamma_4 = 0.00916281$$

$$\gamma_5 = -0.02057706$$

$$\gamma_6 = 0.02635537$$

$$\gamma_7 = -0.01647633$$

$$\gamma_8 = 0.00392377$$

with the absolute value of the error

$$\epsilon < 1.9 \times 10^{-7}$$

A.5.6 $I_1(x)$

In the range $-3.75 \leq x \leq 3.75$, use form A in

$$I_1(x) = xu_a(x)$$

with

$$t = \frac{x}{3.75}$$

and

$$\alpha_0 = 0.50000000$$

$$\alpha_1 = 0.87890594$$

$$\alpha_2 = 0.51498869$$

$$\alpha_3 = 0.15084934$$

$$\alpha_4 = 0.02658733$$

$$\alpha_5 = 0.00301532$$

$$\alpha_6 = 0.00032411$$

with the absolute value of the error

$$\epsilon < 8 \times 10^{-9}$$

In the range $3.75 \leq x < \infty$, use form E

$$I_1(x) = \frac{e^x}{\sqrt{x}} u_e(x)$$

with

$$t = \frac{3.75}{x}$$

and

$$\gamma_0 = 0.39894228$$

$$\gamma_1 = -0.03988024$$

$$\gamma_2 = -0.00362018$$

$$\gamma_3 = 0.00163801$$

$$\gamma_4 = -0.01031555$$

$$\gamma_5 = 0.02282967$$

$$\gamma_6 = -0.02895312$$

$$\gamma_7 = 0.01787654$$

$$\gamma_8 = -0.00420059$$

with the absolute value of the error

$$\epsilon < 2.2 \times 10^{-7}$$

A.5.7 $K_0(x)$

In the range $0 \le x \le 2$, determine $I_0(x)$ and use form A in

$$K_0(x) = -\ln t I_o(x) + u_a(x)$$

with

$$t = \frac{x}{2}$$

and

$$\alpha_0 = -0.57721566$$

$$\alpha_1 = 0.42278420$$

$$\alpha_2 = 0.23069756$$

$$\alpha_3 = 0.03488590$$

$$\alpha_4 = 0.00262698$$

$$\alpha_5 = 0.00010750$$

$$\alpha_6 = 0.00000740$$

with the absolute value of the error

$$\epsilon < 1 \times 10^{-8}$$

In the range $2 \le x < \infty$, use form F with

$$t = \frac{2}{x}$$

and

$$\beta_0 = 1.25331414$$

$$\beta_1 = -0.07832358$$

$$\beta_2 = 0.02189568$$

$$\beta_3 = -0.01062446$$

$$\beta_4 = 0.00587872$$

$$\beta_5 = -0.00251540$$

$$\beta_6 = 0.00053208$$

with the absolute value of the error

$$\epsilon < 1.9 \times 10^{-7}$$

A.5.8 $K_1(x)$

In the range $0 \leq x \leq 2$, determine $I_1(x)$ and $u_a(x)$ and use them in

$$K_1(x) = \ln t\, I_1(x) + \frac{u_a(x)}{x}$$

with

$$t = \frac{x}{2}$$

and

$$\alpha_0 = 1.00000000$$

$$\alpha_1 = 0.15443144$$

$$\alpha_2 = -0.67278579$$

$$\alpha_4 = -0.18156897$$

$$\alpha_5 = -0.01919402$$

$$\alpha_6 = -0.00110404$$

$$\alpha_7 = -0.00004686$$

with the absolute value of the error

$$\epsilon < 8 \times 10^{-9}$$

In the range $2 \leq x < \infty$, use form F with

$$t = \frac{2}{x}$$

and

$$\beta_0 = 1.25331414$$

$$\beta_1 = 0.23498619$$

$$\beta_2 = -0.03655620$$

$$\beta_3 = 0.01504268$$

$$\beta_4 = -0.00780353$$

$$\beta_5 = 0.00325614$$

$$\beta_6 = -0.00068245$$

with the absolute value of the error

$$\epsilon < 2.2 \times 10^{-7}$$

A.6 EXAMPLES OF SOLUTIONS TO SELECTED TEMPERATURE EXCESS EQUATIONS

A.6.1 Longitudinal Fin of Triangular Profile

Compare the equation

$$x\frac{d^2\theta}{dx^2} + \frac{d\theta}{dx} - m^2 b\theta = 0 \tag{A.54}$$

with the generalized Bessel equation

$$\frac{d}{dx}\left(x^p \frac{d\theta}{dx}\right) + (ax^j + bx^k)\theta = 0 \tag{A.13}$$

If the comparison indicates that $p = k + 2$ or $b = 0$, then eq. (A.54) is reducible to Bessel's equation. In this case, $b = 0$ and the solution will be

$$\theta = x^{\beta/\alpha}\left[C_1 I_n\left(\omega x^{1/\alpha}\right) + C_2 K_n\left(\omega x^{1/\alpha}\right)\right] \tag{A.55}$$

if n is an integer and a is negative and where the definition and evaluation of α, β, ω, and n in terms of the values in eq. (A.54) are as follows:

$$\alpha = \frac{2}{2-p+j} = \frac{2}{2-1} = 2$$

$$\beta = \frac{1-p}{2-p-j} = \frac{1-1}{2-1} = 0$$

$$\omega = -\iota\alpha\sqrt{a} = 2m\sqrt{b}$$

$$n = \frac{\sqrt{(1-p^2)-4b}}{2-p+j} = 0$$

The general solution to eq. (A.54) is then

$$\theta = C_1 I_0\left(2m\sqrt{bx}\right) + C_2 K_0\left(2m\sqrt{bx}\right) \tag{A.56}$$

A.6.2 Longitudinal Fin of Convex Parabolic Profile

Compare the equation

$$x^{1/2}\frac{d^2\theta}{dx^2} + \frac{1}{2}x^{-1/2}\frac{d\theta}{dx} - m^2\sqrt{b}\theta = 0 \tag{A.57}$$

with the generalized Bessel equation

$$\frac{d}{dx}\left(x^p\frac{d\theta}{dx}\right)+(ax^j+bx^k)\theta=0 \tag{A.13}$$

If the comparison indicates that $p=k+2$ or $b=0$, then eq. (A.57) is reducible to Bessel's equation. In this case, $b=0$ and the equation is reducible. It will have a solution

$$\theta=x^{\beta/\alpha}\left[C_1I_n\left(\omega x^{1/\alpha}\right)+C_2I_{-n}\left(\omega x^{1/\alpha}\right)\right] \tag{A.58}$$

if n is not an integer and a is negative and where the definition and evaluation of $\alpha,\ \beta,\ \omega$, and n in terms of the values in eq. (A.58) are as follows:

$$\alpha=\frac{2}{2-p+j}=\frac{2}{2-\frac{1}{2}}=\frac{4}{3}$$

$$\beta=\frac{1-p}{2-p+j}=\frac{1-\frac{1}{2}}{2-\frac{1}{2}}=\frac{1}{3}$$

$$\omega=-\iota x\sqrt{a}=\tfrac{4}{3}mb^{1/4}$$

$$n=\frac{\sqrt{(1-p)^2-4b}}{2-p+j}=\frac{\sqrt{(\frac{1}{2})^2}}{\frac{3}{2}}=\frac{1}{3}$$

The general solution to eq. (A.57) is then

$$\theta=x^{1/4}\left[C_1I_{1/3}\left(\frac{4}{3}mb^{1/4}x^{3/4}\right)+C_2I_{-1/3}\left(\frac{4}{3}mb^{1/4}x^{3/4}\right)\right] \tag{A.59}$$

A.6.3 Radial Fin of Hyperbolic Profile

Consider the differential equation

$$\frac{d^2\theta}{dr^2}-M^2r\theta=0 \tag{A.60}$$

where

$$M^2=\frac{m^2}{r_0}$$

This may be compared with the generalized Bessel equation

$$\frac{d}{dr}\left(r^p\frac{d\theta}{dr}\right)+\left(ar^j+br^k\right)\theta=0 \tag{A.13}$$

where

$$p=0 \qquad j=1$$
$$a=-M^2 \qquad k=b=0$$

If the comparison indicates that $p = k + 2$ or $b = 0$, then eq. (A.60) is reducible to Bessel's equation and the solution will be

$$\theta = r^{\beta/\alpha}\left[C_1 I_n\left(\omega r^{1/\alpha}\right) + C_2 I_{-n}\left(\omega r^{1/\alpha}\right)\right] \tag{A.61}$$

if n is not an integer and if a is negative.

In this case, $b = 0$, and the definition and evaluation of the constants α, β, ω, and n proceed as follows:

$$a = \frac{2}{2-p+j} = \frac{2}{2-0+1} = \frac{2}{3}$$

$$\beta = \frac{1-p}{2-p+j} = \frac{1-0}{3} = \frac{1}{3}$$

$$\omega = -\iota x\sqrt{a} = -\iota\left(\frac{2}{3}\right)\sqrt{-M^2} = \frac{2}{3}M$$

$$n = \frac{\sqrt{(1-p)^2 - 4b}}{2-p+j} = \frac{\sqrt{1}}{3} = \frac{1}{3}$$

The general solution is therefore

$$\theta = r^{1/2}\left[C_1 I_{1/3}\left(\frac{2}{3}Mr^{3/2}\right) + C_2 I_{-1/3}\left(\frac{2}{3}Mr^{3/2}\right)\right] \tag{A.62}$$

A.6.4 Convex Parabolic Spine

Consider the equation

$$x\frac{d^2\theta}{dx^2} + \frac{d\theta}{dx} - M^2 x^{1/2}\theta = 0 \tag{A.63}$$

and compare it with Bessel's generalized equation

$$\frac{d}{dx}\left(x^p \frac{d\theta}{dx}\right) + \left(ax^j + bx^k\right)\theta = 0 \tag{A.13}$$

If the comparison indicates that $p = k + 2$ $(b = 0)$, then eq. (A.63) is reducible to Bessel's equation and the solution (if a is negative) will be

$$\theta = x^{\beta/\alpha}\left[C_1 I_n\left(\omega x^{1/\alpha}\right) + C_2 K_n\left(\omega x^{1/\alpha}\right)\right] \tag{A.64a}$$

if n is an integer, or

$$\theta = x^{\beta/\alpha}\left[C_1 I_n\left(\omega x^{1/\alpha}\right) + C_2 I_{-n}\left(\omega x^{1/\alpha}\right)\right] \tag{A.64b}$$

if n is not an integer. For the case at hand,

$$p = 1 \qquad j = \tfrac{1}{2}$$
$$a = -M^2 \qquad b = k = 0$$

and because $b = 0$, eq. (A.63) is reducible to Bessel's equation where evaluation of the constants α, β, ω, and n proceeds as follows:

$$\alpha = \frac{2}{2-p+j} = \frac{2}{2-1+\frac{1}{2}} = \frac{2}{3/2} = \frac{4}{3}$$

$$\beta = \frac{1-p}{2-p+j} = \frac{1-1}{2/3} = 0$$

$$\omega = -\iota\alpha\sqrt{a} = -\iota\frac{4}{3}\sqrt{-M^2} = \frac{4}{3}M$$

$$n = \frac{\sqrt{(1-p)^2-4b}}{2-p+j} = \frac{\sqrt{(1-1)^2-0}}{3/2} = 0$$

The general solution to eq. (A.63) may therefore be set down as

$$\theta = C_1 I_0\left(\tfrac{4}{3}Mx^{3/4}\right) + C_2 K_0\left(\tfrac{4}{3}Mx^{3/4}\right) \tag{A.65}$$

because n is an integer.

A.6.5 Conical Spine

Consider the differential equation

$$x^2\frac{d^2\theta}{dx^2} + 2x\frac{d\theta}{dx} - M^2x\theta = 0 \tag{A.66}$$

and compare it with the generalized Bessel equation

$$\frac{d}{dx}\left(x^p\frac{d\theta}{dx}\right) + \left(ax^j + bx^k\right)\theta = 0 \tag{A.13}$$

If the comparison indicates that $p = k+2$ or $b = 0$, then eq. (A.66) is reducible to Bessel's equation and the solution (if a is negative) will be

$$\theta = x^{\beta/\alpha}\left[C_1 I_n\left(\omega x^{1/\alpha}\right) + C_2 I_{-n}\left(\omega x^{1/\alpha}\right)\right] \tag{A.67a}$$

if n is not an integer, and

$$\theta = x^{\beta/\alpha}\left[C_1 I_n\left(\omega x^{1/\alpha}\right) + C_2 K_n\left(\omega x^{1/\alpha}\right)\right] \tag{A.67b}$$

if n is an integer. In this case,

$$p = 2 \qquad j = 1$$
$$a = -M^2 \qquad b = k = 0$$

and because $b = 0$, eq. (A.66) is reducible to Bessel's equation where evaluation of the constants α, β, ω, and n proceeds as follows:

$$\alpha = \frac{2}{2-p+j} = \frac{2}{2-2+1} = \frac{2}{1} = 2$$

$$\beta = \frac{1-p}{2-p+j} = \frac{1-2}{1} = \frac{-1}{1} = -1$$

$$\omega = -\iota\alpha\sqrt{a} = \iota 2\sqrt{-M^2} = 2M$$

$$n = \frac{\sqrt{(1-p)^2+4b}}{2-p+j} = \frac{\sqrt{(1-2)^2-0}}{1} = \frac{\sqrt{1}}{1} = 1$$

The general solution is therefore

$$\theta = \frac{1}{\sqrt{x}}\left[C_1 I_1\left(2M\sqrt{x}\right) + C_2 K_1\left(2M\sqrt{x}\right)\right] \tag{A.68}$$

A.7 NOMENCLATURE

Roman Letter Symbols

A	constant, dimensionless
a	coefficient in polynomial approximation, dimensionless; coefficient in Frobenius method, dimensionless
B	constant, dimensionless
b	coefficient in polynomial approximation, dimensionless; point, dimensionless
C	constant, dimensionless
d	differential or derivative, dimensionless
f	function, dimensionless
I	modified Bessel function of first kind, dimensionless
ι	complex number, $\sqrt{-1}$
J	Bessel function of first kind, dimensionless
K	modified Bessel function of second kind, dimensionless
M	modified fin performance factor, $m^{-3/2}$
m	fin performance parameter, m^{-1}; counter, dimensionless
n	order of Bessel function; positive number, dimensionless
R	ratio, dimensionless
r	radial coordinate, m
t	transformed variable, dimensionless
u	transformed variable, dimensionless
v	transformed variable, dimensionless
x	variable, dimensionless
Y	Bessel function of second kind, dimensionless

Greek Letter Symbols

α	constant in polynomial approximation, dimensionless
β	constant in polynomial approximation, dimensionless
Γ	gamma function, dimensionless
γ	constant in polynomial approximation, dimensionless

ϵ	error in polynomial approximation, dimensionless
θ	function, dimensionless; constant in polynomial approximation, dimensionless
ω	term in solution to Bessel equation

Roman Letter Subscripts

a	form of polynomial approximation
b	form of polynomial approximation
c	form of polynomial approximation
d	form of polynomial approximation
e	form of polynomial approximation
f	form of polynomial approximation

Roman Letter Superscripts

n	nth derivative
j	exponent in general Bessel equation
k	exponent in general Bessel equation
p	exponent in general Bessel equation; exponent in Frobenius method

Greek Letter Superscripts

α	exponent in solution to general Bessel equation
β	exponent in solution to general Bessel equation

Symbolic Superscripts

$'$	first derivative
$''$	second derivative
$'''$	third derivative

APPENDIX B

MATRICES AND DETERMINANTS

B.1 BASIC CONCEPTS

A system of n linear algebraic equations in n unknowns $x_1, x_2, x_3, \ldots, x_n$ such as

$$\begin{array}{llllll} a_{11}x_1 & +a_{12}x_2 & +a_{13}x_3 & +\cdots & +a_{1n}x_n & = y_1 \\ a_{21}x_1 & +a_{22}x_2 & +a_{23}x_3 & +\cdots & +a_{2n}x_n & = y_2 \\ a_{31}x_1 & +a_{32}x_2 & +a_{33}x_3 & +\cdots & +a_{3n}x_n & = y_3 \\ & & \cdots & \cdots & & \\ a_{n1}x_1 & +a_{n2}x_2 & +a_{n3}x_3 & +\cdots & +a_{nn}x_n & = y_n \end{array}$$

can conveniently be represented by the matrix equation

$$\begin{bmatrix} a_{11} & a_{12} & a_{13} & \cdots & a_{1n} \\ a_{21} & a_{22} & a_{23} & \cdots & a_{2n} \\ a_{31} & a_{32} & a_{33} & \cdots & a_{3n} \\ & \cdots & \cdots & \cdots & \\ a_{n1} & a_{n2} & a_{n3} & \cdots & a_{nn} \end{bmatrix} \begin{bmatrix} x_1 \\ x_2 \\ x_3 \\ \cdots \\ x_n \end{bmatrix} = \begin{bmatrix} y_1 \\ y_2 \\ y_3 \\ \cdots \\ y_n \end{bmatrix}$$

or more simply by

$$\mathbf{AX} = \mathbf{Y}$$

where **A** is a rectangular matrix (in this case square) having elements a_{ij} and where **X** and **Y** are column vectors with elements x_i and y_i, respectively. The foregoing representations imply that

$$\sum_{j=1}^{n} a_{ij}x_i = y_i \qquad i = 1, 2, 3, \ldots, n$$

The matrix **A** is called the *coefficient matrix*. If it is desired to associate the elements of **Y** with the coefficient matrix **A**, one may augment **A** and define an augmented matrix

$$\begin{bmatrix} a_{11} & a_{12} & a_{13} & \cdots & a_{1n} & y_1 \\ a_{21} & a_{22} & a_{23} & \cdots & a_{2n} & y_2 \\ a_{31} & a_{32} & a_{33} & \cdots & a_{3n} & y_3 \\ \cdots & \cdots & \cdots & \cdots & & \\ a_{n1} & a_{n2} & a_{n3} & \cdots & a_{nn} & y_n \end{bmatrix}$$

which has n rows and $n+1$ columns. This matrix may be written more simply as the augmented matrix

$$\mathbf{A}^a = [\mathbf{A}|\mathbf{Y}]$$

where the superscript means *augmented* and where the idea of a partitioned matrix is apparent. For example, in the system of linear algebraic equations

$$\begin{aligned} 6x_1 \quad + 4x_2 \quad + x_3 \quad &= \quad 16 \\ 2x_1 \quad + 7x_2 \quad - 2x_3 \quad &= \quad 12 \\ -4x_1 \quad + x_2 \quad + 8x_3 \quad &= \quad -22 \end{aligned}$$

the matrix

$$\begin{bmatrix} 6 & 4 & 1 \\ 2 & 7 & -2 \\ -4 & 1 & 8 \end{bmatrix}$$

is called the coefficient matrix **A** of the system

$$\mathbf{AX} = \mathbf{B}$$

and the matrix

$$\begin{bmatrix} 6 & 4 & 1 & 16 \\ 2 & 7 & -2 & 12 \\ -4 & 1 & 8 & -22 \end{bmatrix}$$

which contains the constant terms, in addition to the elements of **A**, is called the *augmented matrix* of the system. Moreover, the unknowns and the constant terms form two column vectors **X** and **B**.

In the representation

$$\mathbf{AX} = \mathbf{B}$$

A is said to *premultiply* **X** (**A** is a premultiplier) and **X** is said to *postmultiply* **A** (**X** is a postmultiplier).

B.2 MATRIX AND VECTOR TERMINOLOGY

A matrix of order $m \times n$,

$$\begin{bmatrix} a_{11} & a_{12} & a_{13} & \cdots & a_{1n} \\ a_{21} & a_{22} & a_{23} & \cdots & a_{2n} \\ a_{31} & a_{32} & a_{33} & \cdots & a_{3n} \\ & \cdots & \cdots & \cdots & \\ a_{m1} & a_{m2} & a_{m3} & \cdots & a_{mn} \end{bmatrix}$$

is a rectangular ordered array of a total of mn entries arranged in m rows and n columns. The *order* of this matrix is $m \times n$, which is often written as (m, n).

If $m = n$, the matrix is square of order $n \times n$ (or of n or of nth order)

$$\begin{bmatrix} a_{11} & a_{12} & a_{13} & \cdots & a_{1n} \\ a_{21} & a_{22} & a_{23} & \cdots & a_{2n} \\ a_{31} & a_{32} & a_{33} & \cdots & a_{3n} \\ & \cdots & \cdots & \cdots & \\ a_{n1} & a_{n2} & a_{n3} & \cdots & a_{nn} \end{bmatrix}$$

In both rectangular and square matrices, a_{ij} is called the $(i, j)th$ element of **A**. If the matrix is square and $i = j$, the element is said to *define* and be *located on* the *principal diagonal*. The elements $a_{n1}, a_{(n-1),2}, a_{(n-2),3}, \ldots, a_{1n}$ are located on and constitute the *secondary diagonal*.

All elements where $i \neq j$ are considered to be *off-diagonal: subdiagonal* if $i > j$, and *superdiagonal* if $i < j$. The sum of the elements on the principal diagonal of **A** is called the *trace* of **A**:

$$\mathrm{tr}(\mathbf{A}) = \sum_{k=1}^{n} a_{kk}$$

For example, the matrix

$$\mathbf{A} = \begin{bmatrix} 6 & 3 & 0 & 1 \\ -1 & 4 & 1 & 1 \\ -1 & 1 & 8 & -2 \\ 2 & 5 & 2 & 11 \end{bmatrix}$$

is square and is of fourth order (4×4). The elements 6, 4, 8, and 11 constitute the principal diagonal and the elements 2, 1, 1, and 1 constitute the secondary diagonal. The element 1 is the a_{23} element, which lies at the intersection of the second row and third column. The trace of **A** is

$$\mathrm{tr}(\mathbf{A}) = 6 + 4 + 8 + 11 = 29$$

A vector is a matrix containing a single row or a single column. If it is a $1 \times n$ matrix (a matrix of order $1 \times n$), it is a *row vector*:

$$\mathbf{V} = [v_1 \quad v_2 \quad v_3 \quad \cdots \quad v_n]$$

If the vector is an $m \times 1$ vector (order $m \times 1$), it is a *column vector*:

$$\mathbf{V} = \begin{bmatrix} v_1 \\ v_2 \\ v_3 \\ \vdots \\ v_n \end{bmatrix}$$

This concept and the usual one regarding a vector have certain similarities. These similarities are the reason why the elements of a vector are frequently called *components*. However, caution is necessary because the usual three-dimensional space does not imply that m or n (for column or row vectors, respectively) are limited to an upper bound of 3.

B.3 SOME SPECIAL MATRICES

An $m \times n$ matrix such as the one displayed in Section B.2 is called a *null matrix* if every element in the matrix is identically equal to zero. For example, the 3×4 matrix

$$\begin{bmatrix} 0 & 0 & 0 & 0 \\ 0 & 0 & 0 & 0 \\ 0 & 0 & 0 & 0 \end{bmatrix}$$

is null.

The *transpose* of an $m \times n$ matrix is an $n \times m$ matrix with the rows and columns of the original matrix interchanged. For the 3×4 matrix

$$\mathbf{A} = \begin{bmatrix} 4 & 3 & 1 & -2 \\ -2 & 3 & 0 & 1 \\ 1 & -3 & -4 & 2 \end{bmatrix}$$

the transpose is 4×3

$$\mathbf{A}^{\mathrm{T}} = \begin{bmatrix} 4 & -2 & 1 \\ 3 & 3 & -3 \\ 1 & 0 & -4 \\ -2 & 1 & 2 \end{bmatrix}$$

Note the use of the superscript T to indicate the transpose and recognize that the transpose of the transpose is the original matrix

$$[\mathbf{A}^T]^T = \mathbf{A}$$

The nth-order square matrix

$$\begin{bmatrix} a_{11} & a_{12} & a_{13} & \cdots & a_{1n} \\ a_{21} & a_{22} & a_{23} & \cdots & a_{2n} \\ a_{31} & a_{32} & a_{33} & \cdots & a_{3n} \\ \cdots & \cdots & \cdots & & \\ a_{n1} & a_{n2} & a_{n3} & \cdots & a_{nn} \end{bmatrix}$$

is said to be *diagonal* or a *diagonal matrix* if $a_{ij} = 0$ for all $i \neq j$:

$$\mathbf{A} = \begin{bmatrix} a_{11} & 0 & 0 & \cdots & 0 \\ 0 & a_{22} & 0 & \cdots & 0 \\ 0 & 0 & a_{33} & \cdots & 0 \\ \cdots & \cdots & \cdots & & \\ 0 & 0 & 0 & \cdots & a_{nn} \end{bmatrix}$$

If all a_{ij} are equal for all $i = j$ (that is, $a_{ij} = \alpha; i = j$) and $a_{ij} = 0; i \neq j$, the resulting matrix is said to be a *scalar matrix*, which is a diagonal matrix with all elements (principal diagonal elements) equal:

$$\mathbf{A} = \begin{bmatrix} \alpha & 0 & 0 & \cdots & 0 \\ 0 & \alpha & 0 & \cdots & 0 \\ 0 & 0 & \alpha & \cdots & 0 \\ \cdots & \cdots & \cdots & & \\ 0 & 0 & 0 & \cdots & \alpha \end{bmatrix}$$

If all α in the scalar matrix are equal to unity ($\alpha = 1$), the scalar matrix becomes the *identity matrix*:

$$\mathbf{I} = \begin{bmatrix} 1 & 0 & 0 & \cdots & 0 \\ 0 & 1 & 0 & \cdots & 0 \\ 0 & 0 & 1 & \cdots & 0 \\ \cdots & \cdots & \cdots & & \\ 0 & 0 & 0 & \cdots & 1 \end{bmatrix}$$

B.4 MATRIX EQUALITY

A matrix $\mathbf{A} = [a_{ij}]_{m \times n}$ will be equal to a matrix $\mathbf{B} = [b_{ij}]_{m \times n}$ if and only if $a_{ij} = b_{ij}$ for all i and j. This essentially states that two matrices will be equal if and only if they are of the same order and corresponding elements are equal.

B.5 MATRIX ADDITION AND SUBTRACTION

A matrix $\mathbf{A} = [a_{ij}]_{m\times n}$ may be added to a matrix $\mathbf{B} = [b_{ij}]_{m\times n}$ to form a matrix $\mathbf{C} = [c_{ij}]_{m\times n} = [a_{ij} + b_{ij}]_{m\times n}$. This points out that in order to form the sum of two matrices, the matrices must be of the same order and that the elements of the sum are determined by adding the corresponding elements of the matrices forming the sum.

Example B.1. If

$$\mathbf{A} = \begin{bmatrix} 4 & 3 & 2 \\ -6 & 1 & 5 \end{bmatrix}$$

$$\mathbf{B} = \begin{bmatrix} 3 & -1 & 4 \\ 1 & 0 & 3 \end{bmatrix}$$

and

$$\mathbf{C} = \begin{bmatrix} 3 & 2 \\ 4 & -2 \end{bmatrix}$$

find $\mathbf{A} + \mathbf{B}$ and $\mathbf{A} + \mathbf{C}$.

SOLUTION

$$\mathbf{A} + \mathbf{B} = \begin{bmatrix} 4 & 3 & 2 \\ -6 & 1 & 5 \end{bmatrix} + \begin{bmatrix} 3 & -1 & 4 \\ 1 & 0 & 3 \end{bmatrix}$$

$$= \begin{bmatrix} (4+3) & (3-1) & (2+4) \\ (-6+1) & (1+0) & (5+3) \end{bmatrix} = \begin{bmatrix} 7 & 2 & 6 \\ -5 & 1 & 8 \end{bmatrix}$$

The sum $\mathbf{A} + \mathbf{C}$ does not exist because the order of $\mathbf{C}$ does not equal the order of $\mathbf{A}$.

Matrix addition is both commutative and associative:

$$\mathbf{A} + \mathbf{B} = \mathbf{B} + \mathbf{A}$$

$$\mathbf{A} + (\mathbf{B} + \mathbf{C}) = (\mathbf{A} + \mathbf{B}) + \mathbf{C}$$

In addition, the sum $\mathbf{A} + \mathbf{C}$ is equal to the sum $\mathbf{B} + \mathbf{C}$ if and only if $\mathbf{A} = \mathbf{C}$. This is the *cancellation law for addition.*

The matrix $\mathbf{B} = [b_{ij}]_{m\times n}$ may be subtracted from the matrix $\mathbf{A} = [a_{ij}]_{m\times n}$ to form the matrix $\mathbf{D} = [d_{ij}]_{m\times n}$. This indicates that two matrices of the same order may be subtracted by forming the difference between the corresponding elements of the minuend and the subtrahend. Moreover, it is easy to see that if

$$\mathbf{A} + \mathbf{B} = \mathbf{C}$$

then

$$\mathbf{A} = \mathbf{C} - \mathbf{B}$$

Finally, it may be observed that a square matrix possesses a unique decomposition into a sum of a subdiagonal, a diagonal and a superdiagonal matrix. For example,

$$\mathbf{A} = \begin{bmatrix} 1 & 2 & 3 \\ 4 & 5 & 6 \\ 9 & 8 & 7 \end{bmatrix} = \begin{bmatrix} 0 & 0 & 0 \\ 4 & 0 & 0 \\ 9 & 8 & 0 \end{bmatrix} + \begin{bmatrix} 1 & 0 & 0 \\ 0 & 5 & 0 \\ 0 & 0 & 7 \end{bmatrix} + \begin{bmatrix} 0 & 2 & 3 \\ 0 & 0 & 6 \\ 0 & 0 & 0 \end{bmatrix}$$

B.6 MATRIX MULTIPLICATION

A matrix may be multiplied by a scalar or by another matrix. If $\mathbf{A} = [a_{ij}]$ and α is a scalar, then

$$\alpha\mathbf{A} = [\alpha a_{ij}]$$

This shows that multiplication by a scalar is commutative and that multiplication by a scalar involves the multiplication of each and every element of the matrix by the scalar. In addition, it is easy to see that

$$(\alpha + \beta)\mathbf{A} = \alpha\mathbf{A} + \beta\mathbf{A}$$

$$\alpha(\mathbf{A} + \mathbf{B}) = \alpha\mathbf{A} + \alpha\mathbf{B}$$

and

$$\alpha(\beta\mathbf{A}) = (\alpha\beta)\mathbf{A}$$

Observe that a scalar matrix is equal to the product of the scalar and the identity matrix. For example,

$$\begin{bmatrix} 3 & 0 & 0 \\ 0 & 3 & 0 \\ 0 & 0 & 3 \end{bmatrix} = 3\begin{bmatrix} 1 & 0 & 0 \\ 0 & 1 & 0 \\ 0 & 0 & 1 \end{bmatrix}$$

A modest effort must be expended to use the terminology *multiplication by a scalar* in order to avoid confusion with the process known as scalar multiplication.

The product of a row vector of order $1 \times n$ and a column vector of order $n \times 1$ forms a 1×1 matrix which has no important property that is not possessed by a scalar. This product is therefore called the *scalar* or *dot product* (some sources also use the terminology *inner product*). It is called for through the use of a dot placed between the two matrices in the product; that is, if **A** and **B** are column vectors

$$\mathbf{A} \cdot \mathbf{B} = [a_{ij}]_{1\times n} \cdot [b_{ij}]_{1\times n} = \mathbf{A}\mathbf{B}^{\mathrm{T}} = \mathbf{B}\mathbf{A}^{\mathrm{T}} = \gamma$$

and where γ is a scalar obtained from

$$\gamma = \sum_{k=1}^{n} a_k b_k$$

If the scalar product of two vectors is uniquely equal to zero, the vectors are said to be *orthogonal*.

Example B.2. If

$$\mathbf{A} = \begin{bmatrix} 2 \\ 4 \\ 3 \\ 1 \\ 2 \end{bmatrix}$$

and

$$\mathbf{B} = \begin{bmatrix} -5 \\ 4 \\ -3 \\ 8 \\ 2 \end{bmatrix}$$

what is the dot product $\mathbf{A} \cdot \mathbf{B}$?

SOLUTION

$$\mathbf{A} \cdot \mathbf{B} = 2(-5) + 4(4) + 3(-3) + 1(8) + 2(-2)$$

$$= -10 + 16 - 9 + 8 - 4 = 1$$

In Section B.2, a set of linear simultaneous algebraic equations was shown to be represented by the notation

$$\mathbf{AX} = \mathbf{Y}$$

where $\mathbf{A}$ was the $n \times n$ coefficient matrix and $\mathbf{A}$ and $\mathbf{X}$ were $n \times 1$ column vectors. In order to obtain the original set of equations from a set where $n = 3$,

$$\begin{bmatrix} a_{11} & a_{12} & a_{13} \\ a_{21} & a_{22} & a_{23} \\ a_{31} & a_{32} & a_{33} \end{bmatrix} \begin{bmatrix} x_1 \\ x_2 \\ x_3 \end{bmatrix} \begin{bmatrix} y_1 \\ y_2 \\ y_3 \end{bmatrix}$$

a row by column element product and sum operation is clearly evident:

$$\begin{bmatrix} a_{11}x_1 + a_{12}x_2 + a_{13}x_3 = y_1 \\ a_{21}x_1 + a_{22}x_2 + a_{23}x_3 = y_2 \\ a_{31}x_1 + a_{32}x_2 + a_{33}x_3 = y_3 \end{bmatrix}$$

and it is observed that each element of y is obtained by multiplying the corresponding elements of $\mathbf{A}$ by the elements of $\mathbf{X}$ and adding the results. Notice that the foregoing procedure will not be possible if the number of columns of $\mathbf{A}$ does not equal the number of rows of **X**. In this event there will not always be corresponding elements to multiply. Moreover, it should be noted that **Y** contains the same number of rows as both **A** and **X**.

This suggests a general definition for the multiplication of two matrices. If **A** is $m \times n$ and **B** is $p \times q$, $\mathbf{AB} = \mathbf{C}$ will exist if $n = p$, in which case the matrix **C** will be $m \times q$ with elements given by

$$[c_{ij}]_{m\times q} = \sum_{k=1}^{n=p} a_{ik}b_{kj} \qquad \begin{cases} i = 1, 2, 3, \ldots, m \\ j = 1, 2, 3, \ldots, q \end{cases}$$

When $n = p$, the matrices **A** and **B** are said to be *conformable for multiplication*.

Example B.3. If

$$\mathbf{A} = \begin{bmatrix} 1 & 4 & -2 & 0 \\ 4 & 3 & 2 & 1 \end{bmatrix}$$

$$\mathbf{B} = \begin{bmatrix} -1 & 1 \\ 3 & 2 \\ -2 & 4 \\ 0 & 3 \end{bmatrix}$$

and

$$\mathbf{C} = \begin{bmatrix} 2 & 1 \\ -3 & 4 \end{bmatrix}$$

find **AB**, **BA**, and **AC**.

SOLUTION. The product **AB** exists because **A** is 2×4 and **B** is 4×2. The result **P** will be 2×2:

$$\mathbf{P} = \mathbf{AB} = \begin{bmatrix} -1 & 4 & -2 & 0 \\ 4 & 3 & 2 & 1 \end{bmatrix} \begin{bmatrix} -1 & 1 \\ 3 & 2 \\ -2 & 4 \\ 0 & 3 \end{bmatrix}$$

$$= \begin{bmatrix} (1+12+4+0) & (-1+8-8+0) \\ (-4+9-4+0) & (4+6+8+3) \end{bmatrix} = \begin{bmatrix} 17 & -1 \\ 1 & 21 \end{bmatrix}$$

The product **BA** also exists:

$$\mathbf{BA} = \begin{bmatrix} -1 & 1 \\ 3 & 2 \\ -2 & 4 \\ 0 & 3 \end{bmatrix} \begin{bmatrix} -1 & 4 & -2 & 0 \\ 4 & 3 & 2 & 1 \end{bmatrix}$$

$$\begin{bmatrix} (1+4) & (-4+3) & (2+2) & (0+1) \\ (-3+8) & (12+6) & (-6+4) & (0+2) \\ (2+16) & (-8+12) & (4+8) & (0+4) \\ (0+12) & (0+9) & (0+6) & (0+3) \end{bmatrix} = \begin{bmatrix} 5 & -1 & 4 & 1 \\ 5 & 18 & -2 & 2 \\ 18 & 4 & 12 & 4 \\ 12 & 9 & 6 & 3 \end{bmatrix}$$

Notice that $\mathbf{AB} \neq \mathbf{BA}$, which shows that matrix multiplication, in general, is not commutative. Notice also that the product $\mathbf{AC}$ will not exist because $\mathbf{A}$ and $\mathbf{C}$ are not conformable for multiplication ($\mathbf{A}$ is 2×4 and $\mathbf{C}$ is 2×2).

Although the commutative law does not hold, the multiplication of matrices is associative:

$$(\mathbf{AB})\mathbf{C} = \mathbf{A}(\mathbf{BC})$$

and matrix multiplication is distributive with respect to addition:

$$\mathbf{A}(\mathbf{B} + \mathbf{C}) = \mathbf{AB} + \mathbf{AC}$$

assuming that conformability exists for both addition and multiplication.

If the product $\mathbf{AB}$ is null, that is, $\mathbf{AB} = 0$, it cannot be concluded that either $\mathbf{A}$ or $\mathbf{B}$ is null. Furthermore, if $\mathbf{AB} = \mathbf{AC}$ or $\mathbf{CA} = \mathbf{BA}$, it cannot be concluded that $\mathbf{B} = \mathbf{C}$. This means that, in general, cancellation of matrices is not permissible.

The transpose of a product of matrices is equal to the product of the individual transposes taken in reverse order:

$$(\mathbf{AB})^{\mathrm{T}} = \mathbf{B}^{\mathrm{T}}\mathbf{A}^{\mathrm{T}}$$

B.7 MATRIX DIVISION AND MATRIX INVERSION

Matrix division is not defined. Instead, use is made of a process called *matrix inversion*, which relies on the existence of the identity matrix, which is related to a square matrix $\mathbf{A}$ by

$$\mathbf{AI} = \mathbf{IA} = \mathbf{A}$$

Consider the identity for addition, 0, so that has the property that for all scalars α,

$$\alpha + 0 = 0 + \alpha = \alpha$$

and an identity element for multiplication, 1, so that

$$\alpha 1 = 1\alpha = \alpha$$

The scalar most certainly possesses a reciprocal or multiplicative inverse, $1/\alpha$, which when multiplied by α yields the identity element for scalar multiplication:

$$1/\alpha = \alpha^{-1}\alpha = \alpha$$

This reasoning may be extended to the $n \times n$ matrix **A** and the pair of identity matrices: the $n \times n$ identity matrix for multiplication **I** and the $n \times n$ identity matrix for addition **0** (a null matrix). Thus, as already noted,

$$\mathbf{AI} = \mathbf{IA} = \mathbf{A}$$

and

$$\mathbf{A} + \mathbf{0} = \mathbf{0} + \mathbf{A} = \mathbf{A}$$

If there is an $n \times n$ matrix **A** that pre- and postmultiplies **A** such that

$$\mathbf{A}^{-1}\mathbf{A} = \mathbf{A}\mathbf{A}^{-1} = \mathbf{I}$$

then $\mathbf{A}^{-1}$ is an inverse of **A** with respect to matrix multiplication. The matrix **A** is said to be *invertible* or *nonsingular* if $\mathbf{A}^{-1}$ exists and *singular* if $\mathbf{A}^{-1}$ does not exist.

For example, the 3×3 matrix

$$\mathbf{A} = \begin{bmatrix} 4 & -2 & -1 \\ -2 & 8 & -5 \\ -1 & -5 & 8 \end{bmatrix}$$

can be shown to possess the inverse

$$\mathbf{A}^{-1} = \begin{bmatrix} 13/32 & 7/32 & 3/16 \\ 7/32 & 31/96 & 11/48 \\ 3/16 & 11/48 & 7/24 \end{bmatrix}$$

A simple multiplication will produce the identity matrix:

$$\mathbf{AA}^{-1} = \begin{bmatrix} 4 & -2 & -1 \\ -2 & 8 & -5 \\ -1 & -5 & 8 \end{bmatrix} \begin{bmatrix} 13/32 & 7/32 & 3/16 \\ 7/32 & 31/96 & 11/48 \\ 3/16 & 11/48 & 7/24 \end{bmatrix} = \begin{bmatrix} 1 & 0 & 0 \\ 0 & 1 & 0 \\ 0 & 0 & 1 \end{bmatrix}$$

It can also be verified that the identity is also produced if the product $\mathbf{A}^{-1}\mathbf{A}$ is taken.

The inverse of a product of matrices is equal to the product of the individual inverses taken in reverse order:

$$(\mathbf{AB})^{-1} = \mathbf{B}^{-1}\mathbf{A}^{-1}$$

B.8 DETERMINANTS

B.8.1 Definitions and Terminology

A square matrix of order n (an $n \times n$ matrix)

$$\mathbf{A} = \begin{bmatrix} a_{11} & a_{12} & a_{13} & \cdots & a_{1n} \\ a_{21} & a_{22} & a_{23} & \cdots & a_{2n} \\ a_{31} & a_{32} & a_{33} & \cdots & a_{3n} \\ & \cdots & \cdots & \cdots & \\ a_{n1} & a_{n2} & a_{n3} & \cdots & a_{nn} \end{bmatrix}$$

possesses a uniquely defined scalar (a single number) which is designated as the determinant of **A** or merely the *determinant*:

$$\det \mathbf{A} = |A|$$

where the order of the determinant is the same as the order of the matrix from which it derives. Observe that only square matrices possess determinants, the use of vertical lines and not brackets to designate determinants, and that the elements of the determinant are identical to the elements of the matrix:

$$\det \mathbf{A} = \begin{vmatrix} a_{11} & a_{12} & a_{13} & \cdots & a_{1n} \\ a_{21} & a_{22} & a_{23} & \cdots & a_{2n} \\ a_{31} & a_{32} & a_{33} & \cdots & a_{3n} \\ & \cdots & \cdots & \cdots & \\ a_{n1} & a_{n2} & a_{n3} & \cdots & a_{nn} \end{vmatrix}$$

A determinant of the first order consists of a single element a and has, therefore, the value $\det \mathbf{A} = a$. A determinant of the second order contains four elements in a 2×2 square array with the value

$$\det \mathbf{A} = |A| = \begin{vmatrix} a_{11} & a_{12} \\ a_{21} & a_{22} \end{vmatrix}$$

A determinant of the third order is described in similar fashion. It is a 3×3 square array containing nine elements:

$$\det \mathbf{A} = |A| = \begin{vmatrix} a_{11} & a_{12} & a_{13} \\ a_{21} & a_{22} & a_{23} \\ a_{31} & a_{32} & a_{33} \end{vmatrix}$$

One may deduce that a determinant of nth order consists of a square array of $n \times n$ elements, a_{ij}, and that the total number of elements in an nth-order determinant is n^2. Although this representation of the determinant looks to be purely abstract, the determinant can be proven to be a very rational function which can be evaluated in a number of ways. Moreover, the value of the use of determinants in the taking of matrix inverses and in the solution of simultaneous linear algebraic equations cannot and should not be underemphasized.

B.8.2 Determinant Evaluation

Consider, for example, the following pair of simultaneous linear algebraic equations, which are presumed to be linearly independent:

$$a_{11}x_1 + a_{12}x_2 = b_1 \tag{B.1a}$$

$$a_{21}x_1 + a_{22}x_2 = b_2 \tag{B.1b}$$

and observe that they may also be written in the matrix form $\mathbf{AX} = \mathbf{B}$:

$$\begin{bmatrix} a_{11} & a_{12} \\ a_{21} & a_{22} \end{bmatrix} \begin{bmatrix} x_1 \\ x_2 \end{bmatrix} = \begin{bmatrix} b_1 \\ b_2 \end{bmatrix}$$

In eqs. (B.1), the x's are the unknowns and the a's form the coefficient matrix $\mathbf{A}$. If $\det \mathbf{A} \neq 0$, the equations are said to be *linearly independent* and one method of solving this second-order system is to multiply eq. (B.1a) by a_{22} and eq. (B.1b) by a_{12}:

$$\begin{aligned} a_{22}a_{11}x_1 &+ a_{22}a_{12}x_2 &= a_{22}b_1 \\ a_{12}a_{21}x_1 &+ a_{12}a_{22}x_2 &= a_{12}b_2 \end{aligned}$$

A subtraction then yields

$$(a_{22}a_{11} - a_{12}a_{21})x_1 = a_{22}b_1 - a_{12}b_2$$

and then x_1 is obtained:

$$x_1 = \frac{a_{22}b_1 - a_{12}b_2}{a_{22}a_{11} - a_{12}a_{21}} \tag{B.2a}$$

A similar procedure yields x_2

$$x_2 = \frac{a_{11}b_2 - a_{21}b_1}{a_{22}a_{11} - a_{12}a_{21}} \tag{B.2b}$$

Observe that the denominators of the equations that yield x_1 and x_2 can be represented by the determinant

$$\begin{vmatrix} a_{11} & a_{12} \\ a_{21} & a_{22} \end{vmatrix} = a_{22}a_{11} - a_{12}a_{21}$$

and it is easy to see that the numerators of these equations can be represented by

$$\begin{vmatrix} b_1 & a_{12} \\ b_2 & a_{22} \end{vmatrix} = b_1a_{22} - b_2a_{21}$$

and

$$\begin{vmatrix} a_{11} & b_1 \\ a_{21} & b_2 \end{vmatrix} = a_{11}b_2 - a_{21}b_1$$

This must always hold unless the determinant in the denominators is equal to zero, which is ruled out because the discussion began originally with the statement that the two equations to be solved were linearly independent.

Thus one may write the solutions for x_1 and x_2 in eqs. (B.1) as

$$x_1 = \frac{\begin{vmatrix} b_1 & a_{12} \\ b_2 & a_{22} \end{vmatrix}}{\begin{vmatrix} a_{11} & a_{12} \\ a_{21} & a_{22} \end{vmatrix}} \tag{B.3a}$$

and

$$x_2 = \frac{\begin{vmatrix} a_{11} & b_1 \\ a_{21} & b_2 \end{vmatrix}}{\begin{vmatrix} a_{11} & a_{12} \\ a_{21} & a_{22} \end{vmatrix}} \tag{B.3b}$$

and this is a demonstration of a method of solution of simultaneous linear algebraic equations known as *Cramer's rule*.

The foregoing reasoning applies equally well to a set of n simultaneous algebraic equations. For a set of three equations in three unknowns

$$\begin{aligned} a_{11}x_1 &+ a_{12}x_2 + a_{13}x_3 = b_1 \\ a_{21}x_1 &+ a_{22}x_2 + a_{23}x_3 = b_2 \\ a_{31}x_1 &+ a_{32}x_2 + a_{33}x_3 = b_3 \end{aligned}$$

which are assumed to be linearly independent and which may be written in matrix form as

$$\begin{bmatrix} a_{11} & a_{12} & a_{13} \\ a_{21} & a_{22} & a_{23} \\ a_{31} & a_{32} & a_{33} \end{bmatrix} \begin{bmatrix} x_1 \\ x_2 \\ x_3 \end{bmatrix} = \begin{bmatrix} b_1 \\ b_2 \\ b_3 \end{bmatrix}$$

it can be shown that x_1 can be evaluated from

$$x_1 = \frac{b_1a_{22}a_{33} + b_3a_{12}a_{23} + b_2a_{13}a_{32} - b_3a_{22}a_{13} - b_1a_{32}a_{23} - b_2a_{12}a_{33}}{a_{11}a_{22}a_{33} + a_{12}a_{23}a_{31} + a_{13}a_{21}a_{32} - a_{31}a_{22}a_{13} - a_{32}a_{23}a_{11} - a_{33}a_{21}a_{12}}$$

Both the numerator and denominator can be rearranged by employing a little algebra:

$$x_1 = \frac{b_1(a_{22}a_{33} - a_{32}a_{23}) - b_2(a_{12}a_{33} - a_{32}a_{13}) + b_3(a_{12}a_{23} - a_{22}a_{13})}{a_{11}(a_{22}a_{33} - a_{32}a_{23}) - a_{21}(a_{12}a_{33} - a_{13}a_{32}) + a_{31}(a_{12}a_{23} - a_{13}a_{22})} \tag{B.4}$$

and an inspection of the terms within parentheses shows that the solution for x_1 cannot only be written (Cramer's rule) as the quotient of two determinants, but each of the determinants can be represented in terms of three second-order determinants:

$$x_1 = \frac{\begin{vmatrix} b_1 & a_{12} & a_{13} \\ b_2 & a_{22} & a_{23} \\ b_3 & a_{32} & a_{33} \end{vmatrix}}{\begin{vmatrix} a_{11} & a_{12} & a_{13} \\ a_{21} & a_{22} & a_{23} \\ a_{31} & a_{32} & a_{33} \end{vmatrix}}$$

or

$$x_1 = \frac{b_1 \begin{vmatrix} a_{22} & a_{23} \\ a_{32} & a_{33} \end{vmatrix} - b_2 \begin{vmatrix} a_{12} & a_{13} \\ a_{32} & a_{33} \end{vmatrix} + b_3 \begin{vmatrix} a_{12} & a_{13} \\ a_{22} & a_{23} \end{vmatrix}}{a_{11} \begin{vmatrix} a_{22} & a_{23} \\ a_{32} & a_{33} \end{vmatrix} - a_{21} \begin{vmatrix} a_{12} & a_{13} \\ a_{32} & a_{23} \end{vmatrix} + a_{31} \begin{vmatrix} a_{12} & a_{13} \\ a_{22} & a_{23} \end{vmatrix}} \tag{B.5}$$

This expansion is known as the *Laplace expansion* or *Laplace development*.

The method of evaluating second-order determinants is suggested in eqs. (B.2) and (B.3). The second-order determinant is evaluated as the remainder of the product resulting from the multiplication of the upper left and lower right elements (the principal diagonal elements) minus the product of the lower left and the upper right elements (the secondary diagonal elements). This procedure is demonstrated in Fig. B.1*a*.

The third-order determinant may be evaluated by taking the products and then the sums and differences of the elements shown in Fig. B.1*b*. This procedure may be assisted by rewriting the first two columns of the determinant and then proceding as indicated in Fig. B.1*c*. It is important to note that for this purpose, the diagonals of the third-order determinant are continuous that is, the last column is followed by the first column.

Caution is necessary: Fourth- and higher-order determinants may not be evaluated by the following the procedures displayed in Fig. B.1. The Laplace expansion or *pivotal condensation*, to be discussed presently, must be employed in these cases.

Example B.4. Evaluate the determinants

$$|A| = \begin{vmatrix} 3 & 4 \\ 2 & 5 \end{vmatrix}$$

and

$$|B| = \begin{vmatrix} 4 & 3 & 2 \\ 0 & 3 & 1 \\ 1 & 2 & 1 \end{vmatrix}$$

SOLUTION. The second-order determinant is evaluated by the procedure indicated in Fig. B.1*a*:

$$|A| = \begin{vmatrix} 3 & 4 \\ 2 & 5 \end{vmatrix} = 3(5) - 2(4) = 15 - 8 = 7$$

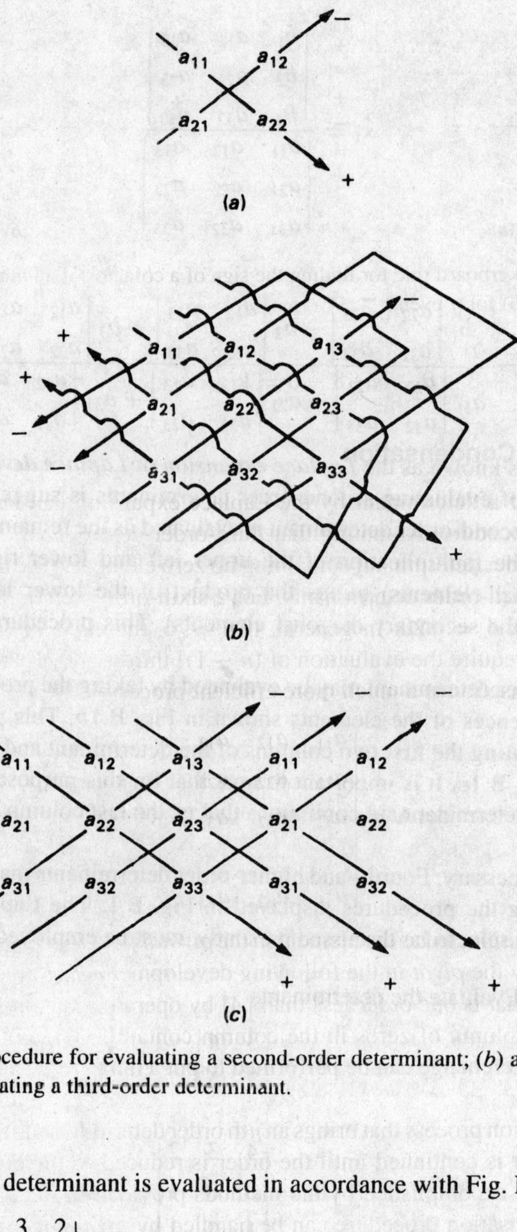

Figure B.1 (*a*) Procedure for evaluating a second-order determinant; (*b*) and (*c*) equivalent procedures for evaluating a third-order determinant.

The third-order determinant is evaluated in accordance with Fig. B.1*b* or *c* as

$$|B| = \begin{vmatrix} 4 & 3 & 2 \\ 0 & 3 & 1 \\ 1 & 2 & 1 \end{vmatrix}$$

$$= 4(3)(1) + 3(1)(1) + 2(0)(2) - 1(3)(2) - 2(1)(4) - 1(0)(3)$$

or

$$\begin{array}{ccccccc}
+ & - & + & - & + & \cdots & + \\
- & + & - & + & - & \cdots & - \\
+ & - & + & - & + & \cdots & + \\
- & + & - & + & - & \cdots & - \\
 & & & \cdots & & & \\
+ & - & + & - & + & \cdots & +
\end{array}
\qquad\qquad
\begin{array}{ccccccc}
+ & - & + & - & + & \cdots & - \\
- & + & - & + & - & \cdots & + \\
+ & - & + & - & + & \cdots & - \\
- & + & - & + & - & \cdots & + \\
 & & & \cdots & & & \\
- & + & - & + & - & \cdots & +
\end{array}$$

(*a*) (*b*)

Figure B.2 Checkerboard rule for finding the sign of a cofactor of an nth-order determinant (*a*) for n odd and (*b*) for n even.

$$B = 12 + 3 + 0 - 6 - 8 - 0 = 1$$

B.8.3 Pivotal Condensation

The evaluation of a determinant by the Laplace expansion can be a long, tedious, and laborious procedure. Assuming that third-order determinants can be evaluated quickly, a fifth-order determinant containing no zero elements requires the evaluation of $5 \times 4 = 20$ third-order determinants. For a sixth-order determinant, this number becomes $6 \times 5 \times 4 = 120$. In general, if $n > 6$, the evaluation of an nth-order determinant can require the evaluation of $(n-1)!$ third-order determinants.

Pivotal condensation is a much more efficient process. Take the determinant

$$\det \mathbf{A} = \begin{vmatrix} a_{11} & a_{12} & a_{13} & \cdots & a_{1n} \\ a_{21} & a_{22} & a_{23} & \cdots & a_{2n} \\ a_{31} & a_{32} & a_{33} & \cdots & a_{3n} \\ & \cdots & \cdots & \cdots & \\ a_{n1} & a_{n2} & a_{n3} & \cdots & a_{nn} \end{vmatrix}$$

The element a_{11} is selected as the element in the *pivotal position*. It is called the *pivotal element* or merely the *pivot* in the following development. The objective is to find a determinant $|B|$ that is one order less than $|A|$ by operating on $|A|$ in such a manner as to produce a column of zeros in the column containing the pivot. If $a_{11} = 0$, a row or column interchange can be performed to put a nonzero element in the pivotal position.

The condensation process that brings an nth order determinant down to an $(n-1)$th-order determinant is continued until the order is reduced to three or two. Then the evaluation can be accomplished by the methods provided in the preceding section. The entire condensation procedure can be handled by the computationally efficient matrix relationship

$$|A| = \frac{1}{a_{11}^{n-2}} \det \left[a_{11} \begin{bmatrix} a_{22} & a_{23} & \cdots & a_{2n} \\ a_{32} & a_{33} & \cdots & a_{3n} \\ & \cdots & \cdots & \\ a_{n2} & a_{n3} & \cdots & a_{nn} \end{bmatrix} \right.$$

$$- \begin{bmatrix} a_{21} \\ a_{31} \\ \vdots \\ a_{n1} \end{bmatrix} [\, a_{12} \quad a_{13} \quad \cdots \quad a_{1n} \,] \Bigg] \tag{B.6}$$

Example B.5. Use pivotal condensation to evaluate the determinant

$$|A| = \begin{vmatrix} 2 & -1 & 1 & 1 & 2 \\ 0 & 2 & 3 & 2 & 1 \\ 0 & 1 & 2 & 1 & 2 \\ 0 & 1 & -1 & -1 & 3 \\ 0 & 2 & 1 & 1 & -2 \end{vmatrix}$$

SOLUTION. By pivotal condensation

$$|A| = \frac{1}{2^3} \det \left[2 \begin{bmatrix} 2 & 3 & 2 & 1 \\ 1 & 2 & 1 & 2 \\ 1 & -1 & -1 & 3 \\ 2 & 1 & 1 & -2 \end{bmatrix} - \begin{bmatrix} 0 \\ 0 \\ 0 \\ 0 \end{bmatrix} [-1 \quad 1 \quad 1 \quad 2] \right]$$

or

$$|A| = \frac{1}{8} \det \begin{bmatrix} 4 & 6 & 4 & 2 \\ 2 & 4 & 2 & 4 \\ 2 & -2 & -2 & 6 \\ 4 & 2 & 2 & -4 \end{bmatrix}$$

Then

$$|A| = \frac{1}{(8)(4)^2} \det \left[4 \begin{bmatrix} 4 & 2 & 4 \\ -2 & -2 & 6 \\ 2 & 2 & -4 \end{bmatrix} - \begin{bmatrix} 2 \\ 2 \\ 4 \end{bmatrix} [6 \quad 4 \quad 2] \right]$$

or

$$|A| = \frac{1}{128} \det \left[\begin{bmatrix} 16 & 8 & 16 \\ -8 & -8 & 24 \\ 8 & 8 & -16 \end{bmatrix} - \begin{bmatrix} 12 & 8 & 4 \\ 12 & 8 & 4 \\ 24 & 16 & 8 \end{bmatrix} \right]$$

$$= \frac{1}{128} \det \begin{bmatrix} 4 & 0 & 12 \\ -20 & -16 & 20 \\ -16 & -8 & -24 \end{bmatrix}$$

The third-order determinant is easily evaluated:

$$|A| = \frac{1}{128}(1536 + 0 + 1920 - 3072 - 0 + 640) = \frac{1024}{128} = 8$$

B.8.4 Additional Properties

Several rules pertaining to the simplification and manipulation of determinants are presented below without formal proof.

- Interchanging any row (or column) of a determinant with its immediately adjacent row (or column) alters the sign of the determinant.
- The multiplication of any single row (column) of a determinant by a scalar constant is equivalent to the multiplication of the entire determinant by the scalar. Observe that this differs from the multiplication of a matrix by a scalar; the multiplication of a matrix by a scalar results in the multiplication of each and every element of the matrix by the scalar.
- If every element in an nth-order determinant is multiplied by the same scalar, α, the value of the determinant is multiplied by α^n.
- If any two rows (columns) of a determinant are identical, the value of the determinant is zero and the matrix from which the determinant derives is said to be *singular*.
- If any row (or column) of a determinant contains nothing but zeros, the value of the determinant is zero.
- If any two rows (columns) of a determinant are proportional, the determinant is equal to zero. In this case, the two rows (columns) are said to be *linearly dependent*.
- If the elements of any row (column) of a determinant are added to or subtracted from the corresponding elements of another row (column), the value of the determinant is unchanged.
- If the elements of any row (column) of a determinant are multiplied by a constant and then added to or subtracted from the corresponding elements of another row (column), the value of the determinant is unchanged.
- The value of the determinant of a diagonal matrix is equal to the product of the terms on the diagonal.
- The value of the determinant of a matrix is equal to the value of the determinant of the transpose of the matrix.
- The determinant of the product of two matrices is equal to the product of the determinants of the two matrices.
- If the determinant of the product of two square matrices is zero, then at least one of the matrices is singular, that is, the value of its determinant is equal to zero.

B.9 MINORS AND COFACTORS

Consider the nth-order determinant

$$\det \mathbf{A} = \begin{vmatrix} a_{11} & a_{12} & a_{13} & \cdots & a_{1n} \\ a_{21} & a_{22} & a_{23} & \cdots & a_{2n} \\ a_{31} & a_{32} & a_{33} & \cdots & a_{3n} \\ \cdots & & & & \\ a_{n1} & a_{n2} & a_{n3} & \cdots & a_{nn} \end{vmatrix} \tag{B.7}$$

which will be used for proposing two useful quantities.

The $(n-1)$th-*order minor* of an nth-order determinant $|A|$ is the determinant formed by deleting one row and one column from $|A|$. The minor, designated by $|M|_{ij}$, is the determinant formed by deleting the ith row and the jth column from $|A|$.

The *cofactor*, designated as A_{ij} without vertical rules and with a double subscript, is the *signed* $(n-1)$th-order minor formed from the nth-order determinant. If the minor has been formed by deleting the ith row and the jth column from $|A|$, then

$$A_{ij} = (-1)^{i+j} M_{ij} \tag{B.8}$$

The sign of the cofactor can be determined from eq. (B.8) or from the *checkerboard rule* summarized in Fig. B.2.

Example B.6. Consider the fourth-order determinant

$$|A| = \det \begin{bmatrix} 1 & 3 & -1 & 2 \\ 4 & 1 & 1 & 3 \\ 3 & 1 & -2 & 1 \\ 1 & 3 & 2 & 5 \end{bmatrix}$$

What is the minor and cofactor formed by deleting the third row and fourth column?

SOLUTION

$$|M|_{34} = \det \begin{bmatrix} 1 & 3 & -1 \\ 4 & 1 & 1 \\ 1 & 3 & 2 \end{bmatrix} = 2 + 3 - 12 + 1 - 3 - 24 = -33$$

The cofactor is the *signed* minor. By the checkerboard rule of Fig. B.2 or by eq. (B.8),

$$A_{34} = (-1)^{3+4}(-33) = -(-33) = 33$$

B.10 COFACTOR MATRIX

A square nth-order matrix

$$\mathbf{A} = \begin{bmatrix} a_{11} & a_{12} & a_{13} & \cdots & a_{1n} \\ a_{21} & a_{22} & a_{23} & \cdots & a_{2n} \\ a_{31} & a_{32} & a_{33} & \cdots & a_{3n} \\ \cdots & \cdots & \cdots & & \\ a_{n1} & a_{n2} & a_{n3} & \cdots & a_{nn} \end{bmatrix}$$

possesses a cofactor matrix with elements indicated by capital letters with double subscripts:

$$\mathbf{A}^c = \begin{bmatrix} A_{11} & A_{12} & A_{13} & \cdots & A_{1n} \\ A_{21} & A_{22} & A_{23} & \cdots & A_{2n} \\ A_{31} & A_{32} & A_{33} & \cdots & A_{3n} \\ \cdots & \cdots & \cdots & & \\ A_{n1} & A_{n2} & A_{n3} & \cdots & A_{nn} \end{bmatrix}$$

Example B.7. Determine the cofactor matrix for the third-order symmetrical matrix

$$\begin{bmatrix} 3 & -2 & 0 \\ -2 & 4 & -1 \\ 0 & -1 & 6 \end{bmatrix}$$

SOLUTION. The nine cofactors with signs determined by eq. (B.8) or from the checkerboard rule in Fig. B.2 are formed from the nine possible second-order minors.

$$A_{11} = +|M|_{11} = \begin{vmatrix} 4 & -1 \\ -1 & 6 \end{vmatrix} = 24 - 1 = 23$$

$$A_{12} = -|M|_{12} = \begin{vmatrix} -2 & 1 \\ 0 & 6 \end{vmatrix} = -(-12) = 12$$

$$A_{13} = +|M|_{13} = \begin{vmatrix} -2 & 4 \\ 0 & -1 \end{vmatrix} = 2$$

$$A_{21} = -|M|_{21} = \begin{vmatrix} -2 & 0 \\ -1 & 6 \end{vmatrix} = -(-12) = 12$$

$$A_{22} = +|M|_{22} = \begin{vmatrix} 3 & 0 \\ 0 & 6 \end{vmatrix} = 18$$

$$A_{23} = -|M|_{23} = \begin{vmatrix} 3 & -2 \\ 0 & -1 \end{vmatrix} = -(-3) = 3$$

$$A_{31} = +|M|_{31} = \begin{vmatrix} -2 & 0 \\ 4 & -1 \end{vmatrix} = 2$$

$$A_{32} = -|M|_{32} = \begin{vmatrix} 3 & 0 \\ -2 & -1 \end{vmatrix} = -(-3) = 3$$

$$A_{33} = +|M|_{33} = \begin{vmatrix} 3 & -2 \\ -2 & 4 \end{vmatrix} = 12 - 4 = 8$$

Thus

$$\mathbf{A}^c = \begin{bmatrix} 23 & 12 & 2 \\ 12 & 18 & 3 \\ 2 & 3 & 8 \end{bmatrix}$$

and this confirms that symmetrical matrices possess symmetrical cofactor matrices.

B.11 LAPLACE EXPANSION

In the denominator of eq. (B.5), the third-order determinant of a matrix **A** was shown to be equal to some function of three second-order determinants

$$\begin{vmatrix} a_{11} & a_{12} & a_{13} \\ a_{21} & a_{22} & a_{23} \\ a_{31} & a_{32} & a_{33} \end{vmatrix} = a_{11} \begin{vmatrix} a_{22} & a_{23} \\ a_{32} & a_{33} \end{vmatrix} - a_{21} \begin{vmatrix} a_{12} & a_{13} \\ a_{32} & a_{33} \end{vmatrix} + a_{31} \begin{vmatrix} a_{12} & a_{13} \\ a_{22} & a_{23} \end{vmatrix} \tag{B.9}$$

Notice that each of the second-order determinants is a second-order minor of **A**. This means that three cofactors exist, and hence eq. (B.9) gives a rule for the evaluation of a third-order determinant which can be extended to an nth-order determinant. For the ith row,

$$|A| = \sum_{j=1}^{j=n} (-1)^{i+j} a_{ij} |M|_{ij}$$

or

$$|A| = \sum_{j=1}^{j=n} a_{ij} A_{ij} \tag{B.10a}$$

and for the jth column,

$$|A| = \sum_{i=1}^{i=n} (-1)^{i+j} a_{ij} |M|_{ij}$$

or

$$|A| = \sum_{i=1}^{i=n} a_{ij} A_{ij} \tag{B.10b}$$

Equations (B.10) describe a procedure known as the *Laplace development* or *Laplace expansion*.

Example B.8. Evaluate the determinant

$$|A| = \begin{vmatrix} 1 & 2 & 3 & 4 \\ 1 & 4 & 3 & 2 \\ 2 & 0 & 3 & 1 \\ 1 & 0 & 2 & 3 \end{vmatrix}$$

SOLUTION. Expand using the second column to reduce the labor (two zeros occur in this column):

$$|A| = a_{21}|M|_{21} + a_{22}|M|_{22}$$

The cofactors derive from the appropriate minors with their sign determined from eq. (B.8) or from the checkerboard rule illustrated in Fig. B.2.

$$A_{21} = -|M|_{21} = \begin{vmatrix} 1 & 3 & 2 \\ 2 & 3 & 1 \\ 1 & 2 & 3 \end{vmatrix} = -(9+3+8-6-2-18) = -(20-26) = 6$$

and

$$A_{22} = +|M|_{22} = \begin{vmatrix} 1 & 3 & 4 \\ 2 & 3 & 1 \\ 1 & 2 & 3 \end{vmatrix} = (9+3+16-12-2-18) = 28-32 = -4$$

The value of the determinant is

$$|A| = a_{21}A_{21} + a_{22}A_{22} = 2(6) + 4(-4) = 12 - 16 = -4$$

It should be noted that if the elements of a row or column of a determinant are multiplied by cofactors of the corresponding elements of a different row or column, the resulting sum of these products is zero:

$$\sum_{i=1}^{i=n} (-1)^{k+j} a_{kj}|M|_{kj} \qquad (i \neq k) \tag{B.11a}$$

and

$$\sum_{j=1}^{j=n} (-1)^{i+k} a_{ik}|M|_{ik} \qquad (j \neq k) \tag{B.11b}$$

B.12 MATRIX INVERSION

An nth-order set of simultaneous linear algebraic equations in n unknowns, x_1, x_2, x_3, . . ., x_n, can be represented conveniently by the matrix equation

$$\mathbf{AX} = \mathbf{Y} \tag{B.12}$$

where **A**, as indicated in Section B.2, is a square matrix of coefficients having elements a_{ij} and where **X** and **Y** are $n \times 1$ column vectors with elements x_i and y_i, respectively. Because division of matrices is not permitted, one method for the solution of matrix equations, such as the one shown in eq. (B.12), is called *matrix inversion*.

If eq. (B.12) is premultiplied by an $n \times n$ square matrix **B** so that

$$\mathbf{BAX} = \mathbf{BY}$$

a solution for the unknowns **X** will evolve if the product **BA** is equal to the identity matrix **I**:

$$\mathbf{BAX} = \mathbf{IX} = \mathbf{BY}$$

or

$$\mathbf{X} = \mathbf{BY} \tag{B.13}$$

If

$$\mathbf{BA} = \mathbf{AB} = \mathbf{I}$$

the matrix **B** is said to be the inverse of **A**:

$$\mathbf{B} = \mathbf{A}^{-1} \tag{B.14a}$$

and, of course, the inverse of the inverse is the matrix itself:

$$\mathbf{A} = \mathbf{B}^{-1} \tag{B.14b}$$

or

$$(\mathbf{A}^{-1})^{-1} = \mathbf{A}$$

It may be recalled that in general, matrix multiplication is not commutative. The multiplication of a matrix by its inverse is one specific case where matrix multiplication is commutative:

$$\mathbf{AA}^{-1} = \mathbf{A}^{-1}\mathbf{A} = \mathbf{I}$$

B.12.1 Properties of the Inverse

The inverse of a product of two matrices is the product of the inverses taken in reverse order. This is easily proved. Consider the product **AB** and postmultiply by **BA**. Because matrix multiplication is associative, this product can be taken with a rearrangement of the parentheses and then by straightforward application of the definition of the matrix inverse:

$$\mathbf{AB}(\mathbf{B}^{-1}\mathbf{A}^{-1}) = \mathbf{A}(\mathbf{BB}^{-1}\mathbf{A}^{-1}) = \mathbf{AIA}^{-1} = \mathbf{AA}^{-1} = \mathbf{I}$$

In addition, the inverse of the transpose of a matrix is equal to the transpose of its inverse:

$$(\mathbf{A}^T)^{-1} = (\mathbf{A}^{-1})^T$$

negative powers of a matrix are related to its inverse:

$$\mathbf{A}^{-n} = (\mathbf{A}^{-1})^n$$

and the determinant of the product of a matrix and its inverse must be equal to unity:

$$\det(\mathbf{A}\mathbf{A}^{-1}) = \det \mathbf{I} = 1$$

If a matrix does not possess an inverse, it is said to be *singular*, but if a matrix does possess an inverse, the inverse is *unique*.

The inverse of a product of matrices is equal to the product of the inverses taken in reverse order:

$$(\mathbf{AB})^{-1} = \mathbf{B}^{-1}\mathbf{A}^{-1}$$

B.12.2 Adjoint Matrix

The *adjoint matrix*, which is sometimes called the *adjugate matrix* and which here will be referred to merely as the *adjoint*, applies only to a square matrix and is the transpose of the cofactor matrix:

$$\text{adj } \mathbf{A} = \mathcal{A} = ((\mathbf{A}^c)^T \tag{B.15}$$

and because symmetrical matrices possess symmetrical cofactor matrices, the adjoint of a symmetrical matrix is the cofactor matrix itself.

The matrix that is of nth order,

$$\mathbf{A} = \begin{bmatrix} a_{11} & a_{12} & a_{13} & \cdots & a_{1n} \\ a_{21} & a_{22} & a_{23} & \cdots & a_{2n} \\ a_{31} & a_{32} & a_{33} & \cdots & a_{3n} \\ & \cdots & \cdots & \cdots & \\ a_{n1} & a_{n2} & a_{n3} & \cdots & a_{nn} \end{bmatrix}$$

has been observed to possess a cofactor matrix,

$$\mathbf{A}^c = \begin{bmatrix} A_{11} & A_{12} & A_{13} & \cdots & A_{1n} \\ A_{21} & A_{22} & A_{23} & \cdots & A_{2n} \\ A_{31} & A_{32} & A_{33} & \cdots & A_{3n} \\ & \cdots & \cdots & \cdots & \\ A_{n1} & A_{n2} & A_{n3} & \cdots & A_{nn} \end{bmatrix}$$

and this cofactor matrix has an adjoint,

$$\text{adj } \mathbf{A} = (\mathbf{A}^c)^T = \begin{bmatrix} A_{11} & A_{21} & A_{31} & \cdots & A_{n1} \\ A_{12} & A_{22} & A_{32} & \cdots & A_{n2} \\ A_{13} & A_{23} & A_{33} & \cdots & A_{n3} \\ & \cdots & \cdots & \cdots & \\ A_{1n} & A_{2n} & A_{3n} & \cdots & A_{nn} \end{bmatrix}$$

B.12.3 One Method for the Determination of the Inverse

Suppose that an $n \times n$ matrix $\mathbf{A}$ is postmultiplied by its adjoint and that the product is designated as $\mathbf{P}$:

$$\mathbf{A}(\text{adj } \mathbf{A}) = \begin{bmatrix} a_{11} & a_{12} & a_{13} & \cdots & a_{1n} \\ a_{21} & a_{22} & a_{23} & \cdots & a_{2n} \\ a_{31} & a_{32} & a_{33} & \cdots & a_{3n} \\ \cdots & \cdots & \cdots & & \\ a_{n1} & a_{n2} & a_{n3} & \cdots & a_{nn} \end{bmatrix} \begin{bmatrix} A_{11} & A_{21} & A_{31} & \cdots & A_{n1} \\ A_{12} & A_{22} & A_{32} & \cdots & A_{n2} \\ A_{13} & A_{23} & A_{33} & \cdots & A_{n3} \\ & \cdots & \cdots & \cdots & \\ A_{1n} & A_{2n} & A_{3n} & \cdots & A_{nn} \end{bmatrix} = \mathbf{P}$$

The elements of $\mathbf{P}$ may be divided into two categories: those that lie on its principal diagonal of which p_{22} is typical and those that do not. For the principal diagonal element p_{22},

$$p_{22} = a_{21}A_{21} + a_{22}A_{22} + a_{23}A_{23} + \cdots + a_{2n}A_{2n}$$

and by eq. (B.10*a*), it is seen that

$$p_{22} = |A|$$

For the off-diagonal element of which p_{13} is typical,

$$p_{13} = a_{11}A_{31} + a_{12}A_{32} + a_{13}A_{33} + \cdots + a_{1n}A_{3n}$$

and by eq. (B.11*a*), it is seen that

$$p_{13} = 0$$

Thus the product of $\mathbf{A}$ and its adjoint is

$$\mathbf{A}(\text{adj } \mathbf{A}) = \begin{bmatrix} |A| & 0 & 0 & \cdots & 0 \\ 0 & |A| & 0 & \cdots & 0 \\ 0 & 0 & |A| & \cdots & 0 \\ & \cdots & \cdots & \cdots & \\ 0 & 0 & 0 & \cdots & |A| \end{bmatrix} = |A|\mathbf{I}$$

If this is put into the form

$$\mathbf{A}\frac{\text{adj } \mathbf{A}}{\det \mathbf{A}} = \mathbf{A}$$

and compared with

$$\mathbf{AA}^{-1} = \mathbf{I}$$

it becomes evident that the inverse of the matrix **A** is equal to its adjoint divided by its determinant:

$$\mathbf{A}^{-1} = \frac{\text{adj } \mathbf{A}}{\det \mathbf{A}} \tag{B.16}$$

Observe that if det **A** = 0, the inverse of **A** cannot exist and is therefore *singular*. Thus the necessary condition for the matrix **A** to be singular is for det **A** = 0.

Example B.9. Determine the inverse of the third-order symmetrical matrix

$$\mathbf{A} = \begin{bmatrix} 3 & -2 & 0 \\ -2 & 4 & -1 \\ 0 & -1 & 6 \end{bmatrix}$$

SOLUTION. In Example B.7 it was shown that the given matrix possesses a cofactor matrix:

$$\mathbf{A}^c = \begin{bmatrix} 23 & 12 & 2 \\ 12 & 18 & 3 \\ 2 & 3 & 8 \end{bmatrix}$$

and the reader may verify that the given matrix has a determinant det **A** = 45.

The given matrix is symmetrical, as is the cofactor matrix. The adjoint (the transpose of the cofactor matrix) is also symmetrical and is equal to the cofactor matrix. Thus, by eq. (B.16), the inverse is

$$\mathbf{A}^{-1} = \begin{bmatrix} 23/45 & 4/15 & 2/45 \\ 4/15 & 2/5 & 1/15 \\ 2/45 & 1/15 & 8/45 \end{bmatrix}$$

which is also observed to be symmetrical.

It is important to note that symmetrical matrices possess symmetrical transposes, symmetrical cofactor matrices, symmetrical adjoints, and symmetrical inverses.

The evaluation of the inverse can always be concluded with a check on its validity. In the example just concluded,

$$\mathbf{AA}^{-1} = \begin{bmatrix} 3 & -2 & 0 \\ -2 & 4 & -1 \\ 0 & -1 & 6 \end{bmatrix} \begin{bmatrix} 23/45 & 4/15 & 2/45 \\ 4/15 & 2/5 & 1/15 \\ 2/45 & 1/15 & 8/45 \end{bmatrix} = \begin{bmatrix} 1 & 0 & 0 \\ 0 & 1 & 0 \\ 0 & 0 & 1 \end{bmatrix}$$

Example B.10. Determine the inverse of the nonsymmetrical second-order matrix

$$\mathbf{A} = \begin{bmatrix} 4 & -1 \\ 1 & 6 \end{bmatrix}$$

SOLUTION. The matrix has a determinant

$$\det \mathbf{A} = 24 + 1 = 25$$

a cofactor matrix

$$\mathbf{A}^c = \begin{bmatrix} 6 & -1 \\ 1 & 4 \end{bmatrix}$$

and an adjoint

$$\text{adj } \mathbf{A} = \begin{bmatrix} 6 & 1 \\ -1 & 4 \end{bmatrix}$$

Its inverse is

$$\mathbf{A}^{-1} = \frac{\text{adj } \mathbf{A}}{\det \mathbf{A}} = \begin{bmatrix} 6/25 & 1/25 \\ -1/25 & 4/25 \end{bmatrix}$$

This can be verified by the reader, and it is observed that the inverse of a second-order determinant is obtained by swapping the elements that lie on the principal diagonal, changing the sign of the off-diagonal elements, and then dividing all elements by the determinant.

B.13 NOMENCLATURE

Roman Letter Symbols

A	matrix, dimensionless
a	element of **A**, dimensionless
B	vector or matrix, dimensionless
b	element of **B**, dimensionless
C	matrix, dimensionless
c	element of **C**, dimensionless
det	determinant, dimensionless
I	identity matrix, dimensionless
i	element of **I**, dimensionless
j	counter, dimensionless
m	number of rows, dimensionless
n	number of columns, dimensionless; order of square matrix
P	matrix, dimensionless
p	element of **P**, dimensionless
tr	trace of a matrix, dimensionless
V	vector, dimensionless

v	element of **V**, dimensionless
X	vector, dimensionless
x	element of **X**, dimensionless
Y	vector, dimensionless
y	element of **Y**, dimensionless

Greek Letter Symbols

α	scalar, dimensionless
β	scalar, dimensionless
γ	scalar, dimensionless

Roman Letter Superscripts

a	augmented
c	cofactor
T	transpose

Symbolic Superscript

-1	inverse

REFERENCES

Abbott, R. W., Norris, R. H., and Spofford, W. A. (1979). Compact Heat Exchangers for General Electric Products: Sixty Years of Advances in Design and in Manufacturing Technologies, *ASME Symp. Vol. HTD-10*, American Society of Mechanical Engineers, New York.

Abramowitz, M., and Stegun, I. A. (1964). *Handbook of Mathematical Functions*, AMS 55, U.S. Government Printing Office, Washington, DC.

Abramzon, B., and Shmurak, M. (1998). An Accurate Analysis of Heat Transfer Performance of Extended Surface in Moist Air Flow, *Proc. International Conference on Heat Exchangers for Sustainable Development*, Lisbon, Portugal.

Achaichia, A., and Cowell, T. A. (1988). Heat Transfer and Pressure Drop Characteristics of Flat Tube and Louvered Plate Fin Surfaces, *Exp. Therm. Fluid Sci.*, 1, 147–157

Acharya, S., and Patankar, S. V. (1985). Laminar Mixed Convection in a Shrouded Fin Array, *J. Heat Transfer*, 103, 559.

Acharya, S., Brand, K. C., and Attar, A. (1986). Calculation of Fin Efficiency for Condensing Fins, *Int. J. Heat Fluid Flow*, 7, 96–98.

Ackerman, J. W., and Brunsvold, A. R. (1969). Heat Transfer and Draft Loss Performance of Extended Surface Tube Banks, *ASME Pap. 69-WA/HT-5*, American Society of Mechanical Engineers, New York

Acosta, R. E., Muller, R. H., and Tobias, C. W. (1985). Transport Processes in Narrow (Capillary) Channels, *AIChE J.*, 31, 473–482.

Adamek, T. (1981). Bestimming der Kondensationgrossen auf feinge-welten Oberflächen zur Auslegun optimaler Wandprofile, *Waerme und Stoffuebertrag.*, 15, 255.

Adamek, T., and Webb, R. L. (1990). Prediction of Film Condensation on Horizontal and Integral-Fin Tubes, *Int. J. Heat Mass Transfer*, 33, 1721–1735.

Advani, C. G., and Sukhatme, S. P. (1973). Analysis of a Vertical Rectangular Cross-Sectioned Fin Losing Heat by Laminar Free Convection, *Indian J. Technol.*, 11(3), 105.

Agrawal, D. D., and Bhavsar, V. C. (1977). Experimental Investigation of Heat Transfer Characteristics of Air Flow Across Finned Cylinders, *J. Inst. Eng. (India)*, 57, 351.

Ahmadi, G., and Razani, A. (1973). Some Optimization Problems Related to Cooling Fins, *Int. J. Heat and Mass Transfer*, 16(12), 2369.

Aihara, T. (1970). Natural Convection Heat Transfer from Vertical Rectangular Fin Arrays, *Bull. JSME*, 13(64), 1182–1192.

Aihara, T., Maruyama, S., and Kobaychawa, S. (1990). Free Convective and Radiative Heat Transfer from Pin-Fin Arrays with a Vertical Base Plate, *Int. J. Heat and Mass Transfer*, 33(6), 1223.

Al Mujahid, A. M. (1987). Analysis of Performance of a Straight Fin with Oscillating Bare Heat Flux and Environment Temperature, *J. Eng. Sci.*, King Saud University, 14, 25–37.

Al Mujahid, A. M. (1989). A Closed Form Solution for Periodic Heat Transfer in Annular Fins, *Waearme Stoffuebertrag.*, 24, 145–150.

Aoki, H., Shinegawa, T., and Suga, K. (1989). An Experimental Study of the Local Heat Transfer Characteristics in Automotive Louvered Fins, *Exp. Therm. Fluid Sci.*, 2, 293–300.

Aparacido, J. B., and Cotta, R. M. (1990). Improved One-Dimensional Fin Solution, *Heat Transfer Eng.*, 11, 49.

Appl, F. C., and Hung, H. M. (1964). Finned Surfaces with Optimum Fins of Restricted Length, *ASME Symp. Vol. Air Cooled Heat Exchangers*, American Society of Mechanical Engineers, New York.

Arias de Velasco, L., Aroca, S., and Gonzalez, C. (1982). Theoretical Analysis of the Constant-Width Annular Fin with Variable Thermal Conductivity and Internal Heat Generation, *Proc. 16th Southeastern Seminar on Thermal Sciences*, Miami, FL, 84.

Arora, S. C., and Messeh, W. A. (1983). Heat Transfer Experiments in High Aspect Ratio Rectangular Channel with Epoxied Short Pin Fins, *ASME Pap. 83-GT-57*, American Society of Mechanical Engineers, New York.

Arpaci, V. S. (1966). *Conduction Heat Transfer*, Addison-Wesley, Reading, MA.

Assis, E., and Kalman, H. (1993). Transient Temperature Response of Different Fins to Step Initial Conditions, *Int. J. Heat and Mass Transfer*, 36, 4107.

Avrami, M., and Little, J. B. (1942). Diffusion of Heat Through a Rectangular Bar and the Cooling and Insulating Effect of Fins: I. The Steady State, *J. Appl. Phys.*, 13(5), 304.

Azer, N. Z., and Sivakumar, V. (1984). Enhancement of Saturated Boiling Heat Transfer by Internally Finned Tubes, *Trans. ASHRAE*, 90, 58.

Aziz, A. (1975). Periodic Heat Transfer in Annular Fins, *J. Heat Transfer*, 97, 302.

Aziz, A. (1977). Perturbation Solution for Convecting Fin with Internal Heat Generation and Temperature Dependent Thermal Conductivity with Convective Boundary Conditions, *Int. J. Heat and Mass Transfer*, 20, 1253.

Aziz, A. (1978). Optimum Rectangular Convection Fin with Tip Heat Loss and Having a Temperature Dependent Thermal Conductivity, *J. Eng. Sci.* (University of Riyadh, Saudi Arabia), 4, 133.

Aziz, A. (1985a). Optimization of Rectangular and Triangular Fins with Convective Boundary Conditions, *Int. Commun. Heat Mass Transfer*, 12(4), 479.

Aziz, A. (1985b). Graduate Course in Fin Heat Transfer, *Mech. Eng. News*, 22, 2.

Aziz, A. (1992). Optimum Dimensions of Extended Surfaces Operating in a Convective Environment, *Appl. Mech. Rev.*, 45(5), 155.

Aziz, A. (1993). The Effects of Internal Heat Generation, Anisotropy and Base Temperature Nonuniformity from a Two-Dimensional Rectangular Fin, *Heat Transfer Eng.*, 14, 63.

Aziz, A., and Benzies, J. Y. (1976). Application of Perturbation Techniques to Heat Transfer Problems with Variable Thermal Properties, *Int. J. Heat Mass Transfer*, 19, 271.

Aziz, A., and Enamul-Huq, S. M. (1975). Perturbation Solution for Convective Fin with Variable Thermal Conductivity, *J. Heat Transfer*, 97, 300.

Aziz, A., and Lunardini, V. J. (1994). Analytical and Numerical Modeling of Steady Periodic Heat Transfer in Extended Surfaces. *J. Comp. Mech.*, 14, 387–410.

Aziz, A., and Na, T. Y. (1980). Transient Response of Fins by Coordinate Perturbation Expansion, *Int. J. Heat Mass Transfer*, 23, 1695.

Aziz, A., and Na, T. Y. (1981a). Periodic Heat Transfer in Fins with Variable Thermal Parameters, *Int. J. Heat Mass Transfer*, 24(8), 1397.

Aziz, A., and Na, T. Y. (1981b). Perturbation Analysis for Periodic Heat Transfer in Radiating Fins, *Thermodyn. Fluid Dyn.*, 15, 245.

Aziz, A., and Na, T. Y. (1984). *Perturbation Methods in Heat Transfer*, Hemisphere, New York.

Aziz, A., and Nguyen, H. (1992). Two-Dimensional Effects in a Triangular Convecting Fin, *AIAA J. Thermophys. Heat Transfer*, 6, 165.

Aziz, A., and Nguyen, H. (1993). Two-Dimensional Performance of Conducting–Radiating Fins of Different Profile Shapes, *Waerme und Stoffuebertrag.*, 28, 481.

Aziz, A., and Sofrata, H. (1981). Fin Performance in an Oscillating Temperature Environment, *Appl. Energy*, 9(1), 13.

Aziz, A., and Tesarik, D. R. (1992). Two-Dimensional Heat Transfer from a Rectangular Fin with Asymmetrical Boundary Conditions, *Int. Commun. Heat Mass Transfer*, 19, 229.

Baclic, B. S. (1988). ϵ–N_{tu} Analysis of Complicated Flow Arrangements, in R. K. Shah, A. D. Kraus, and D. Metzger (Eds.), *Compact Heat Exchangers*, Hemisphere, New York, 31.

Baker, J. J. (1958). The Efficiency of Composite Fins, *Nucl. Sci. Eng.*, 3, 300

Baker, K. W., and Lund, K. O. (1992). Effect of Anisotropic Conduction and Heat Pipe Interaction on Minimum Mass Space Radiators, *Solar Eng.*, 2.

Bakhru, N., and Liehhard, J. H. (1972). Boiling from Small Cylinders, *Int. J. Heat Mass Transfer*, 15(11), 2011–1025.

Baliga, B. R., and Azrak, R. R. (1984). Laminar Fully-Developed Flow and Heat Transfer in Triangular Plate Fin Ducts, 22nd National Heat Transfer Conference, Niagara Falls, NY, *ASME Pap. 84-HT-55*, American Society of Mechanical Engineers, New York.

Bansal, N. K., and Garg, H. P. (1985). Experimental Study on a Finned Type and Non-porous Type Solar Air Heater with a Solar Simulator, *Energy Convers. Manage.*, 26(2), 135.

Bar-Cohen, A. (1979). Fin Thickness for an Optimized Natural Convection Array of Rectangular Fins, *J. Heat Transfer*, 101, 564.

Bar-Cohen, A., and Jelinek, M. (1985). Optimum Arrays of Longitudinal, Rectangular Fins in Convective Heat Transfer, *Heat Transfer Eng.*, 6(3), 68.

Bar-Cohen, A., and Rohsenow, W. M. (1983). Thermally Optimum Arrays of Cards and Fins in Natural Convection, *IEEE Trans. Comput. Hybrids Manuf. Technol.*, 6, 154–158.

Bar-Cohen, A., and Rohsenow, W. M. (1984). Thermally Optimum Spacing of Vertical, Natural Convection Cooled, Parallel Plates, *J. Heat Transfer*, 106, 116.

Barker, J. J. (1958). Efficiency of Composite Fins, *Nucl. Sci. Eng.*, 3, 300.

Barnes, J. F., and Jackson, J. D. (1961). Heat Transfer to Air, Carbon Monoxide and Helium Flowing Through Smooth Circular Tubes Under Conditions of Large Surface/Gas Temperature Ratio, *J. Mech. Eng. Sci.*, 3(4) 303–314.

Barrow, H., Mistry, J. and Clayton, D. (1986). Numerical and Exact Mathematical Analyses of Two-Dimensional Rectangular Composite Fins, *Proc. 8th International Heat Transfer Conferece*, San Francisco, 2, 367.

Bartas, J. G., and Sellars, W. H. (1960). Radiation Fin Effectiveness, *J. Heat Transfer*, 82, 73.

Bathelt, A. G., and Viskanta, R. (1981). Heat Transfer and Interface Melting and Solidification Around a Finned Heat Source/Sink, *J. Heat Transfer*, 103, 720.

Beatty, K. O., and Katz, D. L. (1948). Condensation of Vapors on Outside of Finned Tubes, *Chem Eng. Prog.*, 44(1), 908.

Beauvais, F. N., and Nickol, H. A. (1964). A Rational Approach to Improved Fin Performance, *ASME Symp. Vol. Air Cooled Heat Exchangers*, American Society of Mechanical Engineers, New York.

Beck, J. V., Cole, K. D., Haji-Sheikh, A., and Litkouki, B. (1992) *Heat Conduction Using Green's Functions*, Hemisphere, Washington, D.C.

Bejan, A. (1995). *Convection Heat Transfer*, 2nd ed. Wiley, New York.

Bejan, A., Fowler, A., and Stanescu, G. (1995). The Optimal Spacing Between Horizonal Cylinders in a Fixed Volume Cooled by Natural Convection, *Int. J. Heat Mass Transfer*, 38(11), 2047

Bella, B., Cavallini, A., Longo, G. A., and Rossetto, L. (1993). Pure Vapour Condensation of Refrigerants-11 and -113 on a Horizontal Integral Finned Tube at High Vapour Velocity, *J. Enhanced Heat Transfer*, 1, 77–86.

Bemrose, C. R., and Bott, T. R. (1984). Correlations for Gas Side Fouling of Finned Tubes, *Inst. Chem. Eng. Symp. Ser.*, 86(1), 357.

Berenson, P. (1960). Film Boiling Heat Transfer from a Horizontal Surface, *J. Heat Transfer*, 83, 351–357.

Bergles, A. E. (1973). Recent Developments in Convective Heat Transfer Augmentation, *Appl. Mech. Rev.*, 26(6), 675–682.

Bergles, A. E. (1975). Burnout in Boiling Heat Transfer. I. Pool Boiling Systems, *Nucl. Safety*, 16(1), 29–42.

Bergles, A. E., and Rohsenow, W. M. (1964). The Determination of Forced Convection Surface-Boiling Heat Transfer, *J. Heat Transfer*, 86, 365–372.

Bergles, A. E., Bakhru, N., and Shirer, J. W., Jr. (1968). Cooling High-Power Density Computer Components, *Rep. DSR 70712-60*, Massachusetts Institute of Technology, Cambridge, MA.

Bergles, A. E., Webb, R. L., Junkhan, G. H. and Jensen, M. K. (1978). Bibliography on Augmentation of Convective Heat and Mass Transfer, *HTL-19-ISU-ERI*, Iowa State University, Ames, IA.

Bert, C. W. (1963). Nonsymmetric Temperature Distributions in Varying Thickness Circular Fins, *J. Heat Transfer*, 85, 77.

Bettanini, E. (1970). Simultaneous Heat and Mass Transfer on a Vertical Surface, *Int. Inst. Refrig. Bull.*, 70(1), 307–317.

Bhargava, S., and Duffin, R. J. (1972). Network Models for Maximization of Heat Transfer Under Weight Constraints, *Networks*, 2, 285.

Bhargava, S., and Duffin, R. J. (1973). Dual Extremum Principles Relating to Cooling Fins, *Q. Appl. Math.*, 31, 27.

Biery, J. C. (1981). Prediction of Heat Transfer Coefficients in Gas Flow Normal to Finned and Smooth Tube Banks, *J. Heat Transfer*, 103, 705.

Bilenas, J. A., and Jiji, L. M. (1970). Perturbation Solution for Fins with Conduction, Convection and Radiation Interaction, *AIAA J.*, 8, 168.

Bilitsky, A. (1986). The Effect of Geometry on Heat Transfer by Free Convection from a Fin Array, M.S. thesis, Department of Mechanical Engineering, Ben-Gurion University of the Negev, Beer Sheva, Israel.

Biyikli, S. (1985). Optimum Use of Longitudinal Fins of Rectangular Profiles in Boiling Liquids, *J. Heat Transfer*, 107, 968.

Bobco, R. P., and Starkovs, R. P. (1970). Rectangular Thermal Doublers of Uniform Thickness, *AIAA J.*, 23.

Bogomolov, Y. N., and Orlova, V. I. (1984). Analysis of the Overall Efficiency of Cooling of a Perforated Wall of a Stationary Gas Turbine Blade, *Heat Transfer Sov. Res.*, 16(2), 111.

Bondurant, D. L., and Westwater, J. W. (1971). Performance of Transverse Fins for Boiling Heat Transfer, *Chem. Eng. Prog. Symp. Ser.*, 67, 30.

Bott, T. R., and Bemrose, C. R. (1983). Particulate Fouling on the Gas Side of Finned Tube Heat Exchangers, *J. Heat Transfer*, 105, 178.

Brandt, F., and Wehle, W. (1983). Eine zusammendfassende Darstellung des Wärmeübergangs Rohrbündeln mit glatten Rohren und Rippenrohren, *VGB Konferenz*.

Brauer, H. (1964). Compact Heat Exchangers, *Chem. Process. Eng.*, 451.

Braunlich, R. (1941). M.S. thesis, Massachusetts Institute of Technology, Cambridge, MA.

Brebbia, C. A., Telles, J. C. F., and Wrobel, L. C. (1984). *Boundary Element Techniques. Theory and Applications in Engineering*, Springer-Verlag, New York.

Breen, B. P., and Westwater, J. W. (1962). Effect of Diameter of Horizontal Tubes on Film Boiling Heat Transfer, *Chem. Eng. Prog*, 58, 67.

Briggs, D. E., and London, A. L. (1961). The Heat Transfer and Flow Friction Characteristics of Five Offset Rectangular and Six Plain Plate Fin Heat Transfer Surfaces, *Proc. 2nd International Heat Transfer Conference*, Boulder, CO, I, 14.

Briggs, A., and Rose, J. W. (1994). Effect of "Fin Efficiency" on a Model for Condensation Heat Transfer on a Horizontal Integral-Fin Tube, *Int. J. Heat Mass Transfer*, 33(6), 457–463.

Briggs, D. E., and Young, E. H. (1963). Convective Heat Transfer and Pressure Drop of Air Flowing Across Triangular Pitch Banks of Finned Tubes, *Chem. Eng. Prog. Symp. Ser.*, 59(41), 1.

Brigham, B. A. (1984). Effect of Channel Convergence on Heat Transfer in a Passage with Short Pin Fins, *NASA Tech. Memo. 83801*. National Aeronautics and Space Administration, Washington, DC.

Brigham, B. A., and Van Fossen, G. J. (1984). Length to Diameter Ratio and Row Number Effects in Short Pin Fin Heat Transfer, *J. Eng. Gas Turbines Power*, 106(1), 241.

Bromley, L. A., Leroy, N. R., and Robbers, A. (1953). Heat Transfer in Forced Convection Film Boiling, *Ind. Eng. Chem.*, 45, 2369–2646.

Brown, A. (1965). Optimum Dimensions of Uniform Annular Fins, *Int. J. Heat Mass Transfer*, 8, 655.

Bryan, W. L. (1962). Heat and Mass Transfer in Dehumidifying Extended Surface Coils, *ASHRAE J.*, 4, 60–63.

Buccini, A., and Soliman, H. M. (1986). Optimum Dimensions of Annular Fin Assemblies, *J. Heat Transfer*, 108, 459.

Bueche, W., and Schau, F. (1936). *Arch. Waermewirtsch. Dampfkesselwes.*, 17, 67.

Burmeister, L. C. (1979). Triangular Fin Performance by the Heat Balance Integral Method, *J. Heat Transfer*, 101, 562.

Burmeister, L. C. (1982). Vertical Fin Efficiency with Film Condensation, *J. Heat Transfer*, 104, 391.

Callinan, J. P., and Berggren, W. P. (1959). Some Radiator Design Criteria for Space Vehicles, *J. Heat Transfer*, 81, 237.

Campo, A. (1976). Variational Techniques Applied to Radiative-Convective Fins with Steady and Unsteady Conditions, *Waerme und Stoffuebertrag.*, 9, 193.

Campo, A., and Wolko, H. S. (1973). Optimum Rectangular Radiative Fins Having Temperature-Variant Properties, *J. Spacecr. Rockets*, 10, 811.

Campo, A. (1977) Unsteady Heat Transfer from a Circular Fin with Nonlinear Dissipation, *Waerme Stoffuebertrag.*, 10, 201–210.

Carnavos, T. C. (1980). An Experimental Study: Condensing R-11 on Augmented Tubes, 19th National Heat Transfer Conference, Orlando, FL, *ASME Pap. 80-HT-54*, American Society of Mechanical Engineers, New York.

Carnavos, T. C., (1981). Heat Transfer Performance of Internally Finned Tubes in Turbulent Flow, *Heat Transfer Eng.*, 1(4), 32.

Carrier, W. H., and Anderson, S. W. (1944). The Resistance to Heat Flow Through Finned Tubing, *Heat. Piping Air Cond.*, 16(5), 304.

Cash, D. R., Klein, G. J., and Westwater, J. W. (1971). Approximate Optimum Fin Design for Boiling Heat Trans., *J. Heat Transfer*, 93, 19.

Cengel, Y. A., and Ngai, T. H. (1991). Cooling of Vertical Shrouded Fin Arrays of Rectangular Profile by Natural Convection, *Heat Transfer Eng.*, 12(4), 27–39.

Chaddock, J. B. (1970). Free Convection Heat Transfer from Vertical Rectangular Fin Arrays, *ASHRAE J.*, 12(8), 53.

Chambers, R. L. and Somers, E. V. (1959). Radiation Fin Efficiency for One-Dimensional Heat Flow in a Circular Fin, *J. Heat Transfer*, 81, 327.

Chang, H. V. (1985). Optimization of Linearized Solutions of Radiating Annular Fin, 23rd AIAA Aerospace Sciences Meeting, Reno, NV, *AIAA Pap. 85-0067*, American Institute of Aeronautics and Astronautics, New York.

Chang, Y. M., Chen, C. K., and Cleaver, J. W. (1982a). Transient Response of Fins by Optimal Linearization and Variational Embedding Methods, *J. Heat Transfer*, 104, 813.

Chang, Y. M., Wang, T. S., and Chen C. K. (1982b). Optimal Linearization and Variational Embedding Solutions for Convective Fins with Variable Thermal Conductivity, *Lett. Heat Mass Transfer*, 9, 99.

Chapman, A. J. (1959). Transient Heat Conduction in Annular Fins of Uniform Thickness, 2nd ASME–AIChE Heat Transfer Conference, Chicago, *AIChE Pa. 10*, American Institute of Chemical Engineers, New York.

Chapman, A. J. (1984). *Heat Transfer*, 4th ed., Macmillan, New York.

Chen, L. T. (1991). Two-Dimensional Fin Efficiency with Combined Heat and Mass Transfer Between Water-Wetted Surface and Moving Moist Airstream, *Int. J. Heat Fluid Flow*, 22, 71.

Chen, C. C., and Westwater, J. W. (1983). Application of the Local Assumption for the Design of Compact Heat Exchangers for Boiling Heat Transfer, *Proc. ASME–JSME Thermal Engineering Joint Conference*, 1, 471.

Chen, S. Y., and Zyskowski, G. L. (1963). Steady State Heat Conduction in a Straight Fin with Variable Heat Transfer Coefficient, 6th National Heat Transfer Conference, Boston, MA, *ASME Pap. 63-HT-1*, American Society of Mechanical Engineers, New York.

Chen, C. C., Loh, J. V., and Westwater, J. W. (1981). Prediction of Boiling Heat Transfer Duty in a Compact Plate-Fin Heat Exchanger Using the Improved Local Assumption, *Int. J. Heat and Mass Transfer*, 24(12), 1907.

Chen, C. K., Chen, H. T., Barrow, H., and Cleaver, J. W. (1983). Study of Laminar Film Condensation Heat Transfer onto Fin Subject to Superheated Vapor, *Proc. ASME–JSME Thermal Engineering Joint Conference*, 3, 285.

Chen, C. K., Chen, H. T., and Cleaver, J. W. (1984). Laminar Film Condensation on a Downward Vertical Fin, *Proc. 3rd Symposium–Workshop on Multi-phase Flow and Heat Transfer*, Miami Beach, FL, 787.

Chepernoi, M. N., Lomakin, V. N., Shnaider, V. E., and Chepernoi, V. M. (1985). Study of the Process of Frost Formation on Finned Air Coolers, *J. Eng. Phys.*, 48(1), 34.

Chiou, J. P., and Na, T. Y. (1977). Initial Value Method for the Solution of Radiation Heat Transfer Fin Equations, *Ind. Math.*, 27(1), 13.

Choudhury, D., and Patankar, S. V. (1985). Analysis of Developing Laminar Flow and Heat Transfer in Tubes with Radial Internal Fins, *ASME Symp. Vol. HTD-43*, American Society of Mechanical Engineers, New York, 57.

Christopher, D. M., and Pearson, J. T. (1980). Parametric Analysis of Louvered Air-Heating Solar Collectors, *ASME Pap. 80-WA/Sol-38*, American Society of Mechanical Engineers, New York.

Chu, H. S. (1977). Analysis on the Transient Heat Transfer of Two-Dimensional Straight Fins, M.S. thesis, National Chengking Univ., Taiwan.

Chu, H. S., Chen, C. K., and Weng, C. I. (1982). Applications of Fourier Series Techniques to Transient Heat Transfer Problems, *Chem. Eng. Commun.*, 16, 215.

Chu, H. S., Chen, C. K., and Weng, C. I. (1983a). Transient Response of Circular Pins, *J. Heat Transfer*, 105, 205.

Chu, H. S., Weng, C. I., and Chen, C. K. (1983b). Transient Response of a Composite Straight Fin, *J. Heat Transfer*, 105, 307.

Chumakov, V. L. (1974). Concerning the Calculation of Steady-State Heat Transfer for Radiating Fins, *Heat Transfer Sov. Res.*, 6(6), 122.

Chung, B. T. F., and Nguyen, L. D. (1985). Thermal Analysis and Optimum Design for Radiating Spine of Various Geometries, in B. X. Wang (Ed.), *Heat Transfer in Science and Technology*, Hemisphere, New York.

Chung, B. T. F., and Nguyen, L. D. (1986). Optimization of Design Parameters for Radiating Longitudinal Fins of Various Geometries, 24th AIAA Aerospace Sciences Meeting, Reno, NV, *Pap. AIAA-86-0150*, American Institute of Aeronautics and Astronautics, New York.

Churchill, S. W. (1983). Free Convection Around Immersed Bodies, in E. Schlunder (Ed.), *Heat Exchanger Design Handbook*, Hemisphere, New York, Sect. 2.5.7.

Churchill, S. W., and Bernstein, M. (1977). A Correlating Equation for Forced Convection from Gases and Liquids to a Circular Cylinder in Cross Flow, *J. Heat Transfer*, 99, 300.

Churchill, S. W., and Chu, H. H. S. (1975a). Correlating Equations for Laminar and Turbulent Free Convection from a Horizontal Cylinder, *Int. J. Heat Mass Transfer*, 18, 1049.

Churchill, S. W., and Chu, H. H. S. (1975b). Correlating Equations for Laminar and Turbulent Free Convection from a Vertical Flat Plate, *Int. J. Heat Mass Transfer*, 18, 1323.

Churchill, S. W., and Ozoe, H. (1973). Correlations for Laminar Forced Convection in Flow over an Isothermal Flat Plate in Developing and Fully Developing Flow in an Isothermal Tube, *J. Heat Transfer*, 95, 46.

Churchill, S. W., and Usagi, R. (1972). A General Expression for the Correlation of Rates of Heat Transfer and Other Phenomena, *AIChE J.*, 18(6), 1121–1138.

Cobble, M. H. (1964). Nonlinear Fin Heat Transfer, *J. Franklin Inst.*, 277(3), 206.

Cobble, M. H. (1971). Optimum Fin Shape, *J. Franklin Inst.*, 291, 283.

Cohen, B. M. (1969). Heat Transfer by Conduction and Radiation with Temperature Dependent Thermal Conductivity, *J. Heat Transfer*, 91, 21.

Colburn, A. P. (1933) A Method of Correlating Forced Convection Heat Transfer Data and a Comparison with Fluid Friction, *Trans. AIChE*, 29, 174–210.

Cole, R. (1960). Photographic Study of Boiling in the Region of Critical Heat Flux, *AIChe J.*, 6, 533–538.

Cole, R., and Rohsenow, W. M. (1969). Correlation of Bubble Diameters of Saturated Liquids, *Chem. Eng. Prog. Symp. Ser.*, 65, 92, 211–213.

Coleman, H. W., and Lemmon, E. C. (1973). Turbulent Interference Heating on Several Small Fin Configurations, *J. Spacecr. Rockets*, 10, 820.

Colle, S. (1983). General Solution for the Heat Transfer in Arrays of Radiative Fins of Arbitrary Shape, 21st National Heat Transfer Conference, Seattle, WA, *ASME Pap. 83-HT-57*, American Society of Mechanical Engineers, New York.

Collicott, H. E., Fontaine, W. E., and Grosh, R. J. (1965). Free Convection and Radiation Heat Transfer from Cylindrical Fins, *Trans. ASHRAE*, 71(II), 148.

Coney, J. E. R., Kazeminejad, H., and Sheppard, C. G. W. (1989a). Dehumidification of Air on a Vertical Rectangular Fin: A Numerical Study, *Proc. Inst. Mech. Eng.*, 203, 165–175.

Coney, J. E. R., Kazeminejad, H., and Sheppard, C. G. W. (1989b). Dehumidification of Turbulent Air Flow over a Thick Fin: An Experimental Study, *Proc. Inst. Mech. Eng.*, 203, 177–188.

Cooper, G. S., and Buford, D. L. (1984). Comparison of Methods for Calculating Pressure Drops Across Bar and Solid-Finned Tubes in Fired-Heater Convection Sections, *Proc. API Refining Dep.*, 63, 325.

Cooper, J. R., and Rose, J. W. (1981). Condensation Heat Transfer Enhancement by Vapour-Side Geometry Modification, *Proc. 1981 HTFS Research Symposium*, pap. RS402, Oxford, 647.

Cotta, R. M. (1990). Hybrid Numerical/Analytical Approach to Nonlinear Diffusion Problems, *Num. Heat Transfer*, 17, 217.

Cowan, G. H., Dell, F. R., and Stinchcombe, R. A. (1974). Aerodynamic and Heat Transfer Performance at an Industrial Air Cooled Heat Exchanger, *Proc. 5th International Heat Transfer Conference*, 5, 155.

Cox, B., and Jallouk, P. A. (1973). Methods for Evaluating the Performances of Compact Heat Exchanger Surfaces, *J. Heat Transfer*, 95, 464.

Crawford, H. L. (1978). Thermal Radiation from a Nodal Network on Large Circular Fins, *Proc. 6th International Heat Transfer Conference*, Toronto, Ontario, Canada, 3, 341.

Crawford, M. (1982). Heat Transfer in Trapezoidal Straight Fins with a Periodically Varying Base Temperature, *ASME Pap. 82-WA/HT-41*, American Society of Mechanical Engineers, New York.

Cuffe, K. W., Beatenbough, P. K., Daskavitz, M. J., and Flower, R. J. (1978). Plate Fin Regenerators for Industrial Gas Turbines, *ASME Pap. 78-GT-26*, American Society of Mechanical Engineers, New York.

Culham, J. R., Yovanovich, M. M., and Lee, S. (1994). Thermal Modeling of Isothermal Cuboids and Rectangular Heat Sinks Cooled by Natural Convection, *Proc. IEEE Intersociety Conference on Thermal Performance*, 73.

Cumo, M., Pinchera, G. C., and Urbani, G. (1964) Distribuzióne del Coefficiènte di Addizióne Tèrmica Lungo il Profilo di una Alettatura, *Termotecnica*, 18(1), 14.

Cumo, M., Lopez, S., and Pinchera, G. C. (1965a). Numerical Calculation of Extended Surface Efficiency, *Chem. Eng. Prog. Symp. Ser.*, 61(59), 225.

Cumo, M., Lopez, S., and Pinchera, G. C. (1965b). *Modalità di Utilizzo del Còdice: Relazióne Promossa dal Comitato Nazionale per l'Energia Nucleare*, Centro di Studi Nucleari della Casaccia, Rome.

Dalle Donne, M., and Bowditch, P. W. (1963). Experimental Local Heat Transfer and Friction Coefficients for Subsonic Laminar, Transitional and Turbulent Flow of Air or Helium in a Tube at High Temperatures, *Dragon Proj. Rep. 184*, Winfirth, Dorchester, Dorset, England.

Dalle Donne, M., and Meerwald, E. (1970). Heat Transfer from Surface Roughened by Thread-Type Ribs at High Temperature, *Proc. Heat Transfer and Fluid Mechanics Institute*, 354.

Date, A. W. (1974). Prediction of Fully-Developed Flow in a Tube Containing a Twisted Tape, *Int. J. Heat Mass Transfer*, 17(8), 845.

Davenport, C. J. (1983a). Heat Transfer and Fluid Flow in the Louvered-Fin Heat Exchanger Surfaces, in J. Taborek, G. F. Hewitt and N. Afgan (Eds.), *Heat Exhangers: Theory and Practice*, Hemisphere, New York.

Davenport, C. J. (1983b). Correlations for Heat Transfer and Flow Friction Characteristics of Louvered Fins, *AIChE Symp. Ser.*, 79, 19.

Ded, J. S., and Lienhard, J. H. (1972). The Peak Boiling Heat Flux from a Sphere, *AIChE J.*, 18(2), 337–242.

Deissler, R. G. (1951). Analytical Investigation of Fully Developed Laminar Flow in Tubes with Heat Transfer and Fluid Properties Variable Along the Radius, *NACA TN-2410*, National Advisory Committee on Aeronautics, Washington, DC.

Delfour, M. C., Payre, G., and Zolesio, J. P. (1983). Optimal Design of a Radiating Fin for Communications Satellites, *Proc. 4th IFAC Workshop on Applications of Nonlinear Programming to Optimizations and Control*, San Francisco, 45.

DeLorenzo, B., and Anderson, R. B. (1945). Heat Transfer and Pressure Drop of Liquids in Double Pipe Heat Exchangers, *Trans. ASME*, 67, 697.

Dent, J. C. (1968). An Electrical Network Method for Combined Free Convection and Radiative Heat Transfer from Annular Finned Surfaces, *Br. Chem. Eng.*, 13(1), 90.

Despretz, C. M. (1822). *Ann. Chim. Phys.*, 19, 97.

Despretz, C. M. (1828a). *Ann. Chim. Phys.*, 36, 422.

Despretz, C. M. (1828b). *Ann. Phys. Chem.*, 12, 281.

Dittus, F. W., and Boelter, L. M. K. (1930). *Univ. Calif. (Berkeley) Publ. Eng.*, 2, 443.

Dolgii, A. A., Grinberg, Y. N., Vivsik, S. N., Zevin, I. I., Fadeev, E. A., Moskvichev, V. F., Sviridova, N. M., Zholudov, Y. S., and Lipets, A. U. (1973). Finned Heating Surfaces for Boilers, *Therm. Eng.*, 20, 6.

Donaldson, A. B., and Shouman, A. R. (1972). Unsteady-State Temperature Distribution in a Convecting Fin of Constant Area, *Appl. Sci. Res.*, 26(1), 75.

Donovan, R. C., and Rohrer, W. M. (1971). Radiative and Convective Fins on a Plane Wall Including Mutual Irradiation, *J. Heat Transfer*, 93, 41.

Duffin, R. J. (1959). A Variational Problem Relating to Cooling Fins, *J. Math. Mech.*, 8(1), 47.

Dusinberre, G. M. (1956). Fin Efficiency, *Mech. Eng.*, 78, 500.

Dyer, W. F., and Schaedle, G. C. (1970). Application of a Pin Fin Coldplate to Apollo Spacecraft Electronic Equipment Cooling, *ASME Pap. 70/HT/SpT-25*, American Society of Mechanical Engineers, New York.

Eckels, P. W., and Rabas, T. J. (1985a). Heat Transfer and Pressure Drop of Typical Air Cooler Finned Tubes, *J. Heat Transfer*, 107, 198.

Eckels, P. W., and Rabas, T. J. (1985b). Heat Transfer and Pressure Drop Performance of Finned Tube Bundles, *J. Heat Transfer*, 107, 205.

Eckert, E. R. G., and Soehngen, E. (1952). Distribution of Heat Transfer Coefficients Around Circular Cylinders in Crossflow at Reynolds Numbers from 20 to 500, *Trans. ASME*, 74, 343–346.

Eckert, E. R. G., Irvine, T. F., and Sparrow, E. M. (1960). Analytic Formulation for Radiating Fins with Mutual Irradiation, *Amer. Rocket Soc. J.*, 30(7), 644.

Eder, H. (1973). Dimensioning of Fins with General Nonlinear Cooling Laws, *Thermodyn. Fluid Dyn.*, 6, 86.

Edwards, D. K., Gier, K. D., Ayyaswamy, P., and Catton, I. (1973a). Evaporation and Condensation in Circumferential Grooves on Horizontal Tubes, 12th ASME–AIChE National Heat Transfer Conference, Atlanta, GA, *Paper 73-HT-25*, American Society of Mechanical Engineers, New York.

Edwards, F J., Henry, T. A., and Hayward, G. L. (1973b). Characteristics of the Tube and Continuous Plate Fin Type of Compact Heat Exchanger, *Proc. Recent Developments in Compact High Duty Heat Exchange Conference*, Institute of Mechanical Engineers, London.

Edwards, D. K., Balakrishnan, A., and Catton, I. (1974). Power-Law Solutions for Evaporation from a Finned Surface, *J. Heat Transfer*, 96, 423.

Egorov, E. D., Daschyan, A. A., Nekrasov, M. I., Bugorskii, M. I., and Pikus, V. Y. (1984). Research on Heat Transfer in Banks of Double-Wall Finned Pipes, *Sov. Energy Technol.*, 1, 14

Elenbaas, W. (1942). Heat Dissipation of Parallel Plates by Free Convection, *Physica*, 9(1), 1.

Elmahdy, A. H., and Biggs, R. C. (1979). Finned Tube Heat Exchanger: Correlation of Dry Surface Heat Transfer Data, *Trans. ASHRAE*, 85, 117.

Elmahdy, A. H., and Biggs, R. C. (1983). Efficiency of Extended Surfaces with Simultaneous Heat and Mass Transfer, *Trans. ASHRAE*, 89, 135.

El-Refaie, M. E., El-Ridi, M. K., and El-Kady, M. A. (1983). Incorporation of Fin Effect in Predicting the Performance of Cascaded Solar Stills, *Proc. First Arab International Solar Energy Conference*, Kuwait, United Arab Republic, 336.

El-Ridi, M. K., Chumak, I. G., and Kalinin, L. G. (1976). Free Convection Heat and Mass Transfer on Finned Tubes, *Heat Transfer Sov. Res.*, 8(1), 33.

Eno, B. E. (1967). Combined Convection and Radiation from Rectilinear Fins, *Trans. ASHRAE*, 73(1), pap. 2022.

Epstein, N., and Sandhu, K. (1978). Effect of Uniform Fouling Deposit in Total Efficiency of Extended Heat Transfer Surfaces, *Proc. 6th International Heat Transfer Conference*, Toronto, Ontario, Canada, 4, 397

Eriksen, V. L. (1979). Waste Heat Recovery Exchangers, *ASME Symp. Vol. HTD-10*, American Society of Mechanical Engineers, New York.

Eslinger, R. G., and Chung, B. T. F. (1979) Periodic Heat Transfer in Radiating or Convecting Fins or Fin Arrays, *AIAA J.*, 17, 1134.

Evenko, V. I., and Shishkov, V. M. (1969). Generalized Heat Transfer and Resistance Relations for a Tubular Surface Finned with a Corrugated Strip, *Therm. Eng.*, 16, 51.

Evreinova, V. S. (1967). Study of the Heat Transfer and Hydraulic Resistance of Longitudinally Finned Surfaces, *Kholod. Tekhnol.*, 44(9), 38.

Feldman, K., and Lu, D. C. (1976). Heat Pipe Exchanger Design Considerations, *Proc. 11th Intersociety Energy Conversion Engineering Conference*, State Line, NE, 887.

Finlay, I. C. (1971). Air/Water Sprays as Potential Coolants for Pin Finned Cold Plates, *ARL 71-0087*, U. S. Air Force Systems Command, Aerospace Research Laboratory, Fairborn, OH.

Fitzroy, N. D. (1971). Optimum Spacing of Fins Cooled by Free Convection, *J. Heat Transfer*, 93, 462.

Focke, R. (1942). Die Nadel als Kühlelemente, *Forsch. Geb. Ingenieurwes.*, 13, 34.

Forslund, R. P. (1972). An Experimental Technique for Determining the Diffusion Effect of Braze Material on the Thermal Conductance of Thin Copper Fins, *J. Heat Transfer*, 94, 273.

Fortescue, P. (1957). Some Characteristics of Reactor Gas Cooling Systems, *Nucl. Power*, 2(13), 188.

Fourier, J. B. J. (1822). *The Analytic Theory of Heat*, translated by A. Freeman, Cambridge University Press, Cambridge, 1878.

French, R. J., Jr. (1964). Analysis and Experimental Verification of Space Radiator Performance, *SAE-ASME Pap. 845B*, American Society of Mechanical Engineers, New York.

Frost, W., and Eraslan, A. H. (1968). Solution of Heat Transfer in a Fin with Combined Convection and Radiative Interaction Between the Fin and the Surrounding Surfaces, *Proc. Heat Transfer and Fluid Mechanics Institute*, 206.

Fucinari, C. A. (1977). Effect of Fin Geometry and Manufacturing Process on Ceramic Regenerator Thermodynamic Performance, *J. Eng. Power*, 99(4), 638.

Fujie, K., Itoh. N., Innami, T., Kimura, H., Nakayama, N., and Yanagida, T. (1977). Heat Transfer Pipe, U.S. patent 4,044,797.

Fujii, T. (1991). *Theory of Laminar Film Condensation*, Springer-Verlag, New York.

Fujikake, K., Aoki, H., and Mitui, H. (1983). An Apparatus for Measuring the Heat Transfer Coefficients of Finned Tube Exchangers by Use of a Transient Method, *Proc. Jpn. 20th Symposium on Heat Transfer*, 466–468.

Gambill, W. R. (1962). Generalized Prediction of Burnout Heat Flux for Flowing Subcooled, Wetting Liquids, 5th National Heat Transfer Conference, Houston, TX, *AIChE Prepr. 17*, American Institute of Chemical Engineers, New York.

Ganguli, A., Tung, S. S., and Taborek, J. (1985). Parametric Study of Air-Cooled Heat Exchanger Finned Tube Geometry, AICHE Symp. Ser., 81 (245), 122–128.

Gardner, K. A. (1945). Efficiency of Extended Surface, *Trans. ASME*, 67, 621.

Gardner, K. A. (1951). Discussion on paper by M. L. Ghai, *Proc. General Discussion on Heat Transfer*, Institute of Mechanical Engineers, London, 214.

Gardner, K. A. (1964). Rational Design Temperature Limits for Interference-Fit Finned Tubing, *ASME Symp. Vol. Air Cooled Heat Exchangers*, American Society of Mechanical Engineers, New York.

Gardner, K. A., and Carnavos, T. C. (1960) Thermal Contact Resistance in Finned Tubing, *J. Heat Transfer*, 82, 279.

Garg, V. K., and Velusamy, K. (1986). Heat Transfer Characteristics for a Plate Fin, *J. Heat Transfer*, 108, 224.

Gebhart, B. (1993). *Heat Conduction and Mass Diffusion*, McGraw-Hill, New York.

Gebhart, B., Jaluria, Y., Mahajan, R. L., and Sammakia, B. (1988). *Buoyancy Induced Flows and Transport*, Hemisphere, Washington, DC.

Geidt, W. H. (1949). Investigation of Variation of Point Unit Heat Transfer Coefficients Around a Cylinder Normal to an Air Stream, *Trans. ASME*, 71, 375.

Gentry, C. C., and Small, W. M. (1978). Heat Transfer and Power Expenditure for Agitated Vessels Having Bare and Finned Vertical Tube Baffles, *Proc. 5th International Heat Transfer Conference*, Toronto, Ontario, Canada, 4, 13.

Ghai, M. L. (1951). Heat Transfer in Straight Fins, *Proc. General Discussion of Heat Transfer*, Institute of Mechanical Engineers, London, 180.

Gianolio, E., and Cuti, F. (1981). Heat Transfer Coefficients and Pressure Drops of Air Coolers with Different Numbers of Rows Under Induced and Forced Draft, *Heat Transfer Eng.*, 3(1), 38.

Gilliland, E. R., and Sherwood, T. K. (1934). Diffusion of Vapors into Air Streams, *Ind. Eng. Chem.*, 26, 516–523, 761.

Gnielinski, V. (1976). New Equations for Heat and Mass Transfer in Turbulent Pipe and Channel Flow, *Int. Chem. Eng.*, 16, 359.

Gogonin, I. I., and Svorkova, I. N. (1973). Heat Transfer During Boiling of Freon-21 on Finned Surfaces, *Int. Chem. Eng.*, 13(4), 698.

Gogonin, I. I., and Dorokhov, A. R. (1981). Enhancement of Heat Transfer in Horizontal Shell-and-Tube Exchangers, *Heat Transfer Sov. Res.*, 3, 119–126.

Goldberg, N. (1984). Narrow Channel Forced Air Heat Sink, *IEEE Trans. Cumput. Hybrids Manuf. Technol.* 7, 154-159.

Goldstein, L., and Sparrow, E. M. (1976). Experiments on the Transfer Characteristics of a Corrugated Fin and Tube Heat Exchanger Configuration, *J. Heat Transfer*, 98, 26.

Goshdastidar, P. S., and Muukhopadhyay, A. (1989). Transient Heat Transfer from a Straight Composite Fin: A Numerical Solution by ADI, *Int. Commun. Heat Mass Transfer*, 16, 257.

Graetz, L. (1885). Über die Warmeleitungfähigkeit von Flüssigkeiten (On the Thermal Conductivity of Liquids), Part II, *Ann. Phys. Chem.*, 18, 79.

Graham, R. W., and Hendricks, R. C. (1967). Assessment of Convection, Conduction and Evaporation in Nucleate Boiling, *NASA TND-3943*, National Technical Information Service, Springfield, VA.

Grakovich, L. P., and Konev, S. V. (1974). Cooling of Electronic Circuits and Instruments, *Heat Transfer Sov. Res.*, 6(3), 123.

Granet, I., and McIlroy, W. (1961). Optimum Radiant Straight Fin with Exponential Sides, *Amer. Rocket Soc. J.*, 31(1), 80.

Gray, D. L., and Webb, R. L. (1986). Heat Transfer and Friction Correlations for Plate Fin-and-Tube Heat Exchangers Having Plain Fins, *Proc. 8th International Heat Transfer Conference*, San Francisco, Ca, 6, 2745–2750.

Gregorig, R. (1954). Film Condensation on Finely Rippled Surfaces with Consideration of Surface Tension, *Z. Angew. Math. Phys.*, 5, 36.

Gross, S. (1970). Thermal Coupling of Equipment by Interleaving Fins, *J. Spacecr. Rockets*, 7(4), 489.

Guceri, S., and Maday, C. J. (1975). Least Weight Circular Cooling Fin, *J. Eng. Ind.*, 97(4), 1190.

Guillory, J. L., and McQuiston, F. C. (1973). Experimental Investigation of Air Dehumidification in a Parallel Plate Exchanger, *ASHRAE Trans.*, 79(2), 146–151.

Gunter, A. Y., and Shaw, W. A. (1945). A General Correlation of Friction Factors for Various Types of Cross Flow, *Trans. ASME*, 67, 643.

Hagen, K. D. (1988). Perturbation Analysis of Tapered Fins with Nonlinear Thermal Properties, *J. Thermophys. Heat Transfer*, 2(3), 276.

Haley, K. W. (1965). Ph.D. thesis, Dept of Chem. Eng., University of Illinois, Urbana, IL.

Haley, K. W., and Westwater, J. W. (1966). Boiling Heat Transfer from Single Fins, *Proc. 3rd International Heat Transfer Conference*, Chicago, 3, 245.

Han, L. S., and Lefkowitz, S. G. (1960). Constant Cross Section Fin Efficiencies for Nonuniform Surface Heat Transfer Coefficients, *ASME Pap. 60-WA-41*, American Society of Mechanical Engineers, New York.

Hansen, A. G. (1964). *Similarity Analyses of Boundary Value Problems in Engineering*, Prentice Hall, Englewood Cliffs, NJ.

Harper, D. R., and Brown, W. B. (1922). Mathematical Equations for Heat Conduction in the Fins of Air Cooled Engines, *NACA Rep. 158*, National Advisory Committee on Aeronautics, Washington, DC.

Harris, M. J., and Wilson, J. T. (1961). Heat Transfer and Fluid Flow Investigations on Large-Scale Transverse Fins, *Br. Nucl. Energy Conf. J.*, 6(4), 330.

Hatada, T., and Senshu, T. (1984). Experimental Study on Heat Transfer Characteristics of Convex Louver Fins for Air Conditioning Heat Exchangers, 22nd National Heat Transfer Conference, Niagara Falls, NY, *ASME Pap. 84-HT-74*, American Society of Mechanical Engineers, New York.

Hati, S. K., and Rao, S. S. (1983). Determination of Optimum Profile of One-Dimensional Cooling Fins, *J. Vibrat. Acoust. Stress Reliab. Des.*, 105, 317.

Hausen, H. (1940). Wärmeubertragung durch Rippenrohe, *Z. Ver. Dtsch. Ing. Beiheft*, 55.

Hausen, H. (1943). Darstellung des Wärmeauberganges in Rohren Durch Verallgemeinerte Potenzbeziehungen, *Z. Ver. Dtsch. Ing.*, 4, 91.

Hausen, H. (1959). Neue Gleichungen für Wärmeaubertragung bei Frier oder erzwungener Stromung, *Allg. Waermetech.*, 9, 75–79.

Hausen, H. (1974). Extended Equation for Heat Transfer in Tubes at Turbulent Flow, *Waerme Stoffuebertrag.*, 7, 222–225.

Hauser, S. G., Martini, W. R., and Scheffler, W. A. (1981). Experimental Measurements of Enhanced Heat Transfer in a Stirling Engine Model with Interleaving Fins, *Proc. Intersociety Energy Conversion Engineering Conference*, Atlanta, GA, 1925.

Hazeler, L. (1980). Condensation of Nitrogen in Brazed Aluminum Plate Fin Heat Exchangers, 19th Heat Transfer Conference, Orlando, FL., *ASME Pap. 80-HT-57*, American Society of Mechanical Engineers, New York.

Heaslet, M. A., and Lomax, H. (1961). Numerical Predictions of Radiative Interchange Between Conducting Fins with Mutual Irradiations, *NASA Tech. Rep. R-116*, National Aeronautics and Space Administration, Washington, DC.

Heggs, P. J., and Stones, P. R. (1980). The Effects of Dimensions on the Heat Flow Rate Through Extended Surfaces, *J. Heat Transfer*, 101, 180.

Heggs, P. J., and Stones, P. R. (1983). *Two-Dimensional Heat Flow Through Fin Assemblies, Heat Exchangers: Theory and Practice*, Hemisphere, New York, 917.

Heggs, P. J., Ingham, D. B., and Manzoor, M. (1982). The Analysis of Fin Assembly Heat Transfer by a Series Truncation Method, *J. Heat Transfer*, 104, 210.

Hering, R. G. (1966). Radiative Heat Exchange with Specular Reflection, *J. Heat Transfer*, 88, 29.

Hetsroni, G. (Ed) (1982). *Handbook of Multiphase Systems*, Hemisphere, New York.

Hewitt, T. A., Hill, D. C., and McCullough, R. W. (1978) Design of High Efficiency Flat Plate Solar Collectors for Space and Water Heating, *ASME Pap. 78-DE-7*, American Society of Mechanical Engineers, New York.

Hilding, W. E. (1953) Effect of Fin Geometry on Performance of Heat Exchanger Surfaces, *ASME Pap. 53-S-42*, American Society of Mechanical Engineers, New York.

Hilding, W. E., and Coogan, C. H., Jr. (1964). Heat Transfer and Pressure Loss Measurements in Internally Finned Tubes, *ASME Symp. Vol. Air Cooled Heat Exchangers*, American Society of Mechanical Engi- neers, New York.

Hirasawa, S., Hijikata, K., Mori, Y., and Nakayama, W. (1980). Effect of Surface Tension and Condensate Motion in Laminar Film Condensation (Study of Liquid Film in a Small Trough), *Int. J. Heat Mass Transfer*, 23(11), 1471.

Hofmann, H. (1976). Möglichkeiten zur Berechnung des Wärmeübergangs und des Druckverlustes vin Rippenrohrbünden, *Luft Kaeltetech.*, 12, 136.

Hollworth, B. R., and Gero, L. R. (1985). Entrainment Effects on Impingement Heat Transfer: II. Local Heat Transfer Measurements, *J. Heat Transfer*, 107, 910.

Holman, J. P. (1990). *Heat Transfer*, 7th ed., McGraw-Hill, New York.

Honda, H., and Nozu, S. (1987a). A Prediction Method for Heat Transfer During Film Condensation on Horizontal Low Integral-Fin Tubes, *J. Heat Transfer*, 109, 218–225.

Honda, H., and Nozu, S. (1987b). Effect of Drainage Strips on the Condensation Heat Transfer Performance of Horizontal Finned-Tubes, in B. X. Wang (Ed.), *Heat Transfer Science and Technology*, Hemisphere, New York, 455–462.

Honda, H., and Nozu, S. (1987c). A Theoretical Model of Film Condensation on Horizontal Low Finned Tubes, *J. Heat Transfer*, 109, 525-532.

Honda, H., Nozu, S., and Mitsumori, K. (1983). Augmentation of Condensation on Horizontal Finned Tubes by Attaching a Porous Drainage Plate, *Proc. ASME–JSME Thermal Engineering Joint Conference*, 3, 289.

Honda, H., Nozu, S., Uchima, B., and Fujii, T. (1986) Effect of Vapor Velocity of Film Condensation of R-113 on Horizontal Tubes in a Crossflow, *Int. J. Heat Mass Transfer*, 23(11), 1471.

Honda, H., Nozu, S., and Takada, K. (1987a). Flow Characteristics of Condensate on a Vertical Column of Horizontal Low Finned Tubes, *Proc. ASME–JSME Thermal Engineering Joint Conference*, 1, 517–524.

Honda, H., Nozu, S., and Uchima, B. (1987b). A Generalized Prediction Method for Heat Transfer During Film Condensation on a Horizontal Low Finned Tube, *Proc. ASME–JSME Thermal Engineering Joint Conference*, 4, 385–392.

Horn, G., and Atherton, A. (1969). Statistical Techniques for Evaluating Experimental Performance Results in Batch Heat Exchanger Elements, *Inst. Chem. Eng. Trans.*, 47(2), T43.

Houghton, J. M., Ingham, D. B., and Heggs, P. J. (1990). *Oscillatory Heat Transfer in Extended Surfaces*, Heat Transfer 1990, Hemisphere, New York, 4, 111–116.

Houghton, J. M., Ingham, D. B., and Heggs, P. J. (1991). The Boundary Element Method Applied to Oscillatory Heat Transfer in a Fin, *Numerical Methods Thermal Problems*, 7, 51–62.

Houghton, J. M., Ingham, D. B., and Heggs, P. J. (1992). The One Dimensional Analysis of Oscillating Heat Transfer in a Finned Assembly, *J Heat Transfer*, 114, 546–552.

Houghton, J. M., Ingham, D. B., and Heggs, P. J. (1993a). A New Implementation of the Boundary Element Technique for Time-Dependent Pseudo Stationary Heat Transfer, *Engineering Analysis with Boundary Elements*, 11, 55–65.

Houghton, J. M., Ingham, D B., and Heggs, P. J. (1993b). An Application of the Boundary Element Technique to Periodic Heat Transfer in a Fin Assembly, *Num. Heat Transfer*, Part A, 23, 379–398.

Hrymak, A. N., McRae, G. J., and Westerberg, A. W. (1985). Combined Analysis and Optimization of Extended Heat Transfer Surfaces, *J. Heat Transfer*, 107, 527.

Hsu, Y. Y., (1968). Analysis of Boiling on a Fin, *NASA-TNO-4797*, National Aeronautics and Space Administration, Washington, DC.

Hsu, T.-H., and Tsai, S.-Y. (1992). Thermal Response of a Heated Vertical Plate with Natural Convection-Interaction, *Num. Heat Transfer*, Part A, 22, 307.

Hu, M. H., and Chang, Y. P. (1973). Optimization of Finned Tubes for Heat Transfer in Laminar Flow, 13th Heat Transfer Confereence, Atlanta, GA, *ASME Pap. 73-HT-11*, American Society of Mechanical Engineers, New York.

Huang, C. C. (1985). Analysis on the Transient Response of Two-Dimensional Various Shaped Annular Fins, *Model. Simul. Control*, Ser. B, 5(2), 33.

Huang, M. J., and Chen, C. K. (1984). Vertical Circular Pin with Conjugated Conduction-Forced Convection Flow, *J. Heat Transfer*, 196, 658.

Huang, M. J., and Chen, C. K. (1985). Conjugate Mixed Convection and Conduction Heat Transfer Along a Vertical Circular Pin, *Int. J. Heat Mass Transfer*, 28(3), 523.

Huang, L. J., and Shah, R. K. (1992). Assessment of Calculation Methods for Efficiency of Straight Fins of Rectantangular Profile, *Int. J. Heat Fluid Flow*, 13, 282.

Huang, X. S., Briggs, A. R. and Rose, J. W. (1994). Effect of Thermal Conductivity of Tube Material on Condensate Heat Transfer for Integral-Fin Tubes, *Proc. 10th International Heat Transfer Conference*, Brighton, England.

Hughes, D. F., and Slack, M. R. (1958). A Correlation of Heat Transfer for Finned Fuel Elements for Carbon Dioxide Reactors, *Proc. 2nd United Nations International Conference on the Peaceful Uses of Atomic Energy*, 7, 717.

Hung, H. M. (1969). Heat Transfer in Thin Fins with Stochastic Root Temperature, *J. Heat Transfer*, 91, 129.

Hung, H. M., and Appl, F. C. (1967). Heat Transfer of Thin Fins with Tem- perature Dependent Thermal Properties and Internal Heat Generation, *J. Heat Transfer*, 89, 155.

Hunn, B. D. (1974). Effects of External Fouling on Dry Cooling Tower Performance, *Front of Power Technology Conference*, 7th Annual Proc., Oklahoma State University, Stillwater, OK, paper 6.

Hwang, L. T., Turlik, I., and Reisman, A. (1987). A Thermal Module Design for Advanced Packaging, *J. Electron. Mater.*, 16, 347–355.

Incropera, F. P. (1988). Convection Heat Transfer in Electronic Equipment Cooling, *J. Heat Transfer*, 110, 1097.

Incropera, F. P. and Dewitt, D. P. (1990). *Introduction to Heat Transfer*, Wiley, New York.

Ingham, D. B., Manzoor, M., Heggs, P. J., and Stones, P. R. (1981). Two-Dimensional Analysis of Fin Assembly Heat Transfer: A Comparison of Solution Techniques, *Proc. 2nd National Symposium on Numerical Properties and Methodologies in Heat Transfer*, Hemisphere, New York, 135.

Iqbal, M., and Aggarwala, B. D. (1968). Temperature Distribution in a Two-Dimensional Rectangular Solid in Interplanetary Space, *Proc. ASME Annual Aviation and Space Conference*, New York, 625.

Irey, R. K. (1968). Errors in the One-Dimensional Fin Solution, *J. Heat Transfer*, 90, 175.

Ivey, H. J., and Morris, D. J. (1962). On the Relevance of the Vapor–Liquid Exchange Mechanism for Subcooled Boiling Heat Transfer at High Pressure, *UK Rep.*, London, *AEEW-R-137*.

Jaber, M. H., and Webb, R. L. (1994). An Experimental Investigation of Enhanced Tubes for Steam Generators, *Exp. Heat Transfer*, 6(1), 35–54.

Jakob, M. (1949). *Heat Transfer*, Wiley, New York.

Jameson, S. L. (1945). Tube Spacing in Finned Tube Banks, *Trans. ASME*, 67, 633.

Jones, J. K. (1970). Boiling Heat Transfer Enhancement Using Integral Fins Inside Tubes, *SAE Pap. 700115*, Society of Automotive Engineers, Detroit, MI.

Jones, T. V., and Russell, C. M. B. (1978). Heat Transfer Distribution on Annular Fins, *ASME Pap. 78-HT-30*. American Society of Mechanical Engineers, New York.

Jones, C. D., and Smith L. F. (1970). Optimum Arrangement of Rectangular Fins on Horizontal Surfaces for Free Convection Heat Transfer, *J. Heat Transfer*, 92, 6.

Joos, F. M. (1984). Thin Liquid Films on Arbitrary Surfaces with Condensation or Evaporation, *ASME Symp. Vol. HTD-38*, American Society of Mechanical Engineers, New York, 123.

Joshi, H. M., and Webb, R. L. (1987). Prediction of Heat Transfer and Friction in the Offset Strip Fin Array, *Int. J. Heat Mass Transfer*, 30, 69–84.

Joyce, T. F. (1967). Optimization and Design of Fin-Tube Heat Exchangers by Experiment and Computer Technique, *J. Inst. Heat. Vent. Eng.*, 35, 8.

Ju, Y.-H., Chou, Y. S., and Hsiao, C. I. (1989). A New Approach to the Transient Conduction in a 2-D Rectangular Fin, *Int. J. Heat Mass Transfer*, 32, 1657.

Ju, Y.-H., Shih, Y.-P., and Chiu, C.-Y. (1991). Design of a Tri-laminated Rectangular Fin, *Int. J. Heat Mass Transfer*, 34, 1097.

Jucá, P. C. S., and Prata, A. T. (1993). Two-Dimensional Fins Attached to a Thick Wall: Effect of Nonuniform Root Temperature, *Int. J. Heat Mass Transfer*, 36, 233.

Kadle, D. S., and Sparrow, E. M. (1986). Numerical and Experimental Study of Turbulent Heat Transfer and Fluid Flow in Longitudinal Fin Arrays, *J. Heat Transfer*, 108, 16.

Kajino, M., and Hiramatsu, M. (1987). Research and Development of Automotive Heat Exchangers, in W. J. Yang and Y. Mori, (Eds.), *Heat Transfer in High Technology and Power Engineering*, Hemisphere, New York. pp. 420–432.

Kakac, S., and Yener, Y. (1994). *Convective Heat and Mass Transfer*, 2nd ed., CRC Press, Boca Raton, FL.

Kakac, S., Aung, W., and Viskanta, R. (1985). *Natural Convection Fundamentals and Applications*, Hemisphere, New York.

Kakac, S., Shah, R. K., and Aung, W. (1987). *Handbook of Single-Phase Convective Heat Transfer*, Wiley, New York.

Kantorovich, L. V., and Krylov, V. I. (1964). *Approximate Methods in Higher Analysis*, Wiley, New York.

Karagiozis, A. N. (1991). An Investigation of Laminar Free Convection Heat Transfer from Isothermal Finned Surfaces, Ph.D. thesis, University of Waterloo, Waterloo, Ontario, Canada.

Karam, R. D., and Eby, R. J. (1978). Linearized Solution of Conducting-Radiating Fins, *AIAA J.*, 16, 536.

Karkhu, V. A., and Borovkov, V. P. (1971). Film Condensation of Vapor at Finely Finned Horizontal Tubes, *Heat Transfer Sov. Res.*, 3, 183–191.

Karasev, Z. K. (1967). Investigation of the Fluid Dynamics and Heat Transfer in a Channel with Turbulizers on the Heat Emitting Surfaces, *Proc. 2nd All Soviet Union Conference on Heat and Mass Transfer*, 1, 251.

Karlekar, B. V., and Chao, B. T. (1963). Mass Minimization of Radiating Trapezoidal Fins with Negligible Base Cylinder Interaction, *Int. J. Heat Mass Transfer*, 6, 33.

Karvinen, R. (1981). Natural and Forced Convection Heat Transfer from a Plate Fin, *Int. J. Heat Mass Transfer*, 24(5), 881.

Karvinen, R., Karema, H., and Siskanen, P. (1990). Treatment of Moisture Condensation on Fins, *Waerme Stoffuebertrag.*, 25, 27–31.

Katz, D. L., and Geist, J. M. (1948). Condensation of Six Finned Tubes in a Vertical Row, *Trans. ASME*, 70, 907–914.

Kaushik, N., and Azer, N. Z. (1988). Condensation Heat Transfer Enhancement by Doubly Augmented Tubes, *ASHRAE Trans.*, 94(2), 1159–1175.

Kaushik, N., and Azer, N. Z. (1989). An Analytical Heat Transfer Prediction Model for Condensation Inside Longitudinally Finned Tubes, *ASHRAE Trans.*, 95(2), 516–523.

Kaushik, N., and Azer, N. Z. (1990). A General Pressure Drop Correlation for Condensation Inside Internally Finned Tubes, *ASHRAE Trans.*, 96(1), 242–255.

Kays, W. M. (1960). Basic Heat Transfer and Flow Friction Characteristics of Six High Performance Heat Transfer Surfaces, *J. Eng. Power*, 82, 27.

Kays, W. M., and Crawford, M. E. (1993). *Convective Heat Transfer*, 3rd ed., McGraw-Hill, New York.

Kays, W. M., and London, A. L. (1950). Heat Transfer and Flow Friction Characteristics of Some Compact Heat Exchanger Surfaces: I. Test System and Procedure, *Trans. ASME*, 72, 1075.

Kays, W. M., and London, A. L. (1984). *Compact Heat Exchangers*, 3rd Ed., McGraw-Hill, New York.

Kays, W. M., London, A. L., and Johnson, D. W. (1951). *Gas Turbine Plant Heat Exchangers: Basic Heat Transfer and Flow Friction Data*, ASME Research Publication, American Society of Mechanical Engineers, New York.

Kazeminejad, H. (1993). Effect of Vapor Drag on Laminar Film Condensation on a Vertical Rectangular Fin, *Proc. IMech,* 207, 63–69.

Kazeminejad, H., Yaghoubi, M. A., and Sepehri, M. (1993). Effect of Conjugate Forced Convection-Conduction Analysis at the Performance of a Cooling and Dehumidifying Vertical Rectangular Fin, *Int. J. Heat Mass Transfer,* 36, 3625–3631.

Kedzierski, M. A., and Webb, R. L. (1990). Experimental Methods of Condensation on Vertical Plates with Enhanced Fins, in, E. G. Ragi, T. M. Rudy, T. J. Rabas, and J. M. Robertson, (Eds.), *Boiling and Condensation in Heat Transfer Equipment, ASME Symp. Vol. HTD-85*, American Society of Mechanical Engineers, New York, 87–95.

Kelkar, K. M., and Patankar, S. V. (1984). Numerical Prediction of Flow and Heat Transfer in a Parallel Plate Channel with Staggered Fins, 22nd National Heat Transfer Conference, Niagara Falls, NY, *ASME Pap. 84-HT-70*, American Society of Mechanical Engineers, New York.

Keller, H. H., and Holdredge, E. S. (1969). Radiation Heat Transfer for Annular Fins of Trapezoidal Profile, 11th National Heat Transfer Conference, Minneapolis, MN, *ASME Pap. 69-HT-6*, American Society of Mechanical Engineers, New York.

Keller, H. H., and Somers, E. V. (1959). Heat Transfer from an Annular Fin of Constant Thickness, *J. Heat Transfer*, 81, 151.

Kern, D. Q. (1950). *Process Heat Transfer*, McGraw-Hill, New York.

Kern, D. Q. (1958). Mathematical Development of Loading in Horizontal Condensers, *AIChE J.*, 4, 157–160.

Kern, D. Q., and Kraus, A. D. (1972). *Extended Surface Heat Transfer*, McGraw-Hill, New York.

Keyes, R. W. (1984). Heat Transfer in Forced Convection Through Fins, *IEEE Trans. Electron. Devices*, 31, 1218–1221.

Khader, M. S. (1981). Fin Heat Conduction by the Panel Numerical Scheme, *AIAA Pap. AIAA-81-1138*, American Institute of Aeronautics and Astronautics, New York.

Kharku, V. A., and Borovkov, V. P. (1971). Film Condensation of Vapor at Finely-Finned Horizontal Tubes, *Heat Transfer Sov. Res.*, 3, 183.

Kilic, A., and Onat, K. (1981). Optimum Shape for Convecting Rectangular Fin When Condensation Occurs, *Thermodyn. Fluid Dyn.*, 15, 125.

Kim, R. H. (1976). Kantorovich Method in the Variational Formulation to an Unsteady Heat Conduction, *Lett. Heat Mass Transfer*, 3, 73.

Kishimoto, T., and Ohsaki, T. (1986). VLSI Packaging Technique Using Liquid-Cooled Channels, *Proc. 1986 IEEE Electronic Components Conference*, 595–601.

Kishimoto, T., and Sasaki, E. (1987). Cooling Characteristics of Diamond Shaped Interrupted Cooling Fins for High Power LSI Devices, *Electron. Lett.*, 23, 456–457.

Kishimoto, T., Sasaki, E., and Moriya, K. (1984). Gas Cooling Enhancement Technology for Integrated Circuit Chips, *IEEE Trans. Comput. Hybrids Manuf. Technol.* 7(3), 286.

Klein, G. J., and Westwater, J. W. (1971). Heat Transfer from Multiple Spines to Boiling Liquids, *AIChE J.*, 17, 1050.

Klett, D. E., and McCulloch, J. W. (1972). Effect of Thermal Conductivity and Base-Temperature Depression on Fin Effectiveness, *J. Heat Transfer*, 94, 333.

Koh, J. C. Y., and Colony, R. (1986). Heat Transfer of Microstructures for Integrated Circuits, *International Commun. Heat Mass Transfer*, 13, 89–98.

Konopolev, E. I., and Bychkovskii, A. L. (1970). Method of Calculating the Temperature in Heated Tubes with Cross Fins, *Therm. Eng.*, 17, 111.

Koshelyaev, E. M., (1969). Optimal Parameters of Tubular Radiation Emitters with Profiled Fins, *High Temp.*, 7(3), 459.

Kotan, K., and Arnas, O. A. (1965). On the Optimization of the Design Parameters of Parabolic Radiating Fins, ASME Pap. 65-HT-62, 8th National Heat Transfer Conference, Los Angeles, CA, *ASME Pap. 65-HT-62*, American Society of Mechanical Engineers, New York.

Kovarik, M. (1983). Optimal Heat Transfer Assemblies with Thin Straight Fins, *J. Heat Transfer*, 105, 203.

Kovarik, M. (1984). Optimization of Finned Ducts in Laminar Flow, *J. Heat Transfer*, 106, 586.

Kraemer, H. F., and Westwater, J. W. (1954). Digital Computer Solution for Heat Transfer to Temperature Probes, *Ind. Eng. Chem.*, 46, 2035.

Krane, R. J. (1976). Discussion on Perturbation Solution for Convecting Fin with Variable Thermal Conductivity, *J. Heat Transfer*, 98, 585.

Kraus, A. D. (1958). The Efficiency of a Transistor Cap as a Heat Dissipator, 2nd National Heat Transfer Conference, Chicago, IL, *ASME Pap. 58-HT-15*, American Society of Mechanical Engineers, New York.

Kraus, A. D. (1960). Efficiency of Finned Annular Passages, 4th National Heat Transfer Conference, Buffalo, NY, *ASME Pap. 60-HT-3*, American Society of Mechanical Engineers, New York.

Kraus, A. D. (1961). Efficiency of the Cold Plate Heat Exchanger, *Proc. National Aeronautics and Electronics Conference*, Dayton, OH, Institute of Radio Engineers, New York, 381.

Kraus, A. D. (1962). Optimization of the Cold Plate Heat Exchanger, *Proc. National Aeronautics and Electronics Conference*, Dayton, Ohio, Institute of Radio Engineers, New York, 78.

Kraus, A. D. (1982). *Analysis and Evaluation of Extended Surface Thermal Systems*, Hemisphere, New York.

Kraus, A. D., and Morales, H. J. (1983). The Case for the Magnesium Heat Sink, *Proc. National Electronics Packaging Conference*, Anaheim, CA, 102.

Kraus, A. D., and Bar-Cohen, A. (1995). *Design and Analysis of Heat Sinks*, Wiley, New York.

Kraus, A. D., and Snider, A. D. (1980). New Parameterizations for Heat Transfer in Fins and Spines, *J. Heat Transfer*, 102, 415.

Kraus, A. D., and Snider, A. D. (1985). The Choking and Optimization of Extended Surface Arrays, *J. Heat Transfer*, 107, 746.

Kraus, A. D., and Specht, S. J. (1974). The Effect of Bond Resistance on the Overall Efficiency of a Heat Exchanger Core, *Proc. 5th International Heat Transfer Conference*, Tokyo, V, 165.

Kraus, A. D., Snider, A. D., and Doty, L. F. (1978). An Efficient Algorithm for Evaluating Arrays of Extended Surface, *J. Heat Transfer*, 100, 288.

Kraus, A. D., Snider, A. D., and Landis, F. (1982). The Reciprocity of Extended Surface and the Node Analysis of Finned Arrays, *Proc. Seventh International Heat Transfer Conference*, Munich, 6, 223.

Krause, W. B., and Peters, A. R. (1983). Heat Transfer from Horizontal Serrated Finned Tubes in an Air-Fluidized Bed of Uniformly Sized Particles, *J. Heat Transfer*, 105:319.

Kreid, D. K., Johnson, B. M., and Faletti, D. W. (1978). Approximate Analysis of Heat Transfer from the Surface of a Wet Finned Heat Exchanger, *ASME Paper 78-HT-26*, American Society of Mechanical Engineers, New York.

Kreith, F. (1962). *Radiation Heat Transfer*, International Textbook, Scranton, PA.

Kubanek, G. R., and Miletti, D. L. (1979). Evaporative Heat Transfer and Pressure Drop Performace of Internally Finned Tubes and Refrigerant-22, *J. Heat Transfer*, 101, 447.

Kuehn, T. H., Kwon, S. S., and Tolpadi, A. K. (1983). Similarity Solution for Conjugate Natural Convection Heat Transfer from a Long Vertical Plate Fin, *Int. J. Heat Mass Transfer*, 26(11), 1718.

Kumar, S. S., Nayak, V., and Venkateshan, S. P. (1993). Optimum Finned Space Radiators, *Int. J. Heat Fluid Flow*, 4, 191.

Kuntysh, V. B., and Iokhvedov, F. M. (1971). Effect of Relative Interfin Distance on the Thermal Convective Heat Transfer in Finned Tube Bundles and on Augmenting Heat Transfer, *Heat Transfer Sov. Res.*, 3(2), 50.

Kuntysh, V. B., Taryanyan, I. G., and Iokhvedov, F. M. (1974). On the Effect of the Relative Depth of the Interfin Space on Heat Transfer from Bundles of Finned Tubes, *Heat Transfer Sov. Res.*, 6(6), 5.

Kuznetsova, V. V. (1972). Convective Heat Transfer with Flow of a Viscous Fluid in a Horizontal Tube, *Teploenergetika*, 1995, 84–88.

Kwon, S. S., Kuehn, T. H., and Topaldi, A. K. (1984). On Natural Convection from a Short Conducting Plate Fin Below a Heated Horizontal Cylinder, *J. Heat Transfer*, 106, 661.

Lai, F. S., and Hsu, Y. Y. (1967). Temperature Distribution of a Fin Partially Cooled by Nucleate Boiling, *AIChE J.*, 13(4), 817.

Lam, T. T., and Hardt, B. E. (1977). Analytical Solution of Two-Dimensional Heat Conduction for a Thermally Asymmetric Fin, 6th AIAA–ASME Joint Thermophysics and Heat Transfer Conference, Colorado Springs, CO, *AIAA Pap. 94-1977*, American Institute of Aeronautics and Astronautics, New York.

Langhaar, H. L. (1942). Steady Flow in a Transition Length of a Straight Tube, *J. Appl. Mech.*, 9, A55.

Lau, W., and Tan, C. W. (1973). Errors in One-Dimensional Heat Transfer Analysis in Straight and Annular Fins, *J. Heat Transfer*, 95, 549.

Lee, W. C., and Rose, J. W. (1984). Forced Convection Film Condensation on a Horizontal Tube with and Without Non-condensing Gases, *Int. J. Heat Mass Transfer*, 27, 519–528.

Lee, R. S., Huang, H., and Chen, W. Y. (1990). A Thermal Characteristic Study of Extruded Type Heat Sinks in Considering Air Flow Bypass Phenomenon, *Proc. 6th Semiconductor Thermal and Temperature Measurement Symposium*, Phoeniz, AZ, 95–102.

Lee, S., Yovanovich, M. M., and Jafarpur, K. (1991). Effects of Geometry and Orientation on Lamina Natural Convection from Isothermal Bodies, *AIAA J. Thermophys. Heat Transfer*, 5, 208.

Legkii, V. M., Pavlenko, V. P., Makarov, A. S., and Zholudov, Y. S. (1974). Investigation of Local Heat Transfer in a Tube with Annular Fins in Transverse Air Flow, *Heat Transfer Sov. Res.*, 6(6), 101.

Legkii, V. M., Zholudov, Y. S., and Gerashchenko, O. A. (1976). Local Heat Exchange of a Single Transversely Bathed Round Tube with External Circular Fins, *J. Eng. Phys.*, 30(2), 178.

Lemmon, E. C., and Coleman H. W. (1973). Turbulent Heat Transfer to a Fin Leading Edge: Flight Test Results, *AIAA J.*, 11, 571.

Leung, C. W., Probert, S. D., and Shilston, M. J. (1985a). Heat Exchanger Design: Optimal Separation for Vertical Rectangular Fins Protruding from a Rectangular Base, *Appl. Energy*, 19(2), 77.

Leung, C. W., Probert, S. D., and Shilston, M. J. (1985b). Heat Exchanger Design: Thermal Performance of Rectangular Fins Protuding from Vertical or Horizontal Rectangular Bases, *Appl. Energy*, 20(2), 123–128.

Levy, E. K. (1971). Optimum Plate Spacings for Laminar Natural Convection Heat Transfer from Parallel, Vertical Isothermal Flat Plates, *J. Heat Transfer*, 93, 463.

Levitsky, M. J. (1972). The Criterion for the Validity of the Fin Approximation, *Int. J. Heat Mass Transfer*, 15, 1960.

Li, C. H. (1983). Optimum Cylindrical Pin Fin, *Int. J. Heat Mass Transfer*, 29, 1043.

Lieberman, J., and Gebhart, B. (1969). Interactions in Natural Convection from an Array of Heated Elements, *Int. J. Heat Mass Transfer*, 12, 1385.

Lieblein, S. (1959). Analysis of Temperature Distribution and Radiant Heat Transfer Along a Rectangular Fin of Constant Thickness, *NASA TN-D196*, National Aeronautics and Space Administration, Washington, DC.

Lienhard, J. H. (1987). *A Heat Transfer Textbook*, Prentice Hall, Englewood Cliffs, NJ.

Lienhard, J. H., and Dhir, V. K. (1974). Laminar Film Condensation on Nonisothermal and Arbitrary-Heat-Flux Surfaces and on Fins, *J. Heat Transfer*, 96, 197.

Lienhard, J. H., and Funk J. E. (1971). Design of a Perfectly "Effective" Boiling Fin, *J. Basic Eng.*, 93, 328.

Lienhard, J. H., and Keeling, K. B., Jr. (1970). An Induced Convection Effect upon the Peak Boiling Heat Flux, *J. Heat Transfer*, 92, 1–5.

Lipets, A. U., and Zholudov, Y. S. (1970). Calculating the Effectiveness of Internal Finning of Reheaters, *Therm. Eng.*, 4, 50.

Littlefield, J. M., and Cox, J. E. (1974). Optimization of Annular Fins on a Horizontal Tube, *Thermodyn. Fluid Dyn.*, 7, 87.

Liu, C. Y. (1960). On Minimum Weight Rectangular Radiating Fins, *J. Aerosp. Sci.*, 27, 871–877.

Liu, C. Y. (1961). On Optimum Rectangular Cooling Fins, *Q. Appl. Math.* 19, 72–79.

Liu, C. Y. (1962). A Variational Problem with Applications to Cooling Fins, *Soc. Ind. Appl. Math.*, 10(1), 19–27.

London, A. L. (1954). Air Coolers for High Power Vacuum Tubes, *IRE Trans. Electron. Devices*, 1(2), 9.

London, A. L., and Ferguson, C. K. (1949). Test Results of High Performance Heat Exchanger Surfaces Used in Aircraft Coolers and the Significance in Gas Turbine Regenerator Design, *Trans. ASME*, 71,17.

Look, D. C., Jr. (1988). Two-Dimensional Fin Performance: Bi (Top Surface) $\geq$ Bi (Bottom Surface), *J. Heat Transfer*, 110, 780.

Look, D. C., Jr. (1989). Two-Dimensional Fin with Non-Constant Root Temperature, *Int. J. Heat Mass Transfer*, 32, 977

Look, D. C., Jr. (1992). Separation of Variables on a Triangular Geometry, *Q. Appl. Math.*, 1, 141.

Look, D. C., Jr., and Kang, H. S. (1989). Effects of Variation in Root Temperature on Heat Lost from a Thermally Non-symmetric Fin, *Int. J. Heat Mass Transfer*, 34, 1059.

Look, D. C., Jr., and Kang, H. S. (1992). Optimization of a Thermally Non-Symmetric Fin: Preliminary Evaluation, *Int. J. Heat Mass Transfer*, 35, 2097.

Lubny-Gerczyk, A. L., and Bobkova, N. A. (1963). Generalization of the Calculation of Efficiency Coefficient of Fins and Spines, *Inzh. Fiz. Zh.*, 6(1), 118–124

Lubny-Gerczyk, A. L., Konoplev, E. I., and Bychkovskii, A. L. (1973). Calculation of Stresses in a Hot Tube with Transverse Fins, *Therm. Eng.*, 20, 23.

Lund, K. O., and Baker, K. W. (1993). Minimum-Weight Analysis of Anisotropic Plane-Fin Heat-Pipe Space Radiators, *J. Solar Energy Eng.*, 115, 37.

Luu, M., and Bergles, A. E. (1979a). Enhancement of Horizontal In-Tube Condensation of R-113, *ASHRAE Trans.*, 85(2), 293–312.

Luu, M., and Bergles, A. E. (1979b). Augmentation of Horizontal In-Tube Condensation of R-113, *Heat Transfer Lab. Rep. HTL-23, ISU-ERI-AMES-80175*, Iowa State University, Ames, IA.

Luu, M., and Bergles, A. E. (1980). Augmentation of Horizontal In-Tube Condensation of R-113, *ASHRAE Trans.*, 86(1), 293–311.

Ma, S. W., Bebahani, A. I., and Tsuei, Y. G. (1991). Two-Dimensional Rectangular Fin with Variable Heat Transfer Coefficient, *Int. J. Heat Mass Transfer*, 34, 79.

MacArthur, J. W., and Patankar, S. V. (1985). Effects of Swirling Laminar Flow on Heat Transfer Augmentation in Annular Passages with Twisted Radial Fins, *ASME Symp. Vol. HTD-43*, American Society of Mechanical Engineers, New York, NY., 73.

Mackay, D. B. (1960). *Design of Space Power Plants*, Prentice Hall, Englewood Cliffs, NJ.

Mackay, D. B., and Bacha, C. P. (1962). Performance of Extended Surface Radiation Model, U.S. Government *SID 62-433, Tech. Doc. Rep. ASD-TDR-62-NTIA*, Arlington Hall Station, Arlington, VA.

Mackay, D. B., and Leventhal, E. L. (1960). Radiant Heat Transfer from a Flat Plate Uniformly Heated Along the Edge, *AIChE Pap. 23*, American Institute of Chemical Engineers, New York.

Maday, C. J. (1974). Minimum Weight One-Dimensional Straight Cooling Fin, *J. Eng. Ind.*, 16, 161.

Madjeski, J., and Mikielewicz, J. (1971). Liquid Fin. A New Device for Heat-Transfer Equipment, *Int. J. Heat Mass Transfer*, 14(3), 357.

Mahalingham, M. (1985). Thermal Management in Semiconductor Device Packaging, *IEEE Proc.*, 73, 1396–1404.

Mahalingham, M., and Andrews, J. (1987). High Performance Air Cooling for Microelectronics, *Proc. International Symposium Cooling Technology for Electronic Equipment*, 608–625.

Mandel, S. W., Townsend, M. A., and Parrish, T. F., Jr. (1979). Optimal Fin-Side Design of Compact Tube-in-Fin Heat Exchangers with Rippled Fins, *J. Heat Transfer*, 101, 514.

Manglik, R. M., and Bergles, A. E. (1990). The Thermal-Hydraulic Design of the Rectangular Offset-Strip-Fin Compact Heat Exchanger, in R. K. Shah, A. D. Kraus, and D. E. Metzger, (Eds.), *Compact Heat Exchangers*, Hemisphere, New York.

Mantle, P. L., Freeman A. R., and Watts, J. (1971). Conductivity Effects on Ribbed Surface Heat Transfer, *Int. J. Heat Mass Transfer*, 14(11), 1825.

Manzoor, M. (1984). *Heat Flow Through Extended Surface Heat Exchangers*, Springer-Verlag Lecture Notes in Engineering, Springer-Verlag, New York, 5.

Manzoor, M., Ingham, D. B., and Heggs, P. J. (1983a). The One-Dimensional Analysis of Fin Assembly Heat Transfer, *J. Heat Transfer*, 105, 646.

Manzoor, M., Ingham, D. B., and Heggs, P. J. (1983b). Improved Formulations for the Analysis of Convecting and Radiating Finned Surfaces, *AIAA J.*, 21, 120.

Manzoor, M., Ingham, D. B., and Heggs, P. J. (1984). Accuracy of Perfect Contact Fin Analyses, *J. Heat Transfer*, 106, 234.

Mao, J., and Rook, S. (1993). Transient Analyses of Extended Surfaces with Convective Tip, *Int. Commun. Heat Mass Transfer*, 21, 85.

Markowitz, A., Mikic, B. B., and Bergles, A. E. (1972). Condensation on a Downward Facing Ripple Surface, *J. Heat Transfer*, 94, 315.

Marsh, B. L., and Costello, F. A. (1973). Stochastic Optimization of Convective Fin Design, *J. Heat Transfer*, 95, 339.

Martin, B. W., Brown, A., and Garrett, S. E. (1978). Heat Transfer to a PVD Rotor Blade at High Subsonic Passage Throat Mach Numbers, *Heat Fluid Flow*, 8(1), 33.

Marto, P. J. (1986). Recent Progress in Enhancing Film Condensation Heat Transfer on Horizontal Tubes, *Proc. 8th International Heat Transfer Conferance*, San Francisco, 1, 161.

Marto, P. J. (1988), An Evaluation of Film Condensation on Horizontal Integral-Fin Tubes, *J. Heat Transfer*, 110, 1287–1305.

Marto, P. J. (1992). Film Condensation Heat Transfer Measurements on Horizontal Tubes: Problems and Progress Review, *Exp. Therm. Fluid Sci.*, 5, 565–569.

Marto, P. J., and Weigel, H. (1981). Development of Economical Rotating Heat Pipes, *Proc. 4th Internationa Heat Pipe Conferance*, London, 709.

Marto. P. J., Mitrou, E., Wanniarachchi, A. S., and Rose, J. W. (1986). Film Condensation of Steam on Horizontal Finned Tubes, *Proc. 8th International Heat Transfer Conference*, San Francisco, 4, 1695–7000.

Marto, P. J., Zebrowski, D., Wanniarachchi, A. S. and Rose, J. W. (1988). Film Condensation of R-113 on Horizontal Finned Tubes, Proc. 1988 Heat Transfer Cong., *ASME Symp. Vol.*, HTD-96, H. R. Jacobs (Ed.), American Society of Mechanical Engineers, New York, 2, 583–592

Masliyah, J. H., and Nandakumar, K. (1976). Heat Transfer in Internally Finned Tubes, *J. Heat Transfer*, 98, 257.

Masliyah, J. H., and Nandakumar, K. (1977). Fluid Flow and Heat Transfer in Internally Finned Helical Coils, *Can. J. Chem. Eng.*, 55(1), 27.

Mason, J. (1955). Heat Transfer in Cross Flow, *Proc. 2nd U.S. National Congress on Applied Mechanics*, American Society of Mechanical Engineers, New York.

Masuda, H., and Rose, J. W. (1985). An Experimental Study of Condensation of Refrigerant 113 on Low Integral-Fin Tubes, *Proc. International Symposium on Heat Transfer*, Beijing, 2, paper 32.

Masuda, H., and Rose, J. W. (1987). Models for Condensation on Horizontal Low-Finned Tubes, Proc. 4th UK Heat Transfer Conference, *Ind. Mech. Eng.*, 410, 125–139.

McAdams, W. H. (1954). *Heat Transmission*, 3rd ed., McGraw-Hill, New York.

McQuiston, F. C. (1975). Fin Efficiency with Combined Heat and Mass Transfer, *Trans. ASHRAE*, 81(1), 350.

McQuiston, F. C. (1978). Correlation for Heat, Mass, and Momentum Transport Coefficients for Plate-Fin, Tube Heat Transfer Surfaces with Staggered Tubes, *Trans. ASHRAE*, 84(1), 294–309.

McQuiston, F. C., and Tree, D. R. (1972). Optimum Space Envelopes of the Finned Tube Heat Transfer Surface, *Trans. ASHRAE*, 78, 144.

Mehta, R. C. (1978). Computer-Aided Optimum Design of Heat Sinks, *J. Inst. Electron. Telecommun. Eng.*, 24, 298.

Melese, G. B. (1957). Efficacité d'une ailette longitudinale avec variation du coefficient d'échange de chaleur le long de l'ailette, *J. Nucl. Energy*, 5(3–4), 285.

Melese, G. B., and Wilkins, J. E., Jr. (1966). Efficiency of Longitudinal Fins of Arbitrary Shape with Nonuniform Surface Heat Transfer and Internal Heat Generation, *Proc. 3rd International Heat Transfer Conference*, Chicago, 3, 272.

Metzger, D. E., and Haley, S. W. (1982). Heat Transfer Experiments and Flow Visualization for Arrays of Short Pin Fins, 27th International Gas Turbine Conference, London, *ASME Pap. 82-GT-138*, American Society of Mechanical Engineers, New York.

Metzger, D. E., Berry, R. A., and Bronson, J. P. (1982a). Developing Heat Transfer in Rectangular Ducts with Staggered Arrays of Short Pin Fins, *J. Heat Transfer*, 104, 700.

Metzger, D. E., Fan, C. S., and Shepard, W. B. (1982b). Pressure Loss and Heat Transfer Through Multiple Rows of Short Pin Fins, *Proc. 7th International Heat Transfer Conference*, Munich, 3, 137.

Metzger, D. E., Fan, C. S., and Pennington, J. W. (1983). Heat Transfer and Flow Friction Characteristics of Very Rough Transverse Ribbed Surfaces with and Without Pin Fins, *Proc. ASME–JSME Thermal Eng. Joint Conference*, 1, 429.

Metzger, D. E., Fan, C. S., and Haley S. W. (1984). Effects of Pin Shape and Array Orientation on Heat Transfer and Pressure Drop in Pin Fin Arrays, *J. Eng. Gas Turbines Power*, 106, 252.

Michael, A. G., Marto, P. J., Wanniarachchi, A. S., and Rose, J. W. (1989). Effect of Vapor Velocity on Horizontal Smooth and Finned Tubes, *Heat Transfer for Phase Change*, *ASME Symp. Vol. HTD-114*, American Society of Mechanical Engineers, New York, 1–10.

Michiyoshi, I., Takahashi, O., and Yamaguchi, K. (1974). Radiative and Convective Heat Transfer from Tranverse Finned Fuel Rods, *ASME Pap. 74-WA/HT-6*, American Society of Mechanical Engineers, New York.

Migai, V. K. (1963). Effect of Non-uniform Heat Transfer over the Height of a Fin on Its Efficiency, *Inzh. Fiz. Zh.*, 6(3), 51.

Migai, V. K., Zozulya, N. V., and Zhitomirskaya, I. V. (1984). Research of the Heat Transfer of Finned Tubes in a Fluidized Bed, *Sov. Energy Technol.*, 1, 18.

Mikk, I. (1980). Convective Fin of Minimum Mass, *Int. J. Heat Mass Transfer*, 23(11), 1471.

Mikk, I. (1982). Annular Fin Efficiency, *Teploenergetika*, 11, 75.

Mills, A. F. (1992). *Heat Transfer*, Richard D. Irwin, Homewood, IL.

Mills, A. F., Hubbard, G. L., James, R. K., and Tan, C. (1975). Experimental Study of Film Condensation on Horizontal Grooved Tubes, *Desalination*, 16, 121.

Mil'man, O. O., and Shklover, C. G. (1980). On Ways to Increase the Heat Transfer Coefficients in the Condensation of Steam, *Heat Transfer Sov. Res.*, 13(1), 51.

Mineur, J. M., and Dunstan, G. F. (1979). High Performance Finned Tube Heat Exchanger Design, *Appl. Energy*, 5(2), 89.

Minkler, W. S., and Rouleau, W. T. (1960). The Effects of Internal Heat Generation on Heat Transfer in Thin Fins, *Nucl. Sci. Eng.*, 7, 400.

Minkowycz, W., and Sparrow, E. M., (1974). Local Nonsimilarity Solution for Natural Convection on a Vertical Cylinder, *J. Heat Transfer*, 96, 178–183.

Mirkovic, Z. (1974). Heat Transfer and Flow Resistance Correlations for Helically Finned and Staggered Tube Banks in Cross Flow Heat Exchangers, in N. H. Afgan and E. U. Schlunder, (Eds.), *Design and Theory Source Book*, McGraw-Hill, New York.

Mitrou, E. (1986). Film Condensation Heat Transfer on Horizontal Finned Tubes, Thesis, Naval Postgraduate School, Monterey, CA.

Mochalov, B. V. (1970). Optimum Calculation for the Finned Radiators of Semiconductor Triodes, *Electrotechnika*, 11, 11.

Mochizuki, S. and Yagi, Y. (1977). Heat Transfer and Friction Characteristics of Strip Fins, *Heat Transfer Jpn. Res.*, 6(3), 36.

Mondt, J. R. (1977), Effects of Nonuniform Passages on Deepfold Heat Exchanger Performance, *J. Eng. Power*, 99, 657.

Mondt, J. R., and Siegla, D. C. (1974). Performance of Perforated Heat Exchanger Surfaces, *J. Eng. Power*, 96, 81.

Moody, L. F. (1944). Friction Factors for Pipe Flow, *Trans. ASME*, 66, 671–678.

Moszynski, J. R., and Champaneria, N. (1971). Two-Dimensional Temperature Distributions in Radiating Fins, First National Heat and Mass Conference, Madras, India, *Pap., HMT-36-71*, *I 13-I20*.

Mueller, A. C. (1985). Process Heat Exchangers, in J. P. Hartnett, W. M. Rohsenow, and E. N. Ganic, (Eds.), *Handbook of Heat Transfer: Applications*, McGraw-Hill, New York.

Mueller, A. C. (1990). Condensers, in G F Hewitt, (Ed.), *Handbook of Heat Exchanger Design*, Hemisphere, New York, NY, pp. 3.4.1.1–3.4.9.5.

Murray, W. M. (1938). Heat Transfer Through an Annular Disk or Fin of Uniform Thickness, *Trans. ASME, J. Appl. Mech.*, 60, A78.

Murthy, J. Y., and Patankar, S. V. (1983). Numerical Study of Heat Transfer from a Rotating Cylinder with External Longitudinal Fins, 21st National Heat Transfer Conference, Seattle, WA, *ASME Pap. 83-HT-42*, American Society of Mechanical Engineers, New York.

Mutlu, I., and Al-Shemmeri, T. T. (1993). Steady-State and Transient Performance of a Shrouded Longitudinal Fin Array: I., *Commun. Heat Mass Transfer*, 20, 133.

Myers, G. E. (1971). *Analytical Methods in Heat Conduction*, McGraw-Hill, New York.

Na, T. Y. (1979). Turbulent Heat Transfer for Pipe Flow with Uniform Heat Generation, *Waeerme Stoffeubertrag, Thermodyn. Fluid Dyn.*, 12(1), 55–58.

Nader, W. K. (1978). Extended Surface Heat Transfer with Condensation, *Proc. 6th International Heat Transfer Conference*, Toronto, Ontario, Canada, 407.

Nakajima, K. (1992). Iterative Temperature Calculation Method for Rectangular Sandwich Panel Fins, *AIAA J., Thermophys. Heat Transfer*, 6, 693.

Nakamura, H., Miyan, T. R., Hiraoka, S., Nakamori, Y., and Yamada, I. (1981). Solutions of Two Dimensional Heat Conduction Problem in Triangular Fins, *J. Chem. Eng. Jpn.*, 14, 487.

Nakayama, W. (1982). Enhancement of Heat Transfer, *Proc. 7th International Heat Transfer Conference*, Munich, 1, 223.

Nakayama, W., and Xu, L. P. (1983). Enhanced Fins for Air-Cooled Heat Exchangers: Heat Transfer and Friction Factor Correlations, *Proc. ASME–JSME Thermal Engineering Joint Conference*, 1, 495.

Narayana, K. B., and Kumari, S. U. (1988). Two-Dimensional Heat Transfer Analysis of Radiating Plates, *Int. J. Heat Mass Transfer*, 31, 1767.

Nayak, D., Hwang, T., Turlik, I., and Reisman, A. (1987). A High Performance Thermal Module for Computer Packaging, *J. Electron. Mater.*, 16, 357–364.

Netrakanti, M. N., and Huang C. L. D. (1985). Optimization of Annular Fins with Variable Thermal Paramaters by Invariant Imbedding, *J. Heat Tranfer*, 107, 966.

Newhouse, K. N. (1964). Temperature Distributions in Circular Fins of Rectangular Profile, *J. Heat Transfer*, 86, 563.

Nilson, E. N., and Curry, R., Jr. (1960). The Minimum Weight Straight Fin of Rectangular Profile Radiating to Space, *J. Aerosp. Sci.*, 27, 146.

Ning, Z., and Qiliang, M. (1983). Experimental Investigation on Heat Transfer in Helically Welded Fin Tube Bank and in Every Row in It, *Proc. ASME–JSME Thermal Engineering Joint Conference*, 3, 101.

Nir, A. (1991). Heat Transfer and Friction Correlations for Crossflow over Staggered Finned Tube Banks, *Heat Transfer Eng.*; 12(1), 43

Nusselt, W. (1911). Der Wärmeübergang in Kreuzstrom, *Z. Ver. Dtsch. Ing.*, 55, 2021–2024.

Nusselt, W. (1916). Eine Neue Formel für den Wärmedurchgang in Kreuzstrom, *Tech. Mech. Thermodyn.*, 1, 417–422.

Nusselt, W. (1930). The Condensation of Steam on Cooled Surfaces, *Z. Ver. Dtsch. Ing*, 60, 541–546; translated into English by D. Fullerton, *Chem. Eng. Fundam.*, 1(2), 6–9).

O'Brien, V., and Ehrlich, L. W. (1977). Forced Convection Within Straight Noncircular Ducts, *J. Heat Transfer*, 99, 485.

Okamoto, Y. (1964). Temperature Distribution and Efficiency of a Single Sheet of Radiative Fin Accompanied by an Internal Heat Source, *Bull. JSME*, 7(28), 751.

Okamoto, Y. (1966a). Temperature Distribution and Efficiency of a Plate and Annular Fin of Constant Thickness, *Bull. JSME*, 9(33), 143.

Okamoto, Y. (1966b). Thermal Performances of Radiative and Convective Plate Fins with Mutual Irradiation, *Bull. JSME*, 9(33), 150.

Okamoto, Y., and Kameoka, R. (197[illegible]). Study on Convective Heat Transfer from Planar Fin Surfaces, *Heat Transfer Jpn. Res.*, 2(1), 1.

Okamoto, Y., and Kameoka, R. (1974). Non-steady Heat Transfer of Finned Surface, *Proc. 5th International Heat Transfer Conference*, Tokyo, 2, 353.

Okamoto, Y., and Negishi, K. (1969). Transient Response of Finned Tube-Sheet Radiator for Space Power Plants, *Trans. Jpn. Soc. Aerosp. Sci.*, 12, 73.

O'Meara, J. E. (1976). *Design and Construction of a Residential Solar Heating System at Fermilab*, Solar Energy Soceity of Canada, Winnipeg, Manitoba, Canada, 3, 236.

O'Reilly, W. J. (1978). High-Effectiveness Regenerator Design Concept, *ASME Pap. 78-GT-78*, American Society of Mechanical Enginers, New York.

Ornatskii, A. P., Semena, M. G., and Shcherbakov, V. K. (1970). Investigation of Velocity Distribution in Tubes with Internal Longitudinal Fins, *Therm. Eng.*, 17, 108.

Oshima, T., and Iuchi, S. (1974). Calculation Method for Cooler Condensers with Continuous Phase Finned Tubes, *Heat Transfer Jpn. Res.*, 3(3), 1.

Oskay, R., and Kakac, S. (1973). Effect of Viscosity Variations on Turbulent and Laminar Forced Convection in Pipes, *METU J. Pure Appl. Sci.*, 6, 211–230.

Ouchi, M., and Takeyama, T. (1978). Experimental and Analytical Study of a Fin in Boiling Water, *Heat Transfer Jpn. Res.*, 7(3), 63.

Ozisik, M. N. (1993). *Heat Conduction*, 2nd ed., Wiley, New York.

Palmquist, R. W., Moriarity, L. J., and Beckman, W. A. (1970). Transient Response of Fin-Tube Radiators, *J. Heat Transfer*, 92, 204.

Papadopoulos, K. D., Guzman-Garcia, A. G., and Bailey, R. V. (1990). The Response of Straight and Circular Fins to Fluid Temperature Changes, *Int. Commun. Heat Mass Transfer*, 17, 2587.

Parker, K. O., and Coombs, M. G. (1979). New Developments in Compact Plate Fin Heat Exchangers, *ASME Symp. Vol. HTD-10*, American Society of Mechanical Engineers, New York.

Patankar, S. V., and Murthy, J. Y. (1982). Analysis of Heat Transfer from a Rotating Cylinder with Circumferential Fins, *Proc. of International Centre for Heat and Mass Transfer*, Hemisphere, New York, 16, 155.

Patankar, S. V., and Sparrow, E. M. (1979). Condensation on an Extended Surface, *J. Heat Transfer*, 101, 434.

Patankar, S. V., Ivanovic, M., and Sparrow, E. M. (1979). Analysis of Turbulent Flow and Heat Transfer in Internally Finned Tubes and Channels, *J. Heat Transfer*, 101, 29.

Pearson, J. F., and Withers, J. G. (1969). New Finned Tube Configuration Improves Refrigerant Condensing, *ASHRAE J.* 75, 77–82.

Perkins, H. C., and Worsœ-Schmidt, P. (1965). Turbulent Heat and Momentum Transfer for Gases in a Circular Tube at Wall to Bulk Temperature Ratios of Seven, *Int. J. Heat Mass Transfer*, 8, 1011–1031.

Petukhov, B. S. (1970). Heat Transfer and Friction in Turbulent Pipe Flow with Variable Physical Properties, in J. P. Hartnett and T. F. Irvine (Eds.), *Advances in Heat Transfer*, Academic Press, New York, 504.

Petukhov, B. S., and Popov, V. N. (1963). Theoretical Calculation of Heat Exchange and Frictional Resistance in Turbulent Flow in Tubes of Incompressible Fluid with Variable Physical Properties, *High Temp.*, 1(1), 69–73.

Petukhov, B. S., Kovalev, S. A., Zhukov, V. M., and Kazakov, G. M. (1972). Study of Heat Transfer During the Boiling of a Liquid on the Surface of a Single Fin, *Heat Transfer Sov. Res.*, 4(4), 148.

Petukhov, B. S., Kovalev, S. A., and Geshele, V. D. (1981) Heat Transfer in Boiling on Finned Surfaces, *Heat Transfer Sov. Res.*, 13(1), 51.

Phillips, R. J. (1987). Forced Convection, Liquid Cooled, Microchannel Heat Sinks, M.S. thesis, Mechanical Engineering Department, Massachusetts Institute of Technology, Cambridge, MA.

Phillips, R. J. (1988a). Michrochannel Heat Sinks, *Lincoln Lab. J.*, 1, 31–48.

Phillips, R. J. (1988b). Errata, *Lincoln Lab. J.*, 1, 258–259.

Pieper, R. J., and Kraus, A. D. (1995a). Cold Plates with Asymetric Heat Loading: I. The Single Stack, Advances in Electronic Packaging, *EEP-Vol-10-2*, American Society of Mechanical Engineers, New York, 865.

Pieper, R. J., and Kraus, A. D. (1995b). Cold Plates with Asymetric Heat Loading: II. The Double Stack, Advances in Electronic Packaging, *EEP-Vol-10-2*, American Society of Mechanical Engineers, New York, 871.

Ping, Y. (1983). Heat Transfer and Friction Loss Characteristics of Pin-Fin Cooling Configuration, *ASME Pap. 83-GT-123*, American Society of Mechanical Engineers, New York.

Pis'mennyi, E. N., and Legkii, V. M. (1984). Calculating Heat Transfer of Multi-row Staggered Bundles of Tubes with Transverse Finning, *Therm. Eng.*, 31, 349.

Pnueli, D. (1974). Optimization of Inclined Convective Fins, *J. Heat Transfer*, 96, 541.

Pohlhausen, E. (1921). Der Wärmeaustausch Zwischen festen Körpen and Flössigkeiten mit kleiner Reibung und kleiner Wärmeleitung, *Z. Angew. Math. Mech.*, 1, 115–121.

Pop, I., Sunada, J. K., Cheng, P., and Minkowycz, W. J. (1985). Conjugate Free Convection from Long Vertical Plate Fins Embedded in a Porous Medium at High Rayleigh Numbers, *Int. J. Heat Mass Transfer*, 28(9), 1629.

Poulikakos D., (1994). *Conduction Heat Transfer*, Prentice Hall, Englewood Cliffs, NJ.

Poulikakos D., and Bejan A. (1982). Fin Geometry for Minimum Entropy Generation in Forced Convection, *J. Heat Transfer*, 104, 616.

Prakash, C., and Patankar, S. V. (1981). Combined Free and Forced Convection in Vertical Tubes with Radial Internal Fins, *J. Heat Transfer*, 103, 568.

Prakash, C., and Ronzoni, P. (1985). Effect of Buoyancy on Laminar Fully Developed Flow in a Vertical Annular Passage with Radial Internal Fins, *Int. J. Heat Mass Transfer*, 28(5), 995.

Rabas, T. J., and Eckels, P. W. (1975). Heat Transfer and Pressure Drop Performance of Segmented Extended Surface Tube Bundles, 15th National Heat Transfer Conference, San Franscisco, *ASME Pap. 75-HT-45*, American Society of Mechanical Engineers, New York.

Rabas, T. J., Eckels, P. W., and Sabatino, R. A. (1981). The Effect of Fin Density on the Heat Transfer and Pressure Drop Performance of Low Finned Tube Banks, *Chem. Eng. Commun.*, 10(1), 127–147.

Rabenstein, A. L. (1966). *Introduction to Ordinary Differential Equations*, Academic Press, New York.

Razelos, P. (1979). Note on the Heat Transfer in Convective Fins, *Thermodyn. Fluid Dyn.*, 12, 113.

Razelos, P. (1983). The Optimum Dimensions of Convective Pin Fins, *J. Heat Transfer*, 105, 411.

Razelos, P. (1986). Optimum Dimensions of Convective Pin Fins with Internal Heat Dissipation, *J. Franklin Inst.*, 321, 1.

Razelos, P., and Georgiou, E. (1992). Two Dimensional Effects and Design Criteria for Convective Extended Surfaces, *Heat Transfer Eng.*, 13(3), 38.

Razelos, P., and Imre, K. (1980). The Optimum Dimensions of Circular Fins with Variable Thermal Parameters, *J. Heat Transfer*, 102, 420.

Razelos, P., and Imre, K. (1983). Minimum Mass Convective Fins with Variable Heat Transfer Coefficients, *J. Franklin Inst.*, 315, 269.

Reichl, H. (1986). Silicon Substrates for Chip Interconnection, *Hybrid Circuits*, 11, 5–7.

Reisbig, R. L. (1974). Condensing Heat Transfer Augmentation Inside Splined Tubes, *ASME Pap. 74-HT-7*, American Society of Mechanical Engineers, New York.

Reynolds, W. C. (1963). A Design Oriented Optimization Method of Simple Tapered Radiating Fins, *J. Heat Transfer*, 84, 193.

Reynolds, B. L., and Holmes, R. E. (1975). Heat Transfer Between a Finned Tube and Air in the Presence of a Corona Discharge, *ASME Pap. 75-WA/HT-18*, American Society of Mechanical Engineers, New York.

Rich, D. G. (1973) Effect of Fin Spacing on the Heat Transfer and Friction Performance of Multi-row, Smooth Plate Fin-and-Tube Heat Exchangers, *Trans. ASHRAE*, 79(2), 137.

Rifert, V. G. (1980). New Method for Calculating Rates of Condensation on Finned Tubes, *Heat Transfer Sov. Res.*, 12(3), 142.

Rifert, V. G. (1982). Steam Condensation on Profile Surfaces, in A. Lykov, (Ed.), *Heat and Mass Transfer Processes in Porous Media with Phase Transformation*, Minsk Academy of Sciences, Minsk, Belorussia, 149–170.

Robertson, J. M. (1984). Prediction of Convective Boiling Coefficients in Serrated Plate-Fin Passages Using an Interrupted Liquid-Film Flow Model, *ASME Symp. Vol. HTD-34*, American Society of Mechanical Engineers, New York, 163.

Robertson, J. M., and Lovegrove, P. C. (1983). Boiling Heat Transfer with Freon-11 (R11) in Brazed Aluminum Pin Fin Exchangers, *J. Heat Transfer*, 105, 605.

Robinson, K. K., and Briggs, D. E. (1966). Pressure Drop of Air Flowing Across Triangular Pitch Banks of Finned Tubes, *Chem. Eng. Prog. Symp. Ser.*, 62(64), 177.

Rogers, D. G. (1980). Forced Convection Heat Transfer in a Single Phase Flow of Newtonian Fluid in a Circular Pipe, *CSIR Rep. CENG 322*, Council for Scientific and Industrial Research, , Pretoria, South Africa.

Rohsenow, W. M. (1952). A Method of Correlating Heat Transfer Data for Surface Boiling in Liquids, *Trans. ASME*, 74, 969–975.

Rohsenow, W. M., and Choi, H. (1961). *Heat, Mass and Momentum Transfer*, Prentice Hall, Englewood Cliffs, NJ.

Roizen, L. I. (1966). Heat Transfer in Flow Round Straight Transverse Fins, *Inz. Fiz. Zh.*, 11(2), 148.

Rose, J. W. (1994). Approximate Equation for the Vapor Side Heat Transfer Coefficient for Condensation in Low Finned Tubes, *Int. J. Heat Mass Transfer*, 37, 865–875.

Rosman, E. L., Carajilescov, P., and Saboya, F. E. M. (1984) Performance of One- and Two-Row Tube and Plate Fin Heat Exchangers, *J. Heat Transfer*, 106, 627.

Rowley, G. J., and Patankar S. V. (1983). Analysis of Laminar Flow and Heat Transfer in Tubes with Internal Circumferential Fins, 21st National Heat Transfer Conference, Seattle, WA, *ASME Paper 83-HT-40*, American Society of Mechanical Engineers, New York.

Royal, J. H., and Bergles, A. E. (1978). Augmentation of Horizontal In- Tube Condensation by Means of Twisted Tape Inserts and Internally Finned Tubes, *J. Heat Transfer*, 100, 17.

Ruch, M. A. (1976). Industrial Energy Recovery with Heat Pipe Heat Exchangers, *Proc. 16th ASME Symposium on Energy Alternatives*, Albuquerque, NM, 205.

Rudy, T. M. (1982). Condensate Retention of Horizontal Finned Tubing, Ph.D. dissertation, Pennsylvania State University, University Park, PA.

Rudy, T. M., and Webb, R. L. (1981). Condensation Retention of Horizontal Integral-Fin Tubing, *ASME Symp. Vol. HTD-18*, American Society of Mechanical Engineers, New York, 35.

Rudy, T. M., and Webb, R. L. (1983a). An Analytical Model to Predict Condensate Retention on Horizontal Integral-Fin Tubes, *Proc. ASME–JSME Therm. Engineering Joint Conference*, 1, 373–378.

Rudy, T. M., and Webb, R. L. (1983b). Theoretical Model for Condensation on Horizontal Integral-Fin Tubes, *AIChE Symp. Ser*, No. 225, 79, 11–18.

Rudy, T. M., and Webb, R. L. (1985). An Analytical Model to Predict Condensate Retention on Horizontal Integral-Fin Tubes, *J. Heat Transfer*, 107, 361.

Rudy, T. M., Kedzierski, M. A., and Webb, R. L. (1984). Investigation of Integral-Fin Type Condenser Tubes for Process Industry Applications, *AIChE Chem. Eng. Symp. Ser.*, 86(1), 633.

Russell, C. M. B. (1974). Effect of the External Corrosion of Fintubes on Thermal Performance, *Advances in Enhanced Heat Transfer*, American Society of Mechanical Engineers, New York, 95.

Russell, L. D., and Chapman, A. J. (1967). Analytical Solution for Transient Heat Flow of Energy in a One-Dimensional Radial Fin, *AIAA J.*, 6, 90.

Saboya, F. E. M., and Sparrow, E. M. (1974a). Effect of Tube Relocation on the Transfer Capabilities of a Fin and Tube Heat Exchanger, *J. Heat Transfer*, 96, 421.

Saboya, F. E. M., and Sparrow, E. M. (1974b). Local and Average Transfer Coefficients for One-Row Plate Fin and Tube Heat Exchanger Configurations, *J. Heat Transfer*, 96, 265.

Saboya, F. E. M., and Sparrow, E. M. (1976). Experiments on a Three-Row Fin Tube Heat Exchanger, *J. Heat Transfer*, 98, 520.

Said, M. N. A., and Trupp, A. C. (1983). Predictions of Turbulent Flow and Heat Transfer in Internally Finned Tubes, *AIChE Symp. Ser.*, 79(225), 11.

Said, S. A., and Azer, N. Z. (1983) Heat Transfer and Pressure Drop During Condensation Inside Horizontal Finned Tubes, *ASHRAE Transfer*, 89(1), 114–124.

Sane, N. K., and Sukhatme, S. P. (1974). Natural Convection Heat Transfer from Horizontal Rectangular Fin Arrays, *Proc. 5th International Heat Transfer Conference*, Tokyo, 3, 114.

Sarabia, M. F., and Hitchcock, J. E. (1966). Heat Transfer from Gray Fin-Tube Radiators, *J. Heat Transfer*, 88, 338.

Sasaki, S., and Kishimoto, T. (1986). Optimal Structure for Microgrooved Cooling Fin for High Power LSI Devices, *Electron. Lett.*, 22, 1332–1334.

Schenone, C., and Tanda, D. (1990). Forced Convection Heat Transfer from Shrouded Fin Arrays, *Int. Commun. Heat Mass Transfer*, 17, 247–758.

Schlager, L. M., Pate, M. B., and Bergles, A. E. (1989a). Heat Transfer and Pressure Drop Performance of Smooth and Internally Finned Tubes with Oil and L-22 Mixtures, *ASHRAE Trans.*, 95(2), 375–385.

Schlager, L. M., Pate, M. B., and Bergles, A. E. (1989b). Evaporation and Condensation Heat Transfer and Pressure Drop in Horizontal 12.7 mm Micro-fin Tubes with Refrigerant-22, *ASME Symp. Vol. HTD-108*, American Society of Mechanical Engineers, New York, 205–213.

Schlager, L. M., Pate, M. B., and Bergles, A. E. (1990a). Condensation of Refrigerant–Oil Mixtures in Smooth and Augmented Tubes, *Proc. 2nd International Symposium on Condensers and Condensation*, Heat Transfer and Fluid Flow Service, London, UK, 451–460.

Schlager, L. M., Pate, M. B., and Bergles, A. E. (1990b). Oil Quantity Measurements in Smooth and Micro-fin Tubes During Evaporation and Condensation of Refrigerant–Oil Mixtures, *ASHRAE Trans.*, 96(2), 465–469.

Schmidt, E. (1926). Die Wärmeübertragung durch Rippen, *Z. Ver. Dtsch. Inc.*, 70, 885 and 947.

Schmidt, T. E. (1963). Wärmeübergang an Rippenrohe and Berechnung von Rohrbündelnevärmeanstauschenern *Kaltetechnik*, 15(4), 370.

Schnurr, M. N. (1975). Radiation from an Array of Longitudinal Fins of Triangular Profile, *AIAA J.*, 13, 691.

Schnurr, M. N., and Cothran, C. A. (1974). Radiation from an Array of Gray Circular Fins of Trapezoidal Profile, *AIAA J.*, 12, 1476.

Schnurr, N. M., Shapiro, A. B., and Townsend, M. A. (1976). Optimization of Radiating Fin Arrays with Respect to Weight, *J. Heat Transfer*, 98, 643.

Schulenberg, F. J. (1965). Wahl der Bezugslänge zur Darstellengun Wärmübergang und Druckverlust in Wärmetauschern, *Chem. Ing. Techmol.*, 37, 431.

Scott, M. J., and Webb, R. L. (1981). Analytic Prediction of the Friction Factor for Turbulent Flow in Internally Finned Channels, *J. Heat Transfer*, 103, 423.

Senshu, T., Hatada, T., and Ishibane, K. (1979). Surface Heat Transfer Coefficients of Fins Used in Air-Cooled Heat Exchangers, *Heat Transfer Jpn. Res.*, 8(4), 16.

Serebryanyi, G. L. (1985). Influence of Contacts on the Thermal Resistance of a Bundle of Rectangular Fins, *J. Eng. Phys.*, 48(2), 236.

Sfeir, A. A. (1976). The Heat Balance Integral in Steady State Conduction, *J. Heat Transfer*, 98, 466.

Shah, R. K. (1971a). Temperature Effectiveness of Multiple Rectangular Plate-Fin Surfaces, *J. Heat Transfer*, 93, 471.

Shah, R. K. (1971b). Influence of Brazing on Very Compact Heat Exchanger Surfaces, 12th National Heat Transfer Conference, Tulsa, OK, *ASME paper 71-HT-29*, American Society of Mechanical Engineers, New York.

Shah, R. K., (1978). *Laminar Flow Forced Convection in Ducts*, Academic Press, New York.

Shah, R. K. (1982). Advances in Compact Heat Exchanger Technology and Design, *Proc. 7th International Heat Transfer Conference*, Munich, 1, 123.

Shah, R. K., and Bhatti, M. S. (1987). Laminar Convective Heat Transfer in Ducts, in *Handbook of Single Phase Convective Heat Transfer*, Wiley, New York, Chap. 3.

Shah, R. K., and Webb, R. L. (1982). Compact and Enhanced Heat Exchangers, in J. Taborek, G. F. Hewitt and N. Afgan (Eds.), *Heat Exchangers: Theory and Practice*, Hemisphere, New York, 425–468.

Shcherbakov, V. K., and Bosyy, V. V. (1974). Fin Selection Criteria and the Effect of Fins on the Temperature of the Cooling Wall, *Heat Transfer Sov. Res.*, 6(6), 148.

Shcherbakov, V. K., and Semena, M. G. (1969). Determining the Maximum Temperature of the Wall and the Geometrical Dimensions of Tubes with Internal Longitudinal Fins, *Therm. Eng.*, 16, 113.

Sheffield, J. W. (1982). Effects of Double-Connected Finned Surfaces on Heat Transfer During Solidification, *AIAA Pap. 82-0847*, American Institute of Aeronautics and Astronautics, New York.

Sheffield, J. W., Stafford B. D., and Sauer, H. J., Jr. (1985). Finned-Tube Contact Conductance: Investigation of Contacting Surfaces, *Trans. ASHRAE*, 91, 442.

Sherwin, K. (1973). Development of Finned Tubes for Forced-Draught Air-Cooled Heat Exchangers, *Recent Developments in Compact High Duty Heat Exchange Conference*, Institute of Mechanical Engineers, London, 47.

Sherwood, T. K., and Reed, C. E. (1938). *Applied Mathematics for Chemical Engineering*, McGraw-Hill, New York.

Shih, C. C., and Westwater, J. W. (1974). Spheres, Hemispheres and Discs as High-Performance Fins for Boiling Heat Transfer, *Int. J. Heat Mass Transfer*, 17(1), 125.

Shinohara, Y., and Tobe, M. (1985). Development of an Improved Thermutin Tube, *Hitachi Cable Rev.*, 4, 47–50.

Shinohara, Y., Ohizumi, K., Itoh, Y., and Hori, M. (1987). Heat Transfer Tubes with Grooved Inner Surface, U.S. patent 4,658,992.

Shklover, G. G., Mil'man, O. O., Baskov, V. S., and Ankudenov, G. A. (1985). Heat Transfer in Condensation of Steam on Finely Finned Horizontal Surfaces, *Heat Transfer Sov. Res.*, 13, 108.

Shlykov, Y. P., and Ganin, Y. E. (1964). Thermal Resistance of Metallic Contacts, *Int. J. Heat Mass Transfer*, 7, 921.

Shmerkovich, V. M. (1969). New Air-Cooled Systems, *Chem. Pet. Engineering*, 355–359.

Shouman, A. R. (1965). An Exact Solution for the Temperature Distribution and Radiant Heat Transfer Along a Constant Cross Sectional Area Fin with Finite Equivalent Surrounding Sink Temperature, *Proc. 9th Midwestern Mechanics Conference*, Madison, WI, 175.

Shouman, A. R. (1968). An Exact General Solution for the Temperature Distribution and the Composite Radiation and Convection Heat Exchange Along a Constant Cross-Sectional Area Fin, *Q. Appl. Math.*, 25, 458.

Shvets, Y. I., Didenko, O. I., and Lipovetskaya, O. D. (1985). Calculation of Coefficients of Heat Transfer from Arbitrarily Shaped Fins in Longitudinal Flow, Assuming Conjugate Boundary Conditions, *Heat Transfer Sov. Res.*, 17(1), 5.

Sieder, E. N. and Tate, G. E. (1936). Heat Transfer and Pressure Drop of Liquids in Tubes, *Ind. Eng. Chem.*, 28, 1429.

Sikka, S., and Iqbal, M. (1970). Temperature Distribution and Effectiveness of a Two-Dimensional Radiating and Convecting Circular Fin, *AIAA J.*, 11, 377,

Siman-Tov, M. (1970). Analysis and Design of Extended Surfaces in Boiling Liquids, *Chem. Eng. Prog. Symp. Ser.*, 66(102), 174.

Singh, K. D., and Stasa, F. L. (1982). Transient Finite Element Analysis of Coupled Radiation in an Array of Fins, *Proc. of 16th Southeastern Seminar on Thermal Sciences*, Miami, FL., 78.

Sleicher, C. A., and Rouse, M. W. (1975). A Convenient Correlation for Heat Transfer to Constant and Variable Property Fluids in Turbulent Pipe Flow, *Int. J. Heat Mass Transfer*, 18, 677.

Small, R. D., Larson, D. A., and Brode, H. L. (1984). Asymptotically Large Area Fins, *J. Heat Transfer*, 106, 318.

Smirnov, G. F. (1976). Heat Transfer in Condensation of Binary Mixtures on Finned Surfaces, *Heat Transfer Sov. Res.*, 8(1), 45.

Smirnov, G. F., and Lukanev, I. I. (1972). Study of Heat Transfer from Freon-11 Condensing on a Bundle of Finned Tubes, *Heat Transfer Sov. Res.*, 4(3), 51–56.

Smith, D. M. (1934). Mean Temperature Difference in Cross Flow, *Engineering*, 138, 479–481, 606, 607.

Smith, R. K. (1966). Personal communication.

Smith, M. C. (1972). Performance Analysis and Model Experiments for Louvered Fin Evaporator Core Development, *SAE Pap. 720078*, Society of Automotive Engineers, Detroit, MI.

Smith, R. N. (1983). Analytical and Experimental Study of Freezing Adjacent to a Single Pin Fin, 21st National Heat Transfer Conference, Seattle, WA, *ASME Pap. 83-HT-57*, American Society of Mechanical Engineers, New York.

Smith, S. M. (1992). The Use of Electrical Transmission Line Theory to Predict the Performance of Spacecraft Radiators, M.S.E.E. thesis, Naval Postgraduate School, Monterey, CA.

Smith, T. F., and Lee, H. Y. (1979). Radiant Exchange for a Fin and Tube Solar Collector, *J. Heat Transfer*, 101, 185.

Smith, P. J., and Sucec, J. (1969). Efficiency of Circular Fins of Triangular Profile, *J. Heat Transfer*, 91, 181.

Snider, A. D., and Kraus, A. D. (1981). A General Extended Surface Analysis Method, *J. Heat Transfer*, 103, 699.

Snider, A. D., and Kraus, A. D. (1982). Correction for the Variability of the Heat Transfer Coefficient in Extended Surface Analysis, *Proc. 7th Heat Transfer Conference*, Munich, 6, 239.

Snider, A. D., and Kraus, A. D. (1983). Recent Developments in the Analysis and Design of Extended Surface, *J. Heat Transfer*, 105, 302.

Snider, A. D. and Kraus, A. D. (1986). The Quest for the Optimum Longitudinal Fin Profile, *ASME Symp. Vol. PVP-118, HTD-64*, American Society of Mechanical Engineers, New York.

Sobel, N., Landis, F., and Mueller, W. K. (1966) Natural Convection Heat Transfer in Short Vertical Channels, *Proc. 3rd International Heat Transfer Conference*, Chicago, 2, 121.

Soland, J. G., Mack, W. M., Jr., and Rohsenow, W. M. (1978). Performance Ranking of Plate-Fin Heat Exchanger Surfaces, *J. Heat Transfer*, 100, 514.

Soliman, H. M. (1979). Effect of Fin Material on Laminar Heat Transfer Characteristics of Internally Finned Tubes, *Advances in Enhanced Heat Transfer*, American Society of Mechanical Engineers, New York, 95.

Somerscales, E. F. C., and Knudsen, J. G. (1981). *Fouling of Heat Transfer Equipment*, Hemisphere, New York.

Sonn, A., and Bar-Cohen, A. (1981). Optimum Cylindrical Pin Fin, *J. Heat Transfer*, 103, 814.

Sparrow, E. M., and Acharya, S. (1981). A Natural Convection Fin with a Solution-Determined Nonmonotonically Varying Heat Transfer Coefficient, *J. Heat Transfer*, 103, 218.

Sparrow, E. M., and Bahrami, P. A. (1980). Experiments on Natural Convection Heat Transfer on the Fins of a Finned Horizontal Tube, *Int. J. Heat Mass Transfer*, 23(11), 1555.

Sparrow, E. M., and Becky, T. J. (1981). Pressure Drop Characteristics for a Shrouded Longitudinal-Fin Array with Tip Clearance, *J. Heat Transfer*, 103, 393.

Sparrow, E. M., and Charmchi, M. (1980). Laminar Heat Transfer in an Externally Finned Circular Tube, *J. Heat Transfer*, 102, 605.

Sparrow, E. M., and Chyu M. K. (1982). Conjugate Forced Convection-Conduction Analysis of Heat Transfer in a Vertical Fin, *J. Heat Transfer*, 104, 204.

Sparrow, E. M., and Eckert, E. R. G. (1962). Radiant Interaction Between Fin and Base Surfaces, *J. Heat Transfer*, 84, 12.

Sparrow, E. M., and Gregg, J. L. (1956). Laminar Free Convection from a Vertical Plate, *Trans. ASME*, 78, 435.

Sparrow, E. M., and Hennecke, D. K. (1970). Temperature Depression at the Base of a Fin, *J. Heat Transfer*, 92, 204.

Sparrow, E. M., and Hsu, C. F. (1981). Analytically Determined Fin-Tip Heat Transfer Coefficients, *J. Heat Transfer*, 103, 18.

Sparrow E. M., and Kadle D. S. (1984). Heat Transfer from Rods or Fins Which Extend Outward from a Rotating Shaft, *J. Heat Transfer*, 106, 290.

Sparrow, E. M., and Kang, S. S. (1985). Longitudinally Finned Cross-Flow Tube Banks and Their Heat Transfer and Pressure Drop Characteristics, *Int. J. Heat Mass Transfer*, 28(2), 339.

Sparrow, E. M., and Larsen, E. D. (1982). Heat Transfer from Pin Fins Situated in an Oncoming Longitudinal Flow Which Turns to Crossflow, *Int. J. Heat Mass Transfer*, 25(5), 603.

Sparrow, E. M., and Lee, L. (1975). Effects of Fin Base Depression in a Multifin Array, *J. Heat Transfer*, 97, 63.

Sparrow, E. M., and Lin, S. H. (1964). Heat Transfer Characteristics of Polygonal and Plate Fins, *Int. J. Heat Mass Transfer*, 7, 951.

Sparrow, E. M., and Myrum, T. A. (1985). Crossflow Heat Transfer for Tubes with Periodically Interrupted Annular Fins, *Int. J. Heat Mass Transfer*, 28(2), 509.

Sparrow, E. M., and Niewerth, E. R. (1968). Radiating, Conducting and Convecting Fins: Numerical and Linearized Solutions, *Int. J. Heat Mass Transfer*, 11(2), 377.

Sparrow, E. M., and Semle, F. (1981). Measured Heat Transfer Coefficients at and Adjacent to the Tip of a Wall-Attached Cylinder in Cross Flow: Application to Fins, *J. Heat Transfer*, 103, 778.

Sparrow, E. M., and Vemuri, S. B. (1985). Natural Convection/Radiation Heat Transfer from Highly Populated Pin Fin Arrays, *J. Heat Transfer*, 107, 190.

Sparrow, E. M., and Vemuri, S. B. (1986). Orientation Effects on Natural Convection/Radiation Heat Transfer from Pin-Fin Arrays, *Int. J. Heat Mass Transfer*, 29(3), 359.

Sparrow, E. M., Eckert, E. R. G., and Irvine, T. F. (1961a). The Effectiveness of Radiating Fins with Mutual Irradiation, *J. Aeros. Sci.*, 28(10), 763, 778.

Sparrow, E. M., Gregg, J. L., Szel, V., and Mani, P. (1961b). Analysis Results and Intepretation Between Some Simply Arranged Gray Surfaces, *Trans. ASME*, 83, 207–214.

Sparrow, E. M., Miller, G. B., and Jonsson, V. K. (1962). Radiating Effectiveness of Annular-Finned Space Radiators Including Mutual Irradiation between Radiator Elements, *J. Aeros. Sci.*, 29(11), 1291.

Sparrow, E. M., Johnson, K., and Minkowycz, W. J. (1963). Heat Transfer from Fin-Tube Radiators Including Longitudinal Heat Conduction and Radiant Heat Interchange Between

Longitudinal Non-isothermal Surfaces, *NASA TND-3654*, National Aeronautics and Space Administration, Washington, DC.

Sparrow, E. M., Baliga, B. R., and Patankar, S. V. (1978). Forced Convection Heat Transfer from a Shrouded Fin Array with and without Tip Clearance, *J. Heat Transfer*, 100, 572.

Sparrow, E. M., Ramsey, J. W., and Altemani, C. A. C. (1980). Experiments on In-line Pin Fin Arrays and Performance Comparisons with Staggered Arrays, *J. Heat Transfer*, 102, 44.

Sparrow, E. M., Larsen, E. D., and Ramsey J. W. (1981). Freezing on a Finned Tube for Either Conduction-Controlled or Natural Convection Controlled Heat Transfer, *Int. J. Heat Mass Transfer*, 24(2), 273

Sparrow, E. M., Ansari, M. A., Stryker, P. C., and Ruiz, R. (1986). Enhanced Heat Transfer from a Horizontal Finned Tube Situated in a Vertical Channel, *J. Heat Transfer*, 108, 62.

Stachiewicz, J. W. (1969). Effect of Variation of Local Film Coefficients on Fin Performance, *J. Heat Transfer*, 91, 21.

Starner, K. E., and McManus, H. N. (1963). An Experimental Investigation of Free Convection Heat Transfer from Rectangular Fin Arrays, *J. Heat Transfer*, 85, 273.

Stasiulevicius, J., and Survila, V. (1974). Heat Transfer and Hydraulic Drag of Banks of Finned Tubes at High Reynolds Numbers, *Proc. 5th International Heat Transfer Conference*, 2, 249.

Stefan, K. (1959). Warmeübertragung und Druckabfall Beinichtausgebildeter Laminar Störming in Rohren evenen Spalten, *Chem. Ing. Technol.*, 31, 773–778.

Stockman, N. O., and Bittner, E. C. (1965). Two-Dimensional Heat Transfer in Radiating Stainless Steel Clad Copper Fins, *NASA Tech. Note D-3102*, National Aeronautics and Space Administration, Washington, DC.

Stockman, N. O., and Kramer, J. L. (1963). Effect of Variable Thermal Properties on One Dimensional Heat Transfer in Radiating Fins, *NASA Tech. Note D-1878*, National Aeronautics and Space Administration, Washington, DC.

Stockman, N. O., Bittner, E. C., and Sprague, E. L. (1966). Comparison of One- and Two-Dimensional Heat Transfer Calculations in Central Fin Tube Radiators, *NASA Tech. Note D-3645*, National Aeronautics and Space Administration, Washington, DC.

Sturman, J. J. (1968). Single Cooling Fins: The Design of Test Apparatus and the Testing of Selected Fins, Thesis, University of Western Australia, CSIRO, Division of Chemical Engineering, Clayton, Victoria, Australia.

Sturman, J. J. (1973). Measurement of Convection Heat Transfer Coefficient over the Surfaces of Right Circular Cooling Fins, *Proc. 1st Australasia Conference on Heat and Mass Transfer*, 7.

Stynes, S. K., and Meyers, J. E. (1964). Transport from Extended Surfaces, *AIChE J.*, 10(40), 437.

Suga, T., and Aoki, H. (1991). Numerical Study on Heat Transfer and Pressure Drop in Multilouvered Fins, *Proc. ASME–JSME Thermal Engineering Joint Conference*, 4, 361–368.

Sun, K. H., and Lienhard, J. H. (1970). The Peak Boiling Heat Flux on Horizontal Cylinders, *Int. J. Heat Mass Transfer*, 13(9), 1425–1439.

Sunden, B. (1983a). Conjugate Mixed Convection Heat Transfer from a Vertical Rectangular Fin, *Int. Commun. Heat Mass Transfer*, 10(4), 267.

Sunden, B. (1983b). Numerical Investigation of Coupled Conduction–Mixed Convection for Rectangular Fins, *Proc. 3rd International Conferance on Numerical Methods in Laminar and Turbulent Flow*, Seattle WA, 809.

Sunden, B. (1984). Conjugate Laminar and/or Turbulent Forced Convection Heat Transfer from Rectangular Fins, *AIChE Chem. Eng. Symp. Ser.*, 86(1), 725.

Sunden, B. (1985). Effect of Prandtl Number on Conjugate Heat Transfer from Rectangular Fins, *Int. Commun. Heat Mass Transfer*, 12(3), 225.

Sunden, B., and Svantesson, J. (1990). Thermal Hydraulic Performance of New Multilouvered Fins, *Proc. 9th International Heat Transfer Conference*, Jerusalem, Israel, 5, 91–96.

Sur, B., and Azer, N. Z. (1991). Effect of Oil on Heat Transfer and Pressure Drop During Condensation of R-113 Inside Smooth and Internally Finned Tubes, *ASHRAE Trans.*, 97(1), 356–373.

Suryanarayana, N. V. (1975). Transient Response of Straight Fins (Part I), *J. Heat Transfer*, 97, 417.

Suryanarayana, N. V. (1976). Transient Response of Straight Fins (Part II), *J. Heat Transfer*, 98, 324.

Suryanarayana, N. V. (1977). Two-Dimensional Effects on Heat Transfer Rates from an Array of Straight Fins, *J. Heat Transfer*, 99. 129.

Swift, G. , Migliori, A., and Wheatley, J. C. (1985). Microchannel Crossflow Fluid Heat Exchanger and Method for Its Fabrication, U.S. patent 4,516,632.

Takazawa, H., Nishiyama, K., and Kajikawa, T. (1983). Condensing Heat Transfer Enhancement on Vertical Spiral Double Fin Tubes with Drainage Gutters, *Proc. ASME–JSME Thermal Engineering Joint Conference*, 2, 269.

Takeyama, T., Endo, T., and Owada, K. (1974). Analysis of Boiling Heat Transfer on a Fin, *Heat Transfer Jpn. Res.*, 3(2), 10.

Taranyan, I. G., Iokhvedov, F. M., and Kuntysh, V. B. (1972). Investigation of the Effect of Finning Parameters on Heat Transfer and Resistance of Staggered Tube Bundles with Transverse Smooth and Integrated Fins, *High Temperature*, 10(5), 941.

Test, F. L. (1968). Laminar Flow Heat Transfer and Fluid Flow for Liquids with a Temperature Dependent Viscosity, *J. Heat Transfer*, 90, 385–393.

Theoclitus, G., and Eckrich, T. L. (1964). An Experimental Technique for Determining the Effectiveness of Extended Surface, *ASME Symp. Vol. Air Cooled Heat Exchangers*, American Society of Mechanical Engineers, New York.

Thomas, D. G. (1967). Enhancement of Film Condensation Rates on Vertical Tubes by Vertical Wires, *Ind. Eng. Chem. Fundam.*, 6, 97.

Thomas, D. G., and Hayes, P. H. (1970). High Performance Heat Transfer Surfaces, *Ind. Eng. Chem.*, 62(2), 4.

Thomas, D. G., and Young, G. (1970). Thin-Film Evaporation Enhancement by Finned Surfaces, *Ind. Eng. Chem. Process Des. Dev.*, 9(2), 317.

Tien, C. L. (1967). Approximate Solutions in Radiation Exchange Between Conducting Plates with Specular Reflection, *J. Heat Transfer*, 89, 119.

Timoshenko, S., and Goodier, J. N. (1951). *Theory of Elasticity*, McGraw-Hill, New York.

Tolubinsky, V. I., and Zozulya, N. V. (1974). Application of Heat Transfer Finned Surfaces in High Temperature Heat Exchangers, Heat Exch. Des. and Theory Sourcebook, *Proc. 5th Seminar of the International Center for Heat and Mass Transfer*, Scripta, Washington, DC, 585.

Toner, M., Kilic, A., and Onat, K. (1983). Comparison of Rectangular and Triangular Fins when Condensation Occurs, *Thermodyn. Fluid Dyn.*, 17, 65.

Topaldi, A. K., and Kuehn, T. H. (1984). Natural Convection Heat Transfer from a Conducting Fin Above a Heated Horizontal Cylinder, *ASME Symp. Vol. HTD-32*, American Society of Mechanical Engineers, New York, 85.

Truong, H. V., and Mancuso, R. J. (1980). Performance Predictions of Rotating Annular Fins of Various Profile Shapes, 19th National Heat Transfer Conference, Orlando, FL, *ASME Pap. 80-HT-31*, American Society of Mechanical Engineers, New York.

Tseng, J,-K., Chou, Y,-S., and Wu, W.-J. (1993). Analysis of Transient Heat Transfer in Straight Fins of Various Shapes with Bases Subjected to a Decayed Exponential Function of Time in Heat Flux, *Comput. Struct.*, 47, 289.

Tuckerman, D. B. (1984). Heat Transfer Microstructures for Integrated Circuits, Ph.D. thesis, Stanford University, Stanford, CA.

Tuckerman, D. B., and Pease, R. F. (1981a). High Performance Heat Sinking for VLSI, *IEEE Electron Dev. Lett.*, 2, 126–129.

Tuckerman, D. B. and Pease, R. F. (1981b). Errata, *IEEE Electron Dev. Lett.*, 2, 213.

Tuckerman, D. B. and Pease, R. F. (1981c). Ultrahigh Thermal Conductance Microstructures for Cooling Integrated Circuits, *Proc. 32nd Electronics Components Conference*, 145–149.

Ueda, T., and Harada, I. (1964). Experiment of Heat Transfer on the Surfaces with Transverse Fins for Flow Direction, *Bull. JSME*, 7(28), 759.

Ullman, A., and Kalman, H. (1989). Efficiency and Optimized Dimensions of Annular Fins of Different Cross Section Shapes, *Int. J. Heat Mass Transfer*, 32, 1105.

Unal, H. C. (1985). Determination of the Temperature Distribution in an Extended Surface with a Non-uniform Heat Transfer Coefficient, *Int. J. Heat Mass Transfer*, 28(12), 2279.

Unal, H. C. (1986). A Simple Method of Dimensioning Straight Fins for Nucleate Pool Boiling, *Int. J. Heat Mass Transfer*, 29, 640.

Vampola, J. (1966). Heat Transfer and Pressure Drop in Flow of Gases Across Finned Tube Banks, *Strojirenstvi*, 7, 501.

Van de Pol, D. W., and Tierney, J. K. (1973). Free Convection Nusselt Number for Vertical U-Shaped Channels, *J. Heat Transfer*, 87, 439.

Van Fossen, G. J. (1982). Heat Transfer Coefficients for Staggered Arrays of Short Pin Fins, *J. Eng. Power*, 104, 268.

Venkatesh, C. K. (1984). Estimate of Errors in Exchanger Heat Transfer Coefficients, *Hydrocarbon Proc.*, 63(7), 56–58.

Vrable, D. L., Yang, W. J., and Clark, A. (1974). Condensation of Refrigerant-12 Inside Horizontal Tubes with Internal Axial Fins, *Heat Transfer JSME*, 3, 250–254.

Wang, Q., and Chen H. (1984). Condensation on a Circumferential Fin Surface, *Proc. 3rd Symposium–Workshop on Multi-phase Flow and Heat Transfer*, Miami Beach, FL, 793.

Wanniarachchi, A. S., Marto, P. J., and Rose, J. W. (1985). Film Condensation of Steam on Horizontal Finned Tubes: Effect of Fin Spacing, Thickness and Height, in V. K. Dhir, J. C. Chen, and O. C. Jones (Eds.), *Multiphase Flow and Heat Transfer, ASME Symp. Vol., HTD-47*, American Society of Mechanical Engineers, New York, 93–99.

Wanniarachchi, A. S., Marto, P. J., and Rose, J. W. (1986). Film Condensation of Steam on Horizontal Tubes: Effect of Fin Spacing, *J. Heat Transfer*, 106, 960–966.

Ward, D. J., and Young, E. H. (1959). Heat Transer and Pressure Drop of Air in Forced Convection Across Triangular Pitch Banks of Finned Tubes, *Chem. Eng. Prog. Symp. Ser.*, 55(29), 37.

Watkinson, A. P., Miletti, D. L., and Kubanek, G. R. (1975). Heat Transfer and Pressure Drop of Internally Finned Tubes in Laminar Oil Flow, 15th National Heat Transfer Conference, San Francisco, *ASME Pap. 75-HT-41*, American Society of Mechanical Engineers, New York.

Watson, G. N. (1966). *A Treatise on Bessel Functions*, 3rd Ed, Cambridge University Press, New York.

Webb, R. L. (1971). A Critical Evaluation of Analytical Solutions and Reynolds Analogy Equations for Heat and Mass Transfer in Smooth Tubes, *Waerme und Stoffuebertrag.*, 4, 197–204.

Webb, R. L. (1981a). Heat Transfer and Friction Characteristics For Finned Tubes Having Plane Fins, *Proc. 4th NATO Advanced Study Institute on Heat Transfer*, Ankara, Turkey, 431.

Webb, R. L. (1981b). The Use of Enhanced Surface Geometries in Condensers: An Overview, in *Power Condenser Heat Transfer Technology*, Hemisphere, Washington, DC, p. 287.

Webb, R. L. (1984). Shell Side Condensation in Refrigerant Condensers, *ASHRAE Transfer*, 90(1-B), 39–59.

Webb, R. L. (1994). *Principles of Enhanced Heat Transfer*, Wiley, New York.

Webb, R. L., and Joshi, H. M. (1982). Friction Factor Correlation for Offset Strip Fin Matrix, *Proc. 7th International Heat Transfer Conference*, Munich, 6, 257.

Webb, R. L., and Murawski, C. G. (1990). Row Effect of R-11 Condensation on Enhanced Tubes, *J. Heat Transfer*, 112, 768–776.

Webb, R. L., and Scott, M. J. (1980). A Parametric Analysis of the Performance of Internally Finned Tubes for Heat Exchanger Application, *J. Heat Transfer*, 102, 38:

Webb, R. L., Keswani, S. T., and Rudy, T. M. (1982). Investigation of Surface Tension and Gravity Effects on Film Condensation, *Proc. 7th International Heat Transfer Conference*, Munich, 5, 175–181.

Webb, R. L., Rudy, T. M., and Kedzierski, M. (1985). Prediction of the Condensation Coefficient on Horizontal Integral-Fin Tubes, *J. Heat Transfer*, 107, 369.

Weierman, C. (1976). Correlations Ease the Selection of Finned Tubes, *Oil Gas J.*, 74(36), 94.

Welling, J. R., and Wooldridge, C. B. (1965). Free Convection Heat Transfer Coefficients from Rectangular Vertical Fins, *J. Heat Transfer*, 87, 439.

Welty, J. R., Wicks, C. E., and Wilson, R. E. (1986). *Fundamentals of Momentum, Heat and Mass Transfer*, 3rd ed., Wiley, New York.

West, W. E., Jr., and Westwater, J. W. (1953). Radiation-Conduction Correction for Temperature Measurements in Hot Gases, *Ind. Eng. Chem.*, 45, 2152.

White, F. M. (1994). *Fluid Mechanics*, McGraw-Hill, New York. NY.

Wieting, A. R., (1975). Empirical Correlations for Heat Transfer and Flow Friction Characteristics of Rectangular Offset-Fin Plate-Fin Heat Exchangers, *J. Heat Transfer*, 97, 488.

Wilkie, D., (1971). Criteria for Choice of Surface Form for Gas Cooled Reactors, *Nucl. Eng. International*, (16) 178, 215–217.

Wilkins, J. E., Jr. (1960a). Minimum Mass Thin Fins for Space Radiators, *Proc. Heat Transfer and Fluid Mechanics Institute*, 228.

Wilkins, J. E., Jr. (1960b). Minimum Mass Thin Fins Which Transfer Heat Only by Radiation to Surroundings at Absolute Zero, *J. Soc. Ind. Appl. Math.*, 8, 630.

Wilkins, J. E., Jr. (1961). Minimum Mass Thin Fins with Specified Minimum Thickness, *J. Soc. Ind. Appl. Math.*, 9, 194.

Wilkins, J. E., Jr. (1962a). Minimum-Mass Thin Fins and Constant Temperature Gradients, *J. Soc. Ind. Appl. Math.*, 10(1), 62.

Wilkins, J. E., Jr. (1962b). Minimum-Mass Thin Fins with Internal Heat Generation, *Nucl. Sci. Eng.*, 14(2), 203.

Wilkins, J. E., Jr. (1980). Discussion on Condensation on an Extended Surface, *J. Heat Transfer*, 102, 196–197.

Williams, A. G., Nandapurkar, S. S., and Holland, F. A. (1971) Condensation on a Vertical Rotating Finned Tube, *Can. J. Chem. Eng.*, 49(1), 51.

Wimberly D., and Costello, F. A. (1970). Tube Wall Fin Effects in Spacecraft Radiators, *J. Spacecr. Rockets*, 7(2), 214.

Wong, L. T., and Smith, M. C. (1973). Airflow Phenomena in the Louvered-Fin Heat Exchanger, *SAE Pap. 730237*, Society of Automotive Engineers, Detroit, MI.

Worsœ-Schmidt, P. M. (1966). Heat Transfer and Friction for Laminar Flow of Helium and Carbon Dioxide in a Circular Tube at High Heating Rate, *Int. J. Heat Mass Transfer*, 9, 1291–1295.

Yang, K. T. (1962). Laminar Forced Convection of Liquids in Tubes with Variable Viscosity, *J. Heat Transfer*, 8, 353–362.

Yang, J. W. (1972). Periodic Heat Transfer in Straight Fins, *J. Heat Transfer*, 94, 310.

Yao, P. (1984). Heat Transfer and Friction Loss Characteristics of Pin Fin Cooling Configurations, *J. Eng. Gas Turbines Power*, 106, 264.

Yasuda, K., Ohizumi, K., Hori, M., and Kawamata, D. (1990). Developments of Condensing Thermo-Fin-Hex Tube, *Hitachi Cable Rev.*, 9, 27–30.

Yau, K. K., Cooper, J. R., and Rose, J. W. (1985). Effect of Fin Spacing on the Performance of Horizontal Integral-Fin Condenser Tubes, *J. Heat Transfer*, 107, 377.

Yau, K. K., Cooper, J. R., and Rose, J. W. (1986). Horizontal Plain and Low-Finned Condenser Tubes: Effect of Fin Spacing and Drainage Strips on Heat Transfer and Condensate Retention, *J. Heat Transfer*, 108, 946–950.

Yoshi, T., Ueda, K., and Horiguchi, T. (1971). Anti-fouling System by Electrolysis of Sea Water, *Proc. Conference on Civil Engineering in Oceans*, Miami Beach, FL, 1045–1068.

Yoshimura, T., and Campo, A. (1981). Extended Surface Heat Rejection Accounting For Stochastic Sink Temperature, *AIAA J.*, 19, 221.

Yovanovich, M. M. (1981). New Contact and Gap Conductance Correlations for Conforming Rough Surfaces, 16th AIAA Thermophysics Conference, Palo Alto, CA, *AIAA Pap. 1164*, American Institute of Aeronautics and Astronautics, New York.

Yovanovich, M. M. (1983). Non-iterative Control Volume Approach to One-Dimensional Steady Conduction with Convection: Applications to Extended Surfaces, *ASME Symp. Vol. HTD-28*, American Society of Mechanical Engineers, New York, 13.

Yovanovich, M. M. (1987a). Natural Convection from Isothermal Spheroids in the Conductive to Laminar Flow Regimes, *AIAA Pap. AIAA-87-1587*, American Institute of Aeronautics and Astronautics, New York.

Yovanovich, M. M. (1987b). On the Effect of Shape, Aspect Ratio and Orientation upon Natural Convection from Isothermal Bodies, *ASME Symp. Vol. HTD-82*, American Society of Mechanical Engineers, New York, 103.

Yovanovich, M. M., Culham, J. R., and Lemczyk, T. F. (1988). Simplified Solutions to Circular Annular Fins with Contact Resistance and End Cooling, *AIAA Thermophys. Heat Transfer*, 2, 152.

Yudin, V. F., and Tokhtarova, L. S. (1967). Investigation of Heat Dissipation and Resistance of Finned Checkerwork Stacks with Various Fin Shapes, *Proc. 2nd All-Soviet Union Conference on Heat and Mass Transfer*, Minsk, Belorussia, 1, 283.

Zabronsky, H. (1955). Temperature Distribution and Efficiency of Heat Exchangers Using Square Fins on Round Tubes, *J. Appl Mech.*, 22, 119.

Zappa, R. F., and Geiger, G. E. (1971). Effect of Artificial Roughness on Heat Transfer and Pressure Drop for a High Prandtl Number Fluid in Laminar Flow, *ASME Pap. 71-HT-36*, American Society of Mechanical Engineers, New York.

Zhadan, V. A., Pantyukhov, L. L., Sannikov, D. I., Siphailov, G. A., and Shelekov, S. A. (1970). Effect of the Fin Dimensions upon the Heating of the Series AR Electric Motors, *Elektrotekhnika*, 11, 28.

Zhang, Z., and Patankar, S. V. (1984). Influence of Buoyancy on the Vertical Flow and Heat Transfer in a Shrouded Fin Array, *Int. J. Heat Mass Transfer*, 27(1), 137.

Zhukauskas, A. A. (1972). Heat Transfer from Tubes in Cross Flow, in J. P. Hartnett and T. F. Irvine (Eds.), *Advances in Heat Transfer*, Academic Press, New York, pp. 93–160.

Zhukauskas, A. A. (1974). Investigation of Heat Transfer in Different Arrangements of Heat Exchanger Surfaces, *Teploenergetika*, 21(5), 24.

Zhukauskas, A. A., and Ulinskas, R. V. (1979). Developments in the Compact Heat Exchanger Field, *ASME Symp. Vol. HTD-10*, American Society of Mechanical Engineers, New York.

Zhukauskas, A. A., Ulinskas, R. V., and Zinyavichyus, F. V. (1983). Heat Transfer and Resistance of Staggered-Array Banks of Finned Tubes in Transverse Flows, *Power Eng.*, 21(4), 113.

Zozulya, N. V., and Khavin, A. A. (1974). Experimental Determination of the Thermal Engineering Performance of Large Diameter Tubes with Transversely Welded-on Fins, *Heat Transfer Sov. Res.*, 6(6), 82.

Zozulya, N. V. and Korunzhii, Y. G. (1968). Heat Transfer from Finned Tubes Moving Back and Forth in Liquid, *Chem Pet. Eng.*, 13, 830.

Zozulya, N. V., Kalinin, B. L., and Khavin, A. A. (1970a). Influence of the Layout of a Bank of Finned Aluminum Tubes on Heat Transfer, *Teploenergetika*, 17(6), 221.

Zozulya, N. V., Khavin, A. A., and Kalinin, B. L. (1970b). Effect of Perforated Plate Fins on Oval Pipes on the Rate of Heat Transfer, *Heat Transfer Sov. Res.* 2(1), 77.

Zozulya, N. V., Vorob'yev, Y. P, and Khavin, A. A. (1973). Effect of Flow Turbulization on Heat Transfer in a Finned-Tube Bundle, *Heat Transfer Sov. Res.*, 5(1), 154.

Zozulya, N. V., Kharkhu, V. A., and Borovkov, V. P. (1977). An Analytic and Experimental Study of Heat Transfer in Condensation of Vapor on Finned Surfaces, *Heat Transfer Sov. Res.*, 9(2), 18.

Zuber, N. (1958). On the Stability of Boiling Heat Transfer, *J. Heat Transfer*, 80, 711–720.

Zuber, N., and Tribus, M. (1958). Further Remarks on Stability of Boiling Heat Transer, *UCLA Rep. 58-5*, University of California–Los Angeles, Los Angeles, CA.

AUTHOR INDEX

AUTHOR INDEX